Unit Conversions (Equivalents)

Length

1 in = 2.54 cm
1 cm = 0.394 in
1 ft = 30.5 cm
1 m = 39.4 in = 3.28 ft
1 mi = 5280 ft = 1.61 km
1 km = 0.621 mi
1 nautical mile = 6080 ft = 1.85 km
1 fermi = 1 femtometer (fm) = 10^{-15} m
1 angstrom (Å) = 10^{-10} m
1 light-year = 9.46×10^{15} m
1 parsec = 3.26 light-years

Time

1 day = 8.64×10^4 s
1 year = 3.156×10^7 s

Speed

1 mi/h = 1.47 ft/s = 1.61 km/h = 0.447 m/s
1 km/h = 0.278 m/s = 0.621 mi/h
1 ft/s = 0.305 m/s = 0.682 mi/h
1 m/s = 3.28 ft/s = 3.60 km/h
1 knot = 1.151 mi/h = 0.5144 m/s

Angle

1 radian (rad) = 57.30° = 57°18′
1° = 0.01745 rad
1 rev/min (rpm) = 0.1047 rad/s

Mass

1 atomic mass unit (u) = 1.6605×10^{-27} kg
1 slug = 14.6 kg
1 kg = 0.0685 slug
[1 kg has a weight of 2.20 lb where g = 9.80 m/s^2.]

Force

1 lb = 4.45 N
1 N = 10^5 dyne = 0.225 lb

Energy and Work

1 J = 10^7 ergs = 0.738 ft·lb
1 ft·lb = 1.36 J = 1.29×10^{-3} Btu = 3.25×10^{-4} kcal
1 kcal = 4.18×10^3 J = 3.97 Btu
1 Btu = 252 cal = 778 ft·lb = 1054 J
1 eV = 1.602×10^{-19} J
1 kWh = 3.60×10^6 J = 860 kcal

Power

1 W = 1 J/s = 0.738 ft·lb/s
1 hp = 550 ft·lb/s = 746 W

Pressure

1 atm = 1.013 bar = 1.013×10^5 N/m^2
= 14.7 lb/in^2 = 760 torr
1 lb/in^2 = 6.90×10^3 N/m^2
1 Pa = 1 N/m^2 = 1.45×10^{-4} lb/in^2

SI Derived Units and Their Abbreviations

Quantity	*Unit*	*Abbreviation*	*In terms of base units* †
Force	newton	N	kg·m/s^2
Energy and work	joule	J	kg·m^2/s^2
Power	watt	W	kg·m^2/s^3
Pressure	pascal	Pa	kg/(m·s^2)
Frequency	hertz	Hz	s^{-1}
Electric charge	coulomb	C	A·s
Electric potential	volt	V	kg·m^2/(A·s^3)
Electric resistance	ohm	Ω	kg·m^2/(A^2·s^3)
Capacitance	farad	F	A^2·s^4/(kg·m^2)
Magnetic field	tesla	T	kg/(A·s^2)
Magnetic flux	weber	Wb	kg·m^2/(A·s^2)
Inductance	henry	H	kg·m^2/(s^2·A^2)

† kg = kilogram (mass), m = meter (length), s = second (time), A = ampere (electric current).

Metric (SI) Multipliers

Prefix	*Abbreviation*	*Value*
Tera	T	10^{12}
Giga	G	10^9
Mega	M	10^6
Kilo	k	10^3
Hecto	h	10^2
Deka	da	10^1
Deci	d	10^{-1}
Centi	c	10^{-2}
Milli	m	10^{-3}
Micro	μ	10^{-6}
Nano	n	10^{-9}
Pico	p	10^{-12}
Femto	f	10^{-15}

PHYSICS

Principles with Applications

THIRD EDITION

Douglas C. Giancoli

PRENTICE HALL, ENGLEWOOD CLIFFS, NEW JERSEY 07632

Library of Congress Cataloging-in-Publication Data

Giancoli, Douglas C.
Physics, principles with applications / Douglas C. Giancoli. —
3rd ed.

p. cm.
ISBN 0-13-672510-4
1. Physics. I. Title.
QC23.G399 1991
530—dc20 90-33995
CIP

Editorial/production supervision and page layout: Judith Winthrop
Acquistion editor: Tim Bozik; Holly Hodder
Development editor: Dan Schiller
Cover photos: D. Giancoli
Cover design: Anne T. Bonanno
Interior design: Kenny Beck; Anne T. Bonanno
Prepress buyer: Paula Massenaro
Manufacturing buyer: Lori Bulwin
Photo editor: Lorinda Morris-Nantz
Photo research: John Schultz
Illustrations: Network Graphics
Line art coordinator: Nancy Bauer

Acknowledgments appear on page xx which constitutes
a continuation of the copyright page.

Printed in the United States of America

10 9 8 7 6 5 4 3 2 1

ISBN 0-13-672510-4

Prentice-Hall International (UK) Limited, *London*
Prentice-Hall of Australia Pty. Limited, *Sydney*
Prentice-Hall Canada Inc., *Toronto*
Prentice-Hall Hispanoamericana, S.A., *Mexico City*
Prentice-Hall of India Private Limited, *New Delhi*
Prentice-Hall of Japan, Inc., *Tokyo*
Simon & Schuster Asia Pte. Ltd., *Singapore*
Editora Prentice-Hall do Brasil, Ltda., *Rio de Janeiro*

Contents

27 EARLY QUANTUM THEORY AND MODELS OF THE ATOM 719

28 QUANTUM MECHANICS OF ATOMS 747

29 MOLECULES AND SOLIDS 776

30 NUCLEAR PHYSICS AND RADIOACTIVITY 800

Preface

That physics is exciting, that physics is beautiful, is well-known to physicists. But it may not always be so apparent to others. Indeed, the subject can strike terror in the hearts of students who are required to take a physics course. Even students who recognize the brilliance of physics, but wonder about their struggles with a physics course, ask their professors, "Why must I study physics? I'm a biology (pre-med, architecture, . . ., etc.) major." It's a good question. And it deserves an answer. This book, even more in this third edition, is intended to answer that question: by showing the beauty of physics and the depth of its basic principles, by showing explicitly how physics applies to other fields as well as in daily living, and by showing how physics lies at the base of scientific knowledge.

The basic goal of this introductory algebra-based physics textbook remains the same as in earlier editions: to help students see the world through eyes that know physics.

The book is intended to be readable, interesting, and accessible to students, and is meant to give them a thorough understanding of the basic concepts of physics and, by means of many interesting applications, to prepare them to use physics in their own lives and professions. It is particularly appropriate for introductory physics courses taken by students studying biology, (pre)medicine, architecture, technology, earth and environmental sciences, and other disciplines. It is also suitable for use in colleges that offer only one introductory physics course.

But why a new edition?

NEW IN THIS THIRD EDITION

"One picture is worth more than a thousand words," says an old adage. It is also a well-known instruction to students for solving problems ("first draw a picture"). To imagine learning physics without good diagrams is unthinkable today. In this third edition of *Physics: Principles with Applications,* we have taken an extra step: all of the figures have been redrawn to make use of the full four-color printing process. Full-color figures are obviously more attractive, and so pull the student in to examine them carefully. But more importantly, the many different possible colors can be used (and are used in this book) to carefully delineate the different aspects of a physical situation. In particular, different colors are used consistently throughout the book for different vectors, for the various field lines, and for other essentials. For example, velocity vectors are always green, force vectors red, and so on. (Furthermore, force vectors on different bodies are usually shown in different shades of red—to help students use Newton's second and third laws correctly.) Basic physical objects—such as electric conductors, insulators, magnets, lenses—are always represented with the same color. With full color, we can show the physical situation and our analysis of it much more clearly. A detailed outline of how color is used consistently in the figures to illuminate the physics is presented right after this preface.

Also new to this edition are:

An entirely new chapter on Astrophysics and Cosmology (Chapter 33), which includes the latest results.

A completely revised Chapter (29) on Molecules and Solids including band theory of solids, and semiconductors.

Special sections throughout the text on techniques for solving problems; indicated by a blue section number.

Much more detailed explanation in Chapter 4 on how to solve problems using Newton's second and third laws—in particular, many new worked-out Examples, the first very simple and each succeeding one adding one additional complication. This gentler and more careful introduction to Newton's laws and their use is of crucial pedagogic importance. Students who do not master this early material (and it is not so easy for many of them) can become discouraged and may never recover.

Many new worked-out Examples throughout the text, some fairly simple for helping students get started, others more difficult than in previous editions to show students how to attack complicated situations. Some show how to make useful, quick estimates.

General problems, unranked and not arranged by section, in each chapter. They follow the large group of problems in each chapter that are ranked (I, II, III) and arranged by section. There are many new problems; and the old ones have been reworked for increased clarity. All problems have been checked and rechecked by several professors for reasonableness, clarity, and accuracy. (Also the answers!) The total number of problems has been increased by 20%, to more than 1900.

Many new diagrams and many new photographs (in full color). Each chapter opens with a photo and caption (often detailed) that illustrates the physics of that chapter.

Marginal notes, printed in red (more accurately, magenta) to distinguish them from the main text. They serve (1) as a sort of outline, (2) as a means of quickly locating important concepts or equations, (3) to point out hints for solving problems (these have a blue underline), and (4) to point out the great laws and principles of physics (for which the marginal note has a rose-colored rectangle superposed for emphasis).

Expanded Chapter 1 that now includes how to make rough estimates (order-of-magnitude calculations), as well as measurement and uncertainty, and the use of significant figures.

New material on rotating reference frames, inertial forces, and the Coriolis force (Chapter 8).

All areas, from superconductivity to nuclear physics and elementary particles, have been brought completely up to date. All chapters have been rewritten, at least a little, with significant clarifications in many areas.

A second review of vector addition in Chapter 16, where, after the long hiatus since vectors were first treated in mechanics, students need to be reminded of how to manipulate vectors, this time for the electric force.

The chapters on Vibrations and Waves, and on Sound, now precede those on Heat and Thermodynamics, so that they are included as a part of mechanics.

New applications such as: TIA (temporary blood loss to the brain, which is related to Bernoulli's principle, page 254); Coriolis force and the atmosphere; digital information and magnetic read/write heads.

New appendix on Gauss's law.

Expanded appendix on math background includes not only a review of algebra and geometry, but also the binomial expansion (especially useful for estimating), and trigonometric functions and identities. The Table of Isotopes, Appendix D, contains the latest (1990) data.

This edition contains more pages than its predecessor, but few of the extra pages, with the obvious exception of the final chapter on cosmology, translate into increased material to be covered. The increased number of pages is mainly due to the more open format and the increased space devoted to figures.

GENERAL APROACH

This book offers, above all, an in-depth presentation of physics, and retains the basic approach of the earlier editions. Instead of using the common, dry approach of treating topics formally and abstractly first, and only later relating the material to the students' own experience, my approach is to recognize that physics is a description of reality and thus to start each topic with concrete observations and experiences that students can directly relate to. Readers are then led into the more formal and abstract treatment of topics. Not only does this make the material more interesting and easier to understand, but it is closer to the way physics is actually practiced. Historically, we didn't start with the second law of thermodynamics, for example, and then derive all kinds of consequences from it; rather, the law was a generalization of all kinds of phenomena.

It is necessary, I feel, to pay careful attention to detail, especially when deriving an important result. Whether it is a verbal discussion or a mathematical one, I have aimed at including all steps in a derivation so that students don't get bogged down in details and then fail to understand the concept as a whole. I have tried to make clear which equations are general, and which are not, by explicitly stating the limitations of important equations in brackets next to the equation, such as

$$x = x_0 + v_0t + \frac{1}{2}at^2 \qquad \text{[constant acceleration]}$$

Difficult language, too, can hinder understanding: and to put students at their ease, I have tried to write in a relaxed style, avoiding jargon. New or unusual terms are carefully defined when first used.

I have tried to avoid the problem of leaving certain topics "hanging," with students wondering "why did we study that?" Thus I have tried to indicate why each topic is important, and to bring each topic to completion. We study static forces in structures, for example, because real materials are elastic and can fracture; so I have included elasticity and fracture in the statics chapter.

An important feature of this book is the inclusion of a wide range of examples and applications from other fields: biology, medicine, architecture, technology, earth sciences, the environment, and daily life. Some applications serve only as examples of physical principles. Others are treated in depth, with whole sections devoted to them (among these are the study of medical imaging systems, constructing arches and domes, and the effects of radiation). But applications do not dominate the text (this is, after all, a physics book). They have been carefully chosen and integrated into the text so as not to interfere with the development of the physics but rather to illuminate it.

Throughout the book I have endeavored to present the basic concepts of physics in their historical and philosophical context.

PROBLEM SOLVING

Even more attention is given to problem solving in this edition than in earlier ones. As already mentioned, scattered throughout the book are special sections devoted to how to approach the solving of problems. Many of these special sections are found in the early chapters, where students first begin wrestling with problem solving; but many are also found later in the book, when there are new types of problems to be solved.

Some 300 Examples are fully worked out in the text: these help students to fix ideas in their minds, to demonstrate interesting applications, and to help students develop problem-solving skills. These worked-out Examples range from simple to fairly complicated, and some show how to make quick estimates. Many Examples are taken from everyday life and aim at being realistic. There are close to 2700 end-of-chapter exercises, including more than 750 questions that require verbal answers based on an understanding of the concepts, and over 1900 problems involving mathematical calculation. Each chapter contains a large group of problems arranged by section and graded according to difficulty: level I problems are simple, usually plug-in types, designed to give students confidence; level II are normal problems, requiring more thought and often the combination of two different concepts; level III are the most difficult and serve as a challenge to superior students. The arrangement by section number means only that those problems depend on material up to and including that section; earlier sections and chapters are often relied upon, particularly in level II or III problems. The ranking of the problems by difficulty (I, II, III) is intended only as a guide. Level II problems, particularly, are of a very wide range. I suggest that instructors assign a significant number of the level I and level II problems, and reserve level III problems to stimulate the best students. Although most level I problems may seem easy, they help to build self-confidence—an important part of learning, especially in physics. Each chapter also contains a group of "General Problems" which are unranked and not arranged by section number, but otherwise are not necessarily more difficult than the others. Answers to odd-numbered problems are given at the back of the book. Throughout the text, *Système International* (SI) units are used. Other metric and British units are defined for informational purposes.

ORGANIZATION

The general outline of this new edition retains a traditional order of topics: mechanics (Chapters 1 to 12), including fluid mechanics, vibrations, waves, and sound, followed by kinetic theory and thermodynamics (Chapters 13 to 15), electricity and magnetism (Chapters 16 to 22), light (Chapters 23 to 25), and modern physics (Chapters 26 to 33).

Not every chapter should be given equal weight, however. Whereas Chapter 4 or Chapter 21 might require $1\frac{1}{2}$ to 2 weeks of coverage, Chapter 12 or Chapter 22 may need only $\frac{1}{2}$ week.

Nearly all topics customarily taught in introductory physics courses are included here. The tradition of beginning with mechanics is sensible, I believe, since it was developed first, historically, and since so much else in physics depends on it. Within mechanics, there are various ways to order the topics, and this book allows for considerable flexibility. Statics, for example, can be covered either before or after dynamics. I prefer to cover statics after dynamics, partly because many students have trouble with the concept of force without motion. Once they have understood the connection between force and motion, including Newton's third law, they seem to be better able to deal with force without motion. This order also allows full development of the concept of torque (itself difficult to understand in the absence of motion) before it is used in statics. Moreover statics is a special case of dynamics—we study statics so that we can prevent structures from becoming dynamic (falling down)—and that sense of being at the limit of dynamics is intuitively helpful. Nonetheless statics (Chapter 9) can be covered earlier, if desired, before dynamics, after a brief introduction to vectors. Another option is the position of the chapters on light. I have placed them after electricity and magnetism and EM waves, as is typical. However, light could be treated immediately after the chapters on waves (Chapters 11 and 12), thereby keeping the various types of wave motion in one place. Another choice of sequence involves special relativity (Chapter 26), which is located along with the other chapters on modern physics after EM waves and light have been covered. Relativity could be treated, if desired, along with mechanics—say, after Chapter 7—since it depends (except for the optional Section 26-2) mainly on material through Chapter 7.

The book contains more material than can be covered in most one-year courses, so that instructors have flexibility in choice of topics. The wide range of subjects also means that students can learn about many topics even though there is not class time for them. This aspect makes the book more valuable to students as a resource and as a reference book. Sections marked with a star (asterisk) are considered optional. These sections contain slightly more advanced physics material, often material not usually covered in typical courses, and/or interesting applications. They contain no material needed in later chapters (except perhaps in later optional sections). This does not imply that all nonstarred sections must be covered: there still remains considerable flexibility in the choice of material to suit the needs of students and instructors.

For use in a two-semester course, the book can be considered to be divided roughly in half between Chapters 15 and 16. The first half would then include mechanics, heat, and sound, with the second half containing electricity and magnetism, light, and modern physics. (The slightly longer second half has more optional sections.) As mentioned earlier, the order and choice of topics is flexible, so other course outlines are possible. For a brief course, all the optional material could be dropped as well as major parts of Chapters 10, 12, 19, 22, 28, 29, 32, and 33, as well as selected parts of Chapters 7, 8, 9, 15, 21, 24, 25, and 31.

Mathematics can be an obstacle to student understanding. To avoid frightening students with an initial chapter on mathematics, I have instead incorporated many important mathematical tools, such as addition of vectors and trigonometry, directly in the text where first needed. In addition, the appendices contain a review of many mathematical topics such as algebra and geometry, as well as dimensional analysis.

THANKS

The revision of this book has depended to a great extent on the hundreds of instructors who have used the book in class and were kind enough to send me their comments and suggestions for improvement. To all of them I owe a debt of thanks. I also wish to thank the professors who read and checked the manuscript and offered valuable suggestions, both for the previous editions and for this new third edition: these include John Anderson (University of Pittsburgh), Sirus Aryainejad (Eastern Illinois University), Gene Barnes (California State University, Sacramento), Isaac Bass, Paul A. Bender (Washington State University), Joseph Boyle (Miami-Dade Community College), Peter Brancazio (Brooklyn College, CUNY), Warren Deshotels (Marquette University), Frank A. Ferrone (Drexel University), Hershel J. Hausman (Ohio State University), Laurent Hodges (Iowa State University), Gordon Jones (Mississippi State University), Kirby W. Kemper (Florida State University), Sanford Kern (Colorado State University), James R. Kirk (Edinboro University), Peter Landry (McGill University, Montreal), Michael Lieber (University of Arkansas), David Markowitz (University of Connecticut), Robert Messina, David Mills (College of the Redwoods), Ed Nelson (University of Iowa), John Reading (Texas A&M), William Riley (Ohio State

University), Larry Rowan (University of North Carolina), D. Lee Rutledge (Oklahoma State University), K. Y. Shen (California State University, Long Beach), Joseph Shinar (Iowa State University), Paul Urone (CSU, Sacramento), Jearl Walker (Cleveland State University), Gareth Williams (San Jose State University), Thomas A. Weber (Iowa State University), Wendell S. Williams (Case Western Reserve University), Peter Zimmerman (Louisiana State University). I also wish to acknowledge the work of Rex Joyner (Indiana Institute of Technology), Gregor Novak (Indiana University/Purdue University), Kwok Yeung Tsang (Georgia Institute of Technology), and Upindranath Singh (Embry-Riddle), who checked the problems and worked the solutions. Appendix D data was updated with the help of Arthur Ballato and Jagdish K. Tuli (National Nuclear Data Center, Brookhaven National Laboratory). I owe special thanks to Professors John Heilbron, Buford Price, Howard Shugart, and Richard Marrus, for helpful discussions and for their hospitality at the University of California, Berkeley, as well as to Professor Paolo Galluzzi and the staff at the Institute for the History of Science, Florence. Finally, I wish to thank the many people at Prentice Hall who have worked on this project and made it possible, especially Judy Winthrop, Barbara DeVries, Holly Hodder, and Tim Bozik. The responsibility for all errors lies, of course, with me. I welcome comments and corrections.

Douglas C. Giancoli

NOTES TO STUDENTS AND INSTRUCTORS ON THE FORMAT

1. Sections marked with a star (*) are considered optional. They can be omitted without interrupting the main flow of topics. No later material depends on them, except possibly later starred sections.
2. The customary conventions are used: symbols for quantities (such as m for mass) are italicized, whereas units (such as m for meter) are not italicized; boldface (**F**) is used for vectors.
3. Few equations are valid in all situations. Where practical, the limitations of important equations are stated in square brackets next to the equation.
4. The number of significant figures (see Section 1-4) should not be assumed to be greater than given: if a number is stated as (say) 6, with its units, it is meant to be 6 and not 6.0 or 6.00 (see also Section 1-6).
5. At the end of each chapter is a set of questions that students should attempt to answer (to themselves at least). These are followed by problems which are ranked as level I, II, or III, according to estimated difficulty, with level I problems being easiest. These problems are arranged by sections, but problems for a given section may depend on earlier material as well. There follows a group of General Problems, which are not arranged by section nor ranked as to difficulty. Questions and problems that relate to optional sections are starred.
6. Being able to solve problems is a crucial part of learning physics, and provides a powerful means for understanding the concepts and principles of physics. This book contains many aids to problem solving: (a) worked-out Examples and their solutions in the text, which are set off with a vertical blue line in the margin, and should be studied as an integral part of the text; (b) special sections on techniques for solving problems, which are indicated by a blue section number, as well as by their titles; (c) marginal notes (see below), many of which refer to hints for solving problems, in which case they have a thin blue underline; (d) problems themselves, at the end of each chapter, and they have blue numbers. In other words: blue for problems and problem solving.
7. Marginal notes: brief notes in the margin of almost every page are printed in red (really, magenta) and are of three types: (a) ordinary notes (the majority) that serve as a sort of outline of the text and can help you to later locate important concepts and equations; (b) notes that refer to the great laws and principles of physics, and these are superposed with a rose-colored rectangle for emphasis; (c) notes that refer to a problem-solving hint or technique treated in the text, and these have a thin blue underline.
8. This book is printed in full color. But not simply to make it more attractive. As mentioned in points 6 and 7 above, blue lines and numbers are used for problem-solving material and magenta for the marginal notes. But the color is used above all in the figures, to give them greater clarity for our analysis, and to provide easier learning of the physical principles involved. The table on the next page is a summary of how color is used in the figures, and shows which colors are used for the different kinds of vectors, for field lines, and for other symbols and objects. These colors are used consistently throughout the book.

NOTES ON USE OF COLOR

Vectors

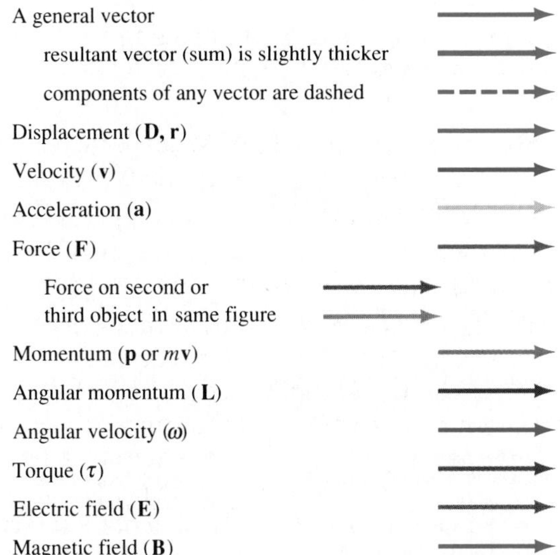

Electricity and magnetism

Electric circuit symbols

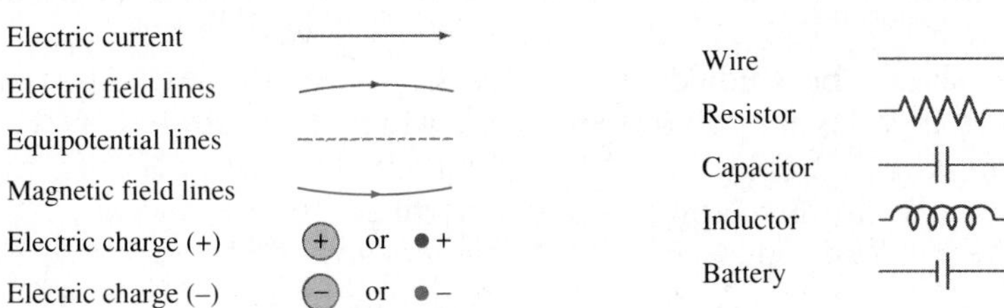

Optics

Other

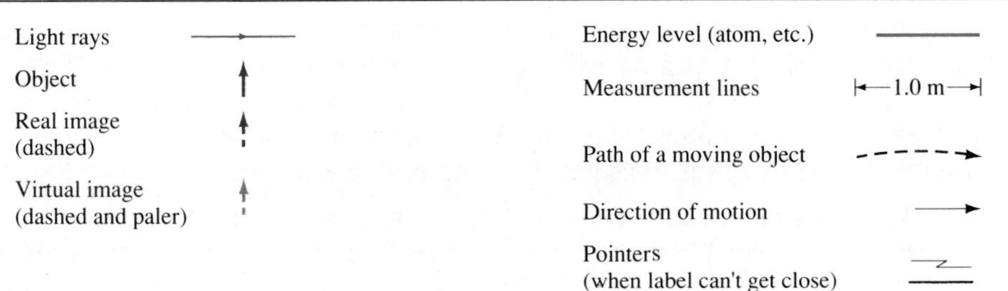

Objects

Real physical objects that recur frequently are shown in the same consistent color for ease of recognition in a complicated diagram. Some examples: magnets (■), iron core of motor or generator (■), electric conductors and plates (■), insulators and dielectrics (■), lenses (■).

Analysis versus real objects

Generally, real objects are shown in natural colors, though often in pale colors when vectors are superimposed so that the latter are prominent.

Additional Photo Credits

CO–1, D. Giancoli; **1–1,** Scala/Art Resource; **1–2, 1–3, 1–4,** Franca Principe-Science Museum, Florence; **1–5(a),** D. Giancoli; **1–5(b),** A. M. Ouzoonian/Weidlinger Associates; **cartoon,** Copyright © 1990 Sidney Harris; **1–6,** D. Giancoli; **1–7,** John Schultz-PAR/NYC, Inc.; **CO–2,** Jon Brenneis/FPG International; **2–9,** Focus on Sports; **2–12,** Scala/Art Resource; **2–13,** Copyright © 1990 Estate of Harold Edgerton. Photo courtesy Palm Press, Inc.; **CO–3,** Berenice Abbott/Commerce Graphics Ltd., Inc.; **3–16,** Copyright © 1989 Globus Brothers/The Stock Market; **3–18,** Educational Development Center, Newton, MA; **3–22(a),** Phil Degginger; **3–22(b),** Stock, Boston; **3–22(c),** Jim Anderson/Woodfin Camp & Associates; **CO–4,** Rick Rickman/Black Star; **4–1,** Jon de Visser/Black Star; **4–2,** Copyright © 1990 Estate of Harold Edgerton. Photo courtesy Palm Press, Inc.; **4–5,** The Granger Collection; **4–6,** Gerard Champlong/The Image Bank; **4–7,** Dr. David Jones/Photo Researchers, Inc.; **4–10,** NASA; **4–13,** Lars Ternblad/The Image Bank; **4–27(a),** Loubat/Vandystadt/Photo Researchers, Inc.; **4–27(b),** Laima Druskis; **4–27(c),** Clyde H. Smith/Peter Arnold, Inc.; **CO–5,** NASA; **5–6(c),** Jay Brousseau/The Image Bank; **5–8,** Guido Alberto Rossi/The Image Bank; **5–16, 5–20,** NASA; **5–21(a),** David Woods/The Stock Market; **5–21(b),** Copyright © 1980 Johan Elbers; **5–21(c),** Morton Beebe/The Image Bank; **5–22,** NASA; **5–23(a)(b),** Copyright © 1990 Barbara Schultz-PAR/NYC, Inc.; **5–27,** D. Giancoli; **5–30,** Daniel L. Feicht/Cedar Point; **CO–6,** Vernon Sigl/Shostal-Superstock; **6–10,** Ric Ferro/Black Star; **6–13,** Copyright © 1990 Estate of Harold Edgerton. Photo courtesy of Palm Press, Inc.; **6–14,** David Madison 1983; **6–17,** Copyright © M. C. Escher Heirs, Cordon Art, Baarn, Holland; **CO–7,** Kim Taylor/Bruce Coleman, Inc.; **7–4,** Copyright © 1990 Estate of Harold Edgerton. Photo courtesy of Palm Press, Inc.; **7–10,** D. J. Johnson; **7–12,** Brookhaven National Laboratory; **7–15,** Copyright © 1962 Berenice Abbott/Photo Researchers, Inc.; **7–17(a),** Wally McNamee/Woodfin Camp & Associates; **CO–8,** G + M Kohler Copyright © 1989/FPG International; **8–3,** Martin Rogers/Stock, Boston; **8–5,** Francesco Ruggeri/The Image Bank; **8–29(c),** NASA; **8–30,** Tim Davis/Photo Researchers, Inc.; **8–31,** Stuart L. Craig/Bruce Coleman, Inc.; **8–33,** Rob Brown/Focus on Sports; **8–45(b),** D. Giancoli; **CO–9,** D. Giancoli; **9–2,** AP/Wide World Photos; **9–25, 9–31, 9–34,** D. Giancoli; **9–36,** National Gallery of Art; **9–37, 9–38,** D. Giancoli; **CO–10,** John S. Shelton; **10–6(a), 10–7,** The Bettmann Archive; **10–10,** C. Falco/Photo Researchers, Inc.; **10–18,** Guy Motil/Westlight; **10–24,** Hans Pfletschinger/Peter Arnold; **10–28,** William Amos/Bruce Coleman, Inc.; **10–30,** Sinclair Stammers-Science Photo Library/Photo Researchers, Inc.; **CO–11,** Comstock, Inc./Bob Grant; **11–9,** D. Giancoli; **11–13(a)-(c),** University of Washington Special Collections; **11–14(a)-(d), 11–27, 11–30(a),** D. Giancoli; **11–32,** John S. Shelton; **CO–12,** Beatriz Schiller; **12–8,** Gregg Mancuso/Stock, Boston; **12–16,** David R. Frazier/Photo Researchers, Inc.; **12–18,** D. Giancoli; **12–21,** A. Macovski, *Medical Imaging Systems,* copyright © 1983 by Prentice-Hall, Inc., Englewood Cliffs, NJ; **12–23,** CNRI-Science Photo Library/Photo Researchers, Inc.; **CO–13,** Comstock, Inc./Tom Grill; **13–2,** Bob Daemmrich/The Image Works; **13–3(a)-(c),** Franca Principe-Science Museum, Florence; **13–7,** Jeff Persons/Stock, Boston; **13–17,** Dick Luria/Photo Researchers, Inc.; **CO–14,** C. Segthters/Photo Researchers, Inc.; **14–1,** Mary Evans Picture Library; **14–11(a)-(b),** Science Photo Library/Photo Researchers, Inc.; **CO–15,** Colin Garratt; **15–14,** Shelly Grossman/Woodfin Camp & Associates; **15–15,** U.S. Dept. of Interior; **15–17(a),** Erich Hartmann/Magnum Photos; **15–17(b),** Sandia National Laboratories; **15–18(b),** Arthur Tress/Woodfin Camp & Associates; **15–20,** Sandia National Laboratories; **CO–16,** Barry L. Runk/Grant Heilman; **16–3,** C. W. Peale-Historical Society of Pennsylvania; **16–11,** AIP Niels Bohr Library-E. Scott Barr Collection; **16–17,** AIP Niels Bohr Library-Zeleny Collection; **CO–17,** Kent Wood/Peter Arnold, Inc.; **CO–18,** Runk/Schoenberger-Grant Heilman; **18–1,** Franca Principe-Science Museum, Florence; **18–2,** Mary Evans Picture Library; **18–3,** Battelle Laboratory; **18–7,** Dan McCoy/Rainbow; **18–8,** Sekai Bunka Photo; **18–9,** Michael Holford; **18–22,** Coco McCoy/Rainbow; **CO–19,** John Halaska/Photo Researchers, Inc.; **19–16, 19–19,** Simpson Electronics Co.; **CO–20,** Manfred Kage/Peter Arnold, Inc.; **20–5,** Silvio Calabi-f/STOP; **20–12,** Comstock, Inc./Russ Kinne; **20–29(b),** AIP Niels Bohr Library; **20–33,** Jon Feingersh/The Stock Market; **CO–21,** Gary Withey/Bruce Coleman, Inc.; **21–17,** Comstock, Inc./Robert Houser; **CO–22,** Copyright © 1990 Bruce Iverson; **22–1,** AIP Niels Bohr Library; **CO–23,** Michael Edrington/The Image Works; **23–2,** Elmer Taylor/AIP Niels Bohr Library; **23–17(a)-(b),** D. Giancoli; **23–23,** Hank Morgan/Photo Researchers, Inc.; **23–34,** D. Giancoli; **CO–24,** AIP Niels Bohr

Library; **24–10,** David Park-Science Photo Library/Photo Researchers, Inc.; **24–15(a),** P. M. Rinard, from *Am. J. Phys.* 44, 1976, p. 70; **24–15(b),** F. W. Sears, *Optics,* copyright © 1949, Addison-Wesley Publishing Co., Inc., Reading, MA Figure 9–8 reprinted with permission; **24–15(c),** M. J. Cagnet, M. Francon, and J. C. Thrierr, *Atlas of Optical Phenomena,* Springer-Verlag, Berlin, 1962; **24–24(a),** Paul Silverman/Fundamental Photographs; **24–24(b),** Martin Rogers/Stock, Boston; **24–24(c),** Diane Schinmo/Fundamental Photographs; **24–26, 24–27(b)-(c),** Bausch & Lomb; **24–36(a)-(b),** D. Giancoli; **24–39,** Sepp Seitz/Woodfin Camp & Associates; **CO–25,** Dorothy Littell/Stock, Boston; **25–3,** Nikon Inc./Martin M. Rotker; **25–12(a)-(b),** Franca Principe-Science Museum, Florence; **25–14(c),** Hale Observatories; **25–25,** Cornell University; **25–30,** Burndy Library, Norwalk, CT; **25–33(b),** National Bureau of Standards; **25–39(a)-(b),** Martin M. Rotker; **CO–26,** Hebrew University of Jerusalem, Israel; **26–1(a),** AIP Niels Bohr Library; **26–1(b),** Ullstein Bilderdienst; **CO–27,** Lawrence Migdale/Stock, Boston; **27–3, 27–8, 27–9,** AIP Niels Bohr Library; **27–10,** Education Development Center, Newton, MA; **27–13(a),** CNRI/Science Photo Library/Photo Researchers, Inc.; **27–13(b),** SECCHI-LECAQUE/ROUSSEL-UCLAF/CNRI/Science Photo Library/Photo Researchers, Inc.; **27–15(a)-(c),** Lawrence Berkeley Laboratory; **27–17,** AIP Niels Bohr Library; **27–20(b)-(c),** Richard Megna/Fundamental Photographs; **CO–28,** Richard Megna/Fundamental Photographs; **28–1,** Niels Bohr Institute/AIP Niels Bohr Library; **28–2,** Segre Collection-AIP Niels Bohr Library; **28–12,** Paul Silverman/Fundamental Photographs; **28–17,** Batelle; **CO–29,** Phillip A. Harrington-Western Electric/Peter Arnold, Inc.; **CO–30,** Michael Collier/Stock, Boston; **30–3,** Burndy Library/Norwalk, CT; **30–5,** AIP Niels Bohr Library; **30–14(a),** Fermilab; **30–14(b),** Brookhaven National Laboratory; **CO–31,** Larry Mulvehill/Science Source-Photo Researchers, Inc.; **31–4,** Argonne National Laboratory; **31–7,** Los Alamos Scientific Laboratory; **31–9(b),** Princeton Plasma Physics Laboratory; **31–10(a),** LLNL/Science Source-Photo Researchers, Inc.; **31–10(b),** Lawrence Livermore National Laboratory; **31–11,** R. Turgeon; **31–12,** J. Van't Hof.; **31–17(b),** SIU-Peter Arnold, Inc.; **CO–32,** Lawrence Berkeley Laboratory/Science Photo Researchers, Inc.; **32–1(b), 32–2,** Brookhaven National Laboratory; **32–4(a)-(b),** Fermilab; **32–6(a),** George Retseck from *Scientific American,* July, 1990. Copyright © 1990 Scientific American, Inc.; **32–6(b),** Universities Research Association; **32–10(a),** David Parker/Science Photo Library-Photo Researchers, Inc.; **32–10(b),** CERN; **32–11(a), 32–13(a)-(b),** Brookhaven National Laboratory; **CO–33,** National Optical Astronomy Observatories; **33–1, 33–3,** U.S. Naval Observatory-Hansen Planetarium; **33–4,** National Optical Astronomy Observatories; **33–5(a),** Copyright © 1979 R. J. Dufour, Rice University-Hansen Planetarium; **33–5(b),** U.S. Naval Observatory/Science Source-Photo Researchers, Inc.; **33–5(c), 33–9, 33–10,** National Optical Astronomy Observatories.

C H A P T E R 1

Introduction

How many aspects of physics can you find in this photograph?

Physics is the most basic of the sciences. It deals with the behavior and structure of matter. The field of physics is usually divided into the areas of motion, fluids, heat, sound, light, electricity and magnetism, and the modern topics of relativity, atomic structure, condensed-matter physics, nuclear physics, elementary particles, and astrophysics. We will cover all these topics in this book, beginning with motion (or mechanics, as it is sometimes called). But before we begin on the physics itself, let us take a brief look at how this activity called "science," including physics, is actually practiced.

1–1 • Science and Creativity

The principal aim of all sciences, including physics, is generally considered to be the ordering of the complex appearances detected by our senses—that is, an ordering of what we often refer to as the "world around us." Many people think of science as a mechanical process of collecting facts and devising theories. This is not the case. Science is a creative activity that in many respects resembles other creative activities of the human mind.

Let's take some examples to see why this is true. One important aspect of science is *observation* of events. But observation requires imagination,

FIGURE 1–1 Aristotle is the central figure on the right at the top of the stairs (the figure next to him is Plato) in this famous Renaissance portrayal of *The School of Athens*, painted by Raphael around 1510. Also in this painting, considered one of the great masterpieces in art, are Euclid (drawing a circle at the lower right), Ptolemy (extreme right with globe), Pythagoras, Sophocles, and Diogenes.

for scientists can never include everything in a description of what they observe. Hence, scientists must make judgments about what is relevant in their observations. As an example, let us consider how two great minds, Aristotle (384–322 B.C.; Fig. 1–1) and Galileo (1564–1642; Fig. 2–12), interpreted motion along a horizontal surface. Aristotle noted that objects given an initial push along the ground (or on a table top) always slow down and stop. Consequently Aristotle believed that the natural state of a body is at rest. Galileo, in his reexamination of horizontal motion in the early 1600s, chose rather to study the idealized case of motion free from resistance. Galileo imagined that if friction could be eliminated, an object given an initial push along a horizontal surface would continue to move indefinitely without stopping. He concluded that for an object to be in motion was just as natural as for it to be at rest. By inventing a new approach, Galileo founded our modern view of motion (more details in Chapters 2, 3, and 4), and he did so with a leap of the imagination.

Theories are creations of humans

Theories are never derived directly from observations—they are *created* to explain observations. They are inspirations that come from the minds of human beings. For example, the idea that matter is made up of atoms (the atomic theory) was certainly not arrived at because someone observed atoms. Rather, the idea sprang from creative minds. The theory of relativity, the electromagnetic theory of light, and Newton's law of universal gravitation were likewise the result of human imagination.

The great theories of science may be compared, as creative achievements, with great works of art or literature. But how does science differ from these other creative activities? One important difference is that science requires *testing* of its ideas or theories to see if their predictions are borne out by experiment. Indeed, careful experimentation is a crucial part of physics.

Although the testing of theories can be considered to distinguish science from other creative fields, it should not be assumed that a theory is "proved" by testing. First of all, no measuring instrument is perfect, so exact confirmation cannot be possible. Furthermore, it is not possible to test a theory in every single possible circumstance. Hence a theory can never be absolutely verified.† In fact theories themselves are generally not perfect—a theory rarely agrees with experiment exactly, within experimental error, in every single case in which it is tested. Indeed, the history of science tells us that theories come and go—that long-held theories are replaced by new ones. The process of one theory replacing another is an important subject in the philosophy of science today; we can discuss it here only briefly.

Testing (which can never be exhaustive)

A new theory is accepted by scientists in some cases because its predictions are quantitatively in much better agreement with experiment than those of the older theory. But in many cases, a new theory is accepted only if it explains a greater *range* of phenomena than does the older one. Copernicus's sun-centered theory of the universe, for example (Fig. 1–2), was no more accurate than Ptolemy's earth-centered theory (Fig. 1–3) for predicting the

Theory acceptance

† Some philosophers therefore emphasize that testing of a theory can be used only to *falsify* it—not to confirm it—and/or to put a limit on its range of validity.

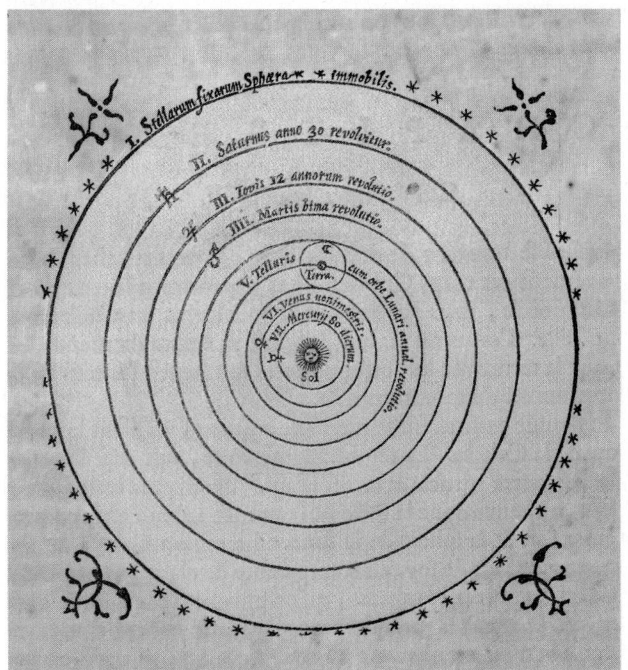

FIGURE 1–2 An early representation of Copernicus's heliocentric view of the universe. (From Jean Blaeu, *Geographie, qui est la Premiere Partie de la Cosmographie Blaviane*, 1667.) (See Chapter 5.)

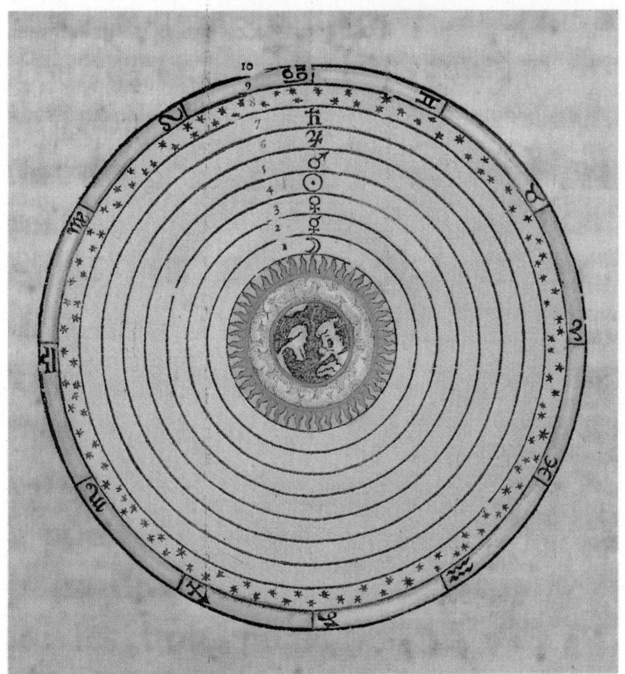

FIGURE 1–3 Ptolemy's geocentric view of the universe. Note at the center the four elements of the ancients: earth, water, air (clouds around the earth), and fire; then the circles, with symbols, for Moon, Mercury, Venus, Sun, Mars, Jupiter, Saturn, the fixed stars, and the signs of the zodiac. (From Jean Blaeu.)

motion of heavenly bodies. But Copernicus's theory had consequences that Ptolemy's did not: for example, it made possible a determination of the order and distance of the planets and predicted the moonlike phases in the appearances of Venus. A simpler (or no more complex) and richer theory, one which unifies and explains a greater variety of phenomena, is more useful and beautiful to a scientist. And this aspect, as well as quantitative agreement, plays a major role in the acceptance of a theory.

An important aspect of any theory is how well it can quantitatively predict phenomena, and from this point of view a new theory may often seem to be only a minor advance over the old one. For example, Einstein's theory of relativity gives predictions that differ very little from the older theories of Galileo and Newton in nearly all everyday situations. Its predictions are better mainly in the extreme case of very high speeds close to the speed of light. In this respect, the theory of relativity might be considered as mere "fine-tuning" of the older theory. But quantitative prediction is not the only important outcome of a theory. Our view of the world is affected as well. As a result of Einstein's theory of relativity, for example, our concepts of space and time have been completely altered, and we have come to see mass and energy as a single entity (via the famous equation $E = mc^2$). Indeed, our view of the world underwent a major change when relativity theory came to be accepted.

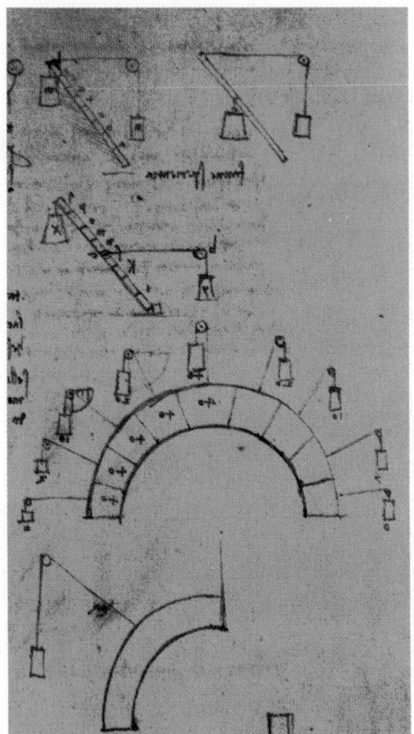

FIGURE 1–4 Studies on the forces in structures by Leonardo da Vinci (1452–1519). (From the Codex Madrid.)

1–2 • Physics and Its Relation to Other Fields

For a long time science was more or less a united whole known as natural philosophy. Not until the last century or two did the distinctions between physics and chemistry and even the life sciences become prominent. Indeed, the sharp distinctions we now see between the arts and the sciences is itself but a few centuries old. It is no wonder then that the development of physics has both influenced and been influenced by other fields. For example, Leonardo da Vinci's notebooks (Fig. 1–4) contain the first references to the forces present within a structure, a subject we consider as physics today; but Leonardo was interested, at least in part, because of the relevance to architecture and building. Early work in electricity that led to the discovery of the electric battery and electric current was done by an eighteenth-century physiologist, Luigi Galvani (1737–1798). He noticed the twitching of frog's legs in response to an electric spark and later that the muscles twitched when in contact with two dissimilar metals (see Chapter 18). At first this phenomenon was known as "animal electricity," but it shortly became clear that electric current itself could exist in the absence of an animal. Later, in the 1930s and 1940s, a number of scientists trained as physicists became interested in applying the ideas and techniques of physics to problems in microbiology. Among the most prominent were Max Delbrück (1906–1981) and Erwin Schrödinger (1887–1961). They hoped, among other things, that studying biological organisms might lead to the discovery of some new unsuspected laws of physics. Alas, this hope has not been realized; but their efforts helped give rise to the field we now call molecular biology and have resulted in a dramatic increase in our understanding of the genetics and structure of living beings.

(a)

(b)

FIGURE 1–5 (a) The Pont du Gard, in southern France, built by the Romans in the early first century A.D., still stands. (b) Collapse of the Hartford Civic Center in 1978 just two years after it was built.

One does not have to be a research scientist in, say, medicine or molecular biology to be able to use physics in his or her work. A zoologist, for example, may find it useful to know how prairie dogs and other animals can live underground without suffocating. A botanist may wonder how water can get to the tops of trees. A physical therapist will do a more effective job if aware of the principles of lever action, center of gravity, and the action of forces within the human body. A knowledge of the operating principles of optical and electronic equipment is helpful in a variety of fields. Life scientists and architects alike will be interested in the nature of heat loss and gain in human beings and the resulting comfort or discomfort. Architects themselves may never have to calculate, for example, the dimensions of the pipes in a heating system or the forces involved in a given structure to determine if it will remain standing (Fig. 1–5). But architects must know the principles behind these systems in order to make realistic designs and to communicate effectively with engineering consultants and other specialists. From the aesthetic or psychological point of view, too, architects must be aware of the forces involved in a structure—for instability, even if only apparent, can be discomforting to those who must live or work in the structure. Indeed, many of the features we admire in the architecture of the past three millennia were introduced not for their decorative effect but rather for practical purposes. For example, the cornices and pediments of windows protected the facades of buildings from the erosive effects of water. The base and capital of a column acted to diffuse the load from the column onto the masonry above and below. The development of the arch as a means to span a space and at the same time support a heavy load will be discussed in Chapter 9. There we will see that the pointed, or Gothic, arch was not originally a decorative device but a technological development of considerable importance.

The list of ways in which physics relates to other fields is extensive. In the chapters that follow we will discuss many such applications as we carry out our principal aim of explaining basic physics.

1–3 • Models, Theories, and Laws

When scientists are trying to understand a particular set of phenomena, they often make use of a **model**. A model, in the scientist's sense, is a kind of analogy or mental image of the phenomena in terms of something we are familiar with. One example is the wave model of light. We cannot see waves of light as we can water waves; but it is valuable to think of light as if it were made up of waves because experiments on light indicate that it behaves in many respects as water waves do.

The purpose of a model is to give us a mental or visual picture—something to hold onto—when we cannot see what actually is happening. Models often give us a deeper understanding: the analogy to a known system (for instance, water waves in the above example) can suggest new experiments to perform and can provide ideas about what other related phenomena might occur.

No model is ever perfect, and scientists are constantly trying to refine their models or to think up new ones when old ones do not seem adequate. The atomic model of matter has gone through many refinements. At one time or another, atoms were imagined to be tiny spheres with hooks on them (to explain chemical bonding), or as tiny billiard balls continually bouncing against each other. More recently, the "planetary model" of the atom visualized the atom as a nucleus with electrons revolving around it, just as the planets revolve about the sun. Yet this model too is oversimplified and fails crucial tests (see Chapters 27 and 28).

Theories are not *really* developed this way.

You may wonder what the difference is between a theory and a model. Sometimes the words are used interchangeably. Usually, however, a model is fairly simple and provides a structural similarity to the phenomena being studied, whereas a **theory** is broader, more detailed, and attempts to solve a set of problems, often with mathematical precision. Often, as a model is developed and modified and corresponds more closely to experiment over a wide range of phenomena, it may come to be referred to as a theory. The atomic theory is an example, as is the wave theory of light.

Models can be very helpful, and they often lead to important theories. But it is important not to confuse a model, or a theory, with the real system or the phenomena themselves.

Scientists give the title **law** to certain concise but general statements about how nature behaves (that momentum is conserved, for example). Sometimes the statement takes the form of a relationship or equation between quantities (such as Einstein's famous equation, $E = mc^2$).

To be called a law, a statement must be found experimentally valid over a wide range of observed phenomena. In a sense, the law brings a unity to many observations. For less general statements, the term **principle** is often used (such as Archimedes' principle). Where to draw the line between laws and principles is, of course, arbitrary, and there is not always complete consistency.

Scientific laws are different from political laws in that the latter are *prescriptive*: they tell us how we must behave. Scientific laws are *descriptive*: they do not say how nature *should* behave, but rather describe how nature *does* behave. As with theories, laws cannot be tested in the infinite variety of cases possible. So we cannot be sure that any law is absolutely true. We use the term "law" when its validity has been tested over a wide range of cases, and when any limitations and the range of validity are clearly understood. Even then, as new information comes in, certain laws may have to be modified or discarded.

Scientists normally do their work as if the accepted laws and theories were true. But they are obliged to keep an open mind in case new information should alter the validity of any given law or theory.

1–4 • Measurement and Uncertainty

In the quest to understand the world around us, scientists seek to find relationships among physical quantities.

We may ask, for example, in what way does the magnitude of a force on an object affect the speed or acceleration of the body? Or by how much does the pressure of gas in a closed container (such as a tire) change if the temperature is raised or lowered? Scientists normally try to express such relationships quantitatively, in terms of equations whose symbols represent the quantities involved. To determine (or confirm) the form of a relationship, careful experimental measurements are required, although creative thinking also plays a role.

Today accurate measurements are an important part of physics. But no measurement is absolutely precise; there is an uncertainty associated with every measurement. Uncertainty arises from different sources. Among the most important, other than blunders, are the limited accuracy of every measuring instrument and the inability to read an instrument beyond some fraction of the smallest division shown. For example, if you were to use a centimeter ruler to measure the width of a board, the result could be claimed to be accurate to about 0.1 cm, the smallest division on the ruler (although half of this value might be a valid claim as well). The reason for this is that it is difficult for the observer to interpolate between the smallest divisions, and the ruler itself has probably not been manufactured to an accuracy much better than this.

Every measurement has an uncertainty

When giving the result of a measurement, it is good practice to state the precision, or **estimated uncertainty**, in the measurement. For example, the width of a board would be written as 23.2 ± 0.1 cm. The ± 0.1 cm ("plus or minus 0.1 cm") represents the estimated uncertainty in the measurement, so that the actual width most likely lies between 23.1 and 23.3 cm. The *percent uncertainty* is simply the ratio of the uncertainty to the measured value, multiplied by 100. For example, if the measurement is 23.2 and the uncertainty about 0.1 cm, the percent uncertainty is

Stating the uncertainty

$$\frac{0.1}{23.2} \times 100 = 0.4\%.$$

Often the uncertainty in a measured value is not specified explicitly.

In such cases, it is generally accepted that the uncertainty is approximately one or two units in the last digit specified. Although this is not as precise as actually specifying the uncertainty, it is often adequate. For example, if a length is given as 23.2 cm, the uncertainty is assumed to be about 0.1 cm (or perhaps 0.2 cm). It is important in this case that you do not write 23.20 cm, for this implies an uncertainty of 0.01 cm; it assumes that the length is probably between 23.19 cm and 23.21 cm when actually you believe it is between 23.1 and 23.3 cm.

Which digits are significant?

The number of reliably known digits in a number is called the number of **significant figures**. Thus there are four significant figures in the number 23.21 and two in the number 0.062 cm (the zeros in the latter are merely place holders). The number of significant figures may not always be clear. Take, for example, the number 80. Are there one or two significant figures? If we say it is *about* 80 km between two cities, there is only one significant figure (the 8) since the zero is merely a place holder. If it is *exactly* 80 km between cities (and not 83 or 77) the 80 has two significant figures.

When making measurements, or when doing calculations, one should not keep more digits in the final answer than the number of significant figures in the least accurate factor in the calculation. For example, to calculate the area of a rectangle 11.3 cm by 6.8 cm, the result of multiplication would be 76.84 cm^2, but this answer is clearly not accurate to 0.01 cm^2, since (using the outer limits of the assumed uncertainty for each measurement) the result could be between $11.2 \times 6.7 = 75.04\ cm^2$ and $11.4 \times 6.9 = 78.66\ cm^2$. At best, we can quote the answer as 77 cm^2, which implies an uncertainty of about 1 or 2 cm^2. The other two digits (in the number 76.84 cm^2) must be dropped since they are not significant. As a general rule, *the final result of a multiplication or division should have only as many digits as the number with the least number of significant figures used in the calculation.* In our example, 6.8 cm has the least number of significant figures, namely two; thus the result 76.84 cm^2 must be rounded off to 77 cm^2. Keep in mind when you use a calculator that all the digits it produces may not be significant. Digits should not be quoted (or written down) unless they are truly significant figures. (However, it is usually good practice to keep an extra significant figure throughout a calculation, if possible, even though it isn't quoted in the final result.)

Problem Solving: Keep only the proper number of significant figures

It is common in physics to write numbers in "powers of ten," or "exponential" notation—for instance, 36,900 as 3.69×10^4, or 0.0021 as 2.1×10^{-3}. (For more details, see Appendices A–2 and A–3.) One advantage of exponential notation is that it allows the number of significant figures to be clearly expressed. For example, it is not clear whether 36,900 has three, four, or five significant figures. With scientific notation this ambiguity can be avoided: if the number is known to an accuracy of three significant figures, we write 3.69×10^4, but if it is known to four, we write 3.690×10^4.

1–5 • Units, Standards, and the SI System

The measurement of any quantity is made relative to a particular standard or unit, and this unit must be specified along with the numerical value of the quantity. For example, we can measure length in units such as inches, feet, or miles, or in the metric system in centimeters, meters, or kilometers.

To specify that the length of a particular object is 18.6 is meaningless. The unit *must* be given; for clearly, 18.6 meters is very different from 18.6 inches or 18.6 millimeters.

Until about 200 years ago, the units of measurement were not standardized, and that made scientific communication difficult. Different people used different units—cubits, leagues, hands, and even the length of the foot varied from place to place. The first real international standard was the standard **meter** (abbreviated m) established by the French Academy of Sciences in the 1790s. In a spirit of rationality, the standard meter was originally chosen to be one ten-millionth of the distance from the earth's equator to either pole,† and a platinum rod to represent this length was made. In 1889, the meter was defined more precisely as the distance between two finely engraved marks on a particular bar of platinum-iridium alloy. Thirty of these rods were manufactured; one was kept at the International Bureau of Weights and Measures near Paris as the international standard, and the others were sent to laboratories around the world. By the end of the nineteenth century it became possible, through the work of the American physicist A. A. Michelson (see Section 24–9), to define the meter in terms of the wavelength of light, and in 1960 the meter was redefined as 1,650,763.73 wavelengths of a particular orange light emitted by the gas krypton 86. In 1983 the meter was again redefined, this time in terms of the speed of light (whose best measured value in terms of the older definition of the meter was 299,792,458 m/s, with an uncertainty of 1 m/s). The new definition reads: "The meter is the length of path traveled by light in vacuum during a time interval of 1/299,792,458 of a second."‡

Standard of length (meter)

British units of length (inch, foot, mile) are now defined in terms of the meter. The inch (in.) is defined as precisely 2.54 centimeters (cm; 1 cm = 0.01 m). Other conversion factors are given in the table on the inside of the front cover of this book.

The standard unit of *time* is the **second** (s). For many years, the second was defined as 1/86,400 of a mean solar day. The standard second is now defined more precisely in terms of the frequency of radiation emitted by cesium atoms when they pass between two particular states. Specifically, one second is defined as the time required for 9,192,631,770 periods of this radiation. There are, of course, 60 s in one minute (min) and 60 minutes in one hour (h).

Standard of time (second)

The definitions of other standard units for other quantities will be given as we encounter them in later chapters.

In the metric system, the larger and smaller units are defined in multiples of 10 from the standard unit, and this makes calculation particularly easy. Thus one centimeter is $\frac{1}{100}$ m, one kilometer (km) is 1000 m, and so on. The prefixes "centi-," "kilo-," and others are listed in Table 1–1 and can be applied not only to units of length, but to units of volume, mass, or any other metric unit. For example, a centiliter (cL) is $\frac{1}{100}$ liter (L) and a kilogram (kg) is 1000 grams (g).

Table 1–1
Metric (SI) Prefixes (Multipliers)

Prefix	Abbreviation	Value
tera	T	10^{12}
giga	G	10^{9}
mega	M	10^{6}
kilo	k	10^{3}
hecto	h	10^{2}
deka	da	10^{1}
deci	d	10^{-1}
centi	c	10^{-2}
milli	m	10^{-3}
micro†	μ	10^{-6}
nano	n	10^{-9}
pico	p	10^{-12}
femto	f	10^{-15}

† μ is the Greek letter "mu"

† Modern measurements of the earth's circumference reveal that the intended length is off by about one-fiftieth of 1 percent.

‡ The new definition of the meter has the effect of giving the speed of light the exact value of 299,792,458 m/s.

The conversion factors between the various units of the British system (for example, 12 inches in a foot) make it unwieldy for calculation. This is one reason why scientists, and nearly all countries of the world, have adopted the metric system.

Problem Solving: Always use a consistent set of units

When dealing with the laws and equations of physics it is very important to use a consistent set of units. To take a very simple example, suppose you want to know how far you can go in your car in 40 min at a speed of 90 km/h. As we shall see in Chapter 2, the distance, x, can be written as the product of the speed, v, and the time, t: $x = vt$. But if you multiply 90 km/h by 40 min, you will get a ridiculous answer. Since the speed, v, is given in terms of hours, the time, t, must also be in hours, and in this case it is $\frac{2}{3}$ h; therefore, $x = (90\text{ km/h})(\frac{2}{3}\text{ h}) = 60$ km. Note how the hour units cancel in this equation; the equals sign in an equation applies not only to the numerical values but also to the units. When dealing with more complicated equations, the necessity of using a consistent set of units is even more crucial.

SI units

Several systems of units have been in use over the years. Today the most important by far is the **Système International** (French for International System), which is abbreviated SI. In SI units, the standard of length is the meter, the standard for time is the second, and the standard for mass is the kilogram. This system used to be called the MKS (meter-kilogram-second) system.

A second metric system is the "cgs system," in which the centimeter, gram, and second are the standard units of length, mass, and time, as abbreviated in the title. The British engineering system takes as its standards the foot for length, the pound for force, and the second for time.

SI units are the principal ones used today in scientific work. We will therefore use SI units almost exclusively in this book, although we will give the cgs and British units for various quantities when introduced, as well as conversion factors.

Table 1–2
SI Base Quantities and Units

Quantity	Unit	Unit Abbreviation
Length	meter	m
Time	second	s
Mass	kilogram	kg
Electric current	ampere	A
Temperature	kelvin	K
Amount of substance	mole	mol
Luminous intensity	candela	cd

Physical quantities can be divided into two categories: *base quantities* and *derived quantities*. (The corresponding units for these quantities are called *base units* and *derived units*.) A **base quantity** must be defined in terms of a standard (as we just discussed for the meter and the second). Scientists, in the interest of simplicity, want the smallest number of base quantities possible consistent with a full description of the physical world. This number turns out to be seven, and for SI are given in Table 1–2. All other quantities can be defined in terms of these seven base quantities,[†] and hence are referred to as **derived quantities**. The selection of which quantities are to be considered base and which derived is somewhat arbitrary. For example, in the British system, force is considered base, whereas mass is derived—just the opposite of SI. An example of a derived quantity is speed, which is defined as distance traveled divided by the time it takes to travel that distance (Chapter 2). The base quantities are not definable in terms of other quantities, which is why they are called base. To define any quantity, whether base or derived, we can specify a rule or procedure, and this is called an **operational definition**.

[†] The only exceptions are for angle (radians—see Chapter 8) and solid angle (steradian). No general agreement has been reached as to whether these are base or derived quantities.

1–6 • Order of Magnitude: Rapid Estimating

We are sometimes interested only in an approximate value for a quantity. This might be because an accurate calculation would take more time than it is worth or would require additional data that are not available. In other cases, we may want to make a rough estimate in order to check an accurate calculation made on a calculator to make sure that no blunders were made when entering the numbers. Also, the correct power of 10 may be lost on a calculator and a rough estimate can be used to obtain it.

Problem Solving: How to make a rough estimate

In general, a rough estimate is made by rounding off all numbers to one significant figure and its power of 10, and after the calculation is made, again only one significant figure is kept. Such an estimate is called an **order-of-magnitude** estimate and can be assumed to be accurate within a factor of 10, and usually better. In fact, the phrase "order of magnitude" is sometimes used to refer simply to the power of 10.

As an example, suppose that a person wants to find out how much water there is in a particular lake (Fig. 1–6), which is roughly circular, about 1 km across, and is estimated to have an average depth of about 10 m. To estimate the volume, we multiply the average depth of the lake times its

FIGURE 1–6 How much water is in this lake? (Carried one step further, we could estimate "How much does a lake like this weigh?" See Sections 4–7 and 10–1.)

surface area as if it were a cylinder. (The volume V of a cylinder is the product of its height times the area of its base: $V = \pi r^2 h$, where r is the radius of the circular base.) We approximate the surface to be a circle, so the area is πr^2, which is approximately $3 \times (5 \times 10^2\ \text{m})^2 \approx 8 \times 10^5\ \text{m}^2$, where the radius r is 500 m and π was rounded off to 3 ($\approx$ means "approximately equal to"). Then the volume is about $(8 \times 10^5\ \text{m}^2) \times (10\ \text{m}) = 8 \times 10^6\ \text{m}^3$, which is on the order of $10^7\ \text{m}^3$. Because of all the estimates that went into this calculation, the order-of-magnitude estimate ($10^7\ \text{m}^3$) is probably better to quote than the $8 \times 10^6\ \text{m}^3$ figure.

Problem Solving: Estimating how many piano tuners there can be

Another example, this one made famous by Enrico Fermi to his students, is to estimate the number of piano tuners in a city, say, Chicago or San Francisco. To get a rough order-of-magnitude estimate of the number of piano tuners in San Francisco, a city of about 700,000 inhabitants, we can proceed by estimating the number of functioning pianos, how often each piano is tuned, and how many pianos each tuner can tune. To estimate the number of pianos in San Francisco, we note that certainly not everyone has a piano. A guess of 1 family in 3 having a piano would correspond to 1 piano per 12 persons, assuming an average family of 4 persons. As an order of magnitude, let's say 1 piano per 10 people. This is certainly more reasonable than 1 per 100 people, or 1 per every person, so let's proceed with the estimate that 1 person in 10 has a piano, or about 70,000 pianos in San Francisco. Now a piano tuner needs an hour or two to tune a piano. So let's estimate that a tuner can tune 4 or 5 pianos a day. A piano ought to be tuned every 6 months or a year—let's say once each year. A piano tuner tuning 4 pianos a day, 5 days a week, 50 weeks a year can tune about 1000 pianos a year. So San Francisco, with its (very) roughly 70,000 pianos needs about 70 piano tuners. This is, of course, only a rough estimate.† It tells us that there must be many more than 10 piano tuners, and surely not as many as 1000. If you were estimating the number of car mechanics, on the other hand, your estimate would be rather different!

1–7 • Mathematics

Physics is sometimes thought of as being a difficult subject. However, it is frequently found that it is the mathematics used that is the source of difficulties rather than the physics itself. The appendices at the end of this book contain a brief summary of simple mathematical techniques, including algebra, geometry, and trigonometry, that will be used in this book. You may find it useful to examine them now to review old topics or learn any new ones. You may also want to reread them later when you need those concepts. Some mathematical techniques, such as vectors as well as trigonometric functions, are treated in the text itself, when we first need them.

† A check of the San Francisco Yellow Pages (done after this calculation) reveals about 50 listings. Each of these listings may employ more than one tuner, but on the other hand, each may also do repairs as well as tuning. In any case, our estimate is reasonable.

SUMMARY[†]

Physics, like other sciences, is a creative endeavor; it is not simply a collection of facts. Important theories are created with the idea of explaining observations. To be accepted, theories are "tested" by comparing their predictions with the results of actual experiments. Note that, in general, a theory cannot be "proved" in an absolute sense.

Physics is related to other fields in important ways. Applications to medicine, biology, earth science, architecture, and technology are of particular interest.

To understand a particular type or range of phenomena, scientists may think up a *model*, which is a kind of picture or analogy that seems to explain the phenomena and thus aids in understanding. A *theory*, often developed from a model, is usually deeper and more complex than a simple model. A scientific *law* is a concise statement, often expressed in the form of an equation, which quantitatively describes a particular range of phenomena over a wide range of cases.

Measurements play a crucial role in physics, but can never be perfectly precise. Thus for any number that results from measurement, it is important to specify the *uncertainty* either by stating it directly using the $\pm$ notation and/or by keeping only the correct number of *significant figures.*

Quantities are specified relative to a particular standard or *unit*, and the unit used should always be specified. The commonly accepted set of units today is the *Système International* (SI) in which the standard units of length, mass, and time are the meter, kilogram, and second, respectively. There are seven independent *base quantities.* All other quantities are called *derived quantities,* since they and their units can be written in terms of the base ones—for example, velocity is distance/time.

Making rough, *order-of-magnitude* estimates is a very useful technique in science as well as in everyday life.

[†] The summaries that appear at the end of each chapter in this book give a brief overview of the main ideas of each chapter. They *cannot* serve to give an understanding of the material, which can be had only by a detailed reading of the chapter.

QUESTIONS

1. It has been said that science is the new religion, complete with high priests and mysteries known only to the select few—the trained scientists. Do you agree? Discuss.
2. Discuss the limitations and strengths of science.
3. Discuss the distinction between science and technology.
4. It is sometimes said that science is responsible for the ills of society. Scientists may respond that their work is of a pure and intellectual nature, and it is technology—the practical application of science—that creates the problems. Discuss.
5. According to the operationalist point of view, a quantity can have meaning in science only if we can describe a set of operations (or procedures) for determining it. What do you see as the advantages and disadvantages of this point of view?
6. It is advantageous that base standards (such as those for length and time) be accessible (easy to compare to), invariable (do not change), indestructible, and reproducible. Discuss why these are advantages and whether any of these criteria can be incompatible with others.
7. What are the merits and drawbacks of using a person's foot as a standard? Discuss in terms of the criteria mentioned in Question 6. Consider both (*a*) a particular person's foot and (*b*) any person's foot.
8. When traveling a highway in the mountains, you may see elevation signs that read "914 m (3000 ft)." Critics of the metric system claim that such numbers show the metric system is more complicated. How would you alter such signs to be more consistent with a switch to the metric system?
9. The SI base units, Table 1–2, contain one quantity that uses a prefix (kilogram). Would there be an advantage to changing this to a nonprefixed unit? How might this be done? Why do you suppose the kilogram is, in fact, used?
10. Suggest a way to measure (*a*) the thickness of a sheet of paper and (*b*) the distance from earth to the sun.
11. Can you set up a complete set of base quantities, as in Table 1–2, that does not include length as one of them?

PROBLEMS

[The problems at the end of each chapter are ranked I, II, or III according to estimated difficulty, with I problems being easiest. The problems are arranged by sections, meaning that the reader should have read up to and including that section, but not only that section—problems often depend on earlier material. Each chapter also has a group of General Problems that are not arranged by Section and not ranked.]

SECTION 1–4

1. (I) How many significant figures do each of the following numbers have: (*a*) 121, (*b*) 81.60, (*c*) 1.63, (*d*) 0.02, (*e*) 0.00265, (*f*) 36, and (*g*) 2700?
2. (I) Write the following numbers in powers of ten notation: (*a*) 56,000, (*b*) 16, (*c*) 0.00066, (*d*) 21.635, and (*e*) 0.2.
3. (I) Write out the following numbers in full: (*a*) 3.6×10^4, (*b*) 1×10^2, (*c*) 5.5×10^{-1}, (*d*) 3.16×10^7, and (*e*) 8.62×10^{-5}.
4. (I) What is the percent uncertainty in the measurement 4.26 ± 0.15 m?
5. (I) What, approximately, is the percent uncertainty for the measurement 5.65?
6. (II) What is the area, and its approximate uncertainty, of a circle of radius 3.4×10^4 cm?
7. (II) What is the percent uncertainty in the volume of a sphere whose radius is $r = 5.15 \pm 0.08$ m?

SECTION 1–5

8. (I) Express the following using the prefixes of Table 1–1: (*a*) 10^6 volts, (*b*) 10^{-6} meters, (*c*) 5×10^2 days, (*d*) 8×10^3 bucks, and (*e*) 3×10^{-9} pieces.
9. (I) Write the following as full (decimal) numbers with standard units: (*a*) 35.6 mm, (*b*) 25 μV, (*c*) 250 mg, (*d*) 500 picoseconds, (*e*) 2.5 femtometers, (*f*) 25 gigavolts.
10. (I) How many kisses is 50 hectokisses? What would you be if you earned a megabuck a year?
11. (I) Determine your height in meters.
12. (I) The moon is 240,000 mi from the earth. How many meters is this? Express (*a*) using powers of ten, and (*b*) using a metric prefix.
13. (I) A typical atom has a diameter of about 1.0×10^{-10} m. What is this in inches?
14. (II) How much longer (percentage) is the 100-m dash than the 100-yd dash?
15. (II) A *light year* is the distance light (speed = 2.998×10^8 m/s) travels in 1 year. (*a*) How many meters are there in 1.00 light year? (*b*) An astronomical unit (AU) is the average distance from the sun to earth, 1.50×10^8 km. How many AU are there in 1.00 light year? (*c*) What is the speed of light in AU/h?

SECTION 1–6

(*Note:* remember that for rough estimates, only round numbers are needed both as input to calculations and as final results.)

16. (II) Estimate how long it would take to row a boat around the world.
17. (II) Estimate the number of times a human heart beats in a lifetime.
18. (II) Make a rough estimate of the volume of your body (in cm^3).
19. (II) Estimate the number of car mechanics (*a*) in San Francisco and (*b*) in your town or city.
20. (II) Estimate the time to drive from Beijing (Peking) to Paris (*a*) today and (*b*) in 1906, when a great car race was run between these two cities.

GENERAL PROBLEMS

21. (*a*) How many seconds are there in 1 year? (*b*) How many nanoseconds are there in 1 year? (*c*) How many years are there in 1 second?
22. If it takes 250 picograms of a certain chemical to poison one drug dealer, how many dealers can you get rid of with one gram of the chemical?
23. Hold a pencil in front of your eye at a position where its tip just blocks out the moon. Make appropriate measurements to estimate the diameter of the moon, given that the earth–moon distance is 3.8×10^5 km.
24. Estimate the number of gallons of gasoline consumed per day by students and employees driving to and from your college.
25. Estimate the number of marbles in the jar shown in Fig. 1–7.

FIGURE 1–7 Problem 25. Estimate the number of marbles in the jar.

C H A P T E R 2

Describing Motion: Kinematics in One Dimension

Bike racers can move with constant velocity (constant speed in a straight line) along a straight road. They may also accelerate–that is, change their velocity.

The motion of objects—baseballs, automobiles, joggers, and even the sun and moon—is an obvious part of everyday life. Motion was undoubtedly the first aspect of the physical world to be thoroughly studied, and this study can be traced back to the ancient civilizations of Asia Minor. Although the ancients acquired significant insight into motion, it was not until comparatively recently, in the sixteenth and seventeenth centuries, that our modern understanding of motion was established. Many contributed to this understanding, but, as we shall soon see, two individuals stand out above the rest: Galileo Galilei (1564–1642) and Isaac Newton (1642–1727).

The study of the motion of objects, and the related concepts of force and energy, forms the field called **mechanics**. Mechanics is customarily divided into two parts: **kinematics**, which is the description of how objects move, and **dynamics**, which deals with why objects move as they do. This chapter and the next deal with kinematics.

We start by discussing objects that move without rotating. Such motion is called **translational motion**. In the present chapter we will mainly be concerned with describing an object that moves along a straight-line path, which we call one-dimensional, or **rectilinear**, motion. In Chapter 3 we will study how to describe motion in two (or three) dimensions.

2–1 • Speed

"Speed" and "velocity" are words we often use interchangeably in everyday language. In physics we make a distinction between them (see Section 2–4). For now, we discuss only speed, which refers to how far an object travels in a given time interval. If a car travels 240 kilometers (km) in 3 hours, we say its average speed was 80 km/h. In general, the **average speed** of an object is defined as *the distance traveled divided by the time it takes to travel this distance*:

Average speed defined

$$\text{average speed} = \frac{\text{distance traveled}}{\text{time elapsed}}.$$

This definition can be written briefly as:

$$\bar{v}_s = \frac{d}{t}, \qquad (2\text{–}1)$$

where d stands for distance, t for the elapsed time, and v_s for the speed. The bar ($^-$) over the v_s is the standard symbol meaning "average."

EXAMPLE 2–1 How far can a cyclist travel in 4.0 h if his average speed is 11.5 km/h?

SOLUTION Since we want to find the distance traveled, we rewrite the equation above as $d = \bar{v}_s t$. Since $\bar{v}_s = 11.5$ km/h and $t = 4.0$ h, then $d = (11.5\ \text{km/h})(4.0\ \text{h}) = 46$ km.

2–2 • Reference Frames and Coordinate Systems

When riding in a train you may observe a bird flying by overhead and remark that it looks as if it is moving at a speed of 20 km/h. But do you mean it is traveling 20 km/h with respect to the train or with respect to the ground?

All measurements are made relative to a frame of reference

Every measurement must be made with respect to a **frame of reference**. For example, while on a train traveling at 80 km/h, you might notice a person walk past you toward the front of the train at a speed of, say, 5 km/h, Fig. 2–1. Of course this is the person's speed with respect to the train as frame of reference. With respect to the ground that person is moving at 85 km/h. It is always important to specify the frame of reference when stating a speed. In everyday life, we usually mean "with respect to the earth" without even thinking about it, but the reference frame should be specified whenever there might be confusion.

The values of other physical quantities also depend on the frame of reference. For example, there is no point in telling you that Yosemite National Park is 300 km away unless I specify 300 km from where. Distances are

FIGURE 2–1 A person walks toward the front of a train at 5 km/h. The train is moving 80 km/h with respect to the ground, so the walking person's speed, relative to the ground, is 85 km/h.

always measured in some frame of reference. Furthermore, when specifying the motion of an object, it is important to specify not only the speed but also the direction of motion. For example, if a friend leaves New York on a jet plane that travels at a speed of 1000 km/h, you would like to know in what direction this person is going—toward Washington, San Francisco, Paris, or wherever. Often we can specify a direction by using the cardinal points, north, east, south, and west, and by "up" and "down." This is not always convenient. So in physics we often draw a set of **coordinate axes**, as shown in Fig. 2–2, to represent a frame of reference. Objects positioned to the right of the origin of coordinates (O) on the x axis are usually said to have an x coordinate with a positive value; points to the left of O then have a negative x coordinate. The position along the y axis is usually considered positive when above and negative when below O, although the reverse convention can be used if convenient. Any point on the plane can be specified by giving its x and y coordinates. In three dimensions, a z axis perpendicular to the x and y axes is also used.

FIGURE 2–2 Standard set of coordinate (Cartesian) axes.

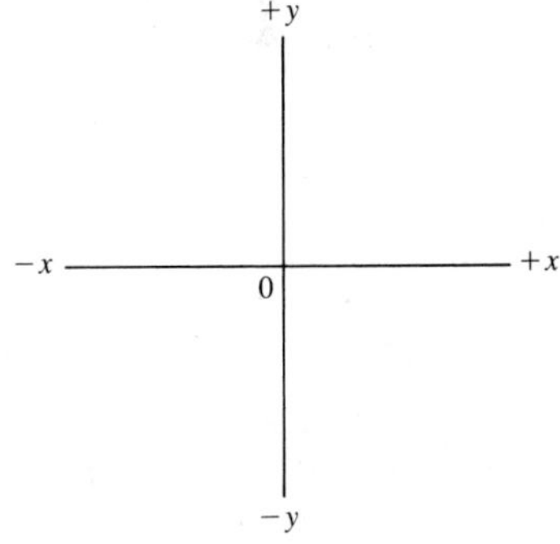

Although most measurements are made in reference frames fixed on the earth, it is important to recognize that reference frames other than the earth are perfectly legitimate. For example, scientific measurements are sometimes made on moving ships, in spacecraft, and on the moon.

2–3 • Changing Units

It is often necessary or useful to change from one set of units to another. For example, it is often preferable to specify the speed of a car in m/s rather than in km/h, as when measuring braking distances, since the times and distances involved are usually seconds and meters rather than hours and kilometers. Also, one sometimes must change units from metric to British, or vice versa.

To determine what a speed of 80 km/h is in m/s, we proceed as follows. There are 1000 m in 1 km, and 3600 s in 1 h. Thus:

Problem Solving: Unit conversions

$$80\ \text{km/h} = \left(\frac{80\ \cancel{\text{km}}}{1\ \cancel{\text{h}}}\right)\left(\frac{1000\ \text{m}}{1\ \cancel{\text{km}}}\right)\left(\frac{1\ \cancel{\text{h}}}{3600\ \text{s}}\right)$$
$$= (80)\left(\frac{1000}{3600}\right)\frac{\text{m}}{\text{s}}$$
$$= (80)\left(\frac{1}{3.6}\right)\frac{\text{m}}{\text{s}} = 22\ \text{m/s}.$$

Notice that in the first line we have multiplied our original number ($80\ \frac{\text{km}}{\text{h}}$) by two conversion factors: $\frac{1000\ \text{m}}{1\ \text{km}} = 1$, and $\frac{1\ \text{h}}{3600\ \text{s}} = 1$. Both of these conversion factors are equal to one and thus do not change the equation. The units of hours and of kilometers cancel out, so we obtain m/s. (Note also that we kept only two significant figures in the final answer, 22 m/s, since the least significant number in the product contained two significant figures; see Section 1–4.)

Problem Solving: Unit conversion is wrong if units do not cancel

When changing units it is often a problem to figure out whether the conversion factor is to go in the numerator or in the denominator. The easiest way to be sure is to check whether the units cancel out as they did above. We would have gotten a wrong result if we had written $(80\,\frac{\text{km}}{\text{h}})(\frac{1000\text{ m}}{1\text{ km}})(\frac{3600\text{ s}}{1\text{ h}})$, and this would have been obvious since the hour units do not cancel.

In the last line in the calculation above we see the factor 1/3.6; this is the conversion factor between m/s and km/h. Note that 1 km/h is the same as 1000 m in 3600 s. That is, $1\,\frac{\text{km}}{\text{h}} = \frac{1000\text{ m}}{3600\text{ s}} = (1/3.6)$ m/s. Thus any speed in km/h can be changed to m/s by dividing by the factor 3.6. Conversely, 1 m/s = 3.6 km/h.

Changes between metric and British units can be done in a similar way, and we leave these calculations as an exercise.

2–4 • Average Velocity and Displacement

Although the terms velocity and speed are used interchangeably in ordinary language, in physics we make a distinction between the two. Most importantly, the term **velocity** is used to signify both the *magnitude* (numerical value) of how fast an object is moving and the *direction* in which it is moving. (Velocity is therefore called a vector, as we shall see shortly.) *Speed*, on the other hand, is a magnitude only. There is a second difference between speed and velocity: namely, the average velocity is defined in terms of "displacement," rather than total distance traveled; that is,

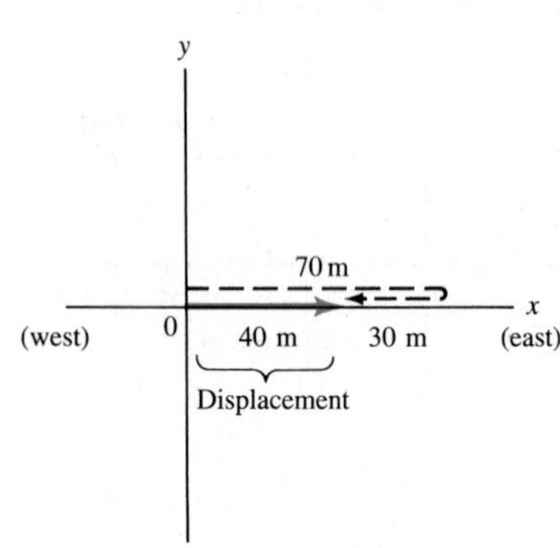

FIGURE 2–3 A person walks 70 m east, then 30 m west. The total distance traveled is 100 m (path is shown in black); but the displacement, shown as a blue arrow, is 40 m to the east.

$$\text{average velocity} = \frac{\text{displacement}}{\text{time elapsed}}.$$

Displacement is defined as *the change in position of an object*. To see the distinction between total distance and displacement, imagine a person walking 70 m to the east and then turning around and walking back (west) a distance of 30 m. The total *distance* traveled is 100 m but the *displacement* is only 40 m since the person is now only 40 m from the starting point (see Fig. 2–3).

Suppose the walk shown in Fig. 2–3 took 80 s; then the average speed was (100 m)/(80 s) = 1.3 m/s. But the magnitude of the average velocity in this case was only (40 m)/(80 s) = 0.50 m/s. This discrepancy between the magnitude of the velocity and the speed occurs in some cases, but only for the *average* values, and we rarely need be concerned with it. We shall see in the next section that the magnitude of the instantaneous velocity and the instantaneous speed are always the same.

FIGURE 2–4 The arrow represents the displacement $x_2 - x_1$.

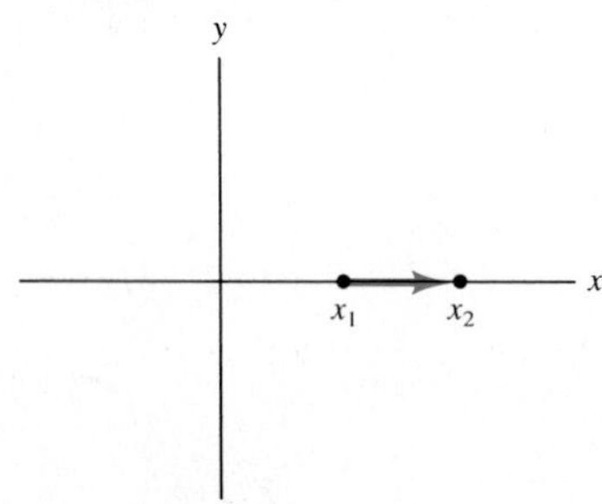

To discuss the motion of an object in general, suppose that at some moment in time, call it t_1, the object is on the x axis at point x_1 in the coordinate system shown in Fig. 2–4. At some later time, t_2, suppose it is at point x_2. The elapsed time is $t_2 - t_1$, and during this time interval the displacement of our object was $x_2 - x_1$ (represented by the arrow pointing to the right in Fig. 2–4). It is convenient to write

$$\Delta x = x_2 - x_1$$

where the symbol Δ (Greek letter delta) means "change in." Then Δx means the displacement or the "change in x." Similarly, we can write the elapsed time (change in time) as $\Delta t = t_2 - t_1$. Then the **average velocity** ($\bar{v}$), defined as *the displacement divided by the elapsed time*, can be written

$$\bar{v} = \frac{x_2 - x_1}{t_2 - t_1} = \frac{\Delta x}{\Delta t}. \qquad (2\text{–}2)$$

Average velocity

Notice that if x_2 is less than x_1, the object is moving to the left, so $\Delta x = x_2 - x_1$ is less than zero. The sign of the displacement, and/or velocity, indicates the direction; the average velocity is positive for an object moving to the right along the x axis and negative when the object moves to the left. Another way to state the definition of velocity is that it is the *rate of change of position in time.*

Problem Solving: + or − sign signifies direction for linear motion

EXAMPLE 2–2 The position of a runner as a function of time is plotted as moving along the x axis of a coordinate system. During a 3.00-s time interval, the runner's position changes from $x_1 = 50.0$ m to $x_2 = 30.5$ m, as shown in Fig. 2–5. What was the runner's average velocity?

SOLUTION Average velocity is the displacement divided by the elapsed time. The displacement is $\Delta x = x_2 - x_1 = 30.5\text{ m} - 50.0\text{ m} = -19.5\text{ m}$. The time interval is $\Delta t = 3.00$ s. Therefore the average velocity is

$$\bar{v} = \frac{\Delta x}{\Delta t} = \frac{-19.5\text{ m}}{3.00\text{ s}} = -6.50\text{ m/s}.$$

The displacement and average velocity are negative, so the runner is moving to the left along the x axis, as indicated by the arrow in Fig. 2–5.

FIGURE 2–5 An object moves from $x_1 = 50.0$ m to $x_2 = 30.5$ m. The displacement is -19.5 m.

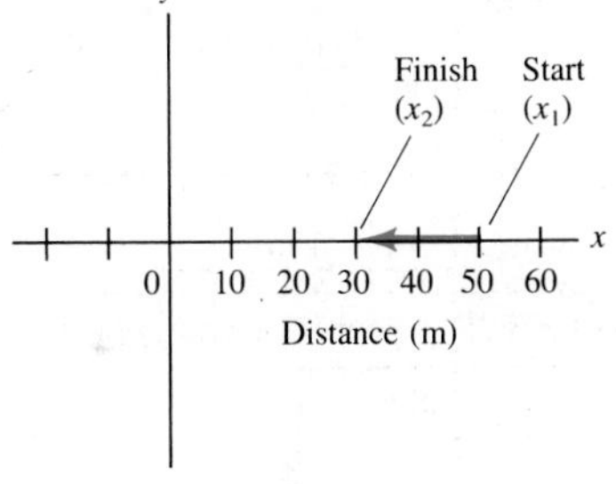

Note that although velocity can be positive or negative, speed is never negative since it represents the rate at which a distance is traveled, not the rate of change of displacement as velocity does.

2–5 • Instantaneous Velocity

If you drive a car along a straight road for 150 km in 2.0 h, the magnitude of your average velocity is 75 km/h. It is unlikely, though, that you were moving at 75 km/h at every instant. To deal with this situation we need the concept of *instantaneous velocity*, which is the velocity at any instant of time. (This is what a speedometer is supposed to indicate.) More precisely, the **instantaneous velocity** at any moment is defined as *the average velocity over an indefinitely short time interval.* That is, Eq. 2–2 is to be evaluated in the limit of Δt being infinitesimally small, approaching zero. We can write the definition of instantaneous velocity, v, for one-dimensional motion as

$$v = \lim_{\Delta t \to 0} \frac{\Delta x}{\Delta t}. \qquad (2\text{–}3)$$

Instantaneous velocity

The notation $\lim_{\Delta t \to 0}$ means the ratio $\Delta x/\Delta t$ is to be evaluated in the limit

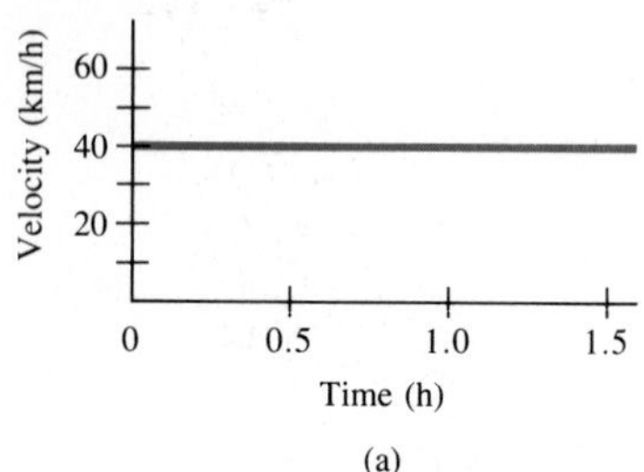

(a)

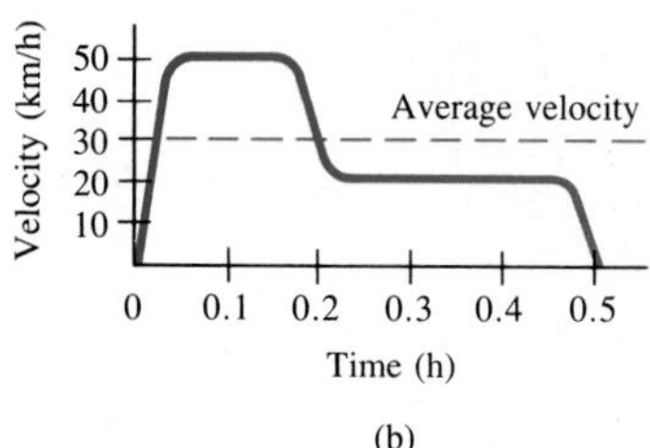

(b)

of Δt approaching zero. But note that we do not simply set $\Delta t = 0$ in this definition, for then Δx would also be zero, and we would have an undefined number. Rather, we are considering the *ratio* $\Delta x/\Delta t$, as a whole. As we let Δt approach zero, Δx approaches zero also. But the ratio $\Delta x/\Delta t$ approaches some definite value, which is the instantaneous velocity at a given instant.†

Note also that for instantaneous velocity we use the symbol v, whereas for average velocity we use $\bar{v}$, with a bar. In the rest of this book, when we use the term "velocity" it will refer to instantaneous velocity. When we want to speak of the average velocity, we will make this clear by including the word "average."

If an object moves at a uniform (or constant) velocity over a particular time interval, then its instantaneous velocity at any instant is the same as its average velocity (see Fig. 2–6a). But in most situations this will not be the case. For example, a car may start from rest, speed up to 50 km/h, remain at that velocity for a time, then slow down to 20 km/h in a traffic jam, and finally stop at its destination after traveling a total of 15 km in 30 min. This trip is plotted on the graph of Fig. 2–6b. Also shown on the graph is the average velocity (dashed line), which is $\bar{v} = \Delta x/\Delta t = 15\text{ km}/0.50\text{ h} = 30\text{ km/h}$.

Finally, note that the instantaneous speed equals the magnitude of the instantaneous velocity under all circumstances. Why? Because in the limit of $\Delta t \to 0$, the displacement also approaches zero, and so there is no way for the magnitude of this infinitesimal displacement to differ from the infinitesimal distance traveled.

2–6 • Vectors and Scalars

Vectors have magnitude and direction

A quantity such as velocity, which has a direction as well as a magnitude, is called a **vector**. Other quantities that are also vectors are displacement, force, and momentum. However, many quantities such as time, temperature, and energy have no direction associated with them. They only have a magnitude. That is, they are specified completely by giving a number (and units, if any). Such quantities are called **scalars**.

Scalars have magnitude only

Drawing a diagram of a particular physical situation is very helpful in physics, and this is especially true when dealing with vectors. On a diagram, each vector is represented by an arrow. The arrow is always drawn so that it points in the direction of the vector it represents. The length of the arrow is drawn proportional to the magnitude of the vector. For example, in Fig. 2–7 arrows have been drawn representing the velocity of a car at various places as it rounds the curve. The magnitude of the velocity at each point can be read off this figure by measuring the length of the corresponding arrow.

When we write the symbol for a vector, we will always use **boldface** type.‡ Thus for velocity we write **v**. If we are only concerned with the magni-

† This material is discussed in more detail later in this chapter, in Section 2–11, which may be considered optional.

‡ The symbol for a vector is usually indicated in handwritten work with an arrow over it, as $\vec{v}$ for velocity, or sometimes with a wiggle underneath, $\underset{\sim}{v}$.

tude we will simply write v. We will deal with vectors more fully in Chapter 3, where we discuss motion in two (or three) dimensions. For now, we deal mainly with motion along a line (rectilinear motion). Vectors which point in one direction along this line will be assigned a positive value, and those pointing in the opposite direction will have a negative sign. In Example 2–2, for instance, the velocity was negative since it pointed to the left along the x axis (as did the displacement—see Fig. 2–5). A velocity or displacement vector pointing to the right (Fig. 2–4) would be considered positive.

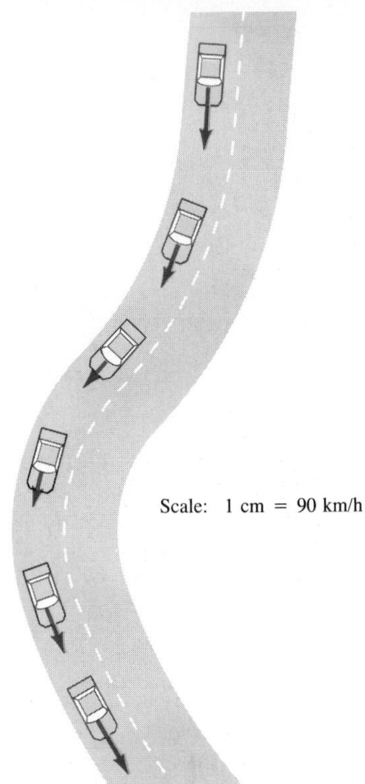

FIGURE 2–7 Car traveling on a road. The green arrows represent the velocity vector at each position.

2–7 • Acceleration

An object whose velocity is changing is said to be accelerating. A car whose velocity increases from zero to 80 km/h is accelerating. If one car can accomplish this change in velocity in less time than another, it is said to undergo a greater acceleration. **Average acceleration** is defined as the *rate of change of velocity*, or the change in velocity divided by the time taken to make this change:

$$\text{average acceleration} = \frac{\text{change of velocity}}{\text{time elapsed}}.$$

In symbols, the average acceleration, $\bar{\mathbf{a}}$, over a time interval $\Delta t = t_2 - t_1$ during which the velocity changes by $\Delta \mathbf{v} = \mathbf{v}_2 - \mathbf{v}_1$, is defined as

Average acceleration

$$\bar{\mathbf{a}} = \frac{\mathbf{v}_2 - \mathbf{v}_1}{t_2 - t_1} = \frac{\Delta \mathbf{v}}{\Delta t}. \tag{2–4}$$

Equation 2–4 is a vector equation. For linear motion, we need only use numerical values for a, v_1, v_2, and Δv, with a plus or minus sign to indicate direction relative to a chosen coordinate system.

The **instantaneous acceleration, a**, can be defined in analogy to instantaneous velocity:

Instantaneous acceleration

$$\mathbf{a} = \lim_{\Delta t \to 0} \frac{\Delta \mathbf{v}}{\Delta t}. \tag{2–5}$$

Here $\Delta \mathbf{v}$ represents the infinitesimally small change in velocity during the infinitesimally short time interval Δt.

The concept of acceleration is often confused with velocity. To help make the distinction, let us take an example.

EXAMPLE 2–3 A car accelerates along a straight road from rest to 60 km/h in 5.0 s. What is the magnitude of its average acceleration?

SOLUTION From Eq. 2–4 the average acceleration is

$$\bar{a} = \frac{60\ \text{km/h} - 0\ \text{km/h}}{5.0\ \text{s}} = 12\ \text{km/h per second} = 12\ \text{km/h/s}.$$

This is read as "twelve kilometers per hour per second" and means that, on the average, the velocity changed by 12 km/h each second. That is, assuming the acceleration was constant, during the first second the car's velocity increased from zero to 12 km/h. During the next second its velocity increased by another 12 km/h up to 24 km/h, and so on. (Of course, if the instantaneous acceleration was not constant, these numbers would be different.)

Caution:
Distinguish carefully between velocity and acceleration

Note carefully that *acceleration is the rate at which the velocity changes,* whereas *velocity is the rate at which position changes.* In this last Example, the calculated acceleration contained two different time units: hours and seconds. We usually prefer to use only seconds. To do so we can change the 60 km/h to $(60\ \text{km/h})(\frac{1}{3.6}\frac{\text{m/s}}{\text{km/h}}) = 17\ \text{m/s}$ (see Section 2–3). Then we get

$$\bar{a} = \frac{17\ \text{m/s} - 0.0\ \text{m/s}}{5.0\ \text{s}} = 3.4\ \text{m/s per second} = 3.4\ \text{m/s}^2.$$

We almost always write these units as m/s^2 (meters per second squared), as done here, instead of m/s/s, which is possible because:

$$\frac{\text{m/s}}{\text{s}} = \frac{\text{m}}{\text{s}\cdot\text{s}} = \frac{\text{m}}{\text{s}^2}.$$

According to the above calculation, the velocity changed on the average by 3.4 m/s during each second, for a total change of 17 m/s over the 5.0 s.

EXAMPLE 2–4 An automobile is moving along a straight highway, which we choose to be to the right along the positive x axis, Fig. 2–8, and the driver puts on the brakes. If the initial velocity is $v_1 = 15.0\ \text{m/s}$ and it takes 5.0 s to slow down to $v_2 = 5.0\ \text{m/s}$, what was the car's average acceleration?

SOLUTION The average acceleration is equal to the change in velocity divided by the elapsed time, Eq. 2–4. Let us say that the initial time was $t_1 = 0$; then $t_2 = 5.0$ s. (Note that our choice of $t_1 = 0$ doesn't matter since only $\Delta t = t_2 - t_1$ appears in Eq. 2–4.) Then

$$\bar{a} = \frac{5.0\ \text{m/s} - 15.0\ \text{m/s}}{5.0\ \text{s}} = -2.0\ \text{m/s}^2.$$

The negative sign appears because the final velocity is less than the initial velocity—that is, the car is decelerating (its speed is decreasing). In this case the direction of the acceleration is to the left—even though the velocity is always pointing to the right.

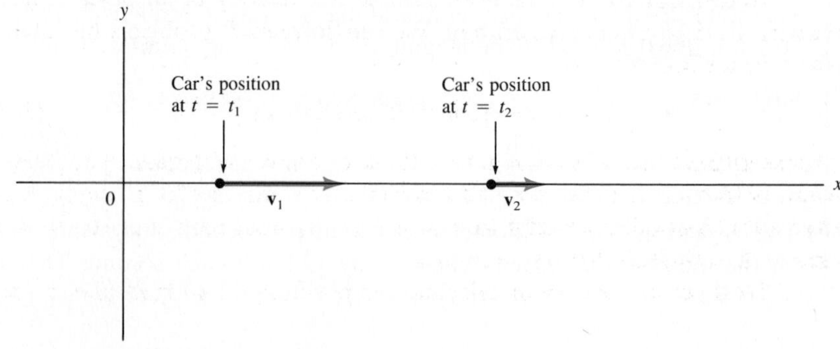

FIGURE 2–8 Example 2–4: graph showing the position of the car at times t_1 and t_2, as well as the car's velocity at t_1 and t_2, which is represented by the arrows labeled $\mathbf{v}_1$ and $\mathbf{v}_2$.

FIGURE 2–9 The racing car in this photo is decelerating with the help of a parachute.

An object is said to be "decelerating" (Fig. 2–9) when its *speed* is decreasing. A deceleration is not necessarily a negative number—the sign of the acceleration depends on the direction. If, however, we choose our reference frame so that the object is moving to the right along the x axis, then the acceleration, a, is positive if the speed is increasing and negative if the speed is decreasing (deceleration), as in Fig. 2–8.

The definition of acceleration involves the change in *velocity*. Thus an acceleration results not only when the *magnitude* of the velocity changes but also if the *direction* changes. For example, a person riding in a car traveling at constant speed around a curve or a child riding on a merry-go-round, will both experience an acceleration because of a change in direction of the velocity. Such accelerations due to change in direction of velocity will be discussed later.

2–8 • Uniformly Accelerated Motion

Let a = constant

Many practical situations occur in which the acceleration is constant. That is, the acceleration doesn't change over time. In many other situations the variation in acceleration is sufficiently small that we are justified in assuming that it is constant. We now treat this situation of **uniformly accelerated motion**: that is, when the magnitude of the acceleration is constant and the motion is in a straight line. In this case, the instantaneous and average accelerations are equal.

To simplify our notation, let us take the initial time in any discussion to be zero: $t_1 = 0$. We can then let $t_2 = t$ be the elapsed time. The initial position (x_1) and initial velocity (v_1) of an object will now be represented by x_0 and v_0, and at time t the position and velocity will be called x and v (rather than x_2 and v_2). The average velocity during the time t will be (from Eq. 2–2)

$$\bar{v} = \frac{x - x_0}{t}.$$

And the acceleration, which is assumed constant in time, will be (from Eq. 2–4)

$$a = \frac{v - v_0}{t}.$$

A common problem is to determine the velocity of an object after a certain time, given its acceleration. We can solve such problems by solving for v in the last equation:

$$v = v_0 + at. \qquad \text{[constant acceleration]} \qquad (2\text{–}6)$$

For example, it may be known that the acceleration of a particular motorcycle is $4.0\ \text{m/s}^2$ and we wish to determine how fast it will be going after, say, 6.0 s. Assuming it starts from rest ($v_0 = 0$), after 6.0 s the velocity will be $v = at = (4.0\ \text{m/s}^2)(6.0\ \text{s}) = 24\ \text{m/s}$.

Next, let us see how to calculate the position of an object after a time

t when it is undergoing constant acceleration. The definition of average velocity (Eq. 2–2) is

$$\bar{v} = \frac{x - x_0}{t},$$

which we can rewrite as

$$x = x_0 + \bar{v}t.$$

Because the velocity increases at a uniform rate, the average velocity, $\bar{v}$, will be midway between the initial and final velocities:

$$\bar{v} = \frac{v_0 + v}{2}. \qquad \text{[constant acceleration]} \qquad (2\text{–}7)$$

(Be careful. This is not generally true if the acceleration is not constant. See, for example, Problems 9 and 59.) We combine the last three equations and find

$$\begin{aligned} x &= x_0 + \bar{v}t \\ &= x_0 + \left(\frac{v_0 + v}{2}\right)t \\ &= x_0 + \left(\frac{v_0 + v_0 + at}{2}\right)t \end{aligned}$$

or

$$x = x_0 + v_0t + \tfrac{1}{2}at^2. \qquad \text{[constant acceleration]} \qquad (2\text{–}8)$$

Equations 2–6, 2–7, and 2–8 are three of the four most useful equations of uniformly accelerated motion. We now derive the fourth equation, which is useful in a situation when, for instance, the acceleration, position, and initial velocity are known, and the final velocity is desired but the time t is not known. To obtain the velocity v at time t in terms of v_0, a, x, and x_0, we begin as above:

$$x = x_0 + \bar{v}t = x_0 + \left(\frac{v + v_0}{2}\right)t$$

where we have substituted in Eq. 2–7. Next we solve Eq. 2–6 for t, obtaining

$$t = \frac{v - v_0}{a},$$

and substituting this into the equation above we have

$$\begin{aligned} x &= x_0 + \left(\frac{v + v_0}{2}\right)\left(\frac{v - v_0}{a}\right) \\ &= x_0 + \frac{v^2 - v_0^2}{2a}. \end{aligned}$$

We solve this for v^2 and obtain the desired relation:

$$v^2 = v_0^2 + 2a(x - x_0). \qquad \text{[constant acceleration]} \qquad (2\text{–}9)$$

We now have four equations relating the various quantities important in uniformly accelerated motion (that is, a is a constant). We collect them here in one place for further reference:

Kinematic equations for constant acceleration (we'll use them a lot)

$$v = v_0 + at \qquad [a = \text{constant}] \qquad (2\text{–}10a)$$

$$x = x_0 + v_0 t + \tfrac{1}{2}at^2 \qquad [a = \text{constant}] \qquad (2\text{–}10b)$$

$$v^2 = v_0^2 + 2a(x - x_0) \qquad [a = \text{constant}] \qquad (2\text{–}10c)$$

$$\bar{v} = \frac{v + v_0}{2}. \qquad [a = \text{constant}] \qquad (2\text{–}10d)$$

These useful equations are not valid unless a is a constant. They can be applied only during time periods when the acceleration is constant. In many cases we can set $x_0 = 0$, and this simplifies the above equations a bit. In the examples that follow, we assume $x_0 = 0$ unless stated otherwise.

EXAMPLE 2–5 Suppose a planner is designing an airport for small planes. One kind of airplane that might use this airfield must reach a speed before takeoff of 100 km/h (27.8 m/s) and can accelerate at 2.0 m/s^2. If the runway is 150 m long, can this airplane reach the proper speed to take off?

SOLUTION To find the velocity, we use Eq. 2–10c with $x_0 = 0$, $v_0 = 0$, $x =$ 150 m, and $a = 2.0\ m/s^2$. Then

$$v^2 = 0 + 2(2.0\ \text{m/s}^2)(150\ \text{m}) = 600\ \text{m}^2/\text{s}^2$$

$$v = \sqrt{600\ \text{m}^2/\text{s}^2} = 24.5\ \text{m/s}.$$

Unfortunately, this runway length is *not* sufficient. By solving Eq. 2–10c for $(x - x_0)$ you can determine how long a runway is needed for this plane.

2–9 • Solving Problems

The solving of problems, such as the examples we have already given, serves two purposes. First, solving problems is useful and practical in itself. Second, solving problems makes you think about the ideas and concepts, and applying the concepts helps you to understand them. But knowing how to do a problem—even to begin it—may not always seem easy. First, and most important, you must read the problem through carefully, and more than once. Spend a moment thinking and trying to understand what physics principles might be involved. Up to this point in the book, we have been concerned mainly with the definitions of velocity and acceleration, and the "kinematic equations for constant acceleration," Eqs. 2–10, that we derived from those definitions. We will discuss problem solving more in Chapter 4, after we have covered more physics. For now it is important to note that physics is *not* a collection of equations to be memorized. (In fact, instead of memorizing the very useful Eqs. 2–10, it might be better to understand how to derive them from the definitions of velocity and acceleration as we did above.) Simply searching for an equation that might work can be very dangerous and can lead you to a wrong result. A much better approach is to follow this (rough) procedure:

Techniques and hints for solving problems

1. Read the whole problem carefully before trying to solve it. Draw a picture of the situation, with coordinate axes where applicable.
2. Write down what quantities are "known" or "given," and then what you *want* to know.
3. Think about what principles, definitions, and/or equations relate the quantities involved. Before using them, be sure their range of validity includes your problem (for example, Eqs. 2–10 are valid only when the acceleration is constant). If you find an applicable equation that involves only known quantities and one desired unknown, solve the equation algebraically for the unknown. In many instances several sequential calculations, and/or a combination of equations, may be needed. It is often preferable to solve algebraically for the desired unknown before putting in numerical values.
4. Think carefully about the result you obtain: Is it reasonable? Does it make sense according to your own intuition and experience?
5. A very important aspect of doing problems is keeping track of units. Note that an equals sign implies the units on each side must be the same, just as the numbers must. If the units do not balance, a mistake has no doubt been made. This can serve as a check on your solution (but it only tells you if you're wrong, not if you're right).†

EXAMPLE 2–6 How long does it take a car to travel 30.0 m if it accelerates from rest at a rate of 2.00 m/s²?

SOLUTION First we make a table, choosing $x_0 = 0$:

Known	Wanted
$x_0 = 0$	t
$x = 30.0$ m	
$a = 2.00\ \mathrm{m/s^2}$	
$v_0 = 0$	

Since a is constant, we can use Eqs. 2–10. Equation 2–10a is not helpful in this case since it contains v, an unknown, as well as the desired unknown t. Equation 2–10c is worse—it contains v but not t. Equation 2–10b is perfect since the only unknown quantity is t. Before we solve for t in this equation, we simplify it by setting $v_0 = 0$ first. Then

$$x = \tfrac{1}{2}at^2$$

$$t^2 = \frac{2x}{a} = \frac{2(30.0\ \mathrm{m})}{2.00\ \mathrm{m/s^2}} = 30.0\ \mathrm{s^2}$$

$$t = \sqrt{30.0\ \mathrm{s^2}} = 5.48\ \mathrm{s}.$$

EXAMPLE 2–7 We consider the stopping distances for a car, which are important for traffic safety and traffic design. The problem is best dealt with in two parts: (1) the time between the decision to apply the brakes and their actual application (the "reaction time"), during which we assume $a = 0$; and (2) the actual braking period when the vehicle decelerates ($a \neq 0$).

† See also Appendix B on dimensional analysis.

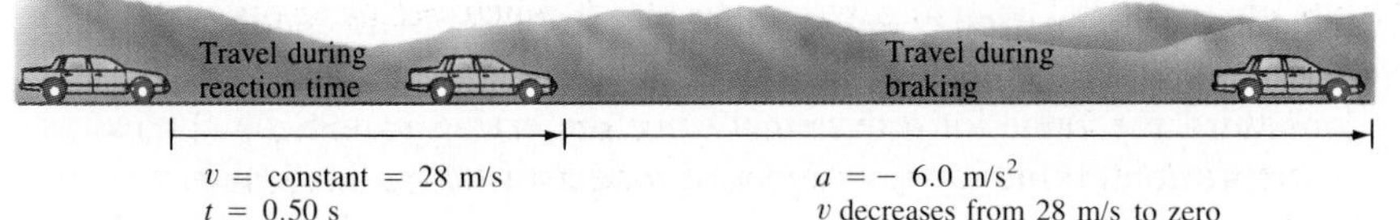

FIGURE 2–10 Example 2–7: stopping distance for a braking car.

The stopping distance depends on the reaction time of the driver, the initial speed of the car (the final speed is zero), and the deceleration of the car. For a dry road, good brakes can decelerate a car at a rate of about 5 m/s^2 to 8 m/s^2. We make the calculation for an initial velocity of 100 km/h (28 m/s) and assume the acceleration of the car is $-6.0\ \text{m/s}^2$ (the minus sign appears because the velocity is taken to be positive and its magnitude is decreasing). Reaction time for normal drivers varies from perhaps 0.3 s to about 1.0 s; we take it to be 0.50 s.

SOLUTION We assume the car is moving to the right along the positive x axis, and we take $x_0 = 0$ for both parts of the problem. For the first part of the problem, the car travels at a constant speed of 28 m/s during the time the driver is reacting (0.50 s); see Fig. 2–10. Thus:

Known	Wanted
$t = 0.50$ s	x
$v_0 = 28$ m/s	
$v = 28$ m/s	
$a = 0$	
$x_0 = 0$	

To find x we can use Eq. 2–10b (note that Eq. 2–10c won't work because x is multiplied by a, which is zero):

$$x = v_0t + 0 = (28\ \text{m/s})(0.50\ \text{s}) = 14\ \text{m}.$$

Now for the second part, during which the brakes are applied and the car is brought to rest:

Known	Wanted
$v_0 = 28$ m/s	x
$v = 0$	
$a = -6.0\ \text{m/s}^2$	

Equation 2–10a doesn't contain x; Eq. 2–10b contains x but also the unknown t. Equation 2–10c is what we want; we solve for x (after setting $x_0 = 0$):

$$v^2 - v_0^2 = 2ax$$

$$x = \frac{v^2 - v_0^2}{2a}$$

$$= \frac{0 - (28\ \text{m/s})^2}{2(-6.0\ \text{m/s}^2)} = \frac{-784\ \text{m}^2/\text{s}^2}{-12\ \text{m/s}^2}$$

$$= 65\ \text{m}.$$

The car traveled 14 m while the driver was reacting and another 65 m dur-

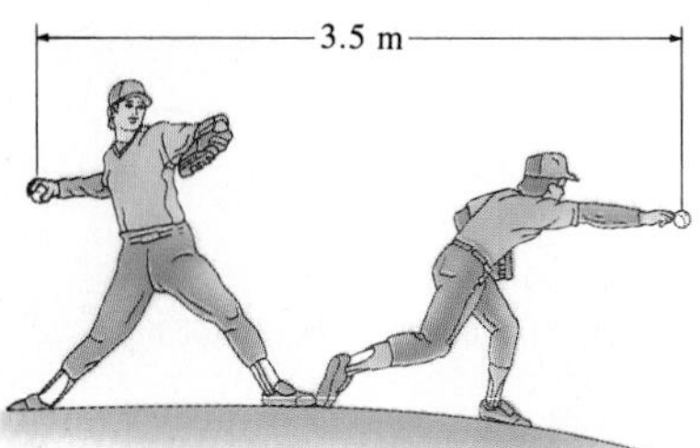

FIGURE 2–11 A baseball pitcher accelerates the ball over a distance of about 3.5 m.

ing the braking period before coming to a stop. The total distance traveled was then 79 m. Under wet or icy conditions, the value of a may be only one-third the value for a dry road since the brakes cannot be applied as hard without skidding, and hence stopping distances are much greater. Note also that stopping distance increases with the *square* of the speed, not just linearly with speed!

EXAMPLE 2–8 A baseball pitcher throws a fastball with a speed of 44 m/s. Estimate the average acceleration of the ball during the throwing motion. It is observed that in throwing the baseball, the pitcher accelerates the ball through a total distance of about 3.5 m from behind his body to the point where it is released (Fig. 2–11).

SOLUTION We want to find the acceleration a given that $x = 3.5$ m, $v_0 = 0$, and $v = 44$ m/s. We use Eq. 2–10c and solve for a:

$$a = \frac{v^2 - v_0^2}{2x}$$

$$= \frac{(44\text{ m/s})^2 - (0\text{ m/s})^2}{2(3.5\text{ m})} = 280\text{ m/s}^2.$$

This is a very large acceleration!

FIGURE 2–12 Galileo Galilei (1564–1642).

The analysis of motion we have been discussing in this chapter is basically algebraic. It is sometimes helpful to use a graphical interpretation as well and this is done in Section 2–11 (optional).

2–10 • Falling Bodies

One of the commonest examples of uniformly accelerated motion is that of an object allowed to fall freely near the earth's surface. That a falling body is accelerating may not be obvious at first. And beware in thinking, as was widely believed until the time of Galileo (Fig. 2–12), that heavier bodies fall faster than lighter bodies and that the speed of fall is proportional to how heavy the object is.

Do heavier bodies fall faster?

Galileo's analysis made use of his new and creative technique of abstraction and simplification—that is, of imagining what would happen in idealized (simplified) cases. For free fall, he postulated that all bodies would fall with the *same constant acceleration* in the absence of air or other resistance. He showed that this postulate predicts that for an object falling from rest, the distance traveled will be proportional to the square of the time (Fig. 2–13); that is, $d \propto t^2$. We can see this from Eq. 2–10b, but Galileo was the first to derive this mathematical relation. In fact, one of Galileo's great contributions to science was to establish such mathematical relations and to insist on their importance. Another great contribution of Galileo was the proposing of a theory with specific experimental consequences that could be quantitatively checked ($d \propto t^2$).

Do falling bodies accelerate?

To support his claim that the speed of falling objects increases as they fall, Galileo made use of the following argument: a heavy stone dropped

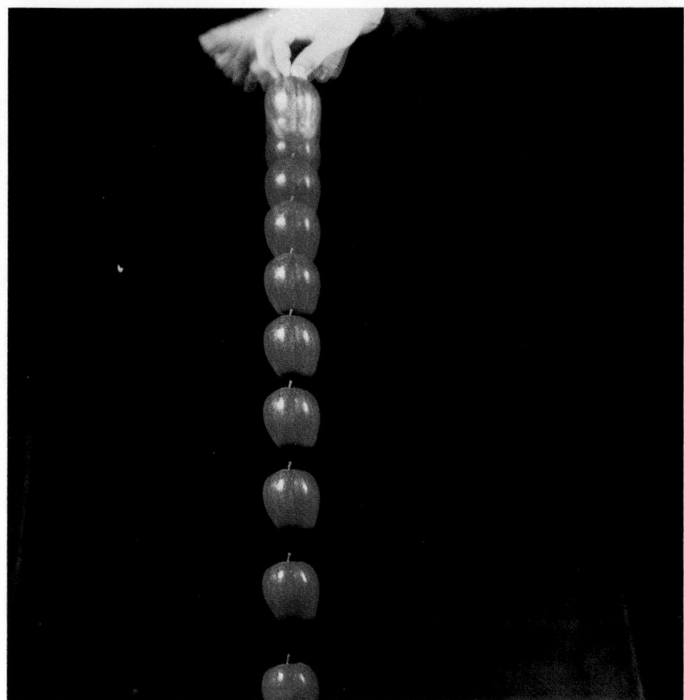

FIGURE 2–13 Multiflash photograph of a falling apple, photographed at equal time intervals. Note that the apple falls further during each successive time interval, which means it is accelerating.

FIGURE 2–14 (a) A ball and a light piece of paper are dropped at the same time. (b) Repeated, with the paper wadded up.

from a height of 2 m will drive a stake into the ground much farther than if the same stone is dropped from a height of only 10 cm. Clearly, the stone must be moving faster in the former case. As we saw, Galileo also claimed that *all* objects, light or heavy, fall with the *same* acceleration, at least in the absence of air. Now common sense may say that the ancients were perhaps closer to the truth. For if you hold a piece of paper horizontally in one hand and a heavier object, say, a baseball, in the other and release them at the same time (see Fig. 2–14a), surely the heavier object will reach the ground first. But if you repeat the experiment, this time crumpling the paper into a small wad (see Fig. 2–14b), you will find that the two objects reach the floor at nearly the same time.

FIGURE 2–15 A rock and a feather are dropped simultaneously (a) in air, (b) in a vacuum.

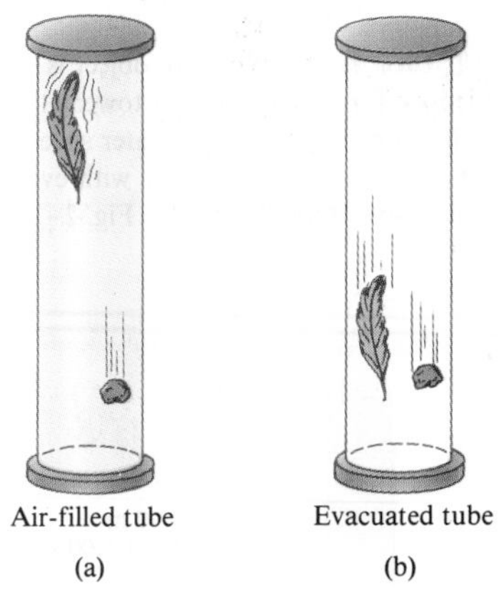

Galileo was sure that air acts as a resistance to very light objects that have a large surface area. But in many ordinary circumstances this air resistance is negligible. In a chamber from which the air has been removed even light objects like a feather or a horizontally held piece of paper will fall with the same acceleration as any other object (see Fig. 2–15). Such a demonstration in vacuum was of course not possible in Galileo's time, which makes Galileo's achievement all the greater. Galileo is often called the "father of modern science," not only for the content of his science (astronomical discoveries, inertia, free fall), but also for his style or approach to science (idealization and simplification, mathematization of theory, prediction of testable consequences that must be checked by experiment).

Galileo's specific contribution to our understanding of the motion of falling objects can be summarized as follows:

Galileo's hypothesis

at a given location on the earth and in the absence of air resistance, all objects fall with the same uniform acceleration.

tance, and (*b*) in the presence of air resistance. [*Hint*: the acceleration due to air resistance is always in a direction opposite to the motion.]

16. An object that is thrown vertically upward will return to its original position with the same speed as it had initially, if air resistance is negligible. If air resistance is appreciable, will this result be altered, and if so, how?

***17.** Describe in words the motion plotted in Fig. 2–23.

***18.** Describe in words the motion of the object graphed in Fig. 2–24.

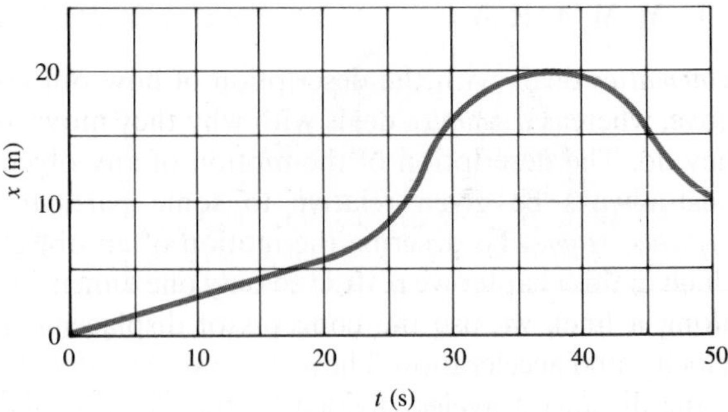

FIGURE 2–23 Question 17, Problems 47, 48, and 53.

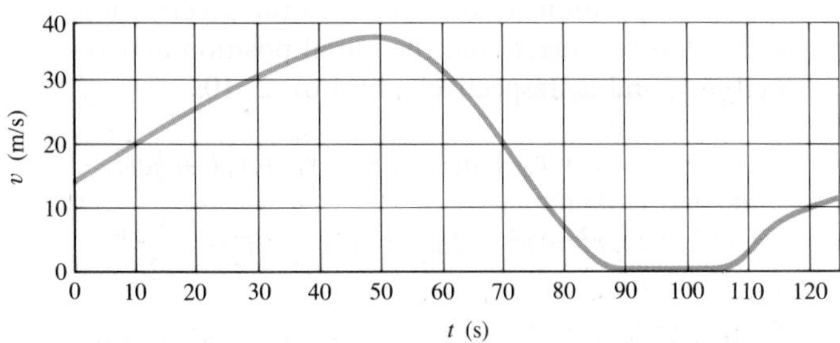

FIGURE 2–24 Question 18, Problems 49, 52, and 54.

PROBLEMS

[The problems at the end of each chapter are ranked I, II, or III according to estimated difficulty, with I problems being easiest. The problems are arranged by sections, meaning that the reader should have read up to and including that section, but not only that section—problems often depend on earlier material. Finally, there is a set of unranked "General Problems" not arranged by section number.]

SECTIONS 2–1 TO 2–6

1. (I) What must be your average speed in order to travel 220 km in 2.25 h?
2. (I) At an average speed of 31.0 km/h, how far will a bicyclist travel in 135 min?
3. (I) A bird can fly 30 km/h. How long does it take to fly 22 km?
4. (I) If you are driving 100 km/h and you look to the side for 2.0 s, how far do you travel during this inattentive period?
5. (I) 55 mph is how many (*a*) km/h, (*b*) m/s, and (*c*) ft/s?
6. (I) Determine the conversion factor between (*a*) km/h and mi/h, (*b*) m/s and ft/s, and (*c*) mi/h and m/s.
7. (II) A person jogs eight complete laps around a quarter-mile track in a total time of 13.5 min. Calculate (*a*) the average speed and (*b*) the average velocity, in m/s.
8. (II) A dog runs 100 m away from its master in a straight line in 8.4 s, and then runs halfway back in one-third the time. Calculate (*a*) its average speed and (*b*) its average velocity.
9. (II) An airplane travels 2100 km at a speed of 1000 km/h. It then encounters a headwind that slows it to 800 km/h for the next 1300 km. What was the average speed of the plane for this trip? [*Hint:* think carefully before using Eq. 2–10d.]
10. (II) Two locomotives approach each other on parallel tracks. Each has a speed of 120 km/h with respect to the earth. If they are initially 8.5 km apart, how long will it be before they pass each other? (See Fig. 2–25.)
11. (III) A ball traveling with constant speed hits the pins placed at the end of a bowling lane 16.5 m long. The bowler heard the sound of the ball hitting the pins 2.50 s after the ball was released from his hands. What was the speed of the ball? The speed of sound is 340 m/s.

SECTION 2–7

12. (I) A sports car accelerates from rest to 100 km/h in 6.6 s. What is its acceleration in m/s^2?
13. (I) At highway speeds, a particular automobile is capable of an acceleration of about 1.7 m/s^2. At this rate,

FIGURE 2–25 Problem 10.

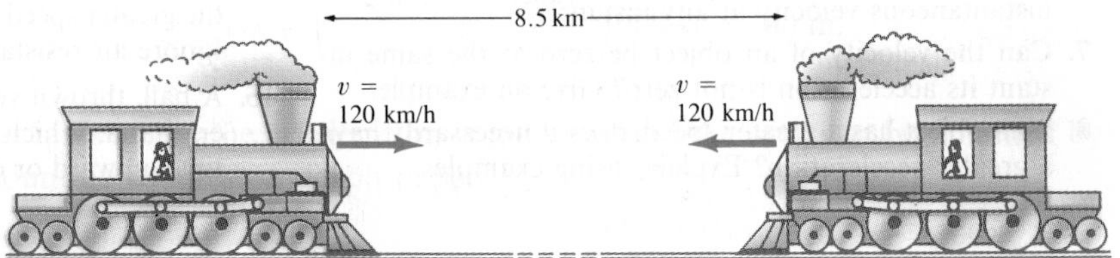

how long does it take to accelerate from 85 km/h to 100 km/h?

14. (II) A sports car is advertised to be able to stop, from a speed of 100 km/h, within 45 m. What is its acceleration in m/s^2? How many *g*'s is this ($g = 9.80\ m/s^2$)?

15. (III) The position of a racing car, which starts from rest at $t = 0$ and moves in a straight line, has been measured as a function of time, as given in the following table. Estimate (*a*) its velocity and (*b*) its acceleration as a function of time. Display each in a table and on a graph.

t (s)	0	0.25	0.50	0.75	1.00	1.50	2.00
x (m)	0	0.11	0.46	1.06	1.94	4.62	8.55

t (s)	2.50	3.00	3.50	4.00	4.50	5.00	5.50	6.00
x (m)	13.79	20.36	28.31	37.65	48.37	60.30	73.26	87.16

SECTIONS 2–8 AND 2–9

16. (I) The principal kinematic equations, 2–10a through 2–10d, become particularly simple if the initial speed is zero. Write down the equations for this special case. (Also put $x_0 = 0$.)

17. (I) A car accelerates from 12 m/s to 25 m/s in 5.0 s. What was its acceleration? How far did it travel in this time? Assume constant acceleration.

18. (I) A car decelerates from a speed of 25 m/s to rest in a distance of 120 m. What was its acceleration, assumed constant?

19. (I) A jet liner must reach a speed of 80 m/s for takeoff. If the runway is 1500 m long, what (constant) acceleration is needed?

20. (II) A car decelerates from a speed of 30.0 m/s to rest in 6.00 s. How far did it travel in that time?

21. (II) In coming to a stop, a car leaves skid marks on the highway 320 m long. Assuming a deceleration of $10\ m/s^2$ (roughly the maximum for rubber tires on dry pavement), estimate the speed of the car just before braking.

22. (II) A car traveling 90 km/h decelerates at a constant $1.6\ m/s^2$. Calculate (*a*) the distance the car goes before it stops, (*b*) the time it takes to stop, and (*c*) the distance it travels during the first and third seconds.

23. (II) A car traveling at 60 km/h strikes a tree; the front end of the car compresses and the driver comes to rest after traveling 0.70 m. What was the average deceleration of the driver during the collision? Express the answer in terms of "*g*'s," where $1.00\ g = 9.80\ m/s^2$.

24. (II) Make up a table of stopping distances for an automobile with an initial speed of 80 km/h and human reaction time of 1.0 s: (*a*) for an acceleration $a = -4.0\ m/s^2$; (*b*) for $a = -8.0\ m/s^2$.

25. (II) Repeat Problem 24, using a reaction time of 0.40 s.

26. (III) Show that the equation for the stopping distance of a car is $d_s = v_0 t_R - v_0^2/(2a)$, where v_0 is the initial speed of the car, t_R is the driver's reaction time, and a is the constant acceleration (and is negative).

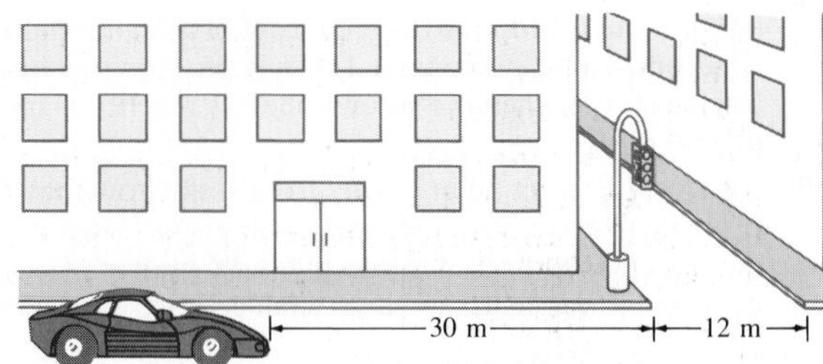

FIGURE 2–26 Problem 27.

27. (III) A person driving her car at 50 km/h approaches an intersection just as the traffic light turns yellow. She knows that the yellow light lasts only 2.0 s before turning to red, and she is 30 m away from the near side of the intersection (Fig. 2–26). Should she try to stop, or should she make a run for it? The intersection is 12 m wide, and her car's maximum deceleration is $-6.0\ m/s^2$. Also her car takes 7.0 s to accelerate from 50 km/h to 70 km/h. Ignore the length of her car and her reaction time.

28. (III) A runner hopes to complete the 5000-m run in less than 13.0 min. After exactly 11.0 min, there are still 800 m to go. The runner must then accelerate at $0.20\ m/s^2$ for how many seconds in order to achieve the desired time?

SECTION 2–10 (Neglect Air Resistance)

29. (I) Calculate the acceleration of the baseball in Example 2–8 in "*g*'s."

30. (I) A stone is dropped from the top of a cliff. It is seen to hit the ground below after 3.5 s. How high is the cliff?

31. (I) (*a*) How long does it take a brick to reach the ground if dropped from a height of 50.0 m? (*b*) What will be its velocity just before it reaches the ground?

32. (II) A baseball is thrown vertically into the air with a speed of 24.0 m/s. (*a*) How high does it go? (*b*) How long does it take to return to the ground?

33. (II) A kangeroo jumps to a vertical height of 2.8 m. How long was it in the air before returning to earth?

34. (II) A ballplayer catches a ball 4.0 s after throwing it vertically upward. With what speed did he throw it, and what height did it reach?

35. (II) Draw graphs of (*a*) the speed and (*b*) the distance fallen, as a function of time, for a body falling under the influence of gravity for $t = 0$ to $t = 5.00$ s.

36. (II) The flea in Example 2–12 actually reached a height of only 3.5 cm because of air resistance. Using this height, and the initial speed of 1.0 m/s, calculate the actual deceleration rate of the flea during its upward flight, assuming it to be constant.

37. (II) A helicopter is ascending vertically with a speed of 6.00 m/s; at a height of 120 m above the earth, a package is dropped from a window. How much time does it take for the package to reach the ground?

38. (II) A stone is dropped from the roof of a high building. A second stone is dropped 1.00 s later. How far apart are the stones when the second one has reached a speed of 15.0 m/s?

39. (II) For an object falling freely from rest, show that the distance traveled during each successive second increases in the ratio of successive odd integers (1, 3, 5, etc.). (This was first shown by Galileo.) See Figs. 2–13 and 2–16.

40. (II) If air resistance is neglected, show that a ball thrown vertically upward with a speed v_0 will have the same speed, v_0, when it comes back down to the starting point.

41. (II) A stone is thrown vertically upward with a speed of 20.0 m/s. (*a*) How fast is it moving when it reaches a height of 16.0 m? (*b*) How long is required to reach this height? (*c*) Why are there two answers to (*b*)?

42. (III) A falling stone takes 0.30 s to travel past a window 2.4 m tall (Fig. 2–27). From what height above the top of the window did the stone fall?

43. (III) A rock is dropped from a seacliff and the sound of it striking the ocean is heard 3.0 s later. If the speed of sound is 340 m/s, how high is the cliff?

44. (III) Suppose you adjust your garden hose nozzle for a hard stream of water. You point the nozzle vertically upward at a height of 1.5 m above the ground (Fig. 2–28). When you quickly move the nozzle away from the vertical, you hear the water striking the ground next to you for 2.0 s. What is the water speed as it leaves the nozzle?

45. (III) A stone is thrown vertically upward with a speed of 10.0 m/s from the edge of a cliff 65 m high (Fig. 2–29). (*a*) How much later does it reach the bottom of the cliff? (*b*) What is its speed just before hitting? (*c*) What total distance did it travel?

46. (III) A baseball is seen to pass upward by a window 30 m above the street with a vertical speed of 12 m/s. If the ball was thrown from the street, (*a*) what was its initial speed, (*b*) what altitude does it reach, (*c*) when was it thrown, and (*d*) when does it reach the street again?

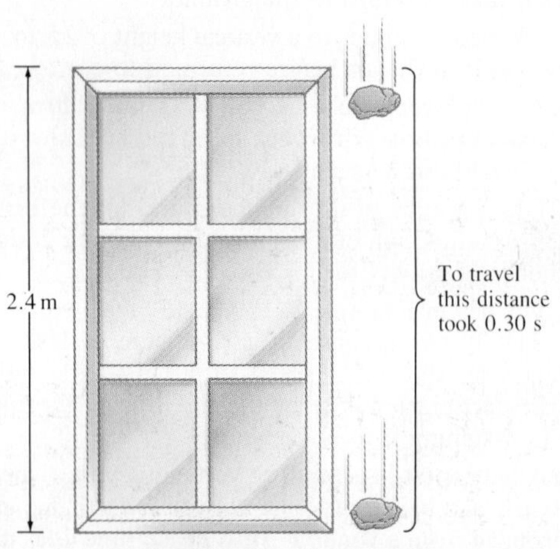

FIGURE 2–27 Problem 42.

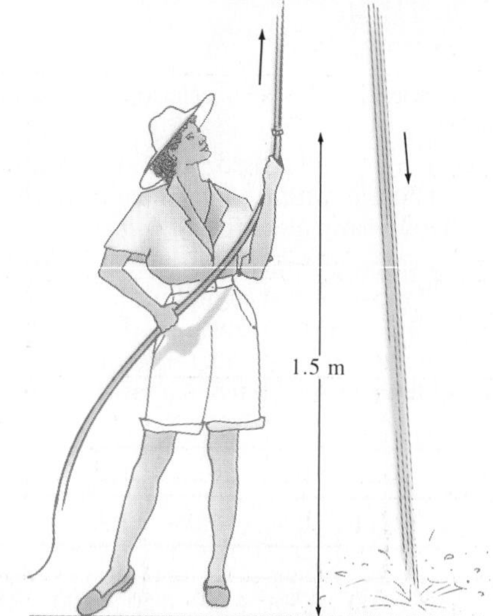

FIGURE 2–28 Problem 44.

*SECTION 2–11

*47. (I) The position of a rabbit along a straight tunnel as a function of time is plotted in Fig. 2–23. What is its instantaneous velocity (*a*) at $t = 10.0$ s and (*b*) at $t = 30.0$ s? What is its average velocity (*c*) between $t = 0$ and $t = 5.0$ s, (*d*) between $t = 25.0$ s and $t = 30.0$ s, and (*e*) between $t = 40.0$ s and $t = 50.0$ s?

*48. (I) In Fig. 2–23, (*a*) during what time periods, if any, is the rabbit's velocity constant? (*b*) At what time is its velocity the greatest? (*c*) At what time, if any, is the velocity zero? (*d*) Does the rabbit run in one direction or in both along its tunnel during the time shown?

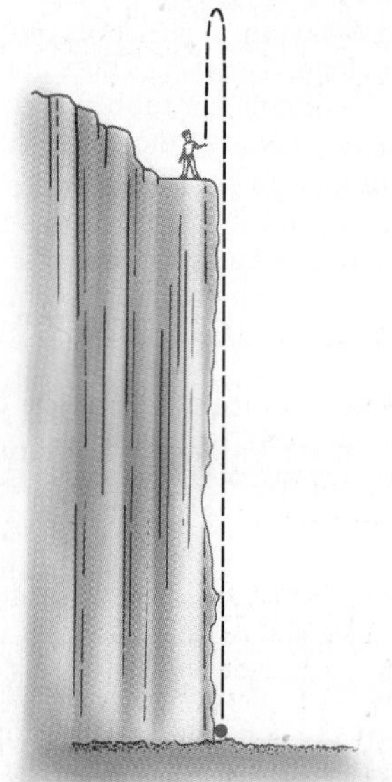

FIGURE 2–29 Problem 45.

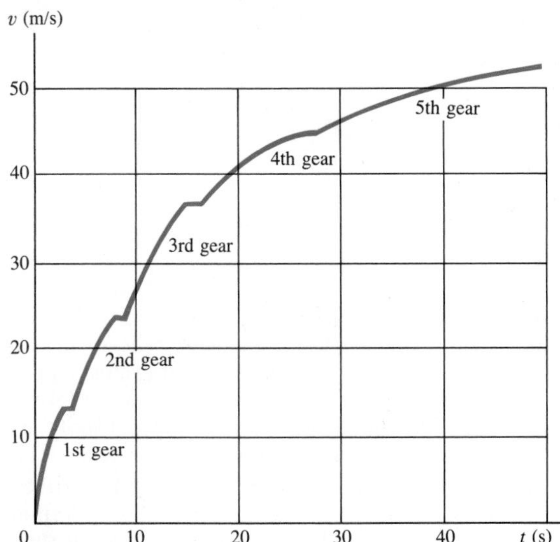

FIGURE 2–30 The velocity of a high-performance automobile as a function of time, starting from a dead stop. The jumps in the curve represent gear shifts. (Problems 50 and 51.)

*49. (I) Figure 2–24 shows the velocity of a train as a function of time. (*a*) At what time was its velocity greatest? (*b*) During what periods, if any, was the velocity constant? (*c*) During what periods, if any, was the acceleration constant? (*d*) When was the magnitude of the acceleration greatest?

*50. (II) A high-performance automobile can accelerate approximately as shown in the velocity-time graph of Fig. 2–30. (The jumps in the curve represent shifting of the gears.) (*a*) Estimate the average acceleration of the car in second gear and in fourth gear. (*b*) Estimate how far the car traveled while in fourth gear.

*51. (II) Estimate the average acceleration of the car in Problem 50 (Fig. 2–30) when it is in (*a*) first, (*b*) third, and (*c*) fifth gear. (*d*) What is its average acceleration through the first four gears?

*52. (II) In Fig. 2–24, estimate the distance the train traveled during (*a*) the first minute and (*b*) the second minute.

*53. (II) Construct the v vs. t graph for the object whose displacement as a function of time is given by Fig. 2–23.

*54. (II) Construct the x vs. t graph for the object whose velocity as a function of time is given by Fig. 2–24.

*55. (II) (*a*) Suppose at times $t_0 = 0$, $t_1 = 1.0$ s, $t_2 = 2.0$ s, and $t_3 = 3.0$ s an object is at the positions $x_0 = 0$, $x_1 = 2.0$ m, $x_2 = 3.5$ m, and $x_3 = 4.5$ m, respectively; plot these points on an x vs. t graph. Compute the average velocity over the time intervals $t_1 - t_0$, $t_2 - t_1$, and $t_3 - t_2$. (*b*) Do the same supposing the object is at positions $x_0 = 10.0$ m, $x_1 = 8.0$ m, $x_2 = 6.5$ m, and $x_3 = 5.5$ m for the respective times. (*c*) Do the same assuming the object was at $x_0 = 0$, $x_1 = 3.0$ m, $x_2 = 6.5$ m, and $x_3 = 10.0$ m at times $t_0 = -2.0$ s, $t_1 = -1.0$ s, $t_2 = 0.0$ s, and $t_3 = 1.0$ s respectively.

*56. (III) Suppose an object's position was measured at seven different moments as given by the following table:

$t_0 = 0.0$ s	$x_0 = 0.000$ m
$t_1 = 1.0$ s	$x_1 = 0.333$ m
$t_2 = 2.0$ s	$x_2 = 1.333$ m
$t_3 = 2.1$ s	$x_3 = 1.470$ m
$t_4 = 2.2$ s	$x_4 = 1.610$ m
$t_5 = 2.5$ s	$x_5 = 2.083$ m
$t_6 = 3.0$ s	$x_6 = 3.000$ m

Plot these points on a graph. Compute the average velocity $\bar{v}$ for the time intervals $t_6 - t_2$, $t_5 - t_2$, $t_4 - t_2$, and $t_3 - t_2$. To what value of the instantaneous velocity at t_2 do these numbers seem to be converging? On the graph, draw a long straight line through each pair of points; for example, draw the first long straight line through the points (t_2, x_2) and (t_6, x_6). The slope of each line equals the average velocity over the corresponding time interval. Note the slope to which the lines are converging as $t \to t_2$. This slope equals the instantaneous velocity at t_2.

GENERAL PROBLEMS

57. The acceleration due to gravity on the moon is about one-sixth what it is on earth. If an object is thrown vertically upward on the moon, how many times higher will it go than it would on earth, assuming the same initial velocity?

58. A person who is properly constrained by an over-the-shoulder seat belt has a good chance of surviving a car collision if the deceleration does not exceed 30 "g's" ($1.00\ g = 9.80\ \text{m/s}^2$). Assuming uniform deceleration at this rate, calculate the distance over which the front end of the car must be designed to collapse if a crash occurs at 100 km/h.

59. A race car driver must average 200 km/h for four laps to qualify for a race. Because of engine trouble, the car averages only 170 km/h over the first two laps. What average speed must be maintained for the last two laps?

60. Calculate the carrying capacity (number of cars passing a given point per hour) on a highway with three lanes (in one direction) using the following assumptions: the average speed is 100 km/h, the average length of a car is 6.0 m, and the average distance between cars should be 80 m.

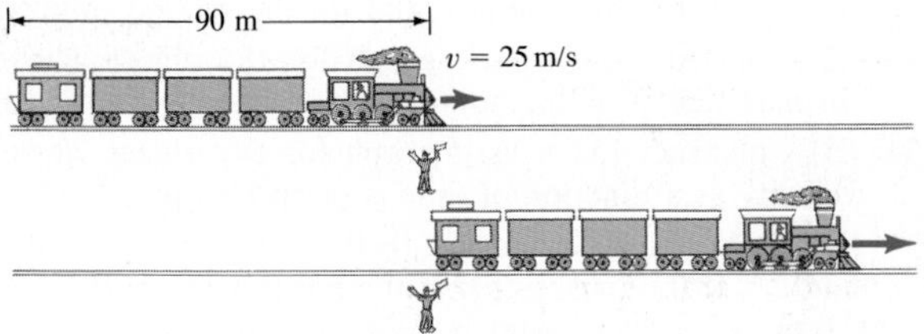

FIGURE 2–31 Problem 62.

61. A car traveling 90 km/h is 100 m behind a truck traveling 60 km/h. How long will it take the car to reach the truck?
62. A 90-m-long train begins accelerating uniformly from rest. The front of the train passes a railway worker, who is standing 200 m from where the front of the train started, at the speed of 25 m/s. What will be the speed of the last car as it passes the worker? (See Fig. 2–31.)
63. In the design of a rapid transit system, it is necessary to balance out the average speed of a train against the distance between stops. The more stops there are, the slower the train's average speed. To get an idea of this problem, calculate the time it takes a train to make a 36-km trip in two situations: (*a*) the stations at which the trains must stop are 0.80 km apart; and (*b*) the stations are 3.0 km apart. Assume that at each station the train accelerates at a rate of 1.1 m/s^2 until it reaches 90 km/h, then stays at this speed until its brakes are applied for arrival at the next station, at which time it decelerates at $-2.0\ m/s^2$. Assume it stops at each intermediate station for 20 s.
64. For the design of a rapid transit system as discussed in the previous problem, derive a general formula for the average speed of a train. Specify the symbols used for all quantities involved such as the acceleration, deceleration, maximum velocity, distance between stations, and time stopped at each station.
65. Pelicans tuck their wings and free-fall straight down when diving for fish. Suppose a pelican starts its dive from a height of 20 m and cannot change its path once committed. If it takes a fish 0.10 s to perform evasive action, at what minimum height must it spot the pelican to escape? Assume the fish is at the surface of the water.
66. In putting, the force with which a golfer strikes a ball is determined so that the ball will stop within some small distance of the cup, say 1.0 m long or short, even if the putt is missed. Accomplishing this from an uphill lie (that is, putting downhill, see Fig. 2–32) is more difficult than from a downhill lie. To see why, assume that on a particular green the ball decelerates constantly at 2.0 m/s^2 going downhill and constantly at 3.0 m/s^2 going uphill. Suppose we have an uphill lie 7.0 m from the cup. Calculate the allowable range of initial velocities we may impart to the ball so that it stops in the range 1.0 m short to 1.0 m long of the cup. Do the same for a downhill lie 7.0 m from the cup. What in your results suggests that the downhill putt is more difficult?

FIGURE 2–32 Problem 66. Golf on Wednesday morning.

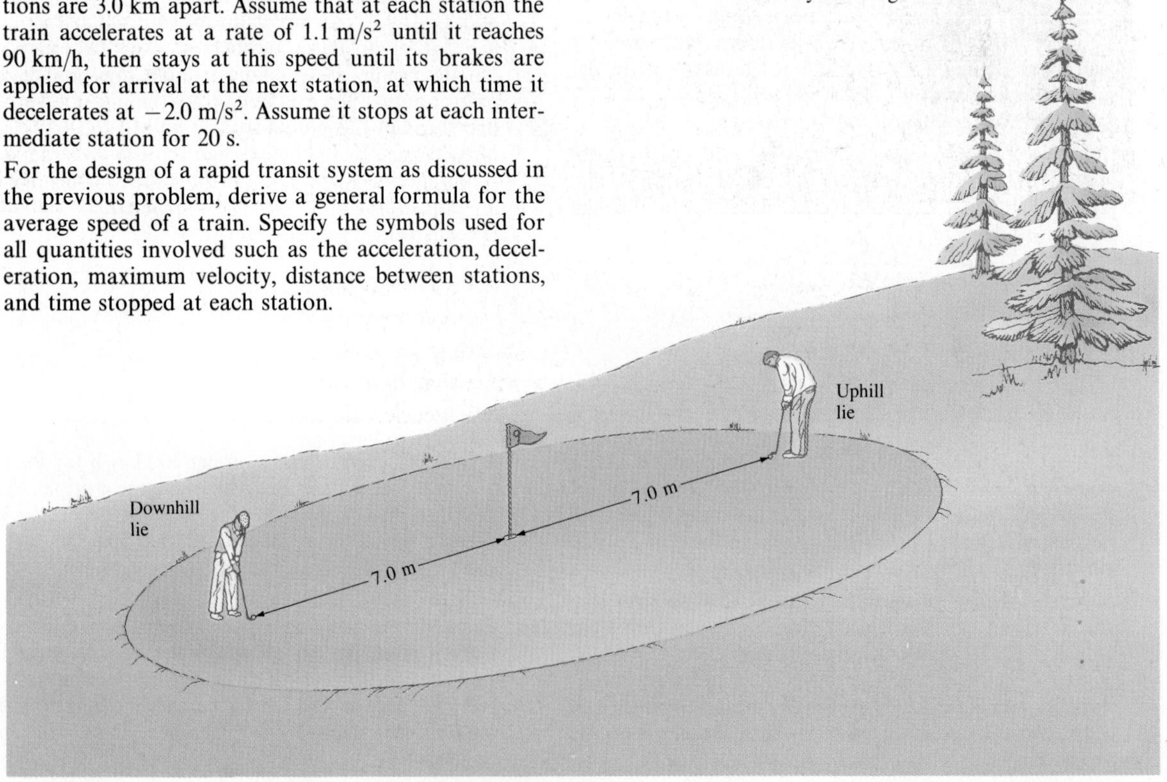

CHAPTER 3

Kinematics in Two or Three Dimensions; Vectors

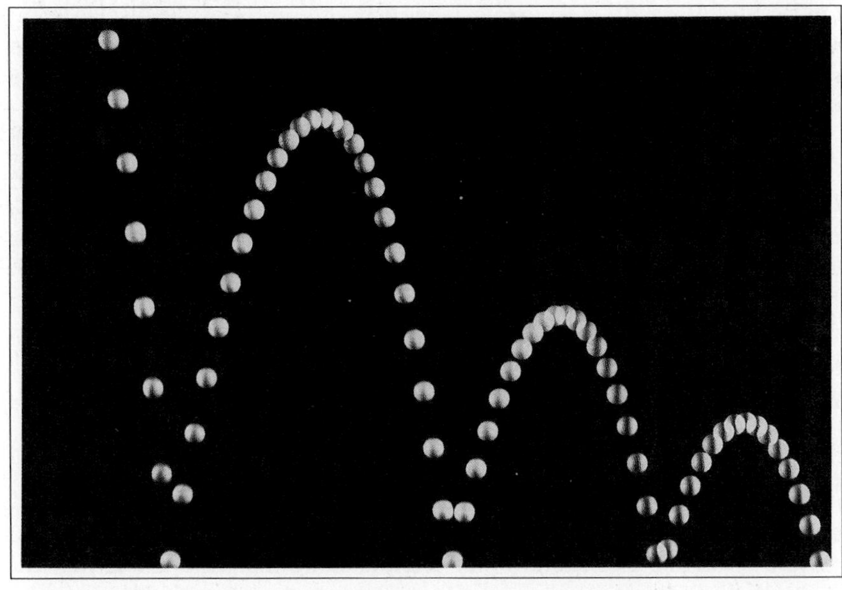

This multiflash photograph of a ball falling through an arc, and making several bounces, is an example of motion in two dimensions. Each arc has the shape of a parabola which is characteristic of "projectile motion." Galileo analyzed projectile motion into its vertical and horizontal components, and arrived at brilliant conclusions.

In Chapter 2 we dealt with motion along a straight line. We now consider the description of the motion of objects that move in paths in two (or three) dimensions. To do so we first need to discuss vectors and how they are added.

As we discussed in Section 2–6, a *vector* quantity is one that has both magnitude and direction. Examples are displacement, velocity, acceleration, and force. *Scalar* quantities, on the other hand, have magnitude only. They are specified completely by giving a number (and units). Examples are temperature and time.

3–1 • Addition of Vectors—Graphical Methods

Because vectors are quantities that have direction as well as magnitude, they must be added in a special way. In this chapter, we will deal mainly with displacement vectors (for which we now use the symbol **D**) and velocity vectors (**v**). But the results will apply for other vectors that we encounter later.

Adding vectors along the same line

We use simple arithmetic for adding scalars, such as time. Simple arithmetic can also be used for adding vectors if they are in the same direction. For example, if a person walks 8 km east one day, and 6 km east the next day, the person will be 8 km + 6 km = 14 km east of the point of origin. We say that the *net* or *resultant* displacement is 14 km to the east. If, on the other hand, the person walks 8 km east on the first day, and 6 km west (in the reverse direction) on the second day, then the person will be 2 km from the origin after the 2 days, so the resultant displacement is 2 km to the east. In this case, the resultant displacement is obtained by subtraction: 8 km − 6 km = 2 km.

Adding vectors that are at right angles

But simple arithmetic cannot be used if the two vectors are not along the same line. For example, suppose a person walks 10.0 km east and then walks 5.0 km north. This motion can be represented on a graph in which the positive y axis points north and the positive x axis points east, Fig. 3–1. On this graph, we draw an arrow, labeled $\mathbf{D}_1$, to represent the displacement vector of the 10.0-km displacement to the east; and a second arrow, $\mathbf{D}_2$, to represent the 5.0-km displacement to the north. Both vectors are drawn to scale.

After taking this walk, the person is now 10.0 km east and 5.0 km north of the point of origin. The *resultant displacement* is represented by the arrow labeled $\mathbf{D}_R$, on the diagram. If you use a ruler and a protractor, you will measure on this diagram that the person is 11.2 km from the origin at an angle of 27° north of east. In other words, the resultant displacement vector has a magnitude of 11.2 km and makes an angle $\theta = 27°$ with the direction of the positive x axis. The magnitude (length) of $\mathbf{D}_R$ can also be obtained using the theorem of Pythagoras, since D_1, D_2, and D_R form a right triangle with D_R as the hypotenuse. Thus $D_R = \sqrt{D_1^2 + D_2^2} = \sqrt{(10.0\text{ km})^2 + (5.0\text{ km})^2} = \sqrt{125\text{ km}^2} = 11.2$ km. You can use the Pythagorean theorem, of course, only when the vectors are *perpendicular* to each other.

The resultant displacement vector, $\mathbf{D}_R$, is the sum of the vectors $\mathbf{D}_1$ and $\mathbf{D}_2$. That is, $\mathbf{D}_R = \mathbf{D}_1 + \mathbf{D}_2$. This is a *vector* equation. An important feature of adding two vectors that are not along the same line is that the magnitude of the resultant vector is not equal to the sum of the magnitudes of the two separate vectors but is smaller than their sum: $D_R < D_1 + D_2$.

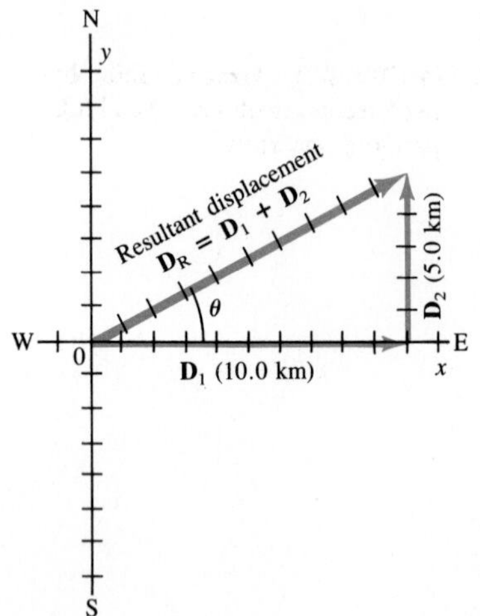

FIGURE 3–1 A person walks 10.0 km east and then 5.0 km north. These two displacements are represented by the vectors $\mathbf{D}_1$ and $\mathbf{D}_2$, which are shown as arrows. The resultant displacement vector, $\mathbf{D}_R$, which is the vector sum of $\mathbf{D}_1$ and $\mathbf{D}_2$, is also shown. Measurement on the graph with ruler and protractor shows that $\mathbf{D}_R$ has magnitude 11.2 km and is at an angle $\theta = 27°$ north of east.

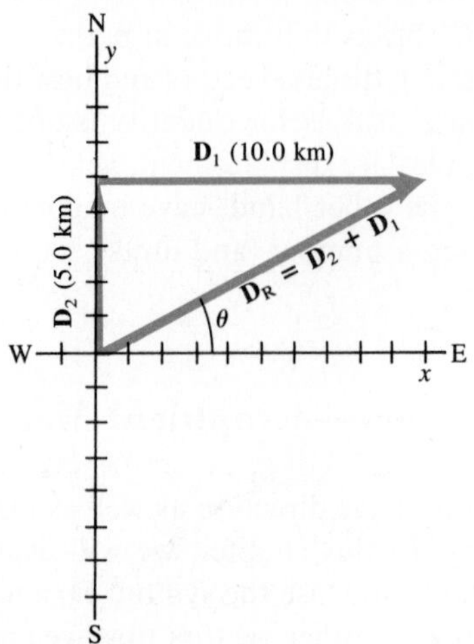

FIGURE 3–2 If the vectors are added in reverse order, the resultant is the same. (Compare Fig. 3–1.)

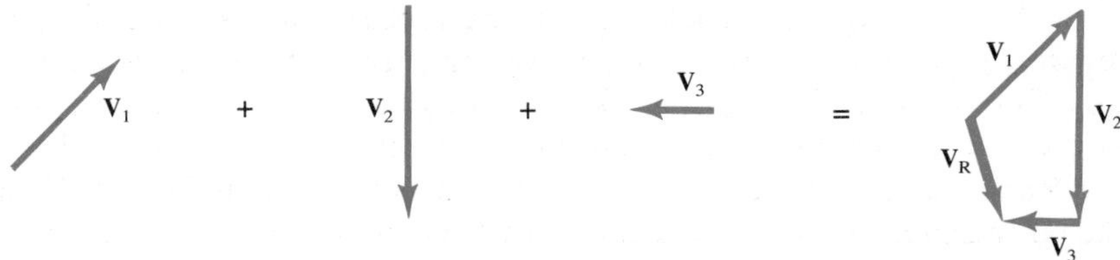

FIGURE 3–3 The resultant of three vectors, $\mathbf{V}_R = \mathbf{V}_1 + \mathbf{V}_2 + \mathbf{V}_3$.

Figure 3–1 illustrates the general rules for graphically adding two vectors together, no matter what angles they make, to get their sum. Specifically, the rules are as follows: (1) on a diagram, draw one of the vectors—call it $\mathbf{V}_1$—to scale; (2) next draw the second vector, $\mathbf{V}_2$, to scale, placing its tail at the tip of the first vector and being sure its direction is correct; (3) the arrow drawn from the tail of the first vector to the tip of the second represents the *sum*, or **resultant**, of the two vectors. The length of the resultant can be measured and compared to the scale. Angles can be measured with a protractor. This method is known as the *tail-to-tip method of adding vectors*. It is, in fact, the definition of how to add vectors.

Tail-to-tip method of adding vectors

Note that it is not important in which order the vectors are taken. For example, a displacement of 5.0 km north, to which is added a displacement of 10.0 km east, yields a resultant of 11.2 km and angle $\theta = 27°$ (see Fig. 3–2), the same as when they were added in reverse order (Fig. 3–1). That is,

$$\mathbf{V}_1 + \mathbf{V}_2 = \mathbf{V}_2 + \mathbf{V}_1.$$

The tail-to-tip method can be extended to three or more vectors. An example is shown in Fig. 3–3; the three vectors could represent displacements (northeast, south, west) or perhaps three forces. Check for yourself that you get the same resultant no matter in which order you add the three vectors.

A second way to add two vectors is the *parallelogram method*; it is equivalent to the tail-to-tip method. In this method, the two vectors are drawn from a common origin and a parallelogram is constructed using these two vectors as adjacent sides. The resultant is the diagonal drawn from the common origin. An example is shown in Fig. 3–4b. In Fig. 3–4a, the tail-to-tip method is shown, and it is clear that both methods yield the same

Parallelogram method of adding vectors

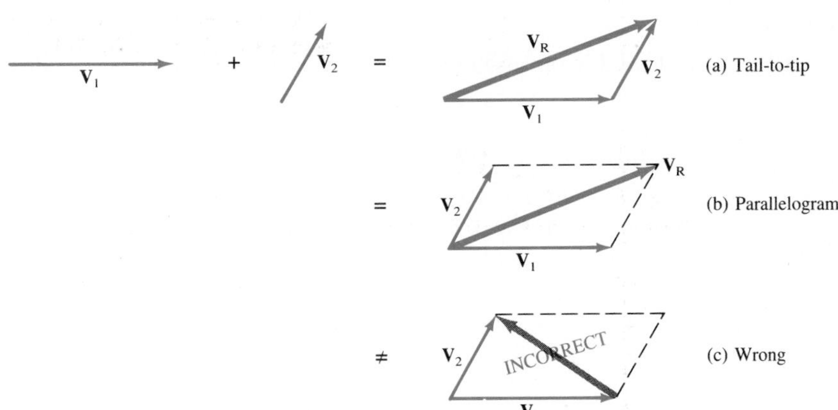

FIGURE 3–4 Vector addition by two different methods, (a) and (b); part (c) is incorrect.

result. It is a common error to draw the sum vector as the diagonal running between the tips of the two vectors, as in Fig. 3–4c. *This is incorrect*: it does not represent the sum of the two vectors (in fact, it represents their difference—see below).

Vectors, when considered solely as mathematical quantities, can be moved about (on paper or in your mind) parallel to themselves as long as their direction and magnitude are not changed. We in fact did this above when adding vectors. (Physically, the position of a vector can be important. For example, the vector representing a force on a body must be placed at the point where the force acts if we are to understand the resulting movement of that body correctly.)

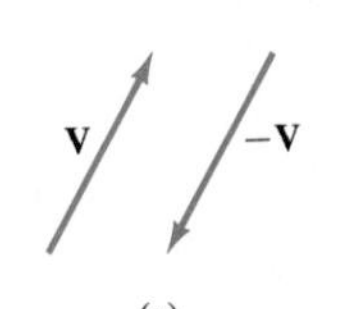

(a)

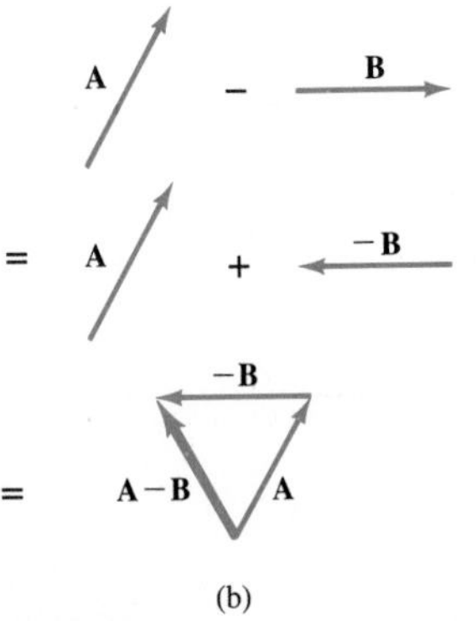

(b)

FIGURE 3–5 (a) The negative of a vector is a vector having the same length but opposite direction. (b) Subtracting two vectors: $\mathbf{A} - \mathbf{B}$.

3–2 • Subtraction of Vectors and Multiplication of a Vector by a Scalar

Given a vector **V**, we define the *negative* of this vector $(-\mathbf{V})$ to be a vector with the same magnitude as **V** but opposite in direction†, Fig. 3–5a. We can use this to define the subtraction of one vector from another: the difference between two vectors, $\mathbf{A} - \mathbf{B}$, is defined as

$$\mathbf{A} - \mathbf{B} = \mathbf{A} + (-\mathbf{B}).$$

That is, the difference between two vectors is equal to the sum of the first plus the negative of the second. Thus our rules for addition of vectors can be applied as shown in Fig. 3–5b using the tail-to-tip method.

A vector **V** can be multiplied by a scalar c. We define this product so that $c\mathbf{V}$ has the same direction as **V** and has magnitude cV. That is, multiplication of a vector by a positive scalar c changes the magnitude of the vector by a factor c but doesn't alter the direction. If c is a negative scalar, the magnitude of the product $c\mathbf{V}$ is still cV (without the minus sign), but the direction is precisely opposite to that of **V**.

3–3 • Analytic Method for Adding Vectors: Components

Adding vectors graphically using a ruler and protractor is often not sufficiently accurate, and is not useful for vectors in three dimensions. We discuss now a more powerful and precise method for adding vectors.

Consider first a vector **V** that lies in a plane. It can be expressed as the sum of two other vectors, called the **components** of the original vector. The components are usually chosen to be along two perpendicular directions. The process of finding the components is known as *resolving the vector into its components*. An example is shown in Fig. 3–6; the vector **V** could be a displacement vector that points at an angle $\theta = 30°$ north of east, where we have chosen the positive x axis to be to the east and the positive y axis north. This vector is resolved into its x and y components by drawing lines from the tip (A) of our given vector that are perpendicular to the x and y axes

Resolving a vector into components

† This is the only sensible way to define the negative of a vector because the sum of a vector and its negative ought to be zero: $\mathbf{V} + (-\mathbf{V}) = 0$.

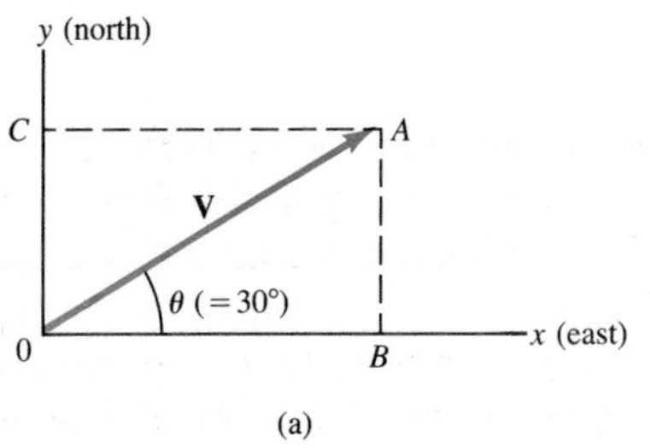

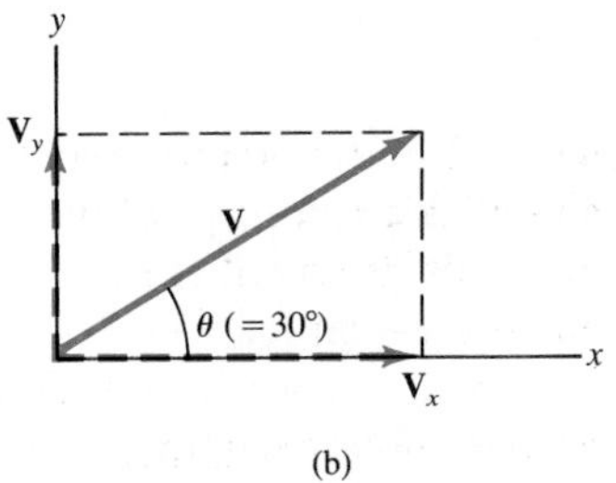

FIGURE 3–6 Resolving a vector **V** into its components along an arbitrarily chosen set of x and y axes. Note that the components, once found, themselves represent the vector. That is, the components contain as much information as the vector itself.

(lines AB and AC). Then the lines OB and OC represent the x and y components of **V**, respectively. These *vector components* are written $\mathbf{V}_x$ and $\mathbf{V}_y$. The magnitudes of $\mathbf{V}_x$ and $\mathbf{V}_y$, namely V_x and V_y, are referred to simply as the *components*, and thus are numbers (with units) that are positive or negative depending on whether they point along the positive or negative x or y axis. As can be seen in Fig. 3–6, $\mathbf{V}_x + \mathbf{V}_y = \mathbf{V}$ by the parallelogram method of adding vectors.

Vector components

Space is made up of three dimensions, and sometimes it is necessary to resolve a vector into components along three mutually perpendicular directions. The components are then called $\mathbf{V}_x$, $\mathbf{V}_y$, and $\mathbf{V}_z$. Although resolution of a vector in three dimensions is merely an extension of the above technique, we will mainly be concerned with situations in which the vectors are in a plane and two components are all that are necessary.

In order to add vectors using the method of components, we need to use the trigonometric functions sine, cosine, and tangent, which we now discuss.

Given any angle, θ, as in Fig. 3–7a, a right triangle can be constructed by drawing a line perpendicular to either of its sides, as in Fig. 3–7b. The longest side of a right triangle, opposite the right angle, is called the hypotenuse, which we label h. The side opposite the angle θ is labeled o, and the side adjacent is labeled a. We let h, o, and a represent the lengths of these sides, respectively. We now define the three trigonometric functions, sine, cosine, and tangent (abbreviated sin, cos, tan), in terms of the right triangle, as follows:

$$\sin\theta = \frac{\text{side opposite}}{\text{hypotenuse}} = \frac{o}{h}$$

$$\cos\theta = \frac{\text{side adjacent}}{\text{hypotenuse}} = \frac{a}{h} \qquad (3\text{–}1)$$

$$\tan\theta = \frac{\text{side opposite}}{\text{side adjacent}} = \frac{o}{a}.$$

Trig. functions defined

Now it is an interesting fact that if we make the triangle bigger, but keep the same angles, then the ratio of the length of one side to the other, or of one side to the hypotenuse, remains the same. That is, in Fig. 3–7c we have

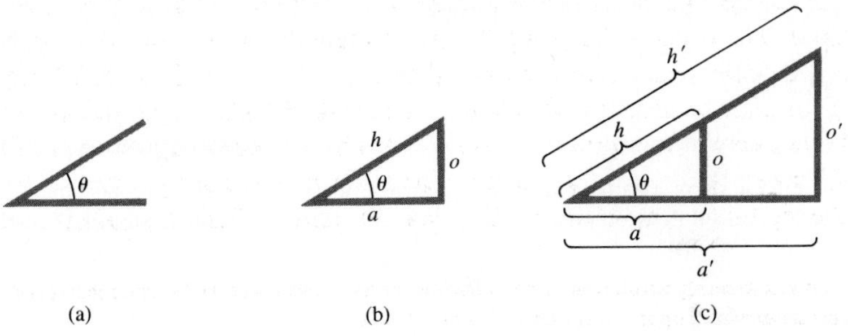

FIGURE 3–7 Starting with an angle θ as in (a), we can construct right triangles of different sizes, (b) and (c), but the ratio of the lengths of the sides does not depend on the size of the triangle.

$a/h = a'/h'$; $o/h = o'/h'$; and $o/a = o'/a'$. Thus the values of sin, cos, and tan do not depend on how big the triangle is. They depend only on the size of the angle. The values of sin, cos, and tan for different angles can be found in the table on the inside back cover of this book, and are given by scientific calculators.

A useful trigonometric identity is

$$\sin^2\theta + \cos^2\theta = 1, \tag{3–2}$$

which follows from the Pythagorean theorem ($o^2 + a^2 = h^2$ in Fig. 3–7):

$$\sin^2\theta + \cos^2\theta = \frac{o^2 + a^2}{h^2} = \frac{h^2}{h^2} = 1.$$

(See also Appendix A for other details on trigonometric functions and identities.) Figure 3–8 shows the signs (+ or −) that the trigonometric functions take on for θ over the range 0° to 360°.

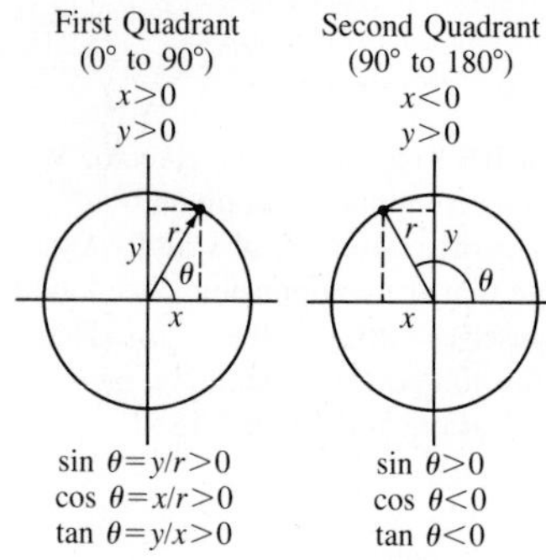

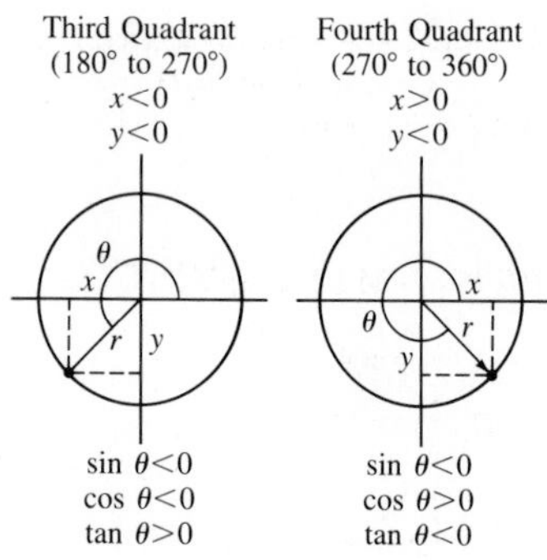

FIGURE 3–8 Signs of the cosine, sine, and tangent for angles θ in the four quadrants (0° to 360°). Note that angles are measured counterclockwise from the x axis as shown; negative angles are measured from *below* the x axis, clockwise: for example, −30° = +330°, and so on.

The use of trigonometric functions for finding the components of a vector is illustrated in Fig. 3–9, where it is seen that a vector and its two components can be thought of as making up a right triangle. If we multiply the definition of $\sin\theta = V_y/V$ by V on both sides, we get

$$V_y = V\sin\theta. \tag{3–3a}$$

Similarly, from the definition of $\cos\theta$, we obtain

$$V_x = V\cos\theta. \tag{3–3b}$$

Note that θ is chosen (by convention) to be the angle that the vector makes with the positive x axis.[†]

Using Eqs. 3–3, we can calculate V_x and V_y for the vector illustrated in Fig. 3–6. Suppose **V** represents a displacement of 500 m in a direction 30° north of east. Then $V = 500$ m. From the trigonometric tables, $\sin 30° = 0.500$ and $\cos 30° = 0.866$. Then $V_x = V\cos\theta = (500\text{ m})(0.866) = 433$ m (east) and $V_y = V\sin\theta = (500\text{ m})(0.500) = 250$ m (north).

Note that there are two ways to specify a vector in a given coordinate system: we can give its components, V_x and V_y; or we can give its magnitude V and the angle θ it makes with the positive x axis. We can shift from one description to the other using Eqs. 3–3, or by using the theorem of Pythagoras[‡] and the definition of tangent

$$V = \sqrt{V_x^2 + V_y^2} \tag{3–4a}$$

$$\tan\theta = \frac{V_y}{V_x} \tag{3–4b}$$

as can be seen in Fig. 3–9.

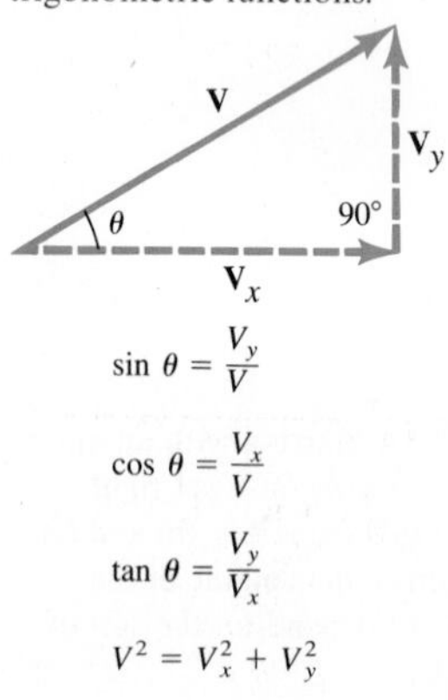

FIGURE 3–9 Finding the components of a vector using trigonometric functions.

We can now discuss how to add vectors analytically by using components. The first step is to resolve each vector into its components. Next we can show (see Fig. 3–10) that the addition of two vectors $\mathbf{V}_1$ and $\mathbf{V}_2$ to give a resultant, $\mathbf{V} = \mathbf{V}_1 + \mathbf{V}_2$, implies that

$$\begin{aligned} V_x &= V_{1x} + V_{2x} \\ V_y &= V_{1y} + V_{2y}. \end{aligned} \tag{3–5}$$

A careful examination of Fig. 3–10 will verify these equations for the x and

[†] Whatever set of axes is used, the component opposite the angle is proportional to the sine, whether we call that component x or y. Normally, we use the convention that it is the y component (Eq. 3–3a).

[‡] In three dimensions, the theorem of Pythagoras becomes $V^2 = \sqrt{V_x^2 + V_y^2 + V_z^2}$, where V_z is the component along the third, or z, axis.

y components. If the magnitude and direction of the resultant vector are desired, they can be obtained using Eqs. 3–4.

The choice of coordinate axes is, of course, always arbitrary. You can often reduce the work involved in adding vectors analytically by a good choice of axes—for example, by choosing one of the axes to be in the same direction as one of the vectors so that vector will have only one nonzero component.

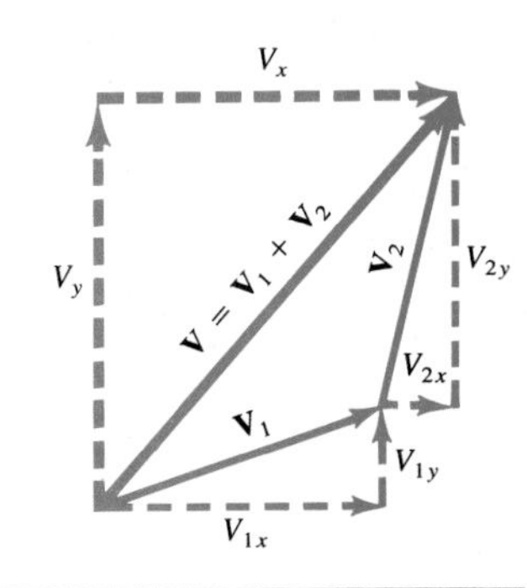

FIGURE 3–10 The components of $\mathbf{V} = \mathbf{V}_1 + \mathbf{V}_2$ are $V_x = V_{1x} + V_{2x}$ and $V_y = V_{1y} + V_{2y}$.

EXAMPLE 3–1 An explorer walks 22.0 km in a northerly direction, and then walks in a direction 60° south of east for 47.0 km (Fig. 3–11a). How far is she from where she started?

SOLUTION We want to find her resultant displacement from the origin, and we choose the positive x axis to be east and the positive y axis north. We resolve each displacement vector into its components (Fig. 3–11b). Since $\mathbf{D}_1$ has magnitude 22.0 km and points north, it has only a y component:

$$D_{1x} = 0, \qquad D_{1y} = 22.0 \text{ km},$$

whereas $\mathbf{D}_2$ has both x and y components:

$$D_{2x} = (47.0 \text{ km})(\cos 60°) = (47.0 \text{ km})(0.500) = 23.5 \text{ km}$$

$$D_{2y} = -(47.0 \text{ km})(\sin 60°) = -(47.0 \text{ km})(0.866) = -40.7 \text{ km}.$$

Notice that D_{2y} is negative because this vector component points along the negative y axis. The resultant vector ($\mathbf{D}$) has components:

$$D_x = D_{1x} + D_{2x} = 0 + 23.5 \text{ km} = 23.5 \text{ km}$$

$$D_y = D_{1y} + D_{2y} = 22.0 \text{ km} - 40.7 \text{ km} = -18.7 \text{ km}.$$

This specifies the resultant vector completely: $D_x = 23.5$ km, $D_y = -18.7$ km. We can also specify it by giving its magnitude and angle:

$$D = \sqrt{D_x^2 + D_y^2} = \sqrt{(23.5 \text{ km})^2 + (-18.7 \text{ km})^2} = 30.0 \text{ km}$$

$$\tan\theta = \frac{D_y}{D_x} = -\frac{18.7 \text{ km}}{23.5 \text{ km}} = -0.796,$$

so $\theta = 38.5°$ south of east. The negative sign for the tangent results because θ is below the x axis, Fig. 3–11c.

FIGURE 3–11 Example 3–1.

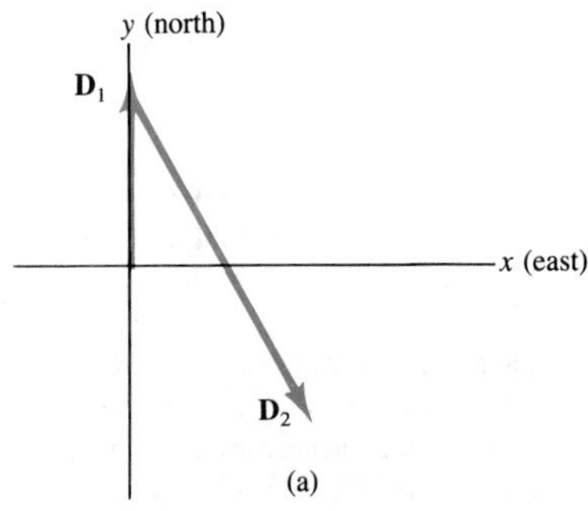

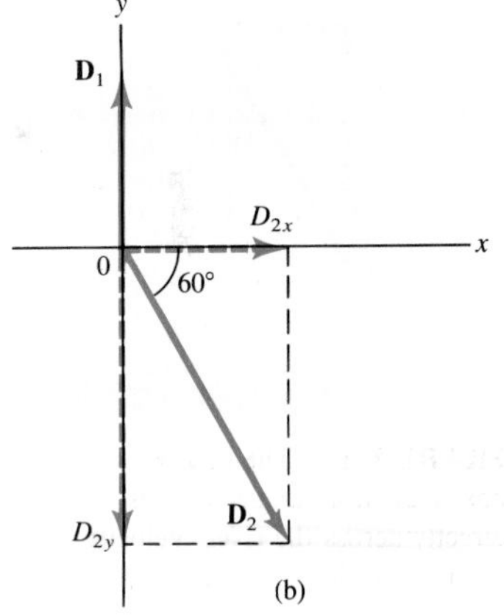

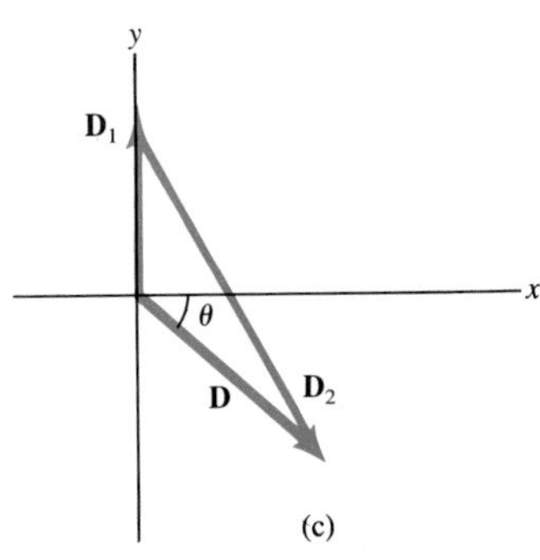

Even though adding vectors analytically is generally more precise than graphical methods, it is important to draw at least rough vector diagrams for given situations since they give you something tangible to look at when analyzing a problem, and can provide a rough check on results.

3–4 • Relative Velocity—Vectors in Problem Solving

A very useful example of adding vectors involves determining *relative velocity*. As a simple example, consider two trains approaching one another, each with a speed of 80 km/h with respect to earth. Then the speed of one train relative to the other is 160 km/h. That is, to an observer on one train, the

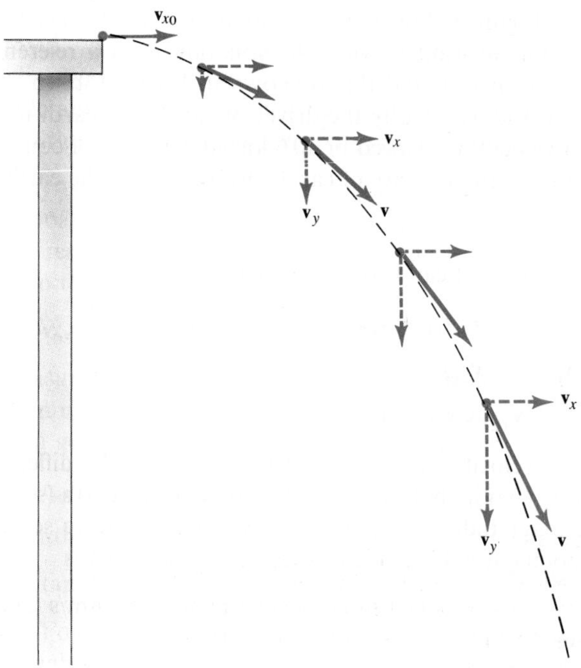

FIGURE 3–17 Projectile motion.

important, in many cases its effect can be ignored, and we will ignore it in the following analysis. We will not be concerned now with the process by which the object is thrown or projected. We consider only its motion after it has been projected and is moving freely through the air under the action of gravity alone. Thus the acceleration of the object is that of gravity, **g**, which acts downward with magnitude $g = 9.80 \text{ m/s}^2$.

Horizontal and vertical motion analyzed separately

It was Galileo (Fig. 2–12) who first accurately described projectile motion; he showed that it could be understood by analyzing the horizontal and vertical components of the motion separately. This was an innovative analysis, not done in this way by anyone prior to Galileo. (It was also idealized in that it did not take into account air resistance.) For convenience, we assume that the motion begins at time $t = 0$ at the origin of an x-y coordinate system (so $x_0 = y_0 = 0$).

FIGURE 3–18 Multiple-exposure photograph showing positions of 2 balls at equal time intervals. One ball was dropped from rest at the same time the other was projected horizontally outward. The vertical position of each ball is seen to be the same.

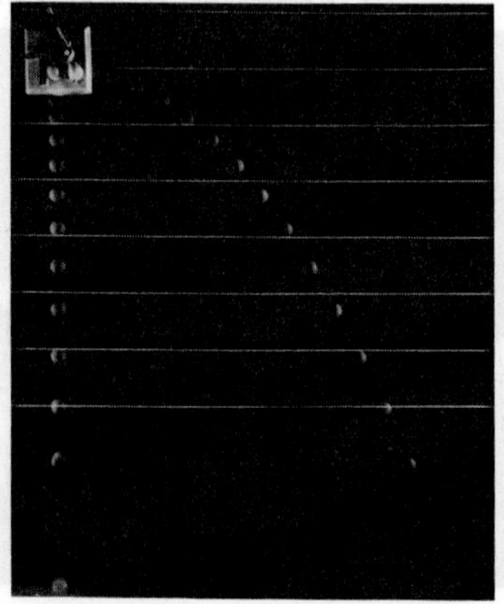

First let us look at a (tiny) ball rolling off the end of a table with an initial velocity v_{x0} in the horizontal (x) direction, Fig. 3–17. The velocity vector **v** at each instant points in the direction of the motion at that instant and is always tangent to the path. Following Galileo's ideas, we treat the horizontal and vertical components of the velocity, v_x and v_y, separately, and we can apply the kinematic equations (2–10a through 2–10d) to each. Once the ball leaves the table (at $t = 0$), it experiences a vertically downward acceleration equal to g, the acceleration of gravity. Thus v_y is continually increasing in the downward direction (until the ball hits the ground). From Eq. 2–10a, we can write $v_y = gt$ since the initial velocity in the vertical direction is zero. The vertical displacement downward, y, is given by $y = \frac{1}{2}gt^2$. Because there is no acceleration in the horizontal direction, the horizontal component of velocity, v_x, remains constant, equal to its initial value, v_{x0}, and thus has the same magnitude at each point on the path. The two vector components, $\mathbf{v}_x$ and $\mathbf{v}_y$, can be added vectorially to obtain the velocity **v** for each point on the path, as shown in Fig. 3–17.

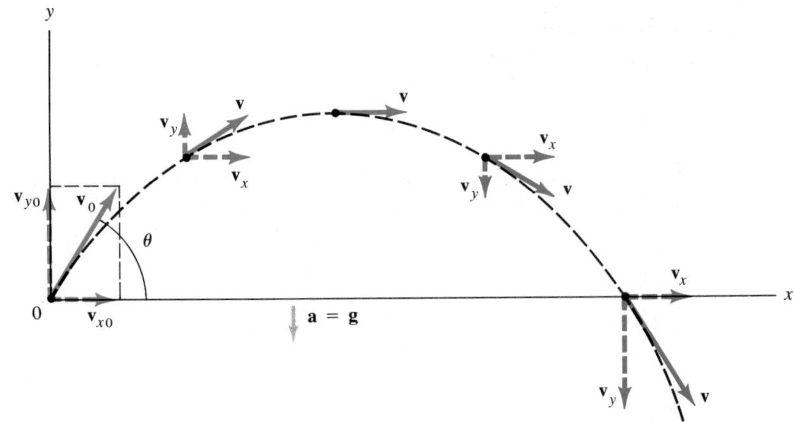

FIGURE 3–19 Path of a projectile fired with initial velocity v_0 at angle θ to the horizontal. Path is shown in black, the velocity vectors are green arrows, and velocity components are dashed.

One result of this analysis, which Galileo himself predicted, is that *an object projected horizontally will reach the ground in the same time as an object dropped vertically.* This is because the vertical motions are the same in both cases. Figure 3–18 is a multiple-exposure photograph of an experiment that confirms this.

If an object is projected at an angle, as in Fig. 3–19, the analysis is essentially the same except that now there is an initial vertical component of velocity, v_{y0}. Because of the downward acceleration of gravity, v_y continually decreases until the object reaches the highest point on its path in Fig. 3–19, at which point $v_y = 0$. Then v_y starts to increase in the downward direction, as shown.

3–6 • Solving Problems Involving Projectile Motion

We now work through several examples of projectile motion quantitatively. We use the kinematic equations (2–10a through 2–10d) separately for the vertical and horizontal components of the motion; we use x or y for the respective displacement. See Table 3–1, which assumes y is positive upward, so $a_y = -9.80\ \text{m/s}^2$. If the y axis is taken positive downward, $a_y = +9.80\ \text{m/s}^2$, and the minus signs in Table 3–1 must be changed to plus signs.

TABLE 3–1
Kinematic Equations for Projectile Motion
(y positive upward)

Horizontal Motion		Vertical Motion†
$v_x = v_{x0}$	(Eq. 2–10a)	$v_y = v_{y0} - gt$
$x = v_{x0}t$	(Eq. 2–10b)	$y = v_{y0}t - \frac{1}{2}gt^2$
	(Eq. 2–10c)	$v_y^2 = v_{y0}^2 - 2gy$

† If y is taken positive downward, the minus (−) signs become + signs.

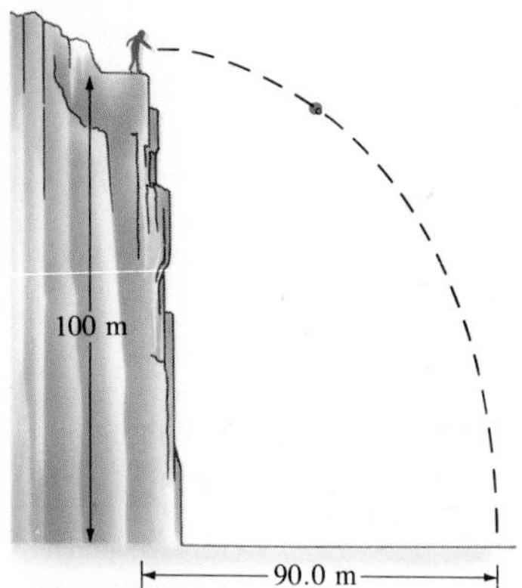

FIGURE 3–20 Example 3–6.

EXAMPLE 3–6 A rock is thrown horizontally from a 100.0-m-high cliff. It strikes level ground 90.0 m from the base of the cliff, Fig. 3–20. At what speed was it thrown? Take the y direction to be positive downward (which means we must change the minus signs in Table 3–1 to plus signs).

SOLUTION First, we find how long it takes the rock to reach the ground below. We use Eq. 2–10b for the vertical (y) direction (which we take as positive downward) with $v_{y0} = 0$ and $a = g = 9.80 \text{ m/s}^2$. Thus

$$y = \tfrac{1}{2}gt^2,$$

$$t = \sqrt{\frac{2y}{g}} = \sqrt{\frac{200.0 \text{ m}}{9.80 \text{ m/s}^2}} = 4.52 \text{ s}.$$

To calculate the initial velocity, v_{x0}, we again use Eq. 2–10b, but this time for the horizontal (x) direction, so $a = 0$.

$$x = v_{x0}t$$

$$v_{x0} = \frac{x}{t} = \frac{90.0 \text{ m}}{4.52 \text{ s}} = 19.9 \text{ m/s}.$$

EXAMPLE 3–7 A football is kicked at an angle $\theta_0 = 37.0°$ with a velocity of 20.0 m/s, as shown in Fig. 3–19. Calculate (*a*) the maximum height, (*b*) the time of travel before it hits the ground, (*c*) how far away it hits the ground, (*d*) the velocity vector at the maximum height, and (*e*) the acceleration vector at maximum height. Assume for simplicity that the ball leaves the foot at ground level.

SOLUTION We take the y direction as positive upward. The components of the initial velocity are

$$v_{x0} = v_0 \cos 37.0° = (20.0 \text{ m/s})(0.799) = 16.0 \text{ m/s}$$

$$v_{y0} = v_0 \sin 37.0° = (20.0 \text{ m/s})(0.602) = 12.0 \text{ m/s}.$$

(*a*) The maximum height is attained where $v_y = 0$, and this occurs (see Eq. 2–10a in Table 3–1) at time $t = v_{y0}/g = (12.0 \text{ m/s})/(9.80 \text{ m/s}^2) = 1.22$ s. From Eq. 2–10b, we have

$$\begin{aligned} y &= v_{y0}t - \tfrac{1}{2}gt^2 \\ &= (12.0 \text{ m/s})(1.22 \text{ s}) - \tfrac{1}{2}(9.80 \text{ m/s}^2)(1.22 \text{ s})^2 \\ &= 7.35 \text{ m}. \end{aligned}$$

Alternatively, we could have used Eq. 2–10c, solved for y, and found

$$y = \frac{v_{y0}^2 - v_y^2}{2g} = \frac{(12.0 \text{ m/s})^2 - (0 \text{ m/s})^2}{2(9.80 \text{ m/s}^2)} = 7.35 \text{ m}.$$

(*b*) To find the time it takes for the ball to return to the ground, we use Eq. 2–10b and set $y = 0$ (ground level):

$$y = v_{y0}t - \tfrac{1}{2}gt^2$$

$$0 = (12.0 \text{ m/s})t - \tfrac{1}{2}(9.80 \text{ m/s}^2)t^2,$$

and therefore

$$t = \frac{2(12.0\ \text{m/s})}{(9.80\ \text{m/s}^2)} = 2.45\ \text{s}.$$

(The solution $t = 0$ also satisfies the equation, but this corresponds to the initial point when y is also zero.)

(*c*) The total distance traveled in the x direction is found by applying Eq. 2–10b with $a = 0$, $v_{x0} = 16.0$ m/s:

$$x = v_{x0}t = (16.0\ \text{m/s})(2.45\ \text{s}) = 39.2\ \text{m}.$$

(*d*) At the highest point, there is no vertical component to the velocity. There is only the horizontal component (which remains constant throughout the flight), so $v = v_{x0} \cos 37.0° = 16.0$ m/s.

(*e*) The acceleration vector is the same at the highest point as it is throughout the flight, $\mathbf{a} = -\mathbf{g}$, 9.80 m/s^2 downward.

Horizontal range of a projectile

EXAMPLE 3–8 (*a*) Derive a formula for the horizontal range R of a projectile, which is defined as the horizontal distance it travels before returning to its original height, $y = 0$ (which is typically the ground). (*b*) Suppose one of Napoleon's cannons had a muzzle velocity, v_0, of 60.0 m/s. At what angle should it have been aimed (ignore air resistance) to strike a target 320 m away?

SOLUTION (*a*) To find a general expression for R, we set $y = 0$ in Eq. 2–10b for the vertical motion (Table 3–1), and we obtain

$$v_{y0}t - \tfrac{1}{2}gt^2 = 0.$$

We solve for t, which gives two solutions: $t = 0$ and $t = 2v_{y0}/g$. The first corresponds to the initial instant of projection and the second ($t = 2v_{y0}/g$) is the time when the projectile returns to $y = 0$. Then the range, using the latter value of t in Eq. 2–10b for the horizontal motion ($x = v_{x0}t$), is given by:

$$R = v_{x0}t = v_{x0}\left(\frac{2v_{y0}}{g}\right) = \frac{2v_{x0}v_{y0}}{g} = \frac{2v_0^2 \sin\theta_0 \cos\theta_0}{g} = \frac{v_0^2 \sin 2\theta_0}{g}.$$

Here we have written $v_{x0} = v_0 \cos\theta_0$ and $v_{y0} = v_0 \sin\theta_0$, and we have used the trigonometric identity $2 \sin\theta \cos\theta = \sin 2\theta$ (Appendix A). We see that the maximum range, for a given initial velocity, v_0, is obtained when the sine takes on its maximum value of 1.0, which occurs for $2\theta_0 = 90°$; so $\theta_0 = 45°$ for maximum range. (When air resistance is important, the range is less for a given v_0, and the maximum range is obtained at an angle smaller than 45°.) Note that the maximum range increases by the square of v_0, so doubling the muzzle speed of a cannon, say, increases its maximum range by a factor of 4.

(*b*) From the equation we just derived:

$$\sin 2\theta_0 = \frac{Rg}{v_0^2} = \frac{(320\ \text{m})(9.80\ \text{m/s}^2)}{(60.0\ \text{m/s})^2} = 0.871.$$

Thus $2\theta_0 = 60.6°$, or $180° - 60.6° = 119.4°$, so that

$$\theta_0 = 30.3° \quad \text{or} \quad 59.7°.$$

Either angle gives the same range.

EXAMPLE 3–9 [A more difficult Example] Suppose the football in Example 3–7 left the punter's foot at a height of 1.00 m above the ground. How far did the football travel before hitting the ground?

SOLUTION We cannot use the range formula from Example 3–8 because $y \neq y_0$. Instead, we take $y_0 = 0$ and the football hits the ground where $y = -1.00$ m (see Fig. 3–21). We can get x from Eq. 2–10b, $x = v_{x0}t$, since we know that $v_{x0} = 16.0$ m/s; but first we must find t, the time at which the ball hits the ground. With $y = -1.00$ m and $v_{y0} = 12.0$ m/s (see Example 3–7), we use the equation

$$y = v_{y0}t - \tfrac{1}{2}gt^2,$$

and obtain

$$-1.00 \text{ m} = (12.0 \text{ m/s})t - (4.90 \text{ m/s}^2)t^2.$$

We rearrange this equation into standard form so we can use the quadratic formula (Appendix A):

$$(4.90 \text{ m/s}^2)t^2 - (12.0 \text{ m/s})t - (1.00 \text{ m}) = 0.$$

Using the quadratic formula gives

$$t = \frac{12.0 \text{ m/s} \pm \sqrt{(12.0 \text{ m/s})^2 + 4(4.90 \text{ m/s}^2)(1.00 \text{ m})}}{2(4.90 \text{ m/s}^2)}$$

$$= 2.53 \text{ s} \quad \text{or} \quad -0.081 \text{ s}.$$

The second solution corresponds to a time previous to the kick, so it doesn't apply here. With $t = 2.53$ s for the time at which the ball touches the ground, the distance the ball traveled is

$$x = v_{x0}t = (16.0 \text{ m/s})(2.53 \text{ s}) = 40.5 \text{ m}.$$

Note that our assumption in Example 3–7 that the ball leaves the foot at ground level results in an underestimate of about 1.3 m in the distance traveled.

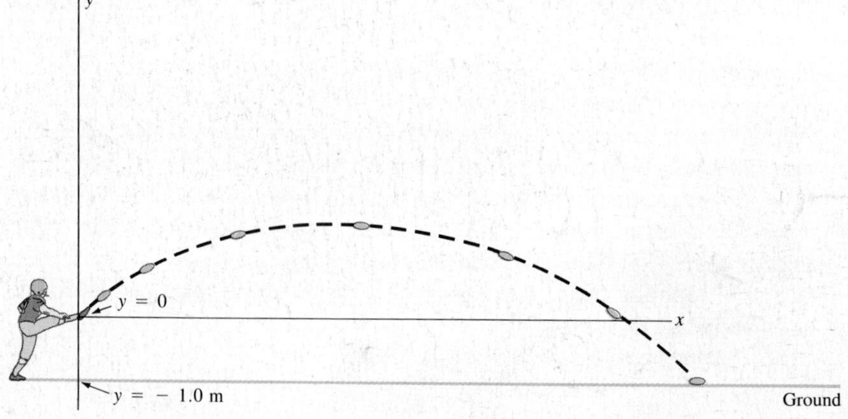

FIGURE 3–21 Example 3–9. The football leaves the punter's foot at $y = 0$, and reaches the ground where $y = -1.00$ m.

We now show that the path followed by any projectile is a parabola, if we can ignore air resistance and can assume that **g** is constant. To do so, we need to find y as a function of x by eliminating the time between the two equations for horizontal and vertical motion (Eq. 2–10b in Table 3–1):

Projectile path is a parabola

$$x = v_{x0}t$$
$$y = v_{y0}t - \tfrac{1}{2}gt^2.$$

From the first equation, we have $t = x/v_{x0}$ and we substitute this into the second one to obtain

$$y = \left(\frac{v_{y0}}{v_{x0}}\right)x - \left(\frac{g}{2v_{x0}^2}\right)x^2.$$

If we write $v_{x0} = v_0 \cos\theta_0$ and $v_{y0} = v_0 \sin\theta_0$, we can also write

$$y = (\tan\theta_0)x - \left(\frac{g}{2v_0^2\cos^2\theta_0}\right)x^2.$$

In either case, we see that y as a function of x has the form

$$y = ax - bx^2,$$

where a and b are constants for any specific projectile motion. This is the well-known equation for a parabola (see Figs. 3–16 and 3–22).

FIGURE 3–22 Examples of projectile motion—sparks (small hot glowing pieces of metal), water, and fireworks. All exhibit the parabolic path characteristic of projectile motion, although the effects of air resistance can be seen to alter the path of some trajectories.

SUMMARY

A quantity that has both a magnitude and a direction is called a *vector*. A quantity that has only a magnitude is called a *scalar*.

Addition of vectors can be done graphically by placing the tail of each successive arrow (representing each vector) at the tip of the previous one. The sum, or *resultant vector*, is the arrow drawn from the tail of the first to the tip of the last. Two vectors can also be added using the parallelogram method. Vectors can be added more accurately using the analytical method of adding their *components* along chosen axes with the aid of trigonometric functions.

The velocity of an object relative to one frame of reference can be found by vector addition if its velocity relative to a second frame of reference, and the relative velocity of the two reference frames, are known.

Projectile motion—that of an object moving in an arc above the earth's surface—can be analyzed as two separate motions if air resistance can be ignored. The horizontal component of motion is at constant velocity, whereas the vertical component is at constant acceleration, g, just as for a body falling vertically under the action of gravity.

QUESTIONS

1. Does the odometer of a car measure a scalar or a vector quantity? What about the speedometer?

2. Can you give several examples of an object's motion in which a great distance is traveled but the displacement is zero?

3. Can the displacement vector for a particle moving in two dimensions ever be longer than the length of path traveled by the particle over the same time interval? Can it ever be less? Discuss.

4. During practice, a baseball player hits a very high fly ball, and then runs in a straight line and catches it. Which had the greater displacement, the player or the ball?

5. If $\mathbf{V} = \mathbf{V}_1 + \mathbf{V}_2$, is V necessarily greater than V_1 and/or V_2? Discuss.

6. Two vectors have length $V_1 = 3.5$ km and $V_2 = 4.0$ km. What are the maximum and minimum magnitudes of their vector sum?

7. Can two vectors, of unequal magnitude, add up to give the zero vector? Can *three* unequal vectors? Under what conditions?

8. Can the magnitude of a vector ever (*a*) equal, or (*b*) be less than, one of its components?

9. Can a vector of magnitude zero have a nonzero component?

10. One car travels due east at 40 km/h, and a second car travels north at 40 km/h. Are their velocities equal? Explain.

11. Two rowers, who can row at the same speed in still water, set off across a river at the same time. One heads straight across and is pulled downstream somewhat by the current. The other one heads upstream at an angle so as to arrive at a point opposite the starting point. Which rower reaches the opposite side first?

12. Two cars with equal speed approach an intersection at right angles to each other. Will they necessarily collide? Show that when the relative velocity of approach is collinear (along the same line) with the relative displacement, we have the nautical maxim "constant bearing means collision."

13. What physical factors are important for an athlete doing the broad jump? What about the high jump?

14. A projectile has the least speed at what point in its path?

15. A child wishes to determine the speed a slingshot imparts to a rock. How can this be done using only a meter stick?

PROBLEMS

SECTIONS 3–1 TO 3–3

1. (I) A car is driven 80 km west and then 30 km southwest. What is the displacement of the car from the point of origin (magnitude and direction)?

2. (I) A delivery truck travels 18 blocks north, 10 blocks east, and 20 blocks south. What is its final displacement from the origin? Assume the blocks are equal length.

3. (I) The three vectors in Fig. 3–3 can be added in six different orders ($\mathbf{V}_1 + \mathbf{V}_2 + \mathbf{V}_3$, $\mathbf{V}_1 + \mathbf{V}_3 + \mathbf{V}_2$, etc.). Show on a diagram that the same resultant is obtained no matter what the order.

4. (I) Show that the vector labeled "wrong" in Fig. 3–4c is actually the difference of the two vectors. Is it $\mathbf{V}_2 - \mathbf{V}_1$ or $\mathbf{V}_1 - \mathbf{V}_2$?

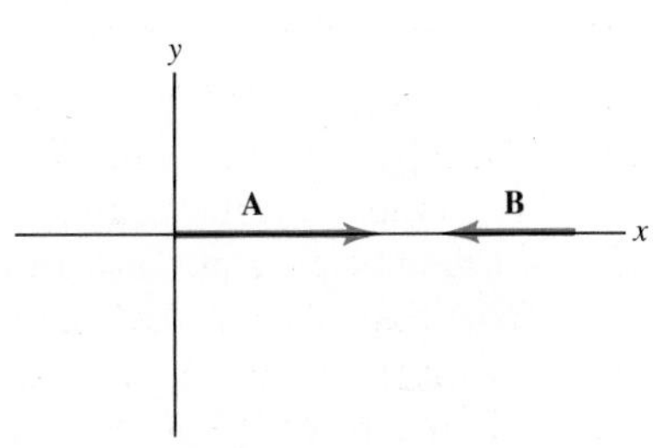

FIGURE 3–23 Problem 5.

5. (I) Figure 3–23 shows two vectors, **A** and **B**, whose magnitudes are $A = 6.3$ units and $B = 3.5$ units. Determine **C** if (*a*) **C** = **A** + **B**, (*b*) **C** = **A** − **B**, (*c*) **C** = **B** − **A**. Give the magnitude and direction for each.
6. (I) If $V_x = -8.50$ units and $V_y = 4.80$ units, determine the magnitude and direction of **V**.
7. (II) Graphically determine the resultant of the following three vector displacements: (1) 20 m, 30° north of east; (2) 16 m, 37° east of north; and (3) 12 m, 30° west of south.
8. (II) Vector $\mathbf{V}_1$ is 6.0 units long and points along the negative *y* axis. Vector $\mathbf{V}_2$ is 4.5 units long and points at +45° to the positive *x* axis. (*a*) What are the *x* and *y* components of each vector? (*b*) Determine the sum of the two vectors (length and angle).
9. (II) An airplane is traveling 900 km/h in a direction 38.5° west of north (Fig. 3–24). (*a*) Find the components of the velocity vector in the northerly and westerly directions. (*b*) How far north and how far west has the plane traveled after 3.00 h?
10. (II) The components of a vector **V** are often written (V_x, V_y, V_z). What are the components and length of a vector that is the sum of the two vectors, $\mathbf{V}_1$ and $\mathbf{V}_2$, whose components are (6, 0, 8) and (3, 4, −3)?
11. (II) Given the vectors $\mathbf{V}_1$ and $\mathbf{V}_2$ from Problem 10, determine a third vector, $\mathbf{V}_3$, such that (*a*) $\mathbf{V}_1 + \mathbf{V}_2 + \mathbf{V}_3 = 0$. (*b*) $\mathbf{V}_1 - \mathbf{V}_2 + \mathbf{V}_3 = 0$.

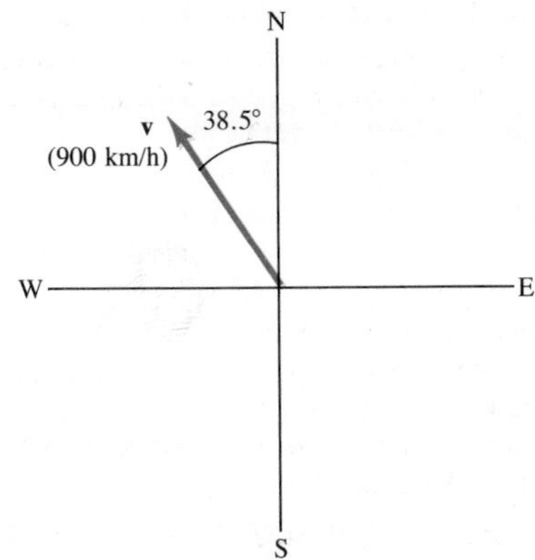

FIGURE 3–24 Problem 9.

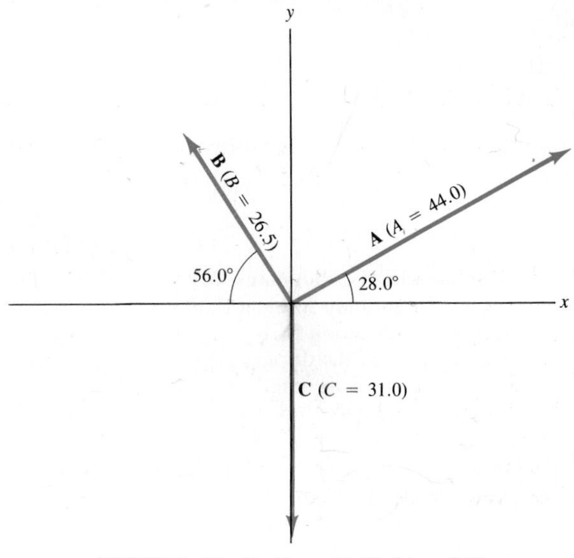

FIGURE 3–25 Problems 12, 13, 14, and 15.

12. (II) Three vectors are shown in Fig. 3–25. Their magnitudes are given in arbitrary units. Determine the sum of the three vectors. Give the resultant in terms of (*a*) components, (*b*) magnitude and angle with *x* axis.
13. (II) Determine the vector **A** − **C** given the vectors **A** and **C** in Fig. 3–25.
14. (II) Given the vectors **A** and **B** shown in Fig. 3–25, determine **B** − **A**.
15. (II) For the vectors given in Fig. 3–25, determine (*a*) **A** − **B** + **C**, and (*b*) **A** + **B** − **C**.
16. (II) (*a*) A skier is accelerating down a 30.0° hill at 3.80 m/s² (Fig. 3–26). What is the vertical component of her acceleration? (*b*) How long will it take her to reach the bottom of the hill, assuming she starts from rest and accelerates uniformly, if the elevation change is 250 m?
17. (II) What is the *y* component of a vector in the *xy* plane whose magnitude is 86.6 and whose *x* component is 35.4? What is the direction of this vector?

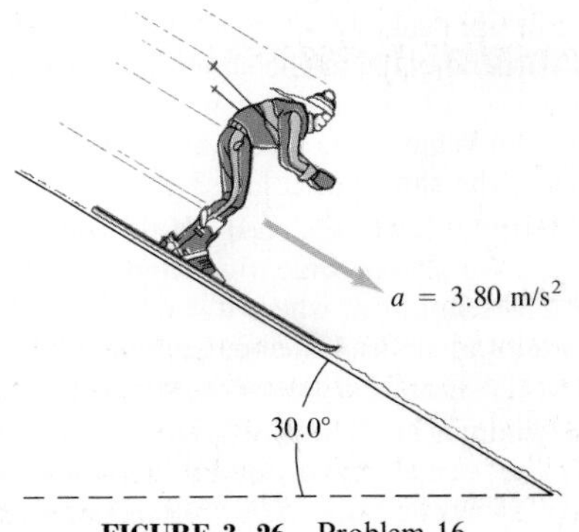

FIGURE 3–26 Problem 16.

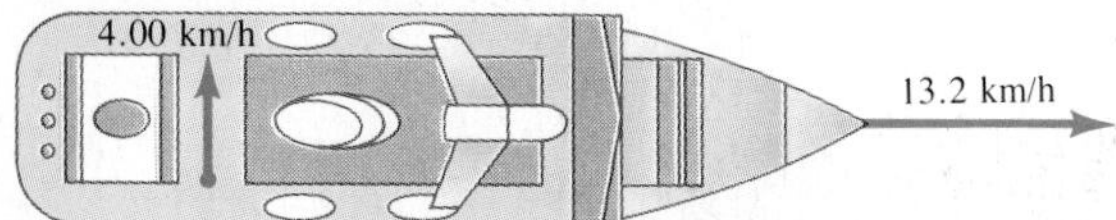

FIGURE 3–27 Problem 20.

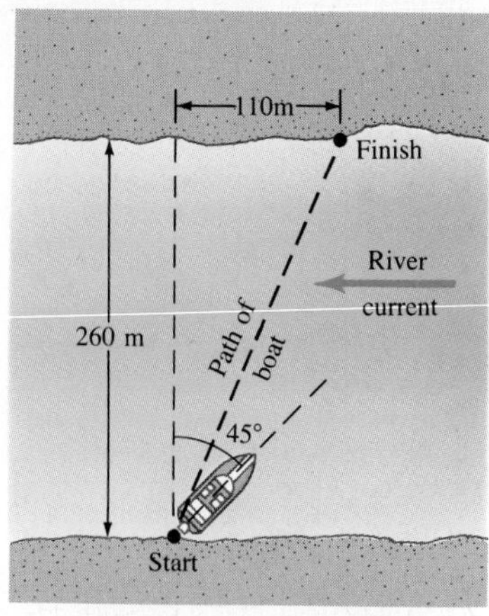

FIGURE 3–28 Problem 26.

18. (II) The summit of a mountain, 2250 m above a camp, is measured on a map to be 4850 m horizontally from the camp in a direction 38.2° west of north. What are the components of the displacement vector from camp to summit? What is its length? Choose the x axis east, y axis north, and z axis up.

SECTION 3–4

19. (I) A cat walks at a speed of 1.8 m/s along the deck toward the front of a boat that is traveling at 6.6 m/s with respect to the water. What is the velocity of the cat with respect to the water? What if the cat were walking toward the back of the boat?
20. (II) A vacationer walks 4.00 km/h directly across a cruise ship whose speed relative to the earth is 13.2 km/h (Fig. 3–27). What is the speed of the vacationer with respect to the earth?
21. (II) A boat can travel 3.10 m/s in still water. (*a*) If the boat points its prow directly across a stream whose current is 1.00 m/s, what is the velocity (magnitude and direction) of the boat relative to the shore? (*b*) What will be the position of the boat, relative to its point of origin, after 4.0 s? (See Fig. 3–14.)
22. (II) Two trains approach each other on parallel tracks. Each has a speed of 80 km/h with respect to the earth. If they are initially 10 km apart, how long will it be before they pass each other?
23. (II) An airplane is heading due north at a speed of 300 km/h. If a wind begins blowing from the southwest at a speed of 50 km/h (average), calculate: (*a*) the velocity (magnitude and direction) of the plane, and (*b*) how far off course it will be after 30 min if the pilot takes no corrective action.
24. (II) Determine the speed of the boat with respect to the shore in Example 3–2.
25. (II) A motorboat whose speed in still water is 3.35 m/s must aim upstream at an angle of 35.5° (with respect to a line perpendicular to the shore) in order to travel directly across the stream. (*a*) What is the speed of the current? (*b*) What is the resultant speed of the boat with respect to the shore? (See Fig. 3–12.)
26. (II) A ferryboat, whose speed in still water is 1.80 m/s, must cross a 260-m-wide river and arrive at a point 110 m upstream from where it starts (Fig. 3–28). To do so, the pilot must head the boat at a 45° upstream angle. What is the speed of the river's current?
27. (II) A swimmer is capable of swimming 1.20 m/s in still water. (*a*) If she aims her body directly across a 200-m-wide river whose current is 0.80 m/s, how far downstream (from a point opposite her starting point) will she land? (*b*) How long will it take her to reach the other side?
28. (II) At what upstream angle must the swimmer in Problem 27 aim if she is to arrive at a point directly across the stream?
29. (II) Two cars approach a street corner at right angles to each other. Car 1 travels at 60 km/h and car 2 at 40 km/h. What is the relative velocity of car 1 as seen by car 2? What is the velocity of car 2 relative to car 1?
30. (II) An airplane, whose air speed is 480 km/h, is supposed to fly in a straight path 28.0° N of E. But a steady 100 km/h wind is blowing from the north. In what direction should the plane head?
31. (III) A motorcycle traveling 90.0 km/h approaches a car traveling in the same direction at 80.0 km/h. When the motorcycle is 50 m behind the car, the rider pushes down on the accelerator and passes the car 10.0 s later. What was the acceleration of the motorcycle?
32. (III) An unmarked police car traveling a constant 90 km/h is passed by a speeder traveling 130 km/h. Precisely 1.00 s after the speeder passes, the policeman steps on the accelerator. If the police car's acceleration is 2.00 m/s^2, how much time elapses after the police car is passed until it overtakes the speeder (assumed moving at constant speed)?
33. (III) Assume in the previous problem that the speeder's speed is not known. If the police car accelerates uniformly as given above and overtakes the speeder after 6.00 s, what was the speeder's speed?

SECTIONS 3–5 AND 3–6 (Neglect air resistance)

34. (I) A diver running 1.8 m/s dives out horizontally from the edge of a vertical cliff and reaches the water below 2.0 s later. How high was the cliff and how far from its base did the diver hit the water?
35. (I) A tiger leaps horizontally from a 15-m-high rock with a speed of 4.0 m/s. How far from the base of the rock will it land?
36. (II) A fire hose held near the ground shoots water at a speed of 13 m/s. At what angle(s) should the nozzle point in order that the water land 16 m away? Why are there two different angles?
37. (II) A ball thrown horizontally at 21 m/s from the roof of a building lands 24 m from the base of the building. How high is the building?

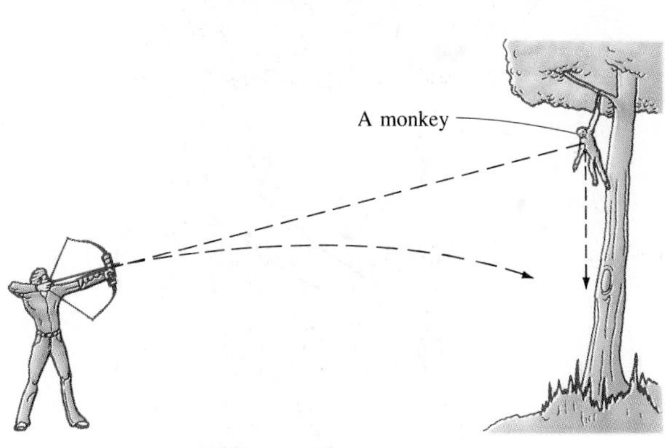

FIGURE 3–29 Problem 44.

38. (II) Show that the speed with which a projectile leaves the ground is equal to its speed just before it strikes the ground at the end of its journey, assuming the firing level equals the landing level.
39. (II) A football is kicked at ground level with a speed of 17.0 m/s at an angle of 40.0° to the horizontal. How much later does it hit the ground? Ignore air resistance.
40. (II) A hunter aims directly at a target (on the same level) 170 m away. If the bullet leaves the gun at a speed of 450 m/s, by how much will it miss the target?
41. (II) A shotputter throws the shot (mass = 7.3 kg) with an initial speed of 14 m/s at a 40° angle to the horizontal. Calculate the horizontal distance traveled by the shot if it leaves the athlete's hand at a height of 2.2 m above the ground.
42. (II) Show that the time required for a projectile to reach its highest point is equal to the time for it to return to its original height.
43. (II) A projectile is fired into the air with an initial speed of 35.0 m/s. Plot on graph paper its trajectory for initial projection angles of $\theta = 15°$, 30°, 45°, 60°, 75°, and 90°. Plot at least 10 points for each curve.
44. (II) A hunter aims his gun directly at a monkey hanging from a high tree branch some distance away (Fig. 3–29). At the instant the gun is fired, the monkey drops from the branch, hoping to avoid the bullet. Show analytically that the monkey made the wrong move. Ignore air resistance.
45. (II) William Tell must split the apple atop his son's head from a distance of 30 m. When he aims directly at the apple, the arrow is horizontal. At what angle must he aim it to hit the apple if the arrow travels at a speed of 35 m/s?

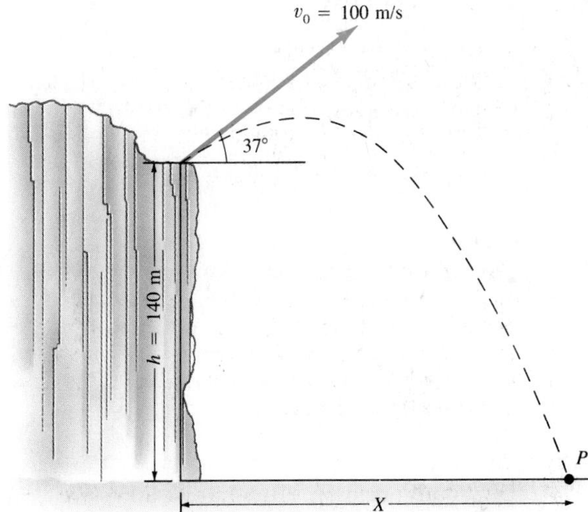

FIGURE 3–30 Problem 47.

46. (II) An athlete executing a long jump leaves the ground at a 30° angle and travels 8.90 m. What was the takeoff speed?
47. (II) A projectile is shot from the edge of a cliff 140 m above ground level with an initial speed of 100 m/s at an angle of 37.0° with the horizontal, as shown in Fig. 3–30. (*a*) Determine the time taken by the projectile to hit point *P* at ground level. (*b*) Determine the range *X* of the projectile as measured from the base of the cliff. At the instant the projectile hits point *P*, find (*c*) the horizontal and the vertical components of its velocity, (*d*) the magnitude of the velocity, and (*e*) the angle made by the velocity vector with the horizontal.
48. (III) A rescue plane wants to drop supplies to isolated mountain climbers on a rocky ridge 200 m below. If the plane is traveling horizontally with a speed of 250 km/h (69 m/s), (*a*) how far in advance of the recipients (horizontal distance) must the goods be dropped (Fig. 3–31a)? (*b*) Suppose, instead, that the plane releases the supplies a horizontal distance of 400 m in advance of the mountain climbers. What vertical velocity (up or down) should the supplies be given so that they arrive precisely at the climbers' position (Fig. 3–31b)? (*c*) With what speed do the supplies land in the latter case?

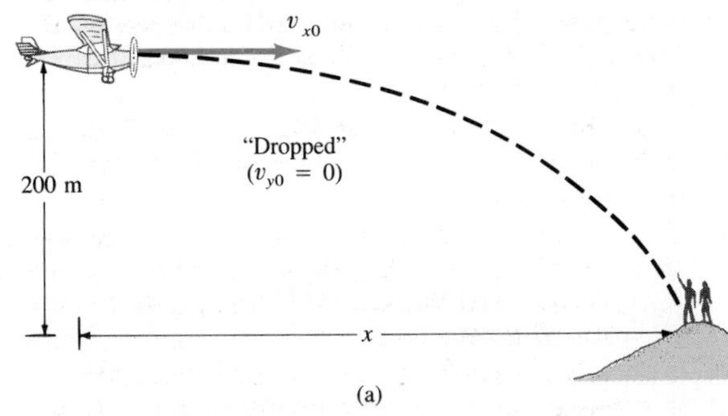

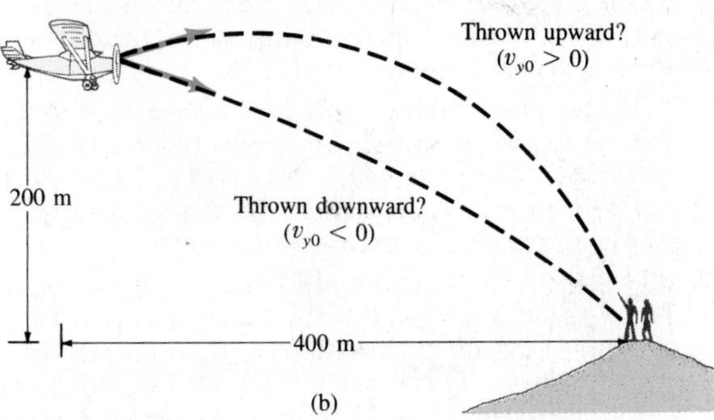

FIGURE 3–31 Problem 48.

49. (III) A ball is thrown horizontally from the top of a cliff with initial speed v_0 (at $t = 0$). At any moment, its direction of motion makes an angle θ to the horizontal (Fig. 3–32). Derive a formula for θ as a function of time, t, as the ball follows a projectile's path.

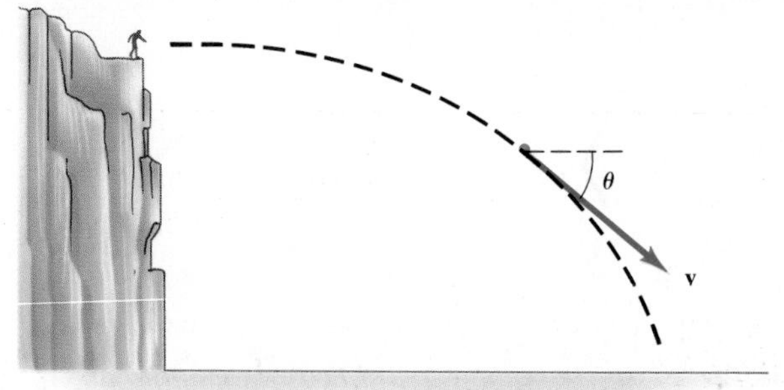

FIGURE 3–32 Problem 49.

GENERAL PROBLEMS

50. Two vectors, $\mathbf{V}_1$ and $\mathbf{V}_2$, add to a resultant $\mathbf{V} = \mathbf{V}_1 + \mathbf{V}_2$. Describe $\mathbf{V}_1$ and $\mathbf{V}_2$ if (*a*) $V = V_1 + V_2$, (*b*) $V^2 = V_1^2 + V_2^2$, (*c*) $V_1 + V_2 = V_1 - V_2$.

51. A deliveryman travels 30 m north, 25 m east, 12 m south, and then takes an elevator 36 m up into a building. What is his final displacement from his starting point? (Give components and also magnitude.)

52. Raindrops make an angle θ with the vertical when viewed through a moving train window (Fig. 3–33). If the speed of the train is v_T, what is the speed of the raindrops in the reference frame of the earth in which they are assumed to fall vertically?

53. A helicopter heads due south with an air speed of 40 km/h. The pilot observes, however, that they have covered 20 km in the previous 45 min in a southwesterly direction. What is the wind speed and direction?

54. An automobile traveling 90.0 km/h overtakes a 1.00-km-long train traveling in the same direction on a track parallel to the road. If the train's speed is 66.0 km/h, how long does it take the car to pass it and how far will the car have traveled in this time? What are the results if the car and train are traveling in opposite directions?

55. An Olympic long jumper is capable of jumping 8.0 m. Assuming his horizontal speed is 9.0 m/s as he leaves the ground, how long is he in the air and how high does he go? Assume that he lands standing upright, that is, the same way he left the ground.

56. Determine how far a person can jump on the moon as compared to the earth if the takeoff speed and angle are the same. The acceleration due to gravity on the moon is one-sixth what it is on earth.

57. When Babe Ruth hit a homer over the 12-m left-field fence 100 m from home plate, roughly what was the minimum speed of the ball when it left the bat? Assume the ball was hit 1.0 m above the ground and its path initially made a 35° angle with the ground.

58. The speed of a boat in still water is v. The boat is to make a round trip in a river whose current travels at speed u. Derive a formula for the time needed to make a round trip of total distance D if the boat makes the round trip by moving (*a*) upstream and back downstream, (*b*) directly across the river and back. We must assume $u < v$; why?

59. Police agents flying a constant 200 km/h horizontally in a low-flying airplane wish to drop an explosive onto a master criminal's automobile traveling 130 km/h on a level highway 88.0 m below. At what angle (with the horizontal) should the car be in their sights when the the bomb is released (Fig. 3–34)?

60. A basketball leaves a player's hands at a height of 2.1 m above the floor. The basket is 2.6 m above the floor. The player likes to shoot the ball at a 35° angle. If the shot is made from a horizontal distance of 12.0 m and must be accurate to ±0.22 m (horizontally), what is the range of initial speeds allowed to make the basket?

FIGURE 3–33 Problem 52.

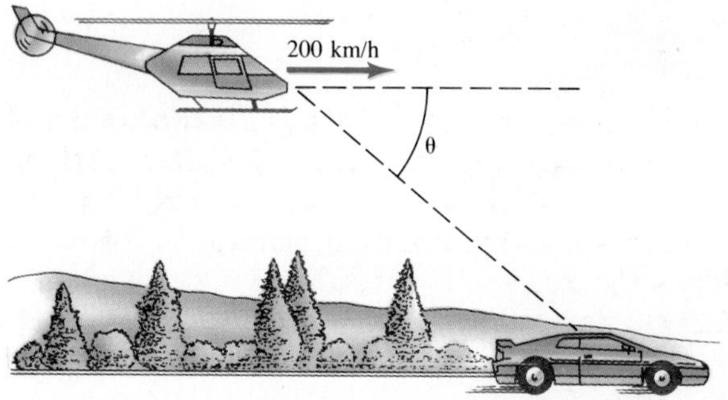

FIGURE 3–34 Problem 59.

C H A P T E R 4

Motion and Force: Dynamics

Acceleration at the start of a race. Can you describe the force that causes an athlete's acceleration?

We have discussed how motion is described in terms of velocity and acceleration. Now we deal with the question of *why* objects move as they do: What makes an object at rest begin to move? What causes a body to accelerate or decelerate? What is involved when an object moves in a circle? We might answer in each case that a force is required. In this chapter, we will investigate the connection between force and motion. Before we delve into this subject of *dynamics*, we first discuss the concept of force in a qualitative way.

4–1 • Force

Intuitively, we can define **force** as any kind of a push or a pull on an object. When you push a grocery cart or a stalled car (Fig. 4–1), you are exerting a force on it. When children pull a wagon, they are exerting a force on the wagon. When a motor lifts an elevator, or a hammer hits a nail, or the wind blows the leaves of a tree, a force is being exerted. We say that an object falls because of the *force of gravity*. Forces do not always give rise to motion. For example, you may push very hard on a heavy desk or sofa and it may not move.

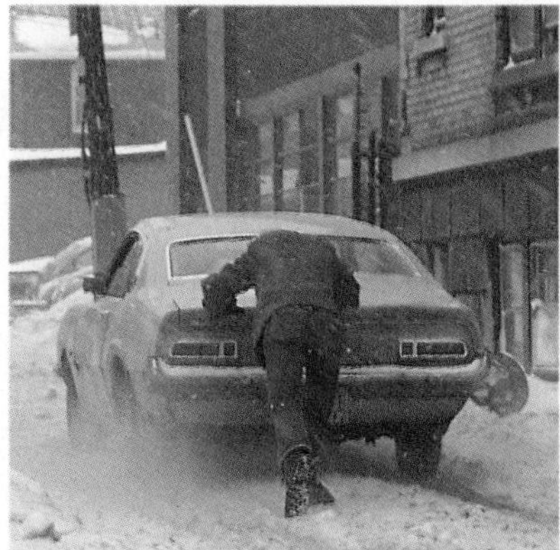

FIGURE 4–1 Exerting a force on a stuck car.

Whether or not an object moves when a force is exerted on it, the object does change shape (Fig. 4–2). This is obvious when you squeeze a balloon, or push on a mattress. You can see the slight deformation of the metal when you push on the side of a refrigerator or the fender of a car. There is always some deformation when a force is exerted, although it may take delicate instruments to detect it for a very rigid object like a heavy steel plate.

FIGURE 4–2 A foot exerts a force on a football. Note the deformation.

One way to quantitatively measure the magnitude (or strength) of a force is to make use of a spring scale (Fig. 4–3). Normally, such a spring scale is used to find the weight of an object. By weight we mean the force of gravity acting on the body (Section 4–7). The spring scale, once calibrated†, can be used to measure other kinds of forces as well, such as the pulling force shown in Fig. 4–3.

A force has direction as well as magnitude and is therefore a vector that follows the rules of vector addition discussed in Chapter 3. We can represent any force on a diagram by an arrow, just as we did with velocity. The direction of the arrow is of course in the direction of the push or pull, and its length is drawn proportional to the strength or magnitude of the force. Although defining a force as a push or a pull is adequate for the moment, a more precise definition is possible, as discussed in Sections 4–4 and 4–5.

4–2 • Newton's First Law of Motion

What is the exact connection between force and motion? Aristotle (384–322 B.C.) believed that a force was required to keep an object moving along a horizontal plane. He would argue that to make a book move across the table, you would have to exert a force on it continuously. To Aristotle, the

† A spring scale is calibrated by hanging from it a series of identical objects of equal weight or mass, say 1 pound or 1 kilogram. The positions of the pointer when 0, 1, 2, 3, . . . units of weight or mass are hung from it are marked. Although the amount the spring stretches is very nearly proportional to the amount of weight hung from it, as long as it isn't stretched too far, we don't have to make use of this fact in this calibration method. We only assume that the pointer rests at the same position when the same weight (force) acts. This could serve as an operational definition (Section 1–5) of force.

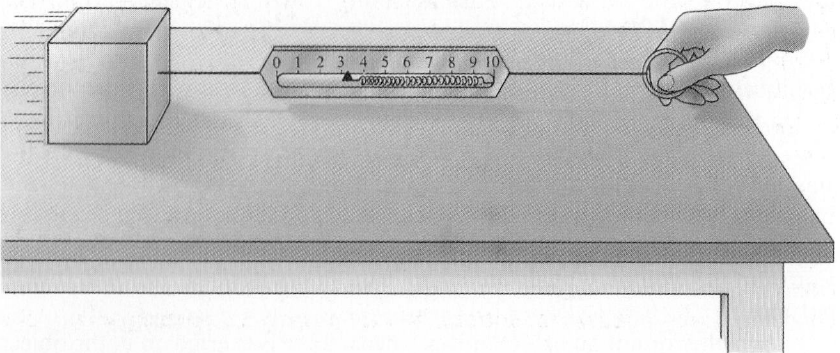

FIGURE 4–3 A spring scale used to measure a force.

natural state of a body was at rest, and a force was believed necessary to keep a body in motion. Furthermore, Aristotle argued, the greater the force on the body, the greater its speed.

Galileo vs. Aristotle

Some 2000 years later, Galileo (Fig. 2–12), skeptical about these Aristotelian views just as he was of those on falling bodies, came to a radically different conclusion. Galileo maintained that it is just as natural for an object to be in horizontal motion with a constant speed as it is to be at rest. To understand Galileo's idea, consider the following observations involving motion along a horizontal plane (where the effects of gravity do not enter). It will take a certain amount of force to push an object with a rough surface along a tabletop at constant speed. To push an equally heavy object with a very smooth surface across the table at the same speed will require less force. Finally, if a layer of oil or other lubricant is placed between the surface of the object and the table, then almost no force is required to move the object. (These observations may well be obvious to you, but if not, you should do these simple experiments for yourself.) Notice that in each successive step, less force is required. As the next step, we can extend the data to a situation in which the object does not rub against the table at all—or there is a perfect lubricant between them—and theorize that once started, the object would move across the table at constant speed with *no* force applied. A steel ball bearing rolling on a hard horizontal surface approaches this situation closely.

It was Galileo's genius to imagine an idealized world—in this case, one where there is no friction—and to see that it could produce a more useful view of the real world. It was this idealization that led him to his remarkable conclusion that if no force is applied to a moving object, it will continue to move with constant speed in a straight line. An object slows down only if a force is exerted on it. Galileo thus interpreted friction as a force akin to ordinary pushes and pulls.

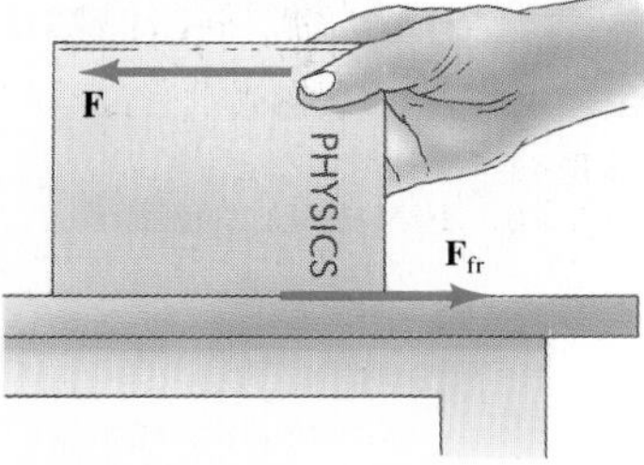

FIGURE 4–4 **F** represents the force applied by the person and $\mathbf{F}_{fr}$ represents the force of friction.

To push an object across a table at constant speed requires a force from your hand only to balance out the force of friction. The pushing force is equal in magnitude to the friction force but they are in opposite directions, so the *net* force on the object (the vector sum of the two forces) is zero, Fig. 4–4. This is consistent with Galileo's viewpoint, for the object moves with constant speed when no net force is exerted on it.

The difference between Aristotle's view and Galileo's is not simply one of right or wrong. Aristotle's view was not really wrong, for our everyday experience indicates that moving objects do tend to come to a stop if not continually pushed. The real difference lies in the fact that Aristotle's view about the "natural state" of a body was essentially a final statement—no further development was possible. Galileo's analysis, on the other hand, could be extended to explain a great many more phenomena. By making the creative leap of imagining the experimentally unattainable situation of no friction and by interpreting friction as a force, Galileo was able to reach his conclusion that an object will continue moving with constant velocity if no force acts to change this motion.

FIGURE 4–5 Isaac Newton (1642–1727).

Upon this foundation, Isaac Newton (Fig. 4–5) built his great theory of motion. Newton's analysis of motion is summarized in his famous "three laws of motion." In his great work, the *Principia* (published in 1687, it contains nearly all his work on motion), Newton readily acknowledged his debt to Galileo. In fact, **Newton's first law of motion** is very close to Galileo's conclusions.[†] It states that

Newton's first law of motion

Every body continues in its state of rest or of uniform speed in a straight line unless it is compelled to change that state by a net force acting on it.

The tendency of a body to maintain its state of rest or of uniform motion in a straight line is called **inertia**. As a result, Newton's first law is often called the **law of inertia**.

4–3 • Mass

Newton's second law (which we come to in the next section) makes use of the concept of mass. Newton himself used the term *mass* as a synonym for *quantity of matter*. This intuitive notion of the mass of a body is not very precise because the concept "quantity of matter" is itself not well defined. More precisely, we can say that **mass** *is a measure of the inertia of a body*. The more mass a body has, the harder it is to change its state of motion. It is harder to start it moving from rest, or to stop it when it is moving, or to change its motion sideways out of a straight-line path. A piano or a truck has much more inertia than a baseball moving at the same speed, and it is much harder to change its state. It therefore has much more mass.

Mass as inertia

To quantify the concept of mass, we must define a standard. In SI units, the unit of mass is the **kilogram** (kg). The actual standard is a particular platinum-iridium cylinder, kept at the International Bureau of Weights and Measures near Paris, whose mass by definition is precisely one kilogram. In cgs units, the unit of mass is the gram (g), where $1\ \mathrm{g} = 10^{-3}\ \mathrm{kg}$. In the British system, the unit of mass is called the slug (see Section 4–4). When dealing with atoms and molecules, we often make use of the **unified atomic mass unit** (u). By definition, the mass of a carbon (^{12}C) atom is assigned an atomic mass of precisely 12 u. In terms of the kilogram

Standard of mass (kg)

Atomic mass unit

$$1\ \mathrm{u} = 1.6605 \times 10^{-27}\ \mathrm{kg}.$$

The terms *mass* and *weight* are often confused with one another, but it is important to distinguish between them. Mass is a property of a body itself (it is a measure of a body's inertia, or its "quantity of matter"). Weight, on the other hand, is a force, the force of gravity acting on a body. To see the difference, suppose we take an object to the moon. The object will weigh only about one-sixth as much as it did on earth, since the force of gravity is weaker, but its mass will be the same. It will have the same amount of matter and it will have just as much inertia—for in the absence of friction, it will be just as hard to start it moving or to stop it once it is moving.

Mass vs. weight

[†] It is not clear in Galileo's works if he adopted "linear" inertia, or "spherical" inertia—that is, natural motion continuing along a spherical surface such as on the surface of the earth. Newton, and Descartes before him, adopted the principle of inertia for motion along a straight line.

4–4 • Newton's Second Law of Motion

FIGURE 4–6 The bobsled accelerates because the team exerts a force.

Newton's first law states that if no net force is acting on a body it remains at rest, or if moving it continues moving with constant speed in a straight line. But what happens if a force is exerted on a body? Newton perceived that the velocity will change (Fig. 4–6). A net force exerted on an object may make its speed increase; or if the net force is in a direction opposite to the motion, it will reduce the speed. If the net force acts sideways on a moving object, the direction as well as the magnitude of the velocity changes. Thus, *a net force gives rise to acceleration.*

What precisely is the relationship between acceleration and force? Ordinary experience can answer this question. Consider the force required to push, say, a cart whose friction is minimal. (If there is friction, consider the net force, which is the force you exert minus the force of friction.) Now if you push with a gentle but constant force for a certain period of time, you will make the cart accelerate from rest up to some speed, say 3 km/h. If you push twice as hard, you will find that it will reach 3 km/h in half the time. That is, the acceleration will be twice as great. If you double the force, the acceleration doubles; if you triple the force, the acceleration is tripled, and so on. Thus, the acceleration of a body is directly proportional to the net applied force. But the acceleration depends on the mass of the object as well. If you push an empty grocery cart with the same force as you push one that is filled with groceries, you will find that the latter accelerates more slowly. The greater the mass, the less the acceleration for a given net force. The mathematical relation, as Newton argued, is that the acceleration of a body is inversely proportional to its mass. These relationships are found to hold in general and can be summarized as follows:

The acceleration of an object is directly proportional to the net force acting on it and is inversely proportional to its mass. The direction of the acceleration is in the direction of the applied net force.

Newton's second law of motion

This is **Newton's second law of motion**. In symbols, we can write it as

$$\mathbf{a} \propto \frac{\mathbf{F}}{m},$$

where **a** stands for acceleration, *m* for the mass, and **F** for the *net force*. By **net force**, we mean the vector sum of *all the forces* acting on the body. To change a proportion into an equation, we merely need to insert a constant of proportionality (see Appendix A–1). The choice of a constant is arbitrary in this case since we are relating quantities with different units. We can therefore choose the unit of force or of mass so that the proportionality constant equals 1. Then $\mathbf{a} = \mathbf{F}/m$. Rearranging, we have the familiar statement of Newton's second law in the form of an equation:

Net force

$$\mathbf{F} = m\mathbf{a}. \tag{4–1}$$

Newton's second law of motion

Newton's second law relates the description of motion to the cause of motion, force. It is one of the most fundamental relationships in physics.[†] From

[†] Newton originally stated his second law of motion in terms of the momentum $\mathbf{p} = m\mathbf{v}$: that is, $\mathbf{F} = \Delta\mathbf{p}/\Delta t$, which for constant mass reduces to $\mathbf{F} = m(\Delta\mathbf{v}/\Delta t) = m\mathbf{a}$. This is discussed in Chapter 7.

Force defined

Newton's second law we can make a more precise definition of *force as an action capable of accelerating an object* (see the next section).

The unit of *force* is chosen so that the proportionality constant in Newton's second law, $\mathbf{F} \propto m\mathbf{a}$, will be 1, and thus $\mathbf{F} = m\mathbf{a}$. With the mass in kilograms, the unit of force is called the **newton** (N). One newton, then, is the force required to impart an acceleration of 1 m/s² to a mass of 1 kg. Thus 1 N = 1 kg·m/s².

In cgs units, the unit of mass is the gram (g) as mentioned earlier. The unit of force is the *dyne*, which is defined as the force required to impart an acceleration of 1 cm/s² to a mass of 1 g; thus 1 dyne = 1 g·cm/s². It is easy to show that 1 dyne = 10^{-5} N. (Do not confuse the abbreviation g for gram with the symbol g for the acceleration due to gravity, which is always italicized or bold faced.)

Table 4–1
Units for Mass and Force

System	Mass	Force (including weight)
SI	kilogram (kg)	newton (N) (=kg·m/s²)
cgs	gram (g)	dyne (=g·cm/s²)
British	slug	pound (lb)

In the British system, the unit of force is the *pound*. The pound is defined as the weight (which is a force) of a body with a mass of 0.45359237 kg at a particular place on the earth where the acceleration due to gravity is $g = 32.1734$ ft/s². The unit of mass is the *slug*, which is defined as that mass which will undergo an acceleration of 1 ft/s² when a force of 1 lb is applied to it. Thus 1 lb = 1 slug·ft/s². It is easy to show that 1 lb ≈ 4.45 N.† Table 4-1 summarizes the units in the three systems.

Problem Solving: Use a consistent set of units

It is very important that only one set of units be used in a given calculation or problem, with the SI being preferred. If the force is given in, say, newtons, and the mass in grams, then before attempting to solve for the acceleration in SI units, the mass must be changed to kilograms. For example, if the force is given as 2.0 N and the mass is 500 g, we change the latter to 0.50 kg and the acceleration will then automatically come out in m/s² when Newton's second law is used:

$$a = \frac{F}{m} = \frac{2.0 \text{ N}}{0.50 \text{ kg}} = 4.0 \text{ m/s}^2.$$

Example 4–1 Estimate the net force needed to accelerate a 1500-kg race car at $\frac{1}{2}g$.

Solution We use Newton's second law, setting $a = 0.50g$:

$$F = ma = (1500 \text{ kg})(0.50)(9.8 \text{ m/s}^2) \approx 7000 \text{ N}.$$

Example 4–2 What net force is required to bring a 1500-kg car to rest from a speed of 100 km/h within 55 m?

Solution We use Newton's second law, $F = ma$, but first we must determine the acceleration a. We assume the motion is along the $+x$ axis. We are given the initial velocity $v_0 = 100$ km/h = 28 m/s, the final velocity $v = 0$, and the distance traveled $x - x_0 = 55$ m. From Eq. 2–10c, we have

$$v^2 = v_0^2 + 2a(x - x_0)$$

† Other systems are occasionally used, and we mention them for reference only. In the "gravitational mks" system, the kilogram is still the mass, but the unit of force is called the kilogram-force (kgf) and 1 kg weighs 1 kgf at a place where $g = 9.8066$ m/s²; in other words, 1 kgf = 9.8066 N. In the British equivalent, the pound remains the unit of force, but the unit of mass is the pound mass (lbm). A mass of 1 lbm weighs 1 lb (at a place where $g = 32.1734$ ft/s²), so 1 slug = 32.1734 lbm.

so

$$a = \frac{v^2 - v_0^2}{2(x - x_0)}$$

$$= \frac{0 - (28 \text{ m/s})^2}{2(55 \text{ m})} = -7.1 \text{ m/s}^2.$$

The net force required is then

$$F = ma = (1500 \text{ kg})(-7.1 \text{ m/s}^2) = -1.1 \times 10^4 \text{ N}.$$

The force must be exerted in the direction opposite to the initial velocity, which is what the negative sign tells us.

We will discuss many more examples of Newton's second law later in this chapter and, in fact, throughout the book.

*4–5 • Laws or Definitions

When is an equation a law and when is it just an equation? To answer this, let us compare Newton's second law, $F = ma$, with the kinematic equations, Eqs. 2–10a, b, c, and d. The kinematic equations, as useful as they are, do not tell us anything startling about nature itself, since they were derived from the definitions of velocity and acceleration and the assumption of constant acceleration. Newton's second law, on the other hand, is based on experiment and tells us something about how things behave in nature. Given the simple definitions of mass, force, and acceleration, there is no obvious relation between them until one actually looks at natural objects and sees how they behave. Because there is actual physical content in the relation $F = ma$, and because it summarizes so much of experience, it is called a law.

However, in this particular case, there is a problem. It is believed by many scientists (the "positivist" or "operationalist" school of philosophy) that the intuitive definition of force as a push or a pull is not a good one because it is too vague. They claim that force must be defined more carefully, and that the only way to do this is via Newton's second law itself. Thus, according to this viewpoint, Newton's second law is not to be considered a law, but rather, as the definition of force.

Similar remarks apply to Newton's first law. It is generally considered to define a particular kind of reference frame, called an **inertial reference frame**. An inertial reference frame is one in which Newton's first law is valid. (The earth for example is, for most purposes, nearly an inertial reference frame.) A **noninertial reference frame** is one in which Newton's first law does not hold. Examples of noninertial reference frames are rotating reference frames (such as a merry-go-round) or accelerating reference frames (as in a freely falling elevator). Newton's first law does not hold in such reference frames (Section 8–10), so it cannot be considered a law; instead it is considered to give the definition of a type of reference frame ("one in which Newton's first law *does* hold").

Inertial reference frames

As for Newton's second law, whether you accept the operationalist viewpoint that $\mathbf{F} = m\mathbf{a}$ is the definition of force, or whether you prefer to think of this relation as a law, has no bearing on the use of this relation in practical situations.

FIGURE 4–7 Multiflash photo of a hammer striking a nail. In accordance with Newton's third law, the hammer exerts a force on the nail and the nail exerts a force back on the hammer. The latter force decelerates the hammer and brings it to rest.

4–6 • Newton's Third Law of Motion

Newton's second law of motion describes quantitatively how forces affect motion. But where, we may ask, do forces come from? Observations suggest that a force applied to any object is always applied *by another object.* A horse pulls a wagon, a person pushes a grocery cart, a hammer pushes on a nail, a magnet attracts an iron nail. In each of these examples, one object exerts the force and the other body feels it: for example, the hammer exerts the force, and the nail feels it.

But Newton realized that things are not so one-sided. True, the hammer exerts a force on the nail (Fig. 4–7). But the nail evidently exerts a force back on the hammer as well, for the hammer's speed is rapidly reduced to zero upon contact. Only a strong force could cause such a rapid deceleration. Thus, said Newton, the two bodies must be treated on an equal basis. The hammer exerts a force on the nail, and the nail exerts a force back on the hammer. This is the essence of **Newton's third law of motion**:

Newton's third law of motion

Whenever one object exerts a force on a second object, the second exerts an equal and opposite force on the first.

Action and reaction act on different objects

This law is sometimes paraphrased as "to every action there is an equal and opposite reaction." To avoid confusion, it is very important to remember that the "action" force and the "reaction" force are acting on *different* objects.

As evidence for the validity of Newton's third law, look at your hand when you push against a grocery cart or against the edge of a desk (Fig. 4–8). Your hand's shape is distorted, clear evidence that a force is being exerted on it. You can *see* the edge of the desk pressing into your hand. You can even *feel* the desk exerting a force on your hand; it hurts. The harder you push against the desk, the harder the desk pushes back on your hand.

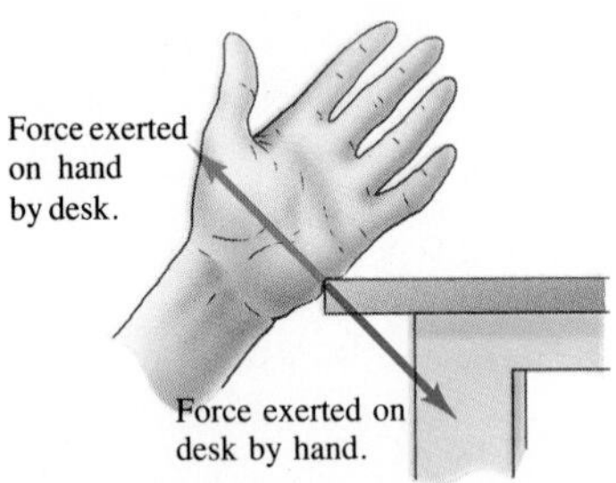

FIGURE 4–8 If your hand pushes against the edge of a desk (the force vector is shown in red), the desk pushes back against your hand (the force vector is shown in orange, to remind us that this force acts on a different object).

Consider next the ice skater in Fig. 4–9. Since there is very little friction between her skates and the ice, she will move freely if a force is exerted on her. She pushes against the railing, and then *she* starts moving backwards. Clearly, there had to be a force exerted on her to make her move. The force she exerts on the railing cannot make *her* move, for that force pushes on the railing and can only affect the railing. Something had to exert a force on her to make her move, and that force could only have been exerted by the railing. The force with which the railing pushes on her is equal and opposite to the force she exerts on the railing.

When a person throws a package out of a boat, the boat moves in the opposite direction. The person exerts a force on the package. The package exerts an equal and opposite force back on the person, and this force propels the person (and the boat) backward slightly. Rockets work on the same principle (Fig. 4–10). A common misconception is that rockets accelerate because the gases rushing out the back of the engine push against the ground or the atmosphere. Actually, a rocket exerts a strong force on the gases, expelling them. The gases exert an equal and opposite force *on the rocket*

How does a rocket accelerate?

FIGURE 4–9 When an ice skater pushes against the railing, the railing pushes back and this force causes her to move away.

FIGURE 4–10 The launch of a rocket.

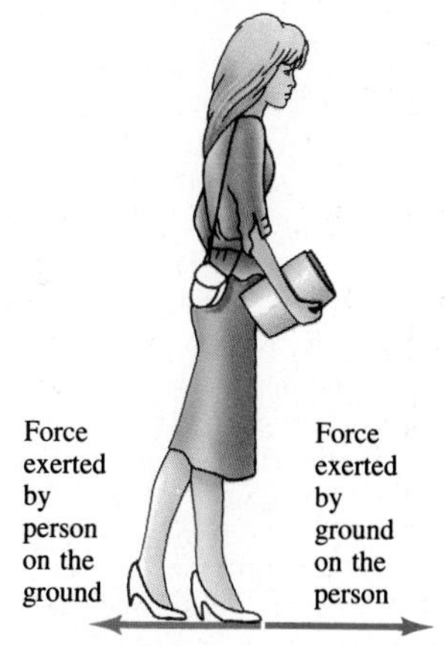

FIGURE 4–11 We can walk forward because the ground pushes forward on our feet when our feet push backward against the ground.

and it is this force that propels the rocket forward. Thus, a space vehicle maneuvers in empty space by firing its rockets in the direction opposite to that in which it wants to accelerate.

How we can walk

Let us consider how we walk. A person begins walking by pushing with the foot against the ground. The ground then exerts an equal and opposite force back on the person (Fig. 4–11) and it is this force, *on* the person, that moves him or her forward. In a similar way, a bird flies forward by exerting a force on the air, but it is the air pushing back on the bird's wings that propels the bird forward. And an automobile moves forward because of the force exerted on it by the ground, which is the reaction to the force exerted on the ground by the tires.

*Problem Solving: Distinguish between forces exerted **by** or **on** a body*

From the above examples, it is clear that it is quite important to remember *on* what object a given force is exerted and *by* what object that force is exerted. The point is that a force influences the motion of an object only when it is applied *on* that object. A force exerted *by* a body does not influence that body; it only influences the other body *on* which it is exerted. Thus, to avoid confusion, the two prepositions *on* and *by* must always be used—and used with care.

Inanimate objects can exert a force

We tend to associate forces with active bodies such as humans, animals, engines, or a moving object like a hammer. It is often difficult to see how an inanimate object at rest, such as a wall or a desk, can exert a force. The explanation lies in the fact that every material, no matter how hard, is elastic, at least to some degree. No one can deny that a stretched rubber band can exert a force on a wad of paper and send it flying across the room. Although other materials may not stretch as easily as rubber, they do stretch when a force is applied to them. And just as a stretched rubber band exerts a force, so does a stretched (or compressed) wall or desk.

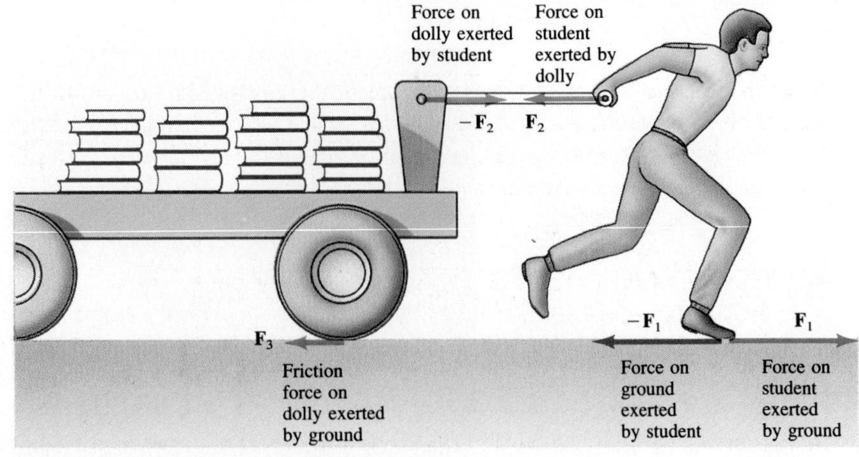

FIGURE 4–12 A student pulls on a dolly. Forces on the student are shown as red (magenta) arrows. Forces on the dolly are shown in orange. Only one of the forces acting on the ground is shown (plum-colored arrow). Action-reaction forces that are equal and opposite are labeled by the same subscript (such as $\mathbf{F}_1$ and $-\mathbf{F}_1$); the two members of a pair act on different objects.

EXAMPLE 4–3 An entering freshman has been assigned the task of pulling a dolly loaded with textbooks (Fig. 4–12). He says to his boss, "When I exert a forward force on the dolly, the dolly exerts an equal and opposite force backward. So how can I move? No matter how hard I pull, the backward reaction force always equals my forward force, so the net force must be zero. I'll never be able to move this load." Is this a case of a little knowledge being dangerous? Explain.

Problem Solving: A study of Newton's second and third laws

SOLUTION Yes. Although it is true that the action and reaction forces are equal, the student has forgotten that they are exerted on different bodies. The forward ("action") force is exerted by the student on the dolly, whereas the backward "reaction" force is exerted by the dolly on the student. To determine if the *student* moves or not, we must consider only the forces *on the student* and then apply $F = ma$, where F is the net force on the student, a is the acceleration of the student, and m is the student's mass. The two forces on the student that affect his forward motion are shown as bright red arrows in Fig. 4–12; they are (1) the horizontal force $\mathbf{F}_1$ exerted on the student by the ground (the harder he pushes back against the ground, the harder the ground pushes forward on him—Newton's third law), and (2) the dolly pulling backward on him ($\mathbf{F}_2$). When the ground pushes forward on the student harder than the dolly pulls backward, the student moves forward (Newton's second law). The cart, on the other hand, moves forward when the student exerts a force on it that is greater than the frictional force acting backwards.

FIGURE 4–13 A tug of war. Describe the forces on each of the people and on the rope.

Is Newton's third law operating in Fig. 4–13? Try to give a description of all forces acting on the people and on the rope.

4–7 • Weight—the Force of Gravity; and the Normal Force

Galileo claimed that objects dropped near the surface of the earth will all fall with the same acceleration, g, if air resistance can be neglected. The force that gives rise to this acceleration is called the force of gravity. We now apply

Newton's second law to the gravitational force and for the acceleration, **a**, we use the acceleration of gravity, **g**. Thus, the force of gravity on a body, $\mathbf{F}_g$, which is also commonly called its **weight** (with symbol **w**, meaning the same as $\mathbf{F}_g$), can be written as

$$\mathbf{F}_g = \mathbf{w} = m\mathbf{g}. \tag{4–2}$$

The direction of this force is down toward the center of the earth.

In SI units, $g = 9.80\ \text{m/s}^2 = 9.80\ \text{N/kg}$,[†] so the weight of a 1.00-kg mass is $1.00\ \text{kg} \times 9.80\ \text{m/s}^2 = 9.80\ \text{N}$. The value of g varies very slightly at different places on the earth's surface, as we saw in Chapter 2, but we usually won't be concerned about this. We will mainly be concerned with the weight of objects on earth, but we note that on the moon, on other planets, or in space, the weight of a given mass will be different. For example, on the moon, g is about one-sixth what it is on earth, and 1 kg weighs only 1.7 N. Although we will not have occasion to use British units, we note that for practical purposes on the earth, a mass of 1 kg weighs about 2.2 lb. (On the moon, 1 kg weighs about 0.4 lb.)

Gravity acts even on a body at rest

The force of gravity acts on an object when it is falling. When an object is at rest on the earth, the gravitational force on it does not disappear, as we know if we weigh it on a spring scale. The same force, given by Eq. 4–2, continues to act. Why, then, doesn't the object move? From Newton's second law, the net force on an object at rest is zero. There must be another force on the object to balance the gravitational force. For an object resting on a table, the table exerts this upward force (see Fig. 4–14a). The table is compressed slightly beneath the object, and due to its elasticity, it pushes up on the object as shown. The force exerted by the table is often called a **contact force**, since it occurs when two objects are in contact. (The force of your hand pushing on a cart is also a contact force.) When a contact force acts perpendicular to the common surface of contact, it is usually referred to as the **normal force** ("normal" means perpendicular); hence it is labeled $\mathbf{F}_N$ in the diagram.

Normal force

Let us examine the forces involved in Fig. 4–14 in more detail, making clear what we can understand from Newton's second law, and what from Newton's third law. The two forces shown in Fig. 4–14a are both acting

[†] Since $1\ \text{N} = 1\ \text{kg·m/s}^2$ (Section 4–4), $1\ \text{m/s}^2 = 1\ \text{N/kg}$.

FIGURE 4–14 (a) The net force on an object at rest is zero according to Newton's second law. Therefore the downward force of gravity ($\mathbf{F}_g$) on an object must be balanced by an upward force (the normal force $\mathbf{F}_N$) exerted by the table in this case. (b) $\mathbf{F}'_N$ is the force exerted on the table by the statue and is the reaction force to $\mathbf{F}_N$ as per Newton's third law. The reaction to $\mathbf{F}_g$ is not shown.

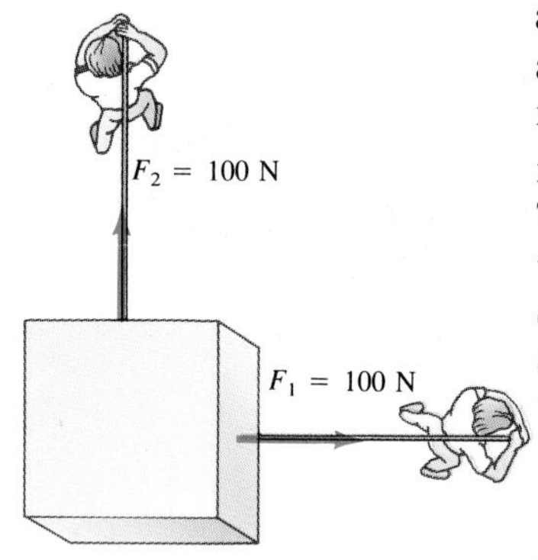

FIGURE 4–16 (a) Two forces, $\mathbf{F}_1$ and $\mathbf{F}_2$, act on an object. (b) The sum, or resultant, of $\mathbf{F}_1$ and $\mathbf{F}_2$ is $\mathbf{F}_R$.

at right angles to each other. Intuitively, we can see that the object will move at a 45° angle and thus the net force acts at a 45° angle. This is just what the rules of vector addition give. The Pythagorean theorem tells us that the resultant force must have magnitude $F_R = \sqrt{(100\ \text{N})^2 + (100\ \text{N})^2} = 141$ N. That this is the correct answer is not intuitively obvious, although clearly the net force will be less than 200 N (since the two persons are pulling at cross purposes to some extent), and it is certainly greater than zero. Carefully done experiments show that two 100-N forces acting at 90° to one another have the same effect as one force of magnitude 141 N acting at 45°, in agreement with our rules for vector addition.

FIGURE 4–17 Two force vectors act on a boat (Example 4–5).

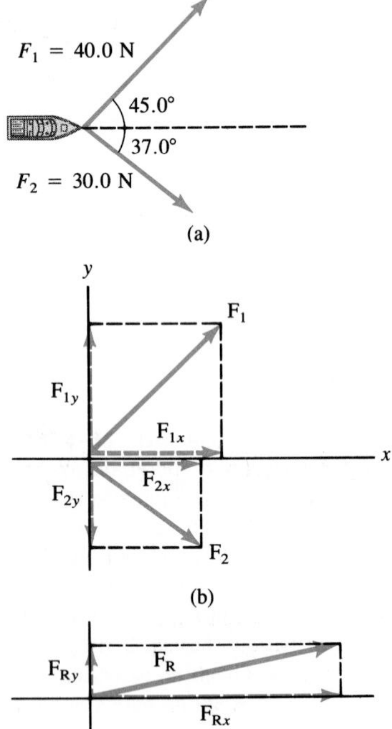

EXAMPLE 4–5 Calculate the sum of the two forces acting on the small boat shown in Fig. 4–17a.

SOLUTION These two forces are shown resolved in Fig. 4–17b. We add the forces using the method of components. The components of $\mathbf{F}_1$ are

$$F_{1x} = F_1 \cos 45.0° = (40.0\ \text{N})(0.707) = 28.3\ \text{N},$$

$$F_{1y} = F_1 \sin 45.0° = (40.0\ \text{N})(0.707) = 28.3\ \text{N}.$$

The components of $\mathbf{F}_2$ are

$$F_{2x} = F_2 \cos 37.0° = (30.0\ \text{N})(0.799) = 24.0\ \text{N},$$

$$F_{2y} = -F_2 \sin 37.0° = -(30.0\ \text{N})(0.602) = -18.1\ \text{N}.$$

F_{2y} is negative since it points along the negative y axis. The components of the resultant force are (see Fig. 4–17c)

$$F_{Rx} = 28.3\ \text{N} + 24.0\ \text{N} = 52.3\ \text{N}$$

$$F_{Ry} = 28.3\ \text{N} - 18.1\ \text{N} = 10.2\ \text{N}.$$

To find the magnitude of the resultant force, we use the Pythagorean theorem:

$$F_R = \sqrt{F_{Rx}^2 + F_{Ry}^2} = \sqrt{(52.3)^2 + (10.2)^2}\ \text{N} = 53.3\ \text{N}.$$

The only remaining question is the angle θ that the net force $\mathbf{F}_R$ makes with the x axis. We use:

$$\tan\theta = \frac{F_{Ry}}{F_{Rx}} = \frac{10.2\ \text{N}}{52.3\ \text{N}} = 0.195,$$

which corresponds to an angle of 11.0°.

EXAMPLE 4–6 Calculate the force required to accelerate the 20-kg cart in Fig. 4–18a from rest to 0.50 m/s in 2.0 s.

SOLUTION If friction is negligible, there are three forces on the cart, which for simplicity are shown in Fig. 4–18b as if all acted at the center of the cart: the forward pushing force exerted by the person, $\mathbf{F}_P$; the downward force of gravity, $\mathbf{F}_g$; and the upward normal force exerted by the floor, $\mathbf{F}_N$ (which is the reaction to the force of the cart pushing down on the floor).

The vertical forces, $\mathbf{F}_g$ and $\mathbf{F}_N$, must add up to zero; if they didn't, the cart would accelerate vertically. So $F_N = mg = (20\text{ kg})(9.8\text{ m/s}^2) = 196\text{ N}$. Then the net force on the cart is simply $\mathbf{F}_P$. To calculate how large F_P must be, we first calculate the acceleration required: $a = \Delta v/\Delta t = (0.50\text{ m/s} - 0)/2.0\text{ s} = 0.25\text{ m/s}^2$. Then the magnitude of the force exerted by the person must be $F_P = ma = (20\text{ kg})(0.25\text{ m/s}^2) = 5.0\text{ N}$.

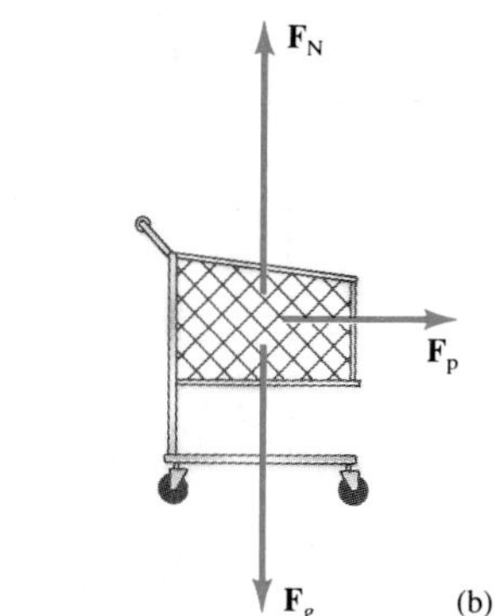

FIGURE 4–18 Forces on a cart (Example 4–6).

This example illustrates several important aspects in the use of Newton's laws. First, we draw a sketch of the situation. Next, if there is only one body involved, we show all the forces acting on that body. If several bodies are involved, we make a diagram of each body *separately*, showing all the forces acting *on that body*. Such a diagram is called a **free-body diagram**. Newton's second law involves vectors, and it is usually important to resolve vectors into components. An x and a y axis should be chosen in a way that simplifies the calculation. Then Newton's second law can be applied to the x and y components separately. That is, the x component of the total force will be related to the x component of the acceleration: $F_x = ma_x$; and similarly for the y direction. (More on problem solving in Section 4–10.)

In the examples that follow, we assume that all surfaces are very smooth so that friction can be ignored. (Friction, and Examples using it, are discussed in the next section.) In the next several Examples, we will deal again with the large candy box on the table, which we first encountered in Example 4–4 (Fig. 4–15). We continue with this system, adding in each successive Example an additional complication so that, step by step, you can see how to approach solving problems. (Because of the many worked-out Examples, this chapter may seem long. However, taking the time to study each Example, leisurely and unrushed, will bring great rewards. Future chapters will be more readable and understandable, and hence more enjoyable. In a way, the present chapter is perhaps the most crucial in the book.)

FIGURE 4–19 Example 4–7; (b) is the free-body diagram.

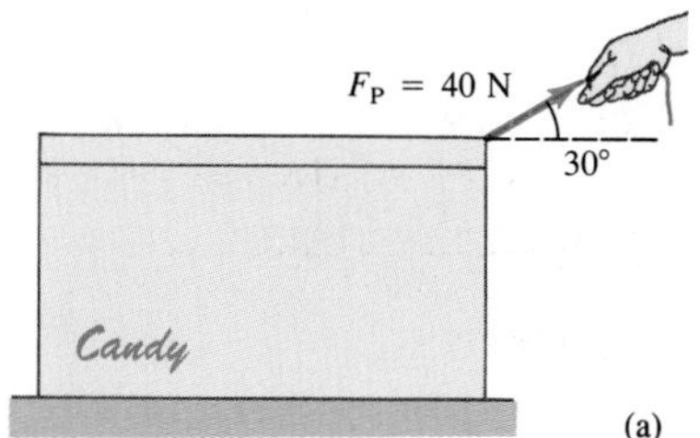

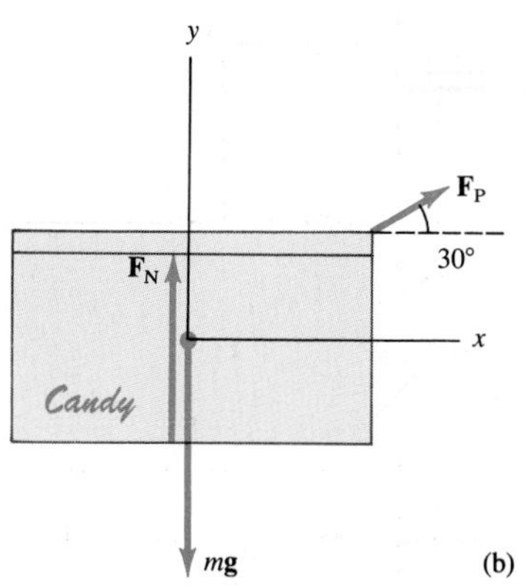

EXAMPLE 4–7 Suppose your friend asks for a piece of candy from the box she gave you (Example 4–4, Fig. 4–15) and you respond, "sure, pull the box over to you." She then pulls the box by the attached ribbon (or string), as shown in Fig. 4–19a, along the smooth surface of the table. The magnitude of the force exerted by the person is $F_P = 40\text{ N}$, and it is exerted at a 30° angle as shown. Calculate (*a*) the acceleration of the box, and (*b*) the magnitude of the upward force F_N exerted by the table on the box. Assume that friction can be neglected.

SOLUTION Figure 4–19b shows the free-body diagram of the box, which means we show *all* the forces acting on the box and *only* the forces acting on the box. They are: the force of gravity $m\mathbf{g}$, the normal force exerted by the table $\mathbf{F}_N$, and the force exerted by the person $\mathbf{F}_P$. With the y axis vertical and the x axis horizontal, the pull of 40 N has components

$$F_{Px} = (40\text{ N})(\cos 30^\circ) = (40\text{ N})(0.866) = 35\text{ N},$$

$$F_{Py} = (40\text{ N})(\sin 30^\circ) = (40\text{ N})(0.50) = 20\text{ N}.$$

(a) In the horizontal (x) direction, $\mathbf{F}_N$ and $m\mathbf{g}$ have zero components; thus the horizontal component of the net force is F_{Px}. From Newton's second law,

$$F_{Px} = ma_x,$$

so

$$a_x = \frac{F_{Px}}{m} = \frac{(35\ \text{N})}{(10\ \text{kg})} = 3.5\ \text{m/s}^2.$$

The acceleration of the box is thus 3.5 m/s^2.

(b) In the vertical (y) direction, we have (using again Newton's second law)

$$F = ma$$

$$F_N - mg + F_{Py} = ma_y.$$

Now $mg = (10\ \text{kg})(9.8\ \text{m/s}^2) = 98\ \text{N}$ and $F_{Py} = 20\ \text{N}$ as we calculated above. Furthermore, we know $a_y = 0$ since the box does not move vertically. Then

$$F_N - 98\ \text{N} + 20\ \text{N} = 0$$

and the normal force is

$$F_N = 78\ \text{N}.$$

Notice that F_N is less than mg. The ground does not push against the full weight of the box since part of the pull exerted by the person is in the upward direction. Compare this to Example 4–4, part c.

Example 4–8 Two boxes, similar to that in the previous example, are connected by a lightweight cord and are resting on a table. The boxes have masses of 12 kg and 10 kg. A horizontal force F_P of 40 N is applied by a person to the 10-kg box, as shown in Fig. 4–20a. Find (a) the acceleration of each box, and (b) the tension in the cord.

SOLUTION (a) The free-body diagram for each of the boxes is shown in Figs. 4–20b and c. The cord is light, so we neglect its mass relative to the mass of the boxes. The force F_P acts on box 1. Box 1 exerts a force T on the connecting cord and the cord exerts a force $-T$ back on box 1 (Newton's third law). These forces on box 1 are shown in Fig. 4–20b. The force T, exerted on the cord by box 1, is transmitted all along the cord up to its attachment at box 2; T is thus called the "tension in the cord", and is the force each piece of cord exerts on the adjacent piece. Because the cord is considered to be massless, the tension at each end is the same.[†]

[†] Since the mass m of the cord is zero, by Newton's second law, the net force on the cord is $F = ma = 0$ no matter what a is; hence the forces pulling on the cord at its two ends must add up to zero (T and $-T$).

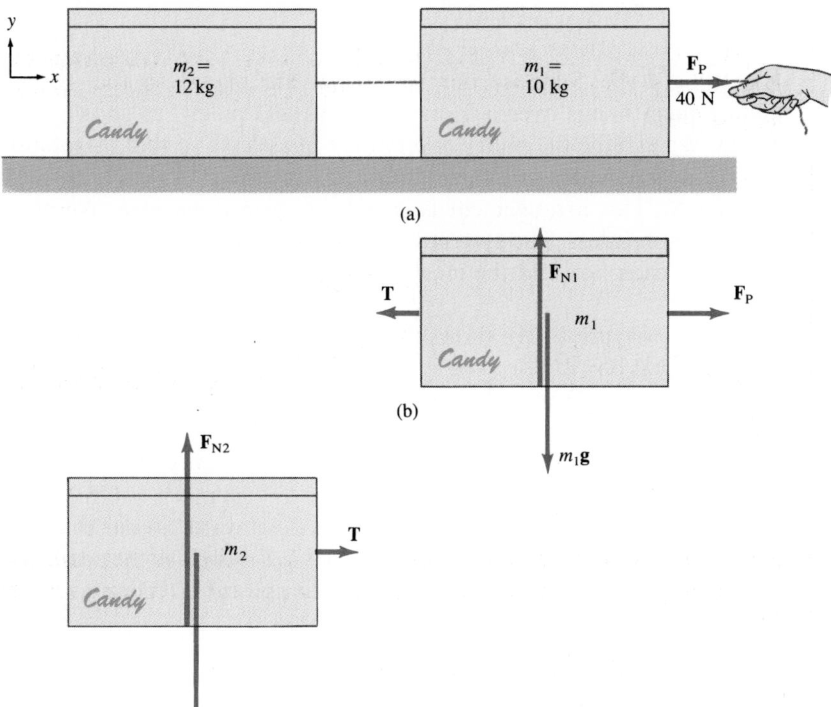

FIGURE 4–20 Example 4–8. (a) Two boxes are connected by a cord. A person pulls horizontally on box 1 with force $\mathbf{F}_P = 40$ N. (b) Free-body diagram for box 1. (c) Free-body diagram for box 2.

Hence the cord exerts a force T on the second box; Fig. 4–20c shows the forces on box 2. There will be only horizontal motion. We take the positive x axis to the right, and we use subscripts 1 and 2 to refer to the two boxes. Applying $F = ma$ to box 1, we have:

$$F_P - T = m_1 a_1.$$

For box 2, the only horizontal force is T, so

$$T = m_2 a_2.$$

The boxes are connected, and if the cord remains taut, then the two boxes will have the same acceleration a. Thus $a_1 = a_2 = a$, and we are given $m_1 = 10$ kg and $m_2 = 12$ kg. We add the two equations above and obtain

$$(m_1 + m_2)a = F_P$$

or

$$a = \frac{F_P}{m_1 + m_2} = \frac{40 \text{ N}}{22 \text{ kg}} = 1.8 \text{ m/s}^2.$$

This is what we sought. Notice that we would have obtained the same result had we considered a single system, of mass $m_1 + m_2$, acted on by a net horizontal force equal to F_P. (The tension forces T are internal to the system as a whole and together make zero contribution to the net force on the *whole* system.)

(*b*) From the equation above for box 2 ($T = m_2 a_2$), the tension in the cord is

$$T = m_2 a = (12 \text{ kg})(1.8 \text{ m/s}^2) = 22 \text{ N}.$$

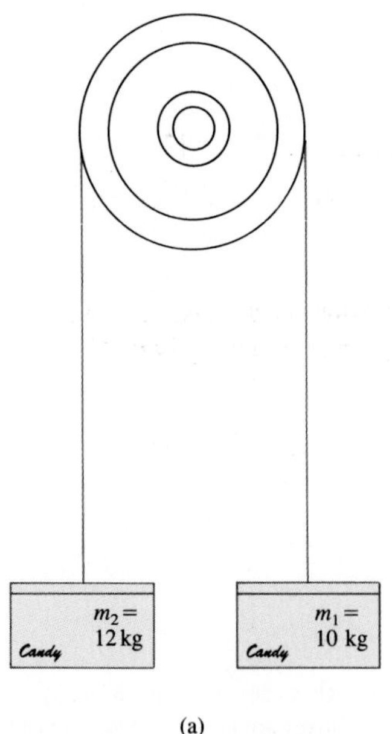

(a)

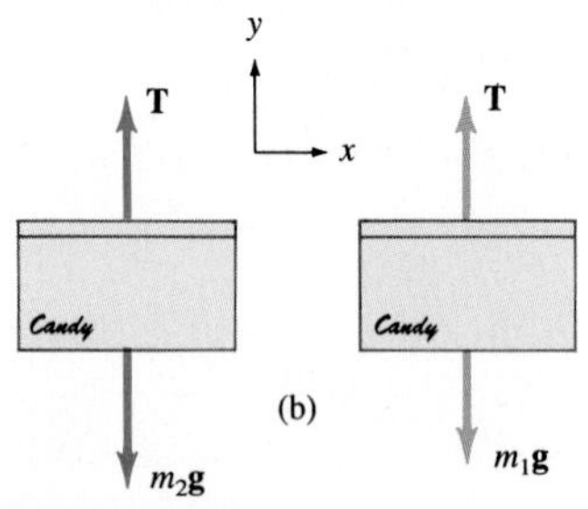

(b)

FIGURE 4–21 Example 4–9. (a) The Atwood machine. (b) The free-body diagrams for each box.

Example 4–9 Suppose our two boxes are placed so that the cord joining them hangs over a frictionless massless pulley, as shown in Fig. 4–21a. We assume the pulley is very light (massless), so that its rotational motion doesn't enter our calculation (we'll see how to deal with that in Chapter 8). This arrangement is called an *Atwood machine.* We assume the cord is massless. The system is released from rest. Calculate the acceleration of each box and the tension in the cord.

SOLUTION Figure 4–21b shows the free-body diagrams for the two boxes. It is clear that box 2, being the heavier, will accelerate downward, and box 1 will accelerate upward; the magnitudes of their accelerations will be equal (as long as the cord doesn't stretch). Since for box 1, $m_1 g = (10\text{ kg})(9.8\text{ m/s}^2) = 98\text{ N}$, we see that T must be greater than 98 N (in order that box 1 will accelerate upwards). For box 2, $m_2 g = (12\text{ kg})(9.8\text{ m/s}^2) = 118\text{ N}$, which must be greater than T since m_2 accelerates downward. So our calculation must give T between 98 N and 118 N. To find T as well as the acceleration a, we apply $F = ma$ to each box, taking the upward direction as positive for both boxes:

$$T - m_2 g = -m_2 a$$

$$T - m_1 g = m_1 a.$$

We subtract the first equation from the second to get

$$(m_2 - m_1)g = (m_1 + m_2)a.$$

We solve this for a:

$$a = \frac{m_2 - m_1}{m_2 + m_1} g = \frac{12\text{ kg} - 10\text{ kg}}{12\text{ kg} + 10\text{ kg}} g = 0.091g = 0.89\text{ m/s}^2.$$

Box 2 accelerates downward (and box 1 upward) at $a = 0.091g = 0.89\text{ m/s}^2$. We can get the tension T from either of the two equations

$$T = m_2 g - m_2 a = (12\text{ kg})(1 - 0.091)(9.8\text{ m/s}^2) = 107\text{ N}$$

$$T = m_1 g + m_1 a = (10\text{ kg})(1 + 0.091)(9.8\text{ m/s}^2) = 107\text{ N},$$

which are consistent.

Problem Solving: Check your result by seeing if it works in situations where the answer is easily guessed

We can check our equation for the acceleration a by noting that if the masses are equal ($m_1 = m_2$), then our equation above gives $a = 0$, as we would expect; also, if one of the masses is zero (say, $m_1 = 0$), then the other mass ($m_2 \neq 0$) would be predicted by our equation to accelerate at $a = g$, again as expected.

To end this section, we take an example of a different type.

How to get out of the mud

Example 4–10 Finding her car stuck in the mud, a clever graduate of a good physics course ties a strong rope to the back bumper of the car and the other end to a tree, as shown in Fig. 4–22a. She pushes at the midpoint of the rope with her maximum effort, which she estimates to be a force $F_P \approx 300$ N. The car just begins to budge with the rope at an angle θ (see the figure), which she estimates to be 5°. With what force is the rope pulling on the car? Neglect the mass of the rope.

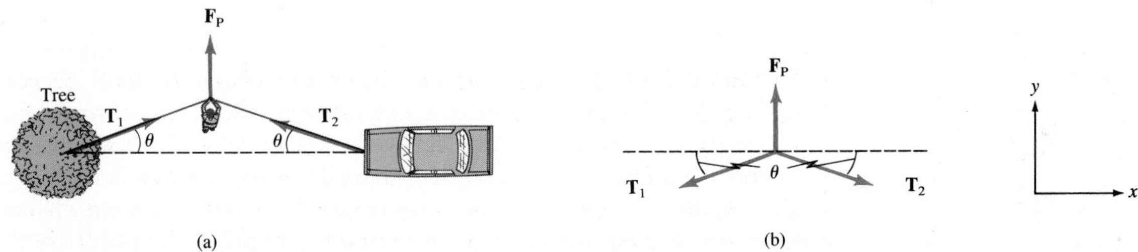

FIGURE 4–22 Example 4–10: getting a car out of the mud.

SOLUTION First, note that the tension in a rope is always along the rope. Any component perpendicular to the rope would cause the rope to bend or buckle (as it does where F_P acts)—in other words, a rope can support a tension force only along its length. Let T_1 and T_2 be the forces the rope exerts on the tree and on the car, as shown in Fig. 4–22a. We ignore the mass of the rope. As our "free body," we use the tiny section of rope where she pushes. The free-body diagram is shown in Fig. 4–22b. F_P has only a y component. $\mathbf{T}_1$ and $\mathbf{T}_2$ have both x and y components. We want to find the force with which the rope pulls on the car, and this equals (by Newton's third law) the force the car exerts on the rope, T_2, the tension in the rope that pulls on the car. At the moment the car budges, the acceleration is still essentially zero. So for the x component of $\mathbf{F} = m\mathbf{a} = 0$, we have

$$T_{2x} - T_{1x} = 0 \quad \text{or} \quad T_1 \cos\theta - T_2 \cos\theta = 0;$$

hence $T_1 = T_2$, and we can write $T = T_1 = T_2$. In the y direction, the forces acting are F_P, and the components of $\mathbf{T}_1$ and $\mathbf{T}_2$ that point in the negative y direction (each equal to $T \sin\theta$). So for the y component of $\mathbf{F} = m\mathbf{a}$, we have

$$F_P - 2T \sin\theta = 0.$$

We solve this for T and insert $F_P \approx 300$ N, which was given:

$$T = \frac{F_P}{2\sin\theta} = \frac{300\text{ N}}{2\sin 5^\circ} \approx 1700\text{ N}.$$

She was able to magnify her effort almost six times using this technique! Notice that the symmetry of the problem ($\mathbf{T}_1$ and $\mathbf{T}_2$ make equal angles with $\mathbf{F}_P$) ensures that $T_1 = T_2$.

Problem Solving: Use any symmetry present to simplify a problem

4-9 • Problems Involving Friction, Inclines

Until now we have ignored friction, but it must be taken into account in most practical situations. Friction exists between two solid surfaces because even the smoothest looking surface is quite rough on a microscopic scale. Even when a body rolls across a surface, there is still some friction, called *rolling friction*, although it is generally much less than when one body slides across a surface. We will be concerned mainly with sliding friction in this section, and it is usually called *kinetic friction* (*kinetic* is from the Greek for "moving").

Kinetic friction

When a body is in motion along a rough surface, the force of kinetic friction acts opposite to the direction of the body's motion. The magnitude of the force of kinetic friction depends on the nature of the two sliding surfaces. For given surfaces, it is proportional to the *normal force* between the two surfaces, which is the force that either object exerts on the other, perpendicular to their common surface of contact (see Fig. 4-23). The normal force on a body sliding on a horizontal surface is equal to the body's weight, mg. The force of friction does not depend appreciably on the total surface area of contact; that is, the friction force on a brick is essentially the same whether it is being slid on its wide face or on its end, as long as the surfaces are the same. We can write the proportionality as an equation by inserting a constant of proportionality, μ_k (Greek lowercase letter mu):

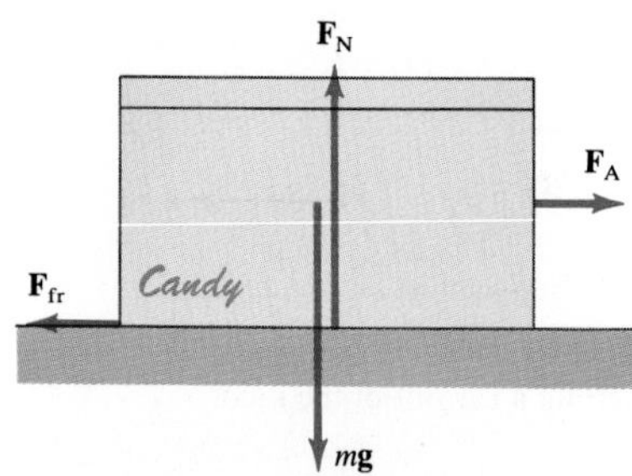

FIGURE 4–23 When an object is pulled by an applied force ($\mathbf{F}_A$) along a surface, the force of friction $\mathbf{F}_{fr}$ opposes the motion. The magnitude of $\mathbf{F}_{fr}$ is proportional to the magnitude of the normal force (F_N).

$$F_{fr} = \mu_k F_N. \qquad \text{[kinetic friction]}$$

This is an approximate (but reasonably accurate and useful) relation. It is not a law; it is a relation between the magnitude of the friction force, F_{fr}, which acts parallel to the two surfaces, and the magnitude of the normal force, F_N, which acts perpendicular to the surfaces. It is not a vector equation since the two forces are perpendicular to one another. The term μ_k is called the *coefficient of kinetic friction* and its value depends on the two surfaces. Measured values for a variety of surfaces are given in Table 4–2. These are only approximate, however, since μ depends on whether the surfaces are wet or dry, on how much they have been sanded or rubbed, if any burrs remain, and other such factors.

Static friction

What we have been discussing up to now is *kinetic friction*, when one body slides over another. There is also *static friction*, which refers to a force parallel to the two surfaces that can arise even when they are not sliding. Suppose an object such as a desk is resting on a horizontal floor. If no horizontal force is exerted on the desk, there also is no friction force. But now, suppose you try to push the desk, but it doesn't move. You are exerting a horizontal force, but the desk isn't moving, so there must be another force on the desk keeping it from moving (the net force is zero). This is the force of *static friction* exerted by the floor on the desk. If you push with a greater force without moving the desk, the force of static friction also has increased.

TABLE 4–2
Coefficients of Friction†

Surfaces	Coefficient of Static Friction, μ_s	Coefficient of Kinetic Friction, μ_k
Wood on wood	0.4	0.2
Ice on ice	0.1	0.03
Metal on metal (lubricated)	0.15	0.07
Steel on steel (unlubricated)	0.7	0.6
Rubber on dry concrete	1.0	0.8
Rubber on wet concrete	0.7	0.5
Rubber on other solid surfaces	1–4	1
Teflon on teflon in air	0.04	0.04
Teflon on steel in air	0.04	0.04
Lubricated ball bearings	<0.01	<0.01
Synovial joints (in human limbs)	0.01	0.01

† Values are approximate and are intended only as a guide.

If you push hard enough, the desk will eventually start to move. At this point, you have exceeded the maximum force of static friction, which is given by $F_{fr} = \mu_s F_N$, where μ_s is the *coefficient of static friction* (Table 4–2). Since the force of static friction varies from zero to this maximum value, we can write

$$F_{fr} \leq \mu_s F_N. \qquad \text{[static friction]}$$

You may have noticed that it is often easier to keep a heavy object (like a desk) moving than it is to start it moving in the first place. This follows from the fact (see Table 4–2) that μ_s is almost always greater than μ_k. (It can never be less. Why?)

EXAMPLE 4–11 A 10-kg box rests on a horizontal floor. The coefficient of static friction is $\mu_s = 0.40$ and the coefficient of kinetic friction is $\mu_k = 0.30$. Determine the force of friction, F_{fr}, acting on the box if a horizontal external applied force F_A is exerted on it of magnitude (*a*) 0, (*b*) 10 N, (*c*) 20 N, (*d*) 38 N, and (*e*) 40 N.

SOLUTION The free-body diagram of the box is shown in Fig. 4–23. In the vertical direction there is no motion, so $F_N - mg = 0$. Hence the normal force for all cases is $F_N = mg = (10\text{ kg})(9.8\text{ m/s}^2) = 98\text{ N}$.

(*a*) Since no force is applied in this first case, the box doesn't move and $F_{fr} = 0$.

(*b*) The force of static friction will oppose any applied force up to a maximum of $\mu_s F_N = (0.40)(98\text{ N}) = 39.2\text{ N}$. The applied force is $F_A = 10\text{ N}$. Thus the box will not move, and since $F_A - F_{fr} = 0$ then $F_{fr} = 10\text{ N}$.

(*c*) An applied force of 20 N is also not sufficient to move the box. Thus $F_{fr} = 20\text{ N}$ to balance the applied force.

(*d*) The applied force of 38 N is still not quite large enough to move the box; so the friction force has now increased to 38 N to keep the box at rest.

(*e*) A force of 40 N will start the box moving since it exceeds the maximum force of static friction, 39.2 N. Instead of static friction, we now have kinetic friction, and its magnitude is $F_{fr} = \mu_k F_N = (0.30)(98\text{ N}) = 29\text{ N}$. There is now a net force (horizontal) on the box of magnitude $F = 40\text{ N} - 29\text{ N} = 11\text{ N}$, so the box will accelerate at a rate $a = F/m = 11\text{ N}/10\text{ kg} = 1.1\text{ m/s}^2$ as long as the applied force is 40 N. Figure 4–24 shows a graph that summarizes this example.

Now we look at some examples involving kinetic friction in a variety of situations. Note that both the normal force and the friction force are forces exerted by one surface on another; one is perpendicular to the contact surfaces (the normal force), and the other is parallel (the friction force).

$F_{fr} \perp F_N$

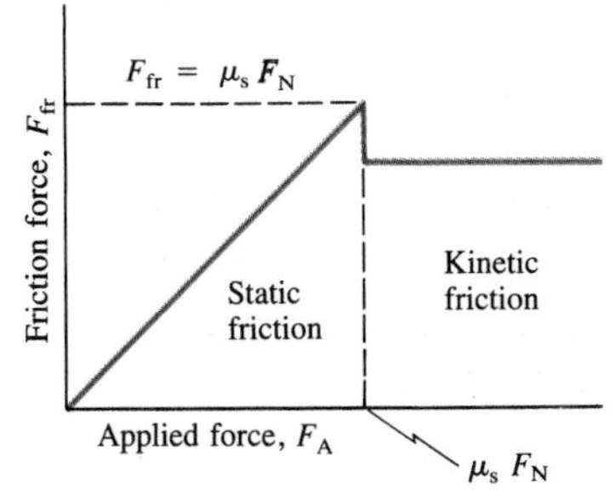

FIGURE 4–24 Magnitude of the force of friction as a function of the external force applied to a body initially at rest. As the applied force is increased in magnitude, the force of static friction increases linearly to just match it, until the applied force equals $\mu_s F_N$. If the applied force increases further, the body will begin to move, and the friction force drops to a constant value characteristic of kinetic friction.

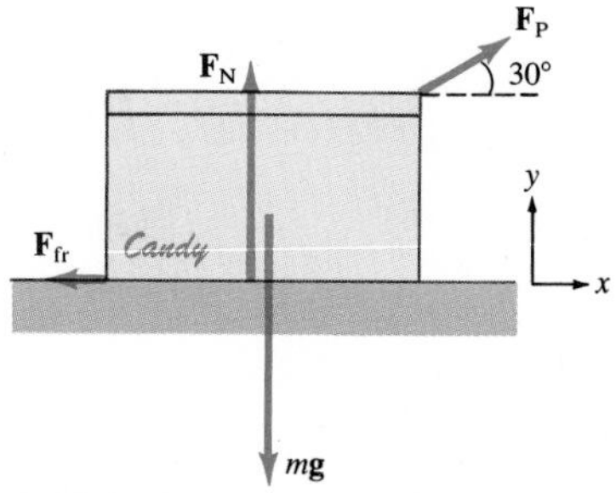

FIGURE 4–25 Example 4–12.

EXAMPLE 4–12 A 10-kg box is pulled along a horizontal surface by a force F_P of 40 N, which is applied at a 30° angle. This is like Example 4–7, except now there is friction, and we assume a coefficient of kinetic friction of 0.30. The free-body diagram is shown in Fig. 4–25. Calculate the acceleration.

SOLUTION The force of kinetic friction acts in the negative x direction, opposite to the direction of motion, and is parallel to the surface of contact. The calculation for the y direction is just the same as before, Example 4–7. Thus the force with which the floor pushes up on the package, the normal force, is still 78 N. For the x direction, we have a new term, the force of friction whose magnitude equals $\mu_k F_N = (0.30)(78\text{ N}) = 23\text{ N}$. Thus

$$ma_x = F_{Px} - \mu_k F_N = 35\text{ N} - 23\text{ N} = 12\text{ N}$$

$$a_x = \frac{12\text{ N}}{10\text{ kg}} = 1.2\text{ m/s}^2.$$

EXAMPLE 4–13 In Fig. 4–26a, two boxes are connected by a cord running over a pulley. The coefficient of kinetic friction between box I and the table is 0.20. (We ignore the mass of the cord and pulley and any friction in the pulley, which means we can assume that a force applied to one end of the cord will have the same magnitude at the other end.) We wish to find the acceleration of the system, which will be the same for both boxes assuming the cord doesn't stretch.

SOLUTION Free-body diagrams are shown for each box in Fig. 4–26b and c. Box I does not move vertically, so the normal force just balances the weight, $F_N = m_I g = (5.0\text{ kg})(9.8\text{ m/s}^2) = 49\text{ N}$. In the horizontal direction, there are two forces on box I: F_c, the pull of the cord (whose value we don't know), and the force of friction $= \mu_k F_N = (0.20)(49\text{ N}) = 9.8\text{ N}$. The horizontal acceleration is what we wish to find, and using Newton's

FIGURE 4–26 Example 4–13.

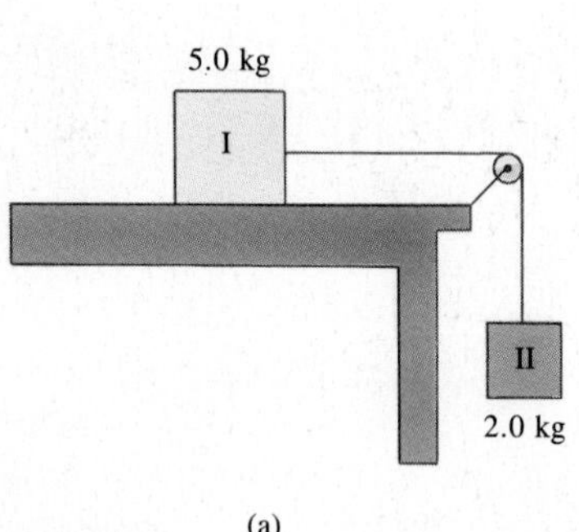

(a)

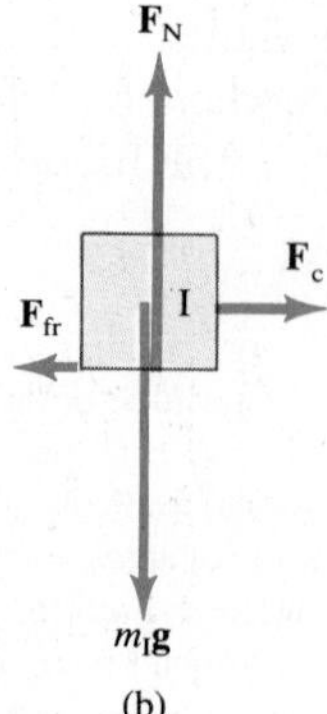

(b)

(c)

second law, $F = ma$, we have (taking the positive direction to the right):

$$F_c - F_{fr} = m_I a.$$

Next consider box II. The force of gravity $F_g = m_{II} g = 19.6$ N pulls downward. And the cord pulls upward with a force F_c (this, the tension in the cord, is the same force the cord exerts on box I since the cord is massless and the tension must be the same at one end as at the other, as we saw in Example 4–8). So we can write Newton's second law for box II (taking the downward direction as positive):

$$m_{II} g - F_c = m_{II} a.$$

We have two unknowns, a and F_c, and we also have two equations. We solve the first for F_c:

$$F_c = F_{fr} + m_I a,$$

and substitute this into the second equation:

$$m_{II} g - F_{fr} - m_I a = m_{II} a.$$

Now we solve for a and put in numerical values (see above):

$$a = \frac{m_{II} g - F_{fr}}{m_I + m_{II}} = \frac{19.6\ \text{N} - 9.8\ \text{N}}{5.0\ \text{kg} + 2.0\ \text{kg}} = 1.4\ \text{m/s}^2.$$

If we wish, we can calculate F_c using the first equation:

$$F_c = F_{fr} + m_I a = 9.8\ \text{N} + (5.0\ \text{kg})(1.4\ \text{m/s}^2) = 17\ \text{N}.$$

We now discuss some examples of objects moving on an incline such as a hill or a ramp. Such situations commonly arise (Fig. 4–27).

FIGURE 4–27 Examples of motion on an incline. [In (b) we are about to see another physics topic—see Section 7–3.]

(a)

(b)

(c)

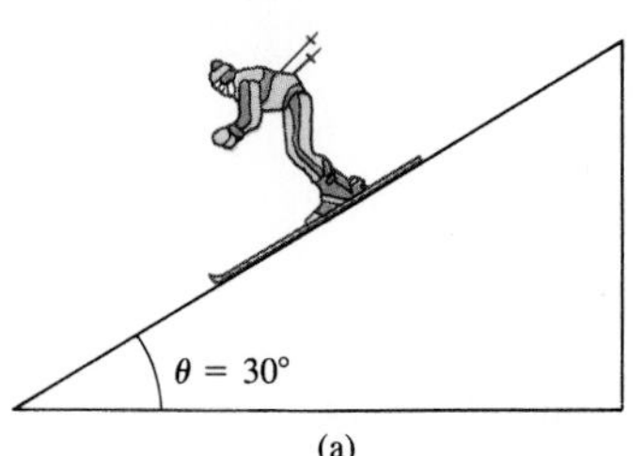

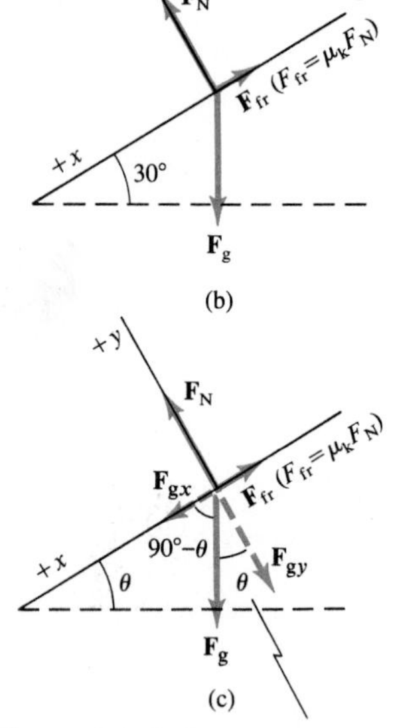

This angle is θ, the same as the slope angle, because the right sides of these two angles are perpendicular to each other and the left sides are perpendicular to each other.

FIGURE 4–28 Example 4–14: a skier descending a slope.

EXAMPLE 4–14 The skier in Fig. 4–28a has just begun descending the 30° slope. Assuming the coefficient of kinetic friction is 0.10, calculate (*a*) her acceleration, and (*b*) the speed she will reach after 6.0 s.

SOLUTION (*a*) The free-body diagram in Fig. 4–28b shows all the forces acting on the skier: her weight ($\mathbf{F}_g = m\mathbf{g}$) downward, and the two forces exerted on her skis by the snow—the normal force perpendicular to the snow's surface, and the friction force parallel to the surface. These three forces are shown acting at one point, for convenience. Also for convenience, we choose the x axis parallel to the snow surface with positive direction downhill, and the y axis perpendicular to the surface. Then we only have to resolve one vector into components, the weight. The components of the weight are shown as dashed lines in Fig. 4–28c; they are given by (notice which angle is θ and the reason given on the diagram):

$$F_{gx} = mg \sin \theta,$$

$$F_{gy} = mg \cos \theta.$$

First we apply Newton's second law to the y direction:

$$F_N - mg \cos \theta = ma_y = 0$$

since there is no motion in the y direction. Thus

$$F_N = mg \cos \theta.$$

In the x direction, $F_x = ma_x$ yields:

$$mg \sin \theta - \mu_k F_N = ma_x$$

$$mg \sin \theta - \mu_k mg \cos \theta = ma_x.$$

There is an m in each term, so they can be canceled out. Thus:

$$\begin{aligned} a_x &= g \sin 30° - \mu_k g \cos 30° \\ &= [0.50 - (0.10)(0.866)]g \\ &= 0.41\ g. \end{aligned}$$

Problem Solving: It is often helpful to put in numbers only at the end

The acceleration is 0.41 times the acceleration of gravity, which in numbers is $a = (0.41)(9.8\ \text{m/s}^2) = 4.0\ \text{m/s}^2$. It is interesting that the mass canceled out and so the acceleration doesn't depend on the mass. That such a cancellation sometimes occurs, and thus saves calculation, is one advantage of working with the algebraic equations and putting in the numbers only at the end.

(*b*) The speed after 6.0 s is found by using Eq. 2–10a:

$$v = v_0 + at = 0 + (4.0\ \text{m/s}^2)(6.0\ \text{s}) = 24\ \text{m/s},$$

where we assumed a start from rest.

A method for determining μ_k

EXAMPLE 4–15 Again consider a skier descending a hill. This time, however, we don't know the coefficient of kinetic friction between the skis and the snow and we want to determine it. We can do so by observing the skier descend slopes of different angles, and noticing at what angle she descends at constant speed (so $a_x = 0$). The $F = ma$ equations for the x

and y components will be the same as in the previous example, except that now $a_x = 0$ and the angle θ will not necessarily be 30°. Thus:

$$F_N - mg \cos \theta = ma_y = 0$$

$$mg \sin \theta - \mu_k F_N = ma_x = 0.$$

From the first equation, we have $F_N = mg \cos \theta$; we substitute this into the second equation:

$$mg \sin \theta - \mu_k(mg \cos \theta) = 0.$$

Now we solve for μ_k:

$$\mu_k = \frac{mg \sin \theta}{mg \cos \theta} = \frac{\sin \theta}{\cos \theta}.$$

From the definitions of sine and cosine, Eqs. 3–1 (see Fig. 3–7):

$$\frac{\sin \theta}{\cos \theta} = \frac{o/h}{a/h} = \frac{o}{a} = \tan \theta.$$

Thus

$$\mu_k = \tan \theta$$

where θ is the angle at which the skier moves at constant speed. For example, if $\theta = 5°$, then $\mu_k = \tan 5° = 0.09$.

EXAMPLE 4–16 In the design of a supermarket, there are to be several ramps connecting different parts of the store. Customers will have to push grocery carts up the ramps and it is obviously desired that this not be too difficult. An engineer has done a survey and found that almost no one complains if the force required is no more than 20 N. Ignoring friction, at what maximum angle θ should the ramps be built assuming a full 20-kg grocery cart?

SOLUTION As shown in Fig. 4–29, the pushing force on the cart must just balance the x component of the weight. For the case of the maximum force of 20 N, we have

$$F_{gx} = F_g \sin \theta = 20 \text{ N}.$$

Since $F_g = mg = (20 \text{ kg})(9.8 \text{ m/s}^2) = 200 \text{ N}$, we have $\sin \theta = 20/200 = 0.10$; so $\theta = 5.7°$. In a real situation, friction would have to be taken into account, particularly friction in the wheels of the carts when they get old, and this would probably be done by experiment.

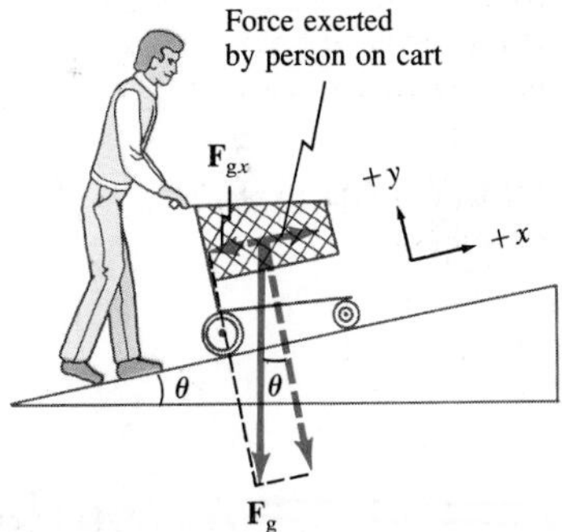

FIGURE 4–29 Example 4–16.

Friction can be a hindrance. It slows down moving objects and causes heating and binding of moving parts in machinery. Friction can be reduced by using lubricants such as oil. More effective in reducing friction between two surfaces is to maintain a layer of air or other gas between them. Devices using this concept, which is not practical for many situations, include air tracks and air tables (or games) in which the layer of air is maintained by forcing air through many tiny holes. Another technique to maintain the air layer is to suspend objects in air using magnetic fields (so-called "magnetic levitation"). On the other hand, friction can be helpful. Our ability to walk

Friction—a help and a hindrance

depends on friction between the soles of our shoes (or feet) and the ground. (Does walking involve static friction or kinetic friction?) The movement of a car and how well it holds the road depend on friction. When friction is low, such as on ice, safe walking and driving become difficult.

4–10 • Problem Solving—A General Approach

A basic part of a physics course is solving problems effectively. Problem solving is important not only because it is valuable in itself, but also because it makes you think about the ideas and concepts, and by applying the concepts, you come to understand them better. We have already studied a range of examples in these first few chapters. It is so important to learn problem solving that we will spend a little time now summarizing how to approach problems.

The solving of problems often requires creativity—each problem is different. Nonetheless, we can outline a general approach that will be of some help in solving problems. But before attacking any problem, it is very important to know the definitions, terminology, and the basic principles and laws that apply.

Techniques and hints for problem solving

1. Read written problems carefully. A common error is to leave out a word or two when reading, which can completely change the sense of a problem.
2. Draw an accurate picture or diagram of the situation. (This is probably the most overlooked, yet most crucial, part of solving a problem.) Use arrows to represent vectors such as velocity or force, and label the vectors with appropriate symbols. Use a separate diagram for different types of vectors: say, one for force and one for velocity, if both types of vector are involved. Make sure to include all forces on a given body, and make clear what forces act on what body (otherwise you may make an error in determining the *net force* on a particular body). A diagram showing all the forces acting on a given body (and only on that body) is called a *free-body diagram*, as discussed in Section 4–8.
3. Usually you will need to choose a convenient x-y coordinate system (choose one that makes your calculations easier). Vectors can be resolved into components along these axes.
4. Determine what the unknowns are—that is, what you are trying to determine.
5. Decide what you need in order to find the unknowns. (*a*) It may help to see if there are one or more relationships (or equations) that relate the unknowns to the knowns. But be sure the relationship(s) is applicable in the given case. Beware of formulas that are not general but apply only in a specific case. (It is dangerous for this reason to thumb through a chapter looking for an equation that will work.) It is very important to know the limitations of each formula or relationship—when it is valid and when not. In this book, the more general equations have been given numbers, but even these can have a limited range of validity. The unnumbered equations are often valid only for very specific cases. (*b*) It is also helpful to determine what information is relevant, and what is irrelevant, in a given situation or problem.

6. Try to solve the problem roughly, to see if it is doable (to check if enough information has been given) and reasonable. Use your intuition, and make rough calculations—see "Order of Magnitude Estimating" in Section 1–6. A rough calculation, or a reasonable guess about what the range of final answers might be, is a very useful tool. And a rough calculation can be checked against the final answer to catch errors in calculation (such as in the powers of 10).
7. Solve the problem, which may include algebraic manipulation of equations and/or numerical calculations. Be sure to keep track of units, for they can serve as a check.
8. Again consider if your answer is reasonable. The use of dimensional analysis, described in Appendix B, can also serve as a check for many problems.

SUMMARY

Newton's *three laws of motion* are the basic classical laws describing motion. Newton's *first law* states that if the net force on an object is zero, an object originally at rest remains at rest, and an object in motion remains in motion in a straight line with constant velocity. The tendency of a body to resist a change in its motion is called *inertia*. *Mass* is a measure of the inertia of a body. *Weight* refers to the force of gravity on a body, and is equal to the product of the body's mass m and the acceleration of gravity $\mathbf{g}$ ($\mathbf{F}_g = m\mathbf{g}$).

Newton's *second law of motion* states that the acceleration of a body is directly proportional to the net force acting on it, and inversely proportional to its mass ($\mathbf{F} = m\mathbf{a}$). Newton's second law is one of the most important and fundamental laws in classical physics. *Force*, which is a vector, can be considered as a push or pull; or, from Newton's second law, force can be defined as an action capable of giving rise to acceleration. The *net* force on an object is the vector sum of all forces acting on it.

Newton's *third law of motion* states that whenever one body exerts a force on a second body, the second body always exerts a force on the first body which is equal in magnitude but opposite in direction.

A consistent set of units must be used when making calculations. The SI is mainly used for scientific work.

When two bodies slide over one another, the force of friction that each body exerts on the other can be written approximately as $F_{fr} = \mu F_N$, where F_N is the normal force (the force each body exerts on the other perpendicular to their contact surface), and μ is the coefficient of kinetic friction if there is relative motion of the bodies; if the bodies are at rest, μ is the coefficient of static friction and F_{fr} is the maximum friction force before motion starts.

For solving problems involving the forces on one or more bodies, it is essential to draw a *free-body diagram* for each body, showing all the forces acting on only that body. Newton's second law can be applied to the vector components for each body.

QUESTIONS

1. Compare the effort (or force) needed to lift a 10-kg object when you are on the moon as compared to being on earth. Compare the force needed to throw a 2-kg object horizontally with a given speed when on the moon as compared to being on earth.
2. Why does a child in a wagon seem to fall backward when you give the wagon a sharp pull?
3. Whiplash sometimes results from an automobile accident when the victim's car is struck violently from the rear. Explain why the head of the victim seems to be thrown backward in this situation. Is it really?
4. When a golf ball is dropped to the pavement, it bounces back up. (*a*) Is a force needed to make it bounce back up? (*b*) If so, what exerts the force?
5. Examine, in the light of Newton's first and second laws, the motion of your leg during one stride while walking.
6. A person wearing a cast on an arm or a leg experiences extra fatigue. Explain this on the basis of Newton's first and second laws.
7. If the acceleration of a body is zero, are no forces acting on it?

8. Why do you push harder on the pedals of a bicycle when first starting out than when moving at constant speed?

9. Only one force acts on an object. Can the object have zero acceleration? Can it have zero velocity?

10. When you are running and want to stop quickly, you must decelerate quickly. (*a*) What is the origin of the force that causes you to stop? (*b*) Estimate (using your own experience) the maximum rate of deceleration of a person running at top speed to come to rest.

11. Why, when you walk on a log floating in water, does the log move in the opposite direction?

12. Why does your foot hurt when you kick a football?

13. When you stand still on the ground, how large a force does the ground exert on you? Why doesn't this force make you rise up into the air?

14. The force of gravity on a 2-kg rock is twice as great as that on a 1-kg rock. Why then doesn't the heavier rock fall faster?

15. A person exerts an upward force of 40 N to hold a bag of groceries. Describe the "reaction" force (Newton's third law) by stating (*a*) its magnitude, (*b*) its direction, (*c*) *on* what body it is exerted, and (*d*) *by* what body it is exerted.

16. According to Newton's third law, each team in a tug-of-war pulls with equal force on the other team. What, then, determines which team will win?

17. Cross-country skiers prefer their skis to have a large coefficient of static friction but a small coefficient of kinetic friction. Explain why.

18. When you must brake your car very quickly, why is it safer if the wheels don't lock?

19. When driving on slick roads, why is it advisable to apply the brakes slowly?

20. Why is the stopping distance of a truck much shorter than for a train going the same speed?

21. Can a coefficient of friction exceed 1.0?

22. Devise a method for using an inclined plane to measure the coefficient of friction between two surfaces.

23. A block is given a push so that it slides up a ramp. When the block reaches its highest point, it slides back down. Why is its acceleration less on the descent than on the ascent?

24. A heavy crate rests on the bed of a flatbed truck. When the truck accelerates, the crate remains where it is on the truck, so it, too, accelerates. What force causes the crate to accelerate?

PROBLEMS

SECTIONS 4–4 TO 4–7

1. (I) How much tension must a rope withstand if it is used to accelerate a 1200-kg car at 0.60 m/s^2? Ignore friction.

2. (I) What force is needed to accelerate a bicycle of mass 80 kg (including a rider) at a rate of 1.85 m/s^2?

3. (I) A net force of 255 N accelerates an object at 4.20 m/s^2. What is the mass of the object?

4. (I) What is your mass in kg, and your weight in newtons?

5. (I) How much force is required to accelerate a 14.0-g object at 10,000 "g's" (say, in a centrifuge)?

6. (I) What is the weight of a 70-kg astronaut (*a*) on earth, (*b*) on the moon ($g = 1.7$ m/s^2), (*c*) on Venus ($g = 8.7$ m/s^2), (*d*) in outer space traveling with constant velocity?

7. (II) What average force is required to stop a 1000-kg car in 6.0 s if it is traveling at 90 km/h?

8. (II) What average force is needed to accelerate a 9.50-g bullet from rest to 500 m/s over a distance of 0.700 m along the barrel of a rifle?

9. (II) A 0.10-g spider is descending on a strand that supports it with a force of 6.6×10^{-4} N. What is the acceleration of the spider? Ignore air resistance.

10. (II) A 0.145-kg baseball traveling 40.0 m/s strikes the catcher's mitt, which, in bringing the ball to rest, recoils backward 12.0 cm. What was the average force applied by the ball on the glove?

11. (II) What is the average force exerted by a shotputter on a 7.0-kg shot if the shot is moved through a distance of 2.8 m and is released with a speed of 13 m/s?

12. (II) An elevator (mass 4750 kg) is to be designed so that the maximum acceleration is 0.0500 g. What are the maximum and minimum forces the motor should exert on the supporting cable?

13. (II) A 40-kg child wants to escape from a third-story window to avoid punishment. Unfortunately, a make-shift rope made of sheets tied together can support a mass of only 30 kg. How can the child use this "rope" to escape? Give quantitative answer.

14. (II) What is the acceleration of a falling 75-kg sky-diver at a moment when air resistance exerts a force of 250 N?

15. (II) The cable supporting a 1650-kg elevator has a maximum strength of 22,500 N. What maximum upward acceleration can it give the elevator without breaking?

16. (III) An exceptional standing jump would raise a person 0.80 m off the ground. To do this, what force must a 70-kg person exert against the ground? Assume the person lowers himself 0.20 m prior to jumping, and thus rises a total distance of 1.00 m.

17. (III) A person jumps from a tower 5.0-m high. When he strikes the ground below, he bends his knees so that his torso decelerates over an approximate distance of 0.70 m. If the mass of his torso (excluding legs) is 50 kg, find (*a*) his velocity just before his feet strike the ground, and (*b*) the average force exerted on his torso by his legs during the deceleration.

18. (III) The 100-m dash can be run by the best sprinters in 10.0 s. (*a*) What is the horizontal component of force exerted on a 75-kg sprinter's feet by the ground during acceleration, which we assume is done roughly uniformly over the first 30 m? (*b*) What is the average speed of the sprinter over the last 70 m of the race?

SECTION 4–8

19. (I) A box weighing 70 N rests on the floor. A rope tied to the box runs vertically upward over a pulley and a weight is hung from the other end. Determine the force that the floor exerts on the box if the weight hanging on the other side of the pulley weighs (*a*) 30 N, (*b*) 60 N, and (*c*) 90 N.

20. (I) A 500-N force acts in a northwesterly direction. A second 500-N force must be exerted in what direction so that the resultant of the two vectors points westward?

21. (II) The two forces $\mathbf{F}_1$ and $\mathbf{F}_2$ shown in Fig. 4–30a and b act on a 10-kg object. If $F_1 = 12$ N and $F_2 = 15$ N, find the net force on the object and its acceleration for each situation, (*a*) and (*b*).

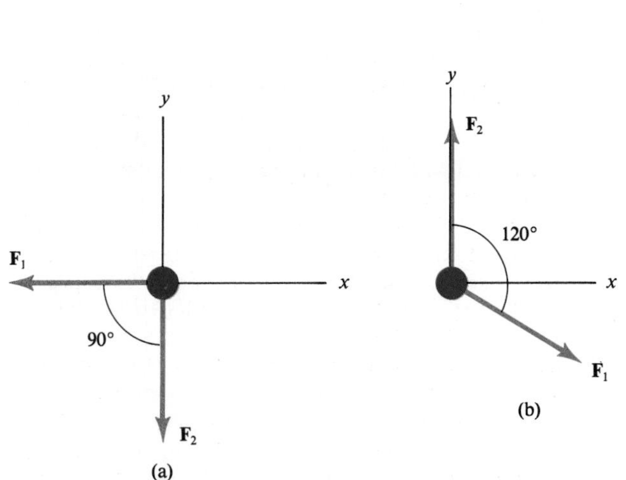

FIGURE 4–30 Problem 21.

FIGURE 4–31 Problem 22.

22. (II) A man pushes a 16-kg lawn mower at constant speed with a force of 90 N directed along the handle, which is at an angle of 45° to the horizontal (Fig. 4–31). Calculate (*a*) the horizontal retarding force on the mower, (*b*) the normal force exerted vertically upward on the mower by the ground, and (*c*) the force the man must exert on the lawn mower to accelerate it from rest to 1.5 m/s in 2.5 seconds. Assume the same retarding force.

23. (II) At the instant a race began, a 60-kg sprinter was found to exert a force of 850 N on the starting block at a 20° angle with respect to the ground. (*a*) What was the horizontal acceleration of the sprinter? (*b*) If the force was exerted for 0.38 s, with what speed did the sprinter leave the starting block?

24. (II) One paint bucket weighing 40.0 N is hanging by a massless cord from another paint bucket also weighing 40.0 N, as shown in Fig. 4–32. The two are being pulled upward with an acceleration of 1.50 m/s^2 by a massless cord attached to the upper bucket. Calculate the tension in each cord.

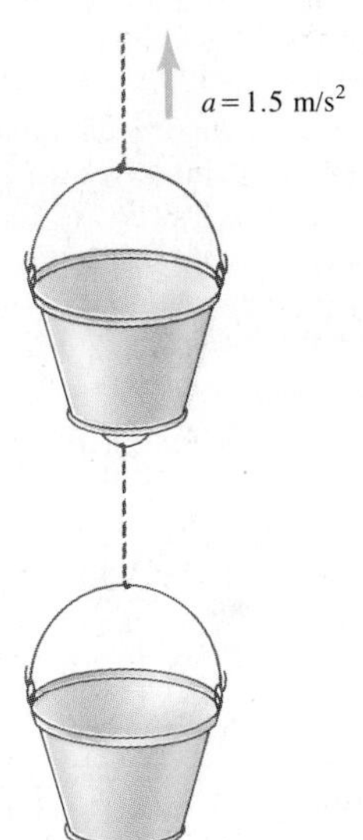

FIGURE 4–32 Problem 24.

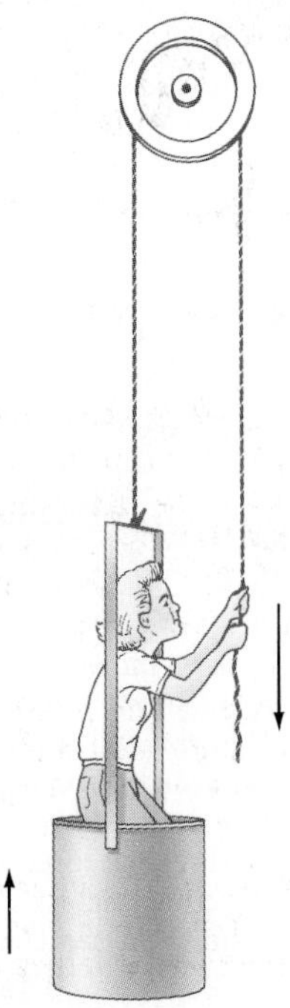

FIGURE 4–33 Problem 27.

25. (II) Redo Example 4–9, but (*a*) set up the equations so that the direction of the acceleration **a** of each box is in the direction of motion of that box. (In Example 4–9, we took **a** as positive upwards for both boxes.) (*b*) Solve the equations to obtain the same answers of Example 4–9.

26. (II) A 5000-kg helicopter accelerates upward at 0.55 m/s^2 while lifting a 1500-kg car. (*a*) What is the lift force exerted by the air on the rotors? (*b*) What is the tension in the cable (ignore its mass) that connects the car to the helicopter?

27. (II) A window washer pulls herself upward using the bucket-pulley apparatus shown in Fig. 4–33. (*a*) How hard must she pull downward to raise herself slowly at constant speed? (*b*) If she increases this force by 10 percent, what will her acceleration be? The mass of the person plus the bucket is 75 kg.

28. (II) A train locomotive is pulling two cars of the same mass behind it. Show that the tension in the coupling between the locomotive and the first car is twice that between the first car and the second car, for any nonzero acceleration of the train.

29. (II) Three blocks on a frictionless horizontal surface are in contact with each other, as shown in Fig. 4–34. If a force **F** is applied to block 1 (mass m_1), determine (*a*) the acceleration of the system (in terms of m_1, m_2, and m_3), (*b*) the net force on each block, and (*c*) the force that each block exerts on its neighbor. (*d*) If $m_1 = m_2 = m_3 = 10$ kg and $F = 100$ N, give numerical answers to (*a*), (*b*), and (*c*). Do your answers make sense intuitively?

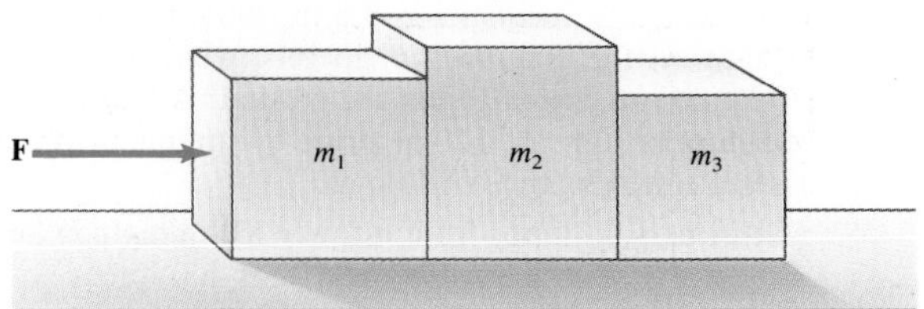

FIGURE 4–34 Problem 29.

30. (III) The two masses shown in Fig. 4–35 are each initially 1.60 m above the ground, and the massless frictionless pulley is 4.8 m above the ground. What maximum height does the lighter object reach after the system is released?

31. (III) Suppose the cord in Example 4–8 and Fig. 4–20 is a heavy rope of mass 1.0 kg. Calculate the acceleration of each box and the tension at each end of the cord, using the free-body diagrams shown in Fig. 4–36. Assume the cord doesn't sag.

SECTION 4–9

32. (I) A force of 400 N is required to start a 40-kg box moving across a concrete floor. What is the coefficient of static friction between the box and the floor?

33. (I) If the coefficient of kinetic friction between a 20-kg crate and the floor is 0.30, what horizontal force is required to move the crate at a steady speed across the floor? What horizontal force is required if μ_k is zero?

34. (II) What is the maximum acceleration a car can undergo if the coefficient of static friction between the tires and level ground is 0.80?

35. (II) The coefficient of friction between hard rubber and normal street pavement is about 0.8. On how steep a hill (maximum angle) can you leave a car parked?

36. (II) A car can decelerate at -5.80 m/s^2 without skidding when coming to rest on a level road. What would its deceleration be if the road were inclined at 12° uphill? Assume the same static friction force.

37. (II) For the system of Fig. 4–26, how large a mass would body I have to have to prevent any motion from occurring? Assume $\mu_s = 0.20$.

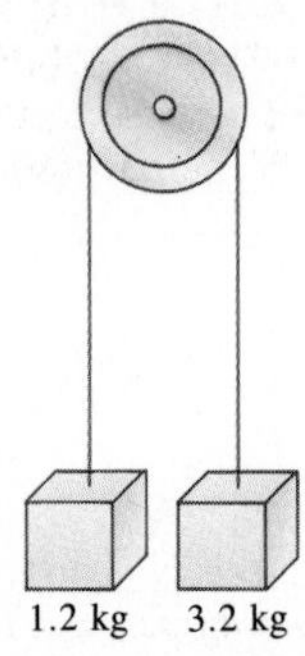

FIGURE 4–35 Problem 30.

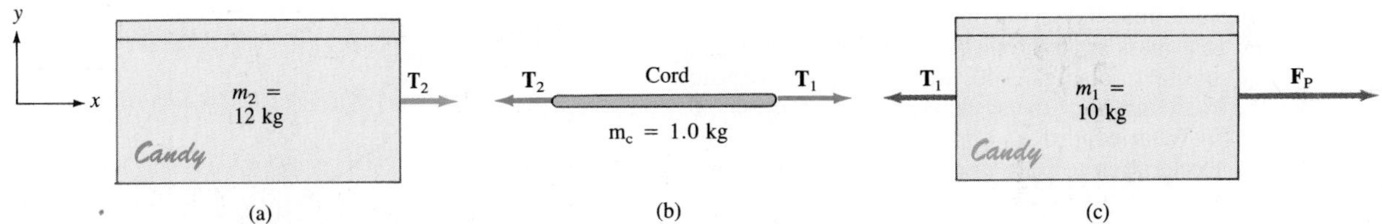

FIGURE 4–36 Problem 31. Free-body diagrams for each of the bodies of the system shown in Fig. 4–20a. Vertical forces, $\mathbf{F}_N$ and $\mathbf{F}_g$, are not shown.

38. (II) A box is given a push so that it slides across the floor. How far will it go, given that the coefficient of kinetic friction is 0.30 and the push imparts an initial speed of 3.0 m/s?
39. (II) Two crates, of mass 80 kg and 110 kg, are in contact and at rest on a horizontal surface (Fig. 4–37). A 650-N force is exerted on the 80-kg crate. If the coefficient of kinetic friction is 0.20, calculate (*a*) the acceleration of the system, and (*b*) the force that each crate exerts on the other.
40. (II) (*a*) Show that the minimum stopping distance for an automobile traveling at speed v is equal to $v^2/2\mu g$, where μ is the coefficient of static friction between the tires and the road, and g is the acceleration of gravity. (*b*) What is this distance for a 1500-kg car traveling 90 km/h if $\mu = 0.85$? (*c*) What would it be if the car were on the moon but all else stayed the same?
41. (II) A flatbed truck is carrying a 3200-kg crate of heavy machinery. If the coefficient of static friction between the crate and the bed of the truck is 0.65, what is the maximum rate at which the driver can decelerate when coming to a stop in order to avoid crushing the cab with the crate?
42. (II) A wet bar of soap slides freely down a ramp 2.0 m long inclined at 6.8°. How long does it take to reach the bottom? Neglect friction.
43. (II) The block shown in Fig. 4–38 has mass $m = 8.0$ kg and lies on a smooth plane tilted at an angle $\theta = 20°$ to the horizontal. (*a*) Determine the acceleration of the block as it slides down the plane. (*b*) If the block starts from rest 12.0 m up the plane from its base, what will be the block's speed when it reaches the bottom of the incline? Ignore friction.
44. (II) A block is given an initial speed of 5.0 m/s up the 20° plane shown in Fig. 4–38. (*a*) How far up the plane will it go? (*b*) How much time elapses before it returns to its starting point? Ignore friction.

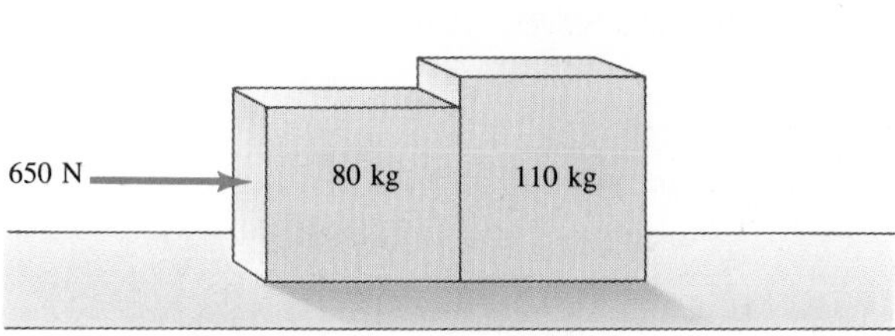

FIGURE 4–37 Problem 39.

45. (II) Repeat (*a*) Problem 43 and (*b*) Problem 44, assuming $\mu_k = 0.20$ between the block and plane.
46. (II) A roller coaster reaches the top of the steepest hill with a speed of 6.0 km/h. It then descends the hill, which is at an average angle of 45° and is 60 m long. What will its speed be when it reaches the bottom? Assume $\mu_k = 0.10$.
47. (II) An 18.0-kg box is released on a 33.0° incline and accelerates down the incline at 0.300 m/s². Find the friction force impeding its motion. How large is the coefficient of friction?
48. (II) A small mass m is set on the surface of a sphere, Fig. 4–39. If the coefficient of static friction is $\mu_s = 0.60$, at what angle θ would the mass no longer resist sliding?

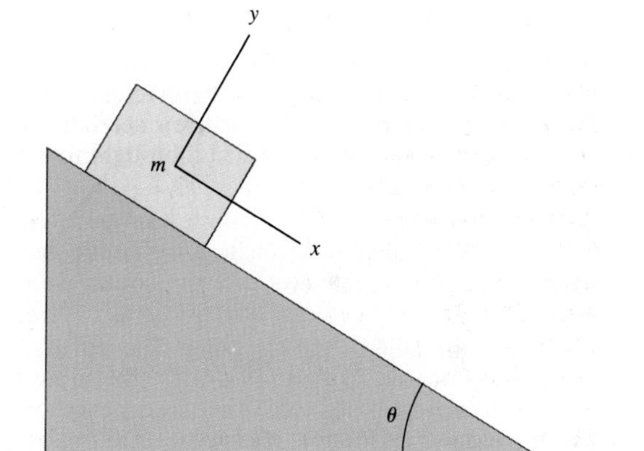

FIGURE 4–38 Block on inclined plane. Problems 43, 44, and 45.

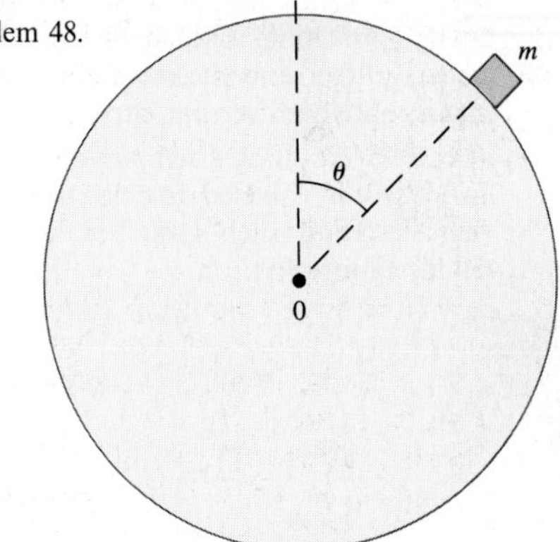

FIGURE 4–39 Problem 48.

49. (III) A block (mass m_1) lying on an inclined plane is connected to a mass m_2 by a massless cord passing over a pulley, as shown in Fig. 4–40. (a) Determine a formula for the acceleration of the system in terms of m_1, m_2, θ, and g. (b) What conditions apply to masses m_1 and m_2 for the acceleration to be in one direction (say, m_1 down the plane), or in the opposite direction?

50. (III) Suppose the coefficient of static friction between m_1 and the plane in Fig. 4–40 is $\mu_k = 0.15$, and that $m_1 = m_2 = 2.0$ kg. Determine the acceleration of m_1 and m_2, given $\theta = 30°$.

51. (III) What minimum value of μ_k will keep the system of Problem 50 from accelerating?

52. (III) If a bicyclist of mass 70 kg (including the bicycle) can coast down a 7.0° hill at a steady speed of 6.0 km/h, how much (average) force must be applied to climb the hill at the same speed?

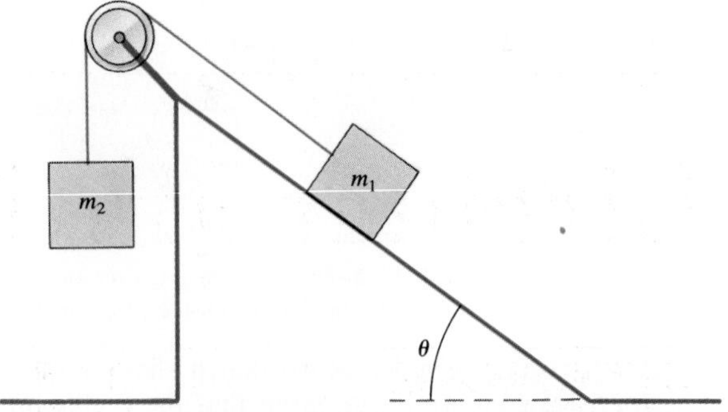

FIGURE 4–40 Problems 49, 50, and 51.

GENERAL PROBLEMS

53. According to a simplified model of a mammalian heart, at each pulse, approximately 20 g of blood is accelerated from 0.25 m/s to 0.35 m/s during a period of 0.10 s. What is the magnitude of the force exerted by the heart muscle?

54. A person has a reasonable chance of surviving an automobile crash if the deceleration is no more than 30 "g's." Calculate the force on a 70-kg person accelerating at this rate. What distance is traveled if brought to rest at this rate from 90 km/h?

55. Suppose that you are standing on a train accelerating at 0.20 g. What minimum coefficient of static friction must exist between your feet and the floor if you are not to slide?

56. A 1000-kg car pulls a 450-kg trailer. The car exerts a horizontal force of 3.5×10^3 N against the ground in order to accelerate. What force does the car exert on the trailer? Assume a friction coefficient of 0.15 for the trailer.

57. Police lieutenants, examining the scene of an accident involving two cars, measure the skid marks of one of the cars, which nearly came to a stop before colliding, to be 80 m long. The coefficient of kinetic friction between rubber and the pavement is about 0.8. Estimate the initial speed of that car.

58. A car starts rolling down a 1-in-4 hill (1-in-4 means that for each 4 m traveled, the elevation change is 1 m). How fast is it going when it reaches the bottom after traveling 50 m? Ignore friction.

59. Repeat Problem 58, assuming an effective coefficient of friction equal to 0.10.

60. A small block of mass m is given an initial speed v_0 up a ramp inclined at angle θ to the horizontal. It travels a distance d up the ramp and comes to rest. (a) Determine a formula for the coefficient of kinetic friction between the block and ramp. (b) What can you say about the value of the coefficient of static friction?

61. A motorcyclist is coasting with the engine off at a steady speed of 20 m/s but enters a sandy stretch where the coefficient of friction is 0.80. Will the cyclist emerge from the sandy stretch without having to start the engine if the sand lasts for 15 m? If so, what will be the speed upon emerging?

62. A city planner is working on the redesign of a hilly portion of a city. An important consideration is how steep the roads can be so that even small cars can get up the hills without slowing down. It is given that a particular small car, with a mass of 1000 kg, can accelerate on the level from rest to 14 m/s (50 km/h) in 8.0 s. Using this figure, calculate the maximum steepness of a hill.

63. A bicyclist can coast down a 4.0° hill at a constant 6.0 km/h. If the force of friction is proportional to the speed v so that $F_{fr} = cv$, calculate (a) the value of the constant c, and (b) the average force that must be applied in order to descend the hill at 20 km/h. The mass of the cyclist plus bicycle is 80 kg.

CHAPTER 5

Circular Motion; Gravitation

The planet Jupiter, showing two of its moons. The moons, held by the force of gravity, revolve around Jupiter. Galileo was the first to observe four of Jupiter's moons, a momentous discovery which he used to argue in favor of the Copernican system.

An object moves in a straight line if the net force on it acts in the direction of motion, or is zero. If the net force acts at an angle to the direction of motion at any moment, then the object moves in a curved path. Projectile motion is an example we have already discussed (Section 3–5). Another important case is that of an object moving in a circle, such as a ball at the end of a string revolving around one's head or the nearly circular motion of the moon about the earth.

In this chapter, we study circular motion of an object, and how Newton's laws of motion apply to it. We will also discuss how Newton conceived of another great law by applying the concepts of circular motion to the motion of the moon and the planets. This new law is the law of universal gravitation, which was the capstone of Newton's analysis of the physical world. Indeed, Newtonian mechanics, with its three laws of motion and the law of universal gravitation, was accepted for centuries (until the early 1900s) as the way the universe works.

5–1 • Kinematics of Uniform Circular Motion

An object that moves in a circle at constant speed v is said to experience **uniform circular motion**. The *magnitude* of the velocity remains constant in this case, but the *direction* of the velocity is continuously changing as the

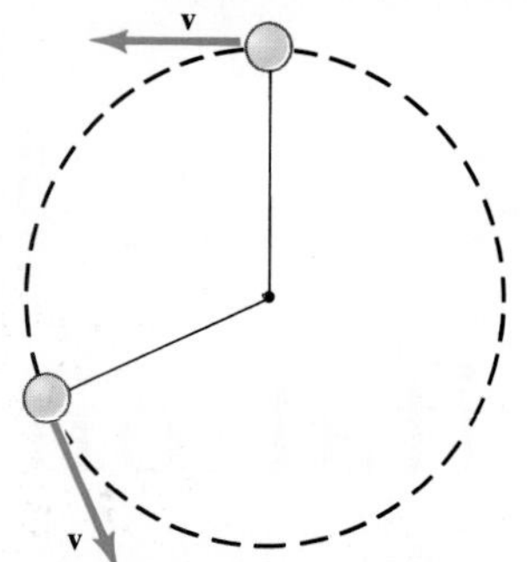

FIGURE 5–1 A particle moving in a circle, showing how the velocity changes. Note that at each point, the instantaneous velocity is in a direction tangent to the circular path.

object moves around the circle, Fig. 5–1. Since acceleration is defined as the rate of change of velocity, a change in direction of velocity constitutes an acceleration just as does a change in magnitude of velocity. Thus, an object experiencing uniform circular motion is continuously accelerating. We now investigate this acceleration quantitatively.

Acceleration is defined as

$$\mathbf{a} = \frac{\mathbf{v} - \mathbf{v}_0}{\Delta t} = \frac{\Delta \mathbf{v}}{\Delta t},$$

where $\Delta \mathbf{v}$ is the change in velocity during the short time interval Δt. We will eventually consider the situation when Δt approaches zero and thus obtain the instantaneous acceleration; but for purposes of making a clear drawing, Fig. 5–2, we consider a nonzero time interval. During the time Δt, the particle in Fig. 5–2a moves from point A to point B, covering a small distance Δl along the arc which subtends a small angle $\Delta\theta$. The change in the velocity vector is $\mathbf{v} - \mathbf{v}_0 = \Delta \mathbf{v}$. If we transfer $\mathbf{v}_0$ to the right side of this equation, we obtain $\mathbf{v} = \mathbf{v}_0 + \Delta \mathbf{v}$. Thus $\Delta \mathbf{v}$ added to $\mathbf{v}_0$ is equal to $\mathbf{v}$, so $\Delta \mathbf{v}$ is the vector shown dashed in Fig. 5–2b. In this diagram, we notice that when Δt is very small (approaching zero), and therefore Δl and $\Delta\theta$ are also very small, $\mathbf{v}$ will be almost parallel to $\mathbf{v}_0$, and $\Delta \mathbf{v}$ will be essentially perpendicular to them. Thus $\Delta \mathbf{v}$ points toward the center of the circle. Since $\mathbf{a}$, by its definition above, is in the same direction as $\Delta \mathbf{v}$, it too must point toward the center of the circle. Therefore, this acceleration is called **centripetal acceleration** ("center-seeking" acceleration) or **radial acceleration** (since it is directed along the radius, toward the center of the circle); henceforth, we denote it by $\mathbf{a}_c$.

Derivation of centripetal acceleration

Now that we have determined the direction of the acceleration, we next determine the magnitude of the centripetal acceleration, a_c. The vectors $\mathbf{v}$, $\mathbf{v}_0$, and $\Delta \mathbf{v}$ in Fig. 5–2b form a triangle that is geometrically similar† to triangle ABC in Fig. 5–2a (the angle between $\mathbf{v}_0$ and $\mathbf{v}$ is equal to $\Delta\theta$, defined as the angle between CA and CB in Fig. 5–2a, because CB is perpendicular to $\mathbf{v}$, and CA is perpendicular to $\mathbf{v}_0$). Thus we can write

$$\frac{\Delta v}{v} \approx \frac{\Delta l}{r}.$$

This is an exact equality when Δt approaches zero, for then the arc length

† Appendix A contains a review of geometry.

FIGURE 5–2 Determining the change in velocity, $\Delta \mathbf{v}$, for a particle moving in a circle.

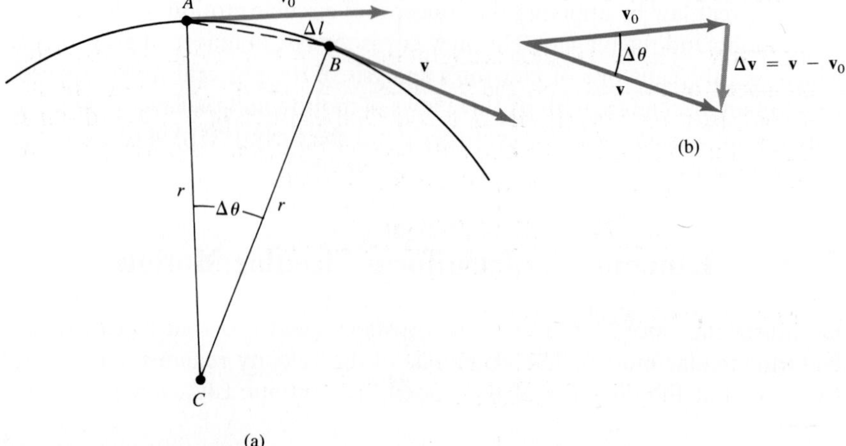

Δl equals the chord length AB. Since we want to find the instantaneous acceleration, which is the case for which Δt approaches zero, we write the above expression as an equality and solve for Δv:

$$\Delta v = \frac{v}{r}\,\Delta l.$$

To get the centripetal acceleration, a_c, we divide Δv by Δt:

$$a_c = \frac{\Delta v}{\Delta t} = \frac{v}{r}\frac{\Delta l}{\Delta t},$$

and since $\Delta l/\Delta t$ is the linear speed, v, of the object:

$$a_c = \frac{v^2}{r}. \tag{5-1}$$

Centripetal acceleration

To summarize, *an object moving in a circle of radius r with constant speed v has an acceleration whose direction is toward the center of the circle and whose magnitude is* $a_c = v^2/r$. It is not surprising that this acceleration depends on v and r. For the greater the speed v, the faster the velocity changes direction; and the larger the radius, the less rapidly the velocity changes direction.

The acceleration vector points toward the center of the circle, as discussed above. But the velocity vector always points in the direction of motion, which is tangential to the circle. Thus the velocity and acceleration vectors are perpendicular to each other at every point for uniform circular motion (see Fig. 5–3). This is another example that illustrates the error in thinking that acceleration and velocity are always in the same direction. For an object falling vertically, **a** and **v** are indeed parallel. But in circular motion, **a** and **v** are not parallel—nor are they in projectile motion (Section 3–5), where the acceleration **a** = **g** is always downward but the velocity vector can have various directions (Figs. 3–17 and 3–19). There is, in fact, *no* fixed general relationship between the directions of **a** and **v**.

Caution:
v and a are not necessarily in the same direction

FIGURE 5–3 For uniform circular motion, **a** is always perpendicular to **v**.

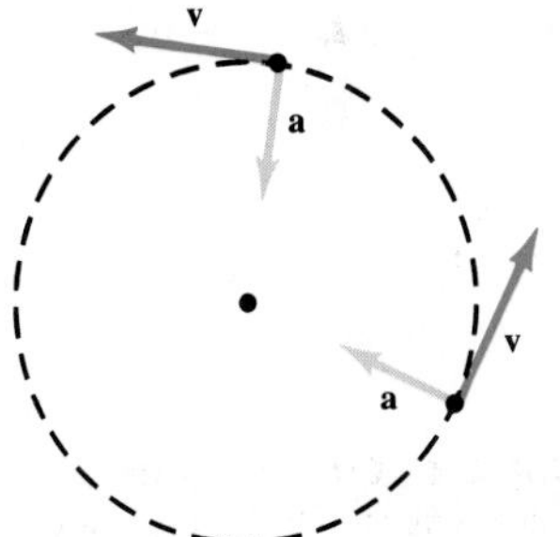

EXAMPLE 5–1 A 150-g ball at the end of a string is swinging in a horizontal circle of radius 0.60 m, as in Fig. 5–1. The ball makes exactly 2.00 revolutions in a second. What is its centripetal acceleration?

SOLUTION The centripetal acceleration is $a_c = v^2/r$. First, we determine the speed of the ball v. If the ball makes two complete revolutions per second, then the ball travels in a complete circle in 0.50 s. The distance traveled in this time is the circumference of the circle, $2\pi r$, where r is the radius of the circle. Therefore, the ball has speed

$$v = \frac{2\pi r}{t} = \frac{2(3.14)(0.60\text{ m})}{(0.50\text{ s})} = 7.5\text{ m/s}.$$

The centripetal acceleration is

$$a_c = \frac{v^2}{r} = \frac{(7.5\text{ m/s})^2}{(0.60\text{ m})} = 95\text{ m/s}^2.$$

EXAMPLE 5–2 The moon's nearly circular orbit about the earth has a radius of about 385,000 km and a period T (the time needed to make one revolution) of 27.3 days. Determine the acceleration of the moon toward the earth.

SOLUTION In orbit around the earth, the moon travels a distance $2\pi r$, where $r = 3.85 \times 10^8$ m is the radius of its circular path. The speed of the moon in its orbit about the earth is then

$$v = \frac{2\pi r}{T} = \frac{(6.28)(3.85 \times 10^8\ \text{m})}{(27.3\ \text{d})(24.0\ \text{h/d})(3600\ \text{s/h})} = 1.02 \times 10^3\ \text{m/s}.$$

Therefore,

$$a_c = \frac{v^2}{r} = \frac{(1.02 \times 10^3\ \text{m/s})^2}{(3.85 \times 10^8\ \text{m})} = 0.00273\ \text{m/s}^2 = 2.73 \times 10^{-3}\ \text{m/s}^2,$$

or about $2.78 \times 10^{-4}\ g$, where $g = 9.80\ \text{m/s}^2$ is the acceleration of gravity at the earth's surface.†

5–2 • Dynamics of Uniform Circular Motion

Force is needed to provide centripetal acceleration

According to Newton's second law ($\mathbf{F} = m\mathbf{a}$), an object that is accelerating must have a net force acting on it. An object moving in a circle, such as a ball on the end of a string, must therefore have a force applied to it to keep it in that circle. That is, a force is necessary to give it a centripetal acceleration. The magnitude of the required force can be calculated using Newton's second law, $F = ma$, where we use the value of the centripetal acceleration, $a_c = v^2/r$, and F must be the total (or net) force:

$$F = ma_c = m\frac{v^2}{r}.$$

Since a_c is directed toward the center of the circle at any moment, the force too must be directed toward the center of the circle. A force is clearly necessary because otherwise, if no force were exerted on the object, it would not move in a circle but in a straight line, as Newton's first law tells us. To pull an object out of its "natural" straight-line path, a force to the side is necessary. For uniform circular motion, this sideways force must act toward the circle's center (see Fig. 5–4). The direction of the force is thus continually changing so that it is always directed toward the center of the circle. This force is sometimes called a centripetal ("aiming toward the center") force. But be aware that "centripetal force" does not indicate some new kind of force. The term merely describes the direction of the force: that the force is directed toward

FIGURE 5–4 A force is required to keep an object moving in a circle. If the speed is constant, the force is directed toward the center of the circle. Thus the direction of the force is continually changing so that it is always pointing toward the center of the circle.

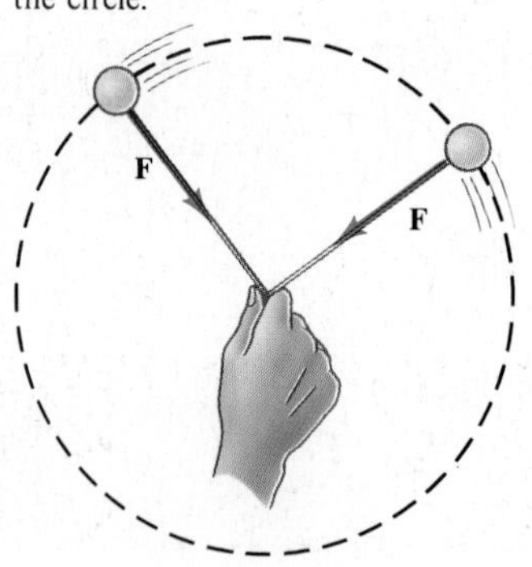

† Note: this acceleration $a = 2.78 \times 10^{-4}\ g$ is *not* the acceleration of gravity for objects at the moon's surface due to the moon's gravity. Rather it is the acceleration due to the *earth's* gravity for any object (such as the moon) that is 385,000 km from the earth.

the circle's center. The force must be applied by other objects. For example, when a person swings a ball in a circle on the end of a string, the person (via the string) exerts the force on the ball. The force on the moon, which gives it its centripetal acceleration so that it revolves around the earth, is the force of gravity (as we shall discuss shortly).

Caution: "Centrifugal force"—a misconception

There is a common misconception that an object moving in a circle has an outward force acting on it, a so-called centrifugal ("center-fleeing") force. Consider, for example, a person swinging a ball on the end of a string around his head (Fig. 5–5). If you have ever done this yourself, you know that you feel a force pulling outward on your hand. The misconception arises when this pull is interpreted as an outward "centrifugal" force pulling on the ball that is transmitted along the string to the hand. But this is not what is happening at all. To keep the ball moving in a circle, the person pulls inwardly on the ball. The ball, then, exerts an equal and opposite force on the hand (Newton's third law), and *this* is the force your hand feels (see Fig. 5–5). The force *on the ball* is the one exerted *inwardly* on it by the person. For even more convincing evidence that a "centrifugal force" does not act on the ball, consider what happens when you let go of the string. If a centrifugal force were acting, the ball would fly outward, as shown in Fig. 5–6a. But it doesn't; the ball flies off tangentially (Fig. 5–6b), in the direction of the velocity it had at the moment it was released, because the inward force no longer acts.

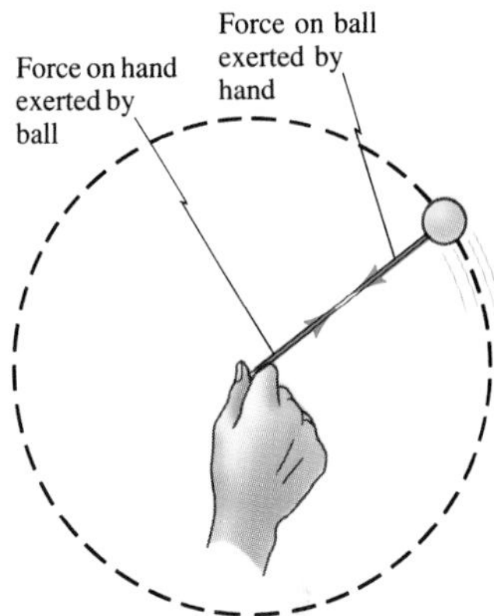

FIGURE 5–5 Swinging a ball on the end of a string. Your hand pulls in on the ball, and the ball pulls outward on your hand (Newton's third law).

An example of centripetal acceleration occurs when a fast-moving automobile rounds a curve. In such a situation, you may feel that you are thrust outward. But there is not some mysterious centrifugal force pulling on you. What is happening is that you tend to move in a straight line, whereas the car has begun to follow a curved path. To make *you* go in the curved path, the back of the seat or the door of the car exerts a force on you (Fig. 5–7).

FIGURE 5–6 If centrifugal force existed, the ball would fly off as in (a) when released. In fact, it flies off as in (b). Similarly, sparks fly in straight lines from the edge of a rotating grinding wheel (c).

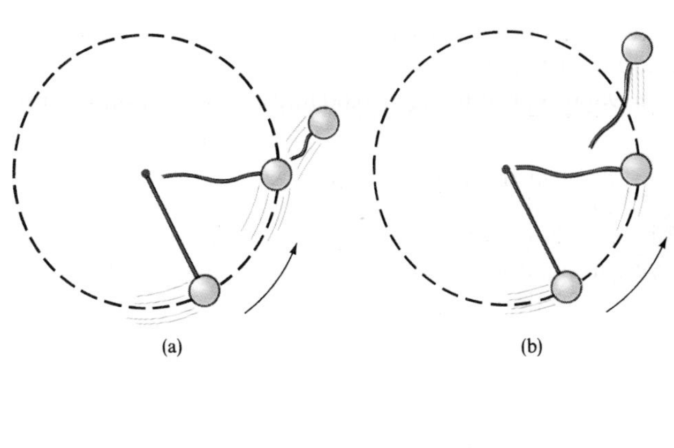

(*b*) For $r = 50$ m and $v = 50$ km/h (or 14 m/s),

$$\tan\theta = \frac{(14\ \text{m/s})^2}{(50\ \text{m})(9.8\ \text{m/s}^2)} = 0.40,$$

so $\theta = 22°$.

*5–3 • Nonuniform Circular Motion

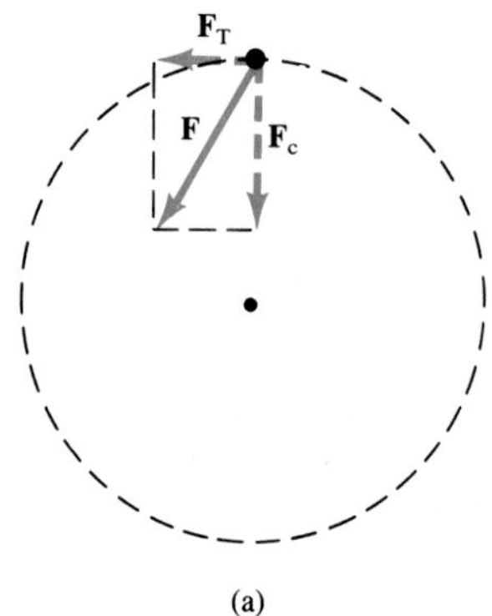

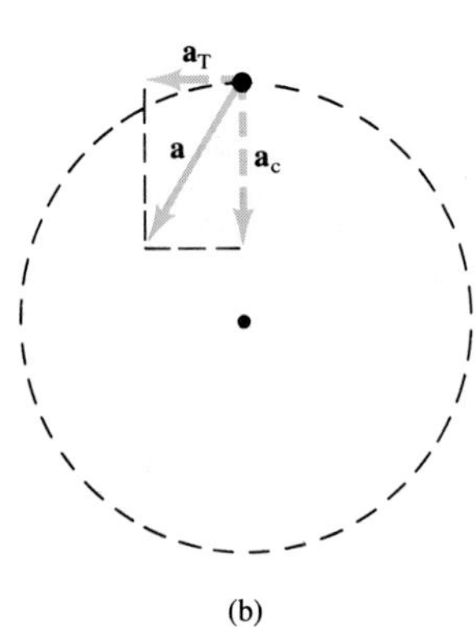

FIGURE 5–11 The speed of a particle moving in a circle changes if the force on it has a tangential component, F_T. Part (a) shows the force **F** and its vector components; part (b) shows the acceleration vector and its vector components.

Circular motion at constant speed occurs when the force on an object is exerted toward the center of the circle. If the force is not directed toward the center but is at an angle, as shown in Fig. 5–11a, the force has two components. The component directed toward the center of the circle, F_c, gives rise to the centripetal acceleration a_c and keeps the object moving in a circle. The component F_T, tangent to the circle, acts to increase (or decrease) the speed, and thus gives rise to a component of the acceleration tangent to the circle, a_T. When an F_T acts, the speed of the object is changed.

When you first start revolving a ball on the end of a string around your head, you must give it tangential acceleration. You do this by pulling on the string with your hand displaced from the center of the circle. In athletics, a hammer thrower accelerates the hammer tangentially in a similar way so that it reaches a high speed before release.

The tangential component of the acceleration, a_T, is equal to the rate of change of the *magnitude* of the velocity of the object:

$$a_T = \frac{\Delta v}{\Delta t}.$$

The centripetal acceleration arises from the change in *direction* of the velocity and, as we have seen, is given by

$$a_c = \frac{v^2}{r}.$$

The tangential acceleration always points in a direction tangent to the circle, and is in the direction of motion (parallel to **v**) if the speed is increasing, as shown in Fig. 5–11b. If the speed is decreasing, $\mathbf{a}_T$ points antiparallel to **v**. In either case, $\mathbf{a}_T$ and $\mathbf{a}_c$ are always perpendicular to each other, and *their directions change* continually as the particle moves along its circular path. The total vector acceleration, **a**, is the sum of these two:

$$\mathbf{a} = \mathbf{a}_T + \mathbf{a}_c.$$

Since $\mathbf{a}_c$ and $\mathbf{a}_T$ are always perpendicular to each other, the magnitude of **a** at any moment is

$$a = \sqrt{a_T^2 + a_c^2}.$$

EXAMPLE 5–5 A racing car starts from rest in the pit area and accelerates at a uniform rate to a speed of 35 m/s in 11 s, moving on a circular track of radius 500 m. Assuming constant tangential acceleration, find (*a*) the tangential acceleration, and (*b*) the centripetal acceleration when the speed is 30 m/s.

SOLUTION (*a*) a_T is constant, of magnitude

$$a_T = \frac{\Delta v}{\Delta t} = \frac{(35 \text{ m/s} - 0 \text{ m/s})}{11 \text{ s}} = 3.2 \text{ m/s}^2.$$

(*b*) $a_c = \dfrac{v^2}{r} = \dfrac{(30 \text{ m/s})^2}{500 \text{ m}} = 1.8 \text{ m/s}^2.$

*5–4 • Centrifugation

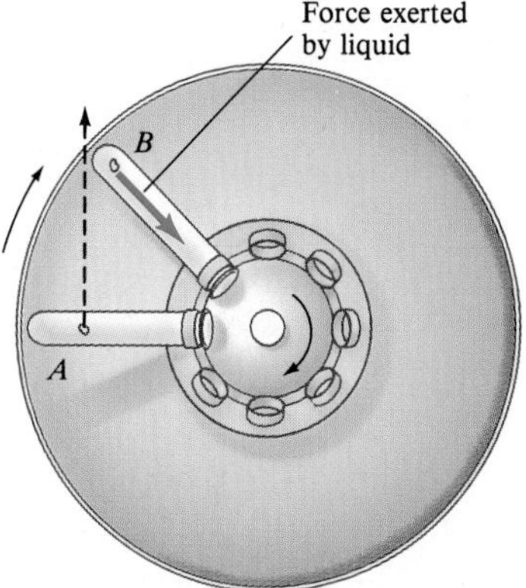

FIGURE 5–12 Rotating test tube in a centrifuge (top view).

A useful device that very nicely illuminates the dynamical aspects of circular motion is the centrifuge, or the very high speed ultracentrifuge. These devices are used to sediment materials quickly or to separate materials with slightly different characteristics. Test tubes or other containers are held in the centrifuge rotor, which is accelerated to very high rotational speeds: see Fig. 5–12, where one test tube is shown in two different positions as the rotor turns. The small circle represents a small particle, perhaps a macromolecule, in a fluid-filled test tube. When the tube is at position *A* and the rotor is turning, the particle has a tendency to move in a straight line in the direction of the dashed arrow. The fluid resists† the motion of the particles, thus exerting a centripetal force that keeps the particles moving in a circle. Usually, the resistance of the fluid (which could be a liquid or a gas, depending on the application) does not quite equal mv^2/r, and the particles eventually reach the bottom of the tube. Then the bottom of the tube exerts a force that keeps the particles moving in a circle. In fact, the bottom of the tube must exert a force on the whole tube of fluid, keeping it moving in a circle. If the tube is not strong enough to exert this force, it will break.

The kinds of materials placed in a centrifuge are those that do not sediment or separate quickly under the action of gravity. The point of a centrifuge is to provide an "effective gravity" much larger than normal gravity because of the high rotational speeds, so that the particles move down the tube more rapidly.

EXAMPLE 5–6 An ultracentrifuge rotor rotates at 50,000 rpm (revolutions per minute). The top of a test tube (perpendicular to the rotation axis) is 6.00 cm, and the bottom of the tube is 10.00 cm, from the axis of rotation. (*a*) Calculate the centripetal acceleration in "*g*'s." (*b*) If the contents of the tube have a total mass of 12.0 g, what force must the bottom of the tube withstand?

SOLUTION We can calculate the centripetal acceleration from $a_c = v^2/r$. (*a*) At the top of the tube, a particle revolves in a circle of circumference $2\pi r$ which is a distance $d = 2(3.14)(0.0600 \text{ m}) = 0.377 \text{ m}$ per revolution. It makes 5.00×10^4 such revolutions each minute, so the time to make one revolution is $t = (60 \text{ s/min})/(50{,}000 \text{ rev/min}) = 1.20 \times 10^{-3} \text{ s/rev}$. The speed of the particle is then

$$v = \frac{d}{t} = \left(\frac{0.377 \text{ m/rev}}{1.20 \times 10^{-3} \text{ s/rev}}\right) = 3.14 \times 10^2 \text{ m/s}.$$

† A type of friction, like air resistance. See Section 10–12.

The centripetal acceleration is

$$a_c = \frac{v^2}{r} = \frac{(3.14 \times 10^2 \text{ m/s})^2}{0.0600 \text{ m}} = 1.64 \times 10^6 \text{ m/s}^2,$$

which, dividing by $g = 9.80 \text{ m/s}^2$, is 1.67×10^5 g's. At the bottom of the tube ($r = 0.1000$ m), the speed is $v = 2\pi(0.1000 \text{ m})(5.00 \times 10^4 \text{ min}^{-1})/(60.0 \text{ s}) = 5.24 \times 10^2$ m/s and $a_c = v^2/r = 2.74 \times 10^6 \text{ m/s}^2$, or 2.79×10^5 g's.

(*b*) Since the acceleration varies with distance from the axis, we estimate the force using the average acceleration $= (1.64 \times 10^6 \text{ m/s}^2 + 2.74 \times 10^6 \text{ m/s}^2)/2 = 2.19 \times 10^6 \text{ m/s}^2$. Then $F = ma = (0.0120 \text{ kg})(2.19 \times 10^6 \text{ m/s}^2) = 2.63 \times 10^4$ N, which is equivalent to the weight of a 2680-kg mass ($m = F/g = 2.63 \times 10^4 \text{ N}/9.80 \text{ m/s}^2 = 2.68 \times 10^3$ kg), or almost 3 tons.

5–5 • Newton's Law of Universal Gravitation

Besides developing the three laws of motion, Sir Isaac Newton also examined the motion of the heavenly bodies—the planets and the moon. In particular, he wondered about the nature of the force that must act to keep the moon in its nearly circular orbit around the earth.

Newton was also thinking about the apparently unrelated problem of gravity. Since falling bodies accelerate, Newton had concluded that they must have a force exerted on them, a force we call the force of gravity. But what, we may ask, *exerts* this force of gravity—for, as we have seen, whenever a body has a force exerted *on* it, that force is exerted *by* some other body. Every object on the surface of the earth feels this force of gravity, and no matter where the object is, the force is directed toward the center of the earth. Newton concluded that it must be the earth itself that exerts the gravitational force on objects at its surface.

Newton's apple

According to an early account, Newton was sitting in his garden and noticed an apple drop from a tree. He is said to have been struck with a sudden inspiration: if gravity acts at the tops of trees, and even at the tops of mountains, then perhaps it acts all the way to the moon! Whether this story is true or not,† it does capture something of Newton's reasoning and inspiration. With this idea that it is terrestrial gravity that holds the moon in its orbit, Newton developed his great theory of gravitation, with considerable help and encouragement from Robert Hooke (1635–1703).

Newton set about determining the magnitude of the gravitational effect that the earth exerts on the moon as compared to the gravitational effect on objects at the earth's surface. At the surface of the earth, the force of gravity accelerates objects at 9.80 m/s^2. But what is the centripetal acceleration of the moon? Since the moon moves with nearly uniform circular

† Many historians of science today believe that Newton may have invented this story to push back the date of discovery to the 1660s, when in fact his correspondence and notebooks indicate he really didn't come to the law of universal gravitation until about 1685. The idea that there might be a connection between terrestrial gravity and the force on the moon, and even the $1/r^2$ dependence, had already been suggested by others, including Hooke. Nonetheless, we still give most of the credit to Newton for the gravitational law since he proposed it in its fullness and gave very strong arguments to support it.

motion, the acceleration can be calculated from $a_c = v^2/r$; we already performed this calculation in Example 5–2 and found that $a_c = 0.00273\ \text{m/s}^2$. In terms of the acceleration of gravity at the earth's surface, g, this is equivalent to

The moon's acceleration

$$a_c \approx \frac{1}{3600}\, g.$$

That is, the acceleration of the moon toward the earth is about $\frac{1}{3600}$ as great as the acceleration of objects at the earth's surface. Now the moon is 385,000 km from the earth, which is about 60 times the earth's radius of 6380 km. That is, the moon is 60 times farther from the earth's center than are objects at the earth's surface. But $60 \times 60 = 3600$—again that number 3600! Newton concluded that the gravitational force exerted by the earth on any object decreases with the square of its distance, r, from the earth's center:

$$\text{force of gravity} \propto \frac{1}{r^2}.$$

The moon, being 60 earth radii away, feels a gravitational force only $\frac{1}{60^2} = \frac{1}{3600}$ times as strong as it would if it were at the earth's surface. Any object placed 385,000 km from the earth would experience the same acceleration due to the earth's gravity as the moon experiences: $0.00273\ \text{m/s}^2$.

Newton realized that the force of gravity on an object depends not only on distance but also on the object's mass. In fact, it is directly proportional to its mass, as we have seen. According to Newton's third law, when the earth exerts its gravitational force on an object, such as the moon, that object exerts an equal and opposite force on the earth (Fig. 5–13). Because of this symmetry, Newton reasoned, the magnitude of the force of gravity must be proportional to *both* the masses. Thus

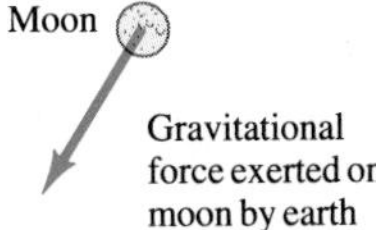

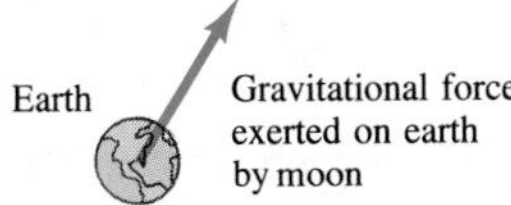

FIGURE 5–13 The gravitational force one body exerts on a second body is directed toward the first body and is equal and opposite to the force exerted by the second body on the first.

$$F \propto \frac{m_e m_o}{r^2}$$

where m_e is the mass of the earth, m_o the mass of the other object, and r the distance from the earth's center to the center of the object.

Newton went a step further in his analysis of gravity. In his examination of the orbits of the planets, he concluded that the force required to hold the different planets in their orbits around the sun seems also to diminish as the inverse square of their distance from the sun. This led him to believe that it is also the gravitational force that acts between the sun and each of the planets to keep them in their orbits. And if gravity acts between these objects, why not between all objects? Thus he proposed his famous **law of universal gravitation**, which we can state as follows:

Every particle in the universe attracts every other particle with a force that is proportional to the product of their masses and inversely proportional to the square of the distance between them. This force acts along the line joining the two particles.

Newton's law of universal gravitation

The magnitude of the gravitational force can be written as

$$F = G\frac{m_1 m_2}{r^2}, \tag{5–2}$$

where m_1 and m_2 are the masses of the two particles, r is the distance between them, and G is a universal constant which must be measured experimentally and has the same numerical value for all objects.

The value of G must be very small, since we are not aware of any force of attraction between ordinary sized objects, such as between two baseballs. The force between two ordinary objects was first measured over 100 years after Newton's publication of his law, by Henry Cavendish in 1798. To detect and measure the incredibly small force, he used an apparatus like that shown in Fig. 5–14. Cavendish not only confirmed Newton's hypothesis that two bodies attract one another and that Eq. 5–2 accurately describes this force, but because he could measure F, m_1, m_2, and r accurately, he was then able to determine the value of the constant G as well. The accepted value today is

Gravitational constant

$$G = 6.67 \times 10^{-11}\ \text{N}\cdot\text{m}^2/\text{kg}^2.$$

[Strictly speaking, Eq. 5–2 gives the magnitude of the gravitational force that one particle exerts on a second particle that is a distance r away. For an extended object (that is, not a point), we must consider how to measure the distance r. You might think that r would be the distance between the centers of the objects. This is not necessarily true, and to do a calculation correctly, each extended body must be considered as a collection of tiny particles; the total force is the sum of the forces due to all the particles. The sum over all these particles is often best obtained using integral calculus, which Newton himself invented. Newton showed, however, that for two uniform spheres, Eq. 5–2 gives the correct force where r is the distance between their centers. Also, when extended bodies are small compared to the distance between them (as for the earth–sun system), little inaccuracy results from considering them as point particles.]

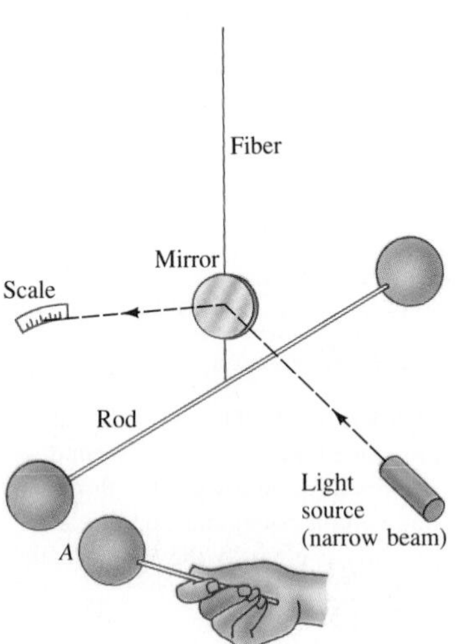

FIGURE 5–14 Schematic diagram of Cavendish's apparatus. Two spheres are attached to a light horizontal rod, which is suspended at its center by a thin fiber. When the sphere labeled A is brought close to one of the suspended spheres, the gravitational force causes the latter to move, and this twists the fiber slightly. The tiny movement is magnified by the use of a narrow light beam directed at a mirror, which is mounted on the fiber. The beam reflects onto a scale. Previous determination of how large a force will twist the fiber a given amount then allows one to determine the magnitude of the gravitational force between two objects.

EXAMPLE 5–7 What is the force of gravity acting on a 2000-kg spacecraft when it orbits two earth radii from the earth's center (that is, 6380 km above the earth's surface)?

SOLUTION The spacecraft is twice as far from the earth's center as when at the surface of the earth. Therefore, since the force of gravity decreases as the square of the distance (and $\frac{1}{2^2} = \frac{1}{4}$), the force of gravity on the spacecraft will be only one-fourth its weight at the surface of the earth. On earth its weight is $mg = (2000\ \text{kg})(9.8\ \text{m/s}^2) = 19{,}600$ N. Thus, when it is in orbit at a height of one earth radius, the force of gravity is $\frac{1}{4}$ as much, or 4900 N. (Note that we could have determined this by plugging all the numbers into Eq. 5–2 directly, but that would have taken more effort.)

EXAMPLE 5–8 Determine the net force on the moon ($m_{\text{m}} = 7.36 \times 10^{22}$ kg) due to the gravitational attraction of both the earth ($m_{\text{e}} = 5.98 \times 10^{24}$ kg) and the sun ($m_{\text{s}} = 1.99 \times 10^{30}$ kg), assuming they are at right angles to each other (Fig. 5–15).

SOLUTION We use Newton's law of universal gravitation, Eq. 5–2. The earth is 3.85×10^5 km $= 3.85 \times 10^8$ m from the moon, so F_{me} (the force on the moon due to the earth) is

$$F_{\text{me}} = \frac{(6.67 \times 10^{-11}\ \text{N}\cdot\text{m}^2/\text{kg}^2)(7.36 \times 10^{22}\ \text{kg})(5.98 \times 10^{24}\ \text{kg})}{(3.85 \times 10^8\ \text{m})^2}$$

$$= 1.98 \times 10^{20}\ \text{N}.$$

The sun is 1.50×10^8 km from the earth, so F_{ms} (the force on the moon due to the sun) is

$$F_{\text{ms}} = \frac{(6.67 \times 10^{-11}\ \text{N}\cdot\text{m}^2/\text{kg}^2)(7.36 \times 10^{22}\ \text{kg})(1.99 \times 10^{30}\ \text{kg})}{(1.50 \times 10^{11}\ \text{m})^2}$$

$$= 4.35 \times 10^{20}\ \text{N}.$$

Since the two forces act at right angles in the case we are considering (Fig. 5–15), the total force is

$$F = \sqrt{(1.98)^2 + (4.35)^2} \times 10^{20}\ \text{N} = 4.79 \times 10^{20}\ \text{N}$$

which acts at an angle $\theta = \tan^{-1}(1.98/4.35) = 24.5°$.

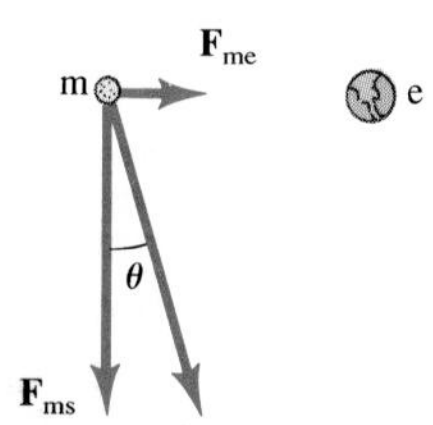

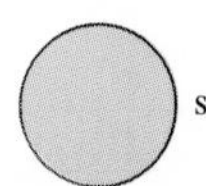

FIGURE 5–15 Orientation of sun (s), earth (e), and moon (m) for Example 5–8 (not to scale).

The law of universal gravitation should not be confused with Newton's second law of motion, $\mathbf{F} = m\mathbf{a}$. The former describes a particular force, gravity, and how its strength varies with the distance and masses involved. Newton's second law, on the other hand, relates the net force on a body (i.e., the vector sum of all the different forces acting on the body) to the mass and acceleration of that body.

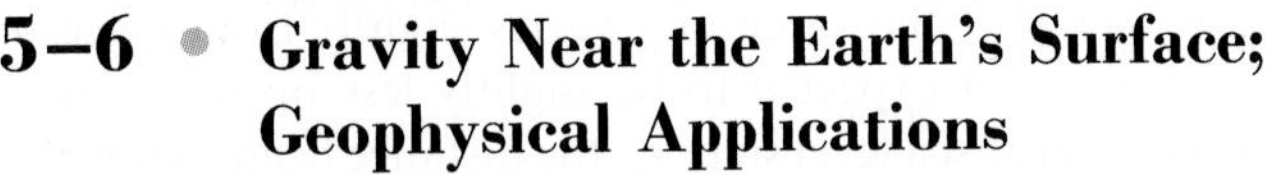

5–6 • Gravity Near the Earth's Surface; Geophysical Applications

When Eq. 5–2 is applied to the gravitational force between the earth and an object at its surface, m_1 becomes the mass of the earth m_{e}, and r becomes the distance from the earth's center,† namely, the radius of the earth r_{e}. This force of gravity due to the earth is the weight of the object, which we have been writing as mg. Thus,

$$mg = G\frac{mm_{\text{e}}}{r_{\text{e}}^2}.$$

Hence

$$g = G\frac{m_{\text{e}}}{r_{\text{e}}^2}. \tag{5–3}$$

g in terms of G

Thus, the acceleration of gravity at the surface of the earth, g, is determined by m_{e} and r_{e}. (Be careful not to confuse G with g; they are very different quantities, but are related by Eq. 5–3.)

Until G was measured, the mass of the earth was not known. But once G was measured, Eq. 5–3 could be used to calculate the earth's mass, and

† That the distance is measured from the earth's center does not imply that the force of gravity somehow emanates from that one point. Rather, all parts of the earth attract gravitationally, but the net effect is a force acting toward the earth's center.

8. Describe how careful measurements of the variation in g in the vicinity of an ore deposit might be used to estimate the amount of ore present.
9. When will your apparent weight be the greatest, as measured by a scale in a moving elevator: when the elevator (*a*) accelerates downward, (*b*) accelerates upward, (*c*) is in free fall, (*d*) moves upward at constant speed? In which case would your weight be the least? When would it be the same as when you are on the ground?
10. The sun is directly below us at midnight, in line with the earth's center. Are we then heavier at midnight, due to the sun's gravitational force on us, than we are at noon? Explain.
11. If you were in a satellite orbiting the earth, how might you cope with walking, drinking, or putting a pair of scissors on a table?
12. An antenna loosens and becomes detached from a satellite in a circular orbit around the earth. Describe the antenna's motion subsequently. If it will land on earth, describe where; if not, describe how it could be made to land on earth.
13. Astronauts who spend long periods in outer space could be adversely affected by weightlessness. One way to simulate gravity is to shape the spaceship like a bicycle wheel that rotates, with the astronauts walking on the inside of the "tire." Explain how this simulates gravity. Consider (*a*) how objects fall, (*b*) the force we feel on our feet, and (*c*) any other aspects of gravity you can think of.
14. People sometimes ask, "What keeps a satellite up in its orbit around the earth?" How would you respond?
15. Explain why a person running experiences "free fall" or "apparent weightlessness" between steps.
*16. The sun's gravitational pull on the earth is much larger than the moon's. Yet the moon is mainly responsible for the tides. Explain.
*17. The earth moves faster in its orbit around the sun in winter than in summer. Is it closer to the sun in summer or in winter? Does this affect the seasons? Explain.

PROBLEMS

SECTIONS 5–1 AND 5–2

1. (I) A jet plane traveling 1800 km/h (500 m/s) pulls out of a dive by moving in an arc of radius 4.00 km. What is the plane's acceleration in g's?

FIGURE 5–27 Problem 8.

2. (I) A child on a merry-go-round is moving with a speed of 1.25 m/s when 11.0 m from the center of the merry-go-round. Calculate (*a*) the centripetal acceleration of the child, and (*b*) the net horizontal force exerted on the child (mass = 25.0 kg).
3. (I) Calculate the centripetal acceleration of the earth in its orbit around the sun and the net force exerted on the earth. What exerts this force on the earth? Assume that the earth's orbit is a circle of radius 1.50×10^{11} m.
4. (I) A horizontal force of 26.0 N is applied to a 0.80-kg stone to keep it rotating uniformly in a horizontal circle of radius 0.50 m. Calculate its speed.
5. (II) What is the maximum speed with which a 1000-kg car can round a turn of radius 85 m on a flat road if the coefficient of friction between tires and road is 0.60? Is this result independent of the mass of the car?
6. (II) How large must the coefficient of friction be between the tires and the road if a car is to round a level curve of radius 68 m at a speed of 55 km/h?
7. (II) A coin is placed 13.0 cm from the axis of a rotating turntable of variable speed. When the speed of the turntable is slowly increased, the coin remains fixed on the turntable until a rate of 42 rpm is reached, at which point the coin slides off. What is the coefficient of static friction between the coin and the turntable?
8. (II) At what minimum speed must a roller coaster be traveling when upside down at the top of a circle (Fig. 5–27) if the passengers are not to fall out? Assume a radius of curvature of 8.0 m.
9. (II) Use dimensional analysis (see Appendix B) to obtain the form for the centripetal acceleration, $a_c = v^2/r$.

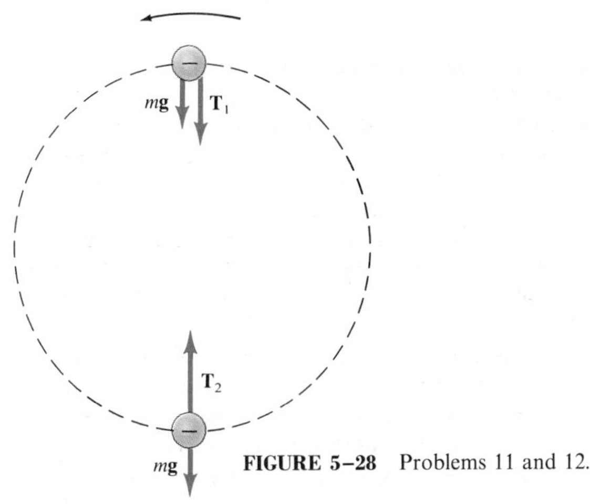

FIGURE 5–28 Problems 11 and 12.

FIGURE 5–30 Problem 14.

10. (II) A 0.35-kg ball, attached to the end of a horizontal cord, is rotated in a circle of radius 1.3 m on a frictionless horizontal surface. If the cord will break when the tension in it exceeds 30 N, what is the maximum speed the ball can have? How would your answer be affected if there were friction?

11. (II) A ball on the end of a string is cleverly revolved at a uniform rate in a vertical circle of radius 85.0 cm, as shown in Fig 5–28. If its speed is 3.25 m/s and its mass is 0.335 kg, calculate the tension in the string when the ball is (*a*) at the top of its path, and (*b*) at the bottom of its path.

12. (II) A ball of mass m is revolved in a vertical circle at the end of a cord of length L. What is the minimum speed v needed at the top of the circle if the cord is to remain taut? [*Hint:* See Fig. 5-28.]

13. (II) Two masses, m_1 and m_2, connected to each other and to a central post by cords, as shown in Fig. 5–29, rotate about the post at a frequency f (revolutions per second) on a frictionless horizontal surface at distances r_1 and r_2, respectively, from the post. Determine the tension in each segment of the cord.

14. (II) In a "Rotor-ride" at a carnival, riders are pressed against the inside wall of a vertical cylinder 2.0 m in radius rotating at a speed of 1.1 revolutions per second when the floor drops out. (See Fig. 5–30.) What minimum coefficient of friction is needed so a person won't slip down? Is this safe?

15. (III) If a curve with a radius of 60 m is properly banked for a car traveling 60 km/h, what must be the coefficient of static friction for a car not to skid when traveling at 90 km/h?

16. (III) A 1000-kg car rounds a curve of radius 65 m banked at an angle of 14°. If the car is traveling at 90 km/h, will a friction force be required? If so, how much and in what direction?

*SECTION 5–3

*17. (I) Determine the tangential and centripetal components of the net force exerted on the car (by the ground) in Example 5-5 when its speed is 30 m/s. The car's mass is 1000 kg.

*18. (II) A car accelerates uniformly from rest to a speed of 40 m/s in one complete revolution of a 100-m-radius circular track. Determine the components of the acceleration (radial and tangential) when the car is halfway around the circle.

*19. (III) A particle revolves in a circle of radius 3.60 m. At a particular instant, its acceleration is 1.25 m/s^2, in a direction that makes an angle of 28.0° to its direction of motion. Determine its speed (*a*) at this moment, and (*b*) 2.00 s later, assuming constant tangential acceleration.

*20. (III) A particle starting from rest revolves with uniformly increasing speed in a clockwise circle in the x-y plane. The center of the circle is at the origin of an x-y coordinate system. At $t = 0$, the particle is at $x = 0.0$, $y = 2.0$ m. At $t = 2.0$ s, it has made one-quarter of a revolution and is at $x = 2.0$ m, $y = 0.0$. Determine (*a*) its speed at $t = 2.0$ s, (*b*) the average velocity vector and (*c*) the average acceleration vector during this interval.

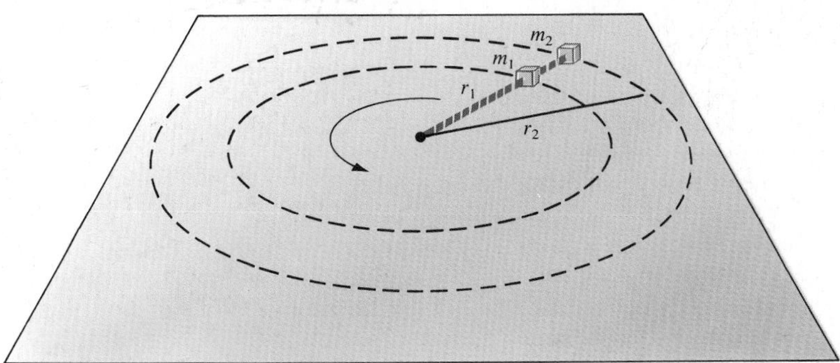

FIGURE 5–29 Problem 13.

SECTIONS 5–5 AND 5–6

21. (I) Calculate the gravitational force between a 60-kg woman and an 80-kg man standing 10 m apart. What if they are practically touching (≈ 0.3 m between their centers)?
22. (I) Calculate the force of gravity on a spacecraft 12,800 km (2 earth radii) above the earth's surface if its mass is 900 kg.
23. (I) Calculate the acceleration due to gravity on the moon. The moon's radius is about 1.7×10^6 m and its mass is 7.4×10^{22} kg.
24. (I) A hypothetical planet has a radius 2.6 times that of the earth, but has the same mass. What is the acceleration due to gravity near its surface?
25. (I) A hypothetical planet has a mass of 2.5 times that of earth, but the same radius. What is g near its surface?
26. (II) Suppose a hypothetical planet has a radius 20.0 times that of earth and a mass 100 times that of earth. What is g near its surface?
27. (II) What is the distance from the earth's center to a point outside the earth where the gravitational acceleration due to the earth is $\frac{1}{25}$ of its value at the earth's surface?
28. (II) At the surface of a certain planet, the gravitational acceleration g has a magnitude of 2.0 m/s^2. A 4.0-kg brass ball is transported to this planet. Find (*a*) the mass of the brass ball on the earth and on the planet, and (*b*) the weight of the brass ball on the earth and on the planet.
29. (II) Suppose the mass of the earth were doubled, but it kept the same density and spherical shape. How would the weight of objects at the earth's surface change?
30. (II) Calculate the effective value of g, the acceleration of gravity, at (*a*) 3200 m, and (*b*) 3200 km, above the earth's surface.
31. (II) Four 8.0-kg spheres are located at the corners of a square of side 0.50 m. Calculate the magnitude and direction of the gravitational force on one sphere due to the other three.
32. (II) Given that the acceleration of gravity at the surface of Mars is 0.38 of what it is on earth and that Mars' radius is 3400 km, determine the mass of Mars.
33. (III) Determine the mass of the sun using the known value for the period of the earth and its distance from the sun.

SECTION 5–7

34. (I) A 15.0-kg monkey hangs from a cord suspended from the ceiling of an elevator. The cord can withstand a tension of 200 N and breaks as the elevator accelerates. What was the elevator's minimum acceleration (magnitude and direction)?
35. (I) Calculate the velocity of a satellite moving in a stable circular orbit about the earth at a height of 3200 km.
36. (II) (*a*) Calculate at what height above the earth's surface a satellite must be placed if it is to remain over the same geographical point on the equator of the earth. (*b*) What is the velocity of such a satellite?
37. (II) What will a spring scale read for the weight of a 54-kg woman in an elevator that moves (*a*) with constant upward speed of 5.0 m/s, (*b*) with constant downward speed of 5.0 m/s, (*c*) with upward acceleration of 0.33 g, (*d*) with downward acceleration 0.33 g, and (*e*) in free fall?
38. (II) A ferris wheel 22.5 m in diameter rotates once every 12.5 s. What is the fractional change in a person's apparent weight (*a*) at the top, and (*b*) at the bottom, as compared to her weight at rest?
39. (II) What is the apparent weight of a 65-kg astronaut 4200 km from the center of the earth's moon in a space vehicle (*a*) moving at constant velocity, and (*b*) accelerating toward the moon at 3.6 m/s^2? State "direction" in each case.
40. (II) Describe a general procedure to determine the mass of a planet from observations on the orbit of one of its satellites.
41. (III) Show that if a satellite orbits very near the surface of a planet with period T, the density of the planet is $\rho = 3\pi/GT^2$.

*SECTION 5–8

*42. (II) (*a*) Calculate the force exerted by the sun ($m = 2.0 \times 10^{30}$ kg) and exerted by the moon ($m = 7.4 \times 10^{22}$ kg) on the earth. (*b*) Explain why the moon has the greater influence on the earth's tides. The sun is 1.5×10^{11} m from earth.

*SECTION 5–9

*43. (I) Use Kepler's laws and the period of the moon (27.4 d) to determine the period of an artificial satellite orbiting near the earth's surface.
*44. (I) The asteroid Icarus, though only a few hundred meters across, orbits the sun like the other planets. Its period is about 410 d. What is its mean distance from the sun?
*45. (I) Neptune is an average distance of 4.5×10^9 km from the sun. Estimate the length of the Neptunian year given that the earth is 1.50×10^8 km from the sun on the average.
*46. (II) Determine the mass of the earth from the known period and distance of the moon.
*47. (II) Our sun rotates about the center of the Galaxy ($M \approx 4 \times 10^{41}$ kg) at a distance of about 3×10^4 light years (1 ly $= 3 \times 10^8$ m/s $\times 3.16 \times 10^7$ s). What is the period of our orbital motion about the center of the Galaxy?
*48. (II) Table 5–2 gives the mean distance, period, and mass for the four largest moons of Jupiter (those discovered by Galileo in 1609). (*a*) Determine the mass of Jupiter using the data for Io. (*b*) Determine the mass of Jupiter using data for each of the other three moons. Are the results consistent?

*49. (II) Determine the mean distance from Jupiter for each of Jupiter's moons using the periods given in Table 5–2. Compare to the values in the table.

*50. (III) (*a*) Use Kepler's second law to show that the ratio of the speeds of a planet at the points of its orbit nearest to and farthest from the sun is equal to the inverse ratio of the near and far distances: $v_N/v_F = d_F/d_N$. (*b*) Given that the earth's distance from the sun varies from 1.47 to 1.52×10^{11} m, determine the minimum and maximum velocities of the earth in its orbit around the sun.

Table 5–2
Principal Moons of Jupiter (Problems 48 and 49)

Moon	Mass (kg)	Period (d)	Mean Distance from Jupiter (km)
Io	8.9×10^{22}	1.77	422×10^3
Europa	4.9	3.55	671
Ganymede	15	7.16	1070
Callisto	11	16.7	1883

GENERAL PROBLEMS

51. How far above the earth's surface will the acceleration of gravity be half what it is on the surface?

52. Tarzan plans to cross a gorge by swinging in an arc from a hanging vine. If his arms are capable of exerting a force of 1200 N on the rope, what is the maximum speed he can tolerate at the lowest point of his swing? His mass is 85 kg; the vine is 4.8 m long.

53. Is it possible to whirl a bucket of water fast enough in a vertical circle so the water won't fall out? If so, what is the minimum speed?

54. Because the earth rotates once per day, the effective acceleration of gravity at the equator is slightly less than it would be if the earth didn't rotate. Estimate the magnitude of this effect. What fraction of g is this?

55. At what distance from the earth will a spacecraft traveling directly from the earth to the moon experience zero net force because the earth and moon pull with equal and opposite forces?

56. A projected space station consists of a circular tube that is set rotating about its center (like a tubular bicycle tire). The circle formed by the tube has a diameter of 1.2 km. (*a*) On which inner wall of the tube will people be able to walk? (*b*) What must be the rotation speed (revolutions per day) if an effect equal to gravity at the surface of the earth (1 g) is to be felt?

57. A jet pilot takes his aircraft in a vertical loop (Fig. 5–31). If the jet is moving at a speed of 700 km/h at the lowest point of the loop, determine (*a*) the minimum radius of the circle so that the acceleration at the lowest point does not exceed 6.0 g's and (*b*) the 80-kg pilot's effective weight (the force with which the seat pushes up on him) at the bottom of the circle.

58. Derive a formula for the mass of a planet in terms of its radius, r, the acceleration due to gravity at its surface, g_p, and the gravitational constant, G.

59. A plumb bob is deflected from the vertical by an angle θ due to a massive mountain nearby (Fig. 5–32). Find an approximate formula for θ in terms of the mass of the mountain, M_M, the distance to its center, D_M, and the radius and mass of the earth.

60. A curve of radius 70 m is banked for a design speed of 90 km/h. If the coefficient of static friction is 0.30 (wet pavement), at what range of speeds can a car safely make the curve?

61. How long would a day be if the earth were rotating so fast that objects at the equator were weightless?

62. Two equal-mass stars maintain a constant distance apart of 8.0×10^{10} m and rotate about a point midway between them at a rate of one revolution every 12.6 yr. (*a*) Why don't the two stars crash into one another due to the gravitational force between them? (*b*) What must be the mass of each star?

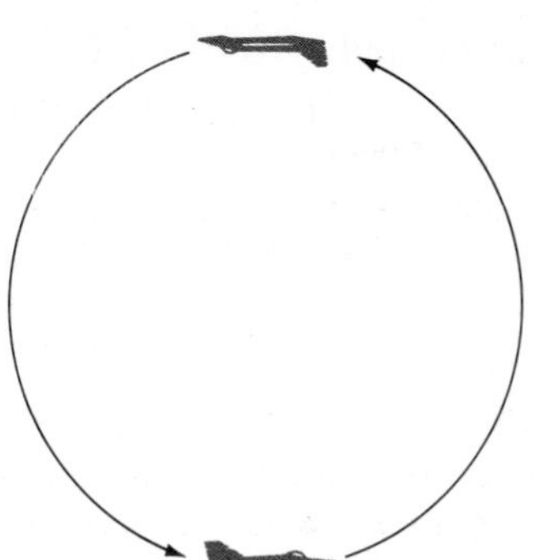

FIGURE 5–31 Problem 57.

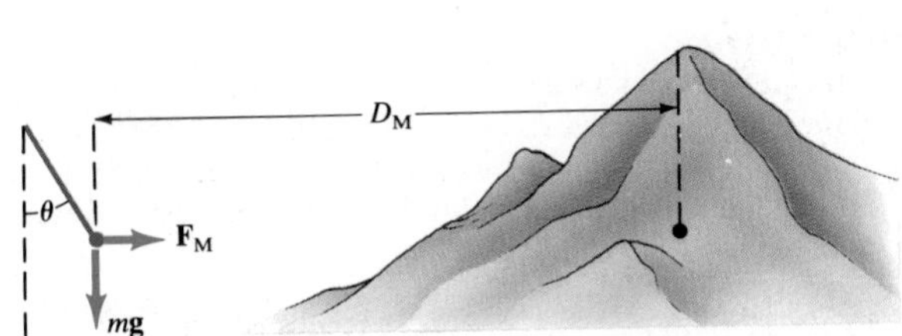

FIGURE 5–32 Problem 59.

CHAPTER 6

Work and Energy

The falling water does work on the water wheel to turn it. Said another way, the potential energy decrease of the water is transformed into kinetic energy of the wheel (and of the grinding stone to which it is connected inside the mill).

Until now we have been studying the motion of a particle in terms of Newton's three laws of motion. In this analysis, *force* has played a central role as the quantity determining the motion. In this chapter and the next, we discuss an alternative analysis of the motion of a particle in terms of the quantities *energy* and *momentum.* The importance of these quantities is that they are *conserved.* That is, in quite general circumstances they remain constant. That such quantities are conserved not only gives us a deeper insight into the nature of the world, but also gives us another way to attack practical problems.

The conservation laws of energy and momentum are especially valuable in dealing with systems of many objects in which a detailed consideration of the forces involved would be difficult.

This chapter is devoted to the very important concept of *energy* and the closely related concept of *work*, which are scalar quantities and have no direction associated with them. Since these two quantities are scalars, they are often easier to deal with than are vector forces. Energy derives its importance from two sources. First, it is a conserved quantity. Second, energy is a concept that is useful not only in the study of motion, but in all areas of physics and other sciences as well. But before discussing energy itself, we first examine the concept of work.

6–1 • Work Done by a Constant Force

The word *work* has a variety of meanings in everyday language. But in physics, work is given a very specific meaning to describe what is accomplished by the action of a force when it makes an object move through a distance. Specifically, the **work** done on a particle by a constant force (constant in both magnitude and direction) is defined to be *the product of the magnitude of the displacement times the component of the force parallel to the displacement.* In equation form, we can write

$$W = F_{\parallel} d$$

where $F_{\parallel}$ is the component of the constant force **F** parallel to the net displacement d. More generally, we can write

$$W = Fd\cos\theta, \qquad (6\text{–}1)$$

Work defined

where F is the constant force, d is the net displacement of the particle, and θ is the angle between the directions of the force and the net displacement. The $\cos\theta$ factor appears in Eq. 6–1 because $F\cos\theta\ (=F_{\parallel})$ is the component of **F** parallel to **d** (Fig. 6–1).

Let's first consider the case in which the motion and the force are in the same direction, so $\theta = 0$ and $\cos\theta = 1$, and then $W = Fd$. For example, if you push a loaded grocery cart a distance of 50 m by exerting a horizontal force of 30 N on the cart, you do 30 N × 50 m = 1500 N·m of work on the cart.

Units for work: the joule

As this example shows, in SI units, work is measured in newton-meters. For convenience, a special name is given to this unit, the **joule** (J): 1 J = 1 N·m. In the cgs system, the unit of work is called the *erg* and is defined as 1 erg = 1 dyne·cm. In British units, work is measured in foot-pounds. It is easy to show that 1 J = 10^7 erg = 0.7376 ft·lb.

Force without work

A force can be exerted on an object and yet do no work. For example, if you hold a heavy bag of groceries in your hands at rest, you do no work on it. You may become tired (and indeed energy is being expended by your muscles), but because the bag is not moved through a distance (the displacement is zero), the work $W = 0$. You also do no work on the bag of groceries if you carry it as you walk horizontally across the floor at constant

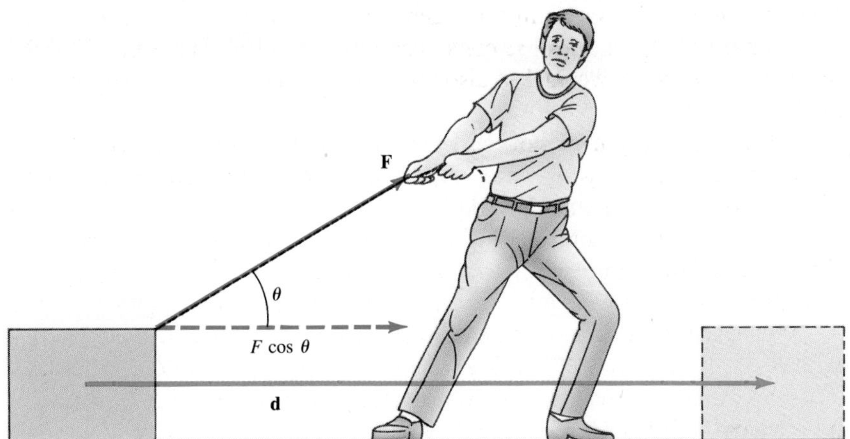

FIGURE 6–1 A person pulling a crate along the floor. The work done by the force **F** is $W = Fd\cos\theta$, where **d** is the displacement.

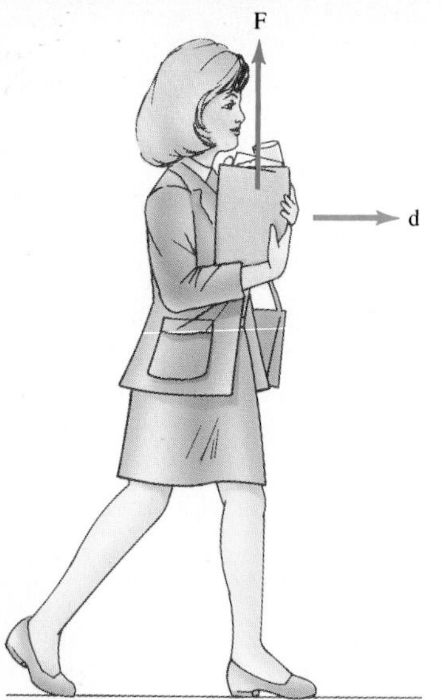

FIGURE 6–2 Work = 0 in this case since **F** is perpendicular to the displacement.

velocity, as shown in Fig. 6–2. No horizontal force is required to move the package at a constant velocity. However, you do exert an upward force **F** on the package equal to its weight. But this upward force is perpendicular to the horizontal motion of the package and thus has nothing to do with that motion. Hence, the upward force is doing no work. This conclusion is consistent with our definition of work, Eq. 6–1: $W = 0$, since $\theta = 90°$ and $\cos 90° = 0$. Thus, when the force is perpendicular to the motion, no work is done by that force.

When dealing with work, as with force, it is necessary to specify whether you are talking about work done *by* a specific object or done *on* a specific object. It is also important to specify whether the work done is due to one particular force or to the total *net force* on the object.

EXAMPLE 6–1 A 50-kg crate is pulled 40 m along a horizontal floor by a constant force exerted by a person, $F_P = 100$ N, which acts at a 37° angle as shown in Fig. 6–3. The floor is rough and exerts a friction force $F_{fr} = 50$ N. Determine the work done by each force acting on the crate, and the net work done on the crate.

SOLUTION We let **x** be the vector that represents the 40-m displacement (that is, along the x axis). There are four forces acting on the crate, as shown in Fig. 6–3: $\mathbf{F}_P$, the friction force $\mathbf{F}_{fr}$, the crate's weight $m\mathbf{g}$, and the normal force $\mathbf{F}_N$ exerted upward by the floor. The work done by the gravitational and normal forces is zero, since they are perpendicular to the displacement **x** ($\theta = 90°$ in Eq. 6–1):

$$W_g = 0$$

$$W_N = 0.$$

The work done by $\mathbf{F}_P$ is

$$W_P = F_P x \cos\theta = (100 \text{ N})(40 \text{ m}) \cos 37° = 3200 \text{ J}.$$

The work done by the friction force is

$$\begin{aligned} W_{fr} &= F_{fr} x \cos 180° \\ &= (50 \text{ N})(40 \text{ m})(-1) = -2000 \text{ J}. \end{aligned}$$

The angle between the displacement **x** and $\mathbf{F}_{fr}$ is 180° because they point in opposite directions. Since $\cos 180° = -1$, we see that the force of friction does negative work on the crate.

Finally, the net work can be calculated in two equivalent ways. (1) The

FIGURE 6–3 Example 6–1: 50-kg crate being pulled along a floor.

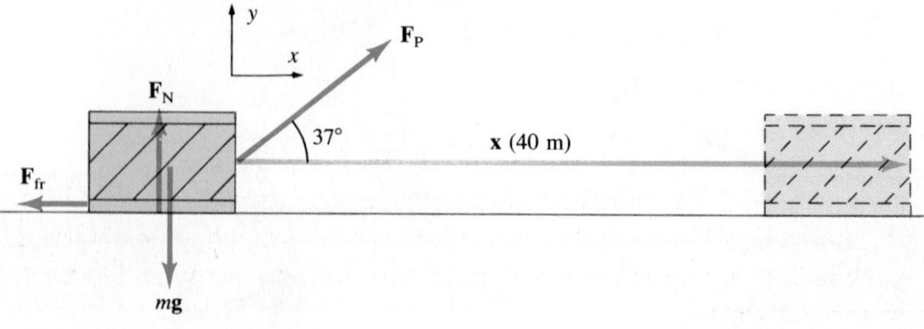

net work done on an object is the algebraic sum of the work done by each force, since work is a scalar:

$$W_{\text{net}} = W_g + W_N + W_P + W_{fr}$$
$$= 0 + 0 + 3200\text{ J} - 2000\text{ J} = 1200\text{ J}.$$

(2) The net work can also be calculated by first determining the net force on the object and then taking its component along the displacement: $(F_{\text{net}})_x = F_P \cos\theta - F_{fr}$. Then the net work is

$$W_{\text{net}} = (F_{\text{net}})_x x = (F_P \cos\theta - F_{fr})x$$
$$= (100\text{ N}\cos 37^\circ - 50\text{ N})(40\text{ m}) = 1200\text{ J}$$

as before.

EXAMPLE 6–2 Determine the work a person must do to carry a 15.0-kg backpack up a hill of height $h = 10.0$ m, as shown in Fig. 6–4a. Determine also the work done by gravity and the net work done on the backpack. For simplicity, assume the motion is smooth and at constant velocity (i.e., there is negligible acceleration).

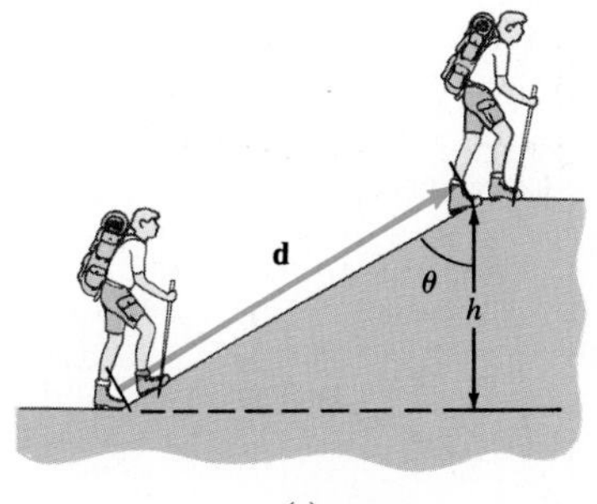

(a)

(b)

FIGURE 6–4 Example 6–2.

SOLUTION The forces on the backpack are shown in Fig. 6–4b: the force of gravity, $m\mathbf{g}$, acting downward, and $\mathbf{F}_P$, the force the person must exert upward to support the pack. Since we assume there is negligible acceleration, horizontal forces are also negligible. By Newton's second law, we have

$$\mathbf{F} = m\mathbf{a}$$

$$F_P - mg = 0,$$

where we have taken the upward direction as positive. Hence, $F_P = mg = (15.0\text{ kg})(9.80\text{ m/s}^2) = 147$ N. To calculate the work, Eq. 6–1 can be written $W = F(d\cos\theta)$, and we note from Fig. 6–4a that $d\cos\theta = h$. So the work done is

$$W_P = F_P h = mgd\cos\theta = mgh$$
$$= (147\text{ N})(10.0\text{ m}) = 1470\text{ J}.$$

Note that the work done depends only on the change in elevation and not on the angle of the hill, θ. The same work would be done to lift the pack vertically to the same height h.

The *net* work done on the backpack is $W_{\text{net}} = 0$, since the net force on the backpack is zero (it is assumed not to accelerate significantly). Finally, since

$$W_{\text{net}} = W_g + W_P = 0,$$

then the work done by gravity is

$$W_g = -W_P = -1470\text{ J}.$$

(Can you think of an independent argument to show $W_g = -1470$ J, with the minus sign?) Note in this example that even though the *net* work on the backpack is zero, the person nonetheless does do work on the backpack equal to 1470 J.

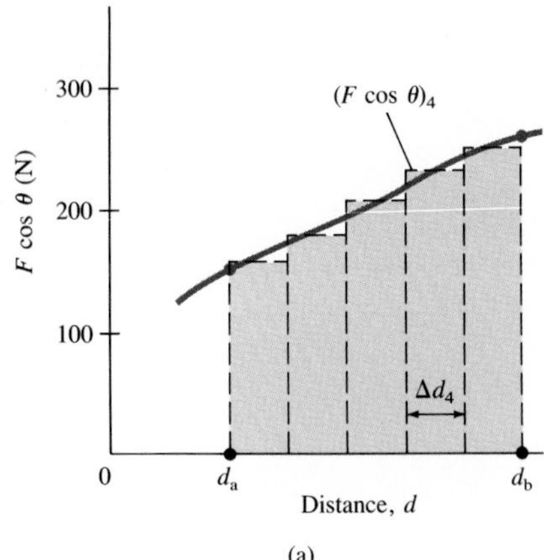

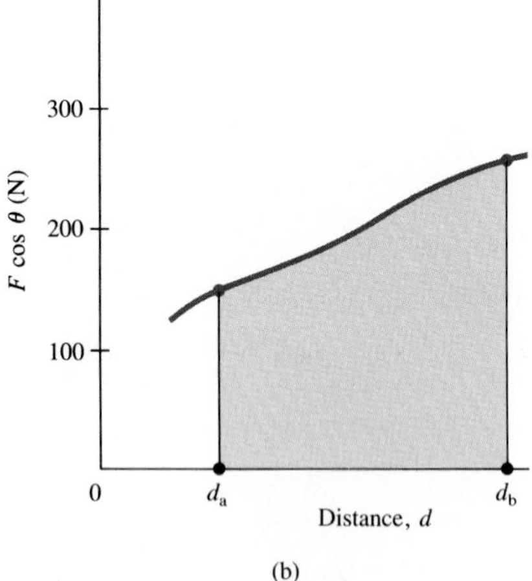

FIGURE 6–5 The work done by a force F can be calculated by taking: (a) the sum of the areas of the rectangles; (b) the area under the curve of $F \cos \theta$ vs. d.

*6–2 • Work Done by a Varying Force

If the force acting on an object is constant, the work done by that force can be calculated using Eq. 6–1. But in many cases, the force varies in magnitude or direction during a process. For example, as a rocket moves away from earth, work is done to overcome the force of gravity, which varies as the inverse square of the distance from the earth's center. Another example is the force exerted by a spring, which increases with the amount of stretch.

The work done by a varying force can be determined graphically. The procedure is like that for determining displacement when the velocity is known as a function of time (Section 2–11). To determine the work done by a variable force, we plot $F \cos \theta$ (the component of **F** parallel to the displacement at any point) as a function of distance d, as in Fig. 6–5a. We divide the distance into small segments Δd. For each segment, we indicate the average of $F \cos \theta$ by a horizontal dashed line. Then the work done for each segment is $\Delta W = (F \cos \theta)(\Delta d)$, which is the area of a rectangle (Δd) wide and $(F \cos \theta)$ high. The total work done to move the object a total distance $d = d_b - d_a$ is the sum of the areas of the rectangles (five in the case shown in Fig. 6–5a). Usually, the average value of $(F \cos \theta)$ for each segment must be estimated, and a reasonable approximation of the work done can then be made. If we subdivide the distance into many more segments, Δd can be made smaller and our estimate more accurate. In the limit as Δd approaches zero, the total area of the many narrow rectangles approaches the area under the curve, Fig. 6–5b. That is, *the work done by a variable force in moving an object between two points is equal to the area under the* $(F \cos \theta)$ *vs. d curve between those two points.*

6–3 • Kinetic Energy and the Work-Energy Theorem

Energy is one of the most important concepts in science. Yet we cannot give a simple but accurate and general definition of energy in only a few words. Energy of various specific types, however, can be defined fairly simply. In this chapter, we define translational kinetic energy and potential energy. In later chapters, we will define other types of energy, such as that related to heat (Chapters 14 and 15). The crucial aspect of all the types of energy is that they can be defined consistently with one another and in such a way that the sum of all types, the *total energy*, is the same after any process occurs as it was before: that is, the quantity "energy" can be defined so that it is a conserved quantity. But more on this later.

For the purposes of this chapter, we can define energy in the traditional way as "the ability to do work." This simple definition is not very precise, nor is it really valid for all types of energy.† It is not incorrect, however, for mechanical energy, which we discuss in this chapter, and it serves to underscore the fundamental connection between work and energy. We now define and discuss one of the basic types of energy, kinetic energy.

† Energy associated with heat is often not available to do work, as we will discuss in detail in Chapter 15.

A moving object can do work on another object it strikes. A flying cannonball does work on a brick wall it knocks down; a moving hammer does work on a nail it strikes. In either case, a moving object exerts a force on a second object and moves it through a distance. An object in motion has the ability to do work and thus can be said to have energy. The energy of motion is called **kinetic energy**, from the Greek word *kinetikos*, meaning "motion."

In order to obtain a quantitative definition for kinetic energy, let us consider a particle of mass m that is moving in a straight line with an initial speed v_1. To accelerate it uniformly to a speed v_2, a constant net force F is exerted on it parallel to its motion over a distance d. Then the work done on the particle is $W = Fd$. We apply Newton's second law, $F = ma$, and Eq. 2–10c, written now as $v_2^2 = v_1^2 + 2ad$, with v_1 as the initial speed and v_2 the final speed. Then we find

$$W = Fd = mad = m\left(\frac{v_2^2 - v_1^2}{2d}\right)d$$

or

$$W = \tfrac{1}{2}mv_2^2 - \tfrac{1}{2}mv_1^2. \qquad (6\text{–}2a)$$

We *define* the quantity $\frac{1}{2}mv^2$ to be the **translational kinetic energy** (KE) of the particle:

Kinetic energy defined

$$\text{KE} = \tfrac{1}{2}mv^2.$$

(We call this "translational" KE to distinguish it from rotational KE, which we will discuss later, in Chapter 8.) We can write Eq. 6–2a as

$$W = \Delta\text{KE}. \qquad (6\text{–}2b)$$

Equation 6–2 in either of its forms is an important result. It can be stated in words:

Work-energy theorem

The net work done on an object is equal to its change in kinetic energy.

This is known as the **work-energy theorem**. Notice, however, that since we made use of Newton's second law, $F = ma$, F must be the *net* force—that is, the sum of all forces acting on the object. Thus, the work-energy theorem is valid only if W is the *net work* done on the object.

The work-energy theorem tells us that if (positive) work W is done on a body, its kinetic energy increases by an amount W. The theorem also holds true for the reverse situation: if negative work W is done on the body, the body's kinetic energy decreases by an amount W. That is, a force exerted on a body opposite to the body's direction of motion reduces its speed and its KE. An example is a moving hammer (Fig. 6–6) striking a nail. The force on the hammer ($-F$ in the figure, where F is assumed constant for simplicity) acts toward the left, whereas the displacement **d** is toward the right; so the work done on the hammer, $W_h = \mathbf{F} \cdot \mathbf{d} = -Fd$, is negative and the hammer's KE decreases. Also note in this example that the hammer, as it slows down, does positive work on the nail: if the nail exerts a force $-F$ on the hammer to slow it down, the hammer exerts a force F on the nail (Newton's third law) through the distance d. Hence the work done on the nail is $W_n = Fd = -W_h$, and W_n is positive. Thus the decrease in KE of the hammer is also equal to

FIGURE 6–6 A moving hammer strikes a nail and comes to rest. The hammer exerts a force F on the nail; the nail exerts a force $-F$ on the hammer (Newton's third law). The work done on the nail is positive ($W_n = Fd > 0$). The work done on the hammer is negative ($W_h = -Fd$).

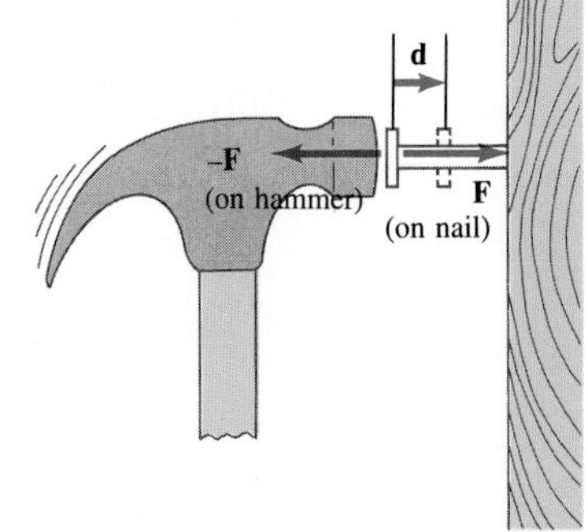

the work the hammer can do on another object (which reinforces what we said earlier about energy being the ability to do work).

Note that whereas the translational KE ($= \frac{1}{2}mv^2$) is directly proportional to the mass of the object, it is proportional to the *square* of the velocity. Thus, if the mass is doubled, the KE is doubled. But if the velocity is doubled, the object has four times as much KE and is therefore capable of doing four times as much work.

Work-energy theorem summarized

To summarize, the connection between work and kinetic energy operates in both directions. If the net work W done on an object is positive, its KE increases, whereas if W is negative, its KE decreases. In the latter case, the object does (positive) work on something else. If the net work done on the object is zero, its KE remains constant.

Energy units: the joule

Because of the direct connection between work and kinetic energy, energy must be measured in the same units as work: joules in SI units, ergs in the cgs, and foot-pounds in the British system. Like work, kinetic energy is a scalar quantity. The kinetic energy of a set of particles is the (scalar) sum of the kinetic energies of the individual particles.

EXAMPLE 6–3 A 145-g baseball is thrown with a speed of 25 m/s. (*a*) What is its KE? (*b*) How much work was done to reach this speed starting from rest?

SOLUTION (*a*) $\text{KE} = \frac{1}{2}mv^2 = \frac{1}{2}(0.145 \text{ kg})(25 \text{ m/s})^2 = 45 \text{ J}$. (*b*) Since the initial KE was zero, the work done is just equal to the final KE, 45 J.

EXAMPLE 6–4 How much work is required to accelerate a 1000-kg car from 20 m/s to 30 m/s?

SOLUTION The work needed is equal to the increase in kinetic energy:

$$\begin{aligned} W &= \tfrac{1}{2}mv^2 - \tfrac{1}{2}mv_0^2 \\ &= \tfrac{1}{2}(1000 \text{ kg})(30 \text{ m/s})^2 - \tfrac{1}{2}(1000 \text{ kg})(20 \text{ m/s})^2 \\ &= 2.5 \times 10^5 \text{ J}. \end{aligned}$$

EXAMPLE 6–5 An automobile traveling 60 km/h can brake to a stop within a distance of 20 m. If the car is going twice as fast, 120 km/h, what is its stopping distance? The force that can be applied to the brakes is approximately independent of speed.

SOLUTION Since the force F is approximately constant, the work needed, Fd, is proportional to the distance traveled. We apply the work-energy theorem, noting that **F** and **d** are in opposite directions and that the final speed of the car is zero:

$$W = -Fd = \Delta\text{KE} = 0 - \tfrac{1}{2}mv^2.$$

Thus the stopping distance increases with the square of the velocity:

$$d \propto v^2.$$

If our car is going twice as fast before braking, the stopping distance is four times as great, or 80 m.

6–4 • Potential Energy

An object can be said to have energy by virtue of its motion, as we have seen. But it can also have **potential energy**, which is the energy associated with the position or configuration of a body (or bodies) and the surroundings. Various types of potential energy (PE) can be defined, and each type is associated with a particular force.

A wound-up clock spring is an example of potential energy. The clock spring acquired its potential energy because work was done *on* it by the person winding the clock. As the spring unwinds, it exerts a force and does work to move the clock hands around.

Perhaps the most common example of potential energy is *gravitational potential energy*. A heavy brick held high in the air has potential energy because of its position relative to the earth. It has the ability to do work, for if it is released, it will fall to the ground due to the gravitational force, and can do work on, say, a stake, driving it into the ground. Let us determine quantitatively the gravitational PE of an object near the surface of the earth. In order to lift an object of mass m vertically, a force at least equal to its weight, mg, must be exerted on it, say, by a person's hand. In order to lift it without acceleration to a height y above the ground, Fig. 6–7, a person must do work equal to the product of the force, mg, and the vertical distance y; that is, $W = mgy$. If we instead allow the object to fall freely under the action of gravity and drive a stake into the ground, the object can do an amount of work equal to mgy on the stake, a fact which can be verified† by using Eqs. 2–10 as we did in Section 6–3 for kinetic energy. Thus, to raise an object of mass m to a height y *requires* an amount of work equal to mgy. And once at height y, the object has the *ability* to do an amount of work equal to mgy. Therefore, we define the *gravitational potential energy* of a body as the product of its weight mg and its height y:

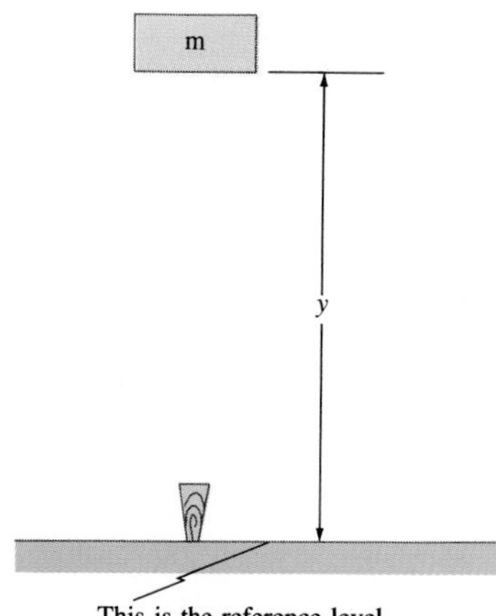

FIGURE 6–7 An object of mass m at height h above the ground can do an amount of work $W = mgh$ when it falls.

Gravitational PE

$$\text{gravitational PE} = mgy. \qquad (6\text{–}3)$$

Notice that the gravitational PE depends on the *vertical height* of the object *above some reference level*, in this case the ground. As we saw earlier in this chapter, the work required to lift an object a height y does not depend on the path taken—that is, on whether one lifts it vertically or goes up a hill as in Fig. 6–4. Similarly, the work it can do after descending does not depend on whether the object falls vertically or by some other path (as down a frictionless incline), but depends only on the vertical height y.

Change in PE *is what is physically meaningful*

In some situations, you may wonder from what point to measure the height y. The gravitational PE of a book held high above a table, for example, depends on whether we measure y from the top of the table, from the floor, or from some other reference point. What is physically important in any situation is the *change* in potential energy because that is what is related to the work done. We can thus choose to measure y from any reference point that is convenient, but we must be consistent throughout any given problem or situation. The change in PE between any two points does not depend on this choice.

† An object that starts from rest and falls a vertical distance y, acquires a velocity given by (see Eq. 2–10c) $v^2 = 2gy$. Just before striking the stake, the falling object's KE will thus be $\frac{1}{2}mv^2 = \frac{1}{2}m(2gy) = mgy$, and by the work-energy theorem, the work done on the stake will be mgy.

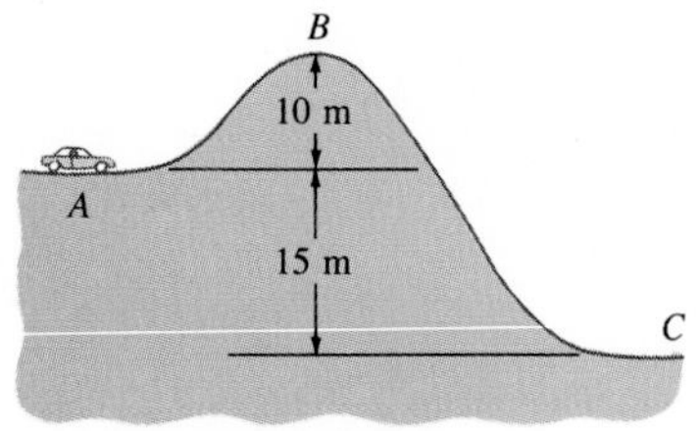

FIGURE 6–8 Example 6–6.

EXAMPLE 6–6 A 1000-kg car moves from point A, Fig. 6–8, to point B and then point C. (*a*) What is its gravitational PE at B and C relative to point A? (*b*) What is the change in potential energy when it goes from B to C? (*c*) Repeat parts (*a*) and (*b*), but take the reference point ($y = 0$) to be at point C.

SOLUTION (*a*) Let us measure the heights from point A, which means initially that the PE is zero. At point B, where $y = 10$ m,

$$\text{PE}_B = mgy = (1000 \text{ kg})(9.8 \text{ m/s}^2)(10 \text{ m}) = 1.0 \times 10^5 \text{ J}.$$

At point C, $y = -15$ m, since C is below A. Therefore,

$$\text{PE}_C = mgy = (1000 \text{ kg})(9.8 \text{ m/s}^2)(-15 \text{ m}) = -1.5 \times 10^5 \text{ J}.$$

(*b*) In going from B to C, the potential energy change is

$$\begin{aligned}\text{PE}_C - \text{PE}_B &= (-1.5 \times 10^5 \text{ J}) - (1.0 \times 10^5 \text{ J}) \\ &= -2.5 \times 10^5 \text{ J}.\end{aligned}$$

The gravitational PE decreases by 2.5×10^5 J.

(*c*) In this instance, the potential energy initially (at A) is equal to $(1000 \text{ kg})(9.8 \text{ m/s}^2)(15 \text{ m}) = 1.5 \times 10^5$ J, since $y = 15$ m. At B, the PE is 2.5×10^5 J, and at C it is zero. But the change in PE going from B to C is $\text{PE}_C - \text{PE}_B = 0 - 2.5 \times 10^5 \text{ J} = -2.5 \times 10^5$ J, which is the same as in part (*b*).

In general, if the object is initially a height y_1 above the reference point, then the change in potential energy when it moves to a different height y_2 is

Change in gravitational PE

$$\text{PE}_2 - \text{PE}_1 = mgy_2 - mgy_1. \qquad [\text{gravitational PE}] \qquad (6\text{–}4)$$

As we already saw, the change in potential energy when an object of mass m moves from height y_1 to height y_2 is equal to the work required by an external force to move the object from height y_1 to height y_2. The change in gravitational PE is also equal to the work done by gravity on the mass if it falls back from y_2 to y_1.

PE *defined in general*

There are other kinds of PE besides gravitational, and each form of PE is associated with a particular force. In general, we define the *change in potential energy associated with a particular force, when an object is moved from one position to another, to be the work that would be done by the force itself in moving the object from the second position back to the first.* Equivalently, we can define the change in PE to be *the work required by some other (external) force to move the object from the first position to the second position, without accelerating it.*

We consider now one other type of PE, that associated with elastic materials. This includes a great variety of practical applications. To take a simple example, consider the simple coil spring shown in Fig. 6–9. The spring has potential energy when compressed (or stretched), for when it is released it can do work on a ball as shown. To hold a spring either stretched or compressed an amount x from its normal (unstretched) length requires a force that is directly proportional to x. That is, $F = kx$, where k is a constant, called the *spring constant*, and is a measure of the stiffness of the particular spring.

The spring itself exerts a force in the opposite direction,

$$F = -kx.$$

Hooke's law

(This is sometimes called a "restoring force" because the spring exerts its force in the direction opposite the displacement (hence the minus sign), acting to return to its normal length.) This **spring equation**, sometimes referred to as **Hooke's law**,† is accurate for springs as long as x is not too great. In order to calculate the potential energy of a compressed spring, we merely need to calculate the work required to compress it (Fig. 6–9b), or the work it does when released (Fig. 6–9c). In either case, the work done is $W = Fx$, where x is the amount it is compressed from its normal length. The force F varies over this distance, becoming greater the more the spring is compressed. We want the average force, and since F varies linearly from zero at the uncompressed position to kx when fully compressed, the average force must be $\frac{1}{2}kx$. The work done is then $W = (\frac{1}{2}kx)(x) = \frac{1}{2}kx^2$. Hence the *elastic potential energy* is proportional to the square of the amount of compression:

$$\text{elastic PE} = \tfrac{1}{2}kx^2. \qquad (6\text{–}5)$$

Elastic PE

If a spring is *stretched* a distance x beyond its normal length, it too has potential energy given by this equation. Thus x can be either the amount compressed or amount stretched from the normal position.

In each of the above examples of potential energy—from a brick held at a height y, to a compressed spring—an object has the capacity or *potential* to do work even though it is not yet actually doing it. That is why we use the term "potential" energy. From these examples, we can also see that energy can be *stored*, for later use, in the form of potential energy. It is also worth noting that although there is a single, universal formula for the kinetic energy of a particle, $\frac{1}{2}mv^2$, there is no single formula for potential energy. Instead, the mathematical form of the PE depends on the force involved.

When dealing with potential energy, we must consider a system, and not a single particle, since potential energy does not exist for an isolated body. Potential energy is associated with a force, and a force on one object is always exerted by some other object. The potential energy is thus a property of the system as a whole. For the system of a mass m on the end of a spring, the potential energy increases by $\frac{1}{2}kx^2$ when the spring is stretched or compressed a distance x from equilibrium, and this is related to the force $F = -kx$ exerted by the spring on the mass. Likewise, when a particle is raised to a height y above the earth's surface, the potential energy change is mgy; the system here is the particle plus the earth.

(a)

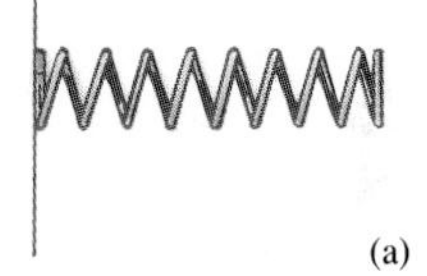

(b)

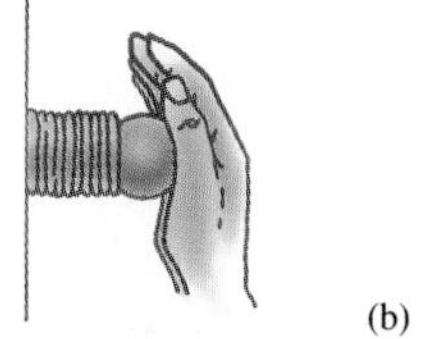

(c)

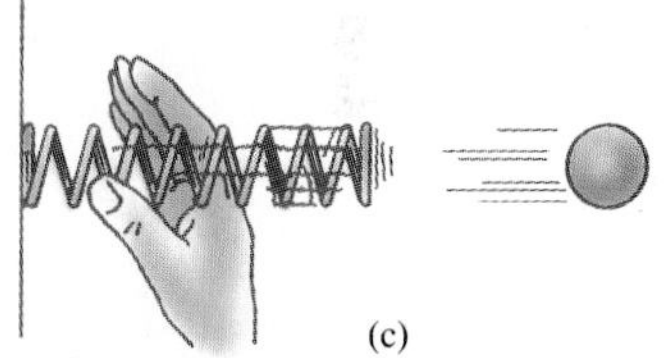

FIGURE 6–9 A spring (a) can store energy (elastic PE) when compressed (b), which can be used to do work (c) when released.

6–5 • Conservative Forces

The work done against gravity in moving an object from one point to another does not depend on the path taken. For example, it takes the same work ($=mgy$) to lift an object of mass m vertically a certain height as to carry it up a hill of the same height as in Fig. 6–4 (see Example 6–2). Forces such as gravity, for which the work done does not depend on the path taken but

† The term "law" applied to this relation is probably not appropriate, since first of all, it is only an approximation, and secondly, it refers only to a limited set of phenomena. Most physicists prefer to reserve the word "law" to those relations which are deeper and more encompassing and precise, such as Newton's laws of motion or the law of conservation of energy (Section 6–7).

only on the initial and final positions, are called **conservative forces.**[†] Friction, on the other hand, is not a conservative force since the work done in, say, pushing a crate across a floor from one point to another depends on whether the path taken is straight, or is curved or zigzag. (In the two latter cases, more work is required, since the distance is greater and, unlike the gravitational force, the friction force is always directed precisely opposite to the direction of motion.)

PE can be defined only for a conservative force

An important fact is that *potential energy can be defined only for a conservative force.* Thus, although potential energy is always associated with a force, we can't write PE for just any force—such as friction, which is a nonconservative force.

We can now extend the **work-energy theorem** (discussed earlier in Section 6–3) to include potential energy. Suppose several forces act on a particle, some of which are conservative, and we can write a potential-energy function for these conservative forces. Then *the work, W′, done by all the other forces acting on a particle is equal to the total change in kinetic and potential energy of the particle*:

General form of work-energy theorem

$$W' = \Delta\text{KE} + \Delta\text{PE}. \qquad (6\text{–}6)$$

It must be emphasized that *all* the forces acting on a body must be included in Eq. 6–6, either in the potential energy term on the right, or in the work term, W', on the left (but not in both!). It is arbitrary whether a given conservative force is considered as doing work (and therefore included in W') or as involved in a change in potential energy, whereas nonconservative forces (such as friction) *must* be included in the work term, W'. We could bring the PE term over to the left side so that all forces are treated as doing work. In this case, the work done is the *net work* since it includes the sum of all the forces. Hence we have

$$\text{net work done} = \Delta\text{KE},$$

which is the original form of the work-energy theorem discussed earlier.

6–6 • Other Forms of Energy; Energy Transformations

Besides the KE and PE of ordinary objects, other forms of energy can be defined as well. These include electric energy, nuclear energy, thermal energy, and the chemical energy stored in food and fuels. With the advent of the atomic theory, these other forms of energy have come to be considered as kinetic or potential energy at the atomic or molecular level. For example, according to the atomic theory, thermal energy is interpreted as the kinetic energy of rapidly moving molecules—when an object is heated, the molecules that make up the object move faster. On the other hand, the energy stored in food and fuel such as gasoline can be regarded as potential energy stored by virtue of the relative positions of the atoms within a molecule due to forces between the atoms (referred to as chemical bonds). For the energy in chemical bonds to be used to do work, it must be released, usually through chemical

[†] The elastic force of, say, a spring (or other elastic material) in which $F = kx$, is also a conservative force.

reactions. This is analogous to a compressed spring which, when released, can do work. Enzymes in our bodies allow the release of energy stored in food molecules. The violent spark of a spark plug in an automobile allows the mixture of gas and air to react chemically, releasing the stored energy and doing work against the piston to propel the car forward. Electric, magnetic, and nuclear energy can also be considered examples of kinetic and potential (or stored) energy. We will deal with these other forms of energy in detail in later chapters.

FIGURE 6–10 Potential energy of a bent bow about to be transformed into kinetic energy of an arrow.

Energy can be transformed from one form to another, and we have already encountered several examples of this. A stone held high in the air has potential energy; as it falls, it loses potential energy, since its height above the ground decreases. At the same time, it gains in kinetic energy, since its velocity is increasing. Potential energy is being transformed into kinetic energy.

Often the transformation of energy involves a transfer of energy from one body to another. The PE stored in the spring of Fig. 6–9b is transformed into KE of the ball, Fig. 6–9c. Water at the top of a dam has potential energy, which is transformed into kinetic energy as the water falls; at the base of the dam, the kinetic energy of the water can be transferred to turbine blades and further transformed into electric energy, as we shall see in a later chapter. The potential energy stored in a bent bow can be transformed into kinetic energy of the arrow (Fig. 6–10).

Work is done when energy is transferred

In each of these examples, the transfer of energy is accompanied by the performance of work. The spring of Fig. 6–9 does work on the ball. Water does work on turbine blades. A bow does work on an arrow. This observation gives us a further insight into the relation between work and energy: *work is done whenever energy is transferred from one object to another.*† A person throwing a ball or pushing a grocery cart provides another example. The work done is a manifestation of energy being transferred from the person (ultimately derived from the chemical energy of food) to the ball or cart.

6–7 • The Law of Conservation of Energy

Whenever energy is transformed, it is found that no energy is gained or lost in the process. As an example, let us consider a stone that is allowed to fall toward the ground (Fig. 6–11) without air resistance. Before it is dropped, it has potential energy equal to mgy (we take the ground as the reference level). As it falls, its PE decreases but its KE increases. Just before hitting the ground, it has only kinetic energy. In fact, its KE at the bottom is exactly equal to the PE it had at the top. We can prove this by using Eq. 2–10c: since $v_0 = 0$ and $a = g$, we have $v^2 = 2gy$, where y is the distance fallen. Thus, the KE after falling a distance y is

$$\tfrac{1}{2}mv^2 = \tfrac{1}{2}m(2gy) = mgy.$$

In other words, the original PE, equal to mgy, has been transformed entirely into KE, with the total amount neither increasing nor decreasing.

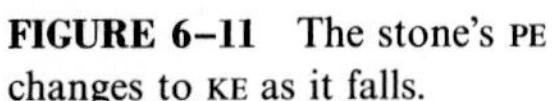

FIGURE 6–11 The stone's PE changes to KE as it falls.

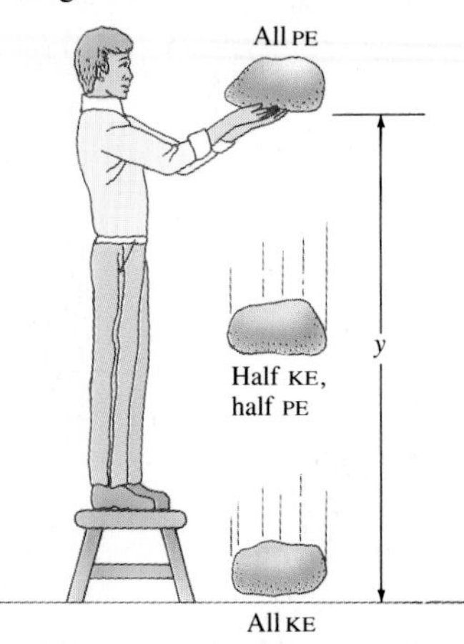

† If the objects are at different temperatures, heat can flow between them instead, or in addition. See Chapters 14 and 15.

This example is an illustration of one of the most important principles in physics, the **law of conservation of energy**:

Law of conservation of energy

The total energy is neither increased nor decreased in any process. Energy can be transformed from one form to another, and transferred from one body to another, but the total amount remains constant.

For mechanical systems involving conservative forces, this law can be derived from Newton's laws and thus is equivalent to them. But in its full generality, the validity of the law of conservation of energy rests on experimental observation, and was not proposed until the middle of the nineteenth century. Even though Newton's laws are found to fail in the submicroscopic world of the atom, the law of conservation of energy has been found to hold in every experimental situation so far tested.

Meaning of the word "conservation"

We need a brief digression here to distinguish the use of the word "conservation" in everyday life from its specific use in physics. In everyday usage, conservation means "saving" or "using wisely"—as when it is said that we should "conserve energy" (what is really meant is that we should conserve *fuel*). In physics the word conservation refers to a quantity that remains strictly *constant*. To be clear, we can say that energy *is* conserved (we don't have to try to conserve it), but we should try to conserve our fuel resources so that we don't use them up too quickly.

Let us now consider some simple illustrations of the law of conservation of energy. For the moment, we consider mechanical systems in which friction (and other nonconservative forces) can be neglected. In this case, the law of conservation of energy can be stated in the following way:

Conservation of mechanical energy

$$\text{KE} + \text{PE} = \text{constant}. \qquad (6\text{–}7)$$

That is, the sum of the kinetic plus potential energies of an object or system of objects, which is called the **total mechanical energy**, remains constant. For example, in the case of the rock falling under gravity, Fig. 6–11, the rock initially has only potential energy. As it falls, its PE decreases, but its KE increases to compensate, so that the sum of the two remains constant. At any point along the path, the total mechanical energy is given by $\frac{1}{2}mv^2 + mgy$, where y is its height above the ground at a given point and v is its velocity at that point. If we let the subscript 1 represent the rock at one point along its path (for example, the initial point), and 2 represent it at some other point, then we can write

$$\text{total energy at point 1} = \text{total energy at point 2}$$

or (see Eq. 6–7)

Conservation of energy when only gravity acts

$$\tfrac{1}{2}mv_1^2 + mgy_1 = \tfrac{1}{2}mv_2^2 + mgy_2. \qquad [\text{grav. PE only}] \qquad (6\text{–}8)$$

To see the practical value of this relation, suppose the original height of the stone in Fig. 6–11 is $y_1 = 3.0$ m, and we wish to calculate its speed when it has fallen to 1.0 m above the ground. Then, since $v_1 = 0$ (the moment of release), $y_2 = 1.0$ m, and $g = 9.8\ \text{m/s}^2$, Eq. 6–8 tells us

$$0 + (m)(9.8\ \text{m/s}^2)(3.0\ \text{m}) = \tfrac{1}{2}mv_2^2 + (m)(9.8\ \text{m/s}^2)(1.0\ \text{m}).$$

The m's cancel out, and solving for v_2^2, we find $v_2^2 = 2[(9.8\ \text{m/s}^2)(3.0\ \text{m}) - (9.8\ \text{m/s}^2)(1.0\ \text{m})] = 39.2\ \text{m}^2/\text{s}^2$, and $v_2 = \sqrt{39.2}\ \text{m/s} = 6.3$ m/s.

6–8 • Problem Solving Using Conservation of Mechanical Energy

Equation 6–8 can be applied to any object moving without friction under the action of gravity. For example, Fig. 6–12 shows a car starting from rest at the top of a hill, and coasting without friction to the bottom and up the hill on the other side. Initially, the car has only PE. As it coasts down the hill, it loses PE and gains in KE, but the sum of the two remains constant. At the bottom of the hill it has its maximum KE, and as it climbs up the other side the KE changes back to PE. When the car comes to rest again, all of its energy will be PE. Since PE is proportional to height and because energy is conserved, the car comes to rest at a height equal to its original height. If the two hills are the same height, the car just barely reaches the top of the second hill when it stops. If the second hill is lower than the first, not all of the car's KE will be transformed to PE and the car continues over the top and down the other side. If the second hill is higher, the car will only reach a height on it equal to its original height on the first hill. This is true (in the absence of friction) no matter how steep the hill is, since PE depends only on the vertical height.

FIGURE 6–12 A car coasting down a hill illustrates the conservation of energy.

EXAMPLE 6–7 Assuming the height of the hill in Fig. 6–12 is 40 m, calculate (*a*) the velocity of the car at the bottom of the hill and (*b*) at what height it will have half this speed.

SOLUTION (*a*) We use Eq. 6–8 with $v_1 = 0$, $y_1 = 40$ m, and $y_2 = 0$. Then

$$0 + (m)(9.8 \text{ m/s}^2)(40 \text{ m}) = \tfrac{1}{2}mv_2^2 + 0.$$

The m's cancel out and we find $v_2 \sqrt{2(9.8 \text{ m/s}^2)(40 \text{ m})} = 28$ m/s.

(*b*) We use the same equation, but now $v_2 = 14$ m/s and y_2 is unknown:

$$0 + (m)(9.8 \text{ m/s}^2)(40 \text{ m}) = \tfrac{1}{2}(m)(14 \text{ m/s})^2 + (m)(9.8 \text{ m/s}^2)(y_2).$$

We cancel the m's and solve for y_2 and find $y_2 = 30$ m. That is, the car has a speed of 14 m/s when it is 30 *vertical* meters above the lowest point, both when descending the left-hand hill and when ascending the right-hand hill.

There are many interesting examples of the conservation of energy in sports, one of which is the pole vault illustrated in Fig. 6–13. We often have to make approximations, but the sequence of events in broad outline for this case is as follows. The KE of the running athlete is transformed into elastic PE of the bending pole and, as the athlete leaves the ground, into gravitational PE. When the vaulter reaches the top and the pole straightens out again, the energy has all been transformed into gravitational PE. (We ignore the vaulter's low speed over the bar.) The pole does not supply any energy, but it acts as a very convenient device to *store* energy and thus aid in the transformation of KE into gravitational PE, which is the net result. The energy required to pass over the bar depends on how high the center of mass† (cm)

† The center of mass (cm) of a body is that point at which the entire mass of the body can be considered as concentrated for purposes of describing its translational motion—that is, so we can consider it as a particle. (This is discussed in Chapter 7.)

FIGURE 6–13 Transformation of energy during a pole vault.

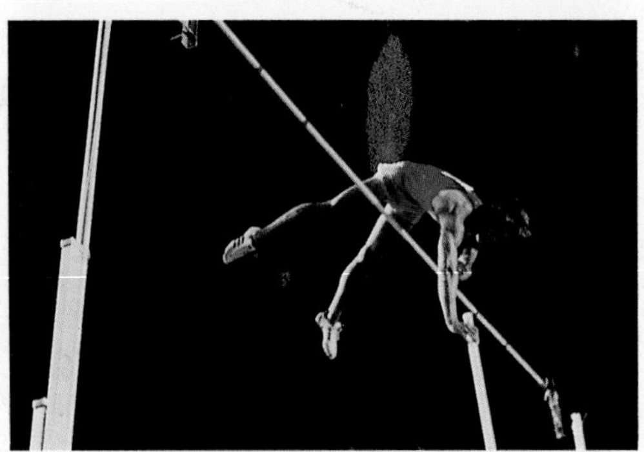

FIGURE 6–14 By bending the body, a pole vaulter can keep his center of mass low so that it may even pass below the bar. By changing his KE (of running) into gravitational PE ($=mgy$) in this way, he can cross over a higher bar than if the change in PE were accomplished without carefully bending the body.

of the vaulter must be raised. By bending their bodies, pole vaulters can keep their cm so low that it actually passes slightly beneath the bar (Fig. 6–14), thus enabling them to cross over a higher bar than would otherwise be possible.

EXAMPLE 6–8 Estimate the kinetic energy and the velocity required for a 70-kg pole vaulter to pass over a bar 5.0 m high. Assume the vaulter's center of mass is initially 0.90 m off the ground and reaches its maximum height at the level of the bar itself.

SOLUTION We equate the total energy just before the vaulter places the end of the pole onto the ground (and the pole begins to bend and store PE) with the vaulter's total energy when passing over the bar (we ignore the small amount of KE at this point). The vaulter's body must be raised a height $y = 5.0\text{ m} - 0.9\text{ m} = 4.1\text{ m}$. Thus

$$\tfrac{1}{2}mv^2 + 0 = 0 + mgy$$

and

$$\begin{aligned} \text{KE} = \tfrac{1}{2}mv^2 &= mgy \\ &= (70\text{ kg})(9.8\text{ m/s}^2)(4.1\text{ m}) = 2.8 \times 10^3\text{ J}. \end{aligned}$$

The velocity is

$$v = \sqrt{\frac{2\text{KE}}{m}} = \sqrt{\frac{2(2800\text{ J})}{70\text{ kg}}} = 8.9\text{ m/s}.$$

This is an approximation because we have ignored such things as the vaulter's speed while crossing over the bar, mechanical energy transformed when the pole is planted in the ground, and work done by the vaulter on the pole.

Notice that the kinematic equations for constant acceleration, Eqs. 2–10, could have been used to treat the simple case of a falling rock (discussed immediately after Eq. 6–8), but they could not be used to solve either Example 6–7 or 6–8 since the acceleration is not constant in these cases. Here we have seen the great power of using the conservation of energy principle. Even though this principle is equivalent to Newton's second law, $\mathbf{F} = m\mathbf{a}$, and can be derived from it,† we have used the conservation of energy to solve problems without having to deal with force and acceleration, which vary in a complicated way for these cases.

Problem Solving: Using energy conservation vs. Newton's laws

As another example of the conservation of mechanical energy, let us consider a mass m connected to a spring whose own mass can be neglected and whose stiffness constant is k. The mass m has speed v at any moment and the potential energy of the system is $\frac{1}{2}kx^2$, where x is the displacement of the spring from its unstretched length. If neither friction nor any other force is acting, the conservation-of-energy principle tells us that

$$\tfrac{1}{2}mv_1^2 + \tfrac{1}{2}kx_1^2 = \tfrac{1}{2}mv_2^2 + \tfrac{1}{2}kx_2^2 \qquad \text{[elastic PE only]}$$

where the subscripts 1 and 2 refer to the velocity and displacement at two different times.

EXAMPLE 6–9 A ball of mass $m = 2.6$ kg, starting from rest, falls a vertical distance $h = 55$ cm before striking a vertical coiled spring, which it compresses (see Fig. 6–15). If the spring has stiffness constant $k = 720$ N/m and negligible mass, what is the maximum compression of the spring? Measure all distances from the point where the ball first touches the uncompressed spring ($y = 0$ at this point).

SOLUTION Since the motion is vertical, we use y instead of x (y positive upward). Let the maximum compression of the spring be called Y ($Y > 0$). Then, initially, the total energy of the system is

$$E_1 = mgh.$$

When the spring has been compressed to its maximum, the total energy is

$$E_2 = \tfrac{1}{2}kY^2 - mgY.$$

The first term on the right is the elastic PE of the compressed spring. The second term is the gravitational PE at the point $y_2 = -Y$ (where $Y > 0$). At both points 1 and 2, the kinetic energy is zero, although at intermediate

† This is done via the work-energy theorem. (See Sections 6–3 and 6–5.)

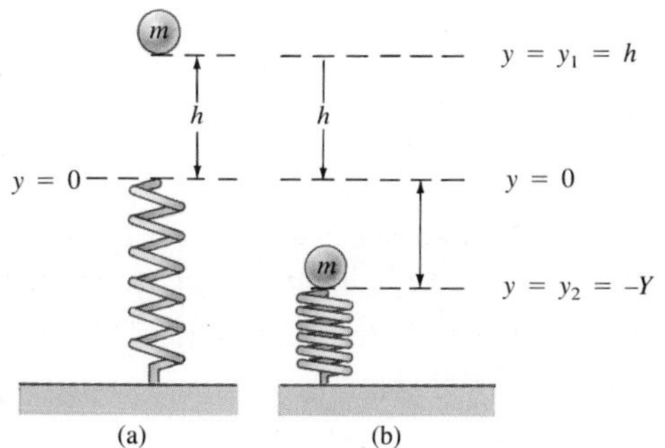

FIGURE 6–15 Example 6–9.

points it is nonzero since the ball is moving. Energy is conserved, so

$$E_1 = E_2$$

$$mgh = \tfrac{1}{2}kY^2 - mgY$$

or

$$\tfrac{1}{2}kY^2 - mgY - mgh = 0.$$

We use the quadratic formula (Appendix A) to find Y:

$$Y = \frac{+mg \pm \sqrt{m^2g^2 + 2mghk}}{k} = 0.24 \text{ m},$$

or 24 cm. We used the root with the plus sign since this gives $Y > 0$, as assumed. The root with the negative sign yields $Y = -0.17$ m, which would correspond to the ball and spring, connected to each other, having sprung back up a distance of 0.17 m above the unstretched ($y = 0$) length.

6–9 • Energy Conservation with Dissipative Forces: Solving Problems

We have neglected friction up to now, but in many situations it cannot be ignored. In a real situation, the car in Fig. 6–12, for example, will not in fact reach the same height on the second hill as it had on the first hill because of friction. In this, and in other natural processes, the mechanical energy (sum of the kinetic and potential energies) does not remain constant but decreases. Because frictional forces reduce the total mechanical energy, they are called **dissipative forces**. Historically, the presence of dissipative forces hindered the formulation of a comprehensive conservation of energy law until well into the nineteenth century. It was not until then that heat, which is always produced when there is friction (try rubbing your hands together), was recognized as energy. Quantitative studies by nineteenth-century scientists (discussed in Chapters 14 and 15) demonstrated that if heat is interpreted as energy (properly called **thermal energy**), then the total energy is conserved in any process. For example, if the car in Fig. 6–12 is subject to frictional forces, then the initial total energy of the car will be equal to the kinetic plus potential energy of the car at any subsequent point along its path plus the amount of thermal energy produced in the process. The thermal energy produced by a constant friction force F_{fr} is equal to the work done against this force: $W = F_{fr}d$, where d is the distance over which the force acts. Equation 6–8 can then be rewritten to take into account friction:

Dissipative forces

Conservation of energy with gravity and friction

$$\tfrac{1}{2}mv_1^2 + mgy_1 = \tfrac{1}{2}mv_2^2 + mgy_2 + F_{fr}d \qquad \begin{bmatrix}\text{gravity and}\\ \text{friction acting}\end{bmatrix} \qquad (6\text{–}9)$$

where d is the distance along the path traveled by the object in going from point 1 to point 2.

EXAMPLE 6–10 The car in Example 6–7 (Fig. 6–12) is found to reach a vertical height of only 25 m on the second hill before coming to a stop. It traveled a total distance of 400 m. Calculate the average friction force on the car. Assume the car has a mass of 1000 kg.

SOLUTION We use Eq. 6–9, taking point 1 to be the instant when the car started coasting and point 2 to be the instant it stopped. Then $v_1 = 0$, $y_1 = 40$ m, $v_2 = 0$, $y_2 = 25$ m, and $d = 400$ m. Thus

$$0 + (1000 \text{ kg})(9.8 \text{ m/s}^2)(40 \text{ m}) = 0 + (1000 \text{ kg})(9.8 \text{ m/s}^2)(25 \text{ m}) + F_{\text{fr}}(400 \text{ m}).$$

We solve this for F_{fr} to find

$$F_{\text{fr}} = 370 \text{ N}.$$

Another example of the transformation of kinetic energy into thermal energy occurs when an object, such as the rock of Fig. 6–11, strikes the ground. Both the rock and the ground will be slightly warmer as a result of their collision. A more apparent example of this transformation of kinetic energy into thermal energy can be observed by vigorously striking a nail several times with a hammer and then gently touching the nail with your finger.

When other forms of energy are involved, such as chemical or electrical energy, the total amount of energy is always found to be conserved. Hence the law of conservation of energy is believed to be universallv valid.

6–10 • Power

Power defined

Power is defined as the *rate at which work is done* or the *rate at which energy is transformed*:

$$\text{power} = \frac{\text{work}}{\text{time}} = \frac{\text{energy transformed}}{\text{time}}.$$

The power of a horse refers to how much work it can do per unit time. The power rating of an engine refers to how much chemical or electrical energy can be transformed into mechanical energy per unit time. In SI units, power is measured in joules per second, and this unit is given a special name, the **watt** (W): 1 W = 1 J/s. We are most familiar with the watt for measuring the rate at which an electric light bulb or heater changes electric energy into light or heat energy, but it is used for other types of energy transformations as well. In the British system, the unit of work is the foot-pound per second (ft·lb/s). For practical purposes, a larger unit is often used, the **horsepower**. One horsepower† (hp) is defined as 550 ft·lb/s, which equals 746 W.

Power and energy distinguished

To see the distinction between energy and power, consider the following example. A person is limited in the work he or she can do, not only by the total energy required, but also by the rate this energy is transformed; that is, by power. For example, a person may be able to walk a long distance or climb many flights of stairs before having to stop because so much energy has been expended. On the other hand, a person who runs very quickly upstairs may fall exhausted after only a flight or two. He or she is limited in this case by power, the rate at which his or her body can transform chemical energy into mechanical energy.

† The unit was first chosen by James Watt (1736–1819) who needed a way to specify the power of his newly developed steam engines. He found by experiment that a good horse can work all day at an average rate of 360 ft·lb/s. So as not to be accused of exaggeration in the sale of his steam engines, he multiplied this by $1\frac{1}{2}$ when he defined the hp.

EXAMPLE 6–11 A 70-kg jogger runs up a long flight of stairs in 4.0 s. The vertical height of the stairs is 4.5 m. Estimate the jogger's power output in watts and horsepower.

SOLUTION The work done is against gravity, so the power P is

$$P = \frac{mgy}{t} = \frac{(70\text{ kg})(9.8\text{ m/s}^2)(4.5\text{ m})}{4.0\text{ s}} = 770\text{ W}.$$

Since there are 750 W in 1 hp, the jogger is doing work at a rate of just over 1 hp. It is worth noting that a human cannot do work at this rate for very long.

Automobiles do work to overcome the force of friction (and air resistance), to climb hills, and to accelerate. A car is limited by the rate it can do work, which is why automobile engines are rated in horsepower. A car needs power most when it is climbing hills and when accelerating. In the next example we will calculate how much power is needed in these situations for a car of reasonable size. Even when a car travels on a level road at constant speed, it needs some power just to do work to overcome the retarding forces of internal friction and air resistance. These forces depend on the conditions and speed of the car, but are typically in the region of 400–1000 N.

It is often convenient to write the power in terms of the net force F applied to an object and its speed v. This is readily done since $P = W/t$ and $W = Fd$ where d is the distance traveled. Then

$$P = \frac{W}{t} = \frac{Fd}{t}$$

$$P = Fv \qquad (6\text{–}10)$$

where $v = d/t$ is the speed of the object.

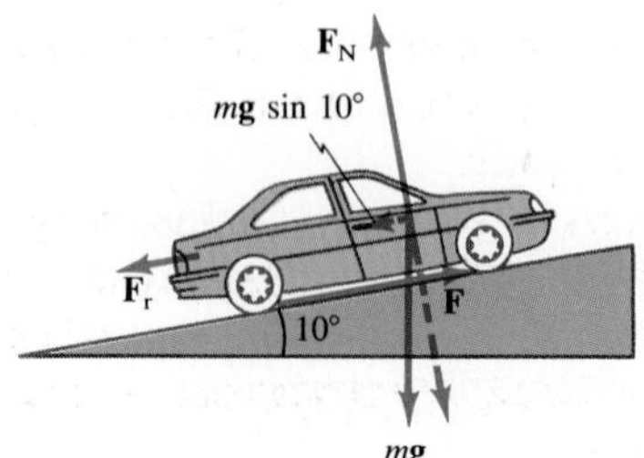

FIGURE 6–16 Calculation of power needed for a car to climb a hill (Example 6–12).

EXAMPLE 6–12 Calculate the power required of a 1400-kg car under the following circumstances: (*a*) the car climbs a 10° hill at a steady 80 km/h; and (*b*) the car accelerates along a level road from 90 to 110 km/h in 6.0 s to pass another car. Assume the retarding force on the car is $F_r = 700$ N. See Fig. 6–16. (Be careful not to confuse $\mathbf{F}_r$, which is due to air resistance and friction that retards the motion, with the force **F** needed to accelerate the car, which is the frictional force exerted by the road on the motor-driven tires—the reaction to the motor-driven tires pushing against the road.)

SOLUTION (*a*) To move at a steady speed up the hill, the car must exert a force equal to the sum of the retarding force, 700 N, and the component of gravity parallel to the hill, $mg \sin 10° = (1400\text{ kg})(9.8\text{ m/s}^2)(0.174) = 2400$ N. Since $v = 80$ km/h $= 22$ m/s and is parallel to **F**, then

$$\begin{aligned} P &= Fv \\ &= (2400\text{ N} + 700\text{ N})(22\text{ m/s}) = 6.8 \times 10^4\text{ W} \\ &= 91\text{ hp}. \end{aligned}$$

(*b*) The car accelerates from 25.0 m/s to 30.6 m/s (90 to 110 km/h). Thus the car must exert a force that overcomes the 700 N retarding force plus that required to give it the acceleration $a = (30.6\text{ m/s} - 25.0\text{ m/s})/6.0\text{ s} = 0.93\text{ m/s}^2$. Since the mass of the car is 1400 kg, the force required for acceleration is $F = ma = (1400\text{ kg})(0.93\text{ m/s}^2) = 1300\text{ N}$, and the total force required is then 2000 N. Since $P = Fv$, the required power increases with speed and the motor must be able to provide a maximum power of

$$\begin{aligned} P &= (2000\text{ N})(30.6\text{ m/s}) \\ &= 6.12 \times 10^4\text{ W} \\ &= 82\text{ hp}. \end{aligned}$$

Even taking into account the fact that only 60 to 80 percent of the engine's power reaches the wheels, it is clear from these calculations that an engine of 100 to 150 hp is quite adequate from a practical point of view even in a reasonably heavy car.

SUMMARY

Work is done on an object by a force when the force moves the object through a distance, *d*. If the direction of a constant force makes an angle θ with the direction of motion, the work done by this force is $W = Fd\cos\theta$.

Energy can be defined as the ability to do work. Both work and energy are measured in *joules* ($1\text{ J} = 1\text{ N}\cdot\text{m}$) in SI units. *Kinetic energy* (KE) is energy of motion; a body of mass m and speed v has translational KE equal to $\frac{1}{2}mv^2$. *Potential energy* (PE) is energy associated with the position or configuration of bodies. Examples include gravitational PE (equal to mgy, where y is the height of the object of mass m above an arbitrary reference point) and elastic PE (equal to $\frac{1}{2}kx^2$ for a compressed spring, where x is the displacement from the unstretched position), as well as chemical, electrical, and nuclear energy. The change in potential energy of an object when it changes position is defined as the work needed to take it from one position to the other.

The *work-energy theorem* states that the *net* work done on a body (by the *net* force) equals the change in kinetic energy of that body.

The law of *conservation of energy* states that energy can be transformed from one type to another, but the total energy remains constant. It is valid even when friction is present since the heat generated can be considered a form of energy.

Power is defined as the rate at which work is done, or the rate at which energy is transformed. The SI unit of power is the *watt* ($1\text{ W} = 1\text{ J/s}$).

QUESTIONS

1. In what ways is the word "work" as used in everyday language the same as defined in physics? In what ways is it different?
2. Can a centripetal force ever do work? Explain.
3. Can the normal force on an object ever do work? Explain.
4. A woman swimming upstream is not moving with respect to the shore. Is she doing any work? If she stops swimming and merely floats, is work done on her?
5. Is the work done by kinetic friction forces always negative? [*Hint:* consider, pulling a tablecloth from under your mom's best china.]
6. Why is it tiring to push hard against a solid wall even though no work is done?
7. You have two springs that are identical except that spring 1 is stiffer than spring 2 ($k_1 > k_2$). On which spring is more work done (*a*) if they are stretched using the same force, (*b*) if they are stretched the same distance?
8. Can kinetic energy ever be negative? Explain.
9. Does the net work done on a particle depend on the choice of reference frame? How does this affect the work-energy theorem?
10. By approximately how much does your gravitational potential energy change when you jump as high as you can?

FIGURE 6–17 Question 11.

11. Describe precisely what is "wrong" physically in the famous Escher drawing shown in Fig. 6–17.

12. Is it true that a rock thrown with a certain speed from the top of a cliff will enter the water below with the same speed whether the rock is thrown horizontally or at any angle? Explain.

13. A coil spring of mass m rests upright on a table. If you compress the spring by pressing down with your hand and then release it, can the spring actually leave the table? Explain using the law of conservation of energy.

14. Does the following statement make sense? "Fiberglass poles led to higher pole vaults because the additional potential energy of bending was converted to gravitational potential energy." Explain.

15. What happens to the gravitational potential energy when water at the top of a waterfall falls to the pool below?

16. Describe the energy transformations when a child hops around on a pogo stick.

17. Describe the energy transformations that take place when a skier starts skiing down a hill, but after a time is brought to rest by striking a snowdrift.

18. An inclined plane has a height h. A body of mass m is released from rest at the top. Does its velocity at the bottom depend on the angle of the plane if (*a*) there is no friction, and (*b*) there is friction?

19. Experienced hikers prefer to step over a fallen log in their path rather than stepping on top and jumping down on the other side. Explain.

20. (*a*) Where does the kinetic energy come from when a car accelerates uniformally starting from rest? (*b*) How is the increase in KE related to the friction force the road exerts on the tires?

21. Analyze the motion of a simple swinging pendulum in terms of energy, (*a*) ignoring friction, and (*b*) taking it into account. Explain why a grandfather clock has to be wound up.

22. When a "superball" is dropped, can it rebound to a height greater than its original height?

23. Suppose you lift a suitcase from the floor to a table. Does the work you do on the suitcase depend on (*a*) whether you lift it straight up or along a more complicated path, (*b*) the time it takes, (*c*) the height of the table, and (*d*) the weight of the suitcase?

24. Repeat the previous question for the *power* needed rather than the work.

25. Why is it easier to climb a mountain via a zigzag trail rather than to climb straight up?

PROBLEMS

SECTION 6–1

1. (I) A 55-kg woman climbs a flight of stairs 5.0 m high. How much work is required?

2. (I) A 650-N crate rests on the floor. How much work is required to move it at constant speed (*a*) 3.0 m along the floor against a friction force of 150 N, and (*b*) 3.0 m vertically?

3. (I) How much work did a horse do that pulled a 200-kg wagon 50 km without acceleration along a level road if the effective coefficient of friction was 0.060?

4. (I) A car does 6.0×10^4 J of work in traveling 2.0 km at constant speed. What was the average force of friction (from all sources) acting on the car?

5. (I) How far must a 200-kg pile driver fall if it is to be capable of doing 13,000 J of work?

6. (II) What is the minimum work needed to push a 1000-kg car 245 m up a 22.5° incline? (*a*) Ignore friction. (*b*) Assume the effective coefficient of friction is 0.30.

7. (II) In pedaling a bicycle, a particular cyclist exerts a downward force of 80 N during each stroke. If the diameter of the circle traced by each pedal is 36 cm, calculate how much work is done in each stroke.

8. (II) Eight bricks, each 6.0 cm thick with mass 1.2 kg, lie flat on a table. How much work is required to stack them one on top of another?

9. (II) A 300-kg piano slides 4.5 m down a 25° incline and is kept from accelerating by a man who is pushing back on it *parallel to the incline*. The effective coefficient of kinetic friction is 0.40. Calculate: (*a*) the force exerted by the man, (*b*) the work done by the man on the piano, (*c*) the work done by the friction force, (*d*) the work done

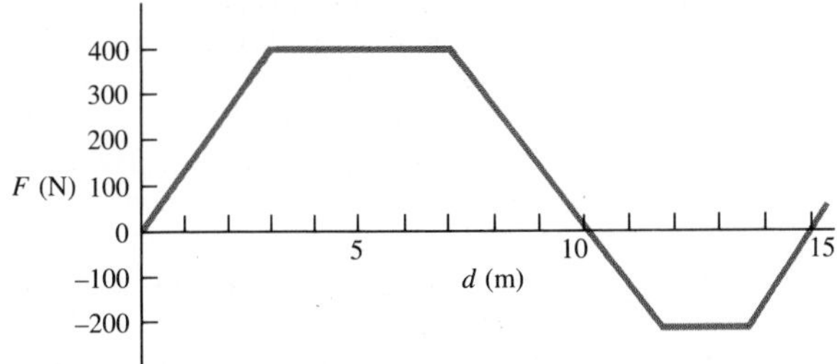

FIGURE 6–18 Problem 12.

by the force of gravity, and (*e*) the net work done on the piano.

10. (III) (*a*) Find the force required to give a helicopter of mass M an acceleration of 0.10 g upward. (*b*) Find the work done by this force as the helicopter moves a distance h upward.

*SECTION 6–2

*11. (II) In Fig. 6–5a, assume the distance axis is linear and that $d_a = 6.0$ m and $d_b = 30.0$ m. Estimate the work done by this force in moving a 2.8-kg object from d_a to d_b.

*12. (II) The force on a particle varies as shown in Fig. 6–18. Determine the work done by this force to move the particle (*a*) from $x = 0.0$ to $x = 10.0$ m; (*b*) from $x = 0.0$ to $x = 15.0$ m.

*13. (II) A spring has $k = 66$ N/m. Use a graph to determine the work needed to stretch it from $x = 3.0$ cm to $x = 5.5$ cm, where x is the displacement from its unstretched length.

*14. (II) The force exerted on a particle increases linearly from zero at $x = 0$ to 24.0 N at $x = 3.0$ m. It remains constant at 24.0 N from $x = 3.0$ m to $x = 8.0$ m and then decreases linearly to zero at $x = 11.0$ m. Determine the work done to move the particle from $x = 0$ to $x = 11.0$ m graphically by determining the area under the F-vs.-x graph.

*15. (III) A 1000-kg space vehicle falls vertically from a height of 2000 km above the earth's surface. Determine how much work is done by the force of gravity in bringing the vehicle to the earth's surface. (First construct an F-vs.-r graph, where r is the distance from the earth's center; then determine the work graphically.)

SECTION 6–3

16. (I) What is the initial KE of a 0.50-mg flea that leaves the ground at a speed of 30 cm/s? (1 mg $= 10^{-3}$ gram.)

17. (I) A carbon atom of mass 1.99×10^{-26} kg has 4.64×10^{-19} J of kinetic energy. How fast is it moving?

18. (I) If the KE of a particle is doubled, by what factor has its speed increased?

19. (I) If the speed of a particle is doubled, by what factor does its KE increase?

20. (I) How much work does it take to accelerate an electron ($m = 9.11 \times 10^{-31}$ kg) from rest to 5.0×10^{6} m/s?

21. (I) How much work must be done to stop a 1000-kg car traveling at 100 km/h?

22. (II) A 1350-kg car rolling on a horizontal surface has speed $v = 40$ km/h when it strikes a horizontal coiled spring and is brought to rest in a distance of 2.5 m. What is the spring constant of the spring? Ignore frictional forces.

23. (II) A baseball ($m = 140$ g) traveling 30 m/s moves a fielder's glove backward 15 cm when the ball is caught. What was the average force exerted by the ball on the glove?

24. (II) If the speed of a car is increased by 50%, by what factor will its minimum braking distance be increased, assuming all else is the same? Ignore the driver's reaction time.

25. (II) In 1955 a paratrooper fell 370 m after jumping from an aircraft without his parachute opening. He landed in a snowbank, creating a crater 1.1 m deep, but survived with only minor injuries. Assuming the paratrooper's mass was 80 kg and his terminal velocity was 50 m/s, estimate: (*a*) the work done by the snow in bringing him to rest; (*b*) the average force exerted on him by the snow to stop him; and (*c*) the work done on him by air resistance as he fell.

26. (III) One car has twice the mass of a second car, but only half as much KE. When both cars increase their speed by 5.0 m/s, they then have the same KE. What were the original speeds of the two cars?

27. (III) A 180-kg load is lifted 23.0 m vertically with an acceleration $a = 0.150\,g$ by a single cable. Determine (*a*) the tension in the cable, (*b*) the net work done on the load, (*c*) the work done by the cable on the load, (*d*) the work done by gravity on the load, and (*e*) the final speed of the load assuming it started from rest.

SECTIONS 6–4 AND 6–5

28. (I) A spring has a spring constant, k, of 380 N/m. How much must this spring be compressed to store 60 J?

29. (I) A 4.3-kg monkey swings from one branch to another 1.3 m higher. What is the change in potential energy?

30. (II) A 1.80-m tall person lifts a 280-g book so it is 2.45 m off the ground. What is the potential energy of the book relative to (*a*) the ground, and (*b*) the top of the person's head? (*c*) How is the work done by the person related to the answers in parts (*a*) and (*b*)?

31. (II) A 65-kg hiker starts at an elevation of 1600 m and climbs to the top of a 2800-m peak. (*a*) What is the hiker's change in potential energy? (*b*) What is the minimum work required of the hiker? (*c*) Can the actual work done be more than this? Explain.

32. (II) (*a*) A spring of spring constant k is initially compressed a distance x_0 from its unstretched length. What is the change in potential energy if it is then compressed by an amount x from its unstretched length? (*b*) The spring is next *stretched* a distance x_0 from the unstretched length. What is the change in potential energy as compared to when it is compressed by an amount x_0 from the unstretched length?

SECTIONS 6–7 TO 6–9

33. (I) Tarzan is running at top speed (6.0 m/s) and grabs a vine hanging vertically from a tall tree in the jungle. How high can he swing upward? Does the length of the vine (or rope) affect your answer?

34. (I) An object, starting from rest, slides down a frictionless 38.0° incline whose vertical height is 14.0 cm. How fast is it going when it reaches the bottom?

35. (II) Two railroad cars, each of mass 4800 kg and traveling 80 km/h, collide head on and come to rest. How much thermal energy is produced in this collision?

36. (II) A 22-kg child descends a slide 5.0 m high and reaches the bottom with a speed of 2.5 m/s. How much thermal energy due to friction was generated in this process?

37. (II) In the high jump, the KE of an athlete is transformed into gravitational potential energy without the aid of a pole. With what minimum speed must the athlete leave the ground in order to lift his center of mass 2.10 m and cross the bar with a speed of 0.80 m/s?

38. (II) A ski starts from rest and slides down an 18° incline 80 m long. (*a*) If the coefficient of friction is 0.080, what is the ski's speed at the base of the incline? (*b*) If the snow is level at the foot of the incline and has the same coefficient of friction, how far will the ski travel along the level? Use energy methods.

39. (II) A 60-kg crate, starting from rest, is pulled across a floor with a constant horizontal force of 100 N. For the first 10 m the floor is frictionless, and for the next 10 m the coefficient of friction is 0.20. What is the final speed of the crate?

40. (II) A roller coaster is shown in Fig. 6–19. Assuming no friction, calculate the speed at points *B*, *C*, *D*, assuming it has a speed of 1.80 m/s at point *A*.

41. (II) The roller coaster in Fig. 6–19 passes point *A* with a speed of 1.20 m/s. If the average force of friction is equal to one-fifth of its weight, with what speed will it reach point *B*? The distance traveled is 67.0 m.

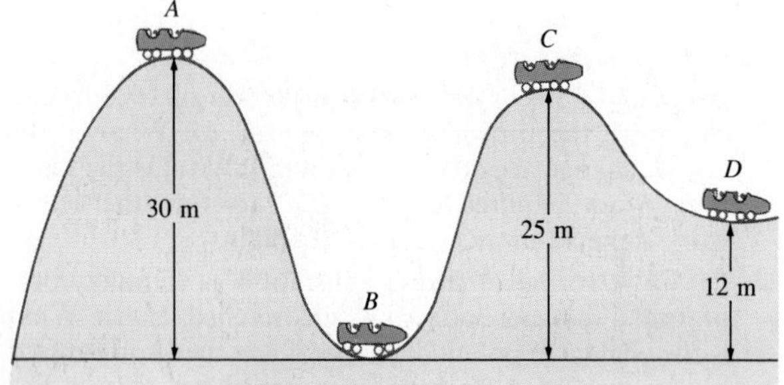

FIGURE 6–19 Problems 40 and 41.

42. (II) A projectile is fired at an upward angle of 45.0° from the top of a 280-m cliff with a speed of 300 m/s. What will be its speed when it strikes the ground at the base of the cliff? (Use conservation of energy.)

43. (II) A vertical spring (ignore its mass), whose spring constant is 850 N/m, stands on a table and is compressed 0.400 m. (*a*) What speed can it give to a 0.300-kg ball when released? (*b*) How high above its original position (spring compressed) will the ball fly?

44. (III) What should be the spring constant k of a spring designed to bring a 1500-kg car to rest from a speed of 90 km/h so that the occupants undergo a maximum acceleration of 5.0 g?

45. (III) A cyclist intends to cycle up a 12.0° hill whose vertical height is 100 m. Assuming the mass of bicycle plus person is 78.0 kg, (*a*) calculate how much work must be done against gravity; (*b*) if each complete revolution of the pedals moves the bike 5.10 m along its path, calculate the average force that must be exerted on the pedals tangent to their circular path. Neglect friction and other losses. The pedals turn in a circle of diameter 36.0 cm.

SECTION 6–10

46. (I) How long will it take a 1500-W motor to lift a 400-kg piano to a sixth-story window 15.0 m above?

47. (I) If a car generates 15 hp when traveling at a steady 100 km/h, what must be the average force exerted on the car due to friction and air resistance?

48. (I) Show that a British horsepower (550 ft·lb/s) is equal to 746 W.

49. (I) Electric energy is often expressed in kilowatt-hours. Show that the kilowatt-hour (kWh) is a unit of energy and is equal to 3.60×10^6 J.

50. (II) A 1200-kg car slows down from 90 km/h to 70 km/h in about 5.0 s on the level when it is in neutral. Approximately what power (watts and hp) is needed to keep the car traveling at a constant 80 km/h?

51. (II) How much work can a 2.0-hp motor do in 1.0 h?

52. (II) A shotputter accelerates a 7.3-kg shot from rest to 15 m/s. If this motion takes 2.0 s, what average power was developed?

53. (II) A pump is to lift 4.00 kg of water per minute through a height of 2.85 m. What output rating (watts) should the pump motor have?

54. (II) How fast must a cyclist climb a 15° hill to maintain a power output of 0.25 hp? Ignore friction and assume the mass of cyclist plus bicycle is 75 kg.

55. (II) A 1000-kg car has a maximum power output of 100 hp. How steep a hill can it climb at a constant speed of 60 km/h if the frictional forces add up to 500 N?

56. (II) A bicyclist coasts down a 6.6° hill at a steady speed of 6.0 m/s. Assuming a total mass of 75 kg (bicycle plus rider), what must be the cyclist's power output to climb the same hill at the same speed?

GENERAL PROBLEMS

57. A 0.25-kg pine cone falls from a branch 20 m above the ground. (*a*) With what speed would it hit the ground if air resistance could be ignored? (*b*) If it actually hits the ground with a speed of 9.0 m/s, what was the average force of air resistance exerted on it?

58. A ball is attached to a horizontal cord of length L whose other end is fixed, Fig. 6–20. (*a*) If the ball is released, what will be its speed at the lowest point of its path? (*b*) A peg is located a distance h directly below the point of attachment of the cord. If $h = 0.75L$, what will be the speed of the ball when it reaches the top of its circular path about the peg?

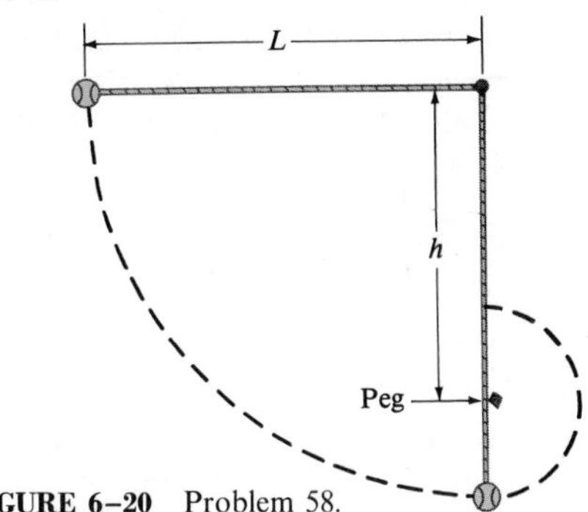

FIGURE 6–20 Problem 58.

59. A skier traveling 13.8 m/s reaches the foot of a steady upward 20° incline and glides 11.4 m up along this slope before coming to rest. What was the average coefficient of friction?

60. A 70-kg hiker climbs to the top of a 4200-m high mountain. The climb is made in 4.0 h starting at an elevation of 2700 m. Calculate (*a*) the work done against gravity, (*b*) the average power output in watts and in horsepower, and (*c*) assuming the body is 15 percent efficient, what rate of energy input was required.

61. A mass m is attached to the end of a spring of spring constant k, Fig. 6–21. The mass is given an initial displacement x_0, after which it oscillates back and forth. Write a formula for the total mechanical energy (ignore friction and mass of the spring) in terms of position x and speed v.

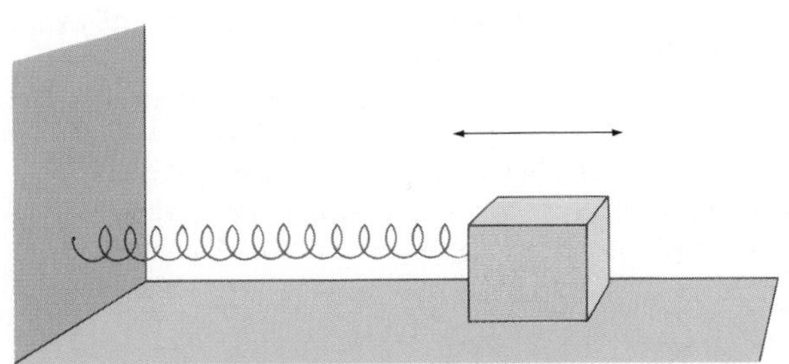

FIGURE 6–21 Problems 61 and 66.

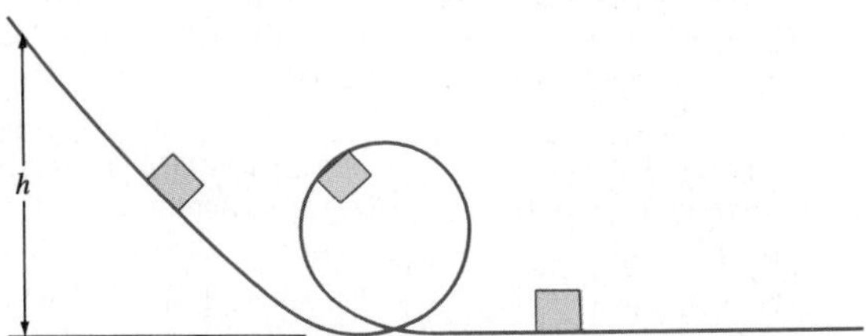

FIGURE 6–22 Problem 62.

62. A small mass m slides without friction along the looped apparatus shown in Fig. 6–22. If the object is to remain on the track, even at the top of the circle (whose radius is r), from what minimum height h must it be released?

63. An elevator cable breaks when a 750-kg elevator is 25 m above a huge spring ($k = 4.0 \times 10^4$ N/m) at the bottom of the shaft. Calculate (*a*) the work done by gravity on the elevator before it hits the spring, (*b*) the speed of the elevator just before striking the spring, and (*c*) the amount the spring compresses (note that work is done by both the spring and gravity in this part).

64. Assume a cyclist of weight w can exert a force on the pedals equal to $0.90w$ on the average. If the pedals rotate in a circle of radius 18 cm, the wheels have a radius of 34 cm, and the front and back sprockets on which the chain runs have 42 and 19 teeth, respectively, determine the maximum steepness of hill the cyclist can climb (Fig. 6–23). Assume the mass of the bike is 12 kg and of the rider is 60 kg. Ignore friction.

65. An engineer is designing a spring to be placed at the bottom of an elevator shaft. If the elevator cable should break at a height h above the top of the spring, calculate the value that the spring constant k should have so that passengers undergo an acceleration of no more than 5.0 g when brought to rest. Let M be the total mass of the elevator and passengers.

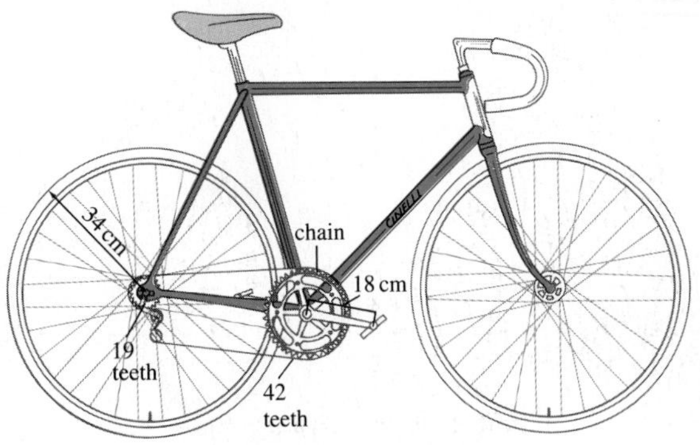

FIGURE 6–23 Problem 64.

66. A 200-g wood block is firmly attached to a horizontal spring, Fig. 6–21. The block can slide along a table where the coefficient of friction is 0.40. A force of 20 N compresses the spring 18 cm. If the spring is released from this position, how far beyond its equilibrium position will it stretch on its first swing?

67. Water flows over a dam at the rate of 800 kg/s and falls vertically 120 m before striking the turbine blades. Calculate (*a*) the velocity of the water just before striking the turbine blades, and (*b*) the rate at which mechanical energy is transferred to the turbine blades. Assume the water loses 80 percent of its speed when striking the blades, and that 12 percent of the energy lost by the water is transformed to thermal energy.

68. A bicyclist of mass 80 kg (including the bicycle) can coast down a 3.4° hill at a steady speed of 9.0 km/h. Pumping hard, the cyclist can descend the hill at a speed of 30 km/h. Using the same power, at what speed can the cyclist climb the same hill? Assume the force of friction is directly proportional to the speed v; that is, $F_{fr} = bv$, where b is a constant.

C H A P T E R 7

Linear Momentum

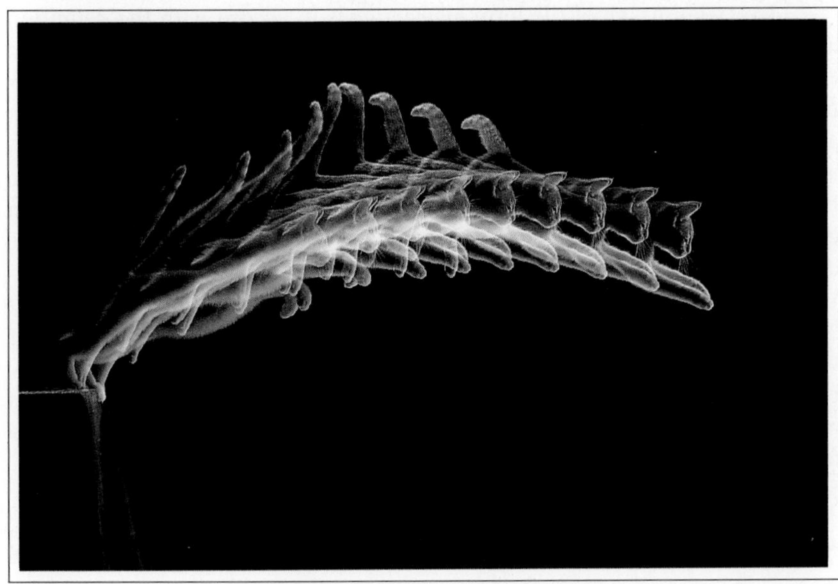

This leaping cat has more momentum than a buzzing bee, but a lot less than a speeding truck. Note that the cat's center of mass moves in a parabolic arc, characteristic of projectile motion.

The law of conservation of energy, which we discussed in the previous chapter, is one of several great conservation laws in physics. Among the other quantities found to be conserved are linear momentum, angular momentum, and electric charge. We will eventually discuss all of these since the conservation laws are among the most important in all of science. In this chapter, we discuss linear momentum (sometimes simply called "momentum") and its conservation. We will then make use of the laws of conservation of linear momentum and of energy to analyze collisions. Indeed, the law of conservation of momentum is particularly useful when dealing with two or more bodies that interact with each other, as in collisions. Our focus up to now has been mainly on the motion of a single particle. But in this chapter we will deal with systems[†] of two or more particles (and with extended bodies which can be considered as collections of particles). An important concept for this study is that of center of mass, which we discuss later in the chapter.

7–1 • Momentum and Its Relation to Force

The **linear momentum** (or "momentum" for short) of a body is defined as the product of its mass and its velocity. Momentum is usually represented by

[†] A *system* is any body or group of bodies we choose to deal with.

the symbol **p**. If we let m represent the mass of a body and **v** represent its velocity, then its momentum **p** is

Linear momentum

$$\mathbf{p} = m\mathbf{v}. \tag{7–1}$$

Since velocity is a vector, momentum is a vector. The direction of the momentum is the direction of the velocity, and the magnitude of the momentum is $p = mv$. Since **v** depends on the reference frame, this frame must be specified. The unit of momentum is just that of mass × velocity, which in SI units is kg·m/s.

Everyday usage of the term *momentum* is in accord with the definition above. For according to Eq. 7–1, a fast-moving car has more momentum than a slow-moving car of the same mass, and a heavy truck has more momentum than a small car moving with the same speed. The more momentum an object has, the harder it is to stop it, and the greater effect it will have if it is brought to rest by impact or collision. A football player is more likely to be stunned if tackled by a heavy opponent running at top speed than by a lighter or slower-moving tackler. A heavy, fast-moving truck can do more damage than a slow-moving small car.

A force is required to change the momentum of an object, whether it is to increase the momentum, to decrease it (such as to bring a moving object to rest), or to change its direction. Newton originally stated his second law in terms of momentum (although he called the product mv the "quantity of motion"). Newton's statement of the **second law of motion**, translated into modern language, is as follows:

The rate of change of momentum of a body is proportional to the net force applied to it.

For a constant force, we can write this as

Newton's second law

$$\mathbf{F} = \frac{\Delta \mathbf{p}}{\Delta t}, \tag{7–2}$$

where **F** is the net force applied to the object and $\Delta\mathbf{p}$ is the resulting momentum change that occurs during the time interval Δt. We can readily derive the familiar form of the second law, $\mathbf{F} = m\mathbf{a}$, from Eq. 7–2 for the case of constant acceleration and constant mass. If $\mathbf{v}_0$ is the initial velocity of an object, and **v** is its velocity after a time Δt has elapsed, then

$$\begin{aligned} \mathbf{F} &= \frac{\Delta \mathbf{p}}{\Delta t} = \frac{m\mathbf{v} - m\mathbf{v}_0}{\Delta t} = \frac{m(\mathbf{v} - \mathbf{v}_0)}{\Delta t} \\ &= m\frac{\Delta \mathbf{v}}{\Delta t} \\ &= m\mathbf{a} \end{aligned}$$

since, by definition, $\mathbf{a} = \Delta\mathbf{v}/\Delta t$. Newton's statement, Eq. 7–2, is actually more general than the more familiar one because it includes the situation in which the mass may change. This is important in certain circumstances, such as for rockets which lose mass as they burn fuel. Another interesting example is the following.

EXAMPLE 7–1 Water leaves a hose at a rate of 1.5 kg/s with a speed of 20 m/s and strikes a wall, which stops it. (That is, we ignore any splashing back.) What is the force exerted by the water on the wall?

SOLUTION In each second, water with a momentum of (1.5 kg)(20 m/s) = 30 kg·m/s is brought to rest. The magnitude of the force (assumed constant) required to change the momentum by this amount is

$$F = \frac{\Delta p}{\Delta t} = \frac{0 - 30\ \text{kg}\cdot\text{m/s}}{1.0\ \text{s}} = -30\ \text{N}.$$

The minus sign indicates that the force on the water is opposite to its original velocity. The wall exerts a force of 30 N to stop the water, so by Newton's third law, the water exerts a force of 30 N on the wall.

7–2 • Conservation of Momentum

The concept of momentum is particularly important because, under certain circumstances, it is a conserved quantity. In the midseventeenth century, shortly before Newton's time, it had been observed that the sum of the momenta of two colliding objects remains constant. Consider, for example, the head-on collision of two billiard balls shown in Fig. 7–1. Although the momentum of each of the two balls changes as a result of the collision, the *sum* of their momenta is found to be the same before as after the collision. If $m_1\mathbf{v}_1$ is the momentum of ball number 1 and $m_2\mathbf{v}_2$ the momentum of ball 2, both measured before the collision, then the total momentum of the two balls before the collision is $m_1\mathbf{v}_1 + m_2\mathbf{v}_2$. After the collision, the balls each have a different velocity and momentum, which we will designate by a "prime" on the velocity: $m_1\mathbf{v}_1'$ and $m_2\mathbf{v}_2'$. The total momentum after the collision is $m_1\mathbf{v}_1' + m_2\mathbf{v}_2'$. No matter what the velocities and masses involved are, it is found that the total momentum before the collision is the same as afterwards, whether the collision is head on or not:

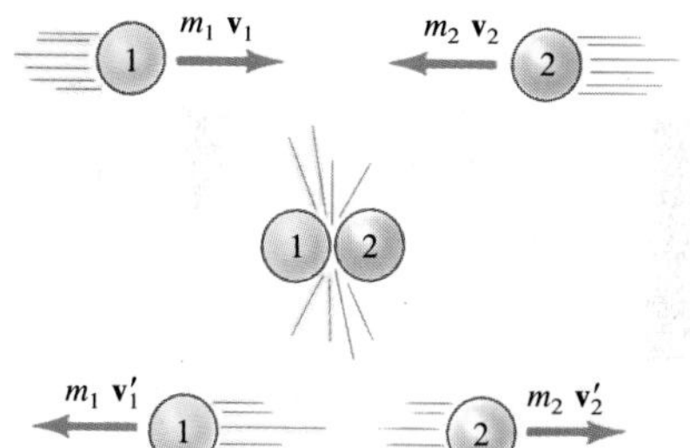

FIGURE 7–1 Momentum is conserved in a collision of two balls.

momentum before = momentum after

$$m_1\mathbf{v}_1 + m_2\mathbf{v}_2 = m_1\mathbf{v}_1' + m_2\mathbf{v}_2'. \tag{7–3}$$

Conservation of momentum (for two bodies, colliding)

That is, the total vector momentum of the two balls is conserved: it stays constant.

Although the conservation-of-momentum principle was discovered experimentally, it is closely connected to Newton's laws of motion and can in fact be derived from them. We will do the derivation for the one-dimensional case illustrated in Fig. 7–1, assuming the force F is constant over the time Δt. We use Newton's second law as expressed in Eq. 7–2 and rewrite it by multiplying both sides by Δt (since the motion is in one dimension, we don't need vector notation):

$$F\,\Delta t = \Delta p. \tag{7–4}$$

We apply Eq. 7–4 to ball 1 in Fig. 7–1, letting v_1 be its velocity before collision and v_1' its velocity after the collision:

$$F\,\Delta t = m_1 v_1' - m_1 v_1.$$

In this relation, F is the force on ball 1 exerted by ball 2, and Δt is the time the two balls are in contact during collision. When we apply Eq. 7–4 to ball 2, we note that by Newton's third law, the force on ball 2 due to ball 1 is $-F$. Thus we have

$$-F\,\Delta t = m_2 v_2' - m_2 v_2.$$

Combining these last two equations, we have

$$m_1 v_1' - m_1 v_1 = -(m_2 v_2' - m_2 v_2)$$

which tells us that any momentum lost by one ball is gained by the other. Thus the total momentum remains constant. This equation is readily rearranged to give Eq. 7–3.

The above derivation can be extended to include any number of interacting bodies. To show this in a simple way, we let $\mathbf{p}$ in Eq. 7–2 represent the total momentum of a system (that is, the vector sum of the momenta of all objects in the system). Then, if the net force $\mathbf{F}$ on the system is zero, $\Delta\mathbf{p} = 0$, so the total momentum doesn't change. Thus the general statement of the **law of conservation of momentum** is

Law of conservation of linear momentum

The total momentum of an isolated system of bodies remains constant.

By a **system**, we simply mean a set of objects that interact with each other. An **isolated system** is one in which the only forces present are those between the objects of the system: that is, there is no net external force. If external forces do act—that is, forces exerted by objects outside the system—then the momentum may not be conserved. However, if the system is redefined so as to include the other objects, then the conservation of momentum principle will apply. For example, if we take as our system a falling rock, it does not conserve momentum since an external force, the force of gravity exerted by the earth, is acting on it and its momentum changes. However, if we include the earth in the system, the total momentum of rock plus earth is conserved. (This of course means that the earth comes up to meet the ball. Since the earth's mass is so great, its upward velocity is very tiny.)

EXAMPLE 7–2 A 10,000-kg railroad car traveling at a speed of 24.0 m/s strikes an identical car at rest. If the cars lock together as a result of the collision, what is their common speed afterward? See Fig. 7–2.

FIGURE 7–2 Example 7–2.

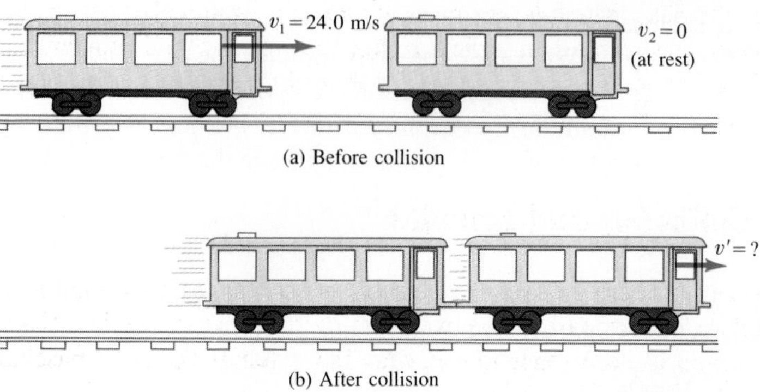

SOLUTION The initial total momentum is

$$m_1 v_1 + m_2 v_2 = (10{,}000\ \text{kg})(24.0\ \text{m/s}) + (10{,}000\ \text{kg})(0\ \text{m/s})$$
$$= 2.40 \times 10^5\ \text{kg}\cdot\text{m/s}.$$

After the collision, the total momentum will be the same but it will be shared by both cars. Since the two cars become attached, they will have the same velocity, call it v'. Then:

$$(m_1 + m_2)v' = 2.40 \times 10^5\ \text{kg}\cdot\text{m/s}$$

$$v' = \frac{2.40 \times 10^5\ \text{kg}\cdot\text{m/s}}{2.00 \times 10^4\ \text{kg}} = 12.0\ \text{m/s}.$$

The law of conservation of momentum is particularly useful when we are dealing with fairly simple systems such as collisions and certain types of explosions. For example, *rocket propulsion*, which we saw in Chapter 4 can be understood on the basis of action and reaction, can also be explained on the basis of the conservation of momentum. Before a rocket is fired, the total momentum of rocket plus fuel is zero. As the fuel burns, the total momentum remains unchanged: the backward momentum of the expelled gases is just balanced by the forward momentum of the rocket itself. Thus, a rocket can accelerate in empty space. There is no need for the expelled gases to push against the earth or the air (as is sometimes erroneously thought), as we already discussed in Chapter 4. Similar examples are the recoil of a gun and the throwing of a package from a boat.

Rocket propulsion

EXAMPLE 7–3 Calculate the recoil velocity of a 4.0-kg rifle that shoots a 0.050-kg bullet at a speed of 280 m/s, Fig. 7–3.

FIGURE 7–3 Example 7–3.

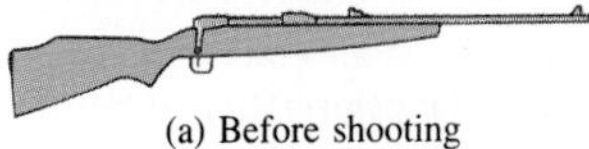

(a) Before shooting

(b) After shooting

SOLUTION The total momentum of the system is conserved. We let the subscripts B represent the bullet and R the rifle; the final velocities are indicated by primes. Then

$$m_B v_B + m_R v_R = m_B v'_B + m_R v'_R$$

$$0 + 0 = (0.050\ \text{kg})(280\ \text{m/s}) + (4.0\ \text{kg})(v'_R)$$

$$v'_R = -\frac{(0.050\ \text{kg})(280\ \text{m/s})}{(4.0\ \text{kg})}$$

$$= -3.5\ \text{m/s}.$$

Since the rifle has a much larger mass, its velocity is much less than that of the bullet. The minus sign indicates that the velocity (and momentum) of the rifle is in the direction opposite to that of the bullet. Notice that it is the *vector sum* of the momenta that is conserved.

7–3 • Collisions and Impulse

Conservation of momentum is a very useful tool for dealing with collision processes, as we already saw in the examples of the previous section. Collisions are a common occurrence in everyday life: a tennis racket, a baseball

FIGURE 7–4 Tennis racket striking a ball. Note the deformation of both ball and racket due to the large force each exerts on the other.

bat or golf club striking a ball, two billiard balls colliding, one railroad car striking another, a hammer hitting a nail. At the atomic and subatomic level, scientists learn about the structure of atoms and nuclei and their constituents, and about the nature of the forces involved, by careful study of collisions between atoms, nuclei and elementary particles.

In a collision of two ordinary objects, both objects are deformed, often considerably, because of the strong forces involved (Fig. 7–4). When the collision occurs, the force normally jumps from zero at the moment of contact to a very large value within a very short time, and then abruptly returns to zero again. A graph of the magnitude of the force one object exerts on the other during a collision, as a function of time, is typically like that shown in Fig. 7–5. The time interval Δt is usually very distinct and usually very small.

From Newton's second law, Eq. 7–2, the *net* force on an object is equal to the rate of change of its momentum:

$$\mathbf{F} = \frac{\Delta \mathbf{p}}{\Delta t}.$$

This equation applies, of course, to *each* of the objects in a collision. If we multiply both sides of this equation by the time interval Δt, we obtain

$$\mathbf{F}\,\Delta t = \Delta \mathbf{p}. \tag{7–5}$$

The quantity on the left, the product of the force **F** times the time Δt over which the force acts, is called the **impulse**. We see that the total change in momentum is equal to the impulse. The concept of impulse is of most help when dealing with forces that act over a short time, as when a bat hits a baseball. The force is generally not constant and often its variation in time is like that graphed in Fig. 7–5. It is often sufficient to approximate such a varying force by an average force $\bar{F}$ acting over a time Δt, as indicated by the dashed line in Fig. 7–6. $\bar{F}$ is chosen so that the area shown shaded in Fig. 7–6 (equal to $\bar{F} \times \Delta t$) is equal to the area under the actual curve of F vs. t (which area represents the impulse).

Example 7–4 (*a*) Calculate the impulse experienced when a 70-kg person lands on firm ground after jumping from a height of 5.0 m. Then estimate the average force exerted on the person's feet by the ground if the landing is (*b*) stiff-legged, and (*c*) with bent legs. In the former case, assume the body moves 1.0 cm during impact, and in the second case, when the legs are bent, about 50 cm.

FIGURE 7–5 Force as a function of time during a typical collision.

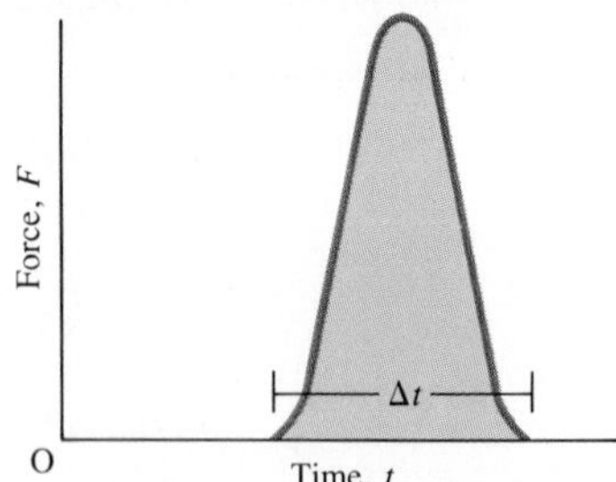

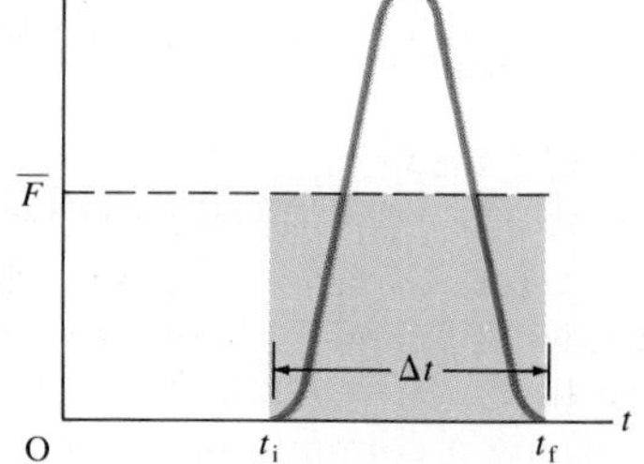

FIGURE 7–6 The average force $\bar{F}$ acting over an interval of time Δt gives the same impulse ($\bar{F}\,\Delta t$) as the actual force.

SOLUTION (*a*) Although we don't know F and thus can't calculate the impulse $F\ \Delta t$ directly, we can use the fact that the impulse equals the change in momentum of the object. We need to determine the velocity of the person just before striking the ground. We can use the appropriate kinematic equation for constant acceleration, Eq. 2–10c, in the vertical (y) direction, with $a = g = 9.8\ \text{m/s}^2$, and initial velocity $v_0 = 0$. After falling 5.0 m, the person's velocity will be

$$v = \sqrt{2a(y - y_0)} = \sqrt{2(9.8\ \text{m/s}^2)(5.0\ \text{m})} = 9.9\ \text{m/s}.$$

As the person strikes the ground, the momentum is quickly brought to zero, Fig. 7–7. The impulse on the person is then

FIGURE 7–7 Period during which impulse acts (Example 7–4).

$$\begin{aligned}\bar{F}\ \Delta t &= \Delta p = p - p_0\\ &= 0 - (70\ \text{kg})(9.9\ \text{m/s}) = -690\ \text{N}\cdot\text{s}.\end{aligned}$$

The negative sign tells us the force must be opposed to the original momentum—that is, it acts upward.

(*b*) In coming to rest, the body decelerates from 9.9 m/s to zero in a distance $d = 1.0\ \text{cm} = 1.0 \times 10^{-2}\ \text{m}$. The average speed during this period is $(9.9\ \text{m/s} + 0\ \text{m/s})/2 = 5.0\ \text{m/s}$, so the time of the collision is $\Delta t = d/\bar{v} = (1.0 \times 10^{-2}\ \text{m})/(5.0\ \text{m/s}) = 2.0 \times 10^{-3}\ \text{s}$. Since the impulse is $\bar{F}\ \Delta t = 690\ \text{N}\cdot\text{s}$, and $\Delta t = 2.0 \times 10^{-3}\ \text{s}$, the average net force $\bar{F}$ is

$$\bar{F} = \frac{690\ \text{N}\cdot\text{s}}{2.0 \times 10^{-3}\ \text{s}} = 3.5 \times 10^5\ \text{N}.$$

The force $\bar{F}$ is the *net* force on the person (we calculated it from Newton's second law). $\bar{F}$ equals the average upward force on the legs exerted by the ground, F_{grd}, minus the downward force of gravity, mg (see Fig. 7–8):

FIGURE 7–8 When the person lands on the ground, the average net force during impact is $\bar{F} = F_{\text{grd}} - mg$, where F_{grd} is the force the ground exerts upward on the person.

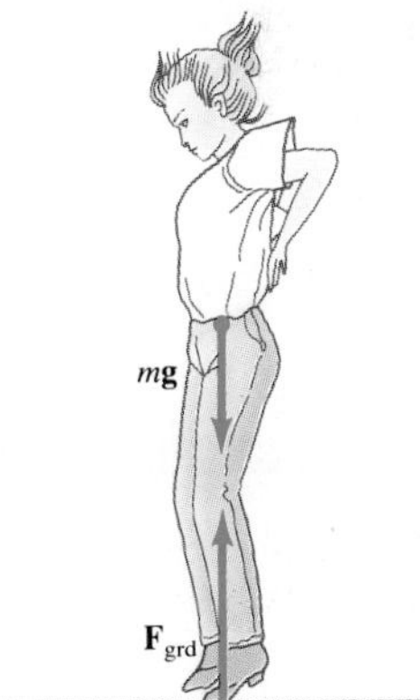

$$\bar{F} = F_{\text{grd}} - mg.$$

Since $mg = (70\ \text{kg})(9.8\ \text{m/s}^2) = 690\ \text{N}$, then

$$F_{\text{grd}} = \bar{F} + mg = 3.5 \times 10^5\ \text{N} + 690\ \text{N} \approx 3.5 \times 10^5\ \text{N}.$$

(*c*) This is just like part (*b*), except $d = 0.50\ \text{m}$, so $\Delta t = (0.50\ \text{m})/(5.0\ \text{m/s}) = 0.10\ \text{s}$, and

$$\bar{F} = \frac{690\ \text{N}\cdot\text{s}}{0.10\ \text{s}} = 6.9 \times 10^3\ \text{N}.$$

The upward force exerted on the person's feet by the earth is, as in part (*b*):

$$F_{\text{grd}} = \bar{F} + mg = 6.9 \times 10^3\ \text{N} + 0.69 \times 10^3\ \text{N} = 7.6 \times 10^3\ \text{N}.$$

Clearly, the force on the feet and legs is much less when the knees are bent. In fact, the ultimate strength of the leg bone (see Chapter 9, Table 9–2) is not great enough to support the force calculated in part (*b*), so the leg would undoubtedly break in such a landing, whereas it probably would not in part (*c*).

7–4 • Conservation of Energy and Momentum in Collisions

During most collisions, we usually don't know how the collision force varies as a function of time. Nonetheless, we can still determine a lot about the motion after a collision, given the initial motion, by making use of the conservation laws for momentum and energy. We saw in Section 7–2 that in the collision of two objects such as billiard balls, the total momentum is conserved. If the two objects are very hard and elastic and no heat is produced in the collision, then kinetic energy is conserved as well. By this we mean that the sum of the kinetic energies of the two objects is the same after the collision as before. Of course, for the brief moment during which the two objects are in contact, some (or all) of the energy is stored momentarily in the form of elastic potential energy. But if we compare the total kinetic energy before the collision with the total after the collision, they are found to be the same. Such a collision, in which the total KE is conserved, is called an **elastic collision.** If we use the subscripts 1 and 2 to represent the two objects, we can write the equation for conservation of total kinetic energy as

KE conserved in elastic collisions

$$\tfrac{1}{2}m_1v_1^2 + \tfrac{1}{2}m_2v_2^2 = \tfrac{1}{2}m_1v_1'^2 + \tfrac{1}{2}m_2v_2'^2, \qquad \text{[elastic collision]} \qquad (7\text{–}6)$$

where primed quantities ($'$) mean after the collision and unprimed mean before the collision.

Although at the atomic level the collisions of atoms and molecules are often elastic, in the "macroscopic" world of ordinary objects, an elastic collision is an ideal that is never quite reached, since at least a little thermal energy (and perhaps sound and other forms of energy) is always produced during a collision. The collision of two hard elastic balls, such as billiard balls, however, is very close to perfectly elastic, and we often treat it as such. Even when the KE is not conserved, the *total* energy is, of course, conserved.

Collisions in which kinetic energy is not conserved are said to be *inelastic.* The kinetic energy that is lost is changed into other forms of energy, often thermal energy, so that the total energy (as always) is conserved. In this case, we can write that $\text{KE}_1 + \text{KE}_2 = \text{KE}_1' + \text{KE}_2' +$ thermal and other forms of energy.

7–5 • Elastic Collisions in One Dimension—Solving Problems Using Energy and Momentum Conservation

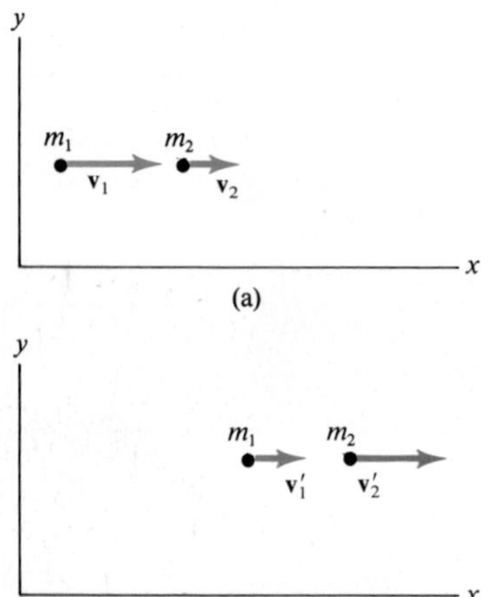

FIGURE 7–9 Two particles, of masses m_1 and m_2, (a) before the collision, and (b) after the collision.

We now apply the conservation laws for momentum and kinetic energy to an elastic collision between two particles that collide head-on, so all the motion is along a line. Let us assume both particles are initially moving with velocities v_1 and v_2 along the x axis, Fig. 7–9a. After the collision, their velocities are v_1' and v_2', Fig. 7–9b. For any $v > 0$, the particle is moving to the right (increasing x), whereas for $v < 0$, the particle is moving to the left (toward decreasing values of x).

From conservation of momentum, we have

$$m_1v_1 + m_2v_2 = m_1v_1' + m_2v_2'.$$

Momentum conservation

Because the collision is assumed to be elastic, kinetic energy is also conserved:

$$\tfrac{1}{2}m_1v_1^2 + \tfrac{1}{2}m_2v_2^2 = \tfrac{1}{2}m_1v_1'^2 + \tfrac{1}{2}m_2v_2'^2.$$

KE *conservation*

We have two equations, and therefore we can solve for two unknowns. If we know the masses and initial velocities, then we can solve these two equations for the velocities after the collision, v_1' and v_2'. Let us take an example.

EXAMPLE 7–5 A particle of mass m moving with speed v collides head on with a second particle of equal mass at rest ($v_2 = 0$). What are the speeds of the two particles after the collision, assuming it is elastic?

SOLUTION Since $v_1 = v$ and $v_2 = 0$, and $m_1 = m_2 = m$, then conservation of momentum and kinetic energy give

$$mv = mv_1' + mv_2'$$

$$\tfrac{1}{2}mv^2 = \tfrac{1}{2}mv_1'^2 + \tfrac{1}{2}mv_2'^2.$$

Since m is the same in all terms, we can divide it out, so these two equations become

$$v = v_1' + v_2'$$

$$v^2 = v_1'^2 + v_2'^2.$$

From the first equation, $v_1' = v - v_2'$, and we substitute this into the second equation:

$$v^2 = (v - v_2')^2 + v_2'^2$$
$$= v^2 - 2v_2'v + 2v_2'^2.$$

Or, after simplifying,

$$0 = v_2'(v - v_2').$$

There are two solutions: $v_2' = 0$ and $v_2' = v$. The first solution, $v_2' = 0$, corresponds to no collision at all; particle 2 is still at rest. The second solution,

$$v_2' = v,$$

is the interesting one (there is a collision). When we substitute this value of v_2' into the momentum equation, $v = v_1' + v_2'$, we find

$$v_1' = v - v_2' = v - v = 0,$$

so particle 1 is brought to rest. In summary, before the collision we have

$$v_1 = v, \qquad v_2 = 0$$

and after the collision

$$v_1 = 0, \qquad v_2 = v.$$

That is, particle 1 is brought to rest, whereas particle 2 acquires the original velocity of particle 1. This result is often observed by billiard players, and is valid only if the two balls have equal masses (and no spin is given to the balls). See Fig. 7–10.

FIGURE 7–10 In this multiflash photo of a head-on collision between two equal mass balls, the white cue ball is accelerated from rest by the cue stick and then strikes the red ball, initially at rest. The white ball stops in its tracks and the (equal mass) red ball moves off, after accelerating quickly, with the same final speed as the white ball had before the collision. See Example 7–5.

*7–6 • Elastic Collisions in Two or Three Dimensions

Conservation of momentum and energy can also be applied to collisions in two or three dimensions, and in such cases the vector nature of momentum is important. One common type of non-head-on collision is that in which a moving particle (called the "projectile") strikes a second particle initially at rest (the "target" particle). This is the common situation in games such as billiards, and for experiments in atomic and nuclear physics (the projectiles, from radioactive decay or accelerated by a high-energy accelerator, strike a stationary target nucleus).

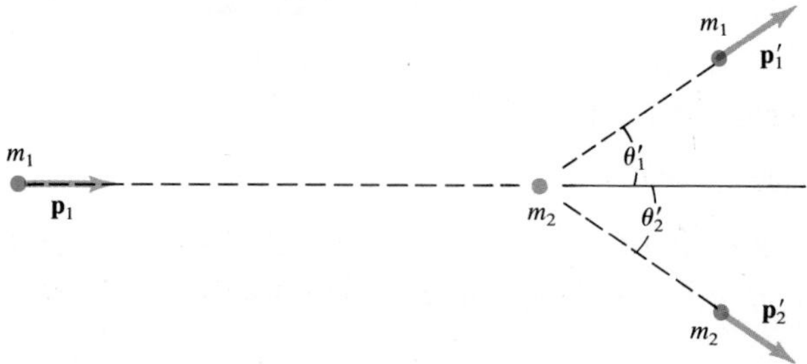

FIGURE 7–11 Particle 1, the projectile, collides with particle 2, the target. They move off, after the collision, with momenta $\mathbf{p}'_1$ and $\mathbf{p}'_2$ at angles θ'_1 and θ'_2.

Figure 7–11 shows particle 1 (the projectile, m_1) heading along the x axis toward particle 2 (the target, m_2), which is initially at rest. If these are, say, billiard balls, m_1 strikes m_2 and they go off at the angles θ'_1 and θ'_2, respectively, which are measured relative to m_1's initial direction (the x axis). If they are electrically charged particles, or nuclear particles, they may begin to deflect even before they touch because of the force (electric or nuclear) between them.[†]

Let us now apply the laws of conservation of momentum and kinetic energy to an elastic collision like that of Fig. 7–11. From conservation of kinetic energy, since $v_2 = 0$, we have

KE conserved

$$\tfrac{1}{2}m_1v_1^2 = \tfrac{1}{2}m_1v_1'^2 + \tfrac{1}{2}m_2v_2'^2. \qquad (7\text{–}7a)$$

We choose the xy plane to be the plane in which lie the initial and final momenta. Since momentum is a vector, and is conserved, its components in the x and y directions remain constant. In the x direction,

p_x conserved

$$m_1v_1 = m_1v'_1\cos\theta'_1 + m_2v'_2\cos\theta'_2. \qquad (7\text{–}7b)$$

Since there is no motion in the y direction initially, the y component of the total momentum is zero:

p_y conserved

$$0 = m_1v'_1\sin\theta'_1 + m_2v'_2\sin\theta'_2. \qquad (7\text{–}7c)$$

We have three independent equations.[‡] This means we can solve for at most three unknowns. If we are given m_1, m_2, v_1 (and v_2, if not zero), we cannot

[†] You might think, for example, of two magnets oriented so that they repel each other: when one moves toward the other, the second moves away before the first one touches it.

[‡] Note that Eqs. 7–7b and c are valid even if the collision is inelastic and KE is not conserved.

uniquely predict the final variables, v'_1, v'_2, θ'_1, and θ'_2, since there are four of them. However, if we measure one of these variables, say θ'_1, then the other three variables (v'_1, v'_2, and θ'_2) are uniquely determined and we can calculate them using the above three equations.

EXAMPLE 7–6 A proton traveling with speed 8.2×10^5 m/s collides elastically with a stationary proton in a hydrogen target. One of the protons is observed to be scattered at a 60° angle. At what angle will the second proton be observed, and what will be the velocities of the two protons after the collision?

SOLUTION Since $m_1 = m_2$, Eqs. 7–7a, b, and c become

$$v_1^2 = v_1'^2 + v_2'^2$$

$$v_1 = v'_1 \cos\theta'_1 + v'_2 \cos\theta'_2$$

$$0 = v'_1 \sin\theta'_1 + v'_2 \sin\theta'_2,$$

where $v_1 = 8.2 \times 10^5$ m/s, and $\theta'_1 = 60°$ are known. In the second and third equations, we take the v'_1 terms to the left side and square both sides of these two equations:

$$v_1^2 - 2v_1v'_1\cos\theta'_1 + v_1'^2\cos^2\theta'_1 = v_2'^2\cos^2\theta'_2$$

$$v_1'^2\sin^2\theta'_1 = v_2'^2\sin^2\theta'_2.$$

We add these two equations and use the fact that $\sin^2\theta + \cos^2\theta = 1$ (Appendix A) to get

$$v_1^2 - 2v_1v'_1\cos\theta'_1 + v_1'^2 = v_2'^2.$$

Into this equation we substitute, from the first equation above, $v_2'^2 = v_1^2 - v_1'^2$, and obtain

$$2v_1'^2 = 2v_1v'_1\cos\theta'_1$$

or

$$v'_1 = v_1\cos\theta'_1 = (8.2 \times 10^5\ \text{m/s})(\cos 60°) = 4.1 \times 10^5\ \text{m/s}.$$

To obtain v'_2, we use the first equation (conservation of KE):

$$v'_2 = \sqrt{v_1^2 - v_1'^2} = \sqrt{(8.2 \times 10^5\ \text{m/s})^2 - (4.1 \times 10^5\ \text{m/s})^2} = 7.1 \times 10^5\ \text{m/s}.$$

Finally, from the third equation:

$$\sin\theta'_2 = -\frac{v'_1}{v'_2}\sin\theta'_1 = -\left(\frac{4.1 \times 10^5\ \text{m/s}}{7.1 \times 10^5\ \text{m/s}}\right)(0.866) = -0.50,$$

so $\theta'_2 = -30°$. (The minus sign means that particle 2 moves at an angle below the x axis if particle 1 is above the axis (as in Fig. 7–11)). An example of such a collision is shown in the bubble chamber photo of Fig. 7–12. Notice that the two trajectories are at right angles to each other after the collision. This can be shown to be true in general for non-head-on collisions of two particles of equal mass, one of which was at rest initially (see the Problems).

FIGURE 7–12 Photo of a proton collision in a hydrogen bubble chamber (a device that makes visible the paths of elementary particles). The many lines represent incoming protons that can strike the protons of the hydrogen in the chamber.

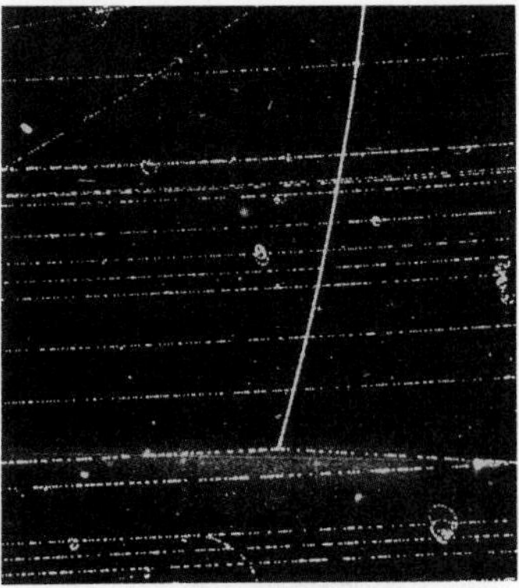

*7–7 • Inelastic Collisions

Collisions in which kinetic energy is not conserved are called **inelastic collisions**. Some of the initial kinetic energy in such collisions is transformed into other types of energy, such as thermal or potential energy, so the total final KE is less than the total initial KE. The inverse can also happen when potential energy (such as chemical or nuclear) is released, and then the total final KE can be greater than the initial KE. Typical macroscopic collisions are inelastic, at least to a small extent, and often to a large extent. If two objects stick together as a result of a collision, the collision is said to be **completely inelastic**. Two colliding balls of putty that stick together, or two railroad cars that couple together when they collide (Example 7–2), are examples. The kinetic energy is not necessarily all transformed to other forms of energy in an inelastic collision. In Example 7–2, for instance, we saw that when a traveling railroad car collided with a stationary one, the coupled cars traveled off with some KE. Even though KE is not conserved in inelastic collisions, the total energy is conserved, and the total vector momentum is also always conserved.

Completely inelastic collision

EXAMPLE 7–7 For the totally inelastic collision we considered in Example 7–2, calculate how much of the initial kinetic energy is transferred to thermal or other forms of energy.

SOLUTION Initially, the total kinetic energy is

$$\tfrac{1}{2}m_1 v_1^2 = \tfrac{1}{2}(10{,}000\text{ kg})(24.0\text{ m/s})^2 = 2.88 \times 10^6\text{ J}.$$

After the collision, the total kinetic energy is

$$\tfrac{1}{2}(20{,}000\text{ kg})(12.0\text{ m/s})^2 = 1.44 \times 10^6\text{ J}.$$

Hence the energy transformed to other forms is

$$2.88 \times 10^6\text{ J} - 1.44 \times 10^6\text{ J} = 1.44 \times 10^6\text{ J}.$$

Ballistic pendulum

EXAMPLE 7–8 The *ballistic pendulum* is a device used to measure the speed of a projectile, such as a bullet. The projectile, of mass m, is fired into a large block (of wood or other material) of mass M, which is suspended like a pendulum. (Usually, M is somewhat greater than m.) As a result of the collision, the center of mass of the pendulum-projectile system swings up to a maximum height h, Fig. 7–13. Determine the relationship between the initial speed of the projectile, v, and the height h.

FIGURE 7–13 Ballistic pendulum.

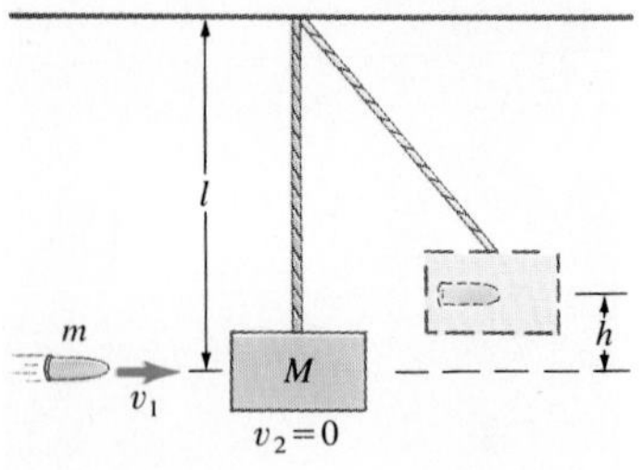

SOLUTION We analyze this process by dividing it into two parts: (1) the collision itself, and (2) the subsequent motion of the pendulum from the vertical hanging position to height h. In part (1), we assume the collision time is very short, and so the projectile comes to rest before the block has moved significantly from its position directly below its support. Thus there is no net external force and momentum is conserved:

$$mv_1 = (m + M)v',$$

where v' is the speed of the block and embedded projectile just after the collision, before they have moved significantly. Once the pendulum begins to move, there will be a net external force (gravity, tending to pull it back to the vertical position). So, for part (2), we cannot use conservation of momentum. But we can use conservation of mechanical energy since the kinetic energy immediately after the collision is changed entirely to gravitational potential energy when the pendulum reaches its maximum height, h. Therefore

$$\tfrac{1}{2}(m + M)v'^2 = (m + M)gh,$$

so $v' = \sqrt{2gh}$. (Why is the work done by the tension force in the supporting cord equal to zero?) We combine these two equations to obtain

$$v_1 = \frac{m + M}{m} v' = \frac{m + M}{m} \sqrt{2gh},$$

which is the final result. To obtain this result, we had to be opportunistic, in that we used what conservation laws we could: in (1) we could use only conservation of momentum, since the collision is inelastic and conservation of mechanical energy is not valid; and in (2), conservation of mechanical energy is valid, but not conservation of momentum. If there were significant motion of the pendulum during the deceleration of the projectile in the block, then there *would* be an external force during the collision—so conservation of momentum would not be valid, and this would have to be taken into account.

7–8 • Center of Mass

Until now, we have been mainly concerned with the motion of single particles. When we have dealt with an extended body (that is, a body that has size), we have assumed that it could be approximated as a point particle or that it underwent only translational motion. Real "extended" bodies, however, can undergo rotational and other types of motion as well. For example, the diver in Fig. 7–14a undergoes only translational motion (all parts of the body follow the same path), whereas the diver in Fig. 7–14b undergoes both translational and rotational motion. We will refer to motion that is not pure translation as *general motion.*

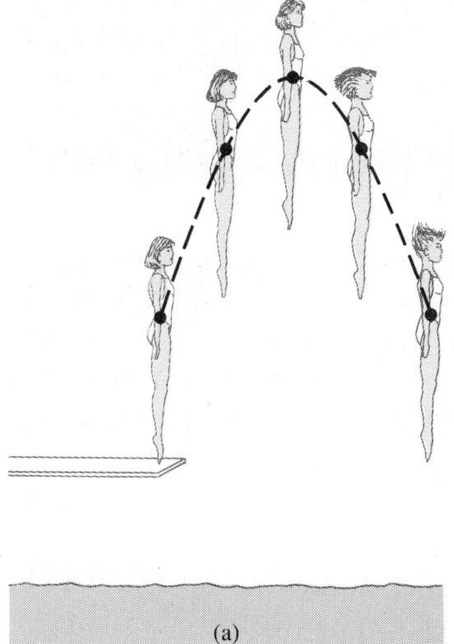

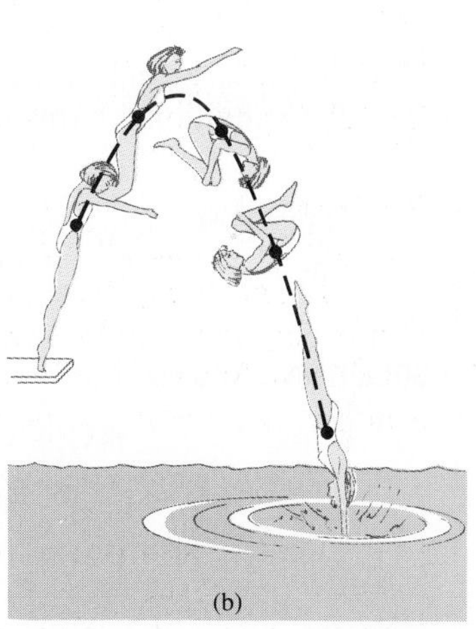

FIGURE 7–14 The motion of the diver is pure translation in (a), but is translation plus rotation in (b).

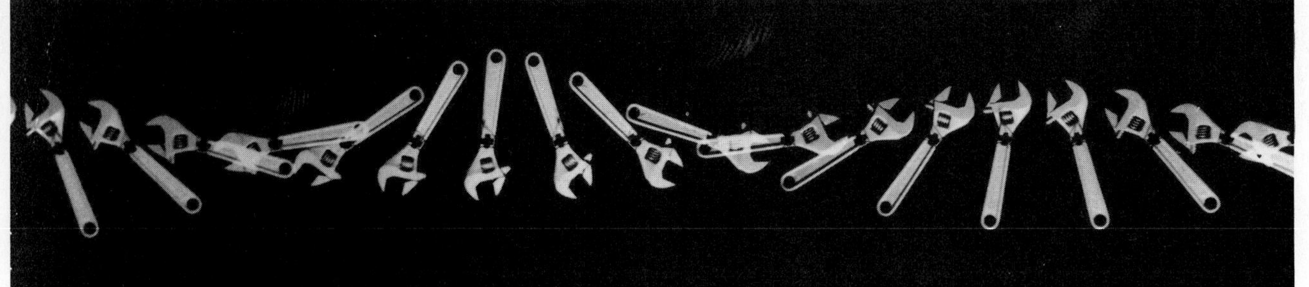

FIGURE 7–15 Translation plus rotation: a wrench moving over a horizontal surface. The cm, marked with an X, moves in a straight line.

Observations of the motion of bodies indicate that even if a body rotates, or there are several bodies that move relative to one another, there is one point that moves in the same path that a particle would if subjected to the same net force. This point is called the **center of mass** (cm). The general motion of an extended body (or system of bodies) can be considered as *the sum of the translational motion of the cm, plus rotational, vibrational, or other types of motion about the cm.*

Center of mass and general motion

As an example, consider the motion of the center of mass of the diver in Fig. 7–14: the cm follows a parabolic path even when the diver rotates, as shown in Fig. 7–14b. This is the same parabolic path that a projected particle follows when acted on only by the force of gravity (that is, projectile motion). Other points in the rotating diver's body follow more complicated paths.

Figure 7–15 shows a wrench translating and rotating along a horizontal surface—note that its cm, marked by a black X, moves in a straight line.

We will show in Section 7–9 (optional) that this important property of the cm follows from Newton's laws if the cm is defined in the following way. We can consider any extended body as being made up of many tiny particles. But first we consider a system made up of only two particles, of mass m_1 and m_2. We choose a coordinate system so that both particles lie on the x axis at positions x_1 and x_2, Fig. 7–16. The center of mass of this system is defined to be at the position, x_{cm}, given by

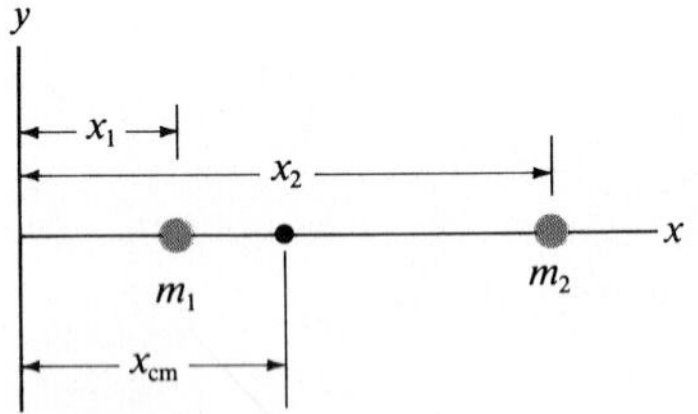

FIGURE 7–16 The center of mass of a two-particle system lies on the line joining the two masses.

Center of mass (x coordinate)

$$x_{cm} = \frac{m_1x_1 + m_2x_2}{m_1 + m_2} = \frac{m_1x_1 + m_2x_2}{M}, \tag{7–8a}$$

where $M = m_1 + m_2$ is the total mass of the system. The center of mass lies on the line joining m_1 and m_2. If the two masses are equal ($m_1 = m_2 = m$), x_{cm} is midway between them, since in this case $x_{cm} = m(x_1 + x_2)/2m = (x_1 + x_2)/2$. If one mass is greater than the other, say, $m_1 > m_2$, then the cm is closer to the larger mass. If there are more than two particles along a line, there will be additional terms in Eq. 7–8a, as the following example shows.

EXAMPLE 7–9 Three particles of equal mass m lie along the x axis at positions $x_1 = 1.0$ m, $x_2 = 5.0$ m, and $x_3 = 6.0$ m. Find the position of the cm.

SOLUTION We use Eq. 7–8a with a third term:

$$x_{cm} = \frac{mx_1 + mx_2 + mx_3}{m + m + m} = \frac{m(x_1 + x_2 + x_3)}{3m}$$

$$= \frac{(1.0\text{ m} + 5.0\text{ m} + 6.0\text{ m})}{3} = \frac{12.0\text{ m}}{3} = 4.0\text{ m}.$$

Table 7–1
Center of Mass of Parts of Typical Male Human Body[†]

Distance of Hinge Points (%)	Hinge Points (•) (Joints)	Center of Mass (×) (% Height Above Floor)		Percent Mass
91.2	Base of skull on spine	Head	93.5	6.9
81.2	Shoulder joint	Trunk and neck	71.1	46.1
	elbow 62.2	Upper arms	71.7	6.6
52.1	Hip; wrist 46.2	Lower arms	55.3	4.2
		Hands	43.1	1.7
		Upper legs (thighs)	42.5	21.5
28.5	Knee			
		Lower legs	18.2	9.6
4.0	Ankle	Feet	1.8	3.4
			58.0	100.0

[†] From *Bioastronautics Data Book*, NASA, Washington, DC.

If the particles are spread out in two or three dimensions, then we need to specify not only the x coordinate of the cm (x_{cm}), but also the y and z coordinates, which will be given by formulas just like Eq. 7–8a. For example, for two particles of mass m_1 and m_2, whose y coordinates are y_1 and y_2, respectively, the y coordinate of their cm, y_{cm}, will be:

y coordinate of center of mass

$$y_{\text{cm}} = \frac{m_1 y_1 + m_2 y_2}{m_1 + m_2} = \frac{m_1 y_1 + m_2 y_2}{M}. \qquad (7\text{–}8b)$$

For more particles, there would be more terms in this formula.

If we have a group of extended bodies, each of whose cm is known, we can find the cm of the group using Eqs. 7–8. As an example, we consider the human body. Table 7–1 indicates the cm and hinge points (joints) for the different components of a "representative" person. Of course, there are wide variations among people, so these data represent only a rough average. Note that the numbers represent a percentage of the total height, which is regarded as 100 units.

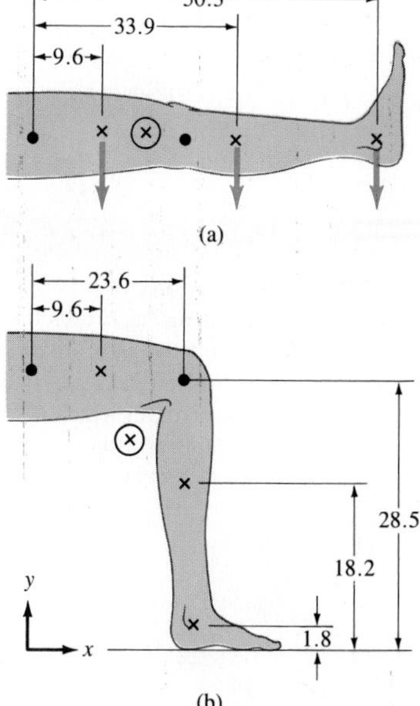

FIGURE 7–17 Example 7–10: finding the cm of a leg in two different positions (ⓧ represents the calculated cm).

Example 7–10 Calculate the cm of a whole leg (*a*) when stretched out, and (*b*) when bent at 90°, as shown in Fig. 7–17.

Solution (*a*) We will use the percentage units; that is, the person has a mass of 100 units and a height of 100 units. We measure the distance from the hip joint using Table 7–1 and obtain the numbers shown in Fig. 7–17a. Using Eq. 7–8a, we obtain

$$x_{\text{cm}} = \frac{(21.5)(9.6) + (9.6)(33.9) + (3.4)(50.3)}{21.5 + 9.6 + 3.4}$$

$$= 20.4 \text{ units.}$$

Thus, the center of mass of the leg and foot is 20.4 units below the hip joint, or $52.1 - 20.4 = 31.7$ units above the base of the foot. (*Note:* If the person is 170 cm tall, this is $31.7 \times 170\text{ cm}/100 = 54$ cm.) (*b*) In this part, we have a two-dimensional problem. We use an xy coordinate system, as shown in Fig. 7–17b. First, we calculate how far to the right of the hip joint the cm lies:

$$x_{\text{cm}} = \frac{(21.5)(9.6) + (9.6)(23.6) + (3.4)(23.6)}{21.5 + 9.6 + 3.4} = 14.9 \text{ units.}$$

Next, we calculate the distance, y_{cm}, of the cm above the floor:

$$y_{\text{cm}} = \frac{(3.4)(1.8) + (9.6)(18.2) + (21.5)(28.5)}{21.5 + 9.6 + 3.4} = 23.1 \text{ units.}$$

Thus, the cm is located 23.1 units above the floor and 14.9 units to the right of the hip joint.

cm *can be outside a body*

Note in this last example that the cm can actually lie *outside* the body. Another example is a doughnut, whose cm is at the center of the hole.

Knowing the cm of the body when it is in various positions is of great use in studying body mechanics. One simple example from athletics is shown in Fig. 7–18. If high jumpers can get into the position shown, their cm can actually pass below the bar, which means that for a particular take-off speed, they can clear a higher bar. This is indeed what they try to do.

FIGURE 7–18 The cm of a high jumper may actually pass beneath the bar.

(a)

(b)

Center of gravity

A concept similar to *center of mass* is **center of gravity** (cg). The cg of a body is that point at which the force of gravity can be considered to act. Of course, the force of gravity actually acts on all the different parts or particles of a body, but for purposes of determining the motion of a body as a whole, we can assume that the entire weight of the body (which is the sum of the weights of all its parts) acts at the cg. Strictly speaking, there is a conceptual difference between the center of gravity and the center of mass, but for practical purposes, they are generally the same point.†

FIGURE 7–19 The force of gravity, considered to act at the cg, causes the body to rotate about the pivot point unless the cg is on a vertical line directly below the pivot, in which case the body remains at rest.

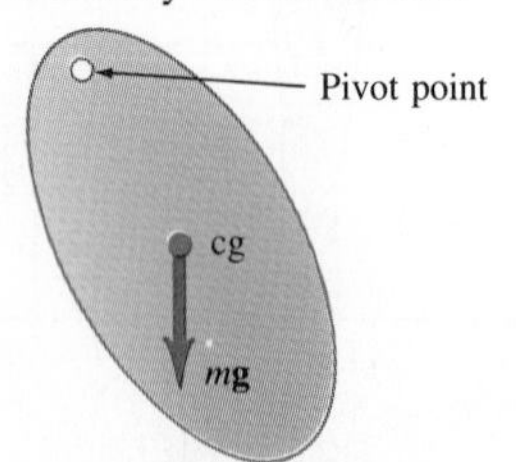

It is often easier to determine the cm or cg of an extended body experimentally rather than analytically. If a body is suspended from any point, it will swing (Fig. 7–19) unless its cm lies on a vertical line directly below the point from which it is suspended. If the object is two-dimensional or has a plane of symmetry, it need only be hung from two different pivot points and the respective vertical (plumb) lines drawn. Then the center of mass will be

† There would be a difference between the two only if a body were large enough so that the acceleration due to gravity was different at different parts of the body.

at the intersection of the two lines, as in Fig. 7–20. If the object doesn't have a plane of symmetry, the cm with respect to the third dimension is found by suspending the object from at least three points whose plumb lines do not lie in the same plane. For symmetrically shaped bodies such as uniform cylinders (wheels), spheres, and rectangular solids, the cm is located at the geometric center of the body.

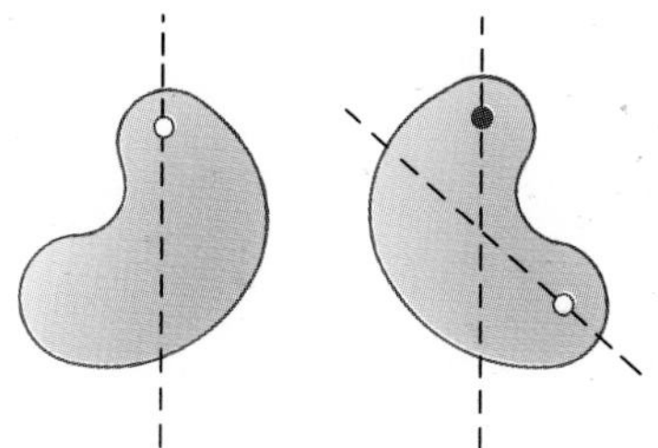

FIGURE 7–20 Finding the cm of an object.

*7–9 • Center of Mass and Translational Motion

As mentioned in the previous section, a major reason for the importance of the concept of center of mass is that the motion of the cm for a system of particles (or an extended body) is directly related to the net force acting on the system as a whole. We now show this, taking the simple case of one-dimensional motion and only three particles, but the extension to more bodies and three dimensions follows the same lines.

Suppose the three particles lie on the x axis and have masses m_1, m_2, m_3, and positions x_1, x_2, x_3. From Eq. 7–8a for the cm, we can write

$$Mx_{\text{cm}} = m_1x_1 + m_2x_2 + m_3x_3,$$

where $M = m_1 + m_2 + m_3$ is the total mass of the system. If these particles are in motion, say with velocities v_1, v_2, and v_3, respectively, along the x axis, then in a short time Δt they each will have traveled a distance:

$$\Delta x_1 = x_1' - x_1 = v_1\,\Delta t$$

$$\Delta x_2 = x_2' - x_2 = v_2\,\Delta t$$

$$\Delta x_3 = x_3' - x_3 = v_3\,\Delta t,$$

where x_1', x_2', and x_3' represent their new positions after a time Δt. The position of the new cm is given by

$$Mx_{\text{cm}}' = m_1x_1' + m_2x_2' + m_3x_3'.$$

If we subtract the two cm equations, we get

$$M\Delta x_{\text{cm}} = m_1\,\Delta x_1 + m_2\,\Delta x_2 + m_3\,\Delta x_3.$$

During the time Δt, the cm will have moved a distance

$$\Delta x_{\text{cm}} = x_{\text{cm}}' - x_{\text{cm}} = v_{\text{cm}}\,\Delta t,$$

where v_{cm} is the velocity of the cm. Into the equation just before the last one, we substitute the relations for all the Δx's:

$$Mv_{\text{cm}}\,\Delta t = m_1v_1\,\Delta t + m_2v_2\,\Delta t + m_3v_3\,\Delta t.$$

We divide out Δt and get

$$Mv_{\text{cm}} = m_1v_1 + m_2v_2 + m_3v_3. \qquad (7\text{–}9)$$

Since $m_1v_1 + m_2v_2 + m_3v_3$ is the sum of the momenta of the particles of the system, it represents the *total momentum* of the system. Thus we see from Eq. 7–9 that *the total (linear) momentum of a system of particles is equal to the product of the total mass M and the velocity of the center of mass of the system.* Or, *the linear momentum of an extended body is the product of the body's mass and the velocity of its cm.*

Total momentum

If there are forces acting on the particles, then the particles may be accelerating. In a short time Δt, each particle's velocity will change by an amount

$$\Delta v_1 = a_1\,\Delta t, \qquad \Delta v_2 = a_2\,\Delta t, \qquad \Delta v_3 = a_3\,\Delta t.$$

If we now use the same reasoning as we did to derive Eq. 7–9, we obtain

$$Ma_{\mathrm{cm}} = m_1 a_1 + m_2 a_2 + m_3 a_3.$$

According to Newton's second law, $m_1 a_1 = F_1$, $m_2 a_2 = F_2$, and $m_3 a_3 = F_3$, where F_1, F_2, and F_3 are the net forces on the three particles, respectively. Thus we get for the system as a whole:

Newton's second law for a system of particles or an extended body

$$Ma_{\mathrm{cm}} = F_1 + F_2 + F_3 = F_{\mathrm{net}}. \tag{7–10}$$

That is, *the sum of all the forces acting on the system is equal to the total mass of the system times the acceleration of its center of mass.* This is **Newton's second law** for a system of particles, and it also applies for an extended body (which can be thought of as a collection of particles). Thus we conclude that the *center of mass of a system of particles (or of an extended body) of total mass M moves like a single particle of mass M which is acted on by the same net external force.* That is, the system moves as if all its mass were concentrated at the cm and all the external forces acted at that point. We can thus treat the translational motion of any body or system of bodies as the motion of a particle (see Figs. 7–14, 7–15, and the photo at start of this Chapter). This theorem clearly simplifies our analysis of the motion of complex systems and extended bodies. Although the motion of various parts of the system may be complicated, we may often be satisfied with knowing the motion of the cm. This theorem also allows us to solve certain types of problems very easily, as illustrated by the following example.

EXAMPLE 7–11 A rocket is fired into the air as shown in Fig 7–21. At the moment it reaches its highest point, a horizontal distance D from its starting point, it separates into two parts of equal mass. Part I falls vertically to earth. Where does part II land? Assume $\mathbf{g}$ = constant.

SOLUTION The path of the cm of the system continues to follow the parabolic trajectory of a projectile acted on by a constant gravitational force. The cm will thus arrive at a point $2D$ from the origin. Since the masses of I and II are equal, the cm must be midway between them. Therefore, II lands a distance $3D$ from the origin. (If part I had been given a kick up or down, instead of merely falling, the answer would have been somewhat more complicated.)

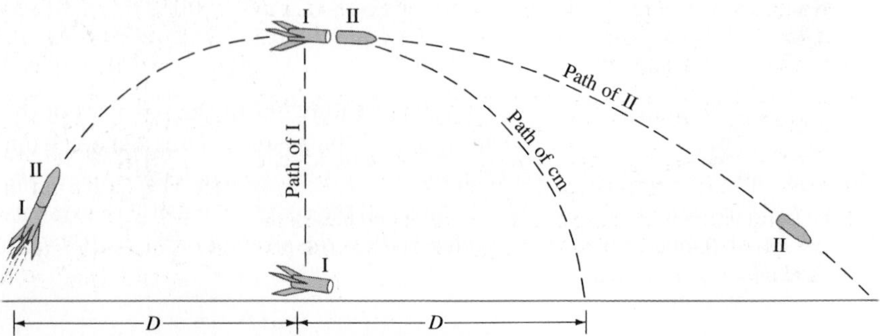

FIGURE 7–21 Example 7–11.

SUMMARY

The *momentum*, $\mathbf{p}$, of a body is defined as its mass times its velocity:

$$\mathbf{p} = m\mathbf{v}.$$

In terms of momentum, Newton's second law can be written

$$\mathbf{F} = \Delta\mathbf{p}/\Delta t.$$

That is, the rate of change of momentum equals the net applied force.

Momentum is a conserved quantity: the law of *conservation of momentum* states that the total momentum of an isolated system of objects remains constant. An isolated system is one on which the net external force is zero.

The law of conservation of momentum is very useful in dealing with *collisions*. In a collision, two (or more) bodies interact with each other for a very short time, and the force between them during this time is very large compared to any other forces acting. The *impulse* of a force on a body is defined as $\mathbf{F}\,\Delta t$, where $\mathbf{F}$ is the average force acting during the (usually short) time Δt. The impulse is equal to the change in momentum of the body: $\mathbf{F}\,\Delta t = \Delta\mathbf{p}$.

Momentum is conserved in any collision. The total energy is also conserved, but this may not be useful unless the only type of energy transformation involves kinetic energy. In such a case, kinetic energy is conserved and the collision is called an *elastic collision*. If kinetic energy is not conserved, the collision is called *inelastic*.

The *center of mass* (cm) of a body (or group of bodies) is that point at which the net force can be considered to act for purposes of determining the translational motion of the body as a whole. The complete motion of a body can be described as the translational motion of its center of mass plus rotation (or other internal motion) about its center of mass.

QUESTIONS

1. We claim that momentum is conserved. Yet most moving objects eventually slow down and stop. Explain.

2. When a person jumps from a tree to the ground, what happens to the momentum of the person upon striking the ground?

3. Explain, on the basis of conservation of momentum, how a fish propels itself forward by swishing its tail back and forth.

4. Why, when you release an inflated, untied balloon, does it fly across the room?

5. It is said that in ancient times a rich man with a bag of gold coins froze to death stranded on the surface of a frozen lake. Because the ice was frictionless, he could not push himself to shore. What could he have done to save himself had he not been so miserly?

6. How can a rocket change direction when it is far out in space and is essentially in a vacuum?

7. According to Eq. 7–5, the shorter the impact time of an impulse, the greater the force must be for the same momentum change, and hence the greater the deformation of the object on which the force acts. Explain on this basis the value of "air bags," which are intended to inflate during an automobile collision and reduce the possibility of fracture or death.

8. Is it possible for a body to receive a larger impulse from a small force than from a large force?

9. A light body and a heavy body have the same kinetic energy. Which has the greater momentum?

10. Is it possible for an object to have momentum without having energy? Can it have energy but no momentum? Explain.

11. In a collision between two cars, which would you expect to be more damaging to the occupants: if the cars collide and remain together, or if the two cars collide and rebound backward? Explain.

12. At a hydroelectric power plant, water is directed at high speed against turbine blades on an axle that turns an electric generator. Do you think the turbine blades should be designed so that the water is brought to a dead stop, or so that the water rebounds?

13. A superball is dropped from a height h onto a hard steel plate (fixed to the earth), from which it rebounds at very nearly its original speed. (*a*) Is the momentum of the ball conserved during any part of this process? (*b*) If we consider the ball and earth as our system, during what parts of the process is momentum conserved? (*c*) Answer part (*b*) for a piece of putty that falls and sticks to the steel plate.

14. If a falling ball were to make a perfectly elastic collision with the floor, would it rebound to its original height? Explain.

***15.** Use conservation of (vector) momentum to show that in a two-particle collision the paths of the particles before and after the collision all lie in one plane if: (*a*) one particle is initially at rest; (*b*) the two particles have momenta in the same or opposite directions.

16. Why do you tend to lean backward when carrying a heavy load in your arms?

17. Why is the cm of a 1-m length of pipe at its midpoint, whereas this is not true for your arm or leg?

18. Show on a diagram how your cm shifts when you change from a lying position to a sitting position.

19. Describe an analytic way of determining the cm of any triangular-shaped, thin uniform plate.

20. Explain why a uniform rectangular brick can be placed so that slightly less than half its length can be suspended over the edge of a table, but no more.

*21. A rocket following a parabolic path through the air suddenly explodes into many pieces. What can you say about the motion of this system of pieces?

*22. Analyze the motion of the spool in Fig. 7–22 if you pull the thread (*a*) directly upward, (*b*) horizontally. Consider the effect on the motion of how hard you pull.

*23. If only an external force can change the momentum of the center of mass of an object, how can the internal force of the engine accelerate a car?

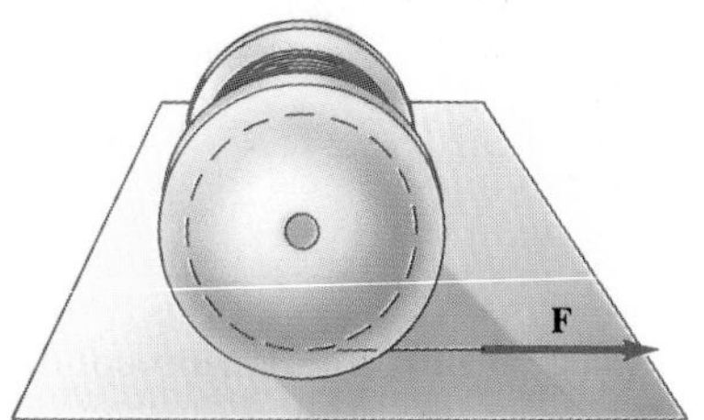

FIGURE 7–22 A spool of thread. (Question 22.)

PROBLEMS

SECTIONS 7–1 AND 7–2

1. (I) What is the momentum of a 15-g sparrow flying with a speed of 12 m/s?

2. (I) Calculate the force exerted on a rocket, given that the propelling gases are expelled at a rate of 1000 kg/s with a speed of 60,000 m/s (at takeoff).

3. (II) A child throws a 3.40-kg package horizontally from a boat with a speed of 10.0 m/s. Calculate the resulting velocity of the boat, assuming it was initially at rest. The mass of the child is 20.0 kg and that of the boat is 60.0 kg.

4. (II) A 9000-kg boxcar traveling at 20 m/s strikes a second car. The two stick together and move off with a speed of 4.0 m/s. What is the mass of the second car?

5. (II) A 15,000-kg railroad car travels alone on a level frictionless track with a constant speed of 18.0 m/s. A 5000-kg additional load is dropped onto the car. What then will be its speed?

6. (II) A 140-kg tackler moving at 3.0 m/s meets head on (and tackles) a 90-kg halfback moving at 6.5 m/s. What will be their mutual speed immediately after the collision?

7. (II) A 15-g bullet strikes and becomes embedded in a 1.24-kg block of wood placed on a horizontal surface just in front of the gun. If the coefficient of kinetic friction between the block and the surface is 0.28, and the impact drives the block a distance of 11.0 m before it comes to rest, what was the muzzle speed of the bullet?

8. (II) A gun is fired vertically into a 2.40-kg block of wood at rest directly above it. If the bullet has a mass of 22.0 g and a speed of 340 m/s, how high will the block rise into the air?

9. (II) An atomic nucleus at rest decays radioactively into an alpha particle and a smaller nucleus. What will be the speed of this recoiling nucleus if the speed of the alpha particle is 3.8×10^5 m/s? Assume the nucleus has a mass 57 times greater than that of the alpha particle.

10. (II) A 10-gram bullet traveling 400 m/s penetrates a 2.0-kg block of wood and emerges going 350 m/s. If the block is stationary on a frictionless surface when hit, how fast does it move after the bullet emerges?

11. (II) A 700-kg two-stage rocket is traveling at a speed of 6.20×10^3 m/s with respect to earth when a predesigned explosion separates the rocket into two sections of equal mass that then move with a relative speed (relative to each other) of 2.45×10^3 m/s along the original line of motion. (*a*) What is the speed and direction of each section (relative to earth) after the explosion? (*b*) How much energy was supplied by the explosion? [*Hint:* what is the change in KE as a result of the explosion?]

12. (III) A rocket of total mass 3300 kg is traveling in outer space with a velocity of 120 m/s toward the sun. It wishes to alter its course by 30.0°, and can do this by firing its rockets briefly in a direction perpendicular to its original motion. If the rocket gases are expelled at a speed of 2200 m/s, how much mass must be expelled?

SECTION 7–3

13. (I) A tennis ball may leave the racket of a top player on the serve with a speed of 65.0 m/s. If the ball's mass is 0.0600 kg and it is in contact with the racket for 0.0300 s, what is the average force on the ball? Would this force be large enough to lift an average-size person?

14. (I) A 0.145-kg baseball pitched at 35.0 m/s is hit on a horizontal line drive straight back toward the pitcher at 50.0 m/s. If the contact time between bat and ball is 5.00×10^{-4} s, calculate the average force between the ball and bat during contact.

15. (II) A 90-kg fullback is running 5.0 m/s and is stopped by a tackler in 1.0 s. Calculate (*a*) the original momentum of the fullback, (*b*) the impulse imparted to the tackler, and (*c*) the average force exerted on the tackler.

16. (II) A tennis ball of mass m and speed v strikes a wall at a 45° angle and rebounds with the same speed at 45° (Fig. 7–23). What is the impulse given the wall?

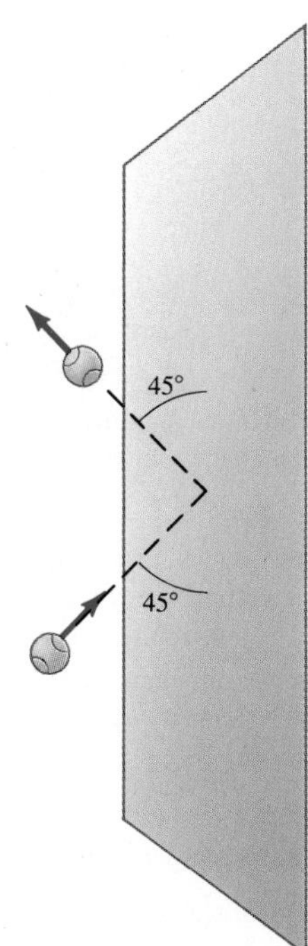

FIGURE 7–23 Problem 16.

17. (II) Suppose the force acting on a tennis ball (mass 0.060 kg) as a function of time is given by the graph of Fig. 7–24. Use graphical methods to estimate (*a*) the total impulse given the ball, and (*b*) the speed of the ball after being struck, assuming the ball is being served so it is nearly at rest initially.

FIGURE 7–24 Problem 17.

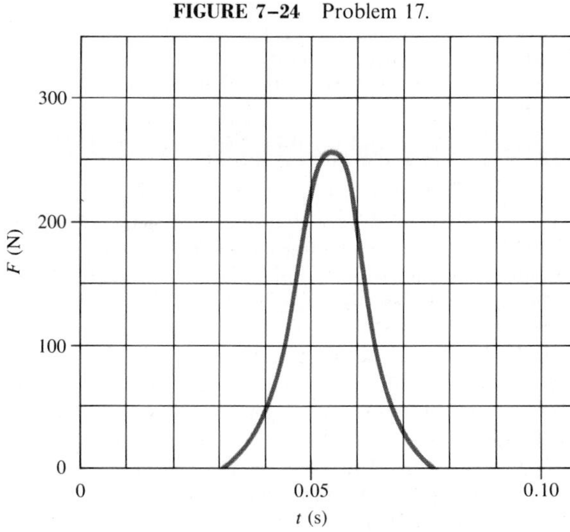

18. (III) From what maximum height can a 60-kg person jump without breaking the lower leg bone, whose cross-sectional area is $3.0 \times 10^{-4}\ \text{m}^2$? Ignore air resistance and assume the cm of the person moves a distance of 0.60 m from the standing to the seated position (that is, in breaking the fall). Assume the breaking strength (force per unit area) of bone is $170 \times 10^6\ \text{N/m}^2$.

SECTIONS 7–4 AND 7–5

19. (II) A ball of mass 0.440 kg moving with a speed of 4.50 m/s collides head-on with a 0.220-kg ball at rest. If the collision is perfectly elastic, what will be the speed and direction of each ball after the collision?
20. (II) A 0.300-kg ball, moving with a speed of 2.50 m/s, has a head-on collision with a 0.600-kg ball initially at rest. Assuming a perfectly elastic collision, what will be the speed and direction of each ball after the collision?
21. (II) Two billiard balls of equal mass undergo a perfectly elastic head-on collision. If the speed of one ball was initially 2.00 m/s, and of the other 3.00 m/s in the opposite direction, what will be their speeds after the collision?
22. (II) Suppose a heavy particle (mass m_1) has an elastic head-on collision with a very light particle of mass m_2 initially at rest. Show that if $m_1 \gg m_2$, the velocity of the projectile (m_1) is practically unchanged, whereas the target particle (m_2) acquires a velocity $v_2' = 2v_1$.
23. (II) Suppose a light particle (m_1) with velocity v_1 strikes head-on a heavy particle (m_2) at rest. If $m_1 \ll m_2$, show that the projectile (m_1) rebounds in the opposite direction with virtually the same speed (that is, $v_1' \approx -v_1$) and that the heavy target (m_2) acquires very little velocity, assuming the collision is elastic.
24. (III) A 0.460-kg ball makes an elastic head-on collision with a second ball initially at rest. The second ball moves off with half the original speed of the first ball. (*a*) What is the mass of the second ball? (*b*) What fraction of the original kinetic energy gets transferred to the second ball?

*SECTION 7–6

*25. (II) A radioactive nucleus at rest decays into a second nucleus, an electron, and a neutrino. The electron and neutrino are emitted at right angle and have momenta of 8.60×10^{-23} kg·m/s and 6.20×10^{-23} kg·m/s, respectively. What is the magnitude and direction of the momentum of the recoiling nucleus?

*26. (II) A billiard ball of mass $m_A = 0.400$ kg moving with speed $v_A = 1.80$ m/s strikes a second ball, initially at rest, of mass $m_B = 0.400$ kg. As a result of the collision, the first ball is deflected off at an angle of 30.0° with a speed $v_A' = 1.10$ m/s. (*a*) Taking the x axis to be the original direction of motion of ball A, write down the equations expressing the conservation of momentum for the components in the x and y directions separately. (*b*) Solve these equations for the speed, v_B', and angle, θ, of ball B. Do not assume the collision is elastic.

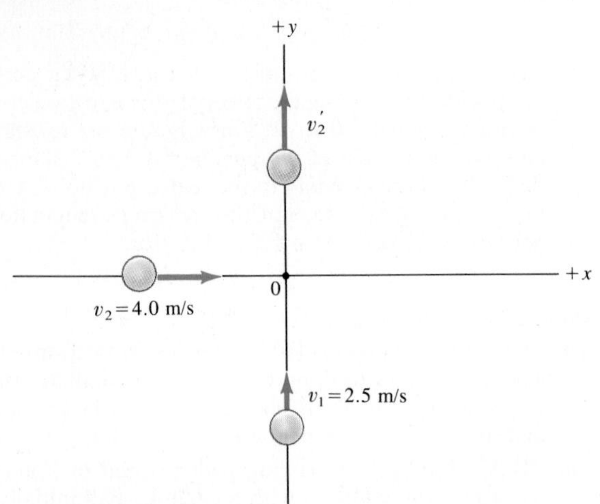

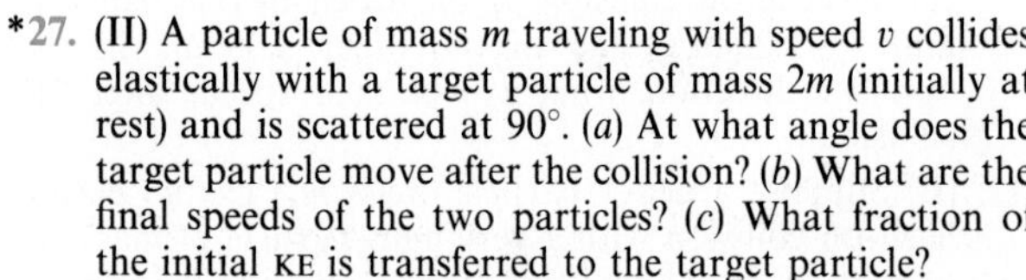

FIGURE 7–25 Problem 28. (Ball 1 after the collision is not shown.)

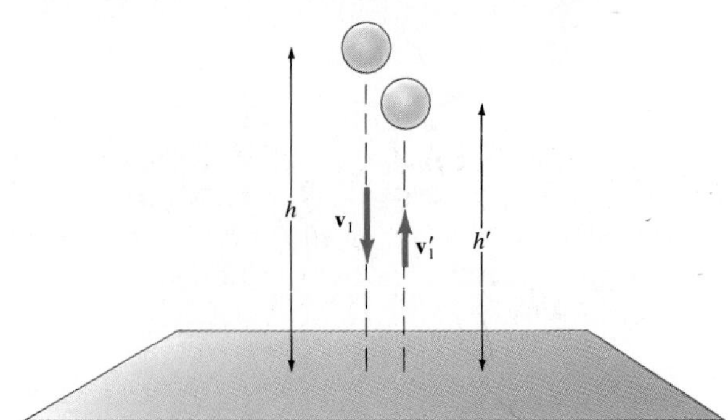

FIGURE 7–26 Problem 35. Measurement of coefficient of restitution.

*27. (II) A particle of mass m traveling with speed v collides elastically with a target particle of mass $2m$ (initially at rest) and is scattered at 90°. (*a*) At what angle does the target particle move after the collision? (*b*) What are the final speeds of the two particles? (*c*) What fraction of the initial KE is transferred to the target particle?

*28. (III) Two billiard balls of equal mass move at right angles and meet at the origin of an xy coordinate system. One is moving upward along the y axis at 2.5 m/s, and the other is moving to the right along the x axis with speed 4.0 m/s. After the collision (assumed elastic), the second ball is moving along the positive y axis (Fig. 7–25). What is the final direction of the first ball, and what are their two speeds?

*29. (III) A billiard ball of mass $m_A = 0.40$ kg strikes a second ball, initially at rest, of mass $m_B = 0.60$ kg. As a result of this collision, ball A is deflected at an angle of 30° and ball B at 53°. What is the ratio of their speeds after the collision?

*30. (III) A neutron collides elastically with a helium nucleus (at rest initially) whose mass is four times that of the neutron. The helium nucleus is observed to rebound at an angle $\theta_2' = 45°$. Determine the angle of the neutron, θ_1', and the speeds of the two particles, v_n' and v_{He}', after the collision. The neutron's initial speed is 4.5×10^5 m/s.

*SECTION 7–7

*31. (II) A 20-gram rifle bullet traveling 250 m/s buries itself in a 3.8-kg pendulum hanging on a 2.3-m-long string. How far does the pendulum swing horizontally?

*32. (II) (*a*) Derive a formula for the fraction of kinetic energy lost, ΔKE/KE, for the ballistic pendulum collision of Example 7–8. (*b*) Evaluate for $m = 10.0$ g and $M = 240$ g.

*33. (II) An explosion breaks an object into two pieces, one of which has 1.5 times the mass of the other. If 4500 J were released in the explosion, how much kinetic energy did each piece acquire?

*34. (II) An eagle ($m_1 = 6.8$ kg) moving with speed $v_1 = 9.6$ m/s is on a collision course with a second eagle ($m_2 = 8.3$ kg) moving at $v_2 = 8.4$ m/s in a direction at right angles to the first. After they collide, they hold onto one another. In what direction, and with what speed, are they moving after the collision?

*35. (II) A measure of inelasticity in a head-on collision of two bodies is the *coefficient of restitution*, e, defined as

$$e = \frac{v_1' - v_2'}{v_2 - v_1},$$

where $v_1' - v_2'$ is the relative velocity of the two bodies after and $v_2 - v_1$ is their relative velocity before the collision. (*a*) Show that for a perfectly elastic collision, $e = 1$, and for a completely inelastic collision, $e = 0$. (*b*) A simple method for measuring the coefficient of restitution for a body colliding with a very hard surface like steel is to drop the body onto a heavy steel plate, as shown in Fig. 7–26. Determine a formula for e in terms of the original height h and the maximum height h' reached after collision.

*36. (II) In the decay $^{232}_{92}\text{U} \rightarrow {}^{228}_{90}\text{Th} + {}^{4}_{2}\text{He}$, the $^{4}_{2}$He nucleus ($m = 4.0$ u) is emitted with a kinetic energy of 5.3 MeV (1 MeV = 1 million electron volts = 1.6×10^{-13} J; see Section 17–4). Assuming the $^{232}_{92}$U nucleus is at rest ($m = 232$ u), determine the KE of the recoiling $^{228}_{90}$Th nucleus ($m = 228$ u) and the total energy released in the decay.

*37. (III) After a completely inelastic collision between two objects of equal mass, each having initial speed, v, the two move off together with speed $v/3$. What was the angle between their initial directions?

*38. (III) A 3.0-kg body moving in the $+x$ direction at 6.0 m/s collides with a 2.0-kg body moving in the $-x$ direction at 4.0 m/s. Find the final velocity of each mass if: (*a*) the bodies stick together; (*b*) the collision is elastic; (*c*) the 3.0-kg body is at rest after the collision; (*d*) the 2.0-kg body is at rest after the collision; (*e*) the 3.0-kg body has a velocity of 4.0 m/s in the $-x$ direction after the collision. Are the results in (*c*), (*d*), and (*e*) "reasonable"? Explain.

SECTION 7–8

39. (I) The distance between a carbon atom ($m = 12$ u) and an oxygen atom ($m = 16$ u) in the CO molecule is 1.13×10^{-10} m. How far from the oxygen atom is the center of mass of the molecule?

40. (I) An empty 1200-kg car has its cm 3.10 m from the front of the car. How far from the front of the car will

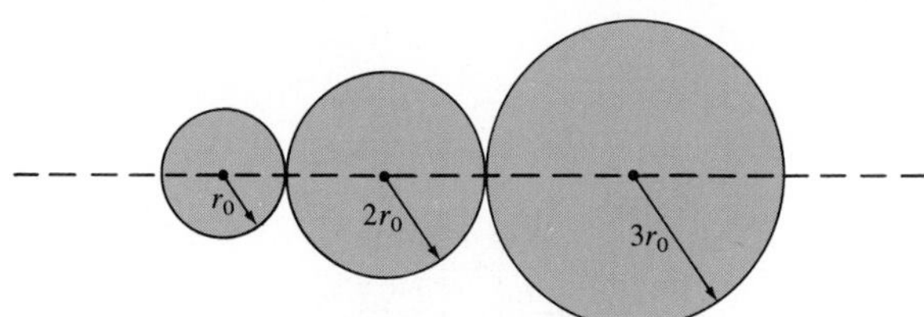

FIGURE 7–27 Problem 45.

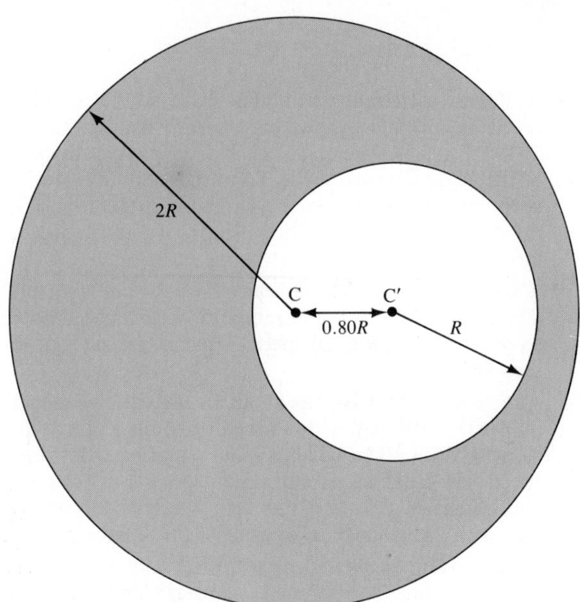

FIGURE 7–28 Problem 47.

the cm be when two people sit in the front seat 2.60 m from the front of the car and three in the back seat 3.85 m from the front? Assume that each person has a mass of 65 kg.

41. (I) Assume that your proportions are the same as those in Table 7–1, and calculate the mass of one of your legs.

42. (I) Determine the cm of an outstretched arm using Table 7–1.

43. (II) Use Table 7–1 to calculate the position of the cm of an arm bent at a right angle. Assume that the person is 170 cm tall.

44. (II) Calculate how far below the torso's median line the cm will be when a jumper is in a position such that his arms and legs are hanging vertically, and the trunk and head are horizontal. Will this be outside the body? Use Table 7–1.

45. (II) Three spheres, of radius r_0, $2r_0$, and $3r_0$, are placed next to one another (in contact) with their centers along a straight line and the $r = 2r_0$ sphere in the center (Fig. 7–27). Where is the cm of this system? Assume the spheres are made of the same uniform material.

46. (III) A square uniform raft, 30 by 30 m, of mass 5400 kg, is used as a ferryboat carrying cars northward across a river. (*a*) If three cars, each of mass 1000 kg, occupy the NE, SE, and SW corners, determine the cm of the loaded ferryboat. (*b*) If the car in the SW corner accelerates northward at 0.80 m/s^2 relative to the raft, where will the cm be after 4.0 s?

47. (III) A uniform circular plate of radius $2R$ has a circular hole of radius R cut out of it. The center of the smaller circle is a distance $0.80R$ from the center of the larger circle, Fig. 7–28. What is the position of the center of mass of the plate? [*Hint:* try subtraction.]

*SECTION 7–9

***48.** (II) The masses of the earth and moon are 5.98×10^{24} kg and 7.36×10^{22} kg, respectively, and their centers are separated by 3.85×10^{8} m. (*a*) Where is the cm of this system located? (*b*) What can you say about the motion of the earth-moon system about the sun, and of the earth and moon separately about the sun?

***49.** (II) A 50-kg girl and a 70-kg boy stand 8.0 m apart on frictionless ice. (*a*) How far from the girl is their cm? (*b*) If they hold on to the two ends of a rope, and the boy pulls on the rope so that he moves 2.0 m, how far from the girl will he be now? (*c*) How far will the boy have moved when he collides with the girl?

***50.** (II) (*a*) Suppose that in Example 7–11 (Fig. 7–21), $m_{II} = 3m_I$. Where then would m_{II} land? (*b*) What if $m_I = 3m_{II}$?

***51.** (III) A helium balloon and its gondola, of mass M, are in the air and stationary with respect to the ground. A passenger, of mass m, then climbs out and slides down a rope with speed v, measured with respect to the balloon. With what speed and direction (relative to earth) does the balloon then move? What happens if the passenger stops?

GENERAL PROBLEMS

52. The air in a 100-km/h wind strikes a 30.0 m × 40.0 m face of a building at a rate of 4.30×10^4 kg/s. Calculate the net force on the building, assuming the air is brought to rest on impact.

53. A 6000-kg open railroad car moves with a constant speed of 15.0 m/s on a level frictionless track. Snow begins to fall vertically and fills the car at a rate of 4.00 kg/min. What is the speed of the car after 80.0 min?

54. A particle of mass m traveling with speed v_0 along the x axis suddenly shoots out one-third of its mass parallel to the y axis with speed $2v_0$. Give the components of the velocity of the remainder of the particle.

55. The decay of a neutron into a proton, an electron, and a neutrino is an example of a three-particle decay process. Use the vector nature of momentum to show that if the neutron is initially at rest, the velocity vectors of the three must be coplanar (that is, all in the same plane). The result is not true for numbers greater than three.

56. A 150-kg astronaut (including space suit) acquires a speed of 2.35 m/s by pushing off with his legs from a 2000-kg space capsule. (*a*) What is the change in speed of the space capsule? (*b*) If the push lasts 0.200 s, what is the average force exerted by each on the other? As the reference frame, use the position of the capsule before the push.

57. A ball of mass m makes a head-on elastic collision with a second ball (at rest) and rebounds with a speed equal to one-third its original speed. What is the mass of the second ball?

58. A meteor whose mass was about 10^8 kg struck the earth ($m = 6.0 \times 10^{24}$ kg) with a speed of about 15 km/s and came to rest in the earth. (*a*) What was the earth's recoil speed? (*b*) What fraction of the meteor's kinetic energy was transformed to KE of the earth? (*c*) By how much did the earth's KE change as a result of this collision?

59. An explosion breaks an object, originally at rest, into two fragments. One fragment acquires twice the kinetic energy of the other. What is the ratio of their masses?

60. The force on a bullet is given by the formula $F = 220 - 1.8 \times 10^5 t$ over the time interval $t = 0$ to $t = 3.0 \times 10^{-3}$ s. In this formula, t is in seconds and F is in newtons. (*a*) Plot a graph of F vs. t for $t = 0$ to $t = 3.0$ ms. (*b*) Estimate, using graphical methods, the impulse given the bullet. (*c*) If the bullet achieves a speed of 250 m/s as a result of this impulse, given to it in the barrel of a gun, what must its mass be?

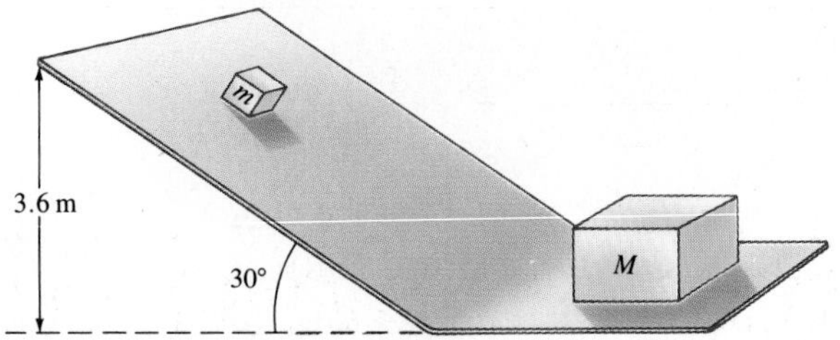

FIGURE 7–29 Problems 61 and 62.

61. A block of mass $m = 1.50$ kg slides down a 30.0° incline which is 3.60 m high. At the bottom, it strikes a block of mass $M = 6.00$ kg which is at rest on a horizontal surface, Fig. 7–29. (Assume a smooth transition at the bottom of the incline.) If the collision is elastic, and friction can be ignored, determine (*a*) the speeds of the two blocks after the collision, and (*b*) how far back up the incline the smaller mass will go.

62. In Problem 61 (Fig. 7–29), what is the upper limit on the mass m if it is to rebound from M and return down the incline and collide with M again?

CHAPTER 8

Rotational Motion

You too can experience rotation—if your stomach can take the high angular velocity and centripetal acceleration. This rotating "Wheel" has rotational KE as well as angular momentum. The riders are in a rotating reference frame.

We have, until now, been concerned with translational motion. In this chapter, we will deal with rotational motion. We will mainly be concerned with rigid bodies. By a **rigid body** we mean a body with a definite shape that doesn't change, so that the particles composing it stay in fixed positions relative to one another. Of course, any real body is capable of vibrating or deforming when a force is exerted on it. But these effects are often very small, so the concept of an ideal rigid body is very useful as a good approximation.

The motion of a rigid body (as mentioned in Chapter 7) can be analyzed as the translational motion of its center of mass, plus rotational motion about its center of mass. We have already discussed translational motion in detail, so now we focus our attention on purely rotational motion. By *purely rotational motion*, we mean that all points in the body move in circles, such as the point P in the rotating wheel of Fig. 8–1, and that the centers of these circles all lie on a line called the **axis of rotation** (which in Fig. 8–1 is perpendicular to the page and passes through point O).

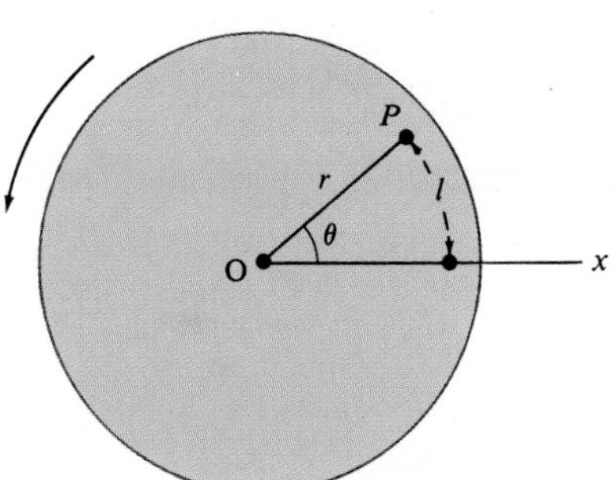

FIGURE 8–1 Looking down on a wheel that is rotating counterclockwise about an axis through the wheel's center at O (perpendicular to the page).

8–1 • Angular Quantities

To describe rotational motion, we make use of angular quantities, such as angular velocity and angular acceleration. These are defined in analogy to the corresponding quantities in linear motion.

Every point in a body rotating about a fixed axis moves in a circle whose center is on the axis and whose radius is r, the perpendicular distance of that point from the axis of rotation. A perpendicular line drawn from the axis to any point sweeps out the same angle θ in the same time. To indicate the position of the body, or how far it has rotated, we specify the angle θ of some particular line in the body with respect to some reference line, such as the x axis (see Fig. 8–1). A point in the body (such as P in Fig. 8–1) moves through an angle θ when it travels the distance l measured along the circumference of its circular path. Angles are commonly measured in degrees, but the mathematics of circular motion is much simpler if we use the *radian* for angular measure. One **radian** (rad) is defined as the angle subtended by an arc whose length is equal to the radius. For example, in Fig. 8–1, point P is a distance r from the axis of rotation and it has moved a distance l along the arc of a circle. If $l = r$, then θ is exactly equal to 1 rad. In general, any angle θ is given by

θ in radians

$$\theta = \frac{l}{r}, \tag{8–1}$$

where r is the radius of the circle, and l is the arc length subtended by the angle θ which is specified in radians. Radians can be related to degrees in the following way. In a complete circle there are 360°, which of course must correspond to an arc length equal to the circumference of the circle, $l = 2\pi r$. Thus $\theta = l/r = 2\pi r/r = 2\pi$ rad in a complete circle, so

$$360° = 2\pi \text{ rad}.$$

1 rad ≈ 57.3°

One radian is therefore $360°/2\pi \approx 360°/6.28 \approx 57.3°$.

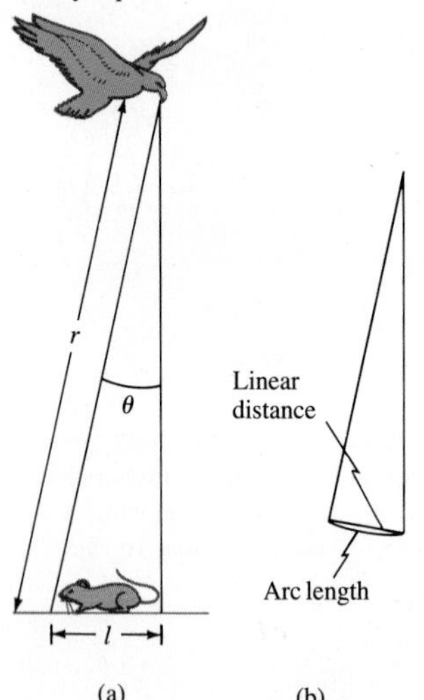

FIGURE 8–2 (a) Example 8–1. (b) For small angles, arc length and the linear distance (the chord) are nearly equal.

EXAMPLE 8–1 A particular bird's eye can just distinguish objects that subtend an angle no smaller than about 3×10^{-4} rad. (How many degrees is this?) How small an object can the bird just distinguish when flying at a height of 100 m (Fig. 8–2a)?

SOLUTION From Eq. 8–1, $l = r\theta$. Strictly speaking, l is the arc length, but for small angles, the linear distance subtended is approximately† the same (Fig. 8–2b). Since $r = 100$ m and $\theta = 3 \times 10^{-4}$ rad, we find

$$l = (100 \text{ m})(3 \times 10^{-4} \text{ rad}) = 3 \times 10^{-2} \text{ m} = 3 \text{ cm}.$$

Had the angle been given in degrees, we would first have had to change it to radians to make this calculation.

† Even for an angle as large as 15°, the error in making this estimate is only 1 percent, but for larger angles the error increases rapidly.

Note in this example that we used the fact that the radian is dimensionless (has no units) since it is the ratio of two lengths.

Angular velocity is defined in analogy with ordinary linear velocity. Instead of distance traveled, we use the angular distance θ. Thus the **average angular velocity** (denoted by ω, the Greek lowercase letter omega) is defined as

$$\bar{\omega} = \frac{\theta}{t}, \tag{8–2a}$$

where θ is the angle through which the body has rotated in the time t. We define the **instantaneous angular velocity** as the small angle, $\Delta\theta$, through which the body turns in the very short time interval Δt:

$$\omega = \frac{\Delta\theta}{\Delta t}. \qquad [\Delta t \text{ very small, approaching zero}] \tag{8–2b}$$

Angular velocity

Angular velocity is generally specified in radians per second. Note that all points in the body rotate with the same angular velocity. This follows since every position in the body moves through the same angle in the same time interval.

Angular acceleration, in analogy to ordinary linear acceleration, is defined as the change in angular velocity divided by the time required to make this change. The **average angular acceleration** (denoted by α, the Greek lowercase letter alpha) is defined as

$$\bar{\alpha} = \frac{\omega - \omega_0}{t}, \tag{8–3a}$$

where ω_0 is the angular velocity initially, and ω is the angular velocity after the time t has passed. **Instantaneous angular acceleration** is defined in the usual way:

$$\alpha = \frac{\Delta\omega}{\Delta t}. \qquad [\Delta t \text{ very small, approaching zero}] \tag{8–3b}$$

Angular acceleration

Since ω is the same for all points of a rotating body, Eq. 8–3 tells us that α also will be the same for all points. Thus, ω and α are properties of the rotating body as a whole. With ω measured in radians per second and t in seconds, α will be expressed as radians per second squared (rad/s^2).

Each particle or point of a rotating rigid body has, at any moment, a linear velocity v and linear acceleration a. We can relate these linear quantities, v and a, of each particle, to the angular quantities, ω and α, of the rotating body as a whole. Consider a particle located a distance r from the axis of rotation, as in Fig. 8–1. If the body rotates with angular velocity ω, any particle will have a linear velocity whose direction is tangent to its circular path. The magnitude of its linear velocity, v, is $v = \Delta l/\Delta t$. From Eq. 8–1, a change in rotation angle $\Delta\theta$ is related to the linear distance traveled by $\Delta l = r\,\Delta\theta$. Hence

$$v = \frac{\Delta l}{\Delta t} = r\frac{\Delta\theta}{\Delta t}$$

$$v = r\omega. \tag{8–4}$$

Linear and angular velocity related

Thus, although ω is the same for every point in the rotating body at any instant, the linear velocity v is greater for points farther from the axis. Note that Eq. 8–4 is valid both instantaneously and on the average.

We can use Eq. 8–4 to show that the angular acceleration α is related to the tangential linear acceleration a_T of a particle by

$$a_T = \frac{\Delta v}{\Delta t} = r\frac{\Delta \omega}{\Delta t}$$

Tangential acceleration

$$a_T = r\alpha. \tag{8–5}$$

In this equation, r is the radius of the circle in which the particle is moving, and the subscript T in a_T stands for "tangential" since the acceleration considered here is along the circle (that is, tangent to it).

The total linear acceleration of a particle is

$$\mathbf{a} = \mathbf{a}_T + \mathbf{a}_c,$$

where the radial† component, $\mathbf{a}_c$, the "centripetal acceleration," points toward the center of the particle's circular path. We saw in Chapter 5 (Eq. 5–1) that $a_c = v^2/r$, and we rewrite this in terms of ω using Eq. 8–4:

Centripetal acceleration

$$a_c = \frac{v^2}{r} = \omega^2 r. \tag{8–6}$$

Equations 8–4, 8–5, and 8–6 relate the angular quantities describing the rotation of a body to the linear quantities for each particle of the body.

It is sometimes useful to relate the angular velocity ω to the frequency of rotation, f. By **frequency**, we mean the number of complete revolutions (rev) per second. One revolution (of, say, a wheel) corresponds to an angle of 2π radians, and thus 1 rev/s = 2π rad/s. Hence, in general, the frequency f is related to the angular velocity ω by

$$f = \frac{\omega}{2\pi}$$

or

$$\omega = 2\pi f. \tag{8–7}$$

The time required for one complete revolution is called the **period**, T, and it is related to the frequency by

$$T = \frac{1}{f}. \tag{8–8}$$

For example, if a particle rotates at a frequency of three revolutions per second, then each revolution takes $\frac{1}{3}$ s.

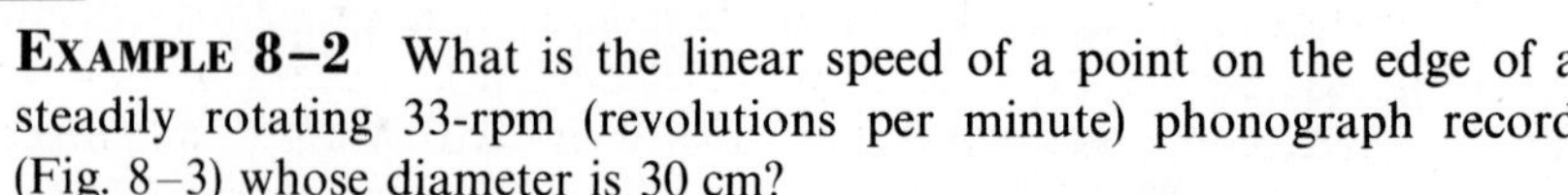

EXAMPLE 8–2 What is the linear speed of a point on the edge of a steadily rotating 33-rpm (revolutions per minute) phonograph record (Fig. 8–3) whose diameter is 30 cm?

FIGURE 8–3 Disc jockey and 33 rpm record. Example 8–2.

†"Radial" means along the radius–that is, toward or away from the center or axis.

SOLUTION First, we find the angular velocity in radians per second: the frequency $f = 33\ \text{rev/min} = 33\ \text{rev}/60\ \text{s} = 0.55\ \text{rev/s}$; then $\omega = 2\pi f = (6.28)(0.55\ \text{rev/s}) = 3.5\ \text{rad/s}$. The radius r is 0.15 m, so the speed v at the edge is

$$v = r\omega = (0.15\ \text{m})(3.5\ \text{rad/s}) = 0.52\ \text{m/s}.$$

EXAMPLE 8–3 What is the magnitude of the acceleration of a speck of dust on the edge of the record in Example 8–2?

SOLUTION From Example 8–2, $\omega = 3.5\ \text{rad/s}$ and $v = 0.52\ \text{m/s}$. Since $\omega = \text{constant}$, then $\alpha = 0$ and the tangential component of the linear acceleration (Eq. 8–5) is

$$a_{\text{T}} = r\alpha = 0.$$

From Eq. 8–6:

$$a_{\text{c}} = \omega^2 r = (3.5\ \text{rad/s})^2(0.15\ \text{m}) = 1.8\ \text{m/s}^2;$$

or $a_{\text{c}} = v^2/r = (0.52\ \text{m/s})^2/(0.15\ \text{m}) = 1.8\ \text{m/s}^2$, which is the same result. What force causes this acceleration? Is it a force of friction exerted by the record?

EXAMPLE 8–4 A centrifuge rotor is accelerated from rest to 20,000 rpm in 5.0 min. What is its average angular acceleration?

SOLUTION Initially, $\omega = 0$. The final angular velocity is

$$\omega = (20{,}000\ \text{rev/min})\left(\frac{2\pi\ \text{rad/rev}}{60\ \text{s/min}}\right) = 2100\ \text{rad/s}.$$

Then, since $\bar{\alpha} = \Delta\omega/\Delta t$, we have

$$\bar{\alpha} = \frac{2100\ \text{rad/s} - 0}{300\ \text{s}} = 7.0\ \text{rad/s}^2.$$

That is, every second it increases its angular velocity by 7.0 rad/s, or by $(7.0/2\pi)$ revolutions per second.

8–2 • Kinematic Equations for Uniformly Accelerated Rotational Motion

In Chapter 2, we derived the important equations (2–10) that relate acceleration, velocity, and distance for the situation of uniform linear acceleration. Those equations were derived from the definitions of linear velocity and acceleration, assuming constant acceleration. The definitions of angular velocity and angular acceleration are the same as for their linear counterparts, except that θ has replaced the linear displacement x, ω has replaced v, and α has replaced a. Therefore, the angular equations for **constant angular acceleration** will be analogous to Eqs. 2–10 with x replaced by θ, v by ω, and a by α, and they can be derived in exactly the same way. We summarize them here, opposite their linear equivalents:

Uniformly accelerated motion

Angular	Linear		
$\omega = \omega_0 + \alpha t$	$v = v_0 + at$	[constant α, a]	(8–9a)
$\theta = \omega_0 t + \frac{1}{2}\alpha t^2$	$x = v_0 t + \frac{1}{2}at^2$	[constant α, a]	(8–9b)
$\omega^2 = \omega_0^2 + 2\alpha\theta$	$v^2 = v_0^2 + 2ax$	[constant α, a]	(8–9c)
$\bar{\omega} = \dfrac{\omega + \omega_0}{2}$	$\bar{v} = \dfrac{v + v_0}{2}$	[constant α, a]	(8–9d)

Note that ω_0 represents the angular velocity at $t = 0$, whereas θ and ω represent the angular position and velocity, respectively, at time t. Since the angular acceleration is constant, $\alpha = \bar{\alpha}$. These equations are valid of course also for constant angular velocity, for which case $\alpha = 0$ and we have $\omega = \omega_0$, $\theta = \omega_0 t$, and $\bar{\omega} = \omega$.

EXAMPLE 8–5 Through how many turns has the centrifuge rotor of Example 8–4 turned during its acceleration period? Assume constant angular acceleration.

SOLUTION We know that $\omega_0 = 0$, $\omega = 2100$ rad/s, $\alpha = \bar{\alpha} = 7.0$ rad/s^2, and $t = 300$ s. We could use either Eq. 8–9b or 8–9c. The former gives

$$\theta = 0 + \tfrac{1}{2}(7.0 \text{ rad/s}^2)(300 \text{ s})^2 = 3.2 \times 10^5 \text{ rad}.$$

To find the total number of revolutions, we divide by 2π and obtain 5.0×10^4 revolutions. (To decide whether to multiply or to divide by 2π, it helps to remember that there are more radians than revolutions, since 2π rad = 1 rev.)

8–3 • Torque

We have so far discussed rotational kinematics—the description of rotational motion in terms of angle, angular velocity, and angular acceleration. Now we discuss the dynamics, or causes, of rotational motion. Just as we found analogies between linear and rotational motion for the description of motion, so rotational equivalents for dynamics exist as well. For example, the rotational equivalent of Newton's first law states that a freely rotating body will continue to rotate with constant angular velocity as long as no net force (or rather, as we shall see shortly, no net torque) acts to change that motion. More difficult is the question of a rotational equivalent for Newton's second law. That is, what gives rise to angular acceleration? To make an object start rotating about an axis clearly requires a force. But the direction of this force, and where it is applied, are also important. Take, for example, an ordinary situation such as the door in Fig. 8–4. If you apply a force $\mathbf{F}_1$ to the door as shown, you will find that the greater the magnitude, F_1, the more quickly the door opens. But now if you apply the same magnitude force at a point closer to the hinge, say $\mathbf{F}_2$ in Fig. 8–4, you will find that the door will not open so quickly. The effect of the force is less.

FIGURE 8–4 Applying the same force with different lever arms, r_1 and r_2.

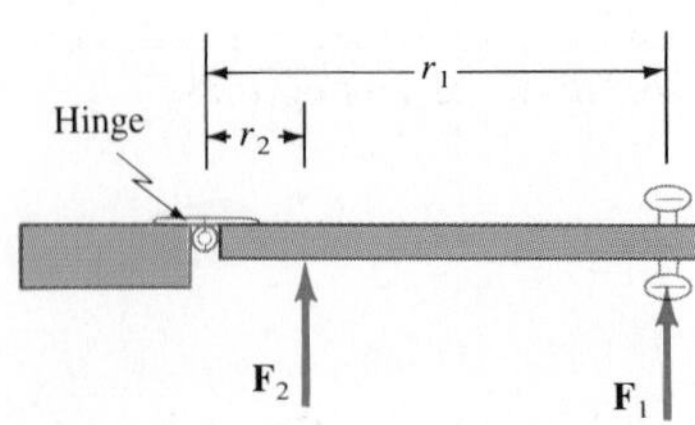

FIGURE 8–5 A plumber can exert greater torque using a wrench with a long lever arm.

So the angular acceleration of the door is proportional not only to the magnitude of the force, but, assuming for the moment that only this one force acts (we ignore friction in the hinges, and so on), it is also proportional to *the perpendicular distance from the axis of rotation to the line along which the force acts.* Thus, if the distance r_1 in Fig. 8–4 is three times larger than r_2, then the angular acceleration of the door will be three times as great, assuming of course that the magnitudes of the forces are the same. To say it another way, if $r_1 = 3r_2$, then F_2 must be three times as large as F_1 to give the same angular acceleration. The distances r_1 and r_2 are called the *lever arms* or *moment arms* of the respective forces. (See Fig. 8–5.)

The angular acceleration, then, will be proportional to the product of the force times the lever arm. This product is called the *moment of the force* about the axis, or, more commonly, it is called the **torque**, and is abbreviated τ (Greek lowercase letter tau). The angular acceleration α of an object is found to be directly proportional to the net applied torque, τ:

Torque defined

$$\alpha \propto \tau.$$

Thus, we see that it is torque that gives rise to angular acceleration. This is the rotational analog of Newton's second law for linear motion, $a \propto F$. (In Section 8–4, we will see what factor is needed to make this proportionality an equation.)

Lever arm

We define the **lever arm** as the *perpendicular* distance of the axis of rotation from the line of action of the force—that is, the distance which is perpendicular both to the axis of rotation and to an imaginary line drawn along the direction of the force. We do this to take into account the effect of forces acting at an angle. It is clear that a force applied at an angle, such as $\mathbf{F}_3$ in Fig. 8–6, will be less effective than the same magnitude force applied straight on, such as $\mathbf{F}_1$ (Fig. 8–6a). And if you push on the end of the door so that the force is directed at the hinge (the axis of rotation), as indicated by $\mathbf{F}_4$, the door will not move at all.

The lever arm for a force such as $\mathbf{F}_3$ is found by drawing a line along the direction of $\mathbf{F}_3$ (this is the "line of action" of $\mathbf{F}_3$), and then drawing another line from the axis of rotation perpendicular to the first line (and also perpendicular to the rotation axis). The length of this second line is the lever arm for $\mathbf{F}_3$, and is labeled r_3 in Fig. 8–6b.

The torque associated with $\mathbf{F}_3$ is then r_3F_3. This short lever arm and the corresponding smaller torque associated with $\mathbf{F}_3$ is consistent with the observation that $\mathbf{F}_3$ is less effective in accelerating the door than is $\mathbf{F}_1$. When the lever arm is defined in this way, experiment shows that the relation $\alpha \propto \tau$ is valid in general. Notice in Fig. 8–6 that the line of action of the force $\mathbf{F}_4$ passes through the hinge and hence its lever arm is zero. Consequently, zero

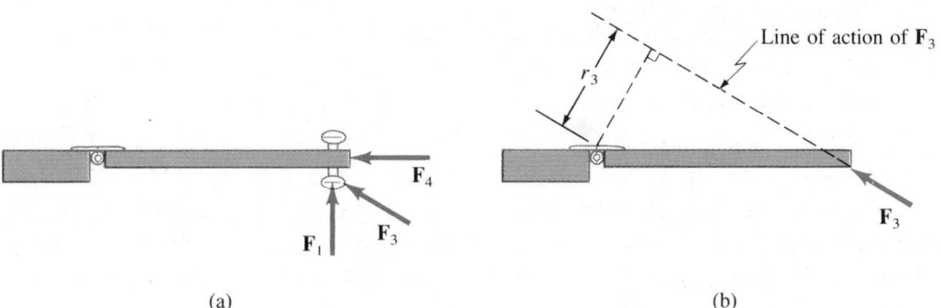

FIGURE 8–6 (a) Forces acting at different angles. (b) The lever arm is defined as the perpendicular distance from the axis of rotation to the line of action of the force.

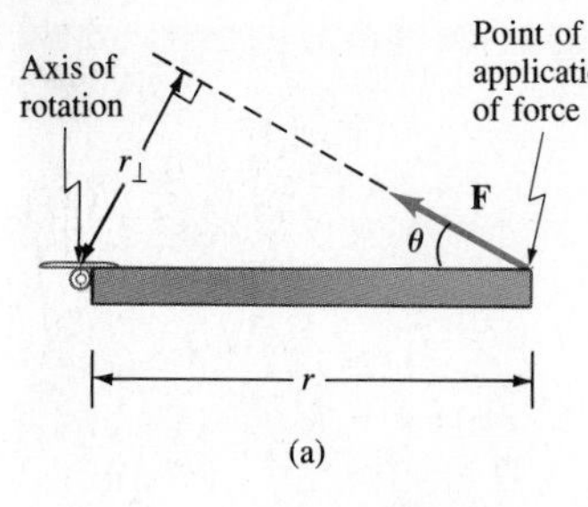

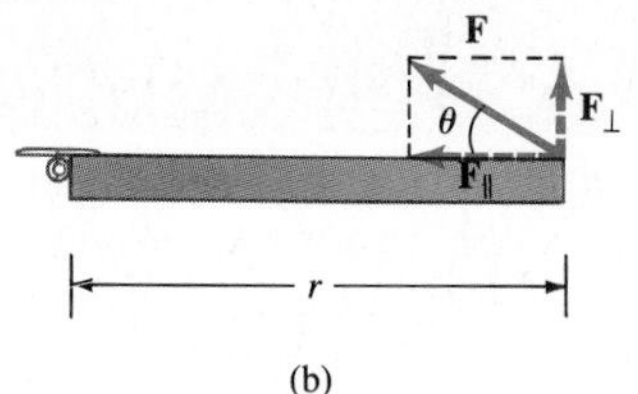

FIGURE 8–7 Torque $= r_\perp F = rF_\perp$.

torque is associated with $\mathbf{F}_4$ and it gives rise to no angular acceleration, in accord with everyday experience.

In general, then, we can write the torque about a given axis as

$$\tau = r_\perp F, \tag{8–10a}$$

where $r_\perp$ is the lever arm, and the perpendicular symbol ($\perp$) reminds us that we must use the distance from the axis of rotation that is perpendicular to the line of action of the force (Fig. 8–7a). An alternate but equivalent way of determining the torque associated with a force is to resolve the force into components parallel and perpendicular to a line joining the point of application of the force to the axis, as shown in Fig. 8–7b. Then the torque will be equal to $F_\perp$ times the distance r from the axis to the point of application of the force:

$$\tau = rF_\perp. \tag{8–10b}$$

That this gives the same result as Eq. 8–10a can be seen from the fact that $F_\perp = F \sin\theta$ and $r_\perp = r \sin\theta$. So

Magnitude of a torque

$$\tau = rF \sin\theta \tag{8–10c}$$

in either case. We can use any of Eqs. 8–10 to calculate the torque, whichever is easiest.

Since torque is a force times a distance, it is measured in units of N·m in SI units,[†] dyne·cm in the cgs system, and lb·ft in the English system.

FIGURE 8–8 Example 8–6.

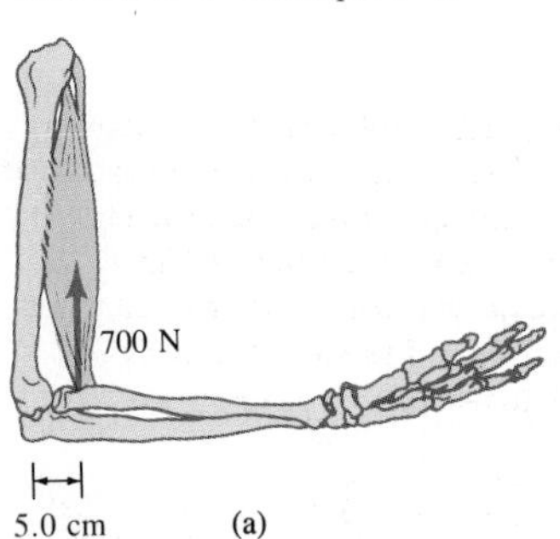

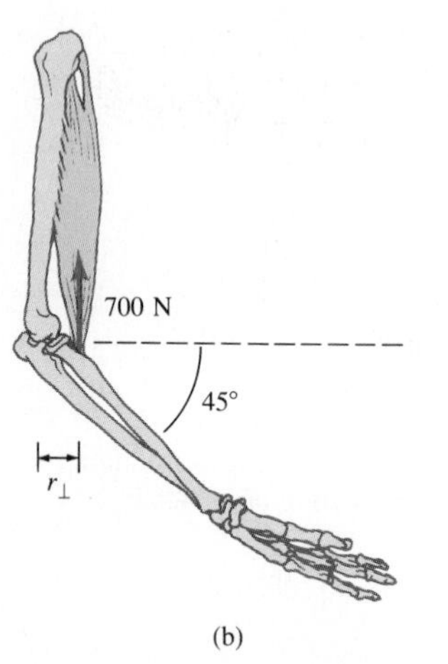

EXAMPLE 8–6 What is the torque applied by the biceps muscle on the lower arm in Figs. 8–8a and b? The axis of rotation is through the elbow joint, and the muscle is attached 5.0 cm from the elbow joint.

SOLUTION (*a*) $F = 700$ N and $r_\perp = 0.050$ m, so $\tau = (0.050 \text{ m})(700 \text{ N}) = 35$ N·m. (*b*) Because the arm is at a 45° angle, the lever arm is shorter (Fig. 8–8b): $r_\perp = (0.050 \text{ m})(\sin 45°)$. F is still 700 N, so

$$\tau = (0.050 \text{ m})(0.71)(700 \text{ N}) = 25 \text{ N·m}.$$

When more than one torque acts on a body, the acceleration α is found to be proportional to the *net* torque. If all the torques acting on a body tend to rotate it in the same direction, the net torque is the sum of the torques. But if, say, one torque acts to rotate a body in one direction, and a second torque acts to rotate the body in the opposite direction (as in Fig. 8–9), the net torque is the difference of the two torques. We assign a positive sign to torques that act to rotate the body in one direction (say counterclockwise) and a negative sign to torques that act to rotate the body in the opposite direction (clockwise).

[†] Note that the unit for torque (N·m in SI) is the same as that for energy. But the two quantities are very different. An obvious difference is that energy is a scalar, whereas torque, as we shall see, is a vector. The special name *joule* (1 J = 1 N·m) is used only for energy (and for work), never for torque.

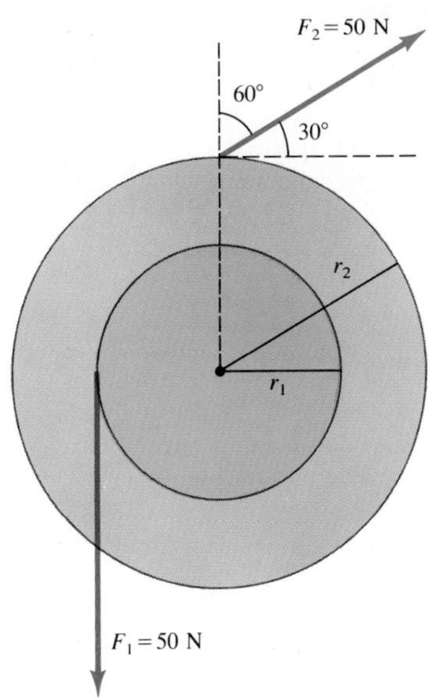

FIGURE 8–9 Example 8–7. The torque due to $\mathbf{F}_1$ tends to rotate the wheel counterclockwise, whereas the torque due to $\mathbf{F}_2$ tends to rotate the wheel clockwise.

EXAMPLE 8–7 Two thin cylindrical wheels, of radii $r_1 = 30$ cm and $r_2 = 50$ cm, are attached to each other on an axle that passes through the center of each, as shown in Fig. 8–9. Calculate the net torque on the wheel due to the two forces shown.

SOLUTION The force $\mathbf{F}_1$ acts to rotate the system counterclockwise, whereas $\mathbf{F}_2$ acts to rotate it clockwise. So the two forces act in opposition to each other. We must choose one direction of rotation to be positive—say, counterclockwise. Then $\mathbf{F}_1$ exerts a positive torque, $\tau_1 = r_1 F_1$, since the lever arm is r_1. $\mathbf{F}_2$, on the other hand, produces a negative (clockwise) torque and does not act perpendicular to r_2, so we must use its perpendicular component to calculate the torque it produces: $\tau_2 = -r_2 F_{2\perp} = -r_2 F_2 \sin\theta$, where $\theta = 60°$. (Note that θ must be the angle between $\mathbf{F}_2$ and a radial line from the axis.) Hence the net torque is

$$\begin{aligned}\tau &= r_1 F_1 - r_2 F_2 \sin 60° \\ &= (0.30\text{ m})(50\text{ N}) - (0.50\text{ m})(50\text{ N})(0.866) = -6.6\text{ N}\cdot\text{m}.\end{aligned}$$

This net torque acts to rotate the wheel clockwise.

[Since we are interested only in rotation about a fixed axis, we consider only forces that act in a plane perpendicular to the axis of rotation. If there is a force (or component of a force) acting parallel to the axis of rotation, it will tend to turn the axis of rotation—the component $\mathbf{F}_{\|}$ in Fig. 8–10 is an example. Since we are assuming the axis remains fixed in direction, either there can be no such forces or else the axis must be an axle (or the like) which is mounted in bearings or hinges that exert a compensating torque to keep the axis fixed. Thus, only a force, or component of a force ($\mathbf{F}_{\perp}$ in Fig. 8–10), in a plane perpendicular to the axis will give rise to rotation about the axis, and it is only these that we consider.]

FIGURE 8–10 Only the component of **F** that acts in the plane perpendicular to the rotation axis, $\mathbf{F}_{\perp}$, acts to turn the wheel about the axis. The component parallel to the axis, $\mathbf{F}_{\|}$, would tend to move the axis itself, which we assume is fixed.

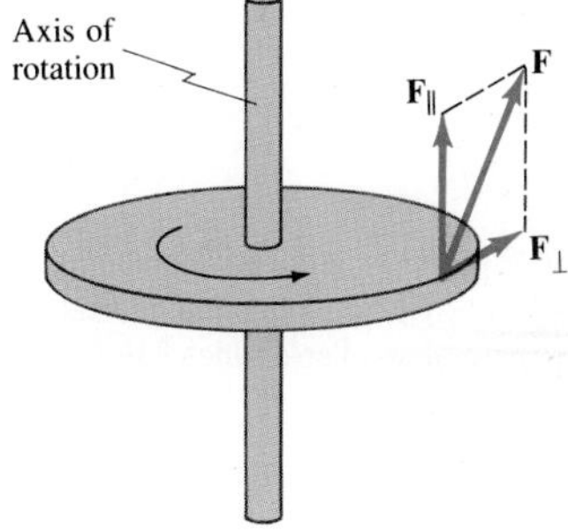

8–4 • Rotational Dynamics; Torque and Rotational Inertia

We have mentioned that the angular acceleration α of a rotating body is proportional to the torque τ applied to it:

$$\alpha \propto \tau.$$

This corresponds to Newton's second law for translational motion, $a \propto F$, where torque has taken the place of force, and, correspondingly, the angular acceleration α takes the place of the linear acceleration a. In the linear case, the acceleration is not only proportional to the net force, but is also inversely proportional to the inertia of the body, which we call its mass, m. Thus we could write $a = F/m$. But what plays the role of mass for the rotational case? That is what we now set out to determine. At the same time, we will see that the relation $\alpha \propto \tau$ follows directly from Newton's second law, $F = ma$.

We first consider a very simple case: a particle of mass m rotating in a circle of radius r at the end of a string or rod whose mass we can ignore

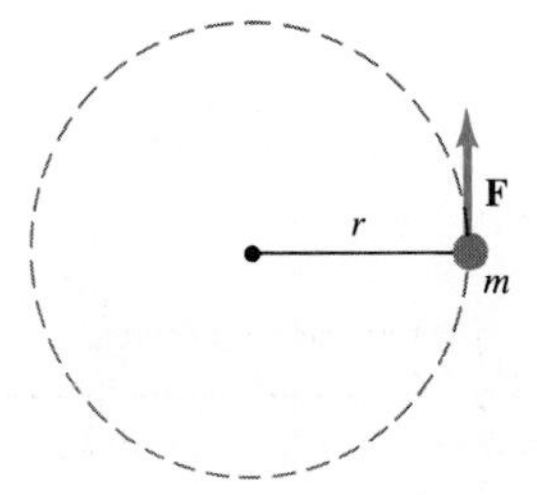

FIGURE 8–11 A mass m rotating in a circle of radius r about a fixed point.

(Fig. 8–11). The torque that gives rise to its angular acceleration is $\tau = rF$. If we make use of Newton's second law for linear quantitites, $F = ma$, and Eq. 8–5 relating the angular acceleration to the tangential linear acceleration, $a_T = r\alpha$, we have

$$F = ma$$
$$= mr\alpha.$$

When we multiply both sides by r, we find that the torque $\tau = rF$ is given by

$$\tau = mr^2\alpha. \qquad \text{[single particle]} \qquad (8\text{–}11)$$

Here at last we have a direct relation between the angular acceleration and the applied torque τ. The quantity mr^2 represents the *rotational inertia* of the particle and is called its *moment of inertia.*

Now let us consider a rotating rigid body, such as a wheel rotating about an axis through its center, such as an axle. We can think of the wheel as consisting of many particles located at various distances from the axis of rotation. We can apply Eq. 8–11 to each particle of the body, and then sum over all the particles. The sum of the various torques is just the total torque, which we now call τ, so we obtain:

$$\tau = (\sum mr^2)\alpha \qquad (8\text{–}12)$$

where we factored out the α since it is the same for all the particles of the body. The $\sum$ (Greek capital letter sigma) means "the sum of." The sum, $\sum mr^2$, represents the sum of the masses of each particle in the body multiplied by the square of the distance of that particle from the axis of rotation. If we give each particle a number (1, 2, 3, . . .), then $\sum mr^2 = m_1r_1^2 + m_2r_2^2 + m_3r_3^2 + \cdots$. This quantity is called the **moment of inertia** (or *rotational inertia*) of the body, I:

Moment of inertia

$$I = \sum mr^2. \qquad (8\text{–}13)$$

Combining Eqs. 8–12 and 8–13, we can write

Newton's second law for rotation

$$\tau = I\alpha. \qquad (8\text{–}14)$$

This is the rotational equivalent of Newton's second law. It is valid for the rotation of a rigid body about a fixed axis. (It can be shown that Eq. 8–14 is valid also when the body is translating with acceleration, but only if I and α are calculated about the center of mass of the body, and the rotation axis through the cm doesn't change direction.)

FIGURE 8–12 A large-diameter wheel has greater rotational inertia than one of smaller diameter but equal mass.

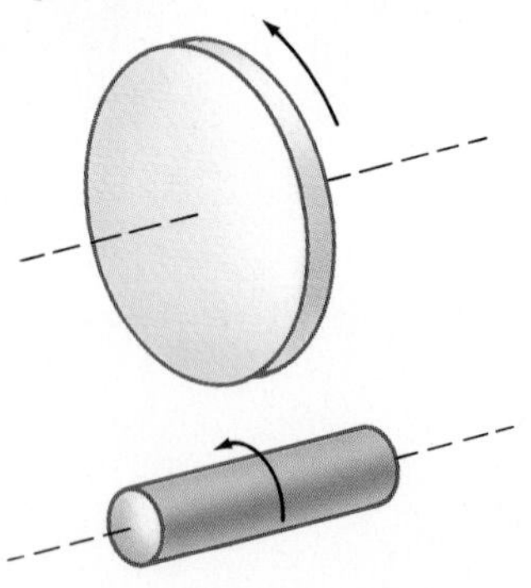

We see that the moment of inertia, I, which is a measure of the rotational inertia of a body, plays the same role for rotational motion that mass does for translational motion. As can be seen from Eq. 8–13, the rotational inertia of an object depends not only on its mass, but also on how that mass is distributed with respect to the axis. For example, a large-diameter cylinder will have greater rotational inertia than one of equal mass but smaller diameter (and therefore greater length), Fig. 8–12. The former will be harder to start rotating, and harder to stop. When the mass is concentrated further from the axis of rotation, the rotational inertia is greater. For rotational motion, the mass of a body *cannot* be considered as concentrated at its center of mass.

8–5 • Solving Problems in Rotational Dynamics

Whenever Eq. 8–14 is used, it must be remembered to give α in rad/s^2 and other quantities in a consistent set of units. The moment of inertia, I, has units of kg·m^2 in SI units.

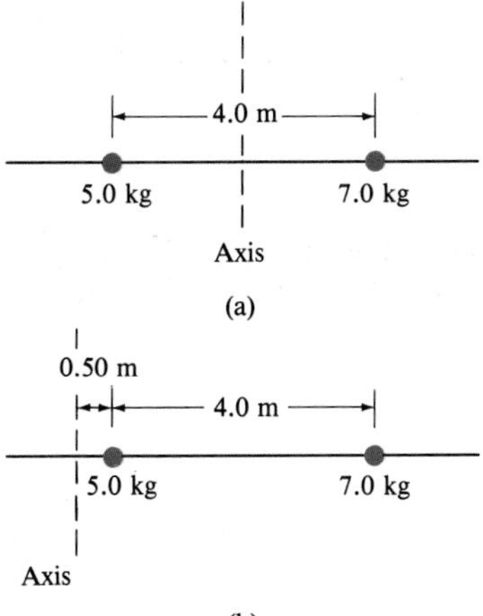

FIGURE 8–13 Example 8–8: Calculating the moment of inertia.

EXAMPLE 8–8 Two particles, of mass 5.0 kg and 7.0 kg, are mounted 4.0 m apart on a light rod (whose mass can be ignored), as shown in Fig. 8–13. Calculate the moment of inertia of the system (*a*) when rotated about an axis passing halfway between the masses, Fig. 8–13a, and (*b*) when the system rotates about an axis located 0.50 m to the left of the 5.0-kg mass (Fig. 8–13b).

SOLUTION (*a*) Both particles are the same distance, 2.0 m, from the axis of rotation. Thus

$$I = \sum mr^2 = (5.0\text{ kg})(2.0\text{ m})^2 + (7.0\text{ kg})(2.0\text{ m})^2 = 48\text{ kg}\cdot\text{m}^2.$$

(*b*) The 5.0-kg mass is now 0.50 m from the axis and the 7.0-kg mass is 4.50 m from the axis. Then

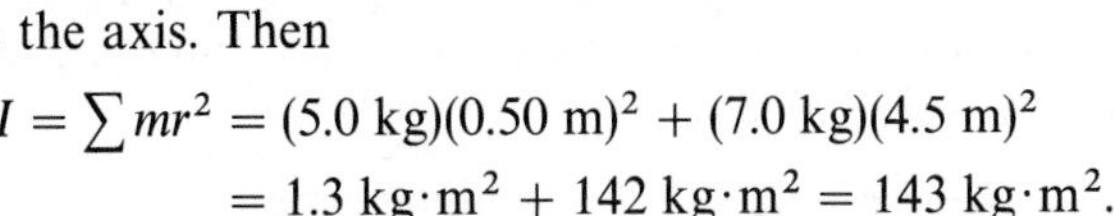

$$I = \sum mr^2 = (5.0\text{ kg})(0.50\text{ m})^2 + (7.0\text{ kg})(4.5\text{ m})^2$$
$$= 1.3\text{ kg}\cdot\text{m}^2 + 142\text{ kg}\cdot\text{m}^2 = 143\text{ kg}\cdot\text{m}^2.$$

I depends on axis of rotation and on distribution of mass

The above Example illustrates two important points. First, the moment of inertia of a given system is different for different axes of rotation. Second, we see in part (*b*) that mass close to the axis of rotation contributes little to the total moment of inertia; in this example, the 5.0-kg object contributed less than 1 percent to the total.

For most ordinary bodies, the mass is distributed continuously and calculation of the moment of inertia, $\sum mr^2$, can be difficult. Expressions can, however, be worked out (using calculus) for the moments of inertia of regularly shaped bodies in terms of their dimensions. Figure 8–14 gives these

	Object	Location of axis		Moment of inertia	Radius of gyration
(a)	Thin ring of radius R	Through center	Axis	MR^2	R
(b)	Uniform cylinder of radius R	Through center	Axis	$\frac{1}{2}MR^2$	$\frac{R}{\sqrt{2}}$
(c)	Uniform sphere of radius R	Through center	Axis	$\frac{2}{5}MR^2$	$\sqrt{\frac{2}{5}}R$
(d)	Long uniform rod of length L	Through center	Axis, L	$\frac{1}{12}ML^2$	$\frac{L}{\sqrt{12}}$
(e)		Through end	L	$\frac{1}{3}ML^2$	$\frac{L}{\sqrt{3}}$

FIGURE 8–14 Moments of inertia for various objects of uniform composition.

expressions for a number of solids rotated about the axes specified. The only one for which the result is obvious is that for the thin hoop rotated about an axis passing through its center perpendicular to the plane of the hoop. For this object, all the mass is concentrated at the same distance from the axis, R. Thus $\sum mr^2 = (\sum m)R^2 = MR^2$, where M is the total mass of the hoop.

When discussing moments of inertia, especially for unusual or irregularly shaped objects, it is often convenient to work with the **radius of gyration**, k, which is a sort of "average radius." In particular, the radius of gyration of an object is defined so that if all the mass of the object were concentrated at this distance from the axis, it would have the same moment of inertia as the original object. For example (see Fig. 8–14), the radius of gyration of a cylinder is $R/\sqrt{2} \approx 0.71R$. This means that a solid cylinder of radius 100 cm has the same moment of inertia as an equal-mass thin hoop of radius 71 cm. The moment of inertia of any object can be written in terms of its radius of gyration as

$$I = Mk^2.$$

FIGURE 8–15 Example 8–9.

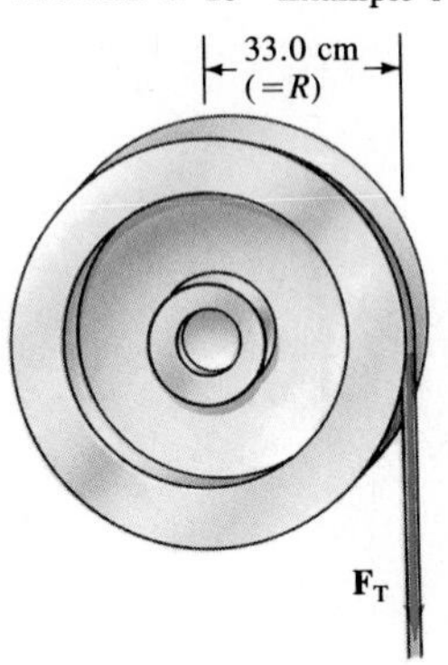

EXAMPLE 8–9 A 15.0-N force (represented by $\mathbf{F}_T$) is applied to a cord wrapped around a wheel of mass $M = 4.00$ kg and radius $R = 33.0$ cm, Fig. 8–15. The wheel is observed to accelerate uniformly from rest to reach an angular speed of 30.0 rad/s in 3.00 s. If there is a frictional torque $\tau_{fr} = 1.10$ N·m, determine the moment of inertia of the wheel and its radius of gyration.

SOLUTION We can calculate the moment of inertia from Eq. 8–14, $\tau = I\alpha$, since from the measurements given we can determine τ and α. The net torque is the applied torque due to $\mathbf{F}_T$ minus the frictional torque:

$$\tau = (0.330 \text{ m})(15.0 \text{ N}) - 1.10 \text{ N·m} = 3.85 \text{ N·m}.$$

The angular acceleration is

$$\alpha = \frac{\Delta\omega}{\Delta t} = \frac{30.0 \text{ rad/s} - 0}{3.00 \text{ s}} = 10.0 \text{ rad/s}^2.$$

Hence

$$I = \frac{\tau}{\alpha} = \frac{3.85 \text{ N·m}}{10.0 \text{ rad/s}^2} = 0.385 \text{ kg·m}^2.$$

The radius of gyration is $k = \sqrt{I/M} = \sqrt{(0.385 \text{ kg·m}^2)/(4.00 \text{ kg})} = 0.310$ m.†

† This result makes sense. We expect the radius of gyration to be less than the outer radius of the wheel. Only if the wheel were a thin ring with all the mass concentrated at the edge would k equal the radius of the wheel. On the other hand, if the wheel were perfectly solid, as in Fig. 8–14b, k would be $k = \sqrt{\frac{1}{2}R^2} = R/\sqrt{2} = 0.233$ m.

EXAMPLE 8–10 Suppose that instead of a constant 15.0-N force being exerted on the cord hanging from the edge of the wheel, as in Fig. 8–15, that a block of weight 15.0 N (and mass $m = 1.53$ kg) hangs from the cord, which we assume not to stretch or slip on the wheel. See Fig. 8–16. (*a*) Calculate the angular acceleration α of the wheel and the linear acceleration a of the mass m. (*b*) Determine the angular velocity ω of the wheel and the linear velocity v of the mass m at $t = 3.00$ s if the wheel starts from rest at $t = 0$.

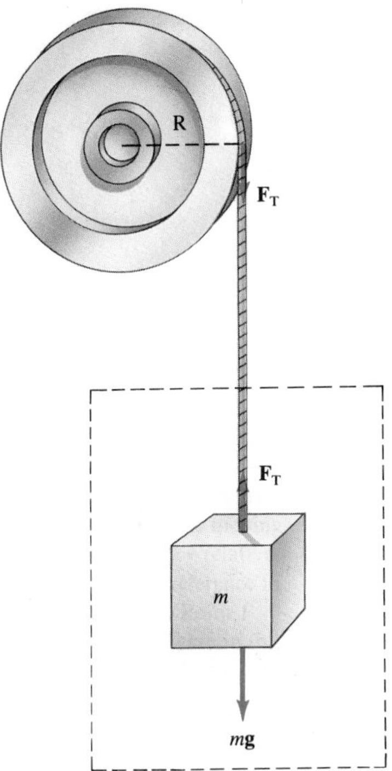

FIGURE 8–16 Example 8–10, with free-body diagram for the falling mass m inside the dashed outline.

SOLUTION (*a*) Let F_T be the tension in the cord. Then a force F_T acts at the edge of the wheel, and we have (see Example 8–9) for the rotation of the wheel:

$$\alpha = \frac{\tau}{I} = \frac{F_T R - \tau_{fr}}{I}.$$

Next we look at the (linear) motion of the block of mass m. Figure 8–16 includes a free-body diagram for the block. Two forces act on the block: the force of gravity mg acts downward, and the tension of the cord F_T pulls upward. So by $F = ma$, we have

$$mg - F_T = ma.$$

Note that the tension F_T, which is the force exerted on the edge of the wheel, is *not* in general equal to the weight of the block ($= mg = 15.0$ N). This is because the block is accelerating. Indeed, by the last equation above, $F_T = mg - ma$. To obtain α, we eliminate F_T between the two equations above and use Eq. 8–5,

$$a = R\alpha,$$

which is valid since the tangential acceleration of a point on the edge of the wheel is the same as the acceleration of the block if the cord doesn't stretch or slip. Substituting $F_T = mg - ma$ into the first equation above, we obtain

$$\alpha = \frac{\tau}{I} = \frac{F_T R - \tau_{fr}}{I} = \frac{(mg - mR\alpha)R - \tau_{fr}}{I} = \frac{mgR}{I} - \frac{mR^2\alpha}{I} - \frac{\tau_{fr}}{I}.$$

Now α appears on both sides of this last relation, so we solve for α:

$$\alpha\left(1 + \frac{mR^2}{I}\right) = \frac{mgR}{I} - \frac{\tau_{fr}}{I}$$

or

$$\alpha(I + mR^2) = mgR - \tau_{fr}.$$

Then, since $I = 0.385\ \text{kg}\cdot\text{m}^2$ (Example 8–9),

$$\alpha = \frac{mgR - \tau_{fr}}{I + mR^2}$$

$$= \frac{(15.0\ \text{N})(0.330\ \text{m}) - 1.10\ \text{N}\cdot\text{m}}{0.385\ \text{kg}\cdot\text{m}^2 + (1.53\ \text{kg})(0.330\ \text{m})^2} = 6.98\ \text{rad/s}^2.$$

friction force because the force and the motion are perpendicular. The reason the rolling sphere in the example above moves down the slope more slowly than if it were sliding is *not* because friction is doing negative work. Rather it is because some of the gravitional PE is converted to rotational KE, leaving less for the translational KE.

8–7 • Angular Momentum and Its Conservation

Throughout this chapter we have seen that if we use the appropriate angular variables, the kinematic and dynamic equations for rotational motion are analogous to those for ordinary linear motion. In the last section (Section 8–6) we saw, for example, that rotational kinetic energy can be written as $\frac{1}{2}I\omega^2$, which is analogous to the translational KE $= \frac{1}{2}mv^2$. In like manner, the linear momentum, $p = mv$, has a rotational analog. It is called **angular momentum**, and for a body rotating about a fixed axis, it is defined as

Angular momentum

$$L = I\omega, \qquad (8\text{–}16)$$

where I is the moment of inertia, and ω is the angular velocity.

We saw in Chapter 7 (Section 7–1) that Newton's second law can be written not only as $F = ma$, but also more generally in terms of momentum (Eq. 7–2), $F = \Delta p/\Delta t$. In a similar way, the rotational equivalent of Newton's second law, which we saw in Eq. 8–14 can be written as $\tau = I\alpha$, can also be written in terms of angular momentum:

Newton's second law for rotation

$$\tau = \frac{\Delta L}{\Delta t}, \qquad (8\text{–}17)$$

where τ is the net torque acting to rotate the body and ΔL is the change in angular momentum in the time Δt. Equation 8–14, $\tau = I\alpha$, is a special case of Eq. 8–17 when the moment of inertia is constant. This can be seen as follows. If a body has angular velocity ω_0 at time $t = 0$, and angular velocity ω at a time Δt later, then its angular acceleration (Eq. 8–3) is

$$\alpha = \frac{\Delta\omega}{\Delta t} = \frac{\omega - \omega_0}{\Delta t}.$$

Then from Eq. 8–17, we have

$$\tau = \frac{\Delta L}{\Delta t} = \frac{I\omega - I\omega_0}{\Delta t} = \frac{I(\omega - \omega_0)}{\Delta t} = I\frac{\Delta\omega}{\Delta t}$$

$$\tau = I\alpha,$$

which is Eq. 8–14.

Angular momentum is an important concept in physics because, under certain conditions, it is a conserved quantity. We can see from Eq. 8–17 that if the net torque τ on a body is zero, then $\Delta L/\Delta t$ equals zero. That is, L does not change. This, then, is the **law of conservation of angular momentum** for a rotating body:

Conservation of angular momentum

The total angular momentum of a rotating body remains constant if the net torque acting on it is zero.

The law of conservation of angular momentum is one of the great conservation laws of physics.

When there is zero net torque acting on a body, and the body is rotating about a fixed axis or about an axis through its cm such that its direction doesn't change, we can write

$$I\omega = I_0\omega_0 = \text{constant}.$$

I_0 and ω_0 are the moment of inertia and angular velocity, respectively, about that axis at some initial time ($t = 0$), and I and ω are their values at some other time. The parts of the body may alter their positions relative to one another, so that I changes. But then ω changes as well and the product $I\omega$ remains constant.

Many interesting phenomena can be understood on the basis of conservation of angular momentum. For example, a skater doing a spin on ice rotates at a relatively low speed when her arms are outstretched. But when she brings her arms in close to her body, she suddenly spins much faster. By remembering the definition of moment of inertia as $I = \sum mr^2$, it is clear that when she pulls her arms in closer to the axis of rotation, her moment of inertia is reduced. Since the angular momentum $I\omega$ remains constant (we ignore the small torque due to friction), if I decreases, then the angular velocity ω must increase. If the skater reduces her moment of inertia by a factor of 2, she will then rotate with twice the angular velocity.

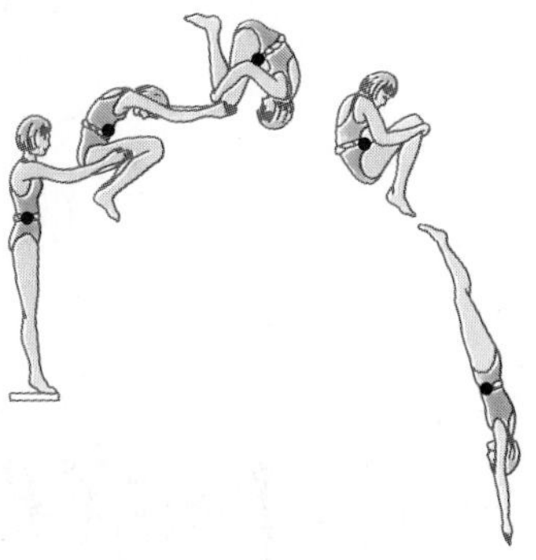

FIGURE 8–20 A diver rotates faster when arms and legs are tucked in than when they are outstretched. Angular momentum is conserved.

A similar example is the diver shown in Fig. 8–20. The push as she leaves the board gives her an initial angular momentum about her cm. When she curls herself into the tuck position, she rotates quickly one or more times. She then stretches out again, increasing her moment of inertia, which reduces the angular velocity to a small value, and then she enters the water. The change in moment of inertia from the straight position to the tuck position can be by a factor of as much as $3\frac{1}{2}$.

Note that for angular momentum to be conserved, the net torque must be zero, but the net force does not necessarily have to be zero. The net force on the diver in Fig. 8–20, for example, is not zero (gravity is acting), but the net torque on her is zero.

EXAMPLE 8–12 A mass m attached to the end of a string revolves in a circle on a frictionless tabletop. The other end of the string passes through a hole in the table (Fig. 8–21). Initially, the ball rotates with a speed $v_1 = 2.4$ m/s in a circle of radius $r_1 = 0.80$ m. The string is then pulled slowly through the hole so that the radius is reduced to $r_2 = 0.48$ m. What is the speed, v_2, of the mass now?

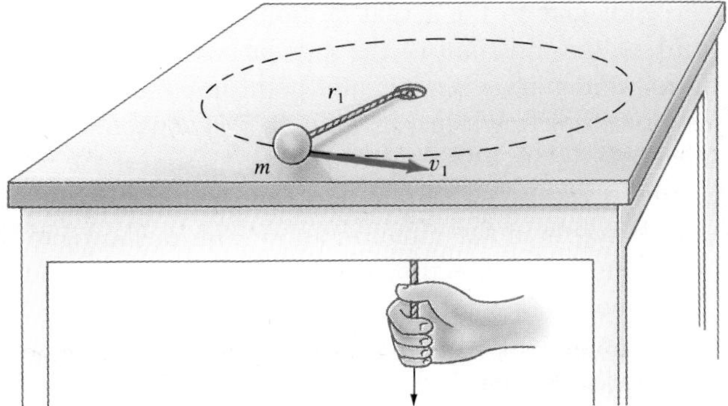

FIGURE 8–21 Example 8–12.

SOLUTION The force exerted by the string on the mass m does not alter its angular momentum about the axis of rotation, because the force is exerted toward the axis so $\tau = 0$. Hence, from conservation of angular momentum:

$$I_1\omega_1 = I_2\omega_2.$$

Since the moment of inertia of a single particle is $I = mr^2$ (Section 8–4, Eq. 8–11), we have

$$mr_1^2\omega_1 = mr_2^2\omega_2,$$

or

$$\omega_2 = \omega_1\left(\frac{r_1^2}{r_2^2}\right)$$

Then, since $v = r\omega$, we can write:

$$v_2 = r_2\omega_2 = r_2\omega_1\left(\frac{r_1^2}{r_2^2}\right) = r_2\frac{v_1}{r_1}\left(\frac{r_1^2}{r_2^2}\right) = v_1\frac{r_1}{r_2}$$

$$= (2.4\ \text{m/s})\left(\frac{0.80\ \text{m}}{0.48\ \text{m}}\right) = 4.0\ \text{m/s}.$$

*8–8 • Vector Nature of Angular Quantities

Just as Newton's second law, $\mathbf{F} = \Delta\mathbf{p}/\Delta t$ or $\mathbf{F} = m\mathbf{a}$, is a vector equation, so too is the rotational equivalent, which we have written only with magnitudes in Eqs. 8–14 ($\tau = I\alpha$) and 8–17 ($\tau = \Delta L/\Delta t$). That is, we can write

$$\boldsymbol{\tau} = I\boldsymbol{\alpha} \tag{8–18a}$$

or

$$\boldsymbol{\tau} = \frac{\Delta\mathbf{L}}{\Delta t} = \frac{\Delta(I\boldsymbol{\omega})}{\Delta t}, \tag{8–18b}$$

where $\boldsymbol{\tau}$, $\boldsymbol{\alpha}$, $\mathbf{L}$, and $\boldsymbol{\omega}$ are vectors. But in what directions do these vectors point? In fact, we have to *define* the directions for rotational quantities, and we take first the angular velocity, $\boldsymbol{\omega}$.

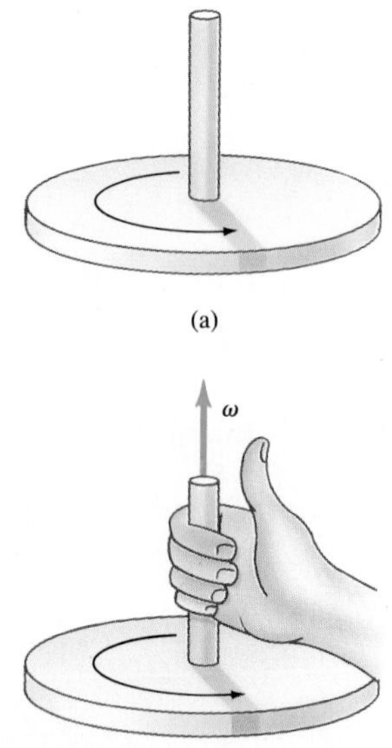

FIGURE 8–22 (a) Rotating wheel. (b) Right-hand rule for obtaining direction of $\boldsymbol{\omega}$.

Consider the rotating wheel shown in Fig. 8–22a. The linear velocities of different particles of the wheel point in all different directions. The only unique direction in space associated with the rotation is along the axis of rotation, perpendicular to the actual motion. We therefore choose the axis of rotation to be the direction of the angular velocity vector, $\boldsymbol{\omega}$. Actually, there is still an ambiguity since $\boldsymbol{\omega}$ could point in either direction along the axis of rotation (up or down in Fig. 8–22a). The convention we use, called the **right-hand rule**, is the following: when the fingers of the right hand are curled around the rotation axis and point in the direction of the rotation, then the thumb points in the direction of $\boldsymbol{\omega}$. This is shown in Fig. 8–22b. Note that $\boldsymbol{\omega}$ points in the direction a right-handed screw would move when turned in the direction of rotation. Thus, if the rotation of the wheel in Fig. 8–22a is counterclockwise, the direction of $\boldsymbol{\omega}$ is upward as shown in Fig. 8–22b. If the wheel rotates clockwise, then $\boldsymbol{\omega}$ points in the opposite direction, downward. Note that no part of the rotating body moves in the direction of $\boldsymbol{\omega}$.

If the axis of rotation is fixed, then $\boldsymbol{\omega}$ can change only in magnitude. Thus $\boldsymbol{\alpha} = \Delta\boldsymbol{\omega}/\Delta t$ must also point along the axis of rotation. If the rotation is counterclockwise as in Fig. 8–22a, and if the magnitude ω is increasing, then $\boldsymbol{\alpha}$ points upward; but if ω is decreasing (the wheel is slowing down), $\boldsymbol{\alpha}$ points downward. If the rotation is clockwise, $\boldsymbol{\alpha}$ will point downward if ω is increasing, and point upward if ω is decreasing.

Since $\boldsymbol{\omega}$ always points along the axis of rotation, if the axis of rotation changes direction, $\boldsymbol{\omega}$ changes direction. In this case, $\boldsymbol{\alpha}$ will not point along the axis of rotation. We will see examples of this in the next section. But for motion about a fixed axis, $\boldsymbol{\omega}$ and $\boldsymbol{\alpha}$ are both along the rotation axis.

The vector nature of torque can be expressed in a similar way. Suppose, for example, the wheel of Fig. 8–22 is given an angular acceleration by a force F applied at the edge of the wheel (and in the plane of the wheel), as shown in Fig. 8–23. The position vector **r** is drawn from a point 0 on the rotation axis to the point where the force **F** acts. The magnitude of the torque, as we saw in Section 8–3, is $\tau = rF \sin\theta$, where θ is the angle between the directions of **r** and **F** (Fig. 8–23a). The *direction* of the torque vector can be defined with the use of another **right-hand rule** as shown in Fig. 8–23b: you orient your hand so that your fingers point along the direction of **r**, and when you bend your fingers, they point along the direction of the force **F** as shown. When your hand is correctly oriented in this way, your thumb will then point in the direction of the torque $\boldsymbol{\tau}$. When we apply this rule to the vectors **r** and **F** of Fig. 8–23, we see that $\boldsymbol{\tau}$ will point outward (toward the reader) along the axis of rotation. This is the same direction that $\boldsymbol{\alpha}$ points, as we discussed above relative to Fig. 8–22. This checks Eq. 8–18a, above,

$$\boldsymbol{\tau} = I\boldsymbol{\alpha},$$

as a vector equation: the directions are the same on the two sides of the equation, as are the magnitudes.

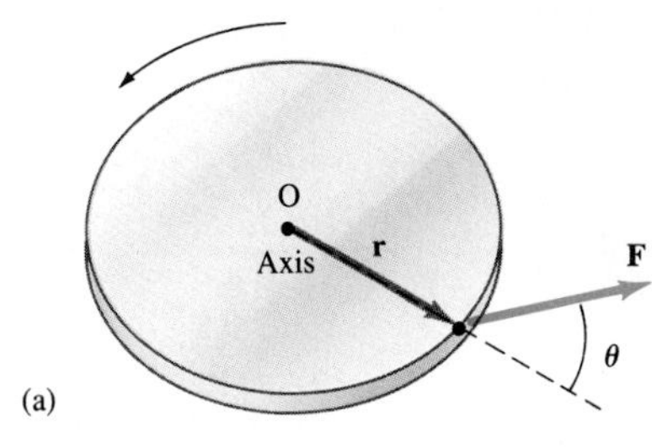

(a)

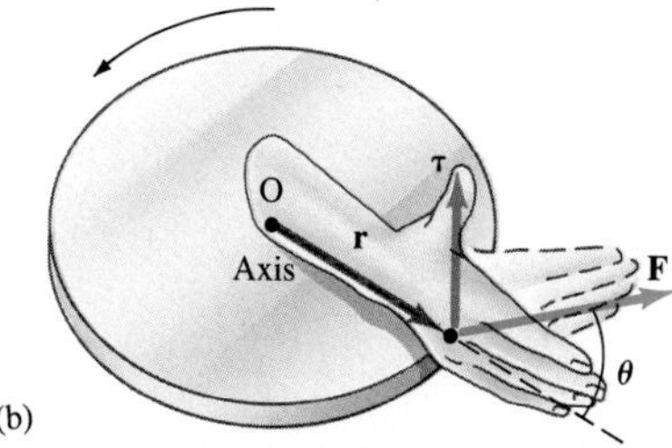

(b)

FIGURE 8–23 (a) The torque due to **F** starts the wheel rotating counterclockwise, so $\boldsymbol{\omega}$ and $\boldsymbol{\alpha}$ point upwards. (b) Determining the direction of the torque vector, $\boldsymbol{\tau}$, using the right-hand rule.

*8–9 • Vector Angular Momentum; a Rotating Wheel

Angular momentum, like linear momentum, is a vector quantity. For a symmetrical body (such as a wheel, cylinder, hoop, sphere) rotating about a symmetry axis,† we can write the vector angular momentum as

$$\mathbf{L} = I\boldsymbol{\omega}. \tag{8–19}$$

The angular velocity vector $\boldsymbol{\omega}$ (and therefore also **L**) points along the axis of rotation in the direction given by the right-hand rule (Fig. 8–22b).

The vector nature of angular momentum can be used to explain a number of interesting (and sometimes surprising) phenomena. For example, consider a person standing at rest on a circular platform capable of rotating without friction about an axis through its center (that is, a simplified merry-

† If the rotation axis is not a symmetry axis of the body, then Eq. 8–19 is valid for the component of **L** along the rotation axis. There may or may not be other components, but we won't consider these more complicated cases here.

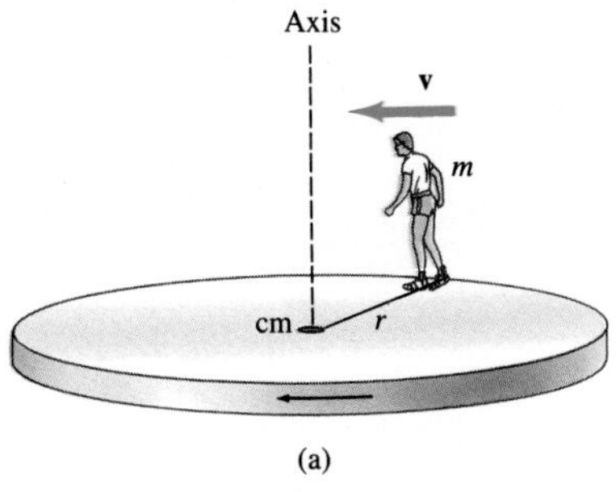

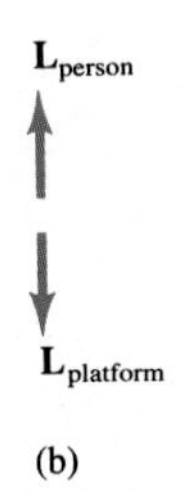

FIGURE 8–24 (a) A person standing on a circular platform, both initially at rest, begins walking along the edge at speed v. The platform, assumed to be mounted on friction-free bearings, begins rotating in the opposite direction, so that the total angular momentum remains zero, as shown in (b).

go-round). If the person now starts to walk along the edge of the platform, Fig. 8–24a, the platform starts rotating in the opposite direction. Why? Well, this is an example of the conservation of angular momentum. If the person starts walking counterclockwise, the person's angular momentum will be pointed upward along the axis of rotation (remember how we defined the direction of $\boldsymbol{\omega}$ using the right-hand rule). The magnitude of the person's angular momentum will be $L = I\omega = (mr^2)(v/r)$, where v is the person's speed (relative to earth, not the platform), r is his distance from the rotation axis, m is his mass, and his moment of inertia is mr^2 if we consider him a particle (mass concentrated at one point). The platform rotates in the opposite direction, so its angular momentum points downward. If the initial total angular momentum was zero (person and platform at rest), it will remain zero after the person starts walking—that is, the upward angular momentum of the person just balances the oppositely directed angular momentum of the platform (Fig. 8–24b), so the total vector angular momentum is still zero. Even though the person exerts a force (and torque) on the platform, and vice versa, these are internal torques (internal to the system consisting of platform plus person). There are no external torques (assuming friction-free bearings of the platform), so from Eq. 8–18b the angular momentum remains constant.

Next we consider an experiment that is easily done, and gives a surprising result which illustrates the vector nature of the relation

$$\boldsymbol{\tau} = \frac{\Delta \mathbf{L}}{\Delta t}.$$

Suppose you are holding a bicycle wheel by a handle connected to its axle, as in Fig. 8–25a. The wheel is spinning rapidly, so its angular momentum **L** points horizontally as shown. Now you suddenly try to tilt the axle upward, as shown by the dashed line in Fig. 8–25a, so that the cm moves vertically. You expect the wheel to go up, but unexpectedly it swerves to the right! To explain this bizarre effect—you may need to do it to believe it—we can use the relation $\boldsymbol{\tau} = \Delta \mathbf{L}/\Delta t$. In the short time Δt, you exert a torque $\boldsymbol{\tau}$ (about an axis through your wrist, Fig. 8–25a) that points along the x axis perpendicular to **L**. The change in **L** is

$$\Delta \mathbf{L} \approx \boldsymbol{\tau}\, \Delta t$$

so $\Delta \mathbf{L}$ must also point (approximately) as $\boldsymbol{\tau}$ does, along the x axis (Fig. 8–25b). Thus the new angular momentum, $\mathbf{L} + \Delta \mathbf{L}$, points to the right of **L** (looking along the axis of the wheel), as shown in Fig. 8–25b. Since the angular momentum is always directed along the axle of the wheel, we see that the axle must move sideways, to the right, in order to lie along the direction of $\mathbf{L} + \Delta \mathbf{L}$, which is what we observe.

Another example of the conservation of angular momentum is the *gy-*

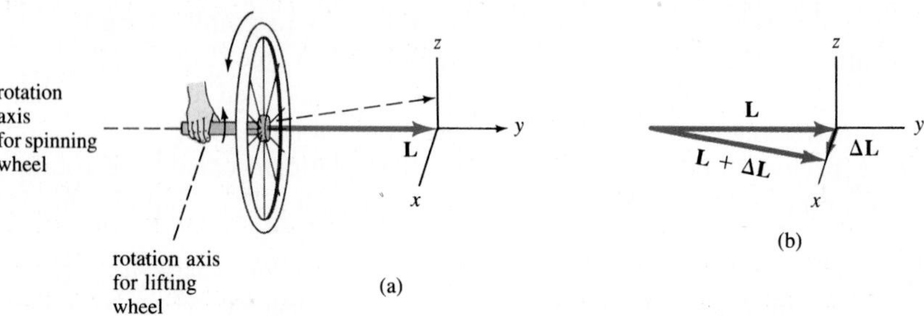

FIGURE 8–25 When you try to tilt a rotating bicycle wheel upward, it swerves to the side.

roscope, used by mariners and as the basis for guidance systems of aircraft. The rapidly spinning wheel is mounted on a complicated set of bearings, so that even when the mount moves, no net torque acts to change the direction of the angular momentum. Thus the axis of the wheel remains pointed in the same direction in space.

*8–10 • Rotating Frames of Reference; Inertial Forces

Up to now, we have examined the motion of bodies, including circular and rotational motion, from the outside, as an observer fixed on the earth. Sometimes it is convenient to place ourselves (in theory, if not physically) into a rotating system. Let us, for example, examine the motion of objects from the point of view, or frame of reference, of a person seated on a rotating platform such as a merry-go-round. It looks to him as if the rest of the world is going around *him*. But let us focus attention on what he observes when he places a tennis ball on the floor of the rotating platform, which we assume is frictionless. If he puts the ball down gently, without giving it any push, he will observe that it accelerates from rest and moves outward as shown in Fig. 8–26a. According to Newton's first law, an object initially at rest should stay at rest if no force acts on it. But, according to the observer on the rotating platform, the ball starts moving even though there is no force applied to it. To an observer on the earth, this is all very clear: the ball has an initial velocity when it is released (because the platform is moving), and it simply continues moving in a straight-line path as shown in Fig. 8–26b, in accordance with Newton's first law.

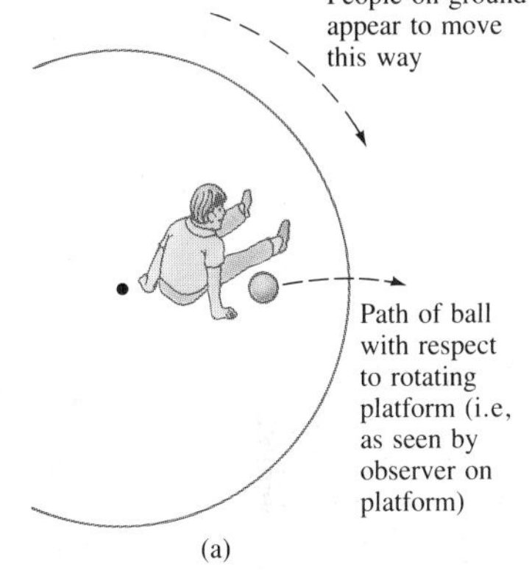

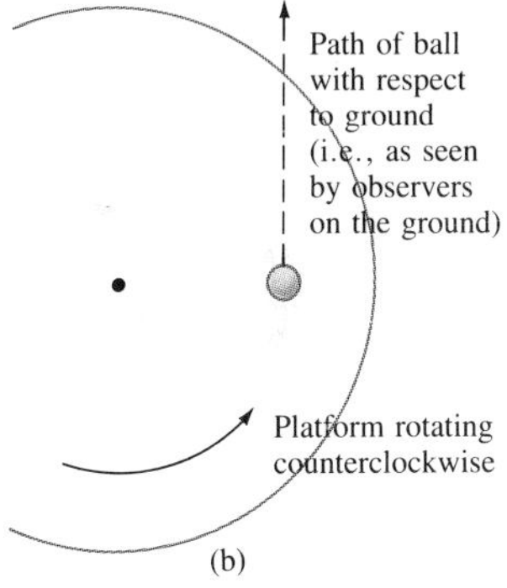

FIGURE 8–26 Path of a ball released on a rotating merry-go-round (a) in the reference frame of the merry-go-round, and (b) in the reference frame of the earth.

But what shall we do about the frame of reference of the person on the rotating platform? Clearly, Newton's first law, the law of inertia, does not hold in this rotating frame of reference. For this reason, such a frame is called a **noninertial reference frame**. An **inertial reference frame** is one in which the law of inertia—Newton's first law—does hold, and so do Newton's second and third laws. In a noninertial reference frame, such as our rotating platform, Newton's second law also does not hold. For instance in the situation described above, there is no net force on the ball; yet, with respect to the rotating platform, the ball accelerates.

Because Newton's laws do not hold when observations are made with respect to a rotating frame of reference, calculation of motion can be complicated. However, we can still apply Newton's laws in such a reference frame if we make use of a trick. We write down the equation $F = ma$ as if a force equal to mv^2/r (or $m\omega^2 r$) were acting radially outward on the object in addition to any other forces that may be acting. This extra force, which might be designated as "centrifugal force" since it seems to act outward, is called a **fictitious force** or **pseudoforce**. It is a pseudoforce ("pseudo" means "false") because there is no object that exerts this force. Furthermore, when viewed from an inertial reference frame, the effect doesn't exist. We have made it up so that we can make calculations in a noninertial frame using the relation $F = ma$. Thus the observer in Fig. 8–26a can determine the motion of the ball by assuming that a force equal to mv^2/r acts on it. These pseudoforces are often called **inertial forces** since they arise only because the reference frame is not an inertial one.

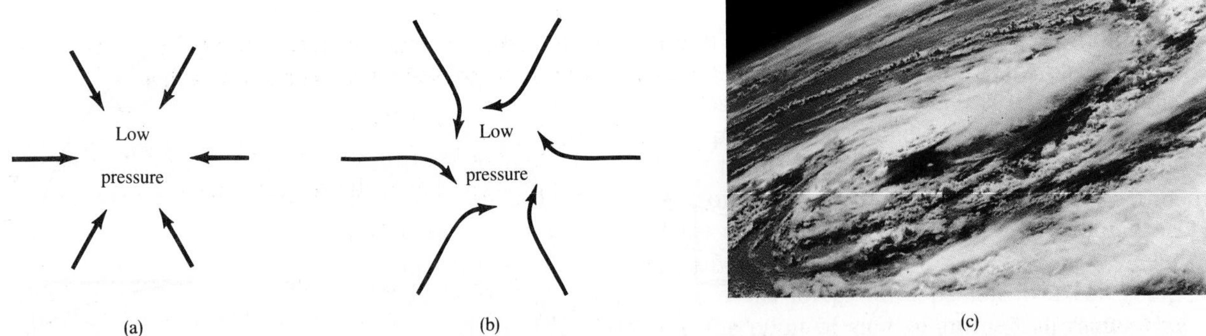

FIGURE 8–29 (a) Winds (moving air masses) would flow directly toward a low-pressure area if the earth did not rotate. (b) and (c): Because of the earth's rotation, the winds are deflected to the right in the Northern Hemisphere (as in Fig. 8–28) as if a fictitious (Coriolis) force were acting.

tion at constant acceleration. For as we saw in Chapter 2 (see Eq. 2–10b), $x = \frac{1}{2}at^2$ for a constant acceleration (with zero initial velocity in the x direction). Thus, if we write Eq. 8–20 in the form $s = \frac{1}{2}a_{\text{cor}}t^2$, we see that the Coriolis acceleration a_{cor} is

$$a_{\text{cor}} = 2\omega v. \tag{8–21}$$

This relation is valid for any velocity in the plane of rotation—that is, in the plane perpendicular to the axis of rotation (in Fig. 8–28, the axis through point O perpendicular to the page).

Because the earth rotates, the Coriolis effect has some interesting manifestations on the earth. It affects the movement of air masses and thus has an influence on weather. In the absence of the Coriolis effect, air would rush directly into a region of low pressure, as shown in Fig. 8–29a. But because of the Coriolis effect, the winds are deflected to the right (Fig. 8–29b), so that there tends to be a counterclockwise wind pattern around a low-pressure area. This is true for the Northern Hemisphere—for which the situation of Fig. 8–28 applies since the earth rotates from west to east. The reverse is true in the Southern Hemisphere. Thus cyclones rotate counterclockwise in the Northern Hemisphere and clockwise in the Southern Hemisphere. The same effect explains the easterly trade winds near the equator: any winds heading south toward the equator will be deflected toward the west (that is, as if coming from the east).

The Coriolis effect also acts on a falling body. A body released from the top of a high tower will not hit the ground directly below the release point, but will be deflected slightly to the east.

SUMMARY

When a rigid body rotates about a fixed axis, each point of the body moves in a circular path. Lines drawn perpendicularly from the rotation axis to various points in the body all sweep out the same angle θ in any given time interval. Angles are conveniently measured in *radians*, where one radian is the angle subtended by an arc whose length is equal to the radius, or

$$2\pi \text{ rad} = 360^\circ$$
$$1 \text{ rad} \approx 57.3^\circ.$$

Angular velocity, ω, is defined as the rate of change of angular position:

$$\omega = \frac{\Delta\theta}{\Delta t}.$$

All parts of a rigid body rotating about a fixed axis have the same angular velocity at any instant. *Angular acceleration*, α, is defined as the rate of change of angular velocity:

$$\alpha = \frac{\Delta\omega}{\Delta t}.$$

The linear velocity v and acceleration a of a point fixed at a distance r from the axis of rotation are related to ω and α by

$$v = r\omega, \qquad a_T = r\alpha, \qquad a_c = \omega^2 r,$$

where a_T and a_c are the tangential and radial components of the linear acceleration, respectively. The frequency f is related to ω by $\omega = 2\pi f$, and to the period T by $T = 1/f$.

The equations describing uniformly accelerated rotational motion (α = constant) have the same form as for uniformly accelerated linear motion:

$$\omega = \omega_0 + \alpha t; \qquad \theta = \omega_0 t + \tfrac{1}{2}\alpha t^2;$$

$$\omega^2 = \omega_0^2 + 2\alpha\theta; \qquad \bar{\omega} = \frac{\omega + \omega_0}{2}.$$

The dynamics of rotation is analogous to the dynamics of linear motion. Force is replaced by *torque*, τ, which is defined as the product of force times lever arm (perpendicular distance from the line of action of the force to the axis of rotation). Mass is replaced by *moment of inertia, I*, which depends not only on the mass of the body, but also on how the mass is distributed about the axis of rotation. Linear acceleration is replaced by angular acceleration. The rotational equivalent of Newton's second law is then

$$\tau = I\alpha.$$

The *rotational kinetic energy* of a body rotating about a fixed axis with angular velocity ω is

$$\text{KE} = \tfrac{1}{2}I\omega^2.$$

For a body both translating and rotating, the total kinetic energy is the sum of the translational KE of the body's cm plus the rotational KE of the body about its cm:

$$\text{KE} = \tfrac{1}{2}Mv_{\text{cm}}^2 + \tfrac{1}{2}I_{\text{cm}}\omega^2$$

as long as the rotation axis is fixed in direction.

The *angular momentum, L*, of a body about a fixed rotation axis is given by

$$L = I\omega.$$

Newton's second law, in terms of angular momentum, becomes

$$\tau = \frac{\Delta L}{\Delta t}.$$

If the net torque on the body is zero, $\Delta L/\Delta t = 0$, so L = constant. This is the *law of conservation of angular momentum* for a rotating body.

QUESTIONS

1. You are standing a known distance from the Statue of Liberty. Describe how you could determine its height using only a meter stick, and without moving from your place.
2. A bicycle odometer (which measures distance traveled) is attached near the wheel hub and is designed for 27-in wheels. What happens if you use it on a bicycle with 24-in wheels?
3. Suppose a record turntable rotates at constant angular velocity. Does a point on the rim have radial and/or tangential acceleration? If the turntable's angular velocity increases uniformly, does the point have radial and/or tangential acceleration? For which cases would the magnitude of either of the components of linear acceleration change?
4. If the angular quantities θ, ω, and α were specified in terms of degrees rather than radians, how would Eqs. 8–9 for uniformly accelerated rotational motion have to be altered?
5. Can a small force exert a greater torque than a larger force? Explain.
6. If a force **F** acts on a body such that its lever arm is zero, does it have any effect on the body's motion?
7. Why is it more difficult to do a situp with your hands behind your head than when they are outstretched in front of you? A diagram may help you to answer this.
8. Expert bicyclists use very lightweight "sew-up" (tubular) tires. They claim that reducing the mass of the tires is far more significant than an equal reduction in mass elsewhere on the bicycle. Explain why this is true.
9. Mammals that depend on being able to run fast have slender lower legs with flesh and muscle concentrated high, close to the body (Fig. 8–30). On the basis of rotational dynamics, explain why this distribution of mass is advantageous.

FIGURE 8–30 A gazelle. Question 9.

FIGURE 8–31 Question 10.

10. Why do tightrope walkers carry a long, narrow beam (Fig. 8–31)?
11. If the net force on a system is zero, is the net torque also zero? If the net torque on a system is zero, is the net force zero?
12. A stick stands vertically on its end on a frictionless surface. Describe the motion of its cm and of each end when it is tipped slightly to one side and falls.
13. A solid sphere, a solid cylinder, and a thin hoop, all of the same diameter, roll down an inclined plane. Which arrives at the bottom first? Which last?
14. Two inclines have the same height but make different angles with the horizontal. The same steel ball is rolled down each incline. On which incline will the speed of the ball at the bottom be greatest? Explain.
15. Two spheres simultaneously start rolling (from rest) down an incline. One sphere has twice the radius and twice the mass of the other. Which reaches the bottom of the incline first? Which has the greater speed there? Which has the greater total kinetic energy at the bottom?
16. A sphere and a cylinder have the same radius and the same mass. They start from rest at the top of an incline. Which reaches the bottom first? Which has the greater speed at the bottom? Which has the greater total kinetic energy at the bottom? Which has the greater rotational KE?
17. A cyclist rides over the top of a hill. Is the bicycle's motion rotational, translational, or a combination of both?
18. Explain how a child "pumps" on a swing to make it go higher.
19. If there were a great migration of people toward the equator, how would this affect the length of the day?
20. We claim that momentum and angular momentum are conserved. Yet most moving or rotating bodies eventually slow down and stop. Explain.
21. A recent roller coaster design, in which riders take a loop upside down, has a teardrop-shaped loop as shown in Fig. 8–32, rather than circular. Explain why the shorter radius of curvature at the top of the loop, and the longer radius near the bottom, increases the safety of riders upside down at the top of the loop.
*22. In what direction is the earth's angular velocity for its daily rotation on its axis?

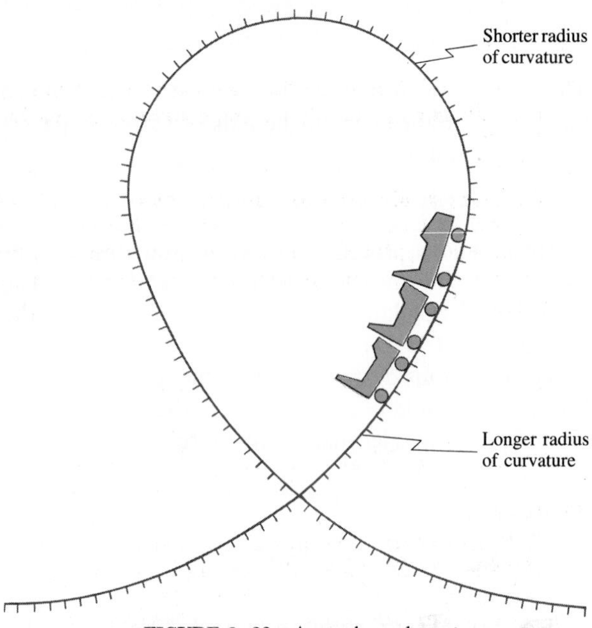

FIGURE 8–32 A teardrop-shaped loop (clothoid) of a roller coaster. (Question 21.)

*23. The angular velocity of a wheel rotating on a horizontal axle points west. In what direction is the linear velocity of a point on the top of the wheel? If the angular acceleration points east, describe the tangential linear acceleration of this point. Is the angular speed increasing or decreasing?
*24. When a motorcyclist leaves the ground on a jump, if the throttle is left on (so that the rear wheel spins), why does the front of the cycle rise up?
*25. A quarterback leaps into the air to throw a forward pass. As he throws the ball, the upper part of his body rotates. If you look quickly you will notice that his hips and legs rotate in the opposite direction (Fig. 8–33). Explain.

FIGURE 8–33 Quarterback in the air, throwing a pass. (Question 25.)

***26.** Look at the face of a clock with a second hand. In what direction is the angular momentum of the second hand?

***27.** On the basis of the law of conservation of angular momentum, discuss why a helicopter must have more than one rotor (or propeller). Discuss one or more ways the second propeller can operate in order to keep the body stable.

***28.** Describe the torque needed if the person in Fig. 8–25 is to tilt the axle of the rotating wheel directly upward with no swerving to the side.

***29.** Suppose you are standing on a turntable that can freely rotate. When you hold a rotating bicycle wheel over your head, with its axis vertical, you are at rest. If you now move the wheel so that its axis is horizontal, what happens to you? What happens if you then point the axis of the wheel downward?

PROBLEMS

SECTION 8–1

1. (I) What are the following angles in radians: (*a*) 30°, (*b*) 90°, and (*c*) 420°? Give as numerical values and as fractions of π.
2. (I) The Eiffel Tower is 300 m tall. When you are standing at a certain place in Paris, it subtends an angle of 6°. How far are you, then, from the Eiffel Tower?
3. (I) A laser beam is directed at the moon, 380,000 km from earth. The beam diverges at an angle θ (Fig. 8–34) of 1.8×10^{-5} rad. How large a spot will it make on the moon?
4. (I) A 20-cm-diameter grinding wheel rotates at 2000 rpm. Calculate its angular velocity in rad/s.
5. (I) What is the linear speed of a point on the edge of the grinding wheel in Problem 4?
6. (I) A 33-rpm phonograph record reaches its rated speed 2.8 s after it is turned on. What was the angular acceleration?
7. (I) A 70-cm-diameter wheel rotating at 1200 rpm is brought to rest in 15 s. Calculate its angular acceleration.
8. (II) A bicycle with 68-cm-diameter tires travels 7.0 km. How many revolutions do the wheels make?
9. (II) Estimate the angle subtended by the moon using a ruler and your finger or other object to just blot out the moon. Describe your measurement and the result obtained and then use it to estimate the diameter of the moon. The moon is about 380,000 km from the earth.

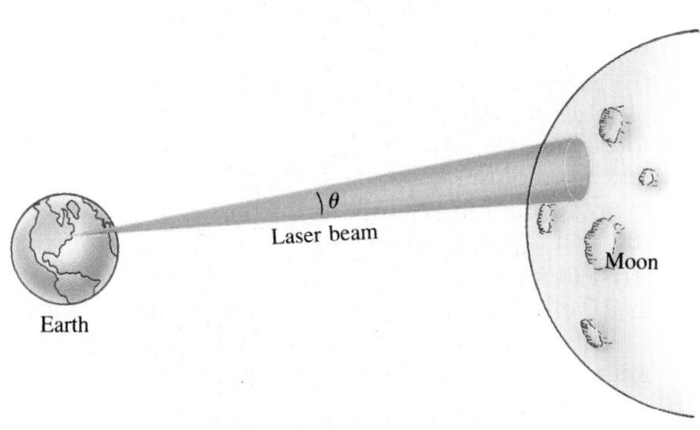

FIGURE 8–34 Problem 3.

10. (II) Calculate the angular velocity of the earth (*a*) in its orbit around the sun, and (*b*) about its axis.
11. (II) What is the linear speed of a point (*a*) on the equator, and (*b*) at a latitude of 60°N, due to the earth's rotation? (See previous problem.)
12. (II) How fast (in rpm) must a centrifuge rotate if a particle 9.0 cm from the axis of rotation is to experience an acceleration of 100,000 *g*'s?
13. (II) A 60-cm-diameter wheel accelerates uniformly from 200 rpm to 300 rpm in 3.0 s. Determine (*a*) its angular acceleration and (*b*) the radial and tangential components of the linear acceleration of a point on the edge of the wheel 2.0 s after it has started accelerating.
14. (II) A record player turntable of radius R_1 is turned by a circular rubber roller of radius R_2 in contact with it at their outer edges. What is the ratio of their angular velocities, ω_1/ω_2?

SECTION 8–2

15. (I) A phonograph turntable reaches its rated speed of 33 rpm after making 2.5 revolutions. What was its angular acceleration?
16. (I) A centrifuge accelerates from rest to 10,000 rpm in 270 s. Through how many revolutions did it turn in this time?
17. (I) An automobile engine slows down from 3600 rpm to 1000 rpm in 5.0 s. Calculate (*a*) its angular acceleration, assumed uniform, and (*b*) the total number of revolutions the engine makes in this time.
18. (II) A 70-cm-diameter wheel accelerates uniformly from 180 rpm to 300 rpm in 5.5 s. How far will a point on the edge of the wheel have traveled in this time?
19. (II) Starting from the definitions of ω and α, derive Eqs. 8–9 assuming constant angular acceleration.
20. (II) Two rubber wheels are mounted next to one another so that their circular edges touch. Wheel 1, of radius $R_1 = 3.0$ cm, accelerates at a rate 0.88 rad/s^2 and drives the second wheel, of radius $R_2 = 5.0$ cm, by contact (without slipping). (*a*) Starting from rest, how long does it take the second wheel to reach an angular speed of 33 rpm? (*b*) What was the angular acceleration of wheel 2?

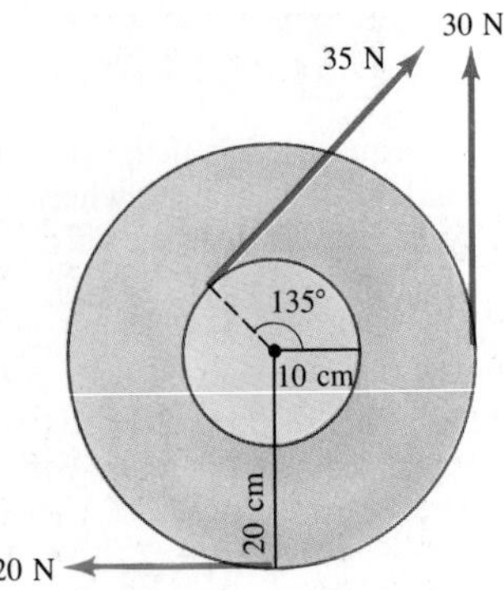

FIGURE 8–35 Problem 25.

21. (II) The tires of a car make 70 revolutions as the car reduces its speed uniformly from 90 km/h to 50 km/h. The tires have a diameter of 1.0 m. (*a*) What was the angular acceleration? (*b*) If the car continues to decelerate at this rate, how much more time is required for it to stop?

22. (III) A wheel, starting from rest, undergoes uniform angular acceleration α about its fixed axle. (*a*) Write the components of the linear acceleration, a_T and a_c, for a point P which is a distance r from the axle in terms of α, r, and time t. (*b*) Let ϕ be the angle between the linear acceleration vector, **a**, and the line drawn between P and the axis. Express ϕ in terms of the total number of revolutions of the wheel, N.

SECTION 8–3

23. (I) What is the maximum torque exerted by a 60-kg person riding a bike if the person puts all her weight on each pedal when climbing a hill? The pedals rotate in a circle of radius 18 cm.

24. (I) A person exerts a force of 28 N on the end of a door 84 cm wide. What is the magnitude of the torque if the force is exerted (*a*) perpendicular to the door, and (*b*) at a 60° angle to the face of the door?

25. (II) Calculate the net torque about the axle of the wheel shown in Fig. 8–35. Assume that a friction torque of 0.30 N·m opposes the motion.

26. (II) The bolts on the cylinder head of certain engines require tightening to a torque of 90 N·m. If a wrench is 20 cm long, what force perpendicular to the wrench must the mechanic exert at its end? If the six-sided bolt head is 15 mm in diameter, estimate the force applied near each of the six points by a socket wrench (Fig. 8–36).

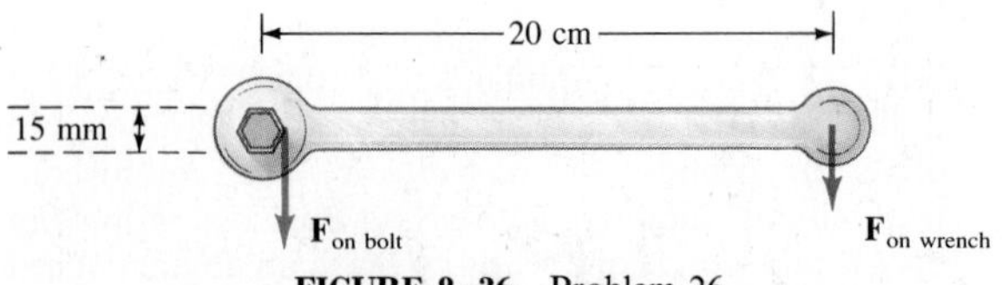

FIGURE 8–36 Problem 26.

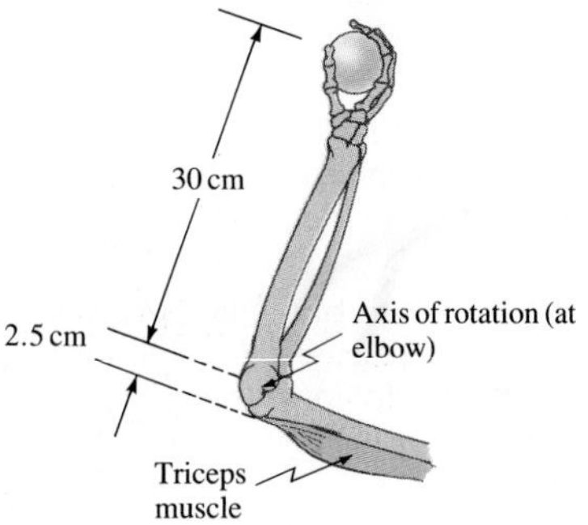

FIGURE 8–37 Problems 36 and 37.

SECTIONS 8–4 AND 8–5

27. (I) Calculate the moment of inertia of a 22.0-kg sphere of radius 0.806 m when the axis of rotation is through its center.

28. (I) Calculate the moment of inertia of a 66.7-cm-diameter bicycle wheel. The rim and tire have a combined mass of 1.25 kg. The mass of the hub can be ignored. (Why?)

29. (II) An oxygen molecule consists of two oxygen atoms whose total mass is 5.3×10^{-26} kg and whose moment of inertia about an axis perpendicular to the line joining the two atoms, midway between them, is 1.9×10^{-46} kg·m^2. Estimate, from these data, the effective distance between the atoms.

30. (II) A small 1.25-kg ball on the end of a light rod is rotated in a horizontal circle of radius 1.20 m. Calculate: (*a*) the moment of inertia of the system about the axis of rotation, and (*b*) the torque needed to keep the ball rotating at constant angular velocity if air resistance exerts a force of 0.0200 N on the ball.

31. (II) A grinding wheel is a uniform cylinder of radius 12.5 cm and mass 0.880 kg. Calculate: (*a*) its moment of inertia, and (*b*) the torque needed to accelerate it from rest to 1200 rpm in 4.00 s if a frictional torque of 0.0145 N·m is also acting.

32. (II) What is the radius of gyration of a 23.6-kg wheel that accelerates from rest to 600 rpm in 10.0 s when a net torque of 3.25 N·m acts on it?

33. (II) A 1.20-m-diameter sphere can be rotated about an axis through its center by a torque of 25 N·m, which accelerates it uniformly from rest through a total of 180 rev in 15 s. What is the mass of the sphere?

34. (II) A merry-go-round accelerates from rest to 1.6 rad/s in 9.0 s. Assuming that the merry-go-round is a uniform disc of radius 8.0 m and mass 31,000 kg, calculate the net torque required to accelerate it.

35. (II) A centrifuge rotor rotating at 10,000 rpm is shut off and is eventually brought to rest by a frictional torque of 1.20 N·m. If the mass of the rotor is 4.80 kg and its radius of gyration is 0.0710 m, through how many revolutions will the rotor turn before coming to rest and how long will it take?

36. (II) The forearm in Fig. 8–37 accelerates a 3.6-kg ball at 7.0 m/s^2 by means of the triceps muscle, as shown. Calculate: (*a*) the torque needed, and (*b*) the force that must be exerted by the triceps muscle. Ignore the mass of the arm.

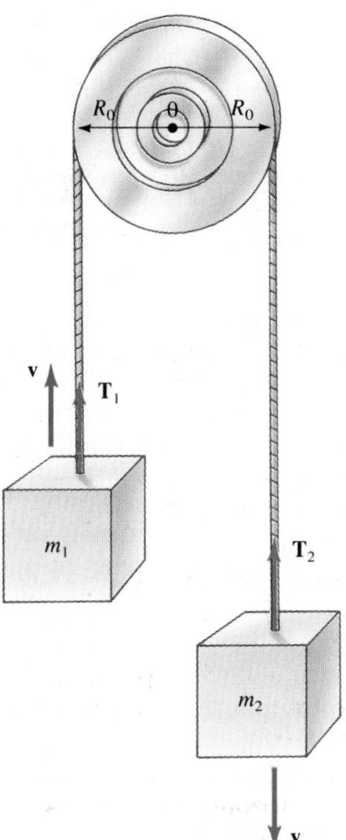

FIGURE 8–38 Atwood's machine. Problems 38 and 45.

37. (II) Assume that a 1.00-kg ball is thrown solely by the action of the forearm, which rotates about the elbow joint under the action of the triceps muscle, Fig. 8–37. The ball is accelerated from rest to 10.0 m/s in 0.220 s, at which point it is released. Calculate: (*a*) the angular acceleration of the arm, and (*b*) the force required of the triceps muscle. Assume that the forearm has a mass of 3.20 kg and rotates like a uniform rod about an axis at its end.

38. (III) An *Atwood machine* consists of two masses, m_1 and m_2, which are connected by a massless inelastic cord that passes over a pulley, Fig. 8–38. If the pulley has radius R_0 and moment of inertia I about its axle, determine the acceleration of the masses m_1 and m_2, and compare to the situation in which the moment of inertia of the pulley is ignored. [*Hint:* the tensions T_1 and T_2 are not necessarily equal.]

39. (III) A hammer thrower accelerates the hammer (mass = 7.30 kg) from rest within four full turns (revolutions) and releases it at a speed of 27.2 m/s. Assuming a uniform rate of increase in angular velocity and a radius of 2.10 m, calculate: (*a*) the angular acceleration, (*b*) the (linear) tangential acceleration, (*c*) the centripetal acceleration just before release, (*d*) the net force being exerted on the hammer by the athlete just before release, and (*e*) the angle of this force with respect to the radius of the circular motion.

SECTION 8–6

40. (I) A centrifuge rotor has a moment of inertia of $4.00 \times 10^{-2}\ \mathrm{kg \cdot m^2}$. How much energy is required to bring it from rest to 10,000 rpm?

41. (II) Calculate the translational speed of a cylinder when it reaches the foot of an incline 11.0 m high. Assume it starts from rest and rolls without slipping.

42. (II) Estimate the kinetic energy of the earth with respect to the sun as the sum of two terms, (*a*) that due to its daily rotation about its axis, and (*b*) that due to its yearly revolution about the sun. [Assume that the earth is a uniform sphere, mass = 6.0×10^{24} kg, radius = 6.4×10^6 m, and is 1.5×10^8 km from the sun.]

43. (II) A merry-go-round has a mass of 1860 kg and a radius of gyration of 18.5 m. How much work is required to accelerate it from rest to a rotation rate of one revolution in 7.10 s?

44. (II) The 1200-kg mass of a car includes four tires, each of mass 30 kg (including wheels) and diameter 0.80 m. Assume that the radius of gyration of the tire and wheel combination is about 0.30 m. Determine (*a*) the total KE of the car when traveling 100 km/h and (*b*) the fraction of the KE in the tires and wheels. (*c*) If the car is initially at rest and is then pulled by a tow truck with a force of 2000 N, what is the acceleration of the car? Ignore frictional losses. (*d*) What percent error would you make in (*c*) if you ignored the rotational inertia of the tires and wheels?

45. (III) Two masses, $m_1 = 22.5$ kg and $m_2 = 33.2$ kg, are connected by a rope that hangs over a pulley (as in Fig. 8–38). The pulley is a uniform cylinder of radius 0.300 m and mass 8.80 kg. Initially, m_1 is on the ground and m_2 rests 2.50 m above the ground. If the system is now released, use conservation of energy to determine the speed of m_2 just before it strikes the ground. Assume the pulley is frictionless.

46. (III) A 5.10-m-long pole is balanced vertically on its tip. What will be the speed of the tip of the pole just before it hits the ground? Assume the lower end of the pole does not slip.

SECTION 8–7

47. (I) What is the angular momentum of a 200-g ball rotating on the end of a string in a circle of radius 1.00 m at an angular speed of 9.45 rad/s?

48. (I) A person stands, hands at the side, on a platform that is rotating at a rate of 1.20 rev/s. If the person now raises her arms to a horizontal position, the speed of rotation decreases to 0.80 rev/s. (*a*) Why does this occur? (*b*) By what factor has the moment of inertia of the person changed?

49. (I) A diver (such as that shown in Fig. 8–20) can reduce his moment of inertia by a factor of about 3.5 when changing from the straight position to the tuck position. If he makes two rotations in 1.5 s when in the tuck position, what is his angular speed (rev/s) when in the straight position?

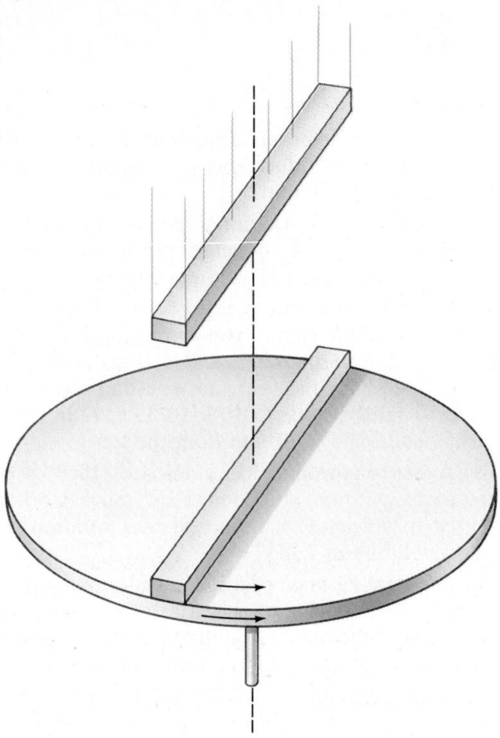

FIGURE 8–39 Problem 53.

50. (II) Determine the angular momentum of the earth (*a*) about its rotation axis (assume the earth is a uniform sphere) and (*b*) in its orbit around the sun (treat the earth as a particle orbiting the sun). The earth has mass $= 6.0 \times 10^{24}$ kg, radius $= 6.4 \times 10^{6}$ m, and is 1.5×10^{8} km from the sun.

51. (II) (*a*) What is the angular momentum of a 2.43-kg uniform cylindrical grinding wheel of radius 12.5 cm when rotating at 1600 rpm? (*b*) How much torque is required to stop it in 8.00 s?

52. (II) A nonrotating cylindrical disk of moment of inertia I is dropped onto an identical disk rotating at angular speed ω. Assuming no external torques, what is the final common angular speed of the two disks?

53. (II) A uniform disk, such as a record turntable, turns at 8.0 rev/s around a frictionless spindle. A nonrotating rod, of the same mass as the disk and length equal to the disk's diameter, is dropped onto the freely spinning disk. They then both turn around the spindle with their centers superposed, Fig. 8–39. What is the angular velocity in rev/s of the combination?

54. (II) A 4.5-m-diameter merry-go-round is rotating freely with an angular velocity of 0.80 rad/s. Its total moment of inertia is 1750 kg·m^2. Four people standing on the ground, each of mass 65 kg, suddenly step onto the edge of the merry-go-round. What is the angular velocity of the merry-go-round now? What if the people were on it initially and then jumped off in a radial direction?

*SECTIONS 8–8 AND 8–9

*55. (II) Suppose a 60-kg person stands at the edge of a 6.0-m-diameter merry-go-round turntable that is mounted on frictionless bearings and has a moment of inertia of 1800 kg·m^2. The turntable is at rest initially, but when the person begins running at a speed of 4.2 m/s (with respect to the turntable) around its edge, the turntable begins to rotate in the opposite direction. Calculate the angular velocity of the turntable.

*56. (II) A person stands on a platform, initially at rest, that can rotate freely without friction. The moment of inertia of the person plus the platform is I_p. The person holds a spinning bicycle wheel with axis horizontal, as in Fig. 8–25. The wheel has moment of inertia I_w and angular velocity ω_w. What will be the angular velocity ω_p of the platform if the person moves the axis of the wheel so that it points (*a*) vertically upward, (*b*) at a 60° angle to the vertical, (*c*) vertically downward. (*d*) What will ω_p be if the person reaches up and stops the wheel in part (*a*)?

*SECTION 8–10

*57. (III) If a plant is allowed to grow from seed on a rotating platform, it will grow at an angle, pointing inward. Calculate what this angle will be (put yourself in the rotating frame) in terms of g, r, and ω. Why does it grow inward rather than outward?

*58. (III) In a rotating frame of reference, Newton's first and second laws remain useful if we assume that a pseudoforce equal to $m\omega^2 r$ is acting. What effect does this assumption have on the validity of Newton's third law?

*SECTION 8–11

*59. (II) Suppose the man at B in Fig. 8–28 throws the ball toward the woman at A. (*a*) In what direction is the ball deflected as seen in the noninertial system? (*b*) Determine a formula for the amount of deflection and for the (Coriolis) acceleration in this case.

*60. (II) A lead ball is dropped from the top of a tall tower. (*a*) Show that its Coriolis acceleration due to the rotation of the earth is $a_{cor} = 2\omega(v \cos \lambda)$ where λ is the latitude (degrees north of the equator), and v is the velocity of the (vertically) falling ball at any moment. [*Hint:* first show that only the component of velocity perpendicular to the earth's axis—that is, $v \cos \lambda$—experiences the Coriolis acceleration.] (*b*) Show that the direction of the displacement due to the Coriolis effect is to the east.

*61. (III) For what directions of velocity would the Coriolis force on a body moving at the earth's equator be zero?

*62. (III) Determine the Coriolis force (magnitude and direction) acting on a 1200-kg race car traveling due north at 500 km/h in Utah (latitude 40°). [*Hint:* use a component of velocity perpendicular to the earth's axis.]

GENERAL PROBLEMS

63. A large spool of rope lies on the ground with the end of the rope lying on the top-edge of the spool. A person grabs the end of the rope and walks a distance L, holding onto it, Fig. 8–40. The spool rolls behind the person without slipping. What length of rope unwinds from the spool? How far does the spool's cm move?

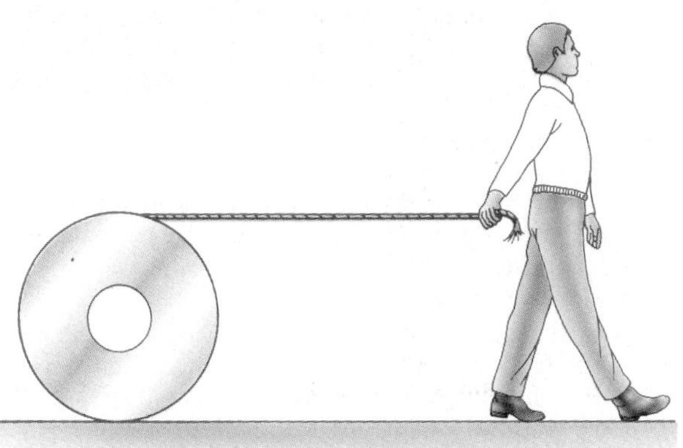

FIGURE 8–40 Problem 63.

64. A cyclist accelerates from rest at a rate of $0.75\ \mathrm{m/s^2}$. How fast will a point on the rim of the tire (diameter = 68 cm) at the top be moving after 3.0 s? [*Hint:* at any moment, the lowest point on the tire is in contact with the ground and hence is at rest—see Fig. 8–41.]

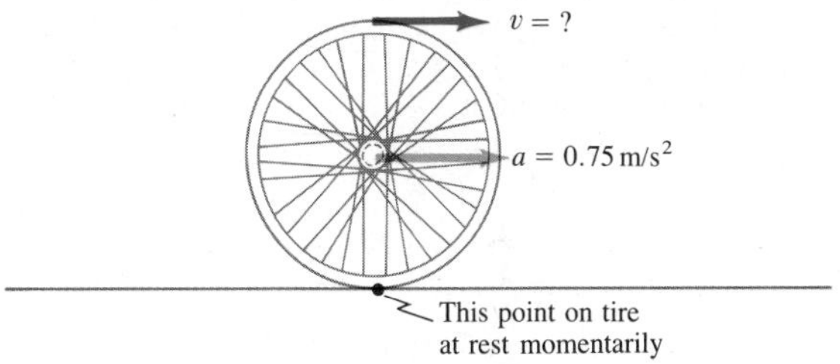

FIGURE 8–41 Problem 64.

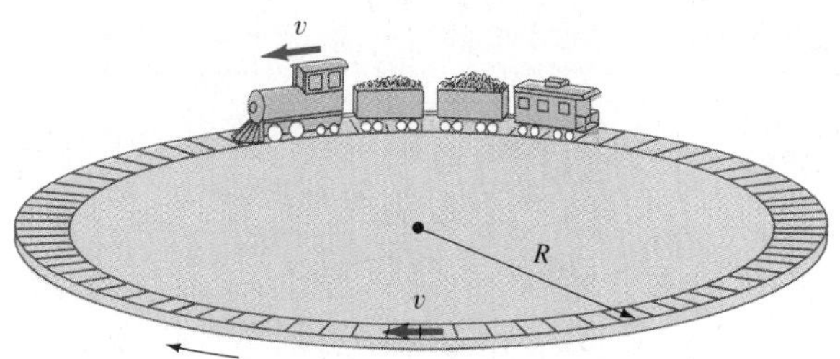

FIGURE 8–42 Problem 67.

65. The moon orbits the earth so that the same side always faces the earth. Determine the ratio of its spin angular momentum (about its own axis) to its orbital angular momentum. (In the latter case, treat the moon as a particle orbiting the earth.)

66. Suppose a star the size of our sun, but of mass 2.0 times as great, were rotating at a speed of 1 revolution every 10 days. If it were to undergo gravitational collapse to a neutron star of radius 10 km, what would its rotation speed be? Assume the star is a uniform sphere at all times.

67. A toy train can run on a horizontal circular track of radius R mounted near the edge of a solid circular board of uniform thickness with mass M, Fig. 8–42. The board is free to rotate, without friction, about a vertical axis through its center. If the system starts from rest, what mass should the train have so that its speed relative to the ground is always equal to (but in the opposite direction from) the speed relative to the ground of a point on the track no matter how fast it travels relative to the ground?

68. One possibility for a low-pollution automobile is for it to use energy stored in a heavy rotating flywheel. Suppose such a car has a total mass of 1400 kg, uses a 1.50-m diameter uniform cylindrical flywheel of mass 240 kg, and should be able to travel 300 km without needing a flywheel "spinup." (*a*) Make reasonable assumptions (average frictional retarding force = 500 N, twenty acceleration periods from rest to 90 km/h, equal uphill and downhill—assuming during downhill, energy can be put back into the flywheel), and show that the total energy needed to be stored in the flywheel is about 1.6×10^8 J. (*b*) What is the angular velocity of the flywheel when it has a full "energy charge"? (*c*) About how long would it take a 150-hp motor to give the flywheel a full energy charge before a trip?

69. A hollow cylinder (hoop) is rolling on a horizontal surface at speed $v = 5.2$ m/s when it reaches a 20° incline. (*a*) How far up the incline will it go? (*b*) How long will it be on the incline before it arrives back at the bottom?

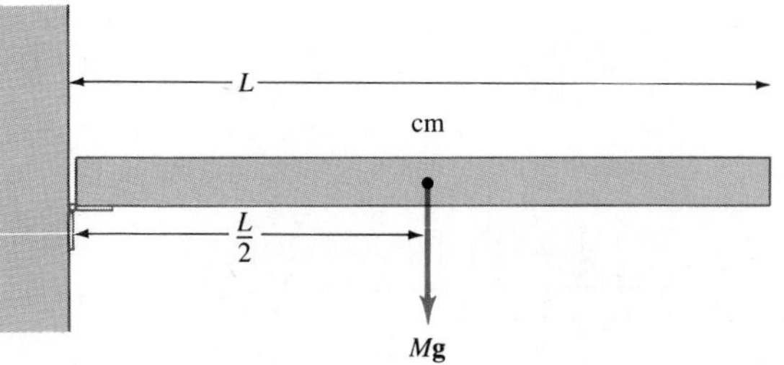

FIGURE 8–43 Problem 70.

70. A uniform rod of mass M and length L can pivot freely (i.e., we ignore friction) about a hinge attached to a wall, as in Fig. 8–43. The rod is held horizontally and then released. At the moment of release, determine (*a*) the angular acceleration of the rod and (*b*) the linear acceleration of the tip of the rod. Assume that the force of gravity acts at the center of mass of the rod, as shown. [*Hint:* see Fig. 8–14e.]

71. A wheel of mass M has radius R. It is standing vertically on the floor, and we want to exert a horizontal force F at its axle so that it will climb a step against which it rests (Fig. 8–44). The step has height h, where $h < R$. What minimum force F is needed?

FIGURE 8–44 Problem 71.

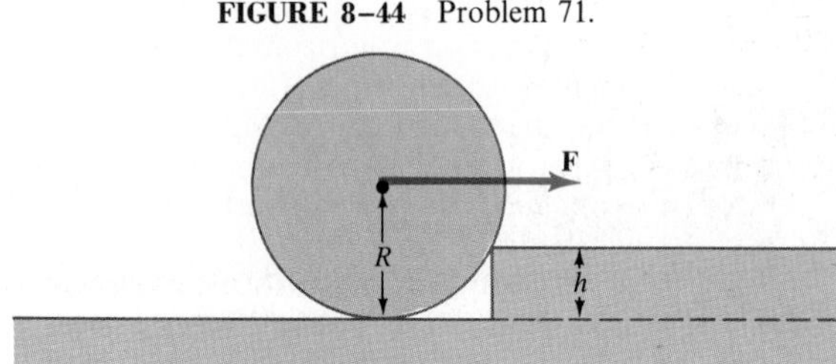

72. A bicyclist traveling with speed $v = 5.4$ m/s on a flat road is making a turn with a radius $r = 4.9$ m. The forces acting on the cyclist and cycle are the normal force ($\mathbf{F}_N$) and friction force ($\mathbf{F}_{fr}$) exerted by the road on the tires, and $m\mathbf{g}$, the total weight of the cyclist and cycle. (*a*) Explain carefully why the angle θ the bicycle makes with the vertical (Fig. 8–45) must be given by $\tan\theta = F_{fr}/F_N$ if the cyclist is to maintain balance. (*b*) Calculate θ for the values given. [*Hint:* consider the "circular" translational motion of the bicycle and rider.] (*c*) If the coefficient of static friction between tires and road is $\mu_s = 0.65$, what is the minimum turning radius?

73. (*a*) Show that the work done by a force F on a rotating body is equal to $W = \tau\,\Delta\theta$, where τ is the torque due to F (assumed constant) and $\Delta\theta$ is the angle through which the body is rotated. (*b*) Show that the rate of work done, or power P, is $P = \tau\omega$, where ω is the angular velocity at any instant. (*c*) An automobile engine develops a torque of 280 N·m at 4000 rpm. What is the horsepower of the engine?

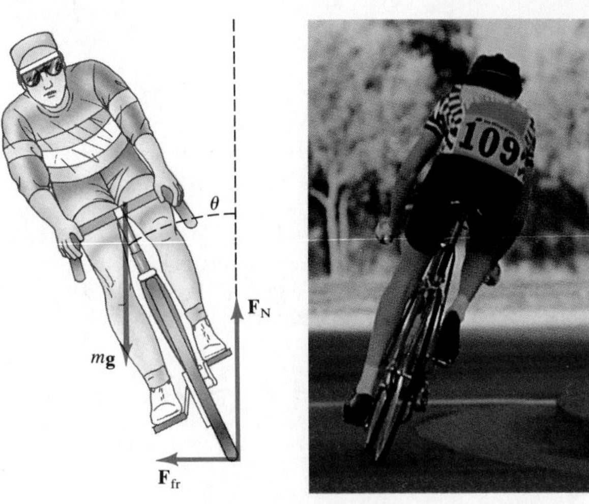

FIGURE 8–45 Problem 72.

74. A marble of mass m and radius r rolls along the looped rough track of Fig. 8–46. What is the minimum value of h if the marble is to reach the highest point of the loop without leaving the track? Assume $r \ll R_0$ and ignore frictional losses.

FIGURE 8–46 Problems 74 and 75.

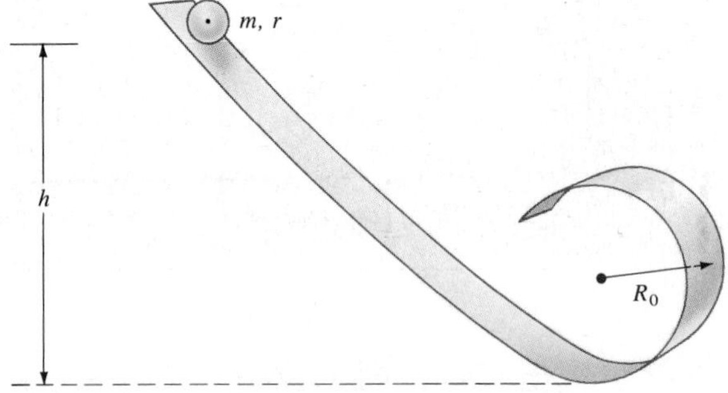

75. Repeat Problem 74, but do not assume $r \ll R_0$.

76. A thin uniform stick of mass M and length L is positioned vertically, with its tip on a frictionless table. It is released and allowed to slip and fall (Fig. 8–47). Determine the speed of its center of mass (cm) just before it hits the table.

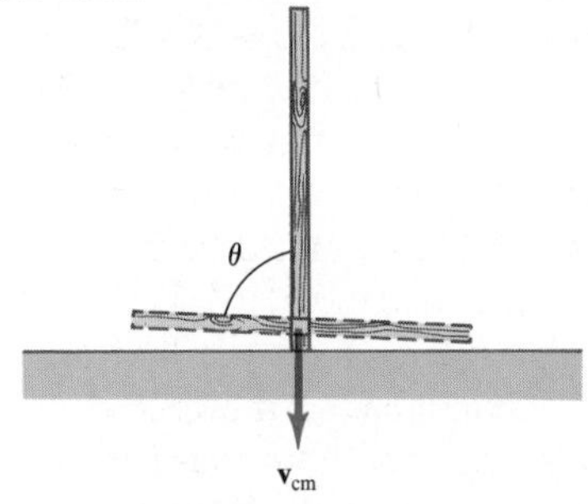

FIGURE 8–47 Problem 76.

CHAPTER 9

Bodies in Equilibrium; Elasticity and Fracture

This Greek temple in Agrigento, Sicily, was built 2500 years ago, and was built to last.

9–1 • Statics—The Study of Forces in Equilibrium

In this chapter, we will study a special case of motion—when the net force and net torque on an object or system of objects are zero, so that the object or system is either at rest, or its center of mass is moving at constant velocity. We will be concerned mainly with the first case, when the object or objects are all at rest. Now you may think that the study of objects at rest is not very interesting since the objects will have neither velocity nor acceleration; and the net force and the net torque will be zero. But it is just this last aspect that makes this subject of **statics** interesting. True, the *net* force on an object at rest is zero, but this does not imply that no forces act on it. In fact it is virtually impossible to find a body on which no forces act at all. Objects within our experience have at least one force acting on them (gravity), and if they are at rest then there must be other forces acting on them as well so that the net force is zero. An object at rest on a table, for example, has two forces acting on it, the downward force of gravity and the force the table exerts upward on it (Fig. 9–1). Since the net force is zero, the upward force

FIGURE 9–1
The book is in equilibrium; the net force on it is zero.

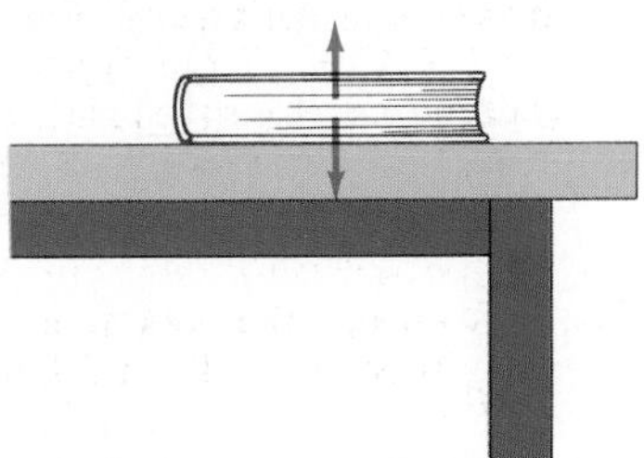

FIGURE 9–2 Elevated walkway collapse in a Kansas City hotel in 1981. How a simple physics calculation could have prevented the tragic loss of 100 lives is considered in Example 9–11.

exerted by the table must be equal in magnitude to the force of gravity acting downward. (Do not confuse these two forces with the equal and opposite forces of Newton's third law which act on different bodies; here both forces act on the same body.) Such a body is said to be in **equilibrium** (Latin for "equal forces") under the action of these two forces.

The subject of statics is concerned with the calculation of the forces acting on and within structures which are in equilibrium. The techniques for doing this can be applied in a wide range of fields. Architects and engineers must be able to calculate the forces on the structural members of buildings, bridges, machines, vehicles, and other structures, since any material will break or buckle if too much force is applied (Fig. 9–2). In the human body, a knowledge of the forces in muscles and joints is of great value for medicine and physical therapy, and is also valuable for the study of athletic activity.

Let us take a simple example of the addition of forces applied to orthodonture.

FIGURE 9–3 Forces on a tooth, Example 9–1.

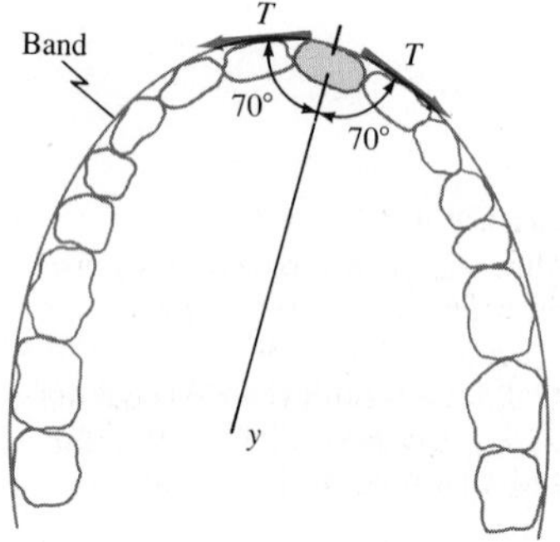

EXAMPLE 9–1 The wire band shown in Fig. 9–3 has a tension T of 2.0 N along it. It therefore exerts forces of 2.0 N on the tooth (to which it is attached) in the two directions shown. Calculate the resultant force on the tooth due to the wire.

SOLUTION Since the two forces are equal, their sum will be directed along the line that bisects the angle between them, which we have labeled the y axis. The x components of the two forces add up to zero. The y component of each force is $(2.0\text{ N})(\cos 70°) = 0.68\text{ N}$ and adding the two together we get a total force of 1.36 N. Note that if the wire is firmly attached to the tooth, the tension to the right, say, can be made larger than that to the left, and the net force would correspondingly be directed more toward the right.

Although it is intended that the tooth in this example should move, it moves very slowly and is essentially in equilibrium. The net force on it is therefore zero. The gums, then, must exert a force of 1.36 N on the tooth in a direction opposite to that exerted by the wire. In a similar way, the force exerted on the leg by the traction apparatus shown in Fig. 9–4 can be calculated to be 320 N. Notice that if the pulleys are essentially frictionless, there is a tension of $20\text{ kg}\,(9.8\text{ m/s}^2) = 200\text{ N}$ all along the cord. Thus there are two 200-N forces acting at 37° angles on the central pulley and on the leg, and the resultant is 320 N acting to the right. The leg is in equilibrium, so there must be another 320-N force acting on the leg to keep it at rest. What exerts this force?

FIGURE 9–4 Traction apparatus exerts force on a leg.

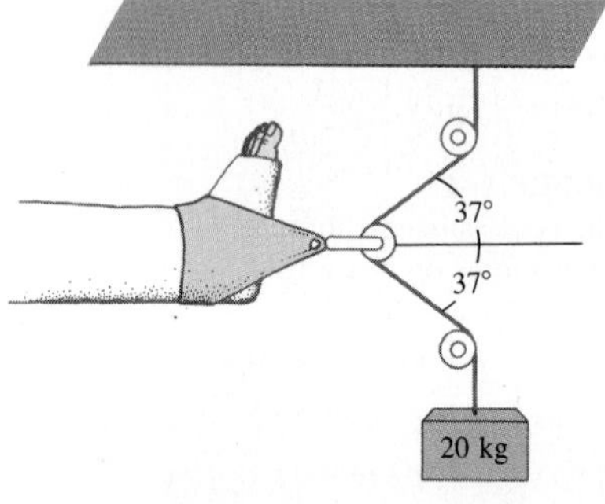

9–2 • The Conditions for Equilibrium

For a body to be at rest, the sum of the forces acting on it must add up to zero. Since force is a vector, the components of the net force must each be zero. Hence, if the forces on the object act in a plane, a condition for equilibrium is that

First condition for equilibrium: the sum of all forces is zero

$$\sum F_x = 0, \qquad \sum F_y = 0. \tag{9–1}$$

If the forces act in three dimensions (a situation we will not consider), then the third component must be zero as well, $\sum F_z = 0$. We must remember that if a particular force component points along the negative x or y axis, it must have a negative sign. Equation 9–1 is called the *first condition* for equilibrium.

EXAMPLE 9–2 Calculate the tensions $\mathbf{F}_1$ and $\mathbf{F}_2$ in the two cords that are connected to the cord supporting the chandelier shown in Fig. 9–5.

SOLUTION The three forces, $\mathbf{F}_1$, $\mathbf{F}_2$, and the 200-kg weight of the chandelier, act at the point where the three cords join. We choose this junction point (it could be a knot) as the object for which we write $\sum F_x = 0$, $\sum F_y = 0$. (We don't bother considering the chandelier itself since only two forces act on it, gravity downward and the equal but opposite force exerted upward by the cord, both of which equal $mg = (200 \text{ kg})(9.8 \text{ m/s}^2) =$ 1960 N.) There are only two unknowns and we can solve for them using Eqs. 9–1. We first resolve $\mathbf{F}_1$ into its horizontal (x) and vertical (y) components. Although we don't know the value of $\mathbf{F}_1$, we can write $F_{1x} = F_1 \cos 60°$ and $F_{1y} = F_1 \sin 60°$. $\mathbf{F}_2$ has only an x component. In the vertical direction, we have only the weight of the chandelier $= (200 \text{ kg})(g)$ acting downward and the vertical component of $\mathbf{F}_1$ upward. Since $\sum F_y = 0$, we have

$$\sum F_y = F_1 \sin 60° - (200 \text{ kg})(g) = 0$$

so

$$F_1 = \frac{(200 \text{ kg})g}{\sin 60°} = (231 \text{ kg})g = 2260 \text{ N}.$$

In the horizontal direction,

$$\sum F_x = F_2 - F_1 \cos 60° = 0.$$

Thus

$$F_2 = F_1 \cos 60° = (231 \text{ kg})(g)(0.500) = (115 \text{ kg})g = 1130 \text{ N}.$$

The magnitudes of $\mathbf{F}_1$ and $\mathbf{F}_2$ determine the strength of cord or wire that must be used. In this case, the wire must be able to hold at least 230 kg. Note in this example that we didn't insert the value of g, the acceleration due to gravity, until the end. In this way we found the magnitude of the force in terms of the number of kilograms (which may be a more familiar quantity than newtons) times g.

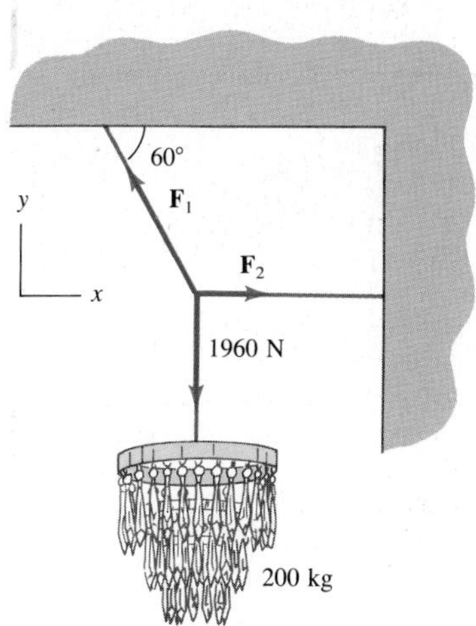

FIGURE 9–5 Example 9–2.

FIGURE 9–6 Although the net force on it is zero, the pencil will move (rotate). A pair of equal forces acting in opposite directions but at different points on a body (as shown here) is referred to as a *couple*.

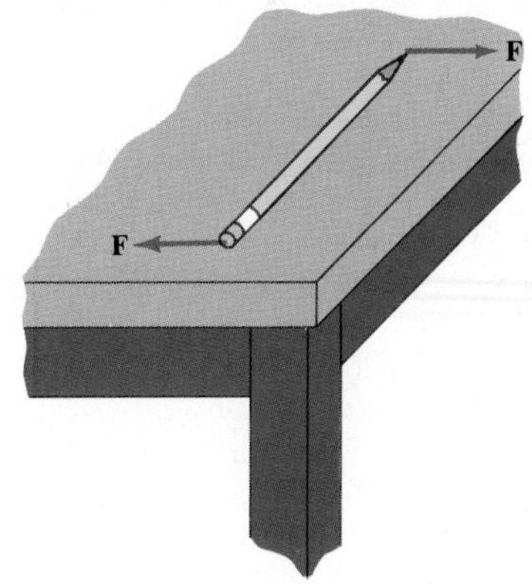

Although Eqs. 9–1 must be true if an object is to be in equilibrium, they are not a sufficient condition. Figure 9–6 shows an object on which the net force is zero. Although the two forces labeled **F** add up to give zero net force on the object, they do give rise to a torque that will rotate the object. Referring to Eq. 8–14, we see that if a body is to remain at rest, the net torque applied to it must be zero. Thus we have the *second condition* for equilibrium: that the sum of the torques acting on a body must be zero:

Second condition for equilibrium: the sum of all torques is zero

$$\sum \tau = 0. \qquad (9\text{–}2)$$

This will assure that the angular acceleration, α, about any axis will be zero.

If the body is not rotating initially ($\omega = 0$) it will not start rotating. Equations 9–1 and 9–2 are the *only* requirements for a body to be in equilibrium.

We will consider cases in which the forces all act in a plane[†] (we call it the xy plane). In this case the torque is calculated about an axis that is perpendicular to the xy plane. The choice of this axis is arbitrary. Since $\tau = I\alpha$ for a rigid body about any fixed axis (Eq. 8–14), $\tau = 0$ about any fixed axis if the body is in equilibrium ($\alpha = 0$). Therefore we can choose any axis that makes our calculation easier.

9–3 • Solving Statics Problems

This subject of statics is important because it allows us to calculate certain forces on (or within) a structure when some of the forces on it are already known. There is no single technique for attacking such statics problems, but the following procedure is helpful. (1) Choose one body at a time for consideration, and make a free-body diagram for it by showing all the forces acting on the body and the points at which these forces act. (2) Choose a convenient coordinate system and resolve the forces into their components. (3) Using letters to represent unknowns, write down equations for $\sum F_x = 0$, $\sum F_y = 0$, and $\sum \tau_z = 0$, and then solve these equations. Undoubtedly the hardest step is (1): *all* the forces *on* the body must be included, but the forces exerted *by* this body on other objects must *not* be included.

Problem-solving hints

It is useful to refer back to Example 9–2 to see how this procedure was followed. Notice that we did not bother to consider the forces on the chandelier (which are gravity downward and the rope pulling upward with an equal magnitude force). Since we wanted to know the tension in the two cords, we chose the junction point (or knot) of the cords as our object. Since the object is essentially a point, there would be no torques, so the torque equation was not used. Of course, one does not have to use all three equations if they are not needed.

In many situations we must make use of the torque equation (9–2). Just as the components of a force are positive or negative depending on their direction, so too are torques. If a torque that tends to rotate the object counterclockwise is considered positive, then a torque that tends to rotate it clockwise must be considered negative. Another way of stating the equilibrium conditions is that the sum of all clockwise torques is equal to the sum of all counterclockwise torques. And for the forces: the sum of the upward forces is equal to the sum of the downward forces, and the sum of horizontal forces to the left is equal to the sum of horizontal forces to the right.

One of the forces that acts on bodies is the force of gravity. Our analysis in this chapter is greatly simplified if we use the concept of center of gravity (cg). As we discussed in Section 7–8, we can consider the force of gravity on the body as a whole as acting at its cg. For uniform symmetrically shaped bodies, the cg is at the geometric center. For more complicated bodies, the cg can be determined as discussed in Section 7–8.

[†] If the forces are not in a plane, the situation is more complicated. There will be three force equations plus three torque equations (torque calculated about three mutually perpendicular axes). We won't usually be concerned with such complicated situations.

EXAMPLE 9–3 A 20-kg board serves as a seesaw for two children, as shown in Fig. 9–7a. One child has a mass of 30 kg and sits 2.5 m from the pivot point (i.e., his cg is 2.5 m from the pivot). At what distance, x, from the pivot must a 25-kg child place herself to balance the seesaw? Assume the board is centered over the pivot.

SOLUTION We draw a free-body diagram for the board, as shown in Fig. 9–7b. The forces acting on the board are the forces exerted downward on it by each child, $\mathbf{F}_1$ and $\mathbf{F}_2$, the upward force exerted by the pivot, $\mathbf{F}_\text{N}$, and the force of gravity (its weight). Let us calculate the torque about the pivot point: the lever arms for F_N and for the weight of the board are then zero. Thus the torque equation will involve only the forces $\mathbf{F}_1$ and $\mathbf{F}_2$, which are equal to the weights of the children. The torque exerted by each child will be mg times the appropriate lever arm. Hence the torque equation is

$$\sum \tau = (30 \text{ kg})(g)(2.5 \text{ m}) - (25 \text{ kg})(g)(x) = 0,$$

where a torque tending to rotate the board counterclockwise has been chosen to be a positive torque, and a torque tending to rotate it clockwise to be negative. The acceleration of gravity, g, appears in both terms and cancels out. We solve for x and find:

$$x = \frac{(30 \text{ kg})(2.5 \text{ m})}{25 \text{ kg}} = 3.0 \text{ m}$$

To balance the seesaw, the second child must sit so that her cg is 3.0 m from the pivot. This makes sense: since she is lighter, she must sit farther from the pivot.

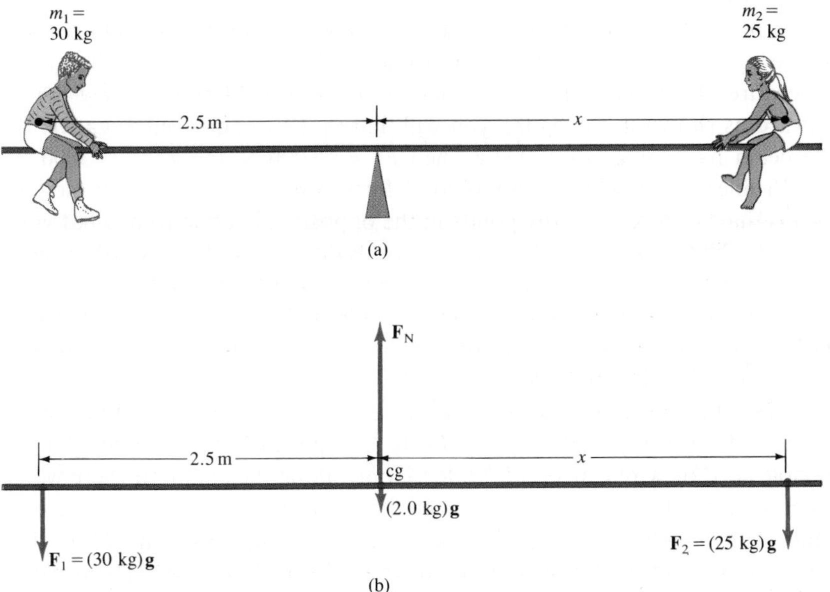

FIGURE 9–7 (a) Two children on a seesaw, Example 9–3. (b) Free-body diagram of the board.

respectively. The values shown are approximations taken from Table 7–1. The distances (in cm) refer to a person 180 cm tall, but are approximately in the same ratio of 1:2:4 for an average person of any height, and the answer below is then independent of the height of the person.

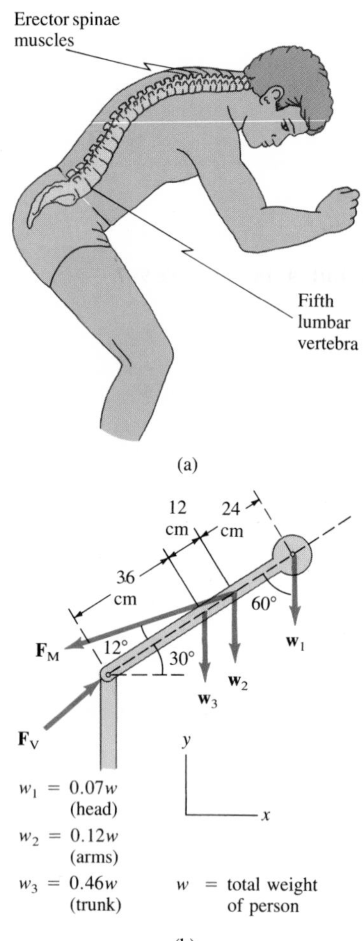

FIGURE 9–14
(a) A person bending over.
(b) Forces required of the back muscles ($\mathbf{F}_M$) and in the vertebrae ($\mathbf{F}_V$) when a person bends over.

EXAMPLE 9–8 Calculate the magnitude and direction of the force $\mathbf{F}_V$ acting on the fifth lumbar vertebra for the example shown in Fig. 9–14b.

SOLUTION First we calculate F_M using the torque equation (Eq. 9–2) about the base of the spine. The torques are calculated using $\tau = rF_\perp = r(F \sin \theta)$ where θ is 12° for $\mathbf{F}_M$ and 60° for $\mathbf{w}_1$, $\mathbf{w}_2$, and $\mathbf{w}_3$:

$$(0.48 \text{ m})(\sin 12°)(F_M) - (0.72 \text{ m})(\sin 60°)(w_1) - (0.48 \text{ m})(\sin 60°)(w_2) - (0.36 \text{ m})(\sin 60°)(w_3) = 0.$$

Putting in the values for w_1, w_2, w_3 given in the figure, we find $F_M = 2.2w$, where w is the total weight of the body. To get the components of F_V we use the x and y components of the force equation:

$$\sum F_y = F_{Vy} - F_M \sin 18° - w_1 - w_2 - w_3 = 0$$

$$F_{Vy} = 1.3w.$$

$$\sum F_x = F_{Vx} - F_M \cos 18° = 0$$

$$F_{Vx} = 2.1w.$$

Then $F_V = \sqrt{F_{Vx}^2 + F_{Vy}^2} = 2.5w$. The angle θ that F_V makes with the horizontal is given by $\tan \theta = F_{Vy}/F_{Vx} = 0.62$, so $\theta = 32°$.

The force on the lowest vertebra is thus $2\frac{1}{2}$ times the body weight! This force is transmitted from the "sacral" bone at the base of the spine, through the fluid-filled and somewhat flexible *intervertebral disc*. The discs at the base of the spine are clearly being compressed under very large forces.

If the person in Fig. 9–14 has a mass of 90 kg, and is holding 20 kg in his hands (this increases w_2 to $0.33w$), then F_V is increased to nearly five times the person's weight ($5w$)! (For this 200-lb person, the force on the disc would be 1000 lb!) With such strong forces acting, it is little wonder that so many people suffer from lower back pain at one time or another in their lives.

*9–5 • Simple Machines: Levers, Pulleys, and Multispeed Bicycles

Simple machines, such as levers, pulleys, gears, and the wheel and axle, are designed to reduce the force needed to lift a heavy load. They can be analyzed using the principles we have discussed in this chapter. Consider, for example, the lever shown in Fig. 9–15. If the stone is lifted slowly, it is essentially in equilibrium, and using the torque rule ($\sum \tau = 0$), with the torques calculated about the pivot point, we see that a force of only (50 kg)(g) = 500 N is needed to lift the stone whose weight is (200 kg)(g) = 2000 N. In a sense, the force has been multiplied by a factor of 4, and we say the lever gives us a *mechanical advantage* of 4. From the geometry of the situation as shown in Fig. 9–15b,

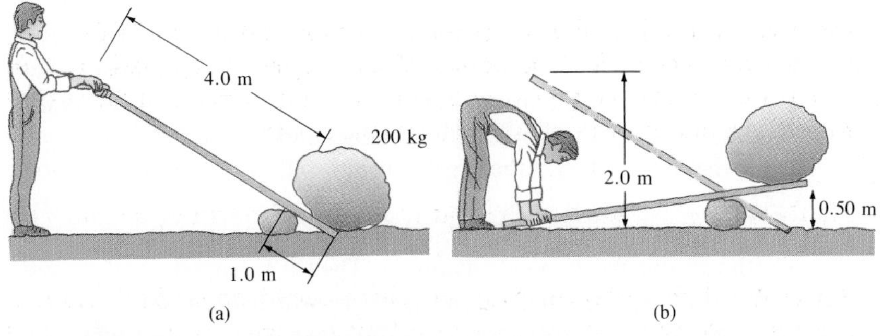

FIGURE 9–15 A simple lever.

the 500-N force must be exerted through a distance of 2.0 m to raise the stone 0.50 m. Thus the work done by the person is (500 N)(2.0 m) = 1000 J, and the work done on the stone is (2000 N)(0.50 m) = 1000 J, which is the same. In other words, the input work equals the output work (as expected from conservation of energy).

This example has been idealized, for in practice it would be found that the force necessary to lift the 2000-N stone would be different from 500 N. This is due to two factors common to all machines: friction and the weight of the machine itself (for example, in Fig. 9–15 we would need to include the weight of the lever in calculating the torques). We therefore define an "ideal" and an "actual" mechanical advantage. The *ideal mechanical advantage* (IMA) is defined as the (output force)/(input force) for the ideal situation, which means the output work = input work, or $F'_i d_i = F'_o d_o$ (the subscript "o" means output and "i" means input and the primes remind us we are talking of the force in the ideal case). From this relation we have $F'_o/F'_i = d_i/d_o$. Thus the ideal mechanical advantage is also the ratio of the distance d_i moved by the input force to the distance d_o moved by the load (the output):

$$\text{IMA} = \frac{F'_o}{F'_i} = \frac{d_i}{d_o}.$$

The *actual mechanical advantage* (AMA) is the ratio of the actual output force to actual input force for the particular machine:

$$\text{AMA} = \frac{F_o}{F_i}.$$

Finally, the efficiency of any machine, be it simple or complex, is defined as (work output)/(work input). This can be written in terms of the IMA and AMA:

$$\begin{aligned}\text{efficiency} &= \frac{F_o d_o}{F_i d_i} \\ &= \left(\frac{F_o}{F_i}\right)\Big/\left(\frac{d_i}{d_o}\right) = \frac{\text{AMA}}{\text{IMA}}.\end{aligned}$$

For example, suppose it actually took a 550-N force to lift the 2000-N stone in Fig. 9–15. The IMA = 2.0 m/0.50 m = 4.0, whereas the AMA = 2000 N/550 N = 3.6. The efficiency is then (3.6)/(4.0) = 0.90, or 90 percent.

Pulleys are often used to mechanical advantage. The pulley system shown in Fig. 9–16 has an IMA of 2.0. This is readily seen by noting that there are two ropes supporting the weight, and in the ideal case each need

FIGURE 9–16 A simple pulley.

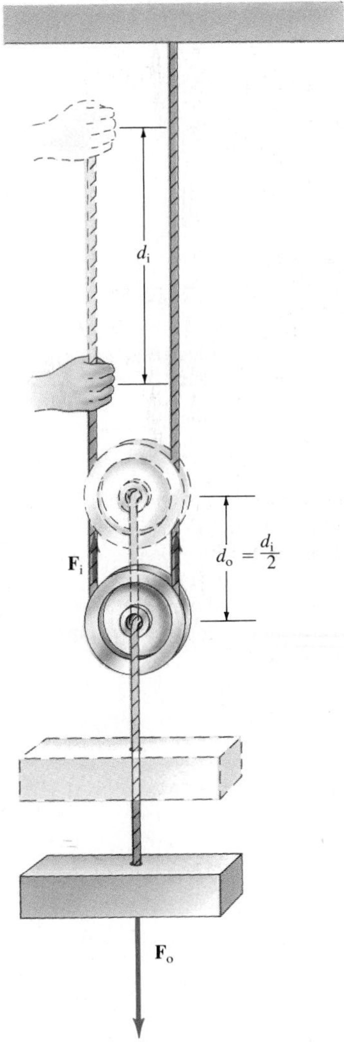

carry only half the weight. This can also be seen from the fact, apparent in Fig. 9–16, that $d_o = d_i/2$. For more complicated systems, such as those shown in Fig. 9–17, similar considerations will reveal that the IMA in general is equal to the number of ropes pulling up on the load. The AMA will always be less than the IMA. Hence the work output will be less than the work input.

10-speed bike

A 10-speed (or 12-, 15-, or 18-speed) bicycle is an example of a simple machine with variable mechanical advantage. The IMA depends on the radius of the wheel (r_w), on the radius of the circle in which the pedals turn (r_p), and on the number of teeth on the front and rear sprockets on which the chain is riding (Fig. 9–18). When you shift gears, you move the chain from one sprocket wheel to another with a different number of teeth, and thus you change the mechanical advantage. Suppose, for example, that the chain rides on a 42-tooth sprocket at the front and a 26-tooth sprocket at the rear (Fig. 9–18). Since the evenly spaced chain links fall sequentially on the teeth, when the pedals make one complete revolution, the chain moves 42 links; then the rear wheel makes (42/26) revolutions. The input force, exerted by the cyclist tangentially to the pedals' circular path, moves through a distance $d_i = 2\pi r_p$ in one revolution. This produces (42/26) revolutions of the rear wheel, which thus moves a distance $d_o = (42/26)\ 2\pi r_w$ along the ground. Hence

$$\text{IMA} = \frac{d_i}{d_o} = \frac{2\pi r_p}{(\frac{42}{26})\ 2\pi r_w} = \left(\frac{26}{42}\right)\frac{r_p}{r_w}.$$

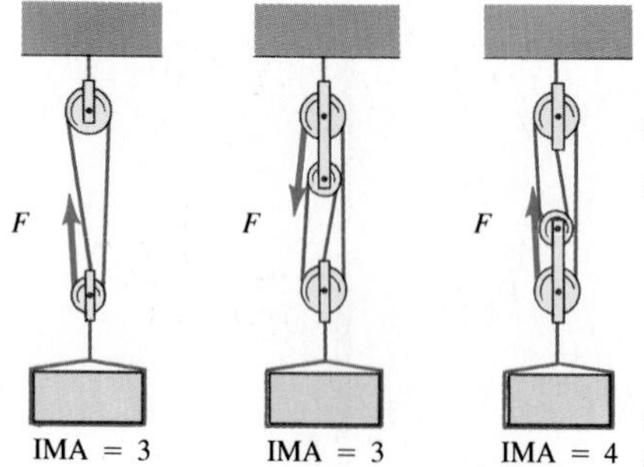

FIGURE 9–17 Various pulley systems.

On a 10-speed racing bike, the pedal arms are 18 cm long, so $r_p = 18$ cm, and the wheels have a diameter of 27 inches or 68 cm, so $r_w = 34$ cm.

In this case

$$\text{IMA} = \left(\frac{26}{42}\right)\left(\frac{18\text{ cm}}{34\text{ cm}}\right) = 0.33,$$

which is less than 1. Indeed, a bicycle is not normally a device to increase force, but rather a device to increase distance traveled (or speed). The IMA just calculated is relatively high and corresponds to a relatively low gear which would be used for climbing hills, where your input force needs to

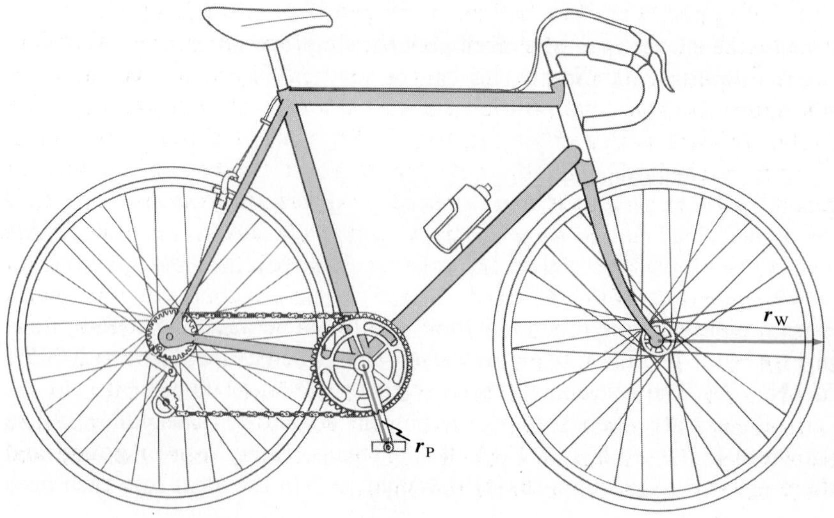

FIGURE 9–18
A multispeed racing bike. Only the sprocket wheels on which the chain is riding at this moment are shown.

oppose the force of gravity. On level ground, where only friction and wind resistance are the opposing forces, you can use a higher gear, corresponding to fewer teeth on the rear sprocket and/or more teeth on the front sprocket. For example, if the front sprocket was still a 42, but you shifted the chain to a rear sprocket with 13 teeth, each revolution of the pedals would correspond to (42/13) revolutions of the rear wheel, twice as many as for the 42:26 ratio. You would thus travel twice as far per pedal revolution. Of course, your maximum output force would be less, since the IMA would be smaller: IMA = (13/42)(18 cm/34 cm) = 0.16. If you also shifted the front sprocket to a 52, your IMA would be even lower and the distance traveled per pedal revolution correspondingly greater.

9–6 • Stability and Balance

A body in static equilibrium, if left undisturbed, will undergo no translational or rotational acceleration since the sum of all the forces and the sum of all the torques acting on it are zero. However, if the object is displaced slightly, three different outcomes are possible: (1) the object returns to its original position, in which case it is said to be in **stable equilibrium**; (2) the object moves even farther from its original position, in which case it is said to be in **unstable equilibrium**; or (3) the object remains in its new position, in which case it is said to be in **neutral equilibrium**.

Stable and unstable equilibrium

Consider the following examples. A ball suspended freely from a string is in stable equilibrium for if it is displaced to one side, it will quickly return to its original position (Fig. 9–19a). On the other hand, a pencil standing on its point is in unstable equilibrium. If its cg is directly over its tip (Fig. 9–19b), the net force and net torque on it will be zero. But if it is displaced ever so slightly—say by a slight vibration or tiny air current—there will be a torque on it and it will continue to fall in the direction of the original displacement. Finally, an example of an object in neutral equilibrium is a sphere resting on a horizontal tabletop. If it is placed slightly to one side, it will remain in its new position.

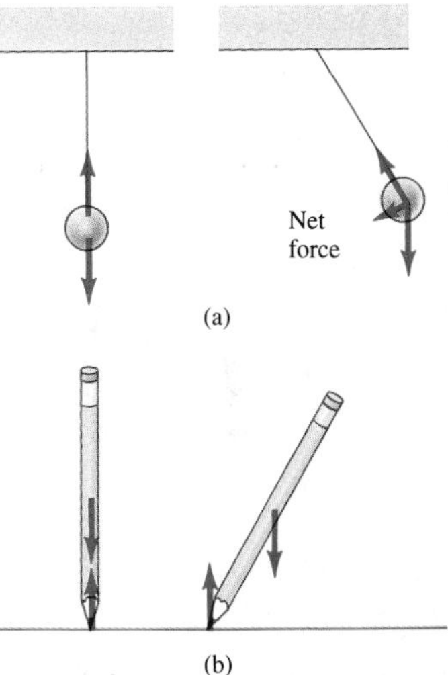

FIGURE 9–19 (a) Stable equilibrium; and (b) unstable equilibrium.

In most situations, such as in the design of structures and in working with the human body, we are interested in maintaining stable equilibrium or *balance*, as we sometimes say. In general, an object whose cg is below its point of support, such as a ball on a string, will be in stable equilibrium. If the cg is above the base of support, we have a more complicated situation. Consider a block standing on its end (Fig. 9–20a). If it is tipped slightly, it will return to its original position due to the torque on it as shown in Fig. 9–20b. But if it is tipped too far, Fig. 9–20c, it will fall over. The critical point is reached when the cg is no longer above the base of support. In general, *a body whose cg is above its base of support will be stable if a vertical line projected downward from its cg falls within the base of support.* This is because the upward force on the object (which balances out gravity) can only be exerted within the area of contact, so that if the force of gravity acts beyond this area, a net torque will act to topple the object. Stability, then, can be relative. A brick lying on its widest face is more stable than a brick standing on its end, for it will take more of an effort to tip it over. In the extreme case of the pencil in Fig. 9–19b, the base is practically a point and the slightest disturbance will topple it. In general, the larger the base and the lower the cg, the more stable the object.

FIGURE 9–20 Equilibrium of a block resting on a surface.

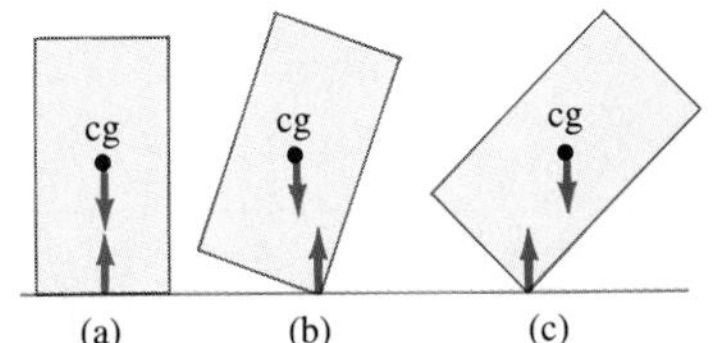

FIGURE 9–21
Humans adjust their posture when carrying loads to achieve stability.

In this sense, humans are much less stable than four-legged mammals, which not only have a larger base of support because of their four legs, but have a lower center of gravity. The human species has had to develop special apparatus, such as certain very strong muscles, in order to deal with the problem of keeping a person upright and at the same time stable. Because of their upright position, humans suffer from numerous ailments such as low back pain due to the large forces involved, as we saw in Example 9–8. When walking and performing other kinds of movement, a person continually shifts the body so that its cg is over the feet, although in the normal adult this requires no conscious thought. Even as simple a movement as bending over requires moving the hips backward so that the cg remains over the feet, and this repositioning is done without thinking about it. To illustrate this, position yourself with your heels and back to a wall and try to touch your toes. Persons carrying heavy loads automatically adjust their posture so that the cg of the total mass is over their feet, as shown in Fig. 9–21.

FIGURE 9–22 Hooke's law: $\Delta L \propto$ applied force.

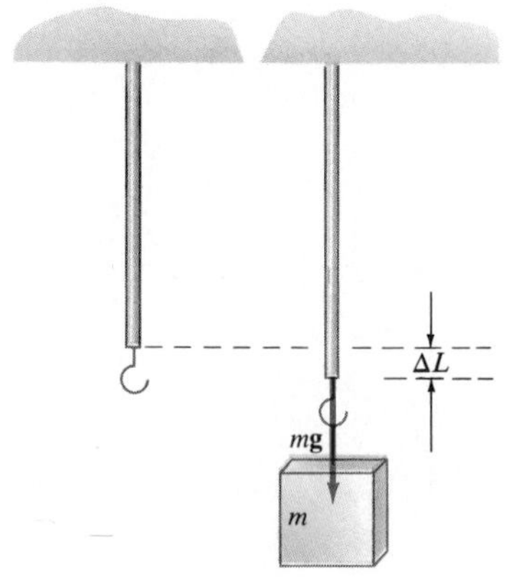

9–7 • Elasticity; Stress and Strain

In the first part of this chapter we studied how to calculate the forces on objects in equilibrium. In this section we study the effects of these forces, for any object changes shape under the action of applied forces. If the forces are great enough, the object will break or *fracture*, and we discuss this in Section 9–8.

If a force is exerted on an object, such as the vertically suspended metal bar shown in Fig. 9–22, the length of the object changes. If the amount of elongation, ΔL, is small compared to the length of the object, experiment shows that ΔL is proportional to the weight or force exerted on the object [a relationship first noted by Robert Hooke (1635–1703)]. This proportionality, as we saw in Section 6–4, can be written as an equation and is sometimes called *Hooke's law*†:

Hooke's law again

$$F = k\,\Delta L. \tag{9–3}$$

Here F represents the force (or weight) pulling on the object, ΔL is the increase in length, and k is a proportionality constant. Equation 9–3 is found to be valid for almost any solid material from iron to bone, but it is valid only up to a point. For if the force is too great, the object stretches excessively and eventually breaks. Figure 9–23 shows a typical graph of elongation versus applied force. Up to a point called the **proportional limit**, Eq. 9–3 is a good approximation for many common materials and the curve is a straight line. Beyond this point, the graph deviates from a straight line and no simple relationship exists between F and ΔL. Nonetheless, up to a point farther along the curve called the **elastic limit**, the object will return to its original length if the applied force is removed. The region from the origin to the elastic limit is called the *elastic region*. If the object is stretched beyond the elastic limit, it enters the *plastic region*: it does not return to the original length upon removal of the external force, but remains permanently deformed. The maxi-

FIGURE 9–23
Applied force vs. elongation for a typical metal under tension.

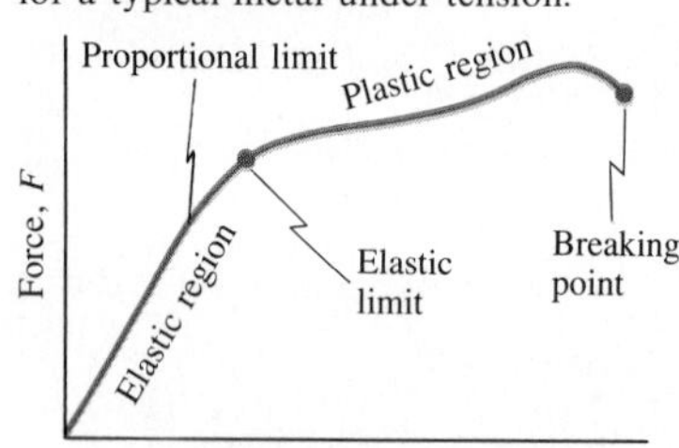

† The term "law" applied to this relation is not really appropriate—see the footnote on this point in Section 6–4.

mum elongation is reached at the *breaking point*. The maximum force that can be applied without breaking is called the **ultimate strength** of the material (discussed in Section 9–8).

The amount of elongation of an object, such as the bar shown in Fig. 9–22, depends not only on the force applied to it, but also on the material from which it is made and on its dimensions. That is, the constant k in Eq. 9–3 can be written in terms of these factors. If we compare bars made of the same material but of different lengths and cross-sectional areas, it is found that for the same applied force, the amount of stretch (again assumed small compared to the total length) is proportional to the original length and inversely proportional to the cross-sectional area. That is, the longer the object, the more it elongates for a given force; and the thicker it is, the less it elongates. These experimental findings can be combined with Eq. 9–3 to yield the relation

$$\Delta L = \frac{1}{E}\frac{F}{A}L_0, \qquad (9\text{–}4)$$

where L_0 is the original length of the object, A is the cross-sectional area, and ΔL is the change in length due to the applied force F. E is a constant of proportionality[†] known as the **elastic modulus**, or **Young's modulus**, and its value depends only on the material. The value of Young's modulus for various materials is given in Table 9–1. Because E is a property only of the

Young's modulus

[†] The fact that E is in the denominator, so that $1/E$ is the actual proportionality constant, is merely a convention.

Table 9–1
Elastic Moduli

Material	Elastic Modulus, E (N/m²)	Shear Modulus, G (N/m²)	Bulk Modulus, B (N/m²)
Solids			
Iron, cast	100×10^9	40×10^9	90×10^9
Steel	200×10^9	80×10^9	140×10^9
Brass	100×10^9	35×10^9	80×10^9
Aluminum	70×10^9	25×10^9	70×10^9
Concrete	20×10^9		
Brick	14×10^9		
Marble	50×10^9		70×10^9
Granite	45×10^9		45×10^9
Wood (pine)			
(parallel to grain)	10×10^9		
(perpendicular to grain)	1×10^9		
Nylon	5×10^9		
Bone (limb)	15×10^9	80×10^9	
Liquids			
Water			2.0×10^9
Alcohol (ethyl)			1.0×10^9
Mercury			2.5×10^9
Gases[†]			
Air, H_2, He, CO_2			1.01×10^5

[†] At normal atmospheric pressure; no variation in temperature during process.

material and is independent of the object's size or shape, Eq. 9–4 is far more useful for practical calculation than Eq. 9–3. From Eq. 9–4, we see that the change in length of an object is directly proportional to the product of the object's length L_0 and the force per unit area F/A applied to it. It is general practice to define the force per unit area as the **stress**:

Stress

$$\text{stress} = \frac{\text{force}}{\text{area}} = \frac{F}{A}.$$

Also, the **strain** is defined to be the ratio of the change in length to the original length:

Strain

$$\text{strain} = \frac{\text{change in length}}{\text{original length}} = \frac{\Delta L}{L_0}.$$

Strain is thus the fractional change in length of the object, and is a measure of how much the bar has been deformed. Equation 9–4 can be rewritten as

$$\frac{F}{A} = E\frac{\Delta L}{L_0} \tag{9–5}$$

or

$$E = \frac{F/A}{\Delta L/L_0} = \frac{\text{stress}}{\text{strain}}.$$

Thus we see that the strain is directly proportional to the stress.

EXAMPLE 9–9 A 1.60-m-long steel piano wire has a diameter of 0.20 cm. How great is the tension in the wire if it stretches 0.30 cm when tightened?

SOLUTION We solve for F in Eq. 9–4 and note that the area $A = \pi r^2 = (3.14)(0.0010\ \text{m})^2 = 3.1 \times 10^{-6}\ \text{m}^2$. Then

$$F = E\frac{\Delta L}{L_0}A$$

$$= (2.0 \times 10^{11}\ \text{N/m}^2)\left(\frac{0.0030\ \text{m}}{1.60\ \text{m}}\right)(3.1 \times 10^{-6}\ \text{m}^2) = 1200\ \text{N},$$

where we obtained the value for E from Table 9–1.

Tension

The bar shown in Fig. 9–22 is said to be under *tension* or *tensile stress.* For not only is there a force pulling down on the bar at its lower end, but since the bar is in equilibrium we know that the support at the top is exerting an equal† upward force on the bar at its upper end, Fig. 9–24a. In fact, this tensile stress exists throughout the material. Consider, for example, the lower half of a suspended bar as shown in Fig. 9–24b. This lower half is in equilibrium, so there must be an upward force on it to balance the downward force at its lower end. What exerts this upward force? It must be the upper part of the bar. Thus we see that external forces applied to an object give rise to internal forces, or stress, within the material itself. (Recall also the discussion of tension in a cord, Example 4–8.)

FIGURE 9–24 Stress exists *within* the material.

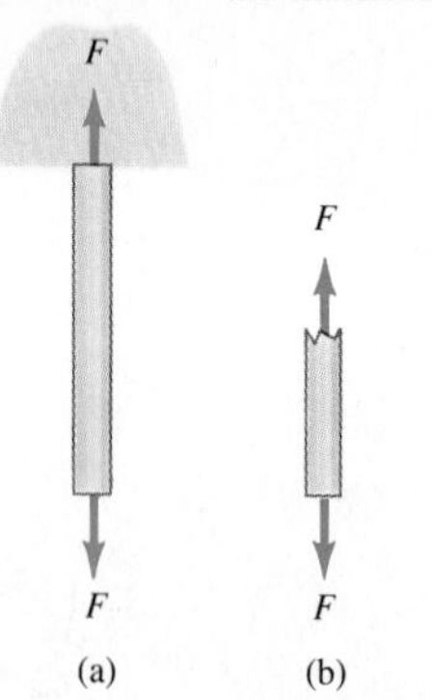

† If we ignore the weight of the bar.

FIGURE 9–25 This Greek temple, in Agrigento, Sicily, shows the post and beam construction. Built 2500 years ago, the temple stands intact today.

Strain or deformation due to tensile stress is but one type of stress to which materials can be subjected. There are two other common types of stress: compressive and shear. *Compressive* stress is the exact opposite of tensile stress. Instead of being stretched, the material is compressed: the forces act inwardly on the body. Any sort of column that supports a weight, such as the columns of a Greek temple (Fig. 9–25) is subjected to a compressive stress. Equations 9–4 and 9–5 apply equally well to compression and tension, and the values for E are usually the same.

Compression

Figure 9–26 compares tensile and compressive stresses as well as the third type, shear stress. An object under *shear* stress has equal and opposite forces applied *across* its opposite faces. An example is a book or brick firmly attached to a tabletop with a force exerted parallel to the top surface. The table exerts an equal and opposite force along the bottom surface. Although the dimensions of the object do not change significantly, the shape of the

Shear

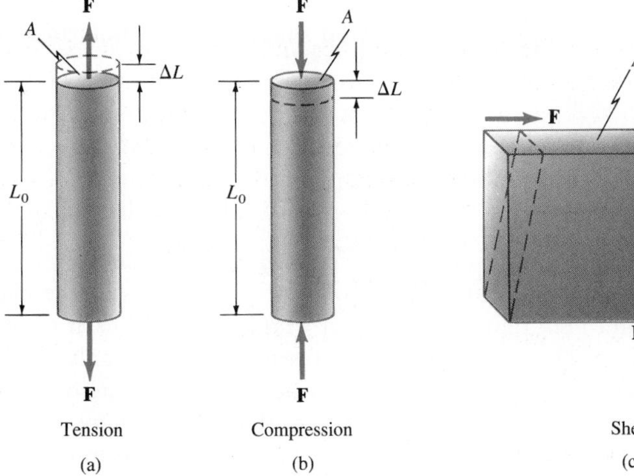

FIGURE 9–26 The three types of stress.

object does change as shown in the figure. An equation similar to 9–4 can be applied to calculate shear strain:

$$\Delta L = \frac{1}{G}\frac{F}{A}L_0 \tag{9–6}$$

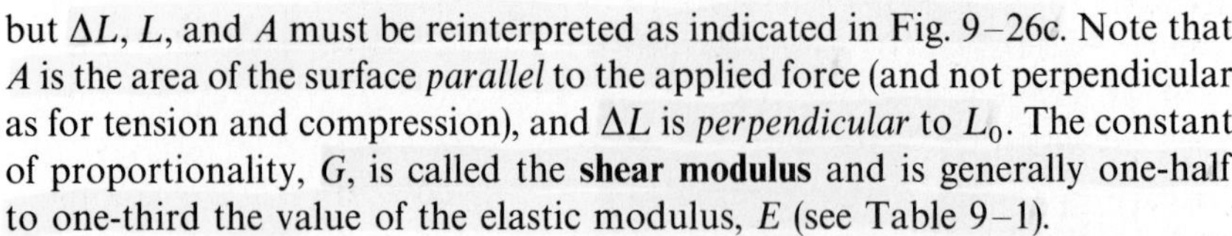

but ΔL, L, and A must be reinterpreted as indicated in Fig. 9–26c. Note that A is the area of the surface *parallel* to the applied force (and not perpendicular as for tension and compression), and ΔL is *perpendicular* to L_0. The constant of proportionality, G, is called the **shear modulus** and is generally one-half to one-third the value of the elastic modulus, E (see Table 9–1).

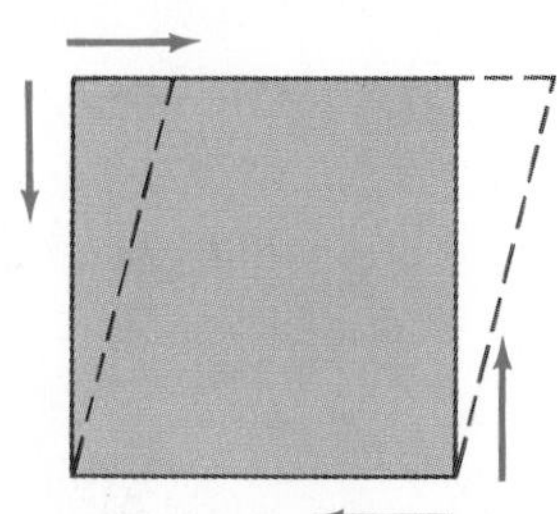

FIGURE 9–27 Balance of forces and torques for shear stress.

The rectangular object undergoing shear in Fig. 9–26c would not actually be in equilibrium under the forces shown, for a net torque would exist. If the object is in fact in equilibrium, there must be two more forces acting on it which balance out this torque. One acts vertically upward on the right, and the other acts vertically downward on the left, as shown in Fig. 9–27. This is generally true of shear forces. If the object is a brick or book lying on a table, these two additional forces can be exerted by the table and whatever exerts the other horizontal force (such as your hand pushing across the top of a book).

If an object is subjected to forces from all sides, its volume will decrease. A common situation is a body submerged in a fluid; for in this case, the fluid exerts a pressure on the object in all directions, as we shall see in Chapter 10. Pressure is defined as force per area, and thus is the equivalent of stress. For this situation the change in volume, ΔV, is found to be proportional to the original volume, V_0, and to the increase in the pressure, ΔP. We thus obtain a relation of the same form as Eq. 9–4 but with a proportionality constant called the **bulk modulus**, B:

$$\frac{\Delta V}{V_0} = -\frac{1}{B}\Delta P \tag{9–7}$$

or

$$B = -\frac{\Delta P}{\Delta V/V_0}.$$

The minus sign is included to indicate that the volume *decreases* with an increase in pressure. Values for the bulk modulus are given in Table 9–1. Since liquids and gases do not have a fixed shape, only the bulk modulus applies to them.

9–8 • Fracture

Ultimate strength

If the stress on a solid object is too great, the object fractures or breaks (Fig. 9–28). In Table 9–2 are listed the ultimate tensile strength, compressive strength, and shear strength for a variety of materials. These give the maximum force per unit area that an object can withstand under each of these three types of stress. They are, however, representative values only and the actual value for a given specimen can differ considerably. It it therefore necessary to maintain a safety factor of from 3 to perhaps 10 or more—that is, the actual stresses on a structure should not exceed one-tenth to one-third

TABLE 9–2
Ultimate Strengths of Materials (force/area)

Material	Tensile Strength (N/m^2)	Compressive Strength (N/m^2)	Shear Strength (N/m^2)
Iron, cast	170×10^6	550×10^6	170×10^6
Steel	500×10^6	500×10^6	250×10^6
Brass	250×10^6	250×10^6	200×10^6
Aluminum	200×10^6	200×10^6	200×10^6
Concrete	2×10^6	20×10^6	2×10^6
Brick		35×10^6	
Marble		80×10^6	
Granite		170×10^6	
Wood (pine)			
(parallel to grain)	40×10^6	35×10^6	5×10^6
(perpendicular to grain)		10×10^6	
Nylon	500×10^6		
Bone (limb)	130×10^6	170×10^6	

of the values given in the table. You may encounter tables of the "allowable stresses" in which appropriate safety factors have already been included.

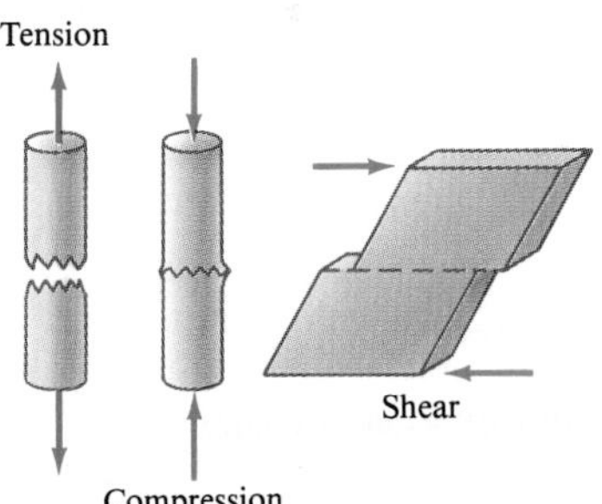

FIGURE 9–28
Fracture as a result of the three types of stress.

EXAMPLE 9–10 (*a*) What minimum cross-sectional area should the supports have to support the beam of Example 9–4 (Fig. 9–8) assuming the supports are made of concrete and a safety factor of 6 is required? (*b*) How much will the chosen supports compress under the given load?

SOLUTION (*a*) The right-hand support receives the larger force, 1.2×10^5 N. It is clearly under compression, and from Table 9–2, we see that the ultimate compressive strength of concrete is $2.0 \times 10^7\ \mathrm{N/m^2}$. Using a safety factor of 6, the maximum allowable stress is $\frac{1}{6}(2.0 \times 10^7\ \mathrm{N/m^2}) = 3.3 \times 10^6\ \mathrm{N/m^2}$. Since $F/A = 3.3 \times 10^6\ \mathrm{N/m^2}$ and $F = 1.2 \times 10^5$ N, we can solve for A and find

$$A = \frac{1.2 \times 10^5\ \mathrm{N}}{3.3 \times 10^6\ \mathrm{N/m^2}} = 3.6 \times 10^{-2}\ \mathrm{m^2} \text{ or } 360\ \mathrm{cm^2}.$$

A support 18 cm × 20 cm will be adequate.

(*b*) We solve for

$$\frac{\Delta L}{L_0} = \frac{1}{E}\frac{F}{A} = \left(\frac{1}{2.0 \times 10^{10}\ \mathrm{N/m^2}}\right)(3.3 \times 10^6\ \mathrm{N/m^2}) = 1.7 \times 10^{-4}.$$

Thus, if the support has a length $L_0 = 5.0$ m, $\Delta L = 0.85 \times 10^{-3}$ m, or about 1 mm. This calculation was for the right-hand support. If the left-hand support is made of the same cross-sectional area, it will compress less and this should be taken into account.

As can be seen in Table 9–2, concrete (like stone and brick) is reasonably strong under compression but extremely weak under tension. Thus concrete can be used as vertical columns placed under compression but is of

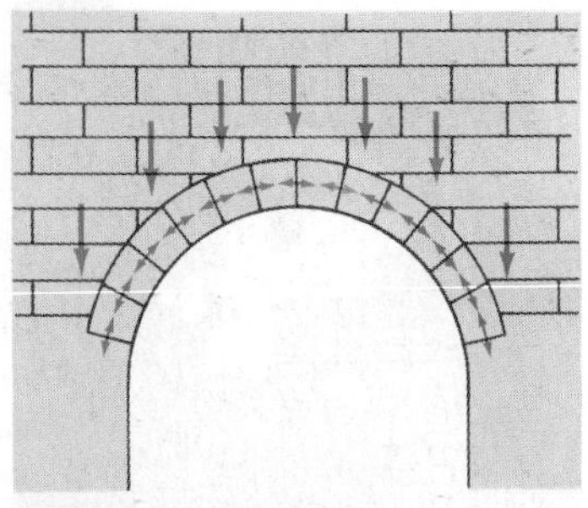

FIGURE 9–33
Stones in a round (or "true") arch are mainly under compression.

FIGURE 9–34 Flying buttresses (on the cathedral of Notre Dame, in Paris).

The pointed arch came into use about A.D. 1100 and soon became the hallmark of the great Gothic cathedrals. It too was an important technical innovation, and it is likely that this was at least part of the reason it was originally introduced: it was first used to support heavy loads such as the tower of a cathedral and as the central arch across the nave. The lesser arches in these "transitional" buildings often remained round. Apparently the builders realized that, because of the steepness of the pointed arch, the forces due to the weight above could be brought down more nearly vertically, so less horizontal buttressing would be needed. The pointed arch reduced the load on the walls, so there could be more openness and light. The smaller amount of buttressing needed was provided on the outside by graceful flying buttresses (Fig. 9–34).

The technical innovation of the pointed arch was achieved not through calculation but through experience and intuition. The beginnings of the science of statics and building appeared during the Renaissance and was advanced by Galileo. But it was not until somewhat later that detailed calculations, such as those presented earlier in this chapter, came into use. To make an accurate analysis of a stone arch is quite difficult in practice. But if we make some simplifying assumptions, we can show why the horizontal component of the force at the base is less for a pointed arch than for a round one. Figure 9–35 shows a round arch and a pointed arch, each with an 8.0-m span. The height of the round arch is thus 4.0 m, whereas that of the pointed arch is larger and has been chosen to be 8.0 m. Each arch supports a weight of 12.0×10^4 N ($= 12{,}000 \text{ kg} \times g$) which, for simplicity, we have divided into two parts (each 6.0×10^4 N) acting on the two halves of each arch as shown. To be in equilibrium, each of the supports must exert an upward force of 6.0×10^4 N. Each support also exerts a horizontal force, F_H, at the base of the arch, and it is this we want to calculate. We focus only on the right half of each arch. We set equal to zero the total torque calculated about the apex of the arch due to the forces exerted on that half arch, as if there were a hinge at the apex. For the round arch, the torque equation is

$$(4.0 \text{ m})(6.0 \times 10^4 \text{ N}) - (2.0 \text{ m})(6.0 \times 10^4 \text{ N}) - (4.0 \text{ m})(F_H) = 0.$$

Thus $F_H = 3.0 \times 10^4$ N. For the pointed arch, the torque equation is

$$(4.0 \text{ m})(6.0 \times 10^4 \text{ N}) - (2.0 \text{ m})(6.0 \times 10^4 \text{ N}) - (8.0 \text{ m})(F_H) = 0.$$

Solving, we find that $F_H = 1.5 \times 10^4$N—only half as much! From this calculation we can see that the horizontal buttressing force required for a pointed

FIGURE 9–35
Forces in a round arch (a), compared with those in a pointed arch (b).

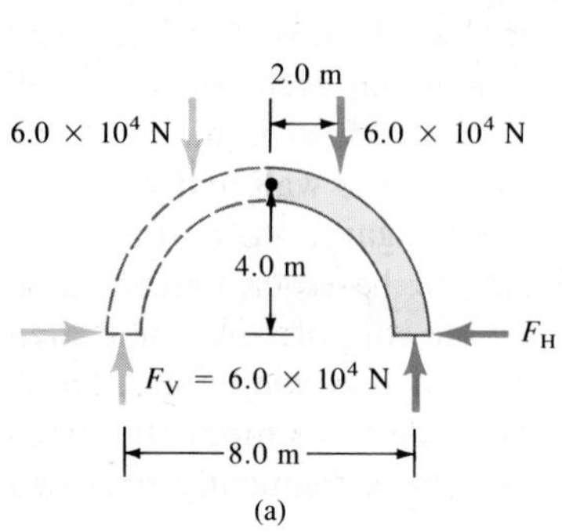

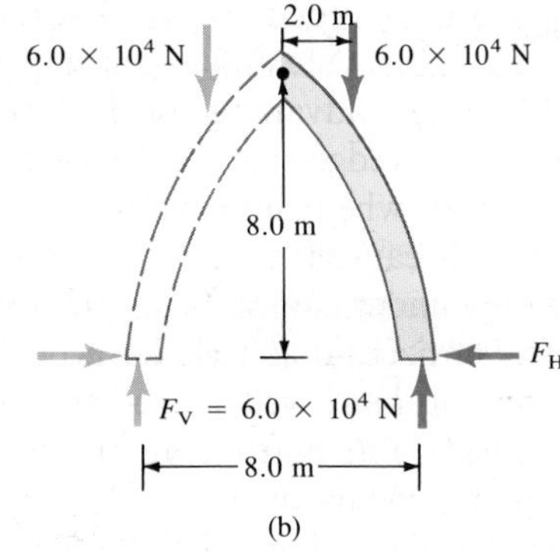

arch is less because the arch is higher, and there is therefore a longer lever arm for this force. Indeed, the steeper the arch, the less the horizontal component of the force needs to be, and hence the more nearly vertical is the force exerted at the base of the arch.

The further development of the arch was one of decline. For the subsequent flattened arches, such as the Tudor arch (Fig. 9–32), were structurally weaker than the simple pointed arch. However, with the coming of advanced methods of calculation in the nineteenth and twentieth centuries, it became possible to calculate the best shape of arch for a given load condition. For example, if the load is uniform across the span, it can be shown that the stresses within the arch will be purely compressive if the arch has a parabolic shape.

Whereas an arch spans a two-dimensional space, a dome—which is basically an arch rotated about a vertical axis—spans a three-dimensional space. The Romans built the first large domes. Their shape was hemispherical and some still stand, such as that of the Pantheon in Rome (Fig. 9–36). By the time of the Renaissance, the technique for constructing large domes seems to have been lost. Indeed, the dome of the Pantheon was a source of wonder to Renaissance architects. (Even today we do not know precisely how it was built.) The problem came to the fore in fifteenth-century Florence with the designing of a new cathedral that was to have a dome 43 m in diameter to rival that of the Pantheon. In 1418, after the cathedral was finished except for the dome, a competition for the design of the dome was held and was won by Brunelleschi. One problem that had to be dealt with was that the dome was to rest on a "drum" that had been completed with no external abutments; and there was no place to put any. Hence the dome must exert the minimum of horizontal force. Brunelleschi solved this by designing a pointed dome (Fig. 9–37), since a pointed dome, like a pointed arch, exerts a smaller side thrust against its base.

The other major problem was how to support the dome during construction. A dome, like an arch, is not stable until all the stones are in place.

FIGURE 9–36
Interior of the Pantheon in Rome, built in the first century. This view, showing the great dome and its central opening for light, was painted about 1740 by Panini. Photographs do not capture its grandeur as well as this painting does.

FIGURE 9–37
The skyline of Florence, showing Brunelleschi's famous dome on the cathedral.

FIGURE 9–38 The dome of the Small Sports Palace in Rome, built for the 1960 Olympics.

It had been the custom to support a dome during construction with a wooden framework. But no trees big enough or strong enough could be found to support the 43-m space required for the cathedral in Florence. Instead of using a wooden framework, Brunelleschi built the dome in horizontal layers. Each layer was bonded to the previous one which held it in place until the last stone of the circle was placed. Each closed ring was then strong enough to support the next layer. It was an amazing feat.

To end this section we will consider the forces necessary to support a modern dome, that of the Small Sports Palace in Rome (Fig. 9–38). A dome, like an arch, is statically most stable when under compression. The 36 buttresses which support the 1.2×10^6-kg dome are positioned at a 38° angle and connect smoothly with the dome.

FIGURE 9–39 Example 9–12.

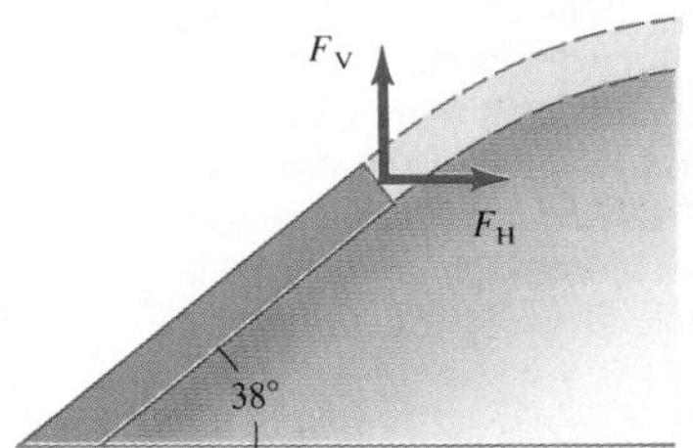

EXAMPLE 9–12 Calculate the components of the force, F_V and F_H, that each buttress exerts on the dome so that the force acts compressively—that is, at a 38° angle (Fig. 9–39).

SOLUTION The vertical load on *each* buttress is 1/36 of the total weight. Thus

$$F_V = \frac{(1.2 \times 10^6\ \text{kg})(9.8\ \text{m/s}^2)}{36} = 3.4 \times 10^5\ \text{N}.$$

The force must act at a 38° angle at the base of the dome in order to be purely compressive. Thus

$$\tan 38° = \frac{F_V}{F_H} = \frac{340{,}000\ \text{N}}{F_H}$$

$$F_H = \frac{340{,}000\ \text{N}}{\tan 38°} = 430{,}000\ \text{N}.$$

In order that each of the buttresses be able to exert this 430,000-N horizontal force, a prestressed-concrete tension ring surrounds the base of the buttresses beneath the ground (see Problem 60 and Fig. 9–59).

SUMMARY

A body at rest, or one in uniform motion at constant velocity, is said to be in *equilibrium*. The subject concerned with the determination of the forces within a structure at rest is called *statics*. The two necessary conditions for a body to be in equilibrium are (1) the vector sum of all the forces on it must be zero, and (2) the sum of all the torques (calculated about any arbitrary axis) must also be zero. It is important when doing statics problems to apply the equilibrium conditions to only one body at a time.

A body in static equilibrium is said to be in (a) *stable*, (b) *unstable*, or (c) *neutral equilibrium*, depending on whether a slight displacement leads to (a) a return to the original position, (b) further movement, or (c) rest in the new position. An object in stable equilibrium is also said to be in *balance*.

Hooke's law applies to many elastic solids, and states that the change in length of an object is proportional to the applied force, $F = k\,\Delta L$. If the force is too great, the object will exceed its *elastic limit*, which means it will no longer return to its original shape when the distorting force is removed. If the force is even greater, the *ultimate strength* of the material can be exceeded and the object fractures. The force per unit area acting on a body is called the *stress*, and the resulting fractional change in length is called the *strain*. The stress on a body is present within the body and can be of three types: *compression*, *tension*, and *shear*. The ratio of stress to strain is called the *elastic modulus* of the material. *Young's modulus* applies for compression and tension, and the *shear modulus* for shear; *bulk modulus* applies to an object whose volume changes as a result of pressure on all sides. All three moduli are constants for a given material when distorted within its elastic region.

QUESTIONS

FIGURE 9–40 Question 2 and Problem 17.

1. Describe several situations where a body is not in equilibrium, even though the net force on it is zero.
2. A bear sling, Fig. 9–40, is used in some national parks for placing backpackers' food out of the reach of bears. Explain why the force needed to pull the backpack up increases as the backpack gets higher and higher. Is it possible to pull the rope hard enough so that it doesn't sag at all?
3. A ladder, leaning against a wall, makes a 60° angle with the ground. When is it more likely to slip: when a person stands near the top or near the bottom?
4. Explain why touching the toes while seated on the floor with outstretched legs produces less stress on the lower spinal column than when touching the toes from a standing position. Use a diagram.

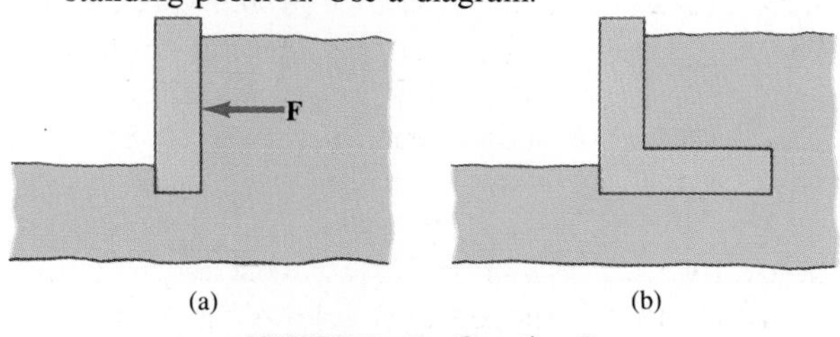

FIGURE 9–41 Question 5.

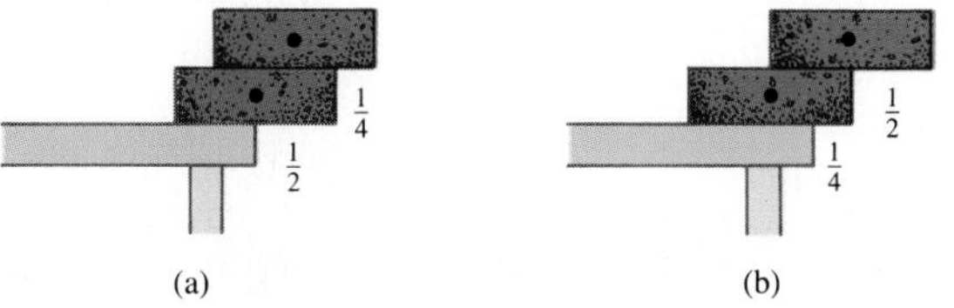

FIGURE 9–42 The dots indicate the cg of each brick. The fractions $\frac{1}{4}$ and $\frac{1}{2}$ indicate what portion of each brick is hanging beyond its support. Question 6.

5. An earth retaining wall is shown in Fig. 9–41a. The earth, particularly when wet, can exert a significant force F on the wall. (*a*) What force produces the torque to keep the wall upright? (*b*) Explain why the retaining wall in Fig 9–41b would be much less likely to overturn.
6. Which of the configurations of brick, (a) or (b) of Fig. 9–42, is the more likely to be stable? Why?
7. Name the type of equilibrium for each position of the ball in Fig. 9–43.
8. In what state of equilibrium is a cube (*a*) when resting on its face, and (*b*) when on its edge?
9. Why do you tend to lean backward when carrying a heavy load in your arms?
10. What purpose does a walking stick serve when you are hiking in rough country? Be specific.
11. Place yourself facing the edge of an open door. Position your feet astride the door with your nose and abdomen touching the door's edge. Try to rise on your tiptoes. Why can't this be done?

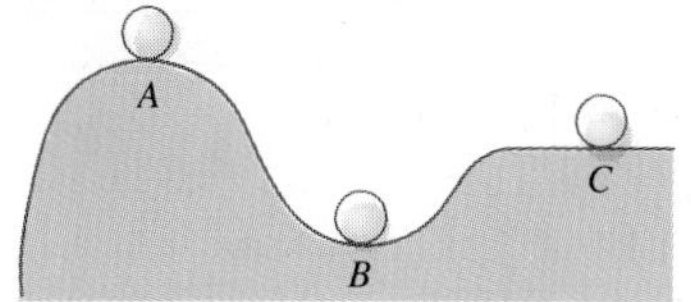

FIGURE 9–43 Question 7.

12. Why is it not possible to sit upright in a chair and rise to one's feet without first leaning forward?
13. Why is it more difficult to do situps when your knees are bent than when your legs are stretched out?
14. Examine how a pair of scissors or shears cuts through a piece of cardboard. Is the name "shears" justified?
15. Materials such as ordinary concrete and stone are very weak under tension or shear. Would it be wise to use such a material for either of the supports of the cantilever shown in Fig. 9–9? If so, which one(s)?

PROBLEMS

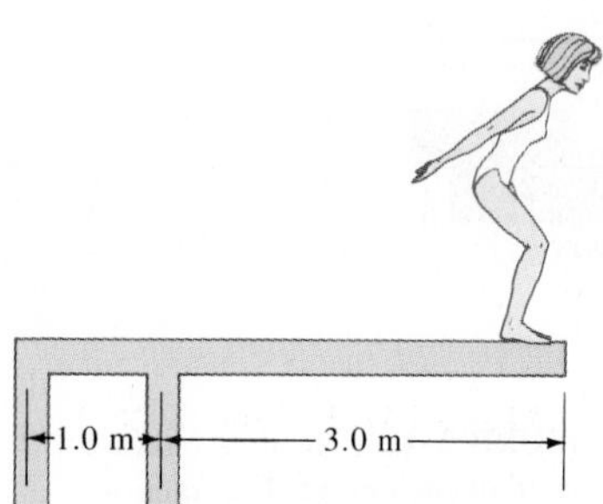

FIGURE 9–44 Problems 2, 11, and 12.

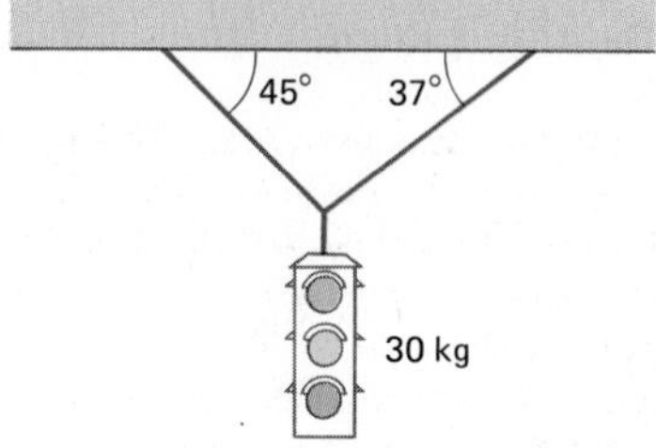

FIGURE 9–46 Problems 7 and 47.

SECTIONS 9–1 TO 9–3

1. (I) What should be the tension in the wire if the net force exerted on the tooth in Fig. 9–3 is to be 0.50 N? Assume that the angle between the two forces is 160° rather than the 140° in the figure.
2. (I) Calculate the torque about the front support on the right in Fig. 9–44 of a diving board exerted by a 70-kg person 3.0 m from that support.
3. (I) Calculate the mass m needed in order to suspend the leg shown in Fig. 9–45. Assume the leg has a mass of 12.0 kg, and its cg is 36.0 cm from the hip joint; the sling is 80.5 cm from the hip joint.
4. (I) Two cords support a chandelier in the manner shown in Fig. 9–5 except that the upper wire makes an angle of 45° with the ceiling. If the cords can sustain a force of 1200 N without breaking, what is the maximum chandelier weight that can be supported?
5. (I) A 180-kg horizontal beam is supported at each end. A 200-kg piano rests a quarter of the way from one end. What is the vertical force on each of the supports?
6. (II) Assume in Example 9–1 that the net force as calculated is 10° to the left of where it should point if the tooth is to move correctly. If the tension to the left is 2.0 N, what should the tension to the right be to make the net force act in the correct direction?
7. (II) Find the tension in the two wires shown in Fig. 9–46.
8. (II) Determine the force F_N that the pivot exerts on the seesaw board of Fig. 9–7.
9. (II) Calculate F_1 and F_2 for the uniform cantilever shown in Fig. 9–9 whose mass is 1200 kg.
10. (II) A door, 2.30 m high and 1.30 m wide, has a mass of 13.0 kg. A hinge 0.40 m from the top and another hinge 0.40 m from the bottom each support half the door's weight (Fig. 9–47). Assume that the center of gravity is at the geometrical center of the door, and determine the horizontal and vertical force components exerted by each hinge on the door.

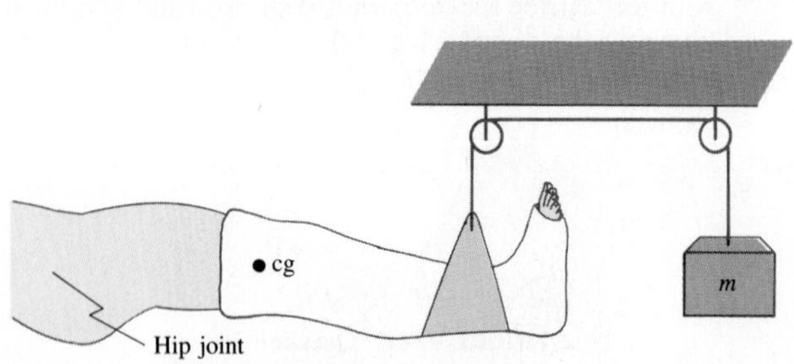

FIGURE 9–45 Problems 3 and 13.

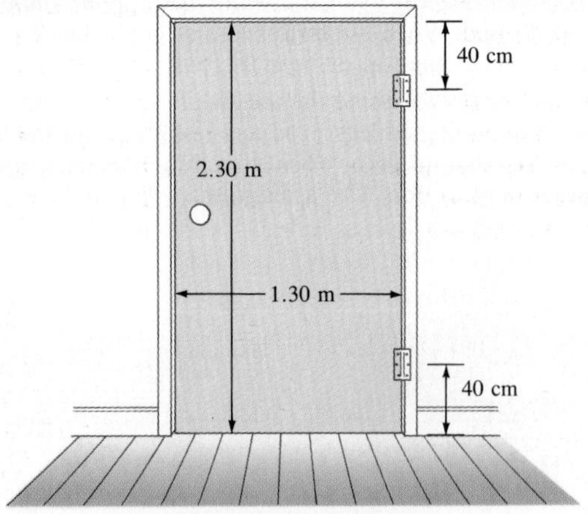

FIGURE 9–47 Problem 10.

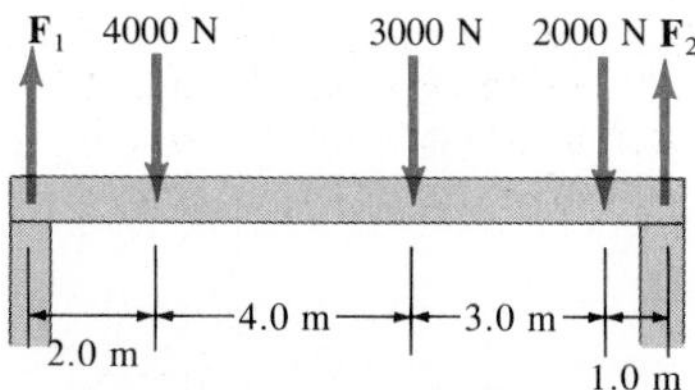

FIGURE 9–48 Problem 14.

11. (II) Calculate the forces F_1 and F_2 that the supports exert on the diving board of Fig. 9–44 when a 50-kg person stands at its tip. Ignore the weight of the board.
12. (II) Repeat the last problem, taking into account the board's mass of 40 kg. Assume the board's cg is at its center.
13. (II) Calculate the mass m required in Fig. 9–45 to support the leg, using the result of Example 7–10 and the values given in Table 7–1, assuming an 88.5-kg person 170 cm tall. The leg pivots about the hip joint and the support acts at the ankle joint.
14. (II) Calculate F_1 and F_2 for the beam shown in Fig. 9–48. Assume it is uniform and has a mass of 280 kg.
15. (II) Calculate the tension F_T in the wire that supports the 20-kg beam shown in Fig. 9–49, and the force $\mathbf{F}_w$ exerted by the wall on the beam (give magnitude and direction).

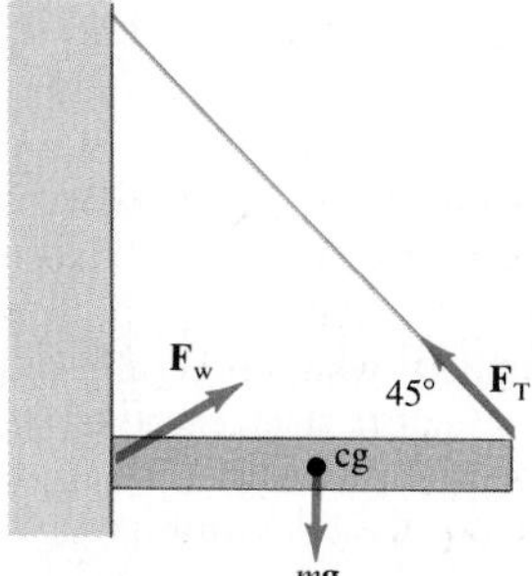

FIGURE 9–49 Problem 15.

16. (II) A 160-cm-tall person lies on a light (massless) board which is supported by two scales, one under the feet and one beneath the top of the head (Fig. 9–50). The two scales read, respectively, 29.4 and 32.8 kg. Where is the center of gravity of this person?
17. (II) The two trees in Fig. 9–40 are 12.5 m apart. Calculate the magnitude of the force **F** a backpacker must exert to hold the 18-kg backpack so that the rope sags at its midpoint by (*a*) 2.0 m, (*b*) 0.20 m.

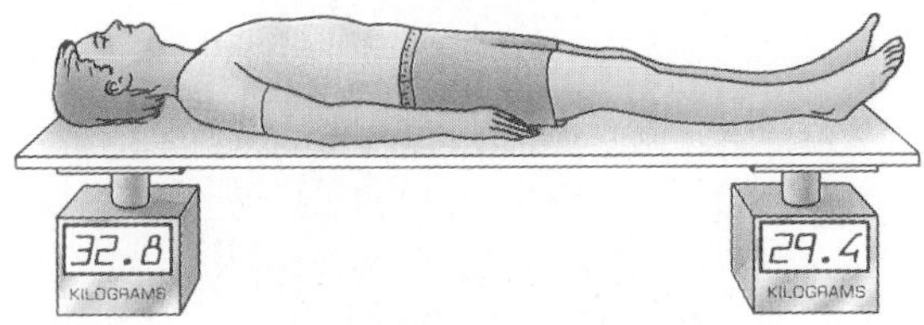

FIGURE 9–50 Problem 16.

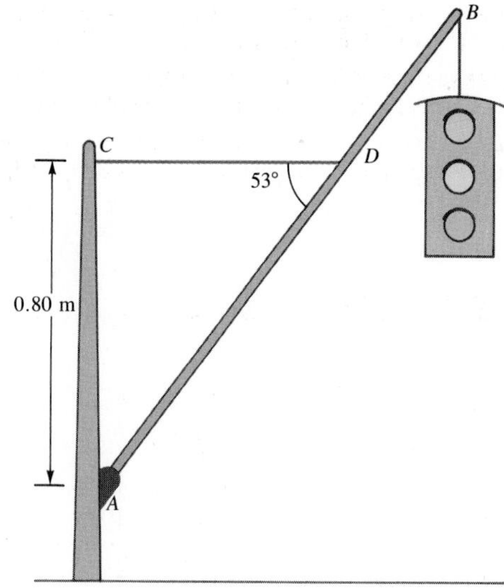

FIGURE 9–51 Problem 18.

18. (II) A traffic light hangs from a structure as shown in Fig. 9–51. The uniform aluminum pole AB is 4.5 m long and weighs 5.0 kg. The weight of the traffic light is 10.0 kg. Determine the tension in the horizontal massless cable CD, and the vertical and horizontal components of the force exerted by the pivot A on the aluminum pole.
19. (II) A uniform ladder of mass m and length L leans at an angle θ against a frictionless wall, Fig. 9–52. If the coefficient of static friction between the ladder and the ground is μ, what is the minimum angle at which the ladder will not slip?
20. (III) A meter stick with a mass of 350 g is supported horizontally by two vertical strings, one at the 0-cm mark and the other at the 90-cm mark. (*a*) What is the tension in the string at 0 cm? (*b*) What is the tension in the string at 90 cm? (*c*) The string at the zero end is cut. What is the tension in the remaining string immediately afterward?

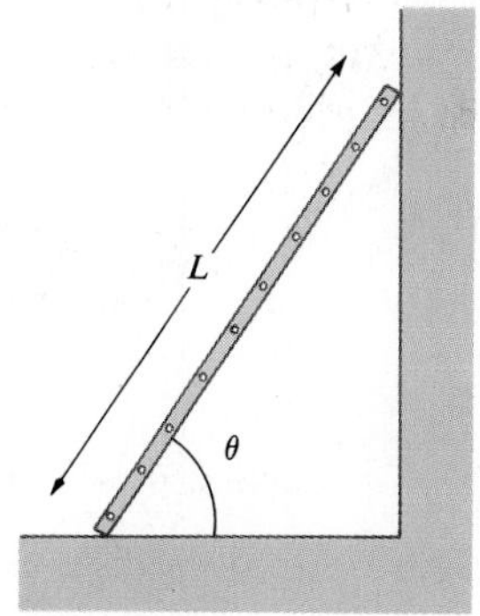

FIGURE 9–52 Problems 19, 72, and 73.

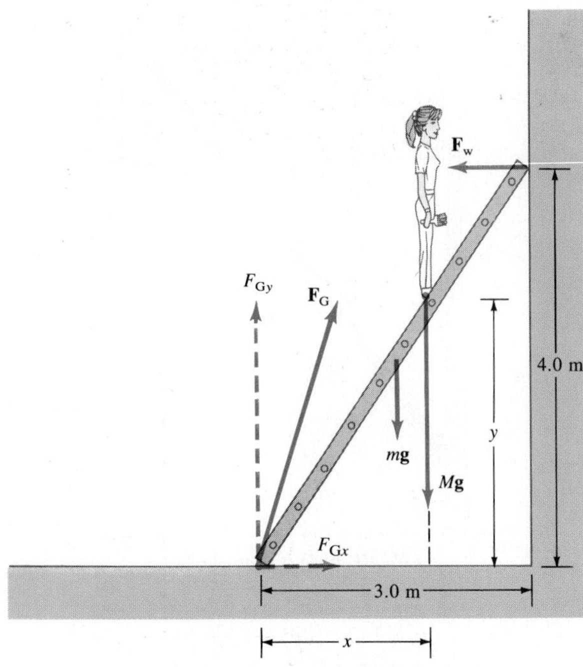

FIGURE 9–53 Problem 21.

21. (III) Consider again the ladder of Example 9–6 (Fig. 9–11). If the coefficient of static friction between the ladder and the ground is 0.40, how high up the ladder can a 58-kg painter climb without the ladder slipping? A free-body diagram is shown in Fig. 9–53.

22. (III) A person wants to push a lamp (mass 9.6 kg) across the floor. (*a*) Assuming the person pushes at a height of 60 cm above the ground and the coefficient of friction is 0.20, determine whether the lamp will slide or tip over (Fig. 9–54). (*b*) Calculate the maximum height above the floor at which the person can push the lamp so that it slides rather than tips.

FIGURE 9–54 Problem 22.

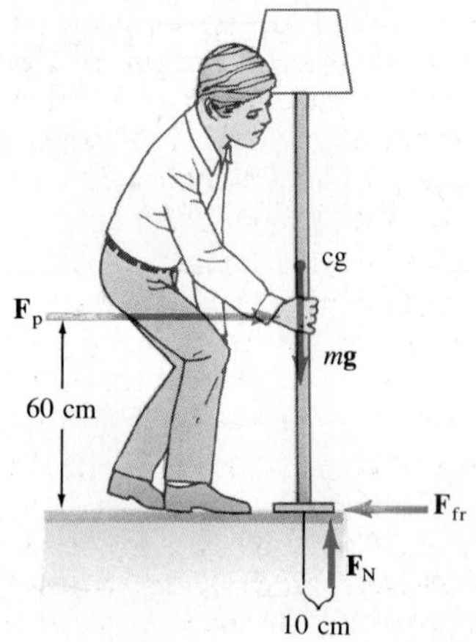

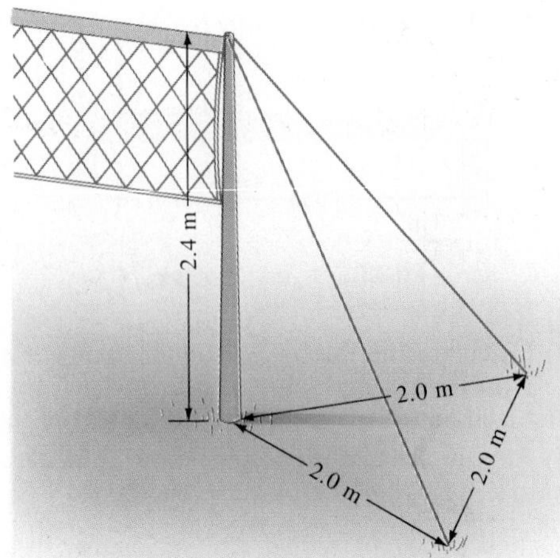

FIGURE 9–55 Problem 23.

23. (III) Two guy wires run from the top of a pole 2.4 m tall that supports a volleyball net. The two wires are anchored to the ground 2.0 m apart and each 2.0 m from the pole (Fig. 9–55). The tension in each wire is 85 N. What is the tension in the net, assumed horizontal and attached at the top of the pole?

*SECTION 9–4

*24. (I) If the point of insertion of the biceps muscle into the lower arm shown in Fig. 9–13a were 6.0 cm, how much mass could the person hold with a muscle exertion of 400 N?

*25. (I) Approximately what force, F_M, must the extensor muscle in the upper arm exert on the lower arm to hold a 7.3 kg shotput (Fig. 9–56)? Assume the lower arm has a mass of 2.8 kg and its cg is 12 cm from the pivot point.

FIGURE 9–56 Problem 25.

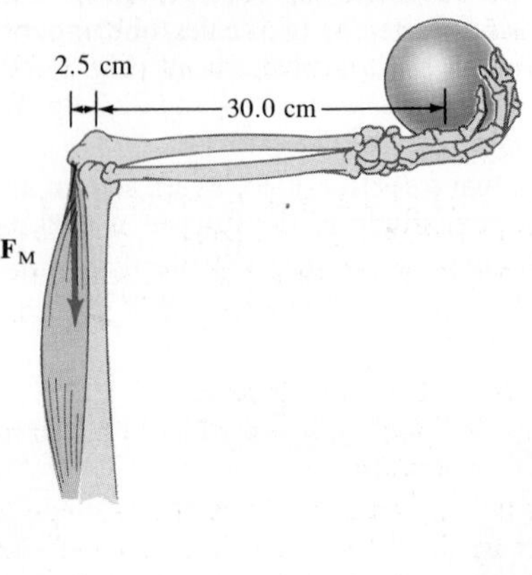

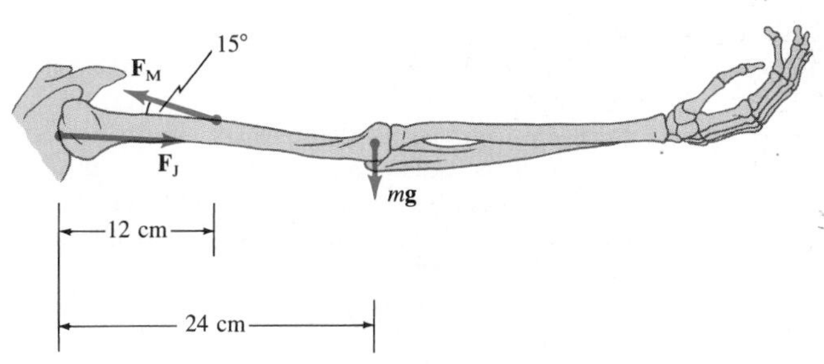

FIGURE 9–57 Problems 26, 27, and 28.

*26. (II) Calculate the force required of the "deltoid" muscle, F_M, to hold up the outstretched arm shown in Fig. 9–57. The total mass of the arm is 2.8 kg.

*27. (II) Suppose the hand in the last problem holds a 10-kg mass. What force, F_M, is required of the deltoid muscle assuming the mass is 50 cm from the shoulder joint?

*28. (II) Calculate the magnitude of the force F_J exerted by the shoulder on the upper arm at the joint for Problems 26 and 27.

*29. (III) Calculate the magnitude of the force at the base of the spine, $\mathbf{F}_V$ in Fig. 9–14b, if a mass of 20 kg is held in the person's hands. Assume the person has a mass of 70 kg.

*30. (III) When walking, a person momentarily puts all the weight on one foot. The cg of the body lies over the supporting foot. Figure 9–58 shows the supporting leg and the forces on it. Calculate the force exerted by the "hip abductor" muscles, $\mathbf{F}_M$, and the x and y components of the force $\mathbf{F}_J$ acting on the joint. Take the whole leg as the object under discussion.

*31. (III) (*a*) Using the information in Problem 30, calculate $\mathbf{F}_M$ and $\mathbf{F}_J$, assuming the person carries a 20-kg suitcase in each hand. Assume the cg of each suitcase lies directly below the edge of the hipbone. (*b*) Calculate $\mathbf{F}_M$ and $\mathbf{F}_J$ assuming the person carries a suitcase only in the hand on the side opposite the supporting leg. [*Hint:* first calculate the common cg of the person plus suitcase; this point will be above the foot, thus shifting the horizontal measurements from those shown in Fig. 9–58.]

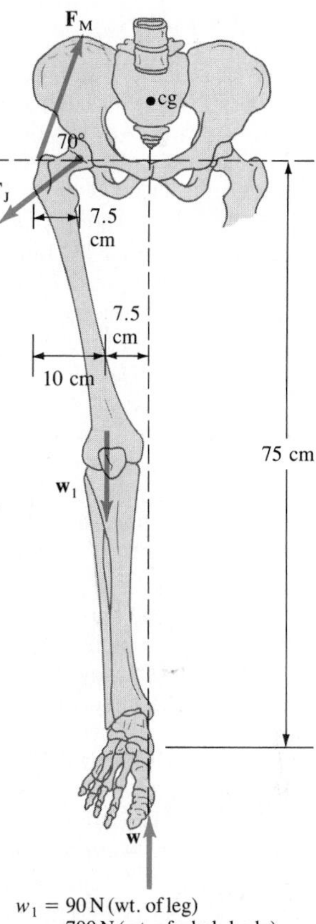

w_1 = 90 N (wt. of leg)
w = 700 N (wt. of whole body)

FIGURE 9–58 Problems 30 and 31.

*SECTION 9–5

*32. (I) One kind of car jack is essentially a lever with a ratchet to hold the lifted object in place while the lever is returned for a new thrust. If the lever is pivoted 2.0 cm from the ratchet and the handle is 35.0 cm long, what is the IMA?

*33. (II) Calculate the force required in Fig. 9–15, taking into account the lever's mass of 12 kg. Assume the cg of the lever is at its center.

*34. (II) Some car jacks work on the principle of a screw. Determine the IMA of a jack in terms of the pitch (distance between threads) P, radius r_1 of the screw, and the turning radius r_2 of the jack handle. Ignore any other effects, such as that due to gears or levers in the jack.

*35. (II) Design a pulley system that can lift 80 kg with a force of (18 kg × g), assuming an efficiency of 75 percent.

*36. (II) A 38-cm-long crank is attached to a spindle with a 2.5-cm radius. Assuming an efficiency of 80 percent, how much weight can be lifted on a cord wrapped around the spindle if the person can exert a steady 50-N force on the crank handle?

*37. (II) An inclined plane (such as used by the Egyptians to build the pyramids) can be considered to be a simple machine because a heavy weight can be raised by the application of a smaller force. (*a*) Calculate the IMA of a plane inclined at an angle of 15° to the horizontal. (*b*) If the coefficient of friction for a particular object on this plane is 0.40, what is the AMA?

*38. (II) Determine formulas for (*a*) the IMA, and (*b*) the road distance traveled per pedal revolution, for any bicycle in terms of the rear wheel radius (r_w), the pedal path radius (r_p), and the number of teeth on the front and rear sprockets (T_F and T_R). See Fig. 9–18.

*39. (II) A 12-speed bicycle has front sprockets with 52 and 42 teeth, and a rear "free-wheel" with sprockets having 13, 15, 17, 20, 24, and 28 teeth. For each of the twelve combinations, determine (*a*) the IMA and (*b*) the distance traveled along the road per pedal revolution. Assume r_p = 18 cm and r_w = 34 cm.

*40. (III) For the bicycle in Problem 39, determine how steep a hill the bicycle can climb for each gear setting. Ignore friction and air resistance. Assume the cyclist can exert, at most, an average force of 0.75 times her weight tangentially to the pedals' circular path. The bike's weight is 0.20 times the cyclist's weight.

*41. (III) Suppose you can climb a 15° hill with your bicycle gears set on sprockets of 42 and 28 teeth. Estimate the average force you exert on the pedals (tangentially) as a factor of mg, your weight. The bicycle's weight is $0.16mg$.

SECTION 9–6

42. (II) The Leaning Tower of Pisa is 55 m tall and about 7.0 m in diameter. The top is 4.5 m off center. Is the tower in stable equilibrium? If so, how much further can it lean before it becomes unstable? Assume the tower is of uniform composition.

43. (III) Four bricks are to be stacked at the edge of a table, each brick overhanging the one below it, so that the top brick extends as far as possible beyond the edge of the table. (*a*) To achieve this, show that successive bricks must extend no more than (starting at the top) 1/2, 1/4, 1/6, and 1/8 of their length beyond the one below. (*b*) Is the top brick completely beyond the base? (*c*) Determine a general formula for the maximum total distance spanned by n bricks if they are to remain stable. (*d*) A builder wants to construct a corbeled arch (Figure 9–32) based on the principle of stability discussed in (*a*) and (*c*) above. What minimum number of bricks, each 0.30 m long, is needed if the arch is to span 1.0 m?

SECTION 9–7

44. (I) A marble column of cross-sectional area 2.0 m^2 supports a mass of 20,000 kg. (*a*) What is the stress within the column? (*b*) What is the strain?

45. (I) By how much is the column in the previous problem shortened if it is 10 m high?

46. (I) How much force is required to stretch a steel piano wire of 0.050 cm diameter by 0.025 percent?

47. (II) If the two wires in Fig. 9–46 are made of steel wire 1.0 mm in diameter, what is the percentage stretch of each because of the load?

48. (II) A 15-cm-long animal tendon was found to stretch 3.7 mm by a force of 13.4 N. The tendon was approximately round with an average diameter of 8.5 mm. Calculate the elastic modulus of this tendon.

49. (II) How much pressure is needed to compress the volume of an iron block by 0.10 percent? Express answer in N/m^2, and compare it to atmospheric pressure ($1.0 \times 10^5\ N/m^2$).

50. (II) One liter of alcohol (1000 cm^3) in a flexible container is carried to the bottom of the sea, where the pressure is $2.4 \times 10^6\ N/m^2$. What will be its volume there?

51. (III) A scallop forces open its shell with an elastic material called abductin, whose elastic modulus is about $2.0 \times 10^6\ N/m^2$. If this piece of abductin is 3.0 mm thick and has a cross-sectional area of 0.50 cm^2, how much potential energy does it store when compressed 1.0 mm?

52. (III) A pole projects horizontally from the front wall of a shop. A 4.5-kg sign hangs from the pole at a point 2.4 m from the wall. (*a*) What is the torque due to this sign calculated about the point where the pole meets the wall? (*b*) If the pole is not to fall off, there must be another torque exerted to balance it. What exerts this torque? Use a diagram to show how this torque must act. (*c*) Discuss whether compression, tension, and/or shear play a role in part (*b*).

SECTION 9–8

53. (I) The femur bone in the leg has an average effective cross section of about 3.0 cm^2 ($=3.0 \times 10^{-4}\ m^2$). How much compressive force can it withstand before breaking?

54. (II) If a compressive force of 3.0×10^4 N is exerted on the end of a 20-cm-long bone of cross-sectional area 3.6 cm^2, (*a*) will the bone break, and (*b*) if not, by how much does it shorten?

55. (II) What is the minimum cross-sectional area required of a vertical steel wire from which is suspended a 280-kg chandelier? Assume a safety factor of 6.0.

56. (II) Assume the supports of the cantilever shown in Fig. 9–9 (mass = 5200 kg) are made of wood. Calculate the minimum cross-sectional area required of each, assuming a safety factor of 8.5.

57. (II) An iron bolt is used to connect two iron plates together. The bolt must withstand shear forces up to about 2000 N. Calculate the minimum diameter for the bolt, based on a safety factor of 7.0.

58. (III) A steel cable is to support an elevator whose total (loaded) mass is not to exceed 2800 kg. If the maximum acceleration of the elevator is 1.0 m/s^2, calculate the diameter of cable required. Assume a safety factor of 6.0.

*SECTION 9–9

*59. (II) How high must a pointed arch be if it is to span a space 8.0 m wide and exert one-third the horizontal force at its base that a round arch would?

*60. (II) The subterranean tension ring that exerts the balancing horizontal force on the abutments for the dome in Fig. 9–38 is 36 sided, so each segment makes a 10° angle with the adjacent one (Fig. 9–59). Calculate the tension F that must exist in each segment so that the required force of 4.3×10^5 N can be exerted at each corner (Example 9–12).

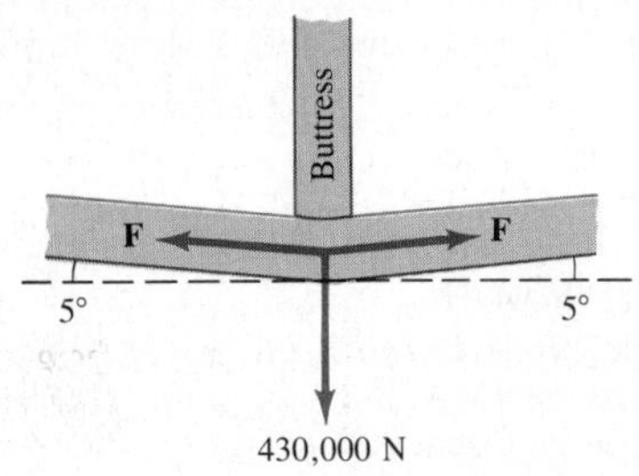

FIGURE 9–59 Problem 60.

GENERAL PROBLEMS

61. A 50-story building is being planned. It is to be 200 m high with a base 40 m by 70 m. Its total mass will be about 1.5×10^7 kg and its weight therefore about 1.5×10^8 N. Suppose a 200-km/h wind exerts a force of 1450 N/m² over the 70-m-wide face (Fig. 9–60). Calculate the torque about the potential pivot point, the rear edge of the building (where $\mathbf{F}_e$ acts in Fig. 9–60), and determine whether the building will topple. Assume the total force of the wind acts at the midpoint of the building's face.

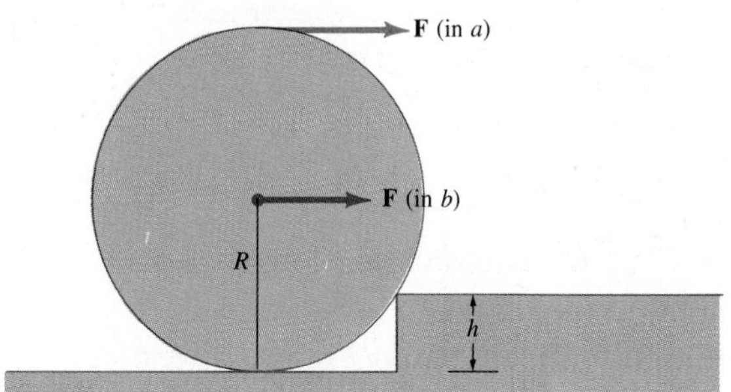

FIGURE 9–61 Problem 64.

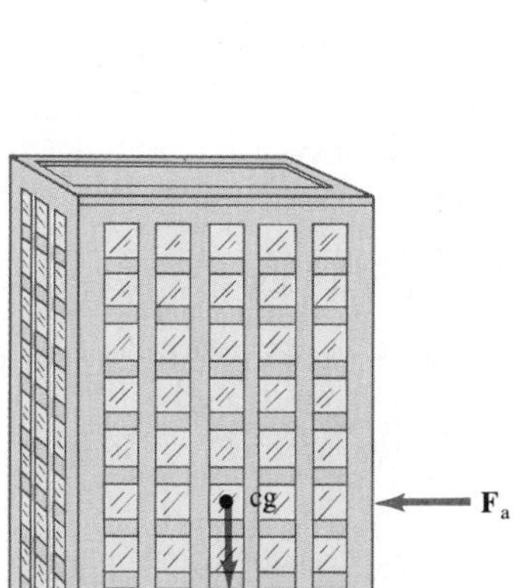

FIGURE 9–60 Force on a building subjected to wind ($\mathbf{F}_a$) and gravity (**w**); $\mathbf{F}_e$ is the force on the building due to the earth in the situation when the building is just about to tip. Problem 61.

62. A tightly stretched "high wire" is 40 m long. It sags 3.0 m when a 60-kg tightrope walker stands at its center. What is the tension in the wire? Is it possible to increase the tension in the wire so that there is no sag?

63. A cubic crate of side $s = 2.0$ m is top-heavy: its cg is 20 cm above its true center. How steep an incline can the crate rest on without tipping over? What would your answer be if the crate were to slide at constant speed down the plane without tipping over?

64. What horizontal force F is needed to pull a wheel of radius R and mass M over a step of height h as shown in Fig. 9–61? (*a*) Assume the force is applied at the top edge as shown. (*b*) Assume the force is applied instead at the wheel's center.

65. In Example 7–4 in Chapter 7, we calculated the impulse and average force on the leg of a person who jumps 5.0 m down to the ground. If the legs are not bent upon landing, so that the body moves a distance d of only 1.0 cm during collision, determine (*a*) the stress in the tibia bone (area $= 3.0 \times 10^{-4}$ m²), and (*b*) whether or not the bone will break. (*c*) Repeat for a bent-knees landing ($d = 50.0$ cm).

66. The roof of a 9.5 m × 10.0 m room in a school is to support a uniformly distributed load of 10,000 kg (133 kg/m²) in addition to its own mass of 4100 kg. The roof is to be supported by "2 × 4s" (actually about 4.0 × 9.0 cm) along the 10.0-m sides. How many supports are required on each side and how far apart must they be? Consider only compression and assume a safety factor of 15.

67. In Fig. 9–62, consider the right-hand (northernmost) section of the Golden Gate Bridge, which has a length $d_1 = 343$ m. Assume the cg of this span is halfway between the tower and anchor. Determine T_1 and T_2 (which act on the northernmost cable) in terms of mg, the weight of the northernmost span, and calculate the tower height h needed for equilibrium. Assume the roadway is supported only by the suspension cable and neglect the mass of the cables. [*Hint:* T_3 does not act on this section.]

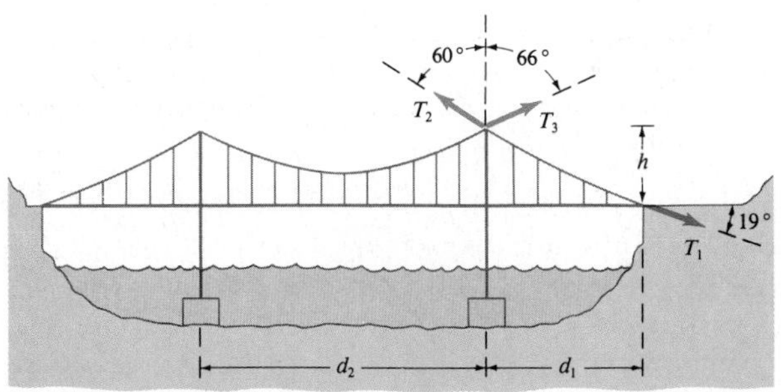

FIGURE 9–62 Problems 67 and 71.

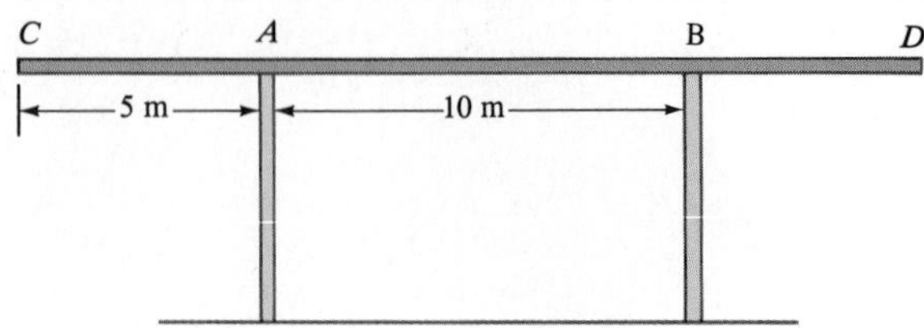

FIGURE 9–63 Problem 68.

68. A 20-m-long uniform beam weighing 500 N is supported on walls A and B, as shown in Fig. 9–63. (*a*) Find the maximum weight a person can be to walk to the extreme end D without tipping the beam over. Find the forces F_A and F_B that the walls A and B exert on the beam when the person is standing: (*b*) at D; (*c*) at a point 2.0 m to the right of B; (*d*) 2.0 m to the right of A.

69. A uniform flexible steel cable of weight mg is suspended between two equal elevation points as shown in Fig. 9–64. Determine the tension in the cable (*a*) at its lowest point, and (*b*) at the points of attachment. (*c*) What is the direction of the tension force in each case?

70. A 30-kg round table is supported by three legs placed equal distances apart on the edge. What minimum mass, placed on the table's edge, will cause the table to overturn?

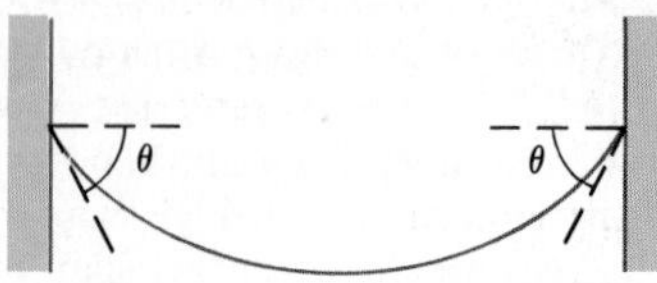

FIGURE 9–64 Problem 69.

71. Assume that a single-span suspension bridge such as the Golden Gate Bridge has the configuration indicated in Fig. 9–62. Assume that the roadway is uniform over the length of the bridge and that each segment of the suspension cable provides the sole support for the roadway directly below it. The ends of the cable are anchored to the ground only, not to the roadway. What must the ratio of d_2 to d_1 be so that the suspension cable exerts no net horizontal force on the towers? Neglect the mass of the cables and the fact that the roadway isn't precisely horizontal.

72. A uniform 8.5-m-long ladder of mass 18.0 kg leans against a smooth wall (so the force exerted by the wall, $\mathbf{F}_w$, is perpendicular to the wall). The ladder makes an angle $\theta = 20°$ with the vertical wall (see Fig. 9–52); and the ground is rough. (*a*) Calculate the components of the force exerted by the ground on the ladder at its base, and (*b*) determine what the coefficient of friction at the base of the ladder must be if the ladder is not to slip when a 75-kg person stands three-fourths of the way up the ladder.

73. If the coefficient of friction between the ladder and the ground in the situation described in the preceding problem is 0.30, how far up the ladder can the person climb before the ladder starts to slip?

74. There is a maximum height of a uniform vertical column made of any material that can support itself without buckling, and it is independent of the cross-sectional area (why?). Calculate this height for (*a*) steel (density 7.8×10^3 kg/m^3), and (*b*) granite (density 2.7×10^3 kg/m^3).

75. From what height must a 1.2-kg rectangular brick 15.0 cm × 6.0 cm × 4.0 cm be dropped above a rigid steel floor in order to break the brick? Assume the brick strikes the floor directly on its largest face, and that the compression of the brick is much greater than that of the steel (that is, ignore compression of the steel). State other simplifying assumptions that may be necessary.

CHAPTER 10

Fluids

The great mass of a glacier's ice moves very slowly, as if it were a very viscous fluid. The dark lines (moraines), consisting of broken rock eroded from the mountain walls by the moving glacier, represent streamlines.

Phases of matter

The three common states, or **phases**, of matter are solid, liquid, and gas. We can distinguish these three phases as follows. A **solid** maintains a fixed shape and a fixed size; even if a large force is applied to a solid, it does not readily change its shape or volume. A **liquid** does not maintain a fixed shape—it takes on the shape of its container—but like a solid it is not readily compressible and its volume can be changed significantly only by a very large force. A **gas** has neither a fixed shape nor a fixed volume—it will expand to fill its container. For example, when air is pumped into an automobile tire, the air does not all run to the bottom of the tire as a liquid would; it fills the whole volume of the tire. Since liquids and gases do not maintain a fixed shape, they each have the ability to flow; they are thus often referred to collectively as **fluids**.

The division of matter into three states is not always simple. How, for example, should butter be classified? Furthermore, a fourth state of matter can be distinguished, the **plasma** state, which occurs only at very high temperatures, as we will discuss in Chapter 31. Some scientists believe that so-called colloids (suspensions of tiny particles in a liquid) should also be considered a separate state of matter. However, for our present purposes we will mainly be interested in the three ordinary states of matter. In Chapter 9 we discussed some of the properties of solid materials. In this chapter we will discuss the properties of fluids.

10–1 • Density and Specific Gravity

It is sometimes said that iron is "heavier" than wood. This cannot really be true since a large log clearly weighs more than an iron nail. What we should say is that iron is *denser* than wood.

The **density**, ρ, of an object is defined as its mass per unit volume:

Density defined

$$\rho = \frac{m}{V}, \tag{10-1}$$

where m is the mass of the object and V its volume. Density is a characteristic property of any pure substance. Objects made of a given pure substance, say pure iron, can have any size or mass, but the density will be the same for each.

Table 10–1
Densities of Substances†

Substance	Density, ρ (kg/m³)
Solids	
Aluminum	2.70×10^3
Iron and Steel	7.8×10^3
Copper	8.9×10^3
Lead	11.3×10^3
Gold	19.3×10^3
Concrete	2.3×10^3
Granite	2.7×10^3
Wood (typical)	$0.3–0.9 \times 10^3$
Glass, common	$2.4–2.8 \times 10^3$
Ice	0.917×10^3
Bone	$1.7–2.0 \times 10^3$
Liquids	
Water (4°C)	1.00×10^3
Blood, plasma	1.03×10^3
Blood, whole	1.05×10^3
Sea water	1.025×10^3
Mercury	13.6×10^3
Alcohol, ethyl	0.79×10^3
Gasoline	0.68×10^3
Gases	
Air	1.29
Helium	0.179
Carbon dioxide	1.98
Water (steam) (100°C)	0.598

† Densities are given at 0°C and 1 atm pressure unless otherwise specified.

The SI unit for density is kg/m³. Sometimes densities are given in g/cm³. Note that since $1\ \text{kg/m}^3 = 1000\ \text{g}/(100\ \text{cm})^3 = 10^{-3}\ \text{g/cm}^3$, then a density given in g/cm³ must be multiplied by 1000 to give the result in kg/m³. Thus the density of aluminum is $\rho = 2.70\ \text{g/cm}^3$, which is equal to 2700 kg/m³. The densities of a variety of substances are given in Table 10–1. The table specifies temperature and pressure because these affect the density of substances (although the effect is slight for liquids and solids).

Example 10–1 What is the mass of a lead sphere of radius 0.500 m?

Solution The volume of the sphere is

$$V = \tfrac{4}{3}\pi r^3 = \tfrac{4}{3}(3.14)(0.500\ \text{m})^3 = 0.523\ \text{m}^3.$$

From Table 10–1, the density of lead is $\rho = 11{,}300\ \text{kg/m}^3$, so we have from Eq. 10–1,

$$m = \rho V = (11{,}300\ \text{kg/m}^3)(0.523\ \text{m}^3) = 5910\ \text{kg}.$$

The **specific gravity** of a substance is defined as the ratio of the density of that substance to the density of water at 4.0°C. Specific gravity (abbreviated SG) is a pure number, without dimensions or units. Since the density of water is $1.00\ \text{g/cm}^3 = 1.00 \times 10^3\ \text{kg/m}^3$, the specific gravity of any substance will be precisely equal numerically to its density specified in g/cm³, or 10^{-3} times its density specified in kg/m³. For example (see Table 10–1), the specific gravity of lead is 11.3 and of alcohol 0.79.

10–2 • Pressure in Fluids

Pressure is defined as force per unit area, where the force F is understood to be acting perpendicular to the surface area A:

Pressure defined

$$\text{pressure} = P = \frac{F}{A}. \tag{10-2}$$

The pascal and other units

The SI unit of pressure is N/m². This unit has the official name **pascal** (Pa): 1 Pa = 1 N/m². However, for simplicity, we will often use N/m². Other units sometimes used are dynes/cm², lb/in² (sometimes abbreviated "psi"), and

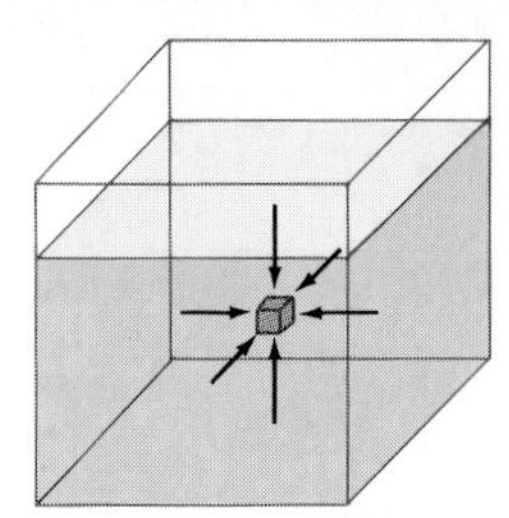

FIGURE 10–1 Pressure is the same in every direction in a fluid at a given depth; if it weren't, the fluid would be in motion.

kg/cm^2 (as if kilograms were a force: that is, 1 kg/cm^2 = 9.8 N/cm^2 = 9.8 × 10^4 N/m^2). The last two are often used on tire gauges. We will meet several other units shortly.

As an example of calculating pressure, a 60-kg person whose two feet cover an area of 500 cm^2 will exert a pressure of $F/A = mg/A = (60\text{ kg})(9.8\text{ m/s}^2)/(0.050\text{ m}^2) = 12 \times 10^3\text{ N/m}^2$ on the ground. If the person stands on one foot, the force is the same but the area will be half, so the pressure will be twice as much: $24 \times 10^3\text{ N/m}^2$.

Fluids exert pressure in all directions

The concept of pressure is particularly useful in dealing with fluids. It is an experimental fact that *a fluid exerts a pressure in all directions.* This is well known to swimmers and divers who feel the water pressure on all parts of their bodies. At a particular point in a fluid at rest, the pressure is the same in all directions. This is illustrated in Fig. 10–1: consider a tiny cube of the fluid which is so small that we can ignore the force of gravity on it. Then the pressure on one side of it must equal the pressure on the opposite side. If this weren't true, the net force on this cube would not be zero, and it would move until the pressure *did* become equal. If the fluid is not flowing, then the pressures must be equal.

Fluid pressure acts perpendicular to the surfaces of their containers

Another important property of a fluid at rest is that the force due to fluid pressure always acts *perpendicularly* to any surface it is in contact with. If there were a component of the force parallel to the surface as shown in Fig. 10–2, then according to Newton's third law, the surface would exert a force back on the fluid that also would have a component parallel to the surface. This component would cause the fluid to flow, in contradiction to our assumption that the fluid is at rest.

Let us now calculate quantitatively how the pressure in a liquid of uniform density varies with depth. Consider a point which is at a depth h below the surface of the liquid (that is, the surface is a height h above this point), as shown in Fig. 10–3. The pressure due to the liquid at this depth h is due to the weight of the column of liquid above it. Thus the force acting on the area is $F = mg = \rho Ahg$, where Ah is the volume of the column, ρ is the density of the liquid (assumed to be constant), and g is the acceleration of gravity. The pressure, P, is then

$$P = \frac{F}{A} = \frac{\rho Ahg}{A}$$

$$P = \rho gh. \qquad \text{[liquid]} \tag{10–3a}$$

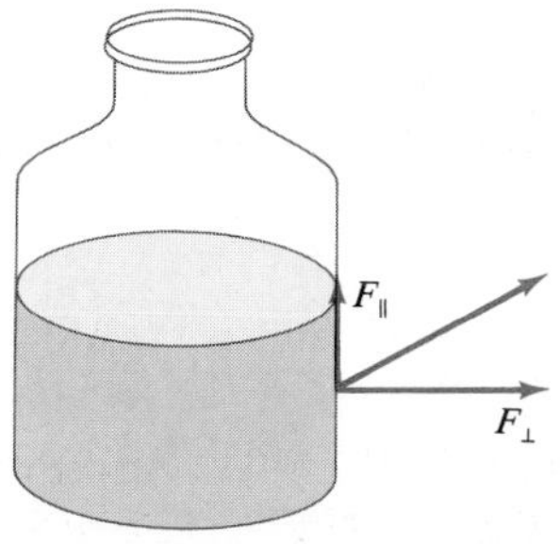

FIGURE 10–2
If there were a component of force parallel to the solid surface, the liquid would move in response to it; for a liquid at rest, $F_{\parallel} = 0$.

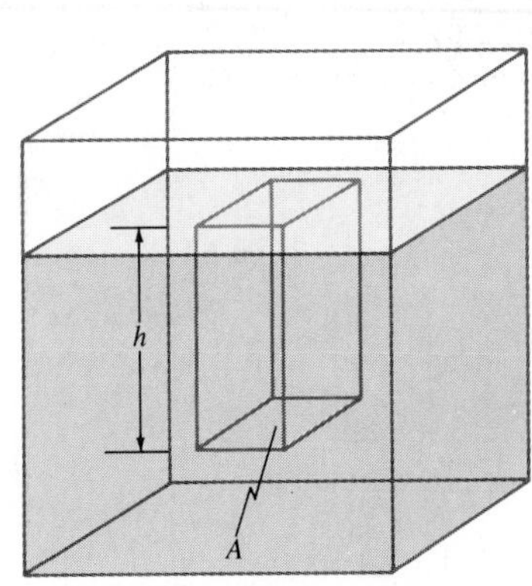

FIGURE 10–3 Calculating the pressure at a depth h in a liquid.

Pressure depends on depth

Thus the pressure is directly proportional to the density of the liquid, and to the depth within the liquid. In general, the *pressure at equal depths within a uniform liquid is the same.* (Equation 10–3a gives the pressure due to the liquid itself. If an external pressure is exerted at the surface of the liquid, this must be taken into account as we will discuss in Section 10–4.)

Equation 10–3a is extremely useful. It is valid for fluids whose density is constant and does not change with depth—that is, if the fluid is *incompressible.* This is usually a good approximation for liquids (although at great depths in the ocean the density of water is increased substantially by compression due to the great weight of water above). Gases, on the other hand, are very compressible and density can vary significantly with depth. If the density varies only slightly, Eq. 10–3a can be used to determine the difference in pressure ΔP at different heights with ρ being the average density:

$$\Delta P = \rho g \, \Delta h. \qquad (10\text{–}3b)$$

EXAMPLE 10–2 The surface of the water in a storage tank is 30 m above a water faucet in the kitchen of a house. Calculate the water pressure at the faucet.

SOLUTION Atmospheric pressure acts both at the surface of the water in the storage tank, and on the water leaving the faucet. The pressure difference between the inside and outside of the faucet is

$$\begin{aligned}\Delta P &= \rho g h = (1.0 \times 10^3 \text{ kg/m}^3)(9.8 \text{ m/s}^2)(30 \text{ m}) \\ &= 2.9 \times 10^5 \text{ N/m}^2.\end{aligned}$$

The height h is sometimes called the **pressure head**. In this example, the head of the water is 30 m.

10–3 • Atmospheric Pressure and Gauge Pressure

The pressure of the earth's atmosphere, as in any fluid, decreases with decreased depth (or increased height). But the earth's atmosphere is somewhat complicated because not only does the density of air vary greatly with altitude, but there is no distinct top surface to the atmosphere from which h (in Eq. 10–3a) can be measured. We can, however, calculate the approximate difference in pressure between two altitudes using Eq. 10–3b.

The pressure of the air at a given place varies slightly according to the weather. At sea level, the pressure of the atmosphere on the average is 1.013×10^5 N/m^2 (or 14.7 lb/in^2). This value is used to define another unit of pressure in common use, the **atmosphere** (abbreviated atm):

One atmosphere

$$1 \text{ atm} = 1.013 \times 10^5 \text{ N/m}^2 = 101.3 \text{ kPa}.$$

The bar

Another unit of pressure sometimes used (in meteorology and on weather maps) is the **bar**, which is defined as 1 bar $= 1.00 \times 10^5$ N/m^2 $= 1.00$ kPa. Thus standard atmospheric pressure is slightly more than 1 bar.

The pressure due to the weight of the atmosphere is exerted on all objects immersed in this great sea of air, including our bodies. How does a human body withstand the enormous pressure on its surface? The answer is that living cells maintain an internal pressure that just balances the external pressure. Similarly, the pressure inside a balloon balances the outside pressure of the atmosphere. An automobile tire, because of its rigidity, can maintain pressures much greater than the external pressure.

One must be careful, however, when determining the pressure in a tire, for tire gauges, and most other pressure gauges as well, register the pressure over and above atmospheric pressure. This is called **gauge pressure**. Thus, to get the absolute pressure P, one must add the atmospheric pressure, P_a, to the gauge pressure, P_G:

Gauge pressure

$$P = P_a + P_G.$$

For example, if a tire gauge registers 220 kPa, the actual pressure within the tire is 220 kPa + 100 kPa = 320 kPa. This is equivalent to about 3.2 atm (2.2 atm gauge pressure).

10–4 • Pascal's Principle

The earth's atmosphere exerts a pressure on all objects with which it is in contact, including other fluids. External pressure acting on a fluid is transmitted throughout that fluid. For instance, according to Eq. 10–3a, the pressure due to the water pressure at a depth of 100 m below the surface of a lake is $P = \rho g h = (1000\ \text{kg/m}^3)(9.8\ \text{m/s}^2)(100\ \text{m}) = 9.8 \times 10^5\ \text{N/m}^2$, or 9.7 atm. However, the total pressure at this point is due to the pressure of water plus the pressure of the air above it. Hence the total pressure (if the lake is near sea level) is 9.7 atm + 1.0 atm = 10.7 atm. This is just one example of a general principle attributed to the French philosopher and scientist Blaise Pascal (1623–1662). **Pascal's principle** states that *pressure applied to a confined fluid increases the pressure throughout by the same amount.*

Pascal's principle

A number of practical devices make use of Pascal's principle. Two examples, hydraulic brakes in an automobile and the hydraulic lift, are illustrated in Fig. 10–4. In the case of a hydraulic lift, a small force can be used to exert a large force by making the area of one piston (the output) larger than the area of the other (the input). This is another example of a simple machine (Chapter 9) and relies on the fact that the pressures on the input and output cylinders are the same at equal heights. (If the difference in height of the two pistons is not too great, this result is not much altered.) Thus, if the input quantities are represented by the subscript "i" and the output by "o," we have

$$P_o = P_i$$

$$\frac{F_o}{A_o} = \frac{F_i}{A_i},$$

or, finally,

$$\frac{F_o}{F_i} = \frac{A_o}{A_i}.$$

The quantity F_o/F_i is the "ideal mechanical advantage" of the hydraulic lift,

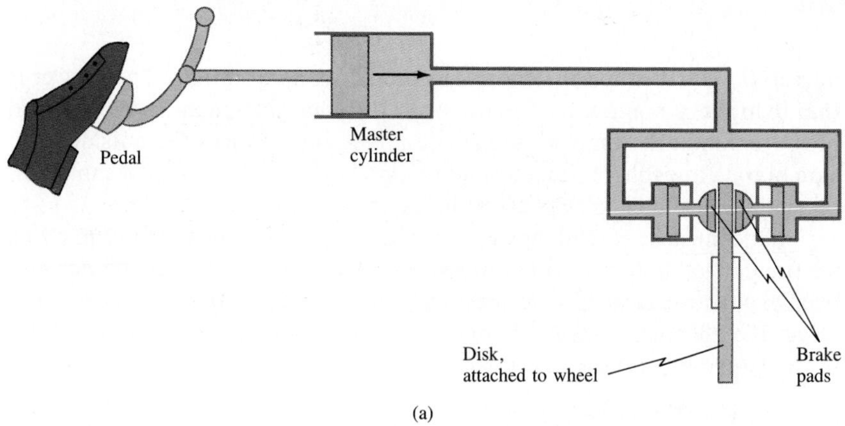

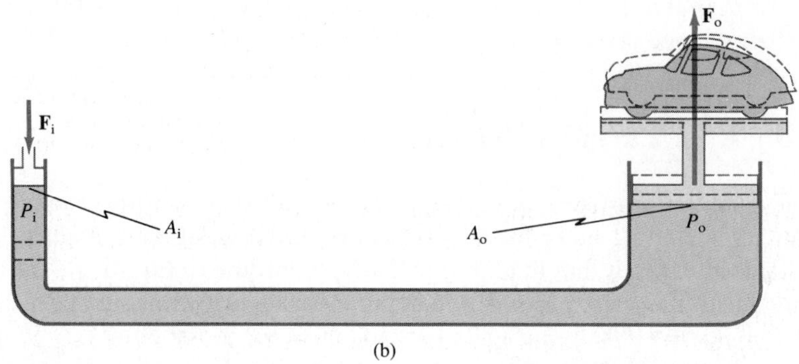

FIGURE 10–4
Applications of Pascal's principle:
(a) hydraulic brakes in a car;
(b) hydraulic lift.

and is equal to the ratio of the areas. For example, if the area of the output piston is 20 times that of the input cylinder, the force is multiplied by a factor of 20; thus a force of 200 lb could lift a 4000-lb car.

A number of living organisms make use of hydraulic pressure. A sea anemone can achieve a variety of shapes by the action of muscles on its sea-water-filled body cavity. Its body is sometimes referred to as a "hydrostatic skeleton." Earthworms move forward by successive contractions of circular muscles along the body axis acting on the hydrostatic skeleton. The legs of spiders have flexor muscles but possess no extensor muscles; their legs are extended by fluid driven into them under pressure.

10–5 • Measurement of Pressure; Gauges and the Barometer

Many devices have been invented to measure pressure, some of which are shown in Fig. 10–5. The simplest is the open-tube *manometer* (Fig. 10–5a) which is a U-shaped tube partially filled with a liquid, usually mercury or water. The pressure P being measured is related to the difference in height of the two levels of the liquid by the relation (see Eq. 10–3b)

$$P = P_0 + \rho g h,$$

where P_0 is atmospheric pressure (acting on the top of the fluid in the left-hand tube), and ρ is the density of the liquid. Note that the quantity $\rho g h$ is the "gauge pressure"—the amount by which P exceeds atmospheric pressure.

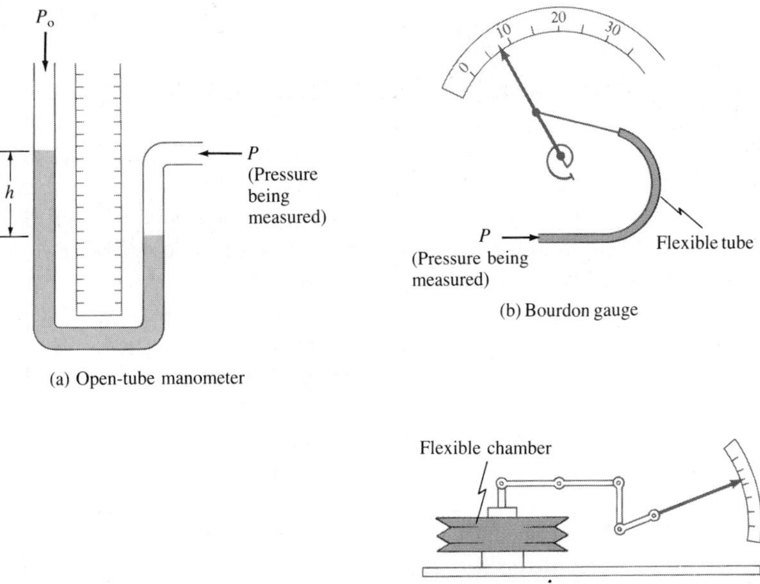

FIGURE 10–5 Pressure gauges: (a) open-tube manometer, (b) Bourdon gauge, (c) aneroid gauge, and (d) common tire pressure gauge.

If the liquid in the left-hand column were lower than that in the right-hand column, this would indicate that P was less than atmospheric pressure (and h would be negative).

Instead of calculating the product $\rho g h$, it is common to merely specify the height h. In fact, pressures are sometimes specified as so many "millimeters of mercury" (mm Hg), and sometimes as so many "mm of water." The unit mm Hg is equivalent to a pressure of 133 N/m^2, since 1.00 mm = 1.00×10^{-3} m and the density of mercury is $13.6 \times 10^3\ kg/m^3$:

$$\begin{aligned}\rho g h &= (13.6 \times 10^3\ kg/m^3)(9.80\ m/s^2)(1.00 \times 10^{-3}\ m)\\ &= 1.33 \times 10^2\ N/m^2.\end{aligned}$$

The unit mm Hg is also called the **torr** in honor of Evangelista Torricelli (1608–1647), who invented the barometer (see below). Conversion factors among the various units of pressure (an incredible nuisance!) are given in Table 10–2. It is important that only N/m^2 = Pa, the proper SI unit, be used in calculations involving other quantities specified in SI units.

Problem Solving:
Use SI unit in calculations: 1 Pa = 1 N/m^2.

TABLE 10–2
Conversion Factors Between Different Units of Pressure

In Terms of 1 Pa = 1 N/m^2	Related to 1 atm
1 atm = $1.013 \times 10^5\ N/m^2$ = 1.013×10^5 Pa = 101.3 kPa	1 atm = $1.013 \times 10^5\ N/m^2$
1 bar = $1.000 \times 10^5\ N/m^2$	1 atm = 1.013 bar
1 $dyne/cm^2$ = 0.1 N/m^2	1 atm = $1.013 \times 10^6\ dyne/cm^2$
1 kg/cm^2 = $9.85 \times 10^4\ N/m^2$	1 atm = 1.03 kg/cm^2
1 lb/in^2 = $6.90 \times 10^3\ N/m^2$	1 atm = 14.7 lb/in^2
1 lb/ft^2 = 47.9 N/m^2	1 atm = $2.12 \times 10^3\ lb/ft^2$
1 cm Hg = $1.33 \times 10^3\ N/m^2$	1 atm = 76 cm Hg
1 mm Hg = 133 N/m^2	1 atm = 760 mm Hg
1 torr = 133 N/m^2	1 atm = 760 torr
1 mm H_2O (4°C) = 9.81 N/m^2	1 atm = 1.03×10^4 mm H_2O (4°C)

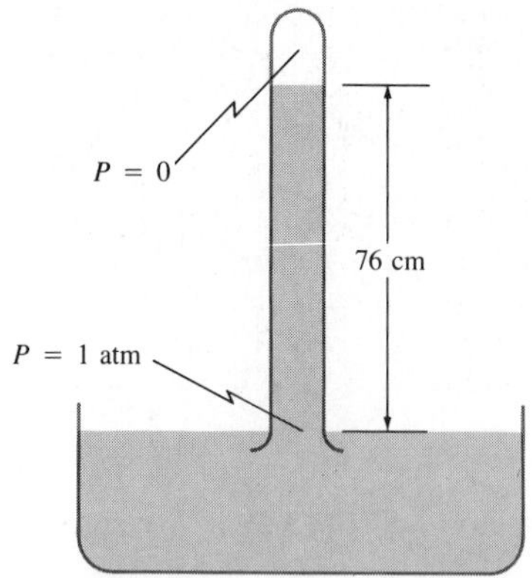

FIGURE 10–6 Diagram of a mercury barometer, when the air pressure is 76 cm Hg.

Other types of pressure gauge are the Bourdon gauge (Fig. 10–5b) in which increased pressure tends to straighten out the delicate tube that is attached to a pointer, and the aneroid gauge (Fig. 10–5c) in which the pointer is connected to the flexible ends of an evacuated thin metal chamber. In a more sophisticated type of gauge, the pressure to be measured is applied to a thin metal diaphragm and the resulting distortion of the diaphragm is detected electrically. How a common tire gauge is constructed is shown in Fig. 10–5d.

Atmospheric pressure is often measured by a modified kind of mercury manometer with one end closed, called a mercury **barometer** (Fig. 10–6). The glass tube is filled with mercury and then inverted into the bowl of mercury. If the tube is long enough, the level of the mercury will drop, leaving a vacuum at the top of the tube, since atmospheric pressure can support a column of mercury only about 76 cm high (exactly 76.0 cm at normal atmospheric pressure). That is, a column of mercury 76 cm high exerts the same pressure as the atmosphere: from the formula $P = \rho g h$, with $\rho = 13.6 \times 10^3\ \text{kg/m}^3$ for mercury and $h = 76.0$ cm, we have

$$\begin{aligned} P &= (13.6 \times 10^3\ \text{kg/m}^3)(9.80\ \text{m/s}^2)(0.760\ \text{m}) \\ &= 1.013 \times 10^5\ \text{N/m}^2 \\ &= 1.00\ \text{atm}. \end{aligned}$$

Household barometers are usually of the aneroid type (Fig. 10–5c).

A calculation similar to that above will show that atmospheric pressure can maintain a column of water 10.3 m high in a tube whose top is under vacuum (Fig. 10–7). A few centuries ago, it was a source of wonder and frustration that no matter how good a vacuum pump was, it could not lift water more than about 10 m. The only way to pump water out of deep mine shafts, for example, was to use multiple stages for depths greater than 10 m. Galileo studied this problem, and his student Torricelli was the first to explain it. The point is that a pump does not really suck water up a tube—it merely reduces the pressure at the top of the tube. It is atmospheric air pressure that pushes the water up the tube if the other end is at low pressure (under a vacuum), just as it is air pressure that pushes (or maintains) the mercury 76 cm high in a barometer.

FIGURE 10–7 A water barometer: water was added at the top and the spigot then closed, but air pressure could not support a column of water more than 10 m high.

10–6 • Buoyancy and Archimedes' Principle

Objects submerged in a fluid appear to weigh less than they do when outside the fluid. For example, a large rock that you would have difficulty lifting off the ground can often be easily lifted from the bottom of a stream. When the rock breaks through the surface of the water, it suddenly seems to be much heavier. Many objects, such as wood, float on the surface of water. These are two examples of *buoyancy*. In each example, the force of gravity is acting downward. But in addition, an upward *buoyant force* is exerted by the liquid.

The buoyant force occurs because the pressure in a fluid increases with depth. Thus the upward pressure on the bottom surface of a submerged object is greater than the downward pressure on its top surface. To see the effect of this consider a cylinder of height h whose top and bottom ends have an area A and which is completely submerged in a fluid of density ρ_f, as shown in Fig. 10–8. The fluid exerts a pressure $P_1 = \rho_f g h_1$ against the top surface of the cylinder. The force due to this pressure on top of the cylinder is $F_1 = P_1 A = \rho_f g h_1 A$, and it is directed downward. Similarly, the fluid exerts an upward force on the bottom of the cylinder equal to $F_2 = P_2 A = \rho_f g h_2 A$. The net force due to the fluid pressure, which is the **buoyant force**, $\mathbf{F}_B$, acts upward and has the magnitude

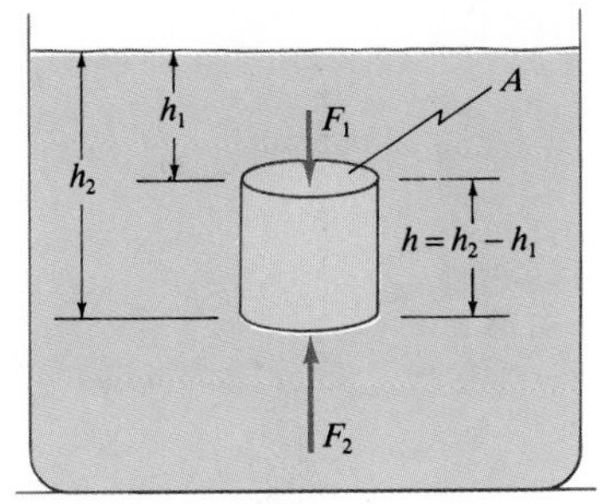

FIGURE 10–8
Determination of buoyant force.

$$\begin{aligned} F_B &= F_2 - F_1 \\ &= \rho_f g A(h_2 - h_1) \\ &= \rho_f g A h \\ &= \rho_f g V, \end{aligned}$$

where $V = Ah$ is the volume of the cylinder. Since ρ_f is the density of the fluid, the product $\rho_f g V = m_f g$ is the weight of fluid which takes up a volume equal to the volume of the cylinder. Thus the buoyant force on the cylinder is equal to the weight of fluid displaced† by the cylinder. This result is valid no matter what the shape of the object. Its discovery is credited to Archimedes (287?–212 B.C.), and it is called **Archimedes' principle**: *the buoyant force on a body immersed in a fluid is equal to the weight of the fluid displaced by that object.*

Archimedes' principle

We can derive Archimedes' principle in general by the following simple but elegant argument. The irregularly shaped object D shown in Fig. 10–9a is acted on by the force of gravity (its weight, $\mathbf{w}$) and the buoyant force, F_B. (If there is no other force acting on the object, such as a hand pulling up on it, the object shown will move downward, since $w > F_B$.) We wish to determine F_B. To do so, we next consider a body, this time made of our same fluid (D' in Fig. 10–9b) with the same shape and size as the original object, and located at the same depth. You might think of this body of fluid as being separated from the rest of the fluid by an imaginary transparent membrane. The buoyant force F_B on this body of fluid will be exactly the same as that on the original object since the surrounding fluid, which exerts F_B, is in exactly the same configuration. Now the body of fluid D' is in equilibrium (the fluid as a whole is at rest); therefore, $F_B = w'$, where w' is the weight of the body of fluid. Hence the buoyant force F_B is equal to the weight of the body of fluid whose volume is equal to the volume of the original submerged object, which is Archimedes' principle.

FIGURE 10–9 Archimedes' principle.

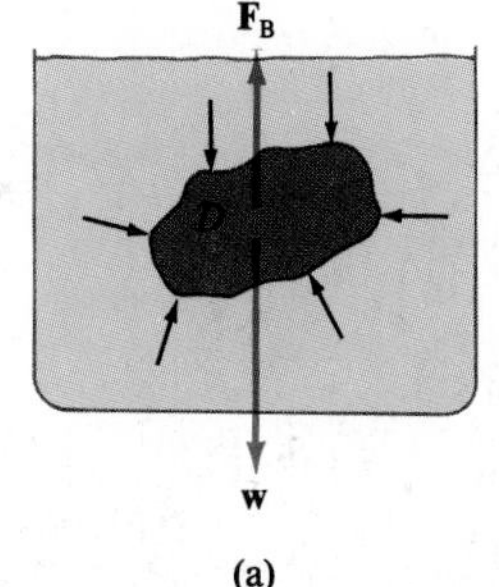

(a)

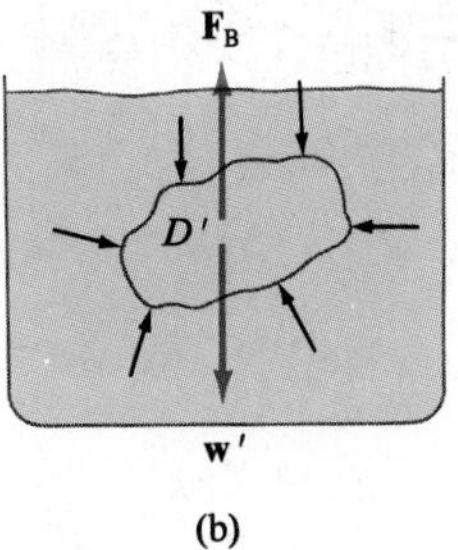

(b)

EXAMPLE 10–3 A 70-kg rock lies at the bottom of a lake. Its volume is $3.0 \times 10^4\ \text{cm}^3$. How much force is needed to lift it?

SOLUTION The buoyant force on the rock due to the water is equal to the weight of $3.0 \times 10^4\ \text{cm}^3 = 3.0 \times 10^{-2}\ \text{m}^3$ of water:

$$\begin{aligned} F_B &= m_{H_2O} g = \rho_{H_2O} g V \\ &= (1.0 \times 10^3\ \text{kg/m}^3)(9.8\ \text{m/s}^2)(3.0 \times 10^{-2}\ \text{m}^3) \\ &= 2.9 \times 10^2\ \text{N}. \end{aligned}$$

† By "fluid displaced," we mean a volume of fluid equal to the volume of the object, or that part of the object submerged if it floats (the fluid that used to be where the object was). If the object is placed in a glass or tub initially filled to the brim with water, the water that flows over the top represents the water displaced by the object.

The weight of the rock is $mg = (70\ \text{kg})(9.8\ \text{m/s}^2) = 6.9 \times 10^2$ N. Hence the force needed to lift it is 690 N − 290 N = 400 N. It is as if the rock had a mass of only $(400\ \text{N})/(9.8\ \text{m/s}^2) = 41$ kg.

Archimedes is said to have discovered his principle in his bath while thinking how he might determine whether the king's new crown was pure gold or a fake. Gold has a specific gravity of 19.3, somewhat higher than that of most metals, but a determination of specific gravity or density is not readily done directly since the volume of an irregularly shaped object is not easily calculated. However, if the object is weighed in air ($=w$) and also "weighed" while it is under water ($=w'$), the density can be determined using Archimedes' principle, as the following example shows.

EXAMPLE 10–4 A 14.7-kg crown has an apparent weight when submerged in water corresponding to 13.4 kg. Is it gold?

SOLUTION The apparent weight of the submerged object, w', equals its actual weight w minus the buoyant force F_B:

$$w' = w - F_B = \rho_o g V - \rho_f g V,$$

where V is the volume of the object, ρ_o its density, and ρ_f the density of the fluid (water in this case). From the above relation, we can see that $w - w' = \rho_f g V$. Then we can write

$$\frac{w}{w - w'} = \frac{\rho_o g V}{\rho_f g V} = \frac{\rho_o}{\rho_f}.$$

[Thus $w/(w - w')$ is equal to the specific gravity of the object if the fluid in which it is submerged is water.] For the crown we have

$$\frac{\rho_o}{\rho_{H_2O}} = \frac{w}{w - w'} = \frac{14.7\ \text{kg}}{1.3\ \text{kg}} = 11.3.$$

This corresponds to a density of 11,300 kg/m³. The crown seems to be made (see Table 10–1) of lead!

Floating objects

Archimedes' principle applies equally well to objects that float, such as wood. In general, *an object floats on a fluid if its density is less than that of the fluid.* At equilibrium—i.e., when floating—the buoyant force on an object has magnitude equal to the weight of the object. For example, a log whose specific gravity is 0.60 and whose volume is 2.0 m³ will have a mass of $(0.60 \times 10^3\ \text{kg/m}^3)(2.0\ \text{m}^3) = 1200$ kg. If the log is fully submerged, it will displace a mass of water $m = \rho V = (1000\ \text{kg/m}^3)(2.0\ \text{m}^3) = 2000$ kg. Hence the buoyant force on the log will be greater than its weight, and it will float to the surface. It will come to equilibrium when it displaces 1200 kg of water, which means that 1.2 m³, or 0.60 of its volume, will be submerged. In general when an object floats, the fraction of the object submerged is given by the ratio of the object's density to that of the fluid.

EXAMPLE 10–5 A **hydrometer** is a simple instrument used to indicate specific gravity by how deeply it sinks in a liquid. A particular hydrometer (Fig. 10–10) consists of a glass tube, weighted at the bottom, which is 25.0 cm long, 2.00 cm^2 in cross-sectional area, and has a mass of 45.0 g. How far from the end should the 1.000 mark be placed?

SOLUTION The hydrometer has a density

$$\rho = \frac{m}{V} = \frac{45.0\ \text{g}}{(2.00\ \text{cm}^2)(25.0\ \text{cm})} = 0.900\ \text{g/cm}^3.$$

Thus, when placed in water, it will come to equilibrium when 0.900 of its volume is submerged. Since it is of uniform cross section, $(0.900)\cdot(25.0\ \text{cm}) = 22.5$ cm of its length will be submerged. Since the specific gravity of water is defined to be 1.000, the mark should be placed 22.5 cm from the end.

FIGURE 10–10 A hydrometer.

Archimedes' principle is useful also in geology. According to the modern theory of plate tectonics and continental drift, the continents can be considered to be floating on a "sea" of slightly deformable rock (mantle rock). Some interesting calculations can be done using very simple models, which we consider here and in the problems.

Continental drift

EXAMPLE 10–6 A simple model (Fig. 10–11) considers a continent as a block (density = 2800 kg/m^3) floating in the mantle rock around it (density = 3300 kg/m^3). Assuming the continent is 35 km thick (the average thickness of earth's crust), estimate the height of the continent above the surrounding rock.

SOLUTION As discussed above, the fraction of continent submerged is equal to the ratio of the densities: $(2800\ \text{kg/m}^3)/(3300\ \text{kg/m}^3) = 0.85$. Hence 0.15 of the continent's height is above the surrounding rock, which is $(0.15)(35\ \text{km}) = 5.3$ km. This result, 5.3 km, represents a rough estimate of the average depth of the ocean bottom relative to the height of the continents. (It is rough since it ignores such things as the weight of the oceans.)

Air is a fluid and it too exerts a buoyant force. Ordinary objects weigh less in air than they do if weighed in a vacuum. Because the density of air is so small, the effect for ordinary solids is slight. There are objects, however, that float in air—helium-filled balloons, for example.

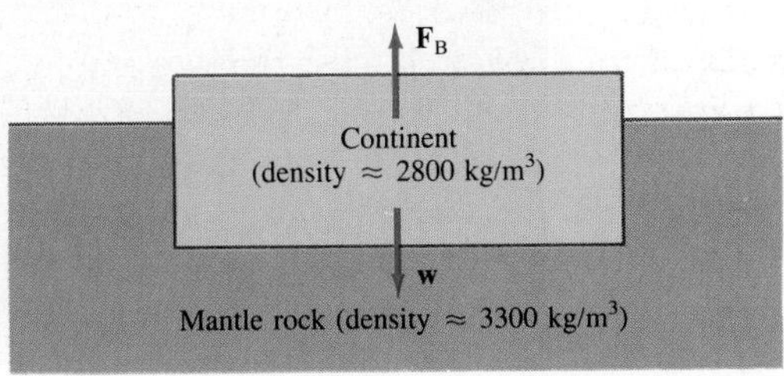

FIGURE 10–11 Model of a continent "floating" in the earth's crust. The buoyant force $\mathbf{F}_B$ has magnitude equal to the weight $\mathbf{w}$ for a body that floats (is in equilibrium).

EXAMPLE 10–7 What volume V of helium is needed if a balloon is to lift a load of 800 kg (including the weight of the empty balloon)?

SOLUTION The buoyant force on the helium, F_B, which is equal to the weight of displaced air, must be at least equal to the weight of the helium plus the load:

$$F_B = (m_{He} + 800\text{ kg})g,$$

where g is the acceleration due to gravity. This equation can be written in terms of density:

$$\rho_{air} Vg = (\rho_{He} V + 800\text{ kg})g.$$

Solving now for V, we find

$$V = \frac{800\text{ kg}}{\rho_{air} - \rho_{He}} = \frac{800\text{ kg}}{(1.29\text{ kg/m}^3 - 0.18\text{ kg/m}^3)}$$

$$= 720\text{ m}^3.$$

This is the volume needed near the earth's surface, where $\rho_{air} = 1.29\text{ kg/m}^3$. To reach a high altitude, a greater volume would be needed since the density of air decreases with altitude.

10–7 • Fluids in Motion; Flow Rate and the Equation of Continuity

We now turn from the study of fluids at rest to the more complex subject of fluids in motion, which is called **hydrodynamics**. Many aspects of fluid motion are still being studied today (for example, turbulence as a manifesta-

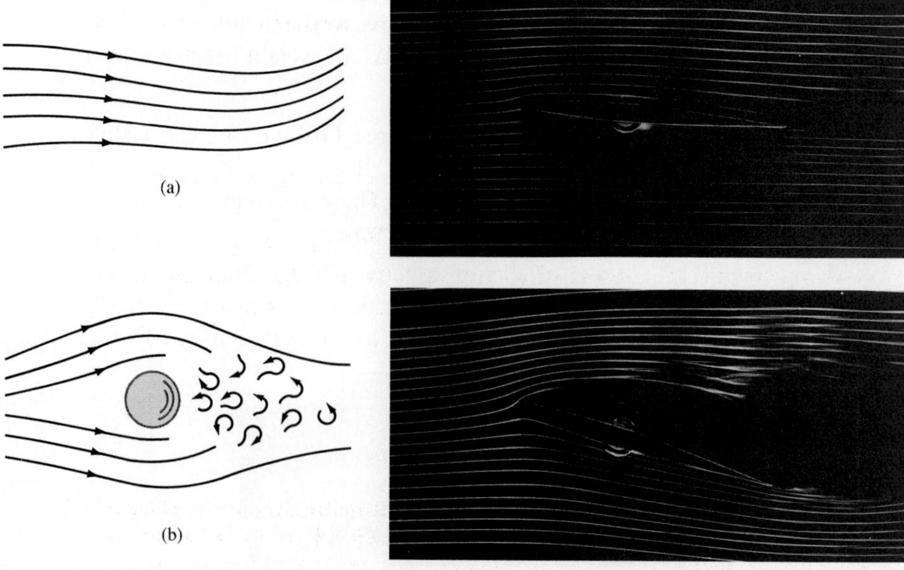

FIGURE 10–12
(a) Streamline or laminar flow; (b) turbulent flow.

tion of chaos is a "hot" topic today). Nonetheless, with certain simplifying assumptions, a good understanding of this subject can be obtained.

To begin with, we can distinguish two main types of fluid flow. If the flow is smooth, such that neighboring layers of the fluid slide by each other smoothly, the flow is said to be **streamline** or **laminar flow**.† In this kind of flow, each particle of the fluid follows a smooth path, and these paths do not cross over one another (Fig. 10–12a). Above a certain speed, which depends on a number of factors, as we shall see later, the flow becomes turbulent. **Turbulent flow** is characterized by erratic, small whirlpool-like circles called eddy currents or eddies (Fig. 10–12b). Eddies absorb a great deal of energy, and although a certain amount of internal friction, called **viscosity**, is present even during streamline flow, it is much greater when the flow is turbulent. A few tiny drops of ink or food coloring dropped into a moving liquid can quickly reveal whether the flow is streamline or turbulent.

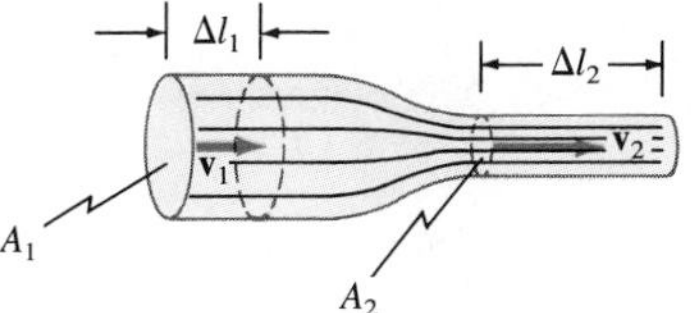

FIGURE 10–13
Fluid flow through a pipe of varying diameter.

Let us consider the steady laminar flow of a fluid through an enclosed tube or pipe as shown in Fig. 10–13. First we determine how the speed of the fluid changes when the size of the tube changes. The mass **flow rate** is defined as the mass Δm of fluid that passes a given point per unit time Δt: mass flow rate $= \Delta m/\Delta t$. In Fig. 10–13, the volume of fluid passing point 1 (that is, through area A_1) in a time Δt is just $A_1 \Delta l_1$, where Δl_1 is the distance the fluid moves in time Δt. Since the velocity‡ of fluid passing point 1 is $v_1 = \Delta l_1/\Delta t$, the mass flow rate $\Delta m/\Delta t$ through area A_1 is

$$\frac{\Delta m}{\Delta t} = \frac{\rho_1 \Delta V_1}{\Delta t} = \frac{\rho_1 A_1 \Delta l_1}{\Delta t} = \rho_1 A_1 v_1,$$

where $\Delta V_1 = A_1 \Delta l_1$ is the volume of mass Δm and ρ_1 is the fluid density. Similarly, at point 2 (through area A_2), the flow rate is $\rho_2 A_2 v_2$. Since no fluid flows in or out the sides, the flow rates through A_1 and A_2 must be equal. Thus

Equation of continuity

$$\rho_1 A_1 v_1 = \rho_2 A_2 v_2.$$

This is called the **equation of continuity**. If the fluid is incompressible, which is an excellent approximation for liquids under most circumstances (and sometimes for gases as well), then $\rho_1 = \rho_2$ and the equation of continuity becomes

$$A_1 v_1 = A_2 v_2. \tag{10–4}$$

Notice that the product Av represents the *volume* rate of flow (volume of fluid passing a given point per second), since $\Delta V/\Delta t = A\ \Delta l/\Delta t = Av$. Equation 10–4 tells us that where the cross-sectional area is large the velocity is small, and where the area is small the velocity is large. That this makes sense can be seen by looking at a river. A river flows slowly and languidly through a meadow where it is broad, but speeds up to torrential speed when passing through a narrow gorge.

† The word laminar means "in layers."

‡ If there were no viscosity, the velocity would be the same across a cross section of the tube. Real fluids have viscosity and this internal friction causes different layers of the fluid to flow at different speeds. In this case v_1 and v_2 represent the average speeds at each cross section.

Human circulatory system

Equation 10–4 can be applied to the flow of blood in the body. Blood flows from the heart into the aorta, from which it passes into the major arteries. These branch into the small arteries (arterioles), which in turn branch into myriads of tiny capillaries. The blood returns to the heart via the veins (Fig. 10–14).

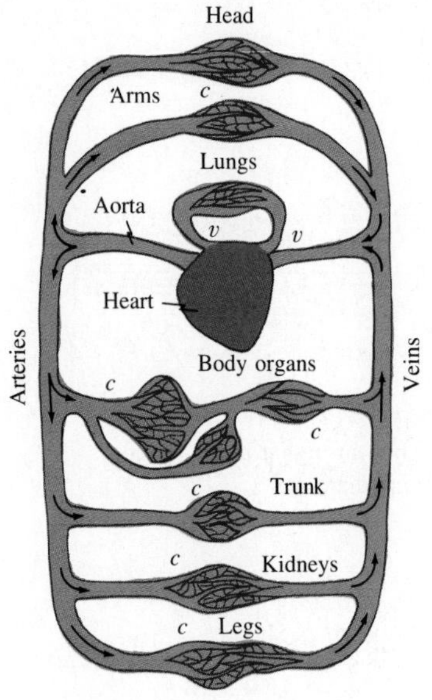

v–valves
c–capillaries

FIGURE 10–14
Human circulatory system.

EXAMPLE 10–8 The radius of the aorta is about 1.0 cm and the blood flowing through it has a speed of about 30 cm/s. Calculate the average speed of the blood in the capillaries given that, although each capillary has a diameter of about 8×10^{-4} cm, there are literally billions of them so that their total cross section is about 2000 cm^2.

SOLUTION The area of the aorta is $A_1 = \pi r^2$, where $r = 0.010$ m. Then the speed of blood in the capillaries is (Eq. 10–4)

$$v_2 = \frac{v_1 A_1}{A_2} = \frac{(0.30 \text{ m/s})(3.14)(0.010 \text{ m})^2}{(2 \times 10^{-1} \text{ m}^2)}$$
$$= 5 \times 10^{-4} \text{ m/s}$$

or 0.5 mm/s.

Another example that makes use of the equation of continuity and the argument leading up to it is the following.

EXAMPLE 10–9 How large must a heating duct be if air moving 3.0 m/s along it can replenish the air in a room of 300-m^3 volume every 15 min? Assume the air's density remains constant.

SOLUTION We can apply Eq. 10–4 if we consider the room (call it point 2) as a large section of the duct. Reasoning in the same way we did to obtain Eq. 10–4 (changing Δt to t), we see that $A_2 v_2 = A_2 l_2 / t = V_2 / t$ where V_2 is the volume of the room. Then $A_1 v_1 = A_2 v_2 = V_2 / t$ and

$$A_1 = \frac{V_2}{v_1 t} = \frac{300 \text{ m}^3}{(3.0 \text{ m/s})(900 \text{ s})} = 0.11 \text{ m}^2.$$

If the duct has a circular cross section, then $A = \pi r^2$ and we find that the radius must be 0.19 m or 19 cm.

10–8 • Bernoulli's Equation

Bernoulli's principle

Have you ever wondered how air can circulate in a prairie dog's burrow, why smoke goes up a chimney, or why a car's convertible top bulges upward at high speeds? These are examples of a principle worked out by Daniel Bernoulli (1700–1782) in the early eighteenth century. In essence, **Bernoulli's principle** states that *where the velocity of a fluid is high, the pressure is low, and where the velocity is low, the pressure is high.* For example, if the pressures at points 1 and 2 in Fig. 10–13 are measured, it will be found that the pressure is lower at point 2, where the velocity is higher, than it is at point 1, where the velocity is lower. At first glance, this might seem strange;

you might expect that the higher speed at point 2 would imply a greater pressure. But this cannot be the case. For if the pressure at point 2 were higher than at 1, this higher pressure would slow the fluid down, whereas in fact it has speeded up in going from point 1 to point 2. Thus the pressure at point 2 must be less than at point 1, which will allow the fluid to accelerate.

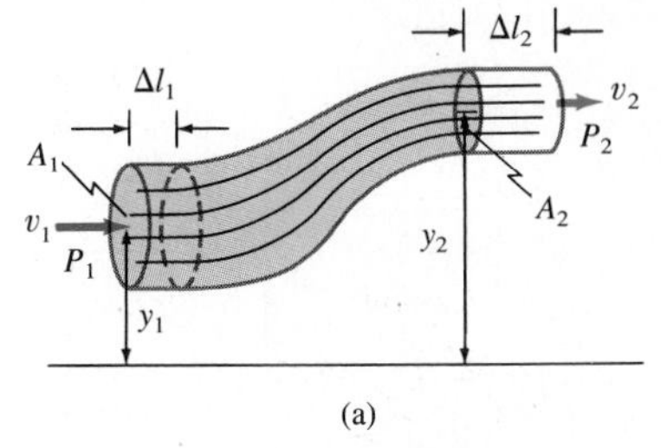

(a)

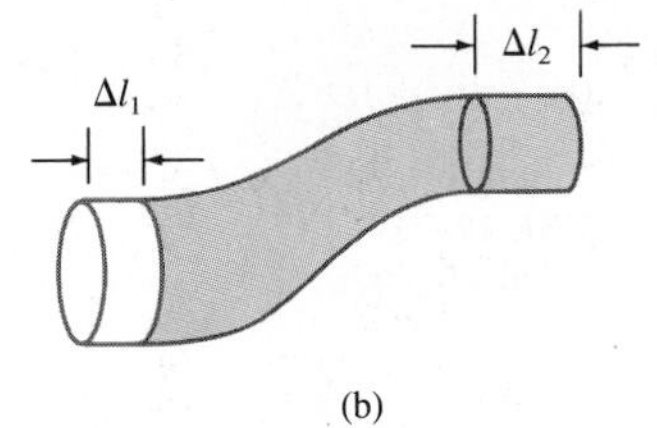

(b)

FIGURE 10–15 Fluid flow: for derivation of Bernoulli's equation.

Bernoulli developed an equation that expresses this principle quantitatively. To derive Bernoulli's equation, we assume the flow is steady and laminar, the fluid is incompressible, and the viscosity is small enough to be ignored. To be general, we assume the fluid is flowing in a tube of nonuniform cross section that varies in height above some reference level, Fig. 10–15. We will consider the amount of fluid shown in color and calculate the work done to move it from the position shown in (a) to that shown in (b). In this process, fluid at point 1 flows a distance Δl_1 and forces the fluid at point 2 to move a distance Δl_2. The fluid to the left of point 1 exerts a pressure P_1 on the fluid and does an amount of work $W_1 = F_1\,\Delta l_1 = P_1 A_1\,\Delta l_1$. At point 2, the work done is $W_2 = -P_2 A_2\,\Delta l_2$; the negative sign is present because the force exerted on the fluid is opposite to the motion (thus the fluid shown in color does work on the fluid to the right of point 2). Work is also done on the fluid by the force of gravity. Since the net effect of the process shown in Fig. 10–16 is to move a mass m of volume $A_1\,\Delta l_1$ ($= A_2\,\Delta l_2$, since the fluid is incompressible) from point 1 to point 2, the work done by gravity is

$$W_3 = -mg(y_2 - y_1),$$

where y_1 and y_2 are heights of the center of the tube above some (arbitrary) reference level. Notice that in the case shown in Fig. 10–15, this term is negative since the motion is uphill against the force of gravity. The net work W done on the fluid is thus:

$$W = W_1 + W_2 + W_3$$

$$W = P_1 A_1\,\Delta l_1 - P_2 A_2\,\Delta l_2 - mgy_2 + mgy_1.$$

According to the work-energy theorem (Section 6–3), the net work done on a system is equal to its change in kinetic energy. Thus

$$\tfrac{1}{2}mv_2^2 - \tfrac{1}{2}mv_1^2 = P_1 A_1\,\Delta l_1 - P_2 A_2\,\Delta l_2 - mgy_2 + mgy_1.$$

The mass m has volume $A_1\,\Delta l_1 = A_2\,\Delta l_2$. Thus we can substitute $m = \rho A_1\,\Delta l_1 = \rho A_2\,\Delta l_2$ and obtain

$$\tfrac{1}{2}\rho(A_1\Delta l_1)v_2^2 - \tfrac{1}{2}\rho(A_1\Delta l_1)v_1^2 = P_1(A_1\Delta l_1) - P_2(A_2\Delta l_2) - \rho(A_1\Delta l_1)(y_2 - y_1)g.$$

Now we can divide through by $A_1\,\Delta l_1 = A_2\,\Delta l_2$ to obtain:

$$\tfrac{1}{2}\rho v_2^2 - \tfrac{1}{2}\rho v_1^2 = P_1 - P_2 - \rho g y_2 + \rho g y_1$$

which we rearrange to get

$$P_1 + \tfrac{1}{2}\rho v_1^2 + \rho g y_1 = P_2 + \tfrac{1}{2}\rho v_2^2 + \rho g y_2. \qquad (10\text{–}5)$$

Bernoulli's equation

This is **Bernoulli's equation**. Since points 1 and 2 can be any two points along a tube of flow, Bernoulli's equation can be written:

$$P + \tfrac{1}{2}\rho v^2 + \rho g y = \text{constant}$$

at every point in the fluid.

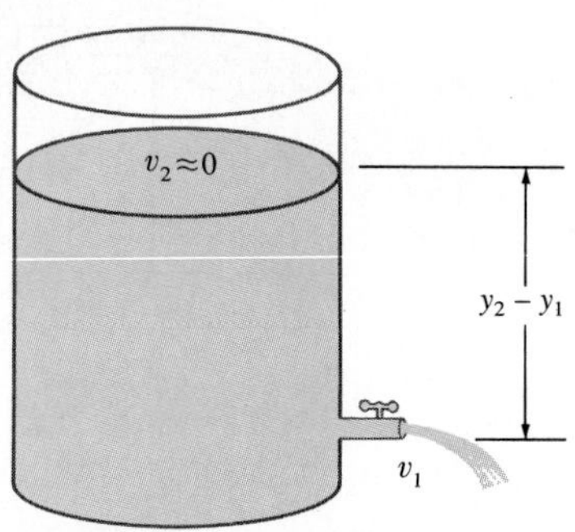

FIGURE 10–16
Torricelli's theorem: $v_1 = \sqrt{2g(y_2 - y_1)}$.

10–9 • Applications of Bernoulli's Principle: From Torricelli to Sailboats, Airfoils and TIA

Bernoulli's equation can be applied to a great many situations. One example is to calculate the velocity, v_1, of a liquid flowing out of a spigot at the bottom of a reservoir, Fig. 10–16. We choose point 2 in Eq. 10–5 to be the top surface of the liquid. Assuming the diameter of the reservoir is large compared to that of the spigot, v_2 will be almost zero. Points 1 (the spigot) and 2 (top surface) are open to the atmosphere so the pressure at both points is equal to atmospheric pressure: $P_1 = P_2$. Then Bernoulli's equation becomes

$$\tfrac{1}{2}\rho v_1^2 + \rho g y_1 = \rho g y_2$$

or

Torricelli's theorem

$$v_1 = \sqrt{2g(y_2 - y_1)}. \tag{10–6}$$

This result is called **Torricelli's theorem**. Although it is seen to be a special case of Bernoulli's equation, it was discovered a century before Bernoulli by Evangelista Torricelli; hence its name. Notice that the liquid leaves the spigot with the same speed that a freely falling object would attain falling the same height. This should not be too surprising since the derivation of Bernoulli's equation relies on the conservation of energy.

Another special case of Bernoulli's equation arises when the fluid is flowing but there is no appreciable change in height; that is, $y_1 = y_2$. Then Eq. 10–5 becomes

$$P_1 + \tfrac{1}{2}\rho v_1^2 = P_2 + \tfrac{1}{2}\rho v_2^2. \tag{10–7}$$

This tells us quantitatively that where the speed is high the pressure is low, and vice versa. It explains many common phenomena, some of which are illustrated in Fig. 10–17. The pressure in the air blown at high speed across the top of the vertical tube of a perfume atomizer (Fig. 10–17a) is less than the normal air pressure acting on the surface of the liquid in the bowl; thus perfume is pushed up the tube because of the reduced pressure at the top. A Ping-Pong ball can be made to float above a blowing jet of air (some

FIGURE 10–17
Examples of Bernoulli's principle.

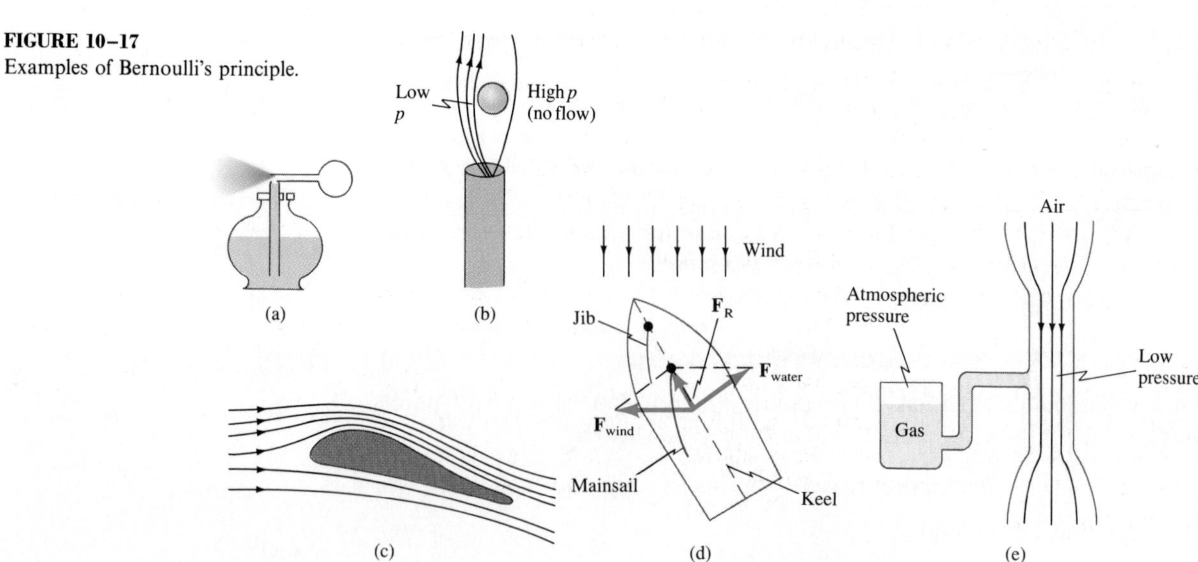

vacuum cleaners can blow air), Fig. 10–17b; if the ball begins to leave the jet of air, the higher pressure outside the jet pushes the ball back in.

Airplane wings and other airfoils are designed to deflect the air so that although streamline flow is largely maintained, the streamlines are crowded together above the wing, Fig. 10–17c. Just as the flow lines are crowded together in a pipe constriction where the velocity is high (see Fig. 10–13), so the crowded streamlines above the wing indicate that the air speed is greater than below the wing. Hence the air pressure above the wing is less than that below and there is thus a net upward force, which is called **dynamic lift**. Actually, Bernoulli's principle is only one aspect of the lift on a wing. Wings are usually tilted slightly upward so that air striking the bottom surface is deflected downward; the change in momentum of the rebounding air molecules results in an additional upward force on the wing. Turbulence also plays an important role.

Airplanes and dynamic lift

A sailboat can move against the wind, Figs. 10–17d and 10–18 and the Bernoulli effect aids in this considerably if the sails are arranged so that the air velocity increases in the narrow constriction between the two sails. The normal atmospheric pressure behind the mainsail is larger than the reduced pressure in front of it (due to the fast moving air in the narrow slot between the sails), and this pushes the boat forward. When going against the wind, the mainsail is set at an angle approximately midway between the wind direction and the boat's axis (keel line) as shown in Fig. 10–17d. The net force on the sail (wind and Bernoulli) acts nearly perpendicular to the sail ($\mathbf{F}_{\text{wind}}$). This would tend to make the boat move sideways if it weren't for the keel beneath—for the water exerts a force ($\mathbf{F}_{\text{water}}$) on the keel nearly perpendicular to the keel. The resultant of these two forces ($\mathbf{F}_{\mathbf{R}}$) is almost directly forward as shown.

Sailing against the wind

FIGURE 10–18
A sailboat moving against the wind makes use of Bernoulli's principle.

A **venturi tube** is essentially a pipe with a narrow constriction (the throat). One example of a venturi tube is the barrel of a carburetor in a car, (Fig. 10–17e). The flowing air speeds up as it passes this constriction (Eq. 10–4) and so the pressure is lower. Because of the reduced pressure, gasoline under atmospheric pressure in the carburetor reservoir is forced into the air stream and mixes with the air before entering the cylinders.

The venturi tube is also the basis of the *venturi meter*, which is used to measure the flow speed of fluids, Fig. 10–19. Venturi meters can be used to measure the flow velocities of gases and liquids and have even been designed to measure blood velocity in arteries. (See the problems.)

FIGURE 10–19 Venturi meter.

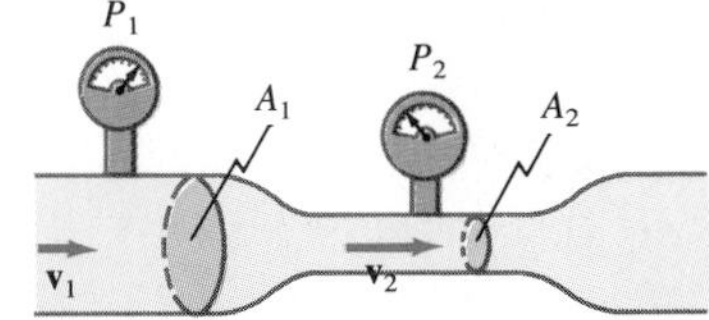

Why does smoke go up a chimney? It's partly because hot air rises (it's less dense and therefore buoyant). But Bernoulli's principle also plays a role. Because wind blows across the top of a chimney, the pressure is less there than inside the house. Hence, air and smoke are pushed up the chimney. Even on an apparently still night there is usually enough ambient air flow at the top of a chimney to assist upward flow of smoke.

If gophers, prairie dogs, rabbits, and other animals that live underground are to avoid suffocation, the air must circulate in their burrows. The burrows are always made to have at least two entrances (Fig. 10–20). The speed of air flow across different holes will usually be slightly different. This results in a slight pressure difference which forces a flow of air through the burrow a la Bernoulli. The flow of air is enhanced if one hole is higher than the other (and this is often contrived by animals) since wind speed tends to increase with height.

FIGURE 10–20 Bernoulli's principle is responsible for air flow in underground burrows.

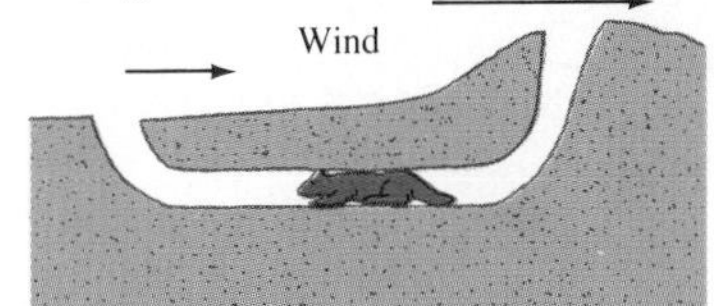

In medicine, one of many applications of Bernoulli's principle is to explain a TIA, a "transient ischemic attack" (meaning a temporary lack of blood supply to the brain), caused by the so-called "subclavian steal syndrome." A person suffering a TIA may experience symptoms such as dizziness, double vision, headache, and weakness of the limbs. A TIA can occur as follows. Blood flows up to the brain at the back of the head via the two vertebral arteries—one going up each side of the neck—which meet to form the basilar artery just below the brain, as shown in Fig. 10–21. The vertebral arteries issue from the subclavian arteries, as shown, before the latter pass to the arms. When an arm is exercised vigorously, blood flow increases to meet the needs of the arm's muscles. If the subclavian artery on one side of the body is partially blocked, however, say by atherosclerosis, the blood velocity will have to be higher on that side to supply the needed blood. (Recall the equation of continuity: smaller area means larger velocity for the same flow rate, Eq. 10–4). The increased blood velocity past the opening to the vertebral artery results in lower pressure (Bernoulli's principle). Thus, blood rising in the vertebral artery on the "good" side at normal pressure can be *diverted down* to the other vertebral artery because of the low pressure on that side (like the Venturi effect), instead of passing upward into the basilar artery and the brain. Hence the blood supply to the brain is reduced due to "subclavian steal syndrome": the subclavian artery "steals" the blood away from the brain. The resulting dizziness or weakness usually causes the person to stop the exertions, followed by a return to normal.

TIA

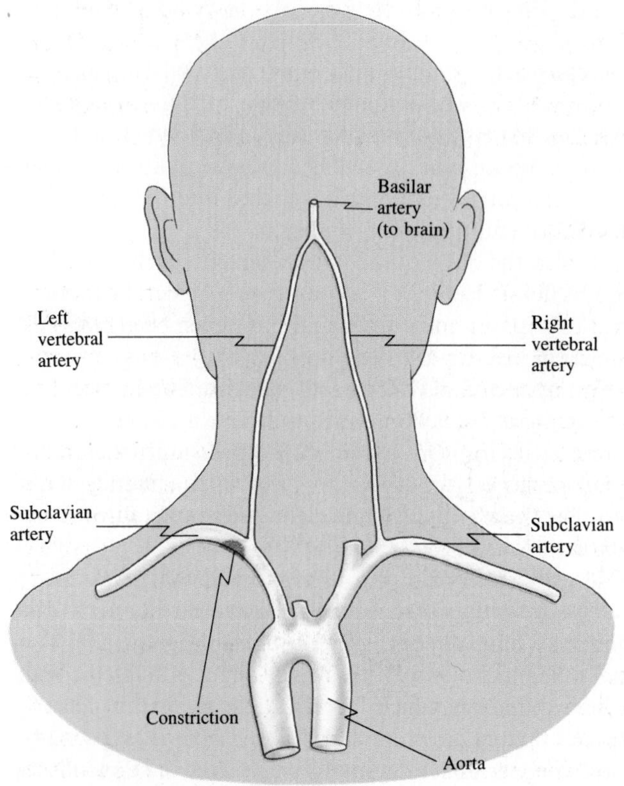

FIGURE 10–21
Rear of the head and shoulders showing arteries leading to brain and to arms. High blood velocity past the constriction in the left subclavian artery causes low pressure in left vertebral artery which can result in reverse (downward) blood flow: so-called "subclavian steal", resulting in a TIA (see text).

Let us next consider a numerical example of Bernoulli's equation.

EXAMPLE 10–10 Water circulates throughout a house in a hot-water heating system. If the water is pumped at a speed of 0.50 m/s through a 4.0-cm-diameter pipe in the basement under a pressure of 3.0 atm, what will be the flow speed and pressure in a 2.6-cm-diameter pipe on the second floor 5.0 m above?

SOLUTION We first calculate the velocity v_2 using the equation of continuity, Eq. 10–4, noting that the areas are proportional to the radii squared ($A = \pi r^2$):

$$v_2 = \frac{v_1 A_1}{A_2} = (0.50 \text{ m/s}) \frac{(0.020 \text{ m})^2}{(0.013 \text{ m})^2} = 1.2 \text{ m/s}.$$

To find the pressure, we use Bernoulli's equation:

$$\begin{aligned} P_2 &= P_1 + \rho g(y_1 - y_2) + \tfrac{1}{2}\rho(v_1^2 - v_2^2) \\ &= (3.0 \times 10^5 \text{ N/m}^2) + (1.0 \times 10^3 \text{ kg/m}^3)(9.8 \text{ m/s}^2)(-5.0 \text{ m}) \\ &\quad + \tfrac{1}{2}(1.0 \times 10^3 \text{ kg/m}^3)[(0.50 \text{ m/s})^2 - (1.2 \text{ m/s})^2] \\ &= 3.0 \times 10^5 \text{ N/m}^2 - 4.9 \times 10^4 \text{ N/m}^2 - 6.0 \times 10^2 \text{ N/m}^2 \\ &= 2.5 \times 10^5 \text{ N/m}^2. \end{aligned}$$

Notice that the velocity term contributes very little in this case.

Bernoulli's equation ignores the effects of friction (viscosity) and the compressibility of the fluid. The energy that is transformed to internal (or potential) energy due to compression and to thermal energy by friction can be taken into account by adding terms to the right side of Eq. 10–5. These terms are difficult to calculate theoretically and are normally determined empirically. We will not pursue it here, but merely note that it does not significantly alter the explanations for the phenomena described above.

*10–10 • Viscosity

As already mentioned, real fluids have a certain amount of internal friction which is called **viscosity**. It exists in both liquids and gases and is essentially a frictional force between different layers of fluid as they move past one another. In liquids, viscosity is due to the cohesive forces between the molecules. In gases, it arises from collisions between the molecules.

Different fluids possess different amounts of viscosity: syrup is more viscous than water; grease is more viscous than engine oil; liquids in general are much more viscous than gases. The viscosity of different fluids can be expressed quantitatively by a *coefficient of viscosity*, η (the Greek lowercase letter eta), which is defined in the following way. A thin layer of fluid is placed between two flat plates. One plate is stationary and the other is made to move, Fig. 10–22. The fluid directly in contact with each plate is held to the surface by the adhesive force between the molecules of the liquid and those of the plate. Thus the upper surface of the fluid moves with the same speed v as the upper plate, whereas the fluid in contact with the stationary plate remains stationary. The stationary layer of fluid retards the flow of the

FIGURE 10–22 Determination of viscosity.

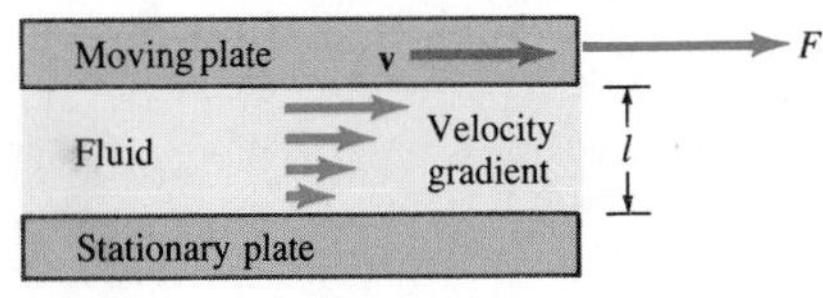

Table 10–3
Coefficient of Viscosity for Various Fluids

Fluid	Temperature (°C)	Coefficient of Viscosity, η (Pa·s)[†]
Water	0	1.8×10^{-3}
	20	1.0×10^{-3}
	100	0.3×10^{-3}
Whole blood	37	$\approx 4 \times 10^{-3}$
Blood plasma	37	$\approx 1.5 \times 10^{-3}$
Ethyl alcohol	20	1.2×10^{-3}
Engine oil (SAE 10)	30	200×10^{-3}
Glycerine	20	1500×10^{-3}
Air	20	0.018×10^{-3}
Hydrogen	0	0.009×10^{-3}
Water vapor	100	0.013×10^{-3}

[†] $1 \text{ Pa·s} = 10 \text{ P} = 1 \times 10^3 \text{ cP}$

layer just above it; this layer retards the flow of the next layer, and so on. Thus the velocity varies linearly from 0 to v, as shown. The increase in velocity divided by the distance over which this change is made—equal to v/l—is called the *velocity gradient*. To move the upper plate requires a force, which you can verify by moving a flat plate across a puddle of syrup on a table. For a given fluid, it is found that the force required, F, is proportional to the area of either plate, A, and to the speed, v, and is inversely proportional to the separation, l, of the plates: $F \propto vA/l$. For different fluids, the more viscous the fluid, the greater is the required force. Hence the proportionality constant for this equation is defined as the *coefficient of viscosity*, η:

$$F = \eta A \frac{v}{l}. \tag{10–8}$$

Solving for η, we find $\eta = Fl/vA$. The SI unit for η is $\text{N·s/m}^2 = \text{Pa·s}$ (pascal·second). In the cgs system, the unit is dyne·s/cm^2 and this unit is called a *poise* (P). Viscosities are often stated in *centipoise* (cP), which is one-hundredth of a poise. Table 10–3 lists the coefficient of viscosity for various fluids. The temperature is also specified, since it has a strong effect—the viscosity of liquids such as motor oil, for example, decreases rapidly as temperature increases.

*10–11 • Flow in Tubes: Poiseuille's Equation, Blood Flow, Reynolds Number

If a fluid had no viscosity, it could flow through a level tube or pipe without a force being applied. Because of viscosity, a pressure difference between the ends of a tube is necessary for the steady flow of any real fluid, be it water or oil in a pipe, or blood in the circulatory system of a human, even when the tube is level.

The rate of flow of a fluid in a round tube depends on the viscosity of the fluid, the pressure difference, and the dimensions of the tube. The French scientist J. L. Poiseuille (1799–1869), who was interested in the physics of blood circulation (and after whom the "poise" is named), determined how the variables affect the flow rate of an incompressible fluid undergoing laminar flow in a cylindrical tube. His result, known as *Poiseuille's equation*, is as follows:

$$Q = \frac{\pi r^4 (P_1 - P_2)}{8\eta L}, \tag{10–9}$$

where r is the inside radius of the tube, L is its length, $P_1 - P_2$ is the pressure difference between the ends, η is the viscosity, and Q is the volume rate of flow (volume of fluid flowing past a given point per unit time). The derivation of Eq. 10–9, which is long and requires the use of calculus, can be found in more advanced textbooks.† Equation 10–9 applies to laminar flow. There is no such simple mathematical relation if the flow is turbulent.

EXAMPLE 10–11 Engine oil (assume SAE 10—Table 10–3) passes through a fine 1.80-mm-diameter tube in a prototype engine. The tube is 5.5 cm long. What pressure difference is needed to maintain a flow rate of 5.6 mL/min?

SOLUTION The flow rate in SI units is $Q = 5.6 \times 10^{-6}\ \text{m}^3/60\ \text{s} = 9.3 \times 10^{-8}\ \text{m}^3/\text{s}$. We solve for $P_1 - P_2$ in Eq. 10–9 and put all terms in SI units:

$$\begin{aligned} P_1 - P_2 &= \frac{8\eta LQ}{\pi r^4} \\ &= \frac{8(2.0 \times 10^{-1}\ \text{N}\cdot\text{s/m}^2)(5.5 \times 10^{-2}\ \text{m})(9.3 \times 10^{-8}\ \text{m}^3/\text{s})}{3.14(0.90 \times 10^{-3}\ \text{m})^4} \\ &= 4.0 \times 10^3\ \text{N/m}^2 \end{aligned}$$

or about 0.040 atm.

Poiseuille's equation tells us that the flow rate Q is directly proportional to the "pressure gradient," $(P_1 - P_2)/L$, and it is inversely proportional to the viscosity of the fluid. This is just what we might expect. It may be surprising, however, that Q also depends on the *fourth* power of the tube's radius. This means that for the same pressure gradient, if the tube radius is halved, the flow rate is decreased by a factor of 16! Thus the rate of flow, or alternately the pressure required to maintain a given flow rate, is greatly affected by only a small change in tube radius.

Blood flow and heart disease

An interesting example of this r^4 dependence is *blood flow* in the human body. Poiseuille's equation is valid only for the streamline flow of an incompressible fluid with constant viscosity η; so it cannot be precisely accurate for blood whose flow is not without turbulence and that contains corpuscles (whose diameter is almost equal to that of a capillary). Hence η depends to a certain extent on the blood flow speed v. Nonetheless, Poiseuille's equation does give a reasonable first approximation. The body controls the flow of blood by means of tiny bands of muscle surrounding the arteries. Contraction of these muscles reduces the diameter of an artery and, because of the r^4

† D. C. Giancoli, *Physics for Scientists and Engineers*, Prentice-Hall, Inc., Englewood Cliffs, New Jersey, 1989, Section 14–5.

in Eq. 10–9, the flow rate is greatly reduced for only a small change in radius. Very small actions by these muscles can thus control precisely the flow of blood to different parts of the body. Another aspect is that the radius of arteries is reduced as a result of arteriosclerosis (hardening of the arteries) and by cholesterol buildup; when this happens, the pressure gradient must be increased to maintain the same flow rate. If the radius is reduced by half, the heart would have to increase the pressure by a factor of about 16 in order to maintain the same blood-flow rate. The heart must work much harder under these conditions, but usually cannot maintain the original flow rate. Thus, high blood pressure is an indication both that the heart is working harder and that the blood-flow rate is reduced.

If the flow velocity is large, the flow through a tube will become turbulent and Poiseuille's equation will no longer hold. When the flow is turbulent, the flow rate Q for a given pressure difference will be less than for laminar flow as given in Eq. 10–9 because friction forces are much greater when turbulence is present.

The onset of turbulence is often abrupt and can be characterized approximately by the so-called *Reynolds number*, *Re*:

$$Re = \frac{2\bar{v}r\rho}{\eta}, \tag{10–10}$$

where $\bar{v}$ is the average speed of the fluid, ρ is its density, η is its viscosity, and r is the radius of the tube in which the fluid is flowing. Experiments show that the flow is laminar if Re has a value less than about 2000, but is turbulent if Re exceeds this value.

Example 10–12 The average speed of blood in the aorta ($r = 1.0$ cm) during the resting part of the heart's cycle is about 30 cm/s. Is the flow laminar or turbulent?

Solution To answer this, we calculate the Reynolds number using the values of ρ and η from Tables 10–1 and 10–3:

$$Re = \frac{(2)(0.30\text{ m/s})(0.010\text{ m})(1.05 \times 10^3\text{ kg/m}^3)}{(4.0 \times 10^{-3}\text{ N}\cdot\text{s/m}^2)} = 1600.$$

The flow will probably be laminar, but is close to turbulence.

Notice in this example that since $1\text{ N} = 1\text{ kg}\cdot\text{m/s}^2$, Re has no units. Hence we see that the Reynolds number is always a *dimensionless* quantity; its value is the same in any consistent set of units. Next we take a geophysical example.

Example 10–13 According to the plate tectonic model, the plates supporting the earth's continents move very slowly on the hot deformable rock below. Show that this flow is laminar using the following data: speed $v = 50$ mm/yr, density and viscosity of deformable rock below are $\rho = 3200\text{ kg/m}^3$ and $\eta = 4 \times 10^{19}$ Pa·s (notice the magnitude!), with thickness ≈ 100 km.

SOLUTION This is not exactly fluid flow in a tube, but it is similar (one edge of the tube moves with the fluid). Also the tube is not round, but Eq. 10–10 can be used as a rough guide, and we set the 100-km thickness equal to $2r = 1 \times 10^5$ m. Then

$$Re = \frac{2\bar{v}r\rho}{\eta}$$

$$= \frac{(5.0 \times 10^{-2}\ \mathrm{m}/3.14 \times 10^{7}\ \mathrm{s})(1 \times 10^{5}\ \mathrm{m})(3200\ \mathrm{kg/m^3})}{4 \times 10^{19}\ \mathrm{Pa \cdot s}} \ll 1.$$

The flow is laminar, due to the low speed of plate movement and the enormous viscosity of the underlying rock.

*10–12 • Object Moving in a Fluid; Sedimentation and Drag

In the previous section we saw how viscosity (and other factors) affect the flow of a fluid through a tube. In this section we will examine a slightly different situation, that of an object moving relative to a fluid. It could be an obstacle that obstructs the flow of a fluid, such as a large rock in a river, or it could be an object moving in a fluid at rest, such as a sky diver, a raindrop, a glider, or a car moving through air, or a fish in water, or a molecule sedimenting in a centrifuge.

When an object moves relative to a fluid, the fluid exerts a frictionlike retarding force on the object. This force, which is referred to as a *drag force*, is due to the viscosity of the fluid and also, at high speeds, to turbulence behind the object.

To characterize the motion of an object relative to a fluid, it is useful to define another *Reynolds number*:

$$Re' = \frac{vL\rho}{\eta},$$

where ρ and η are the density and viscosity of the fluid, v is the object's velocity relative to the fluid, and L is a "characteristic" length of the object. This Reynolds number must be clearly distinguished from the one used for fluid flow in a tube (although the form is similar) since the phenomena are different. When the Reynolds number for our present case is less than about 1, as is typically the case for fairly small objects such as raindrops, pollen grains, and molecules in a centrifuge,[†] the flow around an object is essentially laminar and it is found experimentally that the viscous force F_v is directly proportional to the speed of the object:

$$F_v = kv. \qquad (10\text{–}11)$$

The magnitude of k depends on the size and shape of the object and on the

[†] An object 1 mm long moving at a speed of 1 mm/s through water has a Reynolds number equal to 1. So does an object 2 mm long traveling 7 mm/s in air.

SUMMARY

The three common phases of matter are *solid*, *liquid*, and *gas*. Liquids and gases are collectively called *fluids*, meaning they have the ability to flow. The *density* of a material is defined as its mass per unit volume. *Specific gravity* is the ratio of the density of the material to the density of water (at 4°C).

Pressure is defined as force per unit area. The pressure at a depth h in a liquid is given by $\rho g h$, where ρ is the density of the liquid and g is the acceleration due to gravity. In addition, if an external pressure is applied to a confined fluid, this pressure is transmitted throughout the fluid; this is known as *Pascal's principle*. Pressure is measured using a manometer or other type of gauge. A *barometer* is used to measure atmospheric pressure. Standard atmospheric pressure (average at sea level) is $1.013 \times 10^5\ \mathrm{N/m^2}$. *Gauge pressure* is the total pressure less atmospheric pressure.

Archimedes' principle states that an object submerged wholly or partially in a fluid is buoyed up by a force equal to the weight of fluid it displaces. This principle is used in a method to determine specific gravity, and explains why objects whose density is less than that of a fluid will float in that fluid.

Fluid flow rate is the mass or volume of fluid that passes a given point per unit time. The equation of continuity states that for an incompressible fluid flowing in an enclosed tube, the product of the velocity of flow and the cross-sectional area of the tube remains constant: $Av =$ constant. *Bernoulli's principle* tells us that where the velocity of a fluid is high, the pressure in it is low, and where the velocity is low, the pressure is high. Bernoulli's principle explains many common phenomena.

Fluid flow can be characterized either as *streamline* (sometimes called *laminar*), in which the layers of fluid move smoothly and regularly along paths called streamlines, or as *turbulent*, in which case the flow is not smooth and regular but is characterized by irregularly shaped whirlpools. *Viscosity* refers to friction within a fluid that prevents the fluid from flowing freely and is essentially a frictional force between different layers of fluid as they move past one another.

QUESTIONS

1. Which of the following are fluids at room temperature: gold, mercury, air, glass, alcohol, carbon dioxide?
2. What is the specific gravity of (*a*) gold, (*b*) ice, and (*c*) air?
3. If one material has a higher density than another, does this mean the molecules of the first must be heavier than those of the second? Explain.
4. Airplane travelers often note that their cosmetics bottles and other containers have leaked after a trip. What might cause this?
5. The three containers in Fig. 10–39 are filled with water to the same height and have the same surface area at the base; hence the water pressure, and the total force on the base of each, is the same. Yet the total weight of water is different for each. Explain this "hydrostatic paradox."
6. Consider what happens when you push both a pin and the blunt end of a pen against your skin with the same force. Decide what determines whether your skin suffers a cut—the net force applied to it or the pressure.
7. It is often said that "water seeks its own level." Explain.

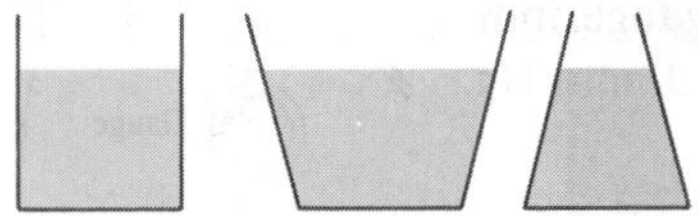

FIGURE 10–39 Question 5.

8. A small amount of water is boiled in a one-gallon gasoline can. The can is removed from the heat and the lid put on. Shortly thereafter the can collapses. Explain.
9. Explain how the tube in Fig. 10–40, known as a siphon, can transfer liquid from one container to a lower one even though the liquid must flow uphill for part of its journey. (Note that the tube must be filled with liquid to start with.) Why doesn't the liquid in each side of the tube flow back into its container?
10. An ice cube floats in a glass of water filled to the brim. As the ice melts, will the glass overflow?

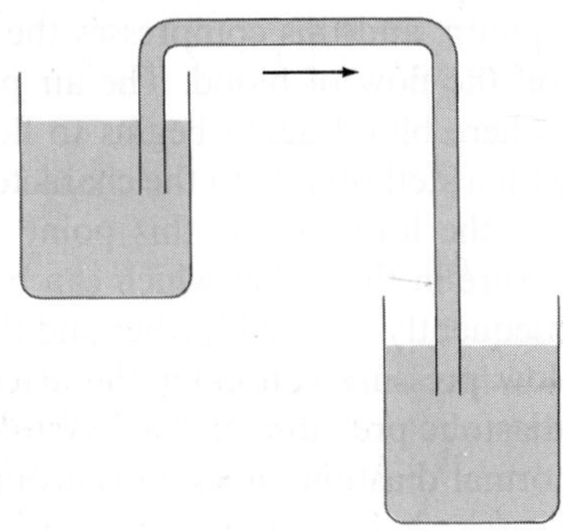

FIGURE 10–40 A siphon. Question 9.

FIGURE 10–41 Question 21.

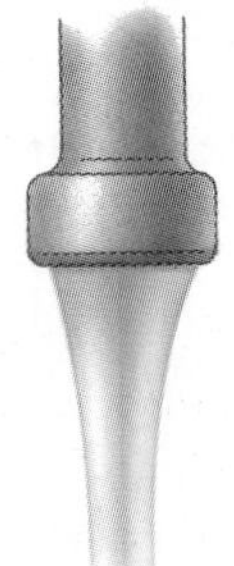

FIGURE 10–42 Water coming from a faucet. Question 29 and Problem 78.

11. A barge filled to overflowing with sand approaches a low bridge over the river and cannot quite pass under it. Should sand be added to, or removed from, the barge?

12. It is harder to pull the plug out of the drain of a bathtub when the tub is full of water than when it is empty. Is this a contradiction of Archimedes' principle? Explain.

13. Will a helium-filled balloon rise indefinitely in the air? Explain.

14. Will an empty balloon have precisely the same apparent weight on a scale as one that is filled with air? Explain.

15. Does the buoyant force on a diving bell deep beneath the ocean have precisely the same value as when the bell is just beneath the surface? Explain.

16. Devise a method for determining the mass of one of your legs, using a swimming pool.

17. (*a*) Show that the buoyant force on a partially submerged object acts at the center of gravity of the fluid before it is displaced. This point is called the *center of buoyancy*. (*b*) For a ship to be stable, should its center of buoyancy be above, below, or at the same point as, its center of gravity? Explain.

18. Why do you float more easily in salt water than in fresh?

19. When measuring blood pressure, why must the jacket be held at the level of the heart?

20. Roofs of houses are sometimes "blown" off (or are they pushed off?) during a tornado or hurricane. Explain, using Bernoulli's principle.

21. If you dangle two pieces of paper vertically, a few inches apart (Fig. 10–41), and blow between them, how do you think the papers will move? Try it and see. Explain.

22. Why does the canvas top of a convertible bulge out when the car is traveling at high speed?

23. Blood corpuscles tend to flow in the center of blood vessels. Explain.

24. Children are told to avoid standing too close to a rapidly moving train because they might get sucked under it. Is this possible? Explain.

25. Why does a sailboat need a keel? In a small sailboat, the keel is removed when the boat is anchored. Why?

26. A tall Styrofoam cup is filled with water. Two holes are punched in the cup near the bottom, and water begins rushing out. If the cup is dropped so it falls freely, will the water continue to flow from the holes? Explain.

27. Why do airplanes normally take off into the wind?

28. Hummingbirds expend 20 times as much energy to hover in front of a flower as they do in normal flight. Explain.

29. Why does the stream of water from a faucet become narrower as it falls (Fig. 10–42)?

***30.** Explain why the speed of wind increases with increased height above the earth's surface. [*Hint:* see Fig. 10–22.] Explain how this affects a mole that builds its burrow with one of the two entrances higher than the other.

***31.** Identical steel ball bearings are dropped into tubs of water at 10°C and 40°C. In which tub will the ball bearing reach the bottom more quickly?

***32.** A duck can float in water because it preens its feathers to apply a layer of grease. Explain how the increased surface tension allows the duck to float.

PROBLEMS

SECTION 10–1

1. (I) The approximate volume of the granite monolith known as El Capitan in Yosemite National Park is about $10^8\ \text{m}^3$. What is its approximate mass?
2. (I) What is the approximate mass of air in a living room $6.0\ \text{m} \times 4.0\ \text{m} \times 2.9\ \text{m}$?
3. (II) A bottle has a mass of 31.20 g when empty and 98.44 g when filled with water. When filled with another fluid, the mass is 88.78 g. What is the SG of the fluid?
4. (II) If 5.0 L of antifreeze solution (SG = 0.80) is added to 4.0 L of water to make a 9.0-L mixture, what is the SG of the mixture?

SECTIONS 10–2 TO 10–5

5. (I) The arm of a record player exerts a force of (1.0 g) $\times$ g on a record. If the diameter of the stylus is 0.0013 cm (= 0.5 mil = 0.5 $\times$ 10^{-3} in), calculate the pressure on the record groove in N/m^2 and in atmospheres.
6. (I) A typical value for systolic blood pressure is 120 mm Hg. Convert this to (*a*) torr, (*b*) N/m^2, (*c*) atm, (*d*) lb/in^2.
7. (I) What is the difference in blood pressure between the top of the head and bottom of the feet of a 1.70-m-tall person standing vertically?
8. (I) (*a*) Calculate the total force of the atmosphere acting on the top of a table that measures 2.0 m $\times$ 2.2 m. (*b*) What is the total force acting upward on the underside of the table?
9. (II) What is the total force and the absolute pressure on the bottom of a swimming pool 8.0 m by 12.0 m whose uniform depth is 2.0 m? What will be the pressure against the *side* of the pool near the bottom?
10. (II) The gauge pressure in each of the four tires of an 1800-kg automobile is 200 kPa. How much area of each tire is in contact with the ground?
11. (II) What is the total force on a rectangular-shaped dam 80 m high and 120 m wide if the water is filled to the top?
12. (II) The maximum gauge pressure in a hydraulic lift is 16 atm. What is the largest size vehicle (kg) it can lift if the diameter of the output line is 20 cm?
13. (II) How high would the level be in an alcohol barometer at normal atmospheric pressure?
14. (II) One arm of a U-shaped tube (open at both ends) contains water and the other alcohol. If the two fluids meet exactly at the bottom of the U, and the alcohol is at a height of 14.5 cm, at what height will the water be?
15. (II) Show that the work done when a pressure P acts to move a volume of fluid ΔV is given by $W = P\,\Delta V$.
16. (II) Determine the minimum gauge pressure needed in the water pipe leading into a building if water is to come out of a faucet on the twelfth floor 30 m above.
17. (III) In working out his principle, Pascal showed dramatically how force can be multiplied with fluid pressure. He placed a long tube of 0.30-cm radius vertically into a 20-cm-radius wine barrel. He found that when the barrel was filled with water and the tube filled to a height of 12 m, the barrel burst. Calculate (*a*) the mass of fluid in the tube, and (*b*) the net force on the lid of the barrel.
18. (III) Estimate the density of the water 10.0 km deep in the sea. (See Section 9–7 and Table 9–1.) By what fraction does it differ from the density at the surface?

SECTION 10–6

19. (I) The hydrometer of Example 10–5 sinks to a depth of 22.8 cm when placed in a fermenting vat. What is the density of the brewing liquid?
20. (I) A geologist finds that a moon rock whose mass is 7.20 kg has an apparent mass of 5.88 kg when submerged in water. What is the density of the rock?
21. (I) What fraction of a piece of aluminium will be submerged when it floats in mercury?
22. (II) A 78-kg person has an effective mass of 54 kg (because of buoyancy) when standing in water that comes up to the hips. Estimate the mass of each leg. Assume the body has SG = 1.00.
23. (II) A 142.0-g cuttlefish is found to have an effective mass of only 0.55 g when submerged in pure water. (*a*) Calculate its density. (*b*) How much would it weigh in seawater (SG = 1.025)?
24. (II) A 0.40-kg piece of wood floats in water but is found to sink in alcohol (SG = 0.79) in which it has an apparent mass of 0.020 kg. What is the SG of the wood?
25. (II) A small animal is found to remain suspended in a mixture of 18.0 percent (by weight) alcohol and 82.0 percent water. What is the density of the animal?
26. (II) Archimedes' principle can be used not only to determine the SG of a solid using a known liquid (Example 10–4). The reverse can be done as well. (*a*) As an example, a 2.80-kg aluminium ball has an apparent mass of 1.90 kg when submerged in a particular liquid; calculate the density of the liquid. (*b*) Derive a simple formula for determining the density of a liquid using this procedure.
27. (II) The specific gravity of ice is 0.917, whereas that for seawater is 1.025. What fraction of an iceberg is above the surface of the water?
28. (III) A 1.84-kg piece of wood (SG = 0.50) floats on water. What minimum mass of lead, hung from it by a string, will cause it to sink?
29. (III) If an object floats in water, its density can be determined by tying a "sinker" on it so that both the object and the sinker are submerged. Show that the specific gravity is given by $w/(w_1 - w_2)$, where w is the weight of the object alone in air, w_1 is the apparent weight when a "sinker" is tied to it and the sinker only is submerged, and w_2 is the apparent weight when both the object and the sinker are submerged.

SECTIONS 10–7 TO 10–9

30. (I) Using the data of Example 10–8, calculate the average speed of blood flow in the major arteries of the body which have a total cross-sectional area of about 2.0 cm^2.
31. (I) A 16-cm-radius air duct is used to replenish the air of a room 10 m $\times$ 6.0 m $\times$ 4.0 m every 10 min. How fast does the air flow in the duct?
32. (I) If wind blows at 22 m/s over your house, what is the net force on the roof if its area is 280 m^2?
33. (II) What gauge pressure in the water mains is necessary if a firehose is to spray water to a height of 20 m?
34. (II) What is the volume rate of flow of water from a 2.0-cm-diameter faucet if the pressure head is 10 m?
35. (II) What is the lift (in newtons) due to Bernoulli's principle on a wing of area 60 m^2 if the air passes over the top and bottom surfaces at speeds of 320 m/s and 290 m/s, respectively?

36. (II) Show that the power needed to drive a fluid through a pipe is equal to the volume rate of flow, Q, times the pressure difference, $P_1 - P_2$.

37. (II) Water at a pressure of 3.5 atm at street level flows into an office building at a speed of 0.60 m/s through a pipe 5.0 cm in diameter. The pipes taper down to 2.6 cm in diameter by the top floor, 25 m above. Calculate the flow velocity and the pressure in such a pipe on the top floor. Ignore viscosity. Pressures are gauge pressures.

38. (III) (*a*) Show that the flow velocity measured by a venturi meter is given by the relation $v_1 = A_2\sqrt{2(P_1 - P_2)/\rho(A_1^2 - A_2^2)}$. See Fig. 10–19. (*b*) A venturi tube is measuring the flow of water; it has a main diameter of 3.0 cm tapering down to a throat diameter of 1.0 cm; if the pressure difference is measured to be 18 mm Hg, what is the velocity of the water?

*SECTION 10–10

*39. (II) A viscometer consists of two concentric cylinders, 10.20 cm and 10.60 cm in diameter. A particular liquid fills the space between them to a depth of 12.0 cm. The outer cylinder is fixed, and a torque of 0.024 N·m keeps the inner cylinder turning at a steady rotational speed of 62 rev/min. What is the viscosity of the liquid?

*SECTION 10–11

*40. (I) A gardener feels it is taking him too long to water a garden with a $\frac{3}{8}$-in-diameter hose. By what factor will his time be cut if he uses a $\frac{5}{8}$-in-diameter hose? Assume nothing else is changed.

*41. (I) During heavy exercise, the flow speed of blood increases by perhaps a factor of 2. Referring to Example 10–12, calculate the Reynolds number and determine what type of flow you would expect in the aorta.

*42. (II) What must be the pressure difference between the two ends of a 1.7-km section of pipe 35 cm in diameter if it is to transport oil ($\rho = 950\ \text{kg/m}^3$, $\eta = 0.20$ Pa·s) at a rate of 400 cm^3/s?

*43. (II) Blood from an animal is placed in a bottle 1.50 m above a 3.8-cm-long needle of inside diameter 0.40 mm from which it flows at a rate of 4.1 cm^3/min. What is the viscosity of this blood?

*44. (II) Calculate the pressure drop per cm along the aorta using the data of Example 10–12 and Table 10–3.

*45. (II) What diameter must a 26-m-long air duct have if the pressure of a ventilation and heating system is to replenish the air in a room 9.0 m × 16 m × 4.0 m every 10 min? Assume the pump can exert a gauge pressure of 0.66×10^{-3} atm.

*46. (II) Assuming a constant pressure gradient, by what factor must a blood vessel decrease in radius if the blood flow is reduced by 80 percent?

*47. (III) What is the (approximate) maximum flow rate, Q, of water in a 15-cm-diameter pipe if turbulence is to be avoided?

*48. (III) A patient is to be given a blood transfusion. The blood is to flow through a tube from a raised bottle to a needle inserted in the vein (Fig. 10–43). The inside diameter of the 4.0-cm-long needle is 0.40 mm and the required flow rate is 4.0 cm^3 of blood per minute. How high should the bottle be placed above the needle? Obtain ρ and η from the tables. Assume the blood pressure is 18 torr above atmospheric pressure.

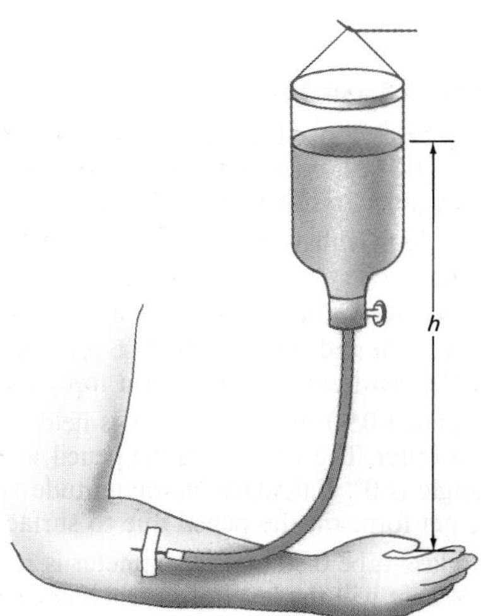

FIGURE 10–43 Problems 48 and 64.

*49. (III) Water shoots 15 m high from a 1.0-cm-diameter pipe in a fountain. What must be the gauge pressure at the pump, which is 4.5 m below the nozzle (in the ground)? Take into account viscosity, but ignore air resistance. Note any simplifying assumptions you make.

*SECTION 10–12

*50. (II) Calculate the magnitude and direction of the terminal velocity of a 1.0-mm-radius air bubble rising in oil of viscosity 0.20 Pa·s and SG 0.90.

*51. (III) If an object requires 35 min to sediment in an ultracentrifuge rotating at 25,000 rpm at an average distance from the axis of 8.0 cm, how long would it take to sediment under gravity in the same tube standing vertically in the lab?

*52. (III) (*a*) Show that the terminal velocity of a small sphere of density ρ_0 falling through a fluid of density ρ_F and viscosity η is

$$v_T = \frac{2}{9}\frac{(\rho_0 - \rho_F)r^2 g}{\eta}.$$

(*b*) What is the terminal velocity of a spherical raindrop of radius $r = 0.020$ cm falling in air?

*53. (III) (*a*) Show that the buoyant force on a small object in the liquid of a centrifuge rotating with angular velocity ω is given by

$$F_B = \rho_f V \omega^2 r,$$

where V is the volume of the object, r is its distance from the axis of rotation, and ρ_f is the density of the liquid. (*b*) Compare the dependence of F_B on position with that for the buoyant force on an object sedimenting under gravity.

is measured from the new (vertical) equilibrium position. (It is easy to show this is valid also when the spring is compressed—see Problem 20.) If the mass is released when in the position shown in Fig. 11–3c, it will oscillate up and down, between $+x$ and $-x$.

SHM

SHO

Any vibrating system for which the restoring force is directly proportional to the negative of the displacement (as in Eq. 11–1, $F = -kx$) is said to exhibit **simple harmonic motion** (SHM).† Such a system is often called a **simple harmonic oscillator** (SHO). We saw in Section 9–7 that most solid materials stretch or compress according to Eq. 11–1 as long as the displacement is not too great. Because of this, many natural vibrations are simple harmonic or close to it.

EXAMPLE 11–1 When a family of four people with a total mass of 200 kg step into their 1200-kg car, the car's springs compress 3.0 cm. (*a*) What is the spring constant of the car's springs, assuming they act as a single spring? (*b*) How far will the car lower if loaded with 300 kg?

SOLUTION (*a*) The added force of (200 kg)(9.8 m/s^2) = 1960 N causes the springs to compress 3.0×10^{-2} m. Therefore, by Eq. 11–1, the spring constant is

$$k = \frac{F}{x} = \frac{1960 \text{ N}}{3.0 \times 10^{-2} \text{ m}} = 6.5 \times 10^4 \text{ N/m}.$$

(*b*) If the car is loaded with 300 kg, $x = F/k = (300 \text{ kg})(9.8 \text{ m/s}^2)/(6.5 \times 10^4 \text{ N/m}) = 4.5 \times 10^{-2}$ m, or 4.5 cm. We could have obtained this answer without solving for k: since x is proportional to F, if 200 kg compresses the spring 3.0 cm, then 1.5 times the force will compress the spring 1.5 times as much, or 4.5 cm.

11–2 • Energy in the Simple Harmonic Oscillator

To stretch or compress a spring, work has to be done. Hence potential energy is stored in a stretched or compressed spring. Indeed, we have already seen in Section 6–4 that the potential energy is given by

$$\text{PE} = \tfrac{1}{2}kx^2.$$

Thus, since the total mechanical energy E is the sum of the kinetic and potential energies, we have

Total energy of SHO

$$E = \tfrac{1}{2}mv^2 + \tfrac{1}{2}kx^2, \tag{11–3}$$

where v is the velocity of the mass m when it is a distance x from the equilibrium position. As long as there is no friction, the total mechanical energy E remains constant. As the mass oscillates back and forth, the energy continuously changes from potential energy to kinetic energy, and back again

† The word "harmonic" refers to the motion being sinusoidal, which we discuss in Section 11–3. It is "simple" when there is pure sinusoidal motion of a single frequency (rather than a mixture of frequencies as, for example, discussed in Section 11–11 and in Chapter 12).

(Fig. 11–4). At the extreme points, $x = A$ and $x = -A$, all the energy is potential energy (and is the same whether the spring is compressed or stretched to the full amplitude). At these extreme points, the mass stops momentarily as it changes direction, so $v = 0$ and:

$$E = \tfrac{1}{2}m(0)^2 + \tfrac{1}{2}kA^2 = \tfrac{1}{2}kA^2. \qquad (11\text{–}4a)$$

Thus, the *total mechanical energy of a SHO is proportional to the square of the amplitude.* At the equilibrium point, $x = 0$, all the energy is kinetic:

Energy of SHO $\propto$ (Amplitude)^2

$$E = \tfrac{1}{2}mv_0^2 + \tfrac{1}{2}k(0)^2 = \tfrac{1}{2}mv_0^2, \qquad (11\text{–}4b)$$

where v_0 represents the *maximum* velocity during the motion (which occurs at $x = 0$). At intermediate points, the energy is part kinetic and part potential. By combining Eq. 11–4a with Eq. 11–3, we can find a useful equation for the velocity as a function of the position x:

$$\tfrac{1}{2}mv^2 + \tfrac{1}{2}kx^2 = \tfrac{1}{2}kA^2.$$

Solving for v^2, we have

$$v^2 = \frac{k}{m}(A^2 - x^2) = \frac{k}{m}A^2\left(1 - \frac{x^2}{A^2}\right).$$

From Eqs. 11–4a and 11–4b, we have $\tfrac{1}{2}mv_0^2 = \tfrac{1}{2}kA^2$, so $v_0^2 = (k/m)A^2$. Inserting this in the equation above and taking the square root, we have

$$v = \pm v_0\sqrt{1 - x^2/A^2}. \qquad (11\text{–}5)$$

This gives the velocity of the object at any position x. The object moves back and forth, so its velocity can be either in the + or − direction, but its magnitude depends only on the magnitude of x.

We can easily show that Eq. 11–3 is valid also for a vertical spring. If we take our reference point for calculating the PE to be the spring's natural length, as in Fig. 11–3a, then the PE at the new (vertical) equilibrium position, Fig. 11–3b, is $\text{PE} = \tfrac{1}{2}kx_0^2 - mgx_0$, where we include both elastic and gravitational PE. (See Section 6–7.) The PE when the spring is stretched an additional distance x, as in Fig. 11–3c, is $\text{PE} = \tfrac{1}{2}k(x + x_0)^2 - mg(x + x_0)$. The difference in these is (remember $x_0 = mg/k$):

$$\begin{aligned}\tfrac{1}{2}k(x + x_0)^2 - mg(x + x_0) - \tfrac{1}{2}kx_0^2 + mgx_0 &= \tfrac{1}{2}kx^2 + kxx_0 - mgx\\ &= \tfrac{1}{2}kx^2 + kx\left(\frac{mg}{k}\right) - mgx\\ &= \tfrac{1}{2}kx^2.\end{aligned}$$

Thus the PE of the system relative to the vertical equilibrium position is $\tfrac{1}{2}kx^2$, and Eq. 11–3 is valid for a vertical spring, as it is for one oscillating horizontally.

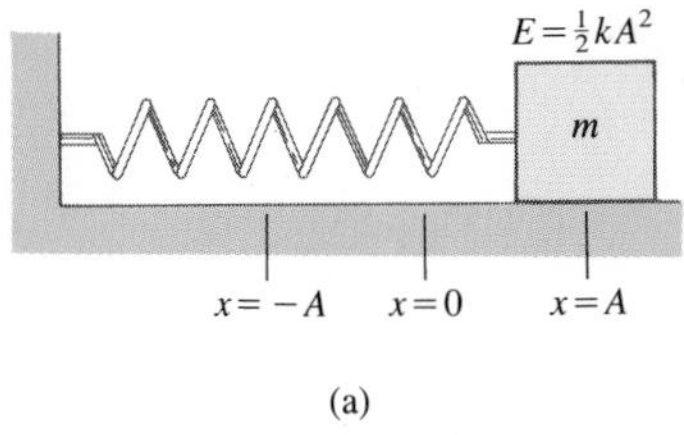

(a)

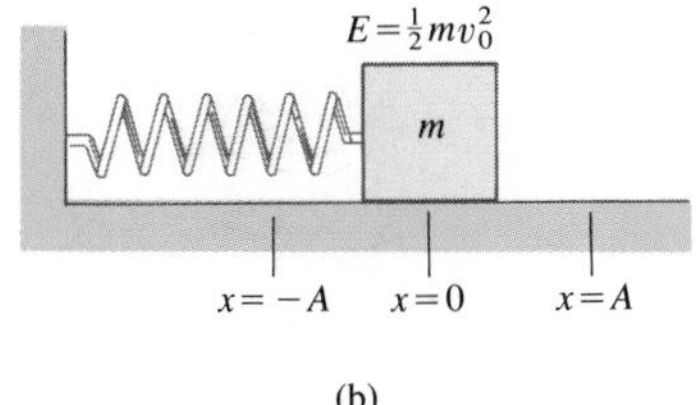

(b)

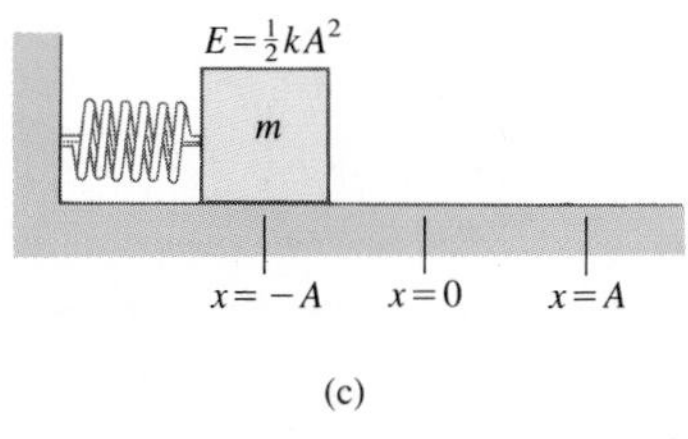

(c)

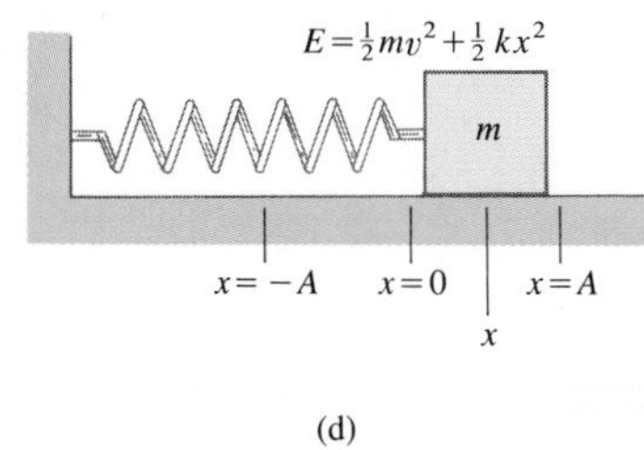

(d)

FIGURE 11–4
Energy changes from kinetic energy to potential energy and back again as the spring oscillates.

EXAMPLE 11–2 A spring stretches 0.150 m when a 0.300-kg mass is hung from it. The spring is then stretched an additional 0.100 m from this equilibrium point and released (as in Fig. 11–3c). Determine (*a*) the spring constant k, (*b*) the amplitude of the oscillation, A, (*c*) the maximum velocity v_0, (*d*) the velocity v when the mass is 0.050 m from equilibrium, and (*e*) the maximum acceleration of the mass.

SOLUTION (*a*) Since the spring stretches 0.150 m when 0.300 kg is hung from it, we find k from Eq. 11–1 to be

$$k = \frac{F}{x} = \frac{mg}{x} = \frac{(0.300 \text{ kg})(9.80 \text{ m/s}^2)}{0.150 \text{ m}} = 19.6 \text{ N/m}.$$

(*b*) Since the spring is stretched 0.100 m from equilibrium (as in Fig. 11–3c) and is given no initial speed, $A = 0.100$ m.

(*c*) The maximum velocity v_0 is attained as the mass passes through the equilibrium point and all the energy is kinetic. By conservation of energy, $\frac{1}{2}mv_0^2 = \frac{1}{2}kA^2$. Solving for v_0, we have

$$v_0 = A\sqrt{\frac{k}{m}} = (0.100 \text{ m})\sqrt{\frac{19.6 \text{ N/m}}{0.300 \text{ kg}}} = 0.808 \text{ m/s}.$$

(*d*) We use Eq. 11–5 and find that

$$v = v_0\sqrt{1 - x^2/A^2}$$

$$= (0.808 \text{ m/s})\sqrt{1 - \frac{(0.050 \text{ m})^2}{(0.100 \text{ m})^2}} = 0.700 \text{ m/s}.$$

(*e*) By Newton's second law, $F = ma$. So the maximum acceleration occurs where the force is greatest—that is, when $x = A = 0.100$ m. Thus

$$a = \frac{kA}{m} = \frac{(19.6 \text{ N/m})(0.100 \text{ m})}{0.300 \text{ kg}} = 6.53 \text{ m/s}^2.$$

EXAMPLE 11–3 For the SHO of Example 11–2, determine (*a*) the total energy, and (*b*) the kinetic and potential energies at half amplitude ($x = \pm A/2$).

SOLUTION (*a*) Since $k = 19.6$ N/m and $A = 0.100$ m, the total energy E from Eq. 11–4a is

$$E = \tfrac{1}{2}kA^2 = \tfrac{1}{2}(19.6 \text{ N/m})(0.100 \text{ m})^2 = 9.80 \times 10^{-2} \text{ J}.$$

(*b*) At $x = A/2 = 0.050$ m, we have

$$\text{PE} = \tfrac{1}{2}kx^2 = 2.5 \times 10^{-2} \text{ J}$$

$$\text{KE} = E - \text{PE} = 7.3 \times 10^{-2} \text{ J}.$$

11–3 • The Reference Circle: the Period and Sinusoidal Nature of SHM

The period of a simple harmonic oscillator is found to depend on the stiffness of the spring and also on the mass m that is oscillating. But—strange as it may seem—the *period does not depend on the amplitude.* You can find this out for yourself by using a watch and timing 10 or 20 cycles of an oscillating spring for a small amplitude and then for a large amplitude.

We can derive a formula for the period of SHM, and this can be done by comparing SHM to an object rotating in a circle. From this same "reference circle" we can obtain a second useful result—a formula for the position of an oscillating mass as a function of time.

Consider now a mass m rotating in a circle of radius A with speed v_0 on top of a table as shown in Fig. 11–5. As viewed from above, the motion is a circle. But a person who looks at the motion from the edge of the table sees an oscillatory motion back and forth, and this corresponds precisely to SHM as we shall now see. What the person sees, and what we are interested in, is the projection of the circulatory motion onto the x axis (Fig. 11–5b). To see that this motion is analogous to SHM, let us calculate the x component of the velocity v_0, which is labeled v in Fig. 11–5. The two triangles shown are similar, so

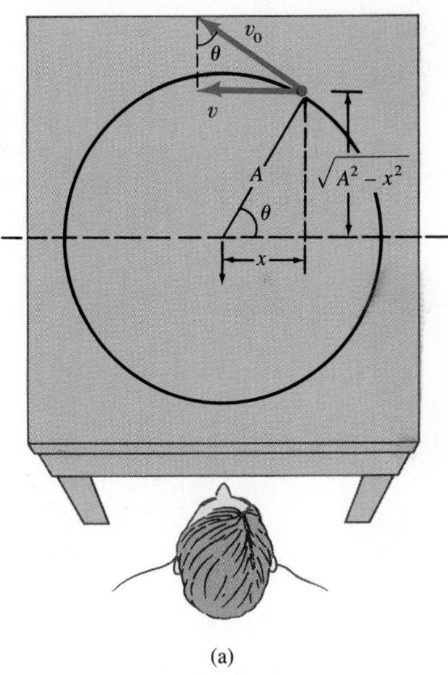

(a)

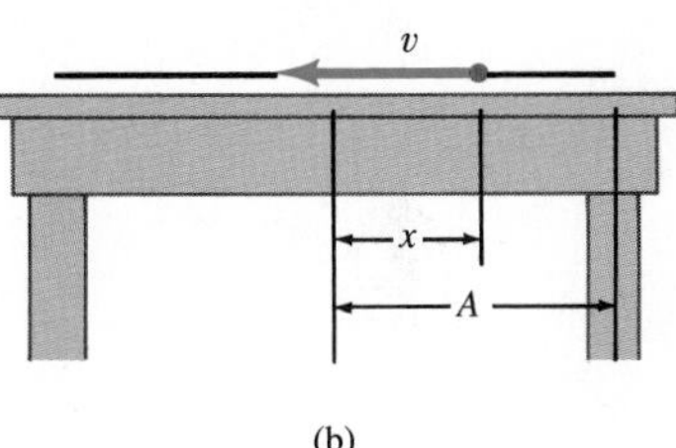

(b)

FIGURE 11–5 Analysis of simple harmonic motion as a side view (b) of circular motion (a).

$$\frac{v}{v_0} = \frac{\sqrt{A^2 - x^2}}{A}$$

or

$$v = v_0\sqrt{1 - \frac{x^2}{A^2}}.$$

This is exactly the equation for the speed of a mass oscillating with SHM (Eq. 11–5). Thus the projection on the x axis of an object rotating in a circle has the same motion as a mass at the end of a spring. The period of the SHO is equal to that of the rotating object making one complete revolution. The time for one revolution is equal to the circumference of the circle (distance) divided by the speed

$$T = \frac{2\pi A}{v_0}.$$

From Eqs. 11–4a and b, we have $\frac{1}{2}kA^2 = \frac{1}{2}mv_0^2$, so $A/v_0 = \sqrt{m/k}$. Thus

$$T = 2\pi\sqrt{\frac{m}{k}}. \qquad (11–6)$$

Period T *of SHO*

This is the formula we were looking for. The period depends on the mass m and the spring constant k, but not on the amplitude. We see from Eq. 11–6 that the greater the mass, the longer the period; and the stiffer the spring, the shorter the period. This makes sense since a greater mass means more inertia and therefore slower response (or acceleration); and larger k means greater force and therefore quicker response. Notice that Eq. 11–6 is not a direct proportion: the period varies as the *square root* of m/k. For example, the mass must be quadrupled to double the period. Equation 11–6 is fully in accord with experiment and is valid not only for a spring, but for all kinds of SHM that obey Eq. 11–1.

EXAMPLE 11–4 What are the period and frequency of the spring in Example 11–2?

SOLUTION From Eq. 11–6,

$$T = 2\pi\sqrt{\frac{m}{k}} = 6.28\sqrt{\frac{0.300\text{ kg}}{19.6\text{ N/m}}} = 0.777\text{ s}.$$

The frequency $f = 1/T = 1.29$ Hz (Eq. 11–2).

EXAMPLE 11–5 A small cockroach of mass 0.30 g is caught in a spider's web. The web vibrates predominantly with a frequency of 15 Hz. (*a*) Estimate the value of the spring constant k for the web. (*b*) At what frequency would you expect the web to vibrate if an insect of mass 0.10 g were trapped?

SOLUTION (*a*) Using Eq. 11–6 and $f = 1/T$, we find

$$f = \frac{1}{2\pi}\sqrt{\frac{k}{m}}.$$

We solve for k and find that

$$k = (2\pi f)^2 m$$
$$= (6.28 \times 15\ \mathrm{s^{-1}})^2(3.0 \times 10^{-4}\ \mathrm{kg}) = 2.7\ \mathrm{N/m}.$$

(*b*) We could substitute $m = 1.0 \times 10^{-4}$ kg in the above equation for f. Instead, we notice that the frequency decreases with the square root of the mass. Since the new mass is one-third the first mass, the frequency increases by a factor of $\sqrt{3}$. That is, $f = (15\ \mathrm{Hz})(\sqrt{3}) = 26$ Hz.

We now use the reference circle to find the position of a mass undergoing SHM as a function of time. From Fig. 11–5, we see that $\cos\theta = x/A$, so the projection of the ball's position on the x axis is

$$x = A\cos\theta.$$

Since the ball is rotating with angular velocity ω, we can write $\theta = \omega t$ (see Section 8–1). Thus

$$x = A\cos\omega t. \qquad (11\text{–}7\text{a})$$

Furthermore, since the angular velocity ω (specified in radians per second) can be written as $\omega = 2\pi f$, where f is the frequency, we also can write

$$x = A\cos 2\pi f t, \qquad (11\text{–}7\text{b})$$

or in terms of the period T:

$$x = A\cos\frac{2\pi t}{T}. \qquad (11\text{–}7\text{c})$$

Notice in Eq. 11–7c that when $t = T$ (that is, after a time equal to the period), we have the cosine of 2π, which is the same as the cosine of zero. This makes sense since the motion repeats itself after a time T.

As we have seen, the projection on the x axis of the rotating object corresponds precisely to the motion of a simple harmonic oscillator. Thus, Eqs. 11–7 give the position of an object oscillating under simple harmonic motion. Since the cosine function varies between 1 and -1, x varies between A and $-A$, as it must. If a pen is attached to a vibrating mass as shown in Fig. 11–6 and a sheet of paper is moved at a steady rate beneath it, a curve will be drawn that accurately follows Eqs. 11–7.

Other equations for SHM are also possible, depending on the initial conditions. For example, if at $t = 0$ the oscillations are begun by giving

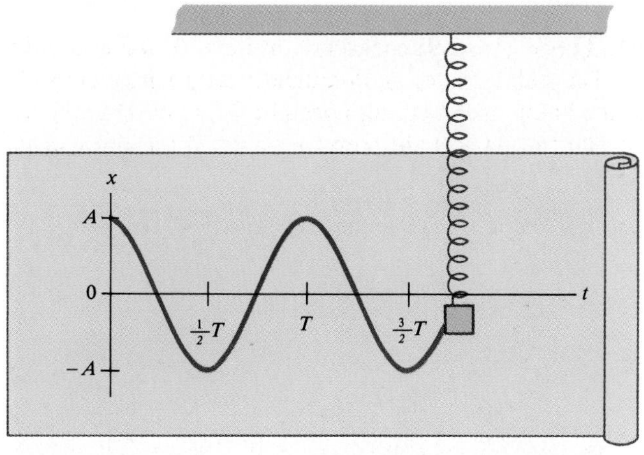

FIGURE 11–6 Sinusoidal nature of SHM as a function of time; in this case, $x = A \cos(2\pi t/T)$.

the mass a push when it is at the equilibrium position, the equation would be

$$x = A \sin \frac{2\pi t}{T}.$$

This curve has precisely the same shape as the cosine curve shown in Fig. 11–6, except it will be shifted to the right by a quarter cycle so that it starts out at $x = 0$ instead of at $x = A$.

SHM is sinusoidal

Both of the curves, sine and cosine, are referred to as being **sinusoidal**. Thus simple harmonic motion† is said to be sinusoidal because the position varies as a sinusoidal function of time. It is easily shown that the velocity and acceleration of SHM are also sinusoidal. This is particularly easy to show for the acceleration, a. From Newton's second law ($F = ma$), the acceleration of the mass m is $a = F/m$, where the force exerted by the spring on the mass is $F = -kx$. Hence

$$a = \frac{F}{m} = \frac{-kx}{m} = -\left(\frac{kA}{m}\right) \sin \frac{2\pi t}{T}.$$

Since the acceleration of a SHO is *not* constant, the equations of uniformly accelerating motion, Eqs. 2–10, do *not* apply to SHM.

EXAMPLE 11–6 (*a*) What is the equation describing the motion of a spring that is stretched 20 cm from equilibrium and then released, and whose period is 1.5 s? (*b*) What will be its displacement after 1.8 s?

SOLUTION (*a*) The amplitude $A = 0.20$ m, and $2\pi/T = 6.28/(1.5\text{ s}) = 4.2\text{ s}^{-1}$. Thus

$$x = 0.20 \cos 4.2t \text{ m}.$$

(*b*) At $t = 1.80$ s, $x = 0.20 \cos 7.56$. The cosine repeats itself after $2\pi = 6.28$ rad, so 7.56 rad gives the same result as $7.56 - 6.28 = 1.28$ rad; thus $x = 0.20 \cos 1.28$. From the tables or a calculator, cos 1.28 is 0.29, so $x = (0.20)(0.29) = 0.058$ m, or 5.8 cm.

† Simple harmonic motion can be *defined* as motion that is sinusoidal. This definition is fully consistent with our earlier definition in Section 11–1.

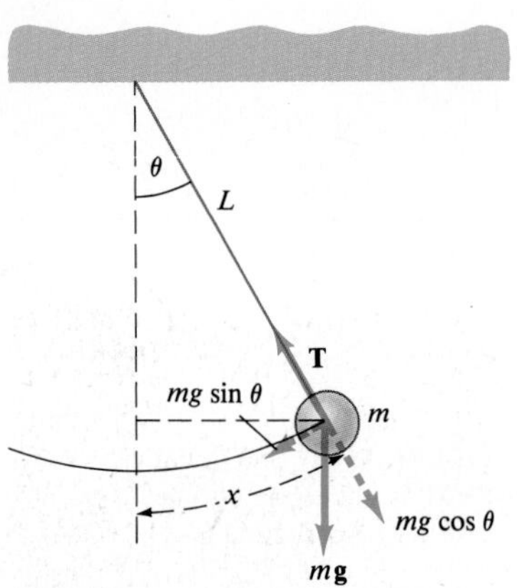

FIGURE 11–7 Simple pendulum.

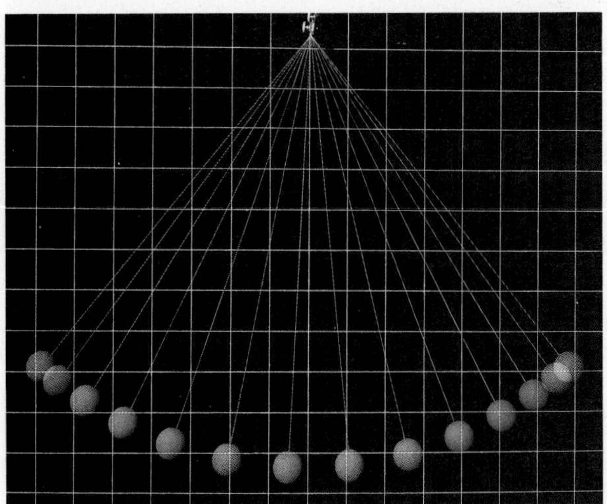

FIGURE 11–8 Strobelight photo of an oscillating pendulum.

11–4 • The Simple Pendulum

A simple pendulum consists of a small object (the pendulum "bob") suspended from the end of a lightweight cord, Fig. 11–7. We assume that the cord doesn't stretch and that its mass can be ignored relative to that of the bob. A simple pendulum moving back and forth (Fig. 11–8) with negligible friction resembles simple harmonic motion: it oscillates along the arc of a circle with equal amplitude on either side of its equilibrium point (where it hangs vertically) and as it passes through the equilibrium point it has its maximum speed. But is it really undergoing SHM? That is, is the restoring force proportional to its displacement? Let us find out.

The displacement of the pendulum along the arc, x, is given by $x = L\theta$, where θ is the angle the cord makes with the vertical and L is the length of the cord (Fig. 11–7). Thus, if the restoring force is proportional to x or to θ, the motion will be simple harmonic. The restoring force is the component of the weight, mg, tangent to the arc:

$$F = -mg \sin \theta.$$

Since F is proportional to the sine of θ and not to θ itself, the motion is *not* SHM. However, if θ is small, then $\sin \theta$ is very nearly equal to θ if the latter is specified in radians. This can be seen by looking at the trigonometry table inside the back cover, or by noting in Fig. 11–7 that the arc length $x\ (= L\theta)$ is nearly the same length as the chord $(= L \sin \theta)$ indicated by the dashed line, *if θ is small.* For angles less than 15°, the difference between θ and $\sin \theta$ is less than 1 percent. Thus, to a very good approximation for small angles,

Pendulum motion is SHO in small-angle approximation

$$F \approx -mg\theta. \qquad [\theta \ll 1 \text{ rad}]$$

Using the fact that $x = L\theta$, we have

$$F \approx -\frac{mg}{L}x.$$

Thus, for small displacements, the motion is essentially simple harmonic, with an effective force constant of $k = mg/L$ (see Eq. 11–1). The period of a simple pendulum can be found using Eq. 11–6, where for k we substitute mg/L:

$$T = 2\pi\sqrt{\frac{m}{mg/L}}$$

$$= 2\pi\sqrt{\frac{L}{g}} \quad (\theta \text{ small}). \qquad (11\text{–}8)$$

Period of pendulum

A surprising result is that the period does not depend on the mass of the pendulum bob! You may have noticed this if you pushed a small child and a large one on the same swing.

We saw in Section 11–3 that the period of any SHM, including a pendulum, does not depend on the amplitude. Galileo is said to have first noted this fact while watching a swinging lamp in the cathedral at Pisa (Fig. 11–9). This discovery led to the pendulum clock,† the first really precise timepiece, which became the standard for centuries.

Because a pendulum does not undergo precisely SHM, the period does depend slightly on the amplitude, mainly for large amplitudes. The accuracy of a pendulum clock would be affected, after many swings, by the decrease in amplitude due to friction; but the mainspring in a pendulum clock (or the falling weight in a grandfather clock) supplies energy to compensate for the friction and to maintain the amplitude constant, so that the timing remains accurate.

The pendulum finds use in geology, for geologists are interested in surface irregularities of the earth and frequently need to measure the acceleration of gravity at a given location very accurately. They often use a carefully designed pendulum to do this, as illustrated in the next example.

FIGURE 11–9 The lamp, hanging by a very long cord from the ceiling of the cathedral at Pisa, whose swinging motion is said to have been observed by Galileo and inspired him to the conclusion that the period of a pendulum does not depend on amplitude.

EXAMPLE 11–7 A geologist's simple pendulum, whose length is 37.10 cm, has a frequency of 0.8190 Hz at a particular location on the earth. What is the acceleration of gravity at this location?

SOLUTION From Eq. 11–8, we have

$$f = \frac{1}{T} = \frac{1}{2\pi}\sqrt{\frac{g}{L}}.$$

Solving for g, we obtain

$$g = (2\pi f)^2 L$$

$$= (6.283 \times 0.8190\ \text{s}^{-1})^2(0.3710\ \text{m})$$

$$= 9.824\ \text{m/s}^2.$$

† It is not known for sure whether or not Galileo actually built a pendulum clock. Christiaan Huygens (1629–1695), more than 10 years after Galileo's death, did build one.

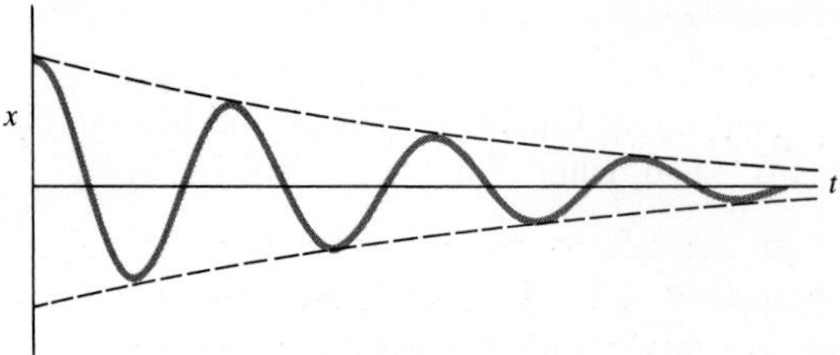

FIGURE 11–10 Damped harmonic motion.

11–5 • Damped Harmonic Motion

The amplitude of any real oscillating spring or swinging pendulum slowly decreases in time until the oscillations stop altogether. Figure 11–10 shows a typical graph of the displacement as a function of time. This is called **damped harmonic motion**. The damping† is generally due to the resistance of air and to internal friction within the oscillating system. The energy that is thus dissipated to thermal energy is reflected in a decreased amplitude of oscillation.

Since natural oscillating systems are damped in general, why do we even talk about (undamped) simple harmonic motion? The answer is that SHM is much easier to deal with mathematically. And if the damping is not large, the oscillations can be thought of as simple harmonic motion on which the damping is superposed—that is, the decrease in amplitude represented by the dashed curves in Fig. 11–10. Although frictional damping does alter the frequency of vibration, the effect is usually small unless the damping is large; thus Eq. 11–6 can still be used in most cases.

Sometimes the damping is so large that the motion no longer resembles simple harmonic motion. Three common cases of heavily damped systems are shown in Fig. 11–11. Curve C represents the **overdamped** situation: when the damping is so large that it takes a long time to reach equilibrium. Curve A represents an **underdamped** situation in which the system makes several swings before coming to rest. Curve B represents **critical damping**; in this case equilibrium is reached the quickest. These terms all derive from the use of practical damped systems such as door-closing mechanisms and shock absorbers in a car. Such devices are usually designed to give critical damping; but as they wear out, underdamping occurs: a door slams or a car bounces up and down several times every time it hits a bump. Needles on analog instruments (voltmeters, ammeters, level indicators on tape recorders) are usually critically damped or slightly underdamped. If they were very underdamped, they would swing back and forth excessively before arriving at the correct value; and if overdamped, they would take too long to reach equilibrium, so rapid changes in the signal (say, recording level) would not be detected.

FIGURE 11–11
Underdamped (A), critically damped (B), and overdamped (C) motion.

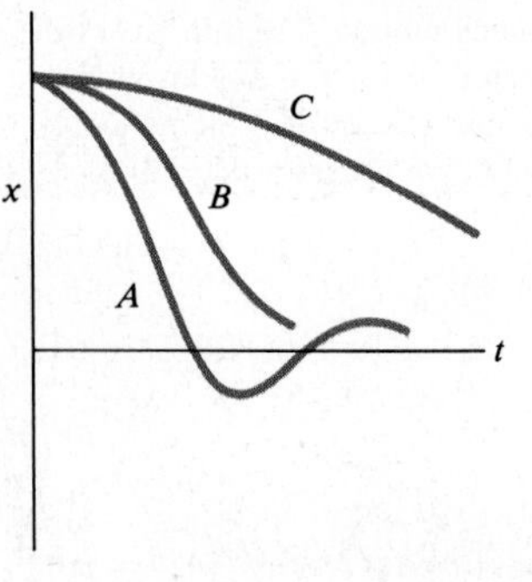

11–6 • Forced Vibrations; Resonance

When a vibrating system is set into motion, it vibrates at its natural frequency. In Sections 11–3 and 11–4 we developed formulas that relate the

† To "damp" means to diminish, restrain, or extinguish, as to "dampen one's spirits."

natural frequency (or period) to the properties of the system for elastic objects (like springs) and pendulums.

However, a system is often not left to merely oscillate on its own, but may have an external force applied to it which itself oscillates at a particular frequency. For example, we might pull the mass on the spring of Fig. 11–1 back and forth at a frequency f. The mass then vibrates at the frequency f of the external force, even if this frequency is different from the **natural frequency** of the spring, which we will now denote by f_0 where (see Eq. 11–6)

$$f_0 = \frac{1}{2\pi}\sqrt{\frac{k}{m}}.$$

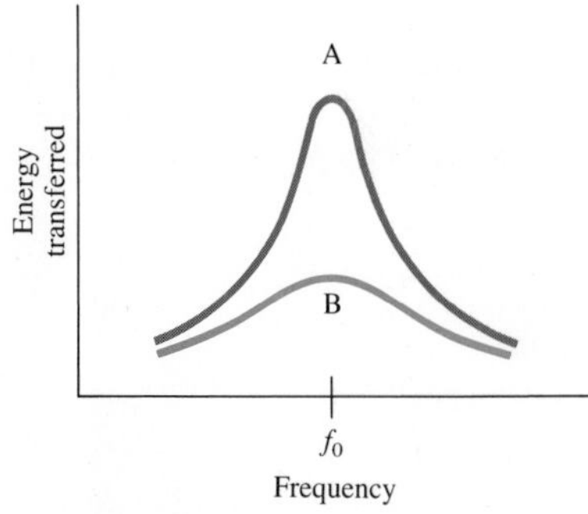

FIGURE 11–12
Resonance for lightly damped (A) and heavily damped (B) systems.

This is an example of **forced vibration**. The amplitude of vibration, and hence the energy transferred to the vibrating system, is found to depend on the difference between f and f_0, and is a maximum when the frequency of the external force equals the natural frequency of the system—that is, when $f = f_0$. The energy transferred to the system (proportional to the amplitude squared—see Eq. 11–4a) is plotted in Fig. 11–12 as a function of the external frequency f. Curve A represents light damping and curve B heavy damping. The amplitude can become large when the driving frequency f is near the natural frequency, $f \approx f_0$, as long as the damping is not too large. When the damping is small, the increase in amplitude near $f = f_0$ is very large (and often dramatic). This effect is known as **resonance**. The natural vibrating frequency f_0 of a system is called its **resonant frequency**.

A simple illustration of resonance is pushing a child on a swing. A swing, like any pendulum, has a natural frequency of oscillation. If you were to close your eyes and push on the swing at a random frequency, the swing would bounce around and reach no great amplitude. But if you push with a frequency equal to the natural frequency of the swing, the amplitude increases greatly. This clearly illustrates that at resonance, relatively little effort is required to obtain a large amplitude.

The great tenor Enrico Caruso was said to be able to shatter a crystal goblet by singing a note of just the right frequency at full voice. This is an example of resonance, for the sound waves emitted by the voice act as a forced vibration on the glass. At resonance, the resulting vibration of the goblet may be large enough in amplitude that the glass exceeds its elastic limit and breaks.

FIGURE 11–13
Large-amplitude oscillations of the Tacoma Narrows Bridge, due to heavy gusty winds that led to its collapse (November 7, 1940).

Since material objects are, in general, elastic, resonance is an important phenomenon in a variety of situations. It is particularly important in building, although the effects are not always to be foreseen. For example, it has been reported that a railway bridge collapsed because a nick in one of the wheels of a passing train set up a resonant vibration in the bridge. Indeed, marching soldiers break step when crossing a bridge to avoid the possibility of a similar catastrophe. And the famous collapse of the Tacoma Narrows Bridge (Fig. 11–13) in 1940 was due in part to resonance of the bridge.

We will meet important examples of resonance later in this chapter and in succeeding chapters. We will also see that vibrating objects often have not one, but many resonant frequencies.

FIGURE 11–14
Water waves spreading outward from a source.

11–7 • Wave Motion

When you throw a stone into a lake or pool of water, circular waves form and move outward, Fig. 11–14. Waves will also travel along a cord (or a "slinky") that is stretched out straight on a table if you vibrate one end back and forth as shown in Fig. 11–15. Water waves and waves on a cord are two common examples of wave motion. We will meet other kinds of wave motion later, but for now we will concentrate on these "mechanical" waves.

If you have ever watched ocean waves moving toward shore, you may have wondered if the waves were carrying water into the beach. This is, in fact, not the case.† Water waves move with a recognizable velocity. But each particle of the water itself merely oscillates about an equilibrium point. This is clearly demonstrated by observing leaves on a pond as waves move by. The leaves (or a cork) are not carried forward by the waves, but simply oscillate about an equilibrium point because this is the motion of the water itself. Similarly, the wave on the rope of Fig. 11–15 moves to the right, but each piece of the rope only vibrates to and fro. This is a general feature of waves: waves can move over large distances, but the medium (the water or the rope) itself has only a limited movement. Thus, although a wave is not matter, the wave pattern can travel in matter. A wave consists of oscillations that move without carrying matter with them.

Waves carry energy from one place to another. Energy is given to a water wave, for example, by a rock thrown into the water, or by wind far out at sea. The energy is transported by waves to the shore. If you have been under an ocean wave when it breaks, you know the energy it carries. The oscillating hand in Fig. 11–15 transfers energy to the rope, which is then

† Do not be confused by the "breaking" of ocean waves, which occurs when the wave interacts with the ground in shallow water and hence is no longer a simple wave.

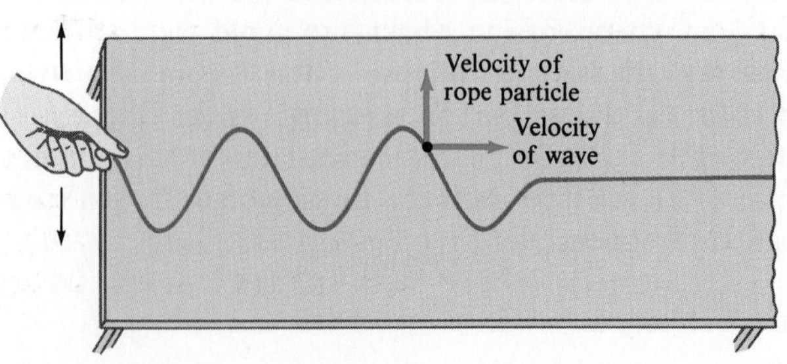

FIGURE 11–15
Wave traveling on a rope. The wave travels to the right along the rope. Particles of the rope oscillate back and forth on the tabletop.

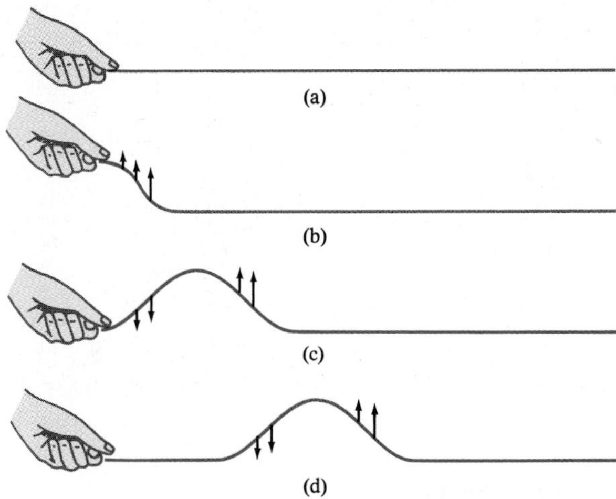

FIGURE 11–16
Motion of a wave pulse. Arrows indicate velocity of rope particles.

transported down the rope and can be transferred to an object at the other end. All forms of wave motion transport energy.

Wave "pulse"

Let us look a little more closely at how a wave is formed and how it comes to "travel." We first look at a single wave bump or **pulse**. A single pulse can be formed on a rope by a quick up-and-down motion of the hand, Fig. 11–16. The hand pulls up on one end of the rope and because the end piece is attached to adjacent pieces, these also feel an upward force and they, too, begin to move upward. As each succeeding piece of rope moves upward, the wave crest moves outward along the rope. Meanwhile, the end piece of rope has been returned to its original position by the hand, and as each succeeding piece of rope reaches its peak position, it, too, is pulled back down again. Thus the source of a traveling wave pulse is a disturbance, and cohesive forces between adjacent pieces of rope cause the pulse to travel outward. Waves in other media are created and propagate outward in a similar fashion.

Periodic wave

A **continuous** or **periodic wave**, such as that shown in Fig. 11–15, has as its source a disturbance that is continuous and oscillating; that is, the source is a *vibration* or *oscillation.* In Fig. 11–15, a hand oscillates one end of the rope. Water waves may be produced by any vibrating object placed at the surface, such as your hand; or the water itself is made to vibrate when wind blows across it or a rock is thrown into it. A vibrating tuning fork or drum membrane gives rise to sound waves in air. And we will see later that oscillating electric charges give rise to light waves. Indeed, almost any vibrating object sends out waves.

The source of any wave, then, is a vibration. And it is the *vibration* that propagates outward and thus constitutes the wave. If the source vibrates sinusoidally in SHM, then the wave itself—if the medium is perfectly elastic—will have a sinusoidal shape both in space and in time. That is, if you take a picture of the wave spread throughout space at a given instant of time, the wave will have the shape of a sine or cosine function. On the other hand, if you look at the motion of the medium at one place over a long period of time—for example, if you look between two closely spaced posts of a pier or out of a ship's porthole as water waves pass by—the up-and-down motion of that small segment of water will be simple harmonic motion—the water moves up and down sinusoidally in time.

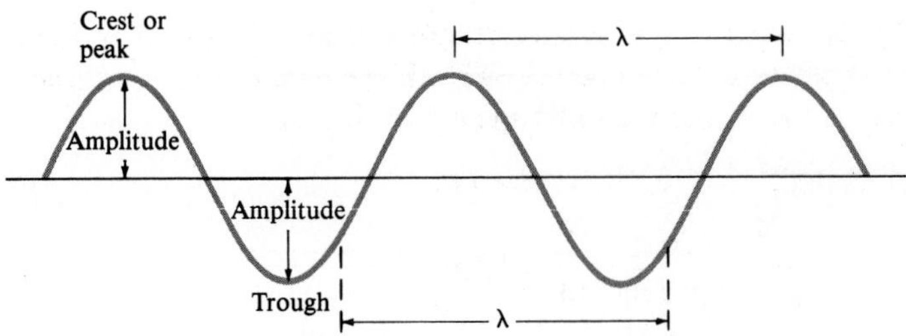

FIGURE 11–17 Characteristics of a single-frequency continuous wave.

Amplitude

Wavelength, λ

Frequency, f

Some of the important quantities used to describe a periodic sinusoidal wave are shown in Fig. 11–17. The high points on a wave are called crests, the low points troughs. The **amplitude** is the maximum height of a crest, or depth of a trough, relative to the normal (or equilibrium) level. The total swing from a crest to a trough is twice the amplitude. The distance between two successive crests is called the **wavelength**, λ (the Greek letter lambda). The wavelength is also equal to the distance between *any* two successive identical points on the wave. The **frequency**, f, is the number of crests—or complete cycles—that pass a given point per unit time. The period, T, of course, is just $1/f$, and is the time elapsed between two successive crests passing by the same point in space.

Wave velocity (don't confuse with the velocity of a particle)

The **wave velocity**, v, is the velocity at which wave crests (or any other part of the waveform) move. The wave velocity must be distinguished from the velocity of a particle of the medium itself. For example, for a wave traveling along a string as in Fig. 11–15, the wave velocity is to the right, along the string, whereas the velocity of a particle of the string is perpendicular to it.

A wave crest travels a distance of one wavelength, λ, in one period, T. Thus the wave velocity v is equal to λ/T, or

$v = \lambda f$ (For all sinusoidal waves)

$$v = \lambda f. \tag{11–9}$$

For example, suppose a wave has a wavelength of 5 m and a frequency of 3 Hz. Since three crests pass a given point per second, and the crests are 5 m apart, the first crest (or any other part of the wave) must travel a distance of 15 m during the 1 s. So its speed is 15 m/s.

The velocity of a wave depends on the properties of the medium in which it travels. The velocity of a wave on a stretched string, for example, depends on the tension in the string, F_T, and on the string's mass per unit length, m/L. For waves of small amplitude, the relationship is

$$v = \sqrt{\frac{F_T}{m/L}}. \quad \text{[wave on a string]} \tag{11–10}$$

This formula makes sense qualitatively on the basis of Newtonian mechanics. That is, we expect the tension to be in the numerator and the mass per unit length in the denominator. Why? Because when the tension is greater, we expect the velocity to be greater since each segment of string is in tighter contact with its neighbor; and the greater the mass per unit length, the more inertia the string has and the more slowly the wave would be expected to propagate.

EXAMPLE 11–8 A wave whose wavelength is 0.30 m is traveling down a 300-m-long wire whose total mass is 30 kg. If the wire is under a tension of 400 N, what is the velocity and frequency of this wave?

SOLUTION From Eq. 11–10, the velocity is

$$v = \sqrt{\frac{400\ \text{N}}{(30\ \text{kg})/(300\ \text{m})}} = 63\ \text{m/s}.$$

The frequency then is

$$f = \frac{v}{\lambda} = \frac{63\ \text{m/s}}{0.30\ \text{m}} = 210\ \text{Hz}.$$

11–8 • Types of Waves

We saw earlier that although waves may travel over long distances, the particles of the medium vibrate only over a limited region of space. When a wave travels down a rope, say from left to right, the particles of the rope vibrate up and down in a direction transverse (or perpendicular) to the motion of the wave itself. Such a wave is called a **transverse wave**. There exists another type of wave known as a **longitudinal wave**. In a longitudinal wave, the vibration of the particles of the medium is along the *same* direction as the motion of the wave. Longitudinal waves are readily formed on a stretched spring or "slinky" by alternately compressing and expanding one end. This is shown in Fig. 11–18b, and can be compared to the transverse wave in Fig. 11–18a. A series of compressions and expansions propagate along the spring. The *compressions* are those areas where the coils are momentarily close together. *Expansions* (sometimes called *rarefactions*) are regions where the coils are momentarily far apart. Compressions and expansions correspond to the crests and troughs of a transverse wave.

Transverse and longitudinal waves

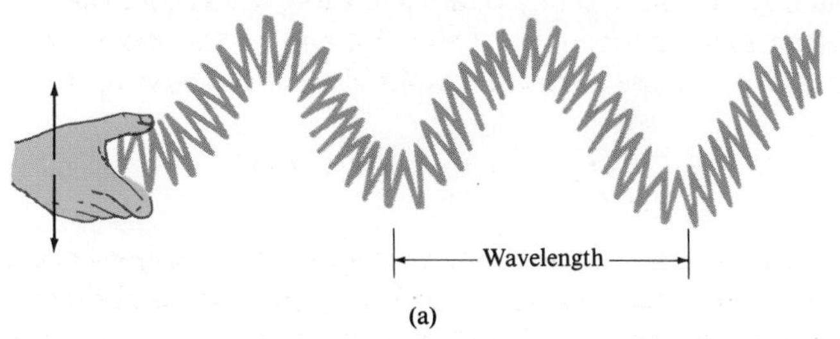

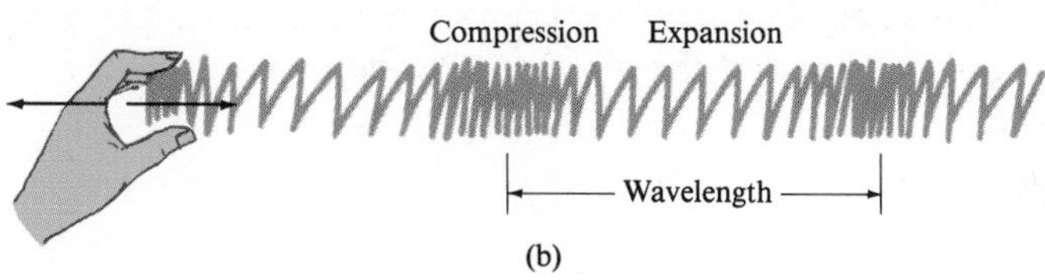

FIGURE 11–18 (a) Transverse wave; (b) longitudinal wave.

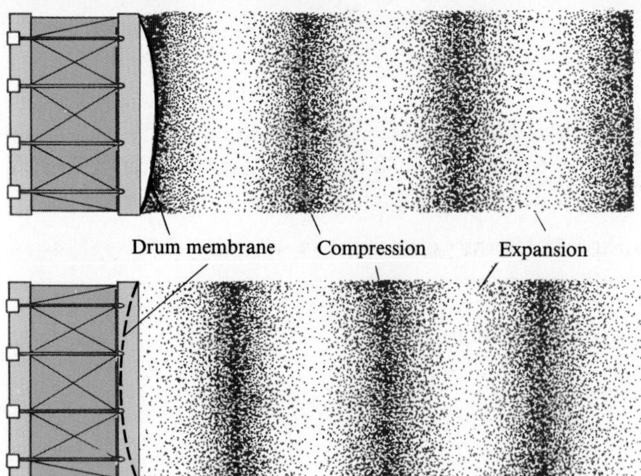

FIGURE 11–19 Production of a sound wave, which is longitudinal.

An important example of a longitudinal wave is a sound wave in air. A vibrating drum head, for example, alternately compresses and rarefies the air and produces a longitudinal wave that travels outward in the air, as shown in Fig. 11–19.

As in the case of transverse waves, each section of the medium in which a longitudinal wave passes oscillates over a very small distance, whereas the wave itself can travel large distances. Wavelength, frequency, and wave velocity all have meaning for a longitudinal wave. The wavelength is the distance between successive compressions (or between successive expansions), and frequency is the number of compressions that pass a given point per second. The wave velocity is the velocity with which each compression appears to move and is equal to the product of wavelength and frequency (Eq. 11–9).

A longitudinal wave can be represented graphically by plotting the density of air molecules (or coils of a slinky) versus position, as shown in Fig. 11–20. We will often use such a graphical representation because it is much easier to illustrate what is happening. Note that the graph looks much like a transverse wave.

The velocity of a longitudinal wave has a form similar to that for a transverse wave on a string (Eq. 11–10); that is

$$v = \sqrt{\frac{\text{elastic force factor}}{\text{inertia factor}}}.$$

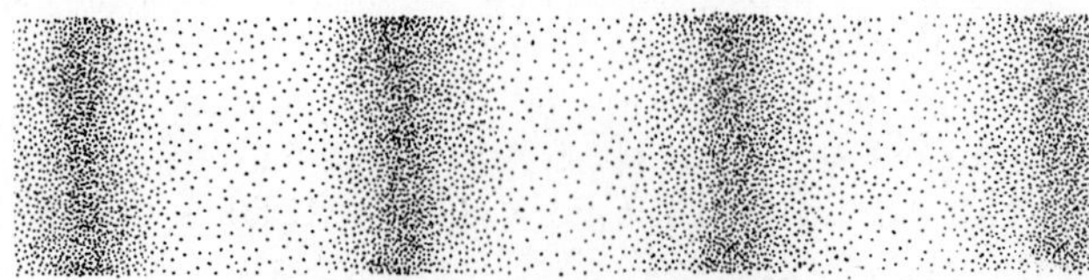

(a)

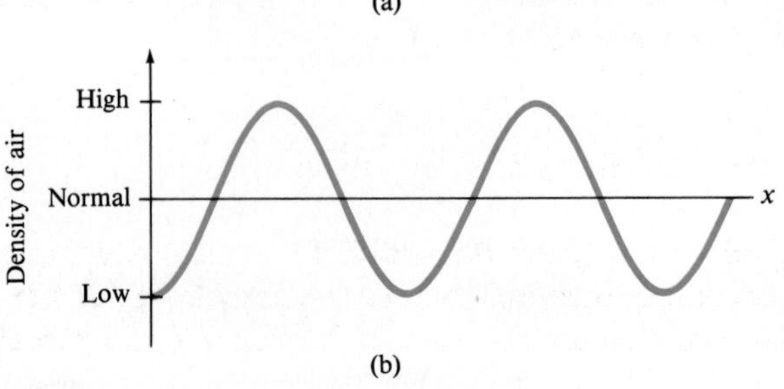

(b)

FIGURE 11–20
(a) A longitudinal wave with (b) its graphical representation.

In particular, for a longitudinal wave traveling down a long solid rod,

$$v = \sqrt{\frac{E}{\rho}}, \tag{11–11a}$$

where E is the elastic modulus (Section 9–7) of the material and ρ is its density. For a longitudinal wave traveling in a liquid or gas,

$$v = \sqrt{\frac{B}{\rho}}, \tag{11–11b}$$

where B is the bulk modulus (Section 9–7) and ρ the density.

EXAMPLE 11–9 You can often hear an approaching train by putting your ear to the track. How long does it take for the wave to travel down the steel track if the train is 1.0 km away?

SOLUTION Referring to Tables 9–1 and 10–1 for the elastic modulus and density of steel, respectively, we have

$$v = \sqrt{\frac{2.0 \times 10^{11}\ \mathrm{N/m^2}}{7.8 \times 10^3\ \mathrm{kg/m^3}}} = 5.1 \times 10^3\ \mathrm{m/s}.$$

Then the time $t = \text{distance/velocity} = (1.0 \times 10^3\ \mathrm{m})/(5.1 \times 10^3\ \mathrm{m/s}) = 0.20\ \mathrm{s}$.

Earthquake waves

Both transverse and longitudinal waves are produced when a disturbance known as an earthquake occurs. The transverse waves that travel through the body of the earth are called S waves and the longitudinal waves are called P waves. Both longitudinal and transverse waves can travel through a solid since the atoms or molecules can vibrate about their relatively fixed positions in any direction. But in a fluid, only longitudinal waves can propagate, because any transverse motion would experience no restoring force since a fluid can flow. This fact was used by geophysicists to infer that the earth's outer core is molten: longitudinal waves are detected diametrically across the earth, but not transverse waves; the only explanation is that the core of the earth must be liquid.

Besides these two types of waves that can pass through the body of the earth (or other substance), there can also be *surface waves* that travel along the boundary between two materials. A wave on water is actually a surface wave that moves on the boundary between water and air. The motion of each particle of water at the surface is circular or elliptical (Fig. 11–21), so it is a combination of transverse and longitudinal motions. Below the surface, there is also transverse plus longitudinal wave motion, as shown. At the bottom, the motion is only longitudinal. (Of course when the water is so shallow that the wave "breaks", the wave motion ceases.) Surface waves are also set up on the earth when an earthquake occurs. The waves that travel along the surface are mainly responsible for the damage caused by earthquakes.

FIGURE 11–21 A water wave is an example of a *surface wave*, which is a combination of transverse and longitudinal wave motions.

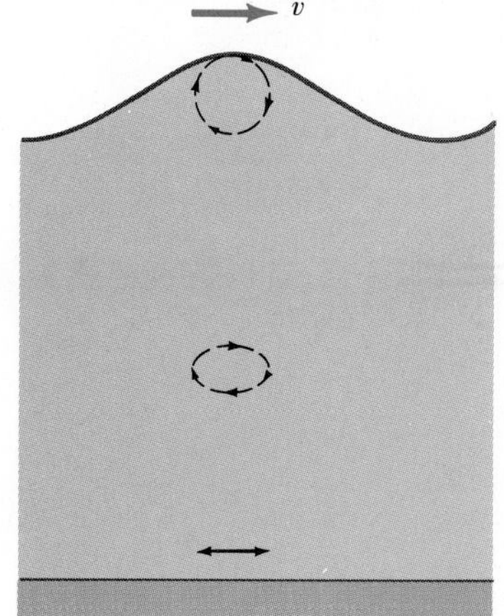

*11–9 • Energy Transmitted by Waves

Waves transmit energy from one place to another. As waves travel through a medium, the energy is transmitted as vibrational energy from particle to particle of the medium. For a sinusoidal wave of frequency f, the particles move in SHM as a wave passes, so each particle has an energy $E = \frac{1}{2}kx_0^2$, where x_0 is the amplitude of its motion, either transversely or longitudinally. (See Eq. 11–4a, in which we have replaced A by x_0.) Using Eq. 11–6, we can write k in terms of the frequency, $k = 4\pi^2 m/T^2 = 4\pi^2 m f^2$, so that

$$E = 2\pi^2 m f^2 x_0^2.$$

The mass $m = \rho V$, where ρ is the density of the medium and V its volume. The volume $V = Al$, where A is the cross-sectional area through which the wave travels. We can write l as the distance the wave travels in a time t as $l = vt$, where v is the speed of the wave. Thus $m = \rho V = \rho Al = \rho Avt$ and

$$E = 2\pi^2 \rho A v t f^2 x_0^2. \tag{11–12}$$

Wave energy $\propto$ (Amplitude)^2

From this equation, we have the important result that the **energy transported by a wave is proportional to the square of the amplitude**. The average *rate* of energy transferred is the power P:

$$\bar{P} = \frac{E}{t} = 2\pi^2 \rho A v f^2 x_0^2. \tag{11–13}$$

Finally, the **intensity** I of a wave is defined as the power transferred across unit area perpendicular to the direction of energy flow:

Intensity

$$I = \frac{\bar{P}}{A} = 2\pi^2 v \rho f^2 x_0^2. \tag{11–14}$$

If a wave flows out from the source in all directions, it is a three-dimensional wave. Examples are sound traveling in the open air, earthquake waves, and light waves. If the medium is isotropic (same in all directions), the wave is said to be a *spherical wave* (Fig. 11–22). As the wave moves outward, it is spread over a larger and larger area since the surface area of a sphere of radius r is $4\pi r^2$. Because energy is conserved, we can see from Eq. 11–12 or 11–13 that as the area A increases, the amplitude x_0 must decrease. That is, at two different distances from the source, r_1 and r_2 (see Fig. 11–22), $A_1 x_{01}^2 = A_2 x_{02}^2$, where x_{01} and x_{02} are the amplitudes of the wave at r_1 and r_2, respectively. Since $A_1 = 4\pi r_1^2$ and $A_2 = 4\pi r_2^2$, we have $(x_{01}^2 r_1^2) = (x_{02}^2 r_2^2)$, or

FIGURE 11–22
Wave traveling outward from source has spherical shape. Two different crests (or compressions) are shown, of radius r_1 and r_2.

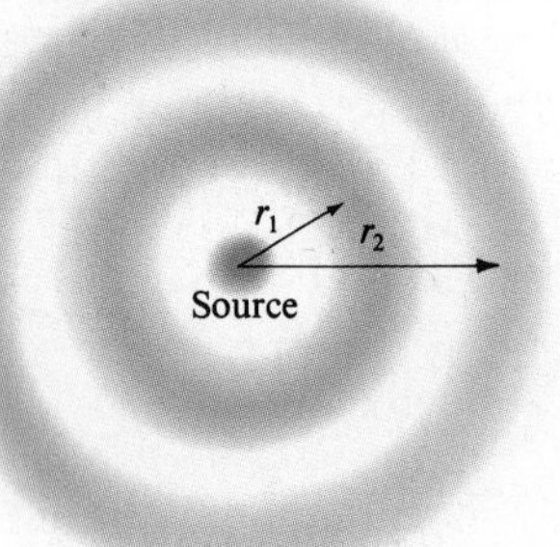

$$\frac{x_{02}}{x_{01}} = \frac{r_1}{r_2}.$$

Thus the amplitude decreases inversely as the distance from the source. When the wave is twice as far from the source, the amplitude is half as large, and so on (ignoring damping due to friction).

$I \propto \frac{1}{r^2}$

The intensity I also decreases with distance. Since I is proportional to x_0^2 (Eq. 11–14), I must decrease as the *square* of the distance from the source. This inverse-square law applies to sound and light and other types of waves. Another way to view this is to consider two points r_1 and r_2 at the same time. If the power output remains constant, the intensity at r_1 is $I_1 = \bar{P}/4\pi r_1^2$ and the intensity at r_2 is $I_2 = \bar{P}/4\pi r_2^2$. Thus

$$\frac{I_2}{I_1} = \frac{r_1^2}{r_2^2}. \tag{11–15}$$

EXAMPLE 11–10 If the intensity of an earthquake P wave 100 km from the source is $1.0 \times 10^6\ \mathrm{W/m^2}$, what is the intensity 400 km from the source?

SOLUTION The intensity decreases as the square of the distance from the source. Therefore, at 400 km, the intensity will be $(\frac{1}{4})^2 = \frac{1}{16}$ of its value at 100 km, or $6.2 \times 10^4\ \mathrm{W/m^2}$. Alternatively, Eq. 11–15 could be used: $I_2 = I_1 r_1^2/r_2^2 = (1.0 \times 10^6\ \mathrm{W/m^2})(100\ \mathrm{km})^2/(400\ \mathrm{km})^2 = 6.2 \times 10^4\ \mathrm{W/m^2}$.

The situation is different for a one-dimensional wave, such as a transverse wave on a string or a longitudinal wave pulse traveling down a uniform metal rod. The area A remains constant, so the amplitude x_0 also remains constant (ignoring friction). Thus the amplitude and the intensity do not decrease with distance.

In practice, frictional damping is generally present and some of the energy is transformed into thermal energy. Thus the amplitude and intensity of a one-dimensional wave decrease with distance from the source, and for a three-dimensional wave the decrease will be greater than that discussed above, although the effect is often small.

11–10 • Behavior of Waves: Reflection, Refraction, Interference, and Diffraction

Reflection. When a wave strikes an obstacle, or comes to the end of the medium it is traveling in, at least a part of the wave is reflected. You have probably seen water waves reflect off a rock or the side of a swimming pool. And you may have heard a shout reflected from a distant cliff—which we call an "echo."

A wave pulse traveling down a rope is reflected as shown in Fig. 11–23.

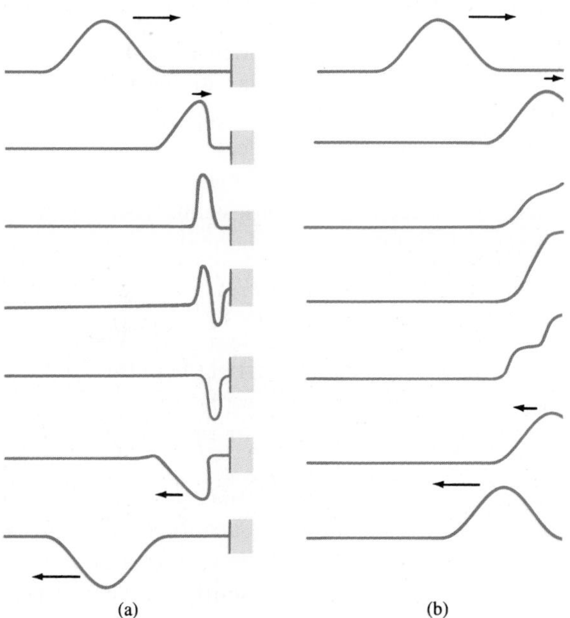

FIGURE 11–23 Reflection of a wave pulse on a rope when the end of the rope is (a) fixed and (b) free.

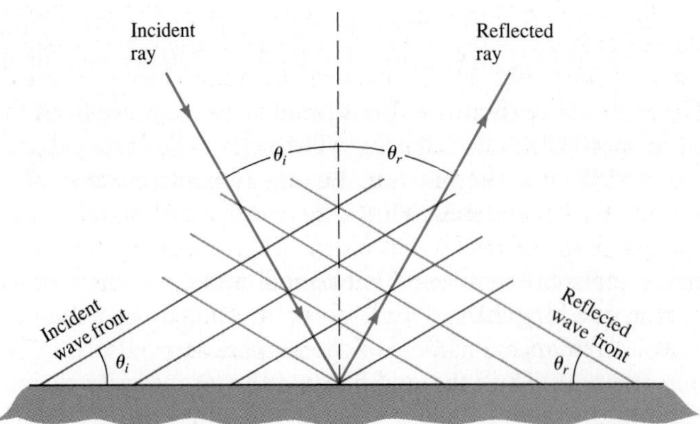

FIGURE 11–24 Law of reflection.

FIGURE 11–25 When a wave pulse reaches a discontinuity, part is reflected and part is transmitted.

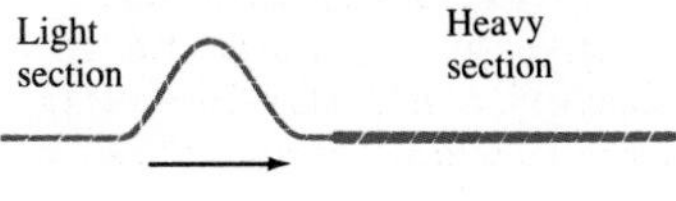

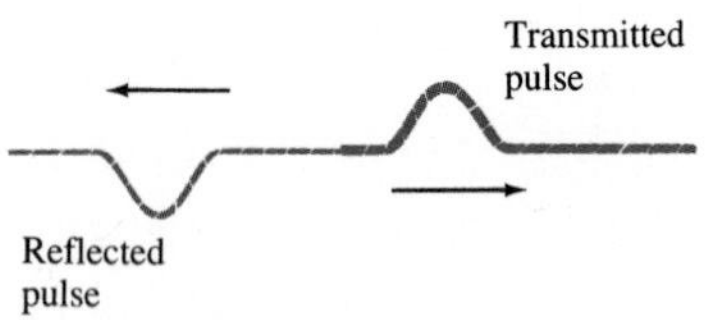

You can observe this for yourself and see that the reflected pulse is inverted as in Fig. 11–23a if the end of the rope is fixed; and returns right side up if the end is free as in Fig. 11–23b. When the end is fixed to a support, as in Fig. 11–23a, the pulse reaching that fixed end exerts a force (upward) on the support. The support exerts an equal but opposite force (Newton's third law) downward on the cord. This downward force on the cord is what "generates" the inverted reflected pulse. In Fig. 11–23b, the free end is constrained by neither a support nor by additional cord. It therefore tends to overshoot—its displacement being momentarily greater than that of the traveling pulse. The overshooting end exerts an upward pull on the cord, and this is what generates the reflected pulse, which is not inverted.

Wave fronts

For a two- or three-dimensional wave, such as a water wave, we are concerned with **wave fronts**, by which we mean the whole width of a wave crest. A line drawn in the direction of motion, perpendicular to the wave front, is called a **ray**. As shown in Fig. 11–24, the angle that the incoming or *incident wave* makes with the reflecting surface is equal to the angle made by the reflected wave. That is, **the angle of reflection equals the angle of incidence**. The "angle of incidence" is defined as the angle the incident ray makes with the perpendicular to the reflecting surface (or the wave front makes with a tangent to the surface), and the "angle of reflection" is the corresponding angle for the reflected wave.

Law of reflection

When the wave pulse in Fig. 11–23a reaches the wall, not all of the energy is reflected. Some of it is absorbed by the wall. Part of the absorbed energy is transformed into thermal energy, and part continues to propagate through the material of the wall. This is more clearly illustrated by considering a pulse that travels down a rope which consists of a light section and a heavy section, as shown in Fig. 11–25. When the wave reaches the boundary between the two sections, part of the pulse is reflected and part is transmitted, as shown. The heavier the second section, the less is transmitted; and when the second section is a wall or rigid support, very little is transmitted.

FIGURE 11–26 Refraction of waves passing a boundary.

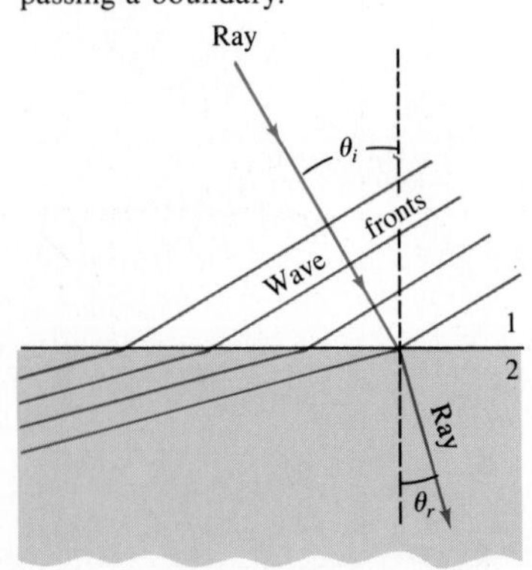

Refraction. When any wave strikes a boundary, some of the energy is reflected and some is transmitted or absorbed. When a two- or three-dimensional wave traveling in one medium crosses a boundary into a medium where its velocity is different, the transmitted wave may move in a different direction than the incident wave, as shown in Fig. 11–26. This phenomenon is known as **refraction**. One example is a water wave; the velocity decreases in shallow water and the waves refract, Fig. 11–27. [When the wave velocity changes gradually, as in Fig. 11–27, without a sharp boundary, the waves change direction (refract) gradually.] In Fig. 11–26, the velocity of the wave in medium 2 is less than in medium 1. In this case, the direction of the wave bends so it moves more nearly perpendicular to the boundary. That is, the *angle of refraction*, θ_r, is less than the *angle of incidence*, θ_i. To see why this is so, and to help us get a quantitative relation between θ_r and θ_i, let us think of each wave front as a row of soldiers. The soldiers are marching from firm ground (medium 1) into mud (medium 2) and hence are slowed down. The soldiers that reach the mud first are slowed down first and the row bends as shown in Fig. 11–28a. Let us consider the wave front (or row of soldiers) labeled A in Fig. 11–28b. In the same time t that A_1 moves a distance $l_1 = v_1 t$, we see that A_2 moves a distance $l_2 = v_2 t$. The two triangles shown have the side labeled a in common. Thus

FIGURE 11–27
Water waves refracting as they approach the shore, where their velocity is less. There is no distinct boundary, as in Fig. 11–26, because the wave velocity changes gradually.

$$\sin\theta_1 = \frac{l_1}{a} = \frac{v_1 t}{a}$$

and

$$\sin\theta_2 = \frac{l_2}{a} = \frac{v_2 t}{a}.$$

Dividing these two equations, we find that

$$\frac{\sin\theta_2}{\sin\theta_1} = \frac{v_2}{v_1}. \qquad (11\text{–}16)$$

Law of refraction

Since θ_1 is the angle of incidence (θ_i), and θ_2 is the angle of refraction (θ_r), Eq. 11–16 gives the quantitative relation between the two. Of course, if the wave were going in the opposite direction, the argument would not be changed. Only θ_1 and θ_2 would change roles: θ_1 would be the angle of re-

FIGURE 11–28 Soldier analogy (a), to derive law of refraction for waves (b).

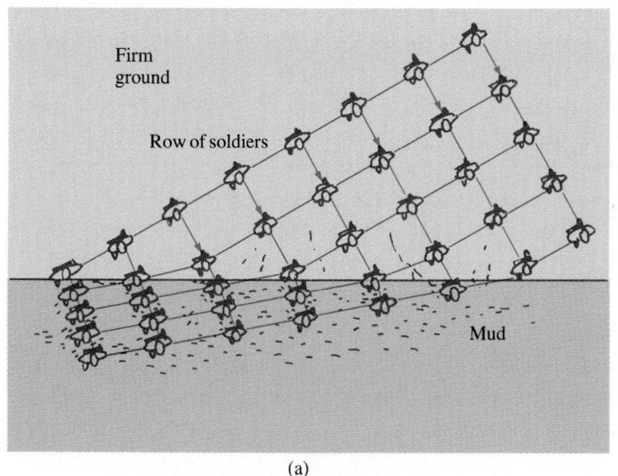

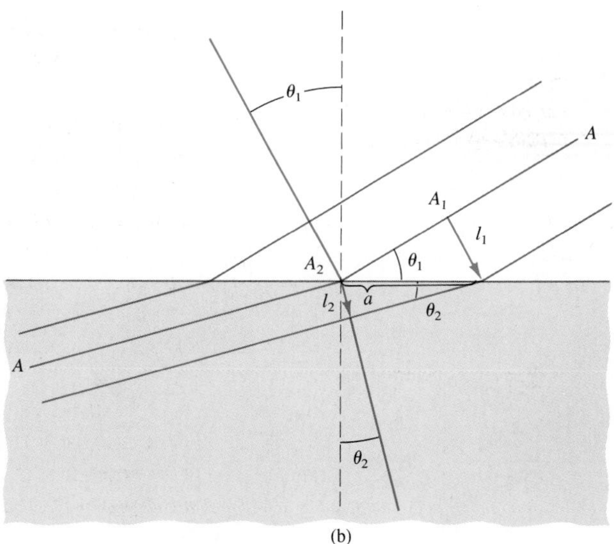

fraction and θ_2 the angle of incidence. Clearly then, if the wave travels into a medium where it can move faster, it will bend in the opposite way, $\theta_r > \theta_i$. We see from Eq. 11–16 that if the velocity increases, the angle increases, and vice versa.

Earthquake waves refract within the earth as they travel through rock of different densities (and therefore the velocity is different) just as water waves do. Light waves refract as well, and when we discuss light we shall find Eq. 11–16 very useful.

EXAMPLE 11–11 An earthquake P wave passes across a boundary in rock where its velocity increases from 6.5 km/s to 8.0 km/s. If it strikes this boundary at 30°, what is the angle of refraction?

SOLUTION Since sin 30° = 0.50, Eq. 11–16 yields

$$\sin\theta_2 = \frac{(8.0\ \text{m/s})}{(6.5\ \text{m/s})}(0.50) = 0.62,$$

so $\theta_2 = 38°$.

Interference. **Interference** refers to what happens when two waves pass through the same region of space at the same time. Consider, for example, the two wave pulses on a string traveling toward each other as shown in Fig. 11–29. In part (a) the two pulses have the same amplitude, but one is a crest and the other a trough; in part (b) they are both crests. In both cases, the waves meet and pass right on by each other. However, in the region where they overlap, the resultant displacement is the *algebraic sum of their separate displacements* (a crest is considered positive and a trough negative). This is called the **principle of superposition**. In Fig. 11–29a, the two waves oppose one another as they pass by and the result is called **destructive inter-**

Superposition principle

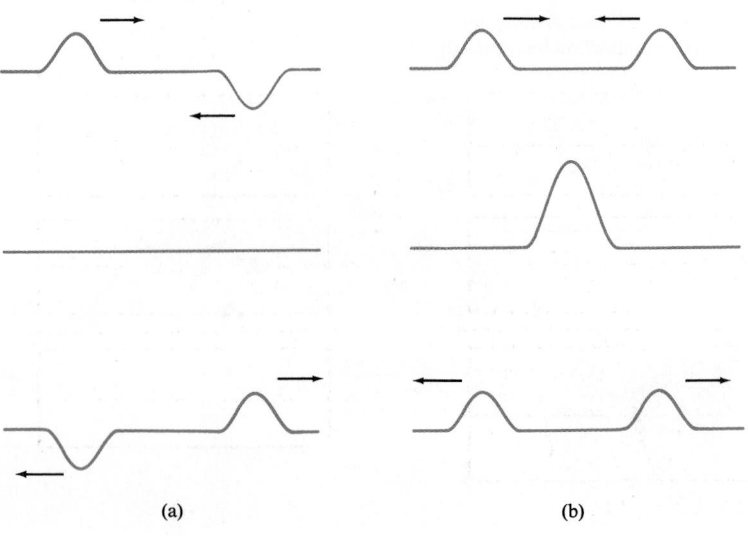

FIGURE 11–29
Two wave pulses pass each other. Where they overlap, inference occurs: (a) destructive, and (b) constructive.

(a)

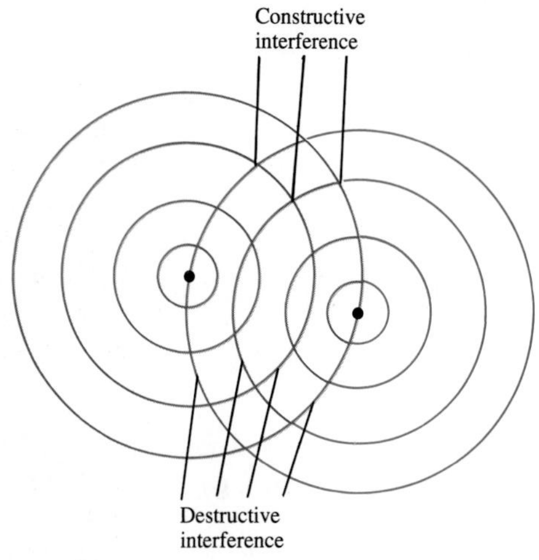

(b)

FIGURE 11–30
Interference of water waves.

Constructive and destructive interference

ference. In Fig. 11–29b, the resultant displacement is greater than that of either pulse and the result is **constructive interference**.

When two rocks are thrown into a pond simultaneously, the two sets of circular waves interfere with one another as shown in Fig. 11–30. In some areas of overlap, crests of one wave meet crests of the other (and troughs meet troughs); this is constructive interference and the water oscillates up and down with greater amplitude than either wave separately. In other areas, destructive interference occurs where the water actually does not move at all—this is where crests of one wave meet troughs of the other, and vice versa. Figure 11–31a shows the displacement of the two waves as a function of time, as well as their sum, for the case of constructive interference. For any two such waves, we use the term **phase** to describe the relative positions of their crests. When the crests and troughs are aligned as in Fig. 11–31a for constructive interference, the two waves are **in phase**. At points where destructive interference occurs—see Fig. 11–31b—crests of one wave meet troughs of the other wave, and the two waves are said to be completely **out of phase** or,

FIGURE 11–31 Two waves interfere: (a) constructively, (b) destructively, and (c) partially destructively.

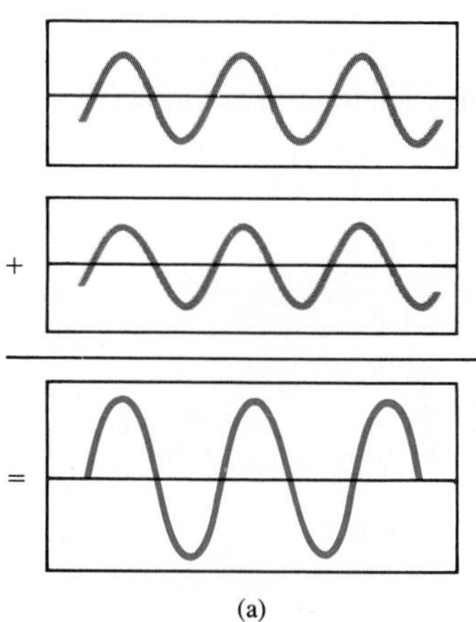

(a)

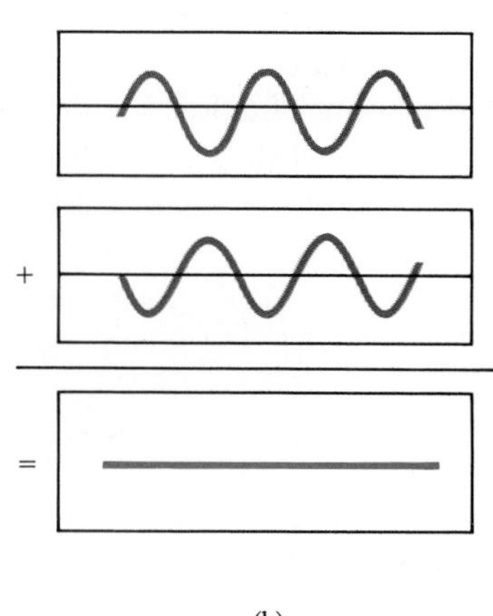

(b)

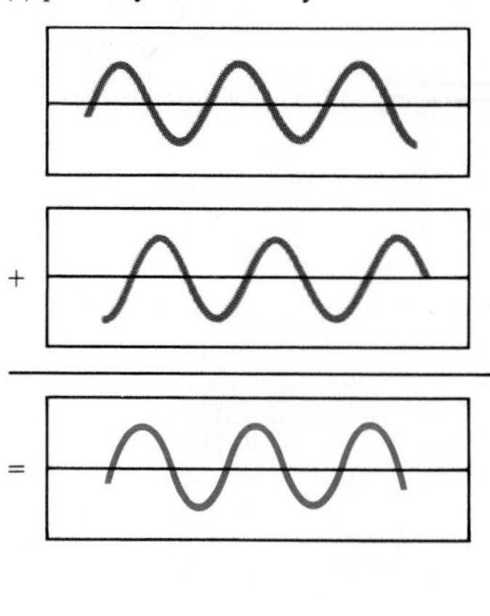

(c)

FIGURE 11–32 Wave diffraction. The waves come from the upper left. Note how the waves, as they pass between the two great rocks, bend around behind each of the rock cliffs. (Each rock can be considered an obstacle in the sense of Fig. 11–33. The two rocks can also be considered as a "slit" through which the waves pass and spread out—compare to Fig. 24–2c for light.)

more precisely, out of phase by one-half wavelength (that is, the crests of one wave occur a half wavelength behind the crests of the other wave). Of course, the relative phase of the two water waves in Fig. 11–30 will at most points be intermediate between these two extremes, which results in *partially* destructive interference, as illustrated in Fig. 11–31c.

Diffraction. Waves spread as they travel, and when they encounter an obstacle they bend around it somewhat and pass into the region behind as shown in Fig. 11–32 for water waves. This phenomenon is called **diffraction**.

The amount of diffraction depends on the wavelength of the wave and on the size of the obstacle, as shown in Fig. 11–33. If the wavelength is much larger than the object, as with the grass blades of Fig. 11–33a, the wave

FIGURE 11–33 Water waves passing objects of various sizes. Note that the larger the wavelength compared to the size of the object, the more diffraction there is into the "shadow region."

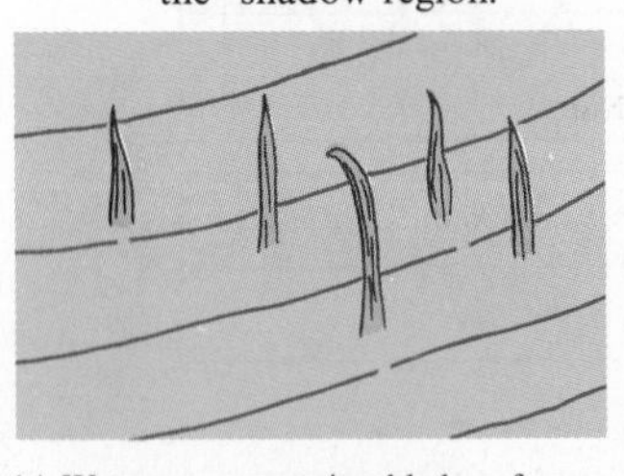

(a) Water waves passing blades of grass

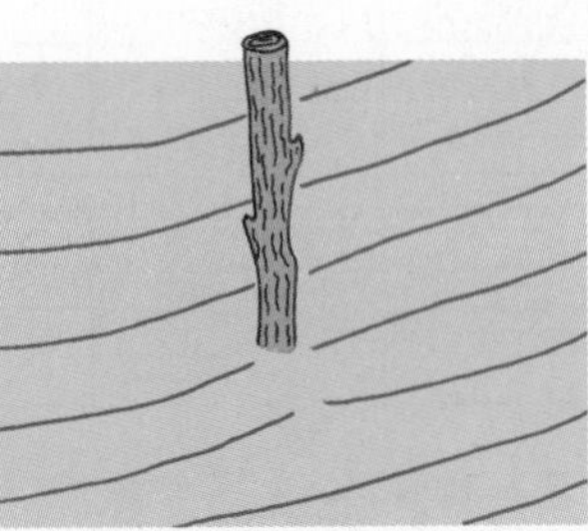

(b) Stick in water

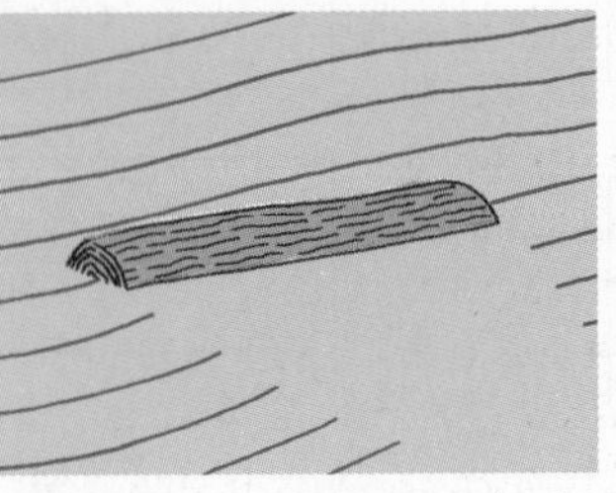

(c) Short-wavelength waves passing log

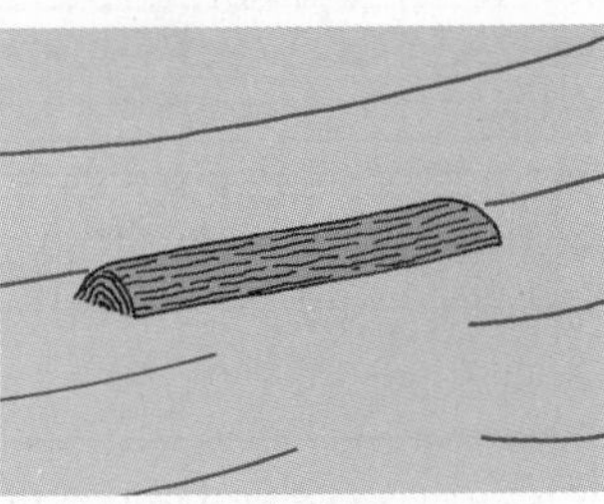

(d) Long-wavelength waves passing log

bends around them almost as if they were not there. For larger objects—parts (b) and (c)—there is more of a "shadow" region behind the obstacle. But notice in part (d), where the obstacle is the same as in part (c) but the wavelength is longer, that there is more diffraction into the shadow region. As a rule of thumb, *only if the wavelength is less than the size of the object will there be a significant shadow region.* It is worth noting that this rule applies to *reflection* from an obstacle as well. Very little of a wave is reflected unless the wavelength is less than the size of the obstacle.

That waves can bend around obstacles, and thus can carry energy to areas behind obstacles, is in clear distinction to energy carried by material particles. A clear example is the following: if you are standing around a corner on one side of a building, you can't be hit by a baseball thrown from the other side, but you can hear a shout or other sound because the sound waves diffract around the edges.

Both interference and diffraction occur only for energy carried by waves and not for energy carried by material particles. This distinction was important for understanding the nature of light, and of matter itself, as we shall see in later chapters.

11–11 • Standing Waves; Resonance

If you shake one end of a rope (or slinky) and the other end is kept fixed, a continuous wave will travel down to the fixed end and be reflected back. As you continue to vibrate the rope, there will be waves traveling in both directions, and the wave traveling down the rope will interfere with the reflected wave coming back. Usually there will be quite a jumble. But if you vibrate the rope at just the right frequency, the two traveling waves will interfere in such a way that a large-amplitude **standing wave** will be produced, Fig. 11–34. It is called a "standing wave" because it doesn't appear to be traveling. The points of destructive interference, called **nodes**, and of constructive interference, called **antinodes**, remain in fixed positions. Standing waves occur at more than one frequency. The lowest frequency of vibration that produces a standing wave gives rise to the pattern shown in Fig. 11–34a. The standing waves shown in parts (b) and (c) are produced at precisely twice and three times the lowest frequency, respectively, assuming the tension in the rope is the same. The rope can also vibrate with four loops at four times the lowest frequency, and so on.

FIGURE 11–34
Standing waves corresponding to three resonant frequencies.

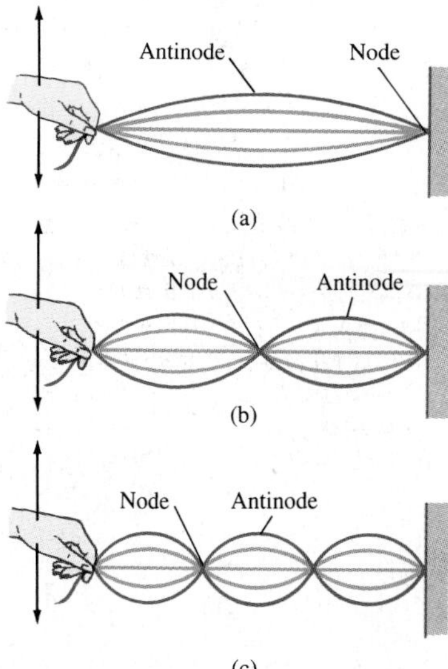

The frequencies at which standing waves are produced are the **natural frequencies** or **resonant frequencies** of the rope, and the different standing wave patterns shown in Fig. 11–34 are different "resonant modes of vibration." For although a standing wave is the result of the interference of two waves traveling in opposite directions, it is also an example of a vibrating object at resonance (Section 11–6). When a standing wave exists on a rope, the rope is vibrating in place; and at the frequencies at which resonance occurs, little effort is required to achieve a large amplitude. Standing waves then represent the same phenomenon as the resonance of a vibrating spring or pendulum, which we discussed earlier. The only difference is that a spring or pendulum has only one resonant frequency, whereas the rope has an infinite number of resonant frequencies, each of which is a whole-number multiple of the lowest resonant frequency.

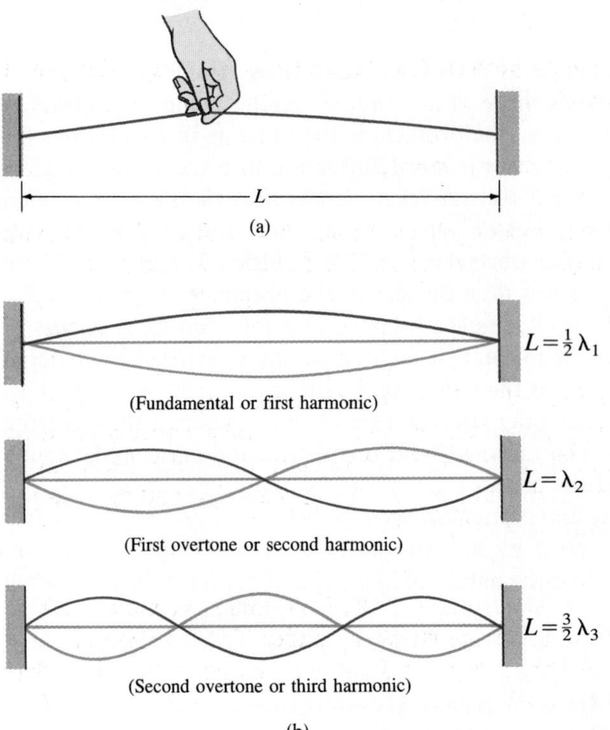

FIGURE 11–35
(a) A string is plucked.
(b) Only standing waves corresponding to resonant frequencies persist for long.

Now let us consider a string stretched between two supports that is plucked like a guitar or violin string, Fig. 11–35a. Waves of a great variety of frequencies will travel in both directions along the string, will be reflected at the ends, and will travel back in the opposite direction. Most of these waves interfere in a random way with each other and quickly die away. However, those waves that correspond to the resonant frequencies of the string will persist. The ends of the string, since they are fixed, will be nodes. There may be other nodes as well. Some of the possible resonant modes of vibration (standing waves) are shown in Fig. 11–35b. Generally, the motion will be a combination of these different resonant modes; but only those frequencies that correspond to a resonant frequency will be present.

Fundamental frequency and harmonics

To determine the resonant frequencies, we first note that the wavelengths of the standing waves bear a simple relationship to the length L of the string. The lowest frequency, called the **fundamental frequency**, corresponds to one antinode (or loop). And as can be seen in Fig. 11–35b, the whole length corresponds to one-half wavelength. Thus $L = \frac{1}{2}\lambda_1$, where λ_1 stands for the wavelength of the fundamental. The other natural frequencies are called **overtones**; when they are integral multiples of the fundamental (as they are for a simple string), they are also called **harmonics**, with the fundamental being referred to as the **first harmonic**.[†] The next mode after the fundamental has two loops and is called the **second harmonic** (or first overtone); the length of the string L corresponds to one complete wavelength: $L = \lambda_2$.

† The term "harmonic" comes from music, because such integral multiples of frequencies "harmonize."

For the third and fourth harmonics, $L = \frac{3}{2}\lambda_3$ and $L = 2\lambda_4$, respectively, and so on. In general, we can write

$$L = \frac{n\lambda_n}{2}, \qquad \text{where} \quad n = 1, 2, 3, \cdots.$$

The integer n labels the number of the harmonic: $n = 1$ for the fundamental, $n = 2$ for the second harmonic, and so on. We solve for λ_n and find

$$\lambda_n = \frac{2L}{n}, \qquad n = 1, 2, 3, \cdots. \qquad (11\text{–}17)$$

In order to find the frequency f of each vibration we use Eq. 11–9, $f = v/\lambda$.

Because a standing wave is equivalent to two traveling waves moving in opposite directions, the concept of wave velocity still makes sense and is given by Eq. 11–10 in terms of the tension F_T in the string and its mass per unit length (m/L)—that is, $v = \sqrt{F_T/(m/L)}$.

EXAMPLE 11–12 A piano string is 1.10 m long and has a mass of 9.00 g. (*a*) How much tension must the string be under if it is to vibrate at a fundamental frequency of 131 Hz? (*b*) What are the frequencies of the first four harmonics?

SOLUTION (*a*) The wavelength of the fundamental is $\lambda = 2L = 2.20$ m (Eq. 11–17). The velocity is then $v = \lambda f = (2.20\text{ m})(131\text{ s}^{-1}) = 288$ m/s. Then, from Eq. 11–10, we have

$$F_T = \frac{m}{L}v^2$$
$$= \left(\frac{9.00 \times 10^{-3}\text{ kg}}{1.10\text{ m}}\right)(288\text{ m/s})^2 = 679\text{ N}.$$

(*b*) The frequencies of the second, third, and fourth harmonics are two, three, and four times the fundamental frequency: 262, 393, and 524 Hz.

A standing wave does appear to be standing in place (and a traveling wave appears to move). The term "standing" wave is also meaningful from the point of view of energy. Since the string is at rest at the nodes, no energy flows past these points. Hence the energy is not transmitted down the string, but "stands" in place in the string.

Standing waves are produced not only on strings, but on any object that is set into vibration. Even when a rock or a piece of wood is struck with a hammer, standing waves are set up that correspond to the natural resonant frequencies of the object. In general, the resonant frequencies depend on the dimensions of the object, just as for a string they depend on its length. For example, a small object does not have as low resonant frequencies as does a large object made of the same material. All musical instruments depend on standing waves to produce their musical sounds, from stringed instruments to wind instruments (in which a column of air vibrates as a standing wave) to drums and other percussion instruments, as we shall discuss in the next chapter.

SUMMARY

A vibrating object undergoes *simple harmonic motion* (SHM) if the restoring force is proportional to the displacement, $F = -kx$. The force constant k is the ratio of restoring force to displacement, $k = F/x$. The maximum displacement is called the *amplitude*. The *period* T is the time required for one complete cycle (back and forth), and the *frequency* f is the number of cycles per second; they are related by $f = 1/T$. The period of vibration for a mass m on the end of a spring is given by $T = 2\pi\sqrt{m/k}$. SHM is *sinusoidal*, which means that the displacement as a function of time follows a sine or cosine curve.

During SHM, the total energy $E = \frac{1}{2}mv^2 + \frac{1}{2}kx^2$ is continually changing from potential to kinetic and back again.

A *simple pendulum* of length L approximates SHM if its amplitude is small and friction can be ignored; its period is then given by $T = 2\pi\sqrt{L/g}$, where g is the acceleration of gravity.

When friction is present (for all real springs and pendulums), the motion is said to be *damped*. The maximum displacement decreases in time and the energy is eventually all transformed to heat.

If an oscillating force is applied to a system capable of vibrating, the amplitude of vibration can be very large if the frequency of the applied force is near the *natural* (or *resonant*) frequency of the oscillator; this is called *resonance*.

Vibrating objects act as sources of waves that travel outward from the source. Waves on water and on a string are examples. The wave may be a *pulse* (a single crest) or it may be continuous (many crests and troughs). The *wavelength* of a continuous sinusoidal wave is the distance between two successive crests. The *frequency* is the number of full wavelengths (or crests) that pass a given point per unit time. The *wave velocity* (how fast a crest moves) is equal to the product of wavelength and frequency, $v = \lambda f$. The *amplitude* of a wave is the maximum height of a crest, or depth of a trough, relative to the normal (or equilibrium) level.

In a *transverse wave*, the oscillations are perpendicular to the direction in which the wave travels; an example is a wave on a string. In a *longitudinal wave*, the oscillations are along (parallel to) the line of travel; sound is an example.

Waves reflect off objects in their path. When the *wave front* (of a two- or three-dimensional wave) strikes an object, the angle of reflection is equal to the angle of incidence. When a wave strikes a boundary between two materials in which it can travel, part of the wave is reflected and part is transmitted. The transmitted wave front may undergo *refraction*, or bending. When two waves pass through the same region of space at the same time, they *interfere*. The resultant displacement at any point and time is the sum of their separate displacements; this can result in *constructive interference*, *destructive interference*, or something in between, depending on the amplitudes and relative phases of the waves. Waves also undergo *diffraction*, meaning they can bend around an obstacle, so there is some wave motion behind the obstacle. The smaller the wavelength relative to the size of the object, the less the diffraction.

Waves traveling on a string (or other medium) of fixed length interfere with waves that have reflected off the end and are traveling back in the opposite direction. At certain frequencies, *standing waves* can be produced in which the waves seem to be standing still rather than traveling. The string (or other medium) is vibrating as a whole. This is a resonance phenomenon and the frequencies at which standing waves occur are called *resonant frequencies*. The points of destructive interference (no vibration) are called *nodes*. Points of constructive interference (maximum amplitude of vibration) are called *antinodes*.

QUESTIONS

1. Give some examples of everyday vibrating objects. Which exhibit SHM, at least approximately?
2. Is the acceleration of a simple harmonic oscillator ever zero? If so, where?
3. Is the motion of a piston in an automobile engine simple harmonic? Explain.
4. Real springs have mass. How will the true period and frequency differ from those given by the equations for a mass oscillating on the end of an idealized massless spring?
5. How could you double the maximum speed of a SHO?

6. If you double the amplitude of a SHO, how does this change the frequency, maximum velocity, maximum acceleration, and total mechanical energy?
7. A 10-kg fish is attached to the hook of a vertical spring scale, and is then released. Describe the scale reading as a function of time.
8. If a pendulum clock is accurate at sea level, will it gain or lose time when taken to high altitude?
9. Why can you make water slosh back and forth in a pan only if you shake the pan at a certain frequency?
10. A tuning fork of natural frequency 264 Hz sits on a table at the front of a room. At the back of the room, two tuning forks, one of natural frequency 260 Hz and one of 420 Hz, are initially silent, but when the tuning fork at the front of the room is set into vibration, the 260-Hz fork spontaneously begins to vibrate, but the 420-Hz fork does not. Explain.
11. Give several everyday examples of resonance.
12. Is a rattle in a car ever a resonance phenomenon? Explain.
13. Over the years, buildings have been able to be built out of lighter and lighter materials. How has this affected the natural vibration frequencies of buildings and the problems of resonance due to passing trucks, airplanes, or natural sources of vibration?
14. Is the frequency of a simple periodic wave equal to the frequency of its source? Why or why not?
15. Explain the difference between the speed of a transverse wave traveling down a rope and the speed of a tiny piece of the rope.
16. Why do the strings used for the lowest-frequency notes on a piano normally have wire wrapped around them?
17. What kind of waves do you think will travel down a horizontal metal rod if you strike its end (*a*) vertically from above and (*b*) horizontally parallel to its length?
18. Since the density of air decreases with an increase in temperature, but the bulk modulus B is nearly independent of temperature, how would you expect the speed of sound waves in air to vary with temperature?
19. The speed of sound in most solids is somewhat greater than in air, yet the density of solids is much greater (10^3–10^4 times). Explain.
20. Give two reasons why circular water waves decrease in amplitude as they travel away from the source.
*21. Two linear waves have the same amplitude and otherwise are identical, except one has half the wavelength of the other. Which transmits more energy? How much more energy does it transfer?
*22. The intensity of a sound in real-life situations does not decrease precisely with the square of the distance from the source as we might expect from Eq. 11–15. Why not?
23. When a sinusoidal wave crosses the boundary between two sections of rope as in Fig. 11–25, the frequency does not change (although the wavelength and velocity do change). Explain why.
24. If we knew that energy was being transmitted from one place to another, how might we determine whether the energy was being carried by particles (material bodies) or by waves?
25. AM radio signals can usually be heard behind a hill, but FM often cannot. That is, AM signals bend more than FM. Explain. (Radio signals, as we shall see, are carried by electromagnetic waves whose wavelength for AM is typically 200 to 600 m and for FM about 3 m.)
26. If a string is vibrating in three segments, are there any places one can touch it with a knife blade without disturbing the motion?
27. When a standing wave exists on a string, the vibrations of incident and reflected waves cancel at the nodes. Does this mean that energy was destroyed?

PROBLEMS

SECTIONS 11–1 TO 11–3

1. (I) A piece of rubber is 35 cm long when a weight of 22 N hangs from it and is 58 cm long when a weight of 60 N hangs from it. What is the "spring" constant of this piece of rubber?
2. (I) When a 70-kg person climbs into a 1200-kg car, the car's springs compress vertically by 2.2 cm. What will be the frequency of vibration when the car hits a bump? Ignore damping.
3. (I) If a particle undergoes SHM with amplitude A, what is the total distance it travels in one period?
4. (I) A fisherman's scale stretches 3.5 cm when a 2.7-kg fish hangs from it. What is the spring constant and what will be the frequency of vibration if the fish is pulled down and released so that it vibrates up and down?
5. (II) Construct a table indicating the position of the mass in Fig. 11–2 at the times $t = 0, \frac{1}{4}T, \frac{1}{2}T, \frac{3}{4}T, T,$ and $\frac{5}{4}T$, where T is the period of oscillation. On a graph of x versus t, plot these six points. Now connect these points with a smooth curve. Does your curve, based on these simple considerations, resemble that of a sine or cosine wave (Fig. 11–6)?
6. (II) A small fly of mass 0.80 g is caught in a spider's web. The web vibrates predominately with a frequency of 10 Hz. (*a*) What is the value of the effective spring constant k for the web? (*b*) At what frequency would you expect the web to vibrate if an insect of mass 0.50 g were trapped?
7. (II) A spring vibrates with a frequency of 2.5 Hz when a weight of 0.40 kg is hung from it. What will its frequency be if only 0.30 kg hangs from it?

8. (II) A 0.26-kg mass at the end of a spring vibrates 2.0 times per second with an amplitude of 0.18 m. Determine (*a*) the velocity when it passes the equilibrium point, (*b*) the velocity when it is 0.10 m from equilibrium, (*c*) the total energy of the system, and (*d*) the equation describing the motion of the mass, assuming that at $t = 0$, x was a maximum.

9. (II) A mass m at the end of a spring vibrates with a frequency of 0.72 Hz. When an additional 700-g mass is added to m, the frequency is 0.48 Hz. What is the value of m?

10. (II) A block of mass m is supported by two identical parallel vertical springs, each with spring constant k. What will be the frequency of vibration?

11. (II) A 0.50-kg mass vibrates according to the equation $x = 0.32 \cos 7.40t$, where x is in meters, and t in seconds. Determine (*a*) the amplitude, (*b*) the frequency, (*c*) the total energy, and (*d*) the kinetic energy and potential energy when $x = 0.26$ m.

12. (II) A mass of 1.80 kg stretches a vertical spring 0.275 m. If the spring is stretched an additional 0.110 m and released, how long does it take to reach the (new) equilibrium position again?

13. (II) It takes a force of 80.0 N to compress the spring of a popgun 0.200 m to "load" a 0.150-kg ball. With what speed will the ball leave the gun?

14. (II) At $t = 0$, a 750-g mass at rest on the end of a horizontal spring ($k = 124$ N/m) is struck by a hammer which gives it an initial speed of 2.76 m/s. Determine (*a*) the period and frequency of the motion, (*b*) the amplitude, (*c*) the maximum acceleration, (*d*) the position as a function of time, and (*e*) the total energy.

15. (II) Show that if a simple harmonic oscillator at rest at $x = 0$ is given a push at $t = 0$, the position x is given by $x = x_0 \sin [(2\pi/T)t]$.

16. (II) A mass on the end of a spring is stretched a distance x_0 from equilibrium and released. At what distance from equilibrium will it have (*a*) velocity equal to half its maximum velocity and (*b*) acceleration equal to half its maximum acceleration?

17. (II) A spring of force constant 160 N/m vibrates with an amplitude of 28.5 cm when 0.380 kg hangs from it. (*a*) What is the equation describing this motion as a function of time? Assume that the mass passes through the equilibrium point at $t = 0$. (*b*) At what times will the spring have its maximum and minimum extensions?

18. (II) A 0.0120-kg bullet strikes a 0.400-kg block attached to a fixed horizontal spring whose spring constant is 6.80×10^3 N/m and sets it into vibration with an amplitude of 17.4 cm. What was the speed of the bullet before impact if the two objects move together after impact?

19. (II) If one vibration has 10 times the energy of a second, but their frequencies and masses are the same, how do their amplitudes compare?

20. (II) If a mass m hangs from a vertical spring, as shown in Fig. 11–3, show that $F = -kx$ holds for compression of the spring, where x is the displacement from the (vertical position) equilibrium point.

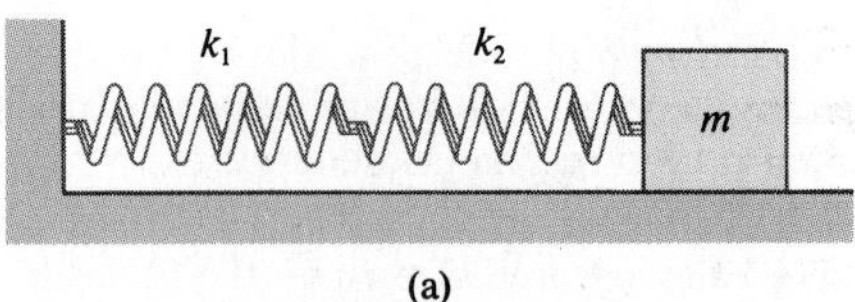

(a)

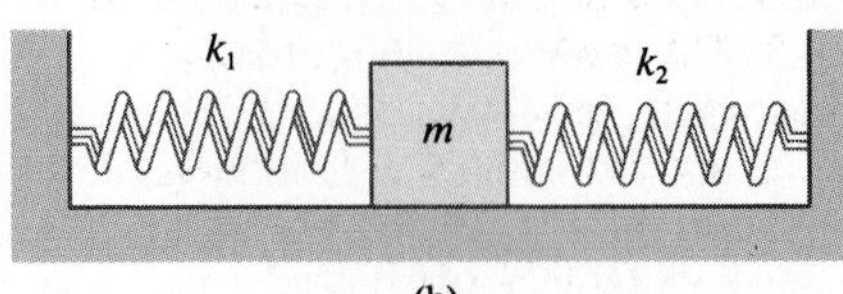

(b)

FIGURE 11–36 Problem 22.

21. (III) Use the reference circle to show that the velocity and acceleration of an object undergoing SHM are given by:

$$v = -v_0 \sin \omega t, \qquad a = -a_0 \cos \omega t,$$

where $v_0 = x_0\sqrt{k/m}$ and $a_0 = kx_0/m$ refer to the maximum speed and acceleration, respectively.

22. (III) A mass m is connected to two springs, with spring constants k_1 and k_2, in two different ways, as shown in Fig. 11–36a and b. Show that the period for the configuration shown in part (a) is given by

$$T = 2\pi \sqrt{m\left(\frac{1}{k_1} + \frac{1}{k_2}\right)}$$

and for that in part (b) is given by

$$T = 2\pi \sqrt{\frac{m}{k_1 + k_2}}.$$

Ignore friction.

SECTION **11–4**

23. (I) A pendulum makes 44 vibrations in 60 s. What is its (*a*) period, and (*b*) frequency?

24. (I) How long must a simple pendulum be if it is to make exactly one complete vibration per second?

25. (II) What is the period of a simple pendulum 80 cm long (*a*) on the earth, and (*b*) when it is in a freely falling elevator?

26. (II) A pendulum oscillates at a frequency of 2.0 Hz. At $t = 0$, it is released starting at an angle of 15°. Ignoring friction, what will be the position (angle) of the pendulum at (*a*) $t = 0.25$ s, (*b*) $t = 1.60$ s, and (*c*) $t = 500$ s?

27. (II) The length of a simple pendulum is 0.36 m and it is released at an angle of 15° to the vertical. (*a*) With what frequency does it vibrate? (*b*) What is the pendulum bob's speed when it passes through the lowest point of the swing? Assume SHM.

28. (II) Derive a formula for the maximum speed v_0 of a simple pendulum bob in terms of g, the length L, and the angle of swing θ_0.

SECTIONS 11–7 AND 11–8

29. (I) A fisherman notices that wave crests pass the bow of his anchored boat every 5.0 s. He measures the distance between two crests to be 12 m. How fast are the waves traveling?

30. (I) A sound wave in air has a frequency of 262 Hz and travels with a speed of 330 m/s. How far apart are the wave crests (compressions)?

31. (I) AM radio signals have frequencies between 550 kHz and 1600 kHz (kilohertz) and travel with a speed of 3.00×10^8 m/s. What are the wavelengths of these signals? On FM, the frequencies range from 88.0 MHz to 108 MHz (megahertz) and travel at the same speed; what are their wavelengths?

32. (I) Calculate the speed of longitudinal waves in (*a*) water, and (*b*) granite.

33. (I) Two solid rods have the same bulk modulus, but one is twice as dense as the other. In which rod will the speed of longitudinal waves be greater, and by what factor?

34. (II) A rope of mass 0.55 kg is stretched between two supports 30 m apart. If the tension in the rope is 150 N, how long will it take a pulse to travel from one support to the other?

35. (II) Determine the wavelength of a 10,000-Hz sound wave traveling along an iron rod. [*Hint:* see Table 9–1.]

36. (II) A sailor strikes the side of his ship just below the surface of the sea. He hears the echo of the wave reflected from the ocean floor directly below 2.0 s later. How deep is the ocean at this point?

37. (II) S and P waves from an earthquake travel at different speeds, and this difference helps in the determination of the earthquake "focus" (where the disturbance took place). (*a*) Assuming typical speeds of 9.0 km/s and 5.0 km/s for P and S waves, respectively, how far away did the earthquake occur if a particular seismic station detects the arrival of these two types of waves 1.8 min apart? (*b*) Is one seismic station sufficient to determine the position of the focus? Explain.

*SECTION 11–9

*38. (I) Two earthquake waves have the same frequency as they travel through the same portion of the earth, but one is carrying twice the energy. What is the ratio of the amplitudes of the two waves?

*39. (II) Leaves on the surface of a pond are observed to move up and down a total vertical distance of 0.16 m as a wave passes. (*a*) What is the amplitude of the wave? (*b*) If the amplitude were changed to 0.12 m, by what factor would the energy in the wave change?

*40. (II) Compare (*a*) the intensities and (*b*) the amplitudes of an earthquake wave as it passes two points 10 km and 20 km from the source.

*41. (II) The intensity of a particular earthquake wave is measured to be $1.0 \times 10^6\ \mathrm{J/m^2 \cdot s}$ at a distance of 100 km from the source. (*a*) What was the intensity when it passed a point only 1.0 km from the source? (*b*) What was the rate energy passed through an area of 5.0 m^2 at 1.0 km?

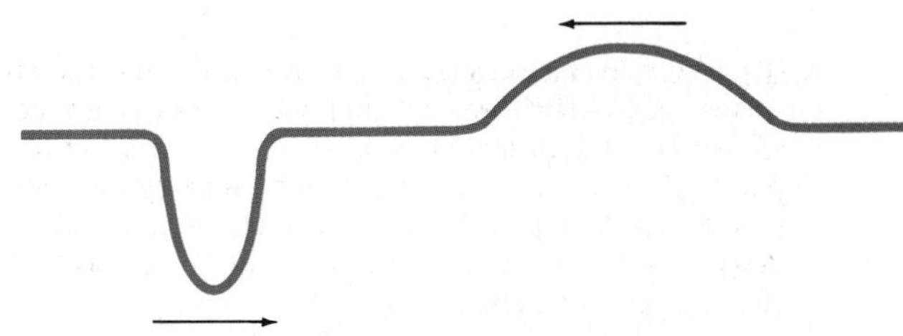

FIGURE 11–37 Problem 43.

*42. (III) Show that the amplitude x_0 of circular water waves decreases as the square root of the distance r from the source: $x_0 \propto 1/\sqrt{r}$. Ignore damping.

SECTION 11–10

43. (I) The two pulses shown in Fig. 11–37 are moving toward each other. (*a*) Sketch the shape of the string at the moment they directly overlap. (*b*) Sketch the shape of the string a few moments later. (*c*) In Fig. 11–29a, at the moment the pulses pass each other, the string is straight. What has happened to the energy at this moment?

44. (I) An earthquake P wave traveling 9.0 km/s strikes a boundary within the earth between two kinds of material. If it approaches the boundary at an incident angle of 47° and the angle of refraction is 27°, what is the speed in the second medium?

45. (I) Water waves approach a shelf where the velocity changes from 2.6 m/s to 2.1 m/s. If the incident waves make a 35° angle with the shelf, what will be the angle of refraction?

46. (II) A longitudinal earthquake wave strikes a boundary between two types of rock at a 20° angle. As it crosses the boundary, the specific gravity of the rock changes from 3.8 to 2.7. Assuming that the elastic modulus is the same for both types of rock, determine the angle of refraction.

47. (II) It is found for any type of wave, say an earthquake wave, that if it reaches a boundary beyond which its speed is increased, there is a maximum incident angle if there is to be a transmitted refracted wave. This maximum incident angle θ_{iM} corresponds to an angle of refraction equal to 90°. If $\theta_i > \theta_{iM}$, all the wave is reflected at the boundary and none is refracted (because this would correspond to $\sin\theta_r > 1$, where θ_r is the angle of refraction, which is impossible); this is referred to as *total internal reflection.* (*a*) Find a formula for θ_{iM} using Eq. 11–16. (*b*) At what angles of incidence will there be only reflection and no transmission for an earthquake P wave traveling 7.5 km/s when it reaches a different kind of rock where its speed is 9.3 km/s?

SECTION 11–11

48. (I) If a violin string vibrates at 294 Hz as its fundamental frequency, what are the frequencies of the first four harmonics?

49. (I) A violin string vibrates at 440 Hz when unfingered. At what frequency will it vibrate if it is fingered one-fourth of the way down from the end?

50. (I) A particular string resonates in four loops at a frequency of 360 Hz. Name at least three other frequencies at which it will resonate.

51. (II) The velocity of waves on a string is 350 m/s. If the frequency of standing waves is 420 Hz, how far apart are the nodes?

52. (II) If two successive overtones of a vibrating string are 360 Hz and 400 Hz, what is the frequency of the fundamental?

53. (II) A guitar string is 90 cm long and has a mass of 3.6 g. From the bridge to the support post ($=L$) is 60 cm, and the string is under a tension of 520 N. What are the frequencies of the fundamental and first two overtones?

54. (II) A particular violin string plays at a frequency of 294 Hz. If the tension is increased 10 percent, what will the frequency be?

55. (II) Show that the frequency of standing waves on a string of length L and linear density μ, which is stretched to a tension F_T, is given by

$$f = \frac{n}{2L}\sqrt{\frac{F_T}{\mu}},$$

where n is an integer.

56. (II) One end of a horizontal string of linear density 4.2×10^{-4} kg/m is attached to a small-amplitude mechanical 60-Hz vibrator. The string passes over a pulley, a distance $L = 1.40$ m away, and weights are hung from this end. What mass must be hung from this end of the string to produce (*a*) one loop, (*b*) two loops, and (*c*) five loops of a standing wave? Assume the string at the vibrator is a node, which is nearly true. Why can the amplitude of the standing wave be much greater than the vibrator amplitude?

57. (III) (*a*) Show that if the tension in a stretched string is changed by an amount ΔF_T, the frequency of the fundamental is changed by an amount $\Delta f = \frac{1}{2}(\Delta F_T/F_T)f$. (*b*) By what percent must the tension in a piano string be increased or decreased to raise the frequency from 438 Hz to 442 Hz? (*c*) Does the formula in part (*a*) apply to the overtones as well?

GENERAL PROBLEMS

58. When you slosh the water back and forth in a tub at just the right frequency, the water alternately rises and falls at each end. Suppose the frequency to produce such a standing wave in a 50-cm-wide tub is 0.60 Hz. What is the speed of the water wave?

59. A 60-kg person jumps from a window to a fire net 15 m below, which stretches the net 1.2 m. Assume that the net behaves like a simple spring, and calculate how much it would stretch if the same person were lying in it. How much would it stretch if the person jumped from 30 m?

60. A simple pendulum oscillates with frequency f. What is its frequency if it accelerates at $\frac{1}{2}g$ (*a*) upward, and (*b*) downward?

61. A 400-kg wooden raft floats on a lake. When a 75-kg man stands on the raft, it sinks 4.5 cm deeper into the water. When the man steps off, the raft vibrates briefly. (*a*) What is the frequency of vibration? (*b*) What is the total energy of vibration (ignoring damping)?

62. The ripples in a certain groove 11.2 cm from the center of a 33-rpm phonograph record have a wavelength of 2.10 mm. What will be the frequency of the sound emitted?

63. Two strings on a musical instrument are tuned to play at 392 Hz (G) and 440 Hz (A). (*a*) What are the first two overtones for each string? (*b*) If the two strings have the same length and are under the same tension, what must be the ratio of their masses (M_G/M_A)? (*c*) If the strings, instead, have the same mass per unit length and are under the same tension, what is the ratio of their lengths (L_G/L_A)? (*d*) If their masses and lengths are the same, what must be the ratio of the tensions in the two strings?

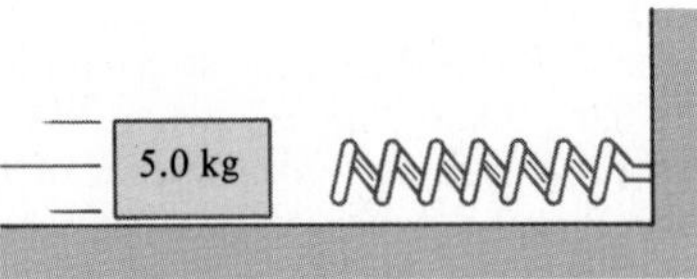

FIGURE 11–38 Problem 64.

64. A 5.0-kg box slides at a speed of 2.5 m/s into a spring of spring constant 85 N/m (Fig. 11–38). (*a*) How far is the spring compressed? (*b*) How long is the box in contact with the spring before it bounces off in the opposite direction?

65. Tides could be considered as a wave. As the moon revolves around the earth once every 25 h, the tides follow (Section 5-8). Consider the two high tides as crests (and the low tides as troughs) and determine the frequency, period, and velocity of this "wave."

66. Consider a sine wave traveling down the stretched two-part cord of Fig. 11–25. Determine a formula (*a*) for the ratio of the speeds of the wave in the two sections, v_2/v_1, and (*b*) for the ratio of the wavelengths in the two sections. (The frequency is the same in both sections. Why?) (*c*) Is the wavelength greater in the heavier cord or the lighter?

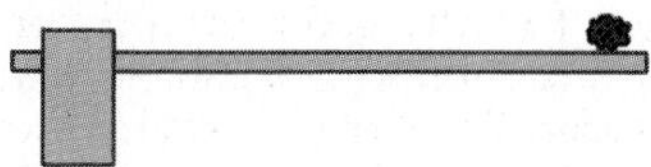

FIGURE 11–39 Problem 67.

67. A diving board oscillates with simple harmonic motion of frequency 6.0 cycles per second. What is the maximum amplitude with which the end of the board can vibrate in order that a pebble placed there (Fig. 11–39) not leave the board during the oscillation?

68. A rectangular block of wood floats in a calm lake. Show that, if friction is ignored, when the block is pushed gently down into the water, it will then vibrate with SHM. Also, determine an equation for the force constant.

69. A pendulum vibrates with an amplitude of 10.0°. What fraction of the time does it spend between +5.0° and −5.0°? Assume SHM.

70. A mass m is gently placed on the end of a freely hanging spring. The mass then falls 30 cm before it stops and begins to rise. What is the frequency of the motion?

71. The pendulum of a clock consists of a small heavy bob at the end of a brass rod. The clock keeps accurate time at 15°C, at which point the pendulum has a period of 0.4820 s. When the temperature is 35°C, will the clock be fast or slow? By how much will it be in error after one day at 35°C? [*Hint*: consult Table 13-1 and Eq. 13-1]

72. The water in a U-shaped tube is displaced an amount Δx from equilibrium. (The level in one side is $2\,\Delta x$ above the level in the other side.) If friction is neglected, will the water oscillate harmonically? Determine a formula for the equivalent spring constant k. Does k depend on the density of the liquid, the cross section of the tube, or the length of the water column?

73. In some diatomic molecules, the force each atom exerts on the other can be approximated by $F = -C/r^2 + D/r^3$, where C and D are positive constants. (*a*) Graph F versus r from $r = 0$ to $r = 2D/C$. (*b*) Show that equilibrium occurs at $r = r_0 = D/C$. (*c*) Let $\Delta r = r - r_0$ be a small displacement from equilibrium, where $\Delta r \ll r_0$. Show that for such small displacements, the motion is approximately simple harmonic, and (*d*) determine the force constant. (*e*) What is the period of such motion? [*Hint:* assume one atom is kept at rest.]

C H A P T E R 12

Sound

If music be the food of physics, play on.

Sound is associated with our sense of hearing and, therefore, with the physiology of our ears and the psychology of our brain which interprets the sensations that reach our ears. The term *sound* also refers to the physical sensation that stimulates our ears: namely, longitudinal waves.

We can distinguish three aspects of any sound. First, there must be a *source* for a sound; and as with any wave, the source of a sound wave is a vibrating object. Second, the energy is transferred from the source in the form of longitudinal sound *waves*. And third, the sound is *detected* by an ear or an instrument. We will discuss sources and detectors of sound later in this chapter, as well as some important applications to fields such as medicine, but now we look at some aspects of sound waves themselves.

12–1 • Characteristics of Sound

We already saw in Chapter 11, Fig. 11–19, how a vibrating drumhead produces a sound wave in air. Indeed, we usually think of sound waves traveling in the air, for normally it is the vibrations of the air that force our eardrums to vibrate. But sound waves can also travel in other materials. Two stones struck together under water can be heard by a swimmer beneath the surface, for the vibrations are carried to the ear by the water. When you put your ear flat against the ground, you can hear an approaching train or truck. In this case the ground does not actually touch your eardrum but the longitudinal wave transmitted by the ground is called a sound wave just the same, for its vibrations cause the outer ear and the air within it to vibrate. Indeed, longitudinal waves traveling in any material medium are often referred to as sound waves. Clearly, sound cannot travel in the absence of matter. For example, a bell ringing inside an evacuated jar cannot be heard.

Table 12–1
Speed of Sound in Various Materials, at 20°C and 1 atm

Material	Speed (m/s)
Air	343
Air (0°C)	331
Helium	1005
Hydrogen	1300
Water	1440
Sea water	1560
Iron and steel	≈5000
Glass	≈4500
Aluminum	≈5100
Hard wood	≈4000

The **speed of sound** is different in different materials. In air at 0°C and 1 atm, sound travels at a speed of 331 m/s. We saw in Eq. 11–11 ($v = \sqrt{B/\rho}$) that the speed depends on the elastic modulus, B, and the density, ρ, of the material. Thus for helium, whose density is much less than that of air but whose elastic modulus is not greatly different, the speed is about three times as large as in air. In liquids and solids, which are much less compressible and therefore have much greater elastic moduli, the speed is larger still. The speed of sound in various materials is given in Table 12–1. The values depend somewhat on temperature, but this is significant mainly for gases. For example, in air, the speed increases approximately 0.60 m/s for each Celsius degree increase in temperature:

Speed of sound in air

$$v \approx (331 + 0.60\,T)\ \text{m/s},$$

where T is the temperature in °C. Unless stated otherwise, we will assume in this chapter that $T = 20°\text{C}$, so that $v = [331 + (0.60)(20)]\ \text{m/s} = 343\ \text{m/s}$.

Two aspects of any sound are immediately evident to a human listener. These are "loudness" and "pitch," and each refers to a sensation in the consciousness of the listener. But to each of these subjective sensations there corresponds a physically measurable quantity. **Loudness** is related to the energy in the sound wave and we shall discuss it in the next section.

Loudness

Pitch

The **pitch** of a sound refers to whether it is high, like the sound of a piccolo or violin, or low, like the sound of a bass drum or string bass. The physical quantity that determines pitch is the frequency, as was first noted by Galileo. The lower the frequency, the lower the pitch, and the higher the frequency, the higher the pitch.† The human ear responds to frequencies in the range from about 20 Hz to about 20,000 Hz. (Recall that 1 Hz is 1 cycle per second.) This is called the **audible range**. These limits vary somewhat from one individual to another. One general trend is that as people age, they are less able to hear the high frequencies, so that the high-frequency limit may be 10,000 Hz or less.

Audible frequency range

Sound waves whose frequencies are outside the audible range may reach the ear, but we are not generally aware of them. Frequencies above 20,000 Hz are called **ultrasonic** (do not confuse with *supersonic*, which is used for an object moving with a speed faster than the speed of sound). Many animals can hear ultrasonic frequencies; dogs, for example, can hear sounds as high as 50,000 Hz and bats can detect frequencies as high as 100,000 Hz. Ultrasonic waves have a number of applications in medicine and other fields, which we will discuss later in this chapter.

† Although pitch is determined mainly by frequency, it also depends to a slight extent on loudness. For example, a very loud sound may seem slightly lower in pitch than a quiet sound of the same frequency.

Sound waves whose frequencies are below the audible range (that is, less than 20 Hz) are called **infrasonic**. Sources of infrasonic waves include earthquakes, thunder, volcanoes, and waves produced by vibrating heavy machinery. This last source can be particularly troublesome to workers, for infrasonic waves—even though inaudible—can cause damage to the human body. These low-frequency waves act in a resonant fashion, causing considerable motion and irritation of internal organs of the body.

12–2 • Intensity of Sound

Like pitch, *loudness* is a sensation in the consciousness of a human being. It too is related to a physically measurable quantity, the **intensity** of the wave. Intensity is defined as the energy transported by a wave per unit time across unit area and, as we saw in the previous chapter (Section 11–9), is proportional to the square of the wave amplitude. Since energy per unit time is power, intensity has units of power per unit area, or watts/meter^2 (W/m^2).

The human ear can detect sounds with an intensity as low as 10^{-12} W/m^2 and as high as 1 W/m^2 (and even higher, although above this it is painful). This is an incredibly wide range of intensity, spanning a factor of 10^{12} from lowest to highest. Presumably because of this wide range, what we perceive as loudness is not directly proportional to the intensity. True, the greater the intensity, the louder the sound. But to produce a sound that sounds about twice as loud requires a sound wave that has about 10 times the intensity. This is roughly valid at any sound level for frequencies near the middle of the audible range. For example, a sound wave of intensity 10^{-2} W/m^2 sounds to an average human being like it is about twice as loud as one whose intensity is 10^{-3} W/m^2, and four times as loud as 10^{-4} W/m^2.

Because of this relationship between the subjective sensation of loudness and the physically measurable quantity "intensity", it is usual to specify sound-intensity levels using a logarithmic scale. The unit on this scale is a **bel**†, or much more commonly, the *decibel* (dB), which is $\frac{1}{10}$ bel (1 dB = 0.1 bel). The **intensity level**, β, of any sound is defined in terms of its intensity, I, as follows:

$$\beta \text{ (in dB)} = 10 \log \frac{I}{I_0}, \qquad (12\text{–}1)$$

where I_0 is the intensity of some reference level, and the logarithm is to the base 10. I_0 is usually taken as the minimum intensity audible to an average person, the "threshold of hearing," which is $I_0 = 1.0 \times 10^{-12}$ W/m^2. The intensity level of a sound, for example, whose intensity is $I = 1.0 \times 10^{-10}$ W/m^2 will be

$$\beta = 10 \log \frac{10^{-10}}{10^{-12}} = 10 \log 100 = 20 \text{ dB},$$

since log 100 is equal to 2.0 (logarithms are discussed in Appendix A). Notice that the intensity level at the threshold of hearing is 0 dB; that is, $\beta = 10 \log (10^{-12}/10^{-12}) = 10 \log 1 = 0$ since $\log 1 = 0$. Notice, too, that an increase in intensity by a factor of 10 corresponds to a level increase of 10 dB.

† After the inventor Alexander Graham Bell (1847–1922).

TABLE 12–2
Intensity of Various Sounds

Source of the Sound	Intensity Level (dB)	Intensity (W/m^2)
Jet plane at 30 m	140	100
Threshold of pain	120	1
Loud indoor rock concert	120	1
Siren at 30 m	100	1×10^{-2}
Auto interior, moving at 90 km/h	75	3×10^{-5}
Busy street traffic	70	1×10^{-5}
Ordinary conversation, at 50 cm	65	3×10^{-6}
Quiet radio	40	1×10^{-8}
Whisper	20	1×10^{-10}
Rustle of leaves	10	1×10^{-11}
Threshold of hearing	0	1×10^{-12}

An increase in intensity by a factor of 100 corresponds to a level increase of 20 dB. Thus a 50-dB sound is 100 times more intense than a 30-dB sound.

The intensities and intensity levels for a number of common sounds are listed in Table 12–2.

EXAMPLE 12–1 A high-quality loudspeaker is advertised to reproduce, at full volume, frequencies from 30 Hz to 18,000 Hz with uniform intensity ± 3 dB. That is, over this frequency range, the intensity level does not vary by more than 3 dB from the average. By what factor does the intensity change for the maximum intensity-level change of 3 dB?

SOLUTION Let us call the average intensity I_1 and the average level β_1. Then the maximum intensity, I_2, corresponds to a level $\beta_2 = \beta_1 + 3$ dB. Thus

$$\beta_2 - \beta_1 = 10 \log \frac{I_2}{I_0} - 10 \log \frac{I_1}{I_0}$$

$$3 \text{ dB} = 10\left(\log \frac{I_2}{I_0} - \log \frac{I_1}{I_0}\right)$$

$$= 10 \log \frac{I_2}{I_1}$$

since $(\log a - \log b) = \log a/b$ (see Appendix A). Then

$$\log \frac{I_2}{I_1} = 0.30.$$

Using a calculator we calculate 10^x with $x = 0.30$, or from a log table, we find what number has a logarithm equal to 0.30. The result is 2.0, so

$$\frac{I_2}{I_1} = 2.0,$$

or I_2 is twice as intense as I_1.

It is worth noting that a sound-level difference of 3 dB (which corresponds to a doubled intensity as we just saw) corresponds to only a very small change in the subjective sensation of apparent loudness. Indeed, the average human can distinguish a difference in level of only about 1 dB.

*12–3 • Intensity Related to Amplitude and Pressure Amplitude

The intensity I of a wave is proportional to the square of the wave amplitude, x_0, as discussed in Section 11–9. Indeed, using Eq. 11–14 we can relate the amplitude quantitatively to the intensity I or level β, as the following example shows.

EXAMPLE 12–2 Calculate the maximum displacement of air molecules for a sound at the threshold of hearing, having a frequency of 1000 Hz.

SOLUTION We use Eq. 11–14 and solve for x_0:

$$x_0 = \frac{1}{\pi f}\sqrt{\frac{I}{2\rho v}}$$

$$= \frac{1}{(3.14)(1.0 \times 10^{3}\ \mathrm{s^{-1}})}\sqrt{\frac{1.0 \times 10^{-12}\ \mathrm{W/m^2}}{(2)(1.29\ \mathrm{kg/m^3})(343\ \mathrm{m/s})}},$$

where we have taken the density of air to be 1.29 kg/m^3 and the speed of sound in air (assumed 20°C) as 343 m/s. When we do the arithmetic, we find that $x_0 = 1.1 \times 10^{-11}$ m.

The result of this example illustrates just how sensitive the human ear is. For it can detect displacements of air molecules that are actually less than the diameter of atoms (about 10^{-10} m)!

We have so far described sound waves in terms of the vibration of the molecules of the medium. But they can also be viewed from the point of view of pressure. Indeed, longitudinal waves are often called **pressure waves**. The pressure variation is usually easier to measure than the displacement. As can be seen in Fig. 11–19, in a wave "compression" (where molecules are closest together), the pressure is higher than normal, whereas in an expansion (or rarefaction) the pressure is less than normal. Figure 12–1 shows a graphical representation of a sound wave in air in terms of (*a*) displacement and (*b*) pressure. Note that the displacement wave is a quarter wavelength out of phase with the pressure wave: where the pressure is a maximum or minimum, the displacement from equilibrium is zero; and where the pressure variation is zero, the displacement is a maximum or minimum. The pressure in a 1000-Hz sound wave at the threshold of hearing differs from atmospheric by at most 3×10^{-5} N/m^2 ($= 3 \times 10^{-10}$ atm). This is called the **pressure amplitude**. Just as the intensity is proportional to the square of the displacement amplitude, x_0, it is also proportional to the square of the pressure amplitude, P_0. Indeed, it can be shown that the intensity is related to the pressure amplitude P_0 by the formula

$$I = \frac{P_0^2}{2\rho v},$$

FIGURE 12–1
Representation of a sound wave in terms of (a) displacement, and (b) pressure.

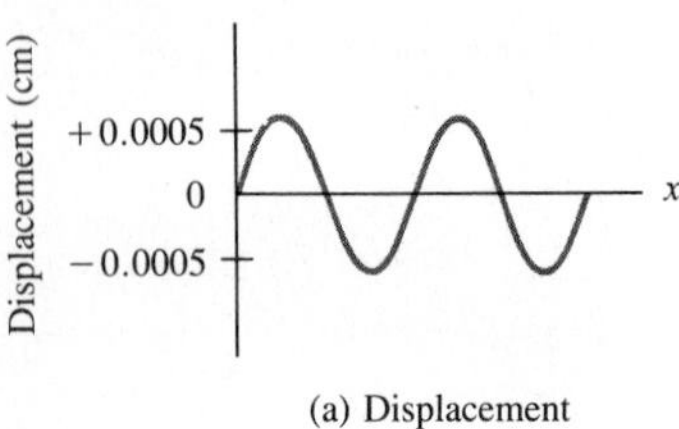

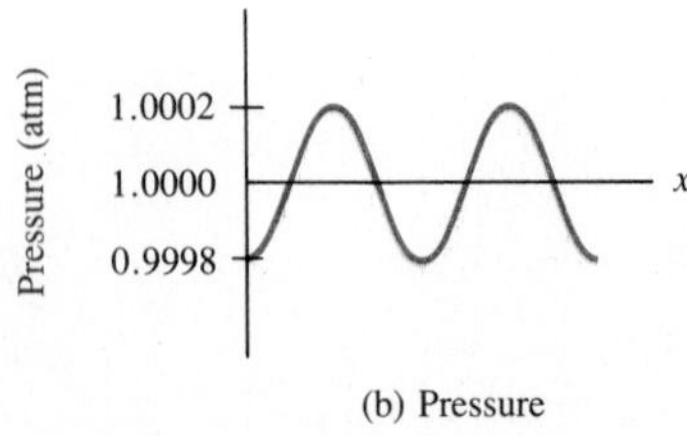

where ρ is the density of the medium and v the velocity of sound in that medium.

Normally, the loudness or intensity of a sound decreases as you get farther from the source of the sound. In interior rooms, this effect is reduced because of reflections from the walls. However, if a source is in the open so that sound can radiate out freely in all directions, the intensity decreases as the inverse square of the distance,

$$I \propto \frac{1}{r^2},$$

as we saw in Eq. 11–15. Of course, if there is significant reflection from structures or the ground, the situation will be more complicated.

EXAMPLE 12–3 The intensity level of the sound from a jet plane at a distance of 30 m is 140 dB. What is the intensity level at 300 m? (Ignore reflections from the ground.)

SOLUTION The intensity I at 30 m is found from Eq. 12–1:

$$140\text{ dB} = 10 \log\left(\frac{I}{10^{-12}\text{ W/m}^2}\right).$$

Reversing the log equation to solve for I we have:

$$10^{14} = \frac{I}{10^{-12}\text{ W/m}^2},$$

so $I = 10^2\text{ W/m}^2$. At 300 m, 10 times as far away, the intensity will be $(\frac{1}{10})^2 = 1/100$ as much, or 1 W/m^2. Hence, the intensity level is

$$\beta = 10 \log\left(\frac{1}{10^{-12}}\right) = 120\text{ dB}.$$

Even at 300 m, the sound is at the threshold of pain. This is why workers at airports wear ear covers to protect their ears from damage.

*12–4 • The Ear and Its Response; Loudness

The human ear, as we have seen, is a remarkably sensitive detector of sound. Mechanical detectors of sound, namely microphones, can barely match the ear in detecting low-intensity sounds.

The function of the ear is to efficiently transform the vibrational energy of sound waves into electrical signals which are carried to the brain by way of nerves. A microphone performs a similar task. Sound waves striking the diaphragm of a microphone set it into vibration, and these vibrations are transformed into an electrical signal with the same frequencies, which can then be amplified and sent to a loudspeaker or tape recorder. We shall discuss the operation of microphones when we study electricity and magnetism in later chapters. Here we shall discuss the structure and response of the ear.

Figure 12–2 is a diagram of the human ear. The ear is conveniently divided into three main divisions; the outer ear, middle ear, and inner ear.

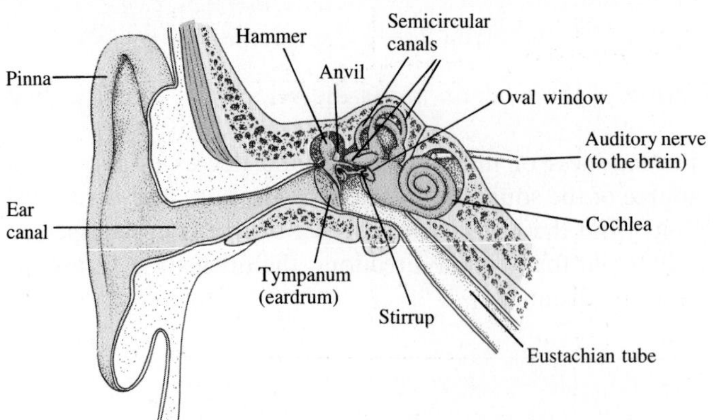

FIGURE 12–2
Diagram of the human ear.

Structure of the ear

In the outer ear, sound waves from the outside travel down the ear canal to the eardrum (the tympanum), which vibrates in response to the impinging waves. The middle ear consists of three small bones known as the hammer, anvil, and stirrup, which transfer the vibrations of the eardrum to the inner ear at the oval window. This delicate system of levers offers a mechanical advantage of about 2. The amplification of the pressure in the wave is much greater than this, however, since the area of the eardrum is about 20 times the area of the oval window. Thus the pressure is amplified by a factor of about 40. The inner ear consists of the semicircular canals, which are important for controlling balance, and the liquid-filled cochlea, where the vibrational energy of sound waves is transformed into electrical energy and sent to the brain. Figure 12–3 is a diagrammatic representation of the cochlea. The sound vibration travels from the oval window down the vestibular canal and back up the tympanic canal. Because of the viscosity of the liquid, considerable damping occurs, but any remaining energy is dissipated at the round window at the end of the tympanic canal. Between these two canals is a third, known as the cochlear duct. On the membrane separating the cochlear duct from the tympanic canal (the basilar membrane) is the "organ of Corti," which contains some 30,000 nerve endings. As a pressure wave passes along the tympanic canal, it causes ripples in the basilar membrane and the attached organ of Corti; this is where the energy is transformed into electrical impulses and sent to the brain by way of the auditory nerve. The basilar membrane is under tension, but is less taut and becomes thicker as it goes from the middle ear to the apex of the cochlea. From our earlier considerations, we might expect that the thicker, less taut end would be more sensitive to low frequencies and the tighter, thinner end more sensitive to the higher frequencies. Careful experiments indicate that this is indeed true, and this fact is important for our sensing of pitch.

FIGURE 12–3 Diagram of cochlea.

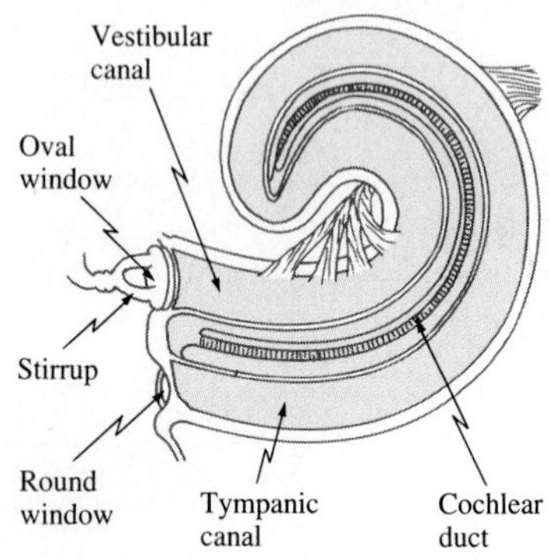

Sensitivity of the ear

The ear is not equally sensitive to all frequencies. The lowest curve in Fig. 12–4 represents the intensity level, as a function of frequency, in the softest sound that is just audible by a very good ear. As can be seen, the ear is most sensitive to sounds of frequency between 2000 and 3000 Hz. Whereas a 1000-Hz sound is audible at a level of 0 dB, a 100-Hz sound must be at least 30 dB to be heard. This lowest (solid) curve represents a very good ear; only about 1 percent of the population, mostly young people, have such a low "threshold of hearing." The middle (dashed) curve represents a more typical curve; 50 percent of the population have a threshold of hearing equal to or better than this. The top curve represents the "threshold of feeling or pain." Sounds above this level can actually be felt and cause pain. As can be seen, it does not vary much with frequency.

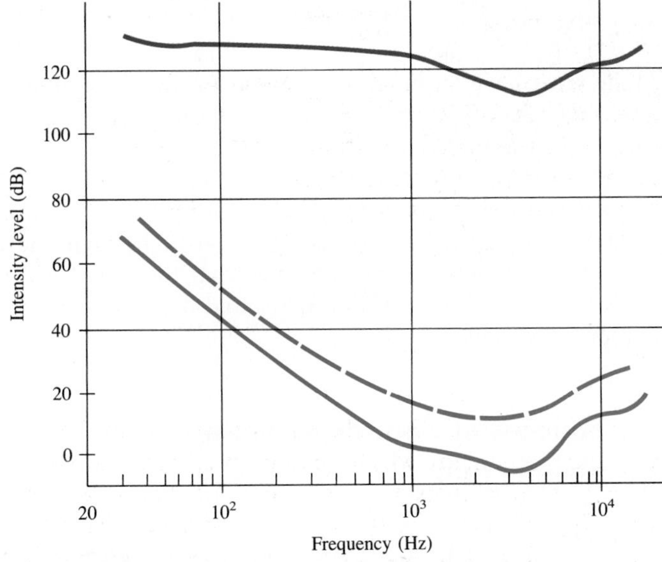

FIGURE 12–4
Sensitivity of the human ear as a function of frequency (see text). Note that the frequency scale is "logarithmic" in order to cover a wide range of frequencies.

Our subjective sensation of loudness obviously depends not only on the intensity, but also on frequency. For example, as seen in Fig. 12–4, an average person will detect a 30-dB sound at 1000 Hz as reasonably loud, but a 30-dB sound at 50 Hz would not be heard at all.

Loudness (in "phons")

To obtain the same loudness for sounds of different frequencies requires different intensities. Studies averaged over large numbers of people have produced the curves shown in Fig. 12–5. On this graph, each curve represents sounds that seemed to be equally loud. The number labeling each curve represents the *loudness level* (the units are called *phons*), which is numerically equal to the intensity level at 1000 Hz for that curve. For example, the curve labeled 40 represents sounds that are heard to have the same loudness as a 1000-Hz sound with an intensity level of 40 dB. From this curve we see that

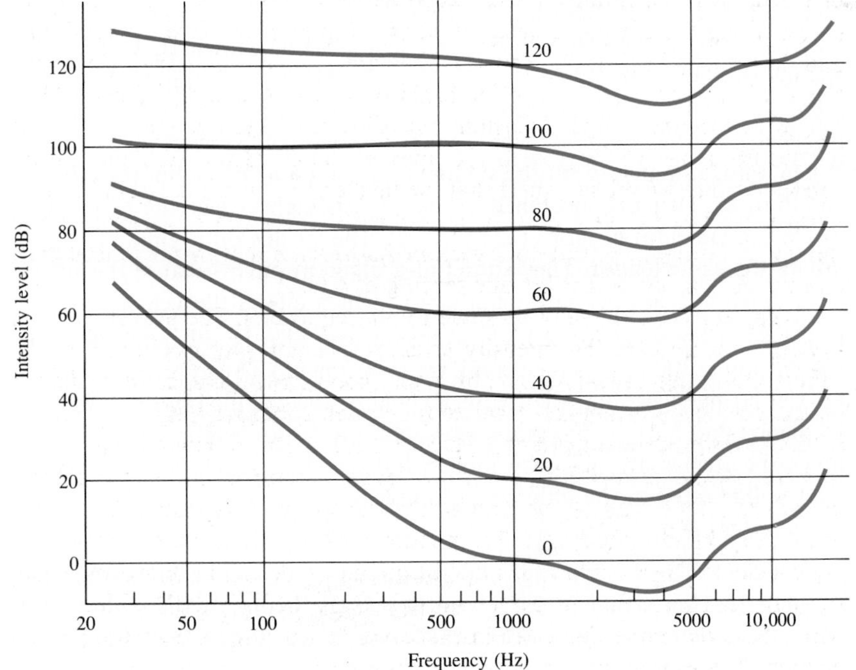

FIGURE 12–5
Loudness-level curves (see text).

TABLE 12–3
Equally Tempered Chromatic Scale†

Note	Frequency (Hz)
C	262
$C^{\#}$ or $D^{\flat}$	277
D	294
$D^{\#}$ or $E^{\flat}$	311
E	330
F	349
$F^{\#}$ or $G^{\flat}$	370
G	392
$G^{\#}$ or $A^{\flat}$	415
A	440
$A^{\#}$ or $A^{\flat}$	466
B	494
C′	524

† Only one octave is included.

a 100-Hz tone must have an intensity of about 62 dB to sound as loud (to an average person) as a 1000-Hz tone of only 40 dB.

It is quite clear that at lower intensity levels, our ears are less sensitive to the high and low frequencies relative to middle frequencies. The "loudness" control on stereo hi-fi systems is intended to compensate for this. As the volume is turned down, the loudness control boosts the high and low frequencies relative to the middle frequencies so that the sound will be more uniform. Many listeners, however, find the sound more pleasing or natural without the loudness control.

12–5 • Sources of Sound: Vibrating Strings and Air Columns

The source of any sound is a vibrating object. Almost any object can vibrate and hence be a source of sound. We now discuss some simple sources of sound, particularly musical instruments. In musical instruments, the source is set into vibration by striking, plucking, bowing, or blowing. Standing waves are produced and the source vibrates at its natural resonant frequencies. A drum has a stretched membrane that vibrates. Xylophones and marimbas have metal or wood bars that can be set into vibration. Bells, cymbals, and gongs also make use of a vibrating metal. The most widely used instruments make use of vibrating strings, such as the violin, guitar, and piano, or make use of vibrating columns of air, such as the flute, trumpet, and pipe organ. (We have already seen that the pitch of a pure sound is determined by the frequency. Typical frequencies for musical notes on the so-called "equally tempered chromatic scale" are given, simply for reference, in Table 12–3 for the octave beginning with middle C.)

Stringed instruments

We saw in Chapter 11, Fig. 11–35, how standing waves are established on a string. This is the basis for all stringed instruments. The pitch is normally determined by the lowest resonant frequency, the **fundamental**, which corresponds to nodes occurring only at the ends. The wavelength of the fundamental is equal to twice the length of the string. Therefore, the fundamental frequency is $f = v/\lambda = v/2L$, where v is the velocity of the wave on the string. When a finger is placed on the string of, say, a guitar or violin, the effective length of the string is shortened; so its pitch is higher since the wavelength of the fundamental is shorter (Fig. 12–6). The strings on a guitar or violin are all the same length. They sound at a different pitch because the strings have different mass per unit length, m/L, which affects the velocity as seen in Eq. 11–10, $v = \sqrt{F_T/(m/L)}$. (The tension may also be different; adjusting the tension is the means for tuning the instrument.) Thus the velocity on a heavier string is less and the frequency will be less for the same wavelength. In pianos and harps, the strings are each of different length. For the lower notes the strings are not only longer, but heavier as well, and the reason why is illustrated in the following example.

FIGURE 12–6 The wavelength of a fingered string (b) is shorter than that of an unfingered string (a). Hence, the frequency of the fingered string is higher. Only the simplest standing wave, the fundamental, is shown.

EXAMPLE 12–4 The highest key on a piano corresponds to a frequency about 150 times that of the lowest key. If the string for the highest note is 5.0 cm long, how long would the string for the lowest note have to be if it had the same mass per unit length and was under the same tension?

SOLUTION The velocity would be the same on each string, so the frequency is inversely proportional to the length L of the string ($f = v/\lambda = v/2L$). Thus

$$\frac{L_l}{L_h} = \frac{f_h}{f_l},$$

where the subscripts l and h refer to the lowest and highest notes, respectively. Thus $L_l = L_h(f_h/f_l) = (5.0\text{ cm})(150) = 750\text{ cm}$, or 7.5 m. This would be ridiculously long for a piano. The longer lower strings are made heavier partly to avoid this, so that even on grand pianos the strings are no longer than about 3 m.

Stringed instruments would not be very loud if they relied on their vibrating strings to produce the sound waves since the strings are simply too thin to compress and expand much air. Stringed instruments therefore make use of a kind of mechanical amplifier known as a *sounding board* (piano) or *sounding box* (guitar, violin) which acts to amplify the sound by putting a greater surface area in contact with the air. When the strings are set into vibration, the sounding board or box is set into vibration as well. Since it has much greater area in contact with the air, it can produce a much stronger sound wave; it thus acts to amplify the sound. On an electric guitar, the sounding box is not so important since the vibrations of the strings are amplified electrically.

Wind instruments

Instruments such as woodwinds, the brasses, and the pipe organ produce sound from the vibrations of standing waves in a column of air within a tube or pipe. Standing waves can occur in the air of any cavity, but the frequencies present are complicated for any but very simple shapes such as a long, narrow tube. Fortunately, and for good reason, this is the situation for most wind instruments. In some instruments, a vibrating reed or the vibrating lip of the player helps to set up vibrations of the air column. In others, a stream of air is directed against one edge of the opening or mouthpiece, leading to turbulence which sets up the vibrations. Because of the disturbance, whatever its source, the air within the tube vibrates with a variety of frequencies; but only certain frequencies persist, which correspond to standing waves.

For a string fixed at both ends, we saw that the standing waves have nodes (no movement) at the two ends, and one or more antinodes (large amplitude of vibration) in between; a node separates successive antinodes. The lowest-frequency standing wave, the *fundamental*, corresponds to a single antinode. The higher-frequency standing waves are called *overtones* or *harmonics*. Specifically, the first harmonic is the fundamental, the second harmonic has twice the frequency of the fundamental,[†] and so on; see Fig. 11–35.

The situation is similar for a column of air, but we must remember that it is now air itself that is vibrating. Thus the air at the closed end of a tube must be a (displacement) node since the air is not free to move there, whereas at the open end of a tube there will be an antinode since the air

[†] When the resonant frequencies above the fundamental (that is, the overtones) are integral multiples of the fundamental, they are called harmonics. But if the overtones are not integral multiples of the fundamental, as is the case for a vibrating drum head, for example, they are not harmonics.

can move freely. The air within the tube vibrates in the form of longitudinal standing waves. The possible modes of vibration for a tube open at both ends (called an **open tube**) and for one that is open at one end but closed at the other (called a **closed tube**) are shown graphically in Fig. 12–7. The graphs represent the displacement amplitude of the vibrating air in the tube; note that the air molecules oscillate horizontally, parallel to the tube length, as shown by the small arrows below the first drawing. The antinodes do not occur precisely at the open ends of tubes. The position of an antinode depends on the diameter of the tube, but if the diameter is small compared to the length, which is the usual case, the antinode occurs very close to the end as shown. We assume this is the case in what follows. (The position of the antinode may also depend slightly on the wavelength and other factors.)

Open and closed tubes

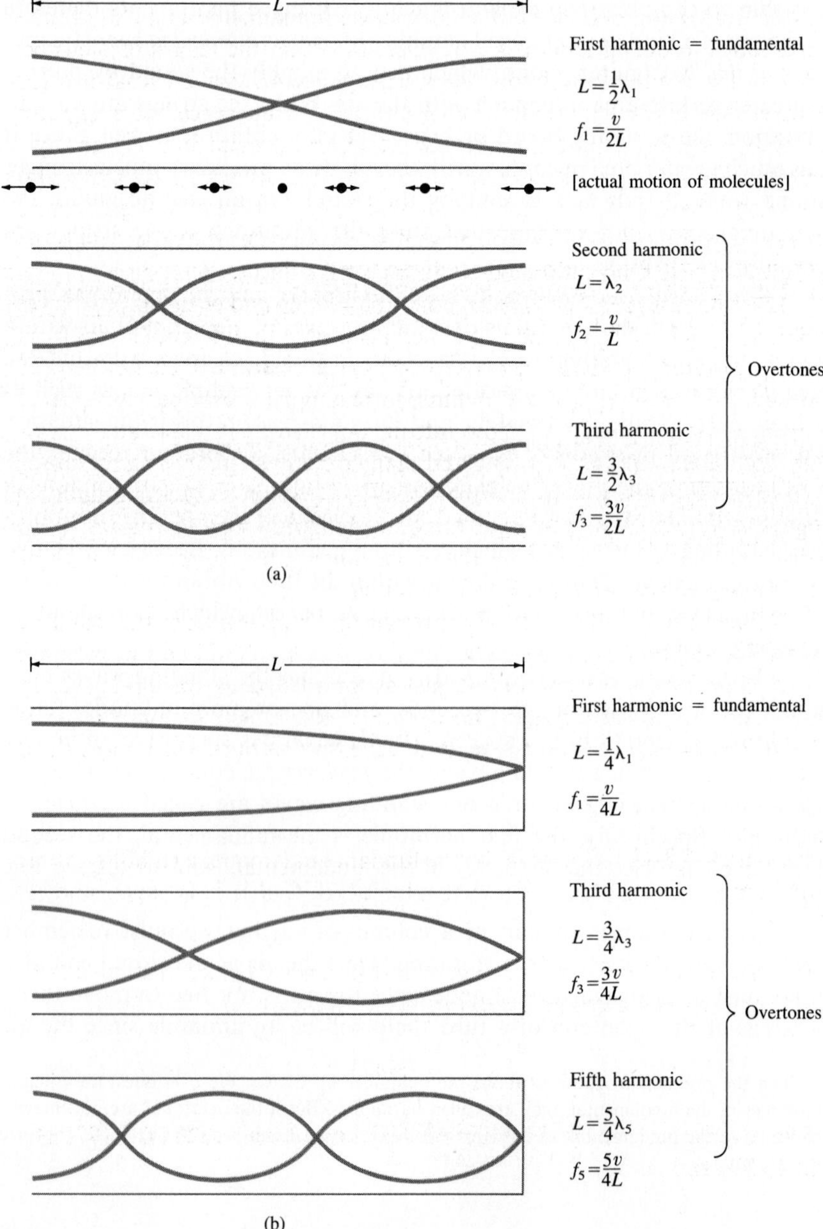

FIGURE 12–7
Modes of vibration (standing waves) for (a) an open tube, and (b) a closed tube. [These graphical representations show the displacement (not pressure) amplitudes.] Although the amplitudes of the graphs are shown vertically, the air molecules themselves oscillate horizontally along the tube. This is shown by the arrows below the first drawing.

Let us look first at the open pipe, Fig. 12–7a. An open pipe has displacement antinodes at both ends. Notice that there must be at least one node within an open pipe if there is to be a standing wave at all. A single node corresponds to the fundamental frequency of the tube. Since the distance between two successive nodes, or between two successive antinodes, is $\frac{1}{2}\lambda$, there is one-half a wavelength within the length of the tube in this case: $L = \frac{1}{2}\lambda$, so the fundamental frequency is $f_1 = v/\lambda = v/2L$, where v is the velocity of sound in air. The standing wave with two nodes is the first overtone or second harmonic and has half the wavelength ($L = \lambda$) and twice the frequency. Indeed, the frequency of each overtone is an integral multiple of the fundamental frequency. This is just what is found for a string.

For a closed tube, Fig. 12–7b, there is always a displacement node at the closed end (because the air is not free to move) and an antinode at the open end (where the air can move freely). Since the distance between a node and the nearest antinode is $\frac{1}{4}\lambda$, we see that the fundamental in a closed tube corresponds to only one-fourth a wavelength within the length of the tube: $L = \lambda/4$. The fundamental frequency is thus $f_1 = v/4L$, or half what it is for an open pipe of the same length. There is another difference, for as we can see from Fig. 12–7b, only the odd harmonics are present in a closed pipe: the overtones have frequencies equal to 3, 5, 7, ... times the fundamental frequency. There is no way for waves with 2, 4, ... times the fundamental frequency to have a node at one end and an antinode at the other, and thus they cannot exist as standing waves in a closed tube.

FIGURE 12–8 A modern pipe organ.

Pipe organs (Fig. 12–8) make use of both open and closed pipes. Notes of different pitch are sounded using different pipes with different lengths from a few centimeters to 5 m or more. Other musical instruments act either like a closed tube or an open tube. A flute, for example, is an open tube, for it is open not only where you blow into it, but also at the opposite end as well. The different notes on a flute and many other instruments are obtained by shortening the length of the tube—that is, by uncovering holes along its length. In a trumpet, on the other hand, pushing down on the valves opens additional lengths of tube. In all these instruments, the longer the length of the vibrating air column, the lower the pitch.

[The diagrams of Fig. 12–7 represent the *displacement* of the standing waves. The amplitude of the *pressure*, on the other hand, will be $\frac{1}{4}$ wavelength out of phase with the displacement, just as for a traveling wave (Fig. 12–1). Thus there is a pressure node at an open end of a tube (which makes sense since it is open to the atmosphere) and a pressure antinode at a closed end of a tube.]

EXAMPLE 12–5 What will be the fundamental frequency and first three overtones for a 26-cm-long organ pipe at 20°C if it is (*a*) open and (*b*) closed?

SOLUTION At 20°C, the speed of sound in air is 343 m/s (Section 12–1). (*a*) For the open pipe, the fundamental frequency is

$$f_1 = \frac{v}{2L} = \frac{343\text{ m/s}}{2(0.26\text{ m})} = 660\text{ Hz}.$$

The overtones, which include all harmonics, are 1320 Hz, 1980 Hz, 2640 Hz, and so on.

(*b*) Referring to Fig. 12–7, we see that

$$f_1 = \frac{v}{4L} = \frac{343 \text{ m/s}}{4(0.26 \text{ m})} = 330 \text{ Hz}.$$

But only the odd harmonics will be present, so the first three overtones will be 990 Hz, 1650 Hz, and 2310 Hz.

EXAMPLE 12–6 A flute is designed to play middle C (264 Hz) as the fundamental frequency when all the holes are covered. Approximately how long should the distance be from the mouthpiece to the far end of the flute? (Note: this is only approximate since the antinode does not occur precisely at the mouthpiece.) Assume the temperature is 20°C.

SOLUTION The speed of sound in air at 20°C is 343 m/s. Then, from Fig. 12–7, the fundamental frequency f_1 is related to the length of the vibrating air column by $f = v/2L$. Solving for L, we find

$$L = \frac{v}{2f} = \frac{343 \text{ m/s}}{2(264 \text{ s}^{-1})} = 0.650 \text{ m}.$$

EXAMPLE 12–7 If the temperature is only 10°C, what will be the frequency of the note played when all the openings are covered in the flute of Example 12–6?

SOLUTION The length L is still 65.0 cm.† But now the velocity of sound is less since it changes by 0.60 m/s per each C°. For a drop of 10 C°, the velocity decreases by 6 m/s to 337 m/s. The frequency will be

$$f = \frac{v}{2L} = \frac{337 \text{ m/s}}{2(0.650 \text{ m})} = 259 \text{ Hz}.$$

This example illustrates why players of wind instruments take time to "warm up" their instruments so they will be in tune. The effect of temperature on stringed instruments is much smaller.

*12–6 • Quality of Sound, and Noise

Whenever we hear a sound, particularly a musical sound, we are aware of its loudness, its pitch, and also of a third aspect called "quality." For example, when a piano and then a flute play a note of the same loudness and the same pitch (say middle C), there is a clear difference in the overall sound. We would never mistake a piano for a flute. This is what is meant by the **quality**‡ of a sound. For musical instruments, the terms *timbre* or *tone color* are also used.

† The length of the flute will change slightly (about 0.01 cm in the present case) due to the temperature change (see Section 13–3), but this is a much smaller effect.

‡ Note that quality in this sense does not refer to the goodness or badness of a sound or to the craftmanship that went into building the instrument.

Sound quality

Just as loudness and pitch can be related to physically measurable quantities, so too can quality. The quality of a sound depends on the presence of overtones—their number and their relative amplitudes. Generally, when a note is played on a musical instrument, the fundamental as well as overtones are present simultaneously. Figure 12–9 illustrates how the superposition of three wave forms, in this case the fundamental and first two overtones (with particular amplitudes), would combine to give a composite *waveform.* Of course, more than two overtones are usually present.

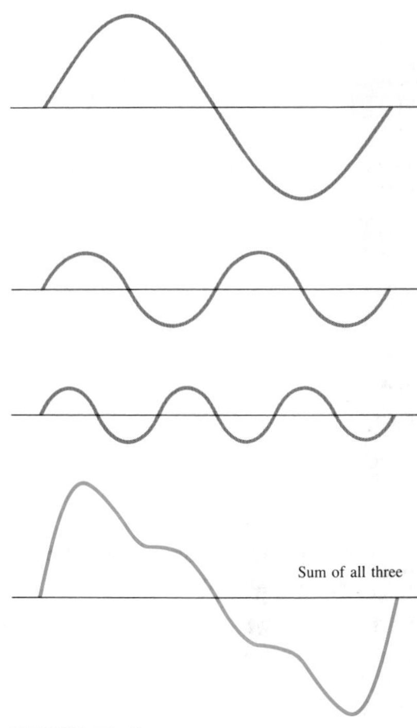

FIGURE 12–9 The amplitudes of the fundamental and first two overtones are added at each point to get the "sum," or composite waveform.

The relative amplitudes of the various overtones are different for different musical instruments, and this is what gives each instrument its characteristic quality or timbre. A graph showing the relative amplitudes of the harmonics produced by an instrument is called a "sound spectrum." Several typical examples for different instruments are shown in Fig. 12–10. Normally, the fundamental has the greatest amplitude and its frequency is what is heard as the pitch.

The manner in which an instrument is played strongly influences the sound quality. Plucking a violin string, for example, makes a very different sound than pulling a bow across it. The sound spectrum at the very start (or end) of a note (as when a hammer strikes a piano string) can be very different from the subsequent sustained tone. This too affects the subjective tone quality of an instrument.

An ordinary sound, like that made by striking two stones together, is a noise that has a certain quality, but a clear pitch is not discernible. A noise such as this is a mixture of many frequencies which bear little relation to one another. If a sound spectrum were made of this noise, it would not show discrete lines like those of Fig. 12–10. Instead it would show a continuous, or nearly continuous, spectrum of frequencies. Such a sound we call "noise" in comparison with the more harmonious sounds which contain frequencies that are simple multiples of the fundamental.

Noise affects us in various ways, particularly psychologically. Sometimes it is a mere annoyance, but loud noise can cause loss of hearing, and this is particularly a problem in factories and other industrial works, where the sound level may be high for long periods. Pop musicians, too, can suffer hearing loss, for levels as high as 120 dB are commonly produced. Hearing loss due to excessive noise levels was recognized by the ancient Romans. Regardless of source, hearing loss due to noise is particularly serious in the frequency range from about 2000 to 5000 Hz, an important region for speech and music.

FIGURE 12–10 Sound spectra for several instruments.

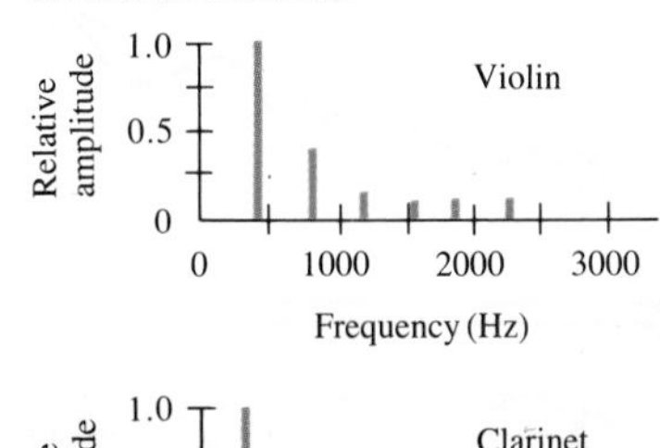

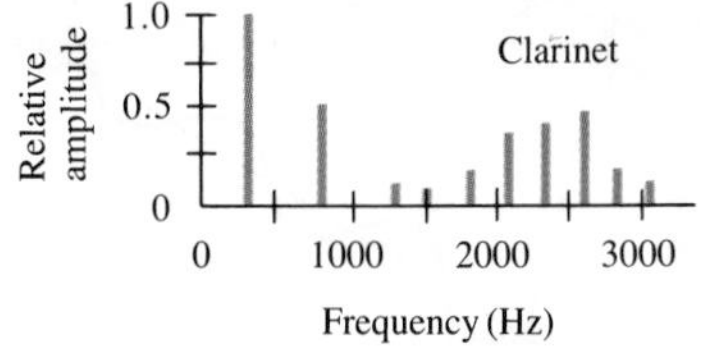

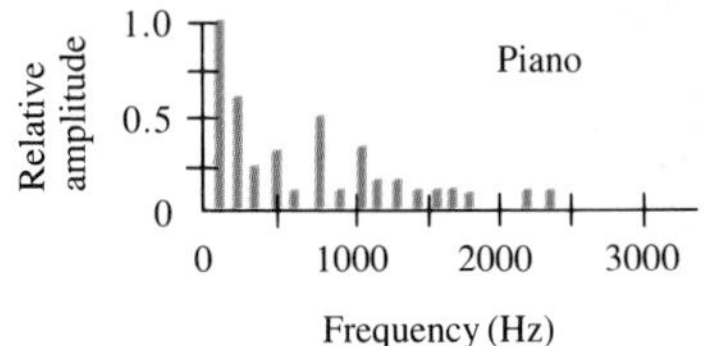

The problem of noise control is a difficult one. Isolation of a source of sound with barriers is helpful but expensive and not always convenient. Attacking the source of the sound is often beneficial. Reducing the area of vibration of machinery is important, for the greater the area, the more air that can be "pushed" and the louder the sound. Constructing surfaces out of stiffer materials, thus reducing the amplitude, or coating surfaces with energy-absorbing material, can be very effective. The placement of machinery is important, for the floor, a wall, or other object may resonate with the machine's vibration, significantly increasing the amplitude. Careful maintenance is important, for lack of lubrication, loose bolts, and worn parts can cause vibrations. The noise produced by jet planes is also a serious problem and can affect a whole community. Much work remains to be done in this area.

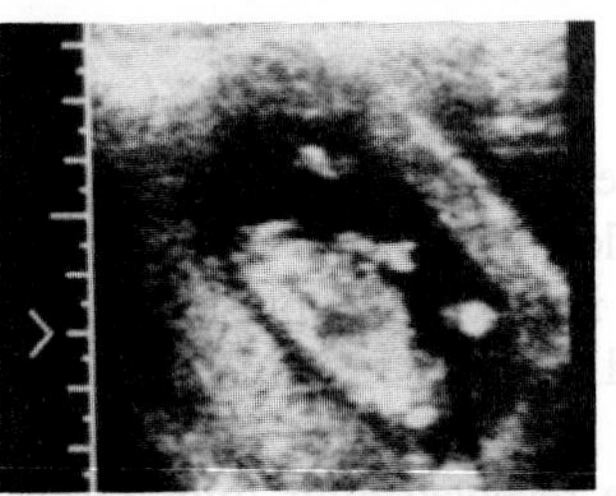

FIGURE 12–23 Ultrasound image of a 14-week old human fetus.

pulse and receives echoes as shown in Fig. 12–22. Each B-scan trace can be plotted, spaced appropriately one below the other, to form an image as shown in Fig. 12–22b. The display can be done on the screen of a cathode-ray tube. Only 10 lines are shown in Fig. 12–22, so the image is crude More lines would give a more precise image.† A photograph of an ultrasound image is shown in Fig. 12–23.

A faster scan can be obtained by using an array of transducers.‡ Since the echo reflections do not simply travel back along the same line (as suggested in Fig. 12–20a), but actually spread out in other directions, all the transducers in an array could not be pulsed at once. The time between the pulse emitted by one transducer and the pulse emitted by the adjacent one must be greater than the time for the farthest echo to return. As we saw earlier, the echo from a surface 25 cm from a transducer takes about 320 μs. Typical spacing between pulses is on the order of 1000 μs or 1 ms, so transducers can be pulsed sequentially at a rate of 1000 per second. This is done electronically and is much faster than moving a single transducer. This speed allows a series of pictures to be taken per second, providing real-time "moving pictures" of moving objects such as the heart. There is a compromise between the number of lines per picture and the number of frames per second. For example, at a pulse rate of 1000 per second, if each image has 100 lines, there can be 10 frames per second; with images of 50 lines, there can be 20 frames per second.

Ultrasound imaging has been an important advance in medicine. It, along with other types of medical imaging, which we will discuss in later chapters (see Chapters 25 and 31), has in many cases replaced exploratory surgery and other risky, painful, and/or costly procedures. There is no evidence of toxic effects with ultrasound imaging, as there is with X-ray imaging. It is thus considered noninvasive. Nonetheless it is not a technique that replaces all others. Beam spreading limits the sharpness of images. And the way sound reflects from matter is not the same as for light or X-rays, so different kinds of information can be obtained from different imaging techniques. A new mode of ultrasound imaging is now being developed based on *transmitted* waves (instead of reflected) making use of computer-assisted tomography techniques (normally done with X rays—see Chapter 25). Another technique, which we already discussed in Section 12–8, employs the Doppler shift of ultrasound echoes to measure velocities within the human body, such as blood flow and fetal heartbeat.

† *Radar* used for aircraft involves a similar pulse-echo technique except that it uses electromagnetic (EM) waves which, like light, travel with a speed of 3×10^8 m/s. This works for the large distances in the sky, but the high speed precludes the use of EM wave pulse-echo techniques in the body since the picosecond times involved [$\approx (0.5\text{ m})/(3 \times 10^8\text{ m/s}) \approx 3 \times 10^{-10}\text{ s} = 300\text{ ps}$] can not at present be separated electronically. Other techniques using EM waves will be discussed later.

‡ Another fast-scanning technique is to rotate a single transducer about a point so it "looks" in a series of different directions through the body. This produces a "sector scan."

SUMMARY

Sound travels as a longitudinal wave in air and other materials. In air, the speed of sound increases with temperature; at 20°C, it is about 343 m/s.

The *pitch* of a sound is determined by the frequency; the higher the frequency, the higher the pitch. The *audible range* of frequencies is roughly 20 to 20,000 Hz (1 Hz = 1 cycle per second). The *loudness* or *intensity* of a sound is related to the amplitude of the wave. Because the human ear can detect sound intensities from 10^{-12} W/m^2 to over 1 W/m^2, intensity levels are specified on a logarithmic scale. The *intensity level* β, specified in decibels, is defined in terms of intensity I as $\beta = 10 \log (I/I_0)$, where the reference intensity I_0 is usually taken to be 10^{-12} W/m^2. An increase in intensity by a factor of 100, for example, corresponds to a level increase of 20 dB.

Musical instruments are simple sources of sound in which standing waves are produced. The strings of a stringed instrument may vibrate as a whole with nodes only at the ends; the frequency at which this occurs is called the *fundamental.* The string can also vibrate at higher frequencies, called *overtones* or *harmonics,* in which there are one or more additional nodes. The frequency of each harmonic is a whole-number multiple of the fundamental. In wind instruments, standing waves are set up in the column of air within the tube. The vibrating air in an open tube (open at both ends) has antinodes at both ends. The fundamental frequency corresponds to a wavelength equal to twice the tube length. The harmonics have frequencies that are 2, 3, 4, . . . times the fundamental frequency. For a closed tube (closed at one end), the fundamental corresponds to a wavelength four times the length of the tube. Only the odd harmonics are present, equal to 1, 3, 5, 7, . . . times the fundamental frequency.

Sound waves from different sources can interfere with each other. If two sounds are at slightly different frequencies, *beats* can be heard at a frequency equal to the difference in frequency of the two sources.

The *Doppler effect* refers to the change in pitch of a sound due to the motion either of the source or of the listener. If they are approaching each other, the pitch is higher; if they are moving apart, the pitch is lower.

QUESTIONS

1. What is the evidence that sound travels as a wave?
2. What is the evidence that sound is a form of energy?
3. Country folk have a rule of thumb that the time delay between seeing lightning and hearing the thunder is an indication of how far away the lightning hit, and in particular, that each 5 seconds corresponds to 1 mile. Explain. What would be the rule for kilometers?
4. When boating on a lake or river at night, one can often hear clearly the voices or radios of people quite a distance away on the shore. Yet this rarely happens in the daytime. This phenomenon can be explained by considering the refraction of sound due to different layers of air having different densities (because of differences in temperature). Draw a diagram and explain this phenomenon and determine if the layer of air nearest the water surface is denser than the one above at nighttime or in the daytime.
5. When a sound wave passes from air into water, do you expect the frequency or wavelength to change?
6. What evidence can you give that the speed of sound in air does not depend significantly on frequency?
7. The voice of a person who has inhaled helium sounds very high pitched. Why?
8. Explain why the larger area of the eardrum, as compared to that of the oval window, leads to an amplification of the pressure.
9. What is the reason that catgut strings on some musical instruments are wrapped with fine wire?
10. Whistle through your lips and describe how the pitch of the whistle is controlled.
11. Explain how a tube might be used as a filter to reduce the amplitude of sounds in various frequency ranges. (An example is a car muffler.)
12. How will the air temperature in a room affect the pitch of organ pipes?
13. Noise control is an important goal today. One mode of attack is to reduce the area of vibration of noisy machinery, for example by keeping it as small as possible or isolating it (acoustically) from the floor and walls. A second method is to make the surface out of a thicker material. Explain how each of these can reduce the noise level.
14. Why are the frets on a guitar spaced closer together as you move down the fingerboard toward the bridge?
15. Standing waves can be said to be due to "interference in space," whereas beats can be said to be due to "interference in time." Explain.

16. Suppose a source of sound moves at right angles to the line of sight of a listener at rest in still air. Will there be a Doppler effect? Explain.

17. If a wind is blowing, will this alter the frequency of the sound heard by a person at rest with respect to the source? Is the wavelength or velocity changed?

*18. A sonic boom sounds much like an explosion. Explain the similarity between the two.

PROBLEMS

[Unless stated otherwise, assume $T = 20°C$ and $v_{sound} = 343$ m/s.]

SECTION 12–1

1. (I) A hiker determines the length of a lake by listening for the echo of her shout reflected by a cliff at the far end of the lake. She hears the echo 1.20 s after shouting. How long is the lake?

2. (I) Ultrasonic waves with frequencies as high as 250,000 Hz are emitted by dolphins. What would be the wavelength of such a wave (*a*) in water, and (*b*) in air?

3. (II) A person sees a heavy stone strike the concrete pavement. A moment later two sounds are heard from the impact: one travels in the air and the other in the concrete, and they are 1.2 s apart. How far away did the impact occur?

SECTION 12–2

4. (I) What is the intensity level of a sound whose intensity is 1.0×10^{-7} W/m^2?

5. (I) What is the intensity of a sound whose intensity level is 60 dB?

6. (II) Human beings can detect a difference in sound intensity level of 1.0 dB. What is the ratio of the amplitudes of two sounds whose levels differ by this amount?

7. (II) A stereo tape recorder is said to have a signal-to-noise ratio of 62 dB. What is the ratio of intensities of the signal and the background noise?

8. (II) If two firecrackers produce an intensity level of 95 dB at a certain place, what will be the intensity level if only one is exploded?

9. (II) A 95-dB sound wave strikes an eardrum whose area is 5.0×10^{-5} m^2. How much energy is absorbed by the eardrum per second?

10. (II) (*a*) Estimate the power output of sound from a person speaking in normal conversation. Use Table 12–2. Assume the sound spreads roughly uniformly over a hemisphere in front of the mouth. (*b*) How many people would produce a total sound output of 100 W of ordinary conversation?

11. (II) What is the resultant intensity level when an 80-dB sound and an 85-dB sound are heard simultaneously?

*SECTION 12–3

*12. (II) Two sound waves have equal displacement amplitudes, but one has twice the frequency of the other. (*a*) Which has the greater pressure amplitude and by what factor is it greater? (*b*) What is the ratio of their intensities?

*13. (II) If the amplitude of a sound wave is tripled, (*a*) by what factor will the intensity increase? (*b*) By how many dB will the intensity level increase?

*14. (II) What would be the intensity level (in dB) of a sound wave in air that corresponds to a displacement amplitude of vibrating air molecules of 1.8 mm at 280 Hz?

*15. (II) Calculate the maximum displacement of air molecules when a 120-Hz sound wave passes whose intensity is at the threshold of pain (120 dB).

*16. (III) The intensity level 10.0 m from a loudspeaker, placed in the open, is 100 dB. What is the acoustic power output (W) of the speaker?

*17. (III) A jet plane emits 3.0×10^5 J of sound energy per second. (*a*) What is the intensity level 40 m away? Air absorbs sound at a rate of about 7.0 dB/km; calculate what the intensity level will be (*b*) 1.0 km and (*c*) 5.0 km away from this jet plane, taking into account air absorption.

*SECTION 12–4

*18. (I) What is the lowest intensity level that can be heard by an average ear (middle curve of Fig. 12–4)?

*19. (I) What are the lowest and highest frequencies that the average ear (middle curve in Fig. 12–4) can hear when the intensity level is 25 dB?

*20. (I) An 8000-Hz tone must have what intensity level to seem as loud as an 80-Hz tone that has a 60-dB intensity level? (See Fig. 12–5.)

SECTION 12–5

21. (I) The G string on a violin has a fundamental frequency of 196 Hz. The length of the vibrating portion is 32 cm and has a mass of 0.50 g. Under what tension must the string be placed?

22. (I) An unfingered guitar string is 0.70 m long and is tuned to play E above middle C (330 Hz). How far from the end of this string must the finger be placed to play A above middle C (440 Hz)?

23. (I) Determine the length of a closed organ pipe that emits middle C (262 Hz) when the temperature is 15°C.

24. (I) How far from the end of the flute in Example 12–6 should the hole be that must be uncovered to play D above middle C at 294 Hz?

25. (I) An organ pipe is 70 cm long. What are the fundamental and first three audible overtones (*a*) if the pipe is closed at one end, and (*b*) if it is open at both ends?

26. (II) An organ is in tune at 20°C. By what fraction will the frequency be off at 0°C?

*27. (II) Draw a diagram, corresponding to each of those in Fig. 12–7, for the standing *pressure* waves in (*a*) an open tube, and (*b*) a closed tube.

28. (II) (*a*) At $T = 20°C$, how long must an open organ pipe be if it is to have a fundamental frequency of 262 Hz? (*b*) If this pipe were filled with helium, what would its fundamental frequency be?

29. (II) A pipe in air at 20°C is to be designed to produce two successive harmonics at 240 Hz and 280 Hz. How long must the pipe be, and is it open or closed?

30. (II) A uniform narrow tube 1.80 m long is open at both ends. It resonates at two successive harmonics of frequency 380 Hz and 456 Hz. What is the speed of sound in the gas in the tube?

31. (II) How many overtones are present within the audible range for a 100-cm-long organ pipe at 20°C (*a*) if it is open, and (*b*) if it is closed?

*SECTION 12–6

*32. (II) Approximately what are the intensities of the first two overtones of a violin compared to the fundamental? How many decibels softer than the fundamental are the first and second overtones? (See Fig. 12–10.)

SECTION 12–7

33. (I) Two horns emitting sounds of frequency 665 and 671 Hz, respectively, produce beats of what frequency?

34. (I) A piano tuner hears one beat every 2.0 s when trying to adjust two strings, one of which is sounding 440 Hz, so that they sound the same tone. How far off in frequency is the other string?

35. (I) What will be the "beat frequency" if middle C (262 Hz) and C# (277 Hz) are played together? Will this be audible? What if each is played two octaves lower (each frequency reduced by a factor of 4)?

36. (II) Two violin strings are tuned to the same frequency, 294 Hz. The tension in one string is then decreased by 3 percent. What will be the frequency of beats heard when the two strings are played together?

37. (II) Two piano strings are supposed to be vibrating at 132 Hz, but a piano tuner hears one beat every 1.5 s when they are played together. (*a*) If one is vibrating at 132 Hz, what must be the frequency of the other (is there only one answer)? (*b*) By how much (in percent) must the tension be increased or decreased to bring them in tune?

38. (II) How many beats will be heard if two identical flutes each try to play middle C (262 Hz), but one is at 0.0°C and the other at 20.0°C?

39. (II) Two loudspeakers are 2.5 m apart. A person stands 3.0 m from one speaker and 3.5 m from the other. (*a*) What is the lowest frequency at which destructive interference will occur at this point? (*b*) Calculate two other frequencies that also result in destructive interference at this point (give the next two highest).

40. (III) A source emits sound of wavelengths 2.80 m and 3.10 m in air. (*a*) How many beats per second will be heard (assume $T = 20°C$)? (*b*) How far apart in space are the regions of maximum intensity?

41. (III) Show that the two speakers in Fig. 12–11 must be separated by at least a distance d equal to one-half the wavelength λ of sound if there is to be any place where complete destructive interference occurs. The speakers are in phase.

SECTION 12–8

42. (I) The predominant frequency of a certain police car's siren is 1800 Hz when at rest. What is the detected frequency if the car is (*a*) moving toward an observer at 30 m/s, and (*b*) if moving away at the same speed?

43. (I) A factory whistle emits a sound at 900 Hz. What frequency will be heard by an observer in an automobile traveling at 70 km/h (*a*) away from the source, and (*b*) toward the source?

44. (II) Sound waves of frequency 1.0×10^6 Hz are directed at the chest of a fetus and travel with a speed in the body of 1.5×10^3 m/s. What will be the expected shift in frequency if the chest of a normal fetus moves at a maximum speed of 0.10 m/s?

45. (II) Two trains emit whistles of the same frequency, 410 Hz. If one train is at rest and the other is traveling at 100 km/h away from an observer at rest, what will the observer detect as the beat frequency?

46. (II) Two automobiles are equipped with the same single-frequency horn. When one is at rest and the other is moving toward an observer at 12 m/s, a beat frequency of 6.0 Hz is heard. What is the frequency the horns emit? Assume $T = 20°C$.

47. (III) The Doppler effect using ultrasonic waves of frequency 2.25×10^6 Hz is used to monitor the heartbeat of a fetus. A (maximum) beat frequency of 600 Hz is observed. Assuming that the speed of sound in tissue is 1.54×10^3 m/s, calculate the maximum velocity of the surface of the beating heart.

48. (III) In Problem 47, the beat frequency is found to appear and then disappear 180 times per minute, which reflects the fact that the heart is beating and its surface changes speed. What is the heartbeat rate?

*SECTION 12–9

*49. (I) How fast is an object moving on the earth if its speed is specified as Mach 0.65?

*50. (II) A boat traveling 8.2 m/s makes a bow wave at an angle of 18° to its direction of motion. What is the speed of the water waves?

*51. (II) Show that the angle θ a sonic boom makes with the path of a supersonic object is given by Eq. 12–4.

*52. (II) An airplane travels at Mach 2.3 where the speed of sound is 310 m/s. (*a*) What is the angle the shock wave makes with the direction of the airplane's motion? (*b*) If the plane is flying at a height of 6500 m, how long after it is directly overhead will a person on the ground hear the shock wave?

*53. (II) A missile travels in air at $-10°C$ near the surface of the earth with a speed of 1000 m/s. (*a*) What is its Mach number? (*b*) What is the apex angle of the shock wave it produces?

*54. (II) A meteorite traveling 8000 m/s strikes the ocean. Determine the shock wave angle it produces (*a*) in the air just before entering the ocean, and (*b*) in the water just after entering. Assume $T = 5°C$.

GENERAL PROBLEMS

55. A stone is dropped from the top of a cliff. The splash it makes when striking the water below is heard 3.0 s later. How high is the cliff?

56. A single mosquito 10 m from a person makes a sound close to the threshold of human hearing (0 dB). What will be the intensity level of 1000 such mosquitoes?

57. Calculate the resonant frequency of the column of air in the outer ear of a human being, which is about 2.5 cm long. Does this correspond to a region of high sensitivity of the ear? Explain.

58. Each string on a violin is tuned to a frequency $1\frac{1}{2}$ times that of its neighbor. If all the strings are to be placed under the same tension, what must be the mass per unit length of each string relative to that of the lowest string?

59. The A string of a violin is 32 cm long between fixed points with a fundamental frequency of 440 Hz and a linear density of 5.0×10^{-4} kg/m. (*a*) What are the wave speed and tension in the string? (*b*) What is the length of the tube of a simple wind instrument (say, an organ pipe) closed at one end whose fundamental is also 440 Hz if the speed of sound is 343 m/s in air? (*c*) What is the frequency of the first overtone of each instrument?

60. A stereo amplifier is rated at 60 W output at 1000 Hz. The output drops by 2 dB at 20 Hz. What is the power output in watts at 20 Hz?

61. A tuning fork is set into vibration above a vertical open tube filled with water. The water level is allowed to drop slowly. As it does so, the air in the tube above the water level is heard to resonate with the tuning fork when the distance from the tube opening to the water level is 0.125 m and again at 0.395 m. What is the frequency of the tuning fork?

62. In audio and communications systems, the *gain*, β, in decibels is defined as

$$\beta = 10 \log \frac{P_{out}}{P_{in}},$$

where P_{in} is the power input to the system and P_{out} is the power output. A particular stereo amplifier puts out 75 W of power for an input of 1 mW. What is its gain in dB?

63. Two loudspeakers face each other at opposite ends of a long corridor. They are connected to the same source which produces a pure tone of 330 Hz. A person walks from one speaker toward the other at a speed of 1.4 m/s. What "beat" frequency does the person hear?

64. A person hears a pure tone coming from two sources that seems to be in the 500–1000 Hz range. The sound is loudest at points equidistant from the two sources. In order to determine exactly what the frequency is, the person moves about and finds that the sound level is minimal at a point 0.22 m farther from one source than the other. What is the frequency of the sound?

65. The frequency of a train whistle as it approaches you is 460 Hz. After it passes you, its frequency is 380 Hz. How fast was the train moving (assume constant velocity)?

66. An 80-cm-long guitar string of mass 1.60 g is placed near a tube open at one end, and also 80 cm long. How much tension should be in the string so that its fourth harmonic has the same frequency as the third harmonic of of the tube?

67. A source of sound waves (wavelength λ) is a distance l from a detector. Sound reaches the detector directly, and also by reflecting off an obstacle, as shown in Fig. 12–24. The obstacle is equidistant from source and detector. When the obstacle is a distance d to the right of the line of sight between source and detector, as shown, the two waves arrive in phase. How much farther to the right must the obstacle be moved if the two waves are to be out of phase by $\frac{1}{2}$ wavelength, so destructive interference occurs? (Assume $\lambda \ll l, d$.)

68. If the velocity of blood flow in the aorta is normally about 0.32 m/s, what beat frequency would you expect if 4.00-MHz ultrasound waves were directed along the flow and reflected from the red blood cells? Assume that the waves travel with a speed of 1.54×10^3 m/s.

69. A factory whistle emits sound of frequency 650 Hz. On a day when the wind velocity is 12.0 m/s from the north, what frequency will observers hear who are located, at rest, (*a*) due north, (*b*) due south, (*c*) due east, and (*d*) due west, of the whistle? What frequency is heard by a cyclist heading (*e*) north, or (*f*) west, toward the whistle at 15.0 m/s? ($T = 20°C$.)

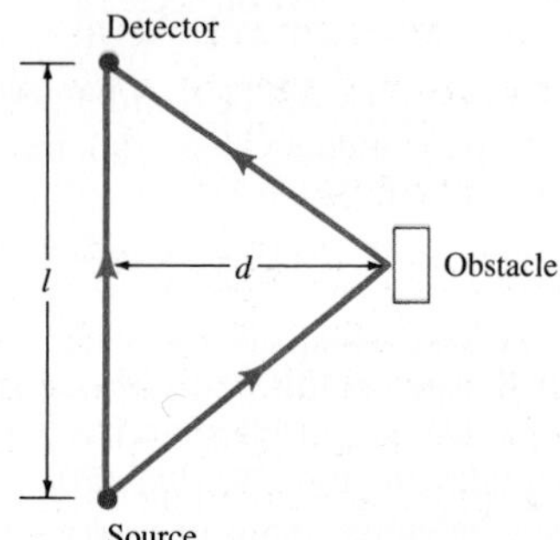

FIGURE 12–24 Problem 67.

C H A P T E R 13

Temperature and Kinetic Theory

Heating the air within a hot-air balloon, and thus raising the air's temperature, forces some air to escape (Charles' law). The air's density (mass of air per unit volume) within the balloon is thus reduced and the balloon can "float" upward.

Much of this chapter will be devoted to an investigation of the theory that matter is made up of atoms and that these atoms are in continuous random motion. This theory is called the *kinetic theory*. ("Kinetic," you may recall, is Greek for "moving.")

In order to investigate this atomic view of matter, we must also discuss the concept of temperature as well as the experimentally measured properties of gases, for these will serve as a foundation for testing the kinetic theory.

13–1 • Atoms

The idea that matter is made up of atoms dates back to the ancient Greeks. According to the Greek philosopher Democritus, if a pure substance—say, a piece of iron—were cut into smaller and smaller bits, eventually a smallest piece of that substance would be obtained which could not be divided further. This smallest piece was called an **atom**, which in Greek means "indivisible."[†]

[†] Today, of course, we don't consider the atom as indivisible, but rather as consisting of a nucleus (containing protons and neutrons) and electrons.

The only real alternative to the atomic theory of matter was the idea that matter is continuous and can be subdivided indefinitely.

Today the atomic theory is generally accepted by scientists. The experimental evidence in its favor, however, came mainly in the eighteenth, nineteenth, and twentieth centuries, and much of it was obtained from the analysis of chemical reactions. A crucial piece of evidence was the *law of definite proportions*, which is a summation of experimental results collected during the half century prior to 1800. It states that when two or more elements[†] combine to form a compound, they always do so in the same proportions by weight. For example, table salt is always formed from 23 parts sodium and 35 parts chlorine; and water is formed from one part hydrogen and eight parts oxygen, by weight. A continuous theory of matter could hardly account for the law of definite proportions, but, as John Dalton (1766–1844) pointed out, the atomic theory could: the weight proportions of each element required to form a compound corresponded to the relative weights of the combining atoms. One atom of sodium (Na), for example, could combine with one atom of chlorine (Cl) to form one molecule of salt (NaCl), and one atom of sodium would have a mass 23/35 times as large as one of chlorine. By measuring the relative amounts of each element needed to form a large variety of compounds, experimenters established the relative weights of atoms. Hydrogen, the lightest atom, was arbitrarily assigned the relative weight of 1. On this scale, carbon was about 12, oxygen 16, sodium 23, and so on.[‡]

Today, we speak of the relative masses of atoms and molecules—what

Atomic mass

we call the **atomic mass** or **molecular mass**, respectively[§]—and these are based on assigning the abundant carbon atom, ^{12}C, the value of exactly 12.0000 atomic mass units(u). The atomic mass of hydrogen is then 1.0078 u, and the values for other atoms are as listed in Appendix D.

Another important piece of evidence for the atomic theory is the so-called **Brownian movement**, named after the biologist Robert Brown, who is credited with its discovery in 1827. While he was observing tiny pollen grains suspended in water under his microscope, Brown noticed that the tiny grains moved about in tortuous paths even though the water appeared to be per-

[†] An *element* is a substance, such as gold, iron, or copper, that cannot be broken down into simpler substances by chemical means. *Compounds* are substances made up of elements, such as carbon dioxide and water. The smallest piece of an element is an atom; the smallest piece of a compound is a molecule. Molecules are made up of atoms; a molecule of water is made up of two atoms of hydrogen and one of oxygen; its chemical formula is H_2O.

[‡] It was not quite so simple, however. For example, from the various compounds oxygen formed, its relative weight was judged to be 16. But this was inconsistent with the weight ratio in water of oxygen to hydrogen (only 8 to 1). This difficulty was explained by assuming two H atoms combine with one O atom to form a water molecule. To make a self-consistent scheme, many other molecules also had to be judged as containing more than one atom of a given type.

[§] The terms *atomic weight* and *molecular weight* are popularly used for these quantities, but properly speaking we are comparing masses.

fectly still. The atomic theory easily explains Brownian movement if the further reasonable assumption is made that the atoms of any substance are continually in motion. Then Brown's tiny pollen grains are jostled about by the vigorous barrage of rapidly moving molecules of water.

In 1905, Albert Einstein examined† Brownian movement from a theoretical point of view and was able to calculate from the experimental data the approximate size and mass of atoms and molecules. His calculations showed that the diameter of a typical atom is about 10^{-10} m.

Macroscopic vs. microscopic properties

At the start of Chapter 10, we distinguished the three common states of matter—solid, liquid, gas—based on **macroscopic**, or "large-scale," properties. Now let us see how these three phases of matter differ from the atomic, or **microscopic**, point of view. Clearly, atoms and molecules must exert attractive forces on each other. For how else could a brick or a piece of aluminum stay together in one piece? The attractive forces between molecules are of an electrical nature (more on this in later chapters). If the molecules come too close together, the force between them becomes repulsive (electric repulsion between their outer electrons). Thus molecules maintain a minimum distance from each other. In a solid material, the attractive forces are strong enough that the atoms or molecules are held in more or less fixed positions, often in an array known as a crystal lattice, as shown in Fig. 13–1a. The atoms or molecules in a solid are in motion—they vibrate about their nearly fixed positions. In a liquid, the atoms or molecules are moving more rapidly, or the forces between them are weaker, so that they are sufficiently free to roll over one another, as in Fig. 13–1b. In a gas, the forces are so weak, or the speeds so high, that the molecules do not even stay close together. They move rapidly every which way, Fig. 13–1c, filling any container and occasionally colliding with one another. On the average, the speeds are sufficiently high in a gas that when two molecules collide, the force of attraction is not strong enough to keep them close together and they fly off in new directions.

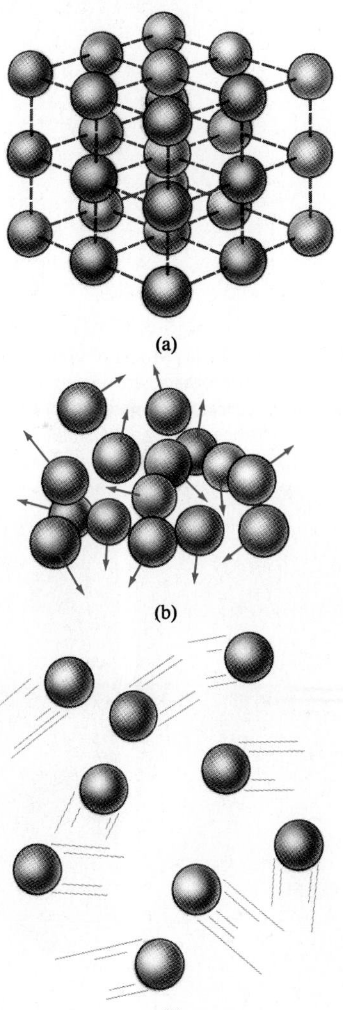

FIGURE 13–1 Atomic arrangements in (a) a crystalline solid, (b) a liquid, and (c) a gas.

13–2 • Temperature

In everyday life, **temperature** is a measure of how hot or cold an object is. A hot oven is said to have a high temperature, whereas a cold tray of ice is said to have a low temperature.

Many properties of matter change with temperature. For example, most‡ materials expand when heated. An iron beam is longer when hot than when cold. Concrete roads and sidewalks expand and contract slightly according to temperature, which is why compressible spacers are placed at

† It is possible that Einstein was unaware of Brown's work, and so independently predicted, from theoretical ideas, the existence of Brownian movement.

‡ Most, but not all, materials expand when their temperature is raised. Water, for example, in the range 0°C to 4°C (see Section 13–3), and certain polymers, contract with an increase in temperature.

FIGURE 13–2
Expansion joint on a bridge.

regular intervals (Fig. 13–2). The electrical resistance of matter changes with temperature (see Chapter 18). So too does the color radiated by objects, at least at high temperatures: you may have noticed that the heating element of an electric stove glows with a red color when hot. At higher temperatures, solids such as iron glow orange or even white. The white light from an ordinary incandescent light bulb comes from an extremely hot tungsten wire.

An instrument designed to measure temperature is called a **thermometer**. There are many kinds of thermometers, but their operation always depends on some property of matter that changes with temperature. Most common thermometers rely on the expansion of a material with an increase in temperature. The first idea for a thermometer (Fig. 13–3a), by Galileo, made use of the expansion of a gas. Common thermometers today consist of a hollow glass tube filled with mercury or with alcohol colored with a red dye, as were the earliest usable thermometers (Fig. 13–3b). Figure 13–3c shows an early clinical thermometer of a different type, also based on a change in density with temperature (see caption). In the common liquid in glass thermometer, the liquid expands more than the glass when the temperature is increased, so the liquid level rises in the tube (Fig. 13–4a). Although metals also expand with temperature, the change in length of, say, a metal rod is generally too small to measure accurately for ordinary changes in temperature. However, a useful thermometer can be made by bonding together two dissimilar metals whose rates of expansion are different (Fig. 13–4b). When the temperature is increased, the different amounts of expansion cause the bimetallic strip to bend. Often the bimetallic strip is in the form of a coil,

FIGURE 13–3 (a) Model of Galileo's original idea for a thermometer. (b) Actual thermometers built by the Accademia del Cimento (1657–67) in Florence, are among the earliest known. These sensitive and exquisite instruments contained alcohol, sometimes colored, like many thermometers today. (c) Clinical thermometers in the shape of a frog, also built by the Accademia del Cimento, could be tied to a patient's wrist. The small spheres suspended in the liquid each have a slightly different density. The number of spheres that would sink was a measure of the patient's fever.

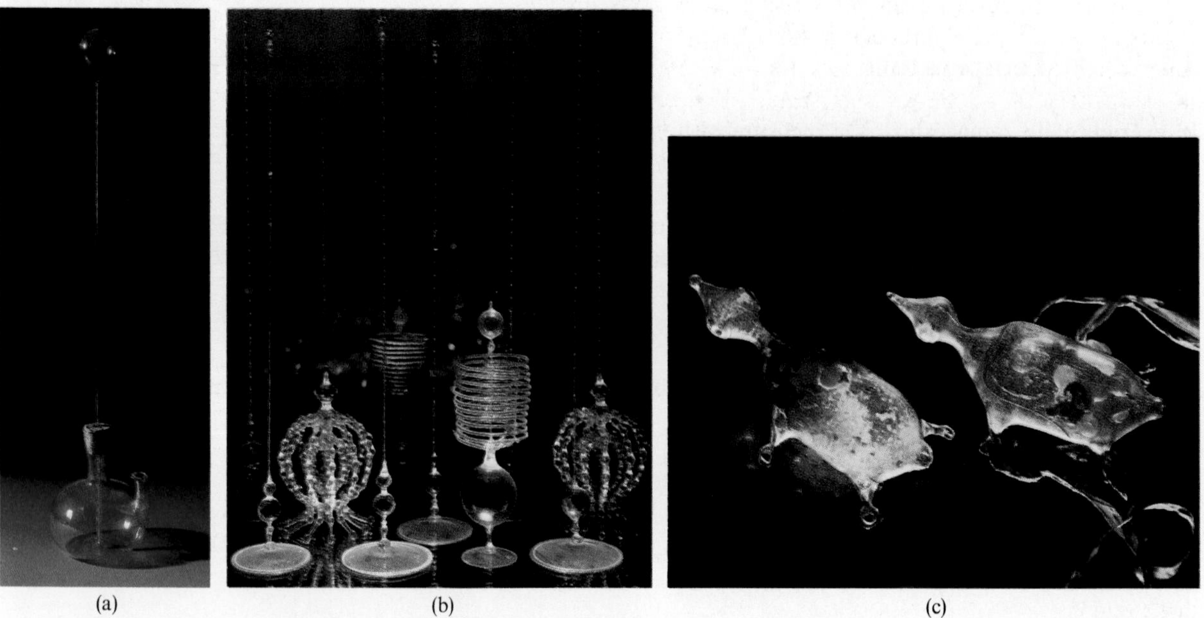

(a) (b) (c)

one end of which is fixed and the other is attached to a pointer (Fig. 13–4c). This kind of thermometer is used as ordinary air thermometers, oven thermometers, in automobiles as the automatic choke, and in thermostats for determining when the heater or air conditioner should go on or off.

In order to measure temperature quantitatively, some sort of numerical scale must be defined. The most common scale today is the **Celsius** scale, sometimes called the **centigrade** scale. In the United States, the **Fahrenheit** scale is also common. The most important scale in scientific work is the absolute, or Kelvin, scale, and it will be discussed later in this chapter.

One way to define a temperature scale is to assign arbitrary values to two readily reproducible temperatures. For both the Celsius and Fahrenheit scales these two fixed points are chosen to be the freezing point and the boiling point† of water, both taken at atmospheric pressure. On the Celsius scale, the freezing point of water is chosen to be 0°C ("zero degrees Celsius") and the boiling point 100°C. On the Fahrenheit scale, the freezing point is defined as 32°F and the boiling point 212°F. A practical thermometer is calibrated by placing it in carefully prepared environments at each of the two temperatures and marking the position of the mercury or pointer. For a Celsius scale, the distance between the two marks is then divided into one hundred equal intervals separated by small marks representing each degree between 0°C and 100°C (hence the name "centigrade scale" meaning "hundred steps"). For a Fahrenheit scale, the two points are labeled 32°F and 212°F and the distance between them is divided into 180 equal intervals. For temperatures below the freezing point of water and above the boiling point of water the scales can be extended using the same equally spaced intervals. However, ordinary thermometers can be used only over a limited temperature range because of their own limitations—for example, the mercury in a mercury-in-glass thermometer solidifies at some point, below which the thermometer will be useless. It is also rendered useless above temperatures where the fluid vaporizes. For very low or very high temperatures, specialized thermometers are required, some of which we will mention later.

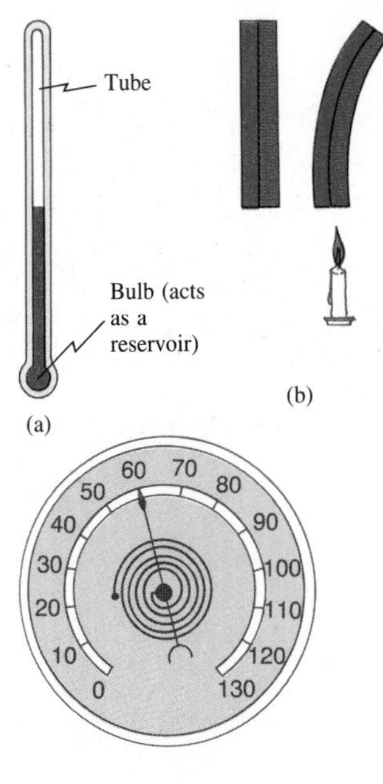

FIGURE 13–4 (a) Mercury or alcohol in glass thermometer; (b) bimetallic strip; (c) bimetallic-strip thermometer.

Every temperature on the Celsius scale corresponds to a particular temperature on the Fahrenheit scale, Fig. 13–5. It is easy to convert from one to the other if you remember that 0°C corresponds to 32°F and that a range of 100° on the Celsius scale corresponds to a range of 180° on the Fahrenheit scale. Thus, one Fahrenheit degree (1 F°) corresponds to $100/180 = \frac{5}{9}$ of a Celsius degree (1 C°). That is, $1\ \text{F}° = \frac{5}{9}\text{C}°$. (Notice that when we refer to a specific temperature, we say "degrees Celsius," as in 20°C; but when we merely refer to a change in temperature or a temperature interval, we say "Celsius degrees," as in "1 C°.") The conversion between the two temperature scales can be written

$$T(°\text{C}) = \tfrac{5}{9}[T(°\text{F}) - 32] \qquad \text{or} \qquad T(°\text{F}) = \tfrac{9}{5}T(°\text{C}) + 32.$$

Rather than memorizing these relations (it would be easy to confuse them), it is simpler to remember that 0°C = 32°F and 5C° = 9F°.

FIGURE 13–5 Celsius and Fahrenheit scales compared.

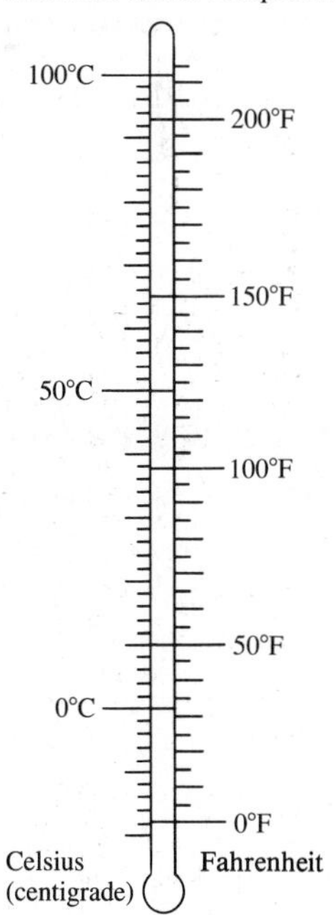

† The freezing point of a substance is defined as that temperature at which the solid and liquid phases coexist in equilibrium—that is, without the liquid changing into the solid or vice versa. Experimentally, this is found to occur at only one definite temperature, for a given pressure. Similarly, the boiling point is defined as that temperature at which the liquid and gas coexist in equilibrium. Since these points vary with pressure, the pressure must be specified (usually it is 1 atm).

EXAMPLE 13–1 Normal body temperature is 98.6°F. What is this on the Celsius scale?

SOLUTION First we note that 98.6°F is $98.6 - 32.0 = 66.6$ F° above the freezing point of water. Since each F° is equal to $\frac{5}{9}$C°, this corresponds to $66.6 \times \frac{5}{9} = 37.0$ Celsius degrees above the freezing point. Since the freezing point is 0°C, the temperature is 37.0°C.

Different materials do not expand in quite the same way over a wide temperature range. Consequently, if we calibrate different kinds of thermometers exactly as described above, they will not usually agree precisely. Because of how we calibrated them, they will agree at 0°C and at 100°C. But because of different expansion properties, they may not agree at intermediate temperatures (remember we arbitrarily divided the thermometer scale into 100 equal divisions between 0°C and 100°C). Thus a carefully calibrated mercury-in-glass thermometer might register a temperature of 52.0°C, whereas a carefully calibrated thermometer of another type might read 52.6°C.

Because of this discrepancy, some standard kind of thermometer must be chosen so that these intermediate temperatures can be precisely defined. The chosen standard for this purpose is the so-called **constant-volume gas thermometer**. As shown in the simplified diagram of Fig. 13–6, this thermometer consists of a bulb filled with a dilute gas connected by a thin tube to a mercury manometer. The volume of the gas is kept constant by raising or lowering the right-hand tube of the manometer so that the mercury in the left tube coincides with the reference mark. An increase in temperature causes a proportional increase in pressure in the bulb. So the tube must be lifted higher to keep the gas volume constant. The height of the mercury in the right-hand column is then a measure of the temperature. This thermometer can be calibrated and gives the same results for all gases in the limit of reducing the gas pressure in the bulb toward zero. The resulting scale is defined as the standard temperature scale.

FIGURE 13-6
Constant-volume gas thermometer.

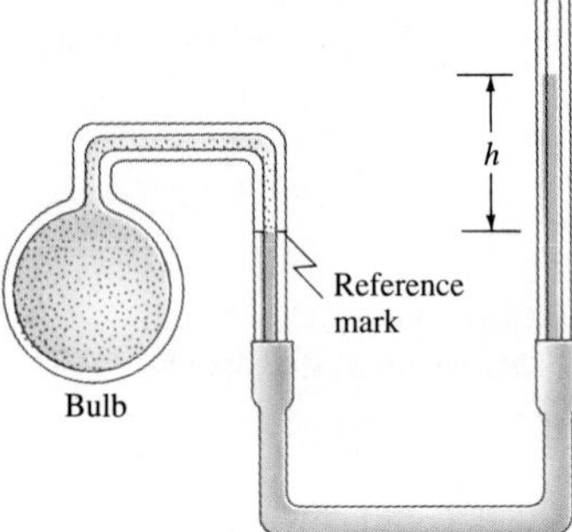

13–3 • Thermal Expansion

Most substances expand when heated and contract when cooled. However, the amount of expansion or contraction varies, depending on the material.

Experiments indicate that the change in length ΔL of almost all solids is, to a very good approximation, directly proportional to the change in temperature ΔT. As might be expected, the change in length is also proportional to the original length of the object, L_0. That is, for the same temperature change, a 4-m-long iron rod will increase twice as much in length as a 2-m-long iron rod. We can write this proportionality as an equation:

$$\Delta L = \alpha L_0 \, \Delta T, \tag{13–1}$$

where α, the proportionality constant, is called the *coefficient of linear expansion* for the particular material and has units of $(\text{C}°)^{-1}$. The values of α for various materials† at 20°C are listed in Table 13–1. It should be noted that

† For certain crystalline materials, α may be different for the three spatial directions. We won't be concerned with this.

Table 13–1
Coefficients of Expansion at 20°C

Material	Coefficient of Linear Expansion, α $(C^\circ)^{-1}$	Coefficient of Volume Expansion, β $(C^\circ)^{-1}$
Solids		
Aluminum	25×10^{-6}	75×10^{-6}
Brass	19×10^{-6}	56×10^{-6}
Iron or steel	12×10^{-6}	35×10^{-6}
Lead	29×10^{-6}	87×10^{-6}
Glass (Pyrex)	3×10^{-6}	9×10^{-6}
Glass (ordinary)	9×10^{-6}	27×10^{-6}
Quartz	0.4×10^{-6}	1×10^{-6}
Concrete and brick	$\approx 12 \times 10^{-6}$	$\approx 36 \times 10^{-6}$
Marble	$1.4–3.5 \times 10^{-6}$	$4–10 \times 10^{-6}$
Liquids		
Gasoline		950×10^{-6}
Mercury		180×10^{-6}
Ethyl alcohol		1100×10^{-6}
Glycerin		500×10^{-6}
Water		210×10^{-6}
Gases		
Air (and most other gases at atmospheric pressure)		3400×10^{-6}

α does vary slightly with temperature (which is why thermometers made of different materials do not agree precisely). However, if the temperature range is not too great, the variation can usually be ignored.

Example 13–2 A steel girder is 200 m long at 20°C. If the extremes of temperature to which it might be exposed are −30°C to +40°C, how much will it contract and expand?

SOLUTION From Table 13–1, we find that $\alpha = 12 \times 10^{-6}(C^\circ)^{-1}$. The increase in length when it is at 40°C will be

$$\begin{aligned}\Delta L &= (12 \times 10^{-6}/C^\circ)(200\text{ m})(40^\circ C - 20^\circ C),\\ &= 4.8 \times 10^{-2}\text{ m},\end{aligned}$$

or 4.8 cm. When the temperature decreases to −30°C, $\Delta T = -50\ C^\circ$. So the change in length is

$$\Delta L = (12 \times 10^{-6}/C^\circ)(200\text{ m})(-50\ C^\circ) = -12.0 \times 10^{-2}\text{ m},$$

or a decrease in length of 12 cm.

Example 13–3 An iron ring is to fit snugly on a cylindrical iron rod. At 20°C, the diameter of the rod is 6.453 cm and the inside diameter of the ring is 6.420 cm. To what temperature must the ring be brought if its hole is to be large enough so it will slip over the rod?

13–5 • The Gas Laws and Absolute Temperature

Equation 13–2 is not very useful for describing the expansion of a gas, partly because the expansion can be so great, and partly because gases generally expand to fill whatever container they are in. Indeed, Eq. 13–2 is meaningful only if the pressure is kept constant. The volume of a gas depends very much on the pressure as well as on the temperature. It is therefore valuable to determine a relation between the volume, the pressure, the temperature, and the mass of a gas. Such a relation is called an **equation of state.**† (By the word *state*, we mean the physical condition of the system.)

If the state of a system is changed, we will always wait until the pressure and temperature have reached the same values throughout. We thus consider only **equilibrium states** of a system—when the variables that describe it (such as temperature and pressure) are the same throughout the system and are not changing in time. We also note that the results of this section are accurate only for gases that are not too dense (the pressure is not too high, on the order of 1 atm or less) and not close to the liquefaction (boiling) point.

For a given quantity of gas it is found experimentally that to a good approximation *the volume of a gas is inversely proportional to the pressure applied to it when the temperature is kept constant.* That is,

Boyle's law

$$V \propto \frac{1}{P}. \qquad [\text{constant } T]$$

For example, if the pressure on a gas is doubled, the volume is reduced to half its original volume. This relation is known as **Boyle's Law**, after Robert Boyle (1627–1691), who first stated it on the basis of his own experiments. Boyle's law can also be written

Boyle's law

$$PV = \text{constant}. \qquad [\text{constant } T]$$

That is, at constant temperature, if either the pressure or volume of the gas is allowed to vary, the other variable also changes so that the product PV remains constant.

Temperature also affects the volume of a gas, but a quantitative relationship between V and T was not found until more than a century after Boyle's work. The Frenchman Jacques Charles (1746–1823) found that when the pressure is not too high and is kept constant, the volume of a gas increases with temperature at a nearly constant rate, as shown in Fig. 13–8a.

† An equation of state can also be sought for solids and liquids since for them, too, the volume depends on the mass, temperature, and external pressure, although the temperature and pressure have much less effect than on gases. However, the situation with solids and liquids is much more complicated.

FIGURE 13–8
Volume of a gas as a function of (a) Celsius temperature, and (b) Kelvin temperature, when the pressure is kept constant.

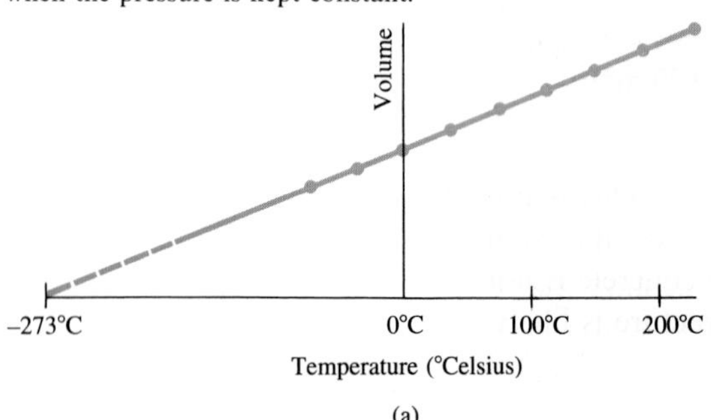

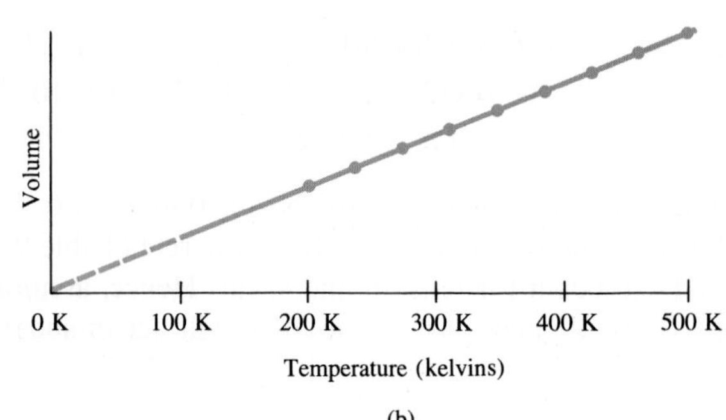

However, all gases liquefy at low temperatures (for example, oxygen liquefies at $-183°C$) and so the graph cannot be extended below the liquefaction point. Nonetheless, the graph is essentially a straight line and if projected to lower temperatures, as shown by the dashed line, it crosses the axis at about $-273°C$.

Such a graph can be drawn for any gas, and the straight line always projects back to $-273°C$ at zero volume. This seems to imply that if a gas could be cooled to $-273°C$ it would have zero volume, and at lower temperatures a negative volume—which makes no sense, of course. It could be argued that perhaps $-273°C$ is the lowest temperature possible, and many other more recent experiments indicate that it is so. This temperature is referred to as the **absolute zero** of temperature. Its value has been measured to be about $-273.15°C$.

Absolute zero

Absolute zero forms the basis of a temperature scale known as the **absolute**, or **Kelvin**, **scale**, and it is used extensively in scientific work. On this scale the temperature is specified as degrees Kelvin or, perferably, simply as kelvins (K) without the degree sign. The intervals are the same as for the Celsius scale, but the zero on this scale (0 K) is chosen as absolute zero itself. Thus the freezing point of water (0°C) is 273.15 K and the boiling point is 373.15 K. Indeed, any temperature on the Celsius scale can be changed to kelvins by adding 273.15 to it:

Kelvin scale

$$T(\text{K}) = T(°\text{C}) + 273.15.$$

Now let us look at Fig. 13–8b, where we see that the graph of the volume of a gas versus absolute temperature is essentially a straight line that passes through the origin. Thus, to a good approximation, *the volume of a given amount of gas is directly proportional to the absolute temperature when the pressure is kept constant.* This is known as **Charles's law**, and can be written

$$V \propto T. \qquad [\text{constant } P]$$

Charles's law

A third gas law, known as **Gay-Lussac's law**, after Joseph Gay-Lussac (1778–1850), states that *at constant volume, the pressure of a gas is directly proportional to the absolute temperature*:

$$P \propto T. \qquad [\text{constant } V]$$

Gay-Lussac's law

A familiar example is that a closed jar or aerosol can thrown into a fire will explode due to the increase in gas pressure inside.

The laws of Boyle, Charles, and Gay-Lussac are not really laws in the sense that we use this term today (precise, deep, wide-ranging validity). They are really only approximations that are accurate for real gases only as long as the pressure and density of the gas are not too high, and the gas is not too close to condensation. The term law applied to these three relationships has become traditional, however, so we have stuck with that usage.

13–6 • The Ideal Gas Law

The gas laws of Boyle, Charles, and Gay-Lussac were obtained by means of a technique that is very useful in science: namely, to hold one or more variables constant in order to see clearly the effects of changing only one of the

Air is saturated with water vapor when the partial pressure of water in the air is equal to the saturated vapor pressure at that temperature. If the partial pressure of water exceeds the saturated vapor pressure, the air is said to be **supersaturated**. This situation can occur when a temperature decrease occurs. For example, suppose the temperature is 30°C and the partial pressure of water is 21 torr, which represents a humidity of 66 percent as we saw above. Suppose now that the temperature falls to, say, 20°C, as might happen at nightfall. From Table 13–4 we see that the saturated vapor pressure of water at 20°C is 17.5 torr. Hence the relative humidity would be greater than 100 percent and the supersaturated air cannot hold this much water. The excess water condenses and appears as dew; this process is also responsible for the formation of fog, clouds, and rain.

When air containing a given amount of water is cooled, a temperature is reached where the partial pressure of water equals the saturated vapor pressure. This is called the **dew point**. Measurement of the dew point is the most accurate means of determining the relative humidity. One method uses a polished metal surface in contact with air, which is gradually cooled down. The temperature at which moisture begins to appear on the surface is the dew point, and the partial pressure of water can then be obtained from saturated vapor pressure tables. If, for example, on a given day the temperature is 20°C and the dew point is 5°C, then the partial pressure of water (Table 13–4) in the original air was 6.54 torr, whereas its saturated vapor pressure was 17.5 torr; hence the relative humidity was $6.54/17.5 = 37$ percent.

A more convenient but less accurate method for measuring relative humidity is the so-called wet-bulb–dry-bulb technique which makes use of two thermometers. One thermometer bulb is fitted with a snug cloth jacket that is soaking wet. The apparatus is usually swung in the air: the lower the humidity, the more evaporation takes place from the wet bulb, causing its temperature reading to be less. A comparison of the temperature readings on the wet-bulb thermometer and the dry (ordinary) thermometer can then be checked against special tables that have been compiled to obtain the relative humidity.

*13–12 • Diffusion

If you carefully place a drop of food coloring in a glass of water, you will find that the color spreads throughout the water. The process may take several hours (assuming you don't shake the glass), but eventually the color will become uniform. This mixing occurs because of the random movement of the molecules, and is called **diffusion**. Diffusion occurs in gases too. Common examples include perfume or smoke (or the odor of something burned on the stove) diffusing in air, although convection often plays a greater role in spreading the odor than does diffusion. Diffusion depends on concentration, by which we mean the number of molecules or moles per unit volume. In general, the diffusing substance moves from a region where its concentration is high to one where its concentration is low.

Diffusion can be readily understood on the basis of kinetic theory and the random motion of molecules. Consider a tube of cross-sectional area A containing molecules in a higher concentration on the left than on the right, Fig. 13–18. We assume the molecules are in random motion. Yet there will be a net flow of molecules to the right. To see why this is true, let us consider the small section of tube of length Δx as shown. Molecules from both regions 1 and 2 cross into this central section as a result of their random motion. The more molecules there are in a region, the more will strike a given area or cross a boundary. Since there is a greater concentration of molecules in region 1 than in region 2, more molecules cross into the central section from region 1 than from region 2. There is, then, a net flow of molecules from left to right, from high concentration toward low concentration. The flow stops only when the concentrations become equal.

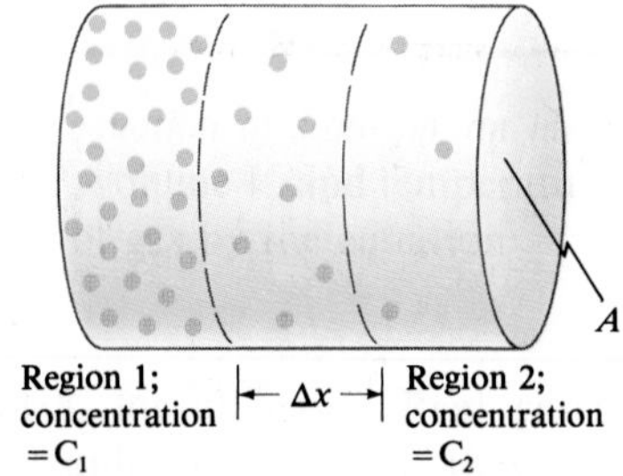

FIGURE 13–18
Diffusion occurs from a region of high concentration to one of lower concentration (only one type of molecule is shown).

You might expect that the greater the difference in concentration, the greater the flow rate. This is indeed the case. In 1855, the physiologist Adolf Fick (1829–1901) determined experimentally that the rate of diffusion (J) is directly proportional to the change in concentration per unit distance $(C_1 - C_2)/\Delta x$ (which is called the **concentration gradient**), and to the cross-sectional area A (see Fig. 13–18):

$$J = DA\frac{C_1 - C_2}{\Delta x}. \qquad (13\text{–}10)$$

Diffusion equation

D is a constant of proportionality called the **diffusion constant**. Equation 13–10 is known as the **diffusion equation**, or **Fick's law**. If the concentrations are given in mol/m^3, then J is the number of moles passing a given point per second; if the concentrations are given in kg/m^3, then J is the mass movement per second (kg/s). The length Δx, of course, is given in meters.

Equation 13–10 applies not only to the simple situation of a gas diffusing as shown in Fig. 13–18, which is called *self-diffusion*, but also to a gas diffusing in a second gas (perfume vapor in air) or to a substance dissolved in a liquid, which are the more common situations. The rate of diffusion will be slower, particularly in liquids, because there will be collisions with the other molecules. Thus, the diffusion constant D will depend on the properties of the substances involved, and also on the temperature and the external pressure. The values of D for a variety of substances are given in Table 13–5.

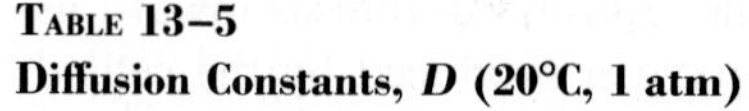

TABLE 13–5
Diffusion Constants, D (20°C, 1 atm)

Diffusing Molecules	Medium	D (m^2/s)
H_2	Air	6.3×10^{-5}
O_2	Air	1.8×10^{-5}
O_2	Water	100×10^{-11}
Blood hemoglobin	Water	6.9×10^{-11}
Glycine (an amino acid)	Water	95×10^{-11}
DNA (molecular mass 6×10^6 u)	Water	0.13×10^{-11}

Example 14–5 How much energy does a refrigerator have to remove from 1.5 kg of water at 20°C to make ice at −12°C?

SOLUTION Heat must flow out to reduce the water from 20°C to 0°C, to change it to ice, and then to lower the ice from 0°C to −12°C:

$$\begin{aligned} Q &= mc_{\text{water}}(20°\text{C} - 0°\text{C}) + ml_{\text{F}} + mc_{\text{ice}}[0° - (-12°\text{C})] \\ &= (1.5\ \text{kg})(4180\ \text{J/kg}\cdot\text{C}°)(20\ \text{C}°) + (1.5\ \text{kg})(3.33 \times 10^5\ \text{J/kg}) \\ &\quad + (1.5\ \text{kg})(2100\ \text{J/kg}\cdot\text{C}°)(12\ \text{C}°) \\ &= 6.6 \times 10^5\ \text{J} = 660\ \text{kJ}. \end{aligned}$$

Example 14–6 A 0.50-kg chunk of ice at −10°C is placed in 3.0 kg of water at 20°C. At what temperature and in what phase will the final mixture be?

SOLUTION In this situation, before we can write down an equation, we must first check to see if the final mixture will be ice, a mixture of ice and water at 0°C, or all water. To bring the 3.0 kg of water at 20°C down to 0°C would require an energy release of $mc\,\Delta T = (3.0\ \text{kg})(4180\ \text{J/kg}\cdot\text{C}°)\cdot(20\ \text{C}°) = 250\ \text{kJ}$. To raise the ice from −10°C to 0°C would require $(0.50\ \text{kg})(2100\ \text{J/kg}\cdot\text{C}°)(10\ \text{C}°) = 10.5\ \text{kJ}$, and to change the ice to water at 0°C would require $(0.50\ \text{kg})(333\ \text{kJ/kg}) = 167\ \text{kJ}$, for a total of 177 kJ. This is not enough energy to bring the 3.0 kg of water at 20°C down to 0°C, so we know that the mixture must end up all water, somewhere between 0°C and 20°C. Now we can determine the final temperature T by writing

$$\begin{pmatrix}\text{heat to raise}\\ \text{0.50 kg of ice}\\ \text{from } -10°\text{C}\\ \text{to } 0°\text{C}\end{pmatrix} + \begin{pmatrix}\text{heat to change}\\ \text{0.50 kg}\\ \text{of ice}\\ \text{to water}\end{pmatrix} + \begin{pmatrix}\text{heat to}\\ \text{raise 0.50 kg}\\ \text{of water}\\ \text{from } 0°\text{C}\\ \text{to } T\end{pmatrix} = \begin{pmatrix}\text{heat lost by}\\ \text{3.0 kg of}\\ \text{water cooling}\\ \text{from } 20°\text{C}\\ \text{to } T\end{pmatrix}$$

Then

$$\begin{aligned} &10.5\ \text{kJ} + 167\ \text{kJ} + (0.50\ \text{kg})(4180\ \text{J/kg}\cdot\text{C}°)(T) \\ &\quad = (3.0\ \text{kg})(4180\ \text{J/kg}\cdot\text{C}°)(20°\text{C} - T) \end{aligned}$$

or

$$\begin{aligned} 14{,}600\ T &= 73{,}800 \\ T &= 5.1°\text{C}. \end{aligned}$$

Example 14–7 The specific heat of mercury is 0.033 kcal/kg·C°. When 1.0 kg of solid mercury at its melting point of −39°C is placed in a 0.50-kg aluminum calorimeter filled with 1.2 kg of water at 20.0°C, the final temperature of the mixture is found to be 16.5°C. What is the heat of fusion of mercury in kcal/kg?

SOLUTION The heat gained by the mercury (Hg) equals the heat lost by the water and calorimeter:

$$m_{Hg}l_{Hg} + m_{Hg}c_{Hg}[16.5°C - (-39°C)]$$
$$= m_w c_w(20°C - 16.5°C) + m_{Al}c_{Al}(20.0°C - 16.5°C)$$

or

$$(1.0 \text{ kg})(l_{Hg}) + (1.0 \text{ kg})(0.033 \text{ kcal/kg}\cdot\text{C°})(55.5 \text{ C°})$$
$$= (1.2 \text{ kg})(1.0 \text{ kcal/kg}\cdot\text{C°})(3.5 \text{ C°})$$
$$+ (0.50 \text{ kg})(0.22 \text{ kcal/kg}\cdot\text{C°})(3.5 \text{ C°}).$$

Thus

$$l_{Hg} = (4.2 + 0.4 - 1.8) \text{ kcal/kg} = 2.8 \text{ kcal/kg}.$$

The latent heat to change a liquid to a gas is needed not only at the boiling point. Water can change from the liquid to the gas phase even at room temperature. This process is called **evaporation** (see also Section 13–11). The value of the heat of vaporization increases slightly with a decrease in temperature: at 20°C, for example, it is 2450 kJ/kg (585 kcal/kg) compared to 2260 kJ/kg (539 kcal/kg) at 100°C. When water evaporates, it cools, since the energy required (the latent heat of vaporization) comes from the water itself; so its internal energy, and therefore its temperature, must drop.[†]

Evaporation of water from the skin is one of the most important methods the body uses to control its temperature. When the temperature of the blood rises slightly above normal, the hypothalamus gland detects this temperature increase and sends a signal to the sweat glands to increase their production. The energy required to vaporize this water comes from the body, and hence the body cools.

We can make use of kinetic theory to see why energy is needed to melt or vaporize a substance. At the melting point, the latent heat of fusion does not increase the kinetic energy (and the temperature) of the molecules in the solid, but instead is used to overcome the potential energy associated with the forces between the molecules. That is, work must be done against these attractive forces to break the molecules loose from their relatively fixed positions in the solid so they can freely roll over one another in the liquid phase. Similarly, energy is required for molecules held close together in the liquid phase to escape into the gaseous phase. This process is a more violent reorganization of the molecules than is melting (the average distance between the molecules is greatly increased) and hence the heat of vaporization is generally much greater than the heat of fusion for a given substance.

14–7 • Heat Transfer: Conduction

Heat is transferred from one place or body to another in three different ways: by *conduction*, *convection*, and *radiation*. We now discuss each of these in turn; but in practical situations, any two or all three may be operating at the same time. We start with conduction.

When a metal poker is put in a hot fire, or a silver spoon is placed in a hot bowl of soup, the exposed end of the poker or spoon soon becomes

[†] According to kinetic theory, evaporation is a cooling process because it is the fastest-moving molecules that escape from the surface (Section 13–11). Hence the average speed of the remaining molecules is less, so by Eq. 13–8 the temperature is less.

2000 years ago the Romans, even in houses in the remote province of Great Britain, made use of hot-water and steam conduits in the floor to heat their houses.

Heating of an object by radiation from the sun cannot be calculated using Eq. 14–5 since this equation assumes a uniform temperature, T_2, of the environment surrounding the object, whereas the sun is essentially a point source. Hence the sun must be treated as an additional source of energy. Heating by the sun is calculated using the fact that about 1350 J of energy strikes the atmosphere of the earth from the sun per second per square meter of area at right angles to the sun's rays. This number, 1350 W/m^2, is called the **solar constant**. The atmosphere may absorb as much as 70 percent of this energy before it reaches the ground, depending on the cloud cover. On a clear day, about 1000 W/m^2 reaches the earth's surface. An object of emissivity e with area A facing the sun absorbs heat at a rate, in watts, of about $1000eA \cos \theta$, where θ is the angle between the sun's rays and a line perpendicular to the area A (Fig. 14–10). That is, $A \cos \theta$ is the "effective" area, at right angles to the sun's rays. The explanations for the seasons, the polar ice caps, and why the sun heats the earth more at midday than at sunrise or sunset, are also related to this $\cos \theta$ factor.

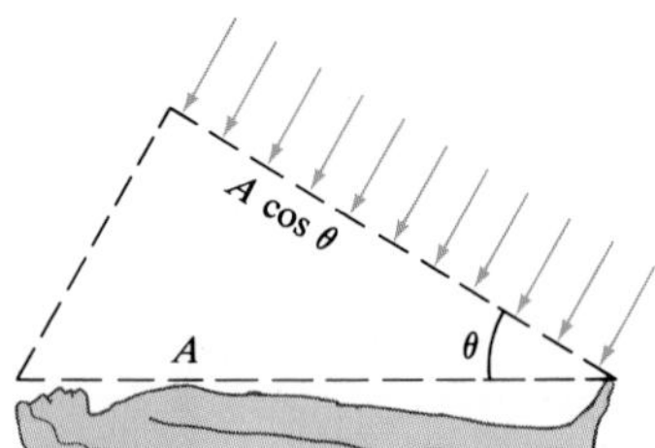

FIGURE 14–10 Radiant energy striking a body at an angle θ.

EXAMPLE 14–10 What is the rate of energy absorption from the sun by a person lying flat on the beach on a clear day if the sun makes a 30° angle with the vertical? Assume that $e = 0.70$, the area of the body exposed to the sun is 0.80 m^2, and that 1000 W/m^2 reaches the earth's surface.

SOLUTION Since $\cos 30° = 0.866$, we have

$$\frac{\Delta Q}{\Delta t} = (1000 \text{ W/m}^2)eA \cos \theta$$

$$= (1000 \text{ W/m}^2)(0.70)(0.80 \text{ m}^2)(0.866)$$

$$= 490 \text{ W}.$$

Notice that if a person wears light-colored clothing, e is much smaller, so the energy absorbed is less.

An interesting application of thermal radiation to diagnostic medicine is **thermography**. A special instrument, the thermograph, scans the body, measuring the intensity of radiation from many points and forming a picture that resembles an X ray (Fig. 14–11). Areas where metabolic activity is high, such as in tumors, can often be detected on a thermogram as a result of their higher temperature and consequent increased radiation.

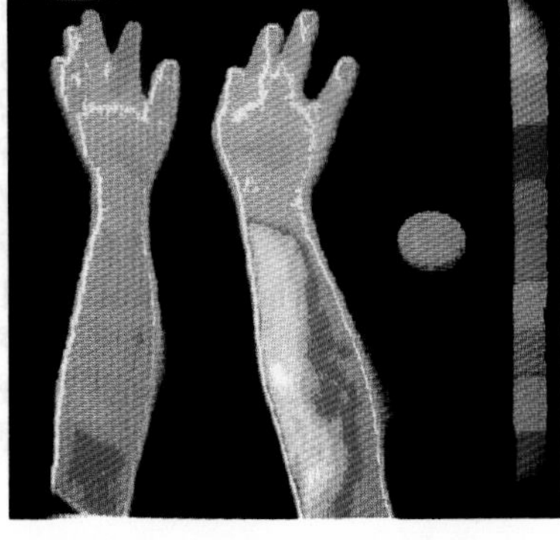

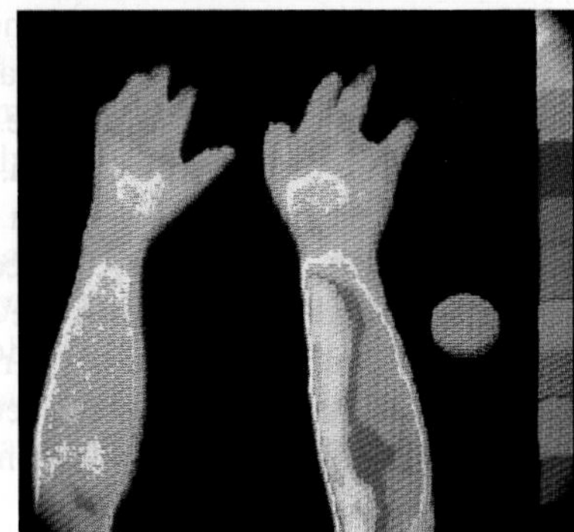

FIGURE 14–11 Thermograms of a person's arms and hands (a) before smoking, (b) after smoking a cigarette, showing temperature decrease due to impaired blood circulation associated with smoking in a healthy individual. The thermograms have been color-coded according to temperature; the scale on the right goes from blue (cold) to white (hot).

SUMMARY

Thermal energy, or *internal energy*, U, refers to the total energy of all the molecules in a body. *Heat* refers to the transfer of energy from one body to another because of a difference of temperature. Heat is thus measured in energy units, such as joules. Heat and thermal energy are also sometimes specified in calories or kilocalories, where 1 cal = 4.18 J is the amount of heat needed to raise the temperature of 1 g of water by 1 C°.

The *specific heat*, c, of a substance is defined as the energy (or heat) required to change the temperature of unit mass of substance by 1 degree; as an equation,

$$Q = mc\,\Delta T,$$

where Q is the heat absorbed or given off, ΔT is the temperature increase or decrease, and m is the mass of the substance. When heat flows within an isolated system, the heat gained by one part of the system is equal to the heat lost by the other part of the system; this is the basis of *calorimetry*, which is the quantitative measurement of heat exchange.

An exchange of energy occurs, without a change in temperature, whenever a substance changes phase; this happens because the potential energy of the molecules changes as a result of the changes in the relative positions of the molecules. The *heat of fusion* is the heat required to melt 1 kg of a solid into the liquid phase; it is also equal to the heat given off when the substance changes from liquid to solid. The *heat of vaporization* is the energy required to change 1 kg of a substance from the liquid to the vapor phase; it is also the energy given off when the substance changes from vapor to liquid.

Heat is transferred from one place (or body) to another in three different ways. In *conduction*, energy is transferred from higher-KE molecules or electrons to lower-KE neighbors when they collide. *Convection* is the transfer of energy by the mass movement of molecules over considerable distances. *Radiation*, which does not require the presence of matter, is energy transfer by electromagnetic waves, such as from the sun. All bodies radiate energy in an amount that is proportional to their surface area and to the fourth power of their Kelvin temperature (T^4). The energy radiated (or absorbed) also depends on the nature of the surface (dark surfaces absorb and radiate more than bright shiny ones), which is characterized by the emissivity, e.

QUESTIONS

1. What happens to the work done when a jar of orange juice is vigorously shaken?
2. Is heat really involved in the Joule experiment of Fig. 14–2? Explain.
3. When a hot object warms a cooler object, does temperature flow between them? Are the temperature changes of the two objects equal?
4. In warm zones where tropical plants grow but the temperature may drop below freezing a few times in the winter, the destruction of sensitive plants due to freezing can be reduced by watering them in the evening. Explain.
5. The specific heat of water is quite large. Explain why this fact makes water particularly good for heating systems (that is, hot-water radiators).
6. Why does water in a canteen stay cooler if the cloth jacket surrounding the canteen is kept moist?
7. Explain why burns caused by steam on the skin are often so severe.
8. Explain, using the concepts of latent heat and internal energy, why water cools (its temperature drops) when it evaporates.
9. Will potatoes cook faster if the water is boiling faster?
10. Does an ordinary electric fan cool the air? Why or why not? If not, why use it?
11. The temperature very high in the earth's atmosphere can be as high as 700°C. Yet an animal there would freeze to death rather than roast. Explain.
12. Why is it important, when hot-air furnaces are used to heat a house, that there be a vent for air to return to the furnace? What happens if this vent is blocked by a bookcase?
13. Why do Bedouins wear several layers of clothing in their desert environment even when the temperature reaches 50°C (122°F) or higher?
14. Down sleeping bags and parkas are often specified as so many inches or centimeters of *loft*, the actual thickness of the garment when it is fluffed up. Explain.
15. Sea breezes are often encountered on sunny days at the shore of a large body of water. Explain in light of the fact that the temperature of the land rises more rapidly than that of the nearby water.
16. The floor of a house on a foundation under which the air can flow is often cooler than a floor that rests directly on the ground (such as a concrete slab foundation). Explain.

17. The earth cools off at night much more quickly when the weather is clear than when cloudy. Why?

18. Why is the liner of a thermos bottle silvered, and why does it have a vacuum between its two walls?

19. Explain why air-temperature readings are always taken with the thermometer in the shade.

20. Why are light-colored clothes more comfortable in hot climates than are dark clothes?

21. A premature baby in an incubator can be dangerously cooled even when the air temperature in the incubator is warm. Explain.

22. In the Northern Hemisphere, the amount of heat required to heat a room where the windows face north is much higher than that required where the windows face south. Explain.

23. Suppose you are designing one of the following (choose one): house, concert hall, medical office building. List as many sources of heat as you can think of. Estimate the heat produced by each.

24. Heat loss occurs through windows by the following processes: (1) ventilation around edges; (2) through the frame, particularly if it is metal; (3) through the glass panes; and (4) radiation. (*a*) For the first three, what is (are) the mechanism(s): conduction, convection, or radiation? (*b*) Heavy curtains reduce which of these heat losses? Explain in detail.

25. A piece of wood lying in the sun absorbs more heat than a piece of shiny metal. Yet the wood feels less hot than the metal when you pick it up. Explain.

PROBLEMS

SECTION 14–1

1. (I) How much heat (joules) is required to raise the temperature of 5.0 kg of water from 0°C to 80°C?

2. (I) How much work must a person do to offset eating a 600-Cal piece of cake?

3. (I) To what temperature will 8500 J of work raise 2.0 kg of water initially at 10.0°C?

4. (II) A British thermal unit (Btu) is a unit of heat in the British system of units. One Btu is defined as the heat needed to raise 1 pound of water by 1 F°. Show that 1 Btu = 0.252 kcal = 1055 J.

5. (II) A water heater can generate 7500 kcal/h. How much water can it heat from 10°C to 50°C per hour?

6. (II) How many kilocalories of heat are generated when the brakes are used to bring a 1000-kg car to rest from a speed of 100 km/h?

SECTIONS 14–4 AND 14–5

7. (I) An automobile cooling system holds 14 L of water. How much heat does it absorb if its temperature rises from 20°C to 80°C?

8. (I) What is the specific heat of a metal substance if 135 kJ of heat is needed to raise 4.5 kg of the metal from 20°C to 42°C?

9. (I) How many kilograms of copper will experience the same temperature rise as 20 kg of water when the same amount of heat is absorbed?

10. (I) What is the specific heat capacity of water in Btu/lb·F°? (See Problem 4.)

11. (I) What is the water equivalent of 2.40 kg of copper?

12. (II) A 25-g glass thermometer reads 18.0°C before it is placed in 110 mL of water. When the water and thermometer come to equilibrium, the thermometer reads 41.6°C. What was the original temperature of the water?

13. (II) The 0.60-kg head of a hammer has a speed of 6.0 m/s just before it strikes a nail and is brought to rest. Estimate the temperature rise of a 12-g iron nail generated by ten such hammer blows done in quick succession. Assume the nail absorbs all the "heat."

14. (II) When a 290-g piece of iron at 180°C is placed in a 100-g aluminum calorimeter cup containing 250 g of glycerin at 10°C, the final temperature is observed to be 38°C. What is the specific heat of glycerin?

15. (II) What will be the equilibrium temperature when a 300-g block of copper at 240°C is placed in a 180-g aluminum calorimeter cup containing 800 g of water at 17.0°C?

16. (II) How long does it take a 600-W coffee pot to bring to a boil 0.70 L of water initially at 10°C? Assume that the part of the pot which is heated with the water is made of 400 g of aluminum, and that no water boils away.

17. (II) When 220 g of a substance is heated to 330°C and then plunged into a 100-g aluminum calorimeter cup containing 150 g of water at 12.5°C, the final temperature, as registered by a 17-g glass thermometer, is 33.8°C. What is the specific heat of the substance?

SECTION 14–6

18. (I) How much heat is needed to melt 16.00 kg of silver that is initially at 25°C?

19. (I) During exercise, a person may give off 180 kcal of heat in 30 min by evaporation of water from the skin. How much water has been lost?

20. (I) If 2.30×10^5 J of energy is supplied to a flask of oxygen at −183°C, how much oxygen will evaporate?

21. (II) An iron boiler of mass 220 kg contains 880 kg of water at 20°C. A heater supplies energy at the rate of 46,000 kJ/h. How long does it take for the water (*a*) to reach the boiling point, and (*b*) to all have changed to steam?

22. (II) What will be the final result when equal amounts of ice at 0°C and steam at 100°C are mixed together?

23. (II) The specific heat of mercury is 138 J/kg·C°. Determine the latent heat of fusion of mercury using the following calorimeter data: 1.00 kg of solid Hg at its melting point of $-39.0°C$ is placed in a 0.620-kg aluminum calorimeter with 0.400 kg of water at 12.80°C; the resulting equilibrium temperature is 5.06°C.

24. (II) A 55.0-kg ice skater moving at 7.5 m/s glides to a stop. Assuming the ice is at 0°C and that 50 percent of the heat generated by friction is absorbed by the ice, how much ice melts?

SECTIONS 14–7 TO 14–9

25. (I) Calculate the rate of heat flow by conduction in Example 14–8 assuming there are strong gusty winds and that the external temperature is $-5°C$.

26. (I) (*a*) How much power is radiated by a tungsten sphere (emissivity $e = 0.35$) of radius 15 cm at a temperature of 20°C? (*b*) If the sphere is enclosed in a room whose walls are kept at $-5°C$, what is the *net* flow rate of energy out of the sphere?

27. (I) Over what distance must there be heat flow by conduction from the blood capillaries beneath the skin to the surface if the temperature difference is 0.50°C? Assume 200 W must be transferred through the whole body's surface area of 1.5 m^2.

28. (I) Approximately how much radiation does a person, of total area 1.5 m^2, absorb per hour from the sun when it makes a 40° angle to the vertical on a clear day and the person lies flat? Assume $e = 0.80$.

29. (II) A 100-W light bulb generates 95 W of heat, which is dissipated through a glass bulb that has a radius of 3.0 cm and is 1.0 mm thick. What is the difference in temperature between the inner and outer surfaces of the glass?

30. (II) A ceramic teapot ($e = 0.70$) and a shiny one ($e = 0.10$) each hold 0.75 L of tea at 95°C. (*a*) Estimate the rate of heat loss from each, and (*b*) estimate the temperature drop (use Eq. 14–2) after 30 min for each. Consider only radiation and assume the surroundings are at 20°C.

31. (II) Write an equation for the total rate of heat flow through the wall of a house if the wall consists of material with thermal conductivity k_1, total area A_1, and thickness l_1, and of windows with thermal conductivity k_2, area A_2, and thickness l_2. The temperature difference is ΔT.

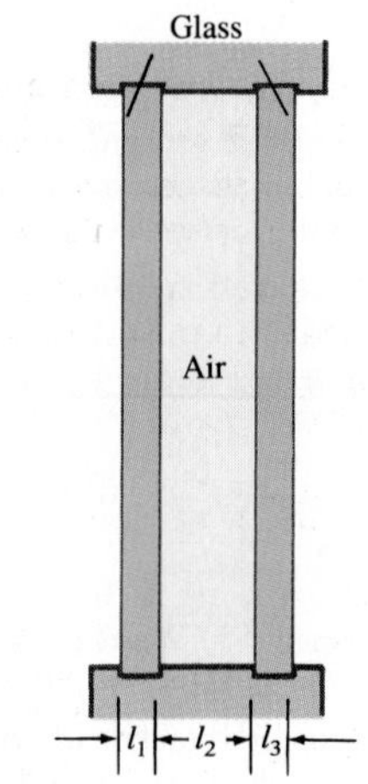

FIGURE 14–12 Problem 32.

32. (III) A double-glazed window is one with two panes of glass separated by an air space, Fig. 14–12. (*a*) Show that the rate of heat flow by conduction is given by

$$\frac{\Delta Q}{\Delta t} = \frac{A(T_2 - T_1)}{l_1/k_1 + l_2/k_2 + l_3/k_3},$$

where k_1, k_2, and k_3 are the thermal conductivities for glass, air, and glass, respectively. (*b*) Generalize this expression for any number of materials placed next to one another.

33. (III) Approximately how long should it take 12 kg of ice at 0°C to melt when it is placed in a carefully sealed Styrofoam "icebox" of dimensions 30 cm × 35 cm × 50 cm whose walls are 1.5 cm thick? Assume that the conductivity of Styrofoam is double that of air and that the outside temperature is 25°C.

34. (III) A house thermostat is normally set to 22°C, but at night it is turned down to 12°C for 8 h. Estimate how much more heat would be needed (state as a percentage of daily usage) if the thermostat were not turned down at night. Assume that the outside temperature averages 0°C for the 8 h at night and 8°C for the remainder of the day, and that the heat loss from the house is proportional to the difference in temperature inside and out. To obtain an estimate from the data, you will have to make other simplifying assumptions; state what these are.

GENERAL PROBLEMS

35. If coal gives off 7000 kcal/kg when it is burned, how much coal will be needed to heat a house that requires 4.2×10^7 kcal for the whole winter? Assume that an additional 30 percent of the heat is lost up the chimney.

36. A 15-g lead bullet traveling 300 m/s is brought to rest when it penetrates a 0.25-kg block of wood. What is the temperature rise of the block and bullet (assumed to have reached thermal equilibrium) as a result of this collision?

37. During light activity, a 70-kg person may generate 200 kcal/h. Assuming that 20 percent of this goes into useful work and the other 80 percent is converted to heat, calculate the temperature rise of the body after 1.00 h if none of this heat were transferred to the environment.

38. A large 200-kg marble rock falls vertically a height of 150 m before striking the ground. Estimate the temperature rise of the rock if 50 percent of the heat generated remains in the rock.

39. The *heat capacity*, C, of an object is defined as the amount of heat needed to raise its temperature by 1 C°. Thus, to raise the temperature by ΔT requires heat Q given by

$$Q = C\,\Delta T.$$

(*a*) Write the heat capacity C in terms of the specific heat, c, of the material. (*b*) What is the heat capacity of 1 kg of water? (*c*) Of 35 kg of water?

40. A very long 1.80-cm-diameter lead rod absorbs 295 kJ of heat. By how much does its length change? What would happen if the rod were only 2.0 cm long?

41. A mountain climber wears down clothing 3.5 cm thick whose total surface area is 1.8 m^2. The temperature at the surface of the clothing is -10°C and at the skin is 34°C. Determine the rate of heat flow by conduction through the clothing (*a*) assuming it is dry and that the thermal conductivity, k, is that of down, and (*b*) assuming the clothing is wet, so that k is that of water and the jacket has matted down to 0.50 cm thickness.

42. How much water would have to evaporate from the skin per minute to take away all the heat generated by the basal metabolism (60 kcal/h) of a 65-kg person?

43. Estimate the rate at which heat can be conducted from the interior of the body to the surface. Assume that the thickness of tissue is 4.0 cm, that the skin is at 34°C and the interior at 37°C, and that the surface area is 1.5 m^2. Compare this to the measured value of about 230 W that must be dissipated by a person working lightly. This clearly shows the necessity of convective cooling by the blood.

44. *Newton's law of cooling* states that for small temperature differences, if a body at a temperature T_1 is in surroundings at a temperature T_2, the body cools at a rate given by

$$\frac{\Delta Q}{\Delta t} = K(T_1 - T_2),$$

where K is a constant. It includes the effects of conduction, convection, and radiation. That this linear relationship should hold is obvious if only conduction is considered. Show that it is also approximately true for radiation by showing that Eq. 14–5 reduces to

$$\begin{aligned}\Delta Q/\Delta t &= 4\sigma e A T_2^3(T_1 - T_2)\\ &= \text{constant} \times (T_1 - T_2)\end{aligned}$$

if $(T_1 - T_2)$ is small.

45. A *thermal transmission coefficient*, U, defined by the equation

$$\frac{\Delta Q}{\Delta t} = AU(T_2 - T_1),$$

is often used in practical work. Note that U is the rate of heat flow per unit area per degree of temperature. (*a*) How is U related to the thermal conductivity coefficient k? (*b*) Is U a characteristic property of a material? If not, what else must be known about the material before U can be determined? (*c*) Write an equation describing the total heat loss by conduction from a room in terms of the values of U for the different materials. Assume the windows are of total area A_1 and have thermal transmission coefficient U_1. The walls have a total area A_2 and consist of a layer of brick (U_2) separated by an air gap (U_3) from a layer of wood (U_4). (*d*) What is the thermal transmission coefficient U for a brick wall 15 cm thick and 7.0 m wide by 2.5 m high?

46. A house has well-insulated walls 20.0 cm thick (assume conductivity of air) and area 380 m^2, a roof of wood 5.5 cm thick and area 250 m^2, and uncovered windows 0.65 cm thick and total area 28 m^2. (*a*) Assuming that the heat loss is only by conduction, calculate the rate at which heat must be supplied to this house to maintain its temperature at 20°C if the outside temperature is -5°C. (*b*) If the house is initially at 10°C, estimate how much heat must be supplied to raise the temperature to 20°C within 30 min. Assume that only the air needs to be heated and that its volume is 700 m^3. (*c*) If natural gas costs \$0.065 per kilogram and its heat of combustion is 5.4×10^7 J/kg, how much is the monthly cost to maintain the house as in part (*a*) for 12 h each day assuming 90 percent of the heat produced is used to heat the house? Take the specific heat of air to be 0.17 kcal/kg·C°.

47. A 15-g lead bullet traveling at 500 m/s passes through a thin iron wall and emerges at a speed of 250 m/s. If the bullet absorbs 50 percent of the heat generated, (*a*) what will be the temperature rise of the bullet? (*b*) If the ambient temperature is 20°C, will any of the bullet melt, and if so, how much?

48. A leaf of area 40 cm^2 and mass 4.5×10^{-4} kg directly faces the sun on a clear day. The leaf has an emissivity of 0.85 and a specific heat of 0.80 kcal/kg·K. (*a*) Estimate the rate of rise of the leaf's temperature. (*b*) Calculate the temperature the leaf would reach if it lost all its heat by radiation (the surroundings are at 20°C). (*c*) In what other ways can the heat be dissipated by the leaf?

49. Use the result of part (a) in the previous problem and take into account radiation from the leaf to calculate how much water must be transpired (evaporated) by the leaf per hour to maintain a temperature of 35°C.

CHAPTER 15

The First and Second Laws of Thermodynamics

This old train locomotive is an example of a steam engine, one type of heat engine whose investigation helped lead to the second law of thermodynamics.

Thermodynamics is the name we give to the study of processes in which energy is transferred as heat and as work.

In Chapter 6 we saw that work is done when energy is transferred from one body to another by mechanical means. In Chapter 14 we saw that heat is a transfer of energy from one body to a second body which is at a lower temperature. Thus, heat is much like work. To distinguish them, *heat* is defined as a *transfer of energy due to a difference in temperature*, whereas work is a transfer of energy that is not due to a temperature difference.

Heat vs. work

In discussing thermodynamics, we shall often refer to particular systems. A **system** is any object or set of objects that we wish to consider. Everything else in the universe we will refer to as its "environment." There are several kinds of systems. A **closed system** is one for which no mass enters or leaves. In an **open system**, mass may enter or leave. Many (idealized) systems we study in physics are closed systems. But many systems, including plants and animals, are open systems since they exchange materials (food, oxygen, waste products) with the environment. A closed system is said to be

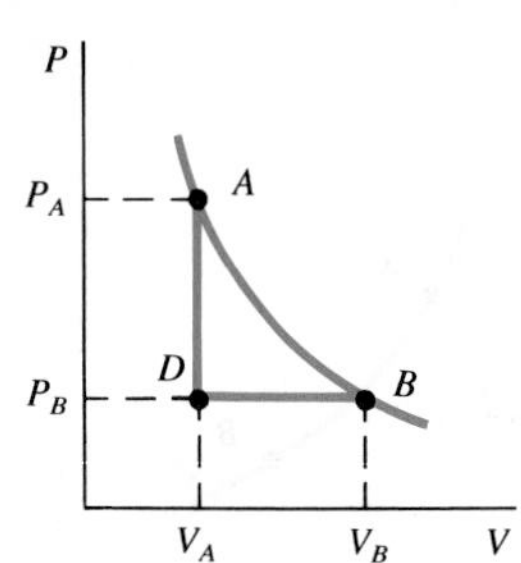

FIGURE 15–4 *PV* diagram for several processes (see the text).

Suppose that a gas, instead of being taken isothermally from point *A* to point *B*, is taken by way of the path *ADB* as shown in Fig. 15–4. In going from *A* to *D*, the gas does no work since the volume does not change. But in going from *D* to *B*, the gas does work equal to $P_B(V_B - V_A)$. This is the total work done in the process.

If the pressure varies during a process, such as for the isothermal process *AB* in Fig. 15–1, Eq. 15–2 cannot be used directly. A rough estimate can be obtained, however, by using an "average" value for *P* in Eq. 15–2. More accurately, the work done is equal to the area under the *PV* curve. This is obvious when the pressure is constant, for as can be seen from Fig. 15–5a, the shaded area is just $P_B(V_B - V_A)$ and this is the work done. Similarly, the work done during an isothermal process is equal to the shaded area shown in Fig. 15–5b. The calculation of work done in this case can be carried out using calculus.

FIGURE 15–5
Work done by a gas is equal to the area under the *PV* curve.

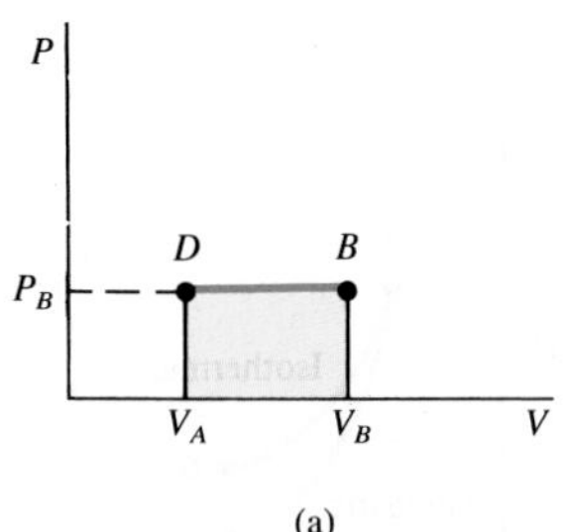

(a)

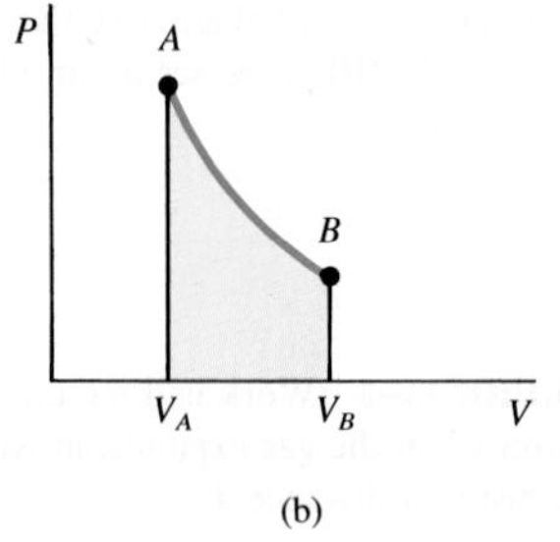

(b)

EXAMPLE 15–1 An ideal gas is slowly compressed at a constant pressure of 2.0 atm from 10.0 L to 2.0 L. (In this process, some heat flows out and the temperature drops.) Heat is then added to the gas, holding the volume constant, and the pressure and temperature are allowed to rise until the temperature reaches its original value. This total process is shown as *BDA* in Fig. 15–4. Calculate (*a*) the total work done by the gas in the process, and (*b*) the total heat flow into the gas.

SOLUTION (*a*) Work is done only in the first part of the compression (*BD*):

$$\begin{aligned} W = P\,\Delta V &= (2.0 \times 10^5\ \mathrm{N/m^2})(2.0 \times 10^{-3}\ \mathrm{m^3} - 10.0 \times 10^{-3}\ \mathrm{m^3}) \\ &= -1.6 \times 10^3\ \mathrm{J}. \end{aligned}$$

(*b*) Since the temperature at the beginning and at the end of the process is the same, there is no change in internal energy: $\Delta U = 0$. From the first law of thermodynamics we have

$$\begin{aligned} Q &= W \\ &= -1.60 \times 10^3\ \mathrm{J}. \end{aligned}$$

Since *Q* is negative, we know that 1600 J of heat flow out of the gas. Note that this is for the whole process, *BDA*.

*15–3 • Human Metabolism and the First Law

Human beings and other animals do work. Work is done when a person walks or runs, or lifts a heavy object. Work requires energy. Energy is also needed for growth—to make new cells, and to replace old cells that have died. A great many energy-transforming processes occur within an organism, and they are referred to as *metabolism*.

We can apply the first law of thermodynamics,

$$\Delta U = Q - W,$$

to an organism, say the human body. Work *W* is done by the body, and this would result in a decrease in the body's internal energy (and tempera-

Table 15–1
Metabolic Rate for a 65-kg Human Being

Activity	Approximate Metabolic Rate	
	kcal/h	watts
Sleeping	60	70
Sitting upright	100	115
Light activity (eating, dressing, household chores)	200	230
Moderate work (tennis, walking)	400	460
Running (15 km/h)	1000	1150
Bicycling (race)	1100	1270

ture), which must be replenished. The body's internal energy is not maintained by a flow of heat Q into the body, however. Normally, the body is at a higher temperature than its surroundings, so heat usually flows *out* of the body. Even on a very hot day when heat is absorbed, the body has no way of utilizing this heat to support its vital processes. What then is the source of energy? It is the internal energy (chemical potential energy) stored in foods. Now in a closed system, the internal energy changes only as a result of heat flow or work done. In an open system, such as an animal, internal energy itself can flow into or out of the system. When we eat food, we are bringing internal energy into our bodies directly, which thus increases the total internal energy U in our bodies. This energy eventually goes into work and heat flow from the body according to the first law.

The metabolic rate is the rate at which internal energy is transformed within the body. It is usually specified in kcal/h or in watts. Typical metabolic rates for a variety of human activities are given in Table 15–1 for an "average" 65-kg adult.

Example 15–2 How much energy is transformed in 24 h by a 65-kg person who spends 8.0 h sleeping, 1.0 h at moderate physical labor, 4.0 h in light activity, and 11.0 h working at a desk or relaxing?

SOLUTION Table 15–1 gives the metabolic rate in watts (J/s). Since there are 3600 s in an hour, the total energy transformed is

$$\begin{bmatrix}(8.0\text{ h})(70\text{ J/s}) + (1.0\text{ h})(460\text{ J/s})\\ + (4.0\text{ h})(230\text{ J/s}) + (11.0\text{ h})(115\text{ J/s})\end{bmatrix}(3600\text{ s/h}) = 1.15 \times 10^7\text{ J}.$$

Since 4.18×10^3 J = 1 kcal, this is equivalent to 2800 kcal; so a food intake of 2800 Cal would compensate for this energy output. A person who wanted to lose weight would have to eat less than 2800 Cal a day, or increase his or her level of activity.

15–4 • The Second Law of Thermodynamics—Introduction

The first law of thermodynamics states that energy is conserved. There are, however, many processes we can imagine that conserve energy but are not observed to occur in nature. For example, when a hot object is placed in contact with a cold object, heat flows from the hotter one to the colder one, never spontaneously the reverse. If heat were to leave the colder object and pass to the hotter one, energy would still be conserved. Yet it doesn't happen. As a second example, consider what happens when you drop a rock and it hits the ground. The initial potential energy of the rock changes to kinetic energy as the rock falls, and when the rock hits the ground this energy in turn is transformed into internal energy of the rock and the ground in the vicinity of the impact; the molecules move faster and the temperature rises slightly. But have you seen the reverse happen—a rock at rest on the ground suddenly rise up in the air because the thermal energy of molecules is transformed into kinetic energy of the rock as a whole? Energy would be conserved in this process, yet we never see it happen.

There are many other examples of processes that occur in nature but whose reverse does not. Here are two more. If you put a layer of salt in a jar and cover it with a layer of similar-sized grains of pepper, when you shake it you get a thorough mixture; no matter how long you shake it, the mixture is unlikely to separate into two layers again. Coffee cups and glasses break spontaneously if you drop them; but they don't go back together spontaneously.

The first law of thermodynamics would not be violated if any of these processes occurred in reverse. To explain this lack of reversibility, scientists in the latter half of the nineteenth century came to formulate a new principle known as the **second law of thermodynamics**. This law is a statement about which processes occur in nature and which do not. It can be stated in a variety of ways, all of which are equivalent. One statement, due to R. J. E. Clausius (1822–1888), is that

Second law of thermodynamics (Clausius statement)

heat flows naturally from a hot object to a cold object; heat will not flow spontaneously from a cold object to a hot object.

Since this statement applies to one particular process, it is not obvious how it applies to other processes. A more general statement is needed that will include other possible processes in a more obvious way.

The development of a general statement of the second law was based partly on the study of heat engines. A **heat engine** is any device that changes thermal energy into mechanical work. We now examine heat engines, both from a practical point of view and to show their importance in developing the second law of thermodynamics.

FIGURE 15–6 Schematic diagram of a heat engine.

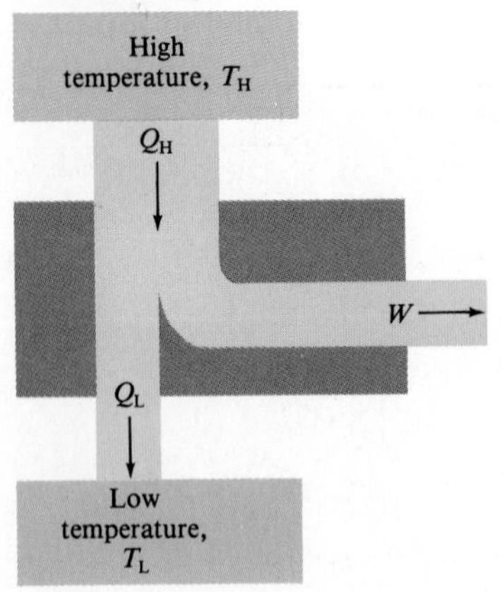

15–5 • Heat Engines and Refrigerators

It is easy to produce thermal energy by doing work—for example, by simply rubbing your hands together briskly, or indeed by any frictional process. But to get work from thermal energy is more difficult, and the invention of a practical device to do this came only about 1700 with the development of the steam engine.

The basic idea behind any heat engine is that mechanical energy can be obtained from thermal energy only when heat is allowed to flow from a high temperature to a low temperature. In the process, some of the heat can then be transformed to mechanical work, as diagrammed in Fig. 15–6. That is, a heat input Q_H at a high temperature T_H is partly transformed into work

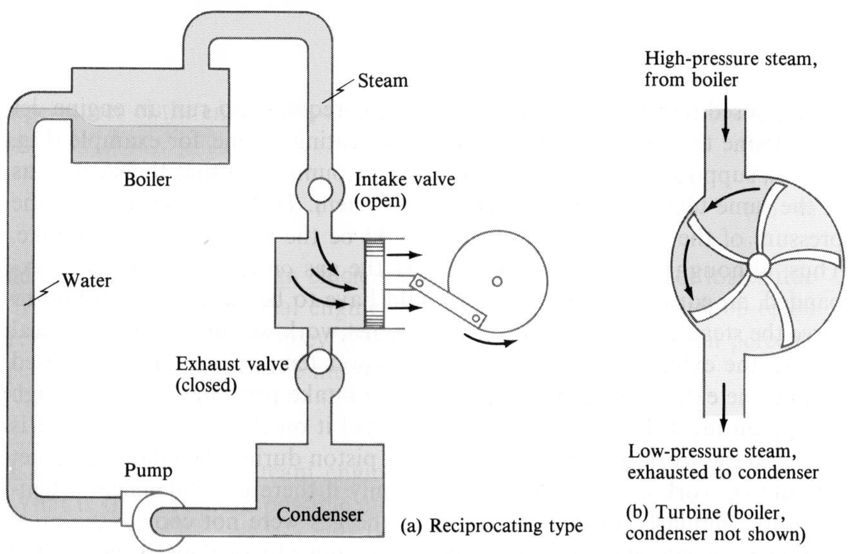

FIGURE 15–7 Steam engines.

W and partly exhausted as heat Q_L at a lower temperature T_L. By conservation of energy, $Q_H = W + Q_L$. The high and low temperatures, T_H and T_L, are called the *operating temperatures* of the engine. We will be interested only in engines that run in a repeating *cycle* (that is, the system returns repeatedly to its starting point) and thus can run continuously.

The operation of two practical engines, the steam engine and the internal combustion engine (used in most automobiles), is illustrated in Figs. 15–7 and 15–8. Steam engines are of two main types. In the so-called reciprocating type, Fig. 15–7a, the heated steam passes through the intake valve and expands against a piston, forcing it to move. As the piston returns to its original position, it forces the gases out the exhaust valve. In a steam turbine, Fig. 15–7b, everything is essentially the same, except that the reciprocating piston is replaced by a rotating turbine that resembles a paddlewheel with many sets of blades. Most of our electricity today is generated using steam turbines.† In a steam engine, the high temperature is obtained by burning coal, oil, or other fuel to heat the steam. In an internal combustion engine, the high temperature is acheived by burning the gasoline–air mixture in the cylinder itself (ignited by the spark plug), Fig. 15–8.

† Even nuclear power plants utilize steam turbines; the nuclear fuel—uranium—merely serves as fuel to heat the steam.

FIGURE 15–8 Four-cycle internal combustion engine: (a) the gasoline–air mixture flows into the cylinder as the piston moves down; (b) the piston moves upward and compresses the gas; (c) firing of the spark plug ignites the gasoline–air mixture, raising it to a high temperature; (d) the gases, now at high temperature and pressure, expand against the piston in this, the power stroke; (e) the burned gases are pushed out to the exhaust pipe; the intake valve then opens, and the whole cycle repeats.

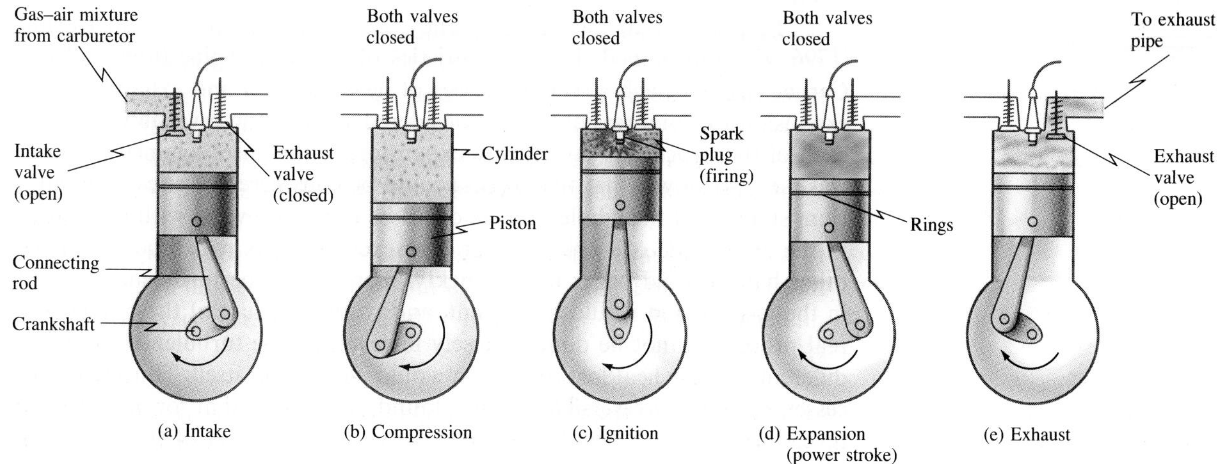

This, then, is the *general statement of the second law of thermodynamics*:

Second law of thermodynamics (general statement)

The total entropy of any system plus that of its environment increases as a result of any natural process.

Although the entropy of one part of the universe may decrease in any process (see Example 15–4), the entropy of some other part of the universe always increases by a greater amount, so the total entropy always increases.

Now that we finally have a quantitative general statement of the second law of thermodynamics, we can see that it is an unusual law. It differs considerably from other laws of physics, which are typically equalities (such as $F = ma$) or conservation laws (such as for energy and momentum). The second law of thermodynamics introduces a new quantity, the entropy, S, but does not tell us it is conserved. Quite the opposite. Entropy is not conserved; it always increases in time.

15–7 • Order to Disorder

The concept of entropy, as we have discussed it so far, may seem rather abstract. To get a feel for the concept of entropy, we can relate it to the concepts of *order* and *disorder*. In fact, the entropy of a system can be considered a *measure of the disorder of the system.* Then the second law of thermodynamics can be stated simply as:

Second law of thermodynamics (general statement)

Natural processes tend to move toward a state of greater disorder.

Exactly what we mean by disorder may not always be clear; so we now consider a few examples. Some of these will show us how this very general statement of the second law actually applies beyond what we usually consider as thermodynamics.

Let us first look at the simple processes mentioned in Section 15–4. A jar containing separate layers of salt and pepper is more orderly than when the salt and pepper are all mixed up. Shaking a jar containing separate layers results in a mixture, and no amount of shaking brings the orderly layers back again. The natural process is from a state of relative order (layers) to one of relative disorder (a mixture), not the reverse. That is, disorder increases. Similarly, a solid coffee cup is a more "orderly" object than the pieces of a broken cup. Cups break when they fall, but they do not spontaneously mend themselves. Again, the normal course of events is an increase of disorder.

When a hot object is put in contact with a cold object, heat flows from the high temperature to the low until the two objects reach the same intermediate temperature. At the beginning of the process we can distinguish two classes of molecules: those with a high average kinetic energy and those with a low average kinetic energy. After the process, all the molecules are in one class with the same average kinetic energy, and we no longer have the more orderly arrangement of molecules in two classes. Order has gone to disorder. Furthermore, note that the separate hot and cold objects could serve as the hot- and cold-temperature regions of a heat engine and thus could be used to obtain useful work. But once the two objects are put in contact and reach the same temperature, no work can be obtained. Disorder has increased, since a system that has the ability to perform work must surely be considered to have a higher order than a system no longer able to do work.

This example illustrates the general concept that an increase in entropy corresponds to an increase in disorder. (We discuss this connection quantitatively in Section 15–10.) In general, we associate disorder with randomness: salt and pepper in layers is more orderly than a random mixture; a neat stack of numbered pages is more orderly than pages strewn randomly about on the floor. We can also say that a more orderly arrangement is one that requires more **information** to specify or classify it. When we have one hot and one cold body, we have two classes of molecules and two pieces of information; when the two bodies come to the same temperature, there is only one class and one piece of information. When salt and pepper are mixed, there is only one (uniform) class; when they are in layers, there are two classes. In this sense, information is connected to order, or low entropy. This is the foundation upon which the modern field of **information theory** is built.

The remaining example of those we discussed earlier was that of a stone falling to the ground, its kinetic energy being transformed to thermal energy; and we noted that the reverse never happens: a stone never rises into the air of its own accord. This is another example of order changing to disorder. For although thermal energy is associated with the disorderly random motion of molecules, the molecules in the falling stone all have the same velocity downward in addition to their own random velocities. Thus, the more orderly kinetic energy of the stone is changed to disordered thermal energy when it strikes the ground. Disorder increases in this process, as it does in all processes that occur in nature.

15–8 • Unavailability of Energy; Heat Death

Let us consider again the natural process of heat conduction from a hot body to a cold one. We have seen that entropy increases and that order goes to disorder. The separate hot and cold objects could serve as the high- and low-temperature regions for a heat engine and thus could be used to obtain useful work. But after the two objects are put in contact and reach the same uniform temperature, no work can be obtained from them. With regard to being able to do useful work, order has gone to disorder in this process.

The same can be said about a falling rock that comes to rest upon striking the ground. Just before hitting the ground, all the kinetic energy of the rock could have been used to do useful work. But once the rock's mechanical kinetic energy becomes internal energy, this is no longer possible.

Both these examples illustrate another important aspect of the second law of thermodynamics—*that in any natural process, some energy becomes unavailable to do useful work.* In any process, no energy is ever lost (it is always conserved). Rather, it becomes less useful—it can do less useful work. As time goes on, **energy is degraded**, in a sense; it goes from more orderly forms (such as mechanical) eventually to the least orderly form, internal or thermal energy.

Degradation of energy

A natural outcome of this is the prediction that as time goes on, the universe will approach a state of maximum disorder. Matter will become a uniform mixture, heat will have flowed from high-temperature regions to low-temperature regions until the whole universe is at one temperature. No work can then be done. All the energy of the universe will have become degraded to thermal energy. All change will cease. This, the so-called **heat death** of the universe, has been much discussed by philosophers. This final state seems an inevitable consequence of the second law of thermodynamics, although it lies very far in the future. Yet it is based on the assumption that the universe is finite, which cosmologists are not really sure of. Furthermore, there is some question as to whether the second law of thermodynamics, as we know it, actually applies in the vast reaches of the universe. The answers are not yet in.

"Heat death"

15–9 • Evolution and Growth; "Time's Arrow"

An interesting example of the increase in entropy relates to biological evolution and to growth of organisms. Clearly, a human being is a highly ordered organism. The process of evolution from the early macromolecules and simple forms of life to *Homo sapiens* is a process of increasing order. So, too, the development of an individual from a single cell to a grown person is a process of increasing order. Do these processes violate the second law of thermodynamics? No, they do not. In the processes of evolution and growth, and even during the mature life of an individual, waste products are eliminated. These small molecules that remain as a result of metabolism are simple molecules without much order. Thus they represent relatively great disorder or entropy. Indeed, the total entropy of the molecules cast aside by organisms during the processes of evolution and growth is greater than the decrease in entropy associated with the order of the growing individual or evolving species.

Another aspect of the second law of thermodynamics is that it tells us in which *direction* processes go. If you were to see a film being run backward, you would undoubtedly be able to tell that it *was* run backward. For you would see odd occurrences, such as a broken coffee cup rising from the floor and reassembling on a table, or a torn balloon suddenly becoming whole again and filled with air. We know these things don't happen in real life; they are processes in which order increases—or entropy decreases. They violate the second law of thermodynamics. When watching a movie (or imagining that time could go backward), we are tipped off to a reversal of time by observing whether entropy is increasing or decreasing. Hence, entropy has been called **time's arrow**, for it tells in which direction time is going.

*15–10 • Statistical Interpretation of Entropy and the Second Law

The ideas of entropy and disorder are made clearer with the use of a statistical or probabilistic analysis of the molecular state of a system. This statistical approach, which was first applied toward the end of the nineteenth century by Ludwig Boltzmann (1844–1906), makes a clear distinction between the "macrostate" and the "microstate" of a system. The **microstate** of a system would be specified when the position and velocity of every particle (or molecule) is given. The **macrostate** of a system is specified by giving the macroscopic properties of the system—the temperature, pressure, number of moles, and so on. In reality, we can know only the macrostate of a system. There are generally far too many molecules in a system to be able to know the velocity and position of every one at a given moment. Nonetheless, it is important to recognize that a great many different microstates can correspond to the *same* macrostate.

Let us take a simple example. Suppose you repeatedly shake four coins in your hand and drop them on the table. Specifying the number of heads and the number of tails that appear on a given throw is the macrostate of this system. Specifying each coin as being a head or a tail is the microstate of the system. In the following table we see the number of microstates that correspond to each macrostate:

Macrostate	**Possible Microstates (H = heads, T = tails)**	**Number of Microstates**
4 heads	H H H H	1
3 heads, 1 tail	H H H T, H H T H, H T H H, T H H H	4
2 heads, 2 tails	H H T T, H T H T, T H H T, H T T H, T H T H, T T H H	6
1 head, 3 tails	T T T H, T T H T, T H T T, H T T T	4
4 tails	T T T T	1

A basic assumption behind the statistical approach is that *each microstate is equally probable.* Thus the number of microstates that give the same macrostate corresponds to the relative probability of that macrostate occurring. The macrostate of two heads and two tails is the most probable one in our case of tossing four coins; out of the total of 16 possible microstates, six correspond to two heads and two tails, so the probability of throwing two heads and two tails is 6 out of 16, or 38 percent. The probability of throwing one head and three tails is 4 out of 16, or 25 percent. The probability of four heads is only 1 in 16, or 6 percent. Of course if you threw the coins 16 times, you might not find that two heads and two tails appear exactly 6 times, or four tails exactly once. These are only probabilities or averages. But if you made 1600 throws, very nearly 38 percent of them would be two heads and two tails. The greater the number of tries, the closer are the percentages to the calculated probabilities.

If we consider tossing more coins, say 100 all at the same time, the relative probability of throwing all heads (or all tails) is greatly reduced. There is only one microstate corresponding to all heads. For 99 heads and 1 tail, there are 100 microstates since each of the coins could be the one tail. The relative probabilities for other macrostates are given in Table 15–2. There are a total of about 10^{30} microstates possible.† Thus the relative probability of finding all heads is 1 in 10^{30}, an incredibly unlikely event! The probability of obtaining 50 heads and 50 tails (see Table 15–2) is $1.0 \times 10^{29}/10^{30} = 0.10$, or 10 percent. The probability of obtaining between 45 and 55 heads is 90 percent.

TABLE 15–2
Probabilities of Various Macrostates for 100 Coin Tosses

Macrostate		**Number of Microstates, W**
Heads	**Tails**	
100	0	1
99	1	1.0×10^{2}
90	10	1.7×10^{13}
80	20	5.4×10^{20}
60	40	1.4×10^{28}
55	45	6.1×10^{28}
50	50	1.0×10^{29}
45	55	6.1×10^{28}
40	60	1.4×10^{28}
20	80	5.4×10^{20}
10	90	1.7×10^{13}
1	99	1.0×10^{2}
0	100	1

† Each coin has two possibilities, heads or tails. Then the possible number of microstates is $2 \times 2 \times 2 \times \cdots = 2^{100}$, which, using a calculator, gives 1.27×10^{30}.

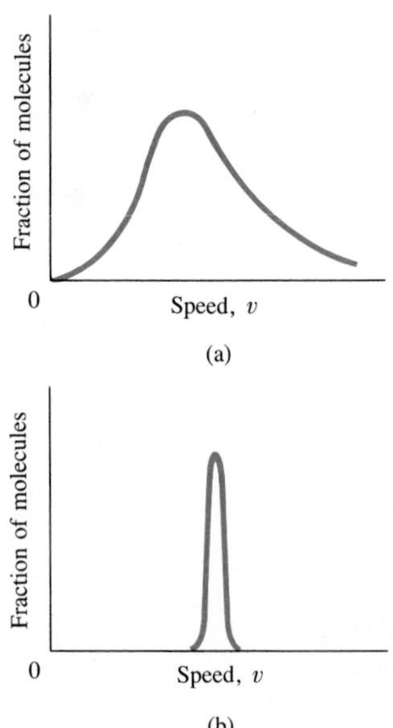

FIGURE 15–12 (a) Most probable distribution of molecular speeds in a gas (Maxwellian, or random); (b) orderly, but highly unlikely, distribution of speeds in which all molecules have nearly the same speed.

Thus we see that as the number of coins increases, the probability of obtaining the most orderly arrangement (all heads or all tails) becomes extremely unlikely. The least orderly arrangement (half heads, half tails) is the most probable and the probability of being within a certain percentage (say, 5 percent) of the most probable arrangement greatly increases as the number of coins increases. These same ideas can be applied to the molecules of a system. For example, the most probable state of a gas (say, the air in a room) is one in which the molecules take up the whole space and move about randomly; this corresponds to the Maxwellian distribution, Fig. 15–12a (and see Chapter 13). On the other hand, the very orderly arrangement of all the molecules located in one corner of the room and all moving with the same velocity (Fig. 15–12b) is extremely unlikely.

From these examples, it is clear that probability is directly related to disorder and hence to entropy. That is, the most probable state is the one with greatest entropy, or greatest disorder and randomness. Boltzmann showed that, consistent with Clausius's definition ($\Delta S = Q/T$), the entropy of a system in a given (macro) state can be written[†]

$$S = 2.3\,k \log W, \qquad (15\text{–}8)$$

where k is Boltzmann's constant ($k = 1.38 \times 10^{-23}$ J/K) and W is the number of microstates corresponding to the given macrostate; that is, W is proportional to the probability of occurrence of that state.

In terms of probability, the second law of thermodynamics—which tells us that entropy increases in any process—reduces to the statement that those processes occur which are most probable. The second law thus becomes a trivial statement. However, there is an additional element now. The second law in terms of probability does not *forbid* a decrease in entropy. Rather, it says the probability is extremely low. It is not impossible that salt and pepper should separate spontaneously into layers, or that a broken tea cup should mend itself. It is even possible that a lake should freeze over on a hot summer day (that is, for heat to flow out of the cold lake into the warmer surroundings). But the probability for such events occurring is extremely small. In our coin examples, we saw that increasing the number of coins from 4 to 100 reduced drastically the probability of large deviations from the average, or most probable, arrangement. In ordinary systems, we are dealing with incredibly large numbers of molecules: in 1 mol alone there are 6×10^{23} molecules. Hence the probability of deviation far from the average is incredibly tiny. For example, it has been calculated that the probability that a stone resting on the ground could transform 1 cal of thermal energy into mechanical energy and rise up into the air is much less likely than the probability that a group of monkeys typing randomly would by chance produce the complete works of Shakespeare.

*15–11 • Energy Resources: Thermal Pollution

When we speak in everyday life of energy usage, we are speaking of the *transformation* of energy from one form to another form that is more useful to

[†] The factor 2.3 comes from the fact that this equation is usually written $S = k \ln W$. Here the "natural logarithm" (ln) to the base e, where $e = 2.718\ldots$, is used instead of the usual base 10. Since $\log e = \log 2.718 = 2.3$, $\ln W = 2.3 \log_{10} W$.

us. For space heating of homes and buildings, the direct burning of fuels such as gas, oil, or coal releases energy stored as potential energy of the molecules. For many transformations, however, a *heat engine* is required, such as to run cars, aircraft, and other vehicles and, very importantly, to generate electricity.

We will discuss various means for producing electricity shortly. But first let us look at two types of pollution associated with heat engines: air pollution and thermal pollution. Air pollution can result from burning any fossil fuels (coal, oil, gas) such as in industrial furnaces for smelting and in electric generating plants. The internal combustion engines of automobiles are especially polluting because the burning takes place so quickly that complete combustion does not take place and more noxious gases are thus produced. To help reduce air pollution, special devices are used (such as catalytic converters). Even when combustion is complete, the CO_2 thrown into the atmosphere absorbs some of the natural infrared radiation emitted by the warm earth that would otherwise escape. The buildup of atmospheric CO_2 and the consequent heating of the atmosphere, referred to as the **greenhouse effect** since a similar effect is partly responsible for warming an enclosed greenhouse, is projected to raise the average temperature of the atmosphere by several degrees C within the next century, causing shifts in rainfall patterns, melting of polar icecaps that would raise sea levels (thus flooding low-lying areas), and turning forests into deserts. An alternative scenario, equally terrible, projects a layer of CO_2 blocking some of the sun's rays and causing a new Ice Age. In either case, experts agree on the urgency for conservation: limit the burning of fossil fuels.

Another type of environmental pollution is **thermal pollution**. Every heat engine, from automobiles to power plants, exhausts heat to the environment (Q_L in Fig. 15–6). Most electricity-producing power plants today make use of a heat engine to transform thermal energy into electricity, and the exhaust heat is generally absorbed by a coolant such as water. If the engine is run efficiently (at best, 30 to 40 percent today), the temperature T_L (see Eq. 15–4) must be kept as low as possible. Hence a great deal of water must flow as coolant through a power plant. The water is usually obtained from a nearby river or lake, or from the ocean. As a result of the transfer of heat to the water, its temperature rises. A rise in temperature of only a few degrees *C* can cause significant damage to aquatic life in the vicinity, in large part because the warmed water holds less dissolved oxygen. The lack of oxygen can adversely affect fish and other organisms and at the same time may encourage excessive growth of other (perhaps alien) organisms, such as algae, thus disrupting the ecology of an area. Another way that waste heat is exhausted is by discharging it into the atmosphere by means of large cooling towers. Unfortunately, this method can also have an environmental effect, for the heated air can alter the weather of a region. Although careful controls may eventually reduce air pollution to an acceptable level, thermal pollution cannot be avoided. What we can do, in light of the second law of thermodynamics, is use less energy and try to build more efficient engines.

We now discuss some of the practical aspects of generating electric power, and various means of doing so. Most of the electricity produced in the United States at the present time makes use of a heat engine coupled with an electric generator. There are a few other possibilities, and we will also discuss them in this section. Electric generators are devices that transform mechanical energy, usually rotational kinetic energy of turbines con-

(a)

(b)

FIGURE 15–17 Two types of windmill: (a) propeller type and (b) "eggbeater" type.

water is released at low tide to drive turbines. At the next high tide, the reservoir is filled again and the inrushing water also turns turbines. A possible future site is in Canada's Bay of Fundy, where the tidal rise reaches 11 m (as compared to an average of 0.3 m elsewhere). This remarkable tide is attributed to a resonance effect (Section 11–6): the resonant frequency for water "sloshing" back and forth within the Bay of Fundy has a period of about 13 hours, as compared to 12.4 h for the tides (which is the driving force—recall from Section 5–8 that the moon's gravitational attraction produces two high tides every 25 h or so). One project for generating electricity proposes a dam to shorten the bay so as to reduce the resonant period slightly and increase the amplitude even more. Other good sites for tidal power (where there is a large difference between high and low tides) are not plentiful and would require large dams across natural or artificial bays. The rather abrupt changes of water level could have an effect on wildlife, but otherwise tidal power would seem to have a minimal environmental effect. Unfortunately, reasonable estimates of available sites indicate that tidal power could at best produce only a small fraction of the world's energy needs.

Wind Power. Windmills were once widely used. Their comeback—on a much grander scale—as a means of turning a generator to produce electricty has already begun. There are already some 16,000 windmills in California alone (Fig. 15–17) capable of producing 1300 MW, enough power for a city of 300,000 people; they are supplying over 1 percent of the state's energy needs. Windmills of various sizes are being produced: from small 3-kW models (say, for a remote house), to large models, each producing several megawatts, with vanes 50 m wide. Windmills are generally "clean,"

FIGURE 15–18 (a) Solar heating system for a house. Water-carrying tubes are located on the roof in contact with a large black surface that absorbs the sun's radiant energy and heats the water. The surface is covered with a piece of glass to prevent loss by convection, and the other sides of the tubes are well insulated to reduce conductive heat losses. The heated water is circulated to a large well-insulated reservoir (kept in the basement perhaps), where it is stored and recirculated to heaters in the house. The reservoir can also serve as the source for the hot-water supply. Either natural convection or forced convection (pumps) can move the water in the two parts of the system. Some form of backup system is needed in many climates to serve when there are prolonged periods of heavy clouds. (b) Solar panels on top of a building.

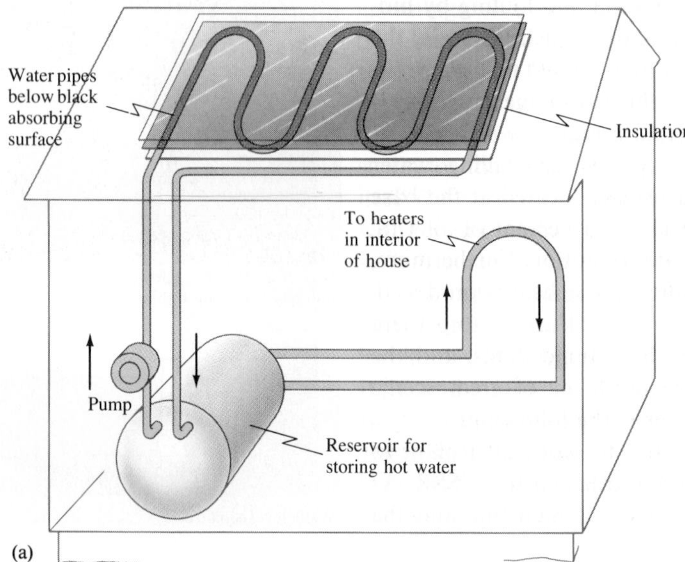

(b)

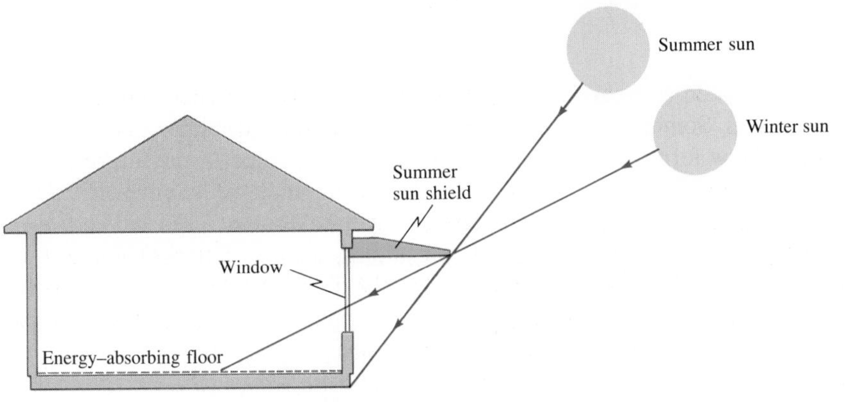

FIGURE 15–19 A passive solar energy system: south-facing windows allow passage of sunlight in winter when the sun is low in the sky. The floor is made of high-heat-capacity material (brick, tile) to store the energy received from the sunlight and release it when the temperature drops. The overhanging sun shield shades the window in summer when the sun is high, helping to keep the interior cooler.

although a large array producing a significant portion of U.S. energy needs might be considered an eyesore. It also might affect the weather, and noise could be a problem.

Solar energy. Many kinds of solar energy are already in use: fossil fuels are the remains of plant life that grew by photosynthesis of light from the sun; hydroelectric power depends on the sun to evaporate water that later comes down as rain; and wind power depends on convection currents produced by the sun heating the atmosphere. *Active solar heating* systems, such as that described in Fig. 15–18, can be used for space heating (keeping interiors warm) and for heating hot water. So-called *passive solar heating* refers to architectural devices to wisely use the sun's energy, such as placing windows along the southern exposure of a building to catch the sun's rays in the winter, but with a sunshade above to keep the sun's rays out in the summer (Fig. 15–19). The sun's rays can also be used to produce electrical energy. For example, an array of large mirrors can focus the sun's rays onto a boiler atop a tall tower, as in a small pilot plant in California (Fig. 15–20). The focused light can heat water to steam to drive a turbine. Such a system could be used for a home generating system, although a backup system would probably be needed for cloudy days. On a larger scale, large areas of land would be needed to collect sufficient sunlight, roughly 1 square mile for 100 MW output. Although the ubiquitous thermal pollution would exist and the climate might be affected, there would be essentially no air or water pollution and no radioactivity, and the technology would not be too difficult. Another direct user of sunlight is the **solar cell**, or more correctly **photovoltaic cell**, which converts sunlight directly into electricity without a heat engine being involved. The efficiency of solar cells, through continuing research, can now be over 30 percent. Solar cells have been very expensive, costing many times per unit of power as ordinary means, but recently the cost has dropped significantly. Solar cells may soon become competitive, making them very desirable, since thermal (and other) pollution would be very low (since no heat engine is involved). Chemical pollution produced in their manufacture in large numbers could, however, be serious. Solar cells might be placed on roofs for home use. On a large scale, they would also require a large land area since the sun's energy is not very concentrated. Another possibility is to place solar cells and concentrators in orbit around the earth. The sun's radiation would be greater there, before the atmosphere absorbs some of it, and the satellites would spend a minimal time in the earth's shadow. The electricity produced might be fed to a microwave generator and transmitted to receivers on earth.

FIGURE 15–20 Array of mirrors focus sunlight on a boiler to produce steam at a solar energy installation near Barstow, California.

It should be clear that all forms of energy production have undesirable side effects. Some are worse than others, and not all problems can be anticipated. New forms of energy production will be needed in the future as old reserves of fuel are used up. It is possible that many different methods—including those we have just discussed—will have to be used in the future to satisfy the needs of our energy-consuming society. The problems posed by any one form of energy production might then not loom as large. Finally, it is clear that conserving our limited fuel supplies, by avoiding wasteful use of energy, should be of great concern to society.

SUMMARY

The *first law of thermodynamics* states that the change in internal energy ΔU of a system is equal to the heat added to the system, Q, minus the work, W, done by the system: $\Delta U = Q - W$. This is a statement of the conservation of energy, and is found to hold for all types of processes. Two simple thermodynamic processes are *isothermal*, which is a process carried out at constant temperature; and *adiabatic*, a process in which no heat is exchanged. The work W done by a gas at constant pressure, P, is given by $W = P\,\Delta V$, where ΔV is the change in volume of the gas.

A *heat engine* is a device for changing thermal energy, by means of heat flow, into useful work. The efficiency of a heat engine is defined as the ratio of the work W done by the engine to the heat input Q_H. Because of conservation of energy, the work output equals $Q_H - Q_L$, where Q_L is the heat exhausted to the environment; hence the efficiency $e = W/Q_H = 1 - Q_L/Q_H$. The upper limit on the efficiency can be written in terms of the higher and lower operating temperatures (in kelvins) of the engine, T_H and T_L, as $e = 1 - T_L/T_H$. The operation of refrigerators and air conditioners is the reverse of that of a heat engine: work is done to extract heat from a cold region and exhaust it to a region at a higher temperature.

The *second law of thermodynamics* can be stated in several equivalent ways: (1) heat flows spontaneously from a hot object to a cold one, but not the reverse; (2) there can be no 100 percent efficient heat engine—that is, one that can change a given amount of heat completely into work; and (3) natural processes tend to move toward a state of greater disorder or greater *entropy*. This last is the most general statement and can be restated as: the total entropy, S, of any system plus that of its environment increases as a result of any natural process: $\Delta S > 0$. Entropy is a quantitative measure of the disorder of a system. As time goes on, energy is degraded to less useful forms—that is, it is less available to do useful work.

QUESTIONS

1. What happens to the internal energy of water vapor in the air that condenses on the outside of a cold glass of water? Is work done or heat exchanged? Explain.

2. Use the conservation of energy to explain why the temperature of a gas increases when it is compressed—say, by pushing down on a cylinder—whereas the temperature decreases when the gas expands.

3. In Fig. 15–2, will more work be done in the isothermal process AB or in the adiabatic process AC? In which process will there be a greater change in internal energy? In which will there be a greater flow of heat?

4. In an isothermal process, 3700 J of work is done by an ideal gas. Is this enough information to tell how much heat has been added to the system? If so, how much?

5. For the processes illustrated in Fig. 15–4, is more work or less work done in the isothermal process AB or in the process ADB?

6. Is it possible for the temperature of a system to remain constant even though heat flows into or out of it? If so, give examples.

7. One liter of air is cooled at constant pressure until its volume is halved. Then it is allowed to expand isothermally back to its original volume. Draw the process on a PV diagram.

8. Discuss how the first law of thermodynamics can apply to metabolism in humans. In particular, note that a person does work W, but very little heat Q is added to the body (rather, it tends to flow out). Why then doesn't the internal energy drop drastically in time?

9. Explain why the temperature of a gas increases when it is adiabatically compressed.

10. Is it possible to cool down a room on a hot summer day by leaving the refrigerator door open?

11. Can mechanical energy ever be transformed completely into heat or internal energy? Can the reverse happen? In each case, if your answer is no, explain why not; if yes, give examples.
12. Would a definition of heat engine efficiency as $e = W/Q_L$ be a useful one? Explain.
13. What plays the role of high-temperature and low-temperature reservoirs in (*a*) an internal combustion engine, and (*b*) a steam engine?
14. Which will give the greater improvement in the efficiency of a Carnot engine, a 10 C° increase in the high-temperature reservoir, or a 10 C° decrease in the low-temperature reservoir?
15. The oceans contain a tremendous amount of thermal energy. Why, in general, is it not possible to put this energy to useful work?
16. A gas is allowed to expand (*a*) adiabatically and (*b*) isothermally. In each process, does the entropy increase, decrease, or stay the same?
17. Give three examples, other than those mentioned in this chapter, of naturally occurring processes in which order goes to disorder. Discuss the observability of the reverse process.
18. Which do you think has the greater entropy, 1 kg of solid iron or 1 kg of liquid iron? Why?
19. What happens if you remove the lid of a bottle containing chlorine gas? Does the reverse process ever happen? Why or why not?
20. Think up several processes (other than those already mentioned) that would obey the first law of thermodynamics, but, if they actually occurred, would violate the second law.
21. Suppose you collect a lot of papers strewn all over the floor and put them in a neat stack; does this violate the second law of thermodynamics? Explain.
22. The first law of thermodynamics is sometimes whimsically stated as, "You can't get something for nothing," and the second law as, "You can't even break even." Explain how these statements could be equivalent to the formal statements.
23. Give three examples of naturally occurring processes that illustrate the degradation of usable energy into internal energy.
24. Entropy is often called "time's arrow" because it tells us in which direction natural processes occur. If a movie film were run backward, name some processes that you might see that would tell you that time was "running backward."

*25. Rank the following five-card hands in order of increasing probability: (*a*) four aces and a king; (*b*) six of hearts, eight of diamonds, queen of clubs, three of hearts, jack of spades; (*c*) two jacks, two queens, and an ace; and (*d*) any hand having no two equal-value cards. Explain ranking using microstates and macrostates.

PROBLEMS

SECTIONS 15–1 AND 15–2

1. (I) One liter of air is cooled at constant pressure until its volume is halved, and then it is allowed to expand isothermally back to its original volume. Draw the process on a *PV* diagram.
2. (I) In Example 15–1, if the heat lost from the gas in the process *BD* is 2.18×10^3 J, what is the change in internal energy of the gas?
3. (II) An ideal gas was slowly compressed at constant temperature to one-half its original volume. In the process, 180 kcal of heat was given off. (*a*) How much work was done (in joules)? (*b*) What was the change in internal energy of the gas? [*Hint:* see Eq. 14–1.]
4. (II) When 539 kcal of heat is added to 1.00 kg of water at 100°C, it is completely changed to steam at 100°C. Calculate (*a*) the work done in this process (the pressure is 1 atm), and (*b*) the change in internal energy of the water. One kilogram of steam occupies 1.67 m^3 at 100°C and 1 atm.
5. (II) An ideal gas is allowed to expand adiabatically to twice its volume. In doing so, the gas does 1860 J of work. (*a*) How much heat flowed into the gas? (*b*) What is the change in internal energy of the gas? (*c*) Did its temperature rise or fall?
6. (II) An ideal gas expands at a constant pressure of 8.0 atm from 380 mL to 630 mL. Heat then flows out of the gas, at constant volume, and the pressure and temperature are allowed to drop until the temperature reaches its original value. Calculate (*a*) the total work done by the gas in the process, and (*b*) the total heat flow into the gas.
7. (II) In the process of taking a gas from state *a* to state *c* along the curved path shown in Fig. 15–21, 80 J of heat leave the system and 55 J of work are done *on* the system. (*a*) Determine the change in internal energy, $U_a - U_c$. (*b*) When the gas is taken along the path *cda*, the work done by the gas is $W = 38$ J. How much heat Q is added to the gas in the process *cda*? (*c*) If $P_a = 2.5P_d$, how much work is done by the gas in the process *abc*? (*d*) What is Q for path *abc*? (*e*) If $U_a - U_b = -10$ J, what is Q for the process *bc*?

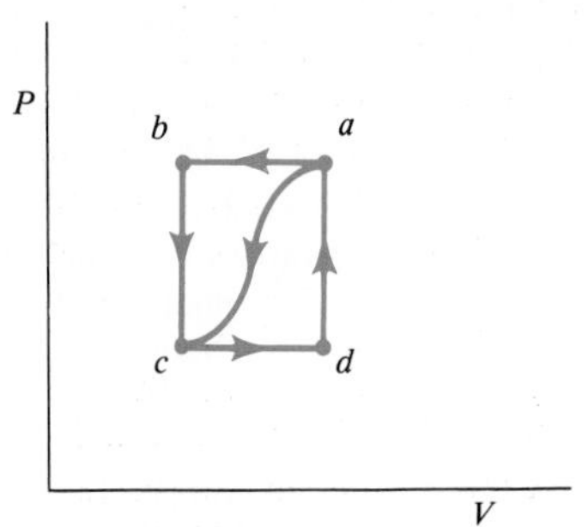

FIGURE 15–21 Problems 7 and 8.

8. (II) When a gas is taken from a to c along the curved path in Fig. 15–21, the work done by the gas is $W = -35$ J and the heat added to the gas is $Q = -63$ J. Along path abc, the work done is $W = -48$ J. (*a*) What is Q for path abc? (*b*) If $P_c = \frac{1}{2}P_b$, what is W for path cda? (*c*) What is Q for path cda? (*d*) What is $U_a - U_c$? (*e*) If $U_d - U_c = 5$ J, what is Q for path da?

*SECTION 15–3

*9. (I) How much energy would the person of Example 15-2 transform if 2.0 h of sleeping were used instead to read?

*10. (I) What is the average power output of a person who spends 10 h each day sleeping, 4 h at light activity, and 10 h watching television or loafing?

*11. (II) A person decides to lose weight by sleeping $\frac{1}{2}$ h less per day, using the time for light activity. How much weight (or mass) can this person expect to lose in 1 year, assuming no change in food intake? Assume that 1 kg of fat stores about 9000 kcal of energy.

SECTION 15–5

12. (I) What is the maximum efficiency of a heat engine whose operating temperatures are 560°C and 340°C?

13. (I) The exhaust temperature of a heat engine is 290°C. What must be the high temperature if the Carnot efficiency is to be 30 percent?

14. (I) A heat engine produces 8200 J of heat while performing 3500 J of useful work. What is the efficiency of this engine?

15. (II) An engine that operates at half its theoretical (Carnot) efficiency operates between 585°C and 340°C while producing work at the rate of 2000 kW. How much heat is discharged per hour?

16. (II) A Carnot engine performs work at the rate of 450 kW while using 960 kcal of heat per second. If the temperature of the heat source is 620°C, at what temperature is the waste heat exhausted?

17. (II) A heat engine utilizes a heat source at 580°C and has an ideal (Carnot) efficiency of 30 percent. To increase the efficiency to 40 percent, what must be the temperature of the heat source?

18. (III) At a steam power plant, steam engines work in pairs, the output of heat from one being the approximate heat input of the second. The operating temperatures of the first are 700°C and 440°C, and of the second 430°C and 310°C. If the heat of combustion of coal is 2.8×10^7 J/kg, at what rate must coal be burned if the plant is to put out 800 MW of power? Assume the efficiency of the engines is 65 percent of the ideal (Carnot) efficiency.

19. (III) Water is used to cool the power plant in Problem 18. If the water temperature is allowed to increase by no more than 7.0 C°, estimate how much water must pass through the plant per hour.

SECTION 15–6

20. (I) What is the change in entropy of 3.0 kg of water at 100°C when it is vaporized to steam at 100°C?

21. (I) What is the change in entropy of 20 kg of water at 0°C when it is frozen to ice at 0°C?

22. (I) One kilogram of water is heated from 0°C to 100°C. Calculate approximately the change in entropy of the water.

23. (II) If the water in Problem 21 were frozen by being in contact with a great deal of ice at −10°C, what would be the total change in entropy of the process?

24. (II) An aluminum rod conducts 2.80 cal/s from a heat source maintained at 280°C to a large body of water at 37°C. Calculate the rate entropy increases per unit time in this process.

25. (II) One kilogram of water at 40°C is mixed with 1 kg of water at 80°C in a well-insulated container. Calculate (approximately) the net change in entropy of the system.

26. (III) A real heat engine working between heat reservoirs at 400 K and 650 K produces 500 J of work per cycle for a heat input of 1500 J. (*a*) Compare the efficiency of this real engine to that of an ideal (Carnot) engine. (*b*) Calculate the total entropy change of the universe per cycle of the real engine. (*c*) Calculate the total entropy change of the universe per cycle of a Carnot engine operating between the same two temperatures.

*SECTION 15–10

*27. (II) Calculate the probabilities, when you throw two dice, of obtaining (*a*) a 2, and (*b*) a 7.

*28. (II) Suppose that you repeatedly shake six coins in your hand and drop them on the table. Construct a table showing the number of microstates that correspond to each macrostate. What is the probability of obtaining (*a*) three heads and three tails, and (*b*) six heads?

*29. (III) (*a*) Suppose you have four coins, all with tails up. You now rearrange them so two heads and two tails are up. What was the change in entropy of the coins? (*b*) Suppose your system is the 100 coins of Table 15–2. What is the change in entropy of the coins if they are mixed randomly initially, 50 heads and 50 tails, and you arrange them so all 100 are heads? (*c*) Compare these entropy changes to ordinary thermodynamic entropy changes.

*30. (II) Estimate the probability that a bridge player will be dealt (*a*) all four aces (among 13 cards), and (*b*) all 13 cards of one suit.

*SECTION 15–11

*31. (I) Solar cells can produce about 40 W of electricity per square meter of surface area if directly facing the sun. How large an area is needed to supply the needs of a house that requires 100 kWh/day? Would this fit on the roof of an average house? (Assume that the sun shines 12 h/day.)

*32. (II) Water falls 70 m over a dam at a rate of 12,000 kg/s. How many megawatts of electric power could be produced by a power plant using this energy?

*33. (II) One way of storing energy for use during peak demand periods is to pump water to a high reservoir when the demand is low, and then release it to drive turbines when needed. Suppose that water is pumped to a lake 100 m above the turbines at a rate of 1.0×10^6 kg/s for 8.0 h during the night. (*a*) How much energy (kWh) is needed to do this each night? (*b*) If all this energy is released during a 16-h day, what is the average power output? Assume that the process is 80 percent efficient.

*34. (II) The basin of the tidal power plant at the mouth of the Rance River in France covers an area of 23 km^2. The average difference in water height between high and low tide (Fig. 15–16) is 8.5 m. Estimate how much work the falling water can do on the turbines per day, assuming there are two high and two low tides per day. Assume the basin is flat, and neglect any KE of water before and after its fall.

GENERAL PROBLEMS

35. To get an idea of how much energy the world's oceans contain, estimate the heat liberated when a cube, 1 km on a side, of ocean water is cooled by 1 K. (Approximate the ocean water by pure water for this estimate.)

36. It has been suggested that a heat engine could be developed that made use of the temperature difference between water at the surface of the ocean and that several hundred meters deep. In the tropics, the temperatures may be 5°C and 25°C, respectively. What is the maximum efficiency such an engine could have? Why might such an engine be feasible in spite of the low efficiency? Can you imagine any adverse environmental effects that might occur?

37. When 4.80×10^4 J of heat are added to a gas enclosed in a cylinder fitted with a light frictionless piston maintained at atmospheric pressure, the volume is observed to increase from 2.0 m^3 to 3.8 m^3. Calculate (*a*) the work done by the gas, and (*b*) the change in internal energy of the gas. (*c*) Graph this process on a *PV* diagram.

38. Two 1000-kg cars are traveling 50 km/h in opposite directions when they collide and are brought to rest. Estimate the change in entropy of the universe as a result of this collision. Assume $T = 20$°C.

39. The burning of gasoline in a car releases about 3.0×10^4 kcal/gal. If a car averages 35 km/gal when driving 90 km/h, which requires 25 hp, what is the efficiency of the engine under those conditions?

40. A 150-g insulated aluminum cup at 20°C is filled with 120 g of water at 40°C. (*a*) What is the final temperature of the mixture? (*b*) Estimate the total change in entropy as a result of the mixing process.

41. A 10-kg box having an initial speed of 3.0 m/s slides along a rough table and comes to rest. Estimate the total change in entropy of the universe. Assume all objects are at room temperature (293 K).

42. A vertical steel I-beam at the base of a building is 7.5 m tall, has a mass of 300 kg, and supports a load of 3.8×10^5 N. If the beam's temperature decreases by 4.0 C°, calculate the change in its internal energy using the facts that for steel, the specific heat is 0.11 kcal/kg·C° and the coefficient of linear expansion is 11×10^{-6} C°$^{-1}$.

43. A "perfect" refrigerator absorbs heat from the freezer compartment at a temperature of −20°C and exhausts it into the room at 20°C. (*a*) How much work must be done by the refrigerator to change 0.60 kg of water at 20°C into ice at −20°C? [*Hint:* think of the refrigerator as a heat engine, and then reverse the process.] (*b*) If the compressor motor does work at the rate of 200 W, what minimum time is needed to take 0.60 kg of 20°C water and freeze it at 0°C?

44. A 40 percent efficient power plant puts out 750 MW (megawatts) of work (electrical energy). Cooling towers are used to take away the exhaust heat. If the air temperature is allowed to rise 7.5 C°, what volume of air (km^3) is heated per day? Will the local climate be heated significantly? If the heated air were to form a layer 200 m thick, how large an area would it cover for 24 h of operation? (The heat capacity of air is about 7.0 cal/mol·C° at constant pressure.)

45. Suppose a power plant delivers energy at 900 MW using steam turbines. The steam goes into the turbines superheated at 520 K and deposits its unused heat in river water at 280 K. Assume that the turbine operates as an ideal Carnot engine. (*a*) If the river flow rate is 45 m^3/s, calculate the average temperature increase of the river water downstream from the power plant. (*b*) What is the entropy increase per kilogram of the downstream river water in J/kg·K?

CHAPTER 16

Electric Charge and Electric Field

This comb has been rubbed by a cloth or paper towel to give it a static electric charge. Because the comb is electrically charged, it induces a separation of charge in all those scraps of paper, and thus attracts them.

The word "electricity" may evoke an image of complex modern technology: computers, lights, motors, electric power. But the electric force plays an even deeper role in our lives, since according to atomic theory, the forces that act between atoms and molecules to hold them together to form liquids and solids are electrical forces. Similarly, the electric force is responsible for the metabolic processes that occur within our bodies. Even ordinary pushes and pulls are the result of the electric force between the molecules of your hand and those of the object being pushed or pulled. Indeed, most of the forces we have dealt with so far, such as elastic forces and the normal force acting on a body, are now considered to be electric forces acting at the atomic level. This does not include gravity, however, which is a separate force.†

† As we discussed in Section 5–10, physicists in this century came to recognize only four different forces in nature: (1) gravitational force, (2) electromagnetic force (we will see later that electric and magnetic forces are intimately related), (3) strong nuclear force, and (4) weak nuclear force. The last two forces operate at the level of the nucleus of an atom. The electromagnetic and weak nuclear forces are now thought to have a common origin known as the electroweak force. We will discuss these forces in later chapters.

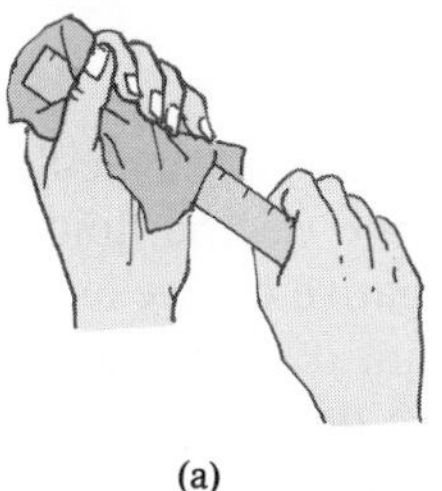

(a)

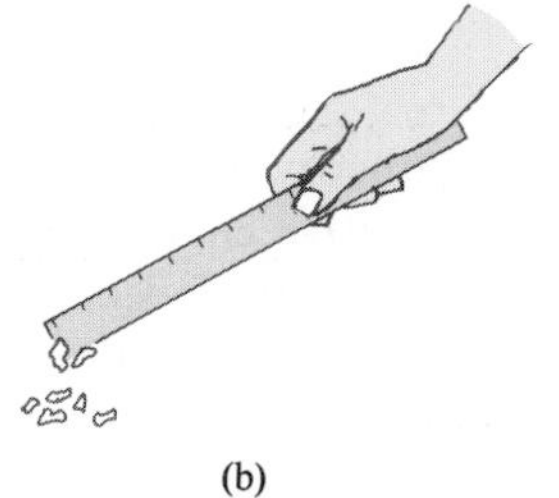

(b)

FIGURE 16–1 Rub a plastic ruler (a) and bring it close to some tiny pieces of paper (b).

The earliest studies on electricity date back to the ancients, but it has been only in the past two centuries that electricity was studied in detail. We will discuss the development of ideas about electricity, including practical devices, as well as the relation to magnetism, in the next seven chapters.

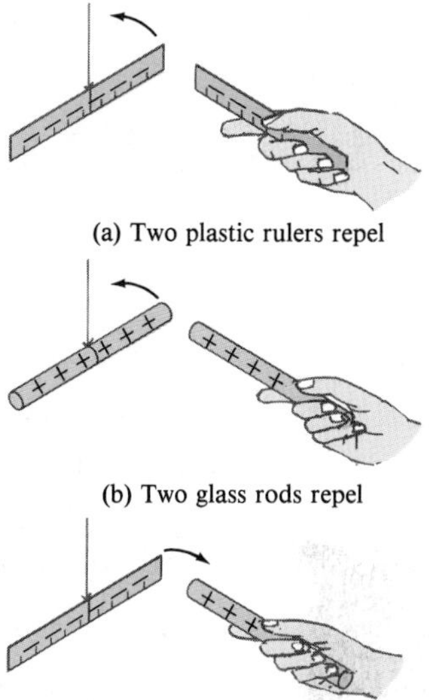

(a) Two plastic rulers repel

(b) Two glass rods repel

(c) Glass rod attracts plastic ruler

FIGURE 16–2 Unlike charges attract, whereas like charges repel one another.

16–1 • Static Electricity; Electric Charge and Its Conservation

The word *electricity* comes from the Greek word *elektron*, which means "amber." Amber is petrified tree resin, and the ancients knew that if you rub an amber rod with a piece of cloth, the amber attracts small pieces of leaves or dust. A piece of hard rubber, a glass rod, or a plastic ruler rubbed with a cloth will also display this "amber effect," or **static electricity** as we call it today. You can readily pick up small pieces of paper with a plastic ruler that you've just vigorously rubbed with even a paper towel, Fig. 16–1. You have probably experienced static electricity when combing your hair or upon taking a synthetic blouse or shirt from a clothes dryer. And you may have felt a shock when you touched a metal door knob after sliding across a car seat or walking across a nylon carpet. In each case, an object becomes "charged" due to a rubbing process and is said to possess an **electric charge**.

Is all electric charge the same, or is it possible that there is more than one type? In fact, there are two types of electric charge, as the following simple experiments show. A plastic ruler is suspended by a thread and rubbed vigorously with a cloth to charge it. When a second ruler, which has also been charged in the same way, is brought close to the first, it is found that the one ruler *repels* the other. This is shown in Fig. 16–2a. Similarly, if a rubbed glass rod is brought close to a second charged glass rod, again a repulsive force is seen to act, Fig. 16–2b. However, if the charged glass rod is brought close to the charged plastic ruler, it is found that they *attract* each other, Fig. 16–2c. The charge on the glass must therefore be different from that on the plastic. Indeed, it is found experimentally that all charged objects fall into one of two categories. Either they are attracted to the plastic and repelled by the glass, just as glass is; or they are repelled by the plastic and attracted to the glass, just as the plastic ruler is. Thus there seems to be two, and only two, types of electric charge. Each type of charge repels the same type but attracts the opposite type. That is: **unlike charges attract**; **like charges repel**.

FIGURE 16–3 Benjamin Franklin.

The two types of electric charge were referred to as *positive* and *negative* by the American statesman, philosopher, and scientist Benjamin Franklin (1706–1790; Fig. 16–3). The choice of which name went with which type of charge was of course arbitrary. Franklin's choice sets the charge on the rubbed glass rod to be positive charge, so the charge on a rubbed plastic ruler (or amber) is called negative charge. We still follow this convention today.

Franklin's theory of electric charge was actually a "single-fluid" theory that viewed a positive charge as an excess of the electric fluid beyond an object's normal content of electricity, and a negative charge as a deficiency. Franklin argued that whenever a certain amount of charge is produced on one body in a process, an equal amount of the opposite type of charge is produced on another body. The names positive and negative are to be taken *algebraically*, so that during any process, the net change in the amount of charge produced is zero. For example, when a plastic ruler is rubbed with a paper towel, the plastic acquires a negative charge and the towel an equal amount of positive charge. The charges are separated, but the sum of the two is zero. This is an example of a law that is now well established: the **law of conservation of electric charge**, which states that

Law of conservation of electric charge

the net amount of electric charge produced in any process is zero.

No violations have ever been found, and this conservation law is as firmly established as those for energy and momentum.

16–2 • Electric Charge in the Atom

Only within the past century has it become clear that electric charge has its origin within the atom itself. In later chapters we will discuss atomic structure and the ideas that led to our present view of the atom in more detail. But it will help our understanding of electricity if we discuss it briefly now.

Today's view, somewhat simplified, shows the atom as having a heavy, positively charged nucleus surrounded by one or more negatively charged electrons (Fig. 16–4). In its normal state, the positive and negative charges within the atom are equal, and the atom is electrically neutral. Sometimes, however, an atom may lose one or more of its electrons, or may gain extra electrons. In this case the atom will have a net positive or negative charge, and is called an **ion**.

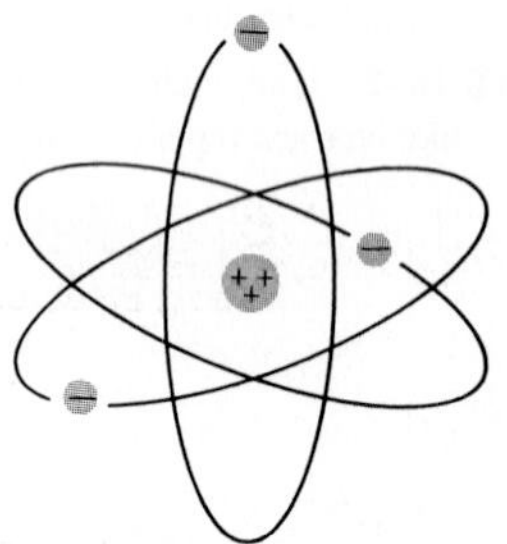

FIGURE 16–4 Simple view of the atom.

The nuclei in a solid material can vibrate, but they tend to remain close to fixed positions, whereas some of the electrons move quite freely. The charging of an object by rubbing is explained by the transfer of electrons or ions from one material to the other. When a plastic ruler becomes negatively charged by rubbing with a paper towel, the transfer of charged particles from the one to the other leaves the towel with a positive charge equal in magnitude to the negative charge acquired by the plastic.

Normally when objects are charged by rubbing, they hold their charge only for a limited time and eventually return to the neutral state. Where does the charge go? In some cases it is neutralized by charged ions in the air (formed, for example, by collisions with charged particles known as cosmic rays that reach the earth from space). Often more importantly, the charge can "leak off" onto water molecules in the air. This is because water molecules are *polar*—that is, even though they are neutral, their charge is not distributed uniformly, Fig. 16–5. Thus the extra electrons on, say, a charged plastic ruler can "leak off" into the air because they are attracted to the positive end of water molecules. A positively charged object, on the other hand, can be neutralized by transfer of loosely held electrons from water molecules in the air. On dry days, static electricity is much more noticeable

FIGURE 16–5 Diagram of a water molecule. Because it has opposite charges on different ends, it is called a "polar" molecule.

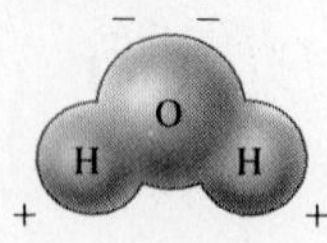

since the air contains fewer water molecules to allow leakage. On humid or rainy days, it is difficult to make any object hold its charge for long.

16–3 • Insulators and Conductors

Suppose we have two metal spheres, one highly charged and the other electrically neutral. If we now place an iron nail so that it touches both the spheres, it is found that the previously uncharged sphere quickly becomes charged. If, however, we had connected the two spheres together with a wooden rod or a piece of rubber, the uncharged ball would not have become noticeably charged. Materials like the iron nail are said to be **conductors** of electricity, whereas wood and rubber are **nonconductors** or **insulators**.

Metals are good conductors

Metals are generally good conductors whereas most other materials are insulators (although even insulators conduct electricity very slightly). It is interesting that nearly all natural materials fall into one or the other of these two quite distinct categories. There are a few materials, however, (notably silicon, germanium, and carbon) that fall into an intermediate (but distinct) category known as *semiconductors.*

From the atomic point of view, the electrons in an insulating material are bound very tightly to the nuclei. In a good conductor, on the other hand, many of the electrons are bound very loosely and can move about freely within the material (although they cannot *leave* the metal easily). When a positively charged object is brought close to or touches a conductor, the free electrons are attracted by this positive charge and move quickly toward it. On the other hand, the free electrons move swiftly away from a negative charge that is brought close. In a semiconductor, there are very few free electrons, and in an insulator, almost none.

16–4 • Induced Charge; the Electroscope

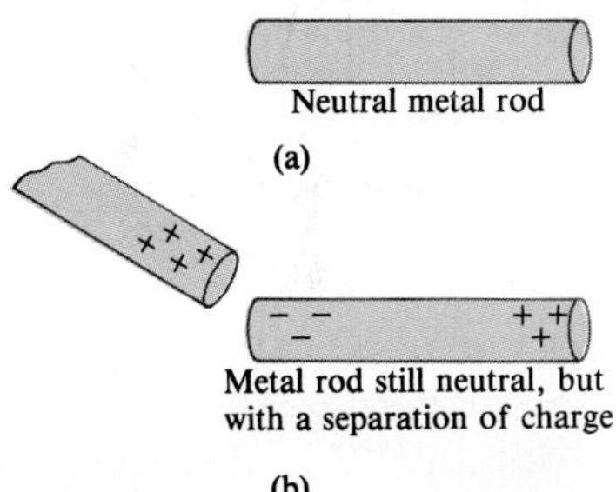

FIGURE 16–6 Charging by induction.

Suppose a positively charged metal object is brought close to a second (neutral) metal object. If the two touch, the free electrons in the neutral one are attracted to the positively charged object and some will pass over to it. Since the second object is now missing some of its negative electrons, it will have a net positive charge. This process is called "charging by conduction."

Now suppose a positively charged object is brought close to a neutral metal rod, but does not touch it. Although the electrons of the metal rod do not leave the rod, they still move within the metal toward the charged object, which leaves a positive charge at the opposite end, Fig. 16–6. A charge is said to have been *induced* at the two ends of the metal rod. Of course no charge has been created: it has merely been *separated.* The net charge on the metal rod is still zero. However, if the metal were divided in two pieces, we could have two charged objects, one charged positively and one charged negatively.

FIGURE 16–7 Inducing a charge on an object connected to ground.

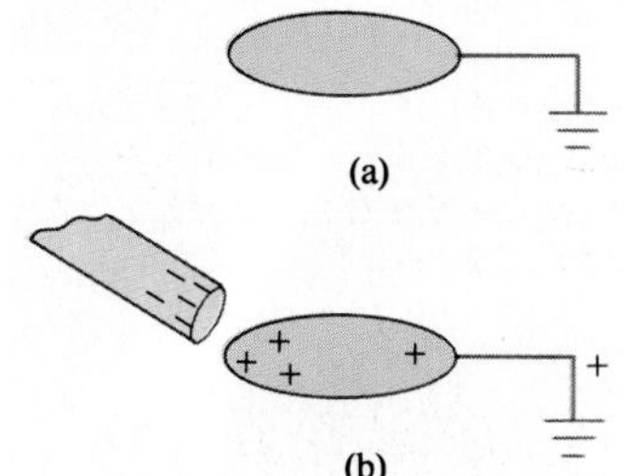

Another way to induce a net charge on a metal object is to connect it with a conducting wire to the ground (or a conducting pipe leading into the ground) as shown in Fig. 16–7a (⏚ means "ground"). The object is then said to be "grounded" or "earthed." Now the earth, since it is so large and can conduct, can easily accept or give up electrons; hence it acts like a reservoir

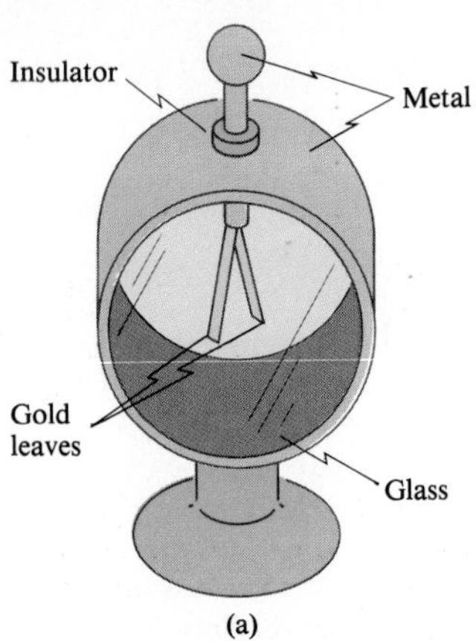

FIGURE 16–8 Electroscope.

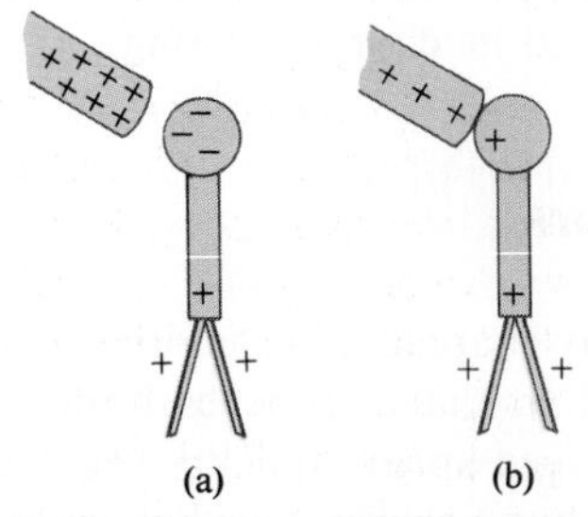

FIGURE 16–9 Electroscope charged (a) by induction, (b) by conduction.

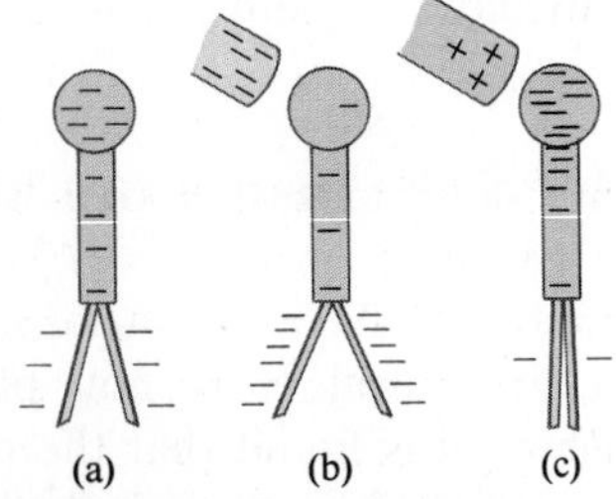

FIGURE 16–10 A previously charged electroscope can be used to determine the sign of a given charge.

for charge. If a charged object—let's say negative this time—is brought up close to the metal, free electrons in the metal are repelled and many of them move down the wire into the earth, Fig. 16–7b. This leaves the metal positively charged. If the wire is now cut, the metal will have a positive induced charge on it. If the wire were cut after the negative object is moved away, the electrons would all have moved back into the metal and it would be neutral.

Electroscope

An **electroscope** (or simple **electrometer**) is a device that can be used for detecting charge. As shown in Fig. 16–8, inside of a case are two movable leaves, often made of gold. (Sometimes only one leaf is movable.) The leaves are connected by a conductor to a metal ball on the outside of the case, but are insulated from the case itself. If a charged object is brought close to the knob, a separation of charge is induced on it, Fig. 16–9a. The two leaves become charged and repel each other as shown. If, instead, the knob is charged by conduction, the whole apparatus becomes charged as shown in Fig. 16–9b. In either case, the greater the amount of charge, the greater the separation of the leaves.

Note however, that you cannot tell the sign of the charge in this way, since a negative charge will cause the leaves to separate just as much as an equal-magnitude positive charge—in either case the two leaves repel each other. An electroscope can, however, be used to determine the sign of the charge if it is first charged by conduction, say negatively, as in Fig. 16–10a. Now if a negative object is brought close, as in Fig. 16–10b, electrons are induced to move farther down into the leaves and they separate further. On the other hand, if a positive charge is brought close, the electrons are induced to flow upward, leaving the leaves less negative and their separation is reduced, Fig. 16–10c.

The electroscope was much used in the early days of electricity. The same principle, aided by some electronics, is used in much more sensitive modern *electrometers*.

FIGURE 16–11 Charles Coulomb.

16–5 • Coulomb's Law

We have seen that an electric charge exerts a force on other electric charges. But how does the magnitude of the charges and other factors affect the magnitude of this force? To answer this, the French physicist Charles Coulomb (1736–1806; Fig. 16–11) investigated electric forces in the 1780s using a torsion balance (Fig. 16–12) much like that used by Cavendish for his studies of the gravitational force (Section 5–5).

Although precise instruments for the measurement of electric charge were not available in Coulomb's time, he was able to prepare small spheres with different amounts of charge in which the *ratio* of the charges was known. He reasoned that if a charged conducting sphere is placed in contact with an identical uncharged sphere, the charge on the first would be shared equally by the two of them because of symmetry. He thus had a way to produce charges equal to $\frac{1}{2}$, $\frac{1}{4}$, and so on, of the original charge. Although he had some difficulty with induced charges, Coulomb was able to argue that the force one tiny charged object exerted on a second tiny charged object is directly proportional to the charge on each of them. That is, if the charge on either one of the objects was doubled, the force was doubled; and if the charge on both of the objects was doubled, the force increased to four times the original value. This was the case when the distance between the two charges remained the same. If the distance between them was allowed to change, he found that the force decreased with the *square of the distance* between them. That is, if the distance was doubled, the force fell to one-fourth of its original value. Thus, Coulomb concluded, the force one tiny charged object (ideally, a **point charge**—one which like an idealized point particle, has no spatial extent) exerts on a second one is proportional to the product of the amount of charge on one, Q_1, times the amount of charge on the other, Q_2, and inversely proportional to the square of the distance r between them. As an equation, we can write **Coulomb's law** as

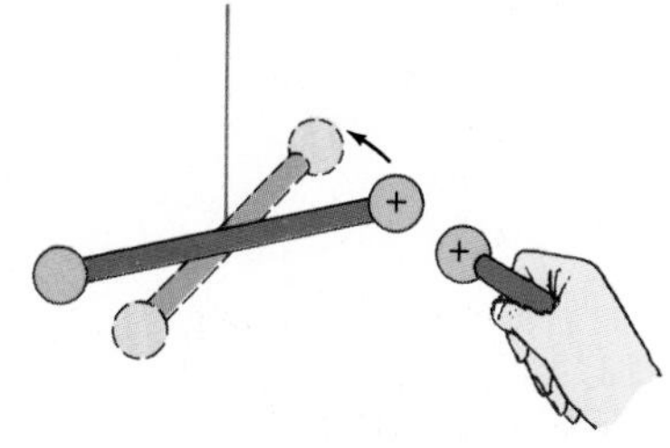

FIGURE 16–12 Schematic diagram of Coulomb's apparatus. It is similar to Cavendish's, which was used to measure the gravitational force. When a charged sphere is placed close to the one on the suspended bar, the bar rotates slightly. The suspending fiber resists the twisting motion and the angle of twist is proportional to the force applied. By the use of this apparatus, Coulomb was able to investigate how the electric force varies as a function of the magnitude of the charges and of the distance between them.

$$F = k\frac{Q_1Q_2}{r^2}, \qquad (16\text{–}1)$$

Coulomb's law

where k is a proportionality constant. Its validity rests on careful experiments that are much more sophisticated than Coulomb's difficult-to-perform experiment. The exponent 2 has been shown to be accurate to 1 part in 10^{16}.

Since we are dealing here with a new quantity (electric charge), we could choose its unit so that the proportionality constant k in Eq. 16–1 would be one. Indeed, such a system of units was once common.† However, the most widely used unit now is the **coulomb** (C), which is the SI unit. The precise definition of the coulomb today is in terms of electric current and magnetic field and will be discussed later (Section 20–13). In SI units, k has the value

$$k = 8.988 \times 10^9 \text{ N}\cdot\text{m}^2/\text{C}^2 \approx 9.0 \times 10^9 \text{ N}\cdot\text{m}^2/\text{C}^2.$$

Thus, 1 C is that amount of charge which, if it exists on each of two point objects placed 1 m apart, will result in each object exerting a force of about $(9.0 \times 10^9 \text{ N}\cdot\text{m}^2/\text{C}^2)(1.0 \text{ C})(1.0 \text{ C})/(1.0 \text{ m})^2 = 9.0 \times 10^9$ N on the other.

Charges produced by rubbing ordinary objects (such as a comb or plastic ruler) are typically a microcoulomb ($1\ \mu\text{C} = 10^{-6}$ C) or less. The magnitude of the charge on one electron, on the other hand, is measured to be about 1.602×10^{-19} C (and its sign is negative). This is the smallest known

† This is a cgs system of units, and the unit of electric charge is called the *electrostatic unit* (esu) or the statcoulomb. One esu is defined as that charge, on each of two point objects 1 cm apart, that gives rise to a force of 1 dyne.

charge,[†] and because of its fundamental nature, it is given the symbol e and is often referred to as the *elementary charge*:

Charge on electron (the elementary charge)

$$e = 1.602 \times 10^{-19}\ \text{C}.$$

Note that e is defined as a positive number, so the charge on the electron is $-e$. (The charge on a proton, on the other hand, is $+e$.) Since an object cannot gain or lose a fraction of an electron, the net charge on any object must be an integral multiple of this charge. Electric charge is thus said to be *quantized* (existing only in discrete amounts). Because e is so small, however, we normally don't notice this discreteness in macroscopic charges (1 μC requires about 10^{13} electrons), which thus seem continuous.

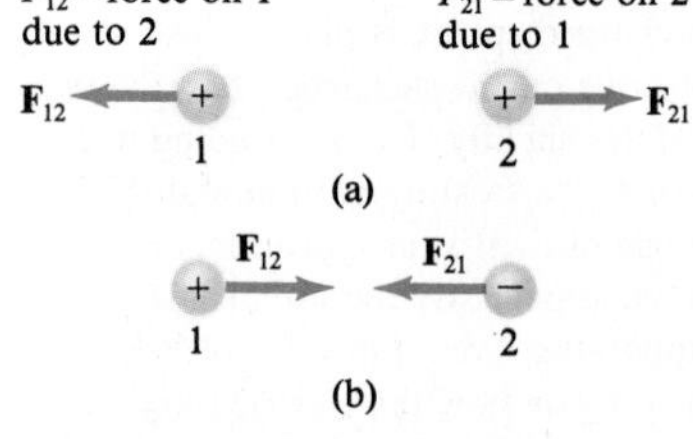

FIGURE 16–13 Direction of force depends on whether the charges have (a) the same sign, or (b) opposite sign.

Equation 16–1 gives the *magnitude* of the force that either object exerts on the other, when the magnitudes of the charges Q_1 and Q_2 are given. The *direction* of this force *is along the line joining the two objects*. If the two charges have the same sign, the force on either object is directed away from the other. If the two objects have opposite charges, the force on one is directed toward the other, Fig. 16–13. If the signs of the two charges are included when using Eq. 16–1, F will be negative for attraction and positive for repulsion. Notice that the force one charge exerts on the second is equal but opposite to that exerted by the second on the first, in accord with Newton's third law.

The constant k in Eq. 16–1 is usually written in terms of another constant, ε_0, called the **permittivity of free space**. It is related to k by $k = 1/4\pi\varepsilon_0$. Coulomb's law can then be written

Coulomb's law (in terms of ε_0)

$$F = \frac{1}{4\pi\varepsilon_0}\frac{Q_1 Q_2}{r^2}, \tag{16–2}$$

where

$$\varepsilon_0 = \frac{1}{4\pi k} = 8.85 \times 10^{-12}\ \text{C}^2/\text{N}\cdot\text{m}^2.$$

Coulomb's law describes the force between two charges when they are at rest. Additional forces come into play when charges are in motion, and these will be discussed in later chapters. In this chapter we discuss only charges at rest, the study of which is called **electrostatics**.

It should be recognized that Eqs. 16–1 and 16–2 apply to objects whose size is much smaller than the distance between them. Ideally, it is precise for point charges. For finite-sized objects, it is not always clear what value to use for r, particularly since the charge may not be distributed uniformly on the objects. If the two objects are spheres and the charge is known to be distributed uniformly on each, then r is the distance between their centers.

It is very important to keep in mind that Eq. 16–1 (or 16–2) gives the force on a charge due to only *one* other charge. If several (or many) charges are present, the *net force on any one of them will be the vector sum of the forces due to each of the others.*

[†] Elementary particle physicists have theorized the existence of smaller particles, called quarks, that would have a smaller charge equal to $\frac{1}{3}$ or $\frac{2}{3}e$. They have not been detected directly, experimentally, and theory indicates that free quarks may not be detectable (see Chapter 32).

16–6 • Solving Problems Involving Coulomb's Law and Vectors

We will now do some examples using Coulomb's law. The electric force between charged particles (sometimes referred to simply as the *Coulomb force*) is, like all forces, a vector: it has both magnitude and direction. When several forces act on an object, the net force $\mathbf{F}_{net}$ on the object is the vector sum of all the forces (call them $\mathbf{F}_1$, $\mathbf{F}_2$, etc.) acting on it:

$$\mathbf{F}_{net} = \mathbf{F}_1 + \mathbf{F}_2 + \cdots.$$

We studied how to add vectors in Chapter 3, and in Chapter 4 we applied the rules for adding vectors to forces. It would be a good idea now to review Sections 3–1, 3–2, 3–3, and 4–8. Here we give a brief review.

Given two vector forces, $\mathbf{F}_1$ and $\mathbf{F}_2$, acting on a body (Fig. 16–14a), they can be added using the tail-to-tip method (Fig. 16–14b) or by the parallelogram method (Fig. 16–14c), as discussed in Section 3–1. These two methods are useful for *understanding* a given problem (for getting a picture in your mind of what is going on), but for *calculating* the direction and magnitude of the resultant sum, it is more precise to use the method of adding components. Figure 16–14d shows the components of our $\mathbf{F}_1$ and $\mathbf{F}_2$ resolved into components along chosen x and y axes (for more details, see Section 3–3). From the definitions of the trigonometric functions (Figs. 3–7 and 3–9), we have

$$F_{1x} = F_1 \cos\theta_1 \qquad F_{2x} = F_2 \cos\theta_2$$

$$F_{1y} = F_1 \sin\theta_1 \qquad F_{2y} = F_2 \sin\theta_2.$$

We add up the x and y components separately to obtain the components of the resultant force $\mathbf{F}$, which are

$$F_x = F_{1x} + F_{2x} = F_1 \cos\theta_1 + F_2 \cos\theta_2,$$

$$F_y = F_{1y} + F_{2y} = F_1 \sin\theta_1 + F_2 \sin\theta_2.$$

The magnitude of $\mathbf{F}$ is

$$F = \sqrt{F_x^2 + F_y^2}.$$

The direction of $\mathbf{F}$ is specified by the angle θ that $\mathbf{F}$ makes with the x axis, which is given by

$$\tan\theta = \frac{F_y}{F_x}.$$

This review has been necessarily brief; a rereading of the appropriate parts of Chapters 3 and 4 is highly recommended.

We now take some examples, the first of which deals only with the magnitude of the Coulomb force.

(a) Two forces acting on an object.

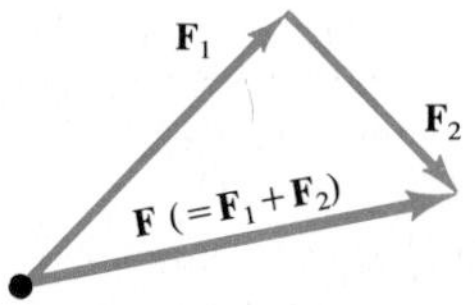

(b) The total, or net, force is $\mathbf{F} = \mathbf{F}_1 + \mathbf{F}_2$ by the tail-to-tip method of adding vectors.

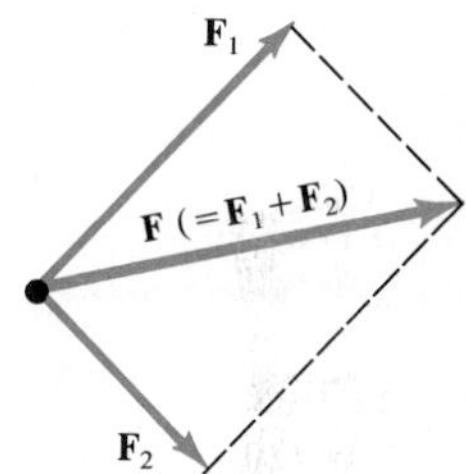

(c) $\mathbf{F} = \mathbf{F}_1 + \mathbf{F}_2$ by the parallelogram method.

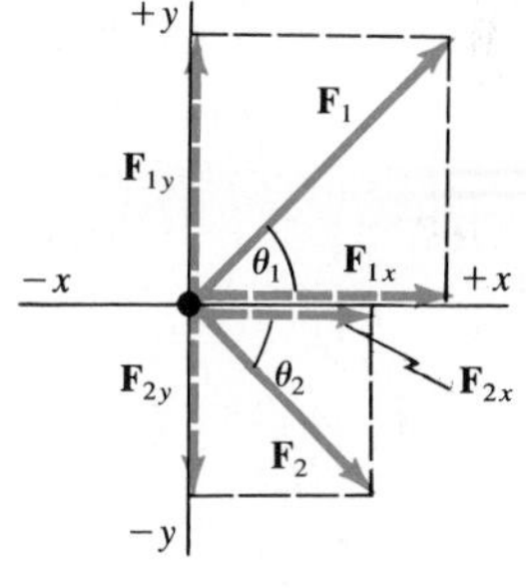

(d) $\mathbf{F}_1$ and $\mathbf{F}_2$ resolved into their x and y components.

FIGURE 16–14 Review of vector addition.

EXAMPLE 16–1 Determine the electric force on the electron of a hydrogen atom exerted by the single proton ($Q_2 = +e$) that is its nucleus, when the electron "orbits" the proton at its average distance of 0.53×10^{-10} m.

SOLUTION We use Eq. 16–1 with $Q_2 = +1.6 \times 10^{-19}$ C, $Q_1 = -Q_2$, and $r = 0.53 \times 10^{-10}$ m:

$$F = \frac{(9.0 \times 10^9\ \text{N}\cdot\text{m}^2/\text{C}^2)(+1.6 \times 10^{-19}\ \text{C})(-1.6 \times 10^{-19}\ \text{C})}{(0.53 \times 10^{-10}\ \text{m})^2}$$

$$= -8.2 \times 10^{-8}\ \text{N}.$$

The minus sign means the force on the electron is toward the proton.

When dealing with several charges, it is often helpful to use subscripts on each of the forces involved. The first subscript refers to the particle *on* which the force acts; the second refers to the particle that exerts the force. For example, if we have three charges, $\mathbf{F}_{31}$ means the force exerted *on* particle 3 *by* particle 1.

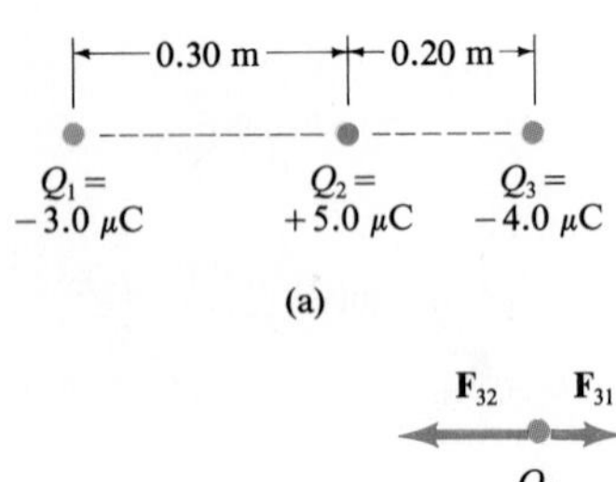

FIGURE 16–15
Diagram for Example 16–2.

EXAMPLE 16–2 Three charged particles are arranged in a line, as shown in Fig. 16–15a. Calculate the net electric force on particle 3 (the $-4.0\ \mu$C on the right) due to the other two charges.

SOLUTION The net force on particle 3 will be the sum of the force $\mathbf{F}_{31}$ exerted by particle 1 and the force $\mathbf{F}_{32}$ exerted by particle 2: $\mathbf{F} = \mathbf{F}_{31} + \mathbf{F}_{32}$. The magnitudes of these two forces are

$$F_{31} = \frac{(9.0 \times 10^9\ \text{N}\cdot\text{m}^2/\text{C}^2)(4.0 \times 10^{-6}\ \text{C})(3.0 \times 10^{-6}\ \text{C})}{(0.50\ \text{m})^2} = 0.43\ \text{N},$$

$$F_{32} = \frac{(9.0 \times 10^9\ \text{N}\cdot\text{m}^2/\text{C}^2)(4.0 \times 10^{-6}\ \text{C})(5.0 \times 10^{-6}\ \text{C})}{(0.20\ \text{m})^2} = 4.5\ \text{N}.$$

Since we were calculating the magnitudes of the forces, we omitted the signs of the charges; but we must be aware of them to get the direction of each force. Let the line joining the particles be the x axis, and we take it positive to the right. Then, because $\mathbf{F}_{31}$ is repulsive and $\mathbf{F}_{32}$ is attractive, the direction of the forces is as shown in Fig. 16–15b: F_{31} points in the positive x direction and F_{32} points in the negative x direction. The net force on particle 3 is then

$$F = F_{32} + F_{31} = -4.5\ \text{N} + 0.4\ \text{N} = -4.1\ \text{N}.$$

The magnitude of the net force is 4.1 N, and it points to the left. (Notice that the charge in the middle (Q_2) in no way blocks the effect of the other charge (Q_1); Q_2 does exert its own force, of course.)

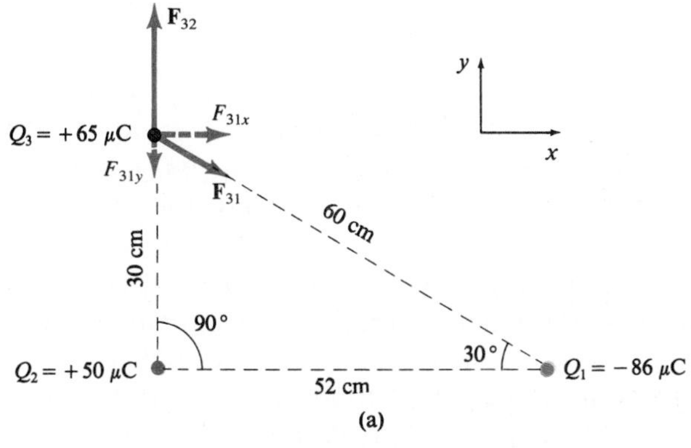

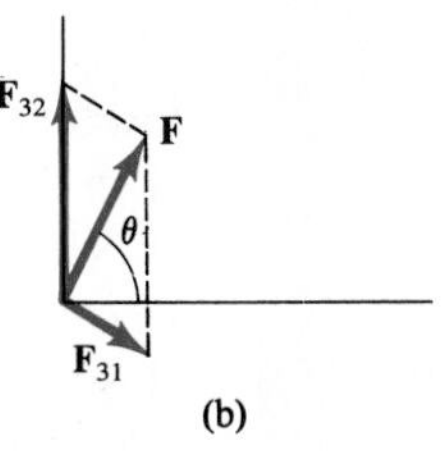

FIGURE 16–16 Determining the forces for Example 16–3.

EXAMPLE 16–3 Calculate the force on charge Q_3 shown in Fig. 16–16a due to the charges Q_1 and Q_2.

SOLUTION The forces $\mathbf{F}_{31}$ and $\mathbf{F}_{32}$ have the directions shown in the diagram since Q_1 exerts an attractive force and Q_2 a repulsive force. The magnitudes of $\mathbf{F}_{31}$ and $\mathbf{F}_{32}$ are (ignoring signs since we know the directions)

$$F_{31} = \frac{(9.0 \times 10^9\ \mathrm{N \cdot m^2/C^2})(6.5 \times 10^{-5}\ \mathrm{C})(8.6 \times 10^{-5}\ \mathrm{C})}{(0.60\ \mathrm{m})^2} = 140\ \mathrm{N},$$

$$F_{32} = \frac{(9.0 \times 10^9\ \mathrm{N \cdot m^2/C^2})(6.5 \times 10^{-5}\ \mathrm{C})(5.0 \times 10^{-5}\ \mathrm{C})}{(0.30\ \mathrm{m})^2} = 330\ \mathrm{N}.$$

We resolve $\mathbf{F}_1$ into its components along the x and y axes, as shown:

$$F_{31x} = F_{31} \cos 30^\circ = 120\ \mathrm{N},$$

$$F_{31y} = -F_{31} \sin 30^\circ = -70\ \mathrm{N}.$$

The force $\mathbf{F}_{32}$ has only a y component. So the net force $\mathbf{F}$ on Q_3 has components

$$F_x = F_{31x} = 120\ \mathrm{N}$$

$$F_y = F_{32} + F_{31y} = 330\ \mathrm{N} - 70\ \mathrm{N} = 260\ \mathrm{N}.$$

Thus the magnitude of the net force is

$$F = \sqrt{F_x^2 + F_y^2} = \sqrt{(120\ \mathrm{N})^2 + (260\ \mathrm{N})^2} = 290\ \mathrm{N};$$

and it acts at an angle θ (see Fig. 16–16b) given by

$$\tan \theta = F_y/F_x = 260\ \mathrm{N}/120\ \mathrm{N} = 2.2,$$

so $\theta = 65^\circ$

16–7 • The Electric Field

Many common forces might be referred to as "contact forces." That is, you exert a force on an object by coming into contact with it: for example, you push or pull on a box, a broom, or a stalled car. Similarly, a tennis racket exerts a force on a tennis ball when they make contact.

On the other hand, both the gravitational force and the electrical force act over a distance: there is a force even when the two objects are not in contact. The idea of a force *acting at a distance* was a difficult one for early

FIGURE 16–17 Michael Faraday, left, with Thomas Henry Huxley, Charles Wheatstone, David Brewster, and John Tyndall.

thinkers. Newton himself felt uneasy with this idea when he published his law of universal gravitation. The conceptual difficulties can be overcome with the idea of the **field**, developed by the British scientist Michael Faraday (1791–1867; Fig. 16–17). In the electrical case, according to Faraday, an *electric field* extends outward from every charge and permeates all of space (Fig. 16–18). When a second charge is placed near the first charge, it feels a force because of the electric field that is there (say, at point P in Fig. 16–18). The electric field at the location of the second charge is considered to interact directly with this charge to produce the force. It must be emphasized, however, that a field is *not* a kind of matter. It is, rather, a concept†—and a very useful one.

We can investigate the electric field surrounding a charge or group of charges by measuring the force on a small positive **test charge**. By a test charge we mean a charge so small that the force it exerts does not significantly alter the distribution of the other charges, the ones that cause the field being measured. The force on a tiny positive test charge q placed at various locations in the vicinity of a single positive charge Q would be as shown in Fig. 16–19. The force at b is less than at a because the distance is greater (Coulomb's law); and the force at c is smaller still. In each case, the force is directed radially outward from Q. The electric field is defined in terms of the force on such a positive test charge. In particular, the **electric field**, **E**, at any point in space is defined as the force **F** exerted on a tiny positive test charge at that point divided by the magnitude of the test charge q:

Definition of electric field

$$\mathbf{E} = \frac{\mathbf{F}}{q}. \qquad (16\text{–}3)$$

Ideally, **E** is defined as the limit of $\mathbf{F}/q$ as q is taken smaller and smaller, approaching zero. From this definition (Eq. 16–3), we see that the direction of the electric field at any point in space is defined as the direction of the force on a positive test charge at that point. And the magnitude of the electric field is the *force per unit charge*. Thus **E** is measured in units of newtons per coulomb (N/C).

† Whether the electric field is "real," and really exists, is a philosophical, even metaphysical, question. In physics it is a very useful idea, a great invention of the human mind.

FIGURE 16–18
An electric field surrounds every charge. P is an arbitrary point.

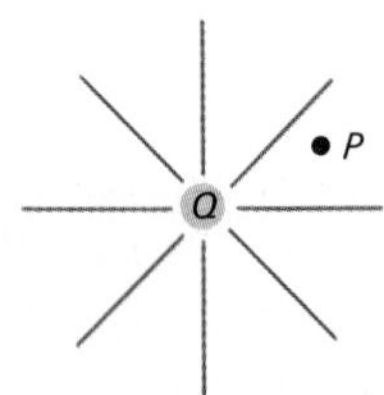

FIGURE 16–19
Force exerted by charge $+Q$ on a small test charge, q, placed at points a, b, and c.

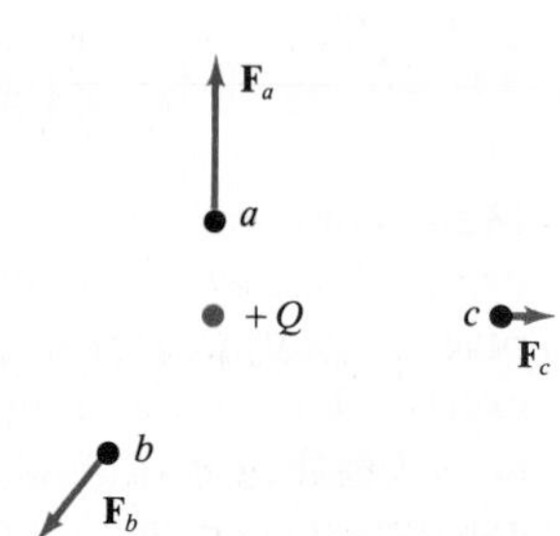

The reason for defining **E** as **F**$/q$ (with $q \to 0$) is so that **E** does not depend on the magnitude of the test charge q. This means that **E** describes only the effect of the charges creating the electric field at that point.

The electric field at any point in space can be measured, based on the definition, Eq. 16–3. For simple situations involving one or several point charges, we can calculate what **E** will be. For example, the electric field at a distance r from a single point charge Q would have magnitude

$$E = k\frac{qQ}{r^2}\frac{1}{q}$$

$$= k\frac{Q}{r^2}; \qquad \text{[single point charge]} \qquad (16\text{–}4a)$$

Electric field due to one point charge

or, in terms of ε_0 (Eq. 16–2)

$$E = \frac{1}{4\pi\varepsilon_0}\frac{Q}{r^2}. \qquad \text{[single point charge]} \qquad (16\text{–}4b)$$

This relation for the electric field due to a single point charge is also (in addition to Eq. 16–1) referred to as Coulomb's law. Notice that E is independent of q—that is, it depends only on the charge Q which produces the field, and not on the value of the test charge q.

EXAMPLE 16–4 Calculate the magnitude and direction of the electric field at a point P which is 30 cm to the right of a point charge $Q = -3.0 \times 10^{-6}$ C.

SOLUTION The magnitude of the electric field is

$$E = k\frac{Q}{r^2} = \frac{(9.0 \times 10^9\ \text{N}\cdot\text{m}^2/\text{C}^2)(3.0 \times 10^{-6}\ \text{C})}{(0.30\ \text{m})^2} = 3.0 \times 10^5\ \text{N/C}.$$

The direction of the electric field is *toward* the charge Q as shown in Fig. 16–20a since we defined the direction as that of the force on a positive test charge. If Q had been positive, the electric field would have pointed away, as in Fig. 16–20b.

FIGURE 16–20 Electric field at point P (a) due to a negative charge Q, and (b) due to a positive charge Q (Example 16–4).

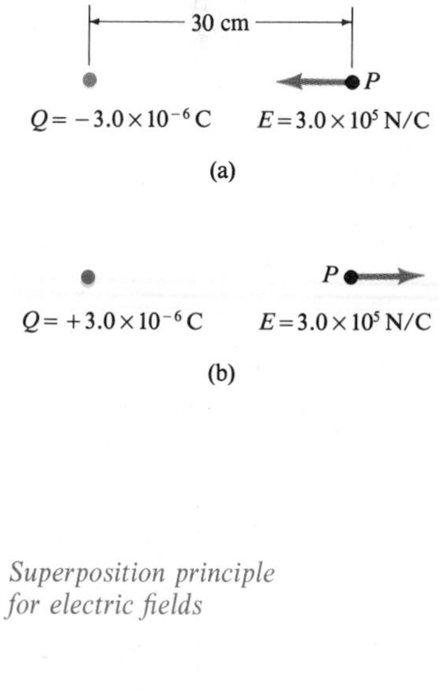

30 cm

P

$Q = -3.0 \times 10^{-6}$ C $\quad E = 3.0 \times 10^5$ N/C

(a)

P

$Q = +3.0 \times 10^{-6}$ C $\quad E = 3.0 \times 10^5$ N/C

(b)

If the field is due to more than one charge, the individual fields (call them $\mathbf{E}_1$, $\mathbf{E}_2$, etc.) due to each charge are added vectorially to get the total field at any point:

$$\mathbf{E} = \mathbf{E}_1 + \mathbf{E}_2 + \cdots.$$

Superposition principle for electric fields

The validity of this **superposition principle** for electric fields is fully confirmed by experiment.

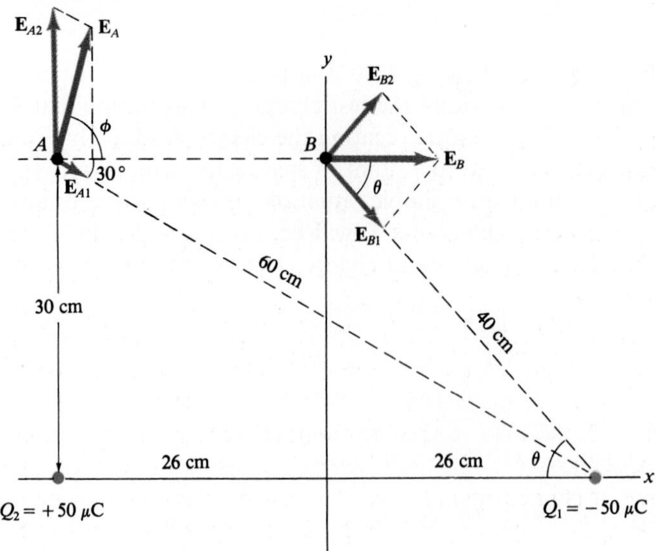

FIGURE 16–21
Calculation of the electric field at points A and B for Example 16–5.

EXAMPLE 16–5 Calculate the total electric field (*a*) at point A and (*b*) at point B in Fig. 16–21 due to both charges, Q_1 and Q_2.

SOLUTION (*a*) The calculation is much like that of Example 16–3, but now we are dealing with electric fields. The electric field at A is the vector sum of the fields $\mathbf{E}_{A1}$ due to Q_1, and $\mathbf{E}_{A2}$ due to Q_2, which have magnitudes:

$$E_{A1} = (9.0 \times 10^9\ \text{N}\cdot\text{m}^2/\text{C}^2)(50 \times 10^{-6}\ \text{C})/(0.60\ \text{m})^2 = 1.25 \times 10^6\ \text{N/C},$$

$$E_{A2} = (9.0 \times 10^9\ \text{N}\cdot\text{m}^2/\text{C}^2)(50 \times 10^{-6}\ \text{C})/(0.30\ \text{m})^2 = 5.0 \times 10^6\ \text{N/C}.$$

The directions are as shown, so the total electric field at A, $\mathbf{E}_A$, has components

$$E_{Ax} = E_{A1} \cos 30° = 1.1 \times 10^6\ \text{N/C},$$

$$E_{Ay} = E_{A2} - E_{A1} \sin 30° = 4.4 \times 10^6\ \text{N/C}.$$

Thus the magnitude of $\mathbf{E}_A$ is

$$E_A = \sqrt{(1.1)^2 + (4.4)^2} \times 10^6\ \text{N/C} = 4.5 \times 10^6\ \text{N/C},$$

and the direction of $\mathbf{E}_A$ is ϕ (see diagram) given by $\tan \phi = 4.4/1.1 = 4.0$, so $\phi = 76°$.

(*b*) Since B is equidistant (40 cm by the Pythagorean theorem) from the two equal charges, the magnitudes of E_{B1} and E_{B2} are the same; that is, $E_{B1} = E_{B2} = (9.0 \times 10^9\ \text{N}\cdot\text{m}^2/\text{C}^2)(5.0 \times 10^{-6}\ \text{N})/(0.40\ \text{m})^2 = 2.8 \times 10^6\ \text{N/C}$. Also, because of the symmetry, the y components are equal and opposite. Hence the total field E_B is horizontal and equals $E_{B1} \cos \theta + E_{B2} \cos \theta = 2E_{B1} \cos \theta$; from the diagram, $\cos \theta = 26\ \text{cm}/40\ \text{cm} = 0.65$. Then

$$E_B = 2E_{B1} \cos \theta = 2(2.8 \times 10^6\ \text{N/C})(0.65) = 3.6 \times 10^6\ \text{N/C},$$

and the direction of $\mathbf{E}_B$ is along the $+x$ direction.

16–8 • Field Lines

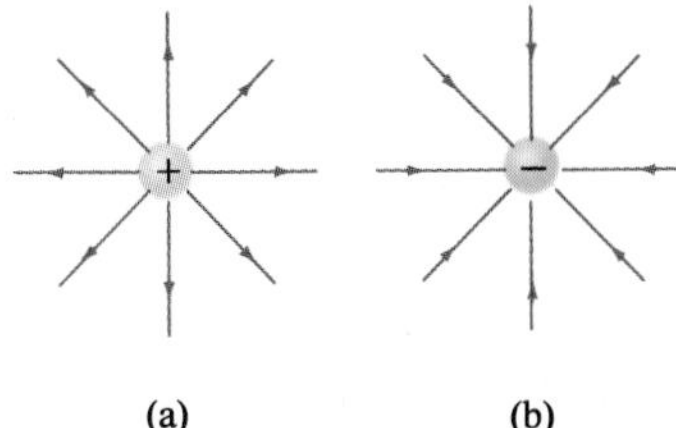

FIGURE 16–22 Electric field lines (a) near a single positive point charge, (b) near a single negative point charge.

Since the electric field is a vector, it is sometimes referred to as a *vector field*. We could indicate the electric field with arrows at various points in a given situation, such as at *a*, *b*, and *c* in Fig. 16–19. The directions of $\mathbf{E}_a$, $\mathbf{E}_b$, and $\mathbf{E}_c$ would be the same as that of the forces already shown, but the lengths (magnitudes) would be different (since we divide by q). However, the relative lengths of $\mathbf{E}_a$, $\mathbf{E}_b$, and $\mathbf{E}_c$ would be the same as for the forces since we divide by the same q each time. However, to indicate the electric field in such a way at many points would result in many arrows, which would appear confusing. To avoid this, we use another technique, that of field lines.

In order to visualize the electric field, we draw a series of lines to indicate the direction of the electric field at various points in space. These **electric field lines**, or (sometimes) *lines of force*, are drawn so that they indicate the direction of the force due to the given field on a positive test charge. The lines of force due to a single positive charge are shown in Fig. 16–22a and for a single negative charge in Fig. 16–22b. In part (a) the lines point radially outward from the charge, and in part (b) they point radially inward toward the charge. The arrows indicate the directions of the force that would be exerted on a positive test charge at any point in the field for each case. Only a few representative lines have been shown. One could just as well draw lines in between those shown since the electric field exists there as well. However, we can always draw the lines so that the *number of lines starting on a positive charge, or ending on a negative charge, is proportional to the magnitude of the charge.* Notice that near the charge, where the force is greatest, the lines are closer together. This is a general property of electric field lines; *the closer the lines are together, the stronger the electric field in that region.* In fact the lines can always be drawn so that the number of lines crossing unit area perpendicular to $\mathbf{E}$ is proportional to the magnitude of the electric field.

FIGURE 16–23 Electric field lines for three arrangements of charges.

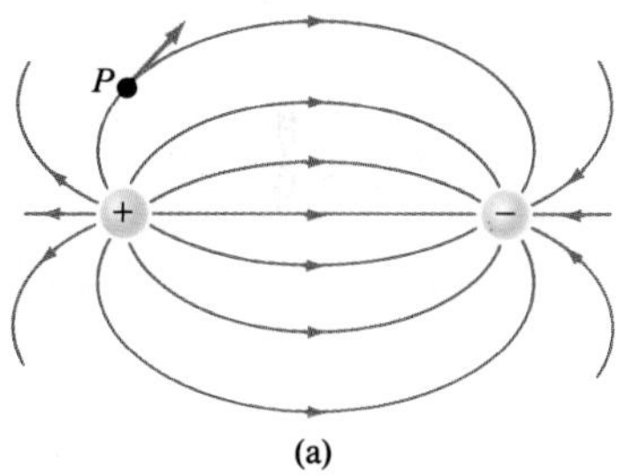

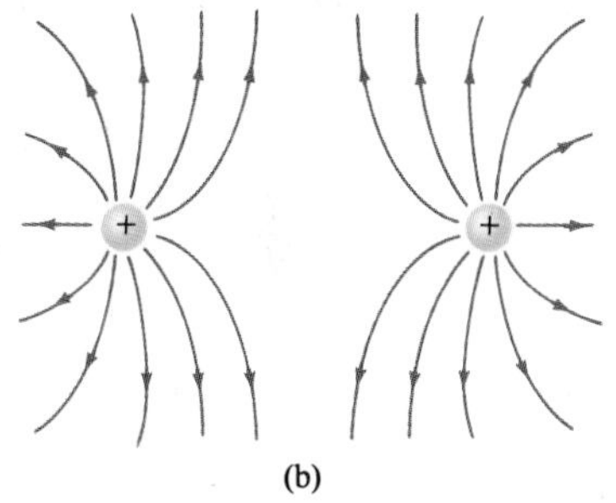

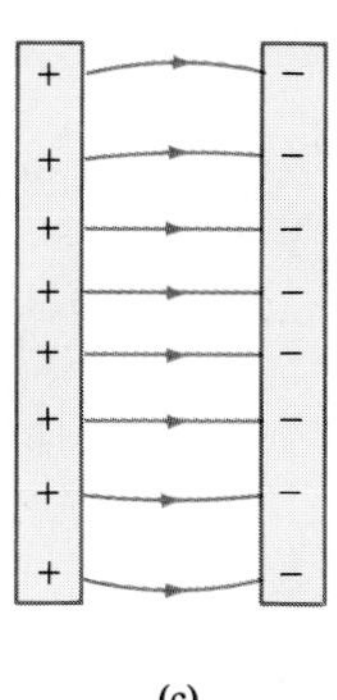

Figure 16–23a shows the electric field lines surrounding two charges of opposite sign. The electric field lines are curved in this case and they are directed from the positive charge to the negative charge. The direction of the field at any point is directed tangentially as shown by the faint arrow at point P. To satisfy yourself that this is the correct pattern for the electric field lines, you can make a few calculations such as those done in Example 16–5 for just this case (see Fig. 16–21). Figures 16–23b and c show the electric field lines surrounding two equal positive charges, and between two oppositely charged parallel plates. Notice that the electric field lines between the two plates are parallel and equally spaced, except near the edges. Thus, in the central region, the electric field has the same magnitude at all points and we can write

$$E = \text{constant.} \qquad \text{[between two closely spaced parallel plates]} \qquad (16\text{–}5)$$

Although the field fringes near the edges (the lines curve), we can often ignore this, particularly if the separation of the plates is small compared to their size. This should be compared to the field of a single point charge where the field decreases as the square of the distance, Eq. 16–4.

We summarize the properties of field lines as follows:

Properties of field lines

1. The field lines indicate the direction of the electric field; the field points in the direction tangent to the field lines at any point.
2. The lines are drawn so that the magnitude of the electric field, E, is proportional to the number of lines crossing unit area perpendicular to the lines. The closer the lines, the stronger the field.
3. Electric field lines start on positive charges and end on negative charges; and the number starting or ending is proportional to the magnitude of the charge.

FIGURE 16–24 The earth's gravitational field.

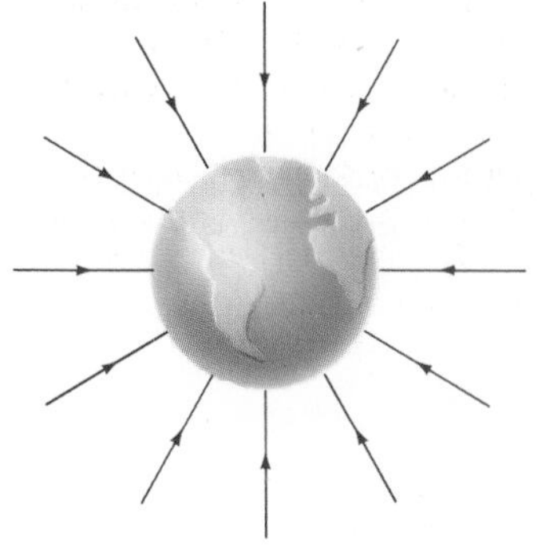

The field concept can also be applied to the gravitational force. Thus we can say that a **gravitational field** exists for every object that has mass. One object attracts another by means of the gravitational field. The earth, for example, can be said to possess a gravitational field (Fig. 16–24) which is responsible for the force on objects that we drop. The *gravitational field intensity* is defined as the *force per unit mass.* The magnitude of the earth's gravitational field intensity at any point is then (GM_e/r^2), where M_e is the mass of the earth, r the distance of the point from the earth's center, and G the gravitational constant (Chapter 5). At the earth's surface, r is simply the radius of the earth and the gravitational field intensity is simply equal to g, the acceleration due to gravity (since $F/m = mg/m = g$). Beyond the earth, the gravitational field intensity can be calculated at any point as a sum of terms due to earth, sun, moon, and other bodies that contribute significantly.

16–9 • Electric Fields and Conductors

$E = 0$ *inside a good conductor*

The electric field inside a good conductor is zero in the static situation—that is, when the charges are at rest. If there were an electric field within a conductor, there would be a force on its free electrons since $\mathbf{F} = q\mathbf{E}$. The electrons would move until they reached positions where the electric field, and therefore the electric force on them, did become zero.

FIGURE 16–25 A charge placed inside a spherical shell. Charges are induced on the conductor surfaces. The electric field exists even beyond the shell but not within the conductor itself.

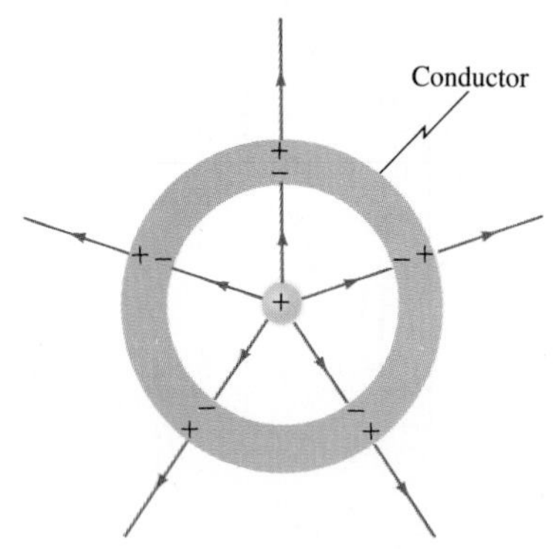

This reasoning has some interesting consequences. For one, *any net charge on a good conductor distributes itself on the outer surface.* For a negatively charged conductor, you can imagine that the negative charges repel one another and race to the surface to get as far away from one another as possible. Another consequence is the following. Suppose that a positive charge Q is surrounded by an isolated uncharged metal conductor whose shape is a spherical shell, Fig. 16–25. Because there can be no field within the metal, the lines leaving the positive charge must end on negative charges on the inner surface of the metal. Thus an equal amount of negative charge, $-Q$, is induced on the inner surface of the spherical shell. Then a positive charge, $+Q$, of the same magnitude must exist on the outer surface of the shell (since the shell is neutral). Thus, although no field exists in the metal itself, an electric field exists outside of it, as shown in Fig. 16–25, as if the metal were not even there.

A related property of static electric fields and conductors is that *the electric field is always perpendicular to the surface outside of a conductor.* If there were a component of **E** parallel to the surface, electrons at the surface would move along the surface in response to this force (Fig. 16–26), until they reached positions where no force was exerted on them—that is, until the electric field was perpendicular to the surface.

E is ⊥ surface outside conductor

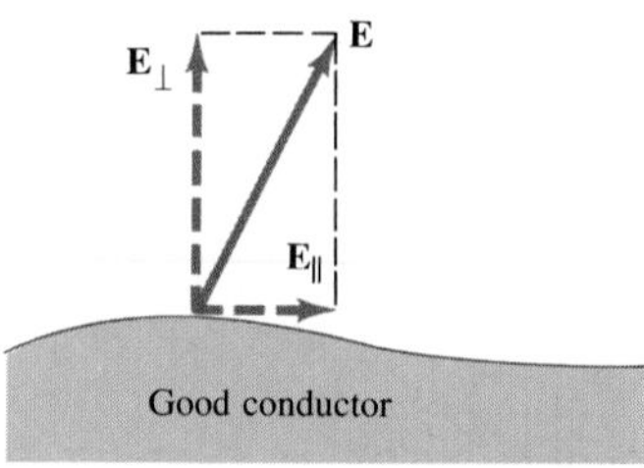

FIGURE 16–26 If the electric field **E** at the surface of a conductor had a component parallel to the surface, $\mathbf{E}_\parallel$, the latter would accelerate electrons into motion. In the static case (no charges are in motion), $\mathbf{E}_\parallel$ must be zero and so the electric field must be perpendicular to the conductor's surface: $\mathbf{E} = \mathbf{E}_\perp$.

These properties pertain only to conductors. Inside a nonconductor, which does not have free electrons, an electric field can exist. And the electric field outside a nonconductor does not necessarily make an angle of 90° to the surface.

*16–10 • Electric Forces in Molecular Biology: DNA Structure and Replication

The study of the structure and functioning of a living cell at the molecular level is known as molecular biology. It is an important area for application of physics. Since the interior of a cell is mainly fluid (mostly water), we can imagine it as a vast sea of molecules continually in motion (as in kinetic theory, Chapter 13), colliding with one another with various amounts of kinetic energy. These molecules interact with one another in various ways—chemical reactions (making and breaking of chemical bonds) and more brief interactions or unions that occur because of *electrostatic attraction* between molecules.

The many activities that occur within the cell, and which help to differentiate living matter from nonliving, are now considered to be the result of random ("thermal") molecular motion plus the ordering effect of the electrostatic force. We now use these ideas to analyze some basic cellular processes involving macromolecules (large molecules) such as DNA and proteins. The picture we present here has not been seen "in action." Rather, it is a model of what happens based on accepted physical theories and a great variety of experimental results.

The genetic information that is passed on from generation to generation in all living objects is contained in the chromosomes, which are made up of genes. Each gene contains the information needed to produce a particular type of protein molecule. A protein consists of one or more chains of small molecules called amino acids. A protein molecule may function as part of a structure (cell wall or muscle fiber, for example), or as an enzyme to catalyze a chemical reaction needed for the growth or survival of the organism. The genetic information contained in a gene is built into the principal molecule of a chromosome, the DNA (deoxyribonucleic acid). A DNA molecule consists of a long chain of many small molecules known as nucleotide bases. There are only four types of bases: adenine (A), cytosine (C), guanine (G), and thymine (T). These are arranged along the molecule according to a code, the "genetic code," which is "translated" into the amino acids that form the protein molecule.

The DNA in a chromosome generally consists of two long DNA chains wrapped about one another in the shape of a "double helix." As shown in

PROBLEMS

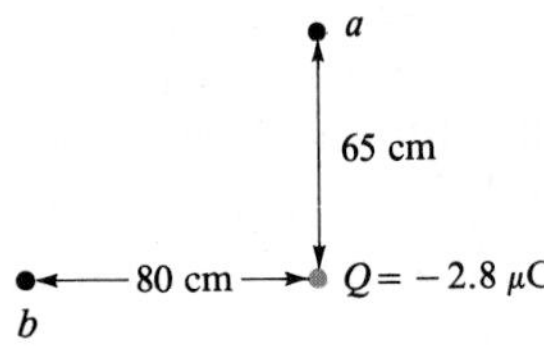

FIGURE 17–10 Problem 15.

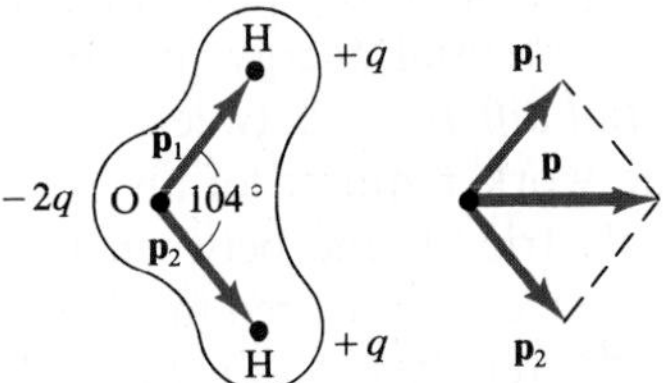

FIGURE 17–12 Problem 20.

SECTIONS 17–1 TO 17–4

1. (I) How much work is needed to move a -8.0-μC charge from ground to a point whose potential is $+75$ V?
2. (I) How much kinetic energy will an electron gain (joules) if it falls through a potential difference of 350 V?
3. (I) An electron acquires 4.2×10^{-16} J of kinetic energy when it is accelerated by an electric field from plate A to plate B. What is the potential difference between the plates, and which plate is at the higher potential?
4. (I) The electric field between two parallel plates connected to a 24-V battery is 600 V/m. How far apart are the plates?
5. (I) How strong is the electric field between two parallel plates 6.0 mm apart if the potential difference between them is 170 V?
6. (I) An electric field of 800 V/m is desired between two parallel plates 16.0 mm apart. How large a voltage should be applied?
7. (I) What potential difference is needed to give a helium nucleus ($Q = 3.2 \times 10^{-19}$ C) 38.0 keV of KE?
8. (II) The work done by an external force to move a -3.0-μC charge from point a to point b is 18.0×10^{-4} J. If the charge was started from rest and had 4.0×10^{-4} J of kinetic energy when it reached point b, what must be the potential difference between a and b?
9. (II) What is the speed of a 950-eV electron?
10. (II) What is the speed of a proton whose kinetic energy is 20 MeV?

SECTION 17–5

11. (I) What is the electric potential 21.0 cm from a 4.85-μC point charge?
12. (I) (*a*) What is the electric potential 0.53×10^{-10} m from a proton (charge $+e$)? (*b*) What is the potential energy of an electron at this point?
13. (II) A $+30$-μC charge is placed 60 cm from an identical $+30$-μC charge. How much work would be required to move a $+0.20$-μC test charge from a point midway between them to a point 10 cm closer to either of the charges?
14. (II) How much work must be done to bring three electrons from a great distance apart to within 1.0×10^{-10} m from one another?
15. (II) Consider point a which is 65 cm north of a -2.8-μC point charge, and point b which is 80 cm west of the charge (Fig. 17–10). Determine (*a*) $V_{ba} = V_b - V_a$, and (*b*) $\mathbf{E}_b - \mathbf{E}_a$ (magnitude and direction).
16. (III) Two equal but opposite charges are separated by a distance d, as shown in Fig. 17–11. Determine a formula for $V_{BA} = V_B - V_A$ for points B and A on the line between the charges situated as shown.

SECTION 17–6

*17. (II) An electron and a proton are 0.53×10^{-10} cm apart. (*a*) What is their dipole moment if they are at rest? (*b*) What is the average dipole moment if the electron revolves about the proton in a circular orbit?

*18. (II) Calculate the electric potential due to a dipole whose dipole moment is 4.8×10^{-30} C·m at a point 1.0×10^{-9} m away if this point is: (*a*) along the axis of the dipole nearer the positive charge; (*b*) 45° above the axis but nearer the positive charge; (*c*) 45° above the axis but nearer the negative charge.

*19. (II) (*a*) In Example 17–5, part (*b*), calculate the electric potential without using the dipole approximation, Eq. 17–4; that is, don't assume $r \gg l$. (*b*) What is the percent error in this case when the dipole approximation is used?

*20. (III) The dipole moment, considered as a vector, points from the negative to the positive charge. The water molecule, Fig. 17–12, has a dipole moment $\mathbf{p}$ which can be considered as the vector sum of the two dipole moments, $\mathbf{p}_1$ and $\mathbf{p}_2$, as shown. The distance between each H and the O is about 0.96×10^{-10} m. The lines joining the center of the O atom with each H atom make an angle of 104°, as shown, and the net dipole moment has been measured to be $p = 6.1 \times 10^{-30}$ C·m. Determine the charge q on each H atom.

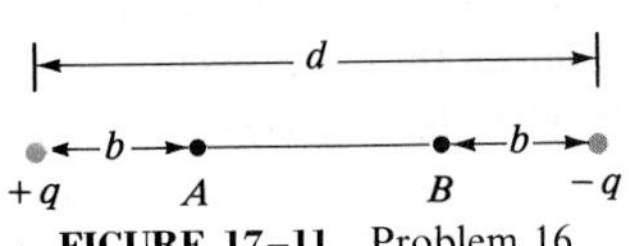

FIGURE 17–11 Problem 16.

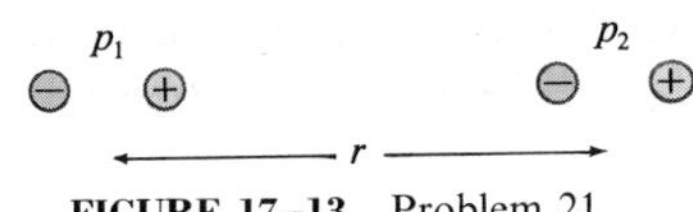

FIGURE 17–13 Problem 21.

*21. (III) Show that if two dipoles with dipole moments p_1 and p_2 are in line with one another (Fig. 17–13), the potential energy of one in the presence of the other (their "interaction energy") is given by

$$\text{PE} = -\frac{2kp_1p_2}{r^3},$$

where r is the distance between the two dipoles. Assume that r is much greater than the length of either dipole.

SECTIONS 17–7 AND 17–8

22. (I) The two plates of a capacitor hold $+2500\ \mu\text{C}$ and $-2500\ \mu\text{C}$ of charge, respectively, when the potential difference is 888 V. What is the capacitance?
23. (I) The potential difference between two parallel wires in air is 20 V. They carry equal and opposite charge of magnitude 65 pC. What is the capacitance of the two wires?
24. (I) A 12,000-pF capacitor holds 18.0×10^{-8} C of charge. What is the voltage across the capacitor?
25. (I) How much charge flows from a 12.0-V battery when it is connected to a 7.50-μF capacitor?
26. (I) A 1.0-F capacitor is desired. What area must the plates have if they are to be separated by a 6.0-mm air gap?
27. (I) What is the capacitance of two square parallel plates 16.0 cm on a side that are separated by 2.3 mm of paraffin?
28. (II) The charge on a capacitor increases by 16 μC when the voltage across it increases from 28 V to 40 V. What is the capacitance of the capacitor?
29. (II) An electric field of 28.0×10^6 V/m is desired between two parallel plates each of area 120 cm^2 and separated by 2.00 cm of air. What charge must be on each plate?
30. (II) How strong is the electric field between the plates of a 30-μF air-gap capacitor if they are 3.0 mm apart and each has a charge of 800 μC?
31. (III) A 5.5-μF capacitor is charged by a 25-V battery and then is disconnected from the battery. When it is then connected to a second (initially uncharged) capacitor, C_2, the voltage on the first drops to 10 V. What is the value of C_2?
32. (III) A 4.0-μF capacitor is charged to 1200 V and a 10.0-μF capacitor is charged to 750 V. The positive plates are now connected to each other and the negative plates are connected to each other. What will be the potential difference across each and the charge on each?

SECTION 17–9

33. (I) 300 V is applied to a 3600-pF capacitor. How much energy is stored?
34. (II) (*a*) How much energy is stored by the electric field between two square plates, 20 cm on a side, separated by a 2.0-mm air gap? The charges on the plates are equal and opposite and of magnitude 800 μC. (*b*) If the gap were filled with mica, how much energy would be stored?
35. (II) A parallel-plate capacitor has a fixed charge Q. The separation of the plates is then doubled. By what factor does the energy stored in the electric field change?
36. (II) How does the energy stored in a capacitor change if (*a*) the potential difference is doubled, (*b*) the charge on each plate is doubled, and (*c*) the separation of the plates is doubled, as the capacitor remains connected to a battery?
37. (III) A 3.0-μF capacitor is charged by a 12-V battery. It is disconnected from the battery and then connected to an uncharged 5.0-μF capacitor. Determine the total stored energy (*a*) before the two capacitors are connected and (*b*) after they are connected. (*c*) What is the change in energy? (*d*) Is energy conserved? Explain why.

GENERAL PROBLEMS

38. There is an electric field near the earth's surface whose intensity is about 150 V/m. How much energy is stored per cubic meter in this field?
39. A lightning flash transfers 3.0 C of charge to earth through a potential difference of 8.5×10^6 V. (*a*) How much energy is dissipated? (*b*) How much water at 0°C could be brought to boiling?
40. What is the KE in electron volts of an oxygen molecule at STP (0°C, 1 atm)?
41. In a television picture tube, electrons are accelerated by thousands of volts through a vacuum. If a television set were laid on its back, would electrons be able to move upward against the force of gravity? What potential difference, acting over a distance of 25 cm, would be needed to balance the downward force of gravity so that an electron would remain stationary? Assume that the electric field is uniform.
42. It takes 8.0 J of energy to move a 3.0-mC charge from one plate of a 6.0-μF capacitor to the other. How much charge is on each plate?
43. How much voltage must be used to accelerate a proton so that it has sufficient energy to penetrate an iron nucleus? An iron nucleus has a charge of $+26e$ and its radius is about 4.0×10^{-15} m. Assume the potential is that for a point charge.
44. An electron starting from rest acquires 2.0 keV of KE in moving from point A to point B. (*a*) How much KE would a proton acquire, starting from rest at B and moving to point A? (*b*) Determine the ratio of their speeds at the end of their respective trajectories.
45. A 7000-pF air-gap capacitor is connected to a 12-V battery. If a piece of mica is placed between the plates, how much charge will then flow from the battery?

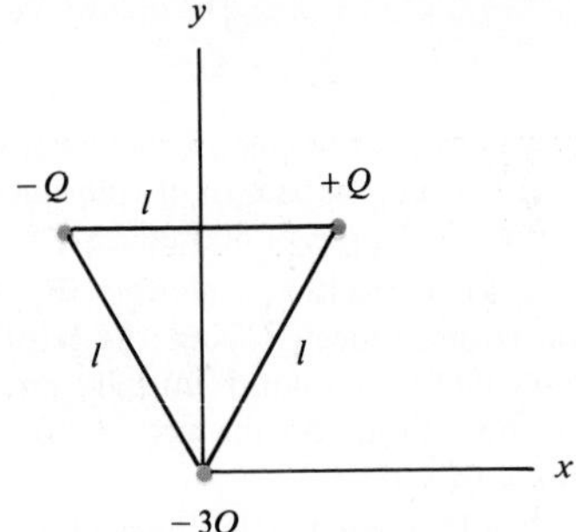

FIGURE 17–14 Problem 48.

46. Dry air will break down if the electric field exceeds 3.0×10^6 V/m. What amount of charge can be placed on a capacitor if the area of each plate is 20 cm^2?

47. A 3.0-μC and a −2.0-μC charge are placed 2.0 cm apart. At what points along the line joining them is (*a*) the electric field zero, and (*b*) the potential zero?

48. Three charges are at the corners of an equilateral triangle (side l) as shown in Fig. 17–14. Determine the potential at the midpoint of each of the sides.

49. A capacitor C_1 carries a charge Q_0. It is then connected directly to a second, uncharged, capacitor C_2. What charge will each carry now? What will be the potential difference across each?

50. An electron is accelerated horizontally from rest in a television picture tube by a potential difference of 15,000 V. It then passes between two horizontal plates 5.0 cm long and 1.2 cm apart that have a potential difference of 250 V (Fig. 17–15). At what angle θ will the electron be traveling after it passes between the plates?

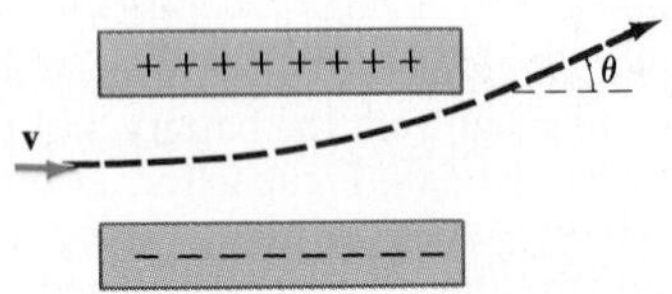

FIGURE 17–15 Problem 50.

51. A thin circular ring of radius R carries a uniformly distributed charge Q. Show that the electric potential at a point P on the axis of the ring a distance x from its center, Fig. 17–16, is given by

$$V = \frac{1}{4\pi\varepsilon_0}\frac{Q}{(x^2 + R^2)^{1/2}}.$$

[*Hint:* assume the ring is made up of tiny point charges, ΔQ (see Fig. 17–16), use Eq. 17–3, and sum over the whole ring.]

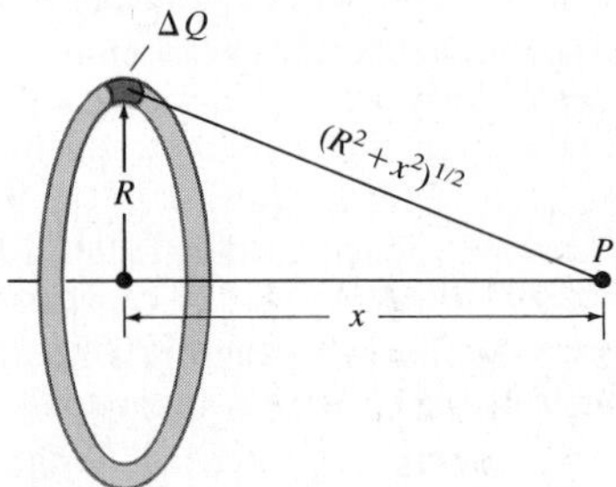

FIGURE 17–16 Problem 51. Calculating the potential at point P, a distance x from the center of a uniform ring of charge.

CHAPTER 18

Electric Currents

The glow of the heating element in a toaster is caused by the electric current passing through it. Electric energy is transformed to thermal energy (via collisions between moving electrons and atoms of the heating element wire), which causes the temperature of the heating element to become so high that it glows. Bread is toasted not by contact but by radiation (see also Section 14–9).

Until the year 1800, the technical development of electricity consisted mainly of producing a static charge by friction. In the preceding century, a number of machines had been built that could produce rather large potentials by frictional means. One type of such apparatus is shown in Fig. 18–1. Large sparks could be produced by these machines, but they had little practical value.

In nature itself there were grander displays of electricity such as lightning and "St. Elmo's fire," which is a glow that appeared around the yardarms of ships during storms. That these phenomena were electrical in origin was not recognized until the eighteenth century. For example, it was only in 1752 that Franklin, in his famous kite experiment, showed that lightning was an electric discharge—a giant electric spark.

Finally, in 1800, an event of great practical importance occurred: Alessandro Volta (1745–1827; Fig. 18–2) invented the electric battery, and with it produced the first steady flow of electric charge—that is, a steady electric current. This discovery opened a new era, which transformed our civilization, for today's electrical technology is based on electric current.

FIGURE 18–1
Early electrostatic generator.

18–1 • The Electric Battery

FIGURE 18–2 Allessandro Volta. In this portrait, Volta exhibits his battery to Napoleon in 1801.

The events that led to the discovery of the battery are interesting. For not only was this an important discovery, but it also gave rise to a famous scientific debate between Volta and Luigi Galvani (1737–1798), eventually involving many others in the scientific world.

In the 1780s, Galvani, a professor at the University of Bologna (thought to be the world's oldest university still in existence), carried out a long series of experiments on the contraction of frog's leg muscle through electricity produced by a static-electricity machine. In the course of these investigations, Galvani found, much to his surprise, that contraction of the muscle could be produced by other means as well: when a brass hook was pressed into the frog's spinal cord and then hung from an iron railing that also touched the frog, the leg muscles again would contract. With further investigation, Galvani found that this strange but important phenomenon also occurred for other pairs of metals.

What was the source of this unusual phenomenon? Galvani believed that the source of the electric charge was in the frog muscle or nerve itself, and that the wire merely transmitted the charge to the proper points; and when he published his work in 1791, he termed it "animal electricity." Many wondered, including Galvani himself, if he had discovered the long-sought "life-force."

Volta, at the University of Pavia 200 km away, was at first skeptical of Galvani's results. At the urging of his colleagues, he soon confirmed and extended those experiments. But Volta doubted Galvani's idea of animal electricity. Instead he came to believe that the source of the electricity was not in the animal iself, but rather in the *contact between the two metals.* Volta made public his views and soon had many followers, although others still sided with Galvani.

Volta was both a strong theoretician and a careful and skillful experimenter. He soon realized that a moist conductor, such as a frog muscle or moisture at the contact point of the two dissimilar metals, was necessary in the circuit if it was to be effective. He also saw that the contracting frog muscle was a sensitive instrument for detecting electric "tension" or "electromotive force" (his words for what we now call potential), in fact more sensitive than the best available electroscopes that he and others had developed. Most important, he recognized that a decisive answer to Galvani could be given only if the sensitive frog leg was replaced by an inorganic detector. That is, to cement his view that it was the contact of two dissimilar metals that caused the frog muscle to contract, he would have to connect the two dissimilar metals directly to an electroscope and observe a separation of the leaves representing a potential difference. This proved difficult since his most sensitive† electroscopes were much less sensitive than the frog muscle. But the eventual success of this experiment vindicated Volta's theory.

Volta's research found that certain combinations of metals produced a greater effect than others, and, using his measurements, he listed them in

† Volta's most sensitive electroscope measured about 40 V per degree (angle of leaf separation). Nonetheless, he was finally able to estimate the potential differences produced by dissimilar metals in contact: for a silver-zinc contact he got about 0.7 V, remarkably close to today's value of 0.78 V.

(a) (b)

FIGURE 18–3 Two types of voltaic battery: (a) a pile; (b) a "crown of cups." Z stands for zinc and A for silver (*argentum* in Latin). Taken from Volta's original publication.

order of effectiveness. (This "electrochemical series" is still used by chemists today.) And he found that carbon could be used in place of one of the metals.

The first battery

Volta then conceived what is his greater contribution to science. Between a disc of zinc and one of silver, he placed a piece of cloth or paper soaked in salt solution or dilute acid and piled a "battery" of such couplings, one on top of another, as shown in Fig. 18–3a. This "pile" or "battery" produced a much increased potential difference. Indeed, when strips of metal connected to the two ends of the pile were brought close, a spark was produced. Volta had designed and built the first electric battery. A second design, known as the "crown of cups" is shown in Fig. 18–3b. Volta made public this great discovery in 1800.

The potential produced by Volta's battery was still weak compared to that produced by the best friction machines of the time, although it could produce considerable charge. (The electrostatic machines were high-potential, low-charge devices.) But it had a great advantage: it was "self-renewing"—it could produce a flow of electric charge continuously for a relatively long period of time. It was not long before even more powerful batteries were constructed.

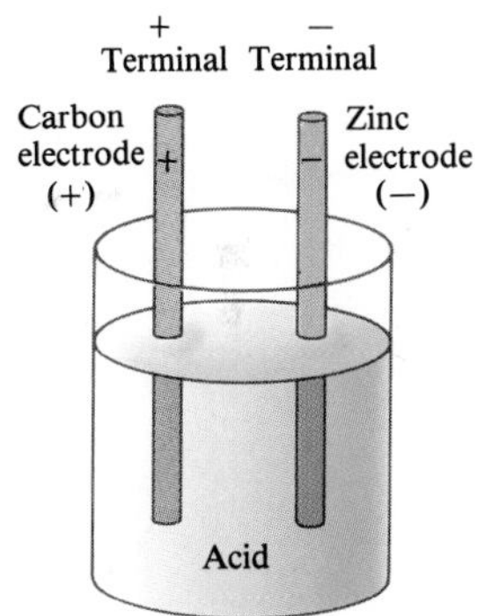

FIGURE 18–4 Simple electric cell.

After Volta's discovery of the electric battery, it was eventually recognized that a battery produces electricity by transforming chemical energy into electrical energy. Today a great variety of electric cells and batteries are available, from flashlight batteries (sometimes called "dry cells") to the storage battery of a car. The simplest batteries contain two plates or rods made of dissimilar metals (one can be carbon) called **electrodes**. The electrodes are immersed in a solution, such as a dilute acid, called the **electrolyte.** (In a dry cell, the electrolyte is absorbed in a powdery paste.) Such a device is properly called an **electric cell**, and several cells connected together is a **battery.** The chemical reactions involved in most electric cells are quite complicated. A detailed description of different electric cells can be found in chemistry textbooks. Here we describe how one very simple cell works, emphasizing the physical aspects.

How a simple battery works

The simple cell shown in Fig. 18–4 uses dilute sulfuric acid as the electrolyte. One of the electrodes is made of carbon, the other of zinc. That part of each electrode remaining outside the solution is called the **terminal**, and connections to wires and circuits are made here. The acid attacks the zinc electrode and tends to dissolve it. But each zinc atom leaves two electrons behind, so it enters the solution as a positive ion. The zinc electrode thus acquires a negative charge. As more zinc ions enter solution, the electrolyte can momentarily become positively charged. Because of this, and through other chemical reactions, electrons are pulled off the carbon electrode. Thus the carbon electrode becomes positively charged. Because there is an opposite charge on the two electrodes, there is a potential difference between the two terminals. In a cell whose terminals are not connected, only a small amount of the zinc is dissolved, for as the zinc electrode becomes increasingly negative, any new positive zinc ions produced are attracted back to the electrode. Thus, a particular potential difference or voltage is maintained between the two terminals. If charge is allowed to flow between the terminals, say, through a wire (or a light bulb), then more zinc can be dissolved. After a time, one or the other electrode is used up and the cell becomes "dead."

The voltage that exists between the terminals of a battery depends on what the electrodes are made of and their relative ability to be dissolved or give up electrons. When two or more cells are connected so that the positive terminal of one is connected to the negative terminal of the next, they are said to be connected in *series* and their voltages add up. Thus, the voltage between the ends of two flashlight batteries so connected is 3.0 V, whereas the six 2-V cells of an automobile storage battery give 12 V.

18-2 • Electric Current

When a continuous conducting path, such as a wire, is connected to the terminals of a battery, we have an electric **circuit**, Fig. 18–5a. On any diagram of a circuit, as in Fig. 18–5b, we will represent a battery by the symbol "⊣⊢"; the longer line on this symbol represents the positive terminal, and the shorter line, the negative terminal. When such a circuit is formed, charge can flow through the circuit from one terminal of the battery to the other. A flow of charge such as this is called an **electric current**.

More precisely, the electric current in a wire is defined as the net amount of charge that passes through it per unit time at any point. Thus, the average current I is defined as

Electric current

$$I = \frac{\Delta Q}{\Delta t}, \tag{18–1}$$

where ΔQ is the amount of charge that passes through a cross-section of conductor at a given point during the time interval Δt. Electric current is measured in coulombs per second; this is given a special name, the **ampere** (abbreviated amp or A), after the French physicist André Ampère (1775–1836). Thus, 1 A = 1 C/s.

The ampere

FIGURE 18–5
(a) Very simple electric circuit.
(b) Schematic drawing of the circuit in part (a).

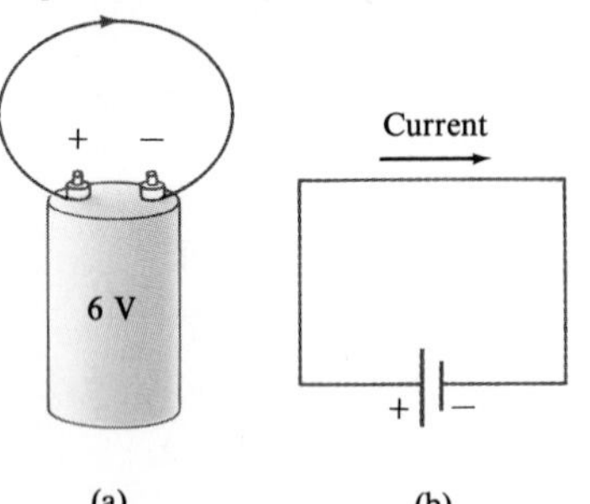

EXAMPLE 18–1 A steady current of 2.5 A flows in a wire connected to a battery. But after 4.0 min, the current suddenly ceases because the wire is disconnected. How much charge passed through the circuit?

SOLUTION Since the current was 2.5 A, or 2.5 C/s, then in 4.0 minutes (= 240 seconds) the total charge that flowed was, from Eq. 18–1,

$$\begin{aligned}\Delta Q &= I\,\Delta t\\ &= (2.5\text{ C/s})(240\text{ s}) = 600\text{ C}.\end{aligned}$$

We saw in Chapter 16 that conductors contain many free electrons. Thus, when a conducting wire is connected to the terminals of a battery as in Fig. 18–5, it is actually the negatively charged electrons that flow in the wire. When the wire is first connected, free electrons at one end of the wire are attracted into the positive terminal. At the same time, electrons leave the negative terminal of the battery and enter the wire at the other end. Thus there is a continuous flow of electrons through the wire that begins as soon as the wire is connected to *both* terminals. However, when the conventions of positive and negative charge were invented two centuries ago, it was assumed positive charge flowed in a wire. Actually, for nearly all purposes, positive charge flowing in one direction is exactly equivalent to negative charge flowing in the opposite direction, Fig. 18–6 (but see Section 20–7). Today, we still use the historical convention of positive current flow when discussing the direction of a current. (It also fits in well with our definition of electric potential in terms of a positive test charge). So when we speak of the current flowing in a circuit, we mean the direction positive charge would flow. This is sometimes referred to as **conventional current**. When we want to speak of the direction of electron flow, we will specifically state it is the electron current. In liquids and gases, both positive and negative charges (ions) can move.

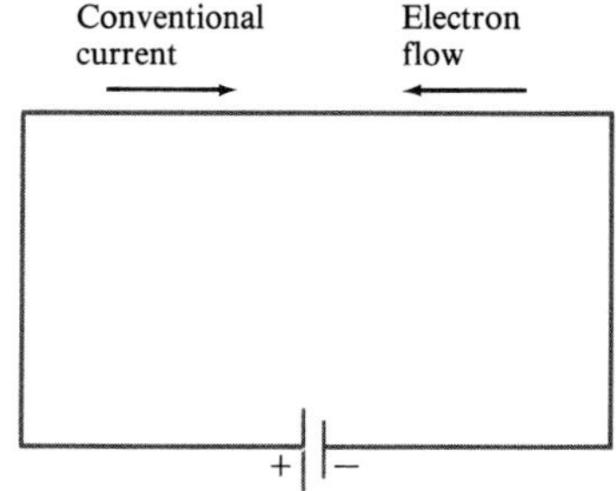

FIGURE 18–6
Conventional current from + to − is equivalent to a negative (electron) current flowing from − to +.

18–3 • Ohm's Law: Resistance and Resistors

To produce an electric current in a circuit, a difference in potential is required. One way of producing a potential difference is by a battery. It was Georg Simon Ohm (1787–1854) who established experimentally that the current in a metal wire is proportional to the potential difference V applied to its ends:

$$I \propto V.$$

This is known as **Ohm's law**. If, for example, we connect a wire to a 6-V battery, the current flow will be twice what it would be if the wire were connected to a 3-V battery.

It is helpful to compare an electric current to the flow of water in a river or a pipe. If the pipe (or river) is nearly level, the flow rate is small. But if one end is somewhat higher than the other, the flow rate—or current—is much greater. The greater the difference in height, the greater the current. We saw in Chapter 17 that electric potential is analogous, in the gravitational case, to the height of a cliff; and this applies in the present case to the height through which the fluid flows. Just as an increase in height causes a greater flow of water, so a greater electric potential difference, or voltage, causes a greater current flow.

Exactly how much current flows in a wire depends not only on the voltage, but also on the resistance the wire offers to the flow of electrons. The walls of a pipe, or the banks of a river and rocks in the middle, offer resistance to the flow of current. Similarly, electrons are slowed down because of interactions with the atoms of the wire. The higher this resistance, the less the current for a given voltage V. We then define resistance so that the current is inversely proportional to the resistance. When we combine this with the above proportion, we have

$$I = \frac{V}{R}, \qquad (18\text{–}2)$$

where R is the **resistance** of a wire or other device, V is the potential difference across the device, and I is the current that flows through it. This relation (Eq. 18–2) is often written

Ohm's law

$$V = IR,$$

and is often referred to as Ohm's law. Actually, it is not a law, but rather is the *definition of resistance*. If we want to call something Ohm's law, it would be the statement that the current through a *metal conductor* is proportional to the applied voltage, $I \propto V$. That is, R is a constant, independent of V, for metal conductors. But this relation does *not* apply generally for other substances and devices such as semiconductors, vacuum tubes, transistors, and so on. Thus "Ohm's law" is not a fundamental law, but rather a description of a certain class of materials (metal conductors). The habit of calling it Ohm's law is so ingrained that we won't quibble with continuing this usage, as long as we keep in mind its limitations. Materials or devices that do not follow Ohm's law are said to be *nonohmic*. The definition of resistance,

$$R = V/I$$

(Eq. 18–2), can be applied also to nonohmic cases. In these cases, R would not be constant and would depend on the applied voltage.

The unit for resistance is called the **ohm** and is abbreviated Ω (Greek capital omega). Because $R = V/I$, we see that 1.0 Ω is equivalent to 1.0 V/A.

FIGURE 18–7 Some resistors. The resistance value of a given resistor is written on the exterior, or may be given as a color code, as shown in the table below: the first two colors represent the first two digits in the value of resistance, the third represents the power of ten that it must be multiplied by, and the fourth is the manufactured tolerance. For example, a resistor whose four colors are red, green, orange, and silver has a resistance of 25,000 Ω (25 kΩ), give or take 10 percent.

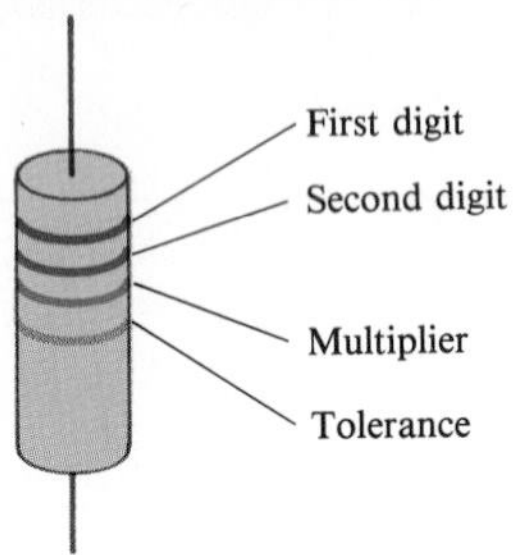

Resistor Color Code

Color	Number	Multiplier	Tolerance (%)
Black	0	1	
Brown	1	10^1	
Red	2	10^2	
Orange	3	10^3	
Yellow	4	10^4	
Green	5	10^5	
Blue	6	10^6	
Violet	7	10^7	
Gray	8	10^8	
White	9	10^9	
Gold		10^{-1}	5%
Silver		10^{-2}	10%
No color			20%

EXAMPLE 18–2 A plate on the bottom of a small tape recorder specifies that it should be connected to 6.0 V and will draw 300 mA. (*a*) What is the net resistance of the recorder? (*b*) If the voltage dropped to 5.0 V, how would the current change?

SOLUTION (*a*) We use Ohm's law, Eq. 18–2, and find

$$R = \frac{V}{I} = \frac{6.0\text{ V}}{0.30\text{ A}} = 20\ \Omega.$$

(*b*) If the resistance stayed constant, the current would be approximately $I = V/R = 5.0\text{ V}/20\ \Omega = 0.25$ A, or a drop of 50 mA. Actually, resistance depends on temperature (Section 18–4), so this is only an approximation.

All electric devices, from heaters to light bulbs to stereo amplifiers, offer resistance to the flow of current. Generally, the connecting wires have very low resistance. In many circuits, particularly in electronic devices, **resistors** are used to control the amount of current. Resistors have resistances from less than an ohm to millions of ohms (see Fig. 18–7). The two main types are "wire-wound" resistors, which consist of a coil of fine wire, and "composition" resistors, which are usually made of the semiconductor carbon.

When we draw a diagram of a circuit, we indicate a resistance with the symbol -\/\/\/-. Wires whose resistance is negligible, however, are shown simply as straight lines.

18–4 • Resistivity

It is found experimentally that the resistance R of a metal wire is directly proportional to its length L and inversely proportional to the cross-sectional area A. That is,

$$R = \rho \frac{L}{A}, \qquad (18\text{–}3)$$

where ρ, the constant of proportionality, is called the **resistivity** and depends on the material used. This relation makes sense since we expect that the resistance of a thick wire would be less than that of a thin one because a thicker wire has more area for the electrons to pass through. And you might expect the resistance to be greater if the length is greater since there would be more obstacles to electron flow.

Typical values of ρ for various materials are given in the first column of Table 18–1. (The values depend somewhat on purity, heat treatment, temperature, and other factors.) Notice that silver has the lowest resistivity and is thus the best conductor, although it is expensive. Copper is not far behind, so it is clear why most wires are made of copper. Aluminum, although it has a higher resistivity, is much less dense than copper; it is thus preferable to copper in some situations, such as transmission lines, because its resistance for the same weight is less than that for copper.

Table 18–1
Resistivity and Temperature Coefficients (at 20°C)

Material	Resistivity, ρ ($\Omega\cdot$m)	Temperature Coefficient, α $(\text{C}^\circ)^{-1}$
Conductors		
Silver	1.59×10^{-8}	0.0061
Copper	1.68×10^{-8}	0.0068
Aluminum	2.65×10^{-8}	0.00429
Tungsten	5.6×10^{-8}	0.0045
Iron	9.71×10^{-8}	0.00651
Platinum	10.6×10^{-8}	0.003927
Mercury	98×10^{-8}	0.0009
Nichrome (alloy of Ni, Fe, Cr)	100×10^{-8}	0.0004
Semiconductors[†]		
Carbon (graphite)	$(3\text{–}60) \times 10^{-5}$	−0.0005
Germanium	$(1\text{–}500) \times 10^{-3}$	−0.05
Silicon	0.1 – 60	−0.07
Insulators		
Glass	$10^{9} - 10^{12}$	
Hard rubber	$10^{13} - 10^{15}$	

[†] Values depend strongly on presence of even slight amounts of impurities.

EXAMPLE 18–3 Suppose you want to connect your stereo set to a remote speaker. If each wire must be 20 m long, what diameter copper wire should you use to keep the resistance less than 0.10 Ω per wire?

SOLUTION We solve Eq. 18–3 for A and use Table 18–1:

$$A = \rho \frac{L}{R} = \frac{(1.7 \times 10^{-8}\ \Omega\cdot\text{m})(20\ \text{m})}{(0.10\ \Omega)} = 3.4 \times 10^{-6}\ \text{m}^2.$$

The cross-sectional area A of a circular wire is related to its diameter d by $A = \pi d^2/4$. The diameter must then be at least $d = \sqrt{4A/\pi} = 2.1 \times 10^{-3}\ \text{m} = 2.1\ \text{mm}$.

The resistivity of a material depends somewhat on temperature. In general, the resistance of metals increases with temperature. This is not surprising, for at higher temperatures, the atoms are moving more rapidly and are arranged in a less orderly fashion. So they might be expected to interfere more with the flow of electrons. If the temperature change is not too great, the resistivity of metals increases nearly linearly with temperature. That is,

$$\rho_T = \rho_0(1 + \alpha\,\Delta T), \qquad (18\text{–}4)$$

where ρ_0 is the resistivity at some reference temperature (such as 0°C or 20°C), ρ_T is the resistivity at a temperature ΔT above the reference temperature (or below, in which case $\Delta T < 0$), and α is the *temperature coefficient of resistivity*. Values for α are given in Table 18–1. Note that the temperature coefficient for semiconductors can be negative. Why? It seems that at higher temperatures, some of the electrons that are not normally free become free and can contribute to the current. Thus, the resistance of a semiconductor can decrease with an increase in temperature, although this is not always the case.

EXAMPLE 18–4 *Resistance thermometer.* The variation in electrical resistance with temperature can be used to make precise temperature measurements. Platinum is usually used since it is relatively free from corrosive effects and has a high melting point. To be specific, suppose at 0°C the resistance of a platinum resistance thermometer is 164.2 Ω. When placed in a particular solution, the resistance is 187.4 Ω. What is the temperature of this solution?

SOLUTION Since the resistance R is directly proportional to the resistivity ρ, we can combine Eq. 18–3 with 18–4 and write, taking 0°C as our reference temperature so $\Delta T = T$ (in °C):

$$R = R_0(1 + \alpha T).$$

Here $R_0 = \rho_0 L/A$ is the resistance of the wire at 0°C. We solve this equation for T and find

$$T = \frac{R - R_0}{\alpha R_0} = \frac{187.4\ \Omega - 164.2\ \Omega}{(3.927 \times 10^{-3}(\text{C}^\circ)^{-1})(164.2\ \Omega)} = 35.9^\circ\text{C}.$$

More convenient for some applications is a *thermistor*, which consists of a metal oxide or semiconductor whose resistance also varies in a repeatable way with temperature. They can be made quite small and respond very quickly to temperature changes.

The value of α in Eq. 18–4 depends on temperature, so it is important to check the temperature range of validity of any value (say, in a handbook of physical data). If the temperature range is wide, Eq. 18–4 is not adequate and terms proportional to the square and cube of the temperature are needed: $\rho_T = \rho_0[1 + \alpha\,\Delta T + \beta(\Delta T)^2 + \gamma(\Delta T)^3]$, where the coefficients β and γ are generally very small, but when ΔT is large, their terms become significant.

*18–5 • Superconductivity

At low temperatures, the resistivity of certain metals and their compounds or alloys becomes zero to within the highest-precision measuring techniques. Materials in such a state are said to be **superconducting**. This phenomenon was first observed by H. K. Onnes (1853–1926) in 1911 when he cooled mercury below 4.2 K. He found that at this temperature, the resistance of mercury suddenly dropped to zero. In general, superconductors become superconducting only below a certain *transition temperature*, which is usually within a few degrees of absolute zero. Current in a ring-shaped superconducting material has been observed to flow for years in the absence of a potential difference, with no measurable decrease.

Much research has been done on superconductivity in recent years to try to understand why it occurs,[†] and to find materials that superconduct at more reasonable temperatures to reduce the cost and inconvenience of refrigeration at the required very low temperature. Before 1986 the highest temperature at which a material was found to superconduct was 23 K, and this required liquid helium or liquid hydrogen to keep the material cold. In 1987, a compound of yttrium, barium, copper, and oxygen was developed that can be superconducting at 90 K. Since this is above the temperature of liquid nitrogen, 77 K, boiling liquid nitrogen is sufficiently cold to keep the material superconducting. This was an important breakthrough since liquid nitrogen is much more easily and cheaply obtained than is liquid helium.

High-temperature superconductors

Applications of superconductivity that once seemed like dreams might become real, although technological development will take some time. A major use of superconductors is for carrying the current in electromagnets (we shall see in Chapter 20 that electric currents produce magnetic fields). In large magnets, an enormous amount of energy is required just to maintain the current, and this energy is wasted as heat (I^2R loss—see Section 18–6). No energy is wasted in maintaining electromagnets using superconductors, although some energy is needed to refrigerate the superconducting material at low temperatures; but the net energy saving is great. Low-temperature

[†] The first successful superconductivity theory was published by Bardeen, Cooper, and Schrieffer (the BCS theory) in 1957. It is based on quantum theory and cannot be explained on the basis of classical physics.

superconducting magnets are already in use at the Fermilab accelerator in Illinois, and in many hospitals for magnetic resonance imaging studies of the body (Section 31–9). If such magnets can be replaced by superconductors at higher temperatures, the energy cost will be reduced significantly.

FIGURE 18–8 An experimental train in Japan, supported by the magnetic field produced by current in coils beneath the tracks (in the red containers).

Although electric motors and generators do not waste a large amount of heat, the use of higher-temperature superconductors would enable motors and generators to be much smaller (perhaps $\frac{1}{10}$ the size today) if superconductors can be developed that can sustain large currents. Transmission of power over long distances using superconductors would similarly require much smaller and less costly transmission lines (that can be placed underground) in place of the grand towers carrying multiple lines that mar the countryside today. And they wouldn't have to carry the power at the high voltage used today (see Section 21–7).

Superconductors could make electric cars more practical, make computers far faster than they are today, and have great potential in devices to store energy for use at peak demand. Superconductors are already being studied for use in high-speed ground transportation: the magnetic fields produced by superconducting magnets would be used to "levitate" vehicles over the tracks so there is essentially no friction (Fig. 18–8). The levitation arises from the repulsive force between the magnet (on the train) and the eddy currents (see Section 21–6) produced in the track below.

18–6 • Electric Power

Electric energy is useful to us because it can be easily transformed into other forms of energy. Motors, whose operations we will examine in Chapter 20, transform electric energy into mechanical work.

In other devices such as electric heaters, stoves, toasters, and hair dryers, electric energy is transformed into thermal energy in a wire resistance known as a "heating element." And in an ordinary light bulb, the tiny wire filament (Fig. 18–9) becomes so hot it glows; only a few percent of the energy is transformed into visible light, and the rest, over 90 percent, into thermal energy. Light-bulb filaments and heating elements in household appliances have resistances typically of a few ohms to a few hundred ohms.

FIGURE 18–9 Incandescent light bulb.

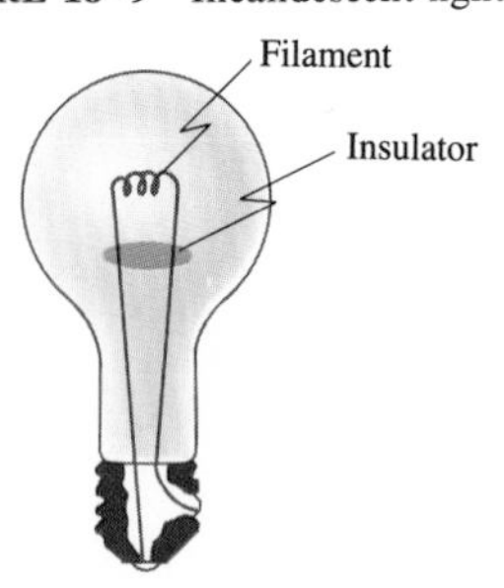

Electric energy is transformed into thermal energy or light in such devices because the current is usually rather large, and there are many collisions between the moving electrons and the atoms of the wire. In each collision, part of the electron's kinetic energy is transferred to the atom with which it collides. As a result, the kinetic energy of the atoms increases and hence the temperature of the wire element increases. The increased thermal energy (internal energy) can be transferred as heat by conduction and convection to the air in a heater or to food in a pan, by radiation to toast in a toaster, or radiated as light.

To find the power transformed by an electric device recall that the energy transformed when a charge Q moves through a potential difference V is QV (Eq. 17–1). Then the power P, which is the rate energy is transformed, is

$$P = \text{power} = \frac{\text{energy transformed}}{\text{time}} = \frac{QV}{t}.$$

The charge that flows per second, Q/t, is simply the electric current I. Thus we have

$$P = IV. \tag{18–5}$$

This general relation gives us the power transformed by any device, where I is the current passing through it and V is the potential difference across it. It also gives the power delivered by a source such as a battery. The SI unit of electric power is the same as for any kind of power, the **watt** (1 W = 1 J/s).

The rate of energy transformation in a resistance R can be written, using Ohm's law ($V = IR$), in two other ways:

Electric power

$$P = IV \tag{18–6a}$$

$$= I(IR) = I^2R \tag{18–6b}$$

$$= \left(\frac{V}{R}\right)V = \frac{V^2}{R}. \tag{18–6c}$$

Equations 18–6b and c apply only to resistors, whereas Eq. 18–6a, $P = IV$, applies to any device.

EXAMPLE 18–5 Calculate the resistance of a 40-W automobile headlight designed for 12 V.

SOLUTION Since we are given $P = 40$ W and $V = 12$ V, we can use Eq. 18–6c and solve for R:

$$R = \frac{V^2}{P} = \frac{(12\text{ V})^2}{(40\text{ W})} = 3.6\ \Omega.$$

This is the resistance when the bulb is burning brightly at 40 W. When the bulb is cold, the resistance is much lower. (Since the current is high when the resistance is low, light bulbs burn out most often when first turned on.)

It is energy, not power, you pay for on your electric bill. Since power is the *rate* energy is transformed, the total energy used by any device is simply its power consumption multiplied by the time it is on. If the power is in watts and the time is in seconds, the energy will be in joules since 1 W = 1 J/s. Electric companies usually specify the energy with a much larger unit, the kilowatt-hour (kWh). One kWh = (1000 W)(3600 s) = 3.60×10^6 J.

EXAMPLE 18–6 An electric heater draws 15 A on a 120-V line. How much power does it use and how much does it cost per month (30 days) if it operates 3.0 h per day and the electric company charges $0.080 per kWh? (For simplicity, assume the current flows steadily in one direction.)

SOLUTION The power is $P = IV = (15\text{ A})(120\text{ V}) = 1800$ W or 1.8 kW. To operate it for (3.0 h/d)(30 d) = 90 h would cost (1.8 kW)(90 h)($0.080) = $13.

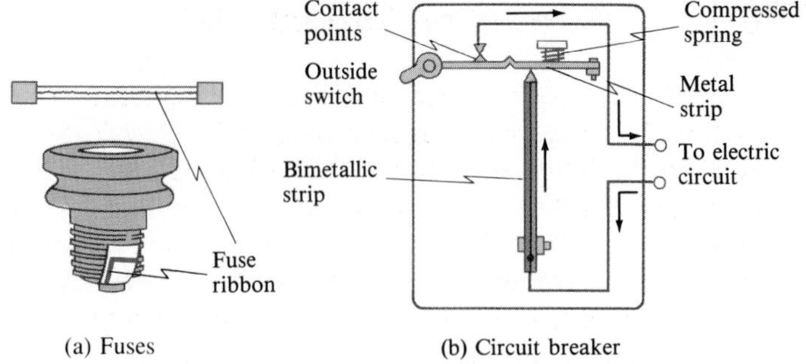

FIGURE 18–10 (a) Fuses. When the current exceeds a certain value, the ribbon melts and the circuit opens. Then the fuse must be replaced. (b) A circuit breaker. Electric current passes through a bimetallic strip. When the current is great enough (i.e., too great to be safe), the bimetallic strip is heated and bends so far to the left that the notch in the spring-loaded metal strip drops down over the end of the bimetallic strip. The circuit then opens at the contact points (one is attached to the metal strip) and the outside switch is also flipped. As the device cools down, it can be reset using the outside switch.

The electric wires that carry electricity to lights and other electric appliances have some resistance, although usually it is quite small. Nonetheless, if the current is large enough, the wires will heat up and produce thermal energy at a rate equal to I^2R, where R is the wire's resistance. One possible hazard is that the current-carrying wires in the wall of a building may become so hot as to start a fire. Thicker wires, of course, have less resistance (see Eq. 18–3) and thus can carry more current without becoming too hot. When a wire carries more current than is safe, it is said to be "overloaded." A building should, of course, be designed with wiring heavy enough for any expected load. To prevent overloading, fuses or circuit breakers are installed in circuits. They are basically switches (Fig. 18–10) that open the circuit when the current exceeds some particular value. A 20-A fuse or circuit breaker, for example, opens when the current passing through it exceeds 20 A. If a circuit repeatedly burns out a fuse or opens a circuit breaker, there are two possibilities: there may be too many devices drawing current in that circuit; or there is a fault somewhere, such as a "short." A short, or "short circuit," means that two wires have crossed (perhaps because the insulation has worn down) so the path of the current is shortened. The resistance of the circuit is then very small, so the current will be very large. Short circuits, of course, should be remedied immediately.

Fuses, circuit breakers, and shorts

FIGURE 18–11
Connections of household appliances.

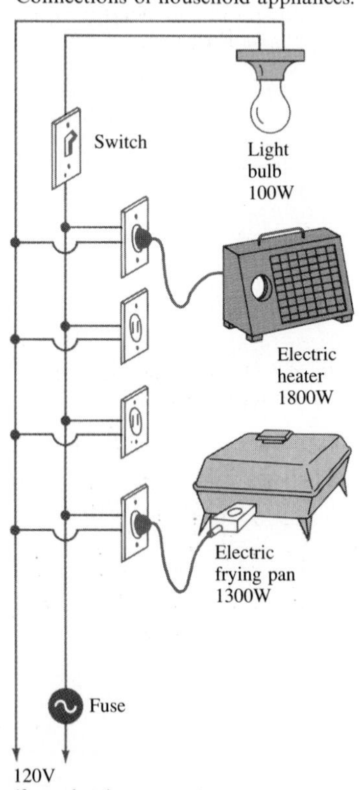

Household circuits are designed with the various devices connected so that each receives the standard voltage (usually 120 V in the US) from the electric company (Fig. 18–11). When a fuse blows or circuit breaker opens, the total current being drawn should be checked first. For example, the circuit in Fig. 18–11 draws the following currents: a light bulb draws $I = P/V = 100\text{ W}/120\text{ V} = 0.8\text{ A}$; a heater, $1800\text{ W}/120\text{ V} = 15.0\text{ A}$; and a frying pan, $1300\text{ W}/120\text{ V} = 10.8\text{ A}$; the total is 26.6 A. If the circuit has a 20-A fuse, you might expect it to blow. If it has a 30-A fuse, it shouldn't blow, so a short may be the problem. (The most likely place is in the cord of one of the appliances.) Proper fuse size is selected according to the wire used to supply the current; a properly rated fuse should never be replaced by a higher-rated one.

In electric circuits, heat dissipation by resistors must be considered. The physical size of a resistor is a rough indicator of the maximum permissible power it can dissipate ($=I^2R$) without appreciable rise in temperature. Common values are $\frac{1}{4}$ W, $\frac{1}{2}$ W, and 1 W; the higher the wattage, the larger the physical size.

18–7 • Alternating Current

When a battery is connected to a circuit, the current flows steadily in one direction. This is called a **direct current**, or dc. Electric generators at electric power plants, however, produce **alternating current**, or ac.[†] An alternating current reverses direction many times per second and is usually sinusoidal, as shown in Fig. 18–12. The electrons in a wire first move in one direction and then in the other. The current supplied to homes and businesses by electric companies is ac throughout virtually the entire world.

DC

AC

Electrons move back and forth in ac

The voltage produced by an ac electric generator is sinusoidal, as we shall see in Chapter 21. The current it produces is thus sinusoidal (Fig. 18–12b). We can write the voltage as a function of time as

$$V = V_0 \sin 2\pi ft.$$

The potential V oscillates between $+V_0$ and $-V_0$. V_0 is referred to as the **peak voltage**. The frequency f is the number of complete oscillations made per second. In most areas of the United States and Canada, f is 60 Hz (cycles per second). In some countries, 50 Hz is used.

From Ohm's law, if a voltage V exists across a resistance R, then the current I is

$$I = \frac{V}{R} = \frac{V_0}{R} \sin 2\pi ft = I_0 \sin 2\pi ft. \qquad (18\text{–}7)$$

The quantity $I_0 = V_0/R$ is the **peak current**. The current is considered positive when the electrons flow in one direction and negative when they flow in the opposite direction. It is clear from Fig. 18–12b that an alternating current is as often positive as it is negative. Thus, the average current is zero. This does not mean, however, that no power is needed or that no heat is produced in a resistor. Electrons do move back and forth, and do produce heat. Indeed, the power delivered to a resistance R at any instant is

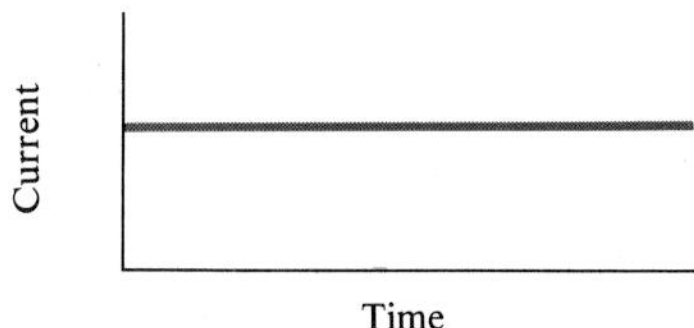

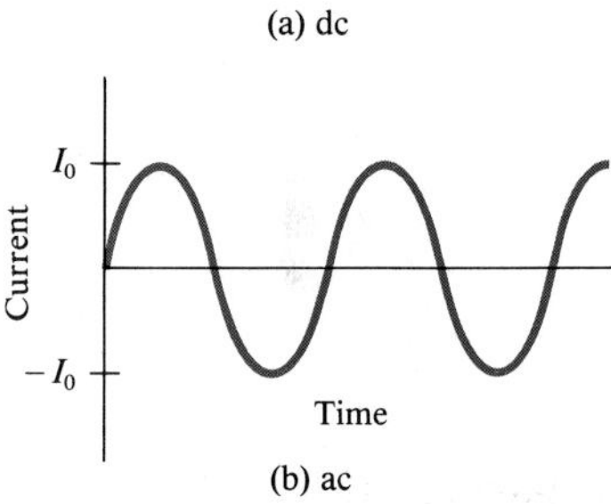

FIGURE 18–12 (a) Direct current. (b) Alternating current.

$$P = I^2R = I_0^2R \sin^2 2\pi ft.$$

Because the current is squared, we see that the power is always positive, Fig. 18–13. The quantity $\sin^2 2\pi ft$ varies between 0 and 1; and it is not too difficult to show that its average value is $\frac{1}{2}$, as can be seen graphically in the figure. Thus, the *average power* developed, $\bar{P}$, is

$$\bar{P} = \tfrac{1}{2}I_0^2R.$$

FIGURE 18–13 Power delivered to a resistor in an ac circuit.

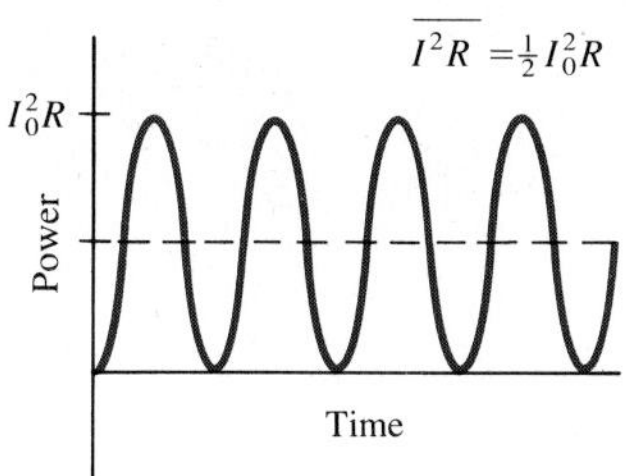

Since power can also be written $P = V^2/R = (V_0^2/R)\sin^2 2\pi ft$, we also have that the average power is

$$\bar{P} = \frac{1}{2}\frac{V_0^2}{R}.$$

The average or mean value of the *square* of the current or voltage is thus what is important for calculating average power: $\overline{I^2} = \frac{1}{2}I_0^2$ and $\overline{V^2} = \frac{1}{2}V_0^2$.

[†] Although it is redundant, we sometimes say "ac current" and "ac voltage," which really mean "alternating current" and "alternating voltage."

The square root of each of these is the **rms** (root-mean-square) value of the current or voltage:

rms current

$$I_{\text{rms}} = \sqrt{\overline{I^2}} = \frac{1}{\sqrt{2}} I_0 = 0.707 I_0, \tag{18–8a}$$

rms voltage

$$V_{\text{rms}} = \sqrt{\overline{V^2}} = \frac{1}{\sqrt{2}} V_0 = 0.707 V_0. \tag{18–8b}$$

The rms values of V and I are sometimes called the "effective values." They are useful because they can be substituted directly into the power formulas, Eqs. 18–6, to get the average power. For example, $\bar{P} = \frac{1}{2} I_0^2 R = I_{\text{rms}}^2 R$. Thus, a direct current whose values of I and V equal the rms values of I and V for an alternating current will produce the same power. Hence it is usually the rms value of current that is specified or measured. For example, in the United States and Canada, standard line voltage[†] is 120 Vac. The 120 V is V_{rms}; the peak voltage V_0 is

$$V_0 = \sqrt{2} V_{\text{rms}} = 170 \text{ V}.$$

In most of Europe the rms voltage is 240 V, so the peak voltage is 340 V.

Example 18–7 Calculate the resistance and the peak current in a 1000-W hair dryer connected to a 120-V line. What happens if it is connected to a 240-V line in Britain?

SOLUTION We can apply Eq. 18–6a using rms values. Then the rms current is

$$I_{\text{rms}} = \frac{\bar{P}}{V_{\text{rms}}} = \frac{1000 \text{ W}}{120 \text{ V}} = 8.33 \text{ A}.$$

Thus $I_0 = \sqrt{2} I_{\text{rms}} = 11.8$ A. From Ohm's law the resistance is

$$R = \frac{V_{\text{rms}}}{I_{\text{rms}}} = \frac{120 \text{ V}}{8.33 \text{ A}} = 14.4 \ \Omega.$$

The resistance could equally well be calculated using peak values: $R = V_0/I_0 = 170 \text{ V}/11.8 \text{ A} = 14.4 \ \Omega$. When connected to a 240-V line, the average power delivered would be

$$\bar{P} = \frac{V_{\text{rms}}^2}{R} = \frac{(240 \text{ V})^2}{(14.4 \ \Omega)} = 4000 \text{ W}.$$

This would undoubtedly melt the heating element or the wire coils of the motor. Be sure your hair dryer (or electric shaver) has a 120/240 V switch before traveling too far from home, or carry a transformer (Section 21–7) for travelers.

This section has given a brief introduction to the simpler aspects of alternating currents. We will discuss ac circuits in more detail in Chapter 21. In Chapter 19 we will deal with the details of dc circuits only.

[†] The line voltage can vary, depending on the total load; the frequency of 60 Hz, however, remains extremely steady.

*18–8 • The Nervous System and Nerve Conduction

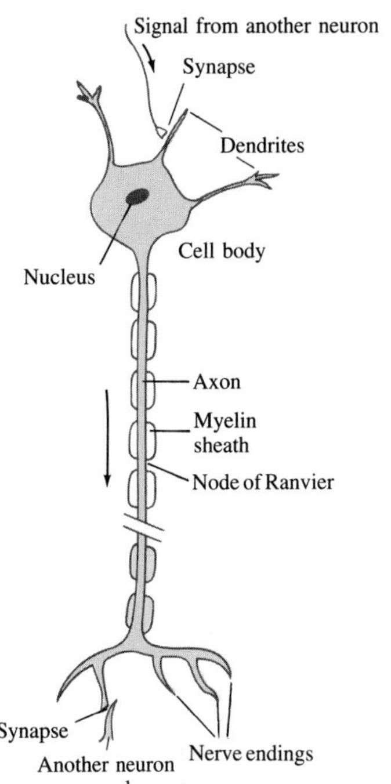

FIGURE 18–14 A neuron.

An interesting example of the flow of electric charge is our remarkable and complex nervous system, which provides us with the means for being aware of the world, for communication within the body, and to control the body's muscles. Although the detailed functioning of the nervous system is still not well understood, we do have a reasonably good understanding of how messages are transmitted within the nervous system: they are electrical signals passing along the basic element of the nervous system, the *neuron.*

Neurons are living cells of unusual shape (Fig. 18–14). Attached to the main cell body are several small appendages known as *dendrites* and a long tail called the *axon.* Signals are received by the dendrites and are propagated along the axon. When a signal reaches the nerve endings, it is transmitted to the next nerve or to a muscle at a connection called a *synapse.* (Some neurons have separate cells, called Schwann cells, wrapped around their axons; they form a layered sheath called a myelin sheath and help to insulate neurons from one another.) Neurons serve in three capacities. "Sensory neurons" carry messages from the eye, ear, skin, and other organs to the central nervous system, which consists of the brain and spinal cord. "Motor neurons" carry signals from the central nervous system to particular muscles and can signal them to contract. These two types of neuron make up the "peripheral system" as distinguished from the central nervous system. The third type of neuron is the interneuron, which transmits signals between neurons. Interneurons are in the brain and spinal column, and often are connected in an incredibly complex array.

A neuron, before transmitting an electrical signal, is in the so-called "resting state." Like nearly all living cells, neurons have a net positive charge on the outer surface of the cell membrane and a negative charge on the inner surface. This was already mentioned in Section 17–10 with regard to heart muscles and the ECG. This difference in charge, or "dipole layer," means that a potential difference exists across the cell membrane. When a neuron is not transmitting a signal, this "resting potential," which is normally stated as

$$V_{\text{inside}} - V_{\text{outside}},$$

is typically -60 to -90 mV, depending on the type of organism. The most common ions in a cell are K^+, Na^+, and Cl^-. There are large differences in the concentrations of these ions inside and outside a cell, as indicated by the typical values given in Table 18–2. Other ions are also present, so the fluids both inside and outside the axon are electrically neutral. Because of the differences in concentration, there is a tendency for ions to diffuse across the membrane (recall diffusion, Section 13–12). However, in the resting state the cell membrane prevents any net flow of Na^+ (through a mechanism of "active pumping" of Na^+ out of the cell). But it does allow the flow of Cl^- ions and less so of K^+ ions, and it is these two ions that produce the dipole charge layer on the membrane. Because there is a greater concentration of K^+ inside the cell than outside, more K^+ ions tend to diffuse outward across the membrane than diffuse inward. A K^+ ion that passes through the membrane becomes attached to the outer surface of the membrane, and leaves behind an equal negative charge that lies on the inner surface of the membrane

TABLE 18–2
Concentrations of Ions Inside and Outside a Typical Axon

	Concentration Inside Axon (mol/m³)	Concentration Outside Axon (mol/m³)
K^+	140	5
Na^+	15	140
Cl^-	9	125

(Fig. 18–15). The fluids themselves remain neutral. Indeed, what keeps the ions on the membrane is their attraction for each other across the membrane. Independent of this process, Cl^- ions tend to diffuse *into* the cell since their concentration outside is higher. Both K^+ and Cl^- diffusion tends to charge the interior surface of the membrane negatively and the outside positively. As charge accumulates on the membrane surface, it becomes increasingly difficult for more ions to diffuse: K^+ ions trying to move outward, for example, are repelled by the positive charge already there. Equilibrium is reached when the tendency to diffuse because of the concentration difference is just balanced by the electrical potential difference across the membrane. The greater the concentration difference, the greater the potential difference across the membrane, which, as mentioned above, is in the range -60 to -90 mV.

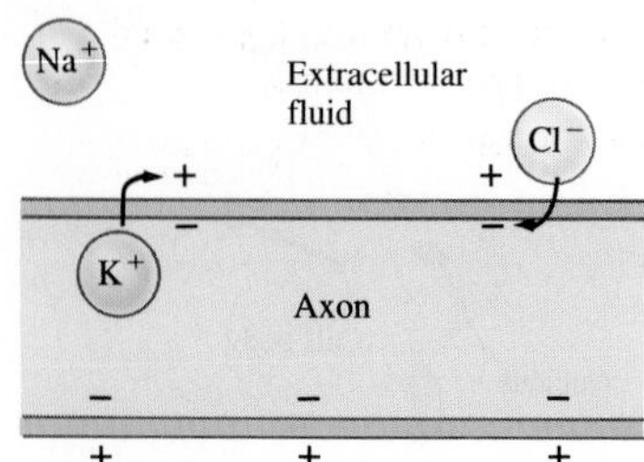

FIGURE 18–15 How a dipole layer of charge forms on a cell membrane.

FIGURE 18–16 Measuring the potential difference between the inside and outside of a nerve.

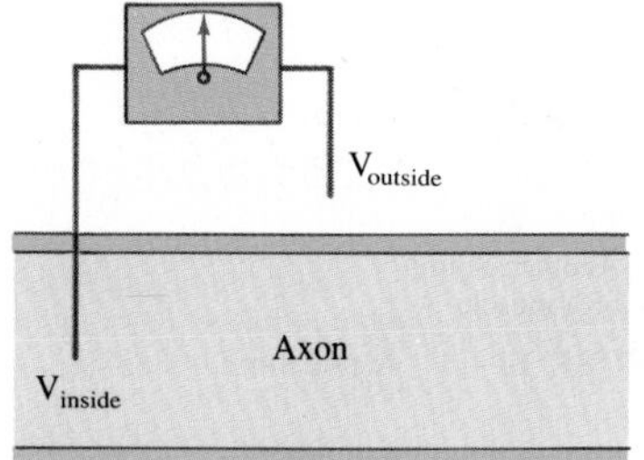

The most important aspect of a neuron is not that it has a resting potential (most cells do), but rather that it can respond to a stimulus and conduct an electrical signal along its length. A nerve can be stimulated in various ways. The stimulus could be thermal (when you touch a hot stove) or chemical (as in taste buds); it could be pressure (as on the skin or at the eardrum), or light (as in the eye); or it could be the electric stimulus of a signal coming from the brain or another neuron. In the laboratory, the stimulus is usually electrical and is applied by a tiny probe at some point on the neuron. If the stimulus exceeds some threshold, a voltage pulse will travel down the axon. This voltage pulse can be detected at a point on the axon using a voltmeter or an oscilloscope connected as in Fig. 18–16. This voltage pulse has the shape shown in Fig. 18–17, and is called an *action potential.* As can be seen, the potential increases from a resting potential of about -70 mV and becomes a positive 30 or 40 mV. The action potential lasts for about 1 ms and travels down an axon with a speed of 30 m/s to 150 m/s.

But what causes the action potential? Apparently, the cell membrane has the ability to alter its permeability properties. At the point where the stimulation occurs, the membrane suddenly becomes much more permeable to Na^+ than to K^+ and Cl^- ions. Thus, Na^+ ions rush into the cell and the inner surface of the wall becomes positively charged, and the potential difference quickly swings positive ($\approx +30$ mV in Fig. 18–17). But then the membrane suddenly returns to its original characteristics: it becomes impermeable to Na^+ and in fact pumps out Na^+ ions. The diffusion of Cl^- and

FIGURE 18–17 Action potential.

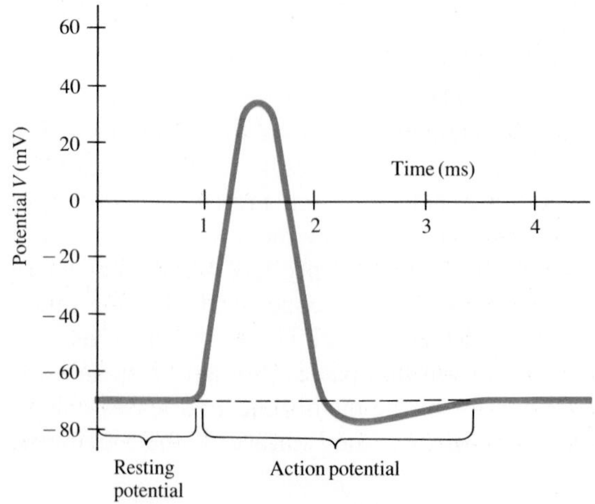

K^+ ions again predominates and the original resting potential is restored (-70 mV in Fig. 18–17).

What causes the action potential to travel along the axon? The action potential occurs at the point of stimulation, as shown in Fig. 18–18a. The membrane momentarily is positive on the inside and negative on the outside at this point. Nearby charges are attracted toward this region, as shown in Fig. 18–18b. The potential in these adjacent regions then drops, causing an action potential there. Thus, as the membrane returns to normal at the original point, nearby it experiences an action potential, so the action potential moves down the axon (Figs. 18–18c and d).

You may wonder if the number of ions that pass through the membrane would significantly alter the concentrations. The answer is no; and we can show why by treating the axon as a capacitor in the following example.

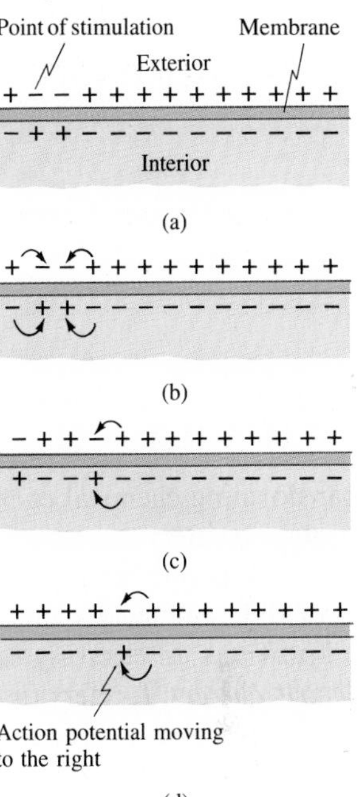

FIGURE 18–18 Propagation of an action potential along an axon membrane.

EXAMPLE 18–8 (*a*) Do an order-of-magnitude estimate for the capacitance of an axon 10 cm long of radius 10 μm. The thickness of the membrane is about 10^{-8} m and the dielectric constant is about 3. (*b*) By what factor does the concentration of Na^+ ions in the cell change as a result of one action potential?

SOLUTION (*a*) The membrane of an axon resembles a cylindrically shaped parallel-plate capacitor, with opposite charges on each side. The separation of the "plates" is the thickness of the membrane, $d \approx 10^{-8}$ m. The area A is the area of a cylinder of radius r and length l: $A = 2\pi rl \approx (6.28)(10^{-5}\text{ m})\cdot(0.1\text{ m}) \approx 6 \times 10^{-6}\text{ m}^2$. From Eq. 17–7, we have

$$C = K\varepsilon_0 \frac{A}{d} \approx (3)(8.85 \times 10^{-12}\ \text{C}^2/\text{N}\cdot\text{m}^2)\frac{6 \times 10^{-6}\ \text{m}^2}{10^{-8}\ \text{m}} \approx 10^{-8}\ \text{F}.$$

(*b*) Since the voltage changes from -70 mV to about $+30$ mV, the total change is about 100 mV. The amount of charge that moves is then $Q = CV \approx (10^{-8}\text{ F})(0.1\text{V}) = 10^{-9}$ C. Each ion carries a charge $e = 1.6 \times 10^{-19}$ C, so the number of ions that flow per action potential is $Q/e = (10^{-9}\text{ C})/(1.6 \times 10^{-19}\text{ C}) \approx 10^{10}$. The volume of our cylindrical axon is $V = \pi r^2 l \approx (3)(10^{-5}\text{ m})^2(0.1\text{ m}) = 3 \times 10^{-11}\text{ m}^3$, and the concentration of Na^+ ions inside the cell (Table 18–2) is $15\text{ mol/m}^3 = 15 \times 6.02 \times 10^{23}$ ions/m$^3 \approx 10^{25}$ ions/m^3. Thus, the cell contains $10^{25} \times (3 \times 10^{-11}) \approx 3 \times 10^{14}$ Na^+ ions. One action potential, then, will change the concentration of Na^+ ions by at most $(10^{10}/3 \times 10^{14}) = 1/3 \times 10^4$ or 1 part in 30,000. This tiny change would not be measurable.

Thus, even 1000 action potentials will not alter the concentration significantly. The sodium pump does not, therefore, have to remove Na^+ ions quickly after an action potential, but can operate slowly over time to maintain a relatively constant concentration.

The propagation of a nerve pulse as described here applies to an unmyelinated axon. Myelinated axons, on the other hand, are insulated from the extracellular fluid by the myelin sheath except at the nodes of Ranvier (see Fig. 18–14). An action potential cannot be regenerated where there is a myelin sheath. Once such a neuron is stimulated, the pulse will still travel along the membrane, but there is resistance and the pulse becomes smaller as it moves down the axon. Nonetheless, the weakened signal can still stimulate a full-fledged action potential when it reaches a node of Ranvier. Thus, the signal is repeatedly amplified at these points. Compare this to an unmyelinated neuron, in which the signal is continually amplified by repeated action potentials all along its length. This naturally requires much more energy. Development of myelinated neurons was a significant evolutionary step, for it meant reliable transmission of nerve pulses with less energy expended. And the pulses travel more quickly, since ordinary conduction is faster than the repeated production of action potentials, whose speed depends on the flow of ions across the membrane.

SUMMARY

By transforming chemical energy into electric energy, an electric batttery serves as a source of potential difference. A simple battery consists of two electrodes made of different metals immersed in a solution or paste known as an electrolyte.

Electric current, I, refers to the rate of flow of electric charge and is measured in *amperes* (A): 1 A equals a flow of 1 C/s past a given point. The direction of current flow is generally taken as being that of positive charge. In a wire, it is actually negatively charged electrons that move, so they flow in a direction opposite to the direction of the *conventional current*. Positive conventional current always flows from a high potential to a low potential.

Ohm's law states that the current in a good conductor is proportional to the potential difference applied to its two ends. The proportionality constant is called the *resistance* R of the material, so $V = IR$. The unit of resistance is the *ohm* (Ω), where $1\ \Omega = 1$ V/A.

The resistance R of a wire is inversely proportional to its cross-sectional area A and directly proportional to its length L and to a property of the material called its resistivity: $R = \rho L/A$. The *resistivity*, ρ, increases with temperature for metals, but for semiconductors it may decrease. A *superconductor* is a material for which the electrical resistance is zero.

The rate at which energy is transformed in a resistance R from electric to other forms of energy (such as heat and light) is equal to the product of current and voltage. That is, the power transformed, measured in watts, is given by $P = IV$ and can be written with the help of Ohm's law as $P = I^2R = V^2/R$. The total electric energy transformed in any device equals the product of power and the time during which the device is operated. In SI units, energy is given in joules ($1\ \text{J} = 1\ \text{W}\cdot\text{s}$), but electric companies use a larger unit, the kilowatt-hour ($1\ \text{kWh} = 3.6 \times 10^6$ J).

Electric current can be *direct* (dc), in which the current is steady in one direction; or it can be *alternating* (ac), in which the current reverses direction at a particular frequency, typically 60 Hz. Alternating currents are often sinusoidal in time, $I = I_0 \sin 2\pi ft$, and are produced by an alternating voltage. The rms values of sinusoidally alternating currents and voltages are given by $I_{\text{rms}} = I_0/\sqrt{2}$ and $V_{\text{rms}} = V_0/\sqrt{2}$, respectively, where I_0 and V_0 are the peak values. The power relationship, $P = IV = I^2R = V^2/R$, is valid for the average power in alternating currents when the rms values are used.

QUESTIONS

1. When you turn on a water faucet, the water usually flows immediately. You don't have to wait for water to flow from the faucet valve to the spout. Why not? Is the same thing true when you connect a wire to the terminals of a battery?
2. Car batteries are often rated in ampere-hours (A·h). What does this rating mean?
3. When an electric cell is connected to a circuit, electrons flow away from the negative terminal in the circuit. But within the cell, electrons flow *to* the negative terminal. Explain.

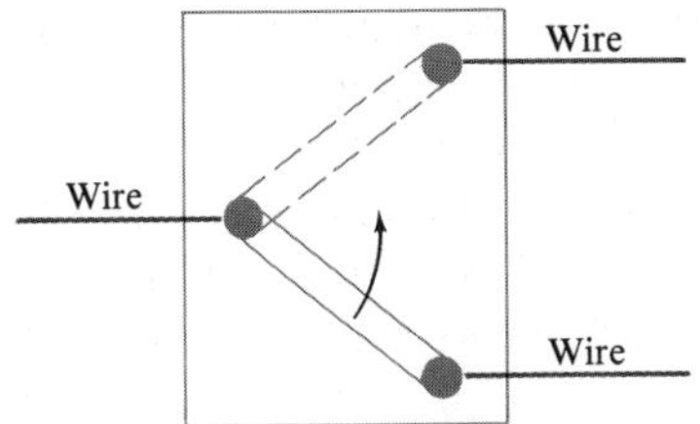

FIGURE 18–19 Question 6.

4. Is current used up in a resistor?
5. Develop an analogy between blood circulation and an electrical circuit. Discuss what plays the role of the heart for the electric case, and so on.
6. Design a circuit in which two different switches of the type shown in Fig. 18–19 can be used to operate the same light bulb from opposite sides of a room.
7. Can a copper wire and an aluminum wire of the same length have the same resistance? Explain.
8. If a large rectangular block of carbon has sides *a*, 2*a*, 3*a*, how would you connect the wires from a battery so as to obtain (*a*) the least resistance, (*b*) the greatest resistance?
9. Describe carefully how you would set up an experiment to measure the resistivity of a large rectangular block of material.
10. In a car, one terminal of the battery is said to be connected to "ground." Since it is not really connected to the ground, what is meant by this expression?
11. The equation $P = V^2/R$ indicates that the power dissipated in a resistor decreases if the resistance is increased, whereas the equation $P = I^2R$ implies the opposite. Is there a contradiction here? Explain.
12. What happens when a light bulb burns out?
13. Which draws more current, a 100-W light bulb or a 75-W bulb?
14. Electric power is transferred over large distances at very high voltages. Explain how the high voltage reduces power losses in the transmission lines.
15. Why is it dangerous to replace a 15-A fuse that blows repeatedly with a 25-A fuse?
16. Electric lights operated on low-frequency ac (say, 10 Hz) flicker noticeably. Why?
17. Suppose Franklin's original convention had been the reverse, so that electrons were considered positive. How would this affect the various results and analyses discussed in this chapter?

PROBLEMS

SECTIONS 18–2 AND 18–3

1. (I) A service station charges a battery using a current of 4.5 A for 7.0 h. How much charge passes through the battery?
2. (I) A current of 1.00 A flows in a wire. How many electrons are flowing past any point in the wire per second? The charge on one electron is 1.60×10^{-19} C.
3. (I) What is the current in amperes if 1000 Na^+ ions were to flow across a cell membrane in 4.0 μs? The charge on the sodium is the same as on an electron, but positive.
4. (I) What voltage will produce 0.15 A of current through a 3000-Ω resistor?
5. (I) What is the resistance of a toaster if 120 V produces a current of 3.5 A?
6. (II) An electrical device draws 3.00 A at 120 V. (*a*) If the voltage drops by 10%, what will be the current, assuming nothing else changes? (*b*) If the resistance of the device were reduced by 10 percent, what current would be drawn at 120 V?
7. (II) A 1.5-V battery is connected to a bulb whose resistance is 1.2 Ω. How many electrons leave the battery per minute?
8. (II) A bird stands on an electric transmission line carrying 1500 A (Fig. 18–20). The line has 1.8×10^{-5} Ω resistance per meter and the bird's feet are 3.0 cm apart. What voltage does the bird feel?

FIGURE 18–20 Problem 8.

9. (III) Copper has approximately 10^{29} free electrons per cubic meter. What is the approximate average velocity of electrons in a 1.0-mm-radius wire carrying 1.0 A?

SECTION 18–4

10. (I) What is the resistance of a 3.0-m length of copper wire 1.5 mm in diameter?
11. (I) What is the diameter of a 30-cm length of tungsten wire whose resistance is 0.20 Ω?
12. (II) A 25.0-m length of wire 1.80 mm in diameter has a resistance of 2.80 Ω. What is the resistance of a 30.0-m length of wire 3.80 mm in diameter made of the same material?
13. (II) Can a 2.0-mm-diameter copper wire have the same resistance as a tungsten wire of the same length? Give numerical details.

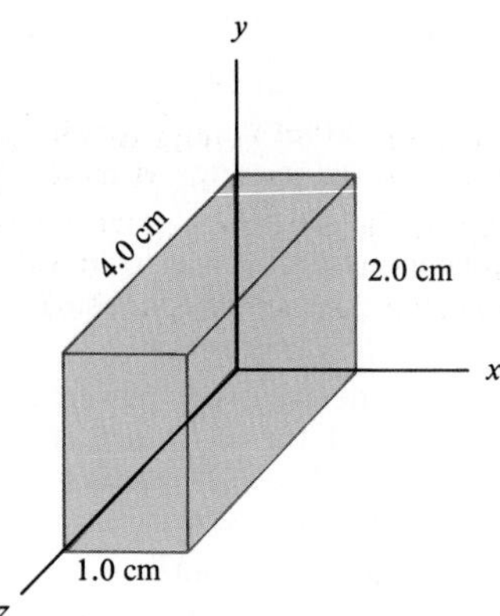

FIGURE 18–21 Problem 17.

14. (II) How much would you have to raise the temperature of a copper wire (originally at 20°C) to increase its resistance by 25%?
15. (II) Estimate at what temperature tungsten will have the same resistivity as copper does at 20°C.
16. (II) A 100-W light bulb has a resistance of about 12 Ω when cold and 140 Ω when on (hot). Estimate the temperature of the filament when "on" assuming an average temperature coefficient of resistivity $\alpha = 0.0060\ (\text{C}^\circ)^{-1}$.
17. (II) A rectangular solid made of carbon has sides lying along the *x*, *y*, and *z* axes, whose lengths are 1.0 cm, 2.0 cm, and 3.0 cm, respectively (Fig. 18–21). Determine the resistance for current that flows through the solid in (*a*) the *x* direction, (*b*) the *y* direction, and (*c*) the *z* direction. Assume the resistivity is $\rho = 3.0 \times 10^{-5}\ \Omega\cdot\text{m}$.
18. (II) A 12.0-m length of wire consists of two 6.0-m portions of equal diameter (2.0 mm), one of copper and one of aluminum, welded together. What voltage is required to produce a current of 8.0 A?
19. (III) For some applications, it is important that the value of a resistance not change with temperature. For example, suppose you made a 2.8-kΩ resistor from a carbon resistor and a nichrome wire-wound resistor connected together so the total resistance is the sum of their separate resistances. What value should each of these resistors have (at 0°C) so that the combination is temperature independent?

SECTION 18–6

20. (I) The element of an electric oven is designed to produce 3.0 kW of heat when connected to a 240-V source. What must be the resistance of the element?
21. (I) What is the current through a 40.0-W light bulb if it is connected to its proper source voltage of 12.0 V?
22. (I) What is the maximum power consumption of a 9.0-V transistor radio that draws a maximum of 300 mA of current?
23. (I) What is the maximum voltage that can be applied to a 300-Ω resistor rated at $\frac{1}{4}$ watt?
24. (I) An automobile starter motor draws 150 A from the 12-V battery. How much power is this?
25. (I) How many kWh does a 1300-W electric frying pan use in 45 min of operation?
26. (II) At $0.080 per kWh, what does it cost to leave a 25-W porch light on all day for a year?
27. (II) What is the total amount of energy stored in a 12-V, 40-A·h car battery when it is fully charged?
28. (II) How many 100-W light bulbs, operated at 120 V, can be used without blowing a 20-A fuse?
29. (II) What is the efficiency of a 0.50-hp electric motor that draws 4.4 A from a 120-V line?
30. (II) A power station delivers 440 kW of power to a factory through 3.0-Ω lines. How much less power is wasted if the electricity is delivered at 40,000 V rather than 12,000 V?
31. (III) The current in an electromagnet connected to a 240-V line is 7.60 A. At what rate must cooling water pass over the coils if the water temperature is to rise by no more than 6.00 C°?
32. (III) A small immersion heater can be used in a car to heat a cup of water for coffee. If the heater can heat 200 mL of water from 5°C to 95°C in 6.0 min, approximately how much current does it draw from the 12-V battery?

SECTION 18–7

33. (I) Calculate the peak current in a 3.2-kΩ resistor connected to a 240-V ac source.
34. (I) An ac voltage, whose peak value is 180 V, is across a 220-Ω resistor. What is the value of the rms and peak currents in the resistor?
35. (I) What is the resistance of an ordinary 60-W, 120-V_{rms} light bulb when it is on?
36. (II) The peak value of an alternating current passing through a 1000-W device is 3.0 A. What is the rms voltage across it?
37. (II) Calculate the peak current passing through a 75-W light bulb connected to a 120-V ac line.
38. (II) What is the maximum instantaneous value of the power dissipated in a 100-W light bulb?
39. (II) A 15-Ω heater coil is connected to a 240-V ac line. What is the average power used? What are the maximum and minimum values of the instantaneous power?

*SECTION 18–8

*40. (I) What is the magnitude of the electric field across an axon membrane 1.0×10^{-8} m thick if the resting potential is −70 mV?

*41. (II) A nerve is stimulated with an electric pulse. The action potential is detected at a point 3.40 cm down the axon 0.0052 s later. When the action potential is detected 7.20 cm from the point of stimulation, the time required is 0.0063 s. What is the speed of the electric pulse along the axon? (Why are two measurements needed instead of only one?)

*42. (III) Estimate how much energy is required to transmit one action potential along the axon of Example 18–8. [*Hint:* one pulse is equivalent to charging and discharging the axon capacitance; see Section 17–9.] What minimum average power is required for 10^4 neurons transmitting 100 pulses per second?

*43. (III) During the action potential, Na^+ ions move into the cell at a rate of about 3×10^{-7} mol/m^2·s. How much power must be produced by the active transport system to produce this flow against a +30-mV potential difference? Assume that the axon is 10 cm long and 20 μm in diameter.

GENERAL PROBLEMS

44. How many coulombs are there in 1.00 ampere-hour?
45. What is the average current drawn by a 0.50-hp 120-V motor?
46. A person accidentally leaves a car with the lights on. If each of the two front lights uses 40 W and each of the two rear lights 6 W, for a total of 92 W, how long will a fresh 12-V battery last if it is rated at 45 A·h? Assume the full 12 V appears across each bulb.
47. The heating element of a 120-V, 1800-W heater is 7.0 m long. If it is made of iron, what must its diameter be?
48. The *conductance* G of an object is defined as the reciprocal of the resistance R; that is, $G = 1/R$. The unit of conductance is a mho (=ohm^{-1}), which is also called the siemens (S). What is the conductance (in siemens) of an object that draws 700 mA of current at 3.0 V?
49. (*a*) A particular household uses a 1.5-kW heater 2.5 h/day ("on" time), six 100-W light bulbs 6.0 h/day, a 3.3-kW electric stove element for a total of 1.2 h, and miscellaneous power amounting to 1.6 kWh/day. If electricity costs $0.090 per kWh, what will be their monthly bill (30 d)? (*b*) How much coal (which produces 7000 kcal/kg) must be burned by a 35-percent-efficient power plant to provide the yearly needs of this household?
50. A length of wire is cut in half and the two lengths are wrapped together side by side to make a thicker wire. How does the resistance of this new combination compare to the resistance of the original wire?
51. A 1200-W hair dryer is designed for 117 V. What will be the percentage change in power output if the voltage drops to 105 V? Assume no change in resistance. How would the actual change in resistivity with temperature affect your answer?
52. The wiring in a house must be thick enough so it doesn't become so hot as to start a fire. What diameter must a copper wire be if it is to carry a maximum current of 40 A and produce no more than 1.8 W of heat per meter of length?
53. Suppose a current is given by the equation $I = 2.00 \sin 180t$, where I is in amperes, and t in seconds. (*a*) What is the frequency? (*b*) What is the rms value of the current? (*c*) If this is the current through a 50.0-Ω resistor, what is the equation that describes the voltage as a function of time?
54. A 1.00-Ω wire is drawn out to 2.00 times its original length. What is its resistance now?
55. 220 V is applied to two different conductors made of the same material. One conductor is twice as long and twice as thick as the second. What is the ratio of the power transformed in the first relative to the second?
56. An electric heater is used to heat a room of volume 45 m^3. Air is brought into the room at 5°C and is changed completely twice per hour. Heat loss through the walls amounts to approximately 550 kcal/h. If the air is to be maintained at 20°C, what minimum wattage must the heater have? (The specific heat of air is about 0.17 kcal/kg·C°.)
57. An electric car (Fig. 18–22) makes use of storage batteries as its source of energy. Suppose that a small 680-kg postman's delivery car is to be powered by ten 70-A·h, 12-V batteries. Assume that the car is driven on the level at an average speed of 30 km/h, and the average friction force is 220 N. Assume 100 percent efficiency and neglect energy used for acceleration. Note that no energy is consumed when the vehicle is stopped since the engine doesn't need to idle. (*a*) Determine the horsepower required. (*b*) After approximately how many kilometers must the batteries be recharged?
58. A 15.0-Ω resistor is made from a coil of copper wire whose total mass is 14.2 g. What is the diameter of the wire and how long is it?

FIGURE 18–22 An electric car. (Problem 57.)

CHAPTER 19

DC Circuits and Instruments

Years ago when one bulb of a certain type of Christmas tree lights burned out, the whole string of lights went out. (Was this a series or parallel circuit?) Today, if one bulb goes out, the rest stay lit. For certain strings, however (usually those with very tiny bulbs), if you *remove* a bulb (burned out or not), all the others go out. What kind of circuit could this be?

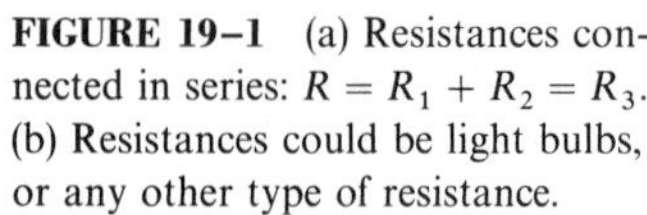

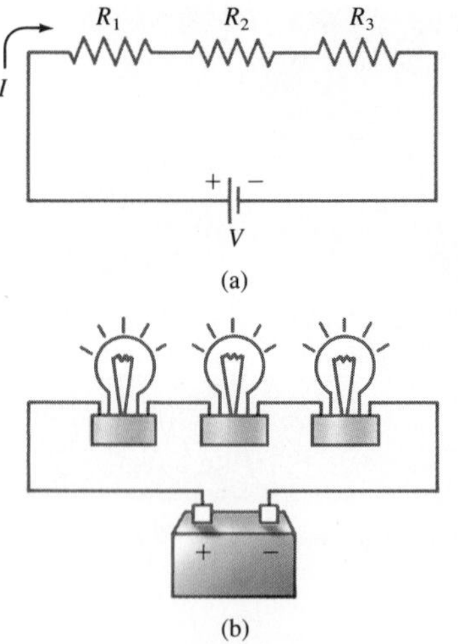

FIGURE 19–1 (a) Resistances connected in series: $R = R_1 + R_2 = R_3$. (b) Resistances could be light bulbs, or any other type of resistance.

In Chapter 18, we discussed the basic principles of current electricity. Now we will apply these principles to analyze dc circuits and to understand the operation of a number of useful instruments.[†]

When we draw a diagram of a circuit, we represent a battery by the symbol ⊣⊢, a capacitor by the symbol ⊣⊢, and a resistor by the symbol -ᴡᴡ-. Wires whose resistance is negligible compared to other resistance in the circuit are drawn simply as straight lines. For the most part in this chapter, except in Section 19–7, we will be interested in circuits operating in their steady state—that is, we won't be looking at a circuit at the moment any change is made in it, such as when a battery or resistor is connected or disconnected, but rather a short time later when the currents have reached their steady values.

19–1 • Resistors in Series and in Parallel

When two or more resistors are connected end to end so the same current passes through each in turn, as in Fig. 19–1, they are said to be connected

[†] Ac circuits that contain only a source of emf and resistors can be analyzed like the dc circuits in this chapter. However, ac circuits in general are more complex—if they contain capacitors, for example—and we discuss them in Chapter 21.

in **series**. The resistors could be like those in the photo of Fig. 18–7, or they could be light bulbs or other resistive electrical devices. On the other hand, if the resistors are connected so that the current from the source splits into separate branches, as shown in Fig. 19–2, the resistors are said to be in **parallel**. In the latter case, the same potential difference exists across each resistor. The wiring in houses and buildings is arranged so that all electric devices are in parallel, as we saw in Fig. 18–11. With parallel wiring, if you disconnect one device, you don't interrupt the current to the other devices, which *does* happen for a series circuit (Fig. 19–1).

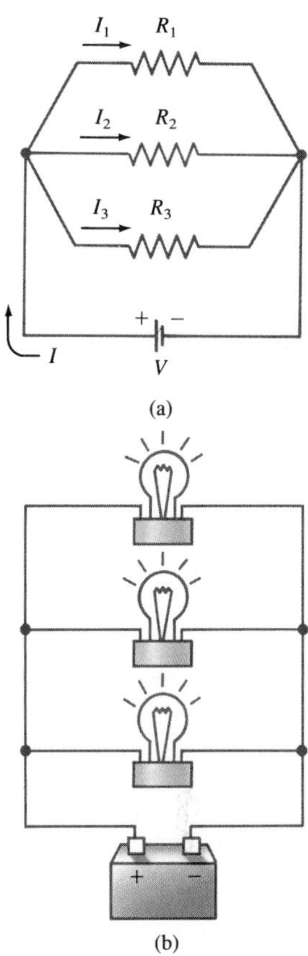

FIGURE 19–2 (a) Resistances connected in parallel: $1/R = 1/R_1 + 1/R_2 + 1/R_3$, which could be (b) light bulbs.

For both series and parallel circuits, we will want to calculate the net resistance of a set of resistors. In other words, we want to find what single resistance R could replace the combination of given resistors without altering the rest of the circuit: that is, the total current I and the potential difference V must stay the same.

First, we consider the series case, Fig. 19–1. The same current I passes through each resistor. If it did not, this would imply that charge was accumulating at some point in the circuit, which does not happen in the steady state. We let V represent the voltage across all three resistors. We assume all other resistance in the circuit can be ignored, and so V equals the voltage of the battery. We let V_1, V_2, and V_3 be the potential differences across each of the resistors, R_1, R_2, and R_3, respectively. By Ohm's law, $V_1 = IR_1$, $V_2 = IR_2$, and $V_3 = IR_3$. Since the resistors are connected end to end, the total voltage V is equal to the sum of the voltages across each resistor, so we have

$$V = V_1 + V_2 + V_3 = IR_1 + IR_2 + IR_3.$$

For the equivalent single resistance R that would draw the same current, we need to have

$$V = IR.$$

We equate these two expressions for V and find

Resistances in series

$$R = R_1 + R_2 + R_3. \qquad \text{[series]} \tag{19–1}$$

This is, in fact, what we expect. When we put several resistances in series, the total resistance is the sum of the separate resistances. This applies to any number of resistances, not simply for three. Clearly, when you add more resistance to the circuit, the current will decrease. If a 12-V battery is connected to a 4-Ω resistor, the current will be 3 A. But if the 12-V battery is connected to three 4-Ω resistors in series, the total resistance is 12 Ω and the current will be only 1 A.

The situation is quite different for the parallel case, Fig. 19–2. Again we want to find what single resistance R is equivalent to the three in parallel. In this situation, the total current I that leaves the battery breaks into three branches. We let I_1, I_2, and I_3 be the currents through each of the resistors, R_1, R_2, and R_3, respectively. Because charge is conserved, the current flowing into a junction must equal the current flowing out, so

$$I = I_1 + I_2 + I_3.$$

The full voltage of the battery is applied to each resistor, so

$$I_1 = \frac{V}{R_1}, \qquad I_2 = \frac{V}{R_2}, \qquad \text{and} \qquad I_3 = \frac{V}{R_3}.$$

Furthermore, for the single resistor R that will draw the same current I as these three in parallel, we must have

$$I = \frac{V}{R}.$$

We now combine these equations:

$$I = I_1 + I_2 + I_3,$$

$$\frac{V}{R} = \frac{V}{R_1} + \frac{V}{R_2} + \frac{V}{R_3}.$$

When we divide out the V from each term, we have

Resistances in parallel

$$\frac{1}{R} = \frac{1}{R_1} + \frac{1}{R_2} + \frac{1}{R_3}. \qquad \text{[parallel]} \qquad (19\text{–}2)$$

For example, if three 30-Ω resistors are put in parallel, the net resistance R offered by this network is

$$\frac{1}{R} = \frac{1}{30\ \Omega} + \frac{1}{30\ \Omega} + \frac{1}{30\ \Omega} = \frac{3}{30\ \Omega} = \frac{1}{10\ \Omega},$$

and so $R = 10\ \Omega$. Thus the net resistance is *less* than that of each single resistance. This may at first seem surprising. But remember that when you put resistors in parallel, you are giving the current additional paths to follow. Hence the net resistance will be less.

FIGURE 19–3 (a) Circuit for Examples 19–1 and 19–2. (b) Equivalent circuit, showing the equivalent resistance of 290 Ω for the two parallel resistors in (a).

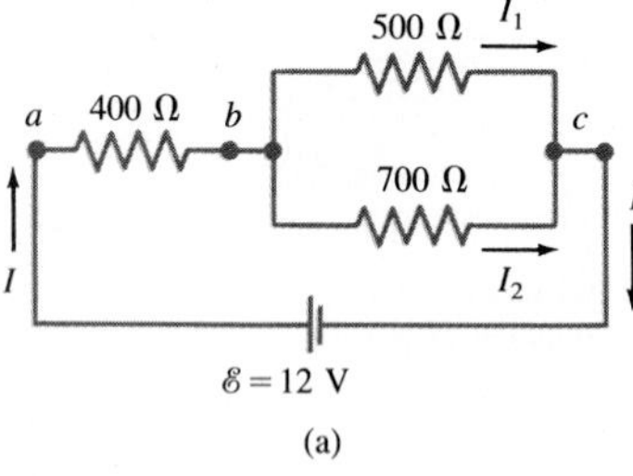

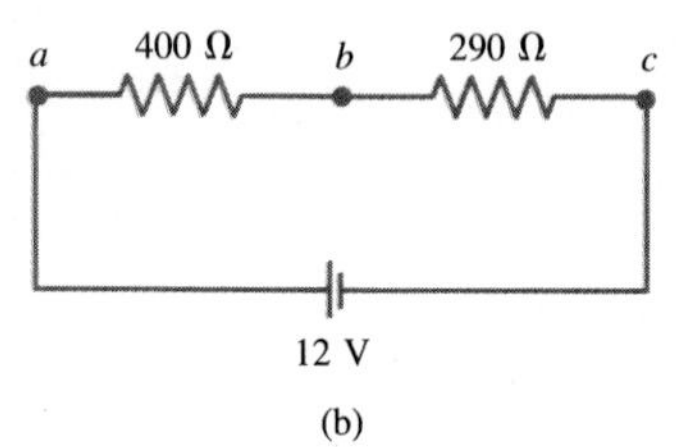

EXAMPLE 19–1 How much current flows from the battery shown in Fig. 19–3a?

SOLUTION First we find the equivalent resistance, R_p, of the 500-Ω and 700-Ω resistors that are in parallel:

$$\frac{1}{R_p} = \frac{1}{500\ \Omega} + \frac{1}{700\ \Omega} = 0.0020\ \Omega^{-1} + 0.0014\ \Omega^{-1}$$

$$= 0.0034\ \Omega^{-1}.$$

Problem Solving: Remember to take the reciprocal

This is $1/R$, so we must take the reciprocal to find R. (It is a common mistake to forget to take this reciprocal. Notice that the units of reciprocal ohms, Ω^{-1}, help to remind us of this.) Thus

$$R_p = \frac{1}{0.0034\ \Omega^{-1}} = 290\ \Omega.$$

This 290 Ω is the equivalent resistance of the two parallel resistors and is in series with the 400-Ω resistor. The equivalent circuit is shown in Fig. 19–3b. To find the total resistance R_T, we add the 400-Ω and 290-Ω resistances together, since they are in series, and find

$$R_T = 400\ \Omega + 290\ \Omega = 690\ \Omega.$$

The total current flowing from the battery is then

$$I = \frac{V}{R} = \frac{12\ \text{V}}{690\ \Omega} = 0.017\ \text{A} = 17\ \text{mA}.$$

EXAMPLE 19–2 What is the current flowing through the 500-Ω resistor in Fig. 19–3a?

SOLUTION To solve this problem, we must find the voltage across the 500-Ω resistor, which is the voltage between points b and c in the diagram and we call it V_{bc}. Once V_{bc} is known, we can apply Ohm's law to get the current. First we find the voltage across the 400-Ω resistor, V_{ab}. Since 17 mA passes through this resistor, the voltage across it can be found using Ohm's law, $V = IR$:

$$V_{ab} = (0.017\ \text{A})(400\ \Omega) = 6.8\ \text{V}.$$

Since the total voltage across the network of resistors is $V_{ac} = 12$ V, then V_{bc} must be 12 V − 6.8 V = 5.2 V. Then Ohm's law tells us that the current I_1 through the 500-Ω resistor is

$$I_1 = \frac{5.2\ \text{V}}{500\ \Omega} = 10\ \text{mA}.$$

This is the answer we wanted. However, we can also calculate the current I_2 through the 700-Ω resistor since the voltage across it is also 5.2 V:

$$I_2 = \frac{5.2\ \text{V}}{700\ \Omega} = 7\ \text{mA}.$$

Notice that when I_1 combines with I_2 to form the total current I (at point c in Fig. 19–3a), their sum is 10 mA + 7 mA = 17 mA. This is, of course, the total current as calculated in Example 19–1.

19–2 • EMF and Terminal Voltage

emf

A device such as a battery or an electric generator that transforms one type of energy (chemical, mechanical, light, and so on) into electric energy is called a **seat** or **source** of **electromotive force** or of **emf**. (Actually, the term "electromotive force" is a misnomer since it does not refer to a "force" that is measured in newtons; hence to avoid confusion we prefer to use the abbreviation, emf.) The potential differences between the terminals of such a source, when no current flows to an external circuit, is called the emf of the source. The symbol $\mathscr{E}$ is usually used for emf (don't confuse it with E for electric field).

Why battery voltage isn't constant

You may have noticed in your own experience that when a current is drawn from a battery, the voltage across its terminals drops below its rated emf. For example, if you start a car with the headlights on, you may notice the headlights dim. This happens because the starter draws a large current and the battery voltage drops as a result. The voltage drop occurs because the chemical reactions in a battery cannot supply charge fast enough to maintain the full emf. For one thing, charge must flow within the electrolyte between the electrodes of the battery, and there is always some hindrance to completely free flow. Thus, a battery itself has some resistance, which is called its **internal resistance**; it is usually designated r. The internal resistance can most simply be represented as if it were in series with the emf, as shown in

Fig. 19–4. Since this resistance r is inside the battery, we can never separate it from the battery. The two points a and b in the diagram represent the two terminals of the battery. What we measure is the **terminal voltage** V_{ab}. When no current is drawn from the battery, the terminal voltage equals the emf, which is determined by the chemical reactions in the battery: $V_{ab} = \mathscr{E}$. However, when a current I flows from the battery, there is an internal drop in voltage equal to Ir. Thus the terminal voltage is†

Terminal voltage

$$V_{ab} = \mathscr{E} - Ir.$$

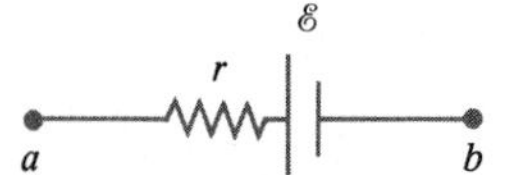

FIGURE 19–4 Diagram for an electric cell or battery.

For example, if a 12-V battery has an internal resistance of 0.1 Ω, then when 10 A flows from the battery, the terminal voltage is 12 V – (10 A)(0.1 Ω) = 11 V. The internal resistance of a battery is usually small. For example, an ordinary flashlight battery when fresh may have an internal resistance of perhaps 0.05 Ω. (However, as it ages and the electrolyte dries out, the internal resistance increases to many ohms.) Car batteries have even lower internal resistance.

EXAMPLE 19–3 A 9.0-V battery whose internal resistance r is 0.50 Ω is connected in the circuit shown in Fig. 19–5a. (*a*) How much current is drawn from the battery? (*b*) What is the terminal voltage of the battery?

FIGURE 19–5 Circuit for Example 19–3, where r is the internal resistance of the battery.

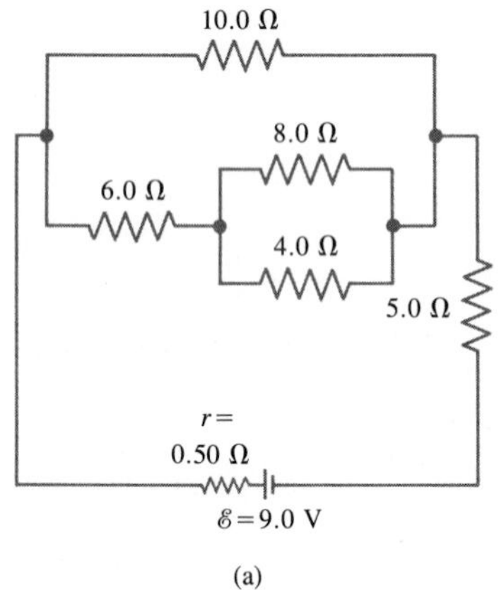

(a)

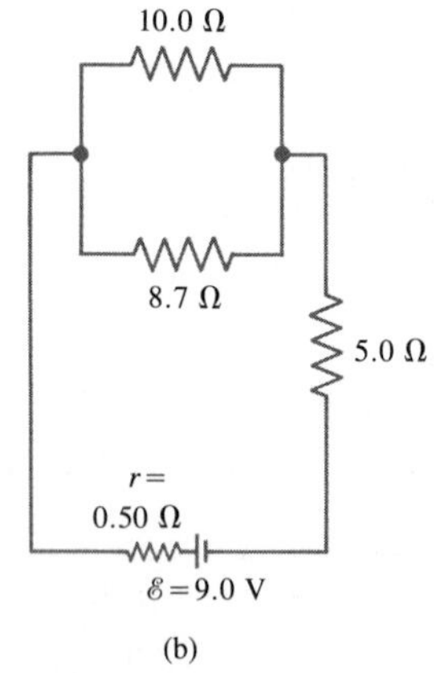

(b)

SOLUTION (*a*) First, we determine the equivalent resistance of the circuit. The 4.0-Ω and 8.0-Ω resistors in parallel have an equivalent resistance R_{I} given by

$$\frac{1}{R_{\mathrm{I}}} = \frac{1}{8.0\,\Omega} + \frac{1}{4.0\,\Omega} = \frac{3}{8.0\,\Omega},$$

so $R_{\mathrm{I}} = 2.7\,\Omega$. This 2.7 Ω is in series with the 6.0-Ω resistor so the net resistance of the lower arm of the circuit is 6.0 Ω + 2.7 Ω = 8.7 Ω, as shown in Fig. 19–5b. The equivalent resistance R_{II} of this 8.7-Ω and the 10.0-Ω resistances in parallel is given by

$$\frac{1}{R_{\mathrm{II}}} = \frac{1}{10.0\,\Omega} + \frac{1}{8.7\,\Omega} = 0.21\,\Omega^{-1}.$$

So $R_{\mathrm{II}} = (1/0.21\,\Omega^{-1}) = 4.8\,\Omega$. This 4.8 Ω is in series with the 5.0-Ω resistor and the 0.50-Ω internal resistance of the battery, so the total resistance R of the circuit is $R = 4.8\,\Omega + 5.0\,\Omega + 0.50\,\Omega = 10.3\,\Omega$. Hence the current drawn is

$$I = \frac{\mathscr{E}}{R} = \frac{9.0\text{ V}}{10.3\,\Omega} = 0.87\text{ A}.$$

(*b*) The terminal voltage of the battery is $V_{ab} = \mathscr{E} - Ir = 9.0\text{ V} - (0.87\text{ A})(0.50\,\Omega) = 8.6\text{ V}$.

† When a battery is being charged, a current is forced to pass through it (this happens in Fig. 19–9b), and we have to write

$$V_{ab} = \mathscr{E} + Ir.$$

See, for example, Problem 19 and Fig. 19–21.

19–3 • Kirchhoff's Rules

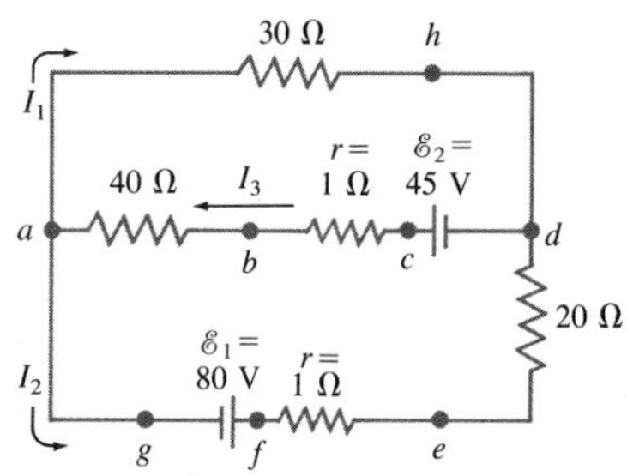

FIGURE 19–6 Currents in the circuit can be calculated using Kirchhoff's rules.

In the last few examples we have been able to find the currents flowing in circuits by combining resistances in series and parallel, and using Ohm's law. This technique can be used for many circuits. However, we sometimes encounter a circuit that is too complicated for this analysis. For example, we cannot find the currents flowing in each part of the circuits shown in Fig. 19–6 simply by combining resistances as we did before.

To deal with such complicated circuits, we use Kirchhoff's rules, invented by G. R. Kirchhoff (1824–1887) in the mid-nineteenth century. There are two of them, and they are simply convenient applications of the laws of conservation of charge and energy. **Kirchhoff's first** or **junction rule** is based on the conservation of charge, and we already used it in deriving the rule for parallel resistors. It states that

Junction rule (conservation of charge)

> **at any junction point, the sum of all currents entering the junction must equal the sum of all currents leaving the junction.**

For example, at point *a* in Fig. 19–6, I_3 is entering whereas I_1 and I_2 are leaving. Thus Kirchhoff's junction rule states that $I_3 = I_1 + I_2$. We already saw an instance of this at the end of Example 19–2, where the currents passing through the 400-Ω and 700-Ω resistors were 10 mA and 7 mA, respectively, and they added at point *c* in Fig. 19–3a to give the outgoing current of 17 mA.

Kirchhoff's junction rule is based on the conservation of charge. Charges that enter a junction must also leave—none is lost or gained. **Kirchhoff's second** or **loop rule** is based on the conservation of energy. It states that

Loop rule (conservation of energy)

> **the algebraic sum of the changes in potential around any closed path of a circuit must be zero.**

To see why this should hold, consider the analogy of a roller coaster on its track. When it starts from the station, it has a particular potential energy. As it climbs the first hill, its potential energy increases and reaches a peak at the top. As it descends the other side, its potential energy decreases and reaches a local minimum at the bottom of the hill. As the roller coaster continues on its path, its potential energy goes through more changes. But when it arrives back at the starting point, it has exactly as much potential energy as it had when it started at this point. Another way of saying this is that there was as much uphill as there was downhill.

The same reasoning can be applied to an electric circuit. As an example, consider the simple circuit in Fig. 19–7. We have chosen it to be the same as the equivalent circuit of Fig. 19–3b already dealt with. The current flowing in this circuit is $I = (12.0\ \text{V})/(690\ \Omega) = 0.017\ \text{A}$, as we calculated in Example 19–1. (Note that we are ignoring the internal resistance of the battery.) The positive side of the battery, point *e* in the figure, is at a high potential compared to point *d* at the negative side of the battery. That is, point *e* is like the top of a hill for a roller coaster. We can now follow the current around the circuit starting at any point we choose. Let us start at point *e* and follow a positive test charge completely around this circuit. As we go, we will note all changes in potential. When the test charge returns to point *e*, the potential there will be the same as when we started, so the total change

FIGURE 19–7 Changes in potential around the circuit in (a) are plotted in (b).

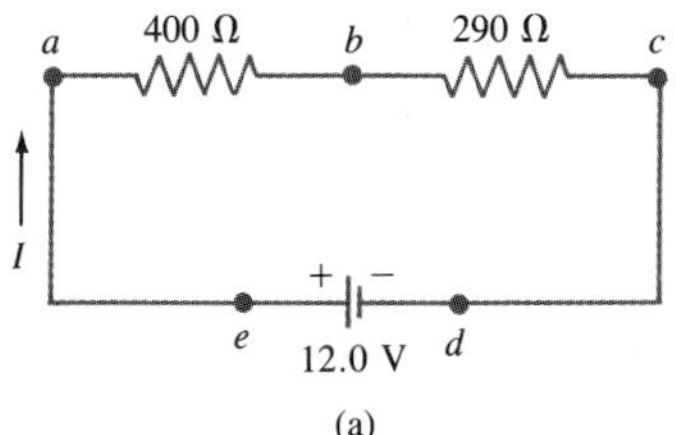

(a)

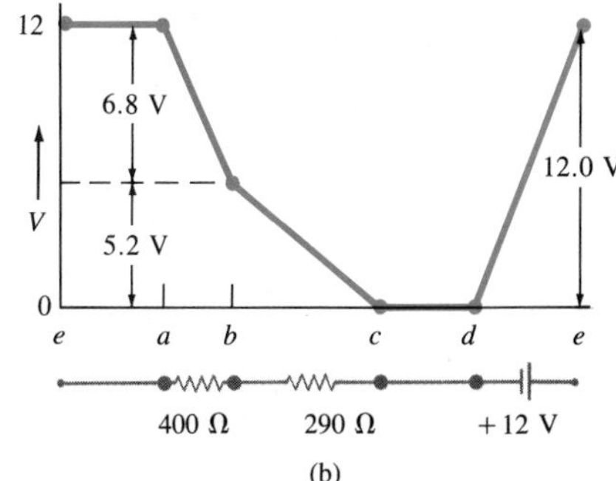

(b)

in potential will be zero. It is helpful to plot the changes in voltage around the circuit as in Fig. 19–7b. (Point d is arbitrarily taken as zero.) As our positive test charge goes from point e to point a, there is no change in potential since there is no source of potential nor any resistance. However, as the charge passes through the 400-Ω resistor to get to point b, there is a decrease in potential of $V = IR = (0.017\text{ A})(400\ \Omega) = 6.8\text{ V}$. In effect, the charge is flowing "downhill" since it is heading toward the negative terminal of the battery. This is indicated in the graph of Fig. 19–7b. The decrease in voltage between the two ends of a resistor ($=IR$) is called a **voltage drop**. Because this is a *decrease* in voltage, we use a negative sign when applying Kirchhoff's loop rule; that is,

Problem Solving: Be consistent with signs

$$V_{ba} = V_b - V_a = -6.8\text{ V}.$$

As the charge proceeds from b to c there is a further voltage drop of $(0.017\text{ A}) \times (290\ \Omega) = 5.2\text{ V}$, and since this is a decrease, we write

$$V_{cb} = -5.2\text{ V}.$$

There is no change in potential as our test charge moves from c to d. But when it moves from d, which is the negative or low potential side of the battery, to point e which is the positive terminal, the voltage *increases* by 12.0 V. That is,

$$V_{ed} = +12.0\text{ V}.$$

The sum of all the changes in potential in going around the circuit of Fig. 19–7 is then

$$-6.8\text{ V} - 5.2\text{ V} + 12.0\text{ V} = 0.$$

And this is exactly what Kirchhoff's loop rule said it would be.

We already knew the details of this circuit, and we gave this example simply to illustrate how the loop rule is applied. In the next section, we will see how to use Kirchhoff's rules to determine the currents in more difficult circuits.

*19–4 • Solving Problems with Kirchhoff's Rules

When using Kirchhoff's rules, we will find it useful to designate the current in each separate branch of the given circuit by a different subscript, such as I_1, I_2, and I_3 in Fig. 19–8 (this is the same circuit as in Fig. 19–6). You do not have to know in advance in which direction these currents actually are moving. You make a guess and calculate the potentials around the circuit as if you were right. If you made the wrong guess for the direction of a current, your answer will merely have a negative sign. This and other details of using Kirchhoff's rules will become clearer in the following example.

Problem Solving: Choose current directions arbitrarily

Example 19–4 Calculate the currents I_1, I_2, and I_3 in each of the branches of the circuit in Fig. 19–8.

SOLUTION We choose the directions of the currents as shown in the figure. Since (positive) current moves away from the positive terminal of a battery, we expect I_2 and I_3 to have the directions shown. It is hard to tell the direction of I_1 in advance, so we arbitrarily chose the direction indicated. We have three unknowns and therefore we need three equations. We first apply Kirchhoff's junction rule to the currents at point a, where I_3 enters and I_2 and I_1 leave. Therefore

$$I_3 = I_1 + I_2. \qquad (a)$$

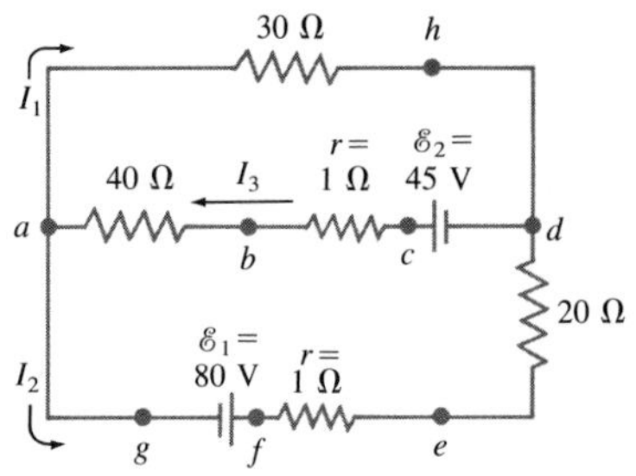

FIGURE 19–8 Currents can be calculated using Kirchhoff's rules. See Example 19–4.

This same equation holds at point d, so we get no new information there. We now apply Kirchhoff's loop rule to two different closed loops. First we apply it to the loop *ahdcba*. From a to h we have a voltage drop $V_{ha} = -(I_1)(30\ \Omega)$. From h to d there is no change, but from d to c the potential increases by 45 V: that is, $V_{cd} = +45\text{V}$. From c to a the voltage drops through the two resistances by an amount $V_{ac} = -(I_3)(40\ \Omega + 1\ \Omega)$. Thus we have $V_{ha} + V_{cd} + V_{ac} = 0$, or

$$-30I_1 - 41I_3 + 45 = 0, \qquad (b)$$

where we have omitted the units. For our second loop, we take the complete circuit *ahdefga*. (We could have just as well taken *abcdefg* instead.) Again we have $V_{ha} = -(I_1)(30\ \Omega)$, and $V_{dh} = 0$. But when we take our positive test charge from d to e, it actually is going uphill, against the natural flow of the current—or at least against the *assumed* direction of the current, which is what counts in this calculation. Thus $V_{ed} = I_2(20\ \Omega)$ has a *positive* sign. Similarly, $V_{fe} = I_2(1\ \Omega)$. From f to g there is a decrease in potential of 80 V since we go from the high potential terminal of the battery to the low. Thus $V_{gf} = -80$ V. Finally, $V_{ag} = 0$ and the sum of the potentials around this loop is then

$$-30I_1 + 21I_2 - 80 = 0. \qquad (c)$$

We now have three equations—labeled (*a*), (*b*), and (*c*)—in three unknowns. From Eq. (*c*) we have

$$I_2 = \frac{80 + 30I_1}{21} = 3.8 + 1.4I_1. \qquad (d)$$

From Eq. (*b*) we have

$$I_3 = \frac{45 - 30I_1}{41} = 1.1 - 0.73I_1. \qquad (e)$$

We substitute these into Eq. (*a*) and solve for I_1:

$$\begin{aligned} I_1 &= I_3 - I_2 \\ &= -2.7 - 2.1I_1, \end{aligned}$$

so

$$\begin{aligned} 3.1I_1 &= -2.7 \\ I_1 &= -0.87\ \text{A}. \end{aligned}$$

I_1 has magnitude 0.87 A. The negative sign indicates that its direction is actually opposite to that initially assumed and shown in the figure. Note that the answer automatically comes out in amperes since all values were in volts and ohms. From Eq. (d) we have

$$I_2 = 3.8 + 1.4I_1 = 2.6 \text{ A},$$

and from (e)

$$I_3 = 1.1 - 0.73I_1 = 1.7 \text{ A}.$$

This completes the solution.

19–5 • EMFs in Series and in Parallel; Charging a Battery

FIGURE 19–9 Batteries in series, (a) and (b), and in parallel, (c).

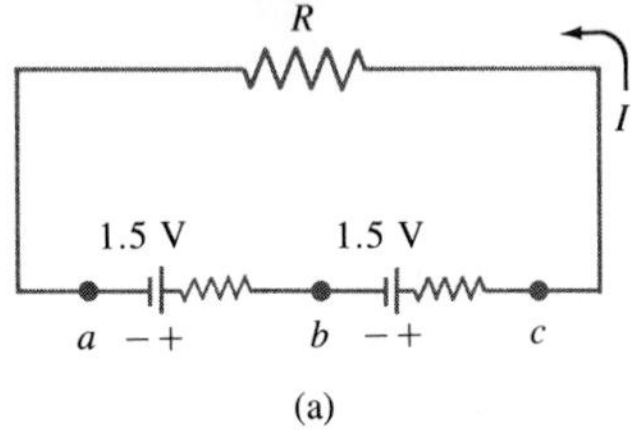

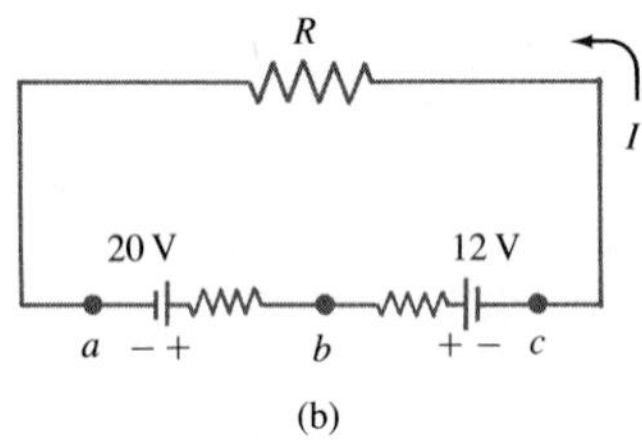

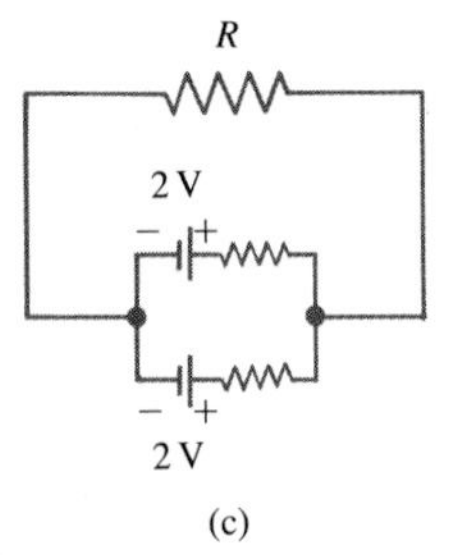

When two or more sources of emf, such as batteries, are arranged in series, the total voltage is the algebraic sum of their respective voltages. For example, if two 1.5-V flashlight batteries are connected as shown in Fig. 19–9a, the voltage V_{ca} across the light bulb, represented by the resistance R, is 3.0 V. (To be absolutely correct, we should also take into account the internal resistance of the batteries, but we assume it to be small.) On the other hand, when a 20-V and a 12-V battery are connected oppositely, as shown in Fig. 19–9b, the net voltage V_{ca} is 8 V. That is, a positive test charge moved from a to b gains in potential by 20 V, but when it passes from b to c it drops by 12 V. So the net change is 20 V − 12 V = 8 V. You might think that connecting batteries in reverse like this would be wasteful. And for most purposes that would be true. But such a reverse arrangement is precisely how a battery charger works. In Fig. 19–9b, the 20-V source is charging up the 12-V battery. Because of its greater voltage, the 20-V source is forcing charge back into the 12-V battery: electrons are being forced into its negative terminal and removed from its positive terminal. An automobile alternator keeps the car battery charged in the same way. A voltmeter placed across the terminals of a (12-V) car battery with the engine running fairly fast can tell you whether or not the alternator is charging the battery. If it is, the voltmeter reads 13 or 14 V; if the battery is not being charged, the voltage will drop below 12 V because the battery is discharging. Car batteries can be recharged, but other batteries may not be rechargable, since the chemical reactions in many cannot be reversed; in such cases, the arrangement of Fig. 19–9b would waste energy.

Sources of emf can also be arranged in parallel, Fig. 19–9c. A parallel arrangement is not used to increase voltage, but rather to provide more energy when large currents are needed. Each of the cells in parallel only has to produce a fraction of the total current, so the loss due to internal resistance is less than for a single cell.

19–6 • Circuits Containing Capacitors in Series and in Parallel

Just as resistors can be placed in series or in parallel in a circuit, so can capacitors (Chapter 17). We first consider a parallel connection as shown in Fig. 19–10a. If a battery of voltage V is connected to points a and b, this voltage exists across each of the capacitors. Each acquires a charge given by

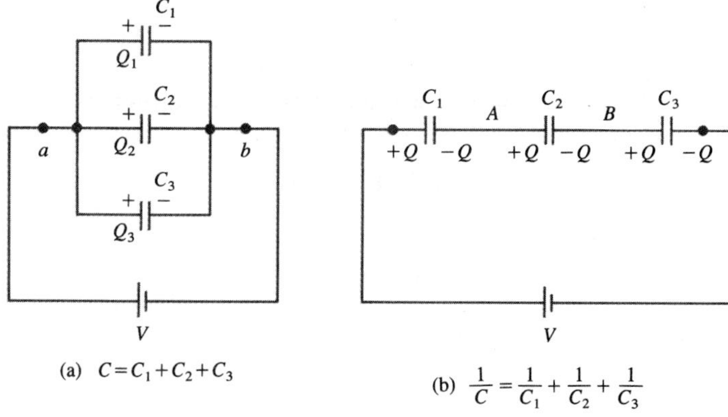

FIGURE 19–10
Capacitors (a) in parallel, (b) in series.

$Q_1 = C_1V$, $Q_2 = C_2V$, and $Q_3 = C_3V$. The total charge Q that must leave the battery is then

$$Q = Q_1 + Q_2 + Q_3 = C_1V + C_2V + C_3V.$$

A single equivalent capacitor that will hold the same charge Q at the same voltage V will have a capacitance C given by

$$Q = CV.$$

Thus we have

$$CV = C_1V + C_2V + C_3V,$$

or

$$C = C_1 + C_2 + C_3. \quad \text{[parallel]} \tag{19–3}$$

Capacitors in parallel

The net effect of connecting capacitors in parallel is thus to increase the capacitance. This is just what we should expect since we are essentially increasing the area of the plates for charge to accumulate on (see Eq. 17–6).

If the capacitors are connected in series, as in Fig. 19–10b, a charge $+Q$ flows from the battery to one plate of C_1, and $-Q$ flows to one plate of C_3. The regions A and B between the capacitors were originally neutral, so the net charge there must still be zero. The $+Q$ on the left plate of C_1 attracts a charge of $-Q$ on the opposite plate; because region A must have a zero net charge, there is thus $+Q$ on the left plate of C_2. The same considerations apply to the other capacitors, so we see that the charge on each capacitor is the same, namely Q. A single capacitor that could replace these three in series without affecting the circuit (that is, Q and V the same) would have a capacitance C given by

$$Q = CV.$$

Now the total voltage V across the three capacitors in series must equal the sum of the voltages across each capacitor:

$$V = V_1 + V_2 + V_3.$$

We also have $Q = C_1V_1$, $Q = C_2V_2$, and $Q = C_3V_3$, so we substitute for V_1, V_2, and V_3 into the last equation and get

$$\frac{Q}{C} = \frac{Q}{C_1} + \frac{Q}{C_2} + \frac{Q}{C_3},$$

or

$$\frac{1}{C} = \frac{1}{C_1} + \frac{1}{C_2} + \frac{1}{C_3}. \quad \text{[series]} \tag{19–4}$$

Capacitors in series

Notice that the forms of the equations for capacitors in series or in parallel are just the reverse of their counterparts for resistance. That is, the formula for capacitors in series resembles the formula for resistors in parallel, and vice versa.

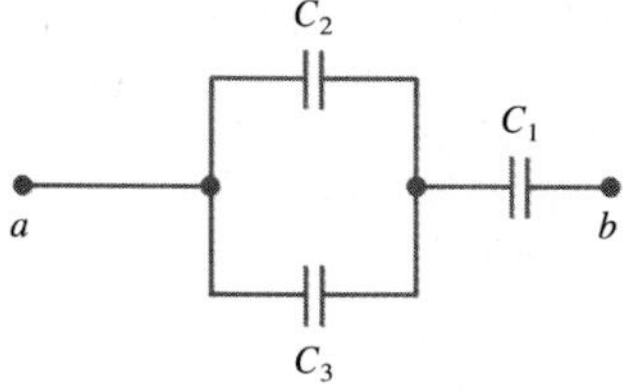

FIGURE 19–11 Example 19–5.

EXAMPLE 19–5 Determine the capacitance of a single capacitor that will have the same effect as the combination shown in Fig. 19–11. Take $C_1 = C_2 = C_3 = C$.

SOLUTION C_2 and C_3 are connected in parallel, so they are equivalent to a single capacitor having capacitance

$$C_{23} = C_2 + C_3 = 2C.$$

C_{23} is in series with C_1, so the equivalent capacitance, C_e, is given by

$$\frac{1}{C_e} = \frac{1}{C_1} + \frac{1}{C_{23}} = \frac{1}{C} + \frac{1}{2C} = \frac{3}{2C}.$$

Hence $C_e = \frac{2}{3}C$, which is the equivalent capacitance of the entire combination.

19–7 • Circuits Containing a Resistor and a Capacitor

RC circuits

Capacitors and resistors are often found together in a circuit. A simple example is shown in Fig. 19–12a. We now analyze this ***RC* circuit**. When the switch S is closed, current immediately begins to flow through the circuit. Electrons will flow out from the negative terminal of the battery, through the resistor R, and accumulate on the upper plate of the capacitor. And electrons will flow into the positive terminal of the battery, leaving a positive charge on the other plate of the capacitor. As charge accumulates on the capacitor, the current is reduced until eventually the voltage across the capacitor equals the emf of the battery, $\mathscr{E}$, and no further current flows. The potential difference across the capacitor, which is proportional to the charge on the capacitor ($V = Q/C$, Eq. 17–5), thus increases gradually, as shown in

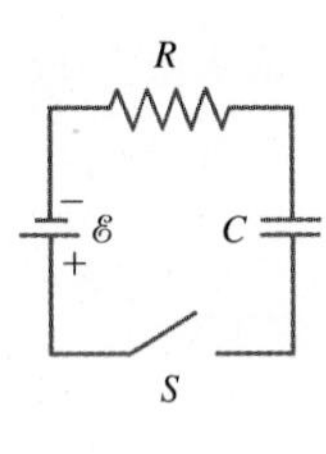

(a)

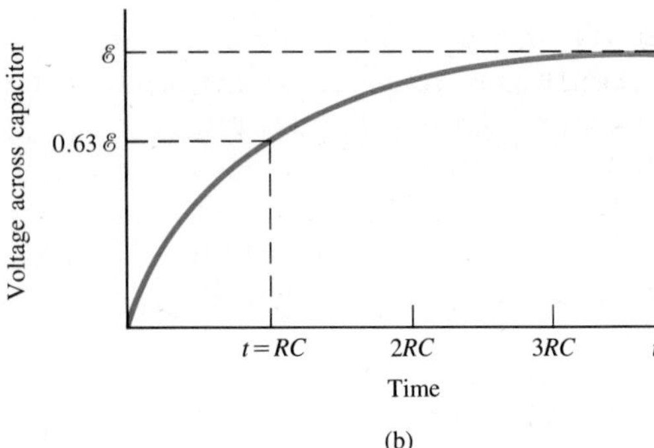

(b)

FIGURE 19–12 For the RC circuit shown in (a), the voltage across the capacitor increases with time, as shown in (b), after the switch S is closed.

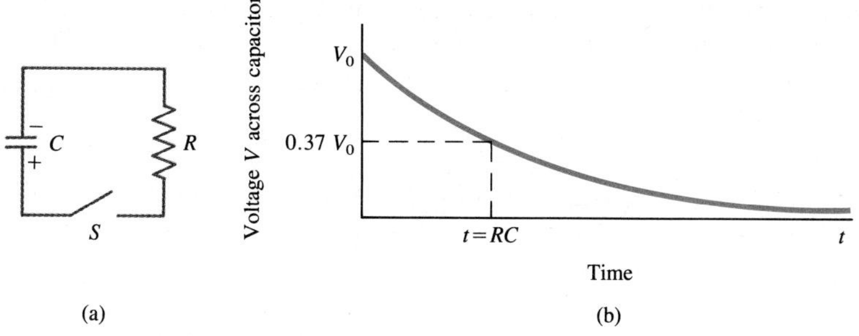

FIGURE 19–13 For the *RC* circuit shown in (a), the voltage *V* (and charge) on the capacitor decreases with time, as shown in (b), after the switch *S* is closed.

Fig. 19–12b. The actual shape of this curve is a type of exponential; it is given by the formula†

$$V = \mathscr{E}(1 - e^{-t/RC}),$$

where V is the voltage across the capacitor as a function of time t. The product of the value of the resistance times that of the capacitor is called the **time constant** τ of the circuit: $\tau = RC$; it is a measure of how quickly the capacitor becomes charged. [The units of RC are $\Omega \cdot \text{F} = (\text{V/A})(\text{C/V}) = \text{C/(C/s)} = \text{s}$.] Specifically, it can be shown that the product RC gives the time required for the capacitor to reach 63 percent of full voltage. This can be checked‡ using any calculator with an e^x key: $e^{-1} = 0.37$, so for $t = RC$, $(1 - e^{-t/RC}) = (1 - e^{-1}) = (1 - 0.37) = 0.63$. In a circuit, for example, where $R = 200\ \text{k}\Omega$ and $C = 3.0\ \mu\text{F}$, the time constant is $(2.0 \times 10^5\ \Omega)(3.0 \times 10^{-6}\ \text{F}) = 0.60$ s. If the resistance is much lower, the time constant is much smaller and the capacitor becomes charged almost instantly. This makes sense, since a lower resistance will retard the flow of charge less. All circuits contain some resistance (if only in the connecting wires), so a capacitor can never be charged instantaneously when connected to a battery.

The circuit just discussed involved the *charging* of a capacitor by a battery through a resistance. Now let us look at another situation: when a capacitor is already charged (say to a voltage V_0), and it is allowed to *discharge* through a resistance R as shown in Fig. 19–13a. (In this case there is no battery.) When the switch S is closed, charge begins to flow through resistor R from one side of the capacitor toward the other side, until it is fully discharged. The voltage across the capacitor decreases, as shown in Fig. 19–13b. This "exponential decay" curve is given by $V = V_0 e^{-t/RC}$, where V_0 is the initial voltage across the capacitor. The voltage falls 63 percent of the way to zero (to 0.37 V_0) in a time $\tau = RC$.

The charging and discharging in an RC circuit can be used to produce voltage pulses at a regular frequency. The charge on the capacitor increases to a particular voltage, and then discharges. A simple way of initiating the discharge is by the use of a gas-filled tube that breaks down when the voltage across it reaches a certain value. After the discharge is finished, the tube no longer conducts current and recharging repeats itself. Figure 19–14 shows a possible circuit, and the "sawtooth" voltage it produces.

FIGURE 19–14 (a) An *RC* circuit, coupled with a gas-filled tube as a switch, can produce a repeating "sawtooth" voltage, as shown in (b). If $R' \ll R$, the time constant for discharging is much smaller than for charging.

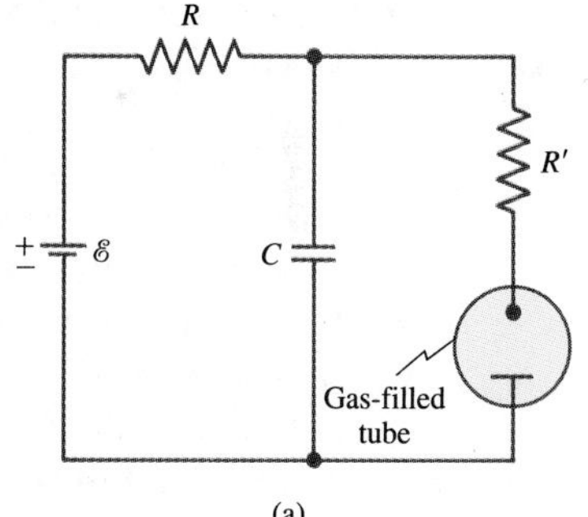

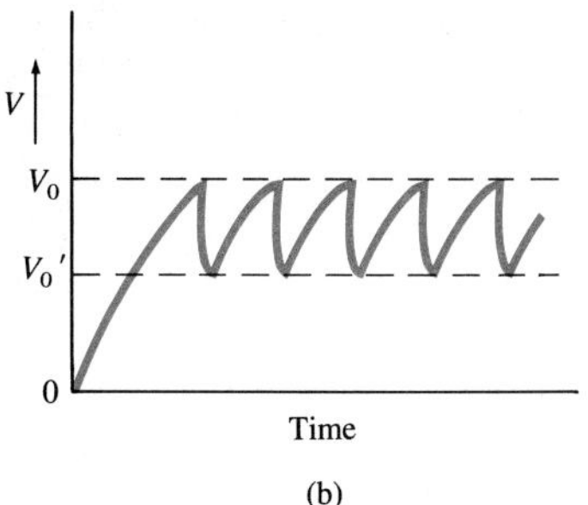

† The derivation is easy, using calculus. The exponential, e, has the value $e = 2.718 \cdots$.

‡ More simply, since $e = 2.718 \cdots$, then $e^{-1} = 1/e = 1/2.718 = 0.37$.

*19–8 • Heart Pacemakers

An interesting use of an RC circuit is the electronic pacemaker, which can make a stopped heart start beating again by applying an electric stimulus through electrodes attached to the chest. The stimulus can be repeated at the normal heartbeat rate if necessary.

The heart itself contains a *pacemaker*, which sends out tiny electric pulses at a rate of 60 to 80 per minute. These are the signals that induce the start of each heartbeat. In some forms of heart disease, the pacemaker cells fail to function properly and the heart loses its beat. *Electronic pacemakers* are now commonplace for people suffering this ailment. They can be external or can be inserted beneath the skin. They produce a regular voltage pulse that starts and controls the frequency of the heartbeat. The "fixed-rate" type produces signals continuously. The "demand" type operates only when the natural pacemaker fails. The electrodes are implanted in or near the heart and the circuit usually contains a capacitor and a resistor. The charge on the capacitor increases to a certain point and then discharges. Then it starts charging again. The pulsing rate depends on the values of R and C. Generally, the power source is a battery that must be replaced or recharged, depending on type. Some pacemakers obtain their energy from the heat produced by a radioactive element; the thermal energy is transformed to electricity by a thermocouple. Another type uses the heart's own contractions as its source; the critical element is a piezoelectric crystal that produces an emf in response to the pressure of the contracting heart.

19–9 • Electric Hazards; Leakage Currents

An electric shock can cause damage to the body or may even be fatal. The severity of a shock depends on the magnitude of the current, how long it acts, and through what part of the body it passes. A current passing through vital organs such as the heart or brain is especially serious for it can interfere with their operation. Electric current heats tissue and can cause burns. A current also stimulates nerves and muscles (whose operation, as we have seen in Sections 17–10 and 18–8, is electrical) and we feel a "shock."

Most people can "feel" a current of about 1 mA. Currents of a few mA cause pain but rarely much damage in a healthy person. However, currents above 10 mA cause severe contraction of the muscles, and a person may not be able to release the source of the current (say, a faulty appliance or wire). Death from paralysis of the respiratory system can occur. Artificial respiration, however, can often revive a victim. If a current above about 70 mA passes across the torso so that a portion passes through the heart for a second or more, the heart muscles will begin to contract irregularly and blood is not properly pumped. This condition is called "ventricular fibrillation." If it lasts for long, death results. Strangely enough, if the current is much larger, on the order of 1 A, the damage may be less and death by heart failure may be less likely† under some conditions.

The seriousness of a shock depends on the effective resistance of the body. Living tissue has quite low resistance since the fluid of cells contains ions that can conduct quite well. However, the outer layers of skin, when dry, offer much resistance. The effective resistance between two points on opposite sides of the body when the skin is dry is in the range of 10^4 to $10^6\ \Omega$. However, when the skin is wet, the resistance may be $10^3\ \Omega$ or less. A person in good contact with the ground who touches a 120-V dc line with wet hands can suffer a current

Beware of wet skin

$$I = \frac{120\ \text{V}}{1000\ \Omega} = 120\ \text{mA}.$$

As we saw above, this could be lethal.

Figure 19–15 shows how the circuit is completed when a person touches an electric wire. One side of a 120-V source is connected to ground by a wire connected to a buried conductor (say, a water pipe). Thus the current passes from the high-voltage wire, through the person, to the ground; it passes through the ground back to the other terminal of the source to complete the circuit. If the person in Fig. 19–15 stands on a good insulator—thick-soled shoes or a dry wood floor—there will be much more resistance in the circuit and consequently much less current will flow. However, if the person stands with bare feet on the ground, or is sitting in a bathtub, there is considerable danger because the resistance is much less. In a bathtub, not only are you wet, but the water is in contact with the drain pipe that leads to the ground. That is why it is strongly recommended not to touch anything electrical in such a situation.

Electricity and a bathtub don't mix

A principal danger comes from touching a bare wire whose insulation has worn off, or from a bare wire inside an appliance when you're tinkering with it. (Always unplug an electrical device before investigating its insides!) Sometimes a wire inside a device breaks or loses its insulation and comes in contact with the case. If the case is metal, it will conduct electricity. A person could then suffer a severe shock merely by touching the case. To prevent an accident, metal cases are supposed to be connected to ground, so they cannot become "hot." Then if a "hot" wire touches the grounded case, a short circuit to ground immediately occurs and the fuse or circuit breaker opens the circuit. (Grounding a metal case is best done by a separate ground wire connected to the third prong of a 3-prong plug, or connected to the larger prong of a so-called "polarized" 2-prong plug.)

Grounding and shocks

The human body acts as if it had capacitance in parallel with its resistance. A dc current can pass through the resistance, but not the capacitance. An ac current, like the changing currents discussed in Section 19–7 (more on this in Chapter 21), can exist also in the capacitive branch. Because of the additional path allowing current flow, the current for a given V_{rms} will be greater than for the same dc voltage. Thus an ac voltage is more dangerous than an equal dc voltage.

FIGURE 19–15 A person receives an electric shock when the circuit is completed.

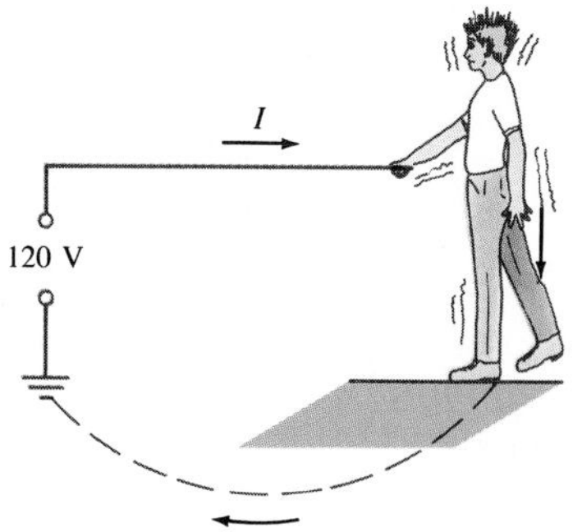

† Apparently, larger currents bring the entire heart to a standstill. Upon release of the current, the heart returns to its normal rhythm. This may not happen when fibrillation occurs since it is often hard to stop once it starts. Fibrillation may also occur as a result of a heart attack or during heart surgery. A device known as a *defibrillator* can apply a brief high current to the heart; this causes complete heart stoppage and is often followed by resumption of normal beating.

Leakage current

Another danger is *leakage current*, by which we mean a current along an unintended path. Leakage currents are often capacitively coupled. For example, a wire in a lamp forms a capacitor with the metal case; charges moving in one conductor attract or repel charge in the other, so there is a current. Typical electrical codes limit leakage currents to 1 mA for any device. A 1-mA leakage current is usually harmless. It can be very dangerous, however, to a hospital patient with implanted electrodes connected to ground through the apparatus. This is because the current can pass directly through the heart as compared to the usual situation where it spreads out through the body. Although 70 mA may be needed to cause heart fibrillation (rapid irregular contractions of muscle fibers) when entering through the hands (very little of it actually passes through the heart), as little as 0.02 mA has been known to cause fibrillation when passing directly to the heart. Thus, a "wired" patient is in considerable danger from leakage current even from as simple an act as touching a lamp.

*19–10 • DC Ammeters and Voltmeters

An **ammeter** is used to measure current, and a **voltmeter** measures potential difference or voltage. The crucial part of an analog ammeter or voltmeter, in which the reading is by a pointer on a scale (Fig. 19–16), is a *galvanometer*. The galvanometer works on the principle of the force between a magnetic field and a current-carrying coil of wire, and will be discussed in Chapter 20. For now, we merely need to know that the deflection of the needle of a galvanometer is proportional to the current flowing through it. The *full-scale current sensitivity*, I_m, of a galvanometer is the current needed to make the needle deflect full scale. If the sensitivity I_m is 50 μA, a current of 50 μA will cause the needle to go to the end of the scale. A current of 25 μA will then make it go only half way. If there is no current, the needle should be on zero and usually there is an adjustment screw to make it so.

Many meters we see in everyday life are galvanometers connected as ammeters or voltmeters. These include the VU meter on a tape recorder, and some of the meters on the dashboards of cars.

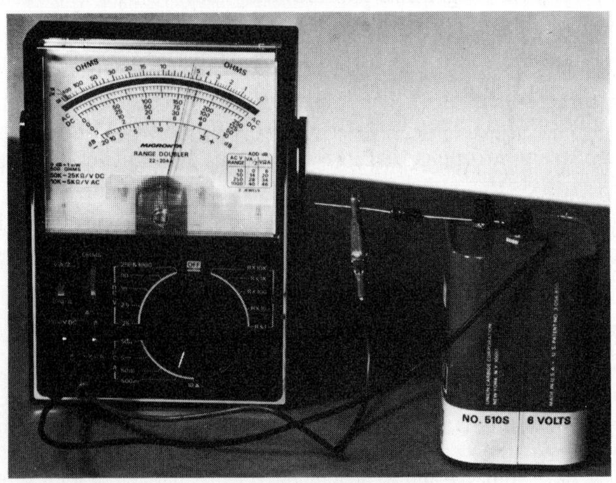

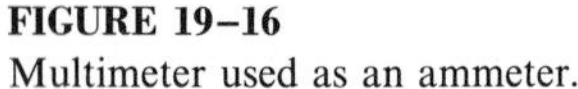
FIGURE 19–16
Multimeter used as an ammeter.

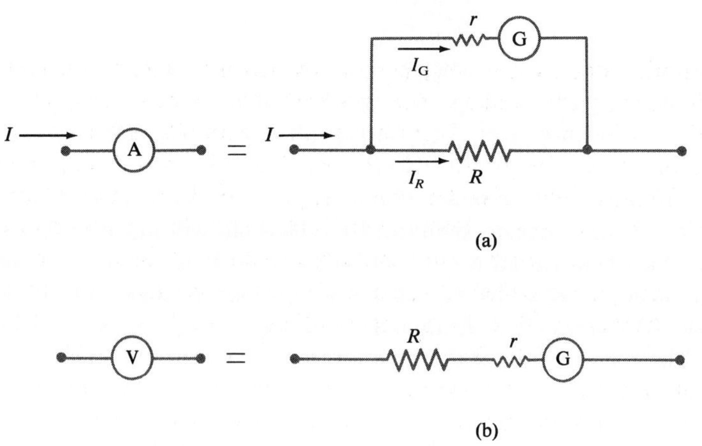

FIGURE 19–17 An ammeter is a galvanometer in parallel with a small (shunt) resistor, R. (b) A voltmeter is a galvanometer in series with a large resistor, R.

A galvanometer can be used directly to measure small dc currents. For example, a galvanometer whose sensitivity I_m is 50 μA can measure currents from about 1 μA (currents smaller than this would be hard to read on the scale) up to 50 μA. To measure larger currents, a resistor is placed in parallel with the galvanometer. Thus, an **ammeter** (represented by the symbol •–Ⓐ–•) consists of a galvanometer (•–Ⓖ–•) in parallel with a resistor called the **shunt resistor**. ("Shunt" is a synonym for "in parallel.") This is shown in Fig. 19–17a. The shunt resistance is R and the resistance of the galvanometer coil (which carries the current) is r. The value of R is chosen according to what full-scale deflection is desired.

Ammeter uses shunt resistor

Suppose we want to design an ammeter to read 1.0 A at full scale using a galvanometer with a full-scale sensitivity of 50 μA and its resistance is $r = 30\ \Omega$. This means that when the total current I entering the ammeter is 1.0 A, we want the current I_G through the galvanometer to be precisely 50 μA (to give full-scale deflection). Thus, when 1.0 A flows into the meter, we want 0.999950 A ($=I_R$) to pass through the shunt resistor R. Since the potential difference across the shunt is the same as across the galvanometer,

$$I_R R = I_G r,$$

then

$$R = \frac{I_G r}{I_R} = \frac{(5.0 \times 10^{-5}\ \text{A})(30\ \Omega)}{(0.999950\ \text{A})} = 1.5 \times 10^{-3}\ \Omega,$$

or 0.0015 Ω. The shunt resistor must thus have a very low resistance so that most of the current passes through it. If the current I into the meter is 0.50 A, this will produce a current in the galvanometer equal to $I_G = I_R R/r = (0.50\ \text{A})(1.5 \times 10^{-3}\ \Omega)/30\ \Omega = 25\ \mu\text{A}$, which gives a deflection half of full scale, as required.

A **voltmeter** (•–Ⓥ–•) also consists of a galvanometer and a resistor. But the resistor R is connected in series, Fig. 19–17b, and it is usually large. Suppose, using the same galvanometer with internal resistance $r = 30\ \Omega$ and full-scale current sensitivity of 50 μA, we want to make a voltmeter that reads from 0 to 15 V. When a potential difference of 15 V exists across the terminals of our voltmeter, we want 50 μA to be passing through it so as to give a full-scale deflection. From Ohm's law we have

Voltmeter uses series resistor

$$15\ \text{V} = (50\ \mu\text{A})(r + R),$$

ordinary analog meters is typically 3 to 4 percent of full-scale deflection. An ammeter also can interfere with a circuit, but the effect is minimal if its resistance is much less than that of the circuit as a whole. For both voltmeters and ammeters, the more sensitive the galvanometer the less effect it will have. A 50,000-Ω/V meter is far better than a 1,000-Ω/V meter.

Electronic meters using transistors, including digital meters, have very high input resistance (usually specified in ohms) in the range 10^6 to 10^8 Ω, and even higher. Hence they have very little effect on most circuits and their readings are reliable for nearly any circuit. The precision of digital meters is typically one part in 10^4 (=0.01 percent) or better. State-of-the-art instruments reach a precision of one part in 10^6.

SUMMARY

When resistances are connected in series, the net resistance is the sum of the individual resistances. When connected in parallel, the reciprocal of the total resistance equals the sum of the reciprocals of the individual resistances. In a parallel connection, the net resistance is less than any of the individual resistances.

A device that transforms one type of energy into electrical energy is called a *seat* or *source* of *emf*. A battery behaves like a source of emf in series with an internal resistance. The emf is the potential difference determined by the chemical reactions in the battery and equals the terminal voltage when no current is drawn. When a current is drawn, the voltage at the battery's terminals is less than its emf by an amount equal to the Ir drop across the internal resistance.

Kirchhoff's rules are helpful in determining the currents and voltages in a complex circuit. Kirchhoff's first, or "junction," rule is based on conservation of electric charge and states that the sum of all currents entering any junction equals the sum of all currents leaving that junction. The second, or "loop," rule is based on conservation of energy and states that the algebraic sum of the voltage changes around any closed path of the circuit must be zero.

When capacitors are connected in parallel, the net capacitance is the sum of the individual capacitances. When connected in series, the reciprocal of the net capacitance equals the sum of the reciprocals of the individual capacitances.

When a circuit containing a resistor R in series with a capacitance C is connected to a dc source of emf, the voltage across the capacitor rises gradually in time characterized by the time constant $\tau = RC$; this is the time it takes for the voltage to reach 63 percent of its maximum value. A capacitor discharging through a resistor is characterized by the same time constant: in a time $\tau = RC$, the voltage across the capacitor drops to 37 percent of its initial value.

Electric shocks are caused by current passing through the body. To avoid shocks, the body must not become part of a circuit by allowing different parts of the body to touch objects at different potentials. Commonly, one part of the body may be touching ground and another part a high or low potential.

QUESTIONS

1. Discuss the advantages and disadvantages of Christmas tree lights connected in parallel versus those connected in series.
2. If all you have is a 120-V line, would it be possible to light several 6-V lamps without burning them out? How?
3. Two light bulbs of resistance R_1 and R_2 ($>R_1$) are connected in series. Which is brighter? What if they are connected in parallel?
4. Describe carefully the difference between emf and potential difference.
5. The internal resistance of an electric cell is not actually constant. Why not?
6. Explain why Kirchhoff's first (junction) rule is equivalent to conservation of electric charge.
7. Explain why Kirchhoff's second (loop) rule is a result of the conservation of energy.

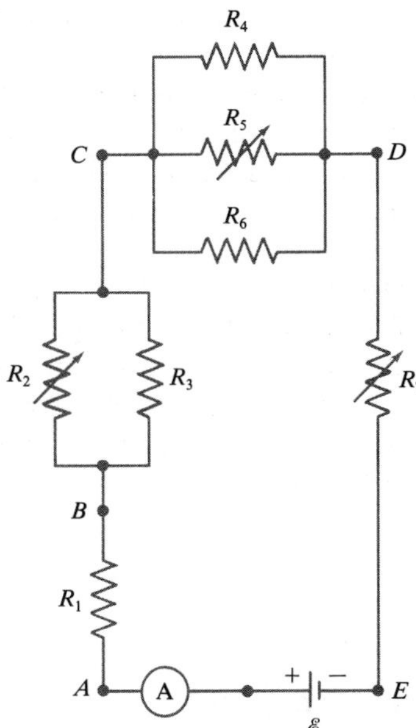

FIGURE 19–20 Question 8.

8. Given the circuit shown in Fig. 19–20 use the words "increases," "decreases," or "stays the same" to complete the following statements:
   (*a*) If R_7 increases, the potential difference between A and E (assume no resistance in Ⓐ and $\mathscr{E}$) ______.
   (*b*) If R_7 increases, the potential difference between A and E (assume Ⓐ and $\mathscr{E}$ have resistance) ______.
   (*c*) If R_7 increases, the voltage drop across R_4 ______.
   (*d*) If R_2 decreases, the current through R_1 ______.
   (*e*) If R_2 decreases, the current through R_6 ______.
   (*f*) If R_2 decreases, the current through R_3 ______.
   (*g*) If R_5 increases, the voltage drop across R_2 ______.
   (*h*) If R_5 increases, the voltage drop across R_4 ______.
   (*i*) If R_2, R_5, and R_7 increase, $\mathscr{E}$ ______.
9. Why are batteries connected in series? Why in parallel? Does it matter if the batteries are nearly identical or not in either case?
10. Describe a situation in which the terminal voltage of a battery is greater than its emf.
11. The 18-V source in Fig. 19–21 is "charging" the 12-V battery. Explain how it does this.

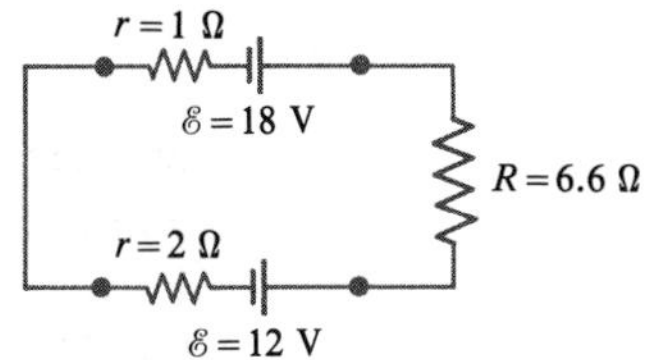

FIGURE 19–21 Question 11 and Problem 19.

12. Explain in detail how you could measure the internal resistance of a battery.
13. Compare and discuss the formulas for resistors and for capacitors when connected in series and in parallel.
14. Suppose that three identical capacitors are connected to a battery. Will they store more energy if connected in series or in parallel?
15. Why is it more dangerous to turn on an electric appliance when you are standing outside in bare feet than when you are inside wearing shoes with thick soles?
16. Figure 19–22 is a diagram of a capacitor (or condenser) *microphone*. The changing air pressure in a sound wave causes one plate of the capacitor C to move back and forth. Explain how a current of the same frequency as the sound wave is produced.

*17. What is the main difference between a voltmeter and an ammeter?

*18. Explain why an ideal ammeter would have zero resistance and an ideal voltmeter infinite resistance.

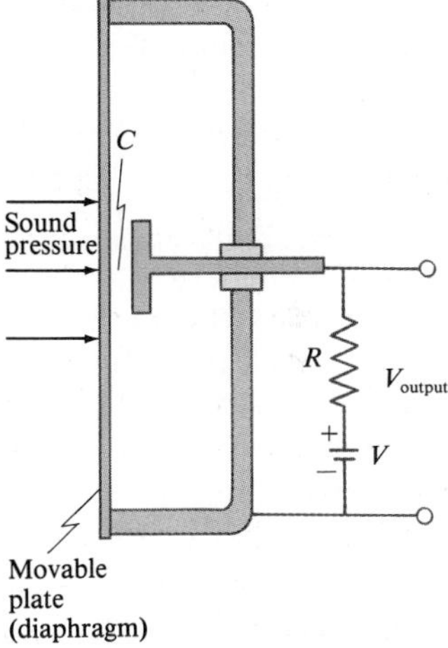

FIGURE 19–22 Diagram of a capacitor microphone. Question 16.

PROBLEMS

SECTION 19–1

In these problems neglect the internal resistance of a battery unless the problem refers to it.

1. (I) Six 80-Ω light bulbs are connected in series. What is the total resistance of the circuit? What is their resistance if they are connected in parallel?
2. (I) Suppose you have a 600-Ω, an 800-Ω, and a 1.30-kΩ resistor. What is (*a*) the maximum, and (*b*) the minimum resistance you can obtain by combining these?
3. (II) Suppose that you have a 6.0-V battery and you wish to apply a voltage of only 2.0 V. Given an unlimited supply of 1.0-Ω resistors, how could you connect them so as to make a "voltage divider" that produced a 2.0-V output for a 6.0-V input?
4. (II) Three 100-Ω resistors can be connected together in four different ways, making combinations of series and/or parallel circuits. What are these four ways and what is the net resistance in each case?

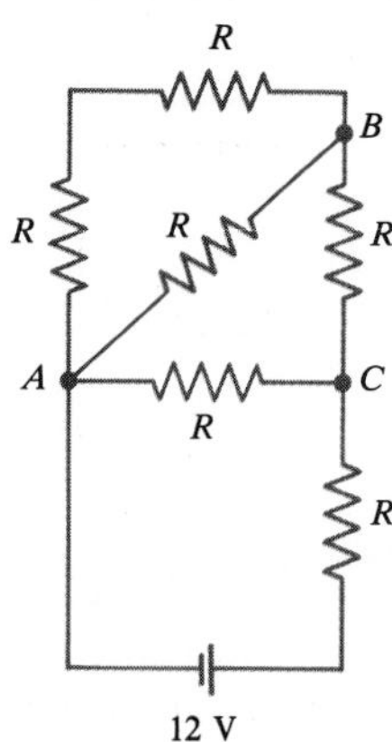

FIGURE 19–23 Problems 5 and 11.

5. (II) What is the net resistance of the circuit connected to the battery in Fig. 19–23? Each resistance has $R =$ 2.2 kΩ.
6. (II) Eight lights are connected in series across a 120-V line. (*a*) What is the voltage across each bulb? (*b*) If the current is 0.60 A, what is the resistance of each bulb and the power dissipated in each?
7. (II) Eight lights are connected in parallel to a 120-V source by two leads of total resistance 2.0 Ω. If 280 mA flows through each bulb, what is the resistance of each, and what fraction of the total power is wasted in the leads?
8. (II) Seven 8.0-W Christmas tree lights are connected in series to a 120-V source. What is the resistance of each bulb?
9. (II) Two resistors when connected in series to a 120-V line use one-fourth the power that is used when they are connected in parallel. If one resistor is 2.8 kΩ, what is the resistance of the other?
10. (II) A 75-W, 120-V bulb is connected in parallel with a 40-W, 120-V bulb. What is the net resistance?
11. (II) Calculate the current through each resistor in Fig. 19–23 if each resistance $R = 2.40$ kΩ. What is the potential difference between points A and B?
12. (III) A 1.3-kΩ and a 1.8-kΩ resistor are connected in parallel; this combination is connected in series with a 1.4-kΩ resistor. If each resistor is rated at $\frac{1}{2}$ W, what is the maximum voltage that can be applied across the whole network?

SECTION 19–2

13. (I) A battery whose emf is 6.00 V and whose internal resistance is 0.60 Ω is connected to a circuit whose net resistance is 7.20 Ω. What is the terminal voltage of the battery?

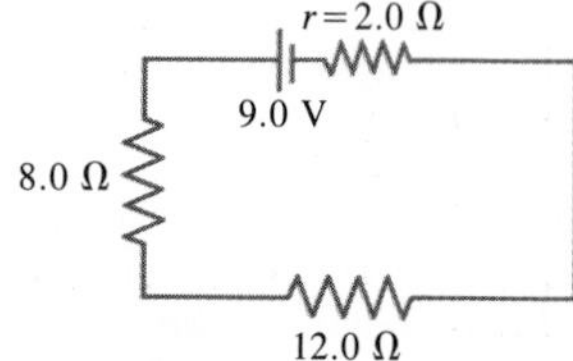

FIGURE 19–24 Problem 18.

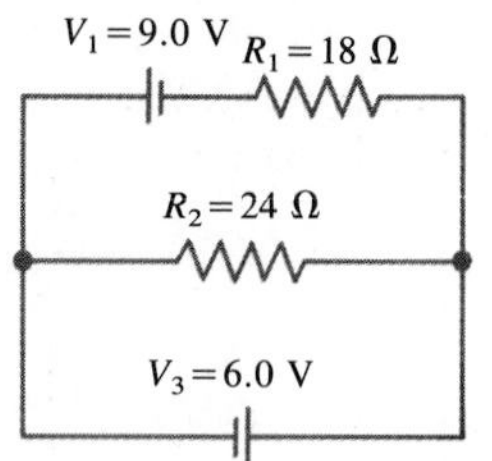

FIGURE 19–25 Problems 22 and 23.

14. (I) Four 1.5-V cells are joined in series to a 6.2-Ω device. The internal resistance of each cell is 0.30 Ω. What current flows to the device?
15. (II) A 1.5-V dry cell can be tested by connecting it to a low-resistance ammeter. It should be able to supply at least 30 A. What is the internal resistance of the cell in this case?
16. (II) What is the internal resistance of a 12.0-V car battery whose terminal voltage drops to 8.8 V when the starter draws 70 A?
17. (II) What is the current in the 8.0-Ω resistor in Fig. 19–5a?

SECTION 19–3

18. (II) Calculate the current in the circuit of Fig. 19–24 and show that the sum of all the voltage changes around the circuit is zero.
19. (II) Determine the terminal voltage of each battery in Fig. 19–21.

*SECTION 19–4

*20. (II) What is the potential difference between points a and d in Fig. 19–8?
*21. (II) What is the terminal voltage of each battery in Fig. 19–8?
*22. (II) Determine the magnitudes and directions of the currents through R_1 and R_2 in Fig. 19–25.
*23. (II) Repeat Problem 22, now assuming that each battery has an internal resistance $r = 1.0$ Ω.
*24. (III) Determine the current through each of the resistors in Fig. 19–26.
*25. (III) If the 30-Ω resistor in Fig. 19–26 were shorted out (resistance = 0), what then would be the current through the 10-Ω resistor?

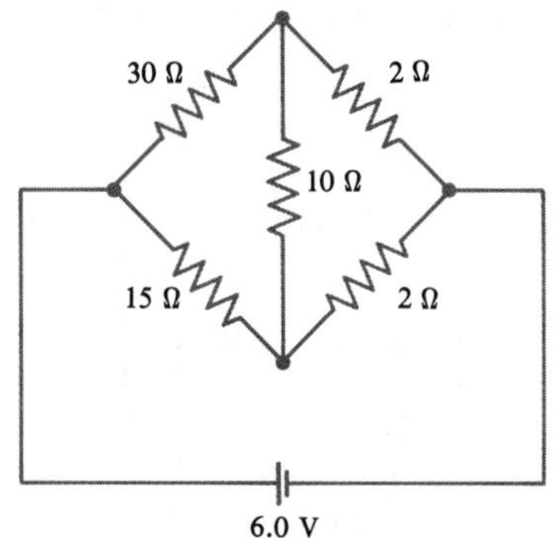

FIGURE 19–26 Problems 24 and 25.

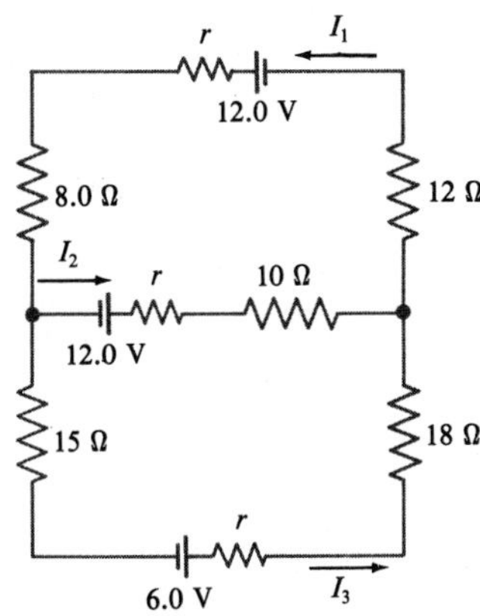

FIGURE 19–27 Problems 26 and 27.

*26. (III) Determine the currents I_1, I_2, and I_3 in Fig. 19–27. Assume the internal resistance of each battery is $r = 1.0\ \Omega$. What is the terminal voltage of the 6.0-V battery?

*27. (III) What would the current I_1 be in Fig. 19–27 if the 12-Ω resistor were shorted out ($r = 1.0\ \Omega$)?

SECTION **19–5**

28. (II) Suppose two batteries are connected as in Fig. 19–9c but their emfs are not equal but are, respectively, 2.0 V and 3.0 V, as shown in Fig. 19–28. If each internal resistance is $r = 0.10\ \Omega$, and $R = 4.0\ \Omega$, what is the voltage across the resistor R?

SECTION **19–6**

29. (I) Six 2.8-μF capacitors are connected in parallel. What is the equivalent capacitance? What is their equivalent capacitance if connected in series?

30. (I) A circuit contains a 3.0-μF capacitor. However, a technician decides 4.8 μF would be better. What size capacitor should be added to the circuit and how should it be connected?

31. (I) The capacitance of a portion of a circuit is to be reduced from 3600 pF to 2000 pF. What capacitance can be added to the circuit to produce this effect without removing anything from the circuit? How should the extra capacitor be connected?

32. (II) Suppose three parallel-plate capacitors, whose plates have areas A_1, A_2, and A_3, and separations d_1, d_2, and d_3, are connected in parallel. Show, using Eq. 17–6, that Eq. 19–3 is valid.

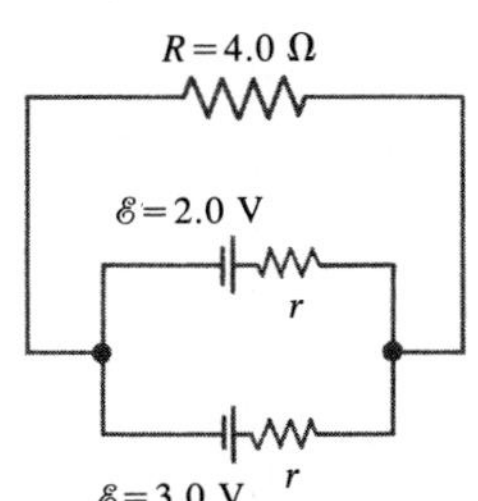

FIGURE 19–28 Problem 28.

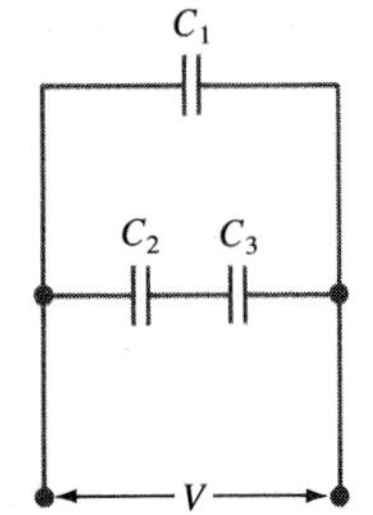

FIGURE 19–29 Problems 33 and 34.

33. (II) (*a*) Determine the equivalent capacitance of the circuit shown in Fig. 19–29. (*b*) If $C_1 = C_2 = 2C_3 = 8.0\ \mu$F, how much charge is stored on each capacitor when $V = 60$ V?

34. (II) In Fig. 19–29, let $V = 200$ V and $C_1 = C_2 = C_3 = 1800$ pF. How much energy is stored in the capacitor network?

35. (II) You have three capacitors, of capacitance 3000 pF, 6000 pF, and 0.010 μF. What is the maximum and minimum capacitance that you can form from these? How do you make the connection in each case?

36. (II) A 0.20-μF and a 0.30-μF capacitor are connected in series to a 9.0-V battery. Calculate (*a*) the potential difference across each capacitor, and (*b*) the charge on each. (*c*) Repeat parts (*a*) and (*b*) assuming the two capacitors are in parallel.

37. (III) A 5.0-μF and a 4.0-μF capacitor are connected in series and this combination is connected in parallel with a 2.0-μF capacitor. (*a*) What is the net capacitance? (*b*) If 60 V is applied across the whole network, calculate the voltage across each capacitor.

SECTION **19–7**

38. (I) Electrocardiographs are often connected as shown in Fig. 19–30. The leads are said to be capacitively coupled. A time constant of 3.0 s is typical and allows rapid changes in potential to be accurately recorded. If $C = 3.0\ \mu$F, what value must R have?

39. (II) Suppose a 3.0-μF capacitor were placed across the 6.0-Ω resistor in Fig. 19–5a. Calculate the charge on the capacitor in the steady state—that is, after the capacitor reaches its maximum charge.

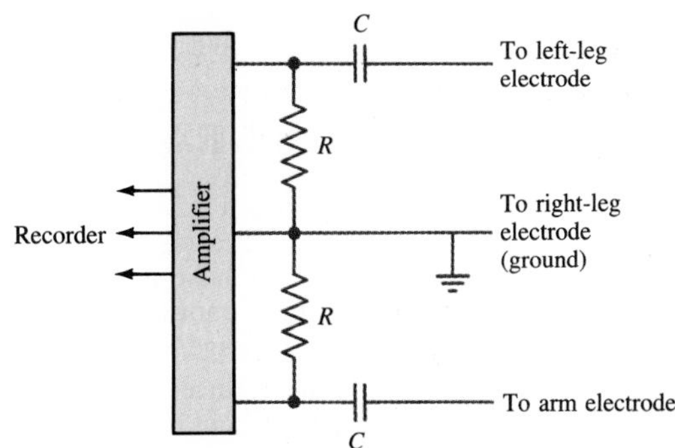

FIGURE 19–30 Problem 38.

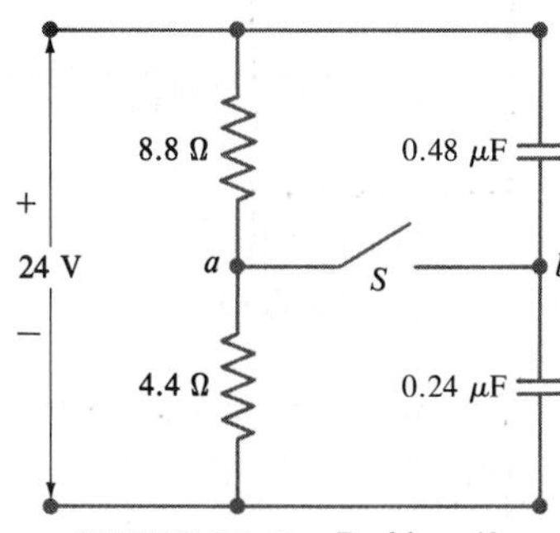

FIGURE 19–31 Problem 42.

40. (II) The capacitance of the circuit of Fig. 19–12a is $C = 0.70\ \mu\text{F}$, the total resistance $R = 22\ \text{k}\Omega$, and the battery emf is $\mathscr{E} = 12$ V. Determine (*a*) the time constant, and (*b*) the time it takes for the voltage across the capacitor to reach 10 V.
41. (II) The *RC* circuit of Fig. 19–13a has $R = 7.5\ \text{k}\Omega$ and $C = 2.0\ \mu\text{F}$. The capacitor is at voltage V_0 at $t = 0$, when the switch is closed. How long does it take the capacitor to discharge to 1.0% of its initial voltage?
42. (III) Two resistors and two uncharged capacitors are arranged as shown in Fig. 19–31. With a potential difference of 24 V across the combination, (*a*) what is the potential at point *a* with *S* open? (Let $V = 0$ at the negative terminal of the source.) (*b*) What is the potential at point *b* with the switch open? (*c*) When the switch is closed, what is the final potential of point *b*? (*d*) How much charge flows through the switch *S* after it is closed?

*SECTION 19–10

*43. (I) What is the resistance of a voltmeter on the 5.0-V scale if the meter sensitivity is 30,000 Ω/V?

*44. (I) An ammeter has a sensitivity of 5000 Ω/V. What current passing through the galvanometer produces full-scale deflection?

*45. (II) How could you make an ammeter with a full-scale deflection of 10 A from a galvanometer that has a 120-μA full-scale deflection when 10 mV is across it?

*46. (II) A galvanometer has an internal resistance of 30 Ω and deflects full scale for a 70-μA current. Describe how to use this galvanometer to make (*a*) an ammeter to read currents up to 15 A, and (*b*) a voltmeter to give a full-scale deflection of 3000 V.

*47. (II) A galvanometer has a sensitivity of 50,000 Ω/V and internal resistance 28.0 Ω. How could you make this into (*a*) an ammeter reading 10.0 mA full scale, (*b*) a voltmeter reading 100 mV full scale?

*48. (II) A milliammeter reads 10 mA full scale. It consists of a 0.20-Ω resistor in parallel with a 30-Ω galvanometer. How can you change this ammeter to a voltmeter giving a full-scale reading of 10 V without taking the ammeter apart? What will be the sensitivity (Ω/V) of your voltmeter?

*SECTION 19–11

*49. (II) A 90-V battery of negligible internal resistance is connected to a 37-kΩ and a 42-kΩ resistor in series. What reading will a voltmeter give when used to measure the voltage across each resistor if its internal resistance is 100 kΩ? What is the percent inaccuracy due to meter resistance for each case?

*50. (II) An ammeter whose internal resistance is 80 Ω reads 3.20 mA when connected in a circuit containing a battery and two resistors in series whose values are 800 Ω and 500 Ω. What is the actual current when the ammeter is absent?

*51. (II) A battery with $\mathscr{E} = 9.0$ V and internal resistance $r = 1.0\ \Omega$ is connected to two 6.0-kΩ resistors in series. An ammeter of internal resistance 0.50 Ω measures the current and at the same time a voltmeter with internal resistance 15 kΩ measures the voltage across one of the 6.0-kΩ resistors in the circuit. What do the ammeter and voltmeter read?

*52. (III) Two 8.9-kΩ resistors are placed in series and connected to a battery. A voltmeter of sensitivity 1000 Ω/V is on the 3.0-V scale and reads 2.0 V when placed across either of the resistors. What is the emf of the battery? (Ignore the battery's internal resistance.)

*53. (III) What internal resistance should the voltmeter have to be in error by less than 3 percent for the situation of Example 19–6?

*54. (III) The voltage across a 120-kΩ resistor in a circuit containing additional resistance (R_2) in series with a battery (V) is measured, by a 20,000-Ω/V meter on the 100-V scale, to be 25 V. On the 30-V scale, the reading is 23 V. What is the actual voltage in the absence of the voltmeter? What is R_2?

*55. (III) A 9.0-V battery (assume the internal resistance = 0) is connected to two resistors in series. A voltmeter whose internal resistance is 10.0 kΩ measures 3.0 V and 4.0 V, respectively, when connected across each of the resistors. What is the resistance of each resistor?

GENERAL PROBLEMS

56. Suppose that you wish to apply a 0.25-V potential difference between two points on the body. The resistance is about 2000 Ω, and you only have a 9.0-V battery. How can you connect up one or more resistors so that you can produce the desired voltage?
57. A three-way light bulb can produce 50 W, 100 W, or 150 W, at 120 V. Such a bulb contains two filaments that can be connected to the 120 V individually or in parallel. Describe how the connections to the two filaments are made to give each of the three wattages. What must be the resistance of each filament?
58. Suppose you want to run some apparatus that is 150 m from an electric outlet. Each of the wires connecting your apparatus to the 120-V source has a resistance per unit length of 0.0065 Ω/m. If your apparatus draws 3.0 A, what will be the voltage drop across the connecting wires and what voltage will be applied to your apparatus?

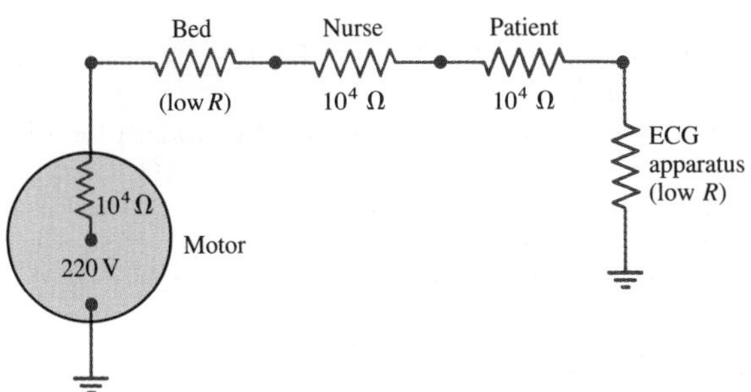

FIGURE 19–32 Problem 59.

59. Electricity can be a hazard in hospitals, particularly to patients who are connected to electrodes, such as an ECG. For example, suppose that the motor of a motorized bed shorts out to the bed frame, and the bed frame's connection to a ground has broken (or was not there in the first place). If a nurse touches the bed and the patient at the same time, she becomes a conductor and a complete circuit can be made through the patient to ground through the ECG apparatus. This is shown schematically in Fig. 19–32. Calculate the current through the patient.
60. How much energy must a 12-V battery expend to fully charge a 0.40-μF and a 0.20-μF capacitor when they are placed (*a*) in parallel, (*b*) in series? (*c*) How much charge flowed from the battery in each case?
61. A heart pacemaker is designed to operate at 70 beats/min using a 4.0-μF capacitor. What value of resistance should be used if the pacemaker is to fire when the voltage reaches 63 percent of maximum?
62. In experiments on nerve axon and muscle fiber cells, two electrodes are inserted into the cell (axon) close to each other. One electrode serves as a signal generator of rectangular pulses. The other serves as a receiver of the cell response to the signal pulses. The signal and response are shown in Figs. 19–33a and b. The electrical characteristics of the cell in such an experiment may be represented by the equivalent circuit illustrated in Fig. 19–33c, where C is the capacitance of the cell membrane, R is the resistance of the cell membrane, the emf is the signal electrode, and the voltmeter is the receiver electrode. (*a*) Use the response graph to estimate the time constant of the cell. (*b*) Use the results of part (*a*) and Example 18–8 to estimate the value of R.
63. The internal resistance of a 1.35-V mercury cell is 0.030 Ω, whereas that of a 1.5-V dry cell is 0.35 Ω. Explain why three mercury cells can more effectively power a 2-W hearing aid that requires 4.0 V than can three dry cells.
64. Suppose that a person's body resistance is 1000 Ω. (*a*) What current passes through the body when the person accidentally is connected to 120 V? (*b*) If there is an alternative path to ground whose resistance is 30 Ω, what current passes through the person? (*c*) If the voltage source can produce at most 1.4 A, how much current passes through the person in case (*b*)?

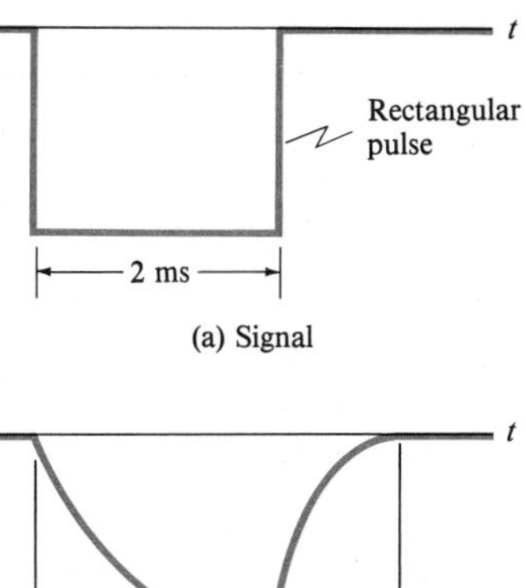

(a) Signal

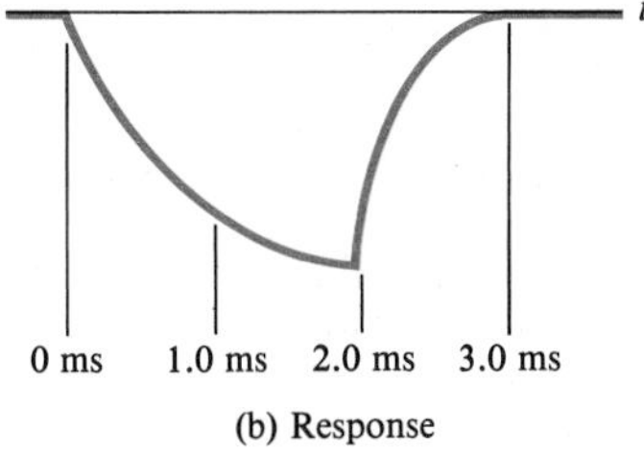

(b) Response

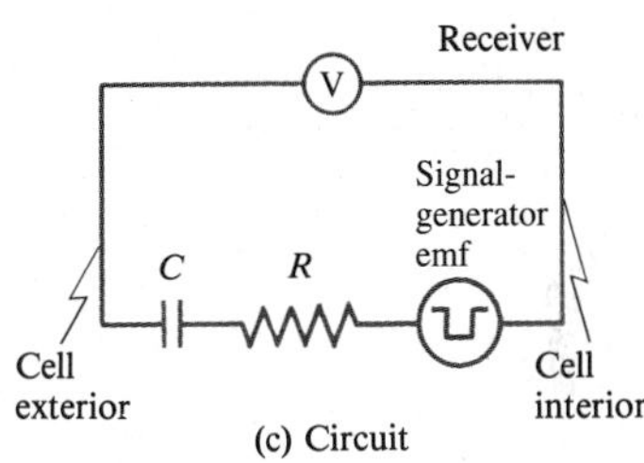

(c) Circuit

FIGURE 19–33 Problem 62.

65. The **Wheatstone bridge** is a circuit used to make precise measurements of resistance. The unknown resistance to be measured, R_x, is placed in the bridge circuit as shown in Fig. 19–34, and the variable resistor R_3 is adjusted until the galvanometer does not deflect when the switch is closed. For this setting, no current flows from B to D, so B and D are at the same potential. (*a*) Show that the unknown resistance R_x is given by

$$R_x = \frac{R_2}{R_1} R_3$$

when R_1, R_2, and R_3 are all accurately known. (*b*) A Wheatstone bridge is balanced when $R_1 = 410\ \Omega$, $R_2 = 848\ \Omega$, and $R_3 = 2.34\ \Omega$. What is the value of the unknown resistance in the fourth arm?

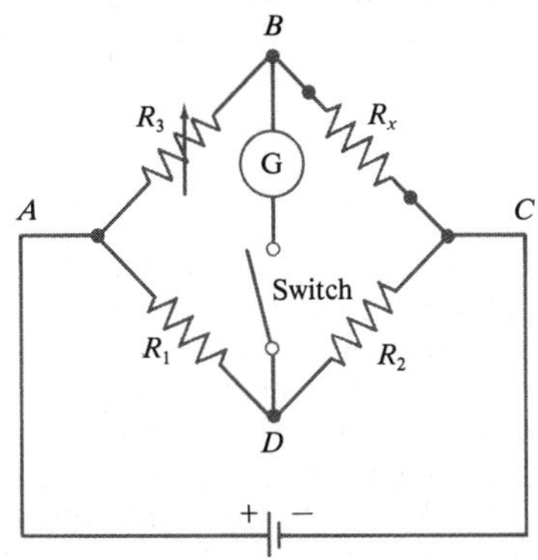

FIGURE 19–34 Wheatstone bridge. Problems 65 and 66.

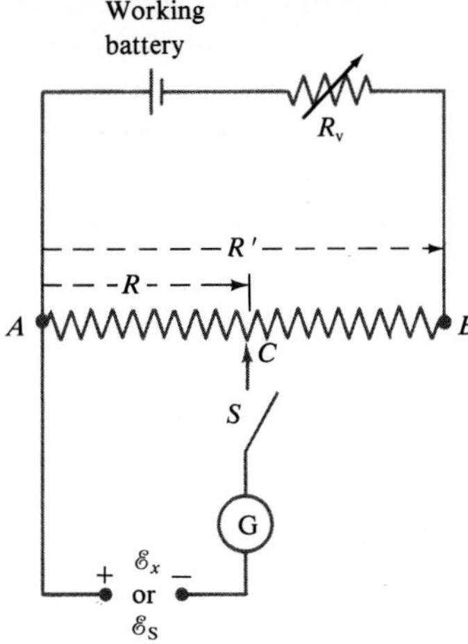

FIGURE 19–35 Potentiometer circuit. Problem 67.

66. An unknown length of platinum wire 1.2 mm in diameter is placed as the unknown resistance in a Wheatstone bridge (see Problem 65). Arms 1 and 2 have resistance of 36.0 Ω and 84.0 Ω, respectively. Balance is achieved when R_3 is 4.76 Ω. How long is the platinum wire?

67. A **potentiometer** is a device to precisely measure potential differences or emf, using a "null" technique. In the simple potentiometer circuit shown in Fig. 19–35, R' represents the total resistance of the resistor from A to B (which could be a long uniform "slide" wire), whereas R represents the resistance of only the part from A to the movable contact at C. When the unknown emf to be measured, $\mathscr{E}_x$, is placed into the circuit as shown, the movable contact C is moved until the galvanometer G gives a null reading (i.e., zero) when the switch S is closed. The resistance between A and C for this situation we call R_x. Next, a standard emf, $\mathscr{E}_s$, which is known precisely, is inserted into the circuit in place of $\mathscr{E}_x$ and again the contact C is moved until zero current flows through the galvanometer when the switch S is closed. The resistance between A and C now is called R_s. (*a*) Show that the unknown emf is given by

$$\mathscr{E}_x = \left(\frac{R_x}{R_s}\right)\mathscr{E}_s$$

where R_x, R_s, and $\mathscr{E}_s$ are all precisely known. The working battery is assumed to be fresh and to give a constant voltage. (*b*) A slide-wire potentiometer is balanced against a 1.0182-V standard cell when the slide wire is set at 31.7 cm out of a total length of 100.0 cm. For an unknown source, the setting is 13.8 cm. What is the emf of the unknown? (*c*) The galvanometer of a potentiometer has an internal resistance of 40 Ω and can detect a current as small as 0.015 mA. What is the minimum uncertainty possible in measuring an unknown voltage? (*d*) Explain the advantage of using this "null" method of measuring emf.

68. The variable capacitance of a radio tuner consists of four plates connected together placed alternately between four other plates, also connected together (Fig. 19–36). Each plate is separated from its neighbor by 1.0 mm of air. One set of plates can move so that the area of overlap varies from 1.0 cm^2 to 4.0 cm^2. (*a*) Are these seven capacitors connected in series or in parallel? (*b*) Determine the range of capacitance values.

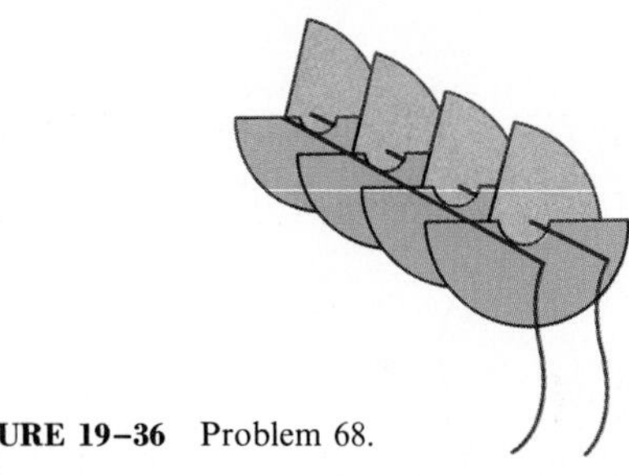

FIGURE 19–36 Problem 68.

69. A battery produces 50.0 V when 6.50 A are drawn from it and 58.2 V when 2.80 A are drawn. What is the emf and internal resistance of the battery?

70. How many $\frac{1}{2}$-W resistors, each of the same resistance, must be used to produce an equivalent 1.8-kΩ, 5-W resistor? What is the resistance of each, and how must they be connected?

71. The current through the 4.0-kΩ resistor in Fig. 19–37 is 3.50 mA. What is the terminal voltage V_{ab} of the "unknown" battery? (There are two answers. Why?) [*Hint:* use conservation of energy or Kirchhoff's rules.]

FIGURE 19–37 Problem 71.

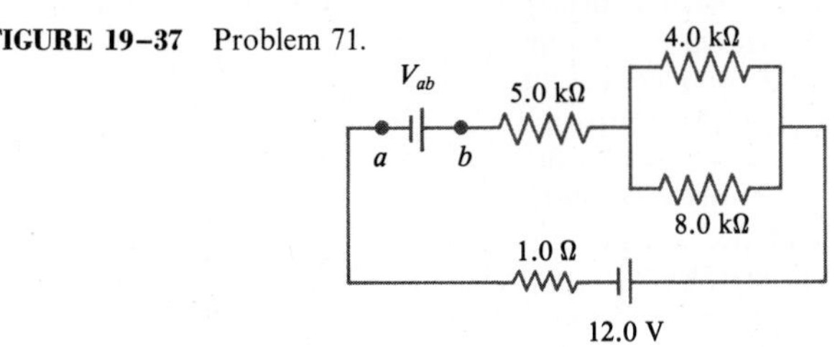

72. One type of *ohmmeter* consists of an ammeter connected to a series resistor and a battery, Fig. 19–38. The scale is different from ammeters and voltmeters. Zero resistance corresponds to full-scale deflection (because maximum current flows from the battery), whereas infinite resistance corresponds to no deflection (no current flow). Suppose a 9.0-V battery is used, and the galvanometer has internal resistance of 30 Ω and deflects full scale for a current of 30 μA. What values of shunt resistance, R_{sh}, and series resistance, R_{ser}, are needed to make an ohmmeter that registers a midscale deflection (that is, half of maximum) for a resistance of 30 kΩ? [Note: an additional series resistor (variable) is also needed so the meter can be zeroed. The zero should be checked frequently by touching the leads together, since the battery voltage can vary. Because of battery voltage variation, such ohmmeters are not precision instruments, but they are useful to obtain approximate values.]

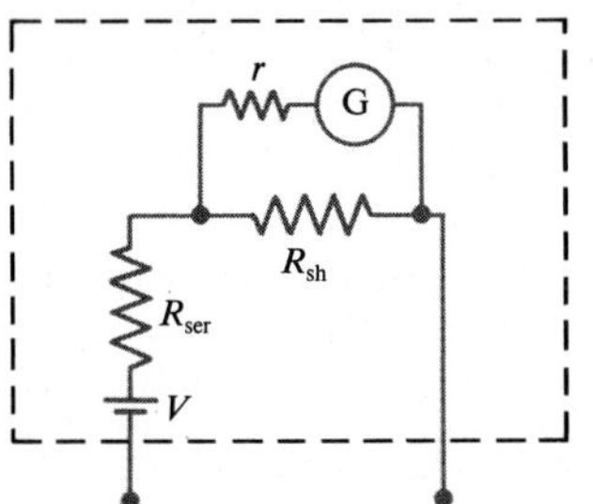

FIGURE 19–38 Ohmmeter, Problem 72.

CHAPTER 20

Magnetism

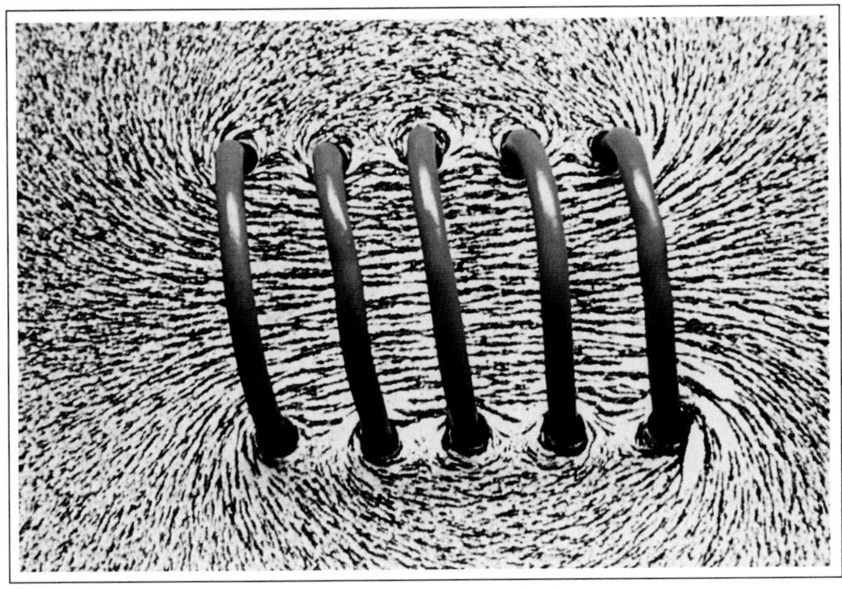

Magnets produce magnet fields, but so do electric currents. An electric current flowing in this coil of wire produces a magnetic field which causes the tiny pieces of iron (iron "filings") to align in the field. Fig. 20–12 shows iron filings in the magnetic field produced by a magnet.

Today it is clear that magnetism and electricity are closely related. This relationship was not discovered, however, until the nineteenth century. The history of magnetism begins much earlier with the ancient civilizations in Asia Minor. It was in a region of Asia Minor known as Magnesia that rocks were found that would attract each other. These rocks were called "magnets" after their place of discovery.

20–1 • Magnets and Magnetic Fields

A magnet will attract paper clips, nails, and other objects made of iron. Any magnet, whether it is in the shape of a bar or a horseshoe, has two ends or faces, called poles, which is where the magnetic effect is strongest. If a magnet is suspended from a fine thread, it is found that one pole of the magnet will always point toward the north. It is not known for sure when this fact was discovered, but it is known that the Chinese were making use of it as an aid to navigation by the eleventh century and perhaps earlier. This is, of course, the principle of a compass. A compass needle is simply a magnet that is supported at its center of gravity so it can rotate freely. That pole of a freely suspended magnet which points toward the north is called the **north pole** of the magnet. The other pole points toward the south and is called the **south pole**.

Poles of a magnet

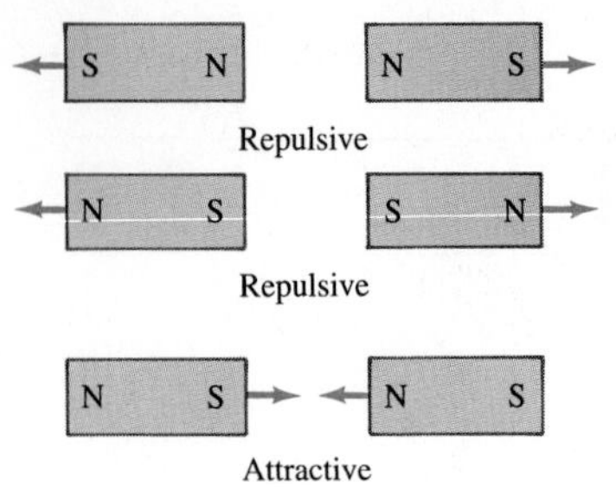

FIGURE 20–1
Like poles of a magnet repel; unlike poles attract.

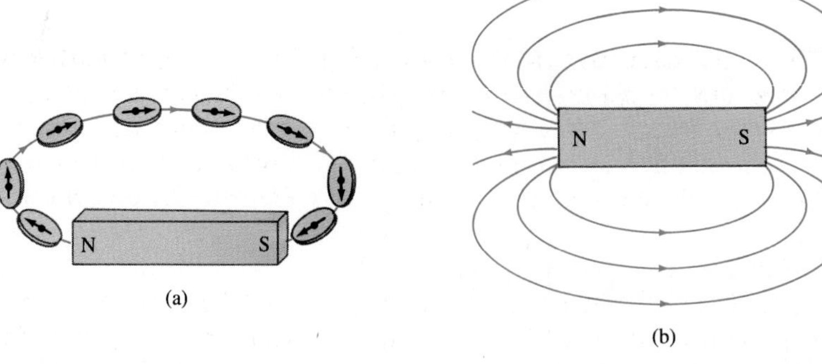

FIGURE 20–2 (a) Plotting a magnetic field line of a bar magnet. (b) Magnetic field lines outside of a bar magnet.

It is a familiar fact that when two magnets are brought near one another, each exerts a force on the other. The force can be either attractive or repulsive and can be felt even when the magnets don't touch. If the north pole of one magnet is brought near the north pole of a second magnet, the force is repulsive. Similarly, if two south poles are brought close, the force is repulsive. But when a north pole is brought near a south pole, the force is attractive, Fig. 20–1. This is reminiscent of the force between electric charges; like poles repel and unlike poles attract. But do not confuse magnetic poles with electric charge. They are not the same thing.

Ferromagnets

Only iron and a few other materials such as cobalt, nickel, and gadolinium show strong magnetic effects. They are said to be **ferromagnetic** (from the Latin word *ferrum* for iron). All other materials show some slight magnetic effect, but it is extremely small and can be detected only with delicate instruments. (We will look in more detail at ferromagnetism in Sections 20–3 and 20–14.)

We found it useful to speak of an electric field surrounding an electric charge. In the same way, we can imagine a **magnetic field** surrounding a magnet. The force one magnet exerts on another can then be described as the interaction between one magnet and the magnetic field of the other. Just as we drew electric field lines, we can also draw **magnetic field lines**. They can be drawn, as for electric field lines, so that (1) the direction of the magnetic field is tangent to a line at any point, and (2) the number of lines per unit area is proportional to the magnitude of the magnetic field.

Magnetic field lines

The *direction* of the magnetic field at a given point can be defined as the direction that the north pole of a compass needle would point when placed at that point. Figure 20–2a shows how one magnetic field line around a bar magnet is found using compass needles. The magnetic field determined in this way for the field outside a bar magnet is shown in Fig. 20–2b. Notice that because of our definition, the lines always point from the north toward the south pole of a magnet (the north pole of a magnetic compass needle is attracted to the south pole of another magnet).

We can define the **magnetic field** at any point as a vector, represented by the symbol **B**, whose direction is as defined above. The *magnitude* of **B** can be defined in terms of the torque exerted on a compass needle when it makes a certain angle with the magnetic field, as in Fig. 20–3. That is, the greater the torque, the greater the magnetic field strength. We can use this (rather rough) definition for now, but a more precise definition will be given in Section 20–5. The terms "magnetic flux density" and "magnetic induction" are often used for **B** rather than our simple term "magnetic field."

FIGURE 20–3
Forces on a compass needle that produce a torque to orient it parallel to the magnetic field lines. The torque will be zero when the needle is parallel to the magnetic field line at that point. (Only the attractive forces are shown; try drawing in the repulsive forces and show that they produce a similar torque.)

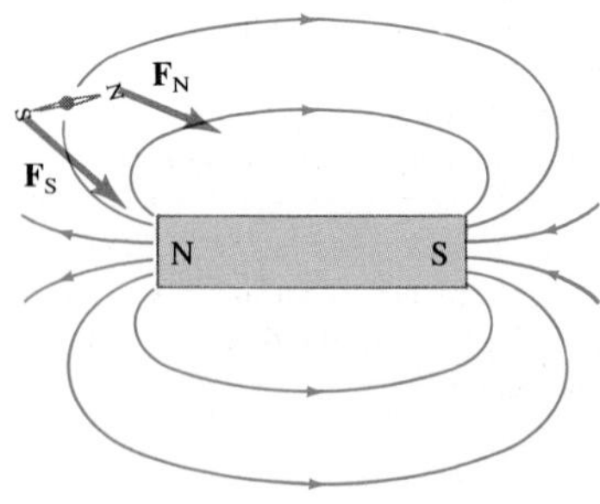

The earth's magnetic field is shown in Fig. 20–4. Since the north pole of a compass needle points north, the magnetic pole which is in the geographic north is magnetically a south pole (remember that the north pole of one magnet is attracted to the south pole of a second). And the earth's magnetic pole near the geographic south pole is magnetically a north pole. The earth's magnetic poles do not, however, coincide with the geographic poles (which are on the earth's axis of rotation). The magnetic south pole, for example, is in northern Canada, about 1500 km from the geographic north pole. This must be taken into account when using a compass (Fig. 20–5). The angular difference between magnetic north and true (geographical) north is called the **magnetic declination**. In the United States it varies from 0° to about 25°, depending on location. Notice in Fig. 20–4 that the earth's magnetic field is not tangent to the earth's surface at all points. The angle **B** makes with the horizontal at any point is referred to as the **angle of dip**.

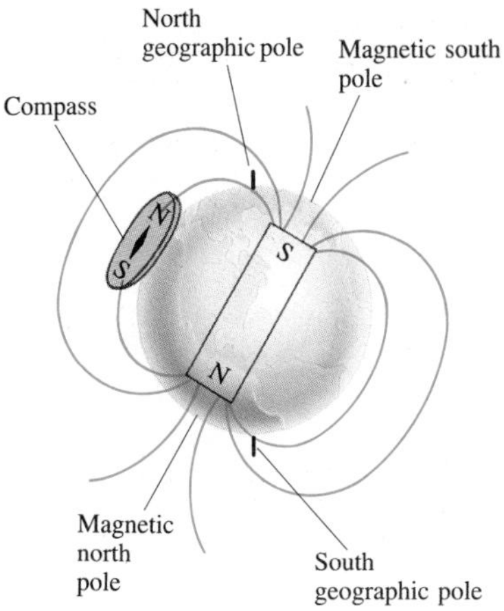

North geographic pole
Magnetic south pole
Compass
S
N
Magnetic north pole
South geographic pole

FIGURE 20–4 The earth acts like a huge magnet with its magnetic south pole near the geographic north pole.

20–2 • Electric Currents Produce Magnetism

FIGURE 20–5 Finding your route using map and compass.

During the eighteenth century, many natural philosophers sought to find a connection between electricity and magnetism. The first to uncover a significant connection was Hans Christian Oersted (1777–1851) in 1820. Oersted had believed for a long time in the unity of nature. Philosophically, he felt there ought to be a connection between magnetism and electricity. However, a stationary electric charge and a magnet had been shown not to have any influence on each other. But Oersted found that when a compass needle is placed near an electric wire, the needle deflects as soon as the wire is connected to a battery and a current flows.

As we have seen, a compass needle can be deflected by a magnetic field. What Oersted found was that **an electric current produces a magnetic field**. He had found a connection between electricity and magnetism.

A compass needle placed near a straight section of current-carrying wire aligns itself so it is tangent to a circle drawn around the wire, Fig. 20–6. Thus, the magnetic field lines produced by a current in a straight wire are in the form of circles with the wire at their center, Fig. 20–7a. The direction

FIGURE 20–6
Deflection of a compass needle near a current-carrying wire, showing the presence and direction of the magnetic field.

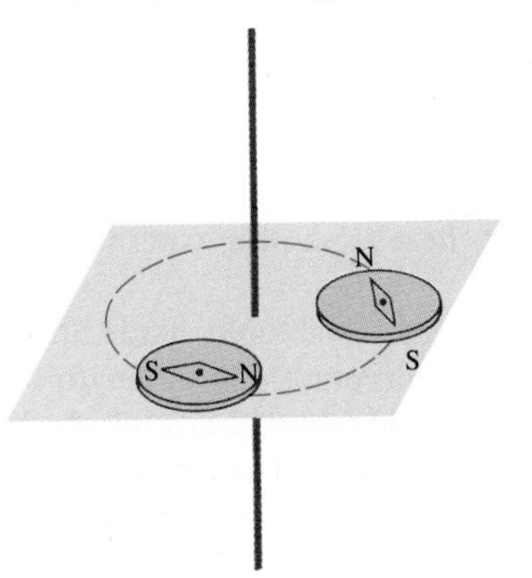

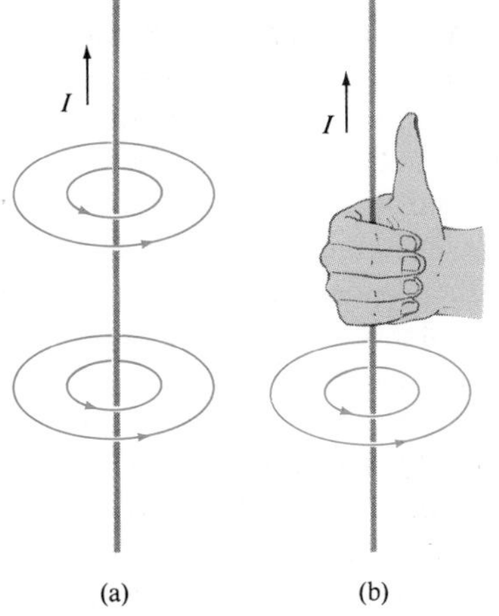

FIGURE 20–7 (a) Magnetic field lines around an electric current in a straight wire. (b) Right-hand-rule for remembering the direction of the magnetic field: when the thumb points in the direction of the conventional current, the fingers wrapped around the wire point in the direction of the magnetic field.

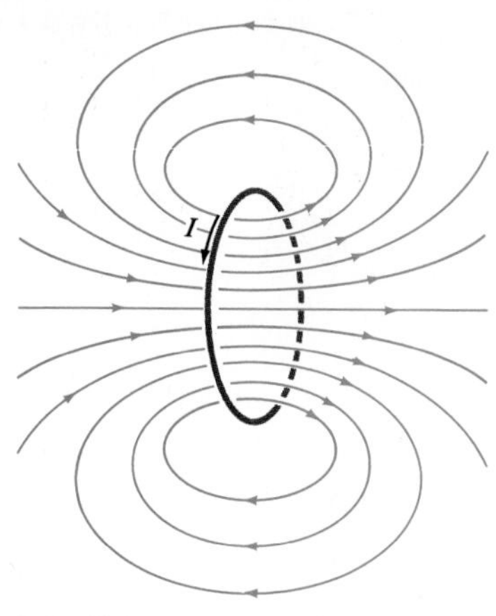

FIGURE 20–8
Magnetic field due to a circular loop of wire.

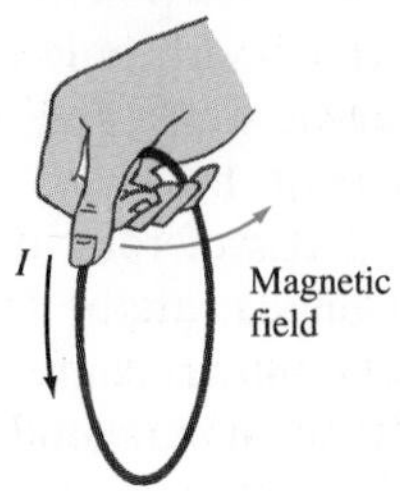

FIGURE 20–9 Right-hand rule for determining the direction of the magnetic field relative to the current.

of these lines is indicated by the north pole of the compass in Fig. 20–6. There is a simple way to remember the direction of the magnetic field lines in this case. It is called a **right-hand rule**: you grasp the wire with your right hand so that your thumb points in the direction of the conventional (positive) current; then your fingers will encircle the wire in the direction of the magnetic field, Fig. 20–7b. The magnetic field lines due to a circular loop of current-carrying wire can be determined in a similar way using a compass. The result is shown in Fig. 20–8. Again the right-hand rule can be used, as shown in Fig. 20–9.

Right-hand rule

20–3 • Ferromagnetism; Domains

We saw in Section 20–1 that iron (and a few other materials) can be made into strong magnets. These materials are said to be **ferromagnetic**. We now look more deeply into the sources of ferromagnetism.

A bar magnet, with its two opposite poles at either end, resembles an electric dipole (equal-magnitude positive and negative charges separated by a distance). Indeed, a bar magnet is sometimes referred to as a "magnetic dipole." There are opposite "poles" separated by a distance. And the magnetic field lines of a bar magnet form a pattern much like that for the electric field of an electric dipole: compare Fig. 16–23a with Fig. 20–2b. One important difference, however, is that a positive or negative electric charge can easily be isolated. But the isolation of a single magnetic pole has proved much more difficult. If a bar magnet is cut in half, you do not obtain isolated north and south poles. Instead, two new magnets are produced, Fig. 20–10. If the cutting operation is repeated, more magnets are produced, each with a north and a south pole. Physicists have tried various ways to isolate a single magnetic pole, and this is an active research field today since certain theories suggest they ought to exist. But so far there is no firm experimental evidence for their existence.

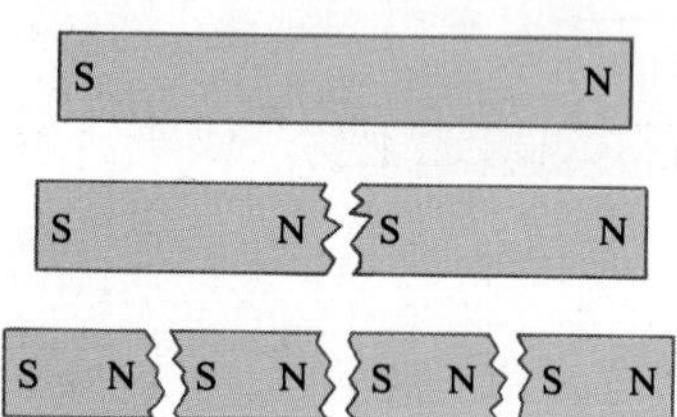

FIGURE 20–10
If you break a magnet in half, you do not obtain isolated north and south poles; instead, two new magnets are produced, each with a north and a south pole.

Magnetic poles not found singly

Microscopic examination reveals that a magnet is actually made up of tiny regions known as **domains**, which are at most about 1 mm in length or width. Each domain behaves like a tiny magnet with a north and a south pole. In an unmagnetized piece of iron, these domains are arranged randomly,

Domains in iron

as shown in Fig. 20–11a. The magnetic effects of the domains cancel each other out, so this piece of iron is not a magnet. In a magnet, the domains are preferentially aligned in one direction as shown in Fig. 20–11b (downward in this case). A magnet can be made from an unmagnetized piece of iron by placing it in a strong magnetic field. (You can make a needle magnetic, for example, by stroking it with one pole of a strong magnet.) Careful observations show in this case that the magnetization of domains may actually rotate slightly so as to be more nearly parallel to the external field. Or, more commonly, the borders of domains move so that those domains whose magnetic orientation is parallel to the external field grow in size at the expense of other domains. This can be seen by comparing Figs. 20–11a and b. This explains how a magnet can pick up unmagnetized pieces of iron like paper clips or bobby pins. The magnet's field causes a slight alignment of the domains in the unmagnetized object so that the object becomes a temporary magnet with its north pole facing the south pole of the permanent magnet, and vice versa; thus, attraction results. In the same way, elongated iron filings will arrange themselves in a magnetic field just as a compass needle does, and will reveal the shape of the magnetic field, Fig. 20–12.

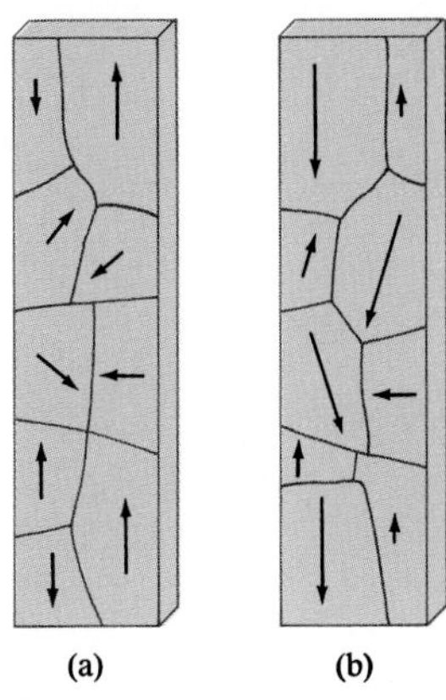

FIGURE 20–11 (a) Unmagnetized piece of iron is made up of domains that are randomly arranged. (b) In a magnet, the domains are preferentially aligned in one direction. (The tip of each arrow represents the north pole of the domain.)

FIGURE 20–12 Iron filings line up along magnetic field lines.

An iron magnet can remain magnetized for a long time, and thus it is referred to as a "permanent magnet." However, if you drop a magnet on the floor or strike it with a hammer, you may jar the domains into randomness. The magnet can thus lose some or all of its magnetism. Heating a magnet too can cause a loss of magnetism, for raising the temperature increases the random thermal motion of the atoms which tends to randomize the domains. Above a certain temperature known as the **Curie temperature** (1043 K for iron), a magnet cannot be made at all.†

There is a striking similarity between the fields produced by a bar magnet and by a loop of electric current (compare Fig. 20–2b with Fig. 20–8). This suggests that the magnetic field produced by a current may have something to do with ferromagnetism, an idea proposed by Ampère in the nineteenth century. According to modern atomic theory, the atoms that make up any material can be roughly visualized as containing electrons that orbit around a central nucleus. Since the electrons are charged, they constitute an electric current and therefore produce a magnetic field. But if there is no external field, the electron orbits in different atoms are arranged randomly, so the magnetic effects due to the many orbits in all the atoms in a material cancel out. However, electrons produce an additional magnetic field, almost as if they and their electric charge were spinning about their own axes. It is the magnetic field due to electron spin‡ that is believed to produce ferromagnetism. In most materials, the magnetic fields due to electron spin cancel out. But in iron and other ferromagnetic materials, a complicated cooperative mechanism operates. The result is that the electrons contributing to the ferro-

† Iron, nickel, cobalt, gadolinium, and certain alloys are ferromagnetic at room temperature; several other elements and alloys have low Curie temperature and thus are ferromagnetic only at low temperatures.

‡ The name "spin" comes from the early suggestion that the additional magnetic field arises from the electron "spinning" on its axis (as well as "orbiting" the nucleus) and this additional motion of the charge was supposed to produce the extra field. However this view of a spinning electron is completely discredited today. See Chapter 28.

magnetism in a domain "spin" in the same direction. Thus the tiny magnetic fields due to each of the electrons add up to give the magnetic field of a domain. And when the domains are aligned, as we have seen, a strong magnet results.

It is believed possible today that *all* magnetic fields are caused by electric currents. This would explain why it has proved difficult to find a single magnetic pole. There is no way to divide up a current and obtain a single magnetic pole. Of course if an isolated pole is found, we will have to alter the idea that all magnetic fields are produced by currents.

The lack of single magnetic poles means that magnetic field lines form closed loops, unlike electric field lines, which begin on + charges and end on − charges.

20–4 • Electromagnets and Solenoids

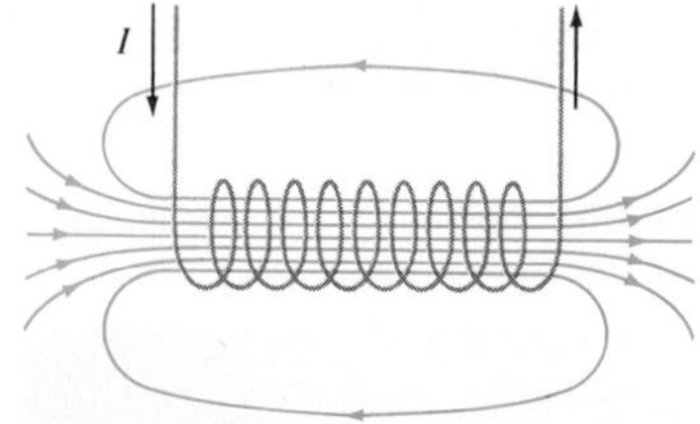

FIGURE 20–13
Magnetic field of a solenoid. The north pole of this solenoid, thought of as a magnet, is on the right, and the south pole is on the left.

A long coil of wire consisting of many loops of wire is called a **solenoid**. The magnetic field within a solenoid can be fairly large since it will be the sum of the fields due to the current in each loop (see Fig. 20–13). The solenoid acts like a magnet; one end can be considered the north pole and the other the south pole, depending on the direction of the current in the loops (use the right-hand rule). Since the magnetic field lines leave the north pole of a magnet, the north pole of the solenoid in Fig. 20–13 is on the right.

If a piece of iron is placed inside a solenoid, the magnetic field is increased greatly because the domains of the iron are aligned by the magnetic field produced by the current. The resulting magnetic field is the sum of that due to the current and that due to the iron, and can be hundreds or thousands of times that due to the current alone (see Section 20–14). This arrangement is called an **electromagnet**. The iron used in electromagnets acquires and loses its magnetism quite readily when the current is turned on or off, and so is referred to as "soft iron." (It is "soft" only in a magnetic sense.) Iron that holds its magnetism even when there is no externally applied field is called "hard iron." Hard iron is used in permanent magnets. Soft iron is usually used in electromagnets so that the field can be turned on and off readily. Whether iron is hard or soft depends on heat treatment and other factors.

FIGURE 20–14
Solenoid used as a doorbell.

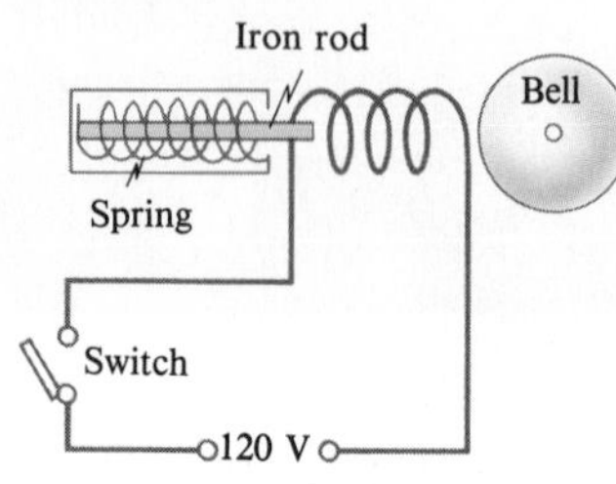

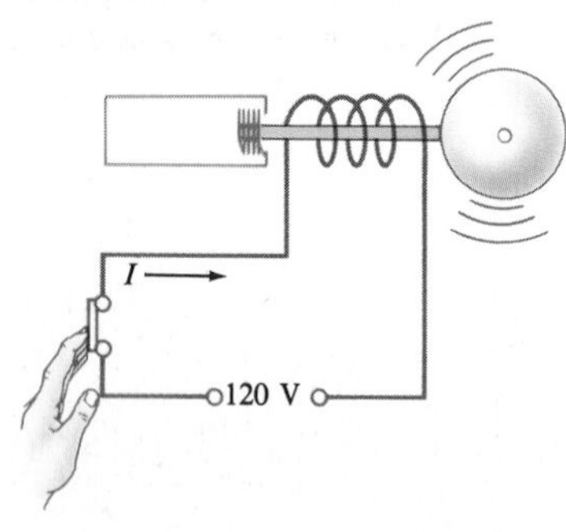

Electromagnets find use in many practical applications, from use in motors and generators to producing large magnetic fields for research. Because the current flows continuously, a great deal of waste heat (I^2R power) is often produced. Cooling coils, which are tubes carrying water, must be used to absorb the heat in bigger installations. For some applications, superconducting magnets are coming into use. The current-carrying wires are made of superconducting material (Section 18–5) kept below the transition temperature. No electric power is needed to maintain large current, which means large savings of electricity. Of course, energy is needed to keep the superconducting coils at the necessary low temperature.

Another useful device consists of a solenoid into which a rod of iron is partially inserted. This combination is also referred to as a **solenoid**. One simple use is as a doorbell (Fig. 20–14). When the circuit is closed by pushing the button, the coil effectively becomes a magnet and exerts a force on the iron rod. The rod is pulled into the coil and strikes the bell. A larger solenoid is used in the starters of cars; when you engage the starter, you are closing a circuit that not only turns the starter motor, but activates a solenoid that first moves the starter into contact with the engine. Solenoids are used as switches in many other devices, such as tape recorders. They have the advantage of moving mechanical parts quickly and accurately.

20–5 • Force on an Electric Current in a Magnetic Field; Definition of B

In the Section 20–2, we saw that an electric current exerts a force on a magnet, such as a compass needle. By Newton's third law, we might expect the reverse to be true as well: we should expect that *a magnet exerts a force on a current-carrying wire.* This is indeed the case and this effect was also discovered by Oersted.

Magnet exerts a force on an electric current

Let us look at the force exerted on a wire in detail. Suppose a straight wire is placed between the pole pieces of a magnet as shown in Fig. 20–15. When a current flows in the wire, a force is exerted on the wire. But this force is *not* toward one or the other poles of the magnet. Instead, the force is directed *at right angles to the magnetic field direction.* If the current is reversed in direction, the force is in the opposite direction. It is found that the direction of the force is always perpendicular to the direction of the current and also perpendicular to the direction of the magnetic field, **B**. This description does not completely describe the direction, however: the force could be either up or down in Fig. 20–15b and still be perpendicular to both the current and to **B**. Experimentally, the direction of the force is given by another right-hand rule, as illustrated in Fig. 20–15c. First you orient your right hand so that the outstretched fingers point in the direction of the (conventional) current; from this position, when you bend your fingers, they should then point in the direction of the magnetic field lines (which point from the N toward the S pole outside a magnet); you may have to rotate your hand and arm about the wrist until they do point along **B** when bent, remembering that straightened fingers must point along the direction of the current. When your hand is oriented in this way, then the extended thumb points in the direction of the force on the wire.

Right-hand rule for force on current due to **B**

This describes the direction of the force. What about its magnitude? It is found experimentally that the magnitude of the force is directly proportional to the current I in the wire, to the length l of wire in the magnetic field (assumed uniform), and to the magnetic field B. The force also depends on the angle θ between the wire and the magnetic field. When the wire is

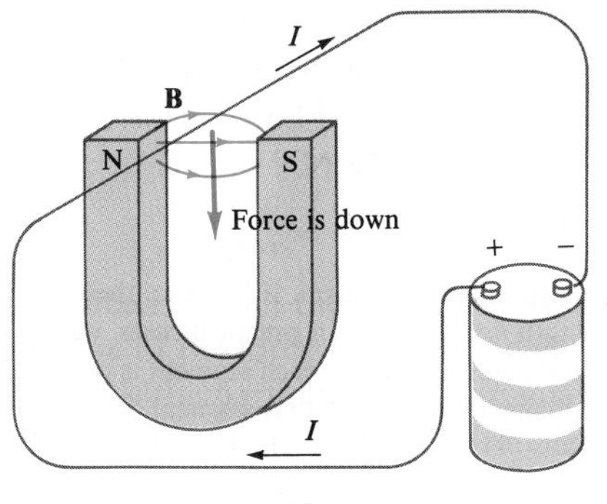

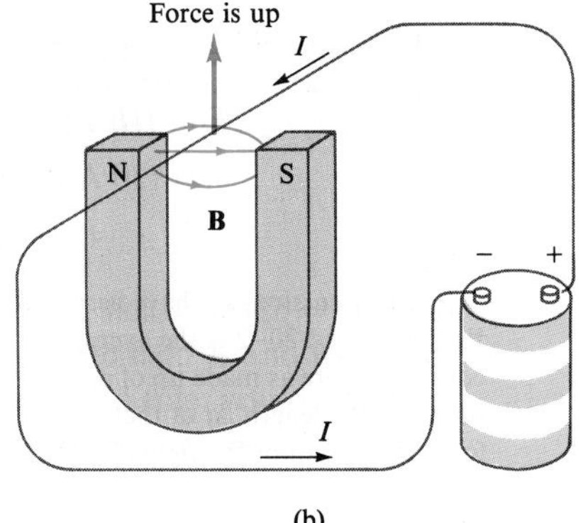

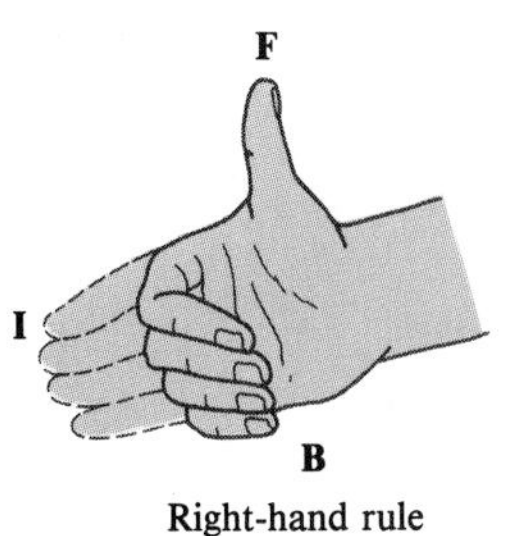

FIGURE 20–15 Force on a current-carrying wire placed in a magnetic field **B**.

vessel and to a magnetic field. The ions in the blood move vertically in the drawing as a result of the electric field. The magnetic field then exerts a sideways force on the ions that causes the fluid to flow along the blood vessel. Do positive and negative ions feel a force in the same direction?

20–8 • Applications: Galvanometers, Motors, Loudspeakers

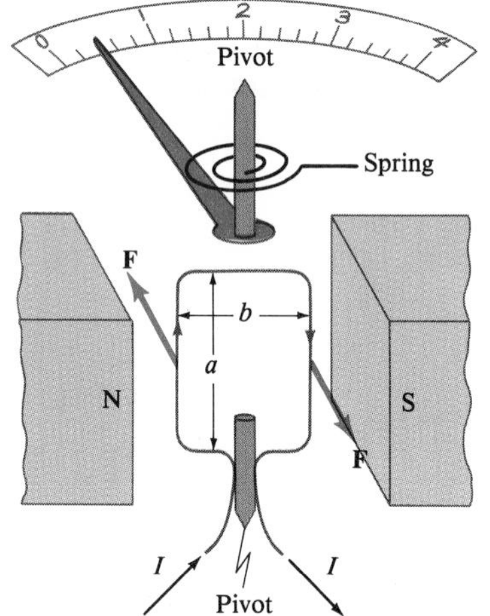

FIGURE 20–22 Galvanometer.

A number of important practical devices make use of the force that exists between a current and a magnetic field. In most of these devices, the current flows in a coil of wire.

As our first example, let us examine a **galvanometer**, the basic component of many meters, including ammeters, voltmeters, and ohmmeters (Chapter 19). As shown in Fig. 20–22, a galvanometer consists of a loop or coil of wire suspended in the magnetic field of a permanent magnet. When current flows through the loop, which we assume is rectangular, the magnetic field exerts a force on the vertical sections of wire as shown. Notice that, by the right-hand rule (Fig. 20–15c), the force on the upward current on the left is inward, whereas that on the descending current on the right is outward. These forces give rise to a net torque that tends to rotate the coil about its vertical axis. The greater the current I, the more the coil and its attached pointer will turn against the resistance of the small spring.

Let us calculate the magnitude of this torque. From Eq. 20–2, the force $F = IaB$, where a is the length of the vertical arm of the coil. The lever arm for each force is $b/2$, where b is the width of the coil and the "axis" is at the midpoint. The total torque is the sum of the torques due to each of the forces, so

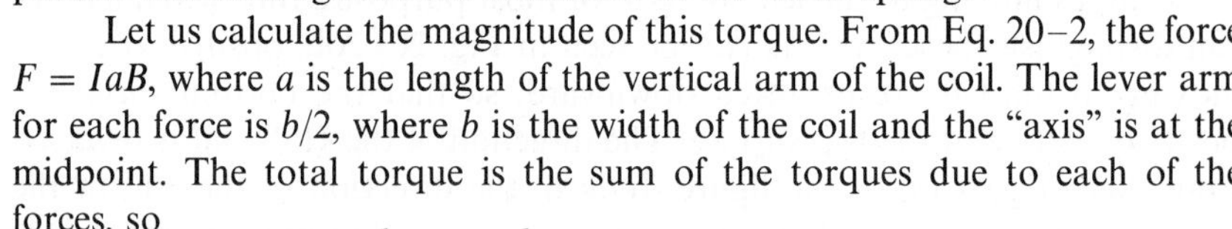

$$\tau = IaB\frac{b}{2} + IaB\frac{b}{2} = IabB = IAB,$$

where $A = ab$ is the area of the coil. If the coil consists of N loops of wire, the current is then NI, so the torque becomes

Torque on galvanometer coil

$$\tau = NIAB. \qquad (20\text{–}5)$$

This formula, derived here for a rectangular coil, is valid for any shape of flat coil. The quantity NIA is referred to as the **magnetic moment** of the coil.

This torque is opposed by the spring. By Hooke's law, the torque τ_s exerted by the spring is proportional to the angle ϕ through which it is turned. That is,

$$\tau_s = k\phi,$$

where k is the stiffness constant of the spring. Thus, the coil and the attached pointer will rotate only to the point where the spring torque balances that due to the magnetic field. From Eq. 20–5, we then have $k\phi = NIAB$, or

$$\phi = \frac{NIAB}{k}. \qquad (20\text{–}6)$$

Thus the angular deflection of the pointer, ϕ, is directly proportional to the current I flowing in the coil. This is what we want for a galvanometer (see Section 19–10).

However, there is a problem. When the loop of Fig. 20–22 is not parallel to the magnetic field, the torque will be less than that given by Eq. 20–5 since the lever arm is less. Indeed, when the loop is rotated 90° from the position shown in Fig. 20–22, the two forces act along the same line and the torque is zero. Thus the torque would depend on the angle, and ϕ would not be proportional to I as in Eq. 20–6. To solve this problem, curved pole pieces are used and the galvanometer coil is wrapped around a cylindrical iron core as shown in Fig. 20–23. The iron tends to concentrate the magnetic field lines so that B always points parallel to the face of the coil at the wire. The force is then always perpendicular to the face of the coil and the torque will not vary with angle. Thus ϕ will be proportional to I, as required.

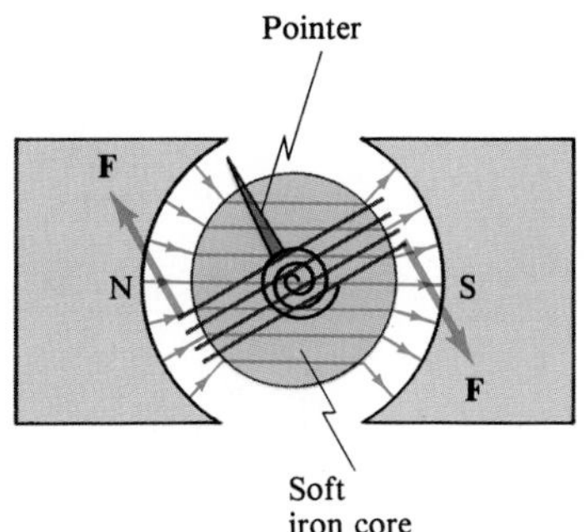

FIGURE 20–23 Galvanometer coil wrapped on an iron core.

A **chart recorder**, in which a pen graphs a signal such as an ECG on a moving roll of paper, is basically a galvanometer. The pen is attached to an arm, which is connected to the galvanometer coil. The instrument could record either voltage or current, just as any galvanometer can be connected as a voltmeter or ammeter.

Electric motor

An **electric motor** changes electric energy into (rotational) mechanical energy. A motor works on the same principle as a galvanometer, except that the coil is larger and is mounted on a large cylinder called the **rotor** or **armature**, Fig. 20–24. Actually, there are several coils, although only one is indicated in the figure. The armature is mounted on a shaft or axle. The permanent magnet is replaced in some motors by an electromagnet. Unlike a galvanometer, a motor must turn continuously in one direction. This presents a problem, for when the coil, which is rotating clockwise in Fig. 20–24, passes beyond the vertical position the forces would then act to return the coil back to vertical. Thus alternation of the current is necessary if a motor is to turn continuously in one direction. This can be achieved in a *dc motor* with the use of **commutators** and **brushes**: as shown in Fig. 20–25, the brushes are stationary contacts that rub against the conducting commutators mounted on the motor shaft. At every half revolution, each commutator changes its connection to the other brush. Thus the current in the coil re-

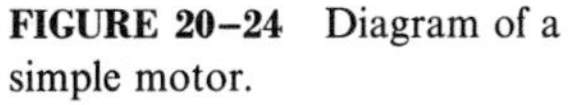

FIGURE 20–24 Diagram of a simple motor.

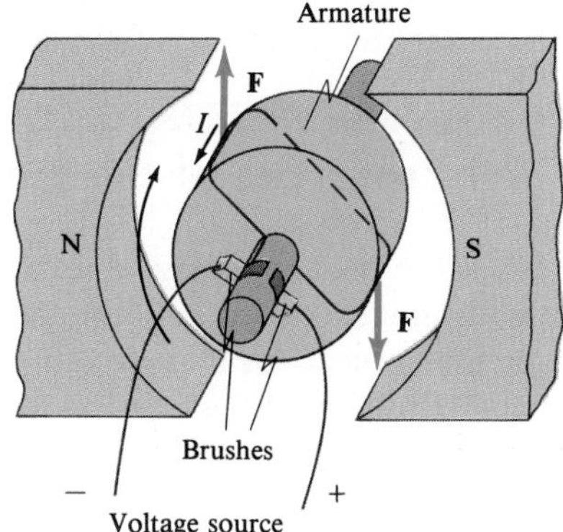

FIGURE 20–25 The commutator-brush arrangement in a dc motor assures alternation of the current in the armature to keep rotation continuous. The commutators are attached to the motor shaft and turn with it, whereas the brushes remain stationary.

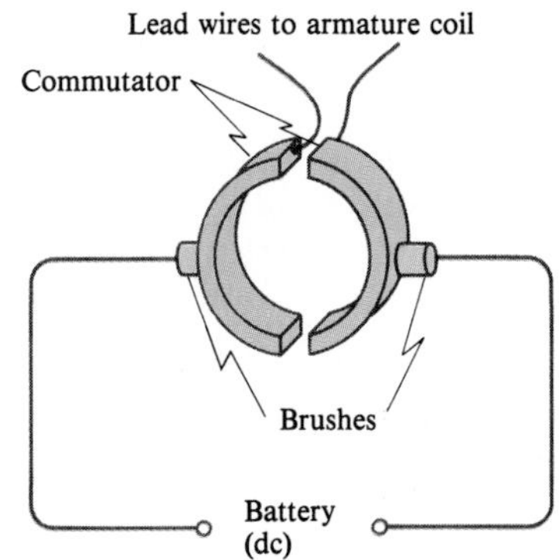

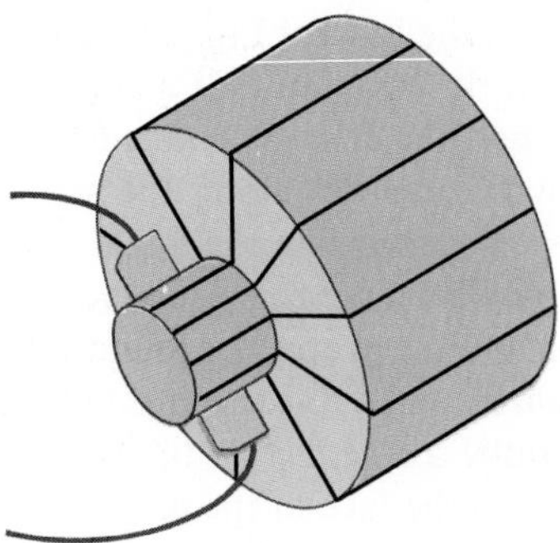

FIGURE 20–26 Motor with many windings.

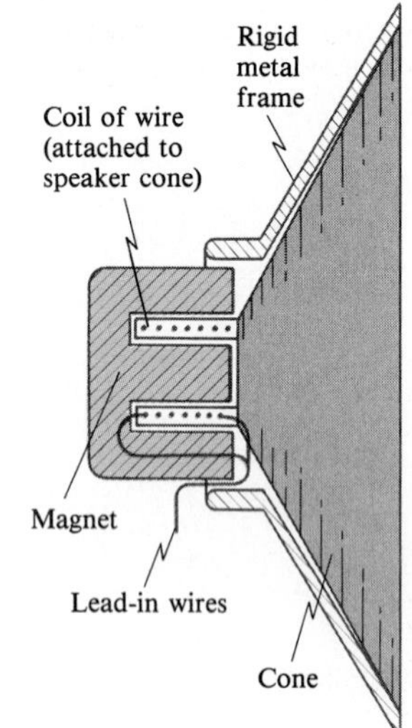

FIGURE 20–27 Loudspeaker.

verses every half revolution as required for continuous rotation. Most motors contain several coils, called "windings," each located in a different place on the armature, Fig. 20–26. Current flows through each coil only during a small part of a revolution, at the time when its orientation results in the maximum torque. In this way, a motor produces a much steadier torque than can be obtained from a single coil. The design of most practicial motors is more complex than described here, but the general principles remain the same.

A **loudspeaker** also works on the principle that a magnet exerts a force on a current-carrying wire. The electrical output of a radio or TV set is connected to the wire leads of the speaker. The speaker leads are connected internally to a coil of wire, which is itself attached to the speaker cone, Fig. 20–27. The speaker cone is usually made of stiffened cardboard and is mounted so that it can move back and forth freely. A permanent magnet is mounted directly in line with the coil of wire. When the alternating current of an audio signal flows through the wire coil, the coil and the attached speaker cone feel a force due to the magnetic field of the magnet. As the current alternates at the frequency of the audio signal, the speaker cone moves back and forth at the same frequency, causing alternate compressions and rarefactions of the adjacent air, and sound waves are produced. A speaker thus changes electrical energy into sound energy, and the frequencies of the emitted sound waves are an accurate reproduction of the electrical input.

20–9 • Discovery and Properties of the Electron

The electron plays a basic role in our understanding of electricity and magnetism today. But its existence was not suggested until the 1890s. We discuss it here because magnetic fields were crucial for measuring its properties.

Toward the end of the nineteenth century, studies were being done on the discharge of electricity through rarefied gases. One apparatus, diagrammed in Fig. 20–28, was a glass tube fitted with electrodes and evacuated so only a small amount of gas remained inside. The negative electrode is called the *cathode*, and the positive one the *anode*.† When a very high voltage

FIGURE 20–28 Discharge tube. In some models, one of the screens is the anode (positive plate).

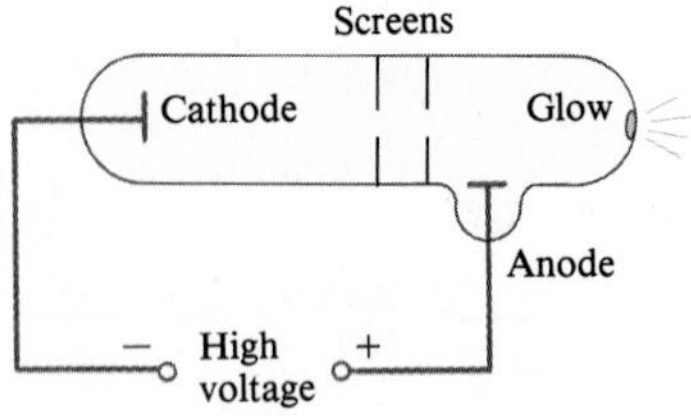

† These terms were coined by Michael Faraday (Fig. 16–17) and came from the Greek words meaning, respectively, "descent" and "a way up."

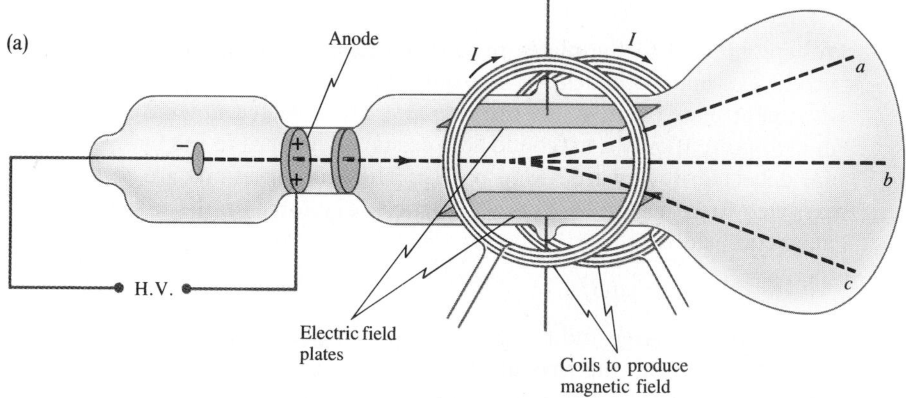

(b)

FIGURE 20–29 (a) Cathode rays deflected by electric and magnetic fields. (b) J. J. Thomson and his cathode ray tube.

was applied to the electrodes, a dark space seemed to extend outward from the cathode toward the opposite end of the tube; and that far end of the tube would glow. If one or more screens containing a small hole were inserted as shown, the glow was restricted to a tiny spot on the end of the tube. It seemed as though something being emitted by the cathode traveled to the opposite end of the tube. They were given the name **cathode rays**.

Observation of cathode rays

There was much discussion at the time about what these rays might be. Some scientists thought they might resemble light. But the observation that the bright spot at the end of the tube could be deflected to one side by an electric or magnetic field suggested that cathode rays could be charged particles; and the direction of the deflection was consistent with a negative charge. Furthermore, if the tube contained certain types of rarefied gas, the path of the cathode rays was made visible by a slight glow.

Estimates of the charge e of the (assumed) cathode-ray particles as well as of their charge-to-mass ratio, e/m, had been made by 1897. But in that year, J. J. Thomson (1856–1940) was able to measure e/m directly, using the apparatus shown in Fig. 20–29. Cathode rays are accelerated by a high voltage and then pass between a pair of parallel plates built into the tube. The

e/m measured

voltage applied to the plates produces an electric field, and a pair of coils produces a magnetic field. When only the electric field is present, say with the upper plate positive, the cathode rays are deflected upward as in path *a* in the figure. If only a magnetic field exists, say inward in the figure, the rays are deflected downward along path *c*. These observations are just what is expected for a negatively charged particle. The force on the rays due to the magnetic field is

$$F = evB,$$

where e is the charge and v is the velocity of the cathode rays. In the absence of an electric field, the rays are bent into a curved path, so we have, from $F = ma$,

$$evB = \frac{mv^2}{r},$$

and thus

$$\frac{e}{m} = \frac{v}{Br}.$$

The radius of curvature r can be measured and so can B. The velocity v is found by applying an electric field in addition to the magnetic field. The electric field E is adjusted so that the cathode rays are undeflected and follow path b in Fig. 20–29. In this situation, the force due to the electric field, $F = eE$, is just balanced by the force due to the magnetic field, $F = evB$. Thus we have $eE = evB$ and

$$v = \frac{E}{B}.$$

Combining this with the above equation, we have

$$\frac{e}{m} = \frac{E}{B^2r}. \qquad (20\text{–}7)$$

The quantities on the right side can all be measured, so that although e and m could not be determined separately, the ratio e/m could be determined. The accepted value today is $e/m = 1.76 \times 10^{11}$ C/kg. Cathode rays soon came to be called **electrons.**

"Discovery" of the electron

It is worth noting that the "discovery" of the electron, like many others in science, is not quite so obvious as discovering gold or oil. Should the discovery of the electron be credited to the person who first saw a glow in the tube? Or to the person who first called them cathode rays? Perhaps neither one, for they had no conception of the electron as we know it today. In fact, the credit for the discovery is generally given to Thomson, but not because he was the first to see the glow in the tube. Rather it is because he believed not only that this phenomenon was due to tiny negatively charged particles and made careful measurements on them. But he also argued that these particles were constituents of atoms, and not ions or atoms themselves as many thought, and he developed an electron theory of matter. His view is close to what we accept today, and this is why Thomson is credited with the "discovery." Note, however, that neither he nor anyone else ever actually saw an electron itself. We discussed this briefly, for it illustrates that discovery in science is not always a clear-cut matter. In fact, some philosophers of science think the word "discovery" is not always appropriate, such as in this case.

Thomson believed that an electron was not an atom, but rather a constituent, or part, of an atom. Convincing evidence for this came soon with the determination of the charge and the mass of the cathode rays. Thomson's student, J. S. Townsend, made the first direct (but rough) measurements of e in 1897. But it was the more refined **oil-drop experiment** of Robert A. Millikan (1868–1953) that yielded a precise value for the charge on the electron and showed that charge comes in discrete amounts. In this experiment, tiny droplets of mineral oil carrying an electric charge were allowed to fall under gravity between two parallel plates, Fig. 20–30. The electric field E between the plates was adjusted until the oil drop was suspended in midair. The downward pull of gravity, mg, was then just balanced by the upward force due to the electric field. Thus, $qE = mg$, so the charge $q = mg/E$. The mass of the droplet was determined by measuring its terminal velocity in the absence of the electric field and using Stoke's equation. (See Section 10–12; the drop is too small to permit direct measurement of its radius.) Sometimes the drop was charged negatively and sometimes positively, suggesting that the drop had acquired or lost electrons (presumably through friction when ejected by the atomizer). Millikan's painstaking observations and analysis presented convincing evidence that any charge was an integral multiple of a smallest charge, e, that was ascribed to the electron, and that the value of e was 1.6×10^{-19} C. (Today's precise value of e, as mentioned in Chapter 16, is $e = 1.602 \times 10^{-19}$ C.) This result, combined with the measurement of e/m (see above), gives the mass of the electron to be $(1.6 \times 10^{-19}\ \text{C})/(1.76 \times 10^{11}\ \text{C/kg}) = 9.1 \times 10^{-31}$ kg. This mass is less than a thousandth the mass of the smallest atom, and thus confirmed the idea that the electron is only a part of an atom. The accepted value today for the mass of the electron is $m_e = 9.11 \times 10^{-31}$ kg.

Millikan oil-drop experiment to determine e

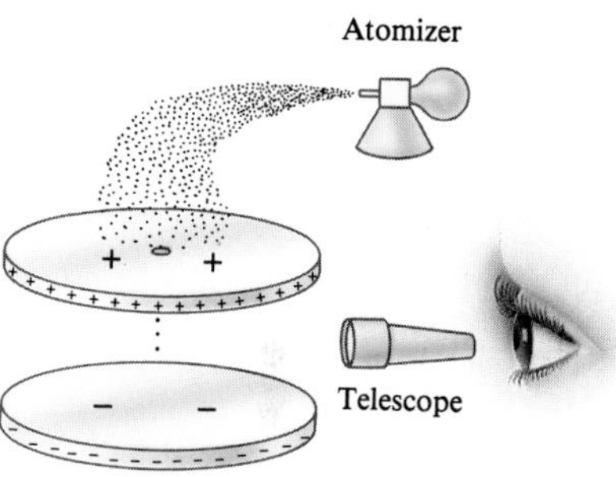

FIGURE 20–30
Millikan's oil-drop experiment.

20–10 • Thermionic Emission and the Cathode-Ray Tube

In the course of experiments on the electric light bulb, Thomas Edison (1847–1931) in 1883 made an interesting discovery. Into an evacuated glass bulb he inserted an electrode or plate in addition to the bulb filament. When a positive potential was applied by a battery to the plate, he found that a current would flow in the external circuit as long as the filament was hot and glowing (heated by a separate current passing through it). However, if a negative potential was applied to the plate, no current was observed to flow. When the filament was cold, no current flowed for either potential. The phenomenon of a current flowing from a heated filament is known as **thermionic emission**.

Although Edison was unaware of the significance of his discovery, it eventually led to the invention of the electronic vacuum tube. Indeed, an explanation for this effect had to await the discovery of the electron over a decade later.

J. J. Thomson, in 1899, measured the value of e/m for the rays emitted in thermionic emission. He obtained the same value as that for the rays in the discharge tube produced by high voltage (Fig. 20–29), and concluded

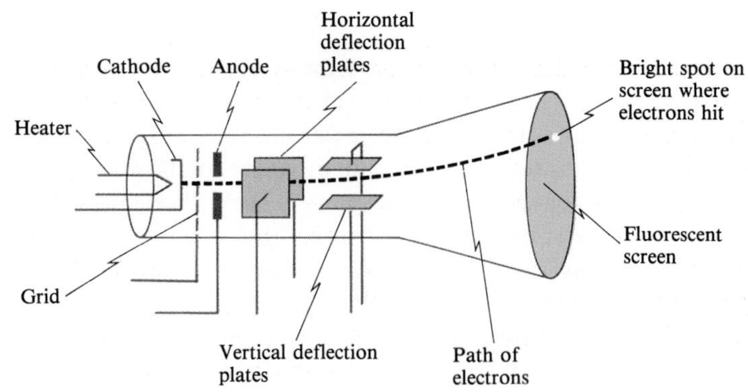

FIGURE 20–31 A cathode-ray tube. Magnetic deflection coils are often used in place of the electric deflection plates. The relative positions of the elements have been exaggerated for clarity.

that it is electrons that are emitted in thermionic emission. Apparently what happens in thermionic emission is that electrons are being "boiled off" the filament when it is hot. When the plate is positive, the electrons are attracted to it and so a current flows. If the plate is negative, the electrons are repelled and no current flows.

FIGURE 20–32 Electron beam sweeps across a television screen in a succession of horizontal lines.

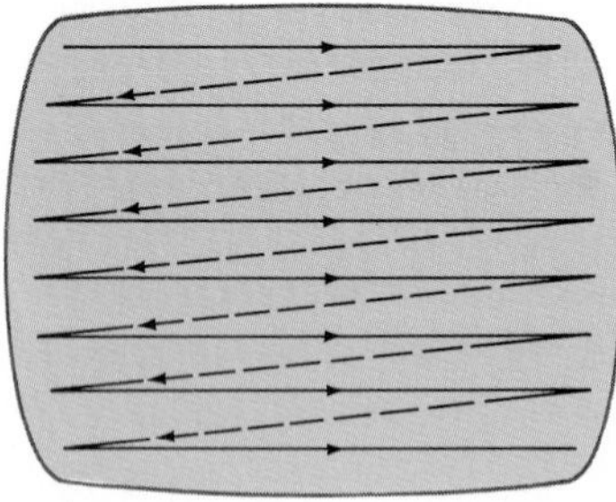

We can understand how electrons might be "boiled off" a hot metal filament if we treat the electrons like molecules in a gas. This makes sense if the electrons are relatively free to move about, which is consistent with the fact that metals are good conductors. However, electrons don't readily escape from the metal. There are forces that keep them in. For example, if an electron were to escape outside the metal surface, a net positive charge would remain behind, and this would attract the electron back. To escape, an electron would have to have a certain minimum kinetic energy, just as molecules in a liquid must have a minimum KE to "evaporate" into the gaseous state. We saw in Chapter 13 that the average kinetic energy ($\overline{\text{KE}}$) of molecules in a gas is proportional to the absolute temperature T. We can apply this idea, but only very roughly, to free electrons in a metal as if they made up an "electron gas." Of course, some electrons have more KE than average and others less. At room temperature, very few electrons would have sufficient energy to escape. At high temperatures, $\overline{\text{KE}}$ is larger and many escape; this is just like evaporation from liquids, which occurs more readily at high temperatures. Thus, significant thermionic emission occurs only at elevated temperatures. And Edison's results are readily understood.

CRT

An important vacuum tube, whose operation depends on thermionic emission, is the **cathode-ray tube** (CRT). It is the picture tube of television sets, oscilloscopes, and computer display terminals. It derives its name from the fact that inside an evacuated glass tube, a beam of cathode rays (electrons) is directed to various parts of a screen to produce a "picture." A simple CRT is diagrammed in Fig. 20–31. Electrons emitted by the heated cathode are accelerated by a high voltage applied to the anode (5000–50,000 V). The electrons pass out of this "electron gun" through a small hole in the anode. The inside of the tube face is coated with a fluorescent material that glows when struck by electrons. A tiny bright spot is thus visible where the electron beam strikes the screen. Two horizontal and two vertical plates deflect the beam of electrons when a voltage is applied to them. The electrons are deflected toward whichever plate is positive. By varying the voltage on the deflection plates, the bright spot can be placed at any point on the screen.

Television

In the picture tube of a television set, magnetic deflection coils are usually used instead of electric plates. The electron beam is made to sweep over the screen in the manner shown in Fig. 20–32. The beam is swept horizontally by the horizontal deflection plates or coils. When the horizontal deflecting field is maximum in one direction, the beam is at one edge of the screen. As the field decreases to zero, the beam moves to the center; and as the field increases to a maximum in the opposite direction, the beam approaches the opposite edge. When the beam reaches this edge, the voltage or current abruptly changes to return the beam to the opposite side of the screen. Simultaneously, the beam is deflected downward slightly by the vertical deflection plates or coils, and then another horizontal sweep is made. In the United States, 525 lines constitutes a complete sweep over the entire screen. (High-definition TV will provide more than double this number of lines, giving greater picture sharpness. Some European systems already provide significantly more lines than the present U.S. standard.) The complete picture of 525 lines is swept out in $\frac{1}{30}$ s. Actually, a single vertical sweep takes $\frac{1}{60}$ s and involves every other line. The lines in between are then swept out over the next $\frac{1}{60}$ s. We see a picture because the image is retained by the fluorescent screen and by our eyes for about $\frac{1}{20}$ s. The picture we see consists of the varied brightness of the spots on the screen. The brightness at any point is controlled by the grid (a "porous" electrode, such as a wire grid, that allows passage of electrons) which can limit the flow of electrons by means of the voltage applied to it: the more negative this voltage, the fewer electrons pass through. The voltage on the grid is determined by the video signal sent out by the station and received by the set. Accompanying this signal are signals that synchronize the grid voltage to the horizontal and vertical sweeps.

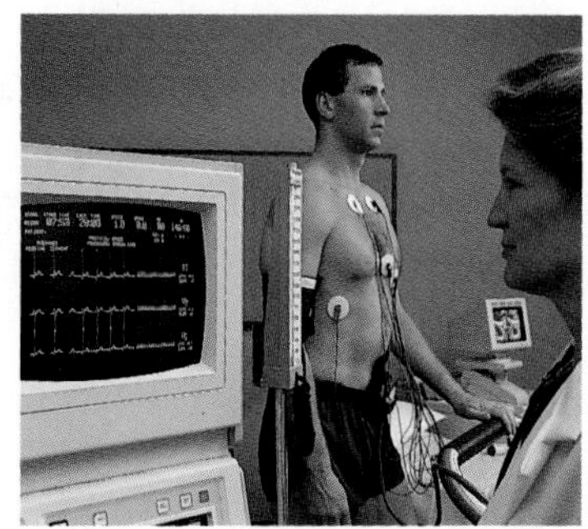

FIGURE 20–33
An electrocardiogram (ECG) trace (see Section 17–10) displayed on a CRT.

Oscilloscope

An **oscilloscope** is a device for amplifying, measuring, and visually observing an electrical signal. It provides a principle means for measuring rapidly changing signals. The signal is displayed on the screen of a CRT. In normal operation, the electron beam is swept horizontally at a uniform rate in time by the horizontal deflection plates. The signal to be displayed is applied, after amplification, to the vertical deflection plates. The visible "trace" on the screen (Fig. 20–33) is thus a plot of the signal voltage (vertically) versus time (horizontally).

*20–11 • Mass Spectrometer

Various methods were developed in the early part of this century to measure the masses of atoms. One of the most accurate was the **mass spectrometer**† of Fig. 20–34. Ions are produced by heating, or by an electric current, in the source S. Those that pass through slit S_1 enter a region where there are crossed electric and magnetic fields as in Thomson's device (Fig. 20–29). Only those ions whose speed is $v = E/B$ will pass through undeflected and emerge through slit S_2. (This arrangement is called a velocity selector; see equation and discussion just before Eq. 20–7.) In this second region, there is only a magnetic field B' and the ions follow a circular path. The radius of their path can be measured because the ions expose the photographic plate where they strike. Since $qvB' = mv^2/r$ and $v = E/B$, then we have

$$m = \frac{qB'r}{v} = \frac{qBB'r}{E}.$$

FIGURE 20–34 Bainbridge mass spectrometer. The magnetic fields B and B' point out of the paper (indicated by the dots).

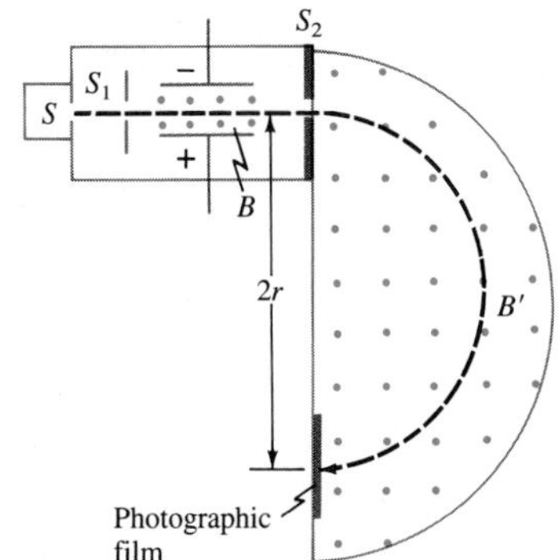

† The term *mass spectrograph* is also used.

All the quantities on the right can be measured, and thus m can be determined. Note that for ions of the same charge, the mass of each is proportional to the radius of its path.

The masses of many atoms were measured in this way. When a pure substance was used, it was sometimes found that two or more closely spaced marks would appear on the film. For example, neon produced two marks whose radii corresponded to atoms of mass 20 and 22 atomic mass units (u). Impurities were ruled out and it was concluded that there must be two types of neon with different mass. These different forms were called **isotopes**. It was soon found that most elements are mixtures of isotopes. We shall see in Chapter 30 that the difference in mass is due to different numbers of neutrons.

Mass spectrometers can be used to separate not only different elements and isotopes, but different molecules as well. They are used in physics and chemistry, and in biological and biomedical laboratories.

EXAMPLE 20–5 Carbon atoms of atomic mass 12.0 u are found to be mixed with another, unknown, element. In a mass spectrometer, the carbon traverses a path of radius 22.4 cm and the unknown's path has a 26.2 cm radius. What is the unknown element? Assume they have the same charge.

SOLUTION Since mass is proportional to the radius, we have

$$\frac{m_x}{m_C} = \frac{26.2\text{ cm}}{22.4\text{ cm}} = 1.17.$$

Thus $m_x = 1.17 \times 12.0\text{ u} = 14.0\text{ u}$. The other element is probably nitrogen (see the periodic table, inside the back cover). However, it could also be an isotope of carbon or oxygen. Further physical or chemical analysis would be needed.

*20–12 • Determination of Magnetic Field Strengths; Ampère's Law

The simplest magnetic field is one that is uniform—it doesn't change from one point to another. A perfectly uniform field over a large area is not easy to produce. But the field between two flat pole pieces of a magnet is nearly uniform if the area of the pole faces is large compared to their separation, Fig. 20–35. At the edges, the field "fringes" out somewhat and is not uniform. The parallel, evenly spaced field lines in the drawing indicate the field is uniform at points not too near the edge. Nonuniform magnetic fields are more common, and it is often important to know how a magnetic field varies from point to point.

FIGURE 20–35 Magnetic field between two large poles of a magnet is nearly uniform except at the edges.

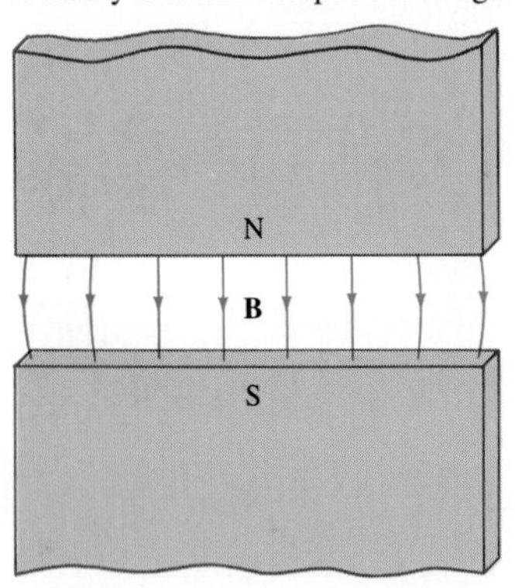

We saw earlier in this chapter (Fig. 20–7) that the magnetic field due to the electric current in a long straight wire is such that the field lines are circles with the wire at the center. You might expect that the field strength at a given point would be greater if the current flowing in the wire were greater; and that the field would be less at points farther from the wire. This is indeed the case. Careful experiments show that the magnetic field B at a point near the wire is directly proportional to the current I in the wire and inversely proportional to the distance r from the wire:

$$B \propto \frac{I}{r}.$$

This relation is valid as long as r, the perpendicular distance to the wire, is much less than the distance to the ends of the wire.

The proportionality constant is written[†] as $\mu_0/2\pi$; thus,

$$B = \frac{\mu_0}{2\pi}\frac{I}{r}. \quad \text{[outside a long straight wire]} \qquad (20\text{–}8)$$

Magnetic field due to current in straight wire

The value of the constant μ_0, which is called the **permeability of free space**, is $\mu_0 = 4\pi \times 10^{-7}\ \text{T}\cdot\text{m/A}$.

EXAMPLE 20–6 A vertical electric wire in the wall of a building carries a dc current of 25 A upward. What is the magnetic field at a point 10 cm due north of this wire?

SOLUTION According to Eq. 20–8:

$$B = \frac{(4\pi \times 10^{-7}\ \text{T}\cdot\text{m/A})(25\ \text{A})}{(2\pi)(0.10\ \text{m})} = 5.0 \times 10^{-5}\ \text{T},$$

or 0.50 G. By the right-hand rule (Fig. 20–7b), the field points to the west at this point. Since this field has about the same magnitude as the earth's, a compass would not point north but in a northwesterly direction.

Equation 20–8 gives the relation between the current in a long straight wire and the magnetic field it produces. This equation is valid only for a long straight wire. The following question arises: Is there a general relation between a current in a wire of whatever shape and the magnetic field around it? The answer is yes: the French scientist André Marie Ampère (1775–1836) proposed such a relation shortly after Oersted's discovery. Consider any (arbitrary) closed path around a current, as shown in Fig. 20–36, and imagine this path as being made up of short segments each of length Δl. First, we take the product of the length of each segment times the component of **B** parallel to that segment. If we now sum all these terms, according to Ampère, the result will be equal to μ_0 times the net current I that passes through the surface enclosed by the path. This is known as **Ampère's law** and can be written mathematically as

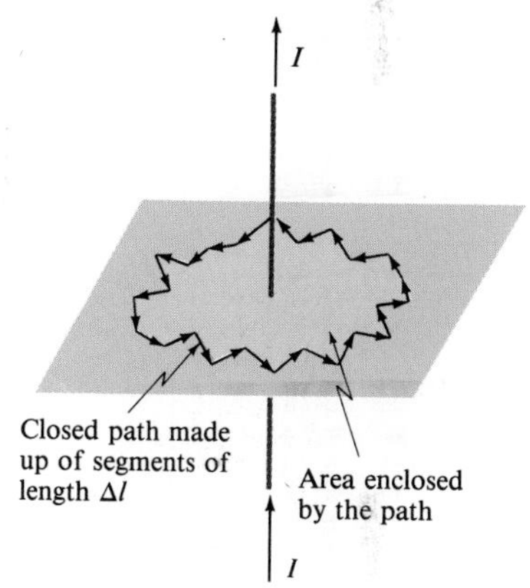

FIGURE 20–36 Arbitrary path enclosing a current, for Ampère's law. The path is broken down into segments of equal length Δl.

$$\sum B_{\parallel}\,\Delta l = \mu_0 I. \qquad (20\text{–}9)$$

Ampère's law

The symbol $\sum$ means "the sum of" and $B_{\parallel}$ means the component of B parallel to that particular Δl. The lengths Δl are chosen so that $B_{\parallel}$ is essentially constant on each length. The sum must be made over a closed path; and I is the net current passing through the surface bounded by this closed path.

We can understand Ampère's law more easily by applying it to a simple

[†] The constant is chosen like this so that Ampère's law—see Eq. 20–9—will have a simple and elegant form.

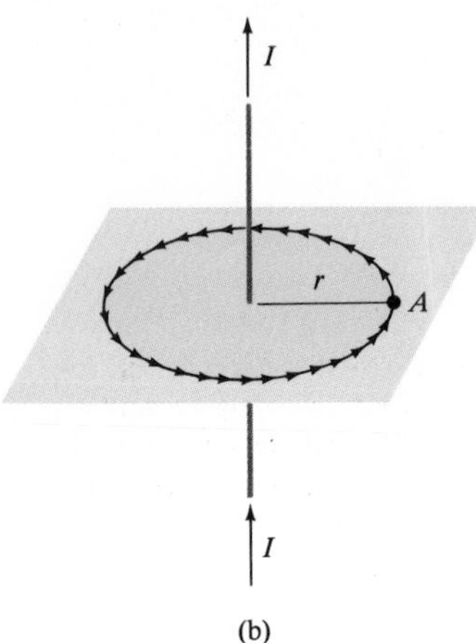

FIGURE 20–37
Circular path of radius r.

case. We take the case of a long, straight wire carrying a current I, which we have already examined and which served as an inspiration for Ampère himself. Suppose that we want to find the magnitude of B at point A, a distance r from the wire in Fig. 20–37. We know that the magnetic field lines are circles with the wire at their center. For the path to be used in Eq. 20–9, we choose a circle of radius r. (The choice of path is ours—so we choose one that will be convenient.) We choose this circular path because at any point on this path, **B** will be tangent to this circle. Thus, for any short segment of the circle (Fig. 20–37), **B** will be parallel to that segment, so $B_{||} = B$. Suppose that we break the circular path down into 100 segments.† Then Ampère's law states that

$$(B\,\Delta l)_1 + (B\,\Delta l)_2 + (B\,\Delta l)_3 + \cdots + (B\,\Delta l)_{100} = \mu_0 I.$$

The dots represent all the terms we did not write down. Since all the segments are the same distance from the wire, we expect B to be the same at each segment. We can then factor out B from the sum:

$$B(\Delta l_1 + \Delta l_2 + \Delta l_3 + \cdots + \Delta l_{100}) = \mu_0 I.$$

The sum of the segment lengths is just the circumference of the circle, $2\pi r$. Thus we have

$$B(2\pi r) = \mu_0 I,$$

or

$$B = \frac{\mu_0 I}{2\pi r}.$$

This is just Eq. 20–8 for the field near a long, straight wire, as discussed earlier.

Importance of Ampère's law

Ampère's law thus works for this simple case. A great many experiments indicate that Ampère's law is valid in general. However, it can be used to calculate the magnetic field mainly for simple cases. Its importance is that it relates the magnetic field to the current in a direct and mathematically elegant way. Ampère's law is thus considered one of the basis laws of electricity and magnetism. It is valid for any situation where the currents and fields are steady and not changing in time.

We now can see why the constant in Eq. 20–8 is written $\mu_0/2\pi$; this is done so that only μ_0 appears in Eq. 20–9 (rather than, say, $2\pi k$ if we had used k in Eq. 20–8). In this way, the more fundamental equation, Ampère's law, has the simpler form.

We now use Ampère's law to calculate the magnetic field inside a long solenoid (Fig. 20–13). If the coils of the solenoid are very closely spaced, the field inside will be parallel to the axis except at the ends (Fig. 20–38). We choose the path *abcd* shown in the figure, far from either end, for applying Ampère's law. We will consider this path as made up of four segments, the sides of the rectangle: *ab*, *bc*, *cd*, *da*. Then the left side of Eq. 20–9 becomes

$$(B_{||}\,\Delta l)_{ab} + (B_{||}\,\Delta l)_{bc} + (B_{||}\,\Delta l)_{cd} + (B_{||}\,\Delta l)_{da}.$$

Now the field outside the solenoid is so small as to be negligible compared to the field inside. Thus the first term will be zero. Furthermore. **B** is perpen-

† Actually, Ampère's law is precisely accurate when there is an infinite number of infinitesimally short segments, but this leads into calculus.

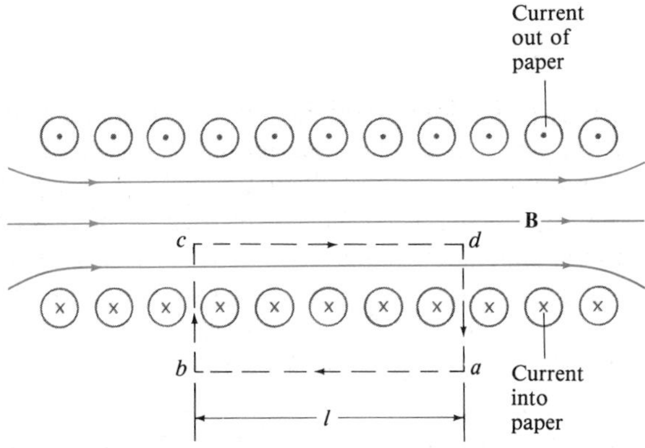

FIGURE 20–38 Magnetic field inside a solenoid is straight except at the ends. Dashed lines indicate the path chosen for use in Ampère's law.

dicular to the segments *bc* and *da* inside the solenoid, and is nearly zero between and outside the coils. Thus these terms are zero. Therefore, the left side of Eq. 20–9 is simply $(B_{\parallel}\,\Delta l)_{cd} = Bl$, where B is the field inside the solenoid, and l is the length *cd*. Now we determine the current enclosed by our chosen rectangular loop, to use for the right side of Eq. 20–9. If a current I flows in the wire of the solenoid, the total current enclosed by our path *abcd* is NI, where N is the number of loops our path encircles (five in Fig. 20–38). Thus Ampère's law gives us

$$Bl = \mu_0 NI.$$

If we let $n = N/l$ be the *number of loops per unit length*, then

Magnetic field inside a solenoid

$$B = \mu_0 nI. \qquad \text{[solenoid]} \tag{20–10}$$

This is the magnitude of the magnetic field within a solenoid. Note that B depends only on the number of loops per unit length, n, and the current I. The field does not depend on the position within the solenoid, so B is uniform. This is strictly true only for an infinite solenoid, but it is a good approximation for real ones for points not close to the ends.

EXAMPLE 20–7 A thin, 10-cm-long solenoid has a total of 400 turns of wire and carries a current of 2.0 A. Calculate the field inside near the center.

SOLUTION The number of turns per unit length is $n = 400/0.10\text{ m} = 4.0 \times 10^3\text{ m}^{-1}$. From Eq. 20–10:

$$B = \mu_0 nI = (12.57 \times 10^{-7}\text{ T}\cdot\text{m/A})(4.0 \times 10^3\text{ m}^{-1})(2.0\text{ A})$$
$$= 1.0 \times 10^{-2}\text{ T}.$$

*20–13 • Force Between Two Parallel Wires; Operational Definition of the Ampere and the Coulomb

You may have wondered how the constant μ_0 could be exactly $4\pi \times 10^{-7}$ T·m/A. With an older definition of the ampere, μ_0 was measured experimentally to be very close to this value. Today, however, μ_0 is *defined* to be exactly $4\pi \times 10^{-7}$ T·m/A. This, of course, could not be done if the ampere were defined independently. The ampere, the unit of current, is now defined in terms of the magnetic field B it produces using the defined value of μ_0.

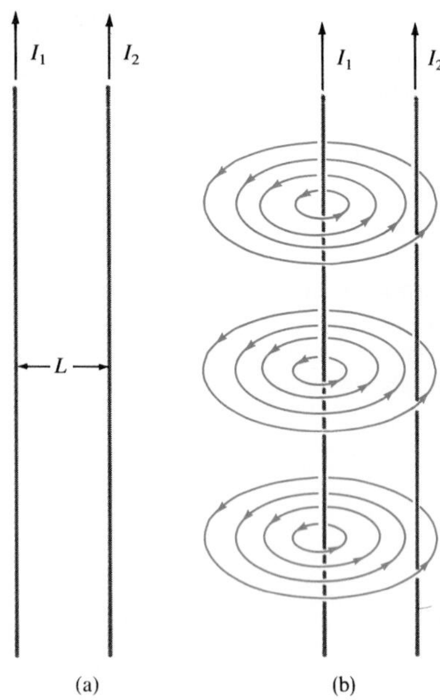

FIGURE 20–39 (a) Two parallel conductors carrying currents I_1, and I_2. (b) Magnetic field produced by I_1.

Let us be more precise. Consider two long parallel conductors separated by a distance L, Fig. 20–39. They carry currents I_1 and I_2, respectively. Each current produces a magnetic field that is "felt" by the other so that each should be expected to exert a force on the other, as Ampère first pointed out. For example, the magnetic field B_1 produced by I_1 is given by Eq. 20–8. At the location of the second conductor, the magnitude of this field is

$$B_1 = \frac{\mu_0}{2\pi}\frac{I_1}{L}.$$

See Fig. 20–39b where the field due *only* to I_1 is shown. According to Eq. 20–2, the force F per unit length l on the conductor carrying current I_2 is

$$\frac{F}{l} = I_2 B_1.$$

Note that the force on I_2 is due only to the field produced by I_1. Of course I_2 also produces a field, but it does not exert a force on itself. We substitute in the above formula for B_1 and find

$$\frac{F}{l} = \frac{\mu_0}{2\pi}\frac{I_1 I_2}{L}. \qquad (20\text{–}11)$$

FIGURE 20–40 (a) Parallel currents in the same direction exert attractive force on each other. (b) Antiparallel currents (in opposite directions) exert repulsive force on each other.

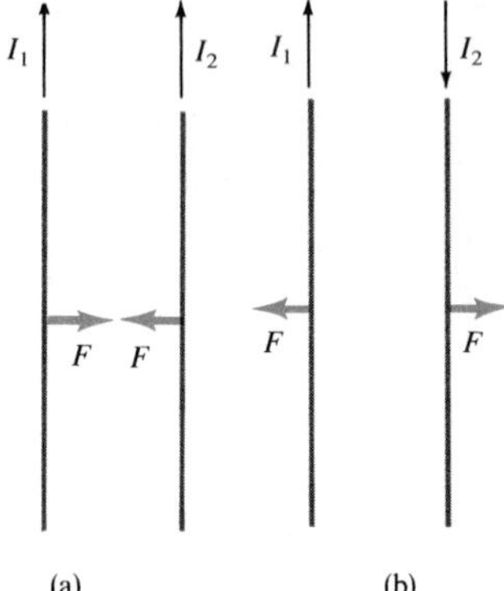

If we use the right-hand rule of Fig. 20–7b, we see that the lines of B_1 are as shown in Fig. 20–39b. Then using the right-hand rule of Fig. 20–15c, we see that the force exerted on I_2 will be to the left in the figure. That is, I_1 exerts an attractive force on I_2 (Fig. 20–40a). This is true as long as the currents are in the same direction. If I_2 is in the opposite direction, the right-hand rule indicates that the force is in the opposite direction. That is, I_1 exerts a repulsive force on I_2 (Fig. 20–40b). Reasoning similar to that above shows that the magnetic field produced by I_2 exerts an equal but opposite force on I_1. We expect this to be true also, of course, from Newton's third law. Thus, as shown in Fig. 20–40, parallel currents in the same directions attract each other, whereas currents in opposite directions repel.

Equation 20–11 is used to define the ampere precisely. If $I_1 = I_2 = 1$ A exactly and the two wires are exactly 1 m apart, then

$$\frac{F}{l} = \frac{(4\pi \times 10^{-7}\ \text{T·m/A})}{(2\pi)}\frac{(1\ \text{A})(1\ \text{A})}{(1\ \text{m})} = 2 \times 10^{-7}\ \text{N/m}.$$

Definitions of ampere and coulomb

Thus, *one* **ampere** *is defined as that current flowing in each of two long parallel conductors 1 m apart, which results in a force of exactly* 2×10^{-7} *N/m of length of each conductor.*

This is the precise definition of the ampere. The **coulomb** is then defined as being *exactly* one ampere-second: 1 C = 1 A·s. The value of k or ε_0 in Coulomb's law (Section 16–5) is obtained from experiment.

This may seem a rather roundabout way of defining quantities. The reason behind it is the desire for **operational definitions** of quantities—that is, definitions of quantities that can actually be measured given a definite set of operations to carry out. For example, the unit of charge, the coulomb, could be defined in terms of the force between two equal charges after defining a value for ε_0 or k in Eqs. 16–1 and 16–2. However, to carry out an actual experiment to measure the force between two charges is very difficult. For one thing, any desired amount of charge is not easily obtained precisely; and charge tends to leak from objects into the air. On the other hand, the amount of current in a wire can be varied accurately and continuously (by putting a variable resistor in a circuit). Thus the force between two current-carrying conductors is far easier to measure precisely. And this is why the ampere is defined first, and then the coulomb in terms of the ampere. At the National Bureau of Standards, Washington, DC, precise measurement of current is made using circular coils of wire rather than straight lengths because it is more convenient and accurate.

Electric and magnetic field strengths are also defined operationally: the electric field in terms of the measurable force on a charge, via Eq. 16–3; and the magnetic field in terms of the force per unit length on a current-carrying wire, via Eq. 20–2.

EXAMPLE 20–8 The two wires of a 2.0-m-long appliance cord are 3.0 mm apart and carry 8.0 A dc. Calculate the force between these wires.

SOLUTION Equation 20–11 gives us

$$F = \frac{(2.0 \times 10^{-7}\ \mathrm{T \cdot m/A})(8.0\ \mathrm{A})^2(2.0\ \mathrm{m})}{(3.0 \times 10^{-3}\ \mathrm{m})} = 8.5 \times 10^{-3}\ \mathrm{N}.$$

Since the currents are in opposite directions, the force would tend to spread them apart.

*20–14 • Magnetic Fields in Magnetic Materials; Hysteresis

The field of a long solenoid is directly proportional to the current. Indeed, Eq. 20–10 tells us that the field B_0 inside a solenoid is given by

$$B_0 = \mu_0 n I.$$

This is valid if there is only air inside the coil. If we put a piece of iron or other ferromagnetic material inside the solenoid, the field will be greatly increased, often by hundreds or thousands of times. This occurs because the domains in the iron become preferentially aligned by the external field. The resulting magnetic field is the sum of that due to the current and that due to the iron. It is sometimes convenient to write the total field in this case as a sum of two terms:

$$\mathbf{B} = \mathbf{B}_0 + \mathbf{B}_M. \qquad (20\text{–}12)$$

Here, $\mathbf{B}_0$ refers to the field due only to the current in the wire (the "external field"). It is the field that would be present in the absence of a ferromagnetic material. Then $\mathbf{B}_M$ represents the additional field due to the ferromagnetic material itself; often $\mathbf{B}_M \gg \mathbf{B}_0$.

The total field inside a solenoid in such a case can also be written by replacing the constant μ_0 in Eq. 20–10 by another constant, μ, characteristic of the material inside the coil:

$$B = \mu n I; \qquad (20\text{–}13)$$

μ is called the **magnetic permeability** of the material. For ferromagnetic materials, μ is much greater than μ_0. For all other materials, its value is very

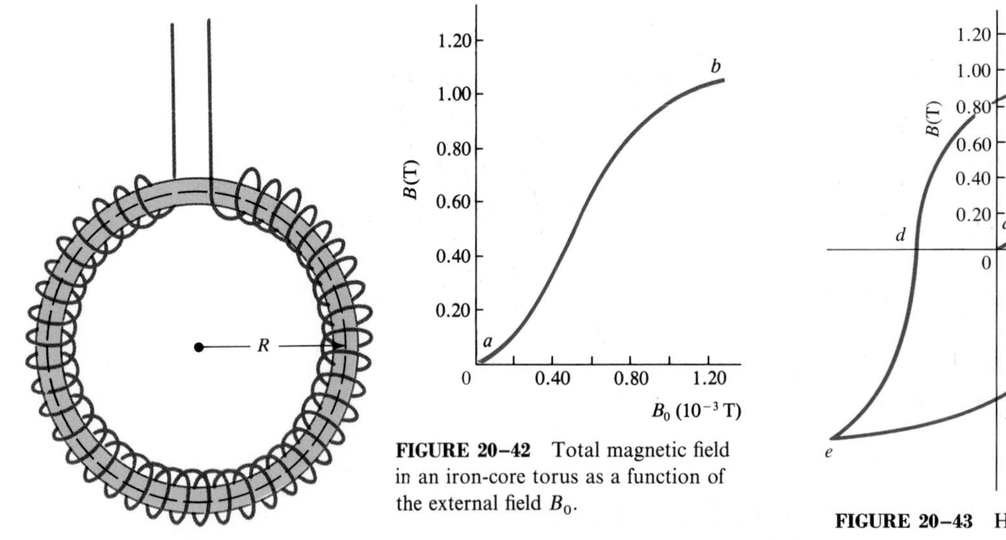

FIGURE 20–41 Iron-core torus.

FIGURE 20–42 Total magnetic field in an iron-core torus as a function of the external field B_0.

FIGURE 20–43 Hysteresis curve.

close to μ_0.† The value of μ, however, is not constant for ferromagnetic materials; it depends on the value of the external field B_0, as the following experiment shows.

Measurements on magnetic materials are generally done using a torus, which is essentially a long solenoid bent into the shape of a circle (Fig. 20–41), so that practically all the lines of **B** remain within the torus. Suppose the torus has an iron core that is initially unmagnetized and there is no current in the windings of the torus. Then the current I is slowly increased, and B_0 increases linearly with I. The total field B also increases, but follows the curved line shown in the graph of Fig. 20–42. (Note the different scales: $B \gg B_0$.) Initially (point *a*), no domains are aligned. As B_0 increases, the domains become more and more aligned until at point *b*, nearly all are aligned. The iron is said to be approaching **saturation**. Point *b* is typically 70 percent of full saturation. (If B_0 is increased further, the curve continues to rise very slowly, and reaches 98 percent saturation only when B_0 reaches a value about a thousandfold above that at point *b*; the last few domains are very difficult to align.) Next, suppose the external field B_0 is reduced by decreasing the current in the coils. As the current is reduced to zero, point *c* in Fig. 20–43, the domains do not become completely unaligned. Some permanent magnetism remains. If the current is then reversed in direction, enough domains can be turned around so $B = 0$ (point *d*). As the reverse current is increased further, the iron approaches saturation in the opposite direction (point *e*). Finally, if the current is again reduced to zero and then increased in the original direction, the total field follows the path *efgb*, again approaching saturation at point *b*.

Notice that the field did not pass through the origin (point *a*) in this cycle. The fact that the curves do not retrace themselves on the same path

† All materials are slightly magnetic. Nonferromagnetic materials fall into two principal classes: **paramagnetic**, in which μ is very slightly larger than μ_0; and **diamagnetic**, in which μ is very slightly less than μ_0. Paramagnetic materials apparently contain atoms that have a net magnetic dipole moment due to orbiting electrons, and these become slightly aligned with an external field just as the galvanometer coil in Fig. 20–22 experiences a torque that tends to align it. Atoms of diamagnetic materials have no net dipole moment. However, in the presence of an external field, electrons revolving in one direction are caused to increase in speed slightly, whereas those revolving in the opposite direction are reduced in speed. The result is a slight net magnetic effect that actually opposes the external field.

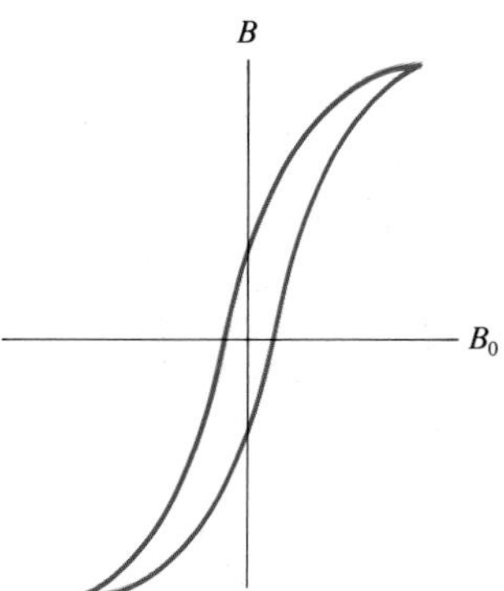

FIGURE 20–44 Hysteresis curve for soft iron.

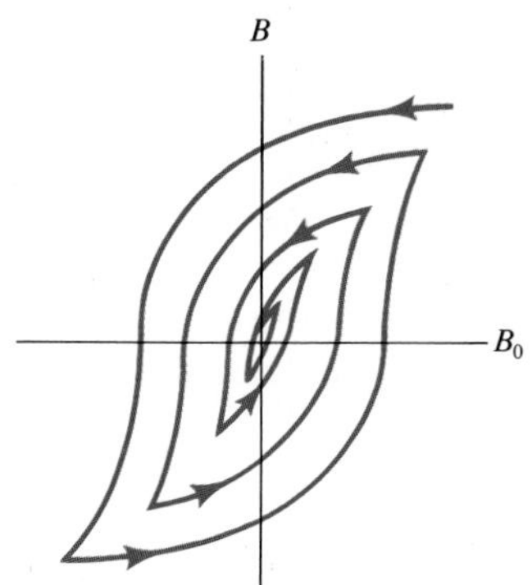

FIGURE 20–45 Successive hysteresis loops during demagnetization.

is called **hysteresis**. The curve *bcdefgb* is called a **hysteresis loop**. In such a cycle, much energy is transformed to thermal energy (friction) due to realigning of the domains. It can be shown that the energy dissipated in this way is proportional to the area of the hysteresis loop.

Hysteresis

At points *c* and *f*, the iron core is magnetized even though there is no current in the coils. These points correspond to a permanent magnet. For a permanent magnet, it is desired that *ac* and *af* be as large as possible. Materials for which this is true are said to have high **retentivity**, and may be referred to as "hard." On the other hand, a hysteresis curve such as that in Fig. 20–44 occurs for so-called "soft iron" (it is soft only from a magnetic point of view). This is preferred for *electromagnets* (Section 20–4) since the field can be more readily switched off, and the field can be reversed with less loss of energy. Whether iron is "soft" or "hard" depends on how it is alloyed, heat treatment, and other factors.

A ferromagnetic material can be demagnetized—that is, made unmagnetized. This can be done by reversing the magnetizing current repeatedly while decreasing its magnitude. This results in the curve of Fig. 20–45. The heads of a tape recorder are demagnetized in this way. The alternating magnetic field acting at the heads due to a demagnetizer is strong when the demagnetizer is placed near the heads and decreases as it is moved slowly away.

SUMMARY

A magnet has two *poles*, north and south. The north pole is that end which points toward the north when the magnet is freely suspended. Unlike poles of two magnets attract each other, whereas like poles repel.

Permanent magnets are made up of tiny *domains*—each a tiny magnet—which are aligned. In an unmagnetized piece of iron, the domains are randomly oriented.

We can imagine that a *magnetic field* surrounds every magnet. The SI unit for magnetic field is the *tesla* (T). The force one magnet exerts on another is said to be an interaction between one magnet and the magnetic field produced by the other.

Electric currents produce magnetic fields. For example, the lines of magnetic field due to a current in a straight wire form circles around the wire and the field exerts a force on magnets placed near it.

A magnetic field exerts a force on an electric current. For a straight wire of length l carrying a current I, the force has magnitude $F = IlB \sin \theta$, where θ is the angle between the magnetic field of strength B and the wire. The direction of the force is perpendicular to the wire and to the magnetic field, and is given by the right-hand rule. Similarly, a magnetic field exerts a force on a charge q moving with velocity v of magnitude $F = qvB \sin \theta$, where θ is the angle between $\mathbf{v}$ and $\mathbf{B}$. The direction of $\mathbf{F}$ is perpendicular to $\mathbf{v}$ and to $\mathbf{B}$. The path of a charged particle moving perpendicular to a uniform magnetic field is a circle.

The force exerted on a current-carrying wire by a magnetic field is the basis for operation of many devices, such as meters, motors, and loudspeakers.

The measurement of the charge-to-mass ratio (e/m) of the electron was done using magnetic and electric fields. The charge e on the electron was measured in the Millikan oil-drop experiment and then its mass was obtained from the measured value of the e/m ratio.

Electric and magnetic fields are employed in the cathode-ray tube (CRT)—which is the picture tube of TV sets, display terminals, and oscilloscopes.

QUESTIONS

1. A compass needle is not always balanced parallel to the earth's surface but one end may dip downward. Explain.

2. Draw the magnetic field lines around a straight section of wire carrying a current horizontally to the left.

3. In what direction are the magnetic field lines surrounding a straight wire carrying a current that is moving directly toward you?

4. The magnetic field due to current in wires in your home can affect a compass. Discuss the problem in terms of currents, depending on whether they are ac or dc.

5. What kind of field or fields surround a moving electric charge?

6. Will a magnet attract any metallic object, or only those made of iron? (Try it and see.) Why is this so?

7. Two iron bars attract each other no matter which ends are placed close together. Are both magnets? Explain.

***8.** Note that the pattern of magnetic field lines surrounding a bar magnet is similar to that of the electric field around an electric dipole. From this fact, predict how the magnetic field will change with distance (*a*) when near one pole of a very long bar magnet, and (*b*) when far from a magnet as a whole.

9. How do you suppose the first magnets found in Magnesia were formed?

10. Why will either pole of a magnet attract an unmagnetized piece of iron?

11. Suppose you have three iron rods, two of which are magnetized but the third is not. How would you determine which two are the magnets without using any additional objects?

12. An unmagnetized nail will not attract an unmagnetized paper clip. However, if one end of the nail is in contact with a magnet, the other end *will* attract a paper clip. Explain.

13. Another type of magnetic switch similar to a solenoid is a **relay**. A relay is an electromagnet (the iron rod inside the coil does not move) that, when activated, attracts a piece of soft iron on a pivot. Design a relay (*a*) to make a doorbell, and (*b*) to close an electrical switch. A relay is used in the latter case when you need to switch on a circuit carrying a very large current but you do not want that large current flowing through the main switch. For example, the starter switch of a car is connected to a relay so that the large currents needed for the starter do not pass to the dashboard switch.

14. How can you make a compass without using iron or other ferromagnetic material?

15. A horseshoe magnet is held vertically with the north pole on the left and south pole on the right. A wire passing perpendicularly between the poles carries a current directly away from you. In what direction is the force on the wire?

16. Can you set a resting electron into motion with a magnetic field? With an electric field?

17. A charged particle is moving in a circle under the influence of a uniform magnetic field. If an electric field that points in the same direction as the magnetic field is turned on, describe the path the charged particle will take.

18. Each of the right-hand rules you learned in this chapter can be changed to *left-hand rules* if you are specifying the direction of movement of *negative* particles, such as electrons in a wire. Show, for each right-hand rule, that the same operations using the left hand give the same results if the direction of charge flow is for negative charges.

19. A charged particle moves in a straight line through a particular region of space. Could there be a nonzero magnetic field in this region? If so, give two possible situations.

20. If a moving charged particle is deflected sideways in some region of space, can we conclude for certain that $\mathbf{B} \neq 0$ in that region?

21. In a particular region of space there is a uniform magnetic field **B**. Outside this region, $B = 0$. Can you inject an electron into the field so it will move in a closed circular path in the field?

22. How could you tell whether moving electrons in a certain region of space are being deflected by an electric field or by a magnetic field (or by both)?

23. A beam of electrons is directed perpendicularly toward a horizontal wire carrying a current from left to right. In what direction is the beam deflected?

24. Charged cosmic ray particles from outside the earth tend to strike the earth more frequently near the poles than at lower latitudes. Explain. (See Fig. 20–4.)

***25.** In the electromagnetic pumping device shown in Fig. 20–21, does the direction of the force, and therefore of the blood flow, depend on the sign of the ions?

26. What factors determine the sensitivity of a galvanometer?

27. Bringing a magnet close to a television screen will distort the picture. Why? (This can damage a color set.)

*28. Two ions have the same mass, but one is singly ionized and the other is doubly ionized. How will their positions on the film of the mass spectrograph of Fig. 20–34 differ?

29. Two long wires carrying equal currents I are at right angles to each other, but don't quite touch. Describe the magnetic force one exerts on the other.

30. A horizontal wire carries a large current. A second wire carrying a current in the same direction is suspended below. Can the current in the upper wire hold the lower wire in suspension? Under what conditions will it be in equilibrium? Under what conditions will the equilibrium be stable?

31. A horizontal current-carrying wire, free to move, is suspended directly above a second, parallel, current-carrying wire. (*a*) In what direction is the current in the lower wire? (*b*) Can the upper wire be held in stable equilibrium due to the magnetic force of the lower wire? Explain.

*32. The presence of iron inside a solenoid can greatly increase **B**. Yet without iron, greater fields could be obtained. Explain.

PROBLEMS

SECTIONS 20–5 AND 20–6

1. (I) (*a*) What is the force per meter on a wire carrying a 3.60-A current when perpendicular to a 1.20-T magnetic field? (*b*) What if the angle between the wire and field is 45.0°?

2. (I) The force on a wire carrying 30.0 A is a maximum of 3.80 N when placed between the pole faces of a magnet. If the pole faces are 25.0 cm in diameter, what is the approximate strength of the magnetic field?

3. (I) How much current is flowing in a wire 3.00 m long if the force on it is 0.900 N when placed in a uniform 0.0800-T field?

4. (I) Calculate the magnetic force on a 180-m length of wire stretched between two towers and carrying a 280-A current. The earth's magnetic field of 5.00×10^{-5} T makes an angle of 60.0° with the wire.

5. (I) Determine the magnitude and direction of the force on an electron traveling 2.84×10^5 m/s horizontally to the east in a vertically upward magnetic field of strength 1.60 T.

6. (I) Describe the path of a proton ($q = e$, $m = 1.67 \times 10^{-27}$ kg) that moves perpendicular to a 0.180-T magnetic field with a speed of 8.25×10^6 m/s. The field points directly toward the observer.

7. (I) Find the direction of the force on a positive charge for each diagram shown in Fig. 20–46, where **v** is the velocity of the charge and **B** is the direction of the magnetic field. (⊗ means the vector points inward. ⊙ means it points outward, toward the viewer.)

8. (I) Determine the direction of **B** for each case in Fig. 20–47, where **F** represents the force on a positively charged particle moving with velocity **v**.

9. (II) An electron experiences the greatest force as it travels 3.4×10^5 m/s in a magnetic field when it is moving southward. The force is upward and of magnitude 6.8×10^{-13} N. What is the magnitude and direction of the magnetic field?

10. (II) A proton moves in a circular path perpendicular to a 1.40-T magnetic field. The radius of its path is 8.45 mm. Calculate the energy of the proton in eV.

11. (II) A particle of charge q moves in a circular path of radius r in a uniform magnetic field B. Show that its momentum is $p = qBr$.

12. (II) For a particle of mass m and charge q moving in a circular path in a magnetic field B, show that its kinetic energy is proportional to r^2, the square of the radius of curvature of its path.

13. (II) An electron is accelerated through a potential difference of 6000 V. What is the strength of the magnetic field if the radius of its path is 8.4 mm?

14. (III) A 4.70-g bullet moves with a speed of 200 m/s perpendicular to the earth's magnetic field of 5.00×10^{-5} T. If the bullet possesses a net charge of 6.80×10^{-9} C, by what distance will it be deflected from its path due to the magnetic field after it has traveled 700 m?

*SECTION 20–7

*15. (II) Show that the emf produced by the Hall effect is given by $\mathcal{E} = vBl$, where v is the speed of the charged particles in the conductor of width l (Fig. 20–20).

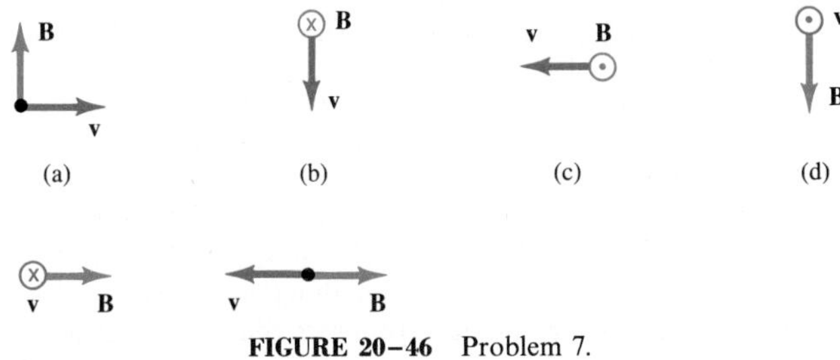

FIGURE 20–46 Problem 7.

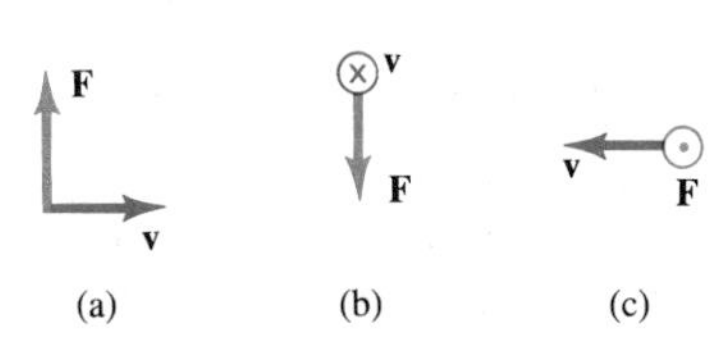

FIGURE 20–47 Problem 8.

*16. (II) The Hall effect can be used to measure blood flow rate because the blood contains ions that constitute an electric current; the apparatus is basically the same as for electromagnetic pumping, Fig. 20–21, except that the blood flows in the artery of its own accord and the external emf is replaced by a voltmeter that measures the Hall emf. (*a*) Does the sign of the ions influence the emf? (*b*) Use the result of Problem 15 to determine the flow velocity in an artery 3.3 mm in diameter if the measured emf is 0.10 mV and B is 0.070 T. (In actual practice, an alternating magnetic field is used.)

SECTION 20–8

17. (I) A galvanometer needle deflects full scale for a 48.0-μA current. What current will give full-scale deflection if the magnetic field weakens to 0.900 of its original value?
18. (I) If the restoring spring of a galvanometer weakens by 15 percent over the years, what current will give full-scale deflection if it originally required 38 μA?
19. (I) If the current to a motor drops by 10 percent, by what factor does the output torque change?
20. (I) A square loop of wire 17.0 cm on a side is placed with its face parallel to the magnetic field between the pole pieces of a large magnet. When 4.50 A flows in the coil, the torque on it is 0.378 N·m. What is the magnetic field strength?
21. (II) Show that the torque acting on the current loop of Fig. 20–22 is given by $\tau = NIAB \cos \theta$, where θ is the angle between the face of the coil and direction of the (uniform) magnetic field.
22. (II) A circular coil 26.0 cm in diameter and containing five loops lies flat on the ground. The earth's magnetic field at this location is 6.10×10^{-5} T; it points into the earth at an angle of 46.0° below a line pointing due north. A 25.0-A clockwise current passes through the coil. (*a*) Determine the torque on the coil. (*b*) Which edge of the coil rises up—north, east, south, or west?

SECTION 20–9

23. (I) What is the value of e/m for a particle that moves in a circle of radius 8.0 mm in a 0.46-T magnetic field if a crossed 200-V/m electric field will make the path straight?
24. (I) Protons move in a circle of radius 5.20 cm in a 0.465-T magnetic field. What value of electric field could make their paths straight? In what direction must it point?
25. (II) What is the velocity of a beam of electrons that go undeflected when passing through crossed electric and magnetic fields of magnitude 8.85×10^3 V/m and 4.50×10^{-3} T, respectively? What is the radius of the electron orbit if the electric field is turned off?
26. (II) An oil drop whose mass is determined to be 3.3×10^{-15} kg is held at rest between two large plates separated by 1.0 cm when the potential difference between them is 340 V. How many excess electrons does this drop have?
27. (III) In Millikan's oil-drop experiment, the mass of the oil drop is obtained by observing the terminal speed v_T of the freely falling drop in the absence of an electric field. Under these circumstances, the "effective" weight equals the viscous force given by Stokes's law (Section 10–12), $F = 6\pi\eta r v_T$, where η is the viscosity of air and r the radius of the drop. Also, the actual weight, $mg = \frac{4}{3}\pi r^3 \rho g$, must be corrected for the buoyant force of the air. This is done by replacing ρ with $\rho - \rho_A$, where ρ is the density of the oil and ρ_A the density of air. With these preliminaries, show that the charge on the drop is given by

$$q = 18\pi \frac{d}{V} \sqrt{\frac{\eta^3 v_T^3}{2(\rho - \rho_A)g}},$$

where d is the separation of the plates (Fig. 20–30) and V is the voltage across them that just keeps the drop stationary. All of the quantities on the right side of this equation are known or can be measured. The terminal velocity v_T is determined by measuring the time it takes the drop to fall a measured distance, which is observed through a small telescope.

SECTION 20–10

28. (I) Use the ideal gas as a model to estimate the rms speed of a free electron in a metal at 300 K, and at 2500 K (the typical temperature of the cathode in a tube).
29. (III) Electrons are accelerated by 10.0 kV in a CRT. The screen is 24 cm wide and is 25 cm from the 2.4-cm-long deflection plates. Over what range must the horizontally deflecting electric field vary to sweep the beam fully across the screen?
30. (III) In a given CRT, electrons are accelerated horizontally by 15 kV. They then pass through a uniform magnetic field B for a distance of 2.8 cm which deflects them upward so they reach the top of the screen 22 cm away, 11 cm above the center. Estimate the value of B.

*SECTION 20–11

*31. (I) In a mass spectrometer, germanium atoms have radii of curvature equal to 21.0, 21.6, 21.9, 22.2, and 22.8 cm. The largest radius corresponds to an atomic mass of 76 u. What are the atomic masses of the other isotopes?
*32. (II) Suppose the electric field between the electric plates in the mass spectrometer of Fig. 20–34 is 2.48×10^4 V/m and the magnetic fields $B = B' = 0.75$ T. The source contains boron isotopes of mass numbers 10 and 11 (to get their masses, multiply by 1.67×10^{-27} kg). How far apart are the lines formed by the singly charged ions of each type on the photographic film?
*33. (II) A mass spectrometer is being used to monitor air pollutants. It is difficult, however, to separate molecules with nearly equal mass such as CO (28.0106 u) and N_2 (28.0134 u). How large a radius of curvature must a spectrometer have if these two molecules are to be separated on the film by 0.33 mm.?
*34. (II) One form of mass spectrometer accelerates ions by a voltage V before they enter a magnetic field B. The ions are assumed to start from rest. Show that the mass of an ion is $m = qB^2R^2/2V$, where R is the radius of the ions' path in the magnetic field and q is their charge.

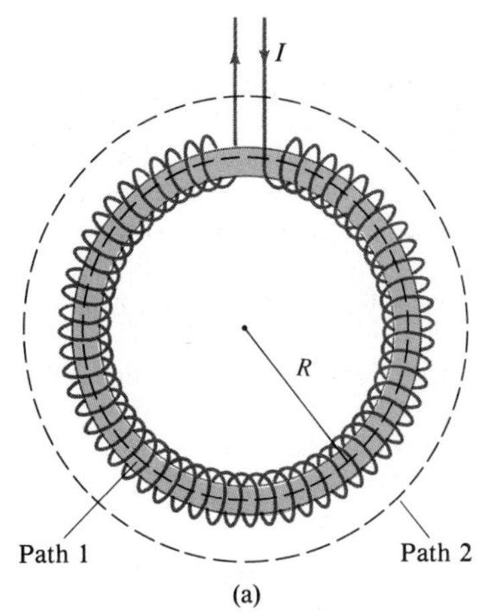

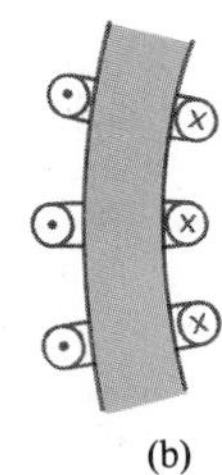

FIGURE 20–48 Problem 44. (a) A torus. (b) A section of the torus showing direction of the current for three loops: ⊙ means current toward viewer, and ⊗ means current away from viewer.

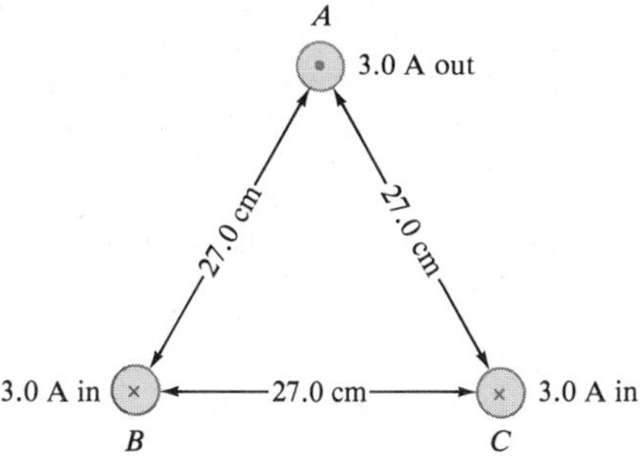

FIGURE 20–49 Problem 50.

*SECTION 20–12

*35. (I) How strong is the magnetic field 8.0 cm from a long straight wire carrying 3.8 A?

*36. (I) If a magnetic field of no more than 1.0×10^{-3} T is to be allowed 30 cm from an electrical wire, what is the maximum current the wire can carry?

*37. (II) What is the acceleration (in g's) of a 300-g model airplane charged to 8.0 C and traveling at 1.6 m/s as it passes within 7.5 cm of a wire, nearly parallel to its path, carrying a 30-A current?

*38. (II) An electron traveling 8.50×10^5 m/s is moving initially at a 45° angle to a straight wire carrying a 25.0-A current. If the electron passes within 10.0 cm of the wire, what maximum force does it feel? Assume the electron's path and the wire are in the same plane.

*39. (II) A horizontal compass is placed 12 cm due south from a straight vertical wire carrying a 20-A current downward. In what direction does the compass needle point at this location? Assume the horizontal component of the earth's field at this point is 0.45×10^{-4} T and the magnetic declination is 0°.

*40. (II) Determine the magnetic field midway between two long straight wires 2.0 cm apart in terms of the current I in one when the other carries 20 A. Assume these currents are (*a*) in the same direction, and (*b*) in opposite directions.

*41. (II) A compass needle in a particular location points 12°E of N outdoors. However, when it is placed 10 cm to the east of a vertical wire inside a building, it points 60°E of N. What is the magnitude and direction of the current in the wire? The earth's field there is 0.50×10^{-4} T and is horizontal.

*42. (II) A 28-cm-long solenoid, 1.0 cm in diameter, is to produce a 0.20-T magnetic field at its center. If the maximum current is 8.8 A, how many turns must the solenoid have?

*43. (II) You have 1.0 kg of copper and want to make a practical solenoid that produces the greatest possible magnetic field. Should you make your copper wire long and thin, short and fat, or what? Consider other variables, such as solenoid diameter, length, and so on.

*44. (II) A torus is a solenoid in the shape of a circle (Fig. 20–48). Use Ampère's law along the circular path, shown dashed in Fig. 20–48a, to determine that the magnetic field (*a*) inside the torus is $B = \mu_0 NI/2\pi R$, where N is the total number of turns, and (*b*) outside the torus is $B = 0$. (*c*) Is the field inside a torus uniform like a solenoid's? If not, how does it vary?

*45. (II) Use Ampère's law to show that a uniform magnetic field, such as between the pole pieces of a magnet, Fig. 20–35, cannot drop abruptly to zero outside the magnet. [*Hint*: take as your path a rectangle with one vertical side inside the field and one vertical side completely outside the field.]

*46. (III) Two long parallel wires 10.0 cm apart carry 25.0-A currents in the same direction. Determine the magnetic field strength at a point 15.0 cm from one wire and 10.0 cm from the other. [*Hint*: make a drawing in a plane containing the field lines and recall the rules for vector addition.]

*47. (III) Suppose that a current I flows uniformly through a long cylindrical conductor of radius r_0. Use Ampère's law to show that the magnetic field inside the conductor at a distance r from the center of the conductor is

$$B = \frac{\mu_0 I r}{2\pi r_0^2} \qquad (r < r_0).$$

Assume that the field lines are circles, just as they are outside the conductor.

*SECTION 20–13

*48. (I) What is the magnitude and direction of the force between two parallel wires 80 m long and 8.0 cm apart, each carrying 35 A in the same direction?

*49. (I) A vertical straight wire carrying a 9.0-A current exerts an attractive force per unit length of 7.0×10^{-4} N/m on a second parallel wire 8.0 cm away. What current (magnitude and direction) flows in the second wire?

*50. (II) Three long parallel wires are 27.0 cm from one another. (Looking along them, they are at three corners of an equilateral triangle.) The current in each wire is 3.0 A, but that in wire A is opposite to that in wires B and C (Fig. 20–49). Determine the magnetic force per unit length on each wire due to the other two.

*51. (III) A long horizontal wire carries a current of 48 A. A second wire, made of 2.5-mm-diameter copper wire and parallel to the first, but 15 cm below it, is held in suspension magnetically, (Fig. 20–50). (*a*) What is the magnitude and direction of the current in the lower wire? (*b*) Is it in stable equilibrium? (*c*) Repeat parts (*a*) and (*b*) if the second wire is suspended 15 cm *above* the first due to the latter's field.

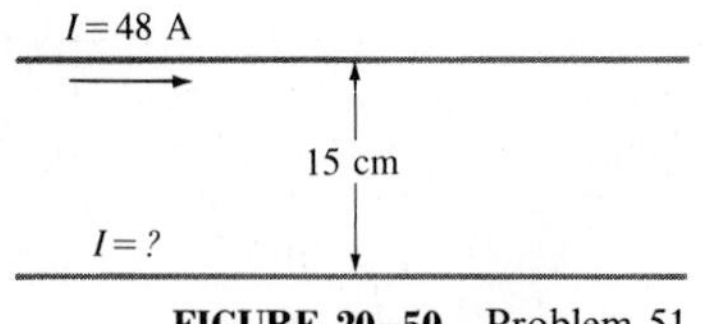

FIGURE 20–50 Problem 51.

*SECTION 20–14

*52. (II) An iron-core solenoid is 36 cm long, 1.5 cm in diameter, and has 600 turns of wire. A magnetic field of 1.8 T is produced when 40 A flows in the wire. What is the permeability μ at this high field strength?

GENERAL PROBLEMS

53. Calculate the force on an airplane which has acquired a net charge of 150 C and moves with a speed of 280 m/s perpendicular to the earth's magnetic field of 5.0×10^{-5} T.

54. A doubly charged helium atom, whose mass is 6.7×10^{-27} kg, is accelerated by a voltage of 1900 V. (*a*) What will be its radius of curvature in a uniform 0.340-T field? (*b*) What is its period of revolution?

55. A straight 2.00-mm-diameter copper wire can just "float" horizontally in air because of the force of the earth's magnetic field **B** which is horizontal and of magnitude 5.00×10^{-5} **T**. What current does the wire carry?

56. Two stiff parallel wires a distance l apart in a horizontal plane act as rails to support a light metal rod of mass m (perpendicular to each rail), Fig. 20–51. A magnetic field **B**, directed vertically upward (outward in diagram), acts throughout. At $t = 0$, wires connected to the rails are connected to a constant current source and a current I begins to flow through the system. Determine the speed of the rod as a function of time (*a*) assuming no friction between the rod and the rails, and (*b*) if the coefficient of friction is μ_k. (*c*) In which direction does the rod move, east or west, if the current through it heads north?

57. Estimate the approximate maximum deflection of the electron beam near the center of a TV screen due to the earth's 5.0×10^{-5}-T field. Assume the screen is 20 cm from the electron gun where the electrons are accelerated (*a*) by 2.0 kV, (*b*) by 30 kV. Note that in color TV sets, the beam must be directed accurately to within less than 1 mm in order to strike the correct phosphor. Because the earth's field is significant here, mu-metal shields are used to reduce the earth's field in the CRT.

58. An electron enters a large solenoid at a 6.0° angle to the axis. If the field is a uniform 0.33 T, determine the radius and pitch (distance between loops) of the electron's helical path if its speed is 3.5×10^5 m/s.

59. The cyclotron (Fig. 20–52) is a device used to accelerate elementary particles such as protons to high speeds. Particles starting at point A with some initial velocity travel in circular orbits in the magnetic field B. The particles are accelerated to higher speeds each time they pass in the gap between the metal "dees," where there is an electric field E. (There is no electric field within the cavity of the metal dees.) The electric field changes direction each half-cycle, owing to an ac voltage $V = V_0 \sin 2\pi f t$, so that the particles are increased in speed at each passage through the gap. (*a*) Show that the frequency f of the voltage must be $f = Bq/2\pi m$, where q is the charge on the particles and m their mass. (*b*) Show that the kinetic energy of the particles increases by $2qV_0$ each revolution, assuming that the gap is small. (*c*) If the radius of the cyclotron is 2.0 m and the magnetic field strength is 0.50 T, what will be the maximum kinetic energy of accelerated protons in MeV? (*d*) How is a cyclotron like a swing?

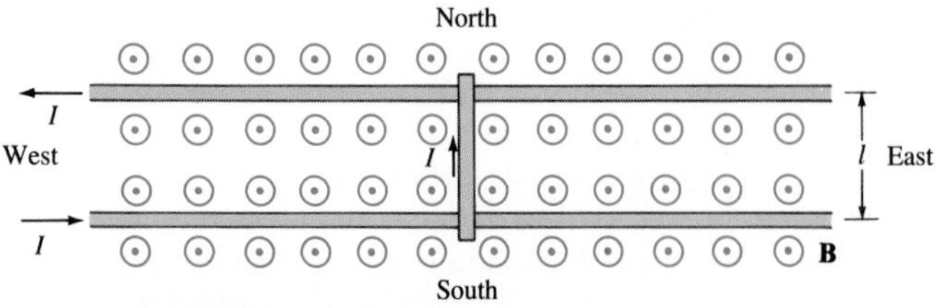

FIGURE 20–51 Looking down on a rod sliding on rails (Problem 56.)

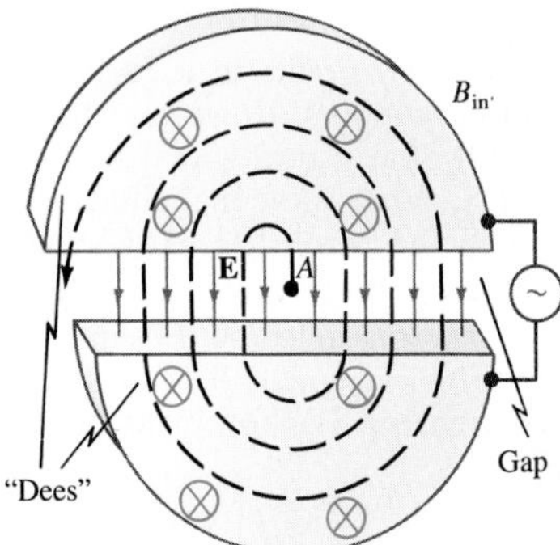

FIGURE 20–52 A cyclotron. Problem 59.

60. A square loop of aluminum wire is 12.5 cm on a side. It is to carry 30 A and rotate in a 2.0-T magnetic field. (*a*) Determine the minimum diameter of the wire so that it will not fracture from tension or shear. Assume a safety factor of 10. (See Table 9–2.) (*b*) What is the resistance of a single loop of this wire?

61. The magnetic field strength B at the center of a circular coil of wire carrying a current I is

$$B = \frac{\mu_0 N I}{2r},$$

where N is the number of loops in the coil and r is its radius. Suppose that an electromagnet uses a coil 1.5 m in diameter made from square copper wire 2.0 mm on a side. The power supply produces 60 V at a maximum power output of 1.0 kW. (*a*) How many turns are needed to run the power supply at maximum power? (*b*) What is the magnetic field strength at the center of the coil? (*c*) If you use a greater number of turns and this same power supply (so the voltage remains at 60 V), will a greater magnetic field strength result? Explain.

CHAPTER 21

Electromagnetic Induction and Faraday's Law; AC Circuits

Transmission lines carry electric power over great distances at very high voltage for greater efficiency. To reduce high voltage to usable voltage, transformers are used, whose operation depends on Faraday's law of induction. Induction is also the basis for electric generators which produce the electric power in the first place.

In Chapter 20, we discussed two ways in which electricity and magnetism are related: (1) an electric current produces a magnetic field; and (2) a magnetic field exerts a force on an electric current or moving electric charge. These discoveries were made in 1820–1821. Scientists then began to wonder: if electric currents produce a magnetic field, is it possible that a magnetic field can produce an electric current? Ten years later the American Joseph Henry (1797–1878) and the Englishman Michael Faraday (1791–1867; see Fig. 16–17) independently found that it was possible. Henry actually made the discovery first. But Faraday published his results earlier and investigated the subject in more detail. We now discuss this phenomenon and some of its world-changing applications.

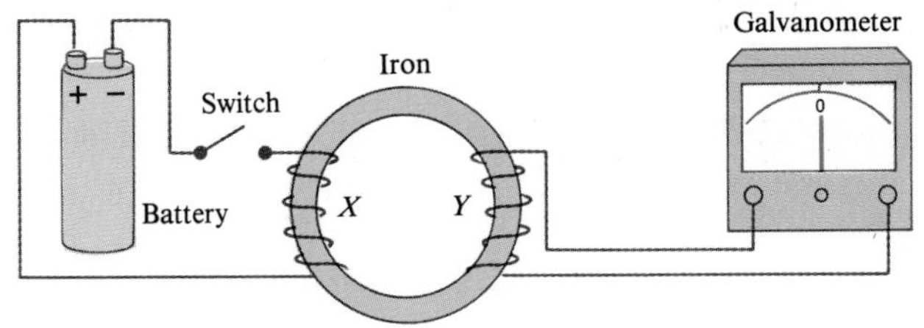

FIGURE 21–1 Faraday's experiment to induce an emf.

21–1 • Induced EMF

In his attempt to produce an electric current from a magnetic field, Faraday used the apparatus shown in Fig. 21–1. A coil of wire, X, was connected to a battery. The current that flowed through X produced a magnetic field that was intensified by the iron core. Faraday hoped that by using a strong enough battery, a steady current in X would produce a great enough magnetic field to produce a current in a second coil Y. This second circuit, Y, contained a galvanometer to detect any current but contained no battery. He met no success with steady currents. But the long-sought effect was finally observed when Faraday saw the galvanometer in circuit Y deflect strongly at the moment he closed the switch in circuit X. And the galvanometer deflected strongly in the opposite direction when he opened the switch. A *steady* current in X had produced *no* current in Y. Only when the current in X was starting or stopping was a current produced in Y.

Faraday concluded that although a steady magnetic field produces no current, a *changing* magnetic field can produce an electric current! Such a current is called an **induced current**. When the magnetic field through coil Y changes, a current flows as if there were a source of emf in the circuit. We therefore say that an

Changing ***B*** *induces an emf*

induced emf is produced by a changing magnetic field.

Faraday did further experiments on **electromagnetic induction**, as this phenomenon is called. For example, Fig. 21–2 shows that if a magnet is moved quickly into a coil of wire, a current is induced in the wire. If the magnet is quickly removed, a current is induced in the opposite direction. Furthermore, if the magnet is held steady and the coil of wire is moved toward or away from the magnet, again an emf is induced and a current

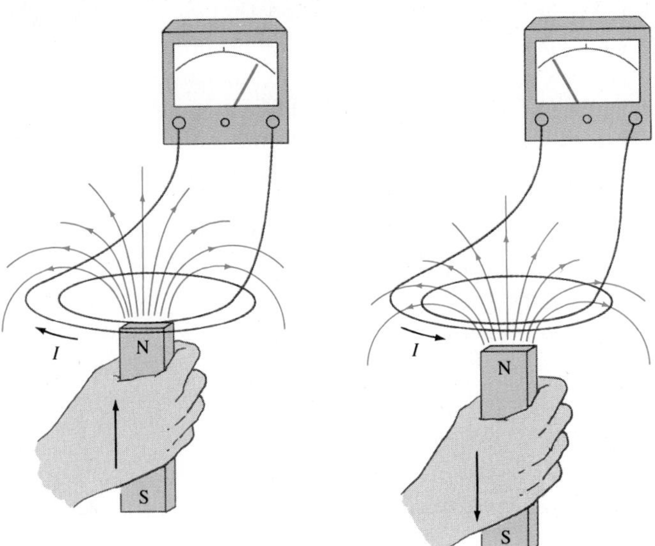

FIGURE 21–2 (a) A current is induced when a magnet is moved toward a coil. (b) The induced current is opposite when the magnet is moved away from the coil. Note that the galvanometer zero is at the center of the scale and deflects left or right, depending on the direction of the current.

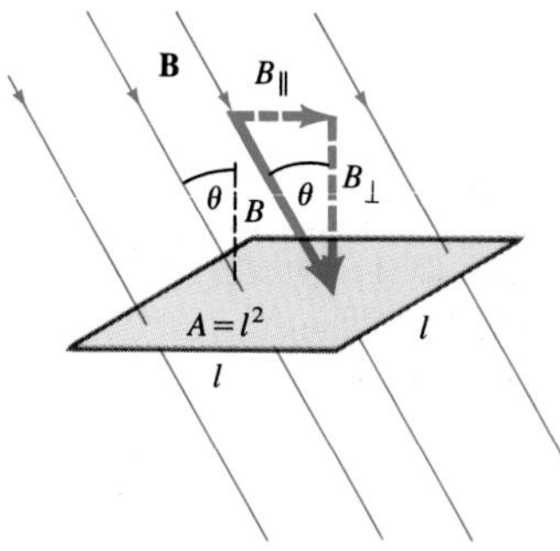

FIGURE 21–3 Determining the flux through a flat loop of wire. This loop is square, of side l and area $A = l^2$.

flows. Motion or change is required to induce an emf. It doesn't matter whether the magnet or the coil moves.

21–2 • Faraday's Law of Induction; Lenz's Law

Faraday investigated quantitatively what factors influence the magnitude of the emf induced. He found first of all that it depends on time: the more rapidly the magnetic field changes, the greater the induced emf. But the emf is not simply proportional to the rate of change of the magnetic field, **B**. Rather it is proportional to the rate of change of the **magnetic flux**, Φ_B, passing through the loop of area A, which is defined as

Magnetic flux defined

$$\Phi_B = B_{\perp}A = BA\cos\theta. \tag{21–1}$$

Here $B_{\perp}$ is the component of the magnetic field **B** perpendicular to the face of the coil, and θ is the angle between **B** and a line drawn perpendicular to the face of the coil. These quantities are shown in Fig. 21–3 for a square coil of side l whose area $A = l^2$. When the face of the coil is parallel to B, $\theta = 90°$ and $\Phi_B = 0$. When B is perpendicular to the coil, $\theta = 0°$ and

$$\Phi_B = BA. \qquad [B \perp \text{coil face}]$$

As we saw in Chapter 20, the lines of **B** can be drawn such that the number of lines per unit area is proportional to the field strength. Then the flux Φ_B can be thought of as being proportional to the *total number of lines passing through the coil.* This is illustrated in Fig. 21–4, where the coil is viewed from the side (on edge). For $\theta = 90°$, no lines pass through the coil and $\Phi_B = 0$, whereas Φ_B is a maximum when $\theta = 0°$. The unit of magnetic flux is the tesla-meter^2; this is called† a **weber**: 1 Wb = 1 T·m^2.

FIGURE 21–4 Magnetic flux Φ_B is proportional to the number of lines of **B** that pass through the loop.

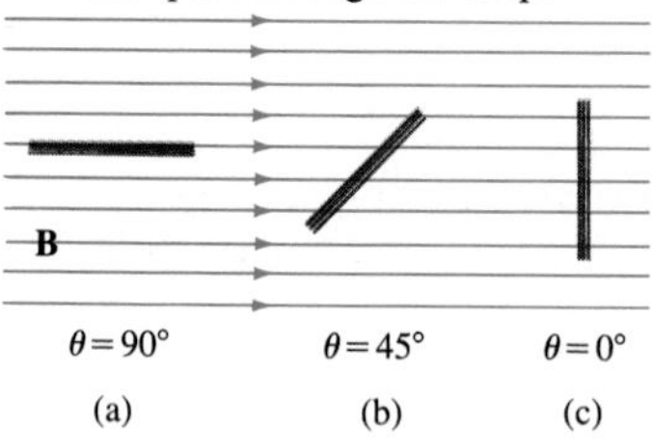

With this definition of the flux, we can now write down the results of Faraday's investigations. If the flux through N loops of wire changes by an amount $\Delta\Phi_B$ during a time Δt, the average induced emf during this time is

Faraday's law of induction

$$\mathscr{E} = -N\frac{\Delta\Phi_B}{\Delta t}. \tag{21–2}$$

This fundamental result is known as **Faraday's law of induction**, and is one of the basic laws of electromagnetism.

The minus sign in Eq. 21–2 is placed there to remind us in which direction the induced emf acts. Experiments show that *an induced emf always gives rise to a current whose magnetic field opposes the original change in flux.* This is known as **Lenz's law**. Let us apply it to the case of relative motion between a magnet and a coil, Fig. 21–2. The changing flux induces an emf, which produces a current in the coil. And this induced current produces its own magnetic field. In Fig. 21–2a the distance between the coil and the magnet decreases. So the magnetic field, and therefore the flux, through the coil increases. The magnetic field of the magnet points upward. To oppose this upward increase, the field produced by the induced current must point *downward.* Thus, Lenz's law tells us that the current must move as shown (use the right-hand rule). In Fig. 21–2b, the flux *decreases* (because the magnet is moved away), so the induced current produces an *upward* magnetic field that is "trying" to maintain the status quo. Thus the current must be as shown.

Lenz's law

† Note also that 1 T = 1 Wb/m^2; this is where the old SI unit for magnetic field, Wb/m^2, comes from. The magnetic field B, since it equals Φ_B/A, is sometimes called the **flux density**.

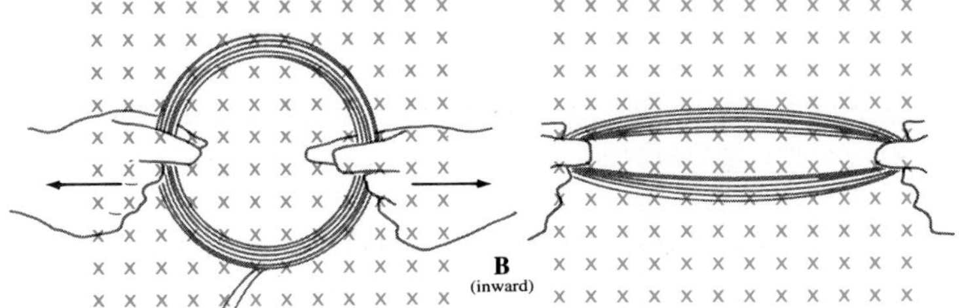

FIGURE 21–5
A current can be induced by changing the area of the coil. In both this case and that of Figure 21–6, the flux through the coil is reduced.

Let us consider what would happen if Lenz's law were not true, but were just the reverse. The induced current in this imaginary situation would produce a flux in the same direction as the original change. This greater change in flux would produce an even larger current followed by a still greater change in flux, and so on. The current would continue to grow indefinitely, producing power ($=I^2R$) even after the original stimulus ended. This would violate the conservation of energy. Such "perpetual motion" devices do not exist. Thus, Lenz's law as stated above (and not its opposite) is consistent with the law of conservation of energy.

It is important to note that an emf is induced whenever there is a change in flux. Since magnetic flux $\Phi_B = BA$, we see that an emf can be induced in two ways: (1) by a changing magnetic field B; or (2) by changing the area of the loop in the field or its orientation θ with respect to the field. Figure 21–1 and 21–2 illustrated case 1. Examples of case 2 are illustrated in Figs. 21–5 and 21–6, and in the following example.

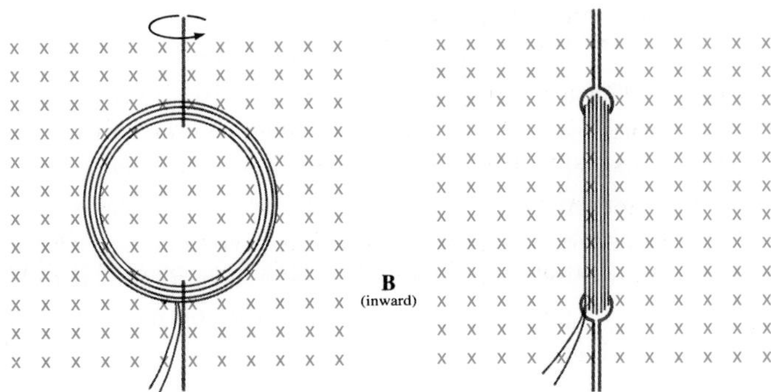

FIGURE 21–6
A current can be induced by rotating a coil in a magnetic field.

EXAMPLE 21–1 A square coil of side 5.0 cm contains 100 loops and is positioned perpendicular to a uniform 0.60-T magnetic field. It is quickly and uniformly pulled from the field (moving perpendicularly to **B**) to a region where B drops abruptly to zero (Fig. 21–7). It takes 0.10 s for the whole coil to reach the field-free region. How much energy is dissipated in the coil if its resistance is 100 Ω?

SOLUTION We need to find out how much the magnetic flux, $\Phi_B = BA$, changes during the time interval $\Delta t = 0.10$ s. The area of the coil is $A = (0.050\text{ m})^2 = 2.5 \times 10^{-3}\text{ m}^2$. The flux is initially $\Phi_B = BA = (0.60\text{ T})(2.5 \times 10^{-3}\text{ m}^2) = 1.5 \times 10^{-3}$ Wb. After 0.10 s, the flux is zero. The rate of change of flux is constant during the 0.10 s, so the emf induced (Eq. 21-2) during this period is

$$\mathscr{E} = -(100)\frac{(0 - 1.5 \times 10^{-3}\text{ Wb})}{(0.10\text{ s})} = 1.5\text{ V}.$$

The current $I = \mathscr{E}/R = 1.5\text{ V}/100\ \Omega = 15$ mA. The total energy dissipated is $I^2Rt = (1.5 \times 10^{-2}\text{ A})^2(100\ \Omega)(0.10\text{ s}) = 2.3 \times 10^{-3}$ J. From the conservation of energy principle, this is just equal to the work needed to pull the coil out of the field.

FIGURE 21–7 Example 21–1. The square coil in a magnetic field $B = 0.60$ T is pulled abruptly to the right to a region where $B = 0$.

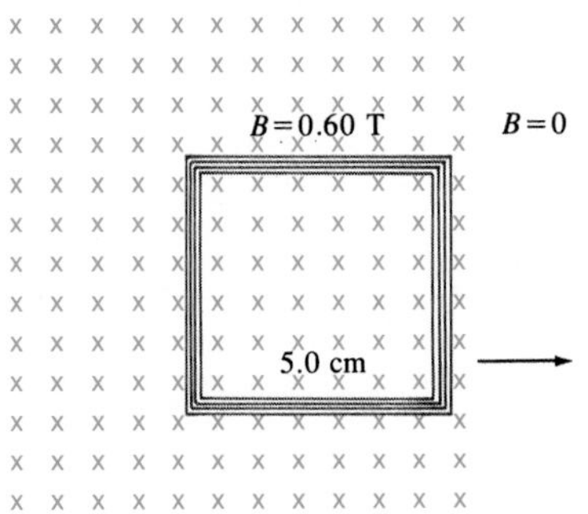

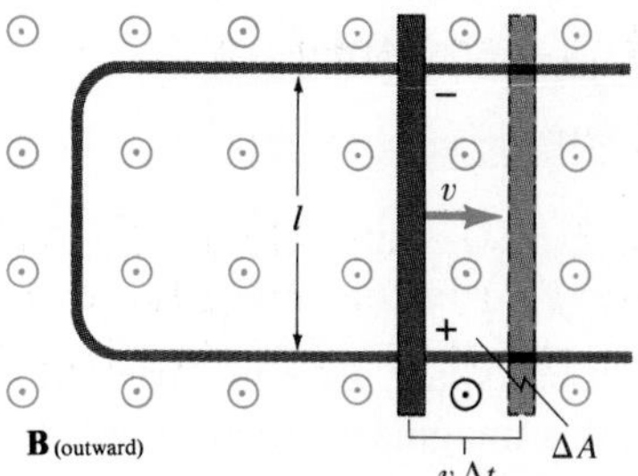

FIGURE 21–8 A conducting rod is moved to the right on a U-shaped conductor in a uniform magnetic field **B** that points out of the paper.

21–3 • EMF Induced in a Moving Conductor

Another way to induce an emf is shown in Fig. 21–8 and this situation helps illuminate the nature of the induced emf. Assume that a uniform magnetic field **B** is perpendicular to the area bounded by the U-shaped conductor and the movable rod resting on it. If the rod is made to move at a speed v, it travels a distance $\Delta x = v\,\Delta t$ in a time Δt. Therefore, the area of the loop increases by an amount $\Delta A = l\,\Delta x = lv\,\Delta t$ in a time Δt. By Faraday's law, there is an induced emf $\mathscr{E}$ whose magnitude is given by

Motional emf

$$\mathscr{E} = \frac{\Delta\Phi_B}{\Delta t} = \frac{B\,\Delta A}{\Delta t} = \frac{Blv\,\Delta t}{\Delta t} = Blv. \tag{21–3}$$

This equation is valid as long as B, l, and v are mutually perpendicular. (If they are not, we use only the components of each that are mutually perpendicular.) An emf induced in this way is sometimes called *motional emf*.

We can also obtain Eq. 21–3 without the use of Faraday's law. We saw in Chapter 20 that a charged particle moving perpendicular to a magnetic field B with speed v experiences a force $F = qvB$. When the rod of Fig. 21–8 moves to the right with speed v, the electrons in the rod move with this same speed. Therefore, each feels a force $F = qvB$, which acts upward in the figure. If the rod were not in contact with the U-shaped conductor, electrons would collect at the upper end of the rod, leaving the lower end positive. There must thus be an induced emf. If the rod does slide on the U-shaped conductor, the electrons will flow into it. There will then be a clockwise (conventional) current flowing in the loop. To calculate the emf, we determine the work W needed to move a charge q from one end of the rod to the other against this potential difference: W = force × distance = $(qvB)(l)$. The emf equals the work done per unit charge, so $\mathscr{E} = qvBl/q = Blv$, just as above.†

EXAMPLE 21–2 An airplane travels 1000 km/h in a region where the earth's field is 5.0×10^{-5} T and is nearly vertical. What is the potential difference induced between the wing tips that are 70 m apart?

SOLUTION Since $v = 1000$ km/h $= 280$ m/s, and $\mathbf{v} \perp \mathbf{B}$, we have $\mathscr{E} = Blv = (5.0 \times 10^{-5}\text{ T})(70\text{ m})(280\text{ m/s}) = 1.0$ V. Not much to worry about.

† This argument, which is basically the same as for the Hall effect, explains this one way of inducing an emf. It does not explain the general case of electromagnetic induction, however.

21–4 • Changing Magnetic Flux Produces an Electric Field

As we just discussed, the electrons in the moving conductor of Fig. 21–8 feel a force. This implies that there is an electric field in the conductor. Since electric field is defined as the force per unit charge, $E = F/q$, the effective field E in the rod must be (since $F = qvB$)

$$E = \frac{F}{q} = \frac{qvB}{q} = vB. \tag{21–4}$$

In the situation in which a changing magnetic field (rather than a moving conductor) induces an emf (as, for example, in Fig. 21–2), there also is an induced current. And again this implies that there is an electric field in the wire. Thus we come to the important conclusion that

a changing magnetic flux produces an electric field.

Electric field is produced by a changing magnetic flux

This applies not only to wires and other conductors, but is a general result that applies to any region in space: an electric field will be produced at any point in space where there is a changing magnetic field.

EXAMPLE 21–3 *Electromagnetic blood-flow measurement.* The rate of blood flow can be measured using the apparatus shown in Fig. 21–9 since blood contains charged ions. Suppose that the blood vessel is 2.0 mm in diameter, the magnetic field is 0.080 T, and the measured emf is 0.10 mV. What is the flow velocity of the blood?

SOLUTION We solve for v in Eq. 21–3, and find that $v = \mathscr{E}/Bl = (1.0 \times 10^{-4}\text{ V})/(0.080\text{ T})(2.0 \times 10^{-3}\text{ m}) = 0.63\text{ m/s}$. (In actual practice, an alternating current is used to produce an alternating magnetic field. The induced emf is then alternating.)

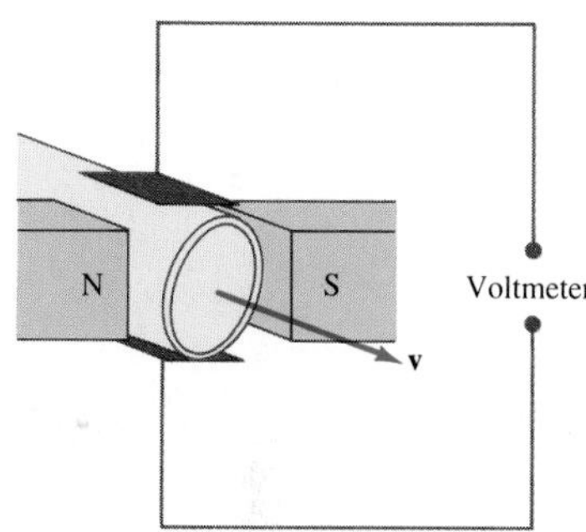

FIGURE 21–9 Measurement of blood velocity from the induced emf.

21–5 • Electric Generators

Probably the most important practical result of Faraday's great discovery was the development of the **electric generator** or **dynamo**. A generator transforms mechanical energy into electric energy. This is just the opposite of what a motor does. Indeed, a generator is basically the inverse of a motor.† A simplified diagram of an ac generator is shown in Fig. 21–10. A generator consists of many coils of wire (only one is shown) wound on an armature that can rotate in a magnetic field. The axle is turned by some mechanical means and an emf is induced in the rotating coil. An electric current is thus the *output* of a generator. In Fig. 21–10 the right-hand rule tells us that, with the armature rotating counterclockwise, the (conventional) current in the wire, labeled L on the armature, is outward; therefore it is outward at

FIGURE 21–10 An ac generator.

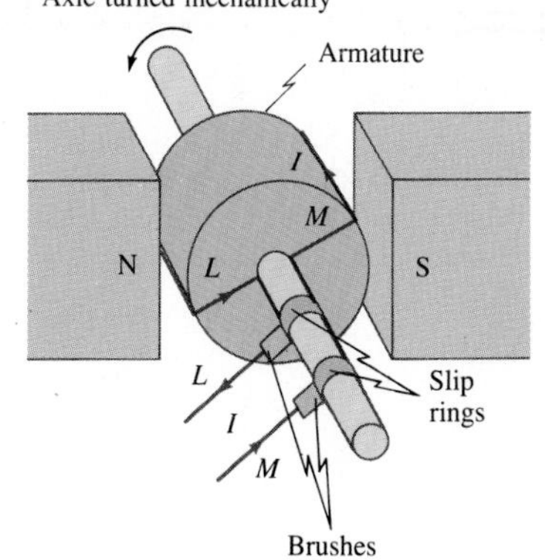

† You can, for example, actually run a car generator backward as a motor by connecting its output terminals to a battery.

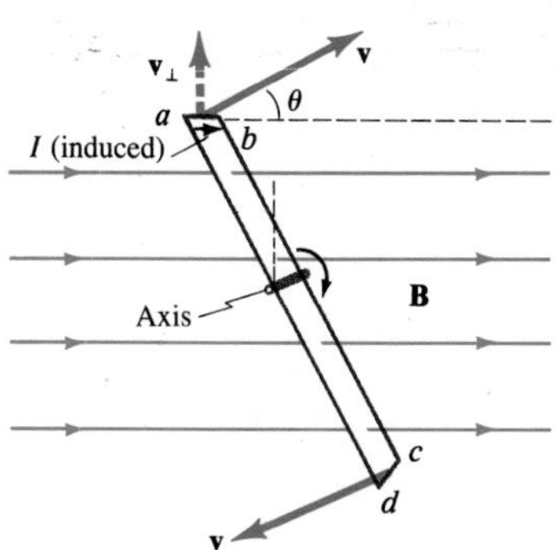

FIGURE 21–11 The emf is induced in the segments *ab* and *cd*, whose velocity components perpendicular to the field **B** are $v \sin \theta$.

brush *L*. (The brushes press against continuous slip rings.) After one-half revolution, wire *L* will be where wire *M* is now in the drawing, and the current then at brush *L* will be inward. Thus the current produced is alternating. Let us look at this in more detail.

In Fig. 21–11, the loop is being made to rotate clockwise in a uniform magnetic field **B**. The velocity of the two lengths *ab* and *cd* at this instant are shown. Although the sections of wire *bc* and *da* are moving, the force on electrons in these sections is toward the side of the wire, not along its length. The emf generated is then due only to the force on charges in the sections *ab* and *cd*. From the right-hand rule, we see that the direction of the induced current in *ab* is from *a* toward *b*. And in the lower section, it is from *c* to *d*; so the flow is continuous in the loop. The magnitude of the emf generated in *ab* is given by Eq. 21–3, except that we must take the component of the velocity perpendicular to *B*:

$$\mathscr{E} = Blv_\perp,$$

where l is the length of *ab*. From the diagram we can see that $v_\perp = v \sin \theta$, where θ is the angle the face of the loop makes with the vertical. The emf induced in *cd* has the same magnitude and is in the same direction. Therefore, they add and the total emf is

$$\mathscr{E} = 2NBlv \sin \theta,$$

where we have multiplied by N, the number of loops in the coil (if there is more than one). If the coil is rotating with constant angular velocity ω, then the angle $\theta = \omega t$. We also have from the angular equations (Chapter 8) that $v = \omega r = \omega(h/2)$, where h is the length of *bc* or *ad*. Thus $\mathscr{E} = 2NlB\omega(h/2) \cdot \sin \omega t$, or

Output emf of a generator

$$\mathscr{E} = NAB\omega \sin \omega t, \tag{21–5}$$

where $A = lh$ is the area of the loop. This equation holds for any shape coil, not just for a rectangle as derived. Thus, the output emf of the generator is sinusoidally alternating (Fig. 21–12 and Section 18–7). Since ω is expressed in radians per second, we can write $\omega = 2\pi f$, where f is the frequency.

Over 99 percent of the electricity used in the United States is produced from generators. The frequency f is 60 Hz for general use in the United States and Canada, although 50 Hz is used in many countries. In electric power generating plants, the armature is mounted on a heavy axle connected to a turbine, which is the modern equivalent of a waterwheel. Water falling over a dam can turn the turbine at a hydroelectric plant. Most of the power generated at present in the United States, however, is done at steam plants,

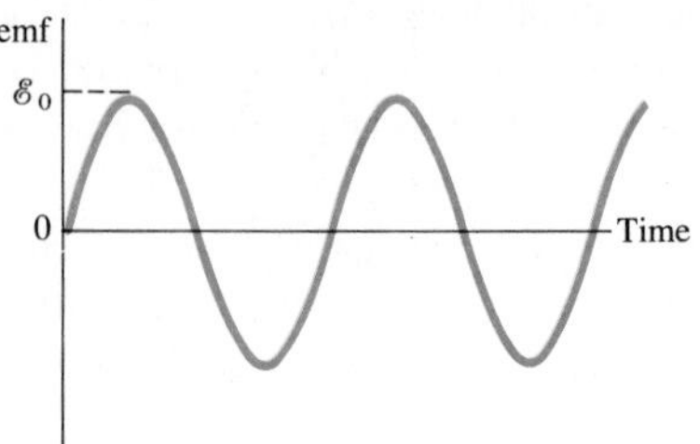

FIGURE 21–12 An ac generator produces an alternating current. The output emf $\mathscr{E} = \mathscr{E}_0 \sin \omega t$, where $\mathscr{E}_0 = \text{NAB}\,\omega$ (Eq. 21–5).

where the burning of fossil fuels (coal, oil, natural gas) boils water to produce high-pressure steam that turns the turbines. Likewise, at nuclear power plants, the released nuclear energy is used to produce steam to turn turbines. Thus, a heat engine (Chapter 15) connected to a generator is the principal means of generating electric power.

EXAMPLE 21–4 The armature of a 60-Hz ac generator rotates in a 0.15-T magnetic field. If the area of the coil is $2.0 \times 10^{-2}\ \mathrm{m}^2$, how many loops must the coil contain if the peak output is to be $\mathscr{E}_0 = 170$ V?

SOLUTION From Eq. 21–5, we see that the maximum emf is $\mathscr{E}_0 = NAB\omega$. Since $\omega = 2\pi f = (6.28)(60\ \mathrm{s}^{-1}) = 377\ \mathrm{s}^{-1}$, we have

$$N = \frac{\mathscr{E}_0}{AB\omega} = \frac{170\ \mathrm{V}}{(2.0 \times 10^{-2}\ \mathrm{m}^2)(0.15\ \mathrm{T})(377\ \mathrm{s}^{-1})} = 150\ \text{turns}.$$

A dc generator is much like an ac generator, except the slip rings are replaced by split-ring commutators, Fig. 21–13a, just as in a dc motor. The output of such a generator is as shown and can be smoothed out by placing a capacitor in parallel with the output. More common is the use of many armature windings, as in Fig. 21–13b, which produces a smoother output.

In the past, automobiles used dc generators. More common now, however, are ac generators or **alternators** (Fig. 21–14). The ac output is changed to dc for charging the battery with the use of diodes (Chapter 29). Most alternators differ from the generator discussed above in that the magnetic field is made to rotate within a stationary armature called a *stator*. The brushes press against a continuous ring instead of a slotted commutator and are on the input instead of the output. Thus large output currents are carried in solid conductors rather than through sliding-ring commutators which are subject to wear and arcing. In a car, the rotation speed can be faster so the battery can be charged even at idling speed and there will still be no problem with electrical arcing across the rings when the car travels at high speeds.

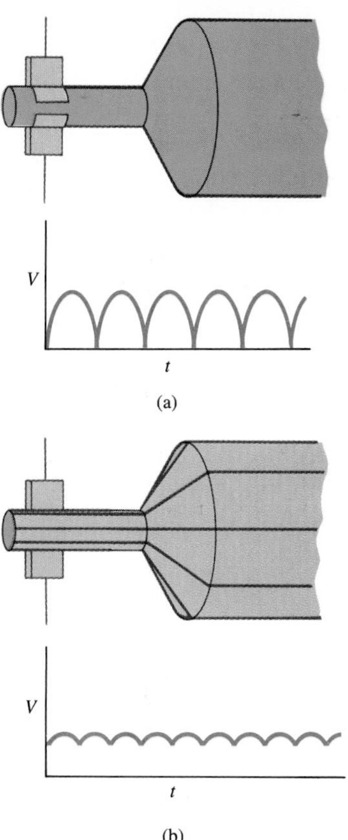

FIGURE 21–13 DC generator (a) with one set of commutators, and (b) with many sets of commutators and windings.

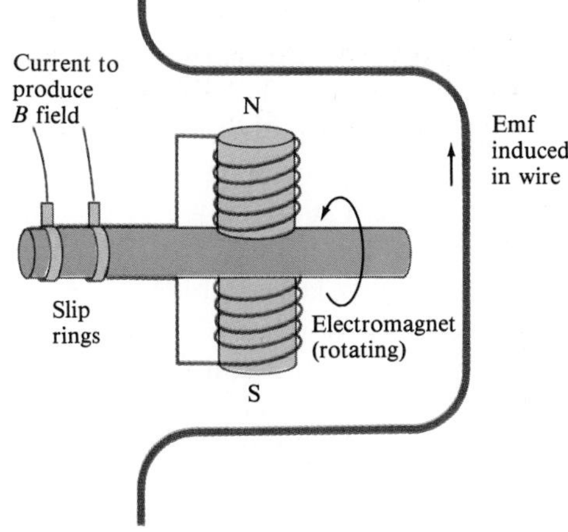

FIGURE 21–14
Simplified diagram of an alternator. The input electromagnet current is connected through continuous slip rings which is preferable to the output of a dc generator connected through many sets of split commutators (Fig. 21–13b), where poor connection or arcing can occur at high speeds. Sometimes the electromagnet is replaced by a permanent magnet.

21–6 • Counter EMF and Torque; Eddy Currents

A motor turns and produces mechanical energy when a current is made to flow in it. From our description in Sec. 20–8 of a simple dc motor, you might expect that the armature would accelerate indefinitely due to the torque on it. However, as the armature of the motor turns, the magnetic flux through the coil changes and an emf is generated. This induced emf acts to oppose the motion (Lenz's law) and is called the **back emf** or **counter emf**. The greater the speed of the motor, the greater the counter emf. A motor normally turns something, but if there were no load, its speed would increase until the counter emf equaled the input voltage. In the normal situation, when there is a mechanical load, the speed of the motor is limited also by the load. The counter emf will then be less than the external voltage. The greater the load, the slower the motor rotates and the lower is the counter emf.

Back emf

EXAMPLE 21–5 The armature windings of a dc motor have a resistance of 5.0 Ω. The motor is connected to a 120-V line and when the motor reaches full speed against its normal load, the counter emf is 108 V. Calculate (*a*) the current into the motor when it is just starting up, and (*b*) the current when it reaches full speed.

SOLUTION (*a*) Initially, the motor is not turning (or turning very slowly), so there is no induced counter emf. Hence, from Ohm's law, the current is

$$I = \frac{V}{R} = \frac{120 \text{ V}}{5.0 \, \Omega} = 24 \text{ A}.$$

(*b*) At full speed, the counter emf is a source of emf that opposes the exterior source. We represent this counter emf as a battery in the equivalent circuit shown in Fig. 21–15. In this case, Ohm's law (or Kirchhoff's rule) gives

$$120 \text{ V} - 108 \text{ V} = I\,(5.0 \, \Omega).$$

Therefore

$$I = 12 \text{ V}/5.0 \, \Omega = 2.4 \text{ A}.$$

FIGURE 21–15 Circuit of a motor showing induced counter emf.

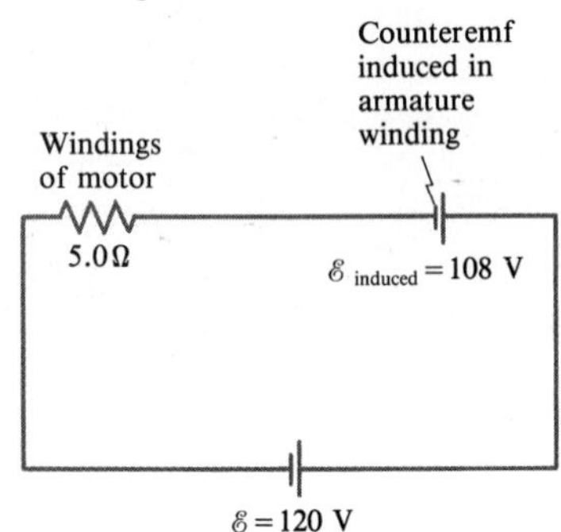

This example illustrates the fact that the current is very high when a motor first starts up. This is why the lights in your house may dim when the motor of the refrigerator (or other large motor) starts up. The large initial current causes the voltage at the outlets to drop (the house wiring has resistance, so there is some voltage drop across it when large currents are drawn). If a motor is overloaded so it turns only at a slow speed, the counter emf will be reduced because of the reduced rotation speed. The current in the motor may then be sufficiently large to burn it out.

In a generator, the situation is the reverse of that for a motor. As we saw, the mechanical turning of the armature induces an emf in the loops, which is the output. If the generator is not connected to an external circuit, the emf exists at the terminals but no current flows. In this case, it takes little effort to turn the armature. But if the generator *is* connected to a device that draws current, then a current flows in the coils of the armature. Because this current-carrying coil is in a magnetic field, there will be a torque exerted on it (as in a motor), and this torque opposes the motion (use the right-hand rule for the force on a wire, in Fig. 21–10). This is called a **counter torque**. The greater the load—that is, the more current that is drawn—the greater will be the counter torque. Hence the external applied torque will have to be greater to keep the generator turning. This of course makes sense from the conservation-of-energy principle. More mechanical-energy input is needed to produce more electrical-energy output.

Counter torque

Induced currents are not always confined to well-defined paths such as in wires. Consider, for example, the rotating metal wheel in Fig. 21–16a. A magnetic field is applied to a limited area as shown and points into the paper. The section of wheel in the magnetic field has an emf induced in it since the conductor is moving (carrying electrons with it). The flow of (conventional) current is upward (Fig. 21–16b) and it follows a downward return path outside the region of the magnetic field. These currents are referred to as **eddy currents** and can be present in any conductor that is moving across a magnetic field or through which the magnetic flux is changing. In Fig. 21–16, the magnetic field exerts a force on the induced currents that opposes (use the right-hand rule) the rotational motion. Eddy currents can be used in this way as a smooth braking device on, say, a rapid-transit car. In order to stop the car, an electromagnet can be turned on that applies its field either to the wheels or to the moving steel rail below. Eddy currents can dampen (reduce) the oscillation of a vibrating system. A common example is in a galvanometer, where induced eddy currents keep the needle from overshooting or oscillating violently. Eddy currents, however, can be a problem. For example, eddy currents induced in the armature of a motor or generator produce heat ($P = I\mathscr{E}$) and waste energy. To reduce the eddy currents, the armatures are *laminated*; that is, they are made of very thin sheets of iron that are well insulated from one another. (See Fig. 21–18 in the next section.) Thus the total path length of the eddy currents is confined to each slab, which increases the total resistance. Hence the current is less and there is less wasted energy.

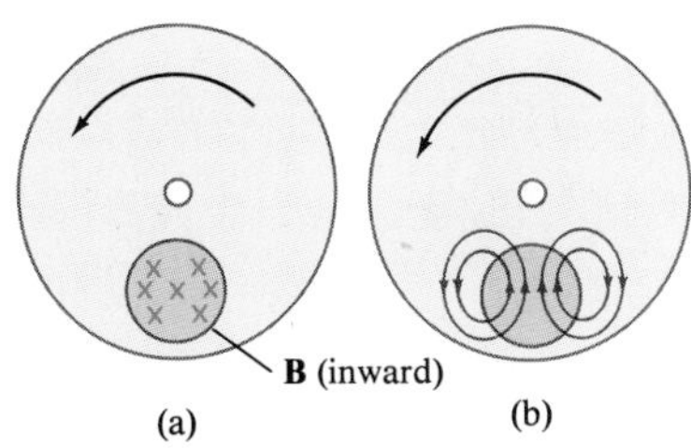

FIGURE 21–16 Production of eddy currents in a rotating wheel.

Eddy currents

FIGURE 21–17 Repairing a transformer on a utility pole.

21–7 • Transformers; Transmission of Power

A transformer is a device for increasing or decreasing an ac voltage. Transformers are found in TV sets to give the high voltage needed for the picture tube, in converters for plugging in a portable "Walkman," on utility poles (Fig. 21–17) to reduce the high voltage from the electric company to that usable in houses (110 V or 220 V), and in many other applications. A **transformer** consists of two coils of wire known as the **primary** and **secondary** coils. The two coils can be interwoven; or they can be linked by a soft iron core (laminated to prevent eddy-current losses), Fig. 21–18. Transformers are designed so that (nearly) all the flux produced by the current in the primary also passes through the secondary coil, and we assume this is true in what follows. We also assume that energy losses in the resistance of the coils and hysteresis in the iron can be ignored—a good approximation for real transformers, which are often better than 99 percent efficient.

When an ac voltage is applied to the primary, the changing magnetic field it produces will induce an ac voltage of the same frequency in the secondary. However, the voltage will be different according to the number

FIGURE 21–18 (below) Step-up transformer ($N_p = 4$, $N_s = 12$).

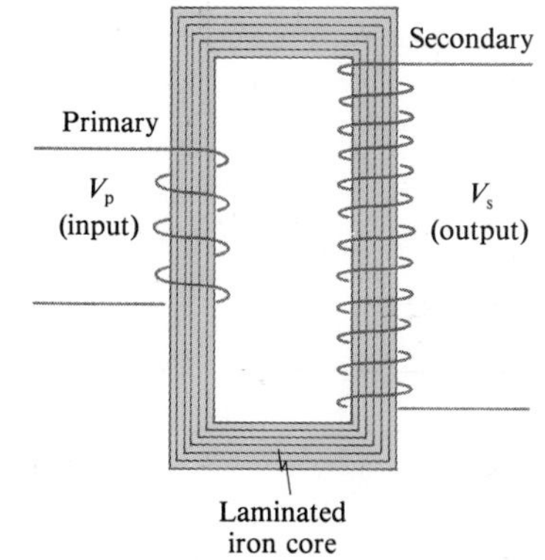

of loops in each coil. From Faraday's law, the voltage or emf induced in the secondary is

$$V_s = N_s \frac{\Delta \Phi_B}{\Delta t},$$

where N_s is the number of turns in the secondary coil, and $\Delta\Phi_B/\Delta t$ is the rate at which the magnetic flux changes. The input primary voltage, V_p, is also related to the rate at which the flux changes

$$V_p = N_p \frac{\Delta \Phi_B}{\Delta t},$$

where N_p is the number of turns in the primary coil.† We divide these two equations to find

Transformer equation

$$\frac{V_s}{V_p} = \frac{N_s}{N_p}. \tag{21–6}$$

This *transformer equation* tells how the secondary (output) voltage is related to the primary (input) voltage; V_s and V_p in Eq. 21–6 can be the rms values for both, or peak values for both.

If N_s is greater than N_p, we have a **step-up transformer**. The secondary voltage is greater than the primary voltage. For example, if the secondary has twice as many turns as the primary, then the secondary voltage will be twice that of the primary. If N_s is less than N_p, we have a **step-down transformer**.

Although voltage can be increased (or decreased) with a transformer, we don't get something for nothing. Energy conservation tells us that the power output can be no greater than the power input. A well-designed transformer can be greater than 99 percent efficient, so little energy is lost to heat. The power input thus essentially equals the power output. Since power $P = VI$ (Eq. 18-5), we have

$$V_p I_p = V_s I_s,$$

or

$$\frac{I_s}{I_p} = \frac{N_p}{N_s}. \tag{21–7}$$

EXAMPLE 21–6 A transformer for a transistor radio reduces 120-V ac to 9.0-V ac. (Such a device also contains diodes to change the 9.0-V ac to dc. See Chapter 29.) The secondary contains 30 turns and the radio draws 400 mA. Calculate: (*a*) the number of turns in the primary; (*b*) the current in the primary; and (*c*) the power transformed.

SOLUTION (*a*) This is a step-down transformer, and from Eq. 21–6 we have

$$N_p = N_s \frac{V_p}{V_s} = \frac{(30)(120\ \text{V})}{(9.0\ \text{V})} = 400\ \text{turns}.$$

† This follows because the changing flux produces a counter emf, $N_p\,\Delta\Phi_B/\Delta t$ in the primary that exactly balances the applied voltage V_p if the resistance of the primary can be ignored (Kirchhoff's rules).

(*b*) From Eq. 21–7:

$$I_{\rm p} = I_{\rm s}\frac{N_{\rm s}}{N_{\rm p}} = (0.40\text{ A})\left(\frac{30}{400}\right) = 0.030\text{ A}.$$

(*c*) The power transformed is

$$P = I_{\rm s}V_{\rm s} = (9.0\text{ V})(0.40\text{ A}) = 3.6\text{ W},$$

which is, assuming 100 percent efficiency, the same as the power in the primary, $P = (120\text{ V})(0.030\text{ A}) = 3.6\text{ W}$.

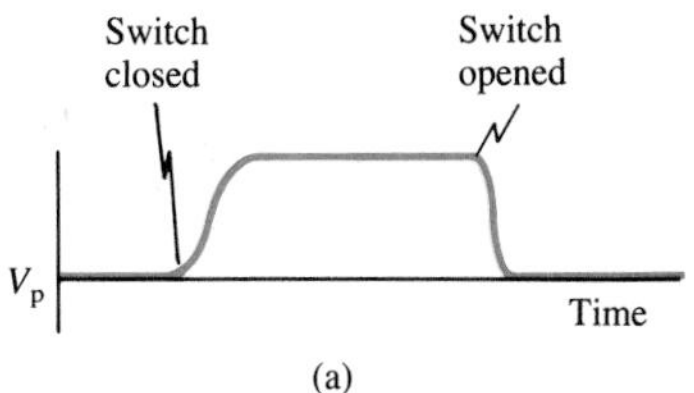

(a)

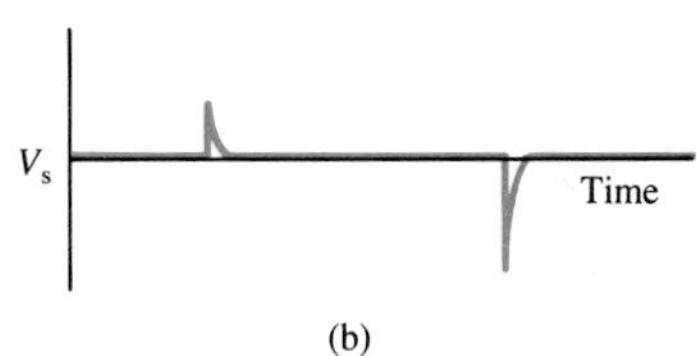

(b)

FIGURE 21–19 A dc voltage turned on and off as shown in (a) produces voltage pulses in the secondary (b). Voltage scales in (a) and (b) are not necessarily the same.

It is important to recognize that a transformer operates only on ac. A dc current in the primary does not produce a changing flux and therefore induces no emf in the secondary. However, if a dc voltage is applied to the primary through a switch, at the instant the switch is opened or closed there will be an induced current in the secondary. For example, if the dc is turned on and off as shown in Fig. 21–19a, the voltage induced in the secondary is as shown in Fig. 21–19b. Notice that the secondary voltage drops to zero when the dc voltage is steady.

Power transmission

Transformers play an important role in the transmission of electricity. Power plants are often situated some distance from metropolitan areas. Hydroelectric plants are located at a dam site and nuclear plants need much cooling water. Fossil-fuel plants too are often situated far from a city because of lack of availability of land or to avoid contributing to air pollution. In any case, electricity must often be transmitted over long distances, and there is always some power loss in the transmission lines. This loss can be minimized if the power is transmitted at high voltage, as the following example shows.

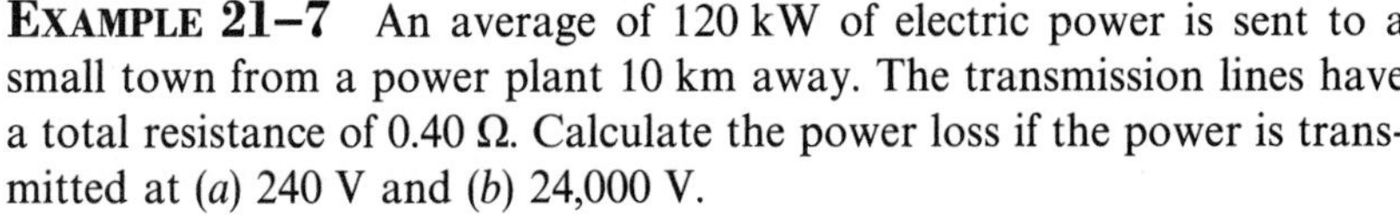

EXAMPLE 21–7 An average of 120 kW of electric power is sent to a small town from a power plant 10 km away. The transmission lines have a total resistance of 0.40 Ω. Calculate the power loss if the power is transmitted at (*a*) 240 V and (*b*) 24,000 V.

SOLUTION For each case we determine the current I in the lines, and then find the power loss from $P = I^2R$. (*a*) If 120 kW is sent at 240 V, the total current will be

$$I = \frac{1.2\times10^5\text{ W}}{2.4\times10^2\text{ V}} = 500\text{ A}.$$

The power loss in the lines, $P_{\rm L}$, is then

$$P_{\rm L} = I^2R = (500\text{ A})^2(0.40\ \Omega) = 100\text{ kW}.$$

Thus, over 80 percent of all the power would be wasted as heat in the power lines!

(*b*) When $V = 24{,}000$ V,

$$I = \frac{P}{V} = \frac{1.2\times10^5\text{ W}}{2.4\times10^4\text{ V}} = 5.0\text{ A}.$$

The power loss is then

$$P_L = I^2R = (5.0\ \text{A})^2(0.40\ \Omega) = 10\ \text{W},$$

which is less than 1/100 of 1 percent.

It should be clear that the greater the voltage, the less the current and thus the less power is wasted in the transmission lines. It is for this reason that power is usually transmitted at very high voltages, as high as 700 kV.

Power is generated at somewhat lower voltages than this, and the voltage in homes and factories is also much lower. The great advantage of ac, and a major reason it is in nearly universal use, is that the voltage can easily be stepped up and down by a transformer. The output voltage of an electric generating plant is stepped up prior to transmission. Upon arrival in a city, it is stepped down in stages at electric substations prior to distribution. The voltage in lines along city streets is typically 2400 V and is stepped down to 240 V or 120 V for home use by transformers (Fig. 21–17).

DC transmission has gained in popularity recently. Although changing voltage with dc is more difficult and expensive, it offers some advantages over ac. A few of these are as follows. AC produces alternating magnetic fields which induce current in nearby wires and reduce transmitted power; this is absent in dc. DC can be transmitted at a higher average voltage than ac since for dc, the rms value equals the peak; and breakdown of insulation or of air is determined by the peak voltage.

*21–8 • Applications of Induction: Magnetic Microphone, Seismograph, Recording Heads and Computers

Microphones

Many microphones work on the principle of induction. In one form, a microphone is just the inverse of a loudspeaker (Section 20–8). A small coil connected to a membrane is suspended close to a small permanent magnet. The coil moves in the magnetic field when sound waves strike the membrane. The frequency of the induced emf will be just that of the impinging sound waves. In a "ribbon" microphone, a thin metal ribbon is suspended between the poles of a permanent magnet. The ribbon vibrates in response to sound waves and the emf induced in the ribbon is proportional to its velocity.

Seismograph

In another field, geophysics, an important device based on electromagnetic induction is one type of *seismograph* or *geophone*. A seismograph is placed in direct contact with the earth and converts the motion of the earth—whether due to an earthquake or to an explosion (such as for mineral prospecting or for detecting a bomb test)—into an electrical signal. A seismograph contains a magnet and a coil of wire, one of which is fixed rigidly to the case, which moves as the earth does where it is planted. The other element is inertial and is suspended from the case by a spring. In the type shown in Fig. 21–20, the coil moves with the earth, and the relative motion of the magnet and coil produces an induced emf in the coil, which is the output of the device. In many geophones, the coil is inertial and the magnet moves with the earth.

FIGURE 21–20 A seismograph or geophone.

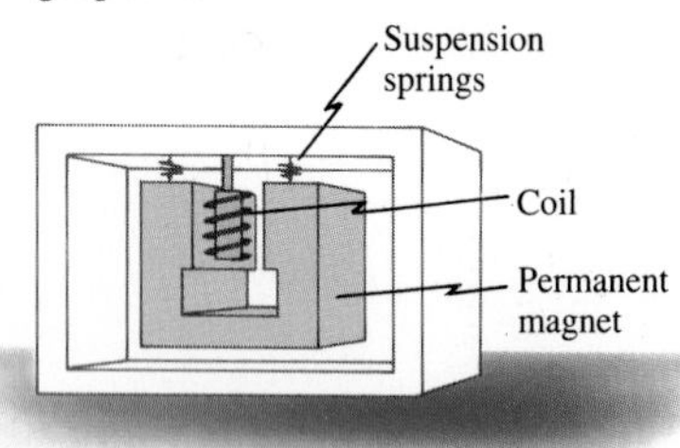

Other common magnetic devices are the *heads* that "read" and "write" on magnetic tapes and disks. Recording tape for use in audio and video tape recorders contains a thin layer of magnetic oxide on a thin plastic tape. During recording, the audio or video signal voltage is sent to the recording head, which acts as a tiny electromagnet (Fig. 21–21) that magnetizes the tiny section of tape passing over the narrow gap in the head at each instant. In playback, the changing magnetism of the moving tape at the gap causes corresponding changes in the magnetic field within the soft-iron head, which in turn induces an emf in the coil (Faraday's law). This induced emf is the output signal that can be amplified and sent to a loudspeaker (or, in the case of a video signal, to a CRT). In audio and video recorders, the signals are usually *analog*—they vary continuously in amplitude over time. The variation in degree of magnetization of the tape at any point reflects the variation in amplitude of the audio or video signal.

Tape heads

Digital information, such as used in the process of reading and writing on computer disks (floppy disks or hard disks) as well as on magnetic computer tape, is managed using heads that are basically the same as just described (Fig. 21–21). The essential difference is that computer signals are not analog, but are digital, and in particular binary, meaning that only two values are possible. The two possible values are usually referred to as 1 and 0. The signal voltage does not vary continuously but rather takes on only two values, say +5 volts and 0 volts, corresponding to the 1 or 0. Thus, information is carried as a series of "bits," each of which can have only one of two values, 1 or 0. We will not go into the details of how computers deal with information coded in this binary fashion. We note only that the magnetism on the storage disk or tape is also digital, and generally is highly saturated (Section 20–14) in one direction, and either not magnetized or highly saturated in the opposite direction, to represent the two states (0 and 1). Thus a magnetic tape, or each of the concentric tracks on a magnetic disk, is a sequence of tiny magnetized spots to represent the sequence of bits. The writing and reading of a disk or tape depends on changing magnetic fields and emfs in the head, as described above for analog recording, Fig. 21–21.

Digital information and computers

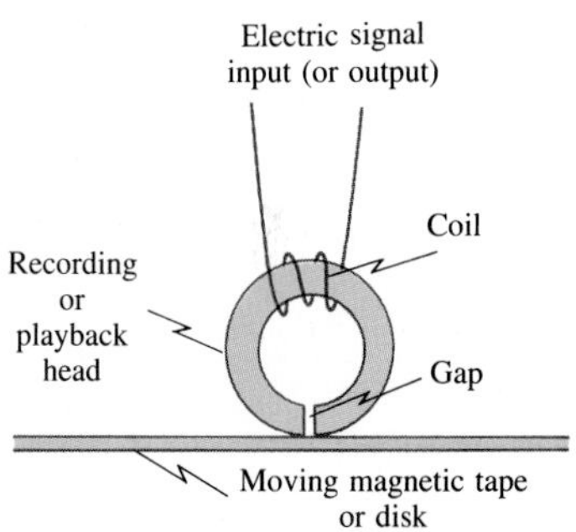

FIGURE 21–21 Recording and/or playback head for tape or disk. In recording (or "writing"), electric input signal to the head, which acts as an electromagnet, magnetizes the passing tape or disk. In playback (or "reading"), the changing magnetic field of the passing tape or disk induces a changing magnetic field in the head, which in turn induces in the coil an emf that is the output signal.

*21–9 • Inductance

Mutual inductance. If two coils of wire are placed near one another, as in Fig. 21–22, a changing current in one will induce an emf in the other. According to Faraday's law, the emf $\mathscr{E}_2$ induced in coil 2 is proportional to the rate of change of flux passing through it. Since the flux is proportional to the current flowing in coil 1, $\mathscr{E}_2$ must be proportional to the rate of change of the current in coil 1, $\Delta I_1/\Delta t$. Thus we can write

$$\mathscr{E}_2 = -M\frac{\Delta I_1}{\Delta t}, \qquad (21\text{–}8a)$$

where the constant of proportionality, M, is called the **mutual inductance**. (The minus sign is simply a reflection of Lenz's law.) Mutual inductance has units of $\mathrm{V \cdot s/A} = \Omega \cdot \mathrm{s}$, which is called the **henry** (H): $1\ \mathrm{H} = 1\ \Omega \cdot \mathrm{s}$. The value of M depends on whether iron is present or not, on the size of the coils, on the number of turns, and on their separation. For example, the closer the two coils are to each other in Fig. 21–22, the more lines of flux will pass through coil 2, so M will be greater. In some arrangements, M can be calculated. More usually it is measured experimentally. If we look at the inverse

FIGURE 21–22 A changing current in one coil will induce a current in the second coil.

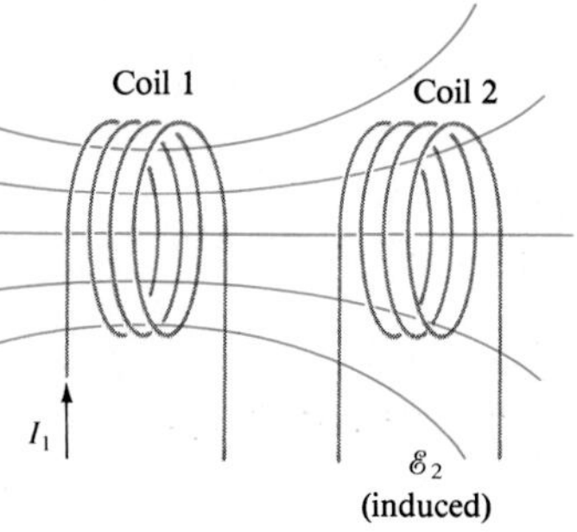

situation—a changing current in coil 2 inducing an emf in coil 1—the proportionality constant, M, turns out to have the same value,

$$\mathscr{E}_1 = -M\frac{\Delta I_2}{\Delta t}. \tag{21–8b}$$

A transformer is an example of mutual inductance in which the coupling is maximized so that nearly all flux lines pass through both coils. However, mutual inductance has other uses as well. For example, some pacemakers, which are used to maintain blood flow in heart patients (Section 19–8), are powered externally. Power in an external coil is transmitted via mutual inductance to a second coil in the pacemaker at the heart. This has the advantage over battery-powered pacemakers in that surgery is not needed to replace a battery when it wears out.

Mutual inductance can sometimes be a problem, however. Any alternating current in a circuit can induce an emf in another part of the same circuit or in a different circuit even though the conductors are not in the shape of a coil. The mutual inductance M is usually small unless multiturn coils and/or iron cores are involved. However, in situations where small signals are present, problems due to mutual inductance often arise. Shielded cable, in which an inner conductor is surrounded by a cylindrical grounded conductor, is often used to reduce the problem.

Self-inductance. The concept of inductance applies also to an isolated single coil. When a changing current passes through a coil or solenoid, a changing magnetic flux is produced inside the coil, and this in turn induces an emf. This induced emf opposes the change in flux (Lenz's law); it is much like the back emf generated in a motor. For example, if the current through the coil is increasing, the increasing magnetic flux induces an emf that opposes the original current and tends to retard its increase. If the current is decreasing in the coil, the decreasing flux induces an emf in the same direction as the current, tending to maintain the original current. In either case, the induced emf $\mathscr{E}$ is proportional to the rate of change in current (and is in the direction opposed to the change):

Self-inductance

$$\mathscr{E} = -L\frac{\Delta I}{\Delta t}. \tag{21–9}$$

The constant of proportionality L is called the **self-inductance**, or simply the **inductance** of the coil. It, too, is measured in henries. The magnitude of L depends on the geometry and on the presence of an iron core.

Inductors

An ac circuit always contains some inductance, but often it is quite small unless the circuit contains a coil of many turns. A coil that has significant self-inductance L is called an **inductor** or a **choke coil**. It is shown on circuit diagrams by the symbol -⌇⌇⌇⌇-. It can serve a useful purpose in certain circuits. Often, inductance is to be avoided in a circuit. Precision resistors are normally wire-wound and thus would have inductance as well as resistance. The inductance can be minimized by winding the wire back on itself so that the current going in the two directions cancels and little magnetic flux is produced; this is called a "noninductive winding."

If an inductor has negligible resistance, it is the inductance, or the back emf, that controls the current. If a source of alternating voltage is applied to the coil, this applied voltage will just be balanced by the induced emf of the coil given by Eq. 21–9. Thus we can see from Eq. 21–9 that, for a given $\mathscr{E}$, if the inductance L is large, the change in the current—and therefore the current itself—will be small. The greater the inductance, the less the current. An inductance thus acts something like a resistance to impede the flow of alternating current. We use the term **impedance** for this quality of an inductor. We shall discuss impedance more fully in Sections 21–12 to 15, and we shall see that it depends not only on L, but also on the frequency. Here we mention one example of its importance. The resistance of the primary in a transformer is usually quite small, perhaps less than 1 Ω. If resistance alone limited the current, tremendous currents would flow when a high voltage was applied. Indeed, a dc voltage applied to a transformer can burn it out. It is the impedance of the coil to an alternating current (or its "back" emf) that limits the current to a reasonable value.

DC can burn out a transformer

EXAMPLE 21–8 (*a*) Determine a formula for the self-inductance L of a solenoid (a long coil) containing N turns of wire in its length l and whose cross-sectional area is A. (*b*) Calculate the value of L if $N = 100$, $l = 5.0$ cm, $A = 0.30\ \text{cm}^2$, and the solenoid is air-filled. (*c*) Calculate L if the solenoid has an iron core with $\mu = 4000\mu_0$.

SOLUTION (*a*) According to Eq. 20–10, the magnetic field inside a solenoid is $B = \mu_0 nI$, where $n = N/l$. From Eqs. 21–2 and 21–9, we have $\mathscr{E} = -N(\Delta\Phi_B/\Delta t) = -L(\Delta I/\Delta t)$. Thus, $L = N(\Delta\Phi_B/\Delta I)$. Since $\Phi_B = BA = \mu_0 NIA/l$, then any change in I causes a change in flux $\Delta\Phi_B = \mu_0 NA\,\Delta I/l$. Thus

$$L = N\frac{\Delta\Phi_B}{\Delta I} = \frac{\mu_0 N^2 A}{l}.$$

(*b*) Since $\mu_0 = 4\pi \times 10^{-7}\ \text{T}\cdot\text{m/A}$,

$$L = \frac{(4\pi \times 10^{-7}\ \text{T}\cdot\text{m/A})(100)^2(3.0 \times 10^{-5}\ \text{m}^2)}{(5.0 \times 10^{-2}\ \text{m})} = 7.5\ \mu\text{H}.$$

(*c*) Here we replace μ_0 by $\mu = 4000\mu_0$, so L will be 4000 times larger: $L = 0.030\ \text{H} = 30\ \text{mH}$.

*21–10 • Energy Stored in a Magnetic Field

In Section 17–9 we saw that the energy stored in a capacitor is equal to $\frac{1}{2}CV^2$. By using a similar argument, we can show that the energy stored in an inductance L, carrying a current I, is

$$\text{energy} = \tfrac{1}{2}LI^2.$$

Just as the energy stored in a capacitor can be considered to reside in the electric field between its plates, so the energy in an inductor can be considered to be stored in its magnetic field.

To write the energy in terms of the magnetic field, let us use the result of Example 21–8 that the inductance of a solenoid is $L = \mu_0 N^2 A/l$. Now the magnetic field B in a solenoid is related to the current I (see Eq. 20–10) by $B = \mu_0 NI/l$. Thus, $I = Bl/\mu_0 N$, and

$$\text{energy} = \frac{1}{2}LI^2 = \frac{1}{2}\left(\frac{\mu_0 N^2 A}{l}\right)\left(\frac{Bl}{\mu_0 N}\right)^2$$

$$= \frac{1}{2}\frac{B^2}{\mu_0}Al.$$

We can think of this energy as residing in the volume enclosed by the windings, which is Al. Then the energy per unit volume, or **energy density**, is

Energy density in magnetic field

$$\text{energy density} = \frac{1}{2}\frac{B^2}{\mu_0}. \tag{21–10}$$

This formula, which was derived for the special case of a solenoid, can be shown to be valid for any region of space where a magnetic field exists. If a ferromagnetic material is present, μ_0 is replaced by μ. This equation is analogous to that for an electric field, $\frac{1}{2}\varepsilon_0 E^2$, Eq. 17–9.

*21–11 • *LR* Circuit

Any inductor will have some resistance. We represent this situation by drawing the inductance L and the resistance R separately, as in Fig. 21–23a. The resistance R could also include a separate resistor connected in series. Now we ask, what happens when a dc source is connected in series to such an LR circuit? At the instant the switch connecting the battery is closed, the current starts to flow. It is, of course, opposed by the induced emf in the inductor. However, as soon as current starts to flow, there is a voltage drop across the resistance. Hence, the voltage drop across the inductance is reduced and there is then less impedance to the current flow from the inductance. The current thus rises gradually, as shown in Fig. 21–23b, and approaches the steady value $I_{max} = V/R$ when all the voltage drop is across the resistance. The actual shape of the curve is

$$I = \left(\frac{V}{R}\right)(1 - e^{-t/\tau}),$$

where e is the exponential function (see Section 19–7) and $\tau = L/R$ is called the **time constant** of the circuit. It is easy to show that τ is the time required to reach $0.63I_{max}$.

If the battery is suddenly removed from the circuit (dashed line in Fig. 21–23a), the current drops off as shown in Fig. 21–23c. This is an exponential decay curve given by $I = I_{max}e^{-t/\tau}$. The time constant τ is the time for the current to drop to 37 percent of the original value, and again equals L/R.

These graphs show that there is always some "reaction time" when an electromagnet, for example, is turned on or off. We also see that an LR circuit has properties similar to an RC circuit (Section 19–7). Unlike the capacitor case, however, the time constant here is *inversely* proportional to R.

FIGURE 21–23 (a) LR circuit; (b) growth of current when connected to a battery; (c) decay of current when the LR circuit is shorted out (battery is out of the circuit).

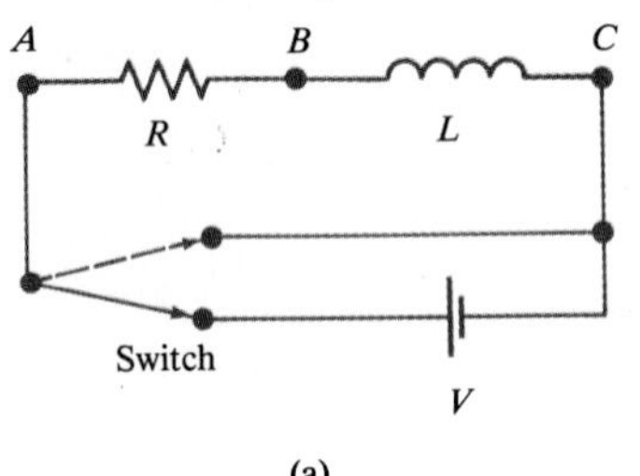

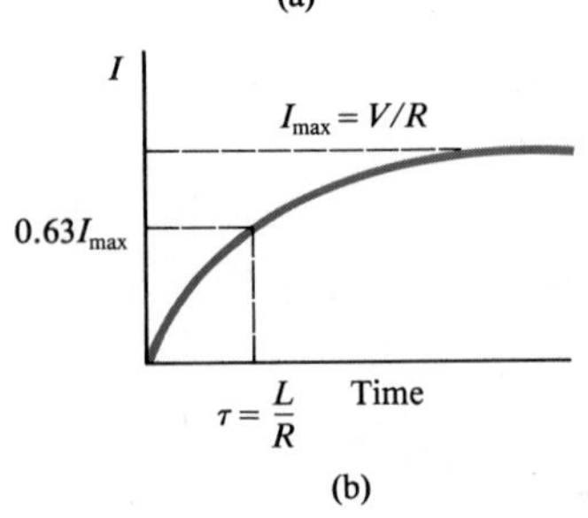

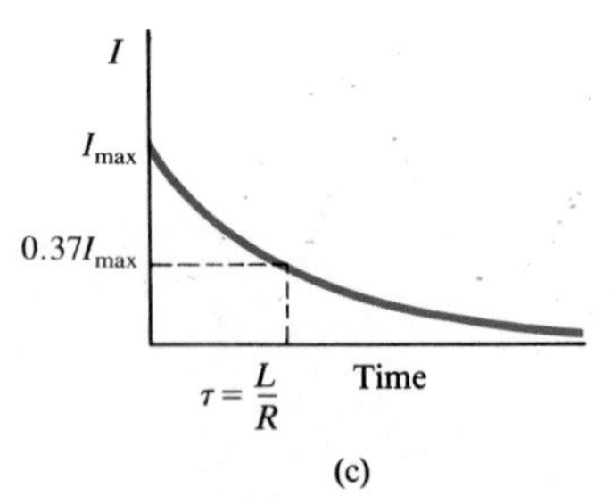

*21–12 • AC Circuits and Impedance

We have previously discussed circuits that contain combinations of resistor, capacitor, and inductor, but only when they are connected to a dc source of emf or to no source (as in the discharge of a capacitor in an *RC* circuit). Now we discuss these circuit elements when connected to a source of alternating emf.

First we examine, one at a time, how a resistor, a capacitor, and an inductor behave when connected to a source of alternating emf, represented by the symbol •–(~)–•, which produces a sinusoidal voltage of frequency f. We assume in each case that the emf gives rise to a current

$$I = I_0 \cos 2\pi ft,$$

where t is time and I_0 is the peak current. We must remember (Section 18–7) that $V_{rms} = V_0/\sqrt{2}$ and $I_{rms} = I_0/\sqrt{2}$ (Eq. 18–8).

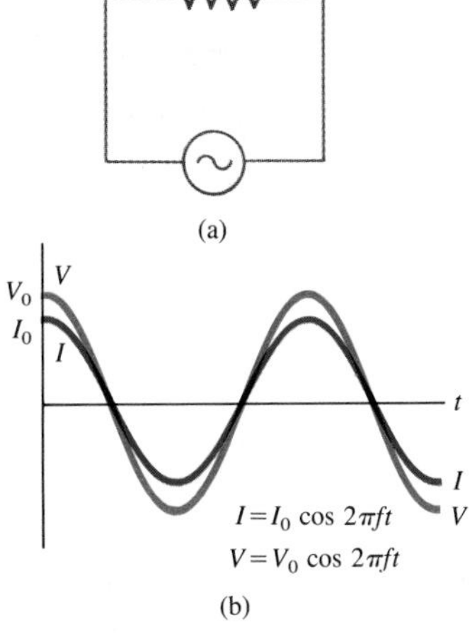

FIGURE 21–24 Resistor connected to ac source. Current is in phase with the voltage.

Resistor. When an ac source is connected to a resistor as in Fig. 21–24a, the current increases and decreases with the alternating emf according to Ohm's law, $I = V/R$. Figure 21–24b shows the voltage (red curve) and the current (blue curve). Because the current is zero when the voltage is zero and the current reaches a peak when the voltage does, we say that the current and voltage are **in phase**.

Resistor: current and voltage are in phase

Energy is transformed into heat, as discussed in Section 18–7, at an average rate $P = \overline{IV} = I^2_{rms}R = V^2_{rms}/R$.

Inductor. In Fig. 21–25a an inductor of inductance L, represented by the symbol –⌇⌇⌇⌇–, is connected to the ac source. We ignore any resistance it might have (it is usually small). The voltage applied to the inductor will be equal to the "back" emf generated in the inductor by the changing current as given by Eq. 21–9. This is because the sum of the emfs around any closed circuit must be zero, as Kirchhoff's rule tells us. Thus

$$V - L\frac{\Delta I}{\Delta t} = 0$$

or

$$V = L\frac{\Delta I}{\Delta t},$$

where V is the sinusoidally varying voltage of the source and $L\,\Delta I/\Delta t$ is the voltage induced in the inductor. According to this equation, I is increasing most rapidly when V has its maximum value, $V = V_0$. And I will be decreasing most rapidly when $V = -V_0$. These two instants correspond to points d and b on the graph of voltage versus time in Fig. 21–25b. At points a and c, $V = 0$. The equation above tells us that $\Delta I/\Delta t = 0$ at these instants, so I is not changing and these points correspond to the maximum and minimum values of the current I, as confirmed by points a and c on the graph. By going point by point in this manner, the curve of I versus t as compared to that for V versus t can be constructed, and they are shown by the blue and red lines, respectively, in Fig. 21–25b. Notice that the current reaches

FIGURE 21–25 Inductor connected to ac source. Current lags voltage by a quarter cycle, or 90°.

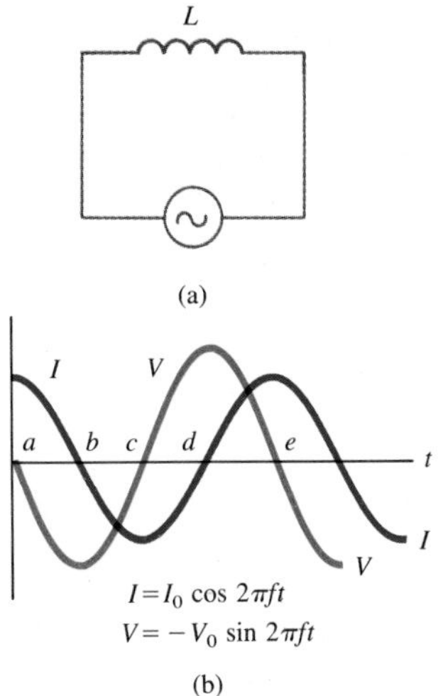

its peaks (and troughs) one-quarter of a cycle after the voltage does. We say that *in an inductor, the current lags the voltage* by 90°. (Remember that 360° corresponds to a full cycle, so 90° is a quarter cycle.) Alternatively, we can say that the voltage leads the current by 90°. (Since we originally chose $I = I_0 \cos 2\pi ft$, we see from the graph that V must vary as $V = -V_0 \sin 2\pi ft$.)

Inductor: current lags voltage

Because the current and voltage are out of phase by 90°, no energy is transformed in an inductor on the average; and no energy is dissipated as thermal energy. This can be seen as follows. From point c to d, the voltage is increasing from zero to its maximum. The current, however, is in the opposite direction to the voltage and is approaching zero. The average power over this interval, VI, is negative. From d to e, however, both V and I are positive so VI is positive; this contribution just balances the negative contribution of the previous quarter cycle. Similar considerations apply to the rest of the cycle. Thus, the average power transformed over one or many cycles is zero. We can see that energy from the source passes into the magnetic field of the inductor, where it is stored temporarily. Then the field decreases and the energy is transferred back to the source. None is dissipated in this process. Compare this to a resistor where the current is always in the same direction as the voltage and energy is transferred out of the source and never back into it. (The product VI is never negative.) This energy is not stored, but is transformed to thermal energy in the resistor.

As mentioned in Section 21–9, the back emf of an inductor impedes the flow of an ac current. Indeed, it is found that the magnitude of the current in an inductor is directly proportional to the applied ac voltage at a given frequency. We can therefore write an equation for a pure inductor (no resistance) that is the equivalent of Ohm's law:

$$V = IX_L, \qquad (21\text{–}11a)$$

Reactance and impedance

where X_L is called the **inductive reactance**, or **impedance**, of the inductor. Normally, we use the term "reactance" to refer solely to the inductive properties. We then reserve the term "impedance" to include the total "impeding" qualities of the coil—its inductance plus any resistance it may have (more on this in the next section). In the absence of resistance, the impedance is the same as the reactance.

The quantities V and I in Eq. 21–11a can refer either to rms or to peak values. Note, however, that although this equation may relate the peak values, the peak current and voltage are not reached at the same time; so Eq. 21–11a is *not* valid at a particular instant, as is the case for Ohm's law, but is valid only on the average.

From the fact that $V = L\,\Delta I/\Delta t$, we see that the larger L is, the less will be the change in current ΔI during the time Δt. Hence I itself will be smaller at any instant for a given frequency. Thus, $X_L \propto L$. The reactance X_L also depends on the frequency. The greater the frequency, the more rapidly the magnetic flux changes in the inductor. If the emf induced by this field is to remain equal to the source emf, as it must, the magnitude of the current must then be less. Hence the greater the frequency, the greater is the reactance X_L, so $X_L \propto fL$. This is also consistent with the fact that if the frequency f is zero (so the current is dc), there is no back emf and no impedance to the flow of charge. Careful calculation (using calculus), as well as experiment, shows that the constant of proportionality is 2π. Thus

$$X_L = 2\pi fL. \qquad (21\text{–}11b)$$

EXAMPLE 21–9 A coil has a resistance $R = 1.00\ \Omega$ and an inductance of 0.300 H. Determine the current in the coil if: (*a*) 120 V dc is applied to it; (*b*) 120 V ac (rms) at 60.0 Hz is applied.

SOLUTION (*a*) There is no inductive impedance ($X_L = 0$ since $f = 0$), so we apply Ohm's law for the resistance:

$$I = \frac{V}{R} = \frac{120\ \text{V}}{1.00\ \Omega} = 120\ \text{A}.$$

(*b*) The inductive reactance in this case is

$$X_L = 2\pi f L = (6.28)(60.0\ \text{s}^{-1})(0.300\ \text{H}) = 113\ \Omega.$$

In comparison to this, the resistance can be ignored. Thus,

$$I_{\text{rms}} = \frac{V_{\text{rms}}}{X_L} = \frac{120\ \text{V}}{113\ \Omega} = 1.06\ \text{A}.$$

(It might be tempting to say that the total impedance is $113\ \Omega + 1\ \Omega = 114\ \Omega$. This might imply that about 1 percent of the voltage drop is across the resistor, or about 1 V; and that across the inductance is 119 V. Although the 1 V_{rms} across the resistor is correct, the other statements are *not* true because of the alteration in phase in an inductor. This will be discussed in the next section.)

Capacitor. When a capacitor is connected to a battery, the capacitor plates quickly acquire equal and opposite charges; but no steady current flows in the circuit. A capacitor prevents the flow of a dc current (unless the capacitor is "leaky," in which case a leakage current would exist across the gap). However, if a capacitor is connected to an alternating source of voltage, as in Fig. 21–26a, an alternating current will flow continuously. This can happen because when the ac voltage is first turned on, charge begins to flow so that one plate acquires a negative charge and the other a positive charge. But when the voltage reverses itself, the charges flow in the opposite direction. Thus, for an alternating applied voltage, an ac current is present in the circuit continuously.

FIGURE 21–26 Capacitor connected to an ac source. Current leads voltage by a quarter cycle, or 90°.

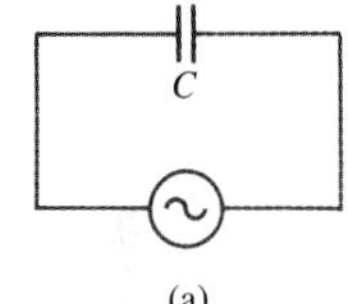

(a)

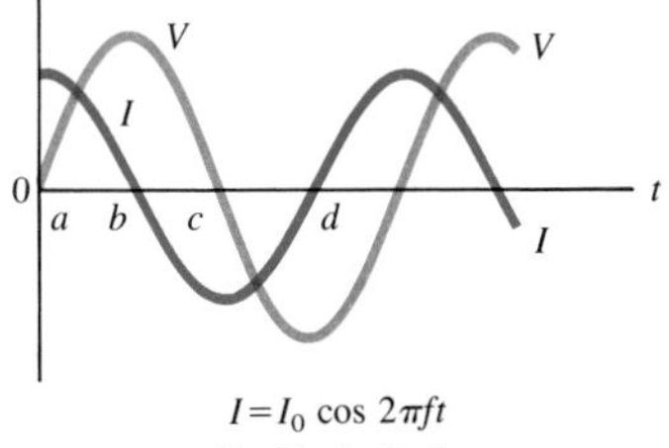

$I = I_0 \cos 2\pi f t$
$V = V_0 \sin 2\pi f t$

(b)

Let us look at this process in more detail. First, we recall that the applied voltage must equal the voltage across the capacitor: $V = Q/C$, where C is the capacitance and Q the charge on the plates. Thus the charge Q on the plates follows the voltage; when the voltage is zero, the charge is zero; when the voltage is a maximum, the charge is a maximum. But what about the current I? At point *a* in Fig. 21–26b, when the voltage starts increasing, the charge on the plates is zero. Thus charge flows readily toward the plates and the current is large. As the voltage approaches its maximum of V_0 (point *b*), the charge that has accumulated on the plates tends to prevent more charge from flowing, so the current drops to zero at point *b*. The charge that has accumulated now starts to flow off the plates and the magnitude of current again increases (blue curve), but in the opposite direction; it reaches a maximum (negatively) when the voltage is at point *c*. Thus the current follows the blue curve in Fig. 21–26b. Like an inductor, the voltage and current are out of phase by 90°. But for a capacitor, the *current leads the voltage by* 90°. (Or the voltage lags the current by 90°.) This is the opposite of what happens

Capacitor: current leads voltage

for an inductor. (Again we have chosen $I = I_0 \cos 2\pi ft$, and we see from the graph that $V = V_0 \sin 2\pi ft$.)

Because the current and voltage are out of phase, the average power dissipated is zero, just as for an inductor. Energy from the source is fed to the capacitor, where it is stored in the electric field between its plates. As the field decreases, the energy returns to the source. Thus, in an ac circuit, *only a resistance will dissipate energy* to thermal energy.

Only R (not C or L) dissipates energy

A relationship between the applied voltage and the current in a capacitor can be written just as for an inductance:

$$V = IX_C, \tag{21–12a}$$

where X_C is the **capacitive reactance** (or **impedance**) of the capacitor. This equation relates the rms or peak values for the voltage and current but is not valid at a particular instant because I and V are out of phase. X_C depends on both the capacitance C and the frequency f. The larger the capacitance, the more charge it can handle, so the less it will retard the flow of an alternating current. Hence, X_C will be inversely proportional to C. It is also inversely proportional to the frequency f, since, when the frequency is higher, there is less time per cycle for the charge to build up on the plates and impede the flow. Again there is a factor of 2π, and

$$X_C = \frac{1}{2\pi fC}. \tag{21–12b}$$

Notice that for dc conditions, $f = 0$ and X_C becomes infinite. This is as it should be, since a capacitor does not pass dc current.

It is interesting to note that the reactance of an inductor increases with frequency, but that of a capacitor decreases with frequency.

EXAMPLE 21–10 What are the peak and rms currents in the circuit of Fig. 21–26a if $C = 1.0\ \mu\text{F}$ and $V_{\text{rms}} = 120$ V? Calculate for (*a*) $f = 60$ Hz, and then for (*b*) $f = 6.0 \times 10^5$ Hz.

SOLUTION (*a*) $V_0 = \sqrt{2}V_{\text{rms}} = 170$ V. Then

$$X_C = \frac{1}{2\pi fC} = \frac{1}{(6.28)(60\ \text{s}^{-1})(1.0 \times 10^{-6}\ \text{F})} = 2.7\ \text{k}\Omega.$$

Thus

$$I_0 = \frac{V_0}{X_C} = \frac{170\ \text{V}}{2.7 \times 10^3\ \Omega} = 63\ \text{mA},$$

$$I_{\text{rms}} = \frac{V_{\text{rms}}}{X_C} = \frac{120\ \text{V}}{2.7 \times 10^3\ \Omega} = 44\ \text{mA}.$$

(*b*) For $f = 6.0 \times 10^5$ Hz, X_C will be $0.27\ \Omega$, $I_0 = 630$ A, and $I_{\text{rms}} = 440$ A. The dependence on f is dramatic.

Uses of capacitors

Capacitors are used for a variety of purposes, some of which have already been described. Two other applications are illustrated in Fig. 21–27.

In Fig. 21–27a, circuit A is said to be capacitively coupled to circuit B. The purpose of the capacitor is to prevent a dc voltage from passing from A to B but allowing an ac signal to pass relatively unimpeded. If C is sufficiently large, the ac signal will not be significantly attenuated. The capacitor in Fig. 21–27b also passes ac but not dc. In this case, a dc voltage can be maintained between circuits A and B. If the capacitance C is large enough, the capacitor offers little impedance to an ac signal leaving A. Such a signal then passes to ground instead of into B. Thus the capacitor in Fig. 21–27b acts like a *filter* when a constant dc voltage is required; any sharp variation in voltage will pass to ground instead of into circuit B. Capacitors used in these two ways are very common in circuits.

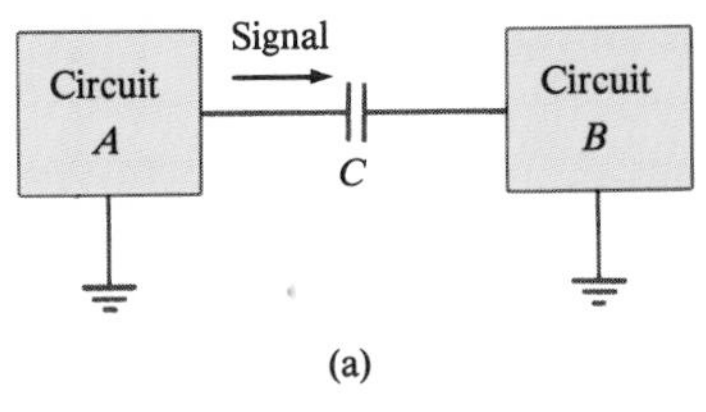

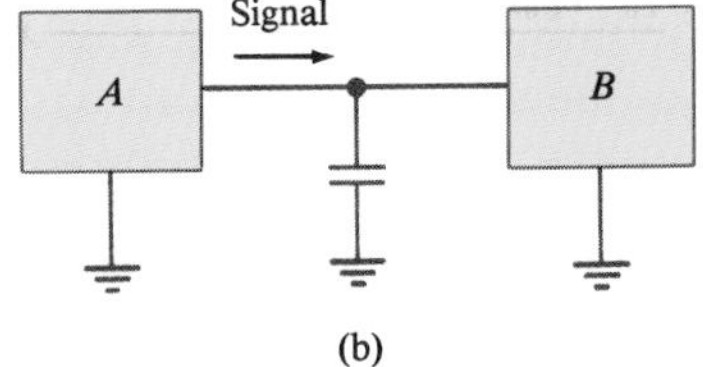

FIGURE 21–27 Two common uses for a capacitor.

*21–13 • *LRC* Series AC Circuit; Problem Solving

We now examine a circuit containing all three elements in series, a resistor R, an inductor L, and a capacitor C (Fig. 21–28). If a given circuit contains only two of these elements, we can still use the results of this section by setting $R = 0$, $L = 0$, or $C = \infty$ (infinity) as needed. We let V_R, V_L, and V_C represent the voltage across each element at a *given instant* in time; and V_{R0}, V_{L0}, and V_{C0} represent the *maximum* (peak) values of these voltages. The voltage across each of the elements will follow the phase relations we discussed in the last section. That is, V_R will be in phase with the current, V_L will lead the current by 90°, and V_C will lag behind the current by 90°. Also, at any instant the total voltage V supplied by the source will be $V = V_R + V_L + V_C$. However, because the various voltages are not in phase, the rms voltages (which is what ac voltmeters usually measure) will not simply add up to give the rms voltage of the source. And V_0 will *not* equal $V_{R0} + V_{L0} + V_{C0}$. Let us now examine the circuit in detail. What we would like to find in particular is the impedance of the circuit as a whole, the rms current that flows, and the phase difference between the source voltage and the current.

FIGURE 21–28 An *LRC* circuit.

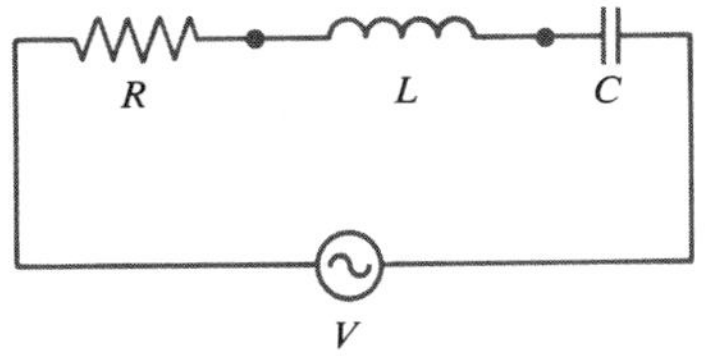

First we note that the current at any instant must be the same at all points in the circuit. Thus, *the currents in each element are in phase with each other, although the voltages are not.* We choose our origin in time, $t = 0$, so that the current I at any time t is $I = I_0 \cos 2\pi ft$.

It is convenient to analyze an *LRC* circuit using a **phasor diagram.** Arrows (acting like vectors) are drawn in an xy coordinate system to represent each voltage. *The length of each arrow represents the magnitude of the peak voltage across each element*:

$$V_{R0} = I_0 R, \qquad V_{L0} = I_0 X_L, \qquad \text{and} \qquad V_{C0} = I_0 X_C.$$

V_{R0} is in phase with the current and is initially ($t = 0$) drawn along the positive x axis, as is the current. Since V_{L0} leads the current by 90°, it leads V_{R0} by 90°, so is initially drawn along the positive y axis. V_{C0} lags the current by 90°, so lags V_{R0} by 90°; hence, V_{C0} is drawn initially along the negative y axis. Such a diagram is shown in Fig. 21–29a. If we let the vector diagram rotate counter-clockwise at frequency f, we get the diagram shown in Fig. 21–29b; after a time, t, each arrow has rotated through an angle $2\pi ft$. Then the *projections of each arrow on the x axis represent the voltages across each element at the instant t.* For example, the projection of

FIGURE 21–29 (below) Phasor diagram for a series *LRC* circuit.

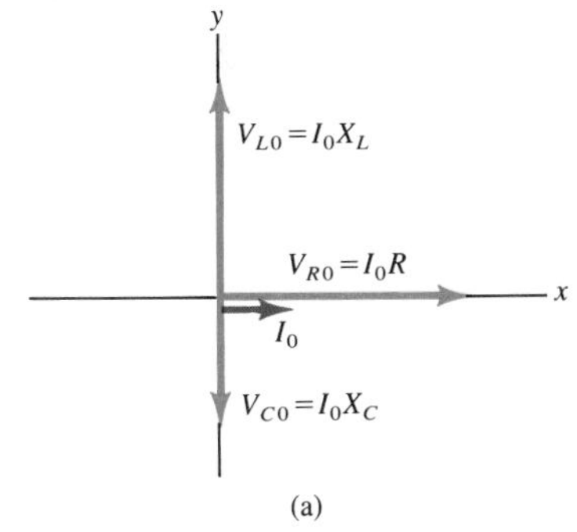

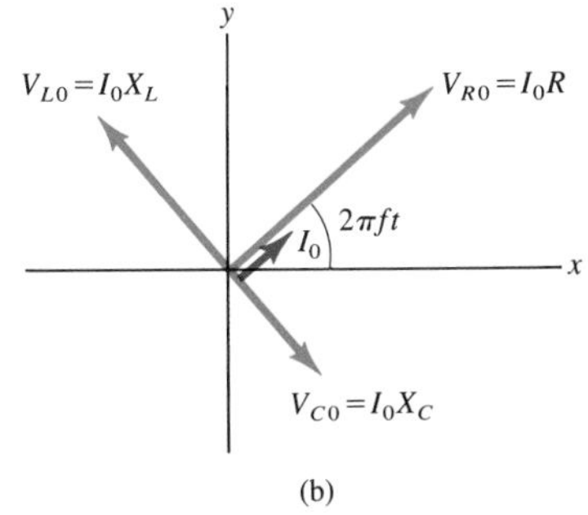

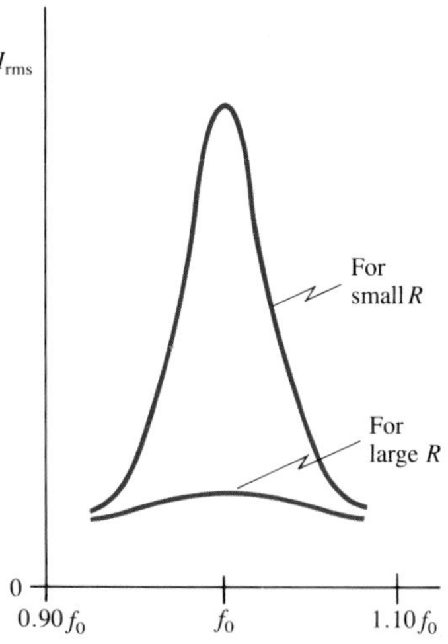

FIGURE 21–31 Current in an LRC circuit as a function of frequency, showing resonance peak at $f = f_0 = (1/2\pi)\sqrt{1/LC}$.

*21–14 • Resonance in AC Circuits; Oscillators

The rms current in an LRC series circuit is given by (see Eqs. 21–13 and 21–14):

$$I_{rms} = \frac{V_{rms}}{Z} = \frac{V_{rms}}{\sqrt{R^2 + \left(2\pi f L - \dfrac{1}{2\pi f C}\right)^2}}. \tag{21–17}$$

Because the impedance of inductors and capacitors depends on the frequency f of the source, the current in an LRC circuit will depend on frequency. From Eq. 21–17 we can see that the current will be maximum at a frequency such that

$$2\pi f L - \frac{1}{2\pi f C} = 0.$$

We solve this for f, and call the solution f_0:

Resonant frequency

$$f_0 = \frac{1}{2\pi}\sqrt{\frac{1}{LC}}. \tag{21–18}$$

This is the **resonant frequency** of the circuit. At this frequency, $X_C = X_L$, so the impedance is purely resistive and $\cos\phi = 1$. A graph of I_{rms} versus f is shown in Fig. 21–31 for particular values of R, L, and C. For smaller R compared to X_L and X_C, the resonance peak will be higher and sharper.

When R is very small, we speak of an ***LC* circuit**. The energy in an LC circuit oscillates, at frequency f_0, between the inductor and the capacitor, with some being dissipated in R (some resistance is unavoidable). To see this in more detail, consider a perfect LC circuit in which $R = 0$. Assume at $t = 0$ that the capacitor C is charged and the switch is closed (Fig. 21–32). The capacitor immediately begins to discharge. As it does so, the current I through the inductor increases. At every instant, the potential difference across the capacitor, $V = Q/C$ (where Q is the charge on the capacitor at that instant), must equal the potential difference across the inductor, which is equal to its emf, $-L(\Delta I/\Delta t)$. At the instant when the charge on the capacitor reaches zero ($Q = 0$), the current I in the inductor has reached its maximum value, but at this instant I is not changing ($-L\,\Delta I/\Delta t = Q/C = 0$). At this moment, the magnetic field B in the inductor is also a maximum. The current next begins to decrease as the flowing charge starts to accumulate on the opposite plate of the capacitor. When the current has dropped to zero, the capacitor has again attained its maximum charge. The capacitor then begins to discharge again, with the current now flowing in the opposite direction. This process of the charge flowing back and forth from one plate of the capacitor to the other, through the inductor, continues to repeat itself. This is called an ***LC* oscillation** or an **electromagnetic oscillation**. Not only does the charge oscillate back and forth, but so does the energy which oscillates between being stored in the electric field of the capacitor and in the magnetic field of the inductor.

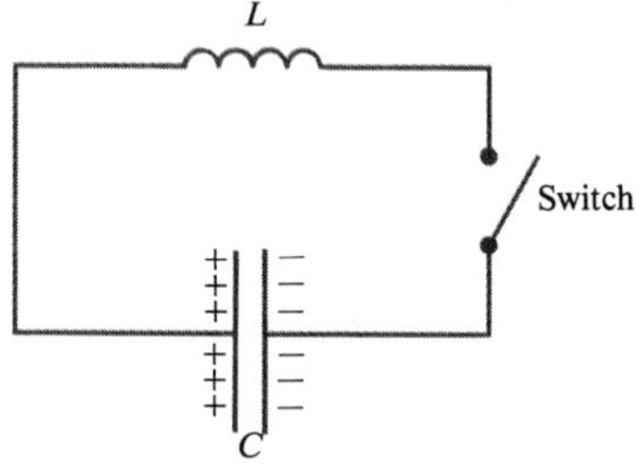

FIGURE 21–32 A pure LC circuit.

EM oscillations

Electric resonance is used in many electronic devices. Radio and television sets, for example, use resonant circuits for tuning in a station. Many frequencies reach the circuit from the antenna, but a significant current flows only for frequencies at or near the resonant frequency. Either L or C is variable so that different stations can be tuned in (more on this in Chapter 22). LC circuits are also used in **oscillators**, which are devices that put out an oscillating signal of particular frequency. Since some resistance is always present, electrical oscillators generally need a periodic input of power to compensate for the energy converted to thermal energy in the resistance.

Electrical resonance is analogous to mechanical resonance, which we discussed in Chapter 11 (see Fig. 11–12). The energy transferred to the system is a maximum at resonance whether it is electrical resonance, the oscillation of a spring, or pushing a child on a swing (Section 11–6). That this is true in the electrical case can be seen from Eqs. 21–14, 15 and 16. At resonance, $z = R$, $\cos\phi = 1$, and I_{rms} is a maximum. For constant V_{rms}, the power is then a maximum at resonance. A graph of power versus frequency looks much like that for the current (Fig. 21–31).

An LRC circuit can have the elements arranged in parallel instead of in series. Resonance will occur in this case, too, but the analysis of such circuits is more involved.

*21–15 • Impedance Matching

It is common to connect one electric circuit to a second circuit. For example, a TV antenna is connected to a TV set; an FM tuner is connected to an amplifier; the output of an amplifier is connected to a speaker; electrodes for an ECG or EEG (electrocardiogram and electroencephalogram—electrical traces of heart and brain signals) are connected to an amplifier or a recorder. In many cases it is important that the maximum power be transferred from one to the other, with a minimum of loss. This can be achieved when the output impedance of the one device matches the input impedance of the second.

To show why this is true, we consider simple circuits that contain only resistance. In Fig. 21–33, the source in circuit 1 could represent a power supply, the output of an amplifier, or the signal from an antenna, a laboratory probe, or a set of electrodes. R_1 represents the resistance of this device and includes the internal resistance of the source. R_1 is called the output impedance (or resistance) of circuit 1. The output of circuit 1 is across the terminals a and b, which are connected to the input of circuit 2. Circuit 2 may be very complicated. By combining the various resistors, we can find an equivalent resistance. This is represented by R_2, the "input resistance" (or input impedance) of circuit 2.

FIGURE 21–33 Output of the circuit on the left is input to the circuit on the right.

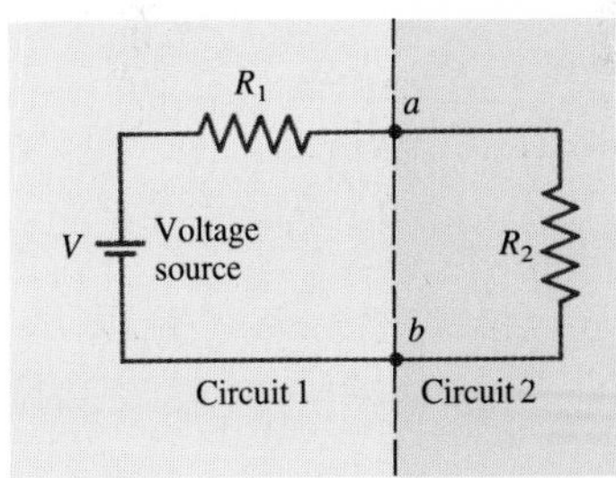

The power delivered to circuit 2 is $P = I^2R_2$, where $I = V/(R_1 + R_2)$. Thus

$$P = I^2R_2 = \frac{V^2R_2}{(R_1 + R_2)^2}.$$

We divide the top and bottom of the right side by R_1 and find

$$P = \frac{V^2}{R_1}\frac{\left(\dfrac{R_2}{R_1}\right)}{\left(1 + \dfrac{R_2}{R_1}\right)^2}. \tag{21–19}$$

The question is, if the resistance of the source is R_1, what value should R_2

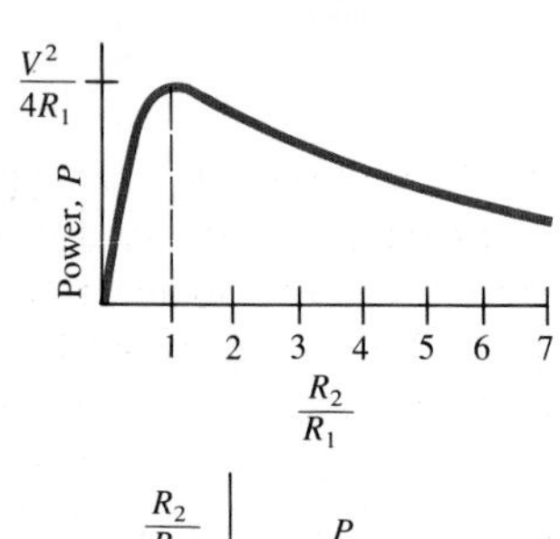

$\frac{R_2}{R_1}$	P
0	0
0.5	0.22 V^2/R_1
1.0	0.25 V^2/R_1
2.0	0.22 V^2/R_1
5.0	0.14 V^2/R_1
100	0.01 V^2/R_1

FIGURE 21–34 Power transferred is at its maximum when $R_2 = R_1$.

have so that the maximum power is transferred to circuit 2? To determine this, we plot a graph of P versus (R_2/R_1). This is shown in Fig. 21–34, where representative values are given in the table. For example, for $R_2/R_1 = 1$, Eq. 21–19 gives $P = V^2/4R_1$; for $R_2/R_1 = 3$, $P = 3V^2/16R_1 = 0.19V^2/R_1$; and so on. As can be seen from the graph, P is a maximum when $R_2 = R_1$.

Thus, the maximum power is transmitted when the *output impedance* of one device *equals the input impedance* of the second. This is called **matching the impedances.**

In an ac circuit that contains capacitors and inductors, the different phases are important and the analysis is more complicated. However, the same result holds: to maximize power transfer it is important to match impedances ($Z_2 = Z_1$). In addition, one must be aware that it is possible to seriously distort a signal. For example, when a second circuit is connected, it may put the first circuit into resonance, or take it out of resonance for a certain frequency.

Without proper consideration of the impedances involved, one can make measurements that are completely meaningless. These considerations are normally examined by engineers when designing an integrated set of apparatus. However, there have been cases when researchers connected several components to one another without regard for impedance matching, and made a "new discovery." Then, after their announcement of the "discovery," it was found that what they had observed was due to impedance mismatch rather than the natural phenomenon they had thought.

Impedance mismatch

In some cases, a transformer is used to alter an impedance, so it can be matched to that of a second circuit. If Z_s is the secondary impedance and Z_p the primary impedance, then $V_s = I_s Z_s$ and $V_p = I_p Z_p$ (I and V are either peak or rms values of current and voltage). Hence,

$$\frac{Z_p}{Z_s} = \frac{V_p I_s}{V_s I_p} = \left(\frac{N_p}{N_s}\right)^2,$$

where we have used Eqs. 21–6 and 21–7 for a transformer. Thus the impedance can be changed with a transformer. A transformer is used for this purpose in some audio amplifiers, which may have several taps corresponding to 4 Ω, 8 Ω, and 16 Ω so the output can be matched to the impedance of any loudspeaker.

Some instruments, such as oscilloscopes, require only a signal voltage but very little power. Maximum power transfer is then not important and such instruments can have a high input impedance. This has the advantage that the instrument draws very little current and disturbs the original circuit as little as possible. This is often desirable in laboratory experiments.

SUMMARY

The *magnetic flux* passing through a loop is equal to the product of the area of the loop times the perpendicular component of the magnetic field strength. If the magnetic flux through a loop of wire changes in time, an emf is induced in the loop; the magnitude of the induced emf equals the time rate of change of the magnetic flux through the loop. This result is called *Faraday's law of induction.* The induced emf produces a current whose magnetic field opposes the original change in flux (*Lenz's law*). Faraday's law also tells us that a changing magnetic field produces an electric field; and that a straight wire of length l moving with speed v perpendicular to a magnetic field of strength B has an emf induced between its ends equal to Blv.

An electric *generator* changes mechanical energy into electrical energy. Its operation is based on Faraday's law: a coil of wire is made to rotate uniformly by mechanical means in a magnetic field, and the changing flux through the coil induces a sinusoidal current, which is the output of the generator.

A motor, which operates in the reverse of a generator, acts like a generator in that a counter emf is induced in its rotating coil; since this counter emf opposes the input voltage, it can act to limit the current in a motor coil. Similarly, a generator acts somewhat like a motor in that a counter torque acts on its rotating coil.

A transformer, which is a device to change the magnitude of an ac voltage, consists of a primary and a secondary coil. The changing flux due to an ac voltage in the primary induces an ac voltage in the secondary. In a 100 percent efficient transformer, the ratio of output to input voltages (V_s/V_p) equals the ratio of the number of turns N_s in the secondary to the number N_p in the primary: $V_s/V_p = N_s/N_p$. The ratio of secondary to primary current is in the inverse ratio of turns: $I_s/I_p = N_p/N_s$.

QUESTIONS

1. What would be the advantage, in Faraday's experiments (Fig. 21–1), of using coils with many turns?
2. What is the difference between magnetic flux and magnetic field? Discuss in detail.
3. Suppose you are holding a circular piece of wire and suddenly thrust a magnet, south pole first, toward the center of the circle. Is a current induced in the wire? Is a current induced when the magnet is held steady within the loop? Is a current induced when you withdraw the magnet? In each case, if your answer is yes, specify the direction.
4. Show, using Lenz's law, that the emf induced in the moving rod in Fig. 21–8 is positive at the bottom and negative at the top, so that the current flows clockwise in the circuit loop on the left.
5. In what direction will the current flow in Fig. 21–8 if the rod moves to the left, which decreases the area of the loop to the left?
6. In situations where a small signal must travel over a distance, a "shielded cable" is used in which the signal wire is surrounded by an insulator and then enclosed by a cylindrical conductor. Why is a "shield" necessary?
7. What is the advantage of placing the two electric wires carrying ac close together?
8. In some early automobiles, the starter motor doubled as a generator to keep the battery charged once the car was started. Explain how this might work.
9. Explain why, exactly, the lights may dim briefly when a refrigerator motor starts up. [*Hint*: consider "terminal voltage."]
10. In older houses, when a refrigerator starts up, the lights may dim briefly. When an electric heater is turned on, the lights may stay dimmed as long as it is on. Explain the difference.
11. Explain what is meant by the statement "a motor acts as a motor and generator at the same time." Can the same be said for a generator?
12. Use Fig. 21–10 and the right-hand rules to show why the counter torque in a generator *opposes* the motion.
13. Will an eddy current brake (Fig. 21–16) work on a copper or aluminum wheel, or must it be ferromagnetic?
14. It has been proposed that eddy currents be used to help sort solid waste for recycling. The waste is first ground into tiny pieces and iron removed with a dc magnet. The waste then is allowed to slide down an incline over permanent magnets. How will this aid in the separation of nonferrous metals (Al, Cu, Pb, brass) from nonmetallic materials?
15. The pivoted metal bar with slots in Fig. 21–35 falls much more quickly through a magnetic field than does a solid bar. Explain in detail.
16. If an aluminum sheet is held between the poles of a large bar magnet, it requires some force to pull it out of the magnetic field even though the sheet is not ferromagnetic and does not touch the pole faces. Explain.
17. A bar magnet falling inside a vertical metal tube reaches a terminal velocity even if the tube is evacuated so that there is no air resistance. Explain.
18. A metal bar, pivoted at one end, oscillates freely in the absence of a magnetic field; but in a magnetic field, its oscillations are quickly damped out. Explain. (This *magnetic damping* is used in a number of practical devices.)
19. An enclosed transformer has four wire leads coming from it. How could you determine the ratio of turns on the two coils without taking the transformer apart? How would you know which wires paired with which?
20. The use of higher-voltage lines in homes, say 600 V or 1200 V, would reduce energy waste. Why are they not used?

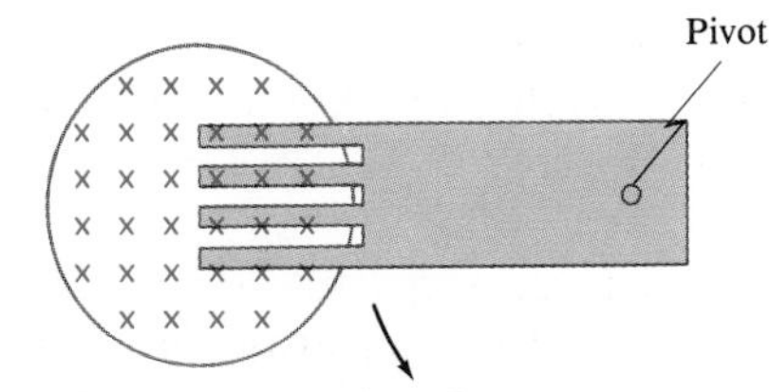

FIGURE 21–35 Question 15.

***21.** Since a magnetic microphone is basically like a loudspeaker, could a loudspeaker (Section 20–8) actually serve as a microphone? That is, could you speak into a loudspeaker and obtain an output signal that could be amplified? Explain. Discuss, in light of your response, how a microphone and loudspeaker differ in construction.

***22.** The primary of a transformer on a telephone pole has a resistance of 0.10 Ω and the input voltage is 2400 V ac. Can you estimate the current that will flow? Will it be 24,000 A? Explain.

***23.** A transformer designed for a 120-V ac input will often "burn out" if connected to a 120-V dc source. Explain. [*Hint*: the resistance of the primary coil is usually very low.]

***24.** How would you arrange two flat circular coils so that their mutual inductance was (*a*) greatest, (*b*) least (without separating them by a great distance)?

***25.** If the two coils in Fig. 21–22 were connected electrically, would there still be mutual inductance?

***26.** If you are given a fixed length of wire, how would you shape it to obtain the greatest self-inductance? The least?

***27.** Does the emf of the battery in Fig. 21–23a affect the time needed for the LR circuit to reach (*a*) a given fraction of its maximum possible current, (*b*) a given value of current?

***28.** How can we justify using phasor diagrams, with the potential differences as vectors, when we know that potential difference is not a vector?

***29.** Can you tell whether the current in an LRC circuit leads or lags the applied voltage from a knowledge of the power factor, $\cos\phi$?

***30.** Under what conditions is the impedance in an LRC circuit a minimum?

***31.** An LC resonance circuit is often called an *oscillator* circuit. What is it that oscillates?

***32.** Compare the oscillations of an LC circuit to the vibration of a mass m on a spring. What do L and C correspond to in the mechanical system?

PROBLEMS

SECTIONS 21–1 TO 21–4

1. (I) A 20-cm-diameter circular loop of wire is in a 0.60-T magnetic field. It is removed from the field in 0.10 s. What is the average induced emf?

2. (I) The rectangular loop shown in Fig. 21–36 is being pulled to the left, out of the magnetic field which points inward as shown. In what direction is the induced current?

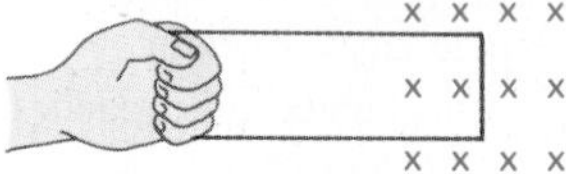

FIGURE 21–36 Problem 2.

3. (I) The magnetic flux through a coil of wire containing two loops changes from −20 Wb to +25 Wb in 0.25 s. What is the emf induced in the coil?

4. (I) If the resistance of the resistor in Fig. 21–37 is slowly increased, what is the direction of the current induced in the small circular loop inside the larger loop?

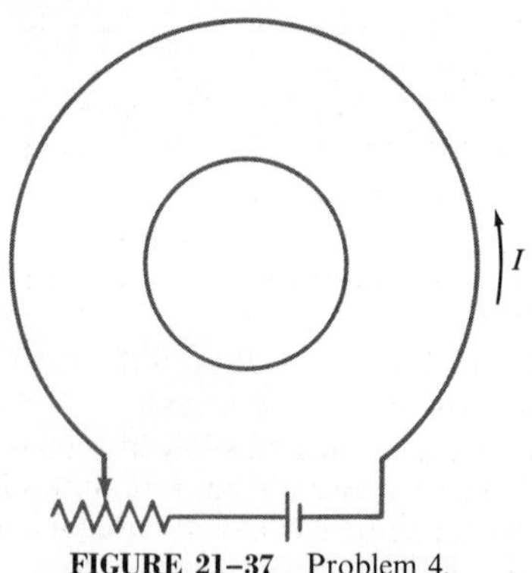

FIGURE 21–37 Problem 4.

5. (I) What is the direction of the induced current in the circular loop due to the current shown in each part of Fig. 21–38?

6. (II) The magnetic field perpendicular to a circular loop of wire 20 cm in diameter is changed from +0.85 T to −0.25 T in 80 ms, where + means the field points away from an observer and − toward the observer. (*a*) Calculate the induced emf. (*b*) In what direction does the induced current flow?

7. (II) The moving rod in Fig. 21–8 is 12.0 cm long and moves with a speed of 15.0 cm/s. If the magnetic field is 0.800 T, calculate (*a*) the emf developed, and (*b*) the electric field in the rod.

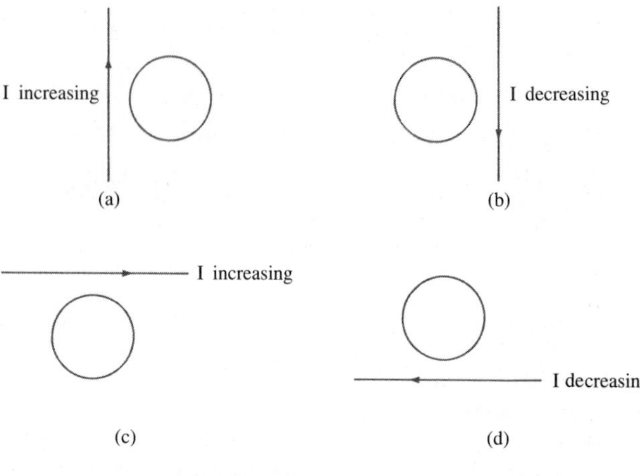

FIGURE 21–38 Problem 5.

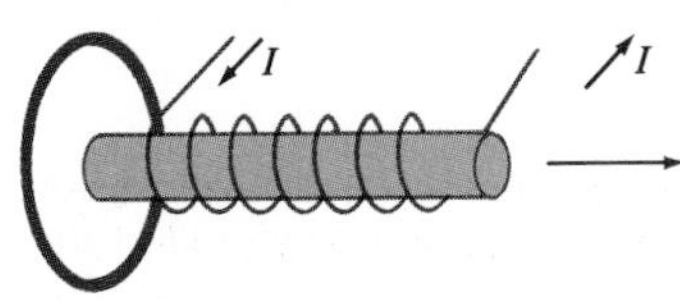

FIGURE 21–39 Problem 8.

8. (II) If the solenoid in Fig. 21–39 is being pulled away from the loop shown, in what direction is the induced current in the part of the loop closest to the viewer?

9. (II) In Fig. 21–8, the rod moves with a speed of 1.80 m/s, is 24.0 cm long, and has negligible resistance. The magnetic field is 0.375 T and the resistance of the U-shaped conductor is 28.5 Ω at a given instant. Calculate (*a*) the emf induced, and (*b*) the current flowing in the U-shaped conductor.

10. (II) A 35.0-cm-diameter coil consists of 30 turns of circular copper wire 2.8 mm in diameter. A uniform magnetic field, perpendicular to the plane of the coil, changes at a rate of 8.65×10^{-3} T/s. Determine (*a*) the current in the loop, and (*b*) the rate at which thermal energy is produced.

11. (II) (*a*) Show that the power $P = Fv$ needed to move the conducting rod to the right in Fig. 21–8 is equal to $B^2l^2v^2/R$, where R is the total resistance of the circuit. (*b*) Show that this equals the power dissipated in the resistance, I^2R.

12. (III) If the U-shaped conductor in Fig. 21–8 has resistivity ρ, whereas that of the moving rod is negligible, derive a formula for the current I as a function of time. Assume the rod has length l, starts at the bottom of the U at $t = 0$, and moves with uniform speed v in the magnetic field B. The cross-sectional area of the rod and all parts of the U is A.

13. (III) The magnetic field perpendicular to a single 18.0-cm-diameter circular loop of copper wire decreases uniformly from 0.850 T to zero. If the wire is 1.75 mm in diameter, how much charge moves past a point in the coil during this operation?

SECTION 21–5

14. (I) A car generator produces 12 V when the armature turns at 800 rev/min. What will be its output at 1800 rev/min, assuming that nothing else changes?

15. (I) Calculate the peak output voltage of a simple generator whose square armature windings are 5.50 cm on a side if the armature contains 125 loops and rotates in a field of 0.400 T at a rate of 120 rev/s.

16. (I) Show that the rms output of an ac generator is $V_{\text{rms}} = NAB\omega/\sqrt{2}$.

17. (II) A 200-loop square armature coil 30 cm on a side rotates at 60 rev/s in a uniform magnetic field. If the rms output is 220 V, what is the strength of the magnetic field?

18. (II) A simple generator has a 300-loop square coil 25.0 cm on a side. How fast must it turn in a 0.550-T field to produce a 120-V peak output?

SECTION 21–6

19. (I) A motor has an armature resistance of 3.40 Ω. If it draws 8.70 A when running at full speed and connected to a 120-V line, how large is the counter emf?

20. (I) The counter emf in a motor is 80 V when operating at 1800 rpm. What would be the counter emf at 2500 rpm if the magnetic field is unchanged?

21. (II) What will be the current in the motor of Example 21–5 if the load causes it to run at half speed?

22. (II) The magnetic field of a "shunt-wound" dc motor is produced by field coils placed in parallel to the armature coils. Suppose that the field coils have a resistance of 66.0 Ω and the armature coils 5.00 Ω. The back emf at full speed is 105 V when the motor is connected to a 115-V line. (*a*) Draw the equivalent circuit for the situations when the motor is just starting and when it is running full speed. (*b*) What is the total current drawn by the motor at start up? (*c*) What is the total current drawn when the motor runs at full speed?

SECTION 21–7

23. (I) A transformer changes 120 V to 15,000 V, and there are 9000 turns in the secondary. How many turns are there in the primary? Assume 100 percent efficiency.

24. (I) A transformer has 280 turns in the primary and 110 in the secondary. What kind of transformer is this and, assuming 100 percent efficiency, by what factor does it change the voltage?

25. (I) A step-up transformer increases 80 V to 220 V. What is the current in the secondary as compared to the primary? Assume 100 percent efficiency.

26. (I) Neon signs require 12 kV for their operation. To operate from a 220-V line, what must be the ratio of secondary to primary turns of the transformer? What would the voltage output be if the transformer were connected backward?

27. (II) A transformer has 1000 primary turns and 150 secondary turns. The input voltage is 120 V and the output current is 8.0 A. What is the secondary voltage and primary current?

28. (II) The output voltage of a 40-W transformer is 25 V and the input current is 15 A. (*a*) Is this a step-up or a step-down transformer? (*b*) By what factor is the voltage multiplied?

29. (II) If 30 MW of power at 45 kV (rms) arrives at a town from a generator via 4.0-Ω transmission lines, calculate (*a*) the emf at the generator end of the lines, and (*b*) the fraction of the power generated that is lost in the lines.

30. (II) Show that the power loss in transmission lines, P_L, is given by $P_L = (P_T)^2R_L/V^2$, where P_T is the power transmitted to the user, V is the delivered voltage, and R_L is the resistance of the power lines.

31. (II) If 50 kW is to be transmitted over two 0.100-Ω lines, estimate how much power is saved if the voltage is stepped up from 120 V to 1200 V and then down again, rather than simply transmitting at 120 V. Assume the transformers are each 99 percent efficient.

32. (III) Design a dc transmission line that can transmit 200 MW of electricity 200 km with only a 2 percent loss. The wires are to be made of aluminum and the voltage is 600 kV.

*SECTION 21–9

*33. (I) If the current in a 420-mH coil changes steadily from 10.0 to 25.0 mA in 250 ms, what is the induced emf?

*34. (I) What is the inductance L of a 0.40-m-long air-filled coil 3.2 cm in diameter containing 10,000 loops?

*35. (I) What is the inductance of a coil if it produces an emf of 8.50 V when the current in it changes from -28.0 mA to $+31.0$ mA in 27.0 ms?

*36. (II) How many turns does an air-filled coil have if it is 1.0 cm in diameter, 9.0 cm long, and its inductance is 0.75 mH? How many turns are needed if it has an iron core and $\mu = 10^3\,\mu_0$?

*37. (II) A 75-V emf is induced in a 0.30-H coil by a current that rises uniformly from zero to I_0 in 2.0 ms. What is the value of I_0?

*38. (II) The wire of a tightly wound solenoid is unwound and used to make another tightly wound solenoid of half the diameter. By what factor does the inductance change?

*39. (II) A coil has 3.00-Ω resistance and 0.550-H inductance. If the current is 8.00 A and is increasing at a rate of 4.50 A/s, what is the potential difference across the coil at this moment?

*40. (II) (*a*) Show that the self-inductance L of a torus (Fig. 20–41) of radius R containing N loops each of radius r is

$$L \approx \frac{\mu_0 N^2 r^2}{2R}$$

if $R \gg r$. Assume the field is uniform inside the toroid; is this actually true? Is this result consistent with L for a solenoid? Should it be? (*b*) Calculate the inductance L of a large toroid if the diameter of the coils is 2.0 cm and the diameter of the whole ring is 60 cm. Assume the field inside the toroid is uniform. There are a total of 800 loops of wire.

*41. (II) (*a*) Ignoring any mutual inductance, what is the equivalent inductance of two inductors connected in series? (*b*) What if they are connected in parallel? (*c*) How does their mutual inductance (their geometrical relationship to each other) affect the results?

*42. (II) A long thin solenoid of length l and cross-sectional area A contains N_1 closely packed turns of wire. Wrapped tightly around it is an insulated coil of N_2 turns, Fig. 21–40. Assume all the flux from coil 1 (the solenoid) passes through coil 2, and calculate the mutual inductance.

*43. (III) A 40-cm-long coil with 1500 loops is wound on an iron core ($\mu = 3000\,\mu_0$) along with a second coil of 300 loops. The loops of each coil have a radius of 2.0 cm. If the current in the first coil drops uniformly from 4.0 A to zero in 8.0 ms, determine (*a*) the emf induced in the second coil, and (*b*) the mutual inductance M.

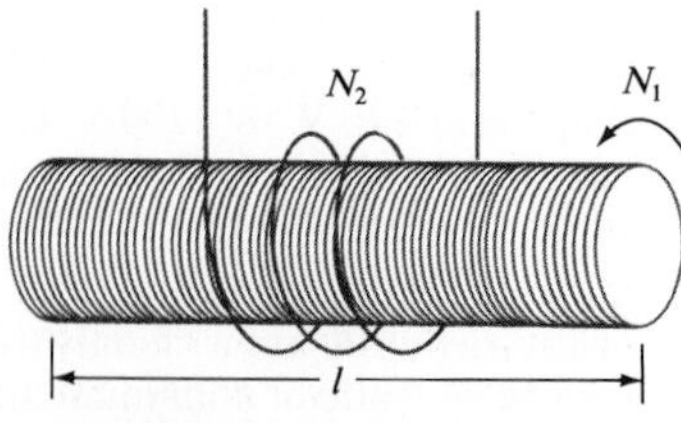

FIGURE 21–40 Problem 42.

*44. (III) The potential difference across a given coil is 22.5 V at an instant when the current is 860 mA and is increasing at a rate of 3.40 A/s. At a later instant, the potential difference is 16.2 V while the current is 700 mA and is decreasing at a rate of 1.80 A/s. Determine the inductance and resistance of the coil.

*SECTION 21–10

*45. (I) The magnetic field inside an air-filled solenoid 20 cm long and 2.0 cm in diameter is 0.60 T. Approximately how much energy is stored in this field?

*46. (I) How much energy is stored in 40.0-mH inductor at an instant when the current is 12.0 A?

*47. (II) Assuming the earth's magnetic field averages about 0.50×10^{-4} T near the surface of the earth, estimate the total energy stored in this field in the first 10 km above the earth's surface.

*48. (II) Typical large values for electric and magnetic fields attained in laboratories are about 1.0×10^4 V/m and 2.0 T. (*a*) Determine the energy density for each field and compare. (*b*) What value electric field would produce the same energy density as the 2.0-T magnetic field?

*SECTION 21–11

*49. (II) Approximately how many time constants does it take for the potential difference across the resistor in an LR circuit (Fig. 21–23a) to drop to 1.0 percent of its original value?

*50. (II) Determine $\Delta I/\Delta t$ at $t = 0$ (when the battery is connected) for the LR circuit of Fig. 21–23a and show that if I continued to increase at this rate, it would reach its maximum value in one time constant.

*51. (II) After how many time constants does the current in Fig. 21–23b reach within (*a*) 10 percent, (*b*) 1.0 percent, and (*c*) 0.1 percent of its maximum value?

*52. (II) It takes 2.66 ms for the current in an LR circuit to increase from zero to half its maximum value. Determine (*a*) the time constant of the circuit, and (*b*) the resistance of the circuit if $L = 350$ mH.

*53. (III) Two tightly wound solenoids have the same length and circular cross-sectional area. But solenoid 1 uses wire that is half as thick as solenoid 2. (*a*) What is the ratio of their inductances? (*b*) What is the ratio of their inductive time constants (assuming no other resistance in the circuits)?

*SECTION 21–12

*54. (I) At what frequency will a 300-mH inductor have a reactance of 2.0 kΩ?

*55. (I) At what frequency will a 7.20-μF capacitor have a reactance of 300 Ω?

*56. (I) Plot a graph of the impedance of a 1.0-μF capacitor as a function of frequency from 10 to 1000 Hz.

*57. (I) Plot a graph of the impedance of a 1.0-mH inductor as a function of frequency from 100 to 10,000 Hz.

*58. (II) Calculate the impedance of, and rms current in, a 100-mH radio coil connected to a 400-V (rms) 8.5-kHz ac line. Ignore resistance.

*59. (II) What is the inductance of a coil whose input is 120 V at 60 Hz and the current drawn is 30 A?

*60. (II) (*a*) What is the impedance of a well-insulated 0.030-μF capacitor connected to a 2.0-kV (rms) 700-Hz line? (*b*) What will be the peak value of the current?

*61. (II) A capacitor is placed in parallel across a load, as in Fig. 21–27b, to filter out stray high-frequency signals, but to allow ordinary 60-Hz ac to pass through with little loss. Suppose that circuit *B* in the figure is a resistance $R = 300\ \Omega$ connected to ground, and that $C = 0.60\ \mu$F. What percent of the incoming current will pass through *C* rather than *R* if (*a*) it is 60 Hz and (*b*) it is 60,000 Hz?

*62. (II) Suppose that circuit *B* in Fig. 21–27a is a resistance $R = 500\ \Omega$, connected to ground, and the capacitance $C = 2.0\ \mu$F. Will this capacitor act to eliminate 60 Hz ac but pass a high-frequency signal of frequency 60,000 Hz? To check this, determine the voltage drop across *R* for a 50-mV signal of frequency (*a*) 60 Hz, and (*b*) 60,000 Hz.

*SECTION 21–13

*63. (I) A 1.0-kΩ resistor and a 2.0-μF capacitor are connected in series to an ac source. Calculate the impedance of the circuit if the source frequency is (*a*) 60 Hz, and (*b*) 2.0×10^4 Hz.

*64. (I) A 20-kΩ resistor is in series with a 0.30-H inductor and an ac source. Calculate the impedance of the circuit if the source frequency is (*a*) 60 Hz, and (*b*) 3.0×10^4 Hz.

*65. (I) For a 120-V rms 60-Hz voltage, a current of 70 mA passing through the body for 1.0 s could be lethal. What would be the impedance of the body for this to occur?

*66. (II) (*a*) What is the rms current in an *RC* circuit if $R = 8.7$ kΩ, $C = 0.70\ \mu$F, and the rms applied voltage is 120 V at 60 Hz? (*b*) What is the phase angle between voltage and current? (*c*) What is the power dissipated by the circuit? (*d*) What are the voltmeter readings across *R* and *C*?

*67. (II) (*a*) What is the rms current in an *RL* circuit when a 60-Hz 120-V rms ac voltage is applied, where $R = 1.2$ kΩ, and $L = 600$ mH? (*b*) What is the phase angle between voltage and current? (*c*) How much power is dissipated? (*d*) What are the rms voltage readings across *R* and *L*?

*68. (II) A voltage $V = 3.5 \sin 754t$ is applied to an *LRC* circuit. If $L = 3.0$ mH, $R = 2.0$ kΩ, and $C = 8.0\ \mu$F, how much power is dissipated in the circuit?

*69. (II) A 60-mH inductor with a 7.0-Ω resistance is connected in series to a 300-μF capacitor and a 60-Hz 55-V (rms) source. For this circuit, calculate the (*a*) rms current, (*b*) phase angle, and (*c*) power dissipated.

*70. (II) What is the total impedance, phase angle, and rms current in an *LRC* circuit connected to a 3.0-kHz, 800-V (rms) source if $L = 8.0$ mH, $R = 4.1$ kΩ, and $C = 3000$ pF?

*71. (II) What is the resistance of a coil if its impedance is 35 Ω and its reactance is 30 Ω?

*72. (II) Show that for the *LRC* circuit of Fig. 21–28, if we have $I = I_0 \cos \omega t$, then $V_R = I_0 R \cos \omega t$, $V_L = I_0 \omega L \cos(\omega t + \pi/2)$, and $V_C = (I_0/\omega C)\cos(\omega t - \pi/2)$, where $\omega = 2\pi f$.

*SECTION 21–14

*73. (I) A 1200-pF capacitor is connected to a 70-μH coil of resistance 3.0 Ω. What is the resonant frequency of this circuit?

*74. (I) The variable capacitor in the tuner of an AM radio has a capacitance of 2500 pF when the radio is tuned to a station at 550 kHz. (*a*) What must be the capacitance for a station at 1600 kHz? (*b*) What is the inductance (assumed constant)?

*75. (II) An *LRC* circuit has $L = 3.6$ mH and $R = 2.0\ \Omega$. (*a*) What value must *C* have to produce resonance at 3600 Hz? (*b*) What will be the maximum current at resonance if the peak external voltage is 50 V?

*76. (II) A 660-pF capacitor is charged to 100 V and then quickly connected to a 75-mH inductor. Determine (*a*) the frequency of oscillation, (*b*) the peak value of the current, and (*c*) the maximum energy stored in the magnetic field of the inductor.

*SECTION 21–15

*77. (I) The output of an ECG amplifier has an impedance of 30 kΩ. It is to be connected to an 8.0-Ω loudspeaker through a transformer. What should be the turns ratio of the transformer?

*78. (I) An audio amplifier has output connections for 4 Ω, 8 Ω, and 16 Ω. If two 8-Ω speakers are to be connected in parallel, to which output terminals should they be connected?

GENERAL PROBLEMS

79. Suppose you are looking along a line through the centers of two circular (but separate) wire loops, one behind the other. A battery is suddenly connected to the front loop, establishing a clockwise current. (*a*) Will a current be induced in the second loop? (*b*) If so, when does this current start? (*c*) When does it stop? (*d*) In what direction is this current? (*e*) Is there a force between the two loops? (*f*) If so, in what direction?

80. A 10.0-cm-diameter circular loop of wire has a resistance of 6.60 Ω. It is initially in a 0.355-T magnetic field, with its plane perpendicular to **B**, but is removed from the field in 70.0 ms. Calculate the electric energy dissipated in this process.

81. A **search coil** for measuring *B* (also called a *flip coil*) is a small coil with *N* turns, each of cross-sectional area *A*. It is connected to a so-called **ballistic galvanometer**,

which is a device to measure the total charge Q that passes through it in a short time. The flip coil is placed in the magnetic field to be measured with its face perpendicular to the field. It is then quickly rotated 180°. Show that the total charge Q that flows in the induced current during this short "flip" time is proportional to the magnetic field strength B; in particular, show that B is given by

$$B = \frac{QR}{2NA},$$

where R is the total resistance of the circuit, including that of the coil and that of the ballistic galvanometer which measures the charge Q.

82. A pair of power transmission lines each have a 0.80-Ω resistance and carry 800 A over 7.0 km. If the rms input voltage is 32 kV, calculate (*a*) the voltage at the other end, (*b*) the power input, (*c*) power loss in the lines, and (*d*) the power output.

83. A small electric car overcomes a 250-N friction force when traveling 30 km/h. The electric motor is powered by ten 12-V batteries connected in series and is coupled directly to the wheels whose diameters are 50 cm. The 300 armature coils are rectangular, 10 cm by 15 cm, and rotate in a 0.60-T magnetic field. (*a*) How much current does the motor draw to produce the required torque? (*b*) What is the back emf? (*c*) How much power is dissipated in the coils? (*d*) What percent of the input power is used to drive the car?

*84. A circuit contains two elements, but it is not known if they are L, R, or C. The current in this circuit when connected to a 120-V 60-Hz source is 6.4 A and lags the voltage by 30°. What are the two elements and what are their values?

*85. A 130-mH coil, whose resistance is 18.0 Ω, is connected to a capacitor C and a 2360-Hz source voltage. If the current and voltage are to be in phase, what value must C have?

*86. What is the inductance L of the primary of a transformer whose input is 110 V at 60 Hz and the current drawn is 1.2 A? Assume no current in the secondary.

*87. A resonant circuit using a 220-pF capacitor is to resonate at 18.0 MHz. The air-core inductor is to be a solenoid with closely packed coils made from 12.0 m of insulated wire 1.1 mm in diameter. How many loops will the inductor contain?

*88. An inductance coil draws 2.5 A dc when connected to a 36-V battery. When connected to a 60-Hz 120-V (rms) source, the current drawn is 3.8 A (rms). Determine the inductance and resistance of the coil.

*89. The ***Q* factor** of a resonant circuit is defined as the ratio of the voltage across the capacitor (or inductor) to the voltage across the resistor, at resonance. The larger the Q factor, the sharper the resonance curve will be and the sharper the tuning. (*a*) Show that the Q factor is given by the equation $Q = (1/R)\sqrt{L/C}$. (*b*) At a resonant frequency $f_0 = 1.0$ MHz, what must be the value of L and R to produce a Q factor of 1000? Assume that $C = 0.010\ \mu\text{F}$.

CHAPTER 22

Electromagnetic Waves

The images we see and the sounds we hear with these devices arrive at the antenna as electromagnetic waves.

The culmination of electromagnetic theory in the nineteenth century was the prediction, and the experimental verification, that waves of electromagnetic fields could travel through space. This achievement opened a whole new world of communication—first the wireless telegraph, then radio and television. And it yielded the spectacular prediction that light is an electromagnetic wave.

The theoretical prediction of electromagnetic waves was the work of the Scottish physicist James Clerk Maxwell (1831–1879; Fig. 22–1), who unified, in one magnificent theory, all the phenomena of electricity and magnetism.

FIGURE 22–1 James Clerk Maxwell.

22-1 • Changing Electric Fields Produce Magnetic Fields; Maxwell's Equations

The development of electromagnetic theory in the early part of the nineteenth century by Oersted, Ampère, and others was not actually done in terms of electric and magnetic fields. The idea of the field was introduced somewhat later by Faraday, and was not generally used until Maxwell showed that all electric and magnetic phenomena could be described using only four equations involving electric and magnetic fields. These equations, known as *Maxwell's equations*, are the basic equations for all electromagnetism. They are fundamental in the sense that Newton's three laws of motion and the law of universal gravitation are for mechanics. In a sense, they are even more fundamental, since they are valid even relativistically (Chapter 26), whereas Newton's laws are not. Because all of electromagnetism is contained in this set of four equations, Maxwell's equations are considered one of the great triumphs of the human mind.

Maxwell's equations

Although we will not present **Maxwell's equations** in mathematical form since they involve calculus, we will summarize them here in words. They are: (1) a generalized form of Coulomb's law known as Gauss's law (Appendix C) that relates electric field to its sources, electric charge; (2) the same for the magnetic field, except that if there are no magnetic monopoles, magnetic field lines are continuous—they do not begin or end (as electric field lines do on charges); (3) an electric field is produced by a changing magnetic field; (4) a magnetic field is produced by an electric current or by a changing electric field.

Changing **E** *produces* **B**

Law number (3) is Faraday's law (see Chapter 21, especially Section 21–4). The first part of (4), that a magnetic field is produced by an electric current, was discovered by Oersted, and the mathematical relation is given by Ampère's law (Section 20-12). But the second part of (4) is an entirely new aspect predicted by Maxwell. Maxwell argued that if a changing magnetic field produces an electric field, as given by Faraday's law, then the reverse might be true as well: **a changing electric field will produce a magnetic field**. This was a *hypothesis* by Maxwell, based on the idea of order in the world. Indeed, the size of the effect in most cases is so small that Maxwell recognized it would be difficult to detect it experimentally.

*22–2 • Maxwell's Fourth Equation; Displacement Current

Maxwell used an indirect argument to back up his idea that a changing electric field would produce a magnetic field. In a simplified fashion it goes something like this. According to Ampère's law (see Section 20–12), $\sum B_{\parallel}\,\Delta l = \mu_0 I$. That is, you divide any closed path you choose into short segments Δl, multiply each segment by the parallel component of the magnetic field B at that segment, and then sum all these products over the complete closed path; this sum will then equal μ_0 times the net current I that passes through the surface bounded by the path. When we applied Ampère's law (Section 20–12) to the field around a straight wire, we imagined the current as passing through the circle enclosed by our circular loop. This would be the flat surface 1 in Fig. 22–2. However, we could just as well use the sack-shaped surface 2 in the figure, since the same current I passes through it. This naturally implies that the current passing through any surface enclosed by the closed path must be the same. Thus the current flowing into the volume enclosed by surfaces 1 and 2 together equals the current that flows out of this volume. This is basically Kirchhoff's point rule, and tells us that the rate charge enters the volume equals the rate at which it leaves.

FIGURE 22–2 Ampere's law applied to two different surfaces bounded by the same closed path.

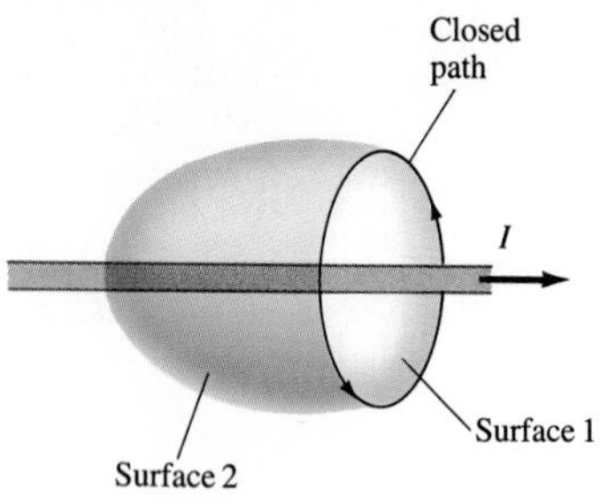

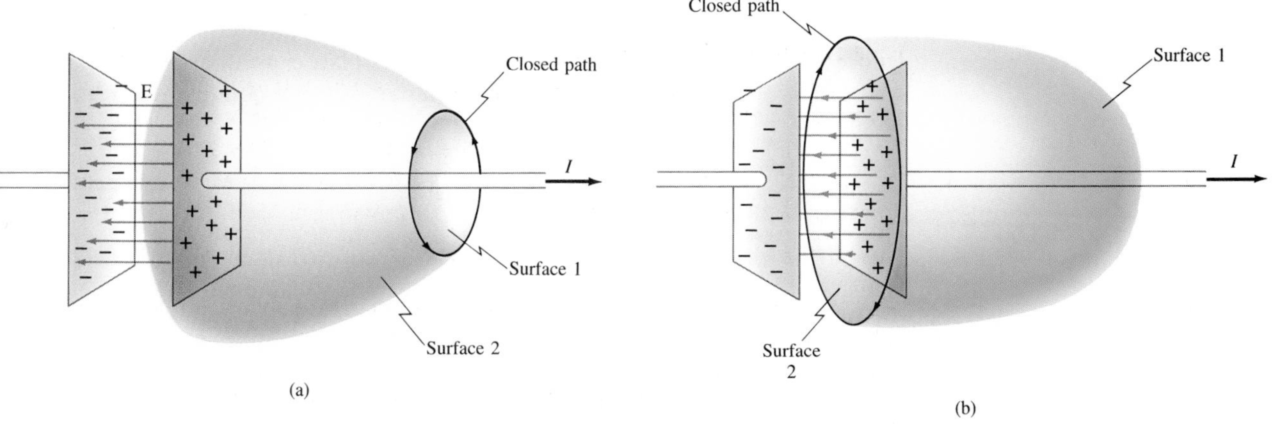

FIGURE 22–3 A capacitor discharging. No conduction current passes through surface 2 in either (a) or (b). An extra term is needed in Ampère's law.

Now consider the closed path for the situation of Fig. 22–3a, where a capacitor is being discharged. Ampère's law works for surface 1, but it does not work for surface 2, since no current passes through surface 2. There is a magnetic field around the wire, so the left side of Ampère's law is not zero; yet for surface 2, the right side *is* zero, since $I = 0$. We seem to have a contradiction of Ampère's law. Similar things can be said for the closed path indicated in Fig. 22–3b, which surrounds the region of the electric field of the plates; there is a current passing through surface 1, but not through surface 2, and there is a magnetic field around the closed path. There is a magnetic field present, however, in Fig. 22–3 only if charge is flowing to or away from the capacitor plates. In this case, the electric field between the plates is changing in time. Maxwell resolved the problem of no current through surface 2 in Figs. 22–3a and b by stating that the changing electric field between the plates is *equivalent to* an electric current. He called this a **displacement current**, I_D. (The name is not especially illuminating today, although historically it had meaning in the context of the then prevalent theory, since discarded.) An ordinary current is then called a **conduction current**, I_C. Ampère's law now becomes

Displacement current defined

$$\sum B_{\parallel}\,\Delta l = \mu_0(I_C + I_D).$$

Ampère's law will now apply even for surface 2 in Figs. 22–3a and b, where the displacement current I_D refers to the changing electric field.†

We can write I_D in terms of the changing electric field. The charge Q on a capacitor of capacitance C is $Q = CV$, where V is the potential difference between the plates. Also recall that $V = Ed$, where d is the (small) separation of the plates and E is the electric field strength between them; we

† Notice that this interpretation of the changing electric field as a displacement current fits in well with our discussion in Chapter 21 where we saw that an alternating current passes through a capacitor. It also means that Kirchhoff's point rule will be valid even at a capacitor plate: when being charged, conduction current flows into the capacitor plate, but no conduction current flows out of the plate. Instead, a "displacement current" flows out of the plate. The introduction of displacement current also resolved other inconsistencies that had theretofore existed.

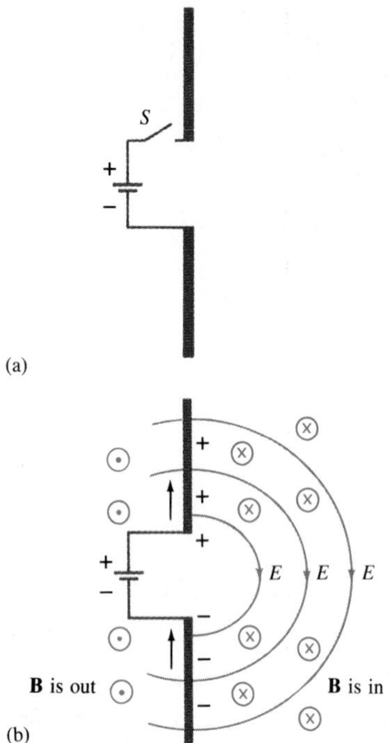

FIGURE 22–4
Fields produced by charge flowing into conductors. It takes time for the **E** and **B** fields to travel outward to distant points.

ignore any fringing of the field. Also, for a parallel-plate capacitor, $C = \varepsilon_0 A/d$, where A is the area of each plate (see Chapter 17). We combine these to obtain

$$Q = CV = \left(\varepsilon_0 \frac{A}{d}\right)(Ed) = \varepsilon_0 AE.$$

Now if the charge on the plate changes at a rate $\Delta Q/\Delta t$, the electric field changes at a proportional rate. Then, from the above expression,

$$\frac{\Delta Q}{\Delta t} = \varepsilon_0 A \frac{\Delta E}{\Delta t}.$$

Now $\Delta Q/\Delta t$ is the rate at which charge accumulates on, or leaves, the capacitor plates; it is therefore equal to the current flowing into or out of the capacitor. If the current flow into the capacitor, $\Delta Q/\Delta t$, is set equal to the displacement current I_D between the plates, then

$$I_D = \frac{\Delta Q}{\Delta t} = \varepsilon_0 A \frac{\Delta E}{\Delta t}.$$

This can be written

$$I_D = \varepsilon_0 \frac{\Delta \Phi_E}{\Delta t},$$

where $\Phi_E = EA$ is the **electric flux**, defined in analogy to magnetic flux (Section 21–2). Then, Ampère's law becomes

Maxwell's extension of Ampère's law

$$\sum B_{\|} \,\Delta l = \mu_0 I_C + \mu_0 \varepsilon_0 \frac{\Delta \Phi_E}{\Delta t}. \tag{22–1}$$

This equation embodies Maxwell's idea that a magnetic field can be caused not only by a normal electric current, but also by a changing electric field or changing electric flux. Eq. 22–1 is essentially Maxwell's fourth equation.†

22–3 • Production of Electromagnetic Waves

According to Maxwell, a magnetic field will be produced in empty space if there is a changing electric field. From this, Maxwell derived another startling conclusion. If a changing magnetic field produces an electric field, the electric field can itself be changing. This changing electric field will, in turn, produce a magnetic field. The latter will be changing and so will produce a changing electric field; and so on. When Maxwell manipulated his equations, he found that the net result of these interacting changing fields was to produce a wave of electric and magnetic fields that can actually propagate (travel) through space. We now examine, in a simplified way, how such **electromagnetic waves** are produced.

How EM waves are produced

Consider two conducting rods that will serve as an "antenna" (Fig. 22–4a). Suppose that these two rods are connected by a switch to the oppo-

† Actually, there is also a third term on the right for the case when a magnetic field is produced by magnetized materials; but we assume in what follows that no magnets are present.

site terminals of a battery. As soon as the switch is closed, the upper rod quickly becomes positively charged and the lower one negatively charged. Electric field lines are formed as indicated by the lines in Fig. 22–4b. While the charges are flowing, a current exists; the direction of conventional current is indicated by the arrows. A magnetic field is therefore produced. The magnetic field lines encircle the wires and therefore, in the plane of the page, **B** points into the page (⊗) on the right and out of the page (⊙) on the left. Now we ask, how far out do these fields extend? In the static case, the fields extend outward indefinitely far. However, when the switch in Fig. 22–4 is closed, the fields quickly appear nearby, but it takes time for them to reach distant points. Both electric and magnetic fields store energy, and this energy cannot be transferred to distant points at infinite speed.

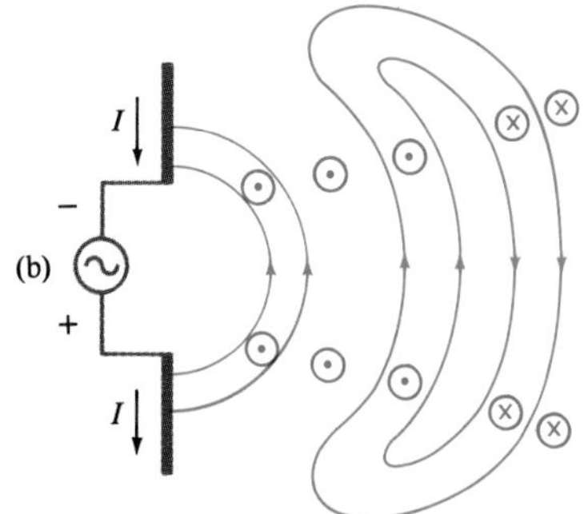

FIGURE 22–5 Sequence showing electric and magnetic fields that spread outward from oscillating charges on two conductors connected to an ac source (see the text).

This example illustrates that it takes time for electric and magnetic fields, once they are formed, to reach distant points. Now we look at a more interesting situation. Suppose our antenna is connected to an ac generator (Fig. 22–5). In Fig. 22–5a, the connection has just been completed. Charge starts building up and fields form just as in Fig. 22–4. The + and − signs indicate the net charge on each rod. The arrows indicate the direction of the current. The electric field is represented by lines in the plane of the page; and the magnetic field, according to the right-hand rule, is into (⊗) or out of (⊙) the page, as shown. In Fig. 22–5b, the emf of the ac generator has reversed in direction; the current is reversed and the new magnetic field is in the opposite direction. The old fields, however, don't suddenly disappear; they are on their way to distant points. But because the new fields have changed direction, the old lines fold back to connect up to some of the new lines and form closed loops as shown.† Actually, the fields not far from the antenna, referred to as the *near field*, become quite complicated, but we are not so interested in them. We are instead mainly interested in the fields far from the antenna (they are generally what we detect), which we refer to as the **radiation field**. The electric field lines form loops, as shown in Fig. 22–6, and continue moving outward. The magnetic field lines also form closed loops, but are not shown since they are perpendicular to the page. Although the lines are shown only on the right of the source, fields also travel in other directions. (The field strengths are greatest in directions perpendicular to the oscillating charges; and they drop to zero along the direction of oscillation—above and below the antenna in Fig. 22–6.) The magnitudes of both **E** and **B** in the radiation field decrease as $1/r$. (Compare this to the static electric field given by Coulomb's law where **E** decreases as $1/r^2$.) The energy carried by the electromagnetic wave is proportional (as for any wave, Chapter 11) to the square of the amplitude, E^2 or B^2, as will be discussed further in Section 22–6, so the intensity of the wave decreases as $1/r^2$.

FIGURE 22–6 The radiation fields (far from the antenna) produced by a sinusoidal signal on a dipole antenna. The closed loops represent electric field lines. The magnetic field lines, perpendicular to the page and represented by ⊗ and ⊙, also form closed loops.

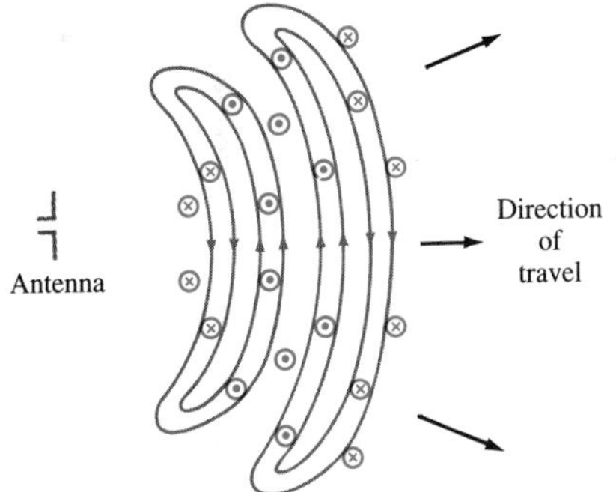

Several things about the radiation field can be noted from Fig. 22–6. First, *the electric and magnetic fields at any point are perpendicular to each other, and to the direction of motion.* Second, we can see that the fields alternate in direction (**B** is into the page at some points and out of the page at others; similarly for **E**). Thus, the field strengths vary from a maximum in one direc-

E ⊥ **B** *in EM waves*

† We are considering waves traveling through empty space, so there are no charges for lines of **E** to start or stop on, so they form closed loops. Magnetic field lines always form closed loops since there are no magnetic poles (as far as we know).

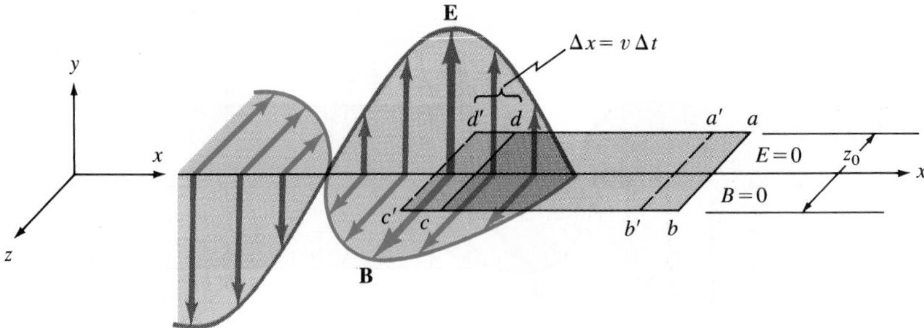

FIGURE 22–9 Rectangle in the xz plane (actually at rest) moves a distance $\Delta x = v\,\Delta t$ relative to a wave traveling to the right.

Now we consider a rectangle in the xz plane, as shown (grey) in Fig. 22–9. Again we show the rectangle moving to the left a distance $\Delta x = v\,\Delta t$ relative to the wave, although actually the rectangle is stationary and the wave is moving to the right; *abcd* is the position of the rectangle initially and *a'b'c'd'* is its position after a time Δt. There is a changing electric flux through this rectangular loop equal to the electric field E (heavy arrow) times the increasing area $\Delta A = z_0\,\Delta x = z_0 v\,\Delta t$ (where z_0 is the width $ab = cd$ of the rectangle). According to Ampère's law, Eq. 22–1 with $I_C = 0$ since there are no conduction currents, we have

$$\sum B_{\parallel}\,\Delta l = \mu_0\varepsilon_0\frac{\Delta\Phi_E}{\Delta t}$$

$$= \mu_0\varepsilon_0\frac{(E)(z_0 v\,\Delta t)}{\Delta t} = \mu_0\varepsilon_0 E z_0 v.$$

The sum of $B_{\parallel}\,\Delta l$ on the sides *ab*, *bc*, and *da* are all zero, because either $B = 0$ or B is perpendicular to these sides. But along side *cd*, the contribution is Bz_0, where B is the magnetic field (heavy arrow) parallel to *cd*. Thus,

$$Bz_0 = \mu_0\varepsilon_0 E z_0 v$$

$$B = \mu_0\varepsilon_0 v E.$$

We combine this with Eq. 21–2 and find that

$$B = \mu_0\varepsilon_0 v(vB) = \mu_0\varepsilon_0 v^2 B.$$

We cancel B on both sides and solve for v:

Speed of light

$$v = \frac{1}{\sqrt{\varepsilon_0\mu_0}}. \qquad (22\text{–}3)$$

When we put in the values for ε_0 and μ_0, we find that

$$v = \frac{1}{\sqrt{\varepsilon_0\mu_0}} = \frac{1}{\sqrt{(8.85 \times 10^{-12}\ \mathrm{C^2/N\cdot m^2})(4\pi \times 10^{-7}\ \mathrm{N\cdot s^2/C^2})}}$$

$$= 3.00 \times 10^8\ \mathrm{m/s}.$$

This is a remarkable result. For this is precisely equal to the measured speed of light!

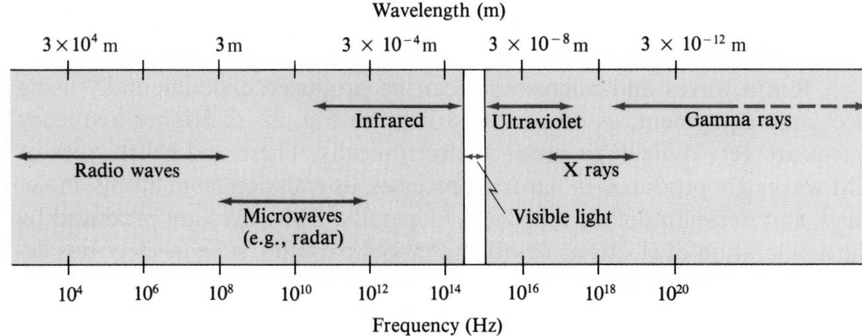

FIGURE 22–10 Electromagnetic spectrum.

22–5 • Light as an Electromagnetic Wave and the Electromagnetic Spectrum

The calculation at the end of the last section gives the result that Maxwell himself determined: that the speed of EM waves is 3.00×10^8 m/s, the same as the measured speed of light.

Light had been shown some 60 years previously to behave like a wave (we'll discuss this in Chapter 24). But nobody knew what kind of wave it was—that is, what is it that is oscillating in a light wave? Maxwell, on the basis of the calculated speed of EM waves, argued that light must be an electromagnetic wave. This soon came to be generally accepted by scientists, but not fully until after EM waves were experimentally detected. EM waves were first generated and detected experimentally by Heinrich Hertz (1857–1894) in 1887, eight years after Maxwell's death. Hertz used a spark-gap apparatus in which charge was made to rush back and forth for a short time, generating waves whose frequency was about 10^9 Hz. He detected them some distance away using a loop of wire in which an emf was produced when a changing magnetic field passed through. These waves were later shown to travel at the speed of light, 3.00×10^8 m/s, and to exhibit all the characteristics of light such as reflection, refraction, and interference. The only difference was that they were not visible. Hertz's experiment was a strong confirmation of Maxwell's theory.

The wavelengths of visible light were measured in the first decade of the nineteenth century, long before anyone imagined that light was an electromagnetic wave. The wavelengths were found to lie between 4.0×10^{-7} m and 7.5×10^{-7} m; or 400 nm to 750 nm (1 nm $= 10^{-9}$ m). The frequencies of visible light rays can be found using Eq. 11–9, which we rewrite here:

$$f\lambda = c, \tag{22–4}$$

where f and λ are the frequency and wavelength, respectively, of the wave. Here, c is the velocity of light, 3.00×10^8 m/s; it gets the special symbol c because of its universality for all EM waves in free space. Equation 22-4 tells us that the frequencies of visible light are between 4.0×10^{14} Hz and 7.5×10^{14} Hz.

But visible light is only one kind of EM wave. As we have seen, Hertz produced EM waves of much lower frequency, about 10^9 Hz. These are called **radio waves**, since frequencies in this range are used today to transmit radio and TV signals. Electromagnetic waves, or EM radiation as we sometimes call it, have been produced or detected over a wide range of frequencies. They are usually categorized as shown in Fig. 22–10. This is known as the **electromagnetic spectrum**.

EM spectrum

Radio waves and microwaves can be produced experimentally using electronic equipment, as discussed earlier; see Fig. 22–5. Higher-frequency waves are very difficult to produce electronically. These and other types of EM waves are produced in natural processes, as emission from atoms, molecules, and nuclei (more on this later). Generally, EM waves are produced by the acceleration of electrons or other charged particles, such as electrons accelerating in the antenna of Fig. 22–5. Another example is X rays, which are produced (see Chapter 28) when fast-moving electrons are rapidly decelerated upon striking a metal target. Even the visible light emitted by an ordinary incandescent light is due to electrons undergoing acceleration within the hot filament. We will meet various types of EM waves later. However, it is worth mentioning here that infrared (IR) radiation (EM waves whose frequency is just less than that of visible light) is mainly responsible for the heating effect of the sun. The sun emits not only visible light but substantial amounts of IR and UV (ultraviolet) as well. The molecules of our skin tend to "resonate" at infrared frequencies, so it is these that are preferentially absorbed and thus warm us up.

EXAMPLE 22–1 Calculate the wavelength: (*a*) of a 60-Hz EM wave, and (*b*) of a 1240-kHz AM radio wave.

SOLUTION (*a*) Since $c = \lambda f$,

$$\lambda = \frac{c}{f} = \frac{3.0 \times 10^8 \text{ m/s}}{60 \text{ s}^{-1}} = 5.0 \times 10^6 \text{ m},$$

or 5000 km. One wavelength stretches all the way across the United States.

(*b*) $$\lambda = \frac{3.0 \times 10^8 \text{ m/s}}{1.24 \times 10^6 \text{ s}^{-1}} = 240 \text{ m}.$$

Electromagnetic waves can travel along transmission lines as well as in empty space. When a source of emf is connected up to a transmission line—be it two parallel wires or a coaxial cable—the electric field within the wire is not set up immediately at all points along the wires. This is based on the same argument we used in Section 22–3 with reference to Fig. 22–5. Indeed, it can be shown that if the wires are separated by air, the electrical signal travels along the wires at the speed $c = 3.0 \times 10^8$ m/s. For example, when you flip a light switch, the light actually goes on a tiny fraction of a second later. If the wires are in a medium whose electric permittivity is ε and magnetic permeability is μ, the speed is not given by Eq. 22–3, but by $v = 1/\sqrt{\varepsilon\mu}$.

EXAMPLE 22–2 When you speak on the telephone from Los Angeles to a friend in New York 4000 km away, how long does it take your voice to travel?

SOLUTION Since speed = distance/time, then time = distance/speed = $(4.0 \times 10^6 \text{ m})/(3.0 \times 10^8 \text{ m/s}) = 1.3 \times 10^{-2}$ s, or about $\frac{1}{100}$ s.

*22–6 • Energy in EM Waves

Electromagnetic waves carry energy from one region of space to another. This energy is associated with the moving electric and magnetic fields. In Section 17–9, we saw that the energy density (J/m^3) stored in an electric field E is $u = \frac{1}{2}\varepsilon_0 E^2$, where u is the energy per unit volume. The energy stored in a magnetic field B, as we discussed in Section 21–10 (Eq. 21–10), is given by $u = \frac{1}{2}B^2/\mu_0$. Thus, the total energy stored per unit volume in a region of space where there is an electromagnetic wave is

$$u = \frac{1}{2}\varepsilon_0 E^2 + \frac{1}{2}\frac{B^2}{\mu_0}. \tag{22–5}$$

In this equation, E and B represent the electric and magnetic field strengths of the wave at any instant in a small region of space. We can write Eq. 22–5 in terms of the E field only, since from Eq. 22–3 we have $\sqrt{\varepsilon_0\mu_0} = 1/c$, and from Eq. 22–2, $B = E/c$. We insert these into Eq. 22–5 to obtain

$$\begin{aligned} u &= \frac{1}{2}\varepsilon_0 E^2 + \frac{1}{2}\frac{\varepsilon_0\mu_0 E^2}{\mu_0} \\ &= \varepsilon_0 E^2. \end{aligned} \tag{22–6a}$$

Notice that the energy density associated with the B field is equal to that associated with the E field, so each contributes half to the total energy. We can also write the energy density in terms of the B field only, or in one term containing both: $u = \varepsilon_0 E^2 = \varepsilon_0 c^2 B^2 = \varepsilon_0 B^2/\varepsilon_0\mu_0$, or

$$u = \frac{B^2}{\mu_0}; \tag{22–6b}$$

and $u = \varepsilon_0 E^2 = \varepsilon_0 EcB = \varepsilon_0 EB/\sqrt{\varepsilon_0\mu_0}$, or

$$u = \sqrt{\frac{\varepsilon_0}{\mu_0}}\, EB. \tag{22–6c}$$

Equations 22–6 give the energy density in any region of space at any instant.

Now let us determine the energy that is transported by the wave per unit time per unit area perpendicular to the wave direction. Let us imagine that the wave is passing through an area A perpendicular to the x axis, as shown in Fig. 22–11. In a short time Δt, the wave moves to the right a

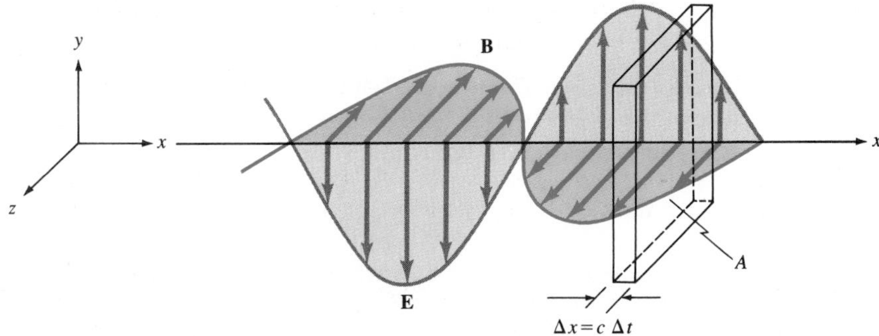

FIGURE 22–11 Electromagnetic wave carrying energy through area A.

distance $\Delta x = c\,\Delta t$. The energy that has passed through A in the time Δt is the energy that now occupies the volume $\Delta V = A\,\Delta x = Ac\,\Delta t$. The energy density u is $u = \varepsilon_0 E^2$, where E is the electric field in this volume at the given instant. So the energy ΔU contained in this volume is the energy density u times the volume: $\Delta U = u\,\Delta V = (\varepsilon_0 E^2)(Ac\,\Delta t)$. Therefore, the energy crossing the area A per time Δt, which we designate† S, is

$$S = \frac{\Delta U}{A\,\Delta t} = \varepsilon_0 c E^2,$$

and is measured in $\mathrm{W/m^2}$. Since $E = cB$ and $c = 1/\sqrt{\varepsilon_0 \mu_0}$, this can also be written

Rate energy is transported by EM waves

$$S = \varepsilon_0 c E^2 = \frac{cB^2}{\mu_0} = \frac{EB}{\mu_0}. \tag{22–7}$$

Equation 22–7 gives the energy transported per unit area per unit time at any *instant*. We often want to know the *average* over an extended period of time. If E and B are sinusoidal, then $\overline{E^2} = E_0^2/2$, just as for electric currents and voltages (Section 18–7), where E_0 is the *maximum* value of E. Thus we can write

$$\bar{S} = \frac{1}{2}\varepsilon_0 c E_0^2 = \frac{1}{2}\frac{c}{\mu_0}B_0^2 = \frac{E_0 B_0}{2\mu_0}, \tag{22–8}$$

where B_0 is the maximum value of B. Equation 22–7 also holds on the average (as well as instantaneously) if for E and B we use the rms values.

EXAMPLE 22–3 Radiation from the sun reaches the earth (above the atmosphere) at a rate of about $1350\ \mathrm{J/s{\cdot}m^2}$. Assume that this is a single EM wave and calculate the maximum values of E and B.

SOLUTION Since $\bar{S} = 1350\ \mathrm{J/s{\cdot}m^2} = \varepsilon_0 c E_0^2/2$, then

$$E_0 = \sqrt{\frac{2\bar{S}}{\varepsilon_0 c}} = \sqrt{\frac{2(1350\ \mathrm{J/s{\cdot}m^2})}{(8.85\times 10^{-12}\ \mathrm{C^2/N{\cdot}m^2})(3.0\times 10^{8}\ \mathrm{m/s})}}$$
$$= 1.01\times 10^{3}\ \mathrm{V/m}.$$

From Eq. 22–2, $B = E/c$, so

$$B_0 = \frac{E_0}{c} = \frac{1.01\times 10^{3}\ \mathrm{V/m}}{3.0\times 10^{8}\ \mathrm{m/s}} = 3.4\times 10^{-6}\ \mathrm{T}.$$

This example illustrates that B has a small numerical value compared to E. This is because of the different units for E and B and the way these units are defined. But, as we saw earlier, B contributes the same energy to the wave as E does.

† The quantity S is called the *Poynting vector*. Its direction is that in which the energy is being transported, which is the direction the wave is traveling.

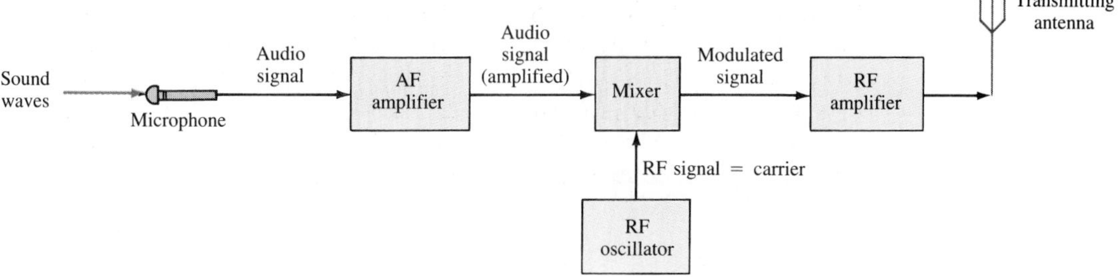

FIGURE 22–12 Block diagram of a radio transmitter.

*22–7 • Radio and Television

Electromagnetic waves offer the possibility of transmitting information over long distances. Among the first to realize this and put it in practice was Guglielmo Marconi (1874–1937), who, in the 1890s invented and developed the wireless telegraph. With it, messages could be sent hundreds of kilometers at the speed of light without the use of wires. The first signals were merely long and short pulses that could be translated into words by a code, such as the "dots" and "dashes" of the Morse code. The next decade saw the development of vacuum tubes. Out of this early work radio and television were born. We now discuss briefly (1) how radio and TV signals are transmitted, and (2) how they are received at home.

Transmission

The process by which a radio station transmits information (words and music) is outlined in Fig. 22–12. The audio (sound) information is changed into an electrical signal of the same frequencies by, say, a microphone, phonograph cartridge, or tape recorder head. This electrical signal is called an audio-frequency (AF) signal, since the frequencies are in the audio range (20 to 20,000 Hz). The signal is amplified† electronically and is then mixed with a radio-frequency (RF) signal. The RF frequency is determined by the values of L and C in a resonant LCR circuit (Section 21–14) and is chosen to produce a particular frequency for each station, called its **carrier frequency**. AM radio stations have carrier frequencies from about 530 to 1600 kHz. For example, "710 on your dial" means a station whose carrier frequency is 710 kHz. FM radio stations have much higher carrier frequencies, between 88 and 108 MHz. The carrier frequencies for TV stations in the United States lie between 54 and 88 MHz for channels 2 through 6, and between 174 and 216 MHz for channels 7 through 13; UHF (ultra-high-frequency) stations have even higher carrier frequencies, between 470 and 890 MHz.

Carrier frequency

AM and FM

The mixing of the audio and carrier frequencies can be done in two ways. In **amplitude modulation** (AM), the amplitude of the audio signal is combined with that of the much higher carrier frequency, as shown in Fig.

† How amplifiers work is discussed in Section 29–8.

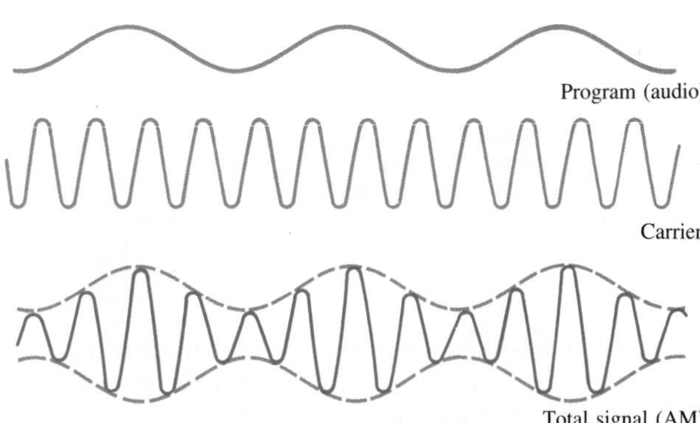

FIGURE 22–13
In amplitude modulation (AM), the amplitudes of the audio and carrier signals are added together.

22–13. It is called "amplitude modulation" because the amplitude of the carrier is altered ("modulate" means to change or alter). In **frequency modulation** (FM), the *frequencies* of the audio and carrier signals are combined. Thus, the *frequency* of the carrier wave is altered by the audio signal, as shown in Fig. 22–14.

The mixed signal is amplified further (since the signal contains radio frequencies, this is called an RF amplifier) and is then sent to the antenna, where the complex mixture of frequencies is sent out in the form of EM waves.

A television transmitter works in a similar way, using frequency modulation, except that both audio and video signals are mixed with carrier frequencies.

Radio and TV receivers

Now let us look at the other end of the process, the reception of radio and TV programs at home. A simple radio receiver is diagrammed in Fig. 22–15. The EM waves sent out by all stations are received by the antenna. One kind of antenna consists of one or more conducting rods; the electric field in the EM waves exert a force on the electrons in the conductor, causing them to move back and forth at the frequencies of the waves. A second type of antenna, often found in AM radios, consists of a tubular coil of wire. This type of antenna detects the magnetic field of the wave, for the changing B field induces an emf in the coil. The signal from the antenna is very small and contains frequencies from many different stations. The receiver selects out a particular RF frequency (actually a narrow range of frequencies) corresponding to a particular station using a resonant LC circuit (Section 21–14) with a variable capacitor or inductor. A simple example is shown in Fig.

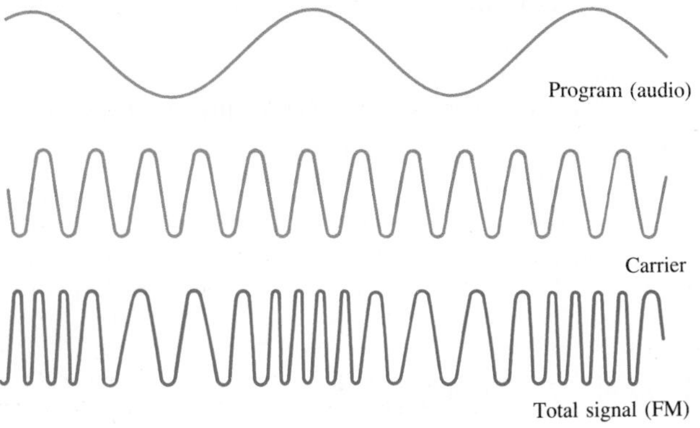

FIGURE 22–14
In frequency modulation (FM), the frequencies of the audio and carrier signals are added together. This method is used by FM radio and television.

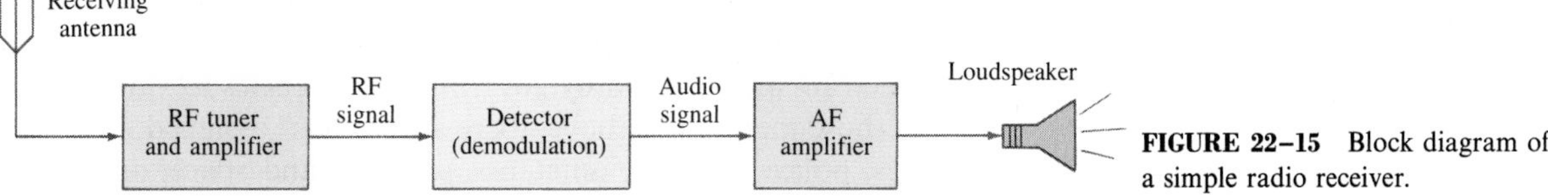

FIGURE 22–15 Block diagram of a simple radio receiver.

22–16. A particular station is "tuned-in" by adjusting L and C so that the resonant frequency of the circuit equals that of the station's carrier frequency. The RF signal may be amplified both before and after the tuning is done. The signal, containing both audio and carrier frequencies, next goes to the *detector* (Fig. 22–15) where "demodulation" takes place—that is, the RF carrier frequency is separated from the audio signal. The audio signal is then amplified and sent to a loudspeaker or headphones.

FIGURE 22–16 Simple tuning stage of a radio.

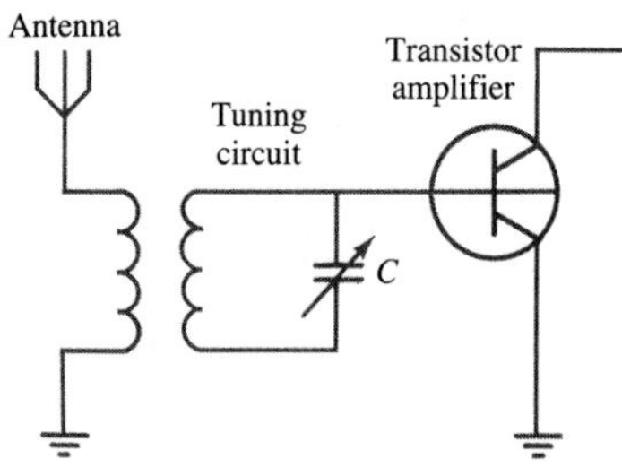

Modern receivers have more stages than those shown. Various means are used to increase the sensitivity and selectivity (ability to detect weak signals and distinguish them from other stations), and to minimize distortion of the original signal.†

A television receiver does similar things to both the audio and the video signals. The audio signal goes finally to the loudspeaker, and the video signal to the picture tube, a *cathode-ray tube* (CRT) whose operation is discussed in Section 20–10.

EXAMPLE 22–4 An FM radio station transmits at 100 MHz. Calculate (*a*) its wavelength, and (*b*) the value of the capacitance in the tuning circuit if $L = 0.40\ \mu\text{H}$.

SOLUTION (*a*) The carrier frequency is $f = 100\ \text{MHz} = 1.0 \times 10^8\ \text{s}^{-1}$. From Eq. 22–4, $\lambda = c/f = (3.0 \times 10^8\ \text{m/s})/(1.0 \times 10^8\ \text{s}^{-1}) = 3.0\ \text{m}$. The wavelengths of other FM signals (88 to 108 MHz) are close to this. FM antennas are typically 1.5 m long, or about a half wavelength. This length is chosen so that the antenna reacts in a resonant fashion and thus is more sensitive.

(*b*) According to Eq. 21–18, the resonant frequency is $f_0 = 1/(2\pi\sqrt{LC})$. Therefore,

$$C = \frac{1}{4\pi^2 f_0^2 L} = \frac{1}{4(3.14)^2(1.0 \times 10^8\ \text{s}^{-1})^2(4.0 \times 10^{-7}\ \text{H})}$$

$$= 6.3\ \text{pF}.$$

Of course, the capacitor or inductor is variable, so other stations can be selected.

† For *FM stereo broadcasting*, two signals are carried by the carrier wave. One of these contains frequencies up to about 17,000 Hz, which includes most audio frequencies. The other signal includes the same range of frequencies, but 21,000 Hz is added to it. A stereo receiver subtracts this 21,000-Hz signal and distributes the two signals to the left and right channels. The first signal actually consists of the sum of left and right channels ($L + R$), so mono radios detect all the sound. The second signal is the difference between left and right ($L - R$). Hence the receiver must add and subtract the two signals to get pure left and right signals for each channel.

Citizens' band (CB) receivers are similar to AM radios, except that the carrier frequencies are in the vicinity of 27 MHz. The various regions of the radio-wave spectrum are assigned by governmental agencies to various purposes. Besides those mentioned above, there are "bands" assigned for use by ships, airplanes, police, military, amateurs, satellites and space, and radar.

SUMMARY

James Clerk Maxwell synthesized an elegant theory in which all electric and magnetic phenomena could be described using four equations, now called *Maxwell's equations.* They are based on earlier ideas, but Maxwell added one more—that a changing electric field produces a magnetic field.

Maxwell's theory predicted that transverse *electromagnetic* (EM) *waves* would be produced by accelerating electric charges, and these waves would propagate through space at the speed of light. The oscillating electric and magnetic fields in an EM wave are perpendicular to each other and to the direction of propagation.

After EM waves were experimentally detected in the late 1800s, the idea that light is an EM wave (of very high frequency) became generally accepted. The *electromagnetic spectrum* includes EM waves of a wide variety of wavelengths, from microwaves and radio waves to visible light to X rays and γ rays, all of which travel through space at a speed $c = 3.0 \times 10^8$ m/s.

QUESTIONS

***1.** What is the direction of the displacement current in Fig. 22–3? (Note: the capacitor is discharging.)

***2.** Suppose you are looking along the same direction as an electric field **E** that is increasing. Will the induced magnetic field be clockwise or counterclockwise? What if **E** points toward you and is decreasing?

3. The electric field in an EM wave traveling north oscillates in an east–west plane. Describe the direction of the magnetic field vector in this wave.

4. Is sound an electromagnetic wave? If not, what kind of wave is it?

5. Can EM waves travel through a perfect vacuum? Can sound waves?

6. How are light and sound alike? How are they different?

7. Are the wavelengths of radio and television signals longer or shorter than those detectable by the human eye?

8. What does the result of Example 22–1 tell you about the phase of a 60-Hz ac current that starts at a power plant as compared to its phase at a house 200 km away?

9. When you connect two loudspeakers to the output of a stereo amplifier, should you be sure the lead wires are equal in length so that there will not be a time lag between speakers? Explain.

10. In the electromagnetic spectrum, what type of EM wave would have a wavelength of 10^3 km? 1 km? 1 m? 1 cm? 1 mm? 1 μm?

***11.** A lost person may signal by flashing a flashlight on and off using Morse code. This is actually a modulated EM wave. Is it AM or FM? What is the frequency of the carrier, approximately?

***12.** Can two radio or TV stations broadcast on the same carrier frequency? Explain.

***13.** If a radio transmitter has a vertical antenna, should a receiver's antenna be vertical or horizontal to obtain best reception?

***14.** The carrier frequencies of FM broadcasts are much higher than for AM broadcasts. On the basis of what you learned about diffraction in Chapter 11, explain why AM signals can be detected more readily than FM signals behind low hills or buildings.

15. Discuss how cordless telephones make use of EM waves.

PROBLEMS

*SECTION 22–2

*1. (I) Calculate the displacement current I_D between the square plates, 2.0 cm on a side, of a capacitor if the electric field is changing at a rate of 2.0×10^6 V/m·s.

*2. (II) At a given instant, a 4.0-A current flows in the wires connected to a parallel-plate capacitor. What is the rate at which the electric field is changing between the plates if the square plates are 1.00 cm on a side?

*3. (II) A 7.2-μF capacitor with parallel circular plates accumulates charge at a rate of 25 mC/s. What will be the magnetic field strength 12 cm radially outward from the center of the plates whose radii are 0.75 cm?

*4. (II) Show that the displacement current through a parallel-plate capacitor can be written $I_D = C\,\Delta V/\Delta t$, where V is the voltage across the capacitor at any instant.

*5. (III) The electric field between two parallel circular capacitor plates (capacitance C) changes at a rate $\Delta E/\Delta t$. (*a*) If the radius of the plates is R, show that the magnetic field B a distance r from the center of the plates, if $r \leqslant R$, is $B = \frac{1}{2}\mu_0\varepsilon_0 r(\Delta E/\Delta t)$; and outside the plates, $r \geqslant R$, show that $B = (\mu_0\varepsilon_0 R^2/2r)(\Delta E/\Delta t)$. (*b*) Show that the magnetic field beyond the edges of the capacitor plates is given by

$$B = \frac{\mu_0 I_D}{2\pi r},$$

where I_D is the displacement current. (*c*) This is the same formula for the field outside a straight wire; explain the similarity.

*SECTION 22–4

*6. (I) If the electric field in an EM wave has a peak of 0.45×10^{-4} V/m, what is the peak value of the magnetic field strength?

*7. (I) In an EM wave traveling west, the B field oscillates vertically and has a frequency of 80.0 kHz and an rms strength of 7.75×10^{-9} T. What is the frequency and rms strength of the electric field and what is its direction?

SECTION 22–5

8. (I) What is the wavelength of a 12.25×10^9 Hz radar signal?

9. (I) What is the frequency of a microwave whose wavelength is 2.50 cm?

10. (II) Who will hear the voice of a singer first—a person in the balcony 50 m away from the stage, or a person 3000 km away at home whose ear is next to the radio? How much sooner? Assume that the microphone is a few centimeters from the singer and the temperature is 20°C.

*SECTION 22–6

*11. (I) The **E** field in an EM wave has a peak of 32 mV/m. What is the average rate at which this wave carries energy across unit area per unit time?

*12. (I) The magnetic field in an EM wave has an rms strength of 1.35×10^{-8} T. How much energy does this wave transport per meter squared per second?

*13. (II) How much energy is transported across a 1.00-cm^2 area per hour by an EM wave whose E field has an rms strength of 21.5 mV/m?

*14. (II) What is the energy contained in a 1.00-m^3 volume near the earth's surface due to radiant energy from the sun? See Example 22–3.

*15. (II) A 15.0-mW laser puts out a narrow beam 2.00 mm in diameter. What are the average (rms) values of E and B in the beam?

*16. (II) Estimate the average power output of the sun, given that about 1350 W/m^2 reaches the upper atmosphere of the earth.

*SECTION 22–7

*17. (I) An FM station broadcasts at 102.1 MHz. What is the wavelength of this wave?

*18. (I) What is the wavelength of an AM station at 810 on the dial?

*19. (I) Compare 940 on the AM dial to 94 on the FM dial. Which has the longer wavelength, and by what factor is it larger?

*20. (I) The variable capacitor in the tuner of an AM radio has a capacitance of 1800 pF when the radio is tuned to a station at 550 kHz. What must the capacitance be for a station at the other end of the dial, 1550 kHz?

*21. (II) The oscillator of a 95.7-MHz FM station has an inductance of 0.85 μH. What value must the capacitance be?

*22. (II) A certain FM radio tuning circuit has a fixed capacitor $C = 780$ pF. Tuning is done by a variable inductance. What range of values must the inductance have to tune stations from 88 to 108 MHz?

*23. (II) An amateur radio operator wishes to build a receiver that can tune a range from 14.0 MHz to 15.0 MHz. A variable capacitor has a minimum capacitance of 92 pF. (*a*) What is the required value of the inductance? (*b*) What is the maximum capacitance used on the variable capacitor?

*24. (II) A 1.40-m-long FM antenna is oriented parallel to the electric field of an EM wave. How large must the **E** field strength be to produce a 1.00-mV (rms) voltage between the ends of the antenna? What is the rate of energy transport per square meter?

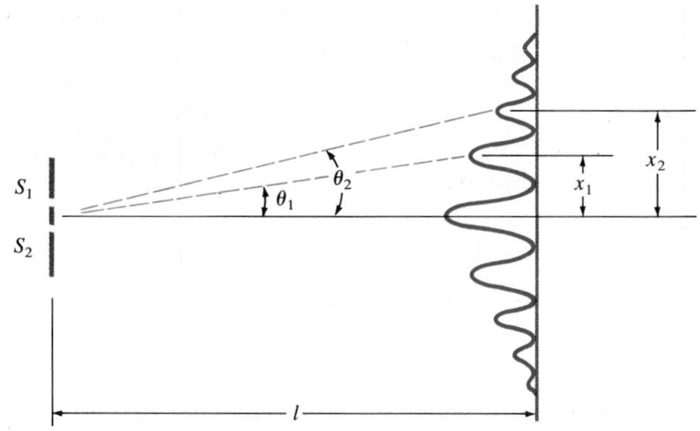

FIGURE 24–8 For small angles, the interference fringes occur at distance $x = \theta l$ above the center ($m = 0$) fringe. θ_1 and x_1 are for the first order ($m = 1$) fringe, θ_2 and x_2 are for $m = 2$.

SOLUTION Since $d = 0.100$ mm $= 1.00 \times 10^{-4}$ m, $\lambda = 500 \times 10^{-9}$ m, and $l = 1.20$ m, the first-order fringe ($m = 1$) occurs at an angle θ given by

$$\sin\theta_1 = \frac{m\lambda}{d} = \frac{(1)(500 \times 10^{-9}\text{ m})}{1.00 \times 10^{-4}\text{ m}} = 5.00 \times 10^{-3}.$$

This is a very small angle, so we can take $\sin\theta = \theta$. The first-order fringe will occur a distance x_1 above the center of the screen (see Fig. 24–8) given by $x_1/l = \theta_1$, so

$$x_1 = l\theta_1 = (1.20\text{ m})(5.00 \times 10^{-3}) = 6.00\text{ mm}.$$

The second fringe ($m = 2$) will occur at

$$x_2 = l\theta_2 = l\frac{2\lambda}{d} = 12.0\text{ mm}$$

above the center, and so on. Thus the fringes are 6.00 mm apart.

From Eqs. 24–2 we can see that, except for the zeroth order fringe at the center, the position of the fringes depends on wavelength. Consequently, when white light falls on the two slits, as Young found in his experiments, the central fringe is white, but the first- (and higher-) order fringes contain a spectrum of colors like a rainbow; θ was found to be smallest for violet light and largest for red. By measuring the position of these fringes, Young was the first to determine the wavelengths of visible light (using Eqs. 24–2). In doing so, he showed that what distinguishes different colors physically is their wavelength, an idea put forward earlier by Grimaldi in 1665.

EXAMPLE 24–2 White light passes through two slits 0.50 mm apart and an interference pattern is observed on a screen 2.5 m away. The first-order fringe resembles a rainbow with violet and red light at either end. The violet light falls about 2.0 mm and the red 3.5 mm from the center of the central white fringe. Estimate the wavelengths of the violet and red lights.

SOLUTION We use Eq. 24–2a with $m = 1$ and $\sin\theta = \theta$. Then for violet light, $x = 2.0$ mm, so

$$\lambda = \frac{d\theta}{m} = \frac{d}{m}\frac{x}{l} = \left(\frac{5.0 \times 10^{-4}\text{ m}}{1}\right)\left(\frac{2.0 \times 10^{-3}\text{ m}}{2.5\text{ m}}\right) = 4.0 \times 10^{-7}\text{ m},$$

or 400 nm. For red light, $x = 3.5$ mm, so λ is about 700 nm.

PROBLEMS

*SECTION 22–2

*1. (I) Calculate the displacement current I_D between the square plates, 2.0 cm on a side, of a capacitor if the electric field is changing at a rate of 2.0×10^6 V/m·s.

*2. (II) At a given instant, a 4.0-A current flows in the wires connected to a parallel-plate capacitor. What is the rate at which the electric field is changing between the plates if the square plates are 1.00 cm on a side?

*3. (II) A 7.2-μF capacitor with parallel circular plates accumulates charge at a rate of 25 mC/s. What will be the magnetic field strength 12 cm radially outward from the center of the plates whose radii are 0.75 cm?

*4. (II) Show that the displacement current through a parallel-plate capacitor can be written $I_D = C\,\Delta V/\Delta t$, where V is the voltage across the capacitor at any instant.

*5. (III) The electric field between two parallel circular capacitor plates (capacitance C) changes at a rate $\Delta E/\Delta t$. (*a*) If the radius of the plates is R, show that the magnetic field B a distance r from the center of the plates, if $r \leqslant R$, is $B = \frac{1}{2}\mu_0\varepsilon_0 r(\Delta E/\Delta t)$; and outside the plates, $r \geqslant R$, show that $B = (\mu_0\varepsilon_0 R^2/2r)(\Delta E/\Delta t)$. (*b*) Show that the magnetic field beyond the edges of the capacitor plates is given by

$$B = \frac{\mu_0 I_D}{2\pi r},$$

where I_D is the displacement current. (*c*) This is the same formula for the field outside a straight wire; explain the similarity.

*SECTION 22–4

*6. (I) If the electric field in an EM wave has a peak of 0.45×10^{-4} V/m, what is the peak value of the magnetic field strength?

*7. (I) In an EM wave traveling west, the B field oscillates vertically and has a frequency of 80.0 kHz and an rms strength of 7.75×10^{-9} T. What is the frequency and rms strength of the electric field and what is its direction?

SECTION 22–5

8. (I) What is the wavelength of a 12.25×10^9 Hz radar signal?

9. (I) What is the frequency of a microwave whose wavelength is 2.50 cm?

10. (II) Who will hear the voice of a singer first—a person in the balcony 50 m away from the stage, or a person 3000 km away at home whose ear is next to the radio? How much sooner? Assume that the microphone is a few centimeters from the singer and the temperature is 20°C.

*SECTION 22–6

*11. (I) The **E** field in an EM wave has a peak of 32 mV/m. What is the average rate at which this wave carries energy across unit area per unit time?

*12. (I) The magnetic field in an EM wave has an rms strength of 1.35×10^{-8} T. How much energy does this wave transport per meter squared per second?

*13. (II) How much energy is transported across a 1.00-cm^2 area per hour by an EM wave whose E field has an rms strength of 21.5 mV/m?

*14. (II) What is the energy contained in a 1.00-m^3 volume near the earth's surface due to radiant energy from the sun? See Example 22–3.

*15. (II) A 15.0-mW laser puts out a narrow beam 2.00 mm in diameter. What are the average (rms) values of E and B in the beam?

*16. (II) Estimate the average power output of the sun, given that about 1350 W/m^2 reaches the upper atmosphere of the earth.

*SECTION 22–7

*17. (I) An FM station broadcasts at 102.1 MHz. What is the wavelength of this wave?

*18. (I) What is the wavelength of an AM station at 810 on the dial?

*19. (I) Compare 940 on the AM dial to 94 on the FM dial. Which has the longer wavelength, and by what factor is it larger?

*20. (I) The variable capacitor in the tuner of an AM radio has a capacitance of 1800 pF when the radio is tuned to a station at 550 kHz. What must the capacitance be for a station at the other end of the dial, 1550 kHz?

*21. (II) The oscillator of a 95.7-MHz FM station has an inductance of 0.85 μH. What value must the capacitance be?

*22. (II) A certain FM radio tuning circuit has a fixed capacitor $C = 780$ pF. Tuning is done by a variable inductance. What range of values must the inductance have to tune stations from 88 to 108 MHz?

*23. (II) An amateur radio operator wishes to build a receiver that can tune a range from 14.0 MHz to 15.0 MHz. A variable capacitor has a minimum capacitance of 92 pF. (*a*) What is the required value of the inductance? (*b*) What is the maximum capacitance used on the variable capacitor?

*24. (II) A 1.40-m-long FM antenna is oriented parallel to the electric field of an EM wave. How large must the **E** field strength be to produce a 1.00-mV (rms) voltage between the ends of the antenna? What is the rate of energy transport per square meter?

GENERAL PROBLEMS

*25. A point source emits light energy uniformly in all directions at an average rate P_0 with a single frequency f. Show that the peak electric field in the wave is given by

$$E_0 = \sqrt{\frac{\mu_0 c P_0}{2\pi r^2}}.$$

*26. What are E_0 and B_0 1.0 m from a 100-W light source? Assume the bulb emits radiation of a single frequency uniformly in all directions.

*27. Suppose a 50-kW radio station emits EM waves uniformly in all directions. (*a*) How much energy per second crosses a 1.0-m^2 area 100 m from the transmitting antenna? (*b*) What is the rms magnitude of the **E** field at this point, assuming the station is operating at full power? (*c*) What is the voltage induced in a 1.0-m-long vertical car antenna?

*28. Repeat Problem 27 for a distance of 20 km from the station.

*29. How large an emf (rms) will be generated in an antenna that consists of a 600-loop circular coil of wire 1.5 cm in diameter if the EM wave has a frequency of 810 kHz and is transporting energy at an average rate of 1.0×10^{-4} W/m^2 at the antenna. [*Hint:* you can use Eq. 21–5, since it could be applied to an observer moving with the coil so that the magnetic field is oscillating with the frequency $f = \omega/2\pi$.]

*30. The variable capacitance of a radio tuner consists of six plates connected together placed alternately between six other plates, also connected together. Each plate is separated from its neighbor by 1.2 mm of air. One set of plates can move so that the area of overlap varies from 1.0 cm^2 to 9.0 cm^2. (*a*) Are these capacitors connected in series or in parallel? (*b*) Determine the range of capacitance values. (*c*) What value of inductor is needed if the radio is to tune AM stations from 550 to 1600 kHz?

CHAPTER 23

Light: Geometric Optics

Reflection from still water, as from a glass mirror, can be analyzed using the ray model of light.

The sense of sight is extremely important to us, for it provides us with a large part of our information about the world. How do we see? What is the something called *light* that enters our eyes and causes the sensation of sight? How does light behave so that we can see the great range of phenomena that we do? The subject of light will occupy us for the next three chapters, and we will return to it in later chapters.

We see an object in one of two ways: (1) the object may be a source, such as a light bulb, a flame, or a star, in which case we see the light emitted directly from the source; or (2), more commonly, we see an object by light reflected from it. In the latter case, the light may have originated from the sun, artificial lights, or some other source. An understanding of how bodies *emit* light was not achieved until the 1920s, and this will be discussed in Chapter 27. How light is *reflected* from objects was understood much earlier, and we will discuss this in Section 23–3.

23–1 • The Ray Model of Light

A great deal of evidence suggests that *light travels in straight lines* under a wide variety of circumstances. For example, a point source of light like the sun casts distinct shadows; and the beam of a flashlight appears to be a straight line. In fact, we infer the positions of objects in our environment by assuming that light moves from the object to our eyes in straight-line paths. Our whole orientation to the physical world is based on this assumption.

Light rays

This reasonable assumption has led to the **ray model** of light. This model assumes that light travels in straight-line paths called light **rays**. Actually, a ray is an idealization: it is meant to represent an infinitely narrow beam of light. When we see an object, light reaches our eyes from each point on the object. Although light rays leave each point in many different directions, normally only a small bundle of these rays can enter an observer's eye, as shown in Fig. 23–1. If the person's head moves to one side, a different bundle of rays will enter the eye from each point.

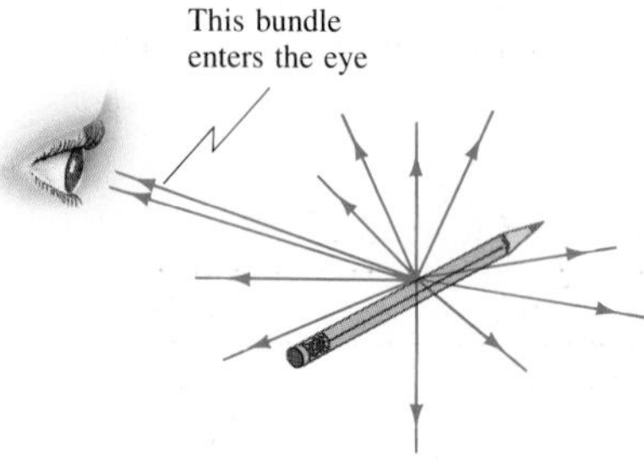

FIGURE 23–1 Light rays emerge from each single point on an object. A small bundle of rays leaving one point is shown entering a person's eye.

We saw in Chapter 22 that light can be considered as an electromagnetic wave. Although the ray model of light does not deal with this aspect of light (we discuss the wave nature of light in Chapter 24), it has been very successful in describing many aspects of light such as reflection, refraction, and the formation of images by mirrors and lenses. Because these explanations involve straight-line rays at various angles, this subject is referred to as **geometric optics**.

23–2 • The Speed of Light and Index of Refraction

Galileo attempted to measure the speed of light by trying to measure the time required for light to travel a known distance between two hilltops. He stationed an assistant on one hilltop and himself on another and ordered the assistant to lift the cover from a lamp the instant he saw a flash from Galileo's lamp. Galileo measured the time between the flash of his lamp and when he received the light from his assistant's lamp. The time was so short that Galileo concluded it merely represented human reaction time, and that the speed of light must be extremely high.

FIGURE 23–2 Albert A. Michelson.

The first successful determination that the speed of light is finite was made by the Danish astronomer Ole Roemer (1644–1710). Roemer had noted that the carefully measured period of one of Jupiter's moons (Io, with an average period of 42.5 h) varied slightly, depending on the relative motion of Earth and Jupiter. When Earth was moving away from Jupiter, the period of the moon was slightly longer, and when Earth was moving toward Jupiter, the period was slightly shorter. He attributed this variation to the extra time needed for light to travel the increasing distance to Earth when Earth is receding, or to the shorter travel time for the decreasing distance when the two planets are approaching one another. Roemer concluded that the speed of light—though great—is finite.

Michelson measures c

Since then a number of techniques have been used to measure the speed of light. Among the most important were those carried out by the American Albert A. Michelson (1852–1931; Fig. 23–2). Michelson used the rotating

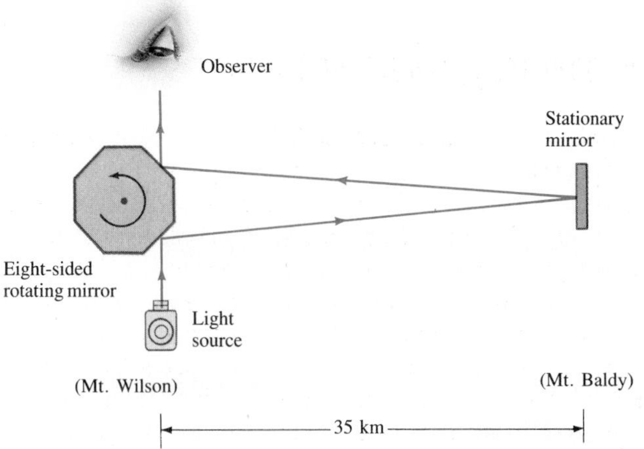

FIGURE 23–3 Michelson's speed-of-light apparatus (not to scale).

mirror apparatus diagrammed in Fig. 23–3 for a series of high-precision experiments carried out from 1880 to the 1920s. Light from a source was directed at one face of a rotating eight-sided mirror. The reflected light traveled to a stationary mirror a large distance away and back again as shown. If the rotating mirror was turning at just the right rate, the returning beam of light would reflect from one face of the mirror into a small telescope through which the observer looked. If the speed of rotation was only slightly different, the beam would be deflected to one side and would not be seen by the observer. From the required speed of the rotating mirror and the known distance to the stationary mirror, the speed of light could be calculated. In the 1920s, Michelson set up the rotating mirror on the top of Mt. Wilson in southern California and the stationary mirror on Mt. Baldy (Mt. San Antonio) 35 km away. He later measured the speed of light in vacuum using a long evacuated tube.

TABLE 23–1
Indices of Refraction†

Material	$n = c/v$
Air (at STP)	1.0003
Water	1.33
Ethyl alcohol	1.36
Glass	
Fused quartz	1.46
Crown glass	1.52
Light flint	1.58
Lucite or Plexiglas	1.51
Sodium chloride	1.53
Diamond	2.42

† $\lambda = 589$ nm

The accepted value today for the speed of light, c, in vacuum is†

Speed of light in vacuum

$$c = 2.99792458 \times 10^8 \text{ m/s}.$$

We usually round this off to

$$3.00 \times 10^8 \text{ m/s}$$

when extremely precise results are not required. In air, the speed is only slightly less. In other transparent materials such as glass and water, the speed is always less than that in vacuum. For example, in water light travels at about $\frac{3}{4}c$. The ratio of the speed of light in vacuum to the speed v in a given material is called the **index of refraction**, n, of that material:

Index of refraction

$$n = \frac{c}{v}. \qquad (23\text{–}1)$$

The index of refraction is never less than 1 (that is, $n \geqslant 1$), and its value for various materials is given in Table 23–1. (As we shall see later, n varies somewhat with the wavelength of the light—except in vacuum—so a particular wavelength is specified, that of yellow light with wavelength $\lambda = 589$ nm.)

† As discussed in Chapter 1 (p. 9), the speed of light is now *defined* to have this value, and the length of the standard meter is defined in terms of this value of c.

For example, since $n = 2.42$ for diamond, the speed of light in diamond is

$$v = \frac{c}{n} = \frac{c}{2.42} = 0.413c$$

or

$$v = \frac{3.00 \times 10^8 \text{ m/s}}{2.42} = 1.24 \times 10^8 \text{ m/s}.$$

23–3 • Reflection; Image Formation by a Plane Mirror

When light strikes the surface of an object, some of the light is reflected. The rest is either absorbed by the object (and transformed to thermal energy) or, if the object is transparent like glass or water, part of it is transmitted through. For a very shiny object such as a silvered mirror, over 95 percent of the light may be reflected.

When a narrow beam of light strikes a flat surface we define the **angle of incidence**, θ_i, to be the angle an incident ray makes with the normal to the surface ("normal" means perpendicular) and the **angle of reflection,** θ_r, to be the angle the reflected ray makes with the normal. For flat surfaces, it is found that *the incident and reflected rays lie in the same plane with the normal to the surface*, and that

Law of reflection

the angle of incidence equals the angle of reflection.

Normal to surface
θ_i Angle of incidence
θ_r Angle of reflection
Light ray

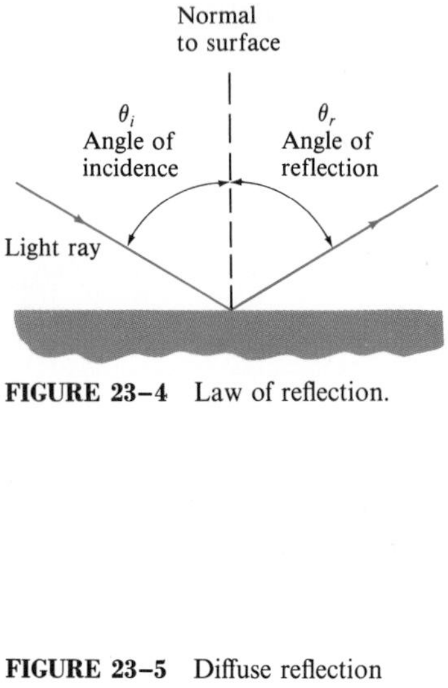

FIGURE 23–4 Law of reflection.

This is the **law of reflection** and is indicated in Fig. 23–4. It was known to the ancient Greeks, and you can confirm it yourself by shining a narrow flashlight beam at a mirror in a darkened room.

When light is incident upon a rough surface, even microscopically rough such as this page, it is reflected in many directions, Fig. 23–5. This is called **diffuse reflection**. The law of reflection still holds, however, at each small section of the surface. Because of diffuse reflection in all directions, an ordinary object can be seen from many different angles. When you move your head to the side, a different bundle of reflected rays reach your eye from each point on the object, Fig. 23–6a. Let us compare diffuse reflection to reflection from a mirror, which is known as *specular* reflection ("speculum" is latin for mirror); when a narrow beam of light is shone on a mirror, the light will not reach your eye unless it is placed at just the right place where the law of reflection is satisfied, as shown in Fig. 23–6b. This is what gives rise to the unusual properties of mirrors. (Galileo, using similar arguments,† showed that the moon must have a rough surface rather than a highly polished surface like a mirror, as some people thought.)

FIGURE 23–5 Diffuse reflection from a rough surface.

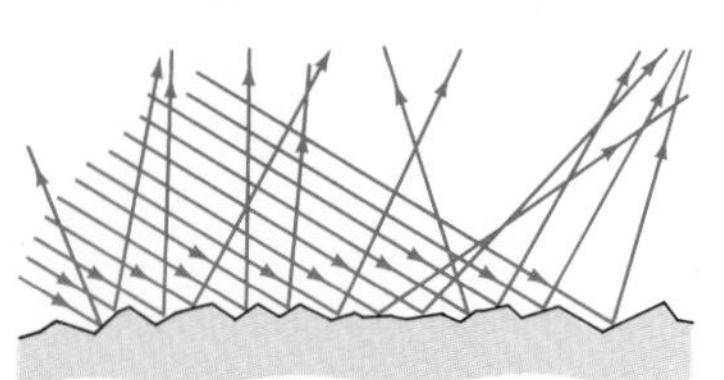

When you look straight in a mirror, you see what appears to be yourself as well as various objects around and behind you. Your face and the other objects look as if they are in front of you, beyond the mirror; but, of course, they are not. What you see in the mirror is an **image** of the objects.

† Galileo Galilei, *Dialogue Concerning the Two Chief World Systems*, trans. Stillman Drake (Berkeley: University of California Press, 1967). See Day 1.

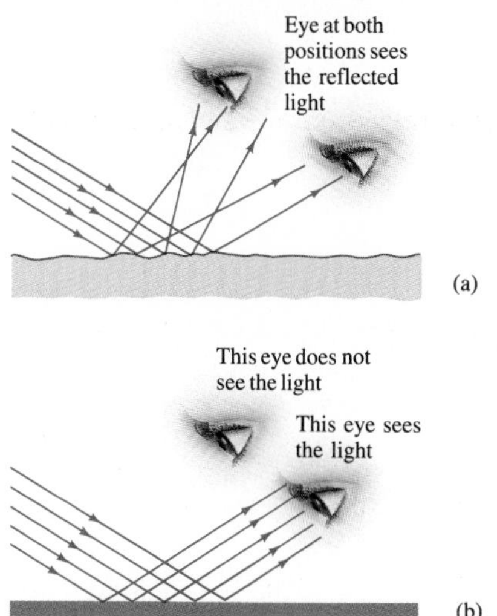

FIGURE 23–6 A beam of light from a flashlight is shined on (a) white paper, and (b) a small mirror. In part (a), you can see the white light reflected at various points because of diffuse reflection. But in part (b), you see the reflected light only when your eye is placed correctly ($\theta_r = \theta_i$); this is known as specular reflection.

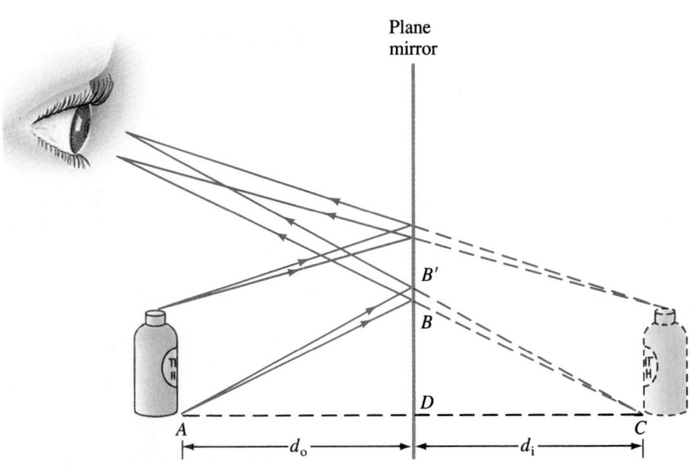

FIGURE 23–7 Formation of a virtual image by a plane mirror.

Figure 23–7 shows how an image is formed by a plane (that is, flat) mirror. Rays from two different points on an object are shown. Rays leave each point on the object going in many directions, but only those that enclose the bundle of rays that reach the eye from the two points are shown. The diverging rays that enter the eye appear to come from behind the mirror as shown by the dashed lines. (Our eyes and brain interpret any rays that enter an eye as having traveled a straight-line path.) The point from which each bundle of rays seems to come is one point on the image. For each point on the object, there is a corresponding image point. Let us concentrate on the two rays that leave the point A on the object and strike the mirror at points B and B'. The angles ADB and CDB are right angles. And angles ABD and CBD are equal because of the law of reflection. Therefore, the two triangles ABD and CDB are congruent and the length $AD = CD$. That is, the image is as far behind the mirror as the object is in front: the **image distance**, d_i (distance from mirror to image, Fig. 23–7), equals the **object distance**, d_o. From the geometry, we also see that the height of the image is the same as that of the object.

Real and virtual images

The light rays do not actually pass through the image itself. It merely *seems* like the light is coming from the image because our brains interpret any light entering our eyes as coming from in front of us. Since the rays do not actually pass through the image, a piece of white paper or film placed at the image would not detect the image. Therefore, it is called a **virtual image**. This is to distinguish it from a **real image** in which the light does pass through the image and which therefore could appear on paper or film placed at the image position. We will see that curved mirrors and lenses can form real images.

EXAMPLE 23–1 A woman 1.60 m tall stands in front of a vertical plane mirror. What is the minimum height of the mirror and how high must its lower edge be above the floor if she is to be able to see her whole body? (Assume her eyes are 10 cm below the top of her head.)

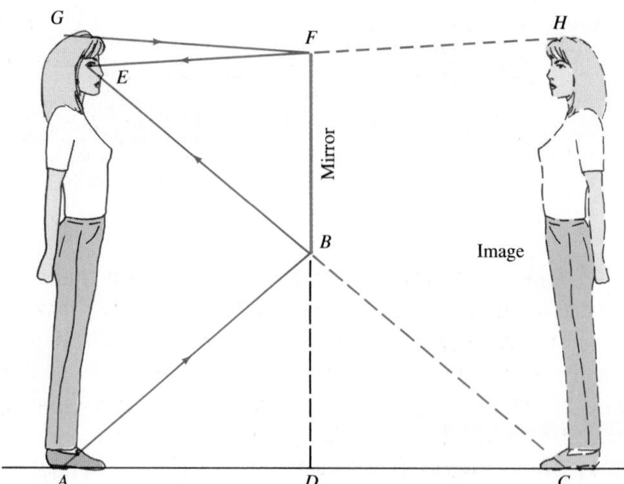

FIGURE 23–8
Seeing oneself in the mirror (Example 23–1).

How tall a mirror is needed to see your whole reflection?

SOLUTION The situation is diagrammed in Fig. 23–8. First consider the ray from the toe, AB, which upon reflection becomes BE and enters the eye E. Since light enters the eye from point A (the toes) after reflecting at B, the mirror needs to extend no lower than B. Because the angle of reflection equals the angle of incidence, the height BD is half of the height AE. Since $AE = 1.60\text{ m} - 0.10\text{ m} = 1.50\text{ m}$, $BD = 0.75\text{ m}$. Similarly, if the woman is to see the top of her head, the top edge of the mirror only needs to reach point F, which is 5 cm below the top of her head (half of $GE = 10\text{ cm}$). Thus, $DF = 1.55\text{ m}$, and the mirror need be only $(1.55\text{ m} - 0.75\text{ m}) = 0.80\text{ m}$ high; and its bottom edge must be 0.75 m above the floor. In general, a mirror need be only half as tall as a person for that person to see all of himself or herself. Does this result depend on the person's distance from the mirror?

*23–4 • Formation of Images by Spherical Mirrors

Convex and concave mirrors

Reflecting surfaces do not have to be flat. The most common *curved* mirrors are *spherical*, which means they form a section of a sphere. A spherical mirror is called **convex** if the reflection takes place on the outer surface of the spherical shape so that the center of the mirror bulges out toward the viewer. A mirror is called **concave** if the reflecting surface is on the inner surface of the sphere so that the center of the mirror bulges away from the viewer. Concave mirrors are used as shaving or makeup mirrors, and convex mirrors are sometimes used on cars and trucks (rearview mirrors) and in shops (to watch for thieves), since they take in a wide field of view.

To see how spherical mirrors form images, we first consider an object that is very far from a concave mirror. In this case, the rays from each point on the object that reach the mirror will be nearly parallel, as shown

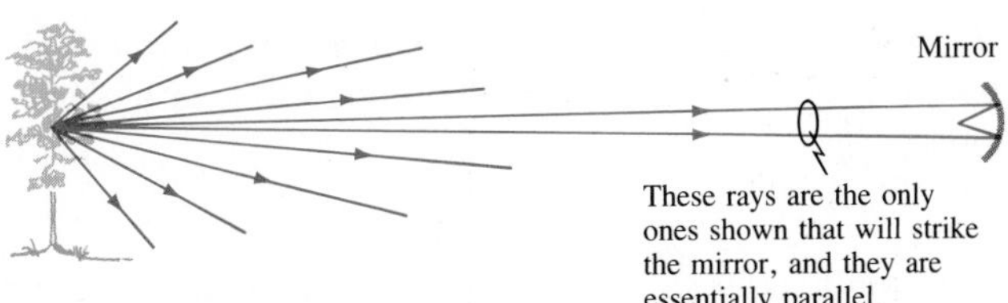

FIGURE 23–9 If the object's distance is large compared to the size of the mirror, the rays are nearly parallel.

in Fig. 23–9. *For an object infinitely far away* (the sun and stars approach this), *the rays would be precisely parallel.* Now consider such parallel rays falling on a concave mirror as in Fig. 23–10. The law of reflection holds for each of these rays at the point each strikes the mirror. As can be seen, they are not all brought to a single point. In order to form a sharp image, the rays must come to a point. Thus a spherical mirror will not make as sharp an image as a plane mirror will. However, if the mirror is small compared to its radius of curvature so that the reflected rays make only a small angle upon reflection, then the rays will cross each other at nearly a single point, or **focus**, as shown in Fig. 23–11. In the case shown, the rays are parallel to the **principal axis**, which is defined as the straight line perpendicular to the curved surface at its center (line CA in the diagram). The point F, where rays parallel to the principal axis come to a focus, is called the **focal point** of the mirror. The distance between F and the center of the mirror, length FA, is called the **focal length**, f, of the mirror. Another way of defining the focal point is to say that it is the *image point for an object infinitely far away* along the principal axis. The image of the sun, for example, would be essentially at F.

FIGURE 23–10 Parallel rays striking a concave spherical mirror do not focus at a single point.

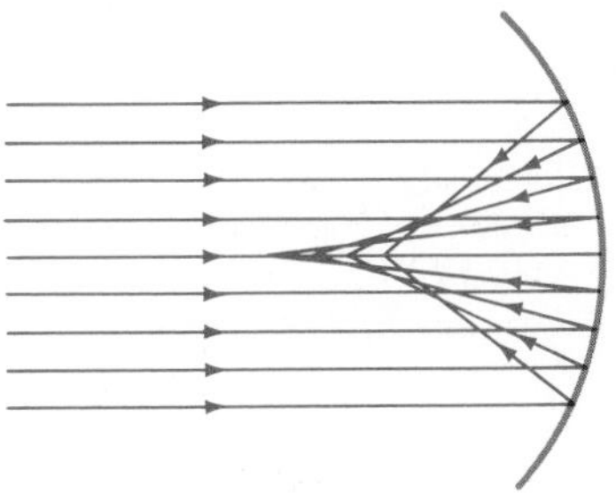

Focal point

Focal length

Now we will show, for a mirror whose reflecting surface is small compared to the radius of curvature, that the rays do indeed meet at a common point, F, and we calculate the focal length f. We consider a ray that strikes the mirror at B in Fig. 23–11. The point C is the center of curvature of the mirror (the center of the sphere of which the mirror is a part). So the dashed line CB is equal to r, the radius of curvature, and CB is normal to the surface at B. From the law of reflection and the geometry, the three angles labeled θ are equal as indicated. The triangle CBF is isosceles because two of its angles are equal. Thus length $CF = BF$. We assume the mirror is small compared to its radius of curvature, so the angles are small and the length FB

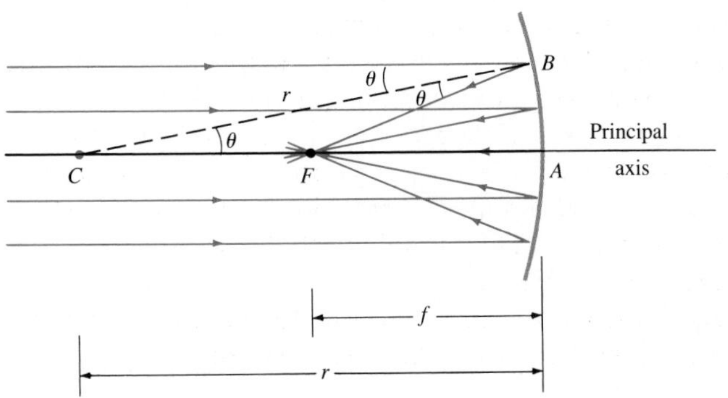

FIGURE 23–11 Rays parallel to the principal axis of a spherical mirror come to a focus at F, called the focal point, as long as the mirror is small in extent as compared to its radius of curvature, r.

is nearly equal to length FA. In this approximation, $FA = FC$. But $FA = f$, the focal length, and $CA = 2FA = r$. Thus the focal length is half the radius of curvature:

Focal length of mirror

$$f = r/2. \tag{23–2}$$

This argument only assumed that the angle θ was small, so the same result applies for all the other rays. Thus all the rays pass through the same point F in this approximation of a mirror small compared to its radius of curvature.

Since it is only approximately true that the rays come to a perfect focus at F, the larger the mirror, the worse the approximation (Fig. 23–10) and the more blurred the image. This "defect" of spherical mirrors is called **spherical aberration**; we will discuss it further with regard to lenses in Chapter 25. A *parabolic* reflector, on the other hand, will reflect the rays to a perfect focus. However, because parabolic shapes are much harder to make and thus much more expensive, spherical mirrors are used for most purposes. We consider here only spherical mirrors and we will assume that they are small compared to their radius of curvature so that the image is sharp and Eq. 23–2 holds.

We saw that for an object at infinity, the image is located at the focal point of a concave spherical mirror, where $f = r/2$. But where does the image lie for an object not at infinity? First consider the object shown in Fig. 23–12, which is placed at point O between F and C. Let us determine where the image will be for a given point O' on the object. To do this, we can draw several rays and make sure these reflect from the mirror such that the reflection angle equals the incidence angle. This can involve much work and our task is simplified if we deal with three particularly simple rays. These are the rays labeled 1, 2, and 3 in Fig. 23–12, and we draw them as follows:

Ray diagramming: finding the image position for a curved mirror

Ray 1 is drawn parallel to the axis; therefore it must pass through F after reflection (as in Fig. 23–11).

Ray 2 is drawn through F; therefore it must reflect so it is parallel to the axis.

Ray 3 is drawn so that it passes through C, the center of curvature, and thus is along a radius of the spherical surface; so it is perpendicular to the mirror and thus will be reflected back on itself.

The point at which these rays cross is the image point I'. All other rays from the same object point will pass through this image point. To find the image

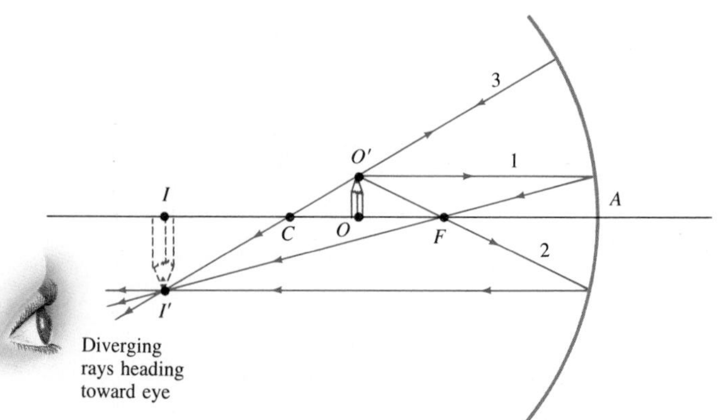

FIGURE 23–12 Rays from point O' on the object form an image at I'.

point for any object point, only these three types of rays need be used. Actually, only two of these rays are needed, but the third serves as a check.

We have only shown the image point in Fig. 23–12 for a single point on the object. Other points on the object are imaged nearby, so a complete image of the object is formed, as shown by the dashed outline. Because the light actually passes through the image itself, this is a **real image**. This can be compared to the virtual image formed by a plane mirror (the light does not actually pass through the image, Fig. 23–7).

Real image

The image in Fig. 23–12 can be seen by the eye when the eye is placed to the left of the image so that some of the rays diverging† from each point on the image (as point I') can enter the eye, as shown. (See also Figs. 23–1 and 23–7.)

Image points can always be determined by drawing the three rays as described above. However, it is possible to derive an equation that gives the image distance if the object distance and radius of curvature of the mirror are known. To do this, we refer to Fig. 23–13. The distance of the object from the center of the mirror, called the **object distance**, is labeled d_o. The **image distance** is labeled d_i. The height of the object OO' is called h_o and the height of the image, $I'I$, is h_i. Two rays are shown, $O'FBI'$ (same as ray 2 in the previous figure) and $O'AI'$. The ray $O'AI'$ obeys the law of reflection, of course, so the triangles $O'AO$ and $I'AI$ are similar. Therefore, we have

$$\frac{h_o}{h_i} = \frac{d_o}{d_i}.$$

For the other ray, $O'FBI'$, the triangles $O'FO$ and AFB are also similar. Therefore,

$$\frac{h_o}{h_i} = \frac{OF}{FA} = \frac{d_o - f}{f},$$

since the length $AB = h_i$ (in our approximation of a mirror that is small compared to its radius) and $FA = f$, the focal length of the mirror. The left

† Why, in order to see the image, the rays must be diverging from each point on the image will be discussed in Section 25–2, but is essentially because we see real objects when diverging rays from each point enter the eye as shown in Fig. 23–1.

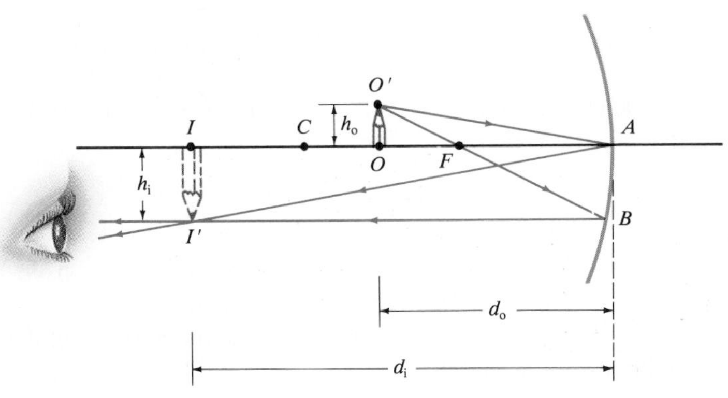

FIGURE 23–13 Diagram for deriving the mirror equation.

sides of the two preceding expressions are the same, so we can equate the right sides:

$$\frac{d_o}{d_i} = \frac{d_o - f}{f}.$$

We now divide both sides by d_o and rearrange to obtain

Mirror equation

$$\frac{1}{d_o} + \frac{1}{d_i} = \frac{1}{f}. \qquad (23\text{–}3)$$

This is the equation we were seeking. It is called the **mirror equation** and relates the object and image distances to the focal length f (where $f = r/2$).

The **lateral magnification**, m, of a mirror is defined as the height of the image divided by the height of the object. From our first set of similar triangles above, we can write:

Lateral magnification for curved mirror

$$m = \frac{h_i}{h_o} = -\frac{d_i}{d_o}, \qquad (23\text{–}4)$$

where the minus sign is inserted as a convention. For consistency, we must be careful about the signs of all quantities in Eq. 23–4. The conventions we use are h_o (or h_i) is considered positive for an object (or image) point above the principal axis, and negative for a point below the axis; d_i and d_o are positive if image and object are on the reflecting side of the mirror (as in Fig. 23–13), but if either image or object is behind the mirror, the corresponding distance is negative (an example is shown in Fig. 23–14, Example 23–3). Thus the magnification (Eq. 23–4) is positive for an upright image and negative for an inverted image.

EXAMPLE 23–2 A 1.50-cm-high object is placed 20.0 cm from a concave mirror whose radius of curvature is 30.0 cm. Determine (*a*) the position of the image, and (*b*) its size.

SOLUTION (*a*) The focal length $f = r/2 = 15.0$ cm. Then, since $d_o = 20.0$ cm, we have from Eq. 23–3 that

$$\frac{1}{d_i} = \frac{1}{f} - \frac{1}{d_o} = \frac{1}{15.0\ \text{cm}} - \frac{1}{20.0\ \text{cm}} = 0.0167\ \text{cm}^{-1}.$$

So $d_i = 1/0.0167\ \text{cm}^{-1} = 60.0$ cm. The image is 60.0 cm from the mirror on the same side as the object. (*b*) From Eq. 23–4, the lateral magnification is $m = -60.0\ \text{cm}/20.0\ \text{cm} = -3.00$. Therefore the image height is $h_i = mh_o = (-3.00)(1.5\ \text{cm}) = -4.5$ cm high; the minus sign tells us the image is inverted.

EXAMPLE 23–3 A 1.00-cm-high object is placed 10.0 cm from a concave mirror whose radius of curvature is 30.0 cm. (*a*) Draw a ray diagram to locate the position of the image. (*b*) Determine the position of the image and the magnification analytically.

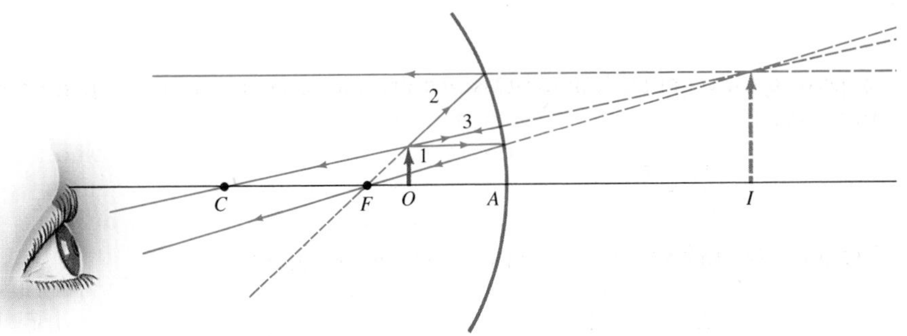

FIGURE 23–14 Object placed within the focal point F (Example 23–3). The image is *behind* the mirror and is *virtual.*

SOLUTION (*a*) Since $f = r/2 = 15.0$ cm, the object is between the mirror and the focal point. We draw the three rays as described earlier (for Fig. 23–12) and this is shown in Fig. 23–14. The rays reflected from the mirror diverge and so never meet at a point. They appear, however, to be coming from a point behind the mirror. The image is thus behind the mirror and is *virtual.* (Why?) (*b*) We use Eq. 23–3 to find d_i where $d_o = 10.0$ cm:

$$\frac{1}{d_i} = \frac{1}{15.0\text{ cm}} - \frac{1}{10.0\text{ cm}} = \frac{2-3}{30.0\text{ cm}} = -\frac{1}{30.0\text{ cm}}.$$

Therefore, $d_i = -30.0$ cm. The minus sign means the image is behind the mirror. The lateral magnification is $m = -d_i/d_o = -(-30.0\text{ cm})/(10.0\text{ cm}) = +3.00$. So the image is 3.00 times larger than the object; the plus sign indicates that the image is upright (which is consistent with the ray diagram, Fig. 23–14).

These examples show that a spherical mirror can produce a magnified image, one that is larger than the object. (There is an old story—maybe fable is a better word—that Julius Caesar spied on the British forces by setting up a curved mirror on the coast of Gaul. Is this reasonable?)

It is useful to compare Figs. 23–12 and 23–14. We can see that if the object is within the focal point, as in Fig. 23–14, the image is virtual, upright, and magnified. This is how a shaving or makeup mirror is used—you must place your head within the focal point if you are to see yourself right side up. If the object is *beyond* the focal point, as in Fig. 23–12, the image is real and inverted. Whether the magnification is greater or less than 1.0 in this case depends on the position of the object relative to the center of curvature, point C

Analysis for convex mirrors

The analysis used for concave mirrors can be applied to *convex* mirrors. Even the mirror equation (Eq. 23–3) holds for a convex mirror, although the quantities involved must be carefully defined. Figure 23–15a shows parallel

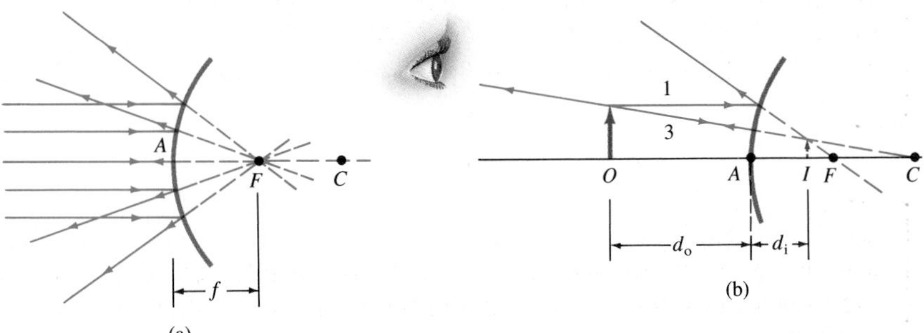

FIGURE 23–15 Convex mirror: (a) the focal point is at F, behind the mirror; (b) the image I of the object at O is virtual, upright, and smaller than the object.

rays falling on a convex mirror. Again spherical aberration will be present, but we assume the mirror's size is small compared to its radius of curvature. The reflected rays diverge, but seem to come from point F behind the mirror. This is the *focal point*, and its distance from the center of the mirror is the *focal length* f. It is easy to show that again $f = r/2$. We see that an object at infinity produces a virtual image in a convex mirror. Indeed, no matter where the object is placed on the reflecting side of the mirror, the image will be virtual and erect, as indicated in Fig. 23–15b. To find the image we draw rays 1 and 3 according to the rules used before on the concave mirror.

The mirror equation, Eq. 23–3, holds for convex mirrors but the focal length f must be considered negative, as must the radius of curvature. The proof is left as a problem. It is also left as a problem to show that Eq. 23–4 for the magnification is also valid.

EXAMPLE 23–4 A convex rearview car mirror has a radius of curvature of 40.0 cm. Determine the location of the image and its magnification for an object 10.0 m from the mirror.

SOLUTION With $r = -40.0$ cm, then $f = -20.0$ cm, and the mirror equation gives

$$\frac{1}{d_i} = \frac{1}{f} - \frac{1}{d_o} = -\frac{1}{0.200\text{ m}} - \frac{1}{10.0\text{ m}} = -\frac{51.0}{10.0\text{ m}}.$$

So $d_i = -10.0\text{ m}/51.0 = -0.196$ m, or 19.6 cm behind the mirror. The lateral magnification is $m = -d_i/d_o = -(-0.196\text{ m})/(10.0\text{ m}) = 0.0196$ or 1/51. So the upright image is reduced by a factor of 51.

We summarize the rules (the **sign conventions**) for applying Eqs. 23–3 and 23–4 to concave and convex mirrors:

Problem Solving: Sign conventions for mirrors

We draw ray diagrams with rays starting on the left and moving to the right. When the object, image, or focal point is on the reflecting side of the mirror (on the left in all our drawings), the corresponding distance is considered positive. If any of these points is behind the mirror (on the right) the corresponding distance is negative.† Object and image heights, h_o and h_i, are positive or negative depending on whether the point is above or below the principal axis, respectively.

It is important to be consistent with these sign conventions.

† We have seen examples where d_i and f are negative. The object distance d_o for any material object is, of course, always positive. However, if the mirror is used in conjunction with a lens or another mirror, the image formed by the first mirror or lens becomes the object for the second mirror. It is then possible for such an "object" to be behind the second mirror, in which case, d_o would be negative. These rules are also consistent with considering the focal length of a concave mirror positive and that of a convex mirror negative.

23–5 • Refraction; Snell's Law

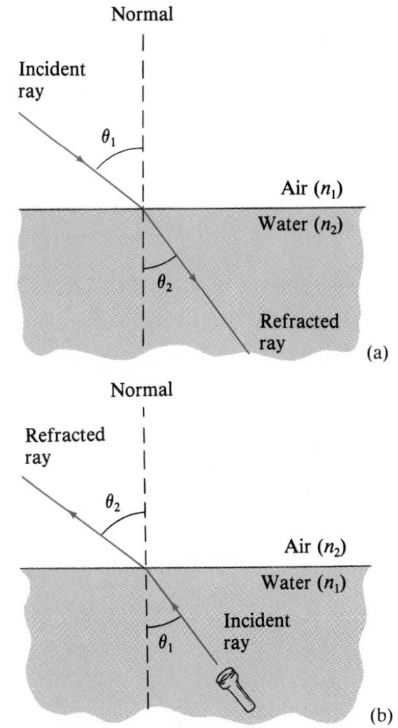

FIGURE 23–16 Refraction.

When light passes from one medium into another, part of the incident light is reflected at the boundary. The remainder passes into the new medium. If a ray of light is incident at an angle to the surface (other than perpendicular), the ray is bent as it enters the new medium. This bending is called **refraction**. Figure 23–16a shows a ray passing from air into water. The angle θ_1 is the **angle of incidence** and θ_2 is the **angle of refraction**. Notice that the ray bends toward the normal when entering the water. This is always the case when the ray enters a medium where the speed of light is less. If light travels from one medium into a second where its speed is greater, the ray bends away from the normal; this is shown in Fig. 23–16b for a ray traveling from water to air.

Refraction is responsible for a number of common optical illusions. For example, a person standing in waist-deep water appears to have shortened legs, Fig. 23–17a. As shown in Fig. 23–17b, the rays leaving the person's foot are bent at the surface. The observer's eye (and brain) assumes the rays to have traveled a straight-line path, and so the feet appear to be higher than they really are. Similarly, when you put a stick in water, it appears to be bent.

The angle of refraction depends on the speed of light in the two media and on the incident angle. An analytical relation between θ_1 and θ_2 was arrived at experimentally about 1621 by Willebrord Snell (1591–1626). It is known as **Snell's law** and is written:

Snell's law (law of refraction)

$$n_1 \sin \theta_1 = n_2 \sin \theta_2; \qquad (23\text{–}5)$$

θ_1 is the angle of incidence and θ_2 is the angle of refraction (both measured with respect to a line perpendicular to the surface between the two media, as shown in Fig. 23–16); n_1 and n_2 are the respective indices of refraction in the materials. The incident and refracted rays lie in the same plane, which

(a)

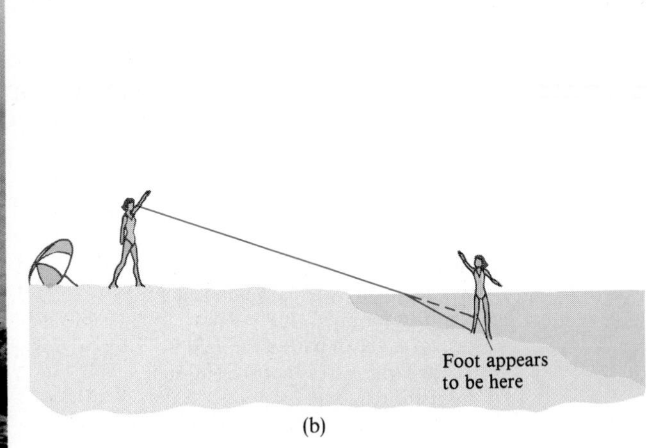

FIGURE 23–17 Because of refraction, when a person is standing in water, his or her legs look shorter.

also includes the perpendicular to the surface. Snell's law† is the basic **law of refraction**.

It is clear from Snell's law that if $n_2 > n_1$, then $\theta_2 < \theta_1$; that is, if light enters a medium where n is greater (and its speed less), then the ray is bent toward the normal. And if $n_2 < n_1$, then $\theta_2 > \theta_1$, so the ray bends away from the normal. This is what we saw in Fig. 23–16.

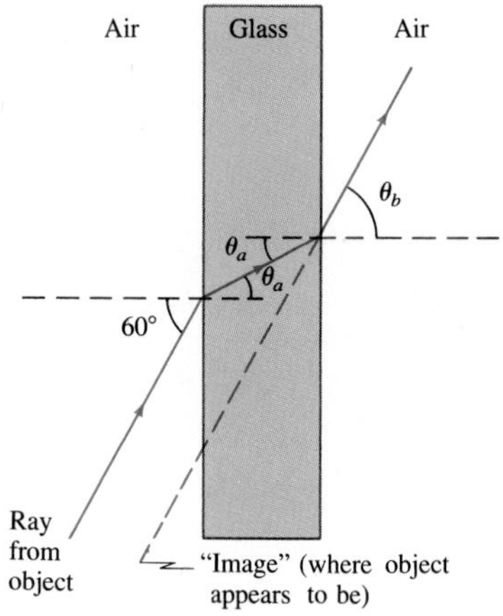

FIGURE 23–18 Light passing through a piece of glass (Example 23–5).

EXAMPLE 23–5 Light strikes a flat piece of glass at an incident angle of 60°, as shown in Fig. 23–18. If the index of refraction of the glass is 1.50, (*a*) what is the angle of refraction θ_a in the glass; (*b*) what is the angle θ_b at which the ray emerges from the glass?

SOLUTION (*a*) We assume the incident ray is in air, so $n_1 = 1.00$ and $n_2 = 1.50$. Then, from Eq. 23–5 we have

$$\sin\theta_a = \frac{1.00}{1.50}\sin 60^\circ = 0.577,$$

so $\theta_a = 35.2^\circ$. (*b*) Since the faces of the glass are parallel, the incident angle in this case is just θ_a, so $\sin\theta_a = 0.577$. This time $n_1 = 1.50$ and $n_2 = 1.00$. Thus, θ_b $(=\theta_2)$ is

$$\sin\theta_b = \frac{1.50}{1.00}\sin\theta_a = 0.866,$$

and $\theta_b = 60.0^\circ$. The direction of the beam is thus unchanged by passing through a plane piece of glass. It should be clear that this is true for any angle of incidence. The ray is displaced slightly to one side, however. You can observe this by looking through a piece of glass (near its edge) at some object and then moving your head to the side so that you see the object directly.

23–6 • Total Internal Reflection; Fiber Optics

When light passes from one material into a second material where the index of refraction is less (say, from water into air), the light bends away from the normal, as for ray A in Fig. 23–19. At a particular incident angle, the angle

† Snell actually was not aware that the index of refraction is related to the speed of light in the particular medium. Only later was it found that the index of refraction could be written as the ratio of the speed of light in vacuum to that in the given material, Eq. 23–1. Snell's law can be derived from the wave theory of light (see the next chapter, Section 24–2), and, in fact, we already did so in Section 11–10 where Eq. 11–16 is just a combination of Eqs. 23–5 and 23–1.

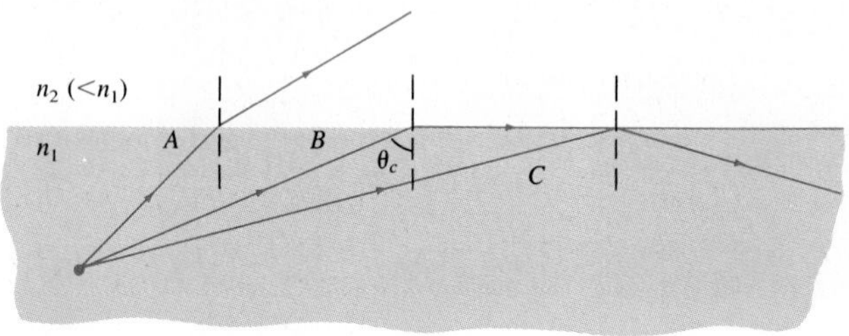

FIGURE 23–19 Since $n_2 < n_1$, light rays are totally internally reflected if $\theta > \theta_c$, as for ray C. If $\theta < \theta_c$, as for ray A, only a part of the light is reflected (this part is not shown), and the rest is refracted.

of refraction will be 90°, and the refracted ray would skim the surface (ray *B*) in this case. The incident angle at which this occurs is called the **critical angle**, θ_c. From Snell's law, θ_c is given by

$$\sin\theta_c = \frac{n_2}{n_1}\sin 90^\circ = \frac{n_2}{n_1}. \qquad (23\text{–}6)$$

Critical angle

For any incident angle less than θ_c there will be a refracted ray, although part of the light will also be reflected at the boundary. However, for incident angles greater than θ_c, Snell's law would tell us that $\sin\theta_2$ is greater than 1.00. Yet the sine of an angle can never be greater than 1.00. In this case there is no refracted ray at all, and *all of the light is reflected*, as for ray *C* in Fig. 23–19. This effect is called **total internal reflection**. But note that total internal reflection can occur only when light strikes a boundary where the medium beyond is optically less dense—that is, it has a lower index of refraction.

Total internal reflection

EXAMPLE 23–6 Describe what a person would see who looked up at the world from beneath the perfectly smooth surface of a lake or swimming pool.

SOLUTION For an air–water interface, the critical angle is given by

$$\sin\theta_c = \frac{1.00}{1.33} = 0.750.$$

Therefore, $\theta_c = 49°$. Thus the person would see the outside world compressed into a circle whose edge makes a 49° angle with the vertical. Beyond this angle, the person would see reflections from the sides and bottom of the pool or lake (Fig. 23–20).

Many optical instruments, such as binoculars, use total internal reflection within a prism to reflect light. The advantage is that very nearly 100 percent of the light is reflected, whereas even the best mirrors reflect somewhat less than 100 percent. Thus the image is brighter. For glass with $n = 1.50$,

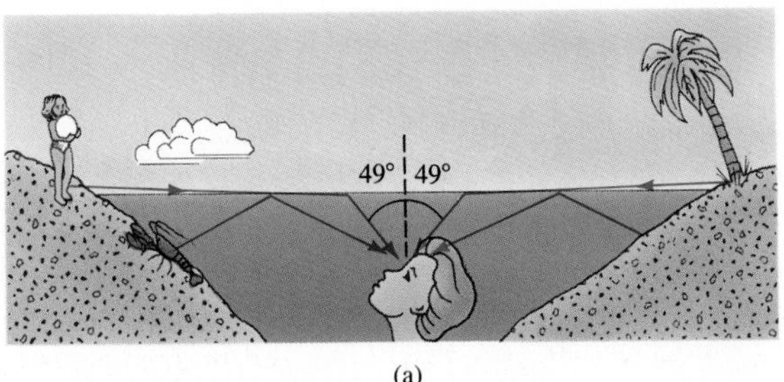

(a)

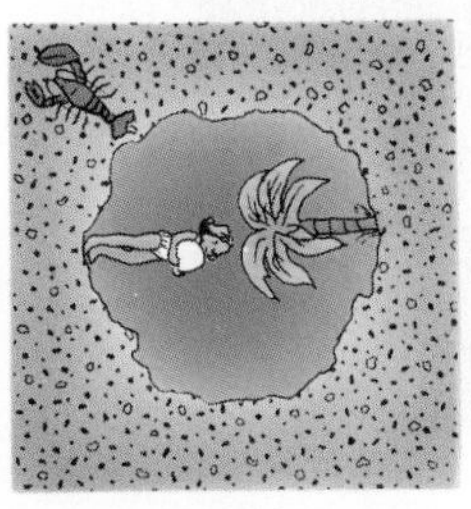
(b)

FIGURE 23–20 View looking upward from beneath the water: the surface of the water is smooth.

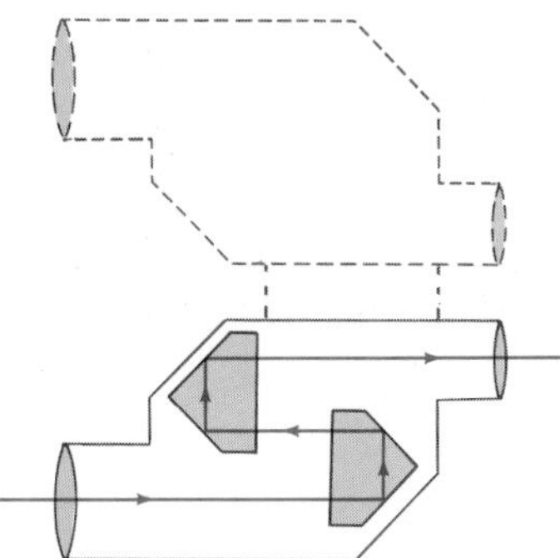

FIGURE 23–21 Total internal reflection of light by prisms in binoculars.

FIGURE 23–22 Light reflected totally at the interior surface of a glass or transparent plastic fiber.

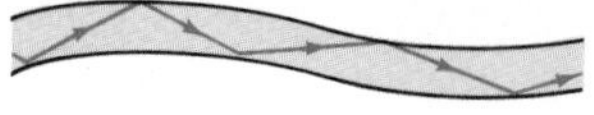

$\theta_c = 41.8°$. Therefore, 45° prisms will reflect all the light internally, as shown in the binoculars of Fig. 23–21.

Total internal reflection is the principle behind **fiber optics**. Very thin glass and plastic fibers can now be made as small as a few micrometers in diameter. A bundle of such tiny fibers is called a **light pipe** and light can be transmitted along it with almost no loss because of total internal reflection. Figure 23–22 shows how light traveling down a thin fiber makes only glancing collisions with the walls so that total internal reflection occurs. Even if the light pipe is bent into a complicated shape, the critical angle still won't (usually) be exceeded, so light is transmitted practically undiminished to the other end (see Fig. 23–23). This effect is used in decorative lamps and to illuminate water streams in fountains. Light pipes can be used to illuminate difficult places to reach, such as inside the human body. They can be used to transmit telephone calls and other communication signals; the signal is a modulated light beam (variable intensity of the light beam) and is transmitted with less loss than an electrical signal in a copper wire. One sophisticated use of fiber optics, particularly in medicine, is to transmit a clear picture, Fig. 23–24. For example, a patient's stomach can be examined by inserting a light pipe through the mouth. Light is sent down one set of fibers to illuminate the stomach. The reflected light returns up another set of fibers. Light directly in front of each fiber travels up that fiber. At the opposite end, a viewer sees a series of bright and dark spots, much like a TV screen—that is, a picture of what lies at the opposite end.† The fibers must be optically insulated from one another, usually by a thin coating of material whose refractive index is less than that of the fiber. The fibers must be arranged precisely parallel to one another if the picture is to be clear. The more fibers there are, and the smaller they are, the more detailed the picture. Such an "endoscope" is useful for observing the stomach or other hard-to-reach places for surgery or searching for lesions without surgery.

† Lenses are used at each end: at the object end to bring the rays in parallel, and at the viewing end a telescope system for viewing.

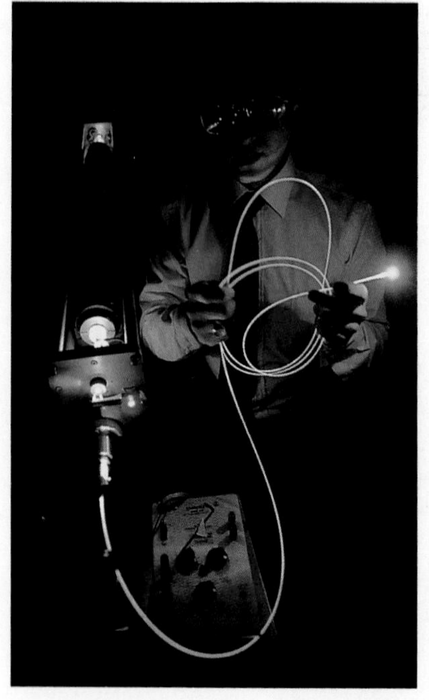

FIGURE 23–23 (left) Total internal reflection within the tiny fibers of this light pipe makes it possible to transmit light in complex paths with minimal loss.

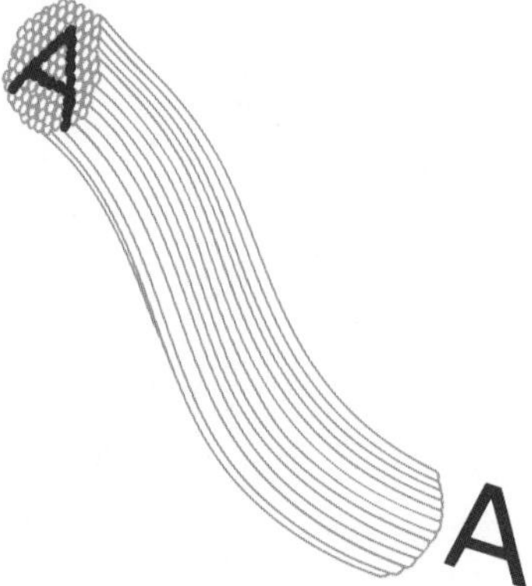

FIGURE 23–24 Fiber-optic image.

23–7 • Thin Lenses; Ray Tracing

Double convex   Planoconvex   Convex meniscus

(a) Converging lenses

Double concave   Planoconcave   Concave meniscus

(b) Diverging lenses

FIGURE 23–25 Types of lenses.

The most important simple optical device is no doubt the thin lens. The development of optical devices using lenses dates to the sixteenth and seventeenth centuries, although the earliest record of eyeglasses dates from the late thirteenth century.† Today we find lenses in eyeglasses, cameras, magnifying glasses, telescopes, binoculars, microscopes, and medical instruments. A thin lens is usually circular, and its two faces are portions of a sphere. (Although cylindrical surfaces are also possible, we will concentrate on spherical.) The two faces can be concave, convex, or plane (Fig. 23–25). The importance of lenses is that they form images of objects.

Consider the rays parallel to the axis of the double convex lens which is shown in cross-section in Fig. 23–26a. We assume the lens is made of glass or transparent plastic, so its index of refraction is greater than that of the air outside. From Snell's law, we can see that each ray is bent toward the axis at both lens surfaces (note the dashed lines indicating the normals to each surface for the top ray). If rays parallel to the principal axis fall on a thin lens, they will be focused to a point called the **focal point**, F. This will not be precisely true for a lens with spherical surfaces. But it will be very nearly true—that is, parallel rays will be focused to a tiny region that is nearly a point—if the diameter of the lens is small compared to the radii of curvature of the two lens surfaces. This criterion is satisfied by a **thin lens**, one that is very thin compared to its diameter, and it is only thin lenses that we consider here.

Thin lens defined

Since the rays from a distant object are essentially parallel (see Fig. 23–9), we can also say that *the focal point is the image point for an object on the principal axis at infinity*. Thus, the focal point of a lens can be found by locating the point where the sun's rays (or those of some other distant object) are brought to a sharp image. The distance of the focal point from the center of the lens is called the **focal length**, f. A lens can be turned around so light can pass through it from the opposite side. The focal length is the same on both sides, as we shall see later, even if the curvatures of the two lens surfaces are different. If parallel rays fall on a lens at an angle, as in

Focal length of lens

† Rounded gem stones used as magnifiers probably date from much earlier.

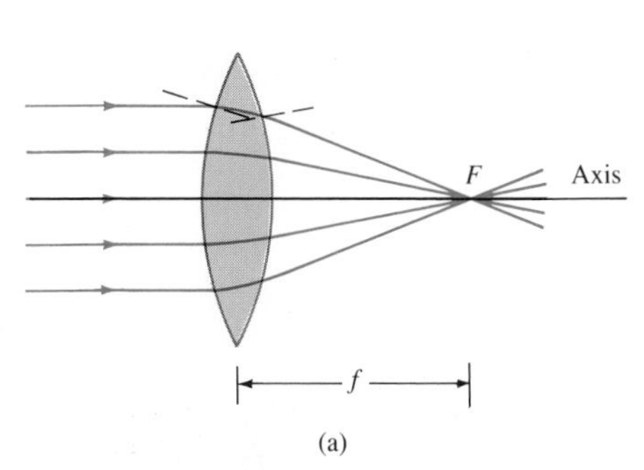

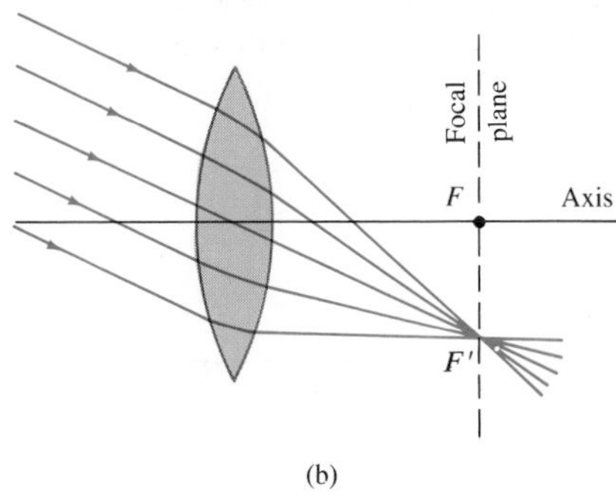

FIGURE 23–26 Parallel rays are brought to a focus by a converging thin lens.

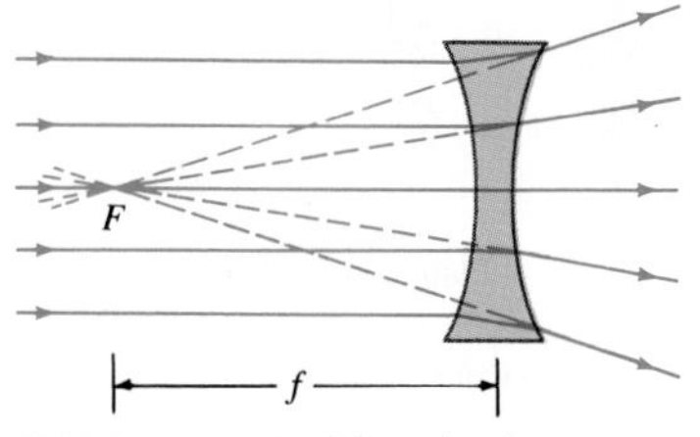

FIGURE 23–27 Diverging lens.

Fig. 23–26b, they focus at a point F'. The plane in which all points such as F and F' fall is called the **focal plane** of the lens.

Any lens[†] that is thicker in the center than at the edges will make parallel rays converge to a point, and is called a **converging lens** (see Fig. 23–25a). Lenses that are thinner in the center than at the edges (Fig. 23–25b) are called **diverging lenses** because they make parallel light diverge, as shown in Fig. 23–27. The *focal point* F of a diverging lens is defined as that point from which refracted rays, originating from parallel incident rays, seem to emerge as shown in the figure. And the distance from F to the lens is called the *focal length*, just as for a converging lens.

Optometrists and ophthalmologists, instead of using the focal length, use the reciprocal of the focal length to specify the strength of eyeglass (or contact) lenses. This is called the **power**, P, of a lens:

Power of lens

$$P = \frac{1}{f}. \tag{23–7}$$

The unit for lens power is the *diopter* (D), which is an inverse meter: 1 D = 1 m^{-1}. For example, a 20-cm-focal-length lens has a power $P = 1/0.20\text{ m} = 5.0\text{ D}$. We will mainly use the focal length here, but will refer again to the power of a lens when we discuss eyeglass lenses in Chapter 25.

The most important parameter of a lens is its focal length f. For a converging lens, f is easily measured by finding the image point for the sun or other distant objects. Once f is known, the image position can be found for any object. To find the image point by drawing rays would be difficult if we had to determine all the refractive angles. Instead, we can do it very simply by making use of certain facts we already know, such as that a ray parallel to the axis of the lens passes (after refraction) through the focal point. In fact, to find an image point, we need consider only the three rays indicated in Fig. 23–28. These three rays, emanating from a single point on the object

[†] We are assuming the lens has an index of refraction greater than that of the surrounding material, such as a glass or plastic lens in air, which is the usual situation.

FIGURE 23–28 Finding the image by ray tracing—converging lens.

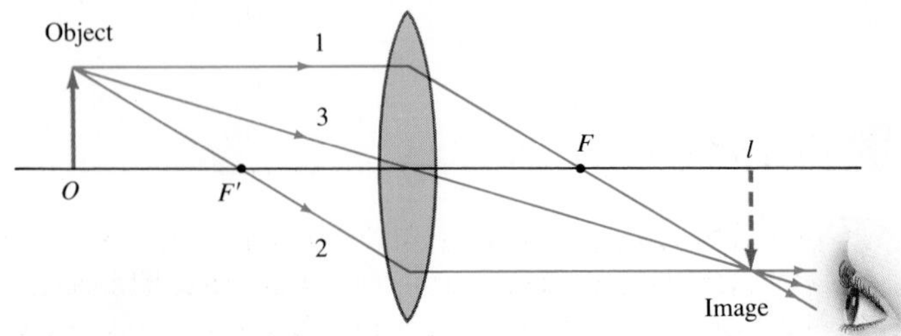

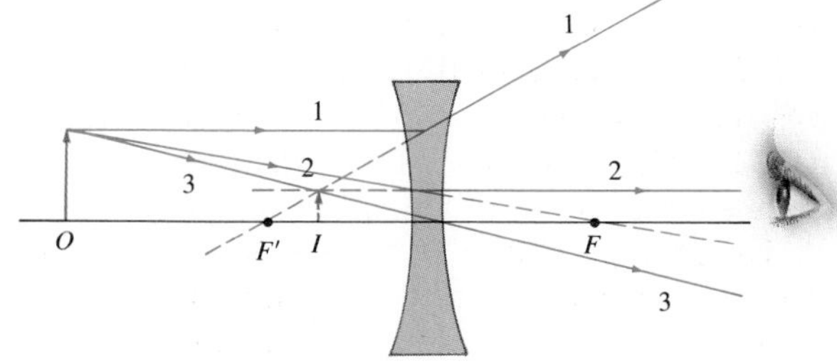

FIGURE 23–29 Finding image I by ray tracing—diverging lens.

and drawn as if the lens were infinitely thin (thus the sharp bend within the lens), are drawn as follows:

Problem solving: Ray diagramming for lenses

Ray 1 is drawn parallel to the axis; therefore it is refracted by the lens so that it passes through the focal point F behind the lens (as in Fig. 23–26a).

Ray 2 is drawn through the focal point F' on the same side of the lens as the object; it therefore emerges from the lens parallel to the axis.

Ray 3 is directed toward the very center of the lens, where the two surfaces are essentially parallel to each other; this ray therefore emerges from the lens at the same angle as it entered; as we saw in Example 23–5, the ray would be displaced slightly to one side, but since we assume the lens is thin, we draw ray 3 straight through as shown.

Actually, any two of these rays will suffice to locate the image point, which is the point where they intersect. Drawing the third can serve as a check.

In this way we can find the image point for one point of the object (the top of the arrow in Fig. 23–28). The image points for all other points on the object can be found similarly to determine the complete image of the object. Because the rays actually pass through the image for the case shown, it is a *real image* (see page 593). The image could be detected by film, or actually seen on a white surface placed at the position of the image. The image can also be seen directly by the eye when the eye is placed behind the image, as shown in Fig. 23–28.

By drawing the same three rays we can determine the image position for a diverging lens, as shown in Fig. 23–29. Note that ray 1 is drawn parallel to the axis, but does not pass through the focal point F behind the lens. Instead it seems to come from the focal point F' in front of the lens (dashed line). Ray 2 is directed toward F and is refracted parallel by the lens. The three refracted rays seem to emerge from a point on the left of the lens. This is the image, I. Since the rays do not pass through the image, it is a *virtual image*.

23–8 • The Lens Equation

We now derive an equation that relates the image distance to the object distance and the focal length of the lens. This will make the determination of image position quicker and more accurate than doing ray tracing. Con-

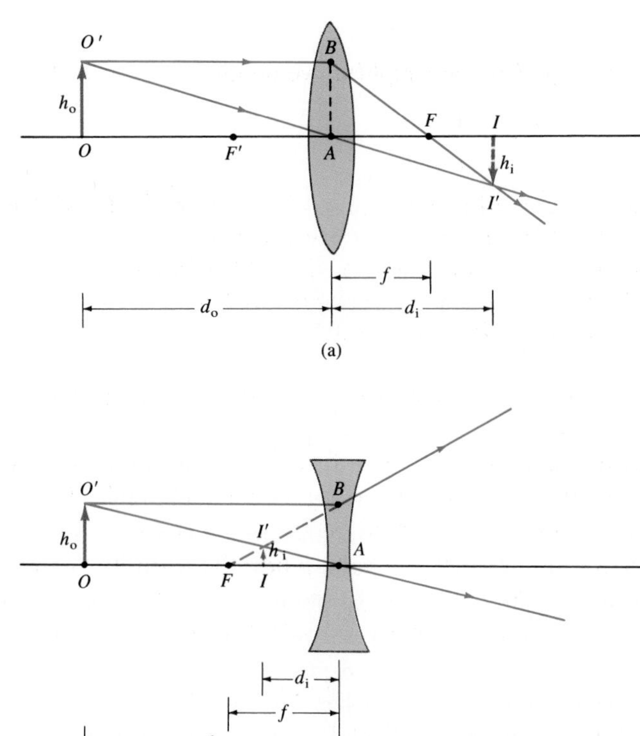

FIGURE 23–30 Deriving the lens equation for (a) converging and (b) diverging lenses.

sider the two rays shown in Fig. 23–30a for a converging lens (assumed to be very thin); h_o and h_i refer to the heights of the object and image, and d_o and d_i are their distances from the lens. The triangles $FI'I$ and FBA are similar, so

$$\frac{h_i}{h_o} = \frac{d_i - f}{f},$$

since length $AB = h_o$. Triangles OAO' and IAI' are similar. Therefore,

$$\frac{h_i}{h_o} = \frac{d_i}{d_o}.$$

We equate these two, divide by d_i, and rearrange to obtain

Lens equation

$$\frac{1}{d_o} + \frac{1}{d_i} = \frac{1}{f}. \tag{23–8}$$

This is called the **lens equation**. It relates the image distance d_i to the object distance d_o and the focal length f. It is the most useful equation in geometric optics. (Interestingly, it is exactly the same as the mirror equation, Eq. 23–3). Note that if the object is at infinity, then $1/d_o = 0$, so $d_i = f$. Thus the focal length is the image distance for an object at infinity, as mentioned earlier.

From the rays in Fig. 23–30b, we can derive the lens equation for a diverging lens. Triangles IAI' and OAO' are similar; and triangles IFI' and AFB are similar. Thus

$$\frac{h_i}{h_o} = \frac{d_i}{d_o} \quad \text{and} \quad \frac{h_i}{h_o} = \frac{f - d_i}{f}.$$

When these are equated and simplifed, we obtain

$$\frac{1}{d_o} - \frac{1}{d_i} = -\frac{1}{f}.$$

This equation becomes the same as Eq. 23–8 if we make f and d_i negative. That is, we take f to be *negative for a diverging lens*, and d_i negative when the image is on the same side of the lens as the light comes from. Thus Eq. 23–8 will be valid for both converging and diverging lenses, and for *all* situations, if we use the following **sign conventions**[†]:

Problem Solving: Sign conventions for lenses

1. The focal length is positive for converging lenses and negative for diverging lenses.
2. The object distance is positive if it is on the side of the lens from which the light is coming (this is normally the case, although when lenses are used in combination, it might not be so); otherwise, it is negative.
3. The image distance is positive if it is on the opposite side of the lens from where the light is coming; if it is on the same side, d_i is negative. Equivalently, the image distance is positive for a real image and negative for a virtual image.
4. Object and image heights, h_o and h_i, are positive for points above the axis, and negative for points below the axis.

The **lateral magnification**, m, of a lens is defined as the ratio of the image height to object height, $m = h_i/h_o$. From Fig. 23–30 and the conventions just stated, we have

Lateral magnification of a lens

$$m = \frac{h_i}{h_o} = -\frac{d_i}{d_o}. \qquad (23\text{–}9)$$

For an upright image the magnification is positive, and for an inverted image m is negative.

From convention 1 above, it follows that the power (Eq. 23–7) of a converging lens, in diopters, is positive, whereas the power of a diverging lens is negative. A converging lens is sometimes referred to as a **positive lens**, and a diverging lens as a **negative lens**.

23–9 • Problem Solving for Lenses

Now we take some examples of using the lens equation to solve problems involving thin lenses.

Example 23–7 What is (a) the position, and (b) the size, of the image of a large 22.4-cm-high flower placed 1.50 m from a +50.0-mm-focal-length camera lens?

[†] These conventions are the same for mirrors, except for 3. See Section 23–4.

SOLUTION (a) First we find the image position using the lens equation, Eq. 23–8. The camera lens is converging, with $f = +5.00$ cm, and $d_o = 150$ cm, so Eq. 23–8 gives

$$\frac{1}{d_i} = \frac{1}{f} - \frac{1}{d_o} = \frac{1}{5.00 \text{ cm}} - \frac{1}{150 \text{ cm}} = \frac{30.0 - 1.0}{150 \text{ cm}}.$$

Then $d_i = 150 \text{ cm}/29.0 = 5.17$ cm, or 51.7 mm behind the lens. Notice that the image is 1.7 mm farther from the lens than would be the image for an object at infinity. This is an example of the fact that when focusing a camera lens, the closer the object is to the camera, the farther the lens must be from the film.

(b) The magnification $m = -d_i/d_o = -5.17 \text{ cm}/150 \text{ cm} = -0.0345$, so $h_i = mh_o = (-0.0345) \cdot (22.4 \text{ cm}) = -0.773$ cm. The image is 7.73 mm high and is inverted as in Fig. 23–30a.

EXAMPLE 23–8 An object is placed 10 cm from a 15-cm-focal-length converging lens. Determine the image position and size (*a*) analytically, and (*b*) using a ray diagram.

SOLUTION (*a*) Since $f = 15$ cm and $d_o = 10$ cm,

$$\frac{1}{d_i} = \frac{1}{15 \text{ cm}} - \frac{1}{10 \text{ cm}} = -\frac{1}{30 \text{ cm}},$$

Problem Solving: Don't forget to take the reciprocal

so $d_i = -30$ cm. (Remember to take the reciprocal!) Since d_i is negative, the image must be virtual and on the same side of the lens as the object. The magnification $m = -(-30 \text{ cm})/(10 \text{ cm}) = 3.0$. So the image is three times as large as the object and is upright. (This is the idea of a simple magnifying glass, which we discuss in more detail in Section 25–3). (*b*) The ray diagram is shown in Fig. 23–31, and confirms the result in part (*a*).

It is a general rule that when an object is placed between a converging lens and its focal point (as in this last Example), the image is virtual.

EXAMPLE 23–9 Where must a small insect be placed if a 25-cm-focal-length diverging lens is to form a virtual image 20 cm in front of the lens?

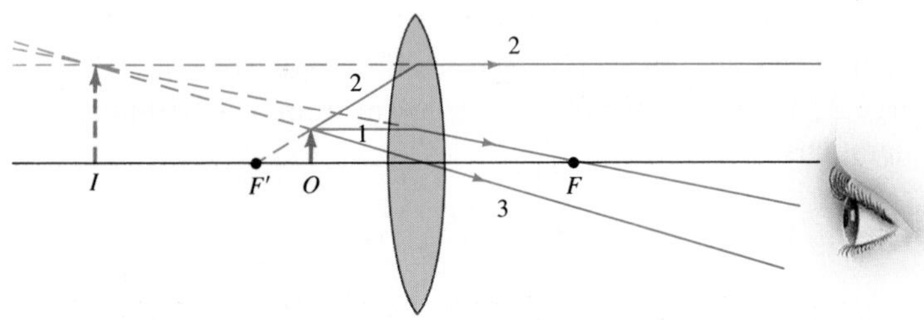

FIGURE 23–31 An object placed within the focal point of a converging lens produces a virtual image (Example 23–8).

SOLUTION Since $f = -25$ cm and $d_i = -20$ cm, then Eq. 23–8 gives

$$\frac{1}{d_o} = -\frac{1}{25\text{ cm}} + \frac{1}{20\text{ cm}} = \frac{1}{100\text{ cm}}.$$

So the object must be 100 cm in front of the lens. The ray diagram is basically that of Fig. 23–30b.

EXAMPLE 23–10 To measure the focal length of a diverging lens, a converging lens is placed in contact with it as in Fig. 23–32. The sun's rays are focused by this combination at a point 28.5 cm behind the lenses as shown. If the converging lens has a focal length f_C of 16.0 cm, what is the focal length f_D of the diverging lens?

SOLUTION Since rays from the sun are focused 28.5 cm behind the combination, the focal length of the total combination is $f_T = 28.5$ cm. If the diverging lens were absent, the converging lens would form the image at its focal point—that is, at a distance $f_C = 16.0$ cm behind it (dashed lines in Fig. 23–32). When the diverging lens is placed next to the converging lens (we assume both lenses are thin and the space between them is negligible), we treat the image formed by the first lens as the *object* for the second (diverging) lens. Since this object lies to the right of the diverging lens, this is a situation where d_o is negative (see the sign conventions, p. 609). Thus, for the diverging lens, the object is virtual and $d_o = -16.0$ cm; and this diverging lens forms the image a distance $d_i = 28.5$ cm away (this was given). Thus,

$$\frac{1}{f_D} = \frac{1}{d_o} + \frac{1}{d_i} = \frac{1}{-16.0\text{ cm}} + \frac{1}{28.5\text{ cm}} = -0.0274\text{ cm}^{-1}.$$

So $f_D = -1/(0.0274\text{ cm}^{-1}) = -36.5$ cm. Note that the converging lens must have a focal length whose magnitude is less than that of the diverging lens if this technique is to work.

This example is our first illustration of how to deal with lenses used in combination. In general, when light passes through several lenses, the image formed by one lens becomes the object for the next lens. The total magnification will be the product of the separate magnification of each lens. We will see more examples in Chapter 25, where we discuss optical instruments such as telescopes and microscopes.

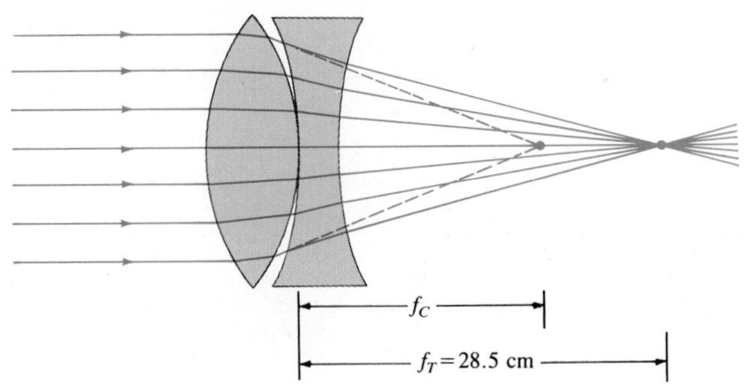

FIGURE 23–32 Determining the focal length of a diverging lens (Example 23–10).

*23–10 • The Lens-Maker's Equation

In this optional section, we will prove that parallel rays are brought to a focus at a *single* point for a thin lens. At the same time, we will also derive an equation that relates the focal length of a lens to the radii of curvature of its two surfaces, which is known as the lens-maker's equation.

In Fig. 23–33, a ray parallel to the axis of a lens is refracted at the front surface of the lens at point A_1 and is refracted at the back surface at point A_2. This ray then passes through point F, which we call the focal point for this ray. Point A_1 is a height h_1 above the axis, and point A_2 is height h_2 above the axis. C_1 and C_2 are the centers of curvature of the two lens surfaces, so the length $C_1A_1 = R_1$, the radius of curvature of the front surface. And $C_2A_2 = R_2$ is the radius of the second surface. The thickness of the lens has been grossly exaggerated so the various angles would be clear. But we will assume that the lens is actually very thin and that angles between the rays and the axis are small. In this approximation, $h_1 \approx h_2$, and the sines and tangents of all the angles will be equal to the angles themselves in radians. For example, $\sin\theta_1 \approx \tan\theta_1 \approx \theta_1$.

To this approximation, then, Snell's law tells us that

$$\theta_1 = n\theta_2$$

$$\theta_4 = n\theta_3$$

where n is the index of refraction of the glass, and we assume that the lens is surrounded by air ($n = 1$). Notice also in Fig. 23–33 that

$$\theta_1 = \sin\theta_1 = \frac{h_1}{R_1}$$

$$\alpha = \frac{h_2}{R_2}$$

$$\beta = \frac{h_2}{f}.$$

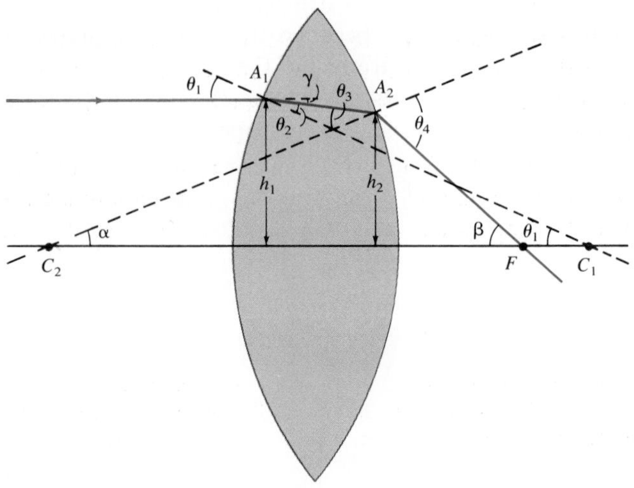

FIGURE 23–33 Diagram of a ray passing through a lens for derivation of the lens-maker's equation.

The last follows because the distance from F to the lens (assumed very thin) is f. From the diagram, the angle γ is defined as

$$\gamma = \theta_1 - \theta_2.$$

A careful examination of Fig. 23–33 shows also that

$$\alpha = \theta_3 - \gamma.$$

This can be seen by drawing a horizontal line to the left from point A_2, which divides the angle θ_3 into two parts. The upper part equals γ and the lower part equals α. (The opposite angles between an oblique line and two parallel lines are equal.) Thus, $\theta_3 = \gamma + \alpha$. Finally, by drawing a horizontal line to the right from point A_2, we divide θ_4 into two parts. The upper part is α and the lower is β. Thus

$$\theta_4 = \alpha + \beta.$$

We now combine all these equations:

$$\alpha = \theta_3 - \gamma = \frac{\theta_4}{n} - (\theta_1 - \theta_2) = \frac{\alpha}{n} + \frac{\beta}{n} - \theta_1 + \theta_2,$$

or

$$\frac{h_2}{R_2} = \frac{h_2}{nR_2} + \frac{h_2}{nf} - \frac{h_1}{R_1} + \frac{h_1}{nR_1}.$$

Since the lens is thin, $h_1 \approx h_2$ and can be canceled from all the numerators. We then multiply through by n and rearrange to find that

$$\frac{1}{f} = (n-1)\left(\frac{1}{R_1} + \frac{1}{R_2}\right). \qquad (23\text{–}10)$$

Lens-maker's equation

This is called the **lens-maker's equation**. It relates the focal length of a lens to the radii of curvature of its two surfaces and its index of refraction. Notice that f does not depend on h_1 or h_2. Thus the position of the point F does not depend on where the ray strikes the lens. Hence, all rays parallel to the axis of a thin lens will pass through the same point F, which we wished to prove.

In our derivation, both surfaces are convex and R_1 and R_2 are considered positive.[†] Equation 23–10 also works for lenses with one or both surfaces concave; but for a concave surface, the radius must be considered *negative*.

Notice in Eq. 23–10 that the equation is symmetrical in R_1 and R_2. Thus, if a lens is turned around so that light impinges on the other surface, the focal length is the same even if the two lens surfaces are different.

[†] Some books use a different convention—for example, R_1 and R_2 are considered positive if their centers of curvature are to the right of the lens, in which case minus signs appear in their equivalent of Eq. 23–10.

EXAMPLE 23–11 A concave meniscus lens (Fig. 23–25b) is made from glass with $n = 1.50$. The radius of curvature of the convex surface is 22.4 cm and that of the concave surface is 46.2 cm. (*a*) What is the focal length? (*b*) Where will it focus an object 2.00 m away?

SOLUTION (*a*) $R_1 = 22.4$ cm and $R_2 = -46.2$ cm; the latter is negative since it refers to the concave surface. Then

$$\frac{1}{f} = (1.50 - 1.00)\left(\frac{1}{22.4\text{ cm}} - \frac{1}{46.2\text{ cm}}\right) = 0.0115\text{ cm}^{-1}.$$

So $f = 1/0.0115\text{ cm}^{-1} = 87$ cm and is converging. Notice that if we turn the lens around so that $R_1 = -46.2$ cm and $R_2 = +22.4$ cm, we get the same result.

(*b*) From the lens equation, with $f = 0.87$ m and $d_o = 2.00$ m, we have

$$\frac{1}{d_i} = \frac{1}{f} - \frac{1}{d_o} = \frac{1}{0.87\text{ m}} - \frac{1}{2.00\text{ m}} = 0.65\text{ m}^{-1},$$

so $d_i = 1/0.65\text{ m}^{-1} = 1.53$ m.

EXAMPLE 23–12 A Lucite planoconcave lens has one flat surface and the other has $R = -18.4$ cm. What is the focal length?

SOLUTION From Table 23–1, n for Lucite is 1.51. A plane surface has infinite radius of curvature; if we call this R_1, then $1/R_1 = 0$. Therefore,

$$\frac{1}{f} = (1.51 - 1.00)\left(-\frac{1}{18.4\text{ cm}}\right).$$

So $f = (-18.4\text{ cm})/0.51 = -36.0$ cm, and the lens is diverging.

SUMMARY

Light appears to travel in straight-line paths, called *rays*, at a speed v that depends on the *index of refraction, n*, of the material; that is, $v = c/n$, where c is the speed of light in vacuum.

When light reflects from a flat surface, the *angle of reflection equals the angle of incidence*. This *law of reflection* explains why mirrors can form *images*. In a plane mirror, the image is virtual, upright, the same size as the object, and is as far behind the mirror as the object is in front.

When light passes from one transparent medium into another, the rays bend or refract. The *law of refraction* (*Snell's law*) states that

$$n_1 \sin\theta_1 = n_2 \sin\theta_2,$$

where n_1 and θ_1 are the index of refraction and angle with the normal to the surface for the incident ray, respectively, and n_2 and θ_2 are for the refracted ray.

When light rays reach the boundary of a material where the index of refraction decreases, the rays will be *totally internally reflected* if the incident angle, θ_1, is such that Snell's law would predict $\sin\theta_2 > 1$; this occurs if θ_1 exceeds the critical angle θ_c given by $\sin\theta_c = n_2/n_1$.

A lens uses refraction to produce a real or virtual image. Parallel rays of light are focused to a point, called the *focal point*, by a converging lens. The distance of the focal point from the lens is called the *focal length f* of the lens. After parallel rays pass through a diverging lens, they appear to diverge from a point, its focal point; and the corresponding focal length is considered negative. The power of a lens, equal to $1/f$, is given in diopters, which are units of inverse meters (m^{-1}). The position and size of the image formed by a lens of a given object can be found by ray tracing. Algebraically, the relation between image and object distances, d_i and d_o, and the focal length is given by the *lens equation*:

$$\frac{1}{d_o} + \frac{1}{d_i} = \frac{1}{f}.$$

The ratio of image height to object height, which equals the magnification, is equal to $-d_i/d_o$.

When using the various equations of geometrical optics, it is important to remember the *sign conventions* for all quantities involved.

QUESTIONS

1. Give arguments to show why the moon must have a rough surface rather than a polished mirrorlike surface.
2. When you look at yourself in a tall plane mirror, you see the same amount of your body whether you are close to the mirror or far away. (Try it and see.) Use ray diagrams to show why this should be true.
3. When you look at the moon's reflection from a ripply sea, it appears elongated (Fig. 23–34). Explain.
4. If a perfect plane mirror reflected 100 percent of the light incident on it, could you see the surface of the mirror?

FIGURE 23–34 Question 3.

5. Although a plane mirror reverses left and right, it doesn't reverse up and down. Explain.

*6. Archimedes is said to have burned the whole Roman fleet in the harbor of Syracuse by focusing the rays of the sun with a huge spherical mirror. Is this reasonable?

*7. Show with diagrams that the magnification of a concave mirror is less than 1 if the object is beyond the center of curvature C and is greater than 1 if it is within this point.

*8. If a concave mirror produces a real image, is the image necessarily inverted?

*9. When you use a concave mirror, you cannot see an inverted image of yourself unless you place your head beyond the center of curvature C. Yet you can see an inverted image of another object placed between C and F, as in Fig. 23–12. Explain. [*Hint*: you can see a real image only if your eye is behind the image, so that the image can be formed.]

*10. Using the rules for the three rays discussed with reference to Fig. 23–12, draw ray 2 for Fig. 23–15b.

*11. What is the focal length of a plane mirror?

*12. What is the magnification of a plane mirror?

*13. Does the mirror equation, Eq. 23–3, hold for a plane mirror? Explain.

14. How might you determine the speed of light in a solid, rectangular, transparent object?
15. When a wide beam of parallel light enters water at an angle, the beam broadens. Explain.
16. What is the angle of refraction when a light ray meets the boundary between two materials perpendicularly?
17. When you look down into a swimming pool or a lake, are you likely to underestimate or overestimate its depth? Explain. How does the apparent depth vary with the viewing angle? (Use ray diagrams.)
18. Draw a ray diagram to show why a stick looks bent when part of it is under water.
19. How are you able to "see" a round drop of water on a table even though the water is transparent and colorless?
20. Can a light ray traveling in air be totally reflected when it strikes a smooth water surface if the incident angle is right?
21. When you look up at an object in air from beneath the surface in a swimming pool, does the object appear to be the same size as when you see it directly in air? Explain.

22. Where must the film be placed if a camera lens is to make a sharp image of an object very far away?

23. Can a diverging lens form a real image under any circumstances? Explain.

24. Show that a real image formed by a thin lens is always inverted, whereas a virtual image is always upright if the object is real.

25. Light rays are said to be "reversible". Is this consistent with the lens equation?

26. Can real images be projected on a screen? Can virtual images? Can either be photographed? Discuss carefully.

27. A thin converging lens is moved closer to a nearby object. Does the real image formed change (*a*) in position, (*b*) in size? If yes, describe how.

28. Why, in Example 23–10, must the converging lens have a shorter focal length than the diverging lens if the latter's focal length is to be determined by combining them?

29. A lens is made of a material with an index of refraction $n = 1.30$. In air, it is a converging lens. Will it still be a converging lens if placed in water? Explain, using a ray diagram.

*30. An unsymmetrical lens (say, planoconvex) forms an image of a nearby object. Does the image point change if the lens is turned around?

*31. The thicker a double convex lens is in the center as compared to its edges, the shorter its focal length for a given lens diameter. Explain.

*32. Does the focal length of a lens depend on the material surrounding it? What about the focal length of a spherical mirror? Explain.

*33. Compare the mirror equation with the lens equation. Discuss similarities and differences, and especially compare the sign conventions for the quantities involved.

PROBLEMS

SECTION 23–2

1. (I) What is the speed of light in (*a*) ethyl alcohol, and (*b*) Lucite?

2. (I) The speed of light in ice is 2.29×10^8 m/s. What is the index of refraction of ice?

3. (I) How long does it take light to reach us from the sun, 1.50×10^8 km away?

4. (I) Our nearest star (other than the sun) is 4.2 light years away. That is, it takes 4.2 years for the light it emits to reach earth. How far away is it in meters?

5. (II) If $n = 1.00030 \pm 0.000010$ for air, what error will this introduce in the value for the speed of light in air?

6. (II) What is the minimum angular speed at which Michelson's eight-sided mirror would have had to rotate in order that light would be reflected into an observer's eye by succeeding mirror faces (Fig. 23–3)?

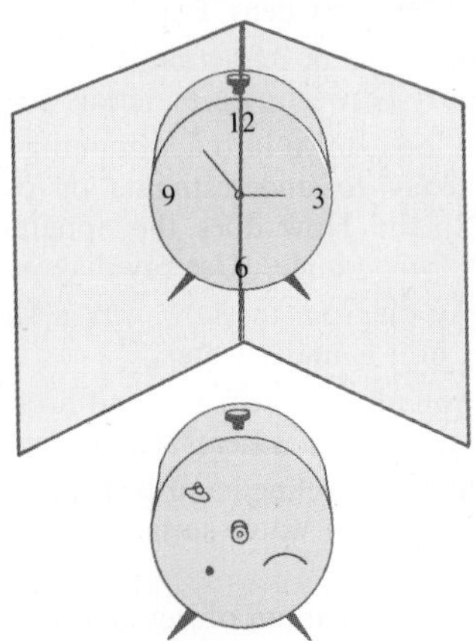

FIGURE 23–35 Problem 8.

SECTION 23–3

7. (I) Suppose that you want to take a photograph of yourself as you look at your image in a flat mirror 2.5 m away. For what distance should the camera lens be focused?

8. (II) Stand up two plane mirrors so they form a right angle as in Fig. 23–35. When you look into this double mirror, you see yourself as others see you, instead of reversed as in a single mirror. Make a ray diagram to show how this occurs.

9. (II) A person whose eyes are 1.48 m above the floor stands 2.40 m in front of a vertical plane mirror whose bottom edge is 40 cm above the floor. What is the horizontal distance to the base of the wall supporting the mirror of the nearest point on the floor that can be seen reflected in the mirror?

10. (II) Suppose you are 70 cm from a plane mirror. What area of the mirror is used to reflect the rays entering one eye from a point on the tip of your nose if your pupil diameter is 5.5 mm?

11. (III) Show that if two plane mirrors meet at an angle ϕ, a single ray reflected (successively) from both mirrors is deflected through an angle of 2ϕ independent of the incident angle. Assume $\phi < 90°$ and that only two reflections, one from each mirror, take place.

12. (III) Suppose a third mirror is placed beneath the two shown in Fig. 23–35, so that all three are perpendicular to each other. (*a*) Show that for such a "corner reflector," any incident ray will return in its original direction after three reflections. (*b*) What happens if it makes only two reflections?

*SECTION 23–4

*13. (I) What is the radius of a concave reflecting surface that brings parallel light to a focus 22.4 cm in front of it?

*14. (I) How far from a concave mirror (radius 18.0 cm) must an object be placed if its image is to be at infinity?

*15. (II) You try to look at yourself in a silvered ball of diameter 64.0 cm when you are 2.70 m away. Where is your image? Is it real or virtual? Can you see yourself clearly?

*16. (II) A dentist wants a small mirror that, when 2.20 cm from a tooth, will produce a 5.5 × upright image. What kind of mirror must be used and what must its radius of curvature be?

*17. (II) A luminous object 3.0 mm high is placed 20 cm from a convex mirror of radius of curvature 20 cm. (*a*) Show by ray tracing that the image is virtual, and estimate the image distance. (*b*) Show that to compute this (negative) image distance from Eq. 23–3, it is sufficient to let the focal length be −10 cm. (*c*) Compute the image size, using Eq. 23–4.

*18. (II) Use the mirror equation to show that the magnitude of the magnification of a concave mirror is less than 1 if the object is beyond the center of curvature C ($d_o > r$), and is greater than 1 if the object is within C ($d_o < r$).

*19. (II) A 2.70-cm-tall object is placed 32.0 cm from a spherical mirror. It produces a virtual image 3.80 cm high. (*a*) What type of mirror is being used? (*b*) Where is the image located? (*c*) What is the radius of curvature of the mirror?

*20. (II) Show, using a ray diagram, that the magnification m of a convex mirror is $m = -d_i/d_o$, just as for a concave mirror. [*Hint*: consider a ray from the top of the object that reflects at the center of the mirror.]

*21. (II) Use ray diagrams to show that the mirror equation, Eq. 23–3, is valid for a convex mirror as long as f is considered negative.

*22. (II) The magnification of a convex mirror is 0.45× for objects 3.0 m away. What is the focal length of this mirror?

*23. (II) (*a*) A plane mirror can be considered a limiting case of a spherical mirror. Specify what this limit is. (*b*) Determine an equation that relates the image and object distances in this limit of a plane mirror. (*c*) Determine the magnification of a plane mirror in this same limit. (*d*) Are your results in parts (*b*) and (*c*) consistent with the discussion of Section 23–3 on plane mirrors?

*24. (III) What is the radius of a concave mirror that gives a 1.8 × magnification of a face 30 cm from it?

SECTION 23–5

25. (I) A flashlight beam strikes the surface of a pane of glass ($n = 1.50$) at a 45° angle. What is the angle of refraction?

26. (I) A diver shines a flashlight upward from beneath the water at a 31.0° angle to the vertical. At what angle does the light leave the water?

27. (I) Rays of the sun are seen to make a 25.0° angle to the vertical beneath the water. At what angle above the horizon is the sun?

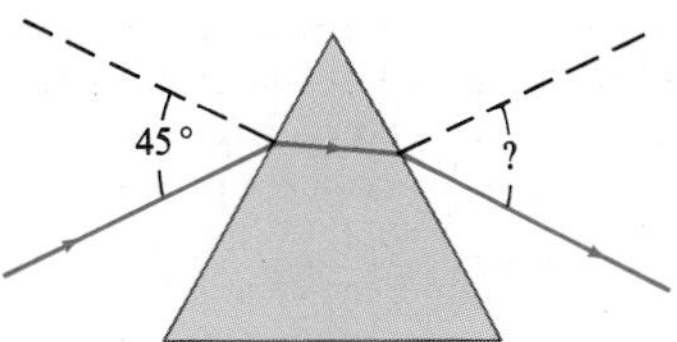

FIGURE 23–36 Problems 28 and 37.

28. (II) Light is incident on an equilateral crown glass prism at a 45.0° angle to one face, Fig. 23–36. Calculate the angle at which light emerges from the opposite face. Assume that $n = 1.52$.

29. (II) A bright light is 2.80 m below the surface of a swimming pool and 1.37 m from one edge of the pool. At what angle does the light leave the water at the edge of the pool? Assume that the water reaches the top edge of the pool.

30. (II) Prove in general that for a light beam incident on a uniform layer of transparent material, as in Fig. 23–18, the direction of the emerging beam is parallel to the incident beam, independent of the incident angle θ and the thickness of the material.

31. (II) An aquarium filled with water has flat glass sides whose index of refraction is 1.52. A beam of light from outside the aquarium strikes the glass at a 43.5° angle to the perpendicular. What is the angle of this light ray when it enters (*a*) the glass and then (*b*) the water? (*c*) What would be the refracted angle if the ray entered the water directly?

32. (III) A light ray is incident on a flat piece of glass as in Fig. 23–18. Show that if the incident angle θ is small, the ray is displaced a distance $d = t\theta(n-1)/n$, where t is the thickness of the glass and θ is in radians.

SECTION 23–6

33. (I) What is the critical angle for a diamond–water interface?

34. (I) The critical angle for a certain liquid–air surface is 46.0°. What is the index of refraction of the liquid?

35. (II) A beam of light is emitted 8.0 cm beneath the surface of a liquid and strikes the surface 7.0 cm from the point directly above the source. If total internal reflection occurs, what can you say about the index of refraction of the liquid?

36. (II) The end faces of a cylindrical glass rod ($n = 1.54$) are perpendicular to the sides. Show that a light ray entering an end face at any angle will be totally internally reflected inside the rod when it strikes the sides. Assume the rod is in air. What if it were in water?

37. (II) Suppose a ray strikes the left face of the prism in Fig. 23–36 at 45° as shown, but is totally internally reflected at the opposite side. If the apex angle (at the top) is $\phi = 75°$, what can you say about the index of refraction of the prism?

38. (III) (*a*) What is the minimum index of refraction for a glass or plastic prism to be used in binoculars (Fig. 23–21) so that total internal reflection occurs at 45°? (*b*) Will binoculars work if its prisms are immersed in water? Assume $n = 1.50$. (*c*) What minimum n is needed if the prisms are immersed in water?

SECTIONS 23–7 TO 23–9

39. (I) A sharp image is located 58.0 mm behind a 50.0-mm-focal-length converging lens. Calculate the object distance.

40. (I) A leaf is placed 88.0 cm in front of a −710-mm-focal-length lens. Where is the image? Is it real or virtual?

41. (I) A certain lens focuses an object 33.5 cm away as an image 45.0 cm on the other side of the lens. What type of lens is it and what is its focal length? Is the image real or virtual?

42. (I) (*a*) What is the power of a 29.0-cm-focal-length lens? (*b*) What is the focal length of a −7.5-diopter lens? Are these lenses converging or diverging?

43. (II) (*a*) An object 28.0 cm in front of a certain lens is imaged 8.10 cm in front of that lens (on the same side as the object). What type of lens is this and what is its focal length? Is the image real or virtual? (*b*) What if the image were located, instead, 35.0 cm in front of the lens?

44. (II) A −6.0-diopter lens is held 22.0 cm from an ant 1.0 mm high. What is the position, type, and height of the image?

45. (II) (*a*) How far from a 50.0-mm-focal-length lens must an object be placed if its image is to be magnified 2.00 × and be real? (*b*) What if the image is to be virtual and magnified 2.00 × ?

46. (II) Repeat Problem 45 for a −50.0-mm-focal-length lens.

47. (II) How large is the image of the sun on the film used in a camera with a 50-mm-focal-length lens? The sun's diameter is 1.4×10^6 km and it is 1.5×10^8 km away.

48. (II) (*a*) A 2.20-cm-high insect is 1.20 m from a 135-mm-focal-length lens. Where is the image, how high is it, and what type is it? (*b*) What if $f = -135$ mm?

49. (II) A diverging lens with $f = -35.5$ cm is placed 12.0 cm behind a converging lens with $f = 16.0$ cm. Where will an object at infinity be focused?

50. (II) Two 27.0-cm-focal-length converging lenses are placed 16.5 cm apart. An object is placed 35.0 cm in front of one. Where will the final image formed by the second lens be located? What is the total magnification?

51. (III) A diverging lens is placed next to a converging lens of focal length f_C, as in Fig. 23–32. If f_T represents the focal length of the combination, show that the focal length of the diverging lens, f_D, is given by

$$\frac{1}{f_D} = \frac{1}{f_T} - \frac{1}{f_C}.$$

52. (III) How far apart are an object and an image formed by a 75-cm-focal-length converging lens if the image is 2.75 × larger than the object and is real?

53. (III) Show that the lens equation can be written in the *Newtonian form*:

$$xx' = f^2,$$

where x is the distance of the object from the focal point on the front side of the lens, and x' is the distance of the image to the focal point on the other side of the lens.

*SECTION 23–10

*54. (I) A double concave lens has surface radii of 25.5 and 20.2 cm. What is the focal length if $n = 1.58$?

*55. (I) Both surfaces of a double convex lens have radii of 28.0 cm. If the focal length is 26.2 cm, what is the index of refraction of the lens material?

*56. (I) Show that if the lens of Example 23–12 is reversed so the light enters the curved face, the focal length is unchanged.

*57. (I) A planoconvex lens is to have a focal length of 23.0 cm. If made from fused quartz, what must be the radius of curvature of the convex surface?

*58. (I) A glass ($n = 1.50$) planoconcave lens has a focal length of −31.6 cm. What is the radius of the concave surface?

*59. (II) An object is placed 100 cm from a glass lens ($n = 1.56$) with one concave surface of radius 21.0 cm and one convex surface of radius 18.5 cm. Where is the final image? What is the magnification?

*60. (III) A glass lens ($n = 1.55$) in air has a power of +4.5 diopters. What would its power be if it were submerged in water?

GENERAL PROBLEMS

61. Light is emitted from an ordinary light bulb filament in wave-train bursts of about 10^{-8} s in duration. What is the length in space of such wave trains?

62. A 31.0-cm-focal-length converging lens is 21.0 cm behind a diverging lens. Parallel light strikes the diverging lens. After passing through the converging lens, the light is again parallel. What is the focal length of the diverging lens?

63. In a slide or move projector, the film acts as the object whose image is projected on a screen. If a 100-mm-focal-length lens is to project an image on a screen 5.0 m away, how far from the lens should the slide be? If the slide is 36 mm wide, how wide will the picture be on the screen?

64. A 35-mm slide (picture size is actually 24 by 36 mm) is to be projected on a screen 1.60 by 2.40 m placed

26.0 m from the projector. What focal-length lens should be used if the image is to cover the screen?

65. Show analytically that the image formed by a converging lens is real and inverted if the object is beyond the focal point ($d_o > f$), and is virtual and upright if the object is within the focal point ($d_o < f$). Describe the image if the object is a virtual image (formed by another lens) for which $-d_o > f$, and for which $0 < -d_o < f$.

66. Show analytically that a diverging lens can never form a real image of a real object. Can you describe a situation in which a diverging lens can form a real image?

67. If the apex angle of a prism is $\phi = 70°$ (see Fig. 23–37), what is the minimum incident angle for a ray if it is to emerge from the opposite side (i.e. not be totally internally reflected), given $n = 1.58$?

68. When light passes through a prism, the angle that the refracted ray makes relative to the incident ray is called the deviation angle δ, Fig. 23–37. This angle is a minimum when the ray passes through the prism symmetrically—that is, for a prism shaped like an isosceles triangle, when the ray inside the prism is parallel to the base.

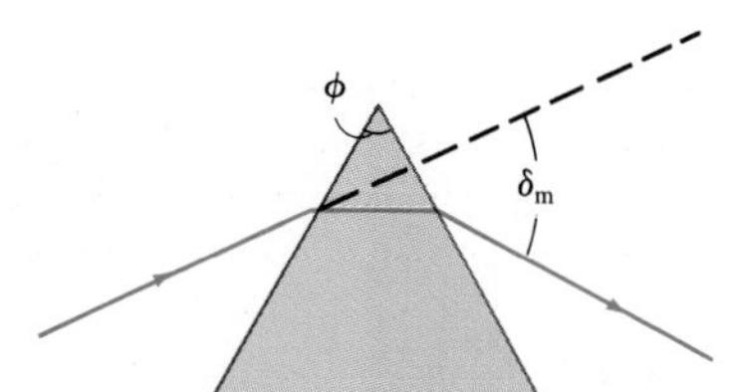

FIGURE 23–37 Problems 67 and 68.

Show that this minimum deviation angle, δ_m, is related to the prism's index of refraction n by

$$n = \frac{\sin \frac{1}{2}(\phi + \delta_m)}{\sin \frac{1}{2}\phi}$$

where ϕ is the apex angle.

69. (*a*) Show that if two lenses of focal lengths f_1 and f_2, respectively, are placed next to each other, the focal length of the combination is given by $f_T = f_1 f_2/(f_1 + f_2)$. (*b*) Show that the power P of the combination of two lenses is the sum of their separate powers, $P = P_1 + P_2$.

70. A bright object is placed on one side of a converging lens of focal length f and a white screen for viewing the image is on the opposite side. The distance $d_T = d_i + d_o$ between the object and the screen is kept fixed, but the lens can be moved. Show that: (*a*) if $d_T > 4f$, there will be *two* positions where the lens can be placed and a sharp image will be produced on the screen; (*b*) if $d_T < 4f$, there will be no lens position where a sharp image is formed. (*c*) Determine the distance between the two lens positions in part (*a*), and the ratio of the image sizes.

*71. A 1.65-m-tall person stands 3.80 m from a convex mirror and notices that he looks precisely half as tall as he does in a plane mirror placed at the same distance. What is the radius of curvature of the convex mirror? (Assume that $\sin \theta \approx \theta$.)

*72. A lens whose index of refraction is n is submerged in a material whose index of refraction is n' ($n' \neq 1$). Derive the equivalent of Eqs. 23–8, 23–9, and 23–10 for this lens.

CHAPTER 24

The Wave Nature of Light

Rainbows, such as this one appearing over Sir Isaac Newton's childhood home in England, are formed by refraction of light within droplets of water. Sunlight is broken into its constituent colors because each wavelength is bent slightly differently by the water.

That light carries energy is obvious to anyone who has focused the sun's rays with a magnifying glass on a piece of paper and burned a hole in it. But how does light travel and in what form is this energy carried? In our discussion of waves in Chapter 11, we noted that energy can be carried from place to place in basically two ways: by particles or by waves. In the first case, material bodies or particles can carry energy, such as a thrown baseball or rushing water. In the second case, water waves and sound waves, for example, can carry energy over long distances even though mass itself does not travel these distances. In view of this, what can we say about the nature of light: does light travel as a stream of particles away from its source; or does it travel in the form of waves that spread outward from the source? Historically, this question has turned out to be a difficult one. For one thing, light does not reveal itself in any obvious way as being made up of tiny particles nor do we see tiny light waves passing by as we do water waves. The evidence seemed to favor first one side and then the other until about 1830, when most physicists had accepted the wave theory. By the end of the nineteenth century, light was considered to be an *electromagnetic wave*

(Chapter 22). In the early twentieth century, light was shown to have a particle nature as well, as we shall discuss in Chapter 27. Nonetheless, the wave theory of light remains valid and has proved very successful. We now investigate the evidence for the wave theory and how it has explained a wide range of phenomena.

24–1 • Waves Versus Particles; Huygens' Principle and Diffraction

The Dutch scientist Christian Huygens (1629–1695), a contemporary of Newton, proposed a wave theory of light that had much merit. Still useful today is a technique he developed for predicting the future position of a wave front when an earlier position is known. This is known as **Huygens' principle** and can be stated as follows: *every point on a wave front can be considered as a source of tiny wavelets that spread out in the forward direction at the speed of the wave itself. The new wave front is the envelope of all the wavelets* (that is, *the tangent to all of them*).

Huygens' principle

As a simple example of the use of Huygens' principle, consider the wave front AB in Fig. 24–1 which is traveling away from a source S. We assume the medium is *isotropic*—that is, the speed v of the waves is the same in all directions. To find the wave front a short time t after it is at AB, tiny circles are drawn whose radius $r = vt$. The centers of these tiny circles are on the original wave front AB and the circles represent Huygens' (imaginary) wavelets. The tangent to all these wavelets, the line CD, is the new position of the wave front.

Huygens' principle is particularly useful when waves impinge on an obstacle and the wave fronts are partially interrupted. Huygens' principle predicts that waves bend in behind an obstacle, as shown in Fig. 24–2. This is just what water waves do, as we saw in Chapter 11 (Figs. 11–32 and 11–33). The bending of waves behind obstacles into the "shadow region" is known as **diffraction**. Since diffraction occurs for waves, but not for particles, it can serve as one means for distinguishing the nature of light.

Does light exhibit diffraction? In the midseventeenth century, a Jesuit priest, Francesco Grimaldi (1618–1663), had observed that when sunlight entered a darkened room through a tiny hole in a screen, the spot on the opposite wall was larger than would be expected from geometric rays; he also observed that the border of the image was not clear but was surrounded by colored fringes. Grimaldi attributed this to the diffraction of light. Newton, who favored a particle theory, was aware of Grimaldi's result. He felt that

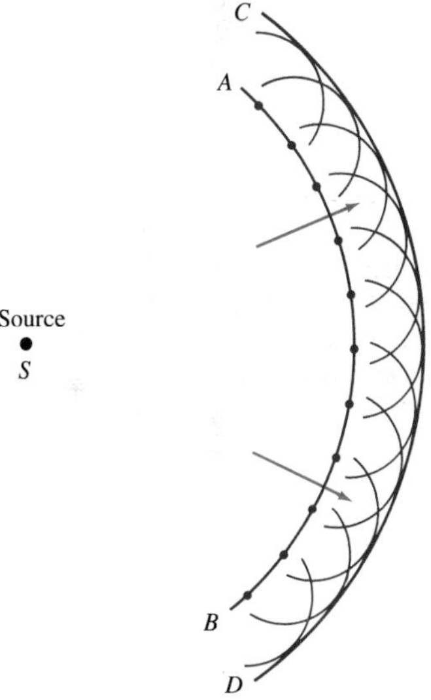

FIGURE 24–1 (above) Huygens' principle used to determine wave front CD when AB is given.

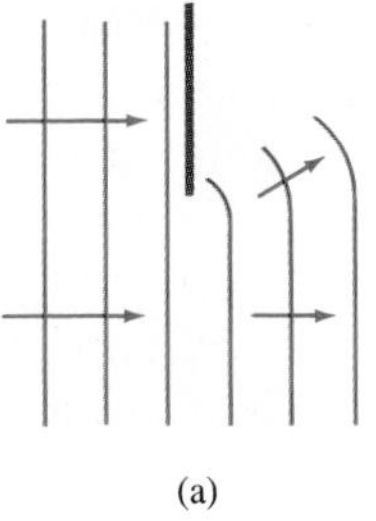
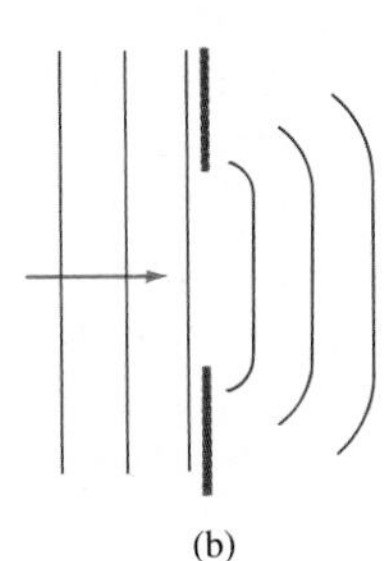
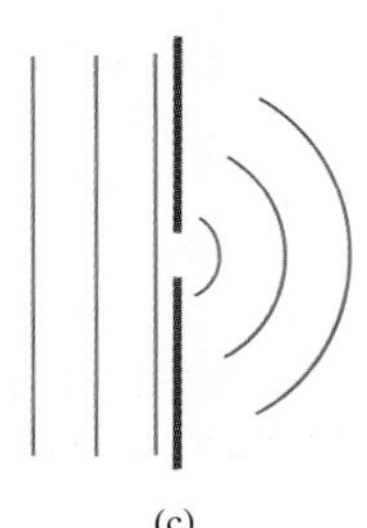

(a) (b) (c)

FIGURE 24–2 (left) Huygens' principle is consistent with diffraction (a) around the edge of an obstacle, (b) through a large hole, (c) through a small hole whose size is on the order of the wavelength of the wave.

Grimaldi's result was due to the interaction of light corpuscles ("little bodies") with the edges of the hole. If light were a wave, he said, the light waves should bend more than that observed. Newton's argument seems reasonable. Yet, as we saw in Chapter 11, diffraction is large only when the size of the obstacle or the hole is on the order of the wavelength of the wave (Fig. 11–33). Newton did not guess that the wavelengths of visible light might be incredibly tiny, and thus diffraction effects would be very small. (Indeed, this is why geometric optics using rays is so successful—normal openings and obstacles are much larger than the wavelength of the light, and so relatively little diffraction or bending occurs.)

24–2 • Huygens' Principle and the Law of Refraction

The laws of reflection and refraction were well known in Newton's time. The law of reflection could not distinguish the two theories. For when waves reflect from an obstacle, the angle of incidence equals the angle of reflection (Fig. 11–24). The same is true of particles—think of a tennis ball without spin striking a flat surface.

The law of refraction is another matter. Consider light entering a medium where it is bent toward the normal, as when it travels from air into water. As shown in Fig. 24–3, this effect can be constructed using Huygens' principle if we assume the speed of light is less in the second medium ($v_2 < v_1$). That is, in time t, the point B on wave front AB goes a distance $v_1 t$ to reach point D. Point A, on the other hand, travels a distance $v_2 t$ to reach point C. Huygens' principle is applied to points A and B to obtain the curved wavelets shown at C and D. The wave front is tangent to these two wavelets, so the new wave front is the line CD. Hence the rays (which are perpendicular to the wave fronts) bend toward the normal if $v_2 < v_1$, as drawn.† Newton's corpuscle theory predicted the opposite result: if the path of light corpuscles

† This is basically the same as the discussion around Fig. 11–28.

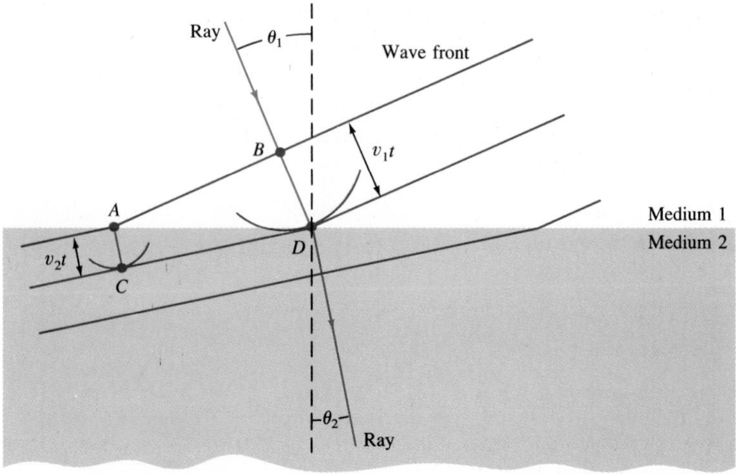

FIGURE 24–3 Refraction explained, using Huygens' principle.

entering a new medium changes direction, it must be because the medium exerts a force on the corpuscles at the boundary. This force was assumed to act perpendicular to the boundary and thus would affect only the perpendicular component of the corpuscles' velocity. When light enters a medium such as water where it is bent toward the normal, the force must accelerate the corpuscles so the perpendicular component of velocity is increased; only in this way will the refracted angle be less than the incident. In Newton's theory, then, the speed of light would be greater in the second medium ($v_2 > v_1$). Thus the wave theory predicts that the speed of light in water (say) is less than in air and Newton's corpuscle theory predicts the reverse. An experiment to actually measure the speed of light in water and confirm the wave-theory prediction was done in 1850 by the French physicist Jean Foucault. By then, however, the wave theory was already fully accepted, as we shall see in the next section.

It is easy to show that Snell's law of refraction follows directly from Huygens' principle, given that the speed of light v in any medium is related to the speed in a vacuum, c, and the index of refraction, n, by Eq. 23–1: that is, $v = c/n$. From the Huygens' construction of Fig. 24–3, angle ADC is equal to θ_2 and angle BAD is equal to θ_1. Then for the two triangles that have the common side AD, we have

$$\sin\theta_1 = \frac{v_1 t}{AD}, \qquad \sin\theta_2 = \frac{v_2 t}{AD}.$$

We divide these two equations and obtain:

$$\frac{\sin\theta_1}{\sin\theta_2} = \frac{v_1}{v_2}.$$

Then, since $v_1 = c/n_1$ and $v_2 = c/n_2$,

$$n_1 \sin\theta_1 = n_2 \sin\theta_2,$$

which is Snell's law of refraction, Eq. 23–5. (The law of reflection can be derived from Huygens' principle in a similar way, and this is given as Problem 1 at the end of the chapter.)

When a light wave travels from one medium to another, its frequency does not change but its wavelength does. This can be seen from Fig. 24–3, where we assume each of the lines representing a wave front corresponds to a crest (peak) of the wave. Then

$$\frac{\lambda_2}{\lambda_1} = \frac{v_2 t}{v_1 t} = \frac{v_2}{v_1} = \frac{n_1}{n_2}.$$

If medium 1 is a vacuum (or air), so $n_1 = 1$, $v_1 = c$, and we call λ_1 simply λ, then the wavelength in another medium of index of refraction n $(=n_2)$ will be

$$\lambda_n = \frac{\lambda}{n}. \qquad (24\text{–}1)$$

λ depends on n

This result is consistent with the frequency f being unchanged since $c = f\lambda$. Combining this with $v = f\lambda_n$ in a medium where $v = c/n$ gives $\lambda_n = v/f = c/nf = f\lambda/nf = \lambda/n$, which checks.

24–3 • Interference—Young's Double-Slit Experiment

In 1801, the Englishman Thomas Young (1773–1829) obtained convincing evidence for the wave nature of light and was even able to measure the wavelengths for visible light. Figure 24–4a shows a schematic diagram of Young's famous double-slit experiment. Light from a single source (Young used the sun) falls on a screen containing two closely spaced slits S_1 and S_2. If light consists of tiny particles, we might expect to see two bright lines on a screen placed behind the slits as in (b). But Young observed instead a series of bright lines as in (c). Young was able to explain this result as a *wave-interference* phenomenon. To see this, imagine plane waves of light of a single wavelength (called **monochromatic**, meaning "one color") falling on the two slits as shown in Fig. 24–5. Because of diffraction, the waves leaving the two small slits spread out as shown. This is equivalent to the interference pattern produced when two rocks are thrown into a lake (Fig. 11–30), or when sound from two loudspeakers interferes (Fig. 12–11).

To see how an interference pattern is produced on the screen, we make use of Fig. 24–6. Waves of wavelength λ are shown entering the slits S_1 and S_2 which are a distance d apart. The waves spread out in all directions after passing through the slits, but they are shown only for three different angles θ. In Fig. 24–6a, the waves reaching the center of the screen are shown ($\theta = 0$). The waves from the two slits travel the same distance, so they are in phase: constructive interference occurs and there is a bright spot at the center of the screen. There will also be constructive interference when the paths of the two rays differ by one wavelength (or any whole number of wavelengths), as shown in Fig. 24–6b. But if one ray travels an extra distance of one-half a wavelength (or $\frac{3}{2}\lambda$, $\frac{5}{2}\lambda$, and so on), the two waves are exactly out of phase when they reach the screen, and so destructive interference occurs and the screen is dark, Fig. 24–6c. Thus, there will be a series of bright and dark lines (or **fringes**) on the viewing screen.

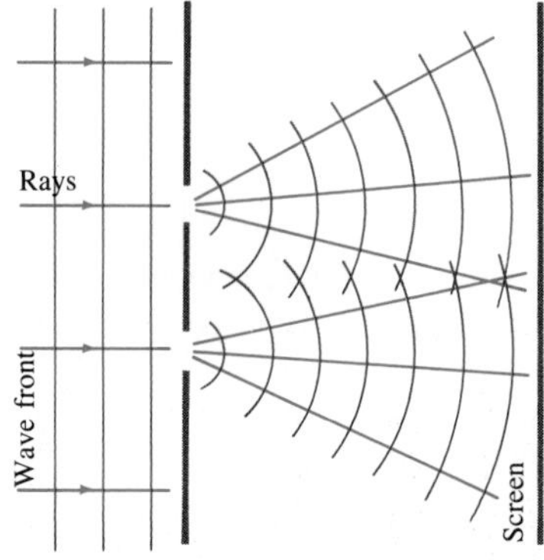

FIGURE 24–5 If light is a wave, light passing through one of two slits should interfere with light passing through the other slit.

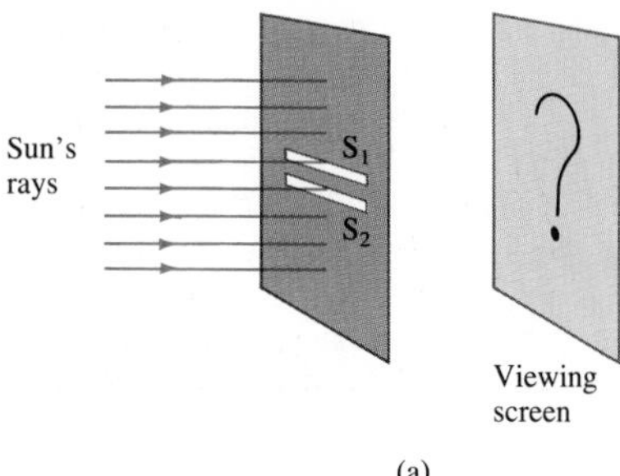

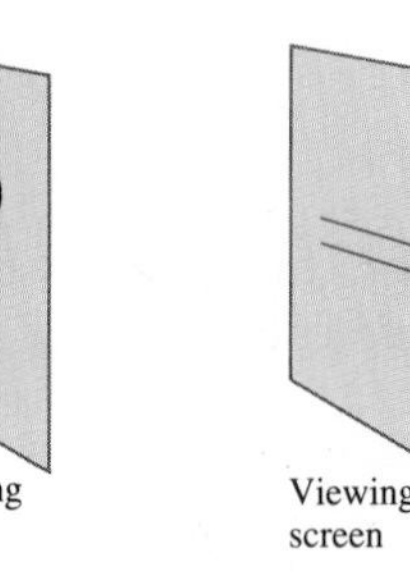

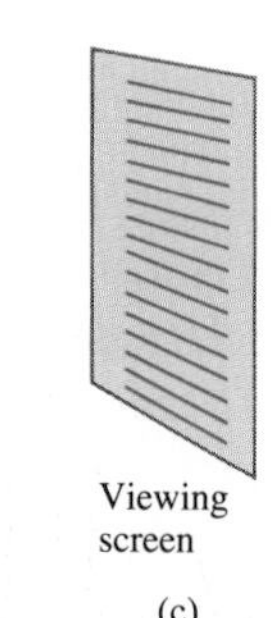

FIGURE 24–4 (a) Young's double-slit experiment. (b) If light consists of particles, we would expect to see two bright lines on the screen behind the slits. (c) Young observed many lines.

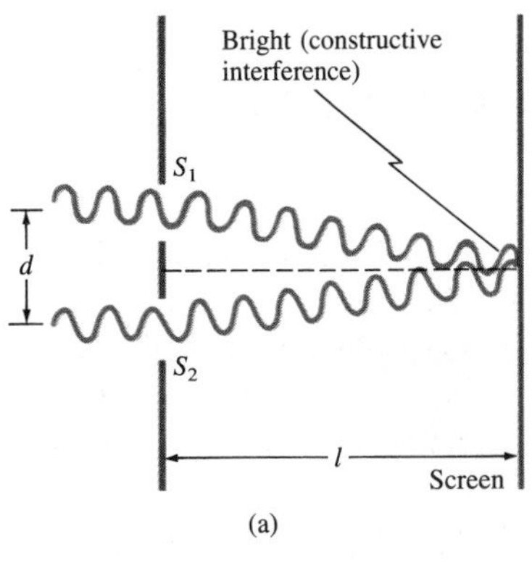

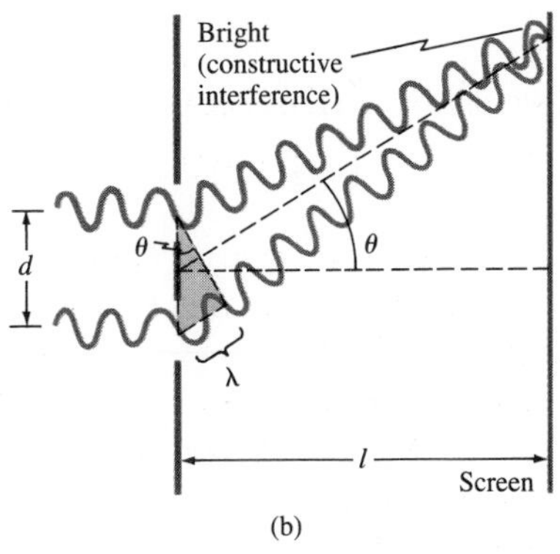

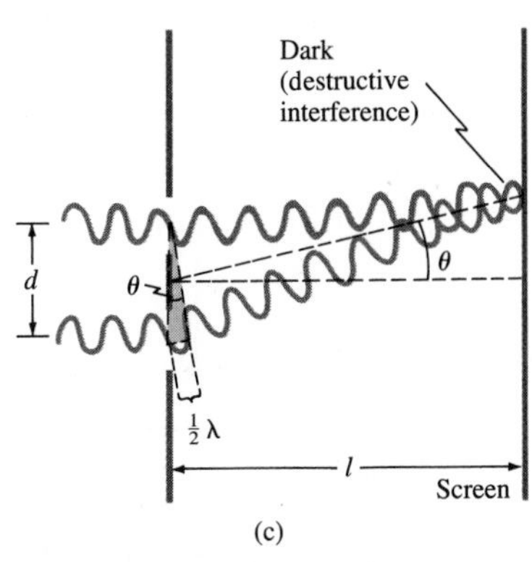

FIGURE 24–6 How the wave theory explains the pattern of lines seen in the double-slit experiment.

To determine exactly where the bright lines fall, first note that Fig. 24–6 is somewhat exaggerated; in real situations, the distance d between the slits is very small compared to the distance l to the screen. The rays from each slit for each case will therefore be essentially parallel and θ is the angle they make with the horizontal. From the shaded right triangle in Fig. 24–6b, we can see that the extra distance traveled by the lower ray is $d \sin \theta$. **Constructive interference** will occur on the screen when $d \sin \theta$ equals a whole number of wavelengths.

Conditions for:

Constructive interference

$$d \sin \theta = m\lambda, \qquad m = 0, 1, 2, \cdots. \qquad \begin{bmatrix}\text{constructive}\\ \text{interference}\end{bmatrix} \qquad (24\text{–}2a)$$

The value of m is called the **order** of the interference fringe. The first order ($m = 1$), for example, is the first fringe on each side of the center one (at $\theta = 0$). **Destructive interference** occurs when the extra distance $d \sin \theta$ is $\frac{1}{2}$, $\frac{3}{2}$, and so on, wavelengths:

Destructive interference

$$d \sin \theta = (m + \tfrac{1}{2})\lambda, \qquad m = 0, 1, 2, \cdots. \qquad \begin{bmatrix}\text{destructive}\\ \text{interference}\end{bmatrix} \qquad (24\text{–}2b)$$

The intensity of the bright fringes is greatest for the center one ($m = 0$) and decreases for higher orders, as shown in Fig. 24–7.

EXAMPLE 24–1 A screen containing two slits 0.100 mm apart is 1.20 m from the viewing screen. Light of wavelength $\lambda = 500$ nm falls on the slits from a distant source. Approximately how far apart will the bright interference fringes be on the screen?

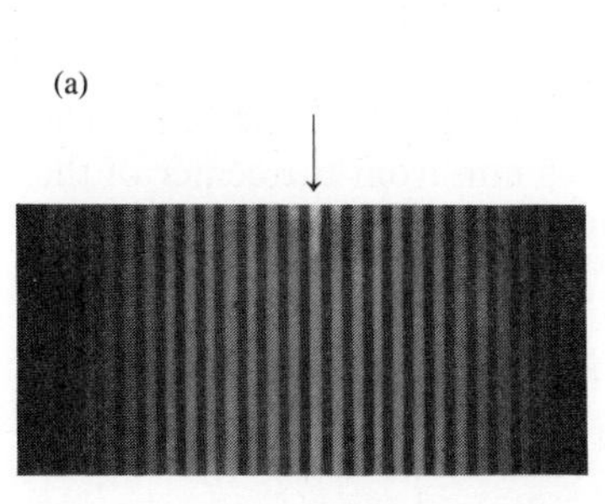

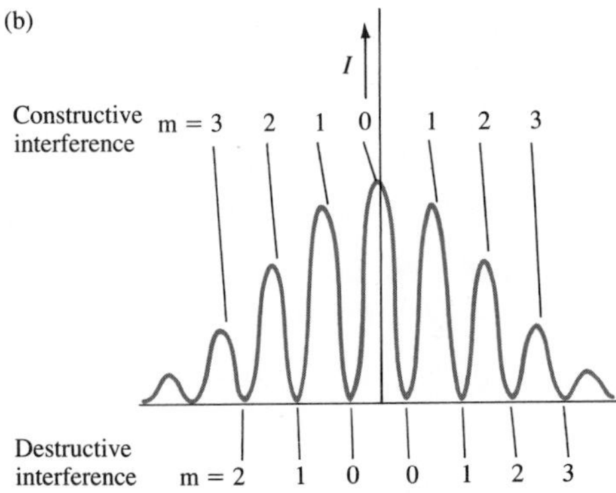

FIGURE 24–7 (a) Interference fringes produced by double-slit experiment and detected by photographic film placed on the viewing screen. The arrow marks the central fringe. (b) Intensity of light in the interference pattern. Also shown are values of m for Eq. 24–2a (constructive interference) and Eq. 24–2b (destructive interference).

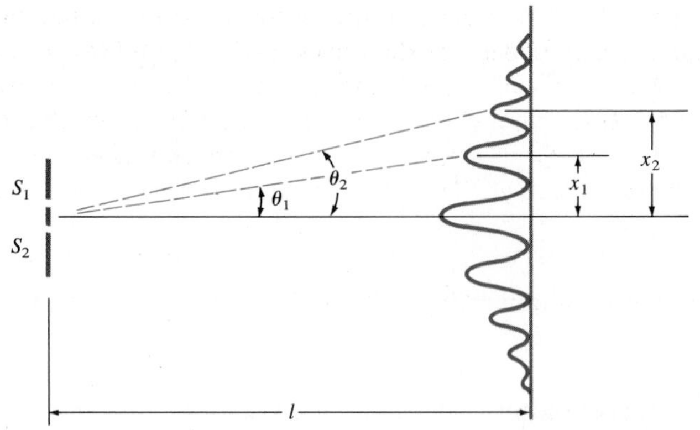

FIGURE 24–8 For small angles, the interference fringes occur at distance $x = \theta l$ above the center ($m = 0$) fringe. θ_1 and x_1 are for the first order ($m = 1$) fringe, θ_2 and x_2 are for $m = 2$.

SOLUTION Since $d = 0.100$ mm $= 1.00 \times 10^{-4}$ m, $\lambda = 500 \times 10^{-9}$ m, and $l = 1.20$ m, the first-order fringe ($m = 1$) occurs at an angle θ given by

$$\sin\theta_1 = \frac{m\lambda}{d} = \frac{(1)(500 \times 10^{-9}\ \text{m})}{1.00 \times 10^{-4}\ \text{m}} = 5.00 \times 10^{-3}.$$

This is a very small angle, so we can take $\sin\theta = \theta$. The first-order fringe will occur a distance x_1 above the center of the screen (see Fig. 24–8) given by $x_1/l = \theta_1$, so

$$x_1 = l\theta_1 = (1.20\ \text{m})(5.00 \times 10^{-3}) = 6.00\ \text{mm}.$$

The second fringe ($m = 2$) will occur at

$$x_2 = l\theta_2 = l\frac{2\lambda}{d} = 12.0\ \text{mm}$$

above the center, and so on. Thus the fringes are 6.00 mm apart.

From Eqs. 24–2 we can see that, except for the zeroth order fringe at the center, the position of the fringes depends on wavelength. Consequently, when white light falls on the two slits, as Young found in his experiments, the central fringe is white, but the first- (and higher-) order fringes contain a spectrum of colors like a rainbow; θ was found to be smallest for violet light and largest for red. By measuring the position of these fringes, Young was the first to determine the wavelengths of visible light (using Eqs. 24–2). In doing so, he showed that what distinguishes different colors physically is their wavelength, an idea put forward earlier by Grimaldi in 1665.

EXAMPLE 24–2 White light passes through two slits 0.50 mm apart and an interference pattern is observed on a screen 2.5 m away. The first-order fringe resembles a rainbow with violet and red light at either end. The violet light falls about 2.0 mm and the red 3.5 mm from the center of the central white fringe. Estimate the wavelengths of the violet and red lights.

SOLUTION We use Eq. 24–2a with $m = 1$ and $\sin\theta = \theta$. Then for violet light, $x = 2.0$ mm, so

$$\lambda = \frac{d\theta}{m} = \frac{d}{m}\frac{x}{l} = \left(\frac{5.0 \times 10^{-4}\ \text{m}}{1}\right)\left(\frac{2.0 \times 10^{-3}\ \text{m}}{2.5\ \text{m}}\right) = 4.0 \times 10^{-7}\ \text{m},$$

or 400 nm. For red light, $x = 3.5$ mm, so λ is about 700 nm.

The two slits in Fig. 24–6 act as if they were sources of radiation. They are called **coherent sources** because the waves leaving them bear the same phase relationship to each other at all times (because ultimately the waves come from a single source to the left of the two slits in Fig. 24–6). An interference pattern is observed only when the sources are coherent. If two tiny light bulbs replaced the two slits (or separate light bulbs illuminated each slit), an interference pattern would not be seen. The light emitted by one light bulb would have a random phase with respect to the second bulb, and the screen would be more or less uniformly illuminated. Two such sources are called **incoherent sources.**

Coherent and incoherent sources

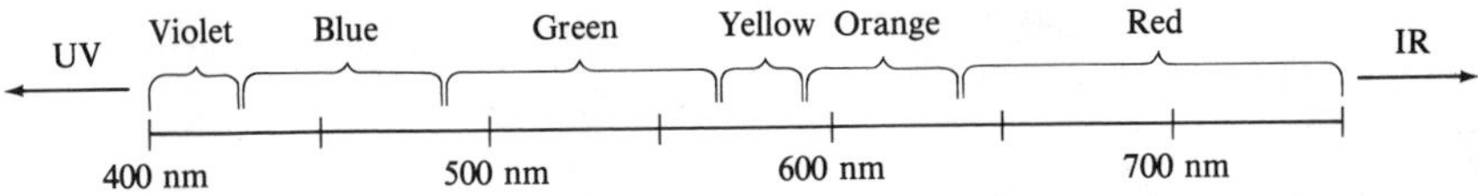

FIGURE 24–9 The spectrum of visible light, showing the range of wavelengths for the various colors.

24–4 • The Visible Spectrum and Dispersion

The two most obvious properties of light are readily describable in terms of the wave theory of light: intensity (or brightness) and color. The *intensity* of light is related to the square of the amplitude of the wave, just as for any wave (see Section 11–9, or Eqs. 22–7 and 22–8). The *color* of the light is related to the wavelength or frequency of the light. Visible light—that to which our eyes are sensitive—falls in the wavelength range of about 400 nm to 750 nm.[†] This is known as the **visible spectrum**, and within it lie the different colors from violet to red, as shown in Fig. 24–9. Light with wavelength shorter than 400 nm is called ultraviolet (UV) and that with wavelength greater than 750 nm is called infrared (IR).[‡] Although human eyes are not sensitive to UV or IR, some types of photographic film do respond to them.

It is a familiar fact that a prism separates white light into a rainbow of colors (Fig. 24–10). This is because the index of refraction of a material depends on the wavelength. This is shown for several materials in Fig. 24–11. White light is a mixture of all visible wavelengths; and when incident on a prism, as in Fig. 24–12, the different wavelengths are bent to varying degrees.

FIGURE 24–10 White light passing through a prism is broken down into its constituent colors.

[†] Sometimes the Angstrom (Å) unit is used when referring to light: 1 Å $= 1 \times 10^{-10}$ m. Then visible light falls in the wavelength range of 4000 Å to 7500 Å.

[‡] The complete electromagnetic spectrum is illustrated in Fig. 22–10.

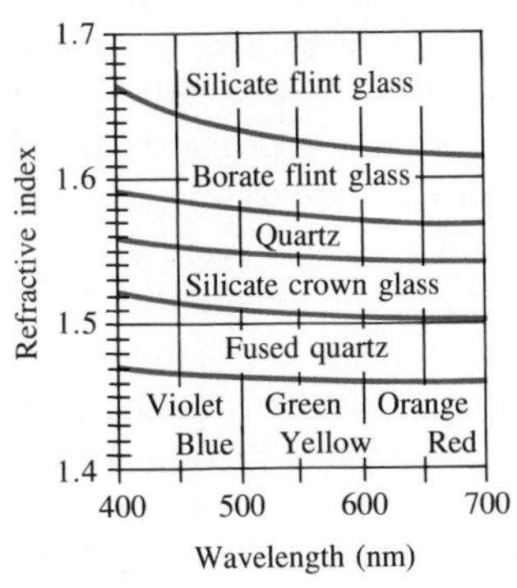

FIGURE 24–11 Index of refraction as a function of wavelength for various transparent solids.

FIGURE 24–12 White light dispersed by a prism into the visible spectrum.

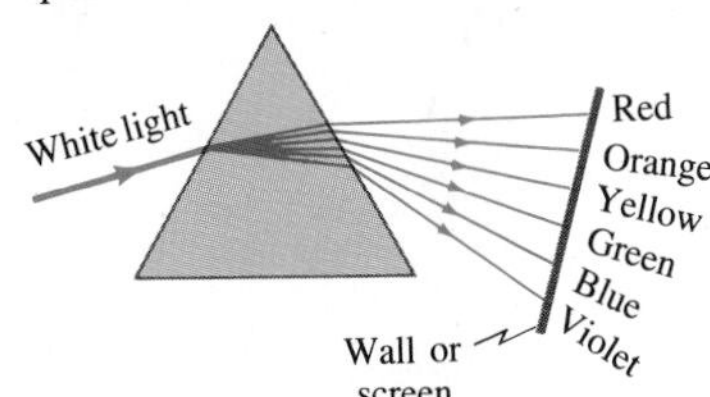

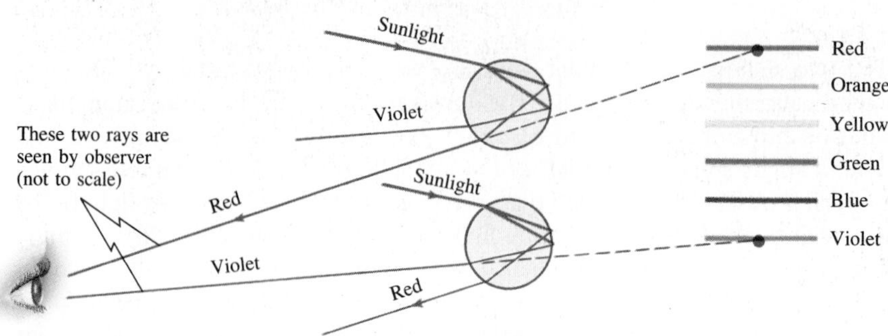

FIGURE 24–13
Formation of a rainbow.

Dispersion

Since the index of refraction is greater for the shorter wavelengths, violet light is bent the most and red the least as indicated. This spreading of white light into the full spectrum is called **dispersion**.

Rainbows

Rainbows (see the photograph at the start of this chapter) are a spectacular example of dispersion (by water in this case). You can see rainbows when you look at falling water with the sun at your back. Figure 24–13 shows how red and violet rays are bent by spherical water droplets and are reflected off the back surface. Red is bent the least and so reaches the observer's eyes from droplets higher in the sky, as shown in the diagram. Thus the top of the rainbow is red.

Diamonds

Diamonds achieve their brilliance from a combination of dispersion and total internal reflection. Since diamonds have a very high index of refraction of about 2.4, the critical angle for total internal reflection is only 25°. Incident light therefore strikes many of the internal surfaces before it strikes one at less than 25° and emerges. After many such reflections, the light has traveled far enough that the colors have become sufficiently separated to be seen individually and brilliantly by the eye after leaving the crystal.

24–5 • Diffraction by a Single Slit or Disk

Young's double-slit experiment put the wave theory of light on a firm footing. But full acceptance came only with studies on diffraction more than a decade later.

We have already discussed diffraction briefly with regard to water waves (Section 11–10) as well as for light (Section 24–1), and we have seen that it refers to the spreading or bending of waves around edges. Now we look at diffraction in more detail.

FIGURE 24–14 If light is a wave, a bright spot will appear at the center of the shadow of a solid disk illuminated by a point source of monochromatic light.

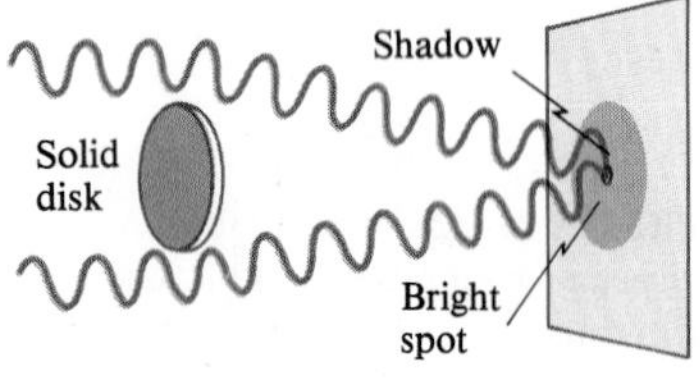

A part of the history of the wave theory of light belongs to Augustin Fresnel (1788–1827) who in 1819 presented to the French Academy a wave theory of light that predicted and explained interference and diffraction effects. Almost immediately Siméon Poisson (1781–1840) pointed out a counterintuitive inference: that according to Fresnel's wave theory, if light from a point source were to fall on a solid disk, then light diffracted around the edges should constructively interfere at the center of the shadow (Fig. 24–14). That prediction seemed very unlikely. But when the experiment was actually carried out by François Arago, the bright spot was seen at the very center of the shadow! This was strong evidence for the wave theory.

The (un)expected diffraction spot

Figure 24–15a is a photograph of the shadow cast by a coin using a (nearly) point source of light (a laser in this case). The bright spot is clearly

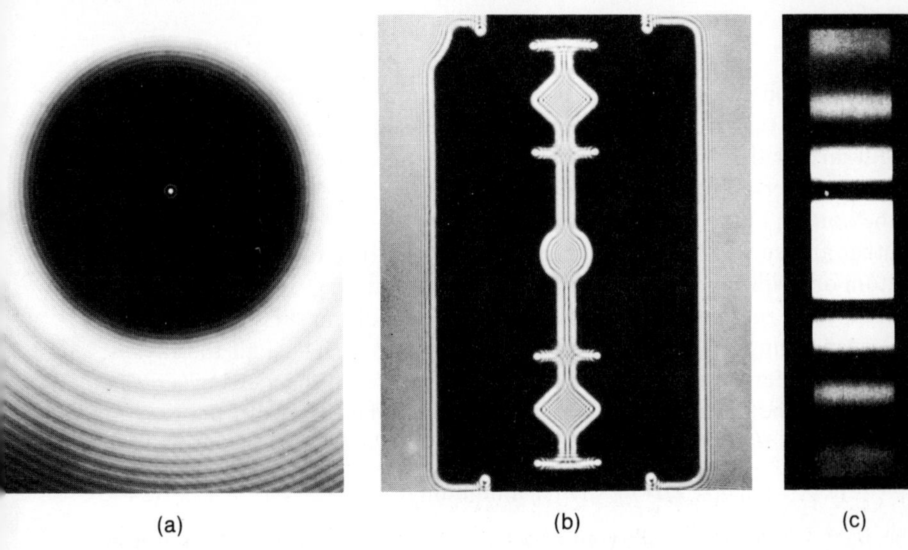

FIGURE 24–15 Diffraction pattern of (a) a penny, (b) a razor blade, (c) a single slit, each illuminated by a nearly point source of monochromatic light.

present at the center. Note that there also are bright and dark fringes beyond the shadow. These resemble the interference fringes of a double slit. Indeed, they are due to interference of waves diffracted around different parts of the disk, and the whole is referred to as a **diffraction pattern**. A diffraction pattern exists around any sharp object illuminated by a point source, as shown in Figs. 24–15b and c. We are not always aware of them because most sources of light in everyday life are not points, so light from different parts of the source washes out the pattern.

To see how a diffraction pattern arises, we will analyze the important case of monochromatic light passing through a narrow slit. We will assume that parallel rays (or plane waves) of light fall on the slit of width D as shown in Fig. 24–16. If the viewing screen is infinitely far away, or a lens is placed behind the slit to focus parallel rays on the screen, the diffraction pattern is called **Fraunhofer diffraction**. If the screen is close and no lenses are used, it is called **Fresnel diffraction**. The analysis in the latter case is rather involved, so we consider only the case of Fraunhofer diffraction. As we know from studying water waves and from Huygens' principle, the waves passing through the slit spread out in all directions. We will now examine how the waves passing through different parts of the slit interfere with each other.

Since the screen is assumed to be very far away, the rays heading for any point are essentially parallel. First we consider rays that pass straight through as in Fig. 24–16a. They are all in phase, so there will be a central

FIGURE 24–16 Analysis of diffraction patterns formed by light passing through a narrow slit.

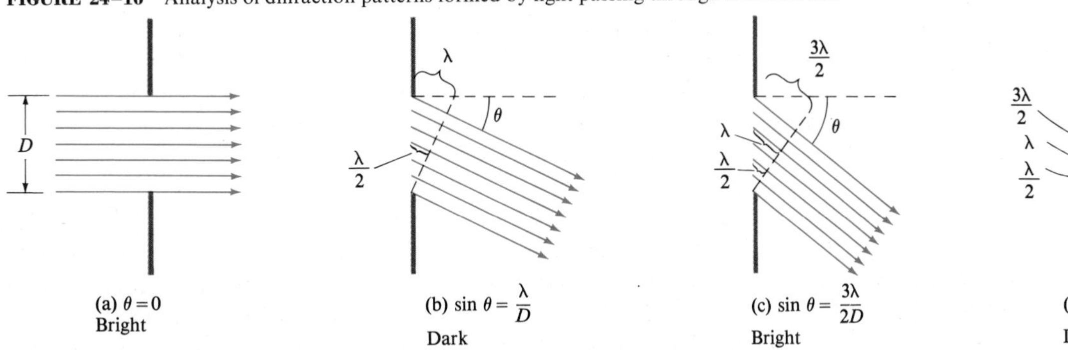

bright spot on the screen. In Fig. 24–16b, we consider rays moving at an angle θ such that the ray from the top of the slit travels exactly one wavelength farther than the ray from the bottom edge. The ray passing through the very center of the slit will travel one-half wavelength farther than the ray at the bottom of the slit. These two rays will be exactly out of phase with one another and so will destructively interfere. Similarly, a ray slightly above the bottom one will cancel a ray that is the same distance above the central one. Indeed, each ray passing through the lower half of the slit will cancel with a corresponding ray passing through the upper half. Thus, all the rays destructively interfere in pairs, and so no light will reach the viewing screen at this angle. The angle θ at which this occurs can be seen from the diagram to be when $\lambda = D \sin \theta$, so

Diffraction equation (angular width of central spot)

$$\sin \theta = \lambda/D. \qquad \text{[first minimum]} \qquad (24\text{–}3)$$

The light intensity is a maximum at $\theta = 0°$ and decreases to a minimum (intensity = zero) at the angle θ given by Eq. 24–3.

Now consider a larger angle θ such that the top ray travels $\frac{3}{2}\lambda$ farther than the bottom ray, as in Fig. 24–16c. In this case, the rays from the bottom third of the slit will cancel in pairs with those in the middle third since they will be $\lambda/2$ out of phase. However, light from the top third of the slit will still reach the screen, so there will be a bright spot, but not nearly as bright as the central spot at $\theta = 0°$. For an even larger angle θ such that the top ray travels 2λ farther than the bottom ray, Fig. 24–16d, rays from the bottom quarter of the slit will cancel with those in the quarter just above it since the path lengths differ by $\lambda/2$. And the rays through the quarter of the slit just above center will cancel with those through the top quarter. At this angle there will again be a minimum of zero intensity in the diffraction pattern. A plot of the intensity as a function of angle is shown in Fig. 24–17. This corresponds well with the photo of Fig. 24–15c. Notice that minima (zero intensity) occur at

FIGURE 24–17 Intensity in the diffraction pattern of a single slit as a function of $\sin \theta$. Note that the central maximum is not only much higher then the maxima to each side, but it is also twice as wide ($2\lambda/D$ wide) as any of the others (only λ/D wide each).

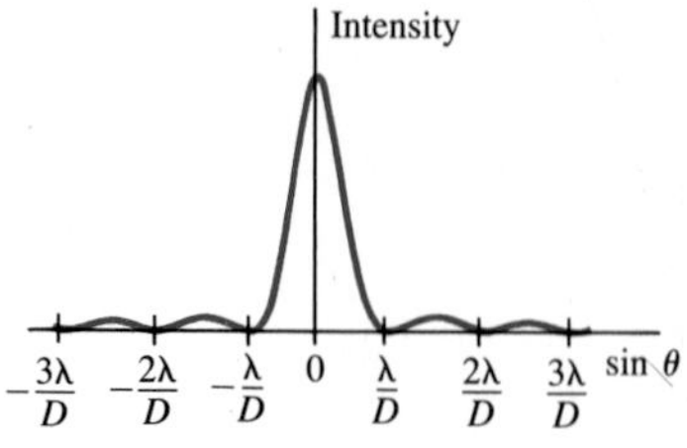

Diffraction minima

$$D \sin \theta = m\lambda, \qquad m = 1, 2, 3, \cdots,$$

but not at $m = 0$ where there is the strongest maximum. Between the minima, smaller brightness maxima occur.

FIGURE 24–18 Example 24–3.

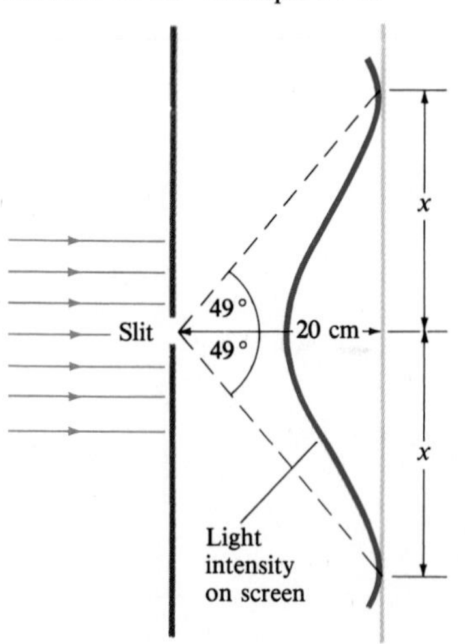

EXAMPLE 24–3 Light of wavelength 750 nm passes through a slit 1.0×10^{-3} mm wide. How wide is the central maximum (*a*) in degrees, and (*b*) in centimeters, on a screen 20 cm away?

SOLUTION (*a*) The first minimum occurs at

$$\sin \theta = \frac{\lambda}{D} = \frac{7.5 \times 10^{-7}\ \text{m}}{1 \times 10^{-6}\ \text{m}} = 0.75.$$

So $\theta = 49°$. This is the angle between the center and the first minimum, Fig. 24–18. The angle subtended by the whole central maximum, between the minima above and below the center, is twice this, or 98°. (*b*) The width of the central maximum is $2x$, where $\tan \theta = x/20$ cm. So $2x = 2(20\ \text{cm}) \cdot (\tan 49°) = 46$ cm. A large width of the screen will be illuminated, but it will not be terribly bright, normally, since the amount of light that passes through such a small slit will be small and it is spread over a large area.

From Eq. 24–3 we can see that the smaller the aperture D, the larger the central diffraction maximum. This is consistent with our earlier study of waves in Chapter 11.

24–6 • Diffraction Grating

A large number of equally spaced parallel slits is called a **diffraction grating**, although the term "interference grating" might be as appropriate. Gratings can be made by using a diamond tip to rule very fine lines on glass. The untouched spaces between the lines serve as the slits. Photographic transparencies of an original grating serve as inexpensive gratings. Photographic reduction can be used to make very fine gratings. Gratings containing 10,000 lines per centimeter are common today, and are very useful for precise measurements of wavelengths. A diffraction grating containing slits is called a **transmission grating**. **Reflection gratings** are also possible; they can be made by ruling fine lines on a metallic or glass surface from which light is reflected and analyzed. The analysis is basically the same as for a transmission grating, and we consider only the latter here.

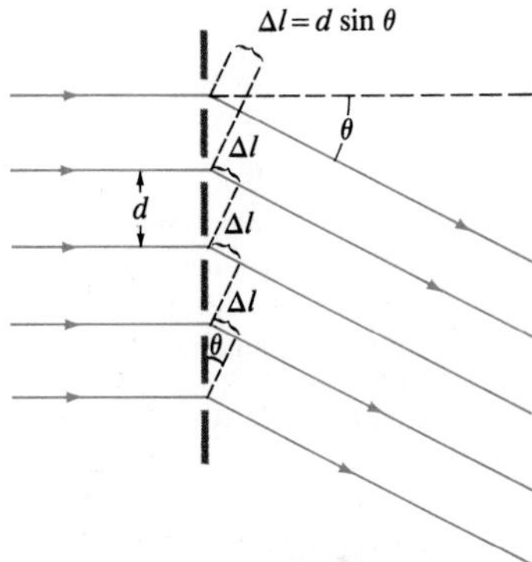

FIGURE 24–19 Diffraction grating.

The analysis of a diffraction grating is much like that of Young's double-slit experiment. We assume parallel rays of light are incident on the grating as shown in Fig. 24–19. We also assume that the slits are narrow enough so that diffraction by each of them spreads light over a very wide angle on a distant screen behind the grating, and interference can occur with light from all the other slits. Light rays that pass through each slit without deviation ($\theta = 0°$) interfere constructively to produce a bright line at the center of the screen. At an angle θ such that rays from adjacent slits travel an extra distance of $\Delta l = m\lambda$ where m is an integer, again constructive interference occurs. Thus, if d is the distance between slits, then $\Delta l = d \sin \theta$, and

Diffraction grating maxima (m = order)

$$\sin \theta = \frac{m\lambda}{d}, \qquad m = 0, 1, 2, \cdots \qquad \text{[principal maxima]} \qquad (24\text{–}4)$$

is the criterion to have a brightness maximum. This is the same equation as for the double-slit situation, and again m is called the **order** of the pattern.

There is an important difference between a double-slit and a multiple-slit pattern, however. The bright maxima are much *sharper* and *narrower* for a grating. Why this happens can be seen as follows. Suppose that the angle θ is increased just slightly beyond that required for a maximum. In the case of only two slits, the two waves will be only slightly out of phase, so nearly full constructive interference occurs. This means the maxima are wide (see Fig. 24–7). For a grating, the waves from two adjacent slits will also not be significantly out of phase. But waves from one slit and those from a second one a few hundred slits away may be exactly out of phase; all or nearly all the light will cancel in pairs in this way. For example, suppose the angle θ is different from its first-order maximum so that the extra path length for a pair of adjacent slits is not exactly λ but rather 1.0010λ. The wave through one slit and another one 500 slits below will be out of phase by 1.5000λ, or exactly $1\frac{1}{2}$ wavelengths, so the two will cancel. A pair of slits, one below each of these, will also cancel. That is, the light from slit 1 cancels with that from slit 501; light from slit 2 cancels with that from slit 502, and so on. Thus even

Why more slits yield sharper peaks

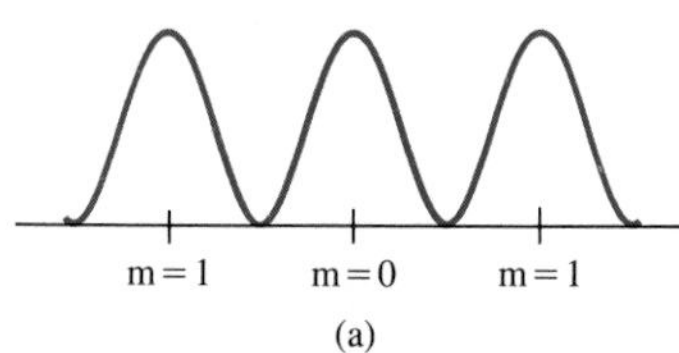

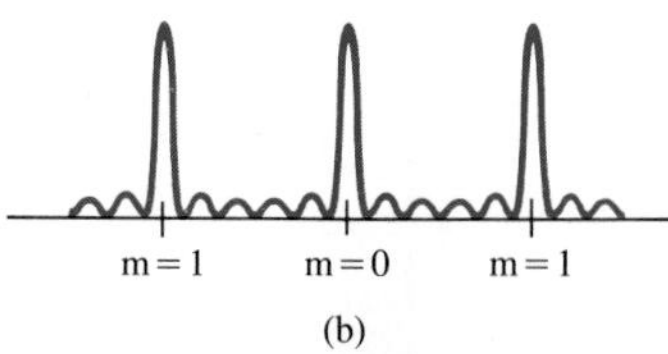

FIGURE 24–20 Intensity as a function of viewing angle θ (or position on screen) for (a) two slits, (b) six slits. For a diffraction grating, the number of slits is very large ($\sim 10^4$) and the peaks are narrower still.

for a tiny angle† corresponding to an extra path length of $\frac{1}{1000}\lambda$, there is much destructive interference, and so the maximum must be very narrow. The more lines there are in a grating, the sharper will be the peaks (see Fig. 24–20). Because a grating produces much sharper (and brighter) lines than two slits alone, it is a far more precise device for measuring wavelengths.

Suppose the light striking a diffraction grating is not monochromatic, but rather consists of two or more distinct frequencies. Then for all orders other than $m = 0$, each wavelength will produce a maximum at a different angle (Fig. 24–21a). If white light strikes a grating, the central ($m = 0$) maximum will be a sharp white peak. But for all other orders, there will be a distinct spectrum of colors spread out over a certain angular width, Fig. 24–21b. Because a diffraction grating spreads out light into its component wavelengths, the resulting pattern is called a **spectrum**.

EXAMPLE 24–4 Calculate the first- and second-order angles for light of wavelength 400 nm and 700 nm if the grating contains 10,000 lines/cm.

SOLUTION Since the grating contains 10^4 lines/cm $= 10^6$ lines/m, the separation between slits is $d = 1/10^6\ \mathrm{m}^{-1} = 1.0 \times 10^{-6}$ m. In first order ($m = 1$), the angles are

$$\sin\theta_{400} = \frac{m\lambda}{d} = \frac{(1)(4.0 \times 10^{-7}\ \mathrm{m})}{1.0 \times 10^{-6}\ \mathrm{m}} = 0.400,$$

$$\sin\theta_{700} = 0.700,$$

so $\theta_{400} = 23.6°$ and $\theta_{700} = 44.0°$. In second order,

$$\sin\theta_{400} = \frac{(2)(4.0 \times 10^{-7}\ \mathrm{m})}{1.0 \times 10^{-6}\ \mathrm{m}} = 0.800,$$

$$\sin\theta_{700} = 1.40,$$

so $\theta_{400} = 53.0°$, but the second order does not exist for $\lambda = 700$ nm since $\sin\theta$ cannot exceed 1. No higher orders will appear.

† Depending on the number of slits, there may or may not be complete cancellation for such an angle, so there will be very tiny peaks between the main maxima (see Fig. 24–20b), but they are usually much too small to be seen.

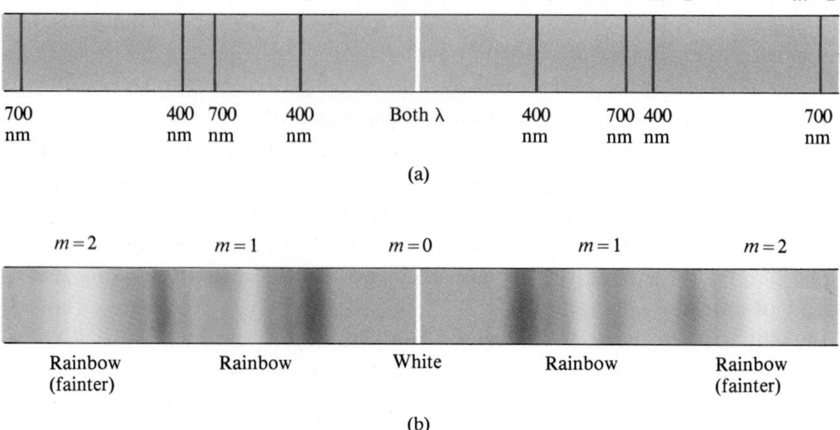

FIGURE 24–21 Spectra produced by a grating: (a) two wavelengths, 400 nm and 700 nm; (b) white light. The second order will normally be dimmer than the first order. (Higher orders are not shown.) If grating spacing is small enough, the second and higher orders will be missing.

EXAMPLE 24–5 White light containing wavelengths from 400 nm to 750 nm strikes a grating containing 4000 lines/cm. Show that the violet of the third-order spectrum overlaps the red of the second order.

SOLUTION The grating spacing is $d = 1/(4000\ \mathrm{cm}^{-1}) = 2.50 \times 10^{-6}$ m. The violet of the third order occurs at an angle θ given by

$$\sin\theta = \frac{(3)(4.00 \times 10^{-7}\ \mathrm{m})}{(2.50 \times 10^{-6}\ \mathrm{m})} = 0.480.$$

Red in second order occurs at

$$\sin\theta = \frac{(2)(7.50 \times 10^{-7}\ \mathrm{m})}{(2.50 \times 10^{-6}\ \mathrm{m})} = 0.600,$$

which is clearly a greater angle.

The diffraction grating is the essential component of a spectroscope, a device for precise measurement of wavelengths, and we discuss it next.

24–7 • The Spectroscope and Spectroscopy

A **spectroscope**, Fig. 24–22, is a device to measure wavelengths accurately using a diffraction grating (discussed in Section 24–6), or a prism, to separate different wavelengths of light. Light from a source passes through a narrow slit S in the collimator. The slit is at the focal point of the lens L, so parallel light falls on the grating. The movable telescope can bring the rays to a focus. Nothing will be seen in the viewing telescope unless it is positioned at an angle θ that corresponds to a diffraction peak (first order is usually used) of a wavelength emitted by the source. The angle θ can be measured† to very high accuracy, so the wavelength of a line can be determined to high accuracy using Eq. 24–4:

$$\sin\theta = \frac{m\lambda}{d},$$

where m is an integer representing the order, and d is the distance between grating lines. (The line you see in a spectroscope corresponding to each wavelength is actually an image of the slit S. So the narrower the slit, the narrower—but dimmer—is the line, and the more precise can be its measurement. If the light contains a continuous range of wavelengths, you will then see a continuous spectrum in the spectroscope.)

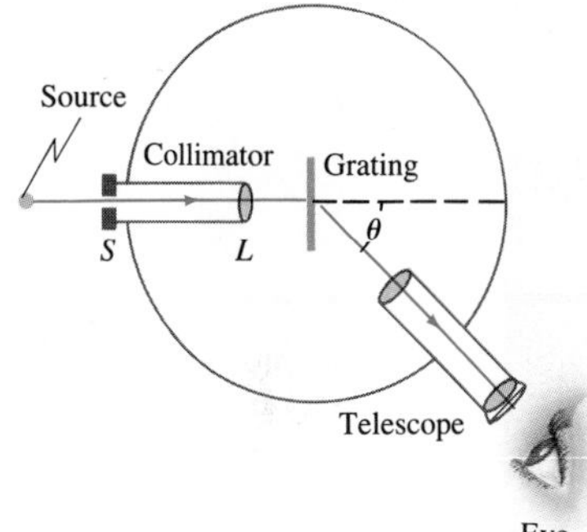

FIGURE 24–22 Spectroscope.

In some spectroscopes, a reflection grating or a prism is used. A prism, because of dispersion (Section 24–4), bends light of different wavelengths into different angles. A prism has the disadvantage that it produces less sharp lines and is less able to separate closely spaced lines. But it has the advantage of deflecting more light (and so is more useful for dim sources) than a normal diffraction grating, since with the latter most of the light passes straight through to the central peak. However, reflection gratings can now be made

† The angle θ for a given wavelength is usually measured on both sides of center because the grating cannot always be aligned precisely; the average of the two values is then taken.

FIGURE 24–23 Line spectra for the gases indicated, and spectrum of the sun showing absorption lines.

that have shaped grooves, so that a large portion of the light is reflected at an appropriate angle to give a strong first-order peak.

If the spectrum of a source is recorded (say, on film) rather than viewed by the eye, the device is called a **spectrometer**, as compared to a **spectroscope**, although these terms are sometimes used interchangeably. Devices that can also measure the intensity of light of a given wavelength are called **spectrophotometers**.

Line spectra

An important use of any of these devices is for the identification of atoms or molecules. When a gas is heated or a large electric current is passed through it, the gas emits a characteristic **line spectrum**. That is, only certain wavelengths of light are emitted, and these are different for different elements and compounds.[†] Figure 24–23 shows the line spectra for a number of elements in the gas state. Line spectra occur only for gases at high temperatures and low pressure. The light from heated solids, such as a light bulb filament, and even from a dense gaseous object such as the sun, produces a **continuous spectrum** including a wide range of wavelengths.

As can be seen in Fig. 24–23, the sun's "continuous spectrum" contains a number of *dark* lines (only the most prominent are shown), called **absorption lines**. Atoms and molecules absorb light at the same wavelengths at which they emit light. The sun's absorption lines are due to absorption by atoms and molecules in the cooler outer atmosphere of the sun, as well as by atoms and molecules in the earth's atmosphere. A careful analysis of all these thousands of lines reveals that at least two-thirds of all elements are present in the sun's atmosphere. The presence of elements in the atmosphere of other planets, in interstellar space, and in stars is also determined by spectroscopy.

Spectroscopy is useful for determining the presence of certain types of molecules in laboratory specimens where chemical analysis would be difficult. For example, biological DNA and different types of protein absorb light in particular regions of the spectrum (such as in the UV). The material to be examined, which is often in solution, is placed in a monochromatic light beam whose wavelength is chosen by placement of a prism or diffraction grating. The amount of absorption, as compared to a standard solution without the specimen, can reveal not only the presence of a particular type of molecule, but also its concentration.

Light emission and absorption also occur outside the visible part of the spectrum, such as in the UV and IR regions. Since glass absorbs light in these regions, reflection gratings and mirrors (in place of lenses) are used. Special types of film, or photocell detectors, are used for detection.

24–8 • Interference by Thin Films

Interference colors from thin oil film on water

Interference of light gives rise to many everyday phenomena such as the bright colors reflected from soap bubbles and from thin oil films on water (Fig. 24–24). In these and other cases, the colors are a result of constructive interference between light reflected from the two surfaces of the thin film. To

[†] Why atoms and molecules emit line spectra was a great mystery for many years and played a central role in the development of modern quantum theory, as we shall see in Chapter 27.

(a)

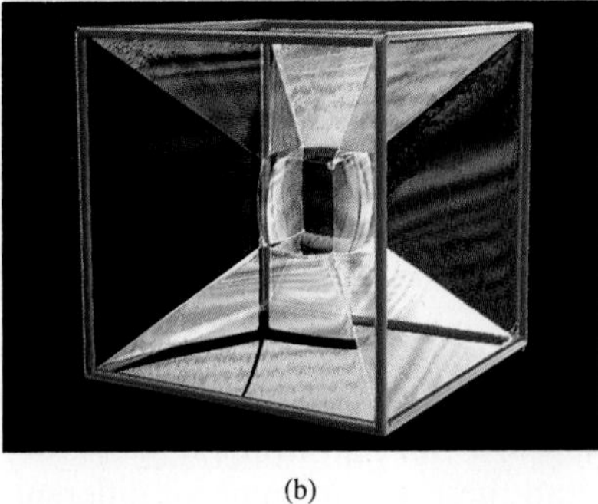

(b)

(c)

FIGURE 24–24 Thin film interference patterns seen in (a) soap bubbles, (b) thin films of soapy water, and (c) a thin layer of oil on the water of a street puddle.

FIGURE 24–25 (above) Light reflected from the upper and lower surfaces of a thin film of oil lying on water.

see how this happens, consider a smooth surface of water on top of which is a thin uniform layer of another substance, Fig. 24–25. Part of the incident light (say, from the sun or street lights) is reflected at A on the top surface, and part of that transmitted is reflected at B on the lower surface. The part reflected at the lower surface must travel the extra distance ABC. If the distance ABC is equal to one or a whole number of wavelengths, the two waves will interfere constructively and the light will be bright.† But if ABC equals $\frac{1}{2}\lambda$, $\frac{3}{2}\lambda$, and so on, the two waves will be exactly out of phase and destructive interference occurs. The wavelength λ is that in the film (see Eq. 24–1).

When white light falls on such a film, the path ABC will equal λ (or $m\lambda$, with m = an integer) for only one wavelength at a given viewing angle. This color will be seen as very bright. For light viewed at a slightly different angle, the path ABC will be longer or shorter and a different color will undergo constructive interference. Thus, for an extended (nonpoint) source emitting white light, a series of bright colors will be seen next to one another. Variations in thickness of the film will also alter the length ABC and therefore affect the color of light that is most strongly reflected.

When a curved glass surface is placed in contact with a flat glass surface, Fig. 24–26, a series of concentric rings is seen when illuminated from

† As is discussed shortly, this is true if the refractive index of the film is less than that of the water.

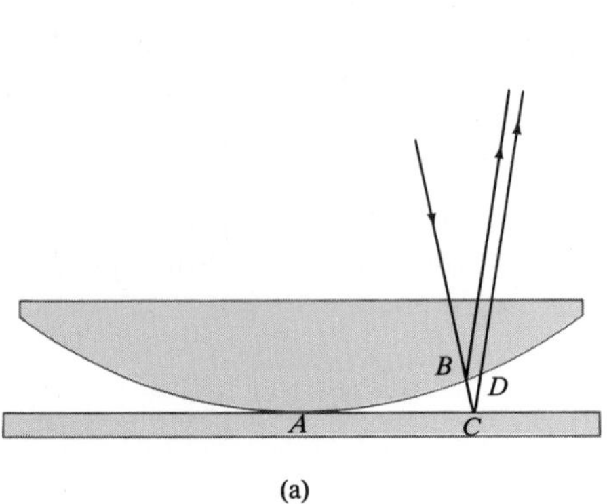

(a)

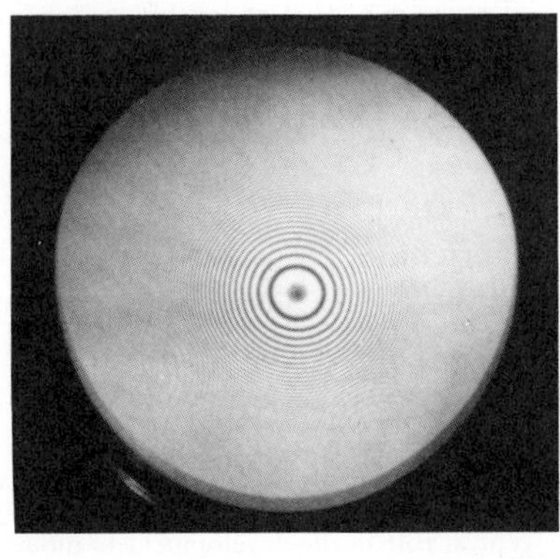

(b)

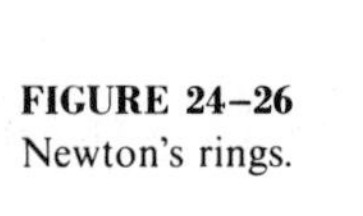

FIGURE 24–26 Newton's rings.

above by monochromatic light. These are called **Newton's rings**† and they are due to interference between rays reflected by the top and bottom surfaces of the *air gap* between the two pieces of glass. Because this gap (which is equivalent to a thin film) increases in width from the central contact point out to the edges, the extra path length for the lower ray (equal to BCD) varies; where it equals 0, $\frac{1}{2}\lambda$, λ, $\frac{3}{2}\lambda$, 2λ, and so on, it corresponds to constructive and destructive interference; and this gives rise to the series of bright and dark lines seen in Fig. 24–26b.

Newton's rings

Change of phase upon reflection

Note that the point of contact of the two glass surfaces (A in Fig. 24–26a) is dark in Fig. 24–26b. Since the path difference is zero here, we expect the rays reflected from each surface to be in phase and this point to be bright. But it is dark, which tells us the two rays must be completely out of phase; this can happen only because one of the waves undergoes a change in phase of 180° upon reflection, corresponding to $\frac{1}{2}$ cycle or $\frac{1}{2}\lambda$. Indeed, this and other experiments reveal that *a beam of light reflected by a material whose index of refraction is greater than that in which it is traveling changes phase by $\frac{1}{2}$ cycle.* If the index is less than that of the material in which it is traveling, no phase change occurs. (This corresponds to the reflection of a wave traveling along a rope when it reaches the end; as we saw in Fig. 11–23, if the end is tied down, the wave changes phase and the pulse flips over, but if the end is free, no phase change occurs.) Thus the ray reflected by the curved surface above the air gap in Fig. 24–26a undergoes no change in phase. That reflected at the lower surface, where the beam in air strikes the glass, undergoes a $\frac{1}{2}\lambda$ phase change. Thus the two rays reflected at the point of contact A of the two glass surfaces (where the air gap approaches zero thickness) will be $\frac{1}{2}\lambda$ out of phase, and a dark spot occurs. Other dark bands will occur when the path difference BCD in Fig. 24–26a is equal to an integral number of wavelengths. Bright bands will occur when the path difference is $\frac{1}{2}\lambda$, $\frac{3}{2}\lambda$, and so on, since the phase change at one surface effectively adds another $\frac{1}{2}\lambda$.

FIGURE 24–27 (a) Light rays reflected from upper and lower surfaces of a thin wedge of air interfere to produce bright and dark bands. (b) Pattern observed when glass plates are optically flat; (c) pattern when plates are not so flat.

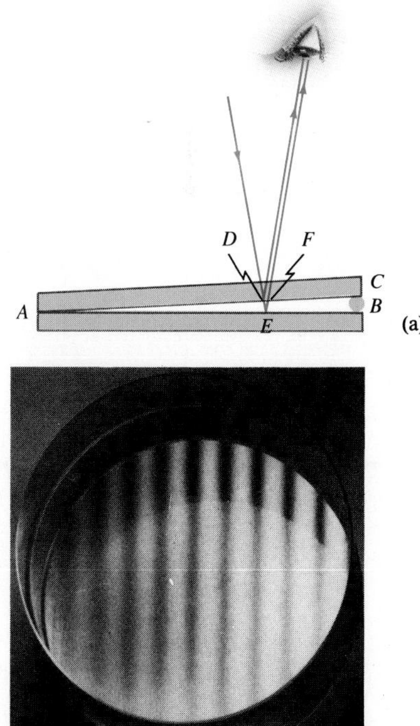

EXAMPLE 24–6 A very fine wire 7.35×10^{-3} mm in diameter is placed between two flat glass plates as in Fig. 24–27a. Light whose wavelength in air is 600 nm falls (and is viewed) perpendicularly to the plates, and a series of bright and dark bands is seen, Fig. 24–27b. How many light and dark bands will there be in this case? Will the area next to the wire be bright or dark?

SOLUTION The thin film is the wedge of air between the two glass plates. Because of the phase change at the lower surface, there will be a dark band when the path difference is 0, λ, 2λ, 3λ, and so on. Since the light rays are perpendicular to the plates, the extra path length equals $2t$, where t is the thickness of the air gap at any point; so dark bands occur where

$$2t = m\lambda, \qquad m = 0, 1, 2, \cdots.$$

Bright bands occur when $2t = (m + \frac{1}{2})\lambda$, where m is an integer. At the position of the wire, $t = 7.35 \times 10^{-6}$ m. At this point there will be $(2)(7.35 \times 10^{-6}\ \text{m})/(6.00 \times 10^{-7}\ \text{m}) = 24.5$ wavelengths. Since this is a

† Although Newton gave an elaborate description of them, they had been first observed and described by his contemporary Robert Hooke. Newton did not realize their significance in support of a wave theory of light.

"half integer," the area next to the wire will be bright. There will be a total of 25 dark lines along the plates, corresponding to path lengths of 0λ, 1λ, 2λ, 3λ, . . . , 24λ, including the one at the point of contact A ($m = 0$). Between them, there will be 24 bright lines plus the one at the end, or 25. The bright and dark bands will be straight only if the glass plates are extremely smooth. If they are not, the pattern is uneven, as in Fig. 24–27c. This is a very precise way of testing a glass surface for smoothness.

If the wedge between the two glass plates of Example 24–6 is filled with some transparent substance other than air—say, water—the pattern shifts because the wavelength of the light changes. In a material where the index of refraction is n, the wavelength is $\lambda_n = \lambda/n$ where λ is the wavelength in vacuum (see Eq. 24–1). For instance, if the thin wedge of Example 24–6 were filled with water, $\lambda_n = 600\text{ nm}/1.33 = 450\text{ nm}$; and instead of 25 dark lines, there would be 33.

When white light (rather than monochromatic light) is incident on the thin wedge of Figs. 24–26a or 24–27a, a colorful series of fringes is seen. This is because constructive interference occurs in the reflected light at different locations along the wedge for different wavelengths. Such a difference in thickness is part of the reason bright colors appear when light is reflected from a soap bubble or a thin layer of oil on a puddle or lake (Fig. 24–24). Which wavelengths appear brightest also depends on the viewing angle, as we saw earlier.

EXAMPLE 24–7 A soap bubble appears green ($\lambda = 540$ nm) at the point on its front surface nearest the viewer. What is its minimum thickness? Assume $n = 1.35$.

SOLUTION The light is reflected perpendicularly from the point on a spherical surface nearest the viewer. Therefore the path difference is $2t$, where t is the thickness of the soap film. Light reflected from the outer surface undergoes a $\frac{1}{2}\lambda$ phase change, whereas that on the inner surface does not. Therefore, green light is bright when the minimum path difference equals $\frac{1}{2}\lambda$. Thus, $2t = \lambda/2n$, so $t = (540\text{ nm})/(4)(1.35) = 100$ nm. This is the minimum thickness. The front surface would also appear green if $2t = 3\lambda/2n$, and so on.

Lens coatings

An important application of thin-film interference is in the coating of glass to make it "nonreflecting," particularly for lenses. A glass surface reflects about 4 percent of the light incident upon it. Good-quality cameras, microscopes, and other optical devices may contain six to ten thin lenses. Reflection from all these surfaces can reduce the light level considerably and multiple reflections produce a background haze that reduces the quality of the image. A very thin coating on the lens surfaces can reduce these problems considerably. The amount of reflection at a boundary depends on the difference in index of refraction between the two materials. Ideally, the coating material should have an index of refraction which is the geometric mean of those for air and glass, so that the amount of reflection at each surface is about equal. Then destructive interference can occur nearly completely for one particular wavelength depending on the thickness of the coating. Nearby wavelengths will at least partially destructively interfere, but it is clear that a single coating cannot eliminate reflections for all wavelengths. Nonetheless, a single coating can reduce total reflection from 4 percent to 1 percent of the incident light. Often the coating is designed to eliminate the center of the reflected spectrum (around 550 nm). The extremes of the spectrum—red and violet—will not be reduced as much. Since a mixture of red and violet produces purple, the light seen reflected from such coated lenses is purple. Lenses containing two or three separate coatings can more effectively reduce a wider range of reflecting wavelengths.

EXAMPLE 24–8 What is the thickness of an optical coating of MgF_2 whose index of refraction is $n = 1.38$ and is designed to eliminate reflected light at wavelengths centered at 550 nm when incident normally on glass for which $n = 1.50$?

SOLUTION Figure 24–28 shows an incoming ray and two rays reflected from the front and rear surfaces of the coating on the lens. The rays are drawn not quite perpendicular to the lens so we can see each of them. To eliminate reflection, we want the reflected rays 1 and 2 to be $\frac{1}{2}$ wavelength out of phase with each other so they destructively interfere. Rays 1 and 2 *both* undergo a change of phase by $\frac{1}{2}\lambda$ when they reflect, respectively, from the front and rear surfaces of the coating. Therefore, we want the extra distance traveled by ray 2 ($=2t$) to be a half integral number of wavelengths. That is, $2t = (m + \frac{1}{2})\lambda_n$, where m is an integer and λ_n is the wavelength inside the MgF_2 coating. The minimum thickness ($m = 0$) is usually chosen because destructive interference will then occur over the widest angle. Then $t = \lambda_n/4 = \lambda/4n = (550\text{ nm})/(4)(1.38) = 99.6\text{ nm}$.

FIGURE 24–28 Incident ray of light is partially reflected at front surface of lens coating (ray 1) and again partially reflected at rear surface of coating (ray 2), with most of the energy passing as the transmitted ray into the glass. Example 24–8.

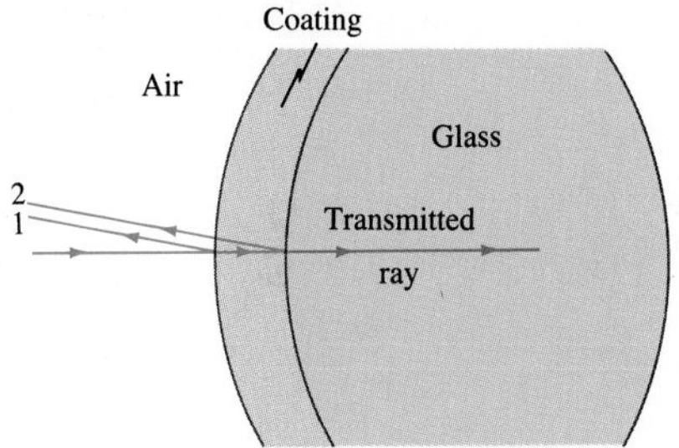

*24–9 • Michelson Interferometer

Interference by thin films is the basis of the **Michelson interferometer** (Fig. 24–29),[†] invented by the American Albert A. Michelson (Section 23–2). Monochromatic light from a single point on an extended source is shown striking a half-silvered mirror M_s (called the **beam splitter**). Half of the beam passes through to a fixed mirror M_2, where it is reflected back. The other half is reflected by M_s up to a mirror M_1 that is movable (by a fine-thread screw), where it is also reflected back. Upon its return, part of beam 1 passes through M_s and reaches the eye; and part of beam 2, on its return, is reflected by M_s into the eye. (A compensator plate C of transparent glass, usually cut from the same plate as M_s, is placed in the path of beam 2 so both beams pass through the same thickness of glass, to within a fraction of a wavelength.) If the two path lengths are identical, the two coherent beams entering the eye constructively interfere and brightness will be seen. If the movable mirror is moved a distance $\lambda/4$, one beam will travel an extra distance equal to $\lambda/2$ (because it travels back and forth over the distance $\lambda/4$). In this case, the two beams will destructively interfere and darkness will be seen. As M_1 is moved farther, brightness will recur (when the path difference is λ), then darkness, and so on.

FIGURE 24–29
Michelson interferometer.

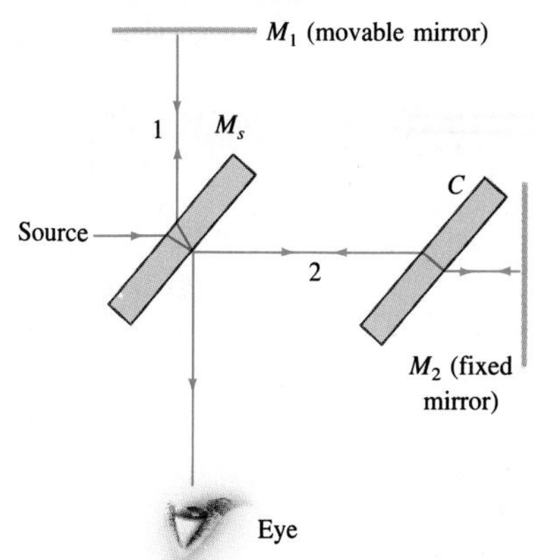

[†] There are other types of interferometer, but Michelson's is the best known.

The eye sees the image of M_2 very close to the position of M_1. In effect, there is a thin layer of air (a "thin film") between M_1 and this image of M_2. If M_1 is tilted slightly, then the air gap is wedge shaped as in Fig. 24–27, and a series of bright and dark lines are seen. When mirror M_1 is then moved, the bright and dark lines move to the left or right. When the path difference is many wavelengths, the observer no longer sees a nearly uniform brightness or darkness when the mirrors M_1 and M_2 are precisely aligned. Because the path difference is different for different angles of view, one sees a series of rings, much like the Newton ring pattern of Fig. 24–26b. When mirror M_1 is moved, dark and bright spots seem to emerge from (or disappear into) the center of the pattern.

Very precise length measurements can be made with an interferometer. The motion of mirror M_1 by only $\frac{1}{4}\lambda$ produces a clear difference between brightness and darkness. For $\lambda = 400$ nm, this means a precision of 100 nm or 10^{-4} mm! By observing the sideways motion of the fringes when the mirrors are not precisely aligned, even greater precision can be obtained. By counting the number of fringes, or fractions thereof, extremely precise length measurements can be made.

Michelson saw that the interferometer could be used to determine the length of the standard meter in terms of the wavelength of a particular light. In 1960, that standard was chosen to be a particular orange–red line in the spectrum of krypton-86 (krypton atoms with atomic mass 86). Careful repeated measurements of the old standard meter (the distance between two marks on a platinum–iridium bar kept in Paris) were made to establish 1 meter as being 1,650,763.73 wavelengths of this light, which was *defined* to be the meter. In 1983, the meter was redefined in terms of the speed of light (Section 1–5).

24–10 • Polarization

An important and useful property of light is that it can be *polarized*. To see what this means, let us examine waves traveling on a rope. A rope can be set into vibration in a vertical plane as in Fig. 24–30a, or in a horizontal plane as in Fig. 24–30b. In either case, the wave is said to be **plane-polarized**—that is, the oscillations are in a plane.

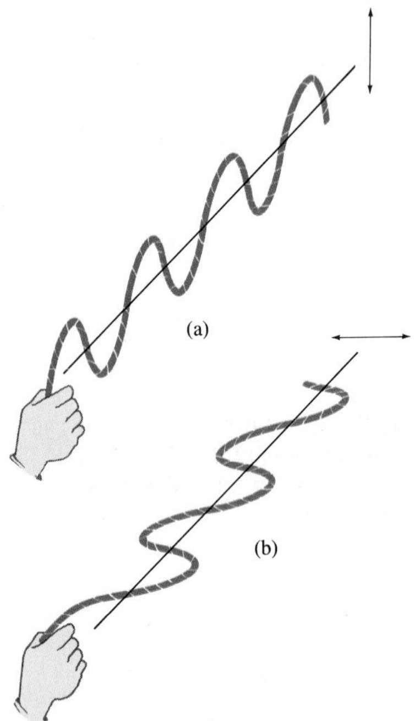

FIGURE 24–30 Transverse waves on a rope polarized (a) in a vertical plane and (b) in a horizontal plane.

If we now place an obstacle containing a vertical slit in the path of the wave, Fig. 24–31, a vertically polarized wave passes through, but a horizontally polarized wave will not. If a horizontal slit were used, the vertically polarized wave would be stopped. If both types of slit were used, both types of wave would be stopped. Note that polarization can exist *only* for *transverse waves*, and not for longitudinal waves. The latter vibrate only along the direction of motion and neither orientation of slit would stop them.

That light can be polarized was recognized only in the nineteenth century. However, even in Newton's time a phenomenon that depends on polarization was already known—namely, that certain types of crystal, such as iceland spar, refract light into two rays; such crystals are called *doubly refracting* (Section 24–12). Certain other crystals, such as tourmaline, would not transmit one or the other of these two rays, depending on the orientation of the crystal. Today we recognize that the two rays in a doubly refracting crystal are plane-polarized in mutually perpendicular directions, and that the tourmaline acts as a "slit" to eliminate one or the other if properly oriented. However, it was not until after Young's and Fresnel's work in the early 1800s that this phenomenon was recognized as evidence that light is a transverse wave. A half century later, Maxwell's theory of light as electromagnetic (EM) waves was fully consistent with the facts of polarization since an EM wave is a transverse wave. The direction of polarization in a plane-polarized EM wave is taken as the direction of the electric field vector (Fig. 22–7).

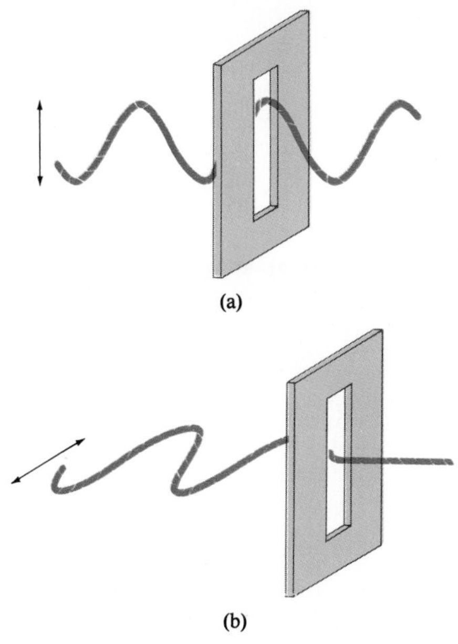

FIGURE 24–31 Vertically polarized wave passes through a vertical slit, but a horizontally polarized wave will not.

Light is not necessarily polarized. It can be **unpolarized**, which means that the source has vibrations in many planes at once, Fig. 24–32. An ordinary incandescent light bulb emits unpolarized light, as does the sun.

Plane-polarized light can be obtained from unpolarized light using certain crystals such as tourmaline. Or, more commonly today, we can use a **Polaroid sheet**. (Polaroid materials were invented in 1929 by Edwin Land.) A Polaroid sheet consists of complicated long molecules arranged parallel to one another. Such a Polaroid acts like a series of parallel slits to allow one orientation of polarization to pass through nearly undiminished (this direction is called the *axis* of the Polaroid), whereas a perpendicular polarization is absorbed almost completely.[†] If a beam of plane-polarized light strikes a Polaroid whose axis is at an angle θ to the incident polarization direction, the beam will emerge plane-polarized parallel to the Polaroid axis and its amplitude will be reduced by $\cos\theta$, Fig. 24–33. Thus, a Polaroid passes only that component of polarization (the electric field vector, **E**) that is parallel to its axis. Since the intensity of a light beam is proportional to the square of the amplitude, we see that the intensity of a plane-polarized beam transmitted by a polarizer is

FIGURE 24–32 Vibration of the electric field vector in unpolarized light. The light is traveling into or out of the page.

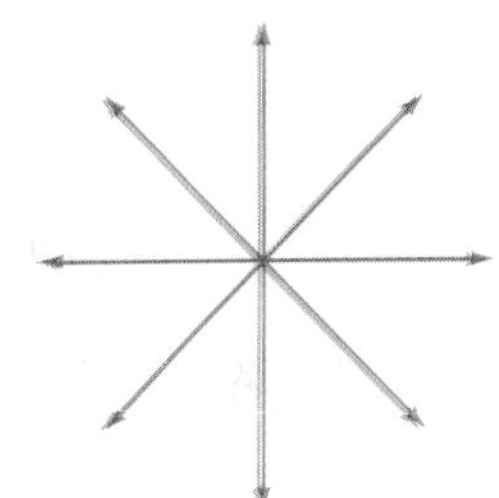

$$I = I_0 \cos^2\theta, \tag{24–5}$$

where θ is the angle between the polarizer axis and the plane of polarization of the incoming wave, and I_0 is the incoming intensity.[‡]

[†] How this occurs can be explained at the molecular level. An electric field **E** that oscillates parallel to the long molecules can set electrons into motion along the molecules, thus doing work on them and transferring energy. Hence, if **E** is parallel to the molecules, it gets absorbed. An electric field **E** perpendicular to the long molecules does not have this possibility of doing work and transfering its energy, and so passes through freely. When we speak of the *axis* of a Polaroid, we mean the direction for which **E** is passed, so a Polaroid axis is perpendicular to the long molecules. (If we want to think of there being slits between the parallel molecules in the sense of Fig. 24–31, then Fig. 24–31 would apply for the **B** field in the EM wave, not the **E** field.)

[‡] Equation 24–5 is often referred to as **Malus' law**, after Etienne Malus, a contemporary of Fresnel.

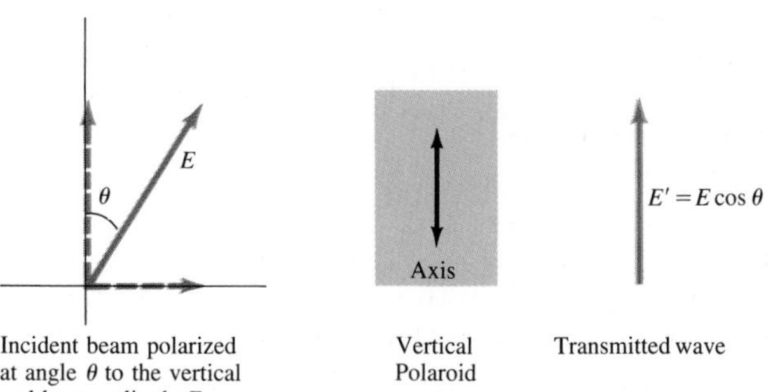

FIGURE 24–33 (left) Vertical Polaroid transmits only the vertical component of a wave incident upon it.

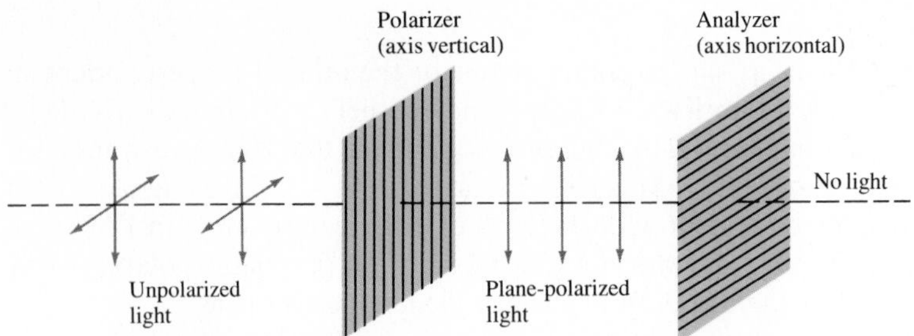

FIGURE 24–34 Crossed Polaroids completely eliminate light.

A Polaroid can be used as a **polarizer** to produce plane-polarized light from unpolarized light, since only the component of light parallel to the axis is transmitted. A Polaroid can also be used as an **analyzer** to determine (1) if light is polarized and (2) what is the plane of polarization. A Polaroid acting as an analyzer will pass the same amount of light independent of the orientation of its axis if the light is unpolarized; try rotating one lens of a pair of Polaroid sunglasses while looking through it at a light bulb. If the light is polarized, however, when you rotate the Polaroid the transmitted light will be a maximum when the plane of polarization is parallel to the Polaroid's axis, and a minimum when perpendicular to it. If you do this while looking at the sky, preferably at right angles to the sun's direction, you will see that skylight is polarized. (Direct sunlight is unpolarized, but don't look directly at the sun, even through a polarizer, for damage to the eye may occur.) If the light transmitted by an analyzer Polaroid falls to zero at one orientation, then the light is 100 percent plane-polarized. If it merely reaches a minimum, the light is *partially polarized.*

Crossed Polaroids

Unpolarized light consists of light with random directions of polarization (electric field vector). Each of these polarization directions can be resolved into components along two mutually perpendicular directions. Thus, an unpolarized beam can be thought of as two plane-polarized beams of equal magnitude perpendicular to one another. When two Polaroids are *crossed*—that is, their axes are perpendicular to one another—unpolarized light can be entirely stopped (or nearly so—Polaroids are not quite perfect). As shown in Fig. 24–34, unpolarized light is made plane-polarized by the first Polaroid (the polarizer). The second Polaroid, the analyzer, then eliminates this component since its axis is perpendicular to the first. You can try this with Polaroid sunglasses. It should be clear that Polaroid sunglasses eliminate 50 percent of unpolarized light because of their polarizing property; they absorb even more because they are colored.

EXAMPLE 24–9 Unpolarized light passes through two Polaroids; the axis of one is vertical and that of the other is at 60° to the vertical. What is the orientation and intensity of the transmitted light?

SOLUTION The first Polaroid eliminates half the light so the intensity is reduced by half: $I_1 = \frac{1}{2}I_0$. The light reaching the second polarizer is vertically polarized and so is reduced in intensity (Eq. 24–5) by

$$I_2 = I_1(\cos 60°)^2 = \tfrac{1}{4}I_1.$$

Thus, $I_2 = \frac{1}{8}I_0$. The transmitted light has an intensity one-eighth that of the original and is plane-polarized at a 60° angle to the vertical.

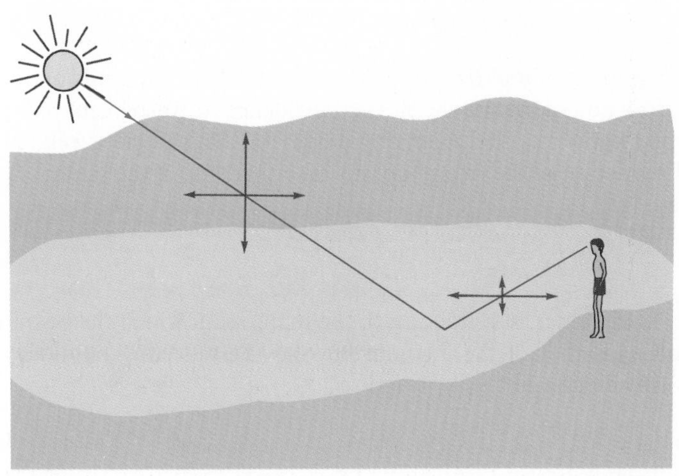

FIGURE 24–35 Light reflected from a nonmetallic surface, such as the smooth surface of water in a lake, is partially polarized parallel to the surface.

Polarization by reflection

Another means of producing polarized light from unpolarized light is by reflection. When light strikes a nonmetallic surface at any angle other than perpendicular, the reflected beam is polarized preferentially in the plane parallel to the surface, Fig. 24–35. In other words, the component with polarization in the plane perpendicular to the surface is preferentially transmitted or absorbed. You can check this by rotating Polaroid sunglasses while looking through them at a flat surface of a lake or road. Since most outdoor surfaces are horizontal, Polaroid sunglasses are made with their axes vertical to eliminate the stronger component, and thus reduce glare. This is well known by fishermen who wear Polaroids to eliminate reflected glare from the surface of a lake or stream and thus see beneath the water more clearly (Fig. 24–36).

(a)

(b)

FIGURE 24–36 Photographs of a river, (a) allowing all light into the camera lens, and (b) using a polarizer which is adjusted to absorb most of the light reflected from the water's surface (the reflected light is strongly polarized). The strong reduction of the light reflected from the surface by the polarizer allows the dimmer light from the bottom of the river, and any fish swimming there, to be seen more readily.

The amount of polarization in the reflected beam depends on the angle, varying from no polarization at normal incidence to 100 percent polarization at an angle known as the **polarizing angle**, θ_p.† This angle is related to the index of refraction of the two materials on either side of the boundary by the equation

$$\tan\theta_p = \frac{n_2}{n_1}, \tag{24–6a}$$

where n_1 is the index of refraction of the material in which the beam is traveling, and n_2 is that of the medium beyond the reflecting boundary. If the beam is traveling in air, $n_1 = 1$ and

$$\tan\theta_p = n. \tag{24–6b}$$

The polarizing angle θ_p is also called **Brewster's angle**, and Eqs. 24–6 *Brewster's law*, after the Scottish physicist David Brewster (1781–1868) who worked it out experimentally in 1812. Equations 24–6 can be derived from the electromagnetic wave theory of light. It is interesting that at Brewster's angle, the reflected and transmitted rays make a 90° angle to each other; that is, $\theta_p + \theta_r = 90°$, Fig. 24–37. This can be seen as follows: we substitute Eq. 24–6a, $n_2 = n_1 \tan\theta_p = n_1 \sin\theta_p/\cos\theta_p$, into Snell's law, $n_1 \sin\theta_p = n_2 \sin\theta_r$, and get $\cos\theta_p = \sin\theta_r$, which can only hold if $\theta_p = 90° - \theta_r$.

EXAMPLE 24–10 (*a*) At what incident angle is sunlight reflected plane-polarized from a lake? (*b*) What is the refraction angle?

SOLUTION (*a*) We use Eq. 24–6b with $n = 1.33$, so $\tan\theta_p = 1.33$ and $\theta_p = 53.1°$.

(*b*) $\theta_r = 90.0° - \theta_p = 36.9°$.

† Only a fraction of the incident light is reflected at the surface of the transparent medium. Although this reflected light is 100% polarized (if $\theta = \theta_p$), the remainder of the light, which is transmitted into the new medium, is only partially polarized.

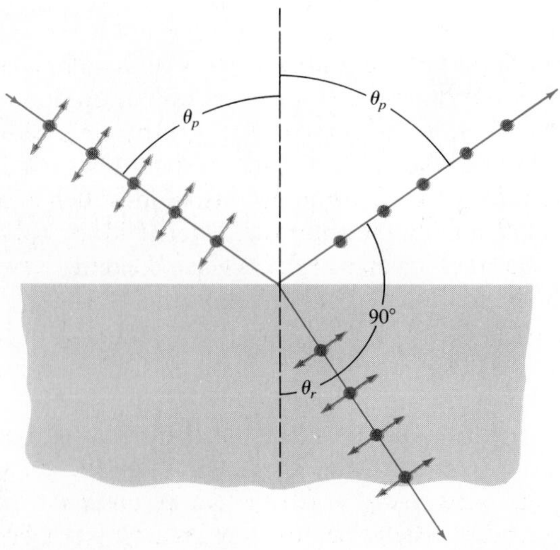

FIGURE 24–37 At θ_p the reflected light is plane-polarized parallel to the surface, and $\theta_p + \theta_r = 90°$, where θ_r is the refraction angle. (The large dots represent vibrations perpendicular to the page.)

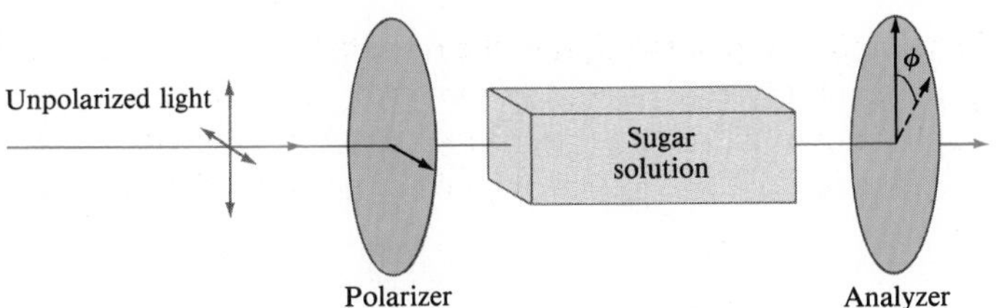

FIGURE 24–38 Sugar solution rotates the plane of polarization of incident light so that it is no longer horizontal but at an angle ϕ. The analyzer thus extinguishes the light when rotated ϕ from the vertical (crossed) position.

*24–11 • Optical Activity

When a beam of plane-polarized light is passed through crystals and solutions, it is found that the plane of polarization is rotated through an angle. For example, Fig. 24–38 shows light passing through a polarizer and then through a sugar solution. The analyzer Polaroid behind the solution does not cut out all the light when placed at 90° to the polarizer. If the analyzer is rotated through an angle ϕ, however, there is no transmitted light; this indicates that the plane of polarization has been rotated through an angle ϕ by the intervening substance. Such substances are said to be **optically active**. Optical activity is due to asymmetry of the molecules, which may have a spiral shape as, for example, some proteins do. Substances that rotate the polarization plane to the right as viewed along the direction of the beam (see Fig. 24–38) are called **dextrorotatory** (also called *right-handed* or *positive*). Substances that rotate it to the left are called **levorotatory** (*left-handed*, or *negative*). The common sugar, dextrose or D-glucose, is dextrorotatory. Most naturally occurring amino acids and proteins are levorotatory.

The angle of rotation ϕ depends on the path length l (meters) through the substance and on the concentration c (kg/m^3) if it is in solution. For dilute solutions, this is a linear relationship and ϕ (in radians) is given by

$$\phi = \alpha l c.$$

The constant α is a property of the substance and is called the specific rotation or the *specific optical rotatory power* (it depends on the temperature and the wavelength of light used). Some typical values are given in Table 24–1.

TABLE 24–1
Specific Optical Rotatory Power, α†

Substance	α ($rad \cdot m^2/kg$)‡
Alanine	$+3.14 \times 10^{-4}$
Leucine	-19.2×10^{-4}
Cysteine	-28.8×10^{-4}
Insulin	-59×10^{-4}
Collagen	-500 to -700×10^{-4}

† Substances in water solution at 20°C for $\lambda = 589$ nm.
‡ + is dextrorotatory; − is levorotatory.

Since ϕ is proportional to the concentration, optical activity is a standard method to measure concentrations of solutions such as for sugars. It is also helpful in determining the three-dimensional shape of large molecules, such as proteins, or their change in shape when conditions are changed. For example, a protein in an α-helix shape will have a large (negative) value for α, but may alter its shape when the pH or temperature is changed, and this will change the value of α. Much information on the shape and properties of molecules has been obtained in this way.

Glass and plastics become optically active when put under stress. The plane of polarization is rotated most where the stress is greatest. Models of bones or machine parts made of plastic can be observed between crossed Polaroids to determine where the points of greatest stress lie. This is called "optical stress analysis." See Fig. 24–39.

FIGURE 24–39 Plastic model of a cross section through the nave of a gothic cathedral, viewed through crossed polarizers to show areas of greatest stress (where lines are closely spaced).

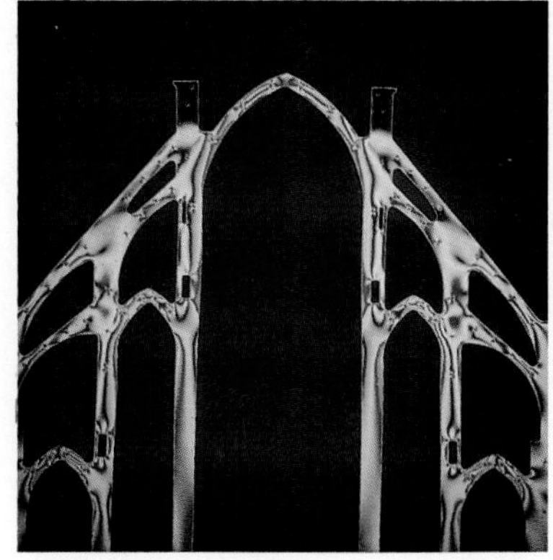

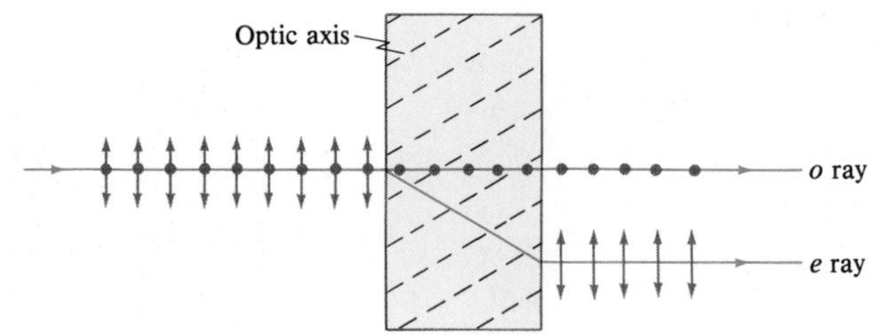

FIGURE 24–40 Unpolarized light, incident normally on a birefringent crystal such as calcite, is separated into two refracted beams. The dashed lines indicate the direction of the optic axis.

*24–12 • Double Refraction: Birefringence and Dichroism

In many transparent materials, the speed of light is the same in all directions. Such materials are said to be *isotropic*. In some crystals and solutions, however, the speed of light is different in different directions, and these are referred to as being *anisotropic*. Such substances are also said to be **doubly refracting** or **birefringent** because of the unusual phenomenon they give rise to.

There is one particular direction in a birefringent crystal, such as calcite, known as its **optic axis** (this is not a single line, but a direction in the crystal). If ordinary light enters a crystal along its optic axis, nothing abnormal is noted. But if unpolarized light falls at an angle to the optic axis, as in Fig. 24–40, a very unusual thing happens: there are two refracted rays. In the case shown, the incident ray is normal to the surface and the optic axis lies in the plane shown. One refracted ray, called the **ordinary ray** (o) passes straight through in a normal way. But the other ray, called the **extraordinary ray** (e), is refracted at an angle.

As can be seen, Snell's law does not hold in this case for the e ray. It does hold, however, for the o ray.

The e ray and o ray are found to be plane-polarized in mutually perpendicular directions. This is indicated in Fig. 24–40 by the dots on the o ray indicating the vibrations are perpendicular to the page, and the polarization of the e ray is indicated by the short arrows.

The phenomenon of double refraction can be explained if we make the following assumption: that the speed of light is different, depending on the orientation of the polarization vector with respect to the crystal's optic axis. As can be seen in Fig. 24–40, the polarization of the o ray is perpendicular to the optic axis. So the speed of the o ray will be the same in all directions, as long as its polarization remains perpendicular to the optic axis. The e ray, on the other hand, has components of polarization both parallel and perpendicular to the optic axis, and it therefore travels with a different speed in different directions. The e ray travels with the same speed as the o ray if its polarization vector is perpendicular to the optic axis. At other angles it has a higher speed (or, in certain crystals, a lower speed), which reaches a maximum (or minimum) when its polarization is parallel to the optic axis. (In what follows, we assume a crystal in which this speed is higher.) Thus the index of refraction n_o for the ordinary ray is the same for all directions, whereas the index of refraction for the extraordinary ray, n_e, depends on angle. Usually n_e is specified for the e ray traveling perpendicular to the optic axis where its polarization direction is parallel to the optic axis. These principal values for n_o and n_e are given in Table 24–2 for several crystals.

TABLE 24–2
Principal Indices of Refraction for Some Doubly Refracting Crystals (λ = 589 nm)

Crystal	n_o	n_e
Ice	1.309	1.313
Quartz	1.544	1.553
Calcite	1.658	1.486
Dolomite	1.681	1.500

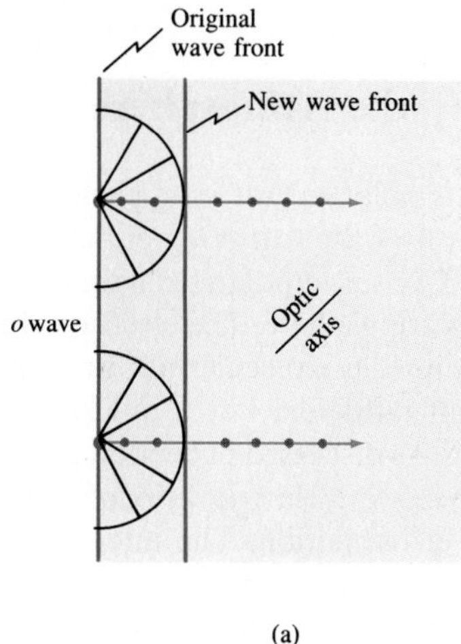

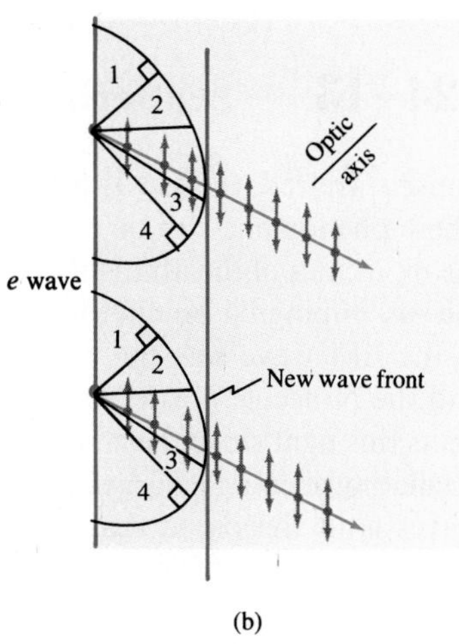

FIGURE 24–41 Huygens' principle used to explain double refraction.

Now we use the above assumptions, plus Huygens' principle, to explain double refraction. In Fig. 24–41a, we examine the ordinary wave. Huygens' wavelets are shown for two points on the wave front just as it has entered the crystal. Since these wavelets travel at the same speed in all directions, they are spheres just as before; and the wave continues moving straight. For the *e* ray, shown in Fig. 24–41b, the Huygens' wavelets are elliptically shaped since the speed depends on direction. Ray 1 on this wavelet has polarization perpendicular to the optic axis, so is traveling at the same speed as the *o* ray. But rays 2, 3, and 4 are traveling at higher speeds, with 4 having the maximum. (This is the case for calcite; for a crystal like ice or quartz, Table 24–2, rays 2, 3, and 4 would be slower.) The advancing wave front, which is tangent to the Huygens' wavelets, will thus be displaced to one side, as shown. And this is the path taken by the *e* ray.

Some birefringent crystals, such as tourmaline, absorb one of the polarized components more strongly than the other (Fig. 24–42). Such a crystal is said to show **dichroism**. If the crystal is sufficiently thick, one component of unpolarized light will be completely eliminated, so the emerging light will be plane-polarized. Dichroism is the basic principle behind Polaroid sheets.

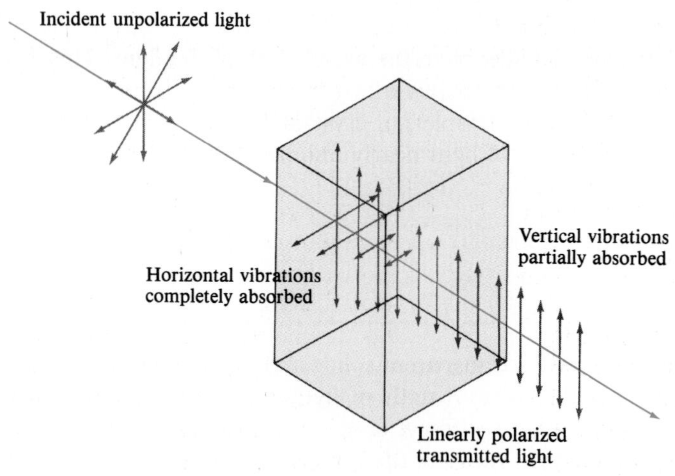

FIGURE 24–42 Dichroic crystal absorbs one polarized component more than the other. If the crystal is thick enough, the transmitted light is plane-polarized.

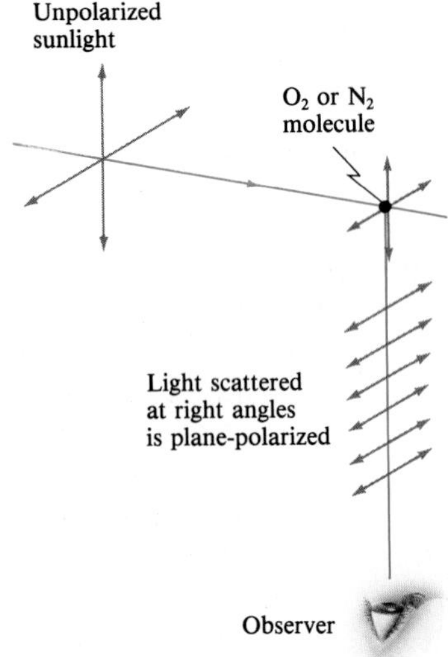

FIGURE 24–43 Unpolarized sunlight scattered by molecules of the air. An observer at right angles sees plane-polarized light, since the component of vibration along the line of sight emits no light along that line.

*24–13 • Scattering of Light by the Atmosphere

Sunsets are red, the sky is blue, and skylight is polarized (at least partially). These phenomena can be explained on the basis of the *scattering* of light by the molecules of the atmosphere. In Fig. 24–43 we see unpolarized light from the sun impinging on a molecule of the earth's atmosphere. The electric field of the EM wave sets the electric charges within the molecule into motion, and the molecule absorbs some of the incident radiation. But it quickly re-emits this light since the charges are oscillating. As discussed in Section 22–3, oscillating electric charges produce EM waves. The electric field of these waves is in the plane that includes the line of oscillation. The intensity is strongest along a line perpendicular to the oscillation, and drops to zero along the line of oscillation (Fig. 22–6). In Fig. 24–43, the motion of the charges is resolved into two components. An observer viewing at right angles to the direction of the sunlight, as shown, will see plane-polarized light since no light is emitted along the line of the other component of the oscillation. (Another way to understand this is to note that when viewing along the line of oscillation, one doesn't see the oscillation, and hence sees no waves made by it.) At other viewing angles, both components will be present; one will be stronger, however, so the light appears partially polarized. Thus, the process of scattering explains the polarization of skylight. [It can also explain complete polarization by reflection, Fig. 24–37. At Brewster's angle, as we saw, the angle between the reflected and refracted ray is 90°. If we think of the molecules of the medium oscillating perpendicular to the direction of the refracted ray, we can see that at 90° to this direction (the direction of the reflected ray) there will be only one component of polarization (Fig. 24–37) just as for scattering (Fig. 24–43).]

Scattering of light by the earth's atmosphere depends on λ. For particles much smaller than the wavelength of light (such as molecules of air), the particles will be less of an obstruction to long wavelengths than to short ones. The scattering decreases, in fact, as $1/\lambda^4$. Red and orange light is thus scattered much less than blue and violet, which is why the sky looks blue. At sunset, the sun's rays pass through a maximum length of atmosphere. Much of the blue has been taken out by scattering. The light that reaches the surface of the earth and reflects off clouds and haze is thus lacking in blue, which is why sunsets appear reddish.

Why the sky is blue

Why sunsets are reddish

The dependence of scattering on $1/\lambda^4$ is valid only if the scattering objects are much smaller than the wavelength of the light. This is valid for oxygen and nitrogen molecules whose diameters are about 0.2 nm. Clouds, however, contain water droplets or crystals that are much larger than λ; they scatter all frequencies of light nearly uniformly. Hence clouds appear white (or gray, if shadowed).

SUMMARY

The wave theory of light is strongly supported by the observations that light exhibits interference and diffraction. Wave theory also explains the refraction of light and the fact that light travels more slowly in transparent solids and liquids than it does in air. The wavelength of light in a medium with index of refraction n is $\lambda_n = \lambda/n$, where λ is the wavelength in vacuum; the frequency is not changed.

Young's double-slit experiment clearly demonstrated the interference of light. The observed bright spots of the interference pattern were explained as constructive interference between the beams coming through the two slits, where they differ in path length by an integral number of wavelengths. The dark areas in between are due to destructive interference when the path lengths differ by $\frac{1}{2}\lambda$, $\frac{3}{2}\lambda$, and so on. The angles θ at which constructive interference occurs are given by $\sin\theta = m\lambda/d$, where λ is the wavelength of the light, d the separation of the slits, and m an integer (0, 1, 2, . . .).

The wavelength of light determines its color. The *visible spectrum* extends from 400 nm (violet) to about 750 nm (red). Glass prisms break down white light into its constituent colors because the index of refraction varies with wavelength, a phenomenon known as *dispersion.*

The formula $\sin\theta = m\lambda/d$ for constructive interference also holds for a *diffraction grating*, which consists of many parallel slits or lines, separated from each other by a distance d. The peaks of constructive interference are much brighter and sharper for a diffraction grating than for the simple two-slit apparatus. A diffraction grating (or a prism) is used in a *spectroscope* to separate different colors or to observe *line spectra*, since for a given order m, θ depends on λ. Precise determination of wavelength can thus be done with a spectroscope by careful measurement of θ.

Diffraction refers to the fact that light, like other waves, bends around objects it passes and spreads out after passing through narrow slits. This bending gives rise to a diffraction pattern due to interference between rays of light that travel different distances. Light passing through a very narrow slit of width D will produce a pattern with a bright central maximum of half-width θ given by $\sin\theta = \lambda/D$, flanked by fainter lines to either side.

Light reflected from the front and rear surfaces of a thin film of transparent material can interfere. Such thin-film interference has many practical applications, such as lens coatings and Newton's rings.

In *unpolarized light*, the electric field vectors vibrate at all angles. If the electric vector vibrates only in one plane the light is said to be *plane-polarized.* Light can also be partially polarized. When an unpolarized light beam passes through a Polaroid sheet or a doubly refracting crystal, the emerging beams are plane-polarized. For a light beam that is polarized, when it passes through a Polaroid, the intensity varies as the Polaroid is rotated. Thus a Polaroid can act as a polarizer or as an analyzer.

Light can also be partially or fully polarized by reflection. If light traveling in air is reflected from a medium of index of refraction n, it will be completely plane-polarized if the incident angle θ_p is given by $\tan\theta_p = n$. The fact that light can be polarized shows that it must be a transverse wave.

QUESTIONS

1. Does Huygens' principle apply to sound waves? To water waves?

2. What is the evidence that light is energy?

3. Why is light sometimes described as rays and sometimes as waves?

4. Two rays of light from the same source destructively interfere if their path lengths differ by how much?

5. If Young's double-slit experiment were submerged in water, how would the fringe pattern be changed?

6. Why was the observation of the double-slit interference pattern more convincing evidence for the wave theory of light than the observation of diffraction?

7. Compare a double-slit experiment for sound waves to that for light waves. Discuss the similarities and differences.

8. How would you produce "beats" with light waves, a well-known phenomenon with sound waves? How might you detect light "beats"?

9. Why doesn't the light from the two headlights of a distant car produce an interference pattern?

10. When white light passes through a flat piece of window glass, it is not broken down into colors as it is by a prism. Explain.

11. For both converging and diverging lenses, discuss how the focal length for red light differs from that for violet light.

12. We can hear sounds around corners, but we cannot see around corners; yet both sound and light are waves. Explain the difference.

13. Hold one hand close to your eye and focus on a distant light source through a narrow slit between two fingers. (Adjust your fingers to get best pattern.) Describe the pattern you see. Is this Fresnel or Fraunhofer diffraction?

14. A rectangular slit is twice as high as it is wide. Will the light spread more in the horizontal or in the vertical plane? Describe the pattern.

15. For diffraction by a single slit, what is the effect of increasing (*a*) the slit width, and (*b*) the wavelength?

16. What is the difference in the interference patterns

formed by two slits 10^{-4} cm apart and by a diffraction grating containing 10^4 lines/cm?

17. Explain why there are tiny peaks between the main peaks produced by a diffraction grating illuminated with monochromatic light. Why are the peaks so tiny?

18. For a diffraction grating, what is the advantage of (*a*) many slits, (*b*) closely spaced slits.

19. White light strikes (*a*) a diffraction grating and (*b*) a prism. A rainbow appears on a wall just below the direction of the horizontal incident beam in each case. What is the color of the top of the rainbow in each case?

20. Why is a diffraction grating preferable to a prism for use in a spectroscope?

21. Why are interference fringes noticeable only for a *thin* film like a soap bubble and not for a thick piece of glass, say?

22. Why are Newton's rings (Fig. 24–26) closer together farther from the center?

23. Some coated lenses appear greenish yellow when seen by reflected light. What wavelengths do you suppose they are designed to eliminate completely?

***24.** Describe how a Michelson interferometer could be used to measure the index of refraction of air.

25. What does polarization tell us about the nature of light?

26. What is the difference between a polarizer and an analyzer?

27. How can you tell if a pair of sunglasses is polarizing or not?

28. Sunlight will not pass through two Polaroids whose axes are at right angles. What happens if a third Polaroid, with axis at 45° to each of the other two, is placed between them?

***29.** For which of the materials in Table 24–2 is the speed of the extraordinary ray less when its polarization vector is parallel to the optic axis than when it is perpendicular?

***30.** What would be the color of the sky if the earth had no atmosphere?

***31.** If the earth's atmosphere were 50 times denser than it is, would sunlight still be white, or would it be some other color?

PROBLEMS

SECTION 24–2

1. (II) Derive the law of reflection—namely, that the angle of incidence equals the angle of reflection from a flat surface—using Huygens' principle for waves.

SECTION 24–3

2. (I) Monochromatic light falling on two slits 0.022 mm apart produces the fourth-order fringe at a 7.2° angle. What is the wavelength of the light used?

3. (I) The second-order fringe when 600-nm light falls on two slits is observed at a 25° angle to the initial beam direction. How far apart are the slits?

4. (II) Monochromatic light falls on two slits 0.030 mm apart. The fringes on a screen 3.00 m away are 9.2 cm apart. What is the wavelength of the light?

5. (II) A parallel beam of 700-nm light falls on two small slits 6.0×10^{-2} mm apart. How far apart are the fringes on a screen 3.0 m away?

6. (II) Light of wavelength 460 nm falls on two slits and produces an interference pattern in which the fourth-order fringe is 35 mm from the central fringe on a screen 1.0 m away. What is the separation of the two slits?

7. (II) The shortest-wavelength visible light falls on two slits 3.30×10^{-2} mm apart. The slits are immersed in water, as is a viewing screen 22.0 cm away. How far apart are the fringes on the screen?

8. (II) If 480-nm and 680-nm light passes through two slits 0.60 mm apart, how far apart are the second-order fringes for these two wavelengths on a screen 2.0 m away?

9. (II) Suppose a thin piece of glass were placed in front of the lower slit in Fig. 24–6 so that the two waves enter the slits 180° out of phase. Describe in detail the interference pattern on the screen.

10. (II) A very thin sheet of plastic ($n = 1.60$) covers one slit of a double-slit apparatus illuminated by 510-nm light. The center point on the screen, instead of being a maximum, is dark. What is the (minimum) thickness of the plastic?

SECTION 24–4

11. (I) By what percent, approximately, does the speed of red light (700 nm) exceed that of violet light (400 nm) in silicate flint glass? (See Fig. 24–11).

12. (II) A light beam strikes a piece of glass at a 60.00° incident angle. The beam contains two wavelengths, 450.0 nm and 700.0 nm, for which the index of refraction of the glass is 1.4820 and 1.4742, respectively. What is the angle between the two refracted beams?

13. (II) A parallel beam of light containing two wavelengths, $\lambda_1 = 400$ nm and $\lambda_2 = 650$ nm, strikes a piece of silicate flint glass at an incident angle of 45.0°. Calculate the angle between the two color beams inside the glass (see Fig. 24–11).

14. (II) Suppose the light beam in Problem 13 is entering the glass of an equilateral prism (Fig. 24–44). At what angle does each beam leave the prism (give angle with normal to the face)?

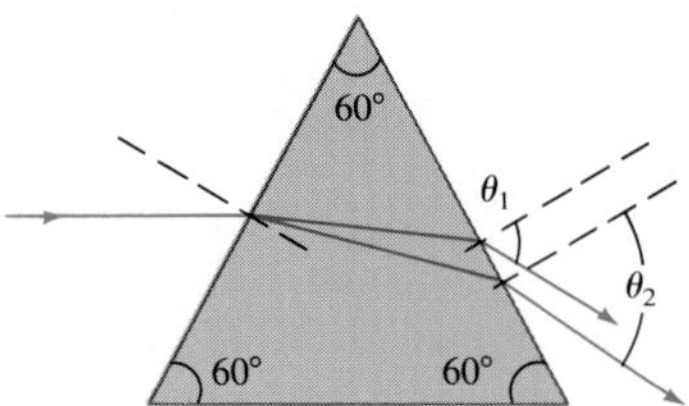

FIGURE 24–44 Problem 14.

*15. (III) A double convex lens whose radii of curvature are both 15.0 cm is made of crown glass. Find the distance between the focal points for 400-nm and 700-nm light. [*Hint:* use the lens-maker's equation, Eq. 23–10, and Fig. 24–11.]

SECTION 24–5

16. (I) If 550-nm light falls on a slit 0.0270 mm wide, what is the angular width of the central diffraction peak?
17. (I) Monochromatic light falls on a slit 3.50×10^{-3} mm wide. If the angle between the first dark fringes on either side of the central maximum is 42.0°, what is the wavelength of the light used?
18. (I) Monochromatic light of wavelength 589 nm falls on a slit. If the angle between the first bright fringes on either side of the central maximum is 38.0°, what is the slit width?
19. (II) How wide is the central diffraction peak on a screen 2.50 m behind a 0.0212-mm-wide slit illuminated by 550-nm light?
20. (II) When 450-nm light falls on a slit, the central diffraction peak on a screen 2.0 m away is 8.0 cm wide. Calculate the slit width.
21. (II) How wide is a slit if it diffracts 690-nm light so that its central diffraction peak is 3.0 cm wide on a screen 2.80 m away?
22. (II) For a given wavelength λ, what is the maximum slit width for which there will be no diffraction minima?

SECTIONS 24–6 AND 24–7

23. (I) The first-order line of 650-nm light falling on a diffraction grating is observed at a 22° angle. How far apart are the slits?
24. (I) At what angle will 670-nm light produce a third-order maximum when falling on a grating whose slits are 1.40×10^{-3} cm apart?
25. (II) Light falling normally on a 10,000-line/cm grating is revealed to contain three lines in the first-order spectrum at angles of 31.2°, 36.4°, and 47.5°. What wavelengths are these?
26. (II) How many lines per centimeter does a grating have if the third-order occurs at a 22.0° angle for 500-nm light?
27. (II) A grating has 5000 lines/cm. How many spectral orders can be seen when it is illuminated by white light?
28. (II) White light containing wavelengths from 400 to 700 nm falls on a grating with 6000 lines/cm. How wide is the first-order spectrum on a screen 2.00 m away?
29. (II) When yellow sodium light, $\lambda = 589$ nm, falls on a diffraction grating, its first-order peak on a screen 35 cm away falls 3.22 cm from the central peak. Another source produces a line 3.81 cm from the central peak. What is its wavelength?
30. (II) Two first-order spectrum lines are measured by a 10,000-line/cm spectroscope at angles, on each side of center, of +26°38′, +41°08′ and −26°48′, −41°19′. What are the wavelengths?
31. (III) Suppose the angles measured in Problem 30 were produced when the spectrometer (but not the source) was submerged in water. What then would be the wavelengths?

SECTION 24–8

32. (I) How far apart are the dark fringes in Example 24–6 if the glass plates are each 20 cm long?
33. (I) If a soap bubble is 120 nm thick, what color will appear at the center when illuminated normally by white light? Assume that $n = 1.34$.
34. (II) A lens appears greenish yellow ($\lambda = 570$ nm is strongest) when white light reflects from it. What minimum thickness of coating ($n = 1.25$) do you think is used on such a (glass) lens, and why?
35. (II) A total of 27 bright and 27 dark Newton's rings (not counting the dark spot at the center) are observed when 550-nm light falls normally on a planoconvex lens resting on a flat glass surface (Fig. 24–26). How much thicker is the center than the edges?
36. (II) A fine metal foil separates one end of two pieces of optically flat glass, as in Fig. 24–27. When light of wavelength 550 nm is incident normally, 32 dark lines are observed (with one at each end). How thick is the foil?
37. (II) How thick (minimum) should the air layer be between two flat glass surfaces if the glass is to appear bright when 640-nm light is incident normally? What if the glass is to appear dark?
38. (II) A thin film of alcohol ($n = 1.36$) lies on a flat glass plate ($n = 1.51$). When monochromatic light, whose wavelength can be changed, is incident normally, the reflected light is a minimum for $\lambda = 520$ nm and a maximum for $\lambda = 640$ nm. What is the thickness of the film?
39. (II) When a Newton's ring apparatus (Fig. 24–26) is immersed in a liquid, the diameter of the eighth dark ring decreases from 2.92 cm to 2.48 cm. What is the refractive index of the liquid?
40. (III) Show that the radius r of the mth dark Newton's ring, as viewed from directly above (Fig. 24–26), is given by $r = \sqrt{m\lambda R}$ where R is the radius of curvature of the curved glass surface and λ is the wavelength of light used. Assume that the thickness of the air gap is much less than R at all points and that $r \ll R$.
41. (III) A planoconvex Lucite lens 5.5 cm in diameter is placed on a flat piece of glass as in Fig. 24–26. When 620-nm light is incident normally, 54 bright rings are observed, the last one right at the edge. What is the radius of curvature of the lens surface, and what is the focal length of the lens?

*SECTION 24–9

*42. (II) What is the wavelength of the light entering an interferometer if 750 fringes are counted when the movable mirror moves 0.255 mm?
*43. (II) A micrometer is connected to the movable mirror of an interferometer. When the micrometer bears on a thin metal foil, the net number of fringes that move, compared to the empty micrometer, is 275. What is the thickness of the foil? The wavelength of light used is 589 nm.

*44. (II) How far must the mirror M_1 in a Michelson interferometer be moved if 1000 fringes of 589-nm light are to pass by a reference line?

*45. (III) One of the beams of an interferometer passes through a small glass container containing a cavity 1.30 cm wide. When a gas is allowed to slowly fill the container, a total of 236 dark fringes are counted to move past a reference line. The light used has a wavelength of 610 nm. Calculate the index of refraction of the gas, assuming that the interferometer is in vacuum.

*46. (III) The yellow sodium D lines have wavelengths of 589.0 and 589.6 nm. When they are used to illuminate a Michelson interferometer, it is noted that the interference fringes disappear and reappear periodically as the mirror M_1 is moved. Why does this happen? How far must the mirror move between one disappearance and the next?

SECTION 24–10

47. (I) Two polarizers are at 45° to one another. Unpolarized light falls on them. What fraction of the light intensity is transmitted?

48. (I) What is Brewster's angle for an air–glass ($n = 1.56$) surface?

49. (I) What is Brewster's angle for a piece of glass ($n = 1.56$) submerged in water?

50. (II) Two Polaroids are aligned so that the light passing through them is a maximum. At what angle should one of them be placed so that the intensity is reduced by half?

51. (II) At what angle should the axes of two Polaroids be placed so as to reduce the intensity of the incident unpolarized light to (*a*) $\frac{1}{3}$, (*b*) $\frac{1}{10}$?

52. (II) Unpolarized light passes through five successive Polaroid sheets, each of whose axis makes a 45° angle with the previous one. What is the intensity of the transmitted beam?

53. (II) Two polarizers are oriented at 38.0° to one another. Light polarized at a 19.0° angle to each polarizer passes through both. What reduction in intensity takes place?

54. (II) What would Brewster's angle be for reflections off the surface of water for light coming from beneath the surface? Compare to the angle for total internal reflection, and to Brewster's angle from above the surface.

55. (III) Describe how to rotate the plane of polarization of a plane-polarized beam of light by 90° and produce only a 10 percent loss in intensity using "perfect" polarizers.

*SECTION 24–11

*56. (II) Calculate the concentration of an insulin solution if the plane of polarization is rotated 18.2° after passing through 11.0 cm of the solution.

*57. (II) A 1.1×10^3 kg/m^3 solution of the sugar fructose 8.0 cm thick produces a rotation in the plane of polarization of incident light by $-117°$. What is the optical rotatory power of fructose?

*SECTION 24–12

*58. (III) A region of a birefringent biological specimen is 1.55×10^{-3} cm thick, with indices of refraction 1.322 and 1.331 for light whose wavelength in air is 590 nm. What phase difference arises between the two rays after passing through this region? That is, by what fraction of a wavelength are the two beams out of step when they exit?

*59. (III) A narrow beam of unpolarized light strikes the face of a flat quartz crystal 2.00 cm thick at a 44.5° angle to the face. The optic axis is parallel to the face and perpendicular to the beam. (*a*) Calculate the distance between the ordinary and extraordinary ray when they emerge from the crystal. (Snell's law is valid for rays that are traveling in the plane perpendicular to the optic axis.) (*b*) What is the polarization state of each emerging beam?

GENERAL PROBLEMS

60. The wings of a certain beetle have a series of parallel lines across them. When normally incident 660-nm light is reflected from the wing, the wing appears bright when viewed at an angle of 48°. How far apart are the lines?

61. How many lines per centimeter must a grating have if there is to be no second-order spectrum for any visible wavelength?

62. Show that the second- and third-order spectra of white light produced by a diffraction grating always overlap.

63. Television and radio waves can reflect from nearby mountains or from airplanes, and the reflections can interfere with the direct signal from the station. (*a*) Determine what kind of interference will occur when 75-MHz television signals arrive at a receiver directly from a distant station, and reflected from an airplane 118 m directly above the receiver. (Assume no change in phase of the signal upon reflection.) (*b*) What kind of interference will occur if the plane is 22 m closer to the station?

64. At what angle above the horizon is the sun when light reflecting off a smooth lake is polarized most strongly?

65. Unpolarized light falls on two polarizer sheets whose axes are at right angles. (*a*) What fraction of the incident light intensity is transmitted? (*b*) What fraction is transmitted if a third polarizer is placed between the first two so that its axis makes a 45° angle with each of their axes? (*c*) What if the third polarizer is in front of the other two?

66. If parallel light falls on a single slit of width D at a 30° angle to the normal, describe the (Fraunhofer) diffraction pattern.

67. Monochromatic light of variable wavelength is incident normally on a thin sheet of plastic film in air. The reflected light is a minimum only for $\lambda = 520$ nm and $\lambda = 650$ nm in the visible spectrum. What is the thickness of the film ($n = 1.60$)?

68. Suppose you viewed the light *transmitted* through a thin film on a flat piece of glass. Draw a diagram, similar to Fig. 24–25, and describe the conditions required for maxima and minima; consider all possible values of index of refraction. Discuss the relative size of the minima compared to the maxima and to zero.

69. Light of wavelength λ strikes a screen containing two slits a distance d apart at an angle θ_i to the normal. Determine the angle θ_m at which the mth-order maximum occurs.

CHAPTER 25

Optical Instruments

These 35 mm single lens reflex cameras are each equipped with a long focal length telephoto lens.

In our discussion of the behavior of light in the two previous chapters, we also described a few instruments such as the spectroscope and the Michelson interferometer. In this chapter, we will discuss some other, more common, instruments. Most of these use lenses, such as the camera, telescope, microscope, and the human eye. To describe their operation, we will use ray diagrams. However, we will see that some aspects of their operation depend on the wave nature of light.

25–1 • The Camera

The basic elements of a camera are a lens, a light-tight box, a shutter to let light pass through the lens only briefly, and a sensitized plate or piece of film (Fig. 25–1). When the shutter is opened, light from external objects in the field of view are focused by the lens as an image on the film. The film contains light-sensitive chemicals that undergo change when light strikes them. In the development process, chemical reactions cause the changed areas to turn black so that the image is recorded on the film.† You can see

† This is called a *negative*, since the black areas correspond to bright objects and vice versa. The same process occurs during printing to produce a black-and-white "positive" picture from the negative. Color film makes use of three dyes corresponding to the primary colors.

an image yourself by removing the camera back and viewing through a piece of tissue or wax paper (on which the image can form) placed at the position of the film with the shutter open.

There are three main adjustments on good-quality cameras: shutter speed, f-stop, and focusing, and we now discuss them. (Although many cameras today make these adjustments automatically, it is valuable to understand these adjustments even to use such automatic cameras effectively. For special or top-quality work, a camera that allows manual adjustments is indispensable.

Shutter speed

Shutter speed. This refers to how long the shutter is open and the film exposed. It may vary from a second or more ("time exposures") to $\frac{1}{1000}$ s or less. To avoid blurring from camera movement, speeds faster than $\frac{1}{100}$ s are normally used. If the object is moving, faster shutter speeds are needed to "stop" the action.

***f*-stop**. The amount of light reaching the film must be carefully controlled to avoid **underexposure** (too little light for any but the brightest objects to show up) or **overexposure** (too much light, so that all bright objects look the same, with a consequent lack of contrast and a "washed-out" appearance). To control the exposure, a "stop" or iris diaphragm, whose opening is of variable diameter, is placed behind the lens (Fig. 25–1). The size of the opening is varied to compensate for bright or dark days, the sensitivity of the film used, and for different shutter speeds. The size of the opening is specified by the ***f*-stop**, defined as

f-stop

$$f\text{-stop} = \frac{f}{D},$$

where f is the focal length of the lens and D is the diameter of the opening. For example, a 50-mm-focal-length lens set at $f/2$ has an opening $D =$ 25 mm; when set at $f/8$, the opening is only $6\frac{1}{4}$ mm. The faster the shutter speed, or the darker the day, the greater the opening must be to get a proper exposure. This corresponds to a smaller f-stop number. The smallest f-number of a lens (largest opening) is referred to as the *speed* of the lens. It is common to find $f/2.0$ lenses today, and even some as fast as $f/1.0$. Fast lenses are expensive to make and require many elements in order to reduce the defects present in simple thin lenses (Section 25–6). The advantage of a fast lens is that it allows pictures to be taken under poor lighting conditions.

FIGURE 25–1 A simple camera.

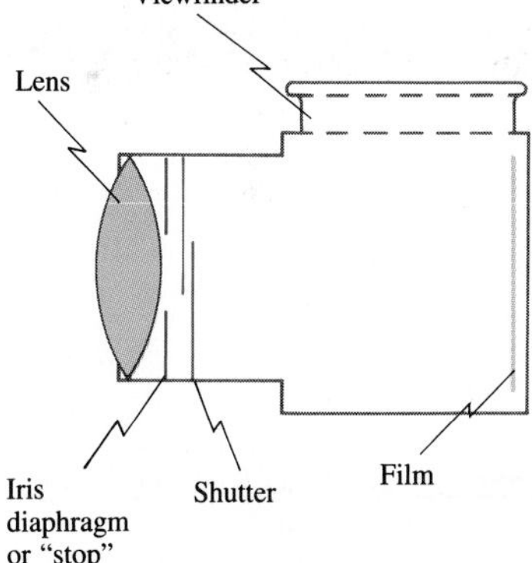

Lenses normally stop down to $f/16$, $f/22$, or $f/32$. Although the lens opening can usually be varied continuously, there are nearly always markings for specific lens openings: the standard f-stop markings are 1.0, 1.4, 2.0, 2.8, 4.0, 5.6, 8, 11, 16, 22, and 32. Notice that each of these corresponds to a diameter of about $\sqrt{2} = 1.4$ times smaller. Since the amount of light reaching the film is proportional to the area of the opening (and therefore proportional to the diameter squared), we see that each standard f-stop corresponds to a factor of 2 in light intensity reaching the film.

Focusing

Focusing. Focusing is the operation of placing the film at the correct position relative to the lens for the sharpest image. The image distance is a minimum for objects at infinity (the symbol ∞ is used for infinity) and is equal to the focal length. For closer objects, the image distance is greater than the focal length, as can be seen from the lens equation, $1/f = 1/d_o + 1/d_i$. To focus on nearby objects, the lens must therefore be moved away from the film, and this is usually done by turning a ring on the lens.

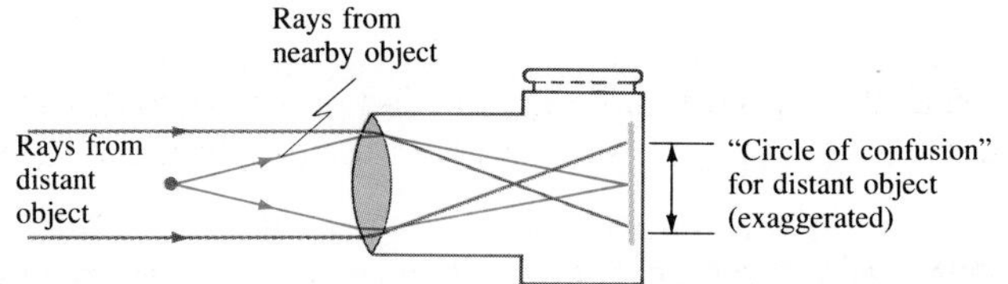

FIGURE 25–2 When the lens is positioned to focus on a nearby object, points on a distant object produce circles and are therefore blurred. (The effect is shown greatly exaggerated.)

If the lens is focused on a nearby object, a sharp image of it will be formed. But then the rays from a point on an object far away will be out of focus—they will form a circle on the film as shown (exaggerated) in Fig. 25–2. The distant object will thus produce an image consisting of overlapping circles and will be blurred. These circles are called **circles of confusion**. If you want to have near and distant objects sharp at the same time, you can try setting the lens focus at an intermediate position. Neither near nor distant objects will then be perfectly sharp, but the circles of confusion may be small enough that the blurriness is not noticeable. For a given distance setting, there is a range of distances over which the circles of confusion will be small enough that the images will be reasonably sharp. This is called the **depth of field**. For a particular circle of confusion diameter (typically taken to be 0.03 mm for 35-mm cameras), the depth of field depends on the lens opening. If the lens opening is smaller, the circles of confusion will be smaller, since only rays through the central part of the lens are accepted, and these form a smaller circle of confusion (Fig. 25–2). Hence, at smaller lens openings, the depth of field is greater.

Depth of field

Other factors also affect the sharpness of the image, such as the graininess of the film, diffraction, and lens aberrations relating to the quality of the lens itself. (Lens quality and diffraction effects will be discussed in Sections 25–6 and 25–7).

Camera lenses are categorized into normal, telephoto, and wide angle, according to focal length and film size. A **normal lens** is one that covers the film with a field of view that corresponds approximately to that of normal vision. A normal lens for 35-mm film has a focal length in the vicinity of 50 mm. A **telephoto lens**, as its name implies, acts like a telescope to magnify images. They have longer focal lengths than a normal lens: as we saw in Chapter 23 (Eq. 23–9), the height of the image for a given object distance is proportional to the image distance, and the image distance will be greater for a lens with longer focal length. For distant objects, the image height is very nearly proportional to the focal length (can you prove this?). Thus a 200-mm telephoto lens for use with a 35-mm camera gives a $4\times$ magnification over the normal 50-mm lens. A **wide-angle lens** has a shorter focal length than normal. A wider field of view is included and objects appear smaller. A **zoom lens** is one whose focal length can be changed so that you seem to zoom up to, or away from, the subject as you change the focal length.

Telephoto and wide angle lenses

Two types of viewing systems are common in cameras today. In a **range-finder** camera, you view through a small window just above the lens as in Fig. 25–1. The name derives from the dual-mirror system usually used to determine the object distance for focusing. In a **single-lens reflex** camera (SLR), you view directly through the lens with the use of prisms and mirrors (Fig. 25–3). A mirror hangs at a 45° angle behind the lens and flips up out of the way just before the shutter opens. SLRs have the great advantage that you can see almost exactly what you will get on film.

FIGURE 25–3 SLR, showing how light is viewed through the lens with the help of a movable mirror and a prism.

EXAMPLE 25–1 How far must a 50-mm-focal-length camera lens be moved from its infinity setting in order to sharply focus an object 3.0 m away?

SOLUTION When focused at infinity, the lens is 50 mm from the film. When focused at $d_o = 3.0$ m, the image distance is given by the lens equation,

$$\frac{1}{d_i} = \frac{1}{f} - \frac{1}{d_o} = \frac{1}{50\text{ mm}} - \frac{1}{3000\text{ mm}}.$$

We solve for d_i and find $d_i = 50.8$ mm, so the lens moves 0.8 mm.

EXAMPLE 25–2 A light meter reads that a lens setting of $f/8$ is correct for a shutter speed of $\frac{1}{250}$ s under certain conditions. What would be the correct lens opening for a shutter speed of $\frac{1}{500}$ s?

SOLUTION The amount of light entering the lens is proportional to the area of the lens opening. Since the exposure is reduced by half (from $\frac{1}{250}$ s to $\frac{1}{500}$ s), the area must be doubled. The area A is proportional to D^2, where D is the diameter of the opening. Since $A(\alpha D^2)$ must be increased by a factor of 2, then D must be increased by $\sqrt{2} = 1.41$. Thus the f stop must be $8.0/1.41 = f/5.6$.

25–2 • The Human Eye; Corrective Lenses

Anatomy of the eye

The human eye resembles a camera in its basic structure (Fig. 25–4). The eye is an enclosed volume into which light passes through a lens. A diaphragm, called the **iris** (the colored part of your eye), adjusts automatically to control the amount of light entering the eye. The hole in the iris through which light passes (the **pupil**) is black because no light is reflected from it (it's a hole), and very little light is reflected back out from the interior of the eye. The **retina**, which plays the role of the film in a camera, is on the curved rear surface. It consists of a complex array of nerves and receptors known as rods and cones which act to change light energy into electrical signals that travel along the nerves. The reconstruction of the image from all these tiny receptors is done mainly in the brain, although some analysis is apparently done in the complex interconnected nerve network at the retina itself. At the center of the retina is a small area called the **fovea**, about 0.25 mm in diameter, where the cones are very closely packed and the sharpest image and best color discrimination are found.

FIGURE 25–4 Diagram of a human eye.

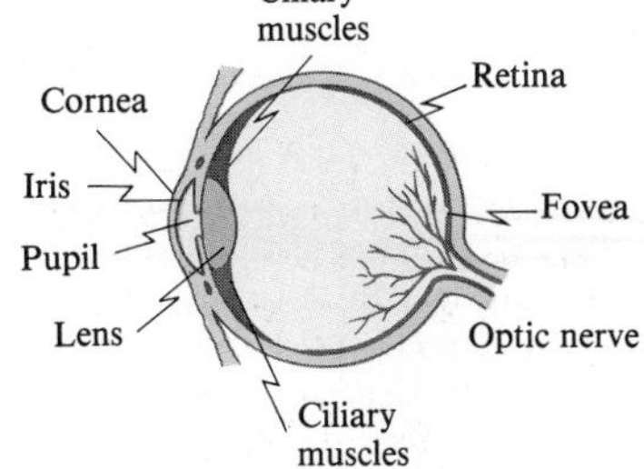

Unlike a camera, the eye contains no shutter. The equivalent operation is carried out by the nervous system, which analyzes the signals to form images at the rate of about 30 per second. This can be compared to motion picture or television cameras, which operate by taking a series of still pictures at a rate of 24 (movies) or 30 (U.S. television) per second. The rapid projection of these on the screen gives the appearance of motion.

The lens of the eye does little of the bending of the light rays. Most of the refraction is done at the front surface of the **cornea** (index of refraction = 1.376), which also acts as a protective covering. The lens acts as a fine adjust-

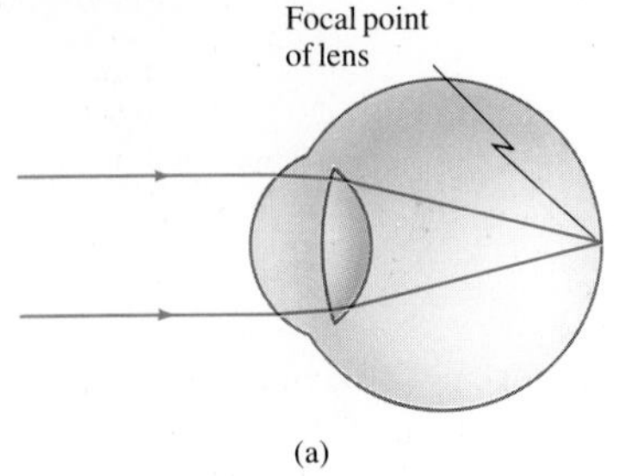

(a)

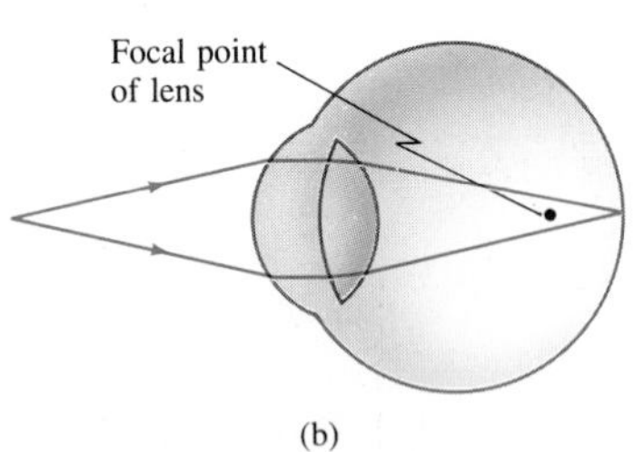

(b)

FIGURE 25–5 Accommodation by the eye: (a) lens relaxed, focused on infinity; (b) lens thickened, focused on nearby object.

ment for focusing at different distances. This is accomplished by the ciliary muscles (Fig. 25–4), which change the curvature of the lens so that its focal length is changed. To focus on a distant object, the muscles are relaxed and the lens is thin, Fig. 25–5a. To focus on a nearby object, the muscles contract, causing the center of the lens to be thicker, Fig. 25–5b, thus shortening the focal length. This focusing adjustment is called **accommodation**.

The closest distance at which the eye can focus clearly is called the **near point** of the eye. For young adults it is typically 25 cm, although younger children can often focus on objects as close as 10 cm. As people grow older, the ability to accommodate is reduced and the near point increases. A given person's **far point** is the farthest distance at which an object can be seen clearly. For some purposes it is useful to speak of a **normal eye** (a sort of average over the population), which is defined as one having a near point of 25 cm and a far point of infinity.

The "normal" eye is more of an ideal than a commonplace. A large part of the population have eyes that do not accommodate within the normal range of 25 cm to infinity, or have some other defect. Two common defects are nearsightedness and farsightedness. Both can be corrected to a large extent with lenses—either eyeglasses or contact lenses.

Nearsightedness

Nearsightedness, or *myopia*, refers to an eye that can focus only on nearby objects. The far point is not infinity but some shorter distance, so distant objects are not seen clearly. It is usually caused by an eyeball that is too long, although sometimes it is the curvature of the cornea that is too great. In either case, images of distant objects are focused in front of the retina. A diverging lens, because it causes parallel rays to diverge, allows the rays to be focused at the retina (Fig. 25–6a) and thus corrects this defect.

Farsightedness

Farsightedness, or *hyperopia*, refers to an eye that cannot focus on nearby objects. Although distant objects are usually seen clearly, the near point is somewhat greater than the "normal" 25 cm, which makes reading difficult. This defect is caused by an eyeball that is too short or (less often) by a cornea that is not sufficiently curved. It is corrected by a converging lens, Fig. 25–6b. Similar to hyperopia is *presbyopia*, which refers to the lessening ability of the eye to accommodate as one ages, and the near point moves out. Converging lenses also compensate for this.

FIGURE 25–6 (a) A nearsighted eye, which cannot focus clearly on distant objects, can be corrected by use of a diverging lens. (b) A farsighted eye, which cannot focus clearly on nearby objects, can be corrected by use of a converging lens.

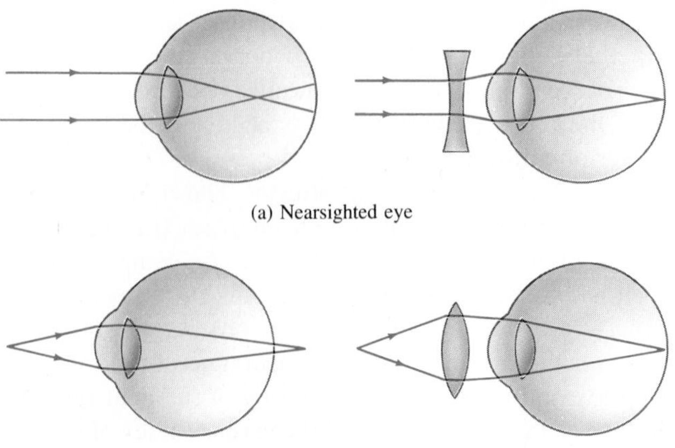

Astigmatism is usually caused by an out-of-round cornea or lens so that point objects are focused as short lines. It is as if the cornea were spherical with a cylindrical section superimposed. As shown in Fig. 25–7, a cylindrical lens focuses a point into a line parallel to its axis. An astigmatic eye focuses rays in a vertical plane, say, at a shorter distance than it does for rays in a horizontal plane. Astigmatism is corrected with the use of a compensating cylindrical lens. Lenses for eyes that are nearsighted or farsighted as well as astigmatic are ground with superimposed spherical and cylindrical surfaces, so that the radius of curvature of the correcting lens is different in different planes. Astigmatism is tested for by looking with one eye at a pattern like that in Fig. 25–8. Sharply focused lines appear dark, whereas those that are spread out slightly appear dimmer or gray.

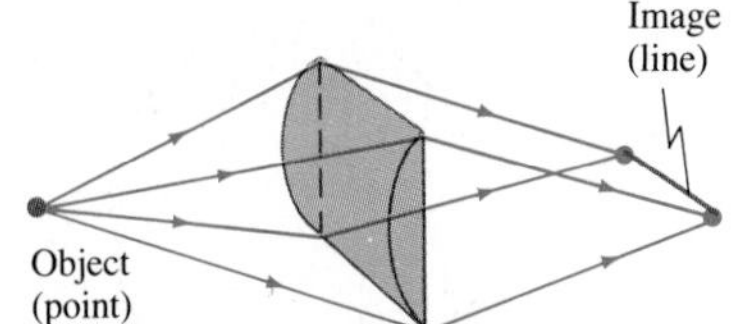

FIGURE 25–7 A cylindrical lens forms a line image of a point object because it is converging in one plane only.

FIGURE 25–8 Test for astigmatism.

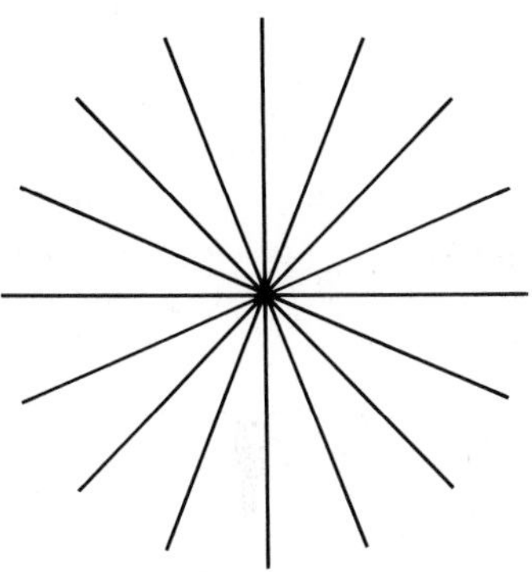

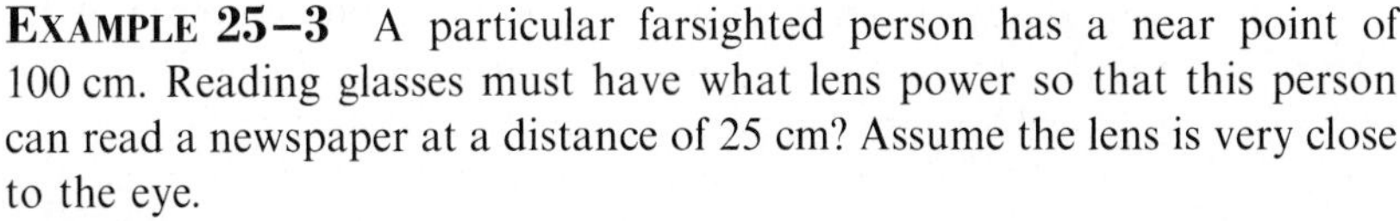

EXAMPLE 25–3 A particular farsighted person has a near point of 100 cm. Reading glasses must have what lens power so that this person can read a newspaper at a distance of 25 cm? Assume the lens is very close to the eye.

SOLUTION When the object is placed 25 cm from the lens, we want the image to be 100 cm away on the *same* side of the lens, and so it will be virtual, Fig. 25–9. Thus, $d_o = 25$ cm, $d_i = -100$ cm, and the lens equation gives

$$\frac{1}{f} = \frac{1}{25\text{ cm}} + \frac{1}{-100\text{ cm}} = \frac{1}{33\text{ cm}}.$$

So $f = 33$ cm $= 0.33$ m. The power P of the lens is $P = 1/f = +3.0$ D. The plus sign indicates that it is a converging lens.

EXAMPLE 25–4 A nearsighted eye has near and far points of 12 cm and 17 cm, respectively. What lens power is needed for this person to see distant objects clearly, and what then will be the near point? Assume that each lens is 2.0 cm from the eye.

SOLUTION First we determine the power of the lens needed to focus objects at infinity, when the eye is relaxed. For a distant object ($d_o = \infty$), the

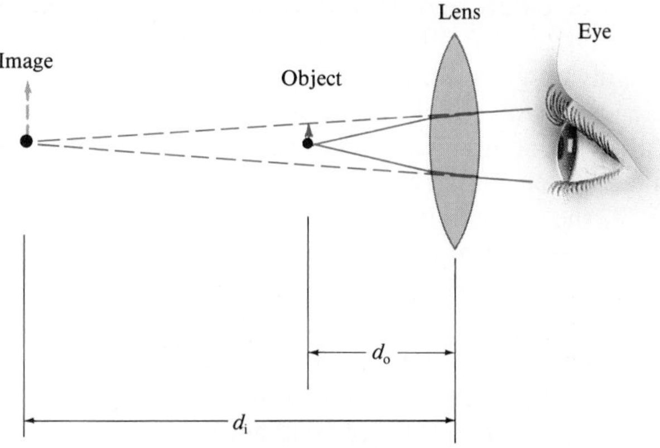

FIGURE 25–9 Lens of reading glasses (Example 25–3).

lens must put the image 17 cm from the eye (its far point), which is 15 cm in front of the lens; hence $d_i = -15$ cm. We use the lens equation to solve for the focal length of the needed lens:

$$\frac{1}{f} = -\frac{1}{15 \text{ cm}} + \frac{1}{\infty} = -\frac{1}{15 \text{ cm}}.$$

So $f = -15$ cm $= -0.15$ m or $P = 1/f = -6.7$ D. The minus sign indicates it must be a diverging lens. To determine the near point when wearing the glasses, we note that the image must be 12 cm from the eye (its near point) or 10 cm from the lens, so $d_i = -0.10$ m and

$$\frac{1}{d_o} = \frac{1}{f} - \frac{1}{d_i} = -\frac{1}{0.15 \text{ m}} + \frac{1}{0.10 \text{ m}} = \frac{1}{0.30 \text{ m}}.$$

So $d_o = 30$ cm, which means the near point when the person is wearing glasses is 30 cm in front of the lens.

Contact lenses could be used to correct the eye in Example 25–4. Since contacts are placed directly on the cornea, we would not subtract out the 2.0 cm for the image distances. That is, for distant objects $d_i = -17$ cm, so $P = 1/f = -5.9$ diopters. Thus we see that a contact lens and an eyeglass lens will require slightly different focal lengths for the same eye because of their different placements relative to the eye.

25–3 • The Magnifying Glass

Much of the remainder of this chapter will deal with optical devices that are used to produce magnified images of objects. We first discuss the **simple magnifier**, or **magnifying glass**, which is simply a converging lens.

How large an object appears, and how much detail we can see on it, depends on the size of the image it makes on the retina. This, in turn, depends on the angle subtended by the object at the eye. For example, a penny held 30 cm from the eye looks twice as high as one held 60 cm away because the angle it subtends is twice as great (Fig. 25–10). When we want to examine detail on an object, we bring it up close to our eyes so that it subtends a greater angle. However, our eyes can accommodate only up to a point (the near point), and we will assume a standard distance of 25 cm as the near point in what follows.

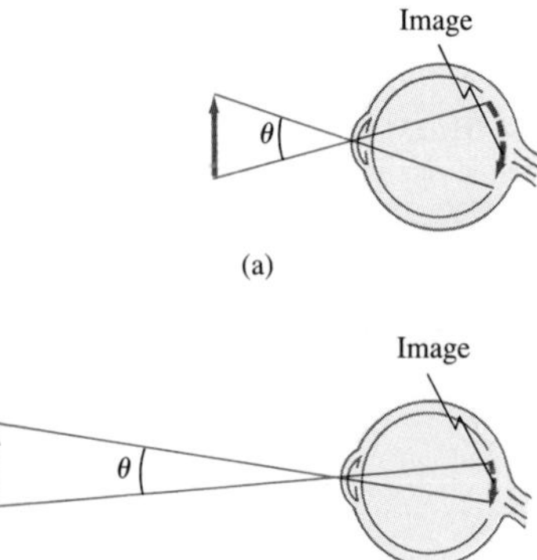

FIGURE 25–10 When the same object is viewed at a shorter distance, the image on the retina is greater, so the object appears larger and more detail can be seen. The angle θ that the object subtends in (a) is greater than in (b).

A magnifying glass allows us to place the object closer to our eye so that it subtends a greater angle. As shown in Fig. 25–11a, the object is placed at the focal point or just within it. Then the converging lens produces a virtual image, which must be at least 25 cm from the eye if the eye is to focus on it. If the eye is relaxed, the image will be at infinity, and in this case the object is exactly at the focal point. (You make this slight adjustment yourself when you "focus" on the object by moving the magnifying glass.)

A comparison of part (a) of Fig. 25–11 with part (b), in which the same object is viewed at the near point with the unaided eye, reveals that the angle the object subtends at the eye is much larger when the magnifier is used. The **angular magnification** or **magnifying power**, M, of the lens is defined as: the ratio of the angle subtended by an object when using the lens, to the

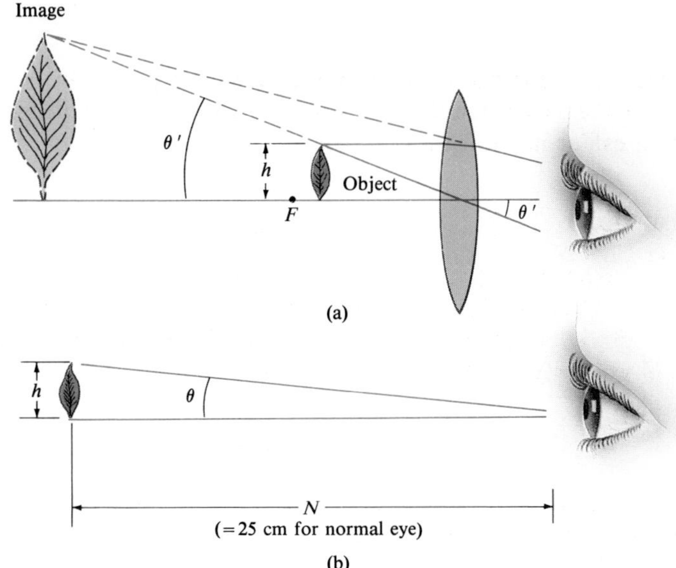

FIGURE 25–11 Leaf viewed (a) through a magnifying glass, and (b) with the unaided eye focused at its near point.

angle subtended using the unaided eye with the object at the near point of the eye (25 cm for the normal eye):

$$M = \frac{\theta'}{\theta},$$

where θ and θ' are shown in Fig. 25–11. This can be written in terms of the focal length f of the lens as follows. Suppose, first, that the image in Fig. 25–11a is at the near point N of the eye: $d_i = -N$, where $N = 25$ cm for the normal eye. Then the object distance d_o is given by

$$\frac{1}{d_o} = \frac{1}{f} - \frac{1}{d_i} = \frac{1}{f} + \frac{1}{N},$$

or $d_o = Nf/(f + N)$. (We see from this equation that $d_o < f$, as shown in Fig. 25–11a, since $N/(f + N)$ must be less than 1.) Let h be the height of an object, and we assume h is small so the angles θ and θ' are approximately equal to their sines and tangents; then $\theta' = h/d_o = (f + N)h/(Nf)$ and $\theta = h/N$. Thus

$$M = \frac{\theta'}{\theta} = \frac{(f + N)h}{Nf}\left(\frac{N}{h}\right),$$

or

$$M = 1 + \frac{N}{f}. \quad \begin{bmatrix}\text{eye focused at near point, } N;\\ N = 25 \text{ cm for normal eye}\end{bmatrix} \quad (25\text{–}1a)$$

Magnification of a simple magnifier

If the eye is relaxed when using the magnifying glass, the image is then at infinity, and the object is then precisely at the focal point. In this case, $\theta' = h/f$, so

$$M = \frac{\theta'}{\theta} = \left(\frac{h}{f}\right)\left(\frac{N}{h}\right) = \frac{N}{f}. \quad \begin{bmatrix}\text{eye focused at } \infty;\\ N = 25 \text{ cm for normal eye}\end{bmatrix} \quad (25\text{–}1b)$$

It is clear that the magnification is slightly greater when the eye is focused at its near point than when relaxed. And the shorter the focal length of the lens, the greater the magnification. It is up to you, of course, whether you relax your eye or not when using a magnifier.

EXAMPLE 25–5 An 8-cm-focal-length converging lens is used as a magnifying glass by a person with normal eyes. Calculate (*a*) the maximum magnification, and (*b*) the magnification when the eye is relaxed.

SOLUTION (*a*) The maximum magnification is obtained when the eye is focused at its near point:

$$M = 1 + \frac{N}{f} = 1 + \frac{25}{8} \approx 4\times.$$

(*b*) With the eye focused at infinity, $M = 25\text{ cm}/8\text{ cm} \approx 3\times$.

25–4 • Telescopes

A telescope is used to magnify objects that are very far away. In most cases, the object can be considered to be at infinity.

Galileo, although he did not invent it,† developed the telescope into a usable and important instrument. He was the first to train the telescope on the heavens (Fig. 25–12), and he made a number of world-shaking discoveries (the moons of Jupiter, the phases of Venus, sunspots, the structure of the moon's surface, that the Milky Way is made up of a huge number of individual stars, among others).

† Galileo built his first telescope in 1609 after having heard of such an instrument existing in Holland. The first telescopes magnified only 3 to 4 times, but Galileo soon made a 30-power instrument. The first Dutch telescope seems to date from about 1604, but there is a reference suggesting it may have been copied from an Italian one from as early as 1590. Kepler (see Chapter 5) gave a ray description (1611) of both the Keplerian and Galilean telescopes (that is, those with two lenses). The former is named for him because he first described it, although he did not build it.

(a)

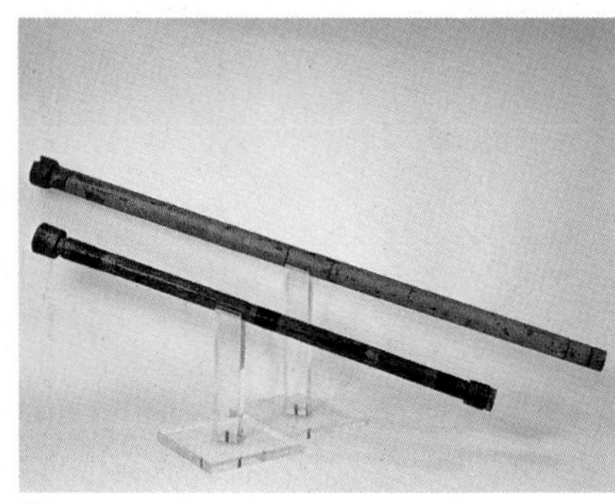

(b)

FIGURE 25–12 (a) Objective lens (mounted now in an ivory frame) from the telescope with which Galileo made his world-shaking discoveries, including the moons of Jupiter, (b) Later telescopes made by Galileo.

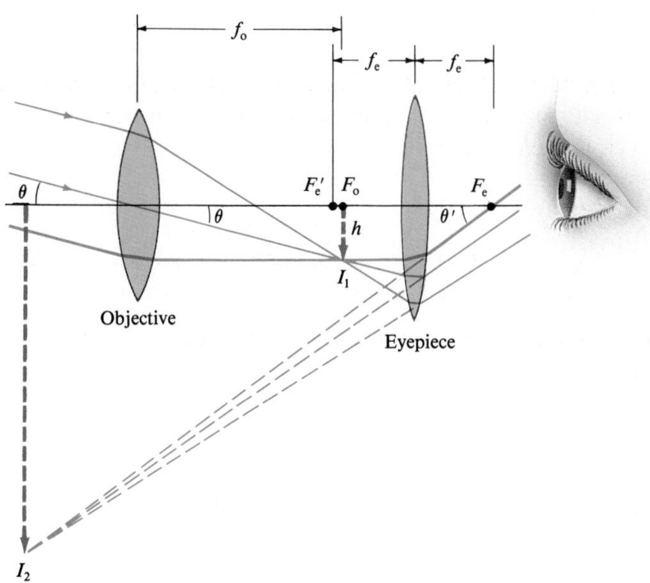

FIGURE 25–13
Astronomical telescope (refracting).

Several types of **astronomical telescope** exist. The common **refracting** type, sometimes called **Keplerian**, contains two converging lens located at opposite ends of a long tube, Fig. 25–13. The lens closest to the object is called the **objective lens** and forms a real image I_1 of the object at its focal point F_o (or near it if the object is not at infinity). Although this image, I_1, is smaller than the original object, it subtends a greater angle and is very close to the second lens, called the **eyepiece**, which acts as a magnifier. That is, the eyepiece magnifies the image produced by the objective to produce a second, greatly magnified image, I_2, which is virtual, and inverted. If the viewing eye is relaxed, the eyepiece is adjusted so the image I_2 is at infinity. Then the real image I_1 is at the focal point F'_e of the eyepiece, and the distance between the lenses is $f_o + f_e$ for an object at infinity.

To find the total magnification of this telescope, we note that the angle an object subtends as viewed by the unaided eye is just the angle θ subtended at the telescope objective. From Fig. 25–13 we can see that $\theta \approx h/f_o$, where h is the height of the image I_1 and we assume θ is small so that $\tan\theta \approx \theta$. Note, too, that the thickest of the three rays drawn in the figure is parallel to the axis before it strikes the eyepiece and therefore passes through the focal point F_e. Thus, $\theta' \approx h/f_e$ and the total magnifying power (angular magnification) of this telescope is

Magnification of telescope

$$M = \frac{\theta'}{\theta} = -\frac{f_o}{f_e}, \tag{25–2}$$

where we have inserted a minus sign to indicate that the image is inverted. To achieve a large magnification, the objective lens should have a long focal length and the eyepiece a short one.

For an astronomical telescope to produce bright images of distant stars, the objective lens must be large to allow in as much light as possible. Indeed, the diameter of the objective (and hence it's "light-gathering power") is the most important parameter for an astronomical telescope, which is why the largest ones are specified by giving the objective diameter. The construction and grinding of large lenses is very difficult. Therefore, the largest telescopes

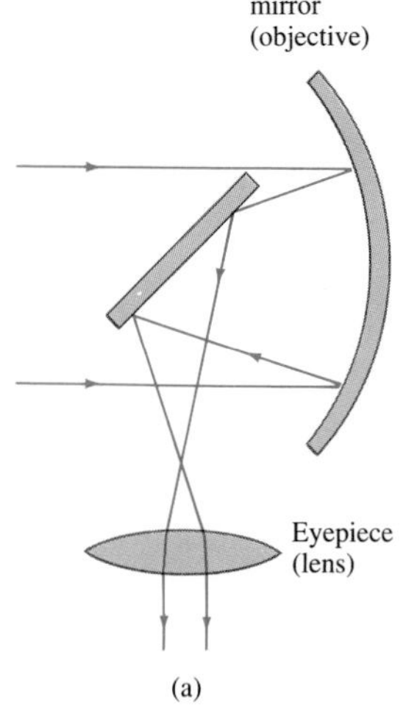

are **reflecting** telescopes using a curved mirror as the objective, Fig. 25–14, since a mirror has only one surface to be ground and can be supported along its entire surface.† (A large lens, supported at its edges, would sag under its own weight.) Normally, the eyepiece lens or mirror (see Fig. 25–14) is removed so that the real image formed by the objective can be recorded on film.

A **terrestrial telescope** (for use in viewing objects on earth), unlike its astronomical counterpart, must provide an upright image. Two designs are shown in Fig. 25–15. The **Galilean** type shown in part (a), which Galileo used for his great astronomical discoveries, has a diverging lens as eyepiece which intercepts the converging rays from the objective lens before they reach a focus, and acts to form a virtual upright image. This design is often used in opera glasses. The tube is reasonably short, but the field of view is small. The second type, shown in Fig. 25–15b, is often called a **spyglass** and makes use of a third lens ("field lens") that acts to make the image upright as shown. A spyglass must be quite long. The most practical design today is the **prism binocular** which was shown in Fig. 23–21. The objective and eyepiece are converging lenses. The prisms reflect the rays by total internal reflection and shorten the physical size of the device, and they also act to produce an upright image. One prism reinverts the image in the vertical plane, the other in the horizontal plane.

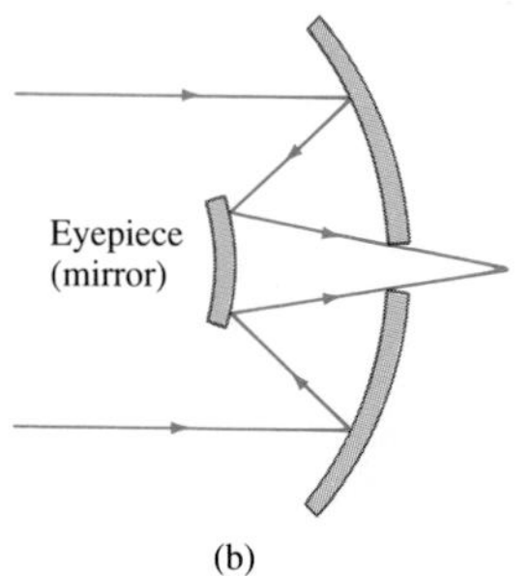

† Another advantage of mirrors is that they exhibit no chromatic aberration since the light doesn't pass through them. Also they can be ground in a parabolic shape to correct for spherical aberration (see Section 25–6). The reflecting telescope was first proposed by Newton.

(c)

FIGURE 25–14 (above) A concave mirror can be used as the objective of an astronomical telescope. Either a lens or a mirror can be used as the eyepiece. Arrangement (a) is called the Newtonian focus and (b) the Cassegrainian focus. Other arrangements are also possible. (c) The 200-inch (Mirror diameter) Hale telescope on Mt. Palomar in California.

FIGURE 25–15 (below) Terrestrial telescopes that produce an upright image: (a) Galilean; (b) spyglass, or field-lens, type.

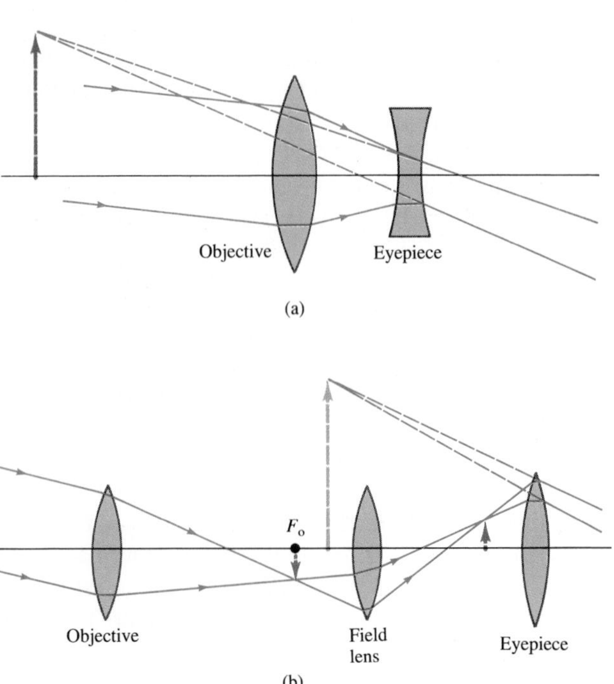

EXAMPLE 25–6 A Galilean telescope has an objective lens whose focal length is 28 cm and an eyepiece with focal length -8.0 cm. What is the magnification?

SOLUTION $M = -f_o/f_e = -(28 \text{ cm})/(-8.0 \text{ cm}) = 3.5\times$.

25–5 • Compound Microscope

The compound microscope, like the telescope, has both objective and eyepiece (or ocular) lenses, Fig. 25–16. The design is different from that for a telescope since a microscope is used to view objects that are very close, so the object distance is very small. The object is placed just beyond the objective's focal point as shown in Fig. 25–16a. The image I_1 formed by the objective lens is real, quite far from the lens, and much enlarged. This image is magnified by the eyepiece into a very large virtual image, I_2, which is seen by the eye and is inverted.

The overall magnification of a microscope is the product of the magnifications produced by the two lenses. The image I_1 formed by the objective is a factor M_o greater than the object itself. From Fig. 25–16a and Eq. 23–9 for the lateral magnification of a simple lens, we have

$$M_o = \frac{d_i}{d_o} = \frac{l - f_e}{d_o},$$

where l is the distance between the lenses (equal to the length of the barrel), and we ignored the minus sign in Eq. 23–9 which only tells us that the image

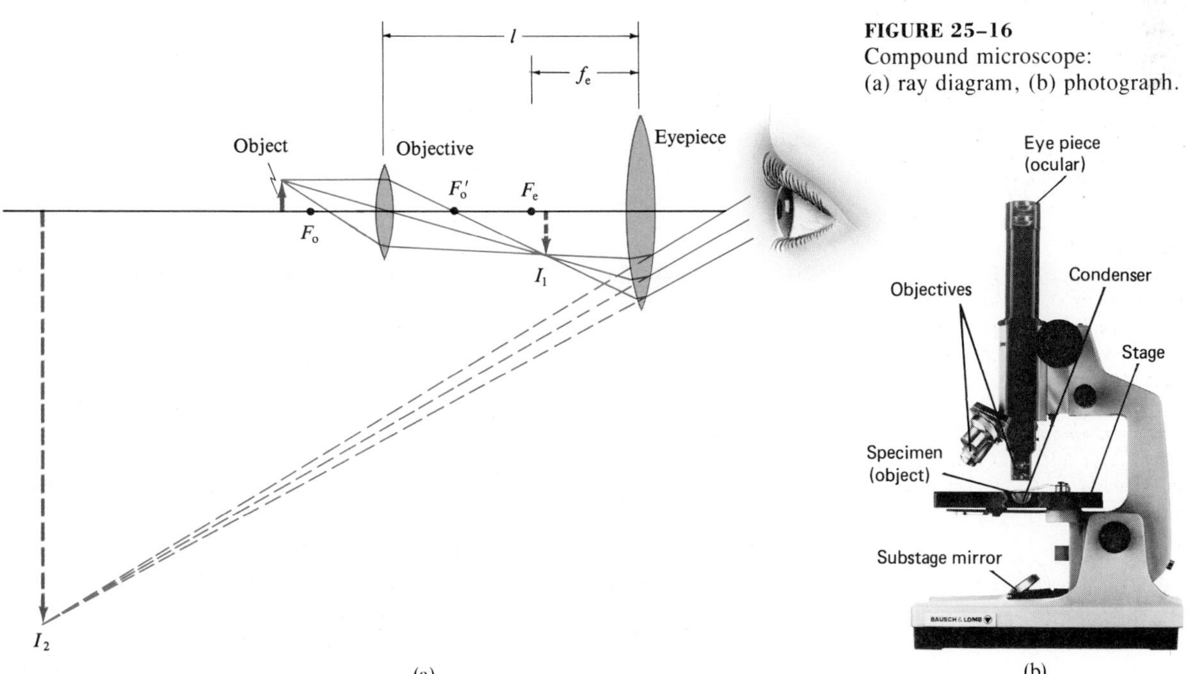

FIGURE 25–16
Compound microscope:
(a) ray diagram, (b) photograph.

is inverted. The eyepiece acts like a simple magnifier. If we assume that the eye is relaxed, its angular magnification M_e is (from Eq. 25–1b)

$$M_e = \frac{N}{f_e},$$

where the near point $N = 25$ cm for the normal eye. Since the eyepiece enlarges the imaged formed by the objective, the overall angular magnification M is

Magnification of microscope

$$M = M_e M_o = \left(\frac{N}{f_e}\right)\left(\frac{l - f_e}{d_o}\right) \quad (25\text{–}3a)$$

$$\approx \frac{Nl}{f_e f_o}. \quad (25\text{–}3b)$$

The approximation, Eq. 25–3b, is accurate when f_e and f_o are small compared to l, so $l - f_e \approx l$ and $d_o \approx f_o$ (Fig. 25–16a). This is a good approximation for large magnifications, since these are obtained when f_o and f_e are very small (they are in the denominator of Eq. 25–3b). In order to make lenses of very short focal length, which can be done best for the objective, compound lenses involving several elements must be used to avoid serious aberrations, as discussed in the next section.

EXAMPLE 25–7 A compound microscope consists of a $10\times$ eyepiece and a $50\times$ objective 17.0 cm apart. Determine (*a*) the overall magnification, (*b*) the focal length of each lens, and (*c*) the position of the object when the final image is in focus with the eye relaxed. Assume a normal eye, so $N = 25$ cm.

SOLUTION (*a*) The overall magnification is $10 \times 50 = 500\times$. (*b*) The eyepiece focal length is (see equation just before Eq. 25–3a) $f_e = N/M_e = 25\text{ cm}/10 = 2.5$ cm. It is easier to next find d_o (part (*c*)) before we find f_o since we can use the equation for M_o on page 665. Solving for d_o, we find $d_o = (l - f_e)/M_o = (17.0\text{ cm} - 2.5\text{ cm})/50 = 0.29$ cm. Then, from the lens equation with $d_i = l - f_e = 14.5$ cm (see Fig. 25–16a),

$$\frac{1}{f_o} = \frac{1}{d_o} + \frac{1}{d_i} = \frac{1}{0.29\text{ cm}} + \frac{1}{14.5\text{ cm}} = 3.52;$$

so $f_o = 0.28$ cm. (*c*) We just calculated $d_o = 0.29$ cm, which is very close to f_o.

Opaque objects are generally illuminated by a source placed above them. If the objects to be viewed are transparent, such as cells or tissue, light is normally passed through the object from a source beneath the microscope stage (see Fig. 25–16b). The illumination system must be carefully designed if maximum sharpness and contrast are to be achieved. Usually, a *condenser* is employed, which is a set of two or three lenses, although inexpensive condensers may be a single lens or mirror. The purpose of the condenser is to gather light over a wide angle from the source, and to "condense" it down to a narrow beam that will illuminate the object strongly and uniformly. A number of different designs are employed. The source is often placed in the focal plane of the condenser so that light from each point on the source is parallel when it passes through the object.

Some specialized microscope types are described in Section 25–10.

25–6 • Lens Aberrations

In Chapter 23, we developed a theory of image formation by a thin lens. We found, for example, that all rays from each point on an object are brought to a single point as the image. This, and other results, were based on approximations such as that all rays make small angles with one another and we can use $\sin\theta \approx \theta$. Because of these approximations, we expect deviations from the simple theory and these are referred to as **lens aberrations**. There are several types of aberration; we will briefly discuss each of them separately but all may be present at one time.

First consider a point object on the axis of a lens. Rays from this point that pass through the outer regions of the lens are brought to a focus at a different point than those that pass through the center of the lens; this is called **spherical abberation**, and is shown exaggerated in Fig. 25–17. Consequently, the image seen on a piece of film (say) will not be a point but a tiny circular patch of light. If the film is placed at the point C, as indicated, the circle will have its smallest diameter, which is referred to as the **circle of least confusion**. Spherical aberration is present whenever spherical surfaces are used. It can be corrected by using nonspherical lens surfaces, but to grind such lenses is very expensive. It can be minimized with spherical surfaces by choosing the curvatures so that equal amounts of bending occur at each lens surface; a lens can only be designed like this for one particular object distance. Spherical aberration is usually corrected (by which we mean reduced greatly) by the use of several lenses in combination.

For object points off the lens axis, additional aberrations occur. Rays passing through different parts of the lens cause spreading of the image that is noncircular. We won't go into the details but merely point out that there are two effects: **coma** (because the image is comet-shaped rather than a circle) and **off-axis astigmatism**.† Furthermore, the image points for objects off the axis but at the same distance from the lens do not fall on a flat plane but on a curved surface—that is, the focal plane is not flat. (We expect this since the points on a flat plane, such as the film in a camera, are not equidistant from the lens.) This aberration is known as **curvature of field** and

† Although the effect is the same as for astigmatism in the eye (Section 25–2), the cause is different. Off-axis astigmatism is no problem in the eye because objects are clearly seen only at the fovea which is on the lens axis.

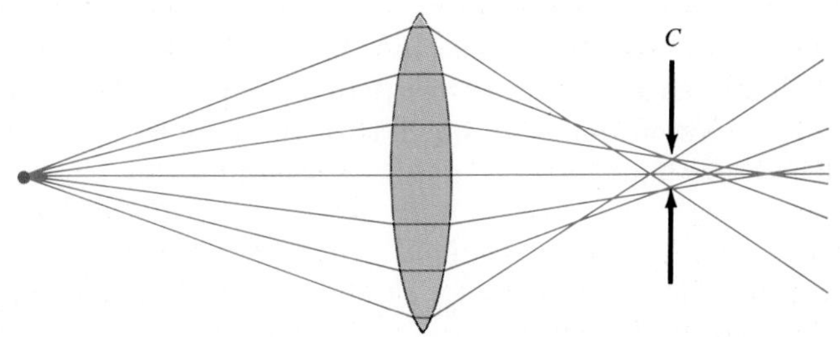

FIGURE 25–17 Spherical aberration (exaggerated). Circle of least confusion is obtained at C.

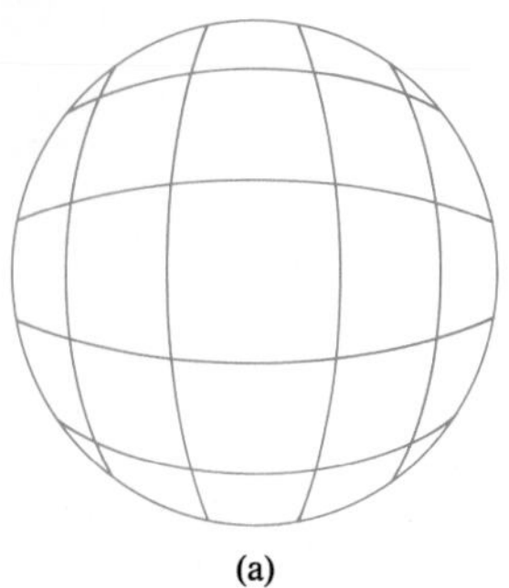

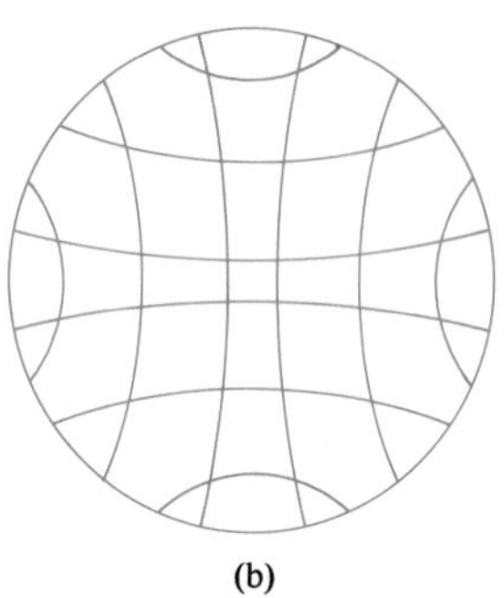

FIGURE 25–18 Distortion. Lenses may image a square grid of perpendicular lines to produce (a) barrel distortion or (b) pincushion distortion.

FIGURE 25–19 Chromatic aberration. Different colors are focused at different points.

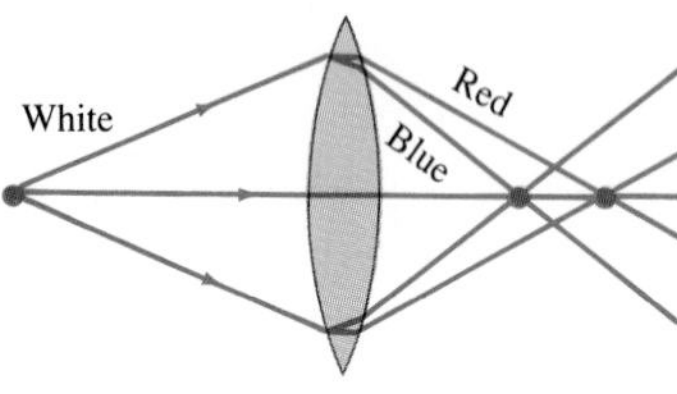

FIGURE 25–20 (below) Achromatic doublet.

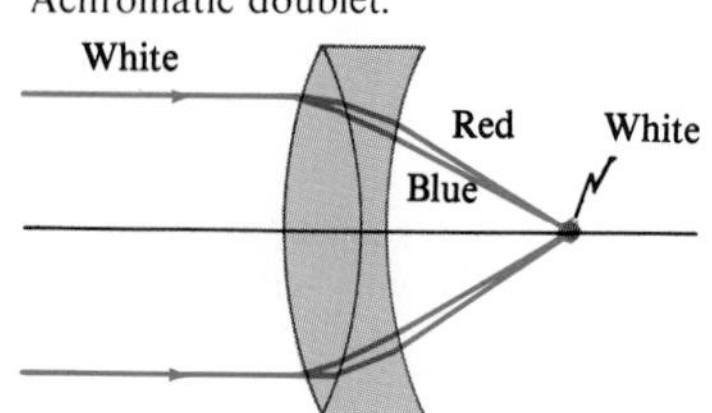

is obviously a problem in cameras and other devices where the film is placed in a flat plane. In the eye, however, the retina is curved, which compensates for this effect. Another aberration, known as **distortion**, is a result of variation of magnification at different distances from the lens axis. Thus a straight line object some distance from the axis may form a curved image. A square grid of lines may be distorted to produce "pincushion distortion" or "barrel distortion," Fig. 25–18. The latter is common in extreme wide-angle lenses.

All the above aberrations occur for monochromatic light and hence are referred to as *monochromatic aberrations*. Normal light is not monochromatic, and there will also be **chromatic aberration**. This aberration arises because of dispersion—the variation of index of refraction of transparent materials with wavelength (Section 24–4). For example, blue light is bent more than red light by glass. So if white light is incident on a lens, the different colors are focused at different points, Fig. 25–19, and there will be colored fringes in the image. Chromatic aberration can be eliminated for any two colors (and reduced greatly for all others) by the use of two lenses made of different materials with different indices of refraction and dispersion. Normally one lens is converging and the other diverging, and they are often cemented together (Fig. 25–20). Such a lens combination is called an **achromatic doublet** (or "color-corrected" lens).

It is not possible to fully correct all aberrations. Combining two or more lenses together can reduce them. High-quality lenses used in cameras, microscopes and other devices are **compound lenses** consisting of many simple lenses (referred to as **elements**). A typical high-quality camera lens may contain six to eight (or more) elements.

For simplicity we will normally indicate lenses in diagrams as if they were simple lenses. But it must be remembered that good-quality lenses are compound.

The human eye is also subject to aberrations. In the course of evolution, however, these have been minimized. Spherical aberration, for example, has been largely corrected since (1) the cornea is less curved at the edges than at the center, and (2) the lens is less dense at the edges than at the center. Both effects cause rays at the outer edges to be bent less strongly, and thus help to reduce spherical aberration. Chromatic aberration is partially compensated for because the lens absorbs the shorter wavelengths appreciably and the retina is less sensitive to the blue and violet wavelengths. This is just the region of the spectrum where dispersion—and thus chromatic aberration—is greatest (Fig. 24–11).

25–7 • Limits of Resolution; the Rayleigh Criterion

The ability of a lens to produce distinct images of two point objects very close together is called the **resolution** of the lens. The closer the two images can be and still be seen as distinct (rather than overlapping blobs), the higher the resolution. The resolution of a camera lens, for example, is often specified as so many lines per millimeter,[†] and can be determined by photographing

[†] This may be specified at the center of the field of view as well as at the edges, where it is usually less because of off-axis aberrations.

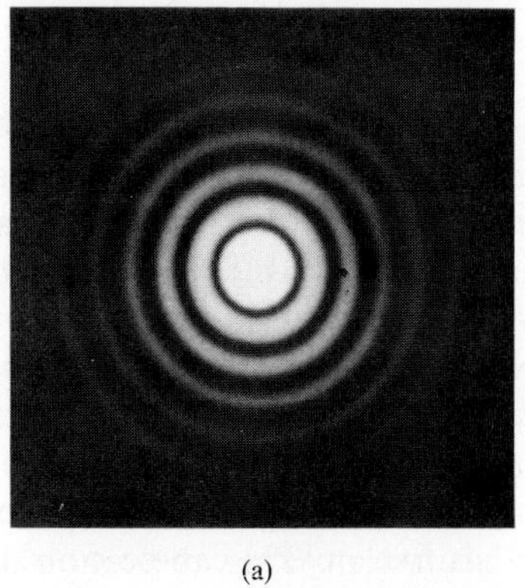
(a)

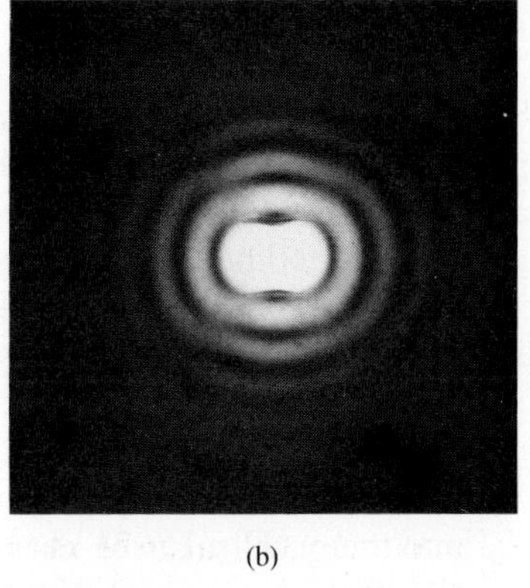
(b)

FIGURE 25–21 Photographs of images (greatly magnified) formed by a lens for: (a) a single point object; (b) two point objects barely resolved.

a standard set of parallel lines on fine-grain film. The minimum spacing of lines distinguishable on film using the lens gives the resolution.

There are two principal factors that limit the resolution of a lens. The first is lens aberrations. As we saw, because of spherical and other aberrations, a point object is not a point on the image but a tiny blob. Careful design of compound lenses can reduce aberrations significantly, but they cannot be eliminated entirely. The second factor that limits resolution is *diffraction*, which cannot be corrected for because it is a natural result of the wave nature of light. We discuss it now.

In Section 24–5 we saw that because light travels as a wave, light from a point source passing through a slit is spread out into a diffraction pattern (Figs. 24–15 and 24–17). A lens, because it has edges, acts like a slit. When a lens forms the image of a point object, the image of that point is actually a tiny diffraction pattern. Thus, an image would be blurred even if aberrations were absent.

In the analysis that follows, we assume that the lens is free of aberrations, so that we can focus our attention on diffraction effects and how much they limit the resolution of a lens. In Fig. 24–17 we saw that the diffraction pattern produced by light passing through a rectangular slit has a central maximum in which most of the light falls. This central peak falls to a minimum on either side of its center at an angle $\theta \approx \sin\theta = \lambda/D$ (this is Eq. 24–3), where D is the diameter of the slit, λ is the wavelength of light used, and we assume θ is small. There are also low-intensity fringes beyond. For a lens, or any circular hole, the image of a point object will consist of a *circular* central peak (called the *diffraction spot* or *Airy disk*) surrounded by faint circular fringes, as in Fig. 25–21a. The central maximum has an angular half width given by

$$\theta = \frac{1.22\lambda}{D}.$$

This differs from that for a slit (Eq. 24–3) by the factor 1.22. This factor comes from the fact that the width of a circular hole is not uniform (like a rectangular slit) but varies from its diameter D to zero. A careful analysis shows that the "average" width is $D/1.22$. Hence we get the equation above rather than Eq. 24–3. The intensity of light in the diffraction pattern of light from a point source passing through a circular opening is shown in Fig. 25–22. (The image for a non-point source would be a superposition of such patterns, thus forming a very complex diffraction pattern.) For most purposes, we need consider only the central spot since the concentric rings are so much dimmer.

FIGURE 25–22 Intensity of light across the diffraction pattern of a circular hole.

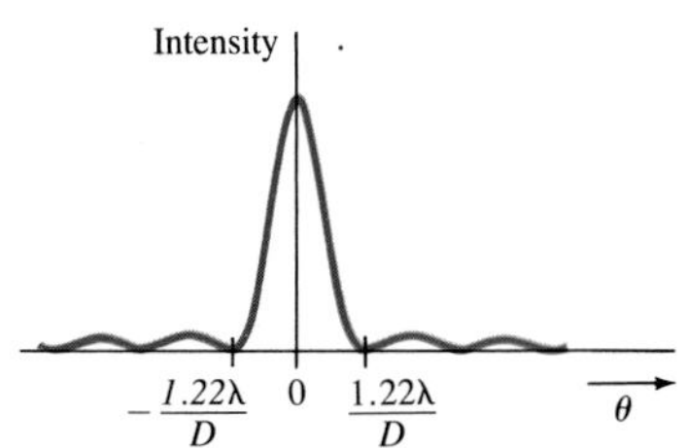

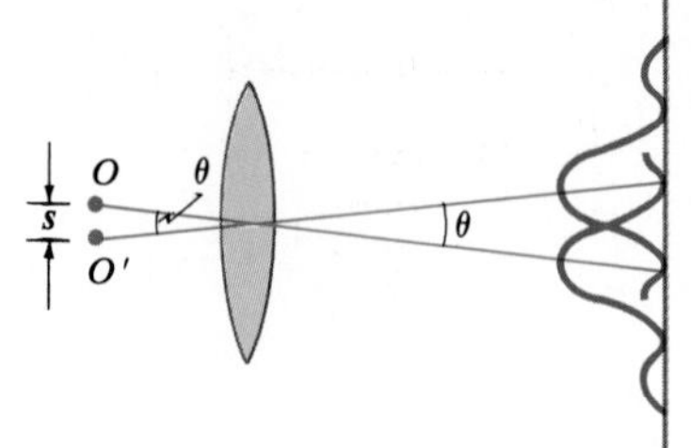

FIGURE 25–23 The *Rayleigh criterion.* Two images are just resolvable when the center of the diffraction peak of one is directly over the first minimum in the diffraction pattern of the other. The two point objects O and O' subtend an angle θ at the lens; one ray only is drawn for each point to indicate the center of the diffraction pattern of its image.

If two point objects are very close, the diffraction patterns of their images will overlap as shown in Fig. 25–21b. As the objects are moved closer, a point is reached where you can't tell if there are two overlapping images or a single image. Where this happens may be judged differently by different observers. However, a generally accepted criterion is one proposed by Lord Rayleigh (1842–1919). This **Rayleigh criterion** states that *two images are just resolvable when the center of the diffraction disk of one is directly over the first minimum in the diffraction pattern of the other.* This is shown in Fig. 25–23. Since the first minimum is at an angle $\theta = 1.22\lambda/D$ from the central maximum, Fig. 25–23 shows us that two objects can be considered just resolvable if they are separated by this angle θ:

$$\theta = \frac{1.22\lambda}{D}. \qquad (25\text{–}4)$$

This is the limit on resolution set by the wave nature of light due to diffraction.

25–8 • Resolution of Telescopes and Microscopes

It might be thought that a microscope or telescope could be designed to produce any desired magnification, depending on the choice of focal lengths. But this is not possible, because of diffraction. An increase in magnification above a certain point merely results in magnification of the diffraction pattern. This would be highly misleading since we might think we are seeing details of an object when we are really seeing details of the diffraction pattern. To examine this, we apply the Rayleigh criterion: two objects (or two nearby points on one object) are just resolvable if they are separated by an angle θ (Fig. 25–23) given by Eq. 25–4:

$$\theta = \frac{1.22\lambda}{D}.$$

This is valid for either a microscope or a telescope, where D is the diameter of the objective lens. For a telescope, the resolution is specified by stating θ as given by this equation.†

For a microscope, it is more convenient to specify the actual distance, s, between two points that are just barely resolvable, Fig. 25–23. Since objects are normally placed near the focal point of the objective, $\theta = s/f$, or $s = f\theta$. If we combine this with Eq. 25–4, we obtain for the **resolving power** (RP),

$$\text{RP} = s = f\theta = \frac{1.22\lambda f}{D}. \qquad (25\text{–}5)$$

This distance s is called the resolving power of the lens because it is the minimum separation of two object points that can just be resolved. The above equation is often written in terms of the angle of acceptance, α, of the ob-

† Telescopes with large-diameter objectives are usually limited not by diffraction but by other effects such as turbulence in the atmosphere. The resolution of a high-quality microscope, on the other hand, is normally limited by diffraction since microscope objectives are complex compound lenses containing many elements of small diameter (since f is small).

jective lens as defined in Fig. 25–24. The derivation is long and we only quote the result:

$$\text{RP} = s = \frac{1.22\lambda}{2\sin\alpha} = \frac{0.61\lambda}{\sin\alpha}.$$

The resolving power can be increased by placing a drop of oil that encloses the object and the front surface of the objective. This is called an **oil-immersion objective**. In the oil, the wavelength of the light is reduced to λ/n (Eq. 24–1), where n is the oil's index of refraction. Thus the resolving power becomes

$$\text{RP} = \frac{0.61\lambda}{n\sin\alpha}. \qquad (25\text{–}6)$$

The oil typically has $n \approx 1.5$, although n may be as great as 1.8. Thus oil immersion increases the resolution by 50 percent or more.

The quantity $(n \sin\alpha)$ is called the **numerical aperture** (NA) of the lens:

$$\text{NA} = n\sin\alpha. \qquad (25\text{–}7)$$

It is usually specified on the objective lens housing along with the magnification. The larger the value of the NA, the finer the resolving power.

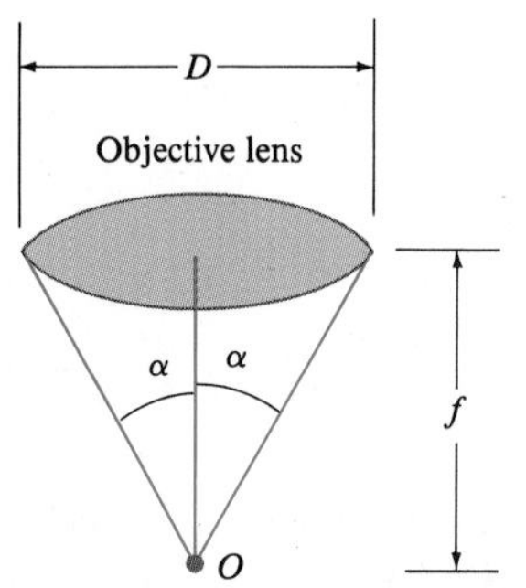

FIGURE 25–24 Objective lens of a microscope, showing the angle of acceptance, α.

EXAMPLE 25–8 What is the theoretical minimum angular separation of two stars that can just be resolved by: (*a*) the 200-inch telescope on Mt. Palomar (Fig. 25–14c); and (*b*) the Arecibo radiotelescope (Fig. 25–25), whose diameter is 300 m and whose radius of curvature is also 300 m. Assume $\lambda = 550$ nm for the visible-light telescope in part (*a*), and $\lambda = 4$ cm (the shortest wavelength at which the radiotelescope has been operated) in part (*b*).

FIGURE 25–25 The 300-meter radiotelescope in Arecibo, Puerto Rico.

SOLUTION (*a*) Since $D = 200$ in $= 5.1$ m, we have from Eq. 25–4 that $\theta = 1.22\lambda/D = (1.22)(5.50 \times 10^{-7}\text{ m})/(5.1\text{ m}) = 1.3 \times 10^{-7}$ rad, or 0.75×10^{-5} deg. This is the limit set by diffraction. In actual fact, the resolution is not this good because of aberrations and, more importantly, turbulence in the atmosphere. In fact, large-diameter objectives are not justified by increased resolution, but by their greater light-gathering ability—they allow more light in, so fainter objects can be seen.

(*b*) Radiotelescopes are not hindered by atmospheric turbulence, and for radio waves with $\lambda = 0.04$ m the resolution is $\theta = (1.22)(0.04\text{ m})/(300\text{ m}) = 1.6 \times 10^{-4}$ rad.

EXAMPLE 25–9 Determine the NA and RP of the best oil-immersion microscopes, where the index of refraction of the oil is $n = 1.8$ and $\sin\alpha \approx 0.90$. Assume that $\lambda = 550$ nm.

SOLUTION The $\text{NA} = n\sin\alpha = 1.6$. The resolving power is $\text{RP} = 0.61\lambda/\text{NA} = (0.61)(5.50 \times 10^{-7}\text{ m})/(1.6) \approx 2 \times 10^{-7}\text{ m} = 200$ nm. This is the best resolution that a visible-light microscope can attain.

Diffraction sets an ultimate limit on the detail that can be seen on any object. In Eq. 25–5 we note that the focal length of a lens cannot be made less than (approximately) the radius of the lens, and even that is very difficult—see the lens-maker's equation (Eq. 23–10). In this best case,† Eq. 25–5 gives, with $f \approx D/2$,

$$\text{RP} \approx \frac{\lambda}{2}. \qquad (25\text{–}8)$$

Thus we can say, to within a factor of 2 or so, that

Resolution limited to λ

it is not possible to resolve detail of objects smaller than the wavelength of the radiation being used.

This is an important and useful rule of thumb.

Compound lenses are now designed so well that the actual limit on resolution is often set by diffraction—that is, by the wavelength of the light used. To obtain greater detail, one must use radiation of shorter wavelength. The use of UV radiation can increase the resolution by a factor of perhaps 2. Far more important, however, was the discovery in the early twentieth century that electrons have wave properties (Chapter 27) and that their wavelengths can be very small. This aspect of electrons is used in the electron microscope (Section 27–6), which can magnify 100 to 1000 times more than a visible-light microscope because of the much shorter wavelengths. X rays, too, have very short wavelengths and are often used to study objects in great detail using special techniques (Section 25–11).

† The same result can be obtained from Eq. 25–6 since $\sin\alpha$ can never exceed 1 and typically is 0.6 to 0.9 at most. With oil immersion, Eq. 25–6 gives, at best, $\text{RP} \approx \lambda/3$, which corresponds to the result of Example 25–9.

25–9 • Resolution of the Human Eye and Useful Magnification

The resolution of the human eye is limited by several factors, all of roughly the same order of magnitude. The resolution is best at the fovea, where the cone spacing is smallest, about 3 μm (= 3000 nm). The diameter of the pupil varies from about 0.1 cm to about 0.8 cm. So for $\lambda = 550$ nm (where the eye's sensitivity is greatest), the diffraction limit is about $\theta \approx 1.22\lambda/D \approx 8 \times 10^{-5}$ rad to 6×10^{-4} rad. Since the eye is about 2 cm long, this corresponds to a resolving power of $s \approx (8 \times 10^{-5}\text{ rad})(2 \times 10^{-2}\text{ m}) \approx 2\ \mu\text{m}$ at best, to about 15 μm at worst (pupil small). Spherical and chromatic aberration also limit the resolution to about 10 μm. The net result is that the eye can resolve objects whose angular separation is about 5×10^{-4} rad at best. This corresponds to objects separated by 1 cm at a distance of about 20 m.

The typical near point of a human eye is about 25 cm. At this distance, the eye can just resolve objects that are $(25\text{ cm})(5 \times 10^{-4}\text{ rad}) \approx 10^{-4}$ m apart. Since the best light microscopes can resolve objects no smaller than about 200 nm (see Example 25–9), the useful magnification [=(resolution by naked eye)/(resolution by microscope)] is limited to about

$$\frac{10^{-4}\text{ m}}{200 \times 10^{-9}\text{ m}} = 500\times.$$

In practice, magnifications of about 1000× are often used to minimize eyestrain. Any greater magnification would simply make visible the diffraction pattern produced by the microscope objective.

*25–10 • Specialty Microscopes and Contrast

All the resolving power a microscope can attain will be useless if the object to be seen cannot be distinguished from the background. The difference in brightness between the image of an object and the image of the surroundings is called **contrast**. Achieving high contrast is an important problem in microscopy and other forms of imaging. The problem arises in biology, for example, because cells consist largely of water and are almost uniformly transparent to light. We now discuss two special types of microscope that can increase contrast: the interference and phase-contrast microscopes.

Interference microscope. The interference microscope makes use of the wave properties of light in a direct way. It is one of the most effective means to increase contrast in a transparent object. To see how it works, let us consider a transparent object—perhaps a bacterium—in a water solution (Fig. 25–26). Light enters uniformly from the left and is coherent (meaning in phase) at all points such as *a* and *b*. If the object is as transparent as the water solution, the beam leaving at *d* will be as bright as that at *c*. There will be no contrast and the object will not be seen. However, if the object's refractive index is slightly different from that of the surrounding medium, the wavelength within the object will be altered as shown. Hence the waves at points *c* and *d* will differ in phase, if not in amplitude. This appears at first to be of no help, since the eye responds only to differences in amplitude or

FIGURE 25–26 Object—perhaps a bacterium—in a water solution.

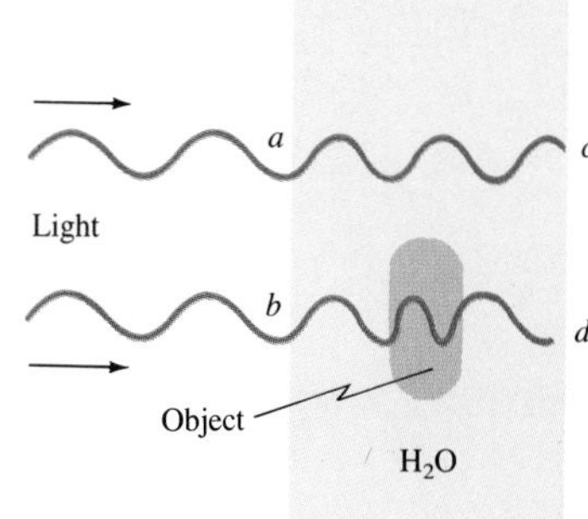

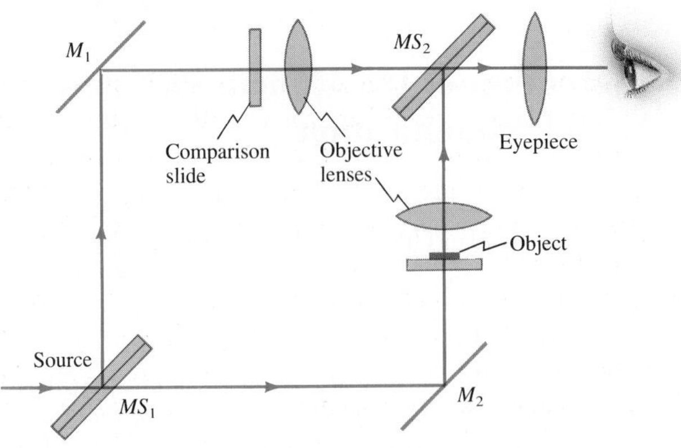

FIGURE 25–27 Diagram of an interference microscope.

brightness, and does not detect this difference in phase. What the interference microscope does is to change this difference in phase into a difference of amplitude. It does so by superimposing the light that passes through the sample onto a reference beam that does not pass through the object, so that they interfere. One way of doing this is shown in Fig. 25–27. Light from a source is split into two equal beams by a half-silvered mirror, MS_1. One beam passes through the object and the second (comparison beam) passes through an identical system without the object. The two meet again and are superposed by the half-silvered mirror MS_2 before entering the eyepiece and the eye. The path length (and amplitude) of the comparison beam is adjustable. It can be adjusted, for example, so that the background is dark; that is, full destructive interference occurs. Light passing through the object (beam *bd* in Fig. 25–26) will also interfere with the comparison beam. But because of its different phase, the interference will not be completely destructive. Thus it will appear brighter than the background. Where the object varies in thickness, the phase difference between beams *ac* and *bd* in Fig. 25–26 will be different; and this will affect the amount of interference. Hence variation in the thickness of the object will appear as variations in brightness in the image. As an example, suppose that the object is a bacterium 1.0 μm thick and has a refractive index of 1.35. Then if yellow light ($\lambda = 550$ nm in air) is used, there will be $(1.0\ \mu\text{m})/(550\ \text{nm}/1.35) = 2.46$ wavelengths in the bacterium and $(1.0\ \mu\text{m})/(550\ \text{nm}/1.33) = 2.42$ wavelengths in the water. Thus the two waves will be out of phase by 0.04 wavelengths, or 14° ($= 0.04 \times 360°$).

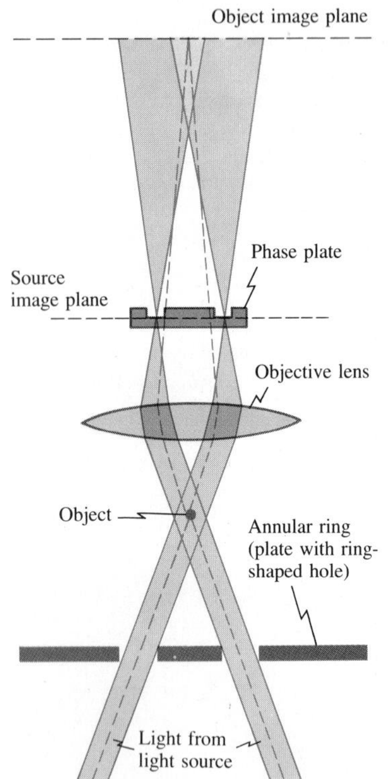

FIGURE 25–28 Phase-contrast microscope. Light beam from the source that is undeviated is shown shaded (pinkish) for clarity. Rays deviated by the object, and which form the image of the object, are shown dashed.

Phase-contrast microscope. The phase-contrast microscope also makes use of interference and differences in phase to produce a high-contrast image. Although it has certain limitations, it is far simpler to construct and operate than an interference microscope. To describe the operation of a phase-contrast microscope in detail, we would have to discuss the diffraction theory of image formation—how the diffraction pattern produced by each point on the object contributes to the final image. This is quite complicated, so we give only a simplified description. Figure 25–28 is a simplified diagram of a phase-contrast microscope. The object to be viewed is illuminated from below as usual. To be specific, we assume that rays from each point on the source are made parallel by a set of condensing lenses (not shown). However, a plate with a ring-shaped hole is placed above the source, so the light can only pass through this annular ring. Light that is not deviated by the object (the beam is shown shaded) is brought into focus by the objective lens in the *source image plane.* If the source is effectively at infinity (because of the condensing lenses), as is assumed here, the source image plane is at the focal point of the lens. Light that strikes the object, on the other hand, is diffracted or scattered by the object. Each point on the object then serves as a source for rays diverging from that point (dashed lines in the figure). These rays are brought to a focus in the *object image plane*, which is behind the source image plane (because the object is so close to the lens). The undeviated light from the source diverges, meanwhile, from its image plane and provides a broad bright background at the object image plane. The object is transparent, however, and the image will not be seen clearly since there will be little contrast. Contrast is achieved by inserting a circular glass *phase plate* at the source image plane. The phase plate has a groove as shown (or a raised portion) in the shape of a ring. This ring is positioned so that all the undeviated rays pass through it. Most of the rays deviated by the object, on the other hand, do not pass through this ring (see Fig. 25–28). Because the rays deviated by the object travel through a different thickness of glass than the undeviated source rays, the two can be out of phase and can interfere destructively at the object image plane. Thus the image of the object will contrast sharply with the background. Actually, because only a small fraction of the light is deviated by the object, the background light will be much stronger and so the contrast will not be great. To compensate for this, the grooved ring on the phase plate is darkened to absorb a good part of the undeviated light so that its intensity is more nearly equal to that of the deviated light. Then nearly complete destructive interference can occur at particular points and the contrast will be very high. The chief limitation of the phase-contrast microscope is that images tend to have "halos" around them as a result of diffraction from the phase-plate opening. Because of this artifact, care must be taken in the interpretation of images.

25–11 • X Rays and X-Ray Diffraction

In 1895, W. C. Roentgen (1845–1923) discovered that when electrons were accelerated by a high voltage in a vacuum tube and allowed to strike a glass (or metal) surface inside the tube, fluorescent minerals some distance away would glow, and film would become exposed. Roentgen attributed these effects to a new type of radiation (different from cathode rays). They were given the name X rays after the algebraic symbol x, meaning an unknown quantity. He soon found that X rays penetrated through some materials better than through others, and within a few weeks he presented the first X-ray photograph (of his wife's hand). The production of X rays today is done in a tube (Fig. 25–29) similar to Roentgen's, using voltages of typically 30 to 150 kV.

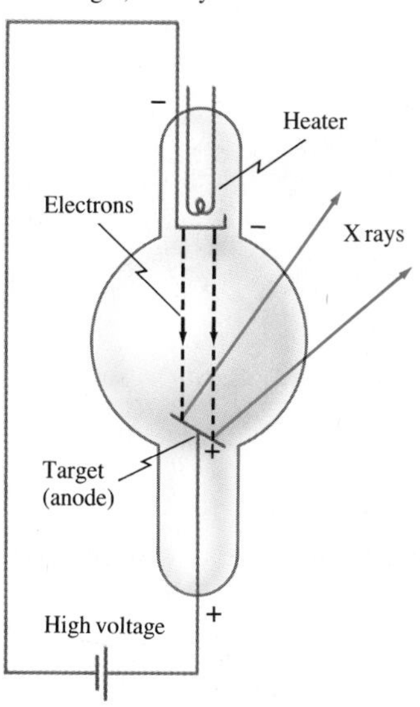

FIGURE 25–29 X-ray tube. Electrons emitted by a heated filament in a vacuum tube are accelerated by high voltage. When they strike the surface of the anode, the "target," X rays are emitted.

Investigations into the nature of X rays indicated they were not charged particles (such as electrons) since they could not be deflected by electric or magnetic fields. It was suggested that they might be a form of invisible light. However, they showed no diffraction or interference effects using ordinary gratings. Of course, if their wavelengths were much smaller than the typical grating spacing of 10^{-6} m ($= 10^3$ nm), no effects would be expected. Around 1912, it was suggested by Max von Laue (1879–1960) that if the atoms in a crystal were arranged in a regular array (see Fig. 13–1a), a theory generally

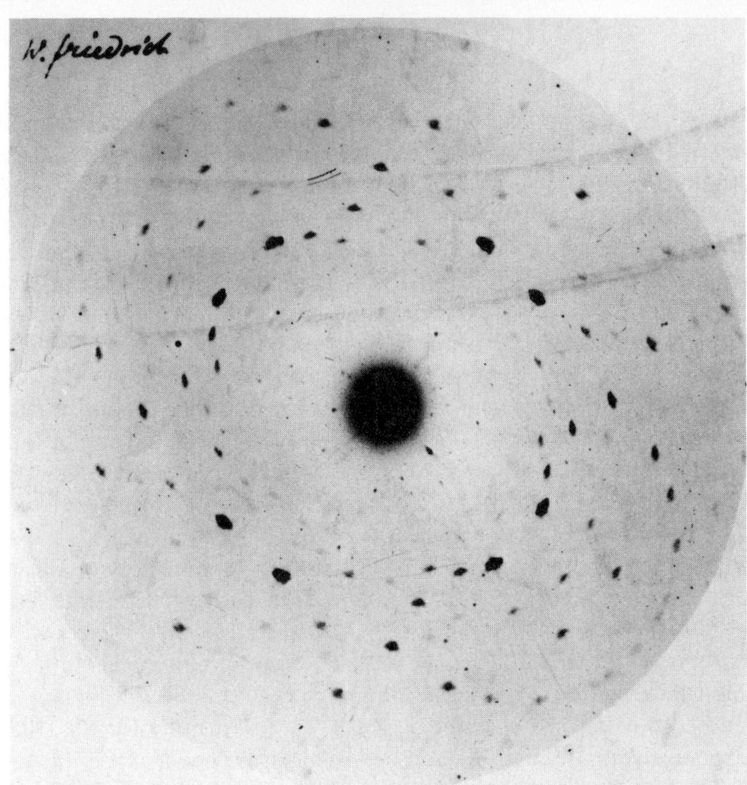

FIGURE 25–30 This X-ray diffraction pattern is one of the first observed by Max von Laue in 1912 when he aimed a beam of X rays at a zinc sulfide crystal. The diffraction pattern was detected directly on a photographic plate.

held by scientists though not then fully tested, such a crystal might serve as a diffraction grating for very short wavelengths on the order of the spacing between atoms, estimated to be about 10^{-10} m ($=10^{-1}$ nm). Experiments soon showed that X rays scattered from a crystal did indeed show the peaks and valleys of a diffraction pattern (Fig. 25–30). Thus it was shown, in a single blow, that X rays have a wave nature and that atoms are arranged in a regular way in crystals. Today, X rays are recognized as electromagnetic radiation with wavelengths in the range of about 10^{-2} nm to 10 nm, the range readily produced in an X-ray tube.

FIGURE 25–31
X-ray diffraction by a crystal.

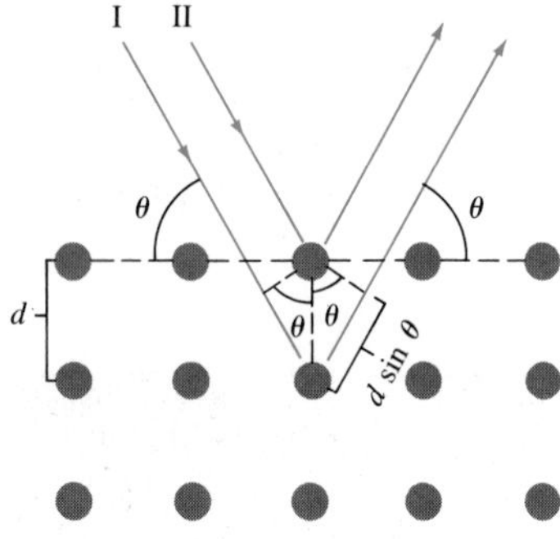

We saw in Sections 25–7 and 25–8 that light of shorter wavelength provides greater resolution when we are examining an object microscopically. Since X rays have much shorter wavelengths than visible light, they should, in principle, offer much greater resolution. However, there seems to be no effective material to use as lenses for the very short wavelengths of X rays. Instead, the clever but complicated technique of **X-ray diffraction** (or **crystallography**) has proved very effective for examining the microscopic world of atoms and molecules. In a simple crystal such as NaCl, the atoms are arranged in an orderly cubical fashion, Fig. 25–31, with atoms spaced a distance d apart. Suppose that a beam of X rays is incident on the crystal at an angle θ to the surface, and that the two rays shown are reflected from two subsequent planes of atoms as shown. The two rays will constructively interfere if the extra distance ray I travels is a whole number of wavelengths farther than what ray II travels. This extra distance is $2d \sin \theta$. Therefore, constructive interference will occur when

Bragg equation

$$m\lambda = 2d \sin \theta, \qquad m = 1, 2, 3, \cdots, \tag{25–9}$$

where m can be any integer. (Notice that θ is *not* the angle with respect to the normal to the surface.) This is called the **Bragg equation** after W. L. Bragg

(1890–1971), who derived it and who, together with his father W. H. Bragg (1862–1942), developed the theory and technique of X-ray diffraction by crystals in 1912–13. Thus, if the X-ray wavelength is known and the angle θ at which constructive interference occurs is measured, d can be obtained. This is the basis for X-ray crystallography.

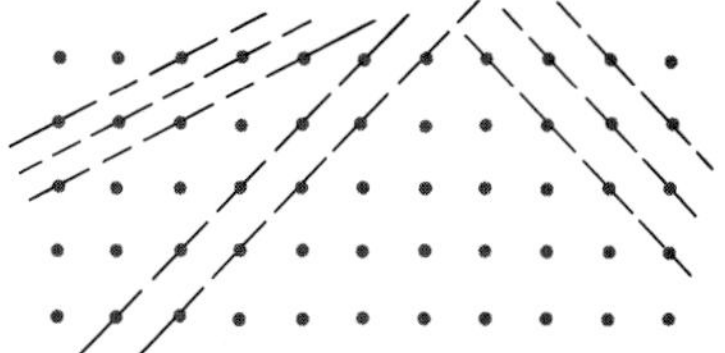

FIGURE 25–32 There are many possible planes existing within a crystal from which X rays can be diffracted.

Actual X-ray diffraction patterns are quite complicated. First of all, a crystal is a three-dimensional object, and X rays can be diffracted from different planes at different angles within the crystal, as shown in Fig. 25–32. Although the analysis is complex, a great deal can be learned about any substance that can be put in crystalline form. If the substance is not a single crystal but a mixture of many tiny crystals—as in a metal or a powder—then instead of a series of spots, as in Fig. 25–30, a series of circles is obtained, Fig. 25–33; each circle corresponds to diffraction of a certain order m (Eq. 25–9), from a particular set of parallel planes.

X-ray diffraction has been very useful in determining the structure of biologically important molecules. Often it is possible to make a crystal of such molecules. The analysis is complex, and it is usually necessary to make various guesses of the structure of the molecule. Predictions of the diffraction patterns for each guessed structure can then be compared to that actually obtained. For larger molecules, such as proteins and nucleic acids, an important innovation has been the "heavy-atom technique." Since very large atoms scatter X rays much more strongly than the ordinary C, N, O, and H atoms of biological molecules, heavy atoms can be used as "markers." The heavy atoms are chemically added to particular spots on the molecule (say, a protein)—hopefully without disturbing its structure significantly. Analysis of the *changes* in the resulting diffraction pattern gives helpful information.

Even when a good crystal cannot be obtained, if the molecule under study has a regularly repeating shape (such as many proteins and DNA have), X-ray diffraction can reveal it. In a sense, each molecule is then like a single crystal and a sample is a collection of such tiny crystals. Indeed, it was with the help of X-ray diffraction that, in 1953, J. D. Watson and F. H. C. Crick worked out the double-helix structure of DNA.

Around 1960, the first detailed structure of a protein molecule was elucidated with the aid of X-ray diffraction; this was for myoglobin, a relative of the important constituent of blood, hemoglobin. Soon the structure of hemoglobin itself was worked out and since then the structures of a great many molecules have been determined with the help of X rays.

FIGURE 25–33 (a) Diffraction of X rays from a polycrystalline substance produces a set of circular rings as in (b), which is for polycrystalline sodium acetoacetate.

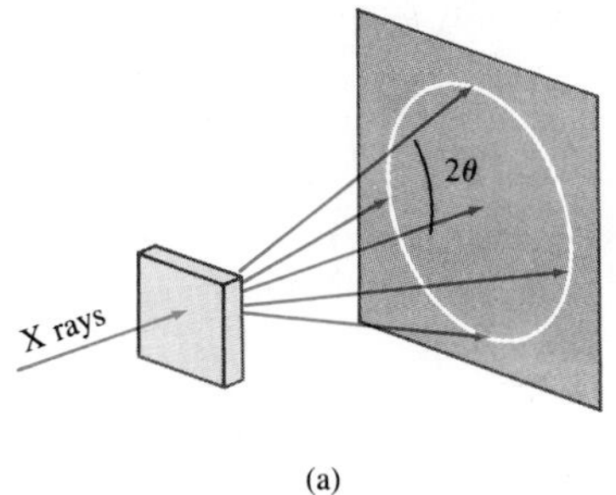

(a)

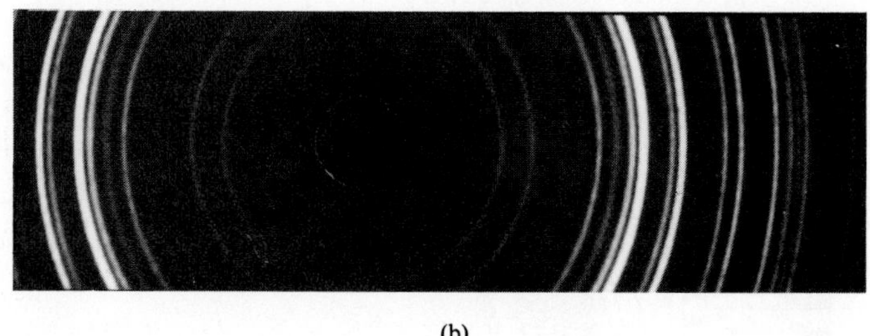

(b)

*25–12 • X-Ray Imaging and Computerized Tomography (CAT Scanning)

For a conventional medical (or dental) X-ray photograph, the X rays emerging from the tube (Fig. 25–29, Section 25–11) pass through the body and are detected on photographic film or a fluorescent screen, Fig. 25–34a. The rays travel in very nearly straight lines through the body with minimal deviation since at X-ray wavelengths there is little diffraction or refraction. There is absorption (and scattering), however; and the difference in absorption by different structures in the body is what gives rise to the image produced by the transmitted rays. The less the absorption, the greater the transmission and the darker the film. The image is, in a sense, a "shadow" of what the rays have passed through. (The X-ray image is *not* produced by focusing rays with lenses as is the case for the instruments discussed earlier in this chapter.)

Within months of Roentgen's 1895 discovery of X rays, they had already become a powerful tool for medical diagnosis, and have remained so to this day. Although many technical advances have been made over the years, the basic principles have not changed significantly. At least not until the 1970s when a revolutionary new technique called **computerized tomography** (CT) was developed.

In conventional X-ray images, the entire thickness of the body is projected onto the film; structures overlap and in many cases are difficult to distinguish. A tomographic image, on the other hand, is an image of a *slice* through the body. (The word *tomography* comes from the Greek: *tomos* = slice, *graph* = picture.) Structures and lesions previously impossible to visualize can now be seen with remarkable clarity. The principle behind CT is shown in Fig. 25–34b: a thin collimated beam of X rays (to "collimate" means to "make straight") passes through the body to a detector that measures the transmitted intensity. Measurements are made at a large number of points as the source and detector are moved past the body together. The apparatus is then rotated slightly about the body axis and again scanned; this is repeated at (perhaps) 1° intervals for 180°. The intensity of the transmitted beam for the many points of each scan, and for each angle, are sent to a computer that reconstructs the image of the slice (more on this in a moment). Note that the imaged slice is perpendicular to the long axis of the

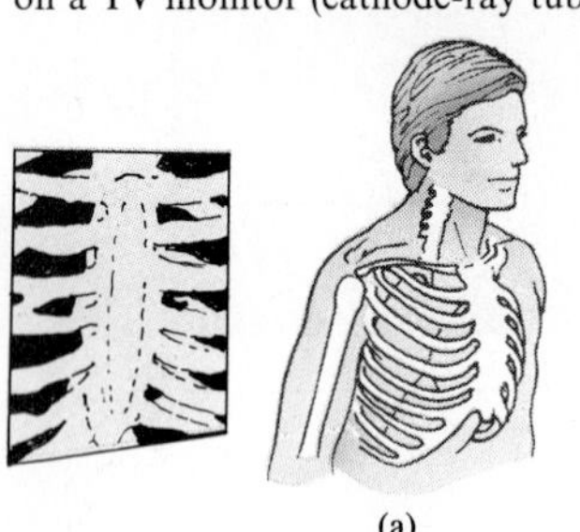

(a)

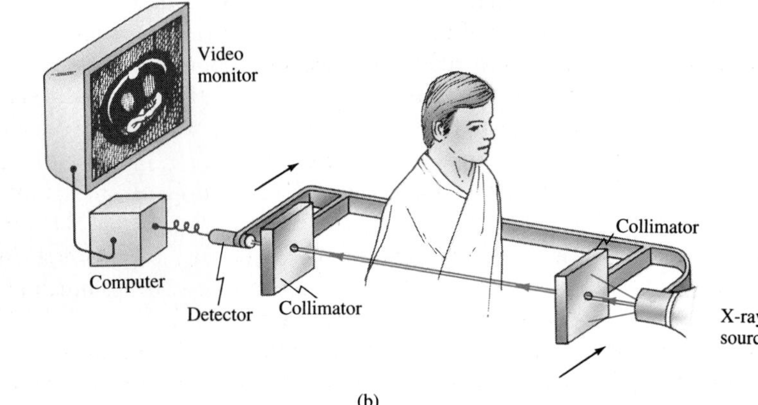

(b)

FIGURE 25–34 (a) Conventional X-ray imaging, which is essentially shadowing. (b) Tomographic imaging: the X-ray source and detector move together horizontally, the transmitted intensity being measured at a large number of points; then the source-detector assembly is rotated slightly (say, 1°) and another scan is made. This is repeated for perhaps 180°. The computer reconstructs the image of the slice and it is presented on a TV monitor (cathode-ray tube).

body. For this reason, CT is sometimes called **computerized axial tomography** (CAT), although the abbreviation CAT, as in CAT scan, can also be read as **computer-assisted tomography**.

The use of a single detector as in Fig. 25–34b requires a few minutes for the many scans needed to form a complete image. Much faster scanners use a fan beam, Fig. 25–35, in which beams through the entire body are detected simultaneously by many detectors. The source and detectors are then rotated about the patient on an apparatus called a gantry. At each of the hundreds of angular positions of the apparatus, several hundred detectors can measure the intensity of transmitted rays simultaneously, so an image requires only a few seconds.

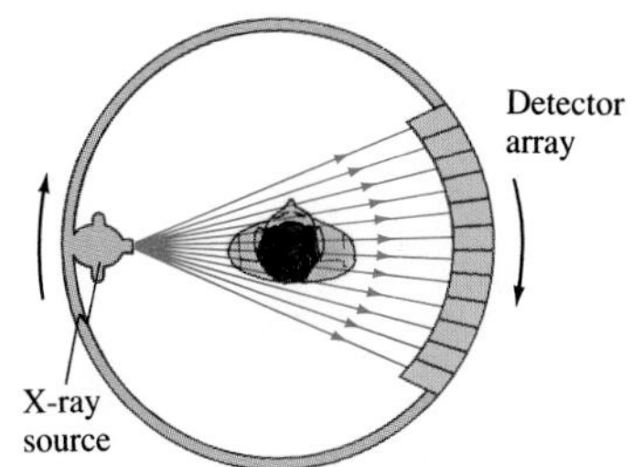

FIGURE 25–35 Fan-beam scanner. Rays through the entire body are measured simultaneously at each angle. The source and detector rotate to take measurements at different angles. In another type of fan-beam scanner, there are detectors around the entire 360° of circle and they remain fixed as the source moves.

But how is the image formed? We can think of the slice to be imaged as being divided into many tiny picture elements (or **pixels**), which could be squares, as was done to make the image shown in Fig. 25–36. For CT, the width of each pixel is chosen according to the width of the detectors and/or the width of the X-ray beams. The pixel size determines the resolution of the image, which is typically about 2 mm. An X-ray detector measures the intensity of the transmitted beam after it has passed through the body. Subtracting this value from the intensity of the beam at the source, we get the total absorption. Note that only the total absorption (called "a projection") along each beam line can be measured (the sum of the absorptions that take place for each of the pixels in a line). To form an image, we need to determine how much radiation is absorbed at *each* pixel. (How that can be done will be discussed in a moment.) We can then assign a "grayness value" to each pixel according to how much radiation was absorbed. The image, then, is made up of tiny spots (pixels) of varying shades of gray, as is a black-and-white television picture. Often the amount of absorption is color coded. The colors in the resulting ("false-color") image have nothing to do, however, with the actual color of the object.

FIGURE 25–36 Example of an image made up of many small squares. This one has rather poor resolution.

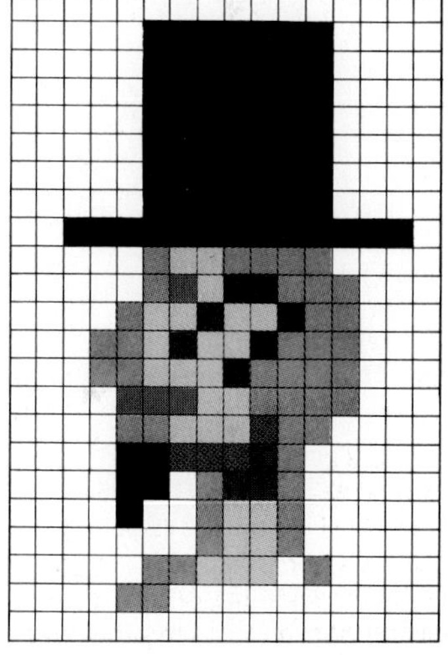

Finally, we must discuss how the "grayness" of each pixel can be determined even though all we can measure is the total absorption along each beam line in the slice. It can be done only by using the many beam scans made at a great many different angles. Suppose the image is to be an array of 100 × 100 elements for a total of 10^4 pixels. If we have 100 detectors and measure the absorption projections at 100 different angles, then we get 10^4 pieces of information. From this information, an image can be reconstructed, but not precisely. If more angles are measured, the reconstruction of the image can be done more accurately.

There are a number of mathematical reconstruction techniques, all of which are complicated and require the use of a computer. To suggest how it is done, we consider a very simple case using the so-called "iterative" technique ("to iterate" is from the Latin "to repeat"). Although this technique is less used now than the more direct "Fourier transform" and "back projection" techniques, it is the simplest to explain. Suppose our sample slice is divided into the simple 2 × 2 pixels as shown in Fig. 25–37. The number in each pixel represents the amount of absorption by the material in that area (say, in tenths of a percent): that is, 4 represents twice as much absorption as 2. But we cannot directly measure these values. All we can measure are the projections—the total absorption along each beam line—and these are shown in the diagram as the sum of the absorptions for the pixels along each line at four different angles. These projections (given at the tip of each

FIGURE 25–37 (below) A simple 2 × 2 image showing true absorption values and measured projections.

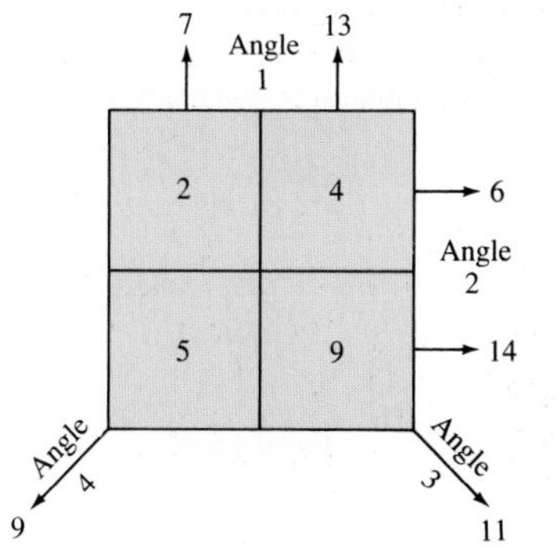

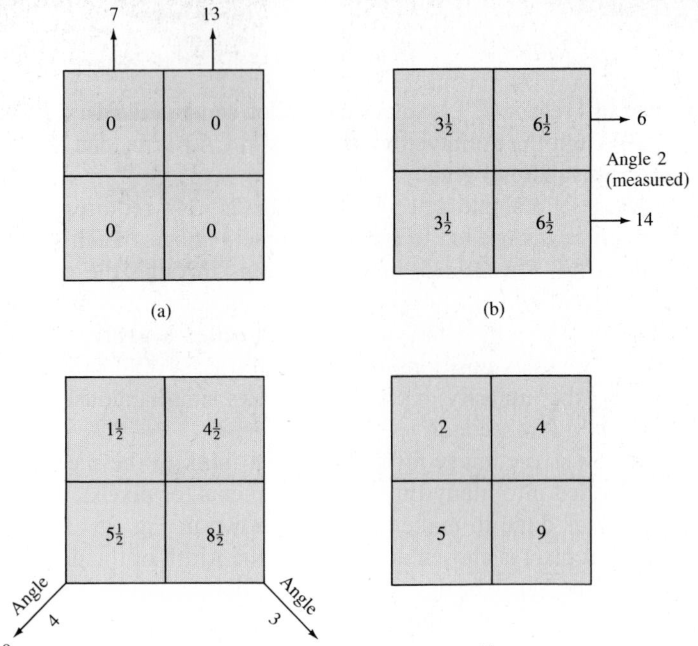

FIGURE 25–38 Reconstructing the image using projections in an iterative procedure.

arrow) are what we assume have been measured, and we now want to work back from them to see how close we can get to the true absorption value for each pixel. We start with each pixel being assigned a zero value, Fig. 25–38a. In the iterative technique, we use the projections to estimate the absorption value in each square, and repeat for each angle. The angle 1 projections are 7 and 13. We divide each of these equally between their two squares: each square in the left column gets $3\frac{1}{2}$ (half of 7), and each square in the right column gets $6\frac{1}{2}$ (half of 13); see Fig. 25–38b. Next we use the projections at angle 2. We calculate the difference between the measured projections at angle 2 (6 and 14) and the projections based on the previous estimate (top row: $3\frac{1}{2} + 6\frac{1}{2} = 10$; same for bottom row). Then we distribute this difference equally to the squares in that row. For the top row, we have

$$3\tfrac{1}{2} + \frac{6 - 10}{2} = 1\tfrac{1}{2} \quad \text{and} \quad 6\tfrac{1}{2} + \frac{6 - 10}{2} = 4\tfrac{1}{2};$$

and for the bottom row,

$$3\tfrac{1}{2} + \frac{14 - 10}{2} = 5\tfrac{1}{2} \quad \text{and} \quad 6\tfrac{1}{2} + \frac{14 - 10}{2} = 8\tfrac{1}{2}.$$

These values are inserted as shown in Fig. 25–38c. Next, the projection at angle 3 gives

$$\text{(upper left) } 1\tfrac{1}{2} + \frac{11 - 10}{2} = 2 \quad \text{and} \quad \text{(lower right) } 8\tfrac{1}{2} + \frac{11 - 10}{2} = 9;$$

and that for angle 4 gives

$$\text{(lower left) } 5\tfrac{1}{2} + \frac{9 - 10}{2} = 5 \quad \text{and} \quad \text{(upper right) } 4\tfrac{1}{2} + \frac{9 - 10}{2} = 4.$$

The result, shown in Fig. 25–38d, corresponds exactly to the true values. (Note that in real situations, the true values are not known, which is why these

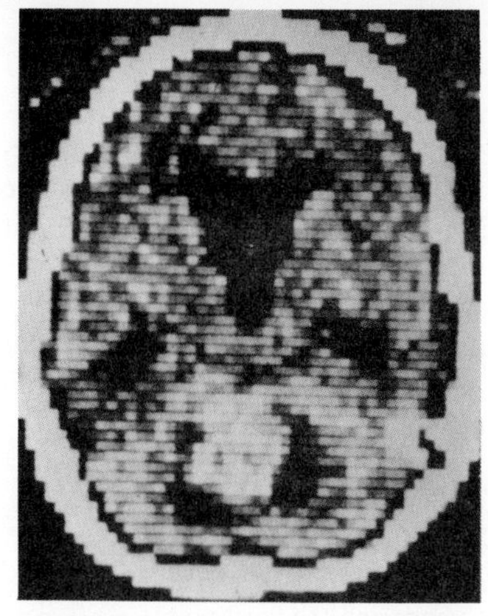

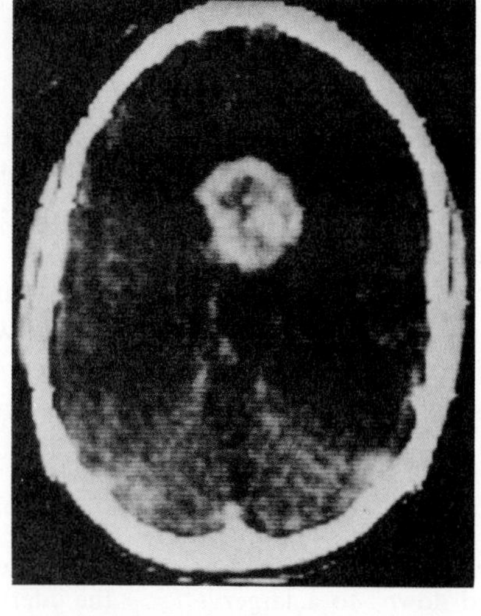

FIGURE 25–39 Two CT images, with different resolutions, each showing a cross section of a brain. The bright area is an unusual growth (a tumor).

computer techniques are required.) To obtain these numbers exactly, we used six pieces of information (two each at angles 1 and 2, one each at angles 3 and 4). For the much larger number of pixels used for actual images, exact values are generally not attained. Many iterations may be needed, and the calculation is considered sufficiently precise when the difference between calculated and measured projections is sufficiently small. The above example illustrates the "convergence" of the process: the first iteration (b to c in Fig. 25–38) changed the values by 2, the last iteration (c to d) by only $\frac{1}{2}$.

Figure 25–39 illustrates what actual CT images look like. It is generally agreed that CT scanning has revolutionized some areas of medicine by providing much less invasive, and/or more accurate, diagnosis.

Computerized tomography can also be applied to ultrasound imaging (Section 12–10) and to emissions from radioisotopes and nuclear magnetic resonance (Sections 31–8 and 31–9).

SUMMARY

A *camera* lens forms an image on film by allowing light in through a shutter. The lens is focused by moving it relative to the film, and its f-stop (or lens opening) must be adjusted for the brightness of the scene and the chosen shutter speed. The f-stop is defined as the ratio of the focal length to the diameter of the lens opening.

The human *eye* also adjusts for the available light—by opening and closing the iris. It focuses not by moving the lens, but by adjusting the shape of the lens to vary its focal length. The image is formed on the retina, which contains an array of receptors known as rods and cones. Diverging eyeglass or contact lenses are used to correct the defect of a nearsighted eye, which cannot focus well on distant objects. Converging lenses are used to correct for defects in which the eye cannot focus on close objects.

A *simple magnifier* is a converging lens that forms a virtual image of an object placed at (or within) the focal point. The *angular magnification*, when viewed by a relaxed normal eye, is $M = N/f$, where f is the focal length of the lens and N is the near point of the eye (25 cm for a "normal" eye).

An *astronomical telescope* consists of an *objective* lens or mirror and an *eyepiece* that magnifies the real image formed by the objective. The *magnification* is equal to the ratio of the objective and eyepiece focal lengths, and the image is inverted. A *terrestrial telescope* uses extra lenses, prisms, or a diverging lens as eyepiece, so that the final image is right side up.

A compound *microscope* also uses objective and eyepiece lenses, and the final image is inverted. The total magnification is the product of the magnifications of the two lenses and is approximately $(N/f_e)(l/f_o)$, where l is the distance between the lenses, N is the nearpoint of the eye, and f_o and f_e are the focal lengths of objective and eyepiece, respectively.

Microscopes, telescopes, and other optical instruments are limited in the formation of sharp images by *lens aberrations*. These include *spherical aberration*, in which rays passing through the edge of a lens are not focused at the same point as those that pass near the center; and *chromatic aberration*, in which different colors are focused at different points. Compound lenses, consisting of several elements, can largely correct for aberrations.

The wave nature of light also limits the sharpness, or *resolution*, of images. Because of diffraction, it is not possible to discern details smaller than the wavelength of the radiation being used. This limits the useful magnification of a light microscope to about 1000×.

X rays are a form of electromagnetic radiation of very short wavelength. They are produced when high-speed electrons, accelerated by high voltage in an evacuated tube, strike a glass or metal target.

QUESTIONS

1. Why is the depth of field greater, and the image sharper, when a camera lens is "stopped down" to a larger f number? Ignore diffraction.

2. Describe how diffraction affects the statement of Question 1. [*Hint:* see Eq. 24-3.]

3. Why must a camera lens be moved farther from the film to focus on a closer object?

4. Why are bifocals needed mainly by older persons and not generally by younger people?

5. Explain why swimmers with good eyes see distant objects as blurry when they are under water. Use a diagram and also show why goggles correct this problem.

6. Will a nearsighted person who wears corrective lenses be able to see clearly underwater when wearing glasses? Use a diagram to show why or why not.

7. The human eye is much like a camera—yet, when a camera shutter is left open and the camera moved, the image will be blurred; but when you move your head with your eyes open, you still see clearly. Explain.

8. Reading glasses use converging lenses. A simple magnifier is also a converging lens. Are reading glasses therefore magnifiers? Discuss the similarities and differences between converging lenses as used for these two different purposes.

9. Is the image formed on the retina of the human eye upright or inverted? Discuss the implications of this for our perception of objects.

10. Complete the ray diagram of Fig. 25–15a by showing the intermediate image and the focal points.

11. Inexpensive microscopes sold for children's use usually produce images that are colored at the edges. Why?

12. Spherical aberration in a thin lens is minimized if rays are bent equally by the two surfaces. If a planoconvex lens is used to form a real image of an object at infinity, which surface should face the object? Use ray diagrams to show why.

13. Which aberrations present in a simple lens are not present (or are greatly reduced) in the human eye?

14. Explain why chromatic aberration occurs for thin lenses but not for mirrors.

15. If monochromatic light were used in a microscope, would the color affect the resolution? Explain.

16. Why can't a light microscope be used to observe molecules in a living cell?

17. Does diffraction limit the resolution of images formed by (*a*) spherical mirrors, (*b*) plane mirrors?

18. Do diffraction effects occur for virtual as well as real images?

19. What are the advantages (give at least two) for the use of large reflecting mirrors in astronomical telescopes?

20. Which color of visible light would give the best resolution in a microscope?

21. Atoms have diameters of about 10^{-8} cm. Can visible light be used to "see" an atom? Why or why not?

PROBLEMS

SECTION 25–1

1. (I) A 200-mm-focal-length lens has f stops ranging from $f/3.5$ to $f/32$. What is the corresponding range of lens diaphragm diameters?

2. (I) A television camera lens has a 14-cm focal length and a lens diameter of 6.0 cm. What is its f number?

3. (I) A properly exposed photograph is taken at $f/16$ and $\frac{1}{125}$ s. What lens opening would be required if the shutter speed were $\frac{1}{1000}$ s?

4. (I) A light meter reports that a camera setting of $\frac{1}{100}$ s at $f/11$ will give a correct exposure. But the photographer wishes to use $f/16$ to increase the depth of field. What should the shutter speed be?

5. (II) A "pinhole" camera uses a tiny pinhole instead of a lens. Show, using ray diagrams, how reasonably sharp images can be formed using such a pinhole camera. In particular, consider two point objects 2.0 cm apart that are 1.0 m from a 1.0-mm-diameter pinhole. Show that on a piece of film 7.0 cm behind the pinhole, each object produces a tiny, easily resolvable spot.

6. (II) Suppose that a correct exposure is $\frac{1}{250}$ s at $f/11$. Under the same conditions, what exposure time would be needed for a pinhole camera if the pinhole diameter is 1.0 mm and the film is 5.0 cm from the hole?

7. (II) A 135-mm-focal-length lens can be adjusted so that

it is 135 to 140 mm from the film. For what range of object distances can it be adjusted?

8. (II) A nature photographer wishes to photograph a 4.5-m-high elephant from a distance of 50 m. What focal-length lens should be used if the image is to fill the 24-mm height of the film?

SECTION 25–2

9. (I) A human eyeball is about 2.0 cm long and the pupil has a maximum diameter of about 5.0 mm. What is the "speed" of this lens?

10. (II) A person's left eye is corrected by a -4.5-diopter lens, 2.0 cm from the eye. (*a*) Is this person near- or farsighted? (*b*) What is this person's far point without glasses?

11. (II) Reading glasses of what power are needed for a person, whose near point is 130 cm, so that he can read at 25 cm? Assume a lens–eye distance of 2.0 cm.

12. (II) A person's left eye can see objects clearly only if they are between 20 and 42 cm away. (*a*) What power of contact lens is required so that objects far away are sharp? (*b*) What, then, will be the near point?

13. (II) About how much longer is the nearsighted eye in Example 25–4 than the 2.0 cm of a normal eye?

14. (II) One lens of a nearsighted person's eyeglasses has a focal length of -25.0 cm and the lens is 1.8 cm from the eye. If the person switches to contact lenses that are placed directly on the eye, what should be the focal length of the corresponding contact lens?

15. (II) What is the focal length of the eye–lens system when viewing an object (*a*) at infinity, and (*b*) 30 cm from the eye? Assume that the lens–retina distance is 2.0 cm.

SECTION 25–3

16. (I) What is the focal length of a magnifying glass of 3.0 × magnification for a relaxed normal eye?

17. (I) A magnifier is rated at 2.5 × for a normal eye focusing on an image at its near point. (*a*) What is its focal length? (*b*) What is its focal length if the 2.5 × refers to a relaxed eye?

18. (II) A 7.0-cm-focal-length lens is used as a simple magnifier. To obtain maximum magnification, where must the object be placed for a normal eye, and what will be the magnification?

19. (II) A 2.30-mm-wide beetle is viewed with a 7.00-cm-focal-length lens. A normal eye views the image at its near point. Calculate (*a*) the angular magnification, (*b*) the width of the image, and (*c*) the object distance from the lens.

20. (II) A small insect is placed 5.20 cm from a $+5.50$-cm-focal-length lens. Calculate (*a*) the position of the image, and (*b*) the angular magnification.

21. (II) A magnifying glass is rated at 3.0 × for a normal eye that is relaxed. What would be the magnification for a relaxed eye whose near point is (*a*) 40 cm, (*b*) 15 cm?

SECTION 25–4

22. (I) What is the magnification of an astronomical telescope whose objective lens has a focal length of 60 cm and whose eyepiece has a focal length of 3.1 cm? What is the overall length of the telescope when adjusted for a relaxed eye?

23. (I) An 8.5 × binocular has 3.0-cm-focal-length eyepieces. What is the focal length of the objective lenses?

24. (II) An astronomical telescope has an objective with focal length 95 cm and a $+40$-D eyepiece. What is the total magnification?

25. (II) What is the magnifying power of an astronomical telescope using a reflecting mirror whose radius of curvature is 5.0 m and an eyepiece whose focal length is 2.8 cm?

26. (II) A Galilean telescope adjusted for a relaxed eye is 33 cm long. If the objective lens has a focal length of 36 cm, what is the magnification?

27. (III) A 60 × astronomical telescope is adjusted for a relaxed eye when the two lenses are 76 cm apart. What is the focal length of each lens?

28. (III) A 7.0 × pair of binoculars has an objective focal length of 26 cm. If the binoculars are focused on an object 4.0 m away (from the objective), what is the magnification? (The 7.0 × refers to objects at infinity; Eq. 25–2 holds only for objects at infinity and not for nearby ones.)

SECTION 25–5

29. (I) An 800 × microscope uses a 0.40-cm-focal-length objective lens. If the tube length is 16.8 cm, what is the focal length of the eyepiece? Assume a normal eye and that the final image is at infinity.

30. (II) A microscope has a 15 × eyepiece and a 40 × objective 16.5 cm apart. Calculate (*a*) the total magnification, (*b*) the focal length of each lens, and (*c*) where the object must be for a normal relaxed eye to see it in focus.

31. (II) The eyepiece of a compound microscope has a focal length of 2.70 cm and the objective has $f = 0.740$ cm. If an object is placed 0.790 cm from the objective lens, calculate (*a*) the distance between the lenses when the microscope is adjusted for a relaxed eye, and (*b*) the total magnification.

32. (II) A microscope has a 2.0-cm-focal-length eyepiece and 0.90-cm objective. Calculate (*a*) the position of the object if the distance between the lenses is 16.0 cm, and (*b*) the total magnification assuming a relaxed normal eye.

33. (II) Repeat Problem 32 assuming that the final image is located 25 cm from the eyepiece (near point of a normal eye).

SECTION 25–6

34. (II) An achromatic lens is made of two very thin lenses placed in contact that have focal lengths $f_1 = -18$ cm and $f_2 = +25$ cm. (*a*) Is the combination converging or diverging? (*b*) What is the net focal length?

SECTIONS 25–7 TO 25–9

35. (I) What is the angle of acceptance α of a microscope oil-immersion objective, and its resolving power, if $n = 1.60$ and the NA $= 1.35$? Use $\lambda = 500$ nm.

36. (I) What is the angular resolution limit set by diffraction for the Mt. Wilson 100-in (mirror diameter) telescope ($\lambda = 500$ nm)?

37. (II) A microscope objective is immersed in oil ($n = 1.60$) and accepts light scattered from the object up to 60° on either side of vertical. (*a*) What is the numerical aperture? (*b*) What is the approximate resolution of the microscope for 550-nm light?

38. (II) Two stars 20 light years away are barely resolved by a 100-in (mirror diameter) telescope. How far apart are the stars? Assume $\lambda = 500$ nm.

39. (II) A certain sea organism has a pattern of dots on its surface with an average spacing of 0.40 μm. If the specimen is viewed using 550-nm light, what minimum value must the numerical aperture be in order that the dots be resolved?

40. (III) What minimum magnification would be required to see the dots on the organism of Problem 39?

41. (III) (*a*) Can a human eye distinguish the two headlights, 1.8 m apart, on a truck 10 km away? Consider only diffraction and assume an eye diameter of 5.0 mm and a wavelength of 500 nm. (*b*) What is the maximum distance at which the two headlights can be resolved?

*SECTION 25–10

*42. (II) Show that the phase difference (in radians) between the two waves *ac* and *bd* in Fig. 25–26 is $\delta = (2\pi/\lambda)(n_2 - n_1)t$, where n_2 and n_1 are the refractive indices of the object and the medium, t the thickness of the object, and λ the wavelength of light used.

SECTION 25–11

43. (II) X rays of wavelength 0.128 nm fall on a crystal whose atoms, lying in planes, are spaced 0.300 nm apart. At what angle must the X rays be directed if the first diffraction maximum is to be observed?

44. (II) First-order Bragg diffraction is observed at 21.6° from a crystal with spacing between atoms of 0.24 nm. (*a*) At what angle will second order be observed? (*b*) What is the wavelength of the X rays?

45. (III) If X-ray diffraction peaks corresponding to the first three orders ($m = 1$, 2, and 3) are measured, can both the X-ray wavelength λ and lattice spacing d be determined? Prove your answer.

*SECTION 25–12

*46. (II) (*a*) Suppose for a conventional X-ray image that the X-ray beam consisted of parallel rays. What would be the magnification of the image? (*b*) Suppose, instead, the X rays came from a point source (as in Fig. 25–34a) that is 15 cm in front of a human body 25 cm thick, and the film is pressed against the person's back. Determine and discuss the range of magnifications that result.

GENERAL PROBLEMS

47. Show that for objects very far away (assume infinity), the magnification of a camera lens is proportional to its focal length.

48. For a camera equiped with a 50-mm-focal-length lens, what is the object distance if the image height equals the object height? How far is the object from the film?

49. A woman can see clearly with her right eye only when objects are between 45 and 220 cm away. Prescription bifocals should have what powers so that she can see distant objects clearly (upper part) and be able to read a book 25 cm away (lower part)? Assume that the glasses will be 2.0 cm from the eye.

50. A child has a near point of 10 cm. What is the maximum magnification the child can obtain using a 7.5-cm-focal-length magnifier? Compare to that for a normal eye.

51. What is the magnifying power of a +16-diopter lens? Assume a relaxed normal eye.

52. A physicist lost in the mountains tries to make a telescope using the lenses from his reading glasses. They have powers of +1.5 and +4.5D, respectively. (*a*) What maximum magnification telescope is possible? (*b*) Which lens should be used as the eyepiece?

53. The normal lens on a 35-mm camera has a focal length of 50 mm. Its aperture diameter varies from a maximum of 25 mm ($f/2$) to a minimum of 3.0 mm ($f/16$). Determine the resolution limit set by diffraction for $f/2$ and $f/16$. Specify as the number of lines per millimeter resolved on the film.

54. A 50-year-old man uses +2.0 diopter lenses to be able to read a newspaper 25 cm away. Ten years later, he finds that he must hold the paper 45 cm away to see clearly with the same lenses. What power lenses does he need now? (Distances are measured from the lens.)

55. Suppose that you wish to construct a telescope that can resolve features 10 km across on the moon, 384,000 km away. You have a 2.2-m-focal-length objective lens whose diameter is 12 cm. What focal-length eyepiece is needed if your eye can resolve objects 0.10 mm apart at a distance of 25 cm? What is the resolution limit set by the size of the objective lens (that is, by diffraction)? Use $\lambda = 500$ nm.

56. Exposure times must be increased for pictures taken at very short distances, because of the increased distance of the lens from the film for a focused image. (*a*) Show that when the object is so close to the camera that the image height equals the object height, the exposure time must be four times longer than when the object is a long distance away (say, ∞), given the same illumination and f-stop. (*b*) Show that if d_o is at least four or five times the focal length f of the lens, the exposure time is increased negligibly relative to the same object being a great distance away.

CHAPTER 26

Special Theory of Relativity

Albert Einstein (1879–1955), one of the great minds of the twentieth century, creator of the special and general theories of relativity, here shown lecturing.

Physics at the end of the nineteenth century looked back on a period of great progress. The theories developed over the preceding three centuries had been very successful in explaining a wide range of natural phenomena. Newtonian mechanics beautifully explained the motion of objects on earth and in the heavens; furthermore, it formed the basis for successful treatments of fluids, wave motion, and sound. Kinetic theory, on the other hand, explained the behavior of gases and other materials. And Maxwell's theory of electromagnetism not only brought together and explained electric and magnetic phenomena, but it predicted the existence of electromagnetic (EM) waves that would behave in every way just like light—so light came to be thought of as an electromagnetic wave. Indeed, it seemed that the natural world, as seen through the eyes of physicists, was very well explained. A few puzzles remained, but it was felt that these would soon be explained using already known principles.

But it did not turn out so simply. Instead, these puzzles were only to be solved by the introduction, in the early part of the twentieth century, of two revolutionary new theories that changed our whole conception of nature: the *theory of relativity* and *quantum theory*.

FIGURE 26–1 Einstein at play.

Classical vs. modern physics

Physics as it was known at the end of the nineteenth century (what we've covered up to now in this book) is referred to as **classical physics**. The new physics that grew out of the great revolution at the turn of the twentieth century is now called **modern physics**. In this chapter, we present the special theory of relativity, which was first proposed by Albert Einstein (1879–1955; Fig. 26–1) in 1905. In the following chapter, we introduce the equally momentous quantum theory.

26–1 • Galilean–Newtonian Relativity

Einstein's special theory of relativity deals with how we observe events, particularly how objects and events are observed from different frames of reference.† This subject had, of course, already been explored by Galileo and Newton. We first briefly discuss these earlier ideas, before seeing (starting in Section 26–3) how the theory of relativity changed them.

The special theory of relativity deals with events that are observed from so-called **inertial reference frames**, which (as mentioned in Chapter 4), are reference frames in which Newton's first law, the law of inertia, is valid. (Newton's first law states that, if an object experiences no net force due to other bodies, the object either remains at rest or remains in motion with constant velocity in a straight line.) It is easiest to deal with events as seen from inertial frames, and the earth, though not quite an inertial frame (it rotates), is close enough that for most purposes we can consider it an inertial frame. Rotating or otherwise accelerating frames of reference are noninertial frames,‡ and Einstein dealt with such complicated frames of reference in his general theory of relativity (Chapter 33).

A reference frame that moves with constant velocity with respect to an inertial frame is itself also an inertial frame, since Newton's laws hold in it as well.

† A reference frame is a set of coordinate axes fixed to some body (or group of bodies) such as the earth, a train, the moon, and so on. See Section 2–2.

‡ On a rotating platform, for example, (say a merry-go-round) an object at rest starts moving outward even though no force is exerted on it. This is therefore not an inertial frame.

FIGURE 26–2 A coin is dropped by a person in a moving car. (a) In the reference frame of the car, the coin falls straight down. (b) In a reference frame fixed on the earth, the coin follows a curved (parabolic) path. The upper views show the movement of the coin's release, and the lower views show it a short time later.

Relativity principle: the laws of physics are the same in all inertial reference frames

Both Galileo and Newton were keenly aware of what we now call the **relativity principle**: that *the basic laws of physics are the same in all inertial reference frames.* You may have recognized its validity in everyday life. For example, objects move in the same way in a smoothly moving (constant-velocity) train or airplane as they do on earth. (This assumes no vibrations or rocking—for they would make the reference frame noninertial.) When you walk, drink a cup of soup, play Ping-Pong, or drop a pencil on the floor while traveling in a train, airplane or ship moving at constant velocity, the bodies move just as they do when you are at rest on earth. Suppose you are in a car traveling rapidly along at constant velocity. If you release a coin from above your head inside the car, how will it fall? It falls straight downward with respect to the car, and hits the floor directly below the point of release, Fig. 26–2a. (If you drop the coin out the car's window, this won't happen because the moving air drags the coin backward relative to the car.) This is just how objects fall on the earth—straight down—and thus our experiment in the moving car is in accord with the relativity principle.

Note in this example, however, that to an observer on the earth, the coin follows a curved path, Fig. 26–2b. The actual path followed by the coin is different as viewed from different frames of reference. This does not violate the relativity principle because this principle states that the *laws* of physics are the same in all inertial frames. The same law of gravity, and the same laws of motion, apply in both reference frames. The difference in Figs. 26–2a and b is that in the earth's frame of reference, the coin has an initial velocity (equal to that of the car). The laws of physics therefore predict it will follow a parabolic path like any projectile. In the car's reference frame, there is no initial velocity and the laws of physics predict that the coin will fall straight down. The laws are the same in both reference frames although the specific paths are different.†

† Galileo, in his great book *Dialogues on the Two Chief Systems of the World*, described a similar experiment and predicted the same results. Galileo's example involved a sailor dropping a knife from the top of the mast of a sailing vessel. If the vessel moves at constant speed, where will the knife hit the deck (ignoring earth's rotation and air resistance)?

Galilean–Newtonian relativity involves certain unprovable assumptions that make sense from everyday experience. It is assumed that the lengths of objects are the same in one reference frame as in another, and that time passes at the same rate in different reference frames. In classical mechanics, then, space and time are considered to be **absolute**: their measurement doesn't change from one reference frame to another. The mass of an object, as well as all forces, are assumed to be unchanged by a change in inertial reference frame.

The position of an object is, of course, different when specified in different reference frames, and so is velocity. For example, a person may walk inside a bus toward the front with a speed of 5 km/h. But if the bus moves 40 km/h with respect to the earth, the person is then moving with a speed of 45 km/h with respect to the earth. The acceleration of a body, however, is the same in any inertial reference frame according to classical mechanics. This is because the change in velocity, and the time interval, will be the same. For example, the person in the bus may accelerate from 0 to 5 km/h in 1.0 seconds, so $a = 5$ km/h/s in the reference frame of the bus. With respect to the earth, the acceleration is (45 km/h − 40 km/h)/(1.0 s) = 5 km/h/s, which is the same.

Since neither F, m, nor a changes from one inertial frame to another, then Newton's second law, $F = ma$, does not change. Thus Newton's second law satisfies the relativity principle. It is easily shown that the other laws of mechanics also satisfy the relativity principle.

All inertial reference frames are equally valid

That the laws of mechanics are the same in all inertial reference frames implies that no one inertial frame is special in any sense. We express this important conclusion by saying that **all inertial reference frames are equivalent** for the description of mechanical phenomena. No one inertial reference frame is any better than another. A reference frame fixed to a car or an aircraft traveling at constant velocity is as good as one fixed on the earth. When you travel smoothly at constant velocity in a car or airplane, it is just as valid to say you are at rest and the earth is moving as it is to say the reverse. There is no experiment one can do to tell which frame is "really" at rest and which is moving. Thus, there is no way to single out one particular reference frame as being at absolute rest.

A complication arose, however, in the last half of the nineteenth century. When Maxwell presented his comprehensive and very successful theory of electromagnetism (Chapter 22), he showed that light can be considered an electromagnetic wave. Maxwell's equations predicted that the velocity of light c would be 3.00×10^8 m/s; and this is just what is measured within experimental error. The question then arose: in what reference frame does light have precisely the value predicted by Maxwell's theory? For it was assumed that light would have a different speed in different frames of reference. For example, if observers were traveling on a rocket ship at a speed of 1.0×10^8 m/s toward a source of light, we might expect them to measure the speed of the light reaching them to be 3.0×10^8 m/s + 1.0×10^8 m/s = 4.0×10^8 m/s. But Maxwell's equations have no provision for relative velocity. They predicted the speed of light to be $c = 3.0 \times 10^8$ m/s. This seemed to imply there must be some special reference frame where c would have this value.

We discussed in Chapters 11 and 12 that waves travel on water and along ropes or strings, and sound waves travel in air and other materials.

Since nineteenth-century physicists viewed the material world in terms of the laws of mechanics, it was natural for them to assume that light too must travel in some *medium*. They called this transparent medium the **ether** and assumed it permeated all space.[†] It was therefore assumed that the velocity of light given by Maxwell's equations must be with respect to the ether.

The "ether"

However, it appeared that Maxwell's equations did *not* satisfy the relativity principle. They were not the same in all inertial reference frames. They were simplest in the frame where $c = 3.00 \times 10^8$ m/s; that is, in a reference frame at rest in the ether. In any other reference frame, extra terms would have to be added to take into account the relative velocity. Thus, although most of the laws of physics obeyed the relativity principle, the laws of electricity and magnetism apparently did not. Instead, they seemed to single out one reference frame that was better than any other—a reference frame that could be considered absolutely at rest.

Scientists soon set out to determine the speed of the earth relative to this absolute frame whatever it might be. A number of clever experiments were designed. The most direct were performed by A. A. Michelson and E. W. Morley in the 1880s. The details of their experiment are discussed in the next section. Briefly, what they did was measure the difference in the speed of light in different directions. They expected to find a difference depending on the orientation of their apparatus with respect to the ether. For just as a boat has different speeds relative to the land when it moves upstream, downstream, or across the stream, so too light would be expected to have different speeds depending on the velocity of the ether past the earth.

Strange as it may seem, they detected no difference at all. This was a great puzzle. A number of explanations were put forth over a period of years, but they led to contradictions or were otherwise not generally accepted.

Then in 1905, Albert Einstein proposed a radical new theory that reconciled these many problems in a simple way. But at the same time, as we shall soon see, it completely changed our ideas of space and time.

*26–2 • The Michelson–Morley Experiment

The Michelson–Morley experiment was designed to measure the speed of the *ether*—the medium in which light was assumed to travel—with respect to the earth. The experimenters thus hoped to find an absolute reference frame, one that could be considered to be at rest.

One of the possibilities nineteenth-century scientists considered was that the ether is fixed relative to the sun, for even Newton had taken the sun as the center of the universe. If this were the case (there was no guarantee, of course), the earth's speed of about 3×10^4 m/s in its orbit around the sun would produce a change of 1 part in 10^4 in the speed of light (3.0×10^8 m/s). Direct measurement of the speed of light to this accuracy was not possible. But A. A. Michelson, later with the help of E. W. Morley, was able to use

[†] The medium for light waves could not be air, since light travels from the sun to earth through nearly empty space. Therefore, another medium was postulated, the ether. The ether was not only transparent, but, because of difficulty in detecting it, was assumed to have zero density.

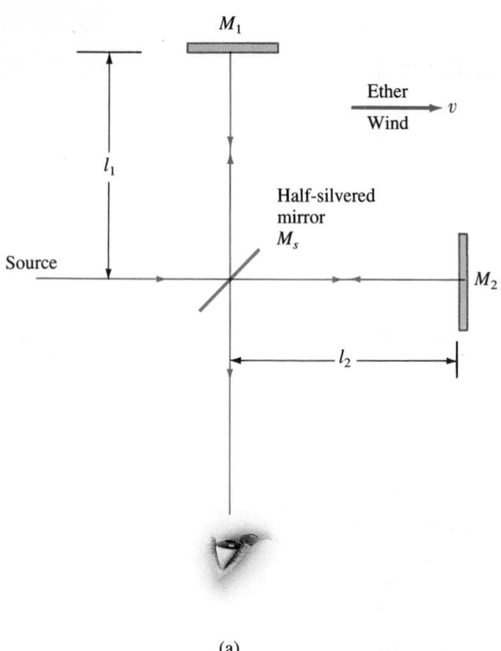

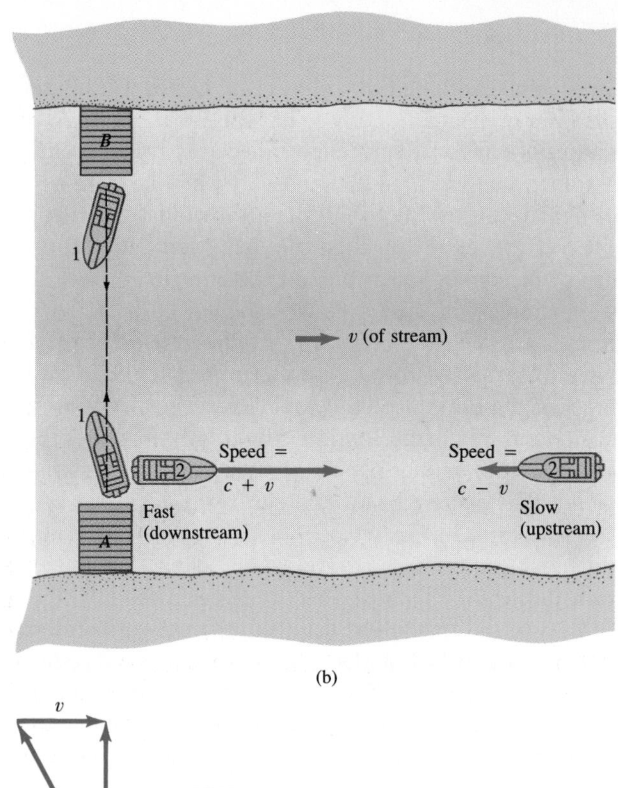

FIGURE 26–3 The Michelson–Morley experiment. (a) Michelson interferometer. (b) Boat analogy: boat 1 goes across the stream and back; boat 2 goes downstream and back upstream. (c) Calculation of the velocity of boat (or light beam) traveling perpendicular to the current (or ether wind).

v

c

$v' = \sqrt{c^2 - v^2}$

(c)

his interferometer (Section 24–9) to measure the difference in the speed of light in different directions to this accuracy. This famous experiment is based on the principle shown in Fig. 26–3. Part (a) is a simplified diagram of the Michelson interferometer, and it is assumed that the "ether wind" is moving with speed v to the right. (Alternatively, the earth is assumed to move to the left with respect to the ether at speed v.) The light from the source is split into two beams by the half-silvered mirror M_s. One beam travels to mirror M_1 and the other to mirror M_2. The beams are reflected by M_1 and M_2 and are joined again after passing through M_s. The now superposed beams interfere with one another and the resultant is viewed by the observer's eye as an interference pattern (discussed in Section 24–9).

Whether constructive or destructive interference occurs at the center of the interference pattern depends on the relative phases of the two beams after they have traveled their separate paths. To examine this let us consider an analogy of a boat traveling up and down, and across, a river whose current moves with speed v as shown in Fig. 26–3b. In still water, the boat can travel with speed c (not the speed of light in this case).

First we consider beam 2 in Fig. 26–3a, which travels parallel to the "ether wind". In its journey from M_s to M_2, we expect the light to travel with speed $c + v$, just as a boat traveling downstream (see Fig. 26–3b) acquires the speed of the river current. Since the beam travels a distance l_2, the time it takes to go from M_s to M_2 is $t = l_2/(c + v)$. To make the return trip from M_2 to M_s, the light must move against the ether wind (like the boat going upstream), so its relative speed is expected to be $c - v$. The time for the return trip is $l_2/(c - v)$. The total time required for beam 2 to travel from M_s to M_2 and back to M_s is

$$t_2 = \frac{l_2}{c + v} + \frac{l_2}{c - v}$$

$$= \frac{2l_2}{c(1 - v^2/c^2)}.$$

The second line was obtained from the first by finding the common denominator and factoring out c^2 in the denominator.

Now let us consider beam 1, which travels crosswise to the ether wind. Here the boat analogy (part b) is especially helpful. The boat is to go from wharf A to wharf B directly across the stream. If it heads directly across, the stream's current will drag it downstream. To reach wharf B, the boat must head at an angle upstream. The precise angle depends on the magnitudes of c and v, but is of no interest to us in itself. Part c of Fig. 26–3 shows how to calculate the velocity v' of the boat relative to earth as it crosses the stream. Since c, v, and v' form a right triangle, we have that $v' = \sqrt{c^2 - v^2}$. The boat has the same velocity when it returns. If we now apply these principles to light beam 1 in Fig. 26–3a, we see that the beam travels with a speed $\sqrt{c^2 - v^2}$ in going from M_s to M_1 and back again. The total distance traveled is $2l_1$, so the time required for beam 1 to make the round trip is $2l_1/\sqrt{c^2 - v^2}$, or

$$t_1 = \frac{2l_1}{c\sqrt{1 - v^2/c^2}}.$$

Notice that the denominator in this equation for t_1 involves a square root, whereas that for t_2 (above) does not.

If $l_1 = l_2 = l$, we see that beam 1 will lag behind beam 2 by an amount

$$\Delta t = t_2 - t_1 = \frac{2l}{c}\left(\frac{1}{1 - v^2/c^2} - \frac{1}{\sqrt{1 - v^2/c^2}}\right).$$

If $v = 0$, then $\Delta t = 0$ and the two beams will return in phase since they were initially in phase. But if $v \neq 0$, then $\Delta t \neq 0$, and the two beams will return out of phase. If this change of phase from the condition $v = 0$ to that for $v = v$ could be measured, then v could be determined. But the earth cannot be stopped. Furthermore, it is not possible to independently assume $l_1 = l_2$.

Michelson and Morley realized that they could detect the difference in phase (assuming that $v \neq 0$) if they rotated their apparatus by 90°, for then the interference pattern between the two beams should change. In the rotated position, beam 1 would now move parallel to the ether and beam 2 perpendicular to it. Thus the roles could be reversed, and in the rotated position the times (designated by primes) would be

$$t_1' = \frac{2l_1}{c(1 - v^2/c^2)} \quad \text{and} \quad t_2' = \frac{2l_2}{c\sqrt{1 - v^2/c^2}}.$$

The time lag between the two beams in the nonrotated position (unprimed) would be

$$\Delta t = t_2 - t_1 = \frac{2l_2}{c(1 - v^2/c^2)} - \frac{2l_1}{c\sqrt{1 - v^2/c^2}}.$$

In the rotated position, the time difference would be

$$\Delta t' = t_2' - t_1' = \frac{2l_2}{c\sqrt{1 - v^2/c^2}} - \frac{2l_1}{c(1 - v^2/c^2)}.$$

When the rotation is made, the fringes of the interference pattern (Section 24–9) will shift an amount determined by the difference:

$$\Delta t - \Delta t' = \frac{2}{c}(l_1 + l_2)\left(\frac{1}{1 - v^2/c^2} - \frac{1}{\sqrt{1 - v^2/c^2}}\right).$$

This expression can be considerably simplified if we assume that $v/c \ll 1$. For in this case we can use the binomial expansion,† so

$$\frac{1}{1 - v^2/c^2} \approx 1 + \frac{v^2}{c^2} \quad \text{and} \quad \frac{1}{\sqrt{1 - v^2/c^2}} \approx 1 + \frac{1}{2}\frac{v^2}{c^2}.$$

Then

$$\begin{aligned}\Delta t - \Delta t' &\approx \frac{2}{c}(l_1 + l_2)\left(1 + \frac{v^2}{c^2} - 1 - \frac{1}{2}\frac{v^2}{c^2}\right)\\ &\approx (l_1 + l_2)\frac{v^2}{c^3}.\end{aligned}$$

Now we take $v = 3.0 \times 10^4$ m/s, the speed of the earth in its orbit around the sun. In Michelson and Morley's experiments, the arms l_1 and l_2 were about 11 m long. The time difference would then be about (22 m)$(3.0 \times 10^4 \text{ m/s})^2/(3.0 \times 10^8 \text{ m/s})^3 \approx 7.0 \times 10^{-16}$ s. For visible light of wavelength $\lambda = 5.5 \times 10^{-7}$ m, say, the frequency would be $f = c/\lambda = (3.0 \times 10^8 \text{ m/s})/(5.5 \times 10^{-7} \text{ m}) = 5.5 \times 10^{14}$ Hz, which means that wave crests pass by a point every $1/(5.5 \times 10^{14} \text{ Hz}) = 1.8 \times 10^{-15}$ s. Thus, with a time difference of 7.0×10^{-16} s, Michelson and Morley should have noted a movement in the interference pattern of $(7.0 \times 10^{-16} \text{ s})/(1.8 \times 10^{-15} \text{ s}) = 0.4$ fringe. They could easily have detected this, since their apparatus was capable of observing a fringe shift as small as 0.01 fringe.

But they found *no significant fringe shift whatever*! They set their apparatus at various orientations. They made observations day and night so that they would be at various orientations with respect to the sun (due to the earth's rotation). They tried at different seasons of the year (the earth at different locations due to its orbit around the sun). Never did they observe a significant fringe shift.

The null result

This "null" result was one of the great puzzles of physics at the end of the nineteenth century. To explain it was a difficult challenge. One possibility was that the ether is not at rest with respect to the sun or other stars, but instead is at rest with respect to the earth. In this case, v would be zero and no fringe shift would be expected. But this implies the earth is somehow a preferred object; only with respect to the earth would the speed of light be c as predicted by Maxwell's equations. This is tantamount to assuming

† The binomial expansion (see Appendix A) states that $(1 \pm x)^n = 1 \pm nx + [n(n-1)/2]x^2 + \cdots$. In our case we have, therefore, $(1 - x)^{-1} \approx 1 + x$, and $(1 - x)^{-1/2} \approx 1 + \frac{1}{2}x$, where only the first term is kept, since $x = v^2/c^2$ is assumed to be small.

that the earth is the central body of the universe, an ancient idea that had been rejected centuries earlier. Another possibility was that the ether was dragged along by the earth and other bodies, so that its speed at the earth's surface would be zero. However, experiments in high-flying balloons, where at least some ether movement might have been expected, also gave a null result. Both possibilities were negated by the observation of "aberration of starlight." Over an observation period of a year, the stars appear to move in tiny ellipses; this phenomenon is explained, if the earth is moving relative to the ether, by assuming a telescope must be pointed not exactly at the star to be observed but rather at an angle as explained in Fig. 26–4. If the ether were at rest with respect to the earth, you could point directly at the star and no "aberration angle" would be present. But the aberration angle is there, so the ether could not be at rest with respect to the earth.

Another theory to explain the null result was put forth independently by G. F. Fitzgerald and H. A. Lorentz (in the 1890s) in which they proposed that any length (including the arm of an interferometer) contracts by a factor $\sqrt{1 - v^2/c^2}$ in the direction of motion through the ether. According to Lorentz, this could be due to the ether affecting the forces between the molecules of a substance, which were assumed to be electrical in nature. This theory, as useful as it was, was eventually replaced by the far more comprehensive theory proposed by Albert Einstein in 1905–the special theory of relativity.

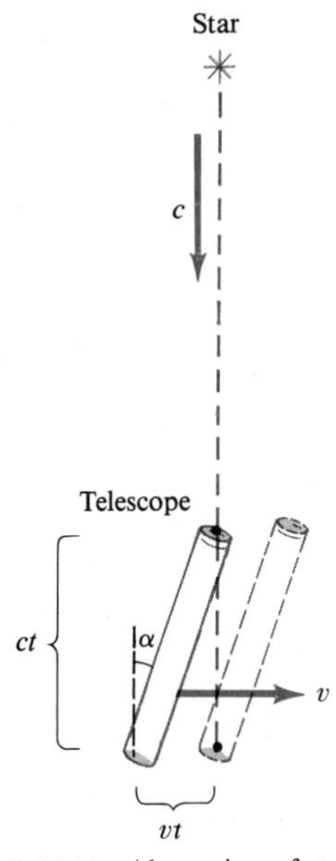

FIGURE 26–4 Aberration of starlight. If a telescope moves with the earth at speed v with respect to the ether, the telescope must be aimed slightly away from the star so that light entering the telescope will travel down the tube (since the tube moves to the right with speed v), rather than hitting the sides. Thus, an overhead star will appear to be at an angle $\alpha = v/c$ to the vertical and, as the earth revolves, will seem to trace a circular path.

26–3 • Postulates of the Special Theory of Relativity

The problems that existed at the turn of the century with regard to electromagnetic theory and Newtonian mechanics were beautifully resolved by Einstein's introduction of the theory of relativity in 1905. Einstein, however, was apparently not influenced directly by the null result of the Michelson–Morley experiment. He had studied and admired the theoretical work of Lorentz. What motivated Einstein were certain questions regarding electromagnetic theory and light waves. For example, he asked himself: "What would I see if I rode a light beam?" The answer was that instead of a traveling electromagnetic wave, he would see alternating electric and magnetic fields at rest whose magnitude changed in space, but did not change in time. Such fields, he realized, had never been detected and indeed were not consistent with Maxwell's electromagnetic theory. He argued, therefore, that it was unreasonable to think that the speed of light relative to any observer could be reduced to zero, or in fact reduced at all. This idea became the second postulate of his theory of relativity.

Einstein concluded that the inconsistencies he found in electromagnetic theory were due to the assumption that an absolute space exists. In his famous 1905 paper, he proposed doing away completely with the idea of the ether and the accompanying assumption of an absolute reference frame at rest. This proposal was embodied in two postulates. The first postulate was an extension of the Newtonian relativity principle to include not only the laws of mechanics but also those of the rest of physics, including electricity and magnetism:

The two postulates of special relativity

> ***First postulate (the relativity principle):* The laws of physics have the same form in all inertial reference frames.**

The second postulate is consistent with the first:

> ***Second postulate (constancy of the speed of light):* Light propagates through empty space with a definite speed c independent of the speed of the source or observer.**

These two postulates form the foundation of Einstein's **special theory of relativity**. It is called "special" to distinguish it from his later "general theory of relativity," which deals with noninertial (accelerating) reference frames (discussed in Chapter 33). The special theory, which is what we discuss here, deals only with inertial frames.

The second postulate seems to be the hardest to accept, for it violates commonsense notions. First of all, we have to think of light traveling through empty space. Giving up the ether is not too hard, however, for after all, it could never be detected. But the second postulate also tells us that the speed of light in vacuum is always the same, 3.00×10^8 m/s, no matter what the speed of the observer or the source. Thus, a person traveling toward or away from a source of light will measure the same speed for that light as someone at rest with respect to the source. This conflicts with our everyday notions, for we would expect to have to add in the velocity of the observer. Part of the problem is that in our everyday experience, we do not measure velocities anywhere near the speed of light. Thus we can't expect our everyday experience to be helpful when dealing with such a high velocity. On the other hand, the Michelson–Morley experiment is fully consistent with the second postulate.[†]

The beauty of Einstein's proposal is evident. For by doing away with the idea of an absolute reference frame, it was possible to reconcile Maxwell's electromagnetic theory with mechanics. The speed of light predicted by Maxwell's equations *is* the speed of light in vacuum in *any* reference frame.

Einstein's theory required giving up commonsense notions of space and time, and in the following sections we will examine some strange but interesting consequences of Einstein's theory. Our arguments for the most part will be simple ones. We will use a technique that Einstein himself did: we will imagine very simple experimental situations in which little mathematics is needed. In this way, we can see many of the consequences of relativity theory without getting involved in detailed calculations. Einstein called these "gedanken" experiments, which is German for "thought" experiments. In Section 26–12 we will look more deeply into the mathematics of relativity.

[†] The Michelson–Morley experiment can also be considered as evidence for the first postulate, for it was intended to measure the motion of the earth relative to an absolute reference frame. Its failure to do so implies the absence of any such preferred frame.

26–4 • Simultaneity

One of the important consequences of the theory of relativity is that we can no longer regard time as an absolute quantity. No one doubts that time flows onward and never turns back. But, as we shall see in this section and the next, the time interval between two events, and even whether two events are simultaneous, depends on the observer's reference frame.

Two events are said to occur simultaneously if they occur at exactly the same time. But how do we know if two events occur precisely at the same time? If they occur at the same point in space—such as two apples falling on your head at the same time—it is easy. But if the two events occur at widely separated places, it is more difficult to know whether the events are simultaneous since we have to take into account the time it takes for the light from them to reach us. Because light travels at finite speed, a person who sees two events must calculate back to find out when they actually occurred. For example, if two events are *observed* to occur at the same time, but one actually took place farther from the observer than the other, then the former must have occurred earlier, and the two events were not simultaneous.

A "thought" experiment

To avoid making calculations, we will now make use of a simple thought experiment. We assume an observer, called O, is located exactly halfway between points A and B where two events occur, Fig. 26–5. The two events may be lightning that strikes the points A and B, as shown, or any other type of events. For brief events like lightning, only short pulses of light will travel outward from A and B and reach O. O "sees" the events when the pulses of light reach point O. If the two pulses reach O at the same time, then the two events had to be simultaneous. This is because the two light pulses travel at the same speed; and since the distance OA equals OB, the time for the light to travel from A to O and B to O must be the same. Observer O can then definitely state that the two events occurred simultaneously. On the other hand, if O sees the light from one event before that from the other, then it is certain the former event occurred first.

The question we really want to examine is this: If two events are simultaneous to an observer in one reference frame, are they also simultaneous to another observer moving with respect to the first? Let us call the observers O_1 and O_2 and assume they are fixed in reference frames 1 and 2 that move with speed v relative to one another. These two reference frames can be thought of as train cars (Fig. 26–6). O_2 says that O_1 is moving to the

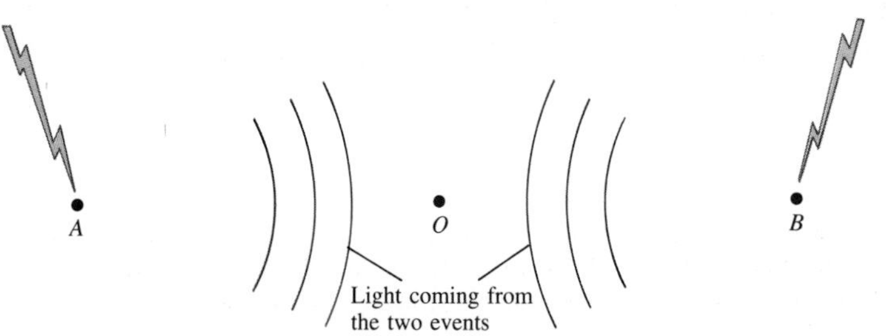

FIGURE 26–5 A moment after lightning strikes points A and B, the pulses of light are traveling toward O, but O "sees" the lightning only when the light reaches him or her.

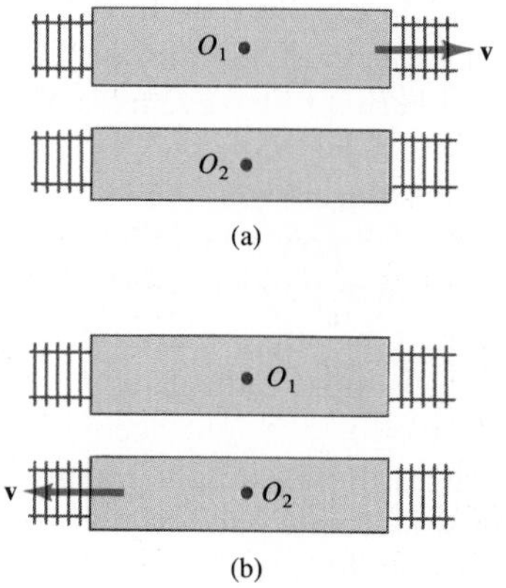

FIGURE 26–6 Observers O_1 and O_2, on two different trains (two different reference frames), are moving with relative velocity v. O_2 says that O_1 is moving to the right (a); O_1 says that O_2 is moving to the left (b). Both viewpoints are legitimate—it all depends on your reference frame.

right with speed v, as in (a); and O_1 says O_2 is moving to the left with speed v, as in (b). Both viewpoints are legitimate according to the relativity principle. (There is, of course, no third point of view which will tell us which one is "really" moving.)

Note necessity of choosing a frame of reference

Now suppose two events occur that are seen by both observers. Let us assume again that the two events are the striking of lightning and that the lightning marks both trains where it struck: at A_1 and B_1 on O_1's train, and at A_2 and B_2 on O_2's train. For simplicity, we assume that O_1 happens to be exactly halfway between A_1 and B_1, and that O_2 is halfway between A_2 and B_2. We must now put ourselves in one reference frame or the other. Let us put ourselves in O_2's reference frame, so we see O_1 moving to the right with speed v. Let us also assume that the two events occur *simultaneously* in O_2's frame, and just at the instant when O_1 and O_2 are opposite each other, Fig. 26–7a. A short time later, Fig. 26–7b, the light reaches O_2 from A_2 and B_2 at the same time (we assumed this). Since O_2 knows (or mea-

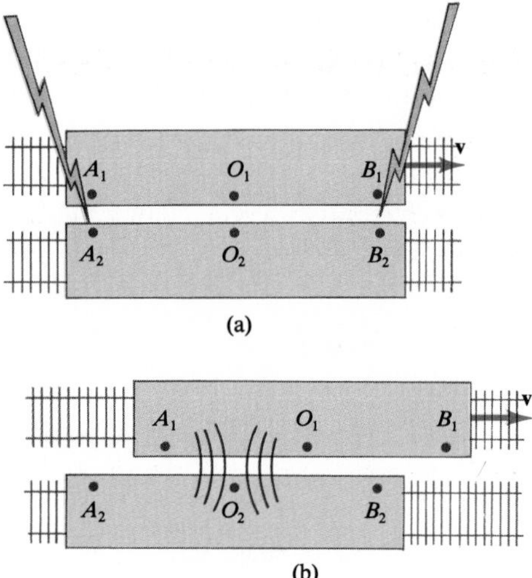

FIGURE 26–7 Thought experiment on simultaneity. To observer O_2, the reference frame of O_1 is moving to the right. In (a), one lightning bolt strikes the two reference frames at A_1 and A_2, and a second lightning bolt strikes at B_1 and B_2. According to observer O_2, the two bolts of lightning strike simultaneously. (b) A moment later, the light from the two events reaches O_2 at the same time (simultaneously). But in O_1's reference frame, the light from B_1 has already reached O_1, whereas the light from A_1 has not yet reached O_1. So in O_1's reference frame, the event at B_1 must have preceded the event at A_1. Time is not absolute.

sures) the distances O_2A_2 and O_2B_2 as equal, O_2 knows the two events are simultaneous in the O_2 reference frame.

But what does observer O_1 see? From our (O_2) reference frame, we can predict what O_1 will see. We see that O_1 moves to the right during the time the light is traveling to O_1 from A_1 and B_1. As shown in Fig. 26–7b, we can see from our O_2 reference frame that the light from B_1 has already passed O_1, whereas the light from A_1 has not yet reached O_1. Therefore, it is clear that O_1 will see the light from B_1 before that from A_1. Now O_1's frame is as good as O_2's. Light travels at the same speed c for O_1 as for O_2 (the second postulate)†; and in the O_1 reference frame, this speed c is of course the same for light traveling from A_1 to O_1 as it is for light traveling from B_1 to O_1. Furthermore the distance O_1A_1 equals O_1B_1. Hence, since O_1 sees the light from B_1 before he sees the light from A_1 (we established this above, looking from the O_2 reference frame, Fig. 26–7b), then observer O_1 can only conclude that the event at B_1 occurred before the event at A_1. The two events are not simultaneous for O_1, even though they are for O_2.

Simultaneity is relative

We thus find that *two events which are simultaneous to one observer are not necessarily simultaneous to a second observer.*

It may be tempting to ask: "Which observer is right, O_1 or O_2?" The answer, according to relativity, is that they are *both* right. There is no "best" reference frame we can choose to determine which observer is right. Both frames are equally good. We can only conclude that simultaneity is not an absolute concept, but is relative. We are not aware of it in everyday life, however, since the effect is noticeable only when the relative speed of the two reference frames is very large (near c), or the distances involved are very large.

Because of the principle of relativity, the argument we gave for the thought experiment of Fig. 26–7 can be done from O_1's reference frame as well. In this case, O_1 will be at rest and will see event B_1 occur before A_1. But O_1 will recognize (by drawing a diagram equivalent to Fig. 26–7) that O_2, who is moving with speed v to the left, will see the two events as simultaneous. The analysis is left as an exercise.

26–5 • Time Dilation and the Twin Paradox

The fact that two events simultaneous to one observer may not be simultaneous to a second observer suggests that time itself is not absolute. Could it be that time passes differently in one reference frame than in another? This is, indeed, just what Einstein's theory of relativity predicts, as the following thought experiment shows.

† Note that O_1 does not see himself catching up with one light beam and running away from the other (that is O_2's viewpoint of what happens in O_1). O_1 sees both light beams traveling the same speed, c.

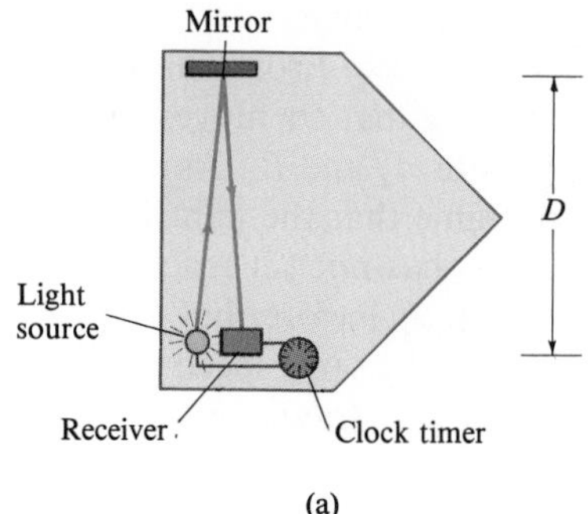

(a)

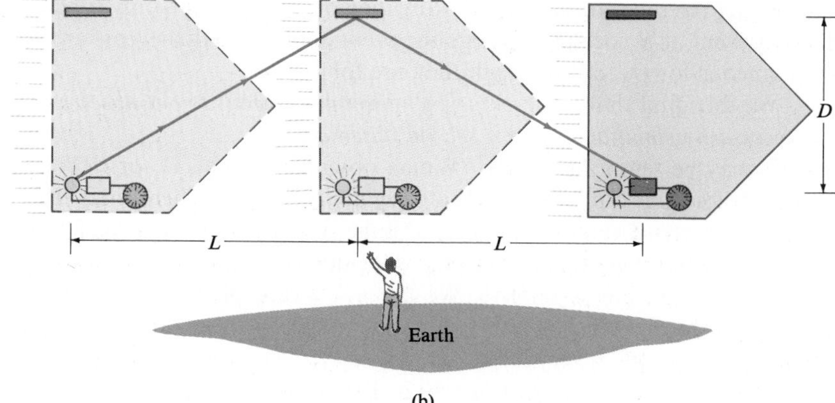

(b)

FIGURE 26–8 Time dilation can be shown by a thought experiment: the time it takes for light to travel over and back on a spaceship is longer for the earth observer (b) than for the observer on the spaceship (a).

Figure 26–8 shows a spaceship traveling past earth at high speed. The point of view of an observer on the spaceship is shown in part (a), and that of an observer on earth in part (b). Both observers have accurate clocks. The person on the spaceship (a) flashes a light and measures the time it takes the light to travel across the spaceship and return after reflecting from a mirror. The light travels a distance $2D$ at speed c, so the time required, which we call Δt_0, is

$$\Delta t_0 = \frac{2D}{c}.$$

This is the time as measured by the observer on the spaceship.

The observer on earth, Fig. 26–8b, observes the same process. But to this observer, the spaceship is moving. So the light travels the diagonal path shown in going across the spaceship, reflecting off the mirror, and returning to the sender. Although the light travels at the same speed to this observer (the second postulate), it travels a greater distance. Hence the time required, as measured by the earth observer, will be greater than that measured by the observer on the spaceship. The time interval, Δt, as observed by the earth observer can be calculated as follows. In the time Δt, the spaceship travels a distance $2L = v\,\Delta t$ where v is the speed of the spaceship (Fig. 26–8b). Thus, the light travels a total distance on its diagonal path of $2\sqrt{D^2 + L^2}$, and therefore

$$c = \frac{2\sqrt{D^2 + L^2}}{\Delta t} = \frac{2\sqrt{D^2 + v^2(\Delta t)^2/4}}{\Delta t}.$$

We square both sides, and then solve for Δt, to find

$$c^2 = \frac{4D^2}{(\Delta t)^2} + v^2,$$

$$\Delta t = \frac{2D}{c\sqrt{1 - v^2/c^2}}.$$

We combine this with the formula above for Δt_0 and find:

$$\Delta t = \frac{\Delta t_0}{\sqrt{1 - v^2/c^2}}. \qquad (26\text{–}1)$$

Time-dilation formula

Since $\sqrt{1 - v^2/c^2}$ is always less than 1, we see that $\Delta t > \Delta t_0$. That is, the time between the two events (the sending of the light, and its reception at the receiver) is *greater* for the earth observer than for the traveling observer. This is a general result of the theory of relativity, and is known as **time dilation**. Stated simply, the time dilation effect says that *clocks moving relative to an observer are measured by the observer to run slowly*. However, we should not think that the clocks are somehow at fault. To the contrary, we assume the clocks are good ones. Time is actually measured to pass more slowly in any moving reference frame as compared to your own. This remarkable result is an inevitable outcome of the two postulates of the theory of relativity.

Time dilation: moving clocks run slowly

The concept of time dilation may be hard to accept, for it violates our commonsense understanding. We can see from Eq. 26–1 that the time-dilation effect is negligible unless v is reasonably close to c. If v is much less than c, then the term v^2/c^2 is much smaller than the 1 in the denominator of Eq. 26–1, and then $\Delta t \approx \Delta t_0$ (see Example 26–2). The speeds we experience in everyday life are very much smaller than c, so it is little wonder we don't ordinarily notice time dilation. Experiments have been done to test the time-dilation effect, and have confirmed Einstein's predictions. In 1971, for example, extremely precise atomic clocks were flown around the world in jet planes. Since the speed of the planes (10^3 km/h) is much less than c, the clocks had to be accurate to nanoseconds (10^{-9} s) in order to detect any time dilation. They were this accurate and confirmed Eq. 26–1 to within experimental error. Time dilation had been confirmed decades earlier, however, by observation on "elementary particles" (see Chapter 32) which have very small masses (typically 10^{-30} to 10^{-27} kg) and so require little energy to be accelerated to speeds close to c. Many of these elementary particles are not stable and decay after a time into smaller particles. One example is the muon, whose mean lifetime is 2.2 μs when at rest. Careful experiments showed that when a muon is traveling at high speeds, its lifetime increases just as predicted by the time-dilation formula.

EXAMPLE 26–1 What will be the mean lifetime of a muon as measured in the laboratory if it is traveling at $v = 0.60c = 1.8 \times 10^8$ m/s with respect to the laboratory? Its mean life at rest is 2.2×10^{-6} s.

SOLUTION If an observer were to move along with the muon (the muon would be at rest to this observer), the muon would have a mean life of

2.2×10^{-6} s. To an observer in the lab, the muon lives longer because of time dilation. From Eq. 26–1 with $v = 0.60c$, we have

$$\Delta t = \frac{\Delta t_0}{\sqrt{1 - \frac{v^2}{c^2}}} = \frac{2.2 \times 10^{-6}\text{ s}}{\sqrt{1 - \frac{0.36c^2}{c^2}}} = \frac{2.2 \times 10^{-6}\text{ s}}{\sqrt{0.64}} = 2.8 \times 10^{-6}\text{ s}.$$

We need to make a comment about the use of Eq. 26–1 and the meaning of Δt and Δt_0. The equation is true only when Δt_0 represents the time interval between the two events in a reference frame where the two events occur at *the same point in space* (as in Fig. 26–8a where the two events are the light flash being sent and being received). This time interval, Δt_0, is called the **proper time**. Then Δt in Eq. 26–1 represents the time interval between the two events as measured in a reference frame moving with speed v with respect to the first. In Example 26–1 above, Δt_0 (and not Δt) was set equal to 2.2×10^{-6} s because it is only in the rest frame of the muon that the two events ("birth" and "decay") occur at the same point in space.

EXAMPLE 26–2 Let's check time dilation for everyday speeds. A car traveling 100 km/h covers a certain distance in 10.00 s according to the driver's watch. What does an observer on earth measure for the time interval?

SOLUTION The car's speed relative to earth is 100 km/h = $(1.00 \times 10^5\text{ m})/(3600\text{ s}) = 27.8$ m/s. We set $\Delta t_0 = 10.00$ s in the time-dilation formula (the driver is at rest in the reference frame of the car), and then Δt is

$$\Delta t = \frac{\Delta t_0}{\sqrt{1 - \frac{v^2}{c^2}}} = \frac{10.00\text{ s}}{\sqrt{1 - \left(\frac{27.8\text{ m/s}}{3.00 \times 10^8\text{ m/s}}\right)^2}} = \frac{10.00\text{ s}}{\sqrt{1 - 8.59 \times 10^{-15}}}.$$

If you put these numbers into a calculator, you will obtain $\Delta t = 10.00$ s, since the denominator differs from 1 by such a tiny amount. Indeed, the time measured by an earth observer would be no different than that measured by the driver, even with the best of today's instruments. A computer that could calculate to a large number of decimal places could reveal a difference between Δt and Δt_0. But we can estimate the difference quite easily using the binomial expansion (see Appendix A), which says that in a formula of the form $(1 \pm x)^n$, if $x \ll 1$, then to a good approximation,

Problem Solving: Use of the binomial expansion

$$(1 \pm x)^n \approx 1 \pm nx.$$

In our time-dilation formula, we have the factor $1/\sqrt{1 - v^2/c^2} = (1 - v^2/c^2)^{-1/2}$. Thus (setting $x = v^2/c^2$ and $n = -\frac{1}{2}$ in the binomial expansion):

$$\begin{aligned}\Delta t = \Delta t_0\left(1 - \frac{v^2}{c^2}\right)^{-\frac{1}{2}} &\approx \Delta t_0\left(1 + \frac{1}{2}\frac{v^2}{c^2}\right)\\ &= 10.00\text{ s}\left[1 + \frac{1}{2}\left(\frac{27.8\text{ m/s}}{3.00 \times 10^8\text{ m/s}}\right)^2\right]\\ &\approx 10.00\text{ s} + 4 \times 10^{-14}\text{ s}.\end{aligned}$$

So the difference between Δt and Δt_0 is predicted to be 4×10^{-14} s, an unmeasurably small amount.

Time dilation has aroused interesting speculation about space travel. According to classical (Newtonian) physics, to reach a star 100 light-years away would not be possible for ordinary mortals (1 light-year is the distance light can travel in 1 year $= 3.0 \times 10^8$ m/s $\times 3.15 \times 10^7$ s $= 9.5 \times 10^{15}$ m). Even if a spaceship could travel at close to the speed of light, it would take over 100 years to reach such a star. But time dilation tells us that the time involved would be less for an astronaut. In a spaceship traveling at $v = 0.999c$, the time for such a trip would be only about $\Delta t_0 = \Delta t\sqrt{1 - v^2/c^2} = (100 \text{ yr})\sqrt{1 - (0.999)^2} = 4.5$ yr. Thus time dilation allows such a trip, but the enormous practical problems of achieving such speeds will not be overcome in the foreseeable future.

Notice, in this example, that whereas 100 years would pass on earth, only 4.5 years would pass for the astronaut on the trip. Is it just the clocks that would slow down for the astronaut? The answer is no. All processes, including life processes, run more slowly for the astronaut according to the earth observer. But to the astronaut, time would pass in a normal way. The astronaut would experience 4.5 years of normal sleeping, eating, reading, and so on. And people on earth would experience 100 years of ordinary activity.

Twin paradox

Not long after Einstein proposed the special theory of relativity, an apparent paradox was pointed out. According to this **twin paradox**, suppose one of a pair of 20-year-old twins takes off in a spaceship traveling at very high speed to a distant star and back again, while the other twin remains on earth. According to the earth twin, the traveling twin will age less. Whereas 20 years might pass for the earth twin, perhaps only 1 year (depending on the spacecraft's speed) would pass for the traveler. Thus, when the traveler returns, the earthbound twin could expect to be 40 years old whereas his twin would be only 21.

This is the viewpoint of the twin on the earth. But what about the traveling twin? If all inertial reference frames are equally good, won't the traveling twin make all the claims the earth twin does, only in reverse? Can't the astronaut twin claim that since the earth is moving away at high speed, time passes more slowly on earth and the twin on earth will age less? This is the opposite of what the earth twin predicts. They cannot both be right, for after all the spacecraft returns to earth and a direct comparison of ages and clocks can be made.

There is, however, not a paradox at all. The consequences of the special theory of relativity—in this case, time dilation—can be applied only by observers in inertial reference frames. The earth is such a frame (or nearly so), whereas the spacecraft is not. The spacecraft accelerates at the start and end of its trip and, more importantly, when it turns around at the far point of its journey. During these acceleration periods, the spacecraft's predictions based on special relativity are not valid. The twin on earth is in an inertial frame and can make valid predictions. Thus, there is no paradox. The traveling twin's point of view expressed above is not correct. The predictions of the earth twin *are* valid, and the prediction that the traveling twin returns having aged less is the proper one.†

† Einstein's general theory of relativity, which deals with accelerating reference frames, confirms this result.

26–6 • Length Contraction

Not only time intervals are different in different reference frames. Space intervals—lengths and distances—are different as well, according to the special theory of relativity, and we illustrate this with a thought experiment.

An observer on Earth watches a spacecraft traveling at speed v from Earth to, say, Neptune, Fig. 26–9a. The distance between the planets, as measured by an Earth observer, is L_0. The time required for the trip, measured from Earth, is $\Delta t = L_0/v$. In Fig. 26–9b we see the point of view of an observer on the spacecraft. In this frame of reference, the spaceship is at rest; Earth and Neptune move with speed v. (We assume v is much greater than the relative speed of Neptune and Earth, so the latter can be ignored.) The time between the departure of Earth and arrival of Neptune (as observed from the spacecraft) is the "proper time" (since the two events occur at the same point in space—i.e., on the spacecraft). Therefore the time interval is less for the spacecraft observer than the earth observer, because of time dilation. From Eq. 26–1, the time for the trip as viewed by the spacecraft is $\Delta t_0 = \Delta t\sqrt{1 - v^2/c^2}$. Since the spacecraft observer measures the same speed but less time between these two events, he must also measure the distance as less. If we let L be the distance between the planets as viewed by the spacecraft observer, then $L = v\,\Delta t_0$. We have already seen that $\Delta t_0 = \Delta t\sqrt{1 - v^2/c^2}$ and $\Delta t = L_0/v$, so we have $L = v\,\Delta t_0 = v\,\Delta t\sqrt{1 - v^2/c^2} = L_0\sqrt{1 - v^2/c^2}$. That is,

Length-contraction formula

$$L = L_0\sqrt{1 - v^2/c^2}. \tag{26–2}$$

This is a general result of the special theory of relativity and applies to lengths of objects as well as to distance. The result can be stated most simply in words as: *the length of an object is measured to be shorter when it is moving than when it is at rest.* This is called **length contraction**. The length L_0 in Eq. 26–2 is called the **proper length**. It is the length of the object (or distance between two points whose positions are measured at the same time)

Length contraction: moving objects are shorter (in the direction of motion)

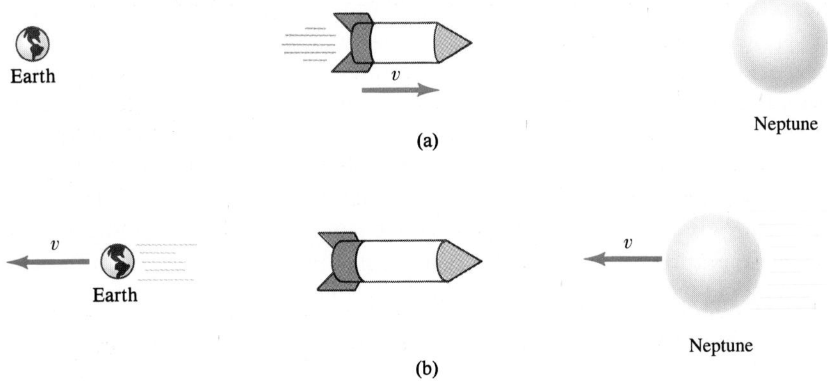

FIGURE 26–9 (a) A spaceship traveling at very high speed from Earth to Neptune, as seen from Earth's frame of reference. (b) As viewed by an observer on the spaceship, Earth and Neptune are moving at the very high velocity v: Earth leaves the spaceship, and a time Δt_0 later planet Neptune arrives at the spaceship. [Note in (b) that each planet does not look shortened because at high speeds we see the trailing edge (as in Fig. 26–10), and the net effect is to leave its appearance as a circle.]

as measured by an observer at rest with respect to it. Equation 26–2 gives the length that will be measured when the object travels past an observer at speed v. It is important to note, however, that length contraction occurs *only along the direction of motion.* For example, the moving spaceship in Fig. 26–9a is shortened in length, but its height is the same as when it is at rest.

Length contraction, like time dilation, is not noticeable in everyday life because the factor $\sqrt{1 - v^2/c^2}$ in Eq. 26–2 differs from 1.00 significantly only when v is very large.

EXAMPLE 26–3 A spaceship passes the earth at speed $v = 0.80c$. Describe the changes in length of a meter stick as it is slowly rotated from vertical to horizontal by a person inside, as viewed (*a*) by another person in the spaceship, and (*b*) by an observer on earth.

SOLUTION (*a*) The meter stick looks 1.0 m long in all orientations since it is at rest. (*b*) The meter stick varies in length from 1.0 m (vertical) to $L = (1.0 \text{ m}) \times \sqrt{1 - (0.80)^2} = 0.60$ m long in the horizontal direction (assuming that is the direction of motion).

Appearance of objects

Equation 26–2 tells us what the length of an object will be *measured* to be when traveling at speed v. The *appearance* of the object is another matter. Suppose, for example, you are traveling to the left past a small building at speed $v = 0.85c$. This is equivalent to the building moving past you to the right at speed v. The building will look narrower (and the same height), but you will also be able to see the side of the building even if you are directly in front of it. This is shown in Fig. 26–10b—part (a) shows the building at rest. The fact that you see the side is not really a relativistic effect, but is due to the finite speed of light. To see how this occurs, we look at Fig. 26–10c which is a top view of the building, looking down. At the instant shown, the observer O is directly in front of the building. Light from points A and B reach O at the same time. If the building were at rest, light from point C could never reach O. But the building is moving at very high speed and does "get out of the way" so that light from C can reach O. Indeed, at the instant shown, light from point C when it was at an earlier location (C' on the diagram) can reach O because the building has moved. In order to reach the observer at the same time as light from A and B, light from C had to leave at an earlier time since it must travel a greater distance. Thus it is light from C' that reaches the observer at the same time as light from A and B. This, then, is how an observer might see both the front and side of an object at

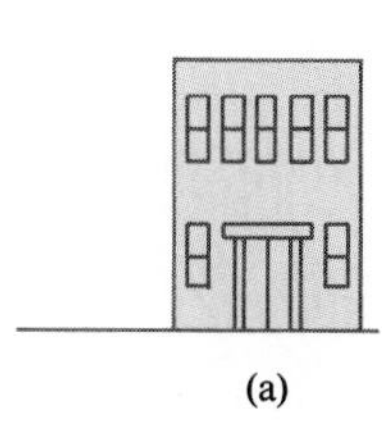
(a)

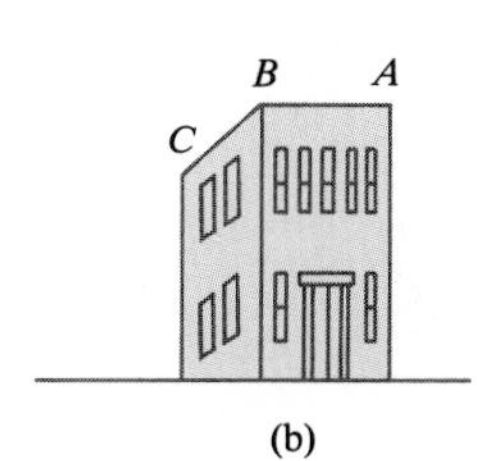

(b)

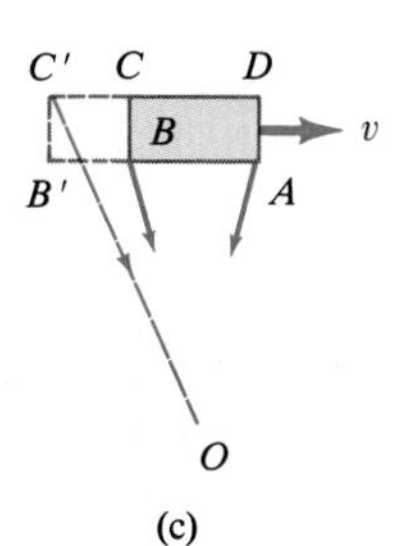

(c)

FIGURE 26–10 Building seen (a) at rest, and (b) moving at high speed. (c) Diagram explains why the side of the building is seen (see the text).

the same time even when directly in front of it.[†] It can be shown, by the same reasoning, that spherical objects will actually still have a circular outline even at high speeds. That is why the planets in Fig. 26–9b are drawn round rather than contracted.

26–7 • Four-Dimensional Space–Time

Let us imagine a person is on a train moving at a very high speed, say $0.65c$, Fig. 26–11. This person begins a meal at 7:00 and finishes at 7:15, according to a clock on the train. The two events, beginning and ending the meal, take place at the same point on the train. So the proper time between these two events is 15 min. To observers on earth, the meal will take longer—20 min according to Eq. 26–1. Let us assume that the meal was served on a 20-cm-diameter plate. To observers on the earth, the plate is only 15 cm wide (length contraction). Thus, to observers on the earth, the meal looks smaller but lasts longer.

In a sense these two effects, time dilation and length contraction, balance each other. When viewed from the earth, what the meal seems to lose in size it gains in length of time it lasts. Space, or length, is exchanged for time.

Considerations like this led to the idea of **four-dimensional space–time**: space takes up three dimensions and time is a fourth dimension. Space and time are intimately connected. Just as when we squeeze a balloon we make one dimension larger and another smaller, so when we examine objects and

[†] It would be an error to think that the building in Fig. 26–10b would look rotated. This is not correct since in that case side A would look shorter than side B. In fact, if the observer is directly in front, these sides appear equal in height. Thus the building looks contracted in its front face, but we also see the side, as described above. Also, though not shown in Fig. 26–10b, the walls of the building would appear curved, because of differing distances from the observer's eye of the various points from top to bottom along a vertical wall.

FIGURE 26–11 According to an accurate clock on a fast-moving train, a person (a) begins dinner at 7:00 and (b) finishes at 7:15. At the beginning of the meal, observers on earth set their watches to correspond with the clock on the train. These observers measure the eating time as 20 minutes.

events from different reference frames, a certain amount of space is exchanged for time, or vice versa.

Although the idea of four dimensions may seem strange, it refers to the idea that any object or event is specified by four quantities—three to describe where in space, and one to describe when in time. The really unusual aspect of four-dimensional space–time is that space and time can intermix: a little of one can be exchanged for a little of the other when the reference frame is changed.

It is difficult for most of us to understand the idea of four-dimensional space–time. Somehow we feel, just as physicists did before the advent of relativity, that space and time are completely separate entities. Yet we have found in our thought experiments that they are not completely separate. Our difficulty in accepting this is reminiscent of the situation in the seventeenth century at the time of Galileo and Newton. Before Galileo, the vertical direction, that in which objects fall, was considered to be distinctly different from the two horizontal dimensions. Galileo showed that the vertical dimension differs only in that it happens to be the direction in which gravity acts. Otherwise, all three dimensions are equivalent, a viewpoint we all accept today. Now we are asked to accept one more dimension, time, which we had previously thought of as being somehow different. This is not to say that there is no distinction whatsoever between space and time. What relativity has shown is that space and time determinations are not independent of one another.

26–8 • Mass Increase

The three basic mechanical quantities are length, time intervals, and mass. The first two have been shown to be relative—their value depends on the reference frame from which they are measured. We might ask if mass, too, is a relative quantity. Indeed, Einstein showed that *the mass of an object increases as its speed increases* according to the formula

$$m = \frac{m_0}{\sqrt{1 - v^2/c^2}}. \qquad (26\text{–}3)$$

Mass increase formula

In this **mass-increase** formula, m_0 is the **rest mass** of the object—the mass it has as measured in a reference frame in which it is at rest; and m is the mass it will be measured to have in a reference frame in which it moves at speed v. Einstein's derivation of Eq. 26–3 was based on the assumption that the law of conservation of momentum is valid.

Relativistic mass increase has been tested countless times on tiny elementary particles (such as muons) and the mass has been found to increase in accord with Eq. 26–3.

EXAMPLE 26–4 Calculate the mass of an electron when it has a speed of (*a*) 4.00×10^7 m/s in the CRT of a television set, and (*b*) $0.98c$ in an accelerator used for cancer therapy.

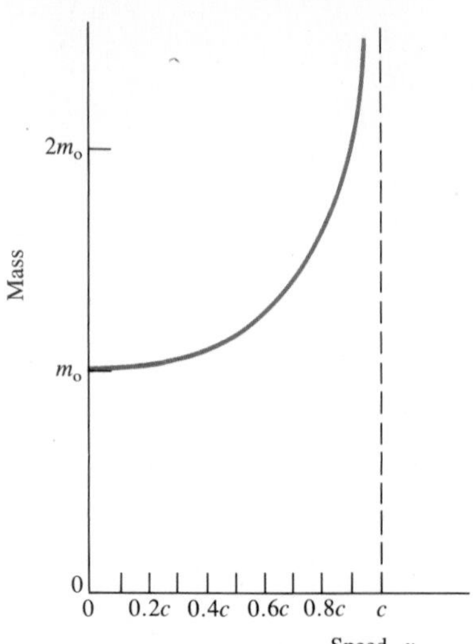

FIGURE 26–12 Mass of a particle (rest mass m_0) as a function of speed v (given as a fraction of c, the speed of light).

SOLUTION The rest mass of an electron is $m_0 = 9.11 \times 10^{-31}$ kg. (*a*) At $v = 4.00 \times 10^7$ m/s, the electron's mass will be

$$m = \frac{m_0}{\sqrt{1 - \dfrac{v^2}{c^2}}} = \frac{9.11 \times 10^{-31}\text{ kg}}{\sqrt{1 - \dfrac{(4.00 \times 10^7\text{ m/s})^2}{(3.00 \times 10^8\text{ m/s})^2}}} = 9.19 \times 10^{-31}\text{ kg}.$$

Even at such a high speed ($v \approx 0.1c$), the electron's mass is only about 1 percent higher then its rest mass. But in (*b*), we have

$$m = \frac{m_0}{\sqrt{1 - \dfrac{v^2}{c^2}}} = \frac{m_0}{\sqrt{1 - \dfrac{(0.98c)^2}{c^2}}} = \frac{m_0}{\sqrt{1 - (0.98)^2}} = 5.0m_0.$$

An electron traveling at 98 percent the speed of light has a mass five times its rest mass!

Figure 26–12 is a graph of mass versus speed for any particle.

26–9 • The Ultimate Speed

A basic result of the special theory of relativity is that the speed of an object cannot equal or exceed the speed of light. That the speed of light is a natural speed limit in the universe can be seen from any one of Eqs. 26–1, 26–2, and 26–3. It is perhaps easiest to see it from Eq. 26–3, the mass-increase formula, $m = m_0/\sqrt{1 - v^2/c^2}$. As an object is accelerated to greater and greater speeds, its mass becomes larger and larger. Indeed, if v were to equal c, the denominator in this equation would be zero and the mass m would become infinite. To accelerate an object up to $v = c$ would thus require infinite energy, and so is not possible.

If v were to exceed c, then the factor $\sqrt{1 - v^2/c^2}$ would be the square root of a negative number, which is imaginary: so lengths, time intervals, and mass would not be real. We conclude, then, that ordinary objects cannot equal or exceed the speed of light. However, as was pointed out in the 1960s, Einstein's equations do not rule out the possibility that objects exist whose speed is *always* greater than c. If such particles exist (the name "tachyon"—meaning "fast"—was proposed), the rest mass m_0 would have to be imaginary; in this way, the mass m would be the ratio of two imaginary numbers for $v > c$, which is real. For such hypothetical particles, c would be a *lower* limit on their speed. In spite of extensive searches for tachyons, none has been found. It seems that the speed of light *is* the ultimate speed in the universe.

26–10 • $E = mc^2$; Mass and Energy

When a steady net force is applied to an object of rest mass m_0, the object increases in speed. Since the force is acting through a distance, work is done on the object and its energy increases. As the speed of the object approaches

c, the speed cannot increase indefinitely since it cannot exceed c. On the other hand, the mass of the object increases with increasing speed. That is, the work done on an object not only increases its speed but also contributes to increasing its *mass*. Since the work done on an object increases its energy, this new twist from the theory of relativity leads to the idea that mass is a form of energy, a crucial part of Einstein's theory.

To find the mathematical relationship between mass and energy, Einstein assumed that the work–energy theorem (Chapter 6) is still valid in relativity. That is, the net work done on a particle is equal to its change in kinetic energy (KE). Using this theorem, Einstein showed that at high speeds the formula $\text{KE} = \frac{1}{2}mv^2$ is not correct. You might think that using Eq. 26–3 for m would give $\text{KE} = \frac{1}{2}m_0v^2/\sqrt{1 - v^2/c^2}$, but this formula, too, is wrong. Instead, Einstein showed that the kinetic energy of a particle is given by

$$\text{KE} = mc^2 - m_0c^2, \tag{26–4}$$

Relativistic kinetic energy

where m is the mass of the particle traveling at speed v and m_0 is its rest mass.

But what does the second term in Eq. 26–4—the m_0c^2—mean? Consistent with the idea that mass is a form of energy, Einstein called m_0c^2 the **rest energy** of the object. We can rearrange Eq. 26–4 to get $mc^2 = m_0c^2 + \text{KE}$. We call mc^2 the *total energy* E of the particle (assuming no potential energy), and we see that the total energy equals the rest energy plus the kinetic energy:

$$E = mc^2, \tag{26–5a}$$

$E = mc^2$, mass related to energy

or

$$E = m_0c^2 + \text{KE}. \tag{26–5b}$$

Here we have Einstein's famous formula $E = mc^2$.

For a particle at rest in a given reference frame, its total energy is $E_0 = m_0c^2$, which we have called its rest energy. This formula mathematically relates the concepts of energy and mass. But if this idea is to have any meaning from a practical point of view, then mass ought to be convertible to energy and vice versa. That is, if mass is just one form of energy, then it should be convertible to other forms of energy just as other types of energy are interconvertible. Einstein suggested that this might be possible, and indeed changes of mass to other forms of energy, and vice versa, have been experimentally confirmed countless times.

Mass and energy interchangeable

The interconversion of mass and energy is most easily detected in nuclear and elementary particle physics. For example, the neutral pion (π^0) of rest mass 2.4×10^{-28} kg is observed to decay into pure electromagnetic radiation (photons). The π^0 completely disappears in the process. The amount of electromagnetic energy produced is found to be exactly equal to that predicted by Einstein's formula, $E = m_0c^2$. The reverse process is also commonly observed in the laboratory: electromagnetic radiation under certain conditions can be converted into material particles such as electrons. On a larger scale, the energy produced in nuclear power plants is a result of the loss in mass of the uranium fuel as it undergoes the process called fission (Chapter 31). Even the radiant energy we receive from the sun is an instance of $E = mc^2$; the sun's mass is continually decreasing as it radiates energy outward.

The relation $E = mc^2$ is now believed to apply to all processes, although the changes are often too small to measure. That is, when the energy of a system changes by an amount ΔE, the mass of the system changes by an amount Δm given by

$$\Delta E = (\Delta m)(c^2).$$

In a chemical reaction where heat is gained or lost, the masses of the reactants and the products will be different. Even when water is heated on a stove, the mass of the water increases very slightly. This example is also easy to understand from the point of view of kinetic theory (Chapter 13), because as heat is added, the temperature and therefore the average speed of the molecules increases; and Eq. 26–3 tells us that the mass also increases.

EXAMPLE 26–5 A π^0 meson ($m_0 = 2.4 \times 10^{-28}$ kg) travels at a speed $v = 0.80c = 2.4 \times 10^8$ m/s. What is its kinetic energy? Compare to a classical calculation.

SOLUTION The mass of the π^0 moving with a speed of $v = 0.80c$ is

$$m = \frac{m_0}{\sqrt{1 - v^2/c^2}} = \frac{2.4 \times 10^{-28}\ \text{kg}}{\sqrt{1 - (0.80)^2}} = 4.0 \times 10^{-28}\ \text{kg}.$$

Thus its KE is

$$\text{KE} = (m - m_0)c^2 = (4.0 \times 10^{-28}\ \text{kg} - 2.4 \times 10^{-28}\ \text{kg})(3.0 \times 10^8\ \text{m/s})^2$$
$$= 1.4 \times 10^{-11}\ \text{J}.$$

Notice that the units of mc^2 are kg·m²/s², which is the joule. A classical calculation would give $\text{KE} = \frac{1}{2}m_0 v^2 = \frac{1}{2}(2.4 \times 10^{-28}\ \text{kg})(2.4 \times 10^8\ \text{m/s})^2 = 6.9 \times 10^{-12}$ J, about half as much, but this is not a correct result.

EXAMPLE 26–6 How much energy would be released if the π^0 meson in the last example is transformed by decay completely into electromagnetic radiation?

SOLUTION The rest energy of the π^0 is

$$E_0 = m_0 c^2 = (2.4 \times 10^{-28}\ \text{kg})(3.0 \times 10^8\ \text{m/s})^2 = 2.2 \times 10^{-11}\ \text{J}.$$

This is how much energy would be released if the π^0 decayed at rest. If it has $\text{KE} = 1.4 \times 10^{-11}$ J, the total energy released would be 3.6×10^{-11} J.

EXAMPLE 26–7 The energy required or released in nuclear reactions and decays comes from a change in mass between the initial and final particles. In one type of radioactive decay (Chapter 30), an atom of uranium ($m = 232.03714$ u) decays to an atom of thorium ($m = 228.02873$ u) plus an atom of helium ($m = 4.00260$ u) where the masses given are in atomic mass units ($1\ \text{u} = 1.6605 \times 10^{-27}$ kg). Calculate the energy released in this decay.

SOLUTION The initial mass is 232.03714 u, and after the decay it is 228.02873 u + 4.00260 u = 232.03133 u, so there is a decrease in mass of 0.00581 u. This mass, which equals $(0.00581\ \text{u})(1.66 \times 10^{-27}\ \text{kg}) = 9.64 \times 10^{-30}$ kg, is changed into energy. By $E = mc^2$, we have

$$E = (9.64 \times 10^{-30}\ \text{kg})(3.0 \times 10^{8}\ \text{m/s})^2 = 8.68 \times 10^{-13}\ \text{J}.$$

Since 1 MeV $= 1.60 \times 10^{-13}$ J, the energy released is 5.4 MeV.

Equation 26–4 for the kinetic energy can be written in terms of the speed v of the object with the help of Eq. 26–3:

$$\text{KE} = m_0c^2\left(\frac{1}{\sqrt{1 - v^2/c^2}} - 1\right). \qquad (26\text{–}6)$$

At low speeds, $v \ll c$, we can expand the square root in Eq. 26–6 using the binomial expansion (see Appendix A or Example 26–2). Then we get

$$\begin{aligned}\text{KE} &\approx m_0c^2\left(1 + \frac{1}{2}\frac{v^2}{c^2} + \cdots - 1\right)\\ &\approx \tfrac{1}{2}m_0v^2,\end{aligned}$$

where the dots in the first expression represent very small terms in the expansion which we have neglected since we assumed that $v \ll c$. Thus at low speeds, the relativistic form for kinetic energy reduces to the classical form, $\text{KE} = \frac{1}{2}m_0v^2$. This is, of course, what must be. Relativity would not be a valuable theory if it did not predict accurate results at low speed as well as at high. Indeed, the other equations of special relativity also reduce to their classical equivalents at ordinary speeds: length contraction, time dilation, and mass increase all disappear for $v \ll c$ since $\sqrt{1 - v^2/c^2} \approx 1$.

A useful relation between the total energy E of a particle and its momentum p can also be derived. Since $E = mc^2$ and $p = mv$, where $m = m_0/\sqrt{1 - v^2/c^2}$, we have

$$\begin{aligned}E^2 = m^2c^4 &= m^2c^2(c^2 + v^2 - v^2)\\ &= m^2c^2v^2 + m^2c^2(c^2 - v^2)\\ &= p^2c^2 + \frac{m_0^2c^4(1 - v^2/c^2)}{1 - v^2/c^2},\end{aligned}$$

or

$$E^2 = p^2c^2 + m_0^2c^4. \qquad (26\text{–}7)$$

Thus the total energy can be written in terms of the momentum p, or in terms of the kinetic energy (Eq. 26–5).

26–11 • Relativistic Addition of Velocities

Consider a rocket ship that travels away from the earth with speed v, and assume that this rocket has fired off a second rocket that travels at speed u' with respect to the first (Fig. 26–13). We might expect that the speed u of rocket 2 with respect to earth is $u = v + u'$, which in the case shown in the figure is $u = 0.60c + 0.60c = 1.20c$. But, as discussed in Section 26–9, no object can travel faster than the speed of light in any reference frame. Indeed, Einstein showed that since length and time are different in different reference

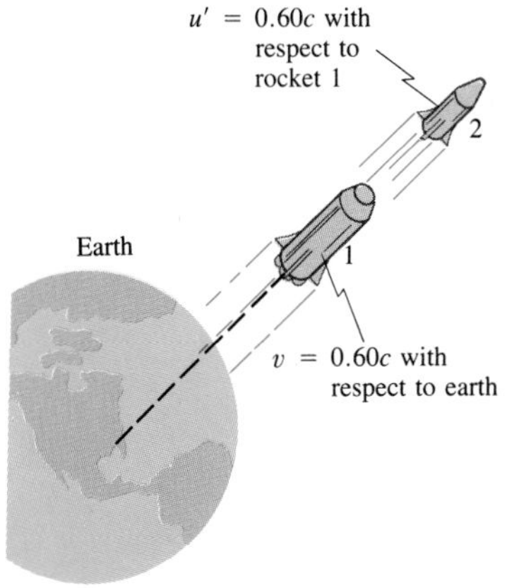

FIGURE 26–13 Rocket 2 is fired from rocket 1 with speed $u' = 0.60c$. What is the speed of rocket 2 with respect to the earth?

For the other components, $u'_y = u_y$ and $u'_z = u_z$, so we have

Galilean velocity transformations

$$u_x = u'_x + v,$$
$$u_y = u'_y, \qquad (26\text{–}10)$$
$$u_z = u'_z.$$

These are known as the **Galilean velocity transformation** equations. We see that the y and z components of velocity are unchanged, but the x components differ by v. This is just what we have used before when dealing with relative velocity. For example, if S' is a train and S the earth, and the train moves with speed v with respect to earth, a person walking toward the front of the train with speed u'_x will have a speed with respect to the earth of $u_x = u'_x + v$.

The Galilean transformations, Eqs. 26–9 and 26–10, are valid only when the velocities involved are much less than c. We can see, for example, that the first of Eqs. 26–10 will not work for the speed of light; for light traveling in S' with speed $u'_x = c$ will have speed $c + v$ in S, whereas the theory of relativity insists it must be c in S. Clearly, then, a new set of transformation equations is needed to deal with relativistic velocities.

We will derive the required equations in a simple way, again looking at Fig. 26–14. We assume the transformation is linear and of the form

$$x = \gamma(x' + vt'), \qquad y = y', \qquad z = z'.$$

That is, we modify the first of Eqs. 26–9 by multiplying by a factor γ which is yet to be determined. But we assume the y and z equations are unchanged since there is no length contraction in these directions. We won't assume a form for t, but will derive it. The inverse equations must have the same form with v replaced by $-v$. (The principle of relativity demands it, since S' moving to the right with respect to S is equivalent to S moving to the left with respect to S'.) Therefore

$$x' = \gamma(x - vt).$$

Now if a light pulse leaves the common origin of S and S' at time $t = t' = 0$, after a time t it will have traveled along the x axis a distance $x = ct$ (in S), or $x' = ct'$ (in S'). Therefore, from the equations for x and x' above,

$$ct = \gamma(ct' + vt') = \gamma(c + v)t',$$
$$ct' = \gamma(ct - vt) = \gamma(c - v)t.$$

We substitute t' from the second equation into the first and find $ct = \gamma(c + v)\gamma(c - v)(t/c) = \gamma^2(c^2 - v^2)t/c$. We cancel out the t on each side and solve for γ to find

$$\gamma = \frac{1}{\sqrt{1 - v^2/c^2}}.$$

Now that we have found γ, we need only find the relation between t and t'. To do so, we combine $x' = \gamma(x - vt)$ with $x = \gamma(x' + vt')$:

$$x' = \gamma(x - vt) = \gamma[\gamma(x' + vt') - vt].$$

We solve for t and find $t = \gamma(t' + vx'/c^2)$. In summary,

$$x = \frac{1}{\sqrt{1 - v^2/c^2}}(x' + vt')$$

$$y = y'$$

$$z = z' \qquad (26\text{–}11)$$

$$t = \frac{1}{\sqrt{1 - v^2/c^2}}\left(t' + \frac{vx'}{c^2}\right).$$

Lorentz transformations

These are called the **Lorentz transformation** equations. They were first proposed, in a slightly different form, by Lorentz in 1904 to explain the null result of the Michelson–Morley experiment and to make Maxwell's equations take the same form in all inertial systems. A year later, Einstein derived them independently based on his theory of relativity. Notice that not only is the x equation modified as compared to the Galilean transformation, but so is the t equation. Indeed, we see directly in this last equation, as well as in the first, how the space and time coordinates mix.

The relativistically correct velocity equations are readily obtained. For example (we let $\gamma = 1/\sqrt{1 - v^2/c^2}$),

$$u_x = \frac{\Delta x}{\Delta t} = \frac{\gamma(\Delta x' + v\,\Delta t')}{\gamma(\Delta t' + v\,\Delta x'/c^2)} = \frac{(\Delta x'/\Delta t') + v}{1 + (v/c^2)(\Delta x'/\Delta t')}$$

$$= \frac{u'_x + v}{1 + vu'_x/c^2}.$$

The others are obtained in the same way and we collect them here:

$$u_x = \frac{u'_x + v}{1 + vu'_x/c^2}$$

$$u_y = \frac{u'_y\sqrt{1 - v^2/c^2}}{1 + vu'_x/c^2} \qquad (26\text{–}12)$$

$$u_z = \frac{u'_z\sqrt{1 - v^2/c^2}}{1 + vu'_x/c^2}.$$

Relativistic velocity transformations

The first of these equations is just Eq. 26–8 of the previous section. As we saw there, velocities do not add in our commonsense (Galilean) way, because of the denominator $(1 + vu'_x/c^2)$. We see, too, that the y and z components of velocity are also altered and that they depend on the x' component of velocity.

EXAMPLE 26–9 Derive the length-contraction formula, Eq. 26–2, from the Lorentz transformation equations.

SOLUTION Let an object of length L_0 be at rest on the x axis in S. The coordinates of its two end points are x_1 and x_2, so that $x_2 - x_1 = L_0$. At any instant in S', the end points will be at x'_1 and x'_2 as given by the Lorentz transformation equations. The length measured in S' is $L = x'_2 - x'_1$. An observer in S' measures this length by measuring x'_2 and x'_1 at the same time (in the S' frame), so $t'_2 = t'_1$. Then, from the first of Eqs. 26–11,

$$L_0 = x_2 - x_1 = \frac{1}{\sqrt{1 - v^2/c^2}}(x_2' + vt_2' - x_1' - vt_1').$$

Since $t_2' = t_1'$, we have

$$L_0 = \frac{1}{\sqrt{1 - v^2/c^2}}(x_2' - x_1') = \frac{L}{\sqrt{1 - v^2/c^2}},$$

or

$$L = L_0\sqrt{1 - v^2/c^2},$$

which is Eq. 26–2.

EXAMPLE 26–10 Derive the time-dilation formula.

SOLUTION The time Δt_0 between two events that occur at the same place ($x_2' = x_1'$) in S' is measured to be $\Delta t_0 = t_2' - t_1'$. Since $x_2' = x_1'$, then from the last of Eqs. 26–11, the time Δt between the events as measured in S is

$$\Delta t = t_2 - t_1 = \frac{1}{\sqrt{1 - v^2/c^2}}\left(t_2' + \frac{vx_2'}{c^2} - t_1' - \frac{vx_1'}{c^2}\right)$$

$$= \frac{1}{\sqrt{1 - v^2/c^2}}(t_2' - t_1') = \frac{\Delta t_0}{\sqrt{1 - v^2/c^2}},$$

which is Eq. 26–1. Notice that we chose S' to be the frame in which the two events occur at the same place, so that $x_1' = x_2'$, and the terms containing x_1' and x_2' cancel out.

26–13 • The Impact of Special Relativity

A great many experiments have been performed to test the predictions of the special theory of relativity. Within experimental error, no contradictions have been found. The vast majority of scientists have therefore accepted relativity as an accurate description of nature.

At speeds much less than the speed of light, the relativistic formulas reduce to the old classical ones, as we have discussed. We would, of course, hope—or rather, insist—that this be true since Newtonian mechanics works so well for objects moving with speeds $v \ll c$. This insistence that a more general theory (such as relativity) give the same results as a more restricted theory (such as classical mechanics which works for $v \ll c$) is called the **correspondence principle**. The two theories must correspond where their realms of validity overlap. Relativity thus does not contradict classical mechanics. Rather, it is a more general theory, of which classical mechanics is now considered to be a special case.

Correspondence principle

The importance of relativity is not simply that it gives more accurate results, especially at very high speeds. Much more than that, it has changed the way we view the world. The concepts of space and time are now seen to be relative, and intertwined with one another, whereas before they were considered absolute and separate. Even our concepts of matter and energy have changed: either can be converted to the other. The impact of relativity extends far beyond physics. It has influenced the other sciences, and even the world of art and literature; it has, indeed, entered the general culture.

From a practical point of view, we do not have much opportunity in our daily lives to use the mathematics of relativity. For example, the factor $\sqrt{1 - v^2/c^2}$, which appears in many relativistic formulas, has a value of 0.995 when $v = 0.10c$. Thus, for speeds even as high as $0.10c = 3.0 \times 10^7$ m/s, the factor $\sqrt{1 - v^2/c^2}$ in relativistic formulas gives a numerical correction of less than 1 percent. For speeds less than $0.10c$, or unless mass and energy are interchanged, we thus don't usually need to use the more complicated relativistic formulas, and can use the simpler classical formulas.

The special theory of relativity we have dealt with in this chapter deals with inertial (nonaccelerating) reference frames. In Chapter 33 we will discuss briefly the more complicated "general theory of relativity" which can deal with noninertial reference frames.

SUMMARY

An *inertial reference frame* is one in which Newton's law of inertia holds. Inertial reference frames can move at constant velocity relative to one another; accelerating reference frames are noninertial.

The *special theory of relativity* is based on two principles: the *relativity principle*, which states that the laws of physics are the same in all inertial reference frames, and the principle of the *constancy of the speed of light*, which states that the speed of light in empty space has the same value in all inertial reference frames.

One consequence of relativity theory is that two events that are simultaneous in one reference frame may not be simultaneous in another. Other effects are *time dilation*: moving clocks are measured to run slowly; *length contraction*: the length of an object is measured to be shorter when it is moving than when it is at rest; *mass increase*: the mass of a body increases with speed; and velocity addition must be done in a special way. All these effects are significant only at high speeds, close to the speed of light, which itself is the ultimate speed in the universe.

The theory of relativity has changed our notions of space and time, and of mass and energy. Space and time are seen to be intimately connected, with time being the fourth dimension in addition to space's three dimensions. Mass and energy are interconvertible. The equation

$$E = mc^2$$

tells how much energy E is needed to create a mass m, or vice versa. Said another way, $E = mc^2$ is the amount of energy an object has because of its mass m. The law of conservation of energy must include mass as a form of energy.

QUESTIONS

1. Give some examples of noninertial reference frames.

2. A woman stands on top of a moving railroad car. She throws a heavy ball straight up (it seems to her). Ignoring air resistance, will the ball land on the car or behind it?

***3.** Discuss a Michelson–Morley type of experiment done with sound waves. Under what conditions would you expect or not expect a null result?

4. According to the principle of relativity, it is just as legitimate for a person riding in a uniformly moving automobile to say that the car is at rest and the earth is moving beneath it, as it is for a person on the ground to say that the car is moving and the earth is at rest. Do you agree, or are you reluctant to accept this? Discuss the reasons for your response.

5. Does the earth really go around the sun? Or is it also valid to say that the sun goes around the earth? Discuss in view of the first principle of relativity (that there is no best reference frame).

6. If you were on a spaceship traveling at $0.5c$ away from a star, at what speed would the starlight pass you?

7. Will two events that occur at the same place and same time for one observer be simultaneous to a second observer moving with respect to the first?

8. (*a*) Explain why two events will be simultaneous to each of two observers moving with respect to each other if the two events occur at the same point in each reference frame. (*b*) Under what other conditions will two events be simultaneous to each observer?
9. Analyze the thought experiment of Section 26–4 from O_1's point of view. (Make a diagram analogous to Fig. 26–7.)
10. The time-dilation effect is sometimes expressed as "moving clocks run slowly." Actually, this effect has nothing to do with motion affecting the functioning of clocks. What then does it deal with?
11. Does time dilation mean that time actually passes more slowly in moving reference frames or that it only *seems* to pass more slowly?
12. Today's subways are said to age people prematurely. Suppose that in the future, subway trains which traveled very close to the speed of light could be designed. How do you think this would affect the aging process?
13. A young-looking woman astronaut has just arrived home from a long trip. She rushes up to an old gray-haired man and in the ensuing conversation refers to him as her son. How might this be possible?
14. If you were traveling away from earth at a speed of $0.5c$, would you notice a change in your heartbeat? Would your mass, height, or waistline change? What would observers on earth using telescopes say about these things?
15. The predictions of the theory of relativity seem to contradict some of our everyday notions about the world and therefore don't seem to make sense. Examine these notions in detail and determine if any measurable contradiction exists.
16. Discuss how our everyday lives would be different if the speed of light were only 25 m/s.
17. Do mass increase, time dilation, and length contraction occur at ordinary speeds, say 90 km/h?
18. Suppose the speed of light were infinite. What would happen to the relativistic predictions of length contraction, time dilation, and mass increase?
19. Explain how the length-contraction and time-dilation formulas might be used to indicate that c is the limiting speed in the universe.
20. Consider an object of mass m to which is applied a constant force for an indefinite period of time. Discuss how its velocity and mass change with time.
21. A white-hot iron bar is cooled to room temperature. Does its mass change?
22. Does the equation $E = mc^2$ conflict with the conservation-of-energy principle? Explain.
23. Does $E = mc^2$ apply to particles that travel at the speed of light? Does it apply only to them?
24. An electron is limited to travel at speeds less than c. Does this put an upper limit on the momentum of an electron? If so, what is this upper limit?
25. If mass is a form of energy, does this mean that a spring has more mass when compressed than when relaxed?
26. It is not correct to say that "matter can neither be created nor destroyed." What must we say instead?
27. A neutrino is an elementary particle whose rest mass is very small, very possibly zero. Could you ever catch up to a neutrino that passed you?

PROBLEMS

SECTIONS 26–5 AND 26–6

1. (I) A beam of a certain type of elementary particle travels at a speed of 2.80×10^8 m/s. At this speed, the average lifetime is measured to be 1.80×10^{-6} s. What is the particle's lifetime at rest?
2. (I) What is the speed of a beam of pions if their average lifetime is measured to be 3.5×10^{-8} s? At rest, their lifetime is 2.6×10^{-8} s.
3. (I) A spaceship passes you at a speed of $0.90c$. You measure its length to be 80 m. How long would it be when at rest?
4. (I) You are sitting in your car when a very fast sports car passes you at a speed of $0.28c$. A person in that car says his car is 6.00 m long and yours is 6.15 m long. What do you measure for these two lengths?
5. (I) Suppose you decide to travel to a star 75 light-years away. How fast would you have to travel so the distance would be only 20 light-years?
6. (I) If you were to travel to a star 60 light-years from earth at a speed of 2.3×10^8 m/s, what would you measure this distance to be?
7. (II) A certain star is 45.0 light-years away. How long would it take a spacecraft traveling $0.970c$ to reach that star from earth, as measured by observers: (*a*) on earth: (*b*) on the spacecraft? (*c*) What is the distance traveled according to observers on the spacecraft? (*d*) What will the spacecraft occupants compute their speed to be from the results of (*b*) and (*c*)?
8. (II) A friend of yours travels by you in her fast sports vehicle at a speed of $0.660c$. It appears to be 5.80 m long and 1.20 m high. (*a*) What will be its length and height at rest? (*b*) How many seconds did you see elapse on your friend's watch when 20.0 s passed on yours? (*c*) How fast did you appear to be traveling to your friend? (*d*) How many seconds did she see elapse on your watch when she saw 20.0 s pass on hers?
9. (III) How fast must a pion be moving to travel 15 m before it decays? The average lifetime, at rest, is 2.6×10^{-8} s.

SECTION 26–8

10. (I) What is the mass of a proton traveling at $v = 0.90\ c$?
11. (I) At what speed will an object's mass be twice its rest mass?
12. (II) At what speed v will the mass of an object be 1 percent greater than its rest mass?
13. (II) Escape velocity from the earth is 40,000 km/h. What would be the percent increase in mass of a 5.5×10^5-kg spacecraft traveling at that speed?
14. (II) (*a*) What is the speed of an electron whose mass is 10,000 times its rest mass? Such speeds are reached in the Stanford Linear Accelerator, SLAC. (*b*) If the electrons travel in the lab through a tube 3.0 km long (as at SLAC), how long is this tube in the electrons' reference frame?

SECTION 26–10

15. (I) How much energy can be obtained from conversion of 1.0 mg of mass? How much mass could this energy raise to a height of 100 m?
16. (I) What is the kinetic energy of an electron whose mass is 3.0 times its rest mass?
17. (I) A certain chemical reaction requires 5.86×10^4 J of energy input for it to go. What is the increase in mass of the products over the reactants?
18. (II) Calculate the rest energy of an electron in joules and in MeV (1 MeV $= 1.60 \times 10^{-13}$ J).
19. (II) Calculate the rest energy of a proton in MeV.
20. (II) (*a*) By how much does the mass of the earth increase each year as a result solely of the sunlight reaching it? (*b*) How much mass does the sun lose per year? (Radiation from the sun reaches the earth at a rate of about 1400 W/m^2 of area perpendicular to the energy flow.)
21. (II) Calculate the kinetic energy and momentum of a proton traveling 9.2×10^7 m/s.
22. (II) What is the momentum of a 500-MeV proton (that is, one with KE = 500 MeV)?
23. (II) What is the speed of a proton accelerated by a potential difference of 250 MV?
24. (II) What is the speed of an electron whose KE is 1.00 MeV?
25. (II) What is the mass and speed of an electron that has been accelerated by a voltage of 200 kV?
26. (II) Calculate the mass of a proton ($m_0 = 1.67 \times 10^{-27}$ kg) whose kinetic energy is half its total energy. How fast is it traveling?
27. (II) Suppose a spacecraft of rest mass 30,000 kg is accelerated to $0.18c$. (*a*) How much kinetic energy would it have? (*b*) If you used the classical formula for KE, by what percentage would you be in error?
28. (II) Calculate the kinetic energy and momentum of a proton ($m_0 = 1.67 \times 10^{-27}$ kg) traveling 9.5×10^7 m/s. By what percentages would your calculations have been in error if you had used classical formulas?
29. (II) What is the speed and momentum of an electron ($m_0 = 9.11 \times 10^{-31}$ kg) whose kinetic energy is half its rest energy?
30. (II) An electron ($m_0 = 9.11 \times 10^{-31}$ kg) is accelerated from rest to speed v by a conservative force. In this process, its potential energy decreases by 5.60×10^{-14} J. Determine the electron's speed, v.
31. (II) Make a graph of the kinetic energy versus momentum for (*a*) a particle of nonzero rest mass, and (*b*) a particle with zero rest mass.
32. (II) What magnetic field intensity is needed to keep 400-GeV protons revolving in a circle of radius 1.0 km (at, say, the Fermilab synchrotron)? Use the relativistic mass. The proton's rest mass is 0.938 GeV/c^2. (1 GeV = 10^9 eV.)
33. (II) Show that the energy of a particle of charge e revolving in a circle of radius r in a magnetic field B is given by E (in eV) $= Brc$ in the relativistic limit ($v \approx c$).
34. (III) Show that the kinetic energy (KE) of a particle of rest mass m_0 is related to its momentum p by the equation $p = \sqrt{(\text{KE})^2 + 2(\text{KE})(m_0c^2)}/c$.

SECTION 26–11

35. (I) A person on a rocket traveling at $0.50c$ (with respect to the earth) observes a meteor come from behind and pass her at a speed she measures as $0.50c$. How fast is the meteor moving with respect to the earth?
36. (II) Two spaceships leave the earth in opposite directions, each with a speed of $0.50c$ with respect to the earth. (*a*) What is the velocity of spaceship 1 relative to spaceship 2? (*b*) What is the velocity of spaceship 2 relative to spaceship 1?
37. (II) A spaceship leaves earth traveling $0.75c$. A second spaceship leaves the first at a speed of $0.86c$ with respect to the first. Calculate the speed of the second ship with respect to earth if it is fired (*a*) in the same direction the first spaceship is already moving, (*b*) directly backward toward earth.

*SECTION 26–12

*38. (I) Suppose in Fig. 26–14 that the origins of S and S' overlap at $t = t' = 0$ and that S' moves at speed $v =$ 50 m/s with respect to S. In S', a person is resting at a point whose coordinates are $x' = 15$ m, $y' = 10$ m, and $z' = 0$. Calculate this person's coordinates in $S(x, y, z)$ at (*a*) $t = 5.0$ s, and (*b*) $t = 10.0$ s. Use the Galilean transformation.

*39. (I) Repeat Problem 38 using the Lorentz transformation and a relative speed $v = 2.0 \times 10^8$ m/s, but calculate at time t' equals (*a*) 5.0 μs and (*b*) 10.0 μs.

*40. (II) In Problem 38, suppose that the person moves with a velocity whose components are $u'_x = u'_y = 30$ m/s. What will be his velocity with respect to S? (Give magnitude and direction.)

*41. (II) In Problem 39, suppose that the person moves with a velocity (in a rocket) whose components are $u'_x = u'_y = 1.8 \times 10^8$ m/s. What will be the person's velocity (magnitude and direction) with respect to S?

*42. (II) A spaceship traveling $0.80c$ away from earth fires a module with a speed of $0.88c$ at right angles to its own direction of travel (as seen by the spaceship). What is the speed of the module, and its direction of travel (relative to the spaceship's direction), as seen by an observer on earth?

*43. (III) In the old West, a marshal riding on a train traveling 50 m/s sees a duel between two men standing on the earth 50 m apart parallel to the train. The marshal's instruments indicate that in his reference frame the two men fired simultaneously. (*a*) Which of the two men, the first one the train passes (A) or the second (B) should be arrested for firing the first shot? That is, in the gunfighters' frame of reference, who fired first? (*b*) How much earlier did he fire? (*c*) Who was struck first?

*44. (III) If a particle moves in the xy plane of system S (Fig. 26–14) in a direction that makes an angle θ with the x axis, show that it makes an angle θ' in S' given by $\tan\theta' = (\sin\theta)\sqrt{1 - v^2/c^2}/(\cos\theta - v/u)$.

GENERAL PROBLEMS

45. The period of a pendulum fixed on the earth is 2.00 s. What would be the pendulum's period when traveling on a spaceship at $v = 2.00 \times 10^8$ m/s as measured (*a*) on earth, and (*b*) on the spaceship?

46. Derive a formula showing how the density of an object changes with speed v relative to an observer.

47. An electron ($m_0 = 9.11 \times 10^{-31}$ kg) enters a uniform magnetic field $B = 1.5$ T, and moves perpendicular to the field lines with a speed $v = 0.60c$. What is the radius of curvature of its path?

48. The nearest star to earth is Proxima Centauri, 4.3 light-years away. (*a*) At what constant velocity must a spacecraft travel from earth if it is to reach the star in 2.5 years, as measured by travelers on the spacecraft? (*b*) How long does the trip take according to earth observers?

49. How many grams of matter would have to be totally destroyed to run a 100-W light bulb for 1 year?

50. What minimum amount of electromagnetic energy is needed to produce an electron and a positron together? A positron is a particle with the same rest mass as an electron, but has the opposite charge. (Note that electric charge is conserved in this process. See Section 27–3.)

51. The sun radiates energy at a rate of about 4×10^{26} W. (*a*) At what rate is the sun's mass decreasing? (*b*) How long does it take for the sun to lose a mass equal to that of earth? (*c*) Estimate how long the sun could last if it radiated constantly at this rate.

52. An airplane travels 1000 km/h around the world, returning to the same place, in a circle of radius essentially equal to that of the earth. Estimate the difference in time to make the trip as seen by earth and airplane observers. [*Hint:* use the binomial expansion, Appendix A.]

53. How much energy would be required to break a helium nucleus into its constituents, two protons and two neutrons? The rest masses of a proton (including an electron), a neutron, and helium are, respectively, 1.00783 u, 1.00867 u, and 4.00260 u. (This is called the *total binding energy* of the $^4_2$He nucleus.)

54. What is the percentage increase in the mass of a car traveling 120 km/h as compared to at rest?

55. A pi meson of rest mass m_π decays at rest into a muon (rest mass m_μ) and a neutrino of zero rest mass. Show that the kinetic energy of the muon is $\text{KE}_\mu = (m_\pi - m_\mu)^2c^2/2m_\pi$.

*56. A farm boy studying physics believes that he can fit a 12-m-long pole into a 10-m-long barn if he runs fast enough (carrying the pole). Can he do it? Explain in detail. How does this fit with the idea that when he is running the barn looks even shorter than 10 m?

CHAPTER 27

Early Quantum Theory and Models of the Atom

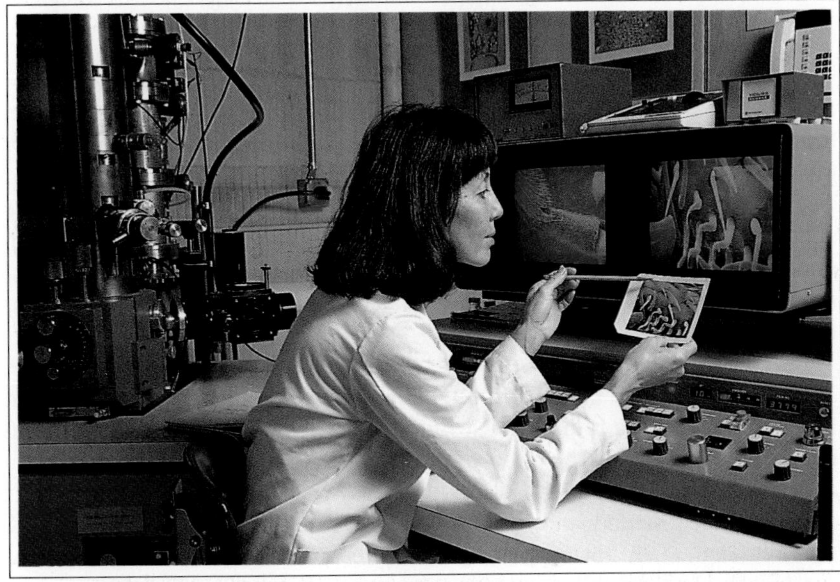

Electron microscopes produce images based on wave properties of electrons. Since the wavelength of electrons can be much smaller than that of visible light, much greater resolution and magnification can be obtained. The operation of this scanning electron microscope allows viewing the image directly on the screen of a CRT.

The second aspect of the revolution that shook the world of physics in the early part of the twentieth century (the first half was Einstein's theory of relativity) was the quantum theory. Unlike the special theory of relativity—whose basic tenets were put forth mainly by one person in a single year—the revolution of quantum theory required almost three decades to unfold, and many scientists contributed to its development. It began in 1900 with Planck's quantum hypothesis, and culminated in the mid-1920s with the theory of quantum mechanics of Schrödinger and Heisenberg which has been so effective in explaining the structure of matter.

27–1 • Planck's Quantum Hypothesis

One of the observations that was unexplained at the end of the nineteenth century was the spectrum of light emitted by hot objects. We saw in Chapter 14 that all objects emit radiation whose total intensity is proportional to the fourth power of the Kelvin temperature (T^4). At normal temperatures, we are not aware of this electromagnetic radiation because of its low intensity.

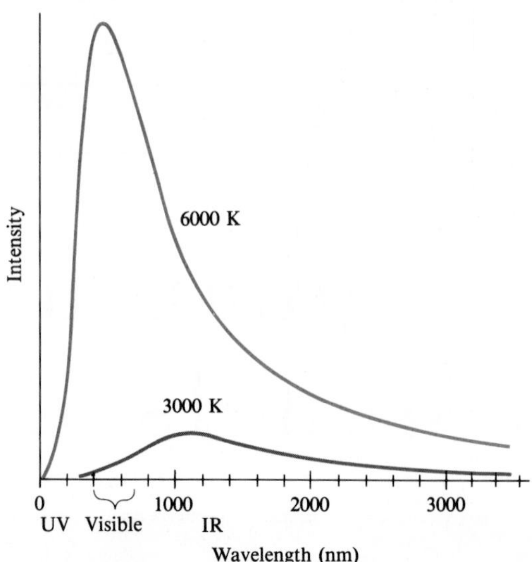

FIGURE 27–1
Spectrum of frequencies emitted by a blackbody at two different temperatures.

At higher temperatures, there is sufficient infrared radiation that we can feel heat if we are close to the object. At still higher temperatures (on the order of 1000 K), objects actually glow, such as a red-hot electric stove burner or element in a toaster (see photo at start of Chapter 18). At temperatures above 2000 K, objects glow with a yellow or whitish color, such as white-hot iron and the filament of a light bulb. As the temperature increases, the electromagnetic radiation emitted by bodies is strongest at higher and higher frequencies.

Blackbody radiation

The spectrum of light emitted by a hot dense object is shown in Fig. 27–1 for an idealized **blackbody**. Such a body would absorb all the radiation falling on it, and the radiation it would emit when hot and luminous, called **blackbody radiation**, is the easiest to deal with. As can be seen, the spectrum contains a continuous range of frequencies. Such a continuous spectrum is emitted by any heated solid or liquid, and even by dense gases. The 6000-K curve in Fig. 27–1, corresponding to the temperature of the surface of the sun, peaks in the visible part of the spectrum. For lower temperatures, the total radiation drops considerably and the peak occurs at higher wavelengths. Hence the blue end of the visible spectrum (and the UV) is relatively weaker. (This is why objects glow with a red color at around 1000 K.) It is found that the wavelength at the peak of the spectrum, λ_p, is related to the Kelvin temperature T by

$$\lambda_p T = 2.90 \times 10^{-3}\ \mathrm{m \cdot K}.$$

This is known as **Wien's displacement law** and for the sun's temperature gives us $\lambda_p = (2.90 \times 10^{-3}\ \mathrm{m \cdot K})/(6.0 \times 10^3\ \mathrm{K}) \approx 500$ nm, which is in the visible region of the spectrum.

A major problem facing scientists in the 1890s was to explain blackbody radiation. Maxwell's electromagnetic theory had predicted that oscillating electric charges produce electromagnetic waves, and the radiation emitted by a hot object could be due to the oscillations of electric charges in the molecules of the material. Although this would explain where the radiation came from, it did not correctly predict the observed spectrum of emitted light. Two important theoretical curves based on classical ideas were those proposed by Wien (1896) and by Rayleigh (1900). The latter was slightly modified later by Jeans and since then has been known as the Rayleigh–Jeans

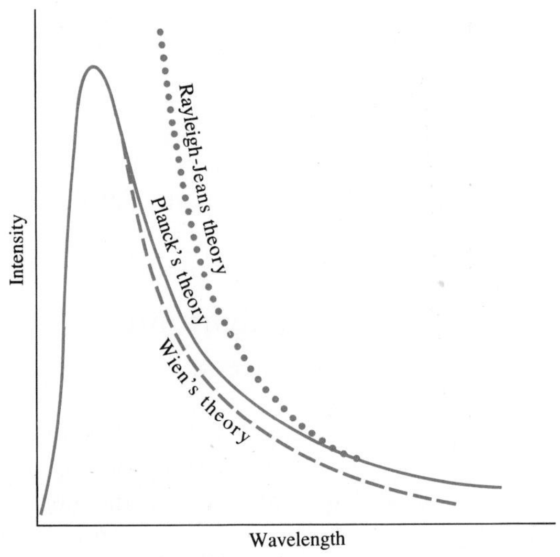

FIGURE 27–2 Comparison of the Wien and the Rayleigh-Jeans theories to that of Planck, which closely follows experiment.

theory. As experimental data came in, it became clear that neither Wien's nor the Rayleigh–Jeans formulations were in accord with experiment. Wien's was accurate at short wavelength but deviated from experiment at longer wavelengths, whereas the reverse was true for the Rayleigh–Jeans theory (see Fig. 27–2).

The break came in late 1900 when Max Planck (1858–1947; Fig. 27–3) proposed an empirical formula that nicely fit the data. He then sought a theoretical basis for the formula and within two months found that he could obtain the formula by making a new and radical (though not so recognized at the time) assumption: that the energy distributed among the molecular oscillators is not continuous but instead consists of a finite number of very small discrete amounts, each related to the frequency of oscillation by

Planck's quantum hypothesis

$$E_{\text{min}} = hf.$$

Here h is a constant, now called **Planck's constant**, whose value was estimated by Planck by fitting his formula for the blackbody radiation curve to experiment. The value accepted today is

$$h = 6.626 \times 10^{-34}\ \text{J}\cdot\text{s}.$$

Planck's assumption suggests that the energy of any molecular vibration could be only some whole number multiple of hf:

$$E = nhf, \qquad n = 1, 2, 3, \cdots. \tag{27–1}$$

This idea is often called **Planck's quantum hypothesis** ("quantum" means "fixed amount") although little attention was brought to this point at the time. In fact, it appears that Planck considered it more as a mathematical device to get the "right answer" rather than as a discovery comparable to those of Newton. Planck himself continued to seek a classical explanation for the introduction of h. The recognition that this was an important and radical innovation did not come until later, after about 1905 when others, particularly Einstein, entered the field.

The quantum hypothesis, Eq. 27–1, as we accept it today, says that the energy of an oscillator can be $E = hf$, or $2hf$, or $3hf$, and so on, but there cannot be vibrations whose energy lies between these values. That is, energy would not be a continuous quantity as had been believed for centuries; rather

FIGURE 27–3 Max Planck.

it is **quantized**—it exists only in discrete amounts. The smallest amount of energy possible (hf) is called the **quantum of energy**. Another way of expressing the quantum hypothesis is that not just any amplitude of vibration is possible. The possible values for the amplitude are related to the frequency f.

27–2 • Photon Theory of Light and the Photoelectric Effect

In 1905, the same year that he introduced the special theory of relativity, Einstein made a bold extension of the quantum idea by proposing a new theory of light. Planck's work had suggested that the vibrational energy of molecules in a radiating object is quantized with energy $E = nhf$, where n is an integer. Einstein argued that therefore, when light is emitted by a molecular oscillator, its energy of nhf must decrease by an amount hf (or by $2hf$, etc.) to another integer times hf, namely $(n - 1)hf$. Then to conserve energy, the light ought to be emitted in packets or quanta, each with an energy

$$E = hf. \tag{27–2}$$

Again h is Planck's constant. Since all light ultimately comes from a radiating source, this suggests that perhaps *light is transmitted as tiny particles*, or **photons**, as they are now called, rather than as waves. This, too, was a radical departure from classical ideas. Einstein proposed a simple test of the quantum theory of light: quantitative measurements on the photoelectric effect.

Photoelectric effect

The **photoelectric effect** is the phenomenon that when light shines on a metal surface, electrons are emitted from the surface. (The photoelectric effect occurs in other materials, but is most easily observed with metals.) It can be observed using the apparatus shown in Fig. 27–4. A metal plate P and a smaller electrode C are placed inside an evacuated glass tube, called a **photocell**. The two electrodes are connected to an ammeter and a source of emf, as shown. When the photocell is in the dark, the ammeter reads zero. But when light of sufficiently high frequency is shone on the plate, the ammeter indicates a current flowing in the circuit. To explain completion of the circuit, we can imagine electrons flowing across the tube from the plate to the "collector" C as shown in the diagram.

FIGURE 27–4
The photoelectric effect.

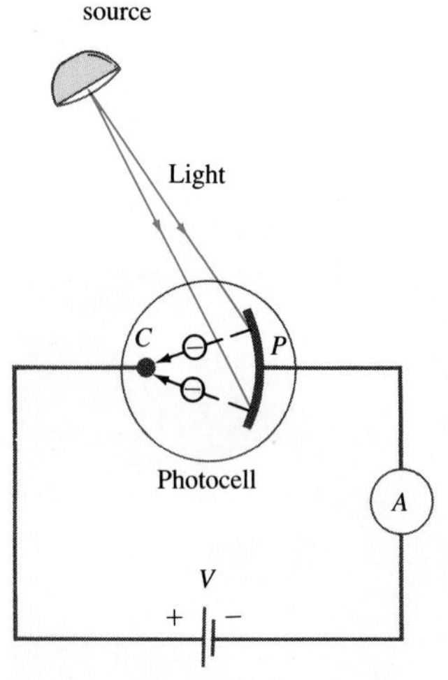

That electrons should be emitted when light shines on a metal is consistent with the electromagnetic (EM) wave theory of light, since the electric field of an EM wave could exert a force on electrons in the metal and thrust some of them out. Einstein pointed out, however, that the wave theory and the photon theory of light give very different predictions on the details of the photoelectric effect. For example, one thing that can be measured with the apparatus of Fig. 27–4 is the maximum kinetic energy (KE_{max}) of the emitted electrons. This can be done by using a variable voltage source and reversing the terminals so that electrode C is negative and P is positive. The electrons emitted from P will be repelled by the negative electrode, but if this reverse voltage is small enough, the fastest electrons will still reach C and there will be a current in the circuit. If the reversed voltage is increased, a point is reached where the current reaches zero—no electrons have sufficient kinetic energy to reach C. This is called the "stopping potential" V_0, and from its measurement, KE_{max} can be determined using conservation of energy (loss of KE = gain in PE):

$$\text{KE}_{\text{max}} = eV_0.$$

Now let us examine the details of the photoelectric effect in view of the wave theory versus Einstein's particle theory. First the wave theory, assuming monochromatic light. The two important properties of a light wave are its intensity and its frequency (or wavelength). When these two quantities are varied, the wave theory makes the following predictions:

1. If the light intensity is increased, the number of electrons ejected and their maximum KE should be increased because the higher intensity means a greater electric-field amplitude, and the greater electric field should thrust electrons out with higher speed.
2. The frequency of the light should not affect the KE of the ejected electrons. Only the intensity should affect KE_{max}.

The photon theory makes completely different predictions. First we note that in a monochromatic beam, all photons have the same energy ($=hf$). Increasing the intensity of the light beam means increasing the number of photons in the beam, but does not affect the energy of each photon as long as the frequency is not changed. According to Einstein's theory, an electron is ejected from the metal by a collision with a single photon. In the process, all the photon energy is transferred to the electron and the photon ceases to exist. Since electrons are held in the metal by attractive forces, some minimum energy W_0 (called the **work function**, which is on the order of a few electron volts for most metals) is required just to get an electron out through the surface. If the frequency f of the incoming light is so low that hf is less than W_0, then the photons will not have enough energy to eject any electrons at all. If $hf > W_0$, then electrons will be ejected and energy will be conserved in the process. That is, the input energy (of the photon), hf, will equal the outgoing KE of the electron plus the energy required to get it out of the metal, W:

$$hf = \text{KE} + W. \qquad (27\text{–}3a)$$

For the least tightly held electrons, W is the work function W_0, and KE in this equation becomes KE_{max}:

$$hf = \text{KE}_{\text{max}} + W_0. \qquad (27\text{–}3b)$$

Many electrons will require more energy than the bare minimum (W_0) to get out of the metal, and thus the KE of such electrons will be less than the maximum.

From these considerations, the photon theory makes the following predictions:

1. An increase in intensity of the light beam means more photons are incident, so more electrons will be ejected; but since the energy of each photon is not changed, the maximum KE of electrons is not changed.

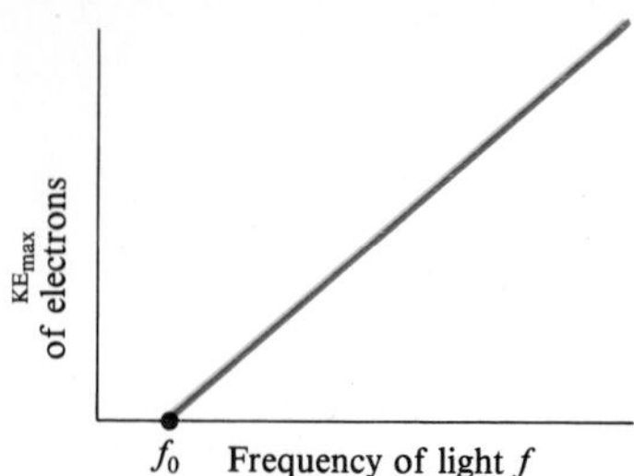

FIGURE 27–5 Photoelectric effect: maximum kinetic energy of ejected electrons increases linearly with frequency of incident light. No electrons are emitted if $f < f_0$.

FIGURE 27–6 The Compton effect.

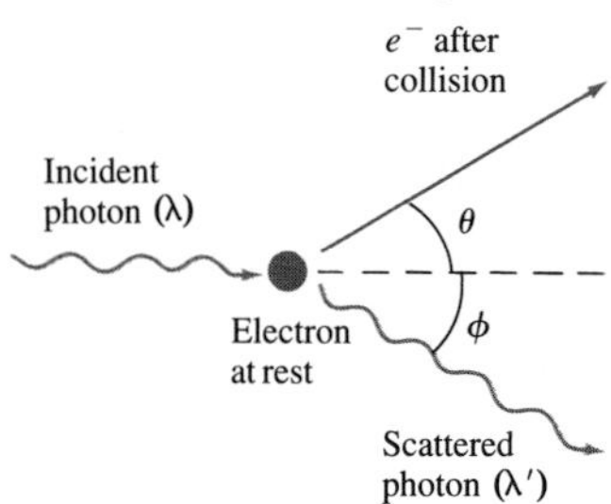

2. If the frequency of the light is increased, the maximum KE of the electrons increases linearly, according to Eq. 27–3b. That is,
$$KE_{max} = hf - W_0.$$
This relationship is plotted in Fig. 27–5.
3. If the frequency f is less than the "cutoff" frequency f_0, where $hf_0 = W_0$, no electrons will be ejected at all, no matter how great the intensity.

These predictions of the photon theory are clearly very different from the predictions of the wave theory. In 1913–1914, careful experiments were carried out by R. A. Millikan. The results were fully in agreement with Einstein's photon theory.

A number of other experiments were carried out in the early twentieth century that also supported the photon theory. One of these was the **Compton effect** (1923), named after its discoverer, A. H. Compton (1892–1962). Compton scattered short-wavelength light (actually X rays) from various materials. He found the scattered light had a slightly lower frequency than did the incident light, indicating a loss of energy. This finding, he showed, could be explained on the basis of the photon theory, as incident photons colliding with electrons of the material, Fig. 27–6. He applied the laws of conservation of energy and momentum to such collisions and found that the predicted energies of scattered photons was in accord with experimental results. Thus the photon theory of light rests on a firm experimental foundation.

EXAMPLE 27–1 Calculate the energy of a photon of blue light, $\lambda = 450$ nm.

SOLUTION Since $f = c/\lambda$, we have
$$E = hf = \frac{hc}{\lambda} = \frac{(6.63 \times 10^{-34}\ \text{J}\cdot\text{s})(3.0 \times 10^{8}\ \text{m/s})}{(4.5 \times 10^{-7}\ \text{m})} = 4.4 \times 10^{-19}\ \text{J},$$
or $(4.4 \times 10^{-19}\ \text{J})/(1.6 \times 10^{-19}\ \text{J/eV}) = 2.7$ eV.

EXAMPLE 27–2 What is the maximum kinetic energy and speed of an electron ejected from a sodium surface whose work function is $W_0 = 2.28$ eV when illuminated by light of wavelength: (*a*) 410 nm; (*b*) 550 nm?

SOLUTION (*a*) For $\lambda = 410$ nm, $hf = hc/\lambda = 4.85 \times 10^{-19}$ J or 3.03 eV. From Eq. 27–3b, $KE_{max} = 3.03\ \text{eV} - 2.28\ \text{eV} = 0.75$ eV, or 1.2×10^{-19} J. Since $KE = \frac{1}{2}mv^2$ where $m = 9.1 \times 10^{-31}$ kg,
$$v = \sqrt{2\,KE/m} = 5.1 \times 10^{5}\ \text{m/s}.$$
Notice that we used the nonrelativistic equation for KE. If v had turned out to be more than about $0.1c$, our calculation would have been inaccurate by more than a percent or so, and we would probably prefer to redo it using the relativistic form (Eq. 26–6).

(*b*) For $\lambda = 550$ nm, $hf = 3.60 \times 10^{-19}\ \text{J} = 2.25$ eV. Since this photon energy is less than the work function, no electrons are ejected.

The photoelectric effect, besides playing an important historical role in confirming the photon theory of light, also has many practical applications.

Burglar alarms and automatic door openers often make use of the photocell circuit of Fig. 27–4. When a person interrupts the beam of light, the sudden drop in current in the circuit activates a switch—often a solenoid—to operate a bell or open the door. Many smoke detectors use the photoelectric effect to detect tiny amounts of smoke that interrupt the flow of light and so alter the electric current. Photographic light meters use this circuit as well. For many applications today, the vacuum-tube photocell of Fig. 27–4 has been replaced by a semiconductor device known as a **photodiode**. In these semiconductors, the absorption of a photon liberates a bound electron, which changes the conductivity of the material, so the current through a photodiode is altered.

The photon theory of light is useful in biology and medicine; here is an example.

EXAMPLE 27–3 In **photosynthesis**, which is the process by which pigments such as chlorophyll in plants capture the energy of sunlight to change CO_2 to useful carbohydrate, about nine photons are needed to transform one molecule of CO_2 to carbohydrate and O_2. Assuming light of wavelength $\lambda = 670$ nm (chlorophyll absorbs most strongly in the range 650 nm to 700 nm), how efficient is the photosynthetic process? The reverse chemical reaction has a heat of combustion of 4.9 eV/molecule of CO_2.

SOLUTION The energy of nine photons, each of energy $hf = hc/\lambda$, is $(9)(6.6 \times 10^{-34}\ \text{J}\cdot\text{s})(3.0 \times 10^{8}\ \text{m/s})/(6.7 \times 10^{-7}\ \text{m}) = 2.7 \times 10^{-18}\ \text{J}$, or 17 eV. Thus the process is (4.9 eV/17 eV) = 29 percent efficient.

27–3 • Photon Interactions; Pair Production

The photon is truly a relativistic particle—it travels at the speed of light. Thus we must use relativistic formulas for dealing with its mass, energy, and momentum. The mass m of any particle is given by $m = m_0/\sqrt{1 - v^2/c^2}$. Since $v = c$ for a photon, the denominator is zero. So the rest mass, m_0, of a photon must also be zero, or its energy $E = mc^2$ would be infinite. Of course a photon is never at rest. The momentum of a photon, from Eq. 26–7 with $m_0 = 0$, is $E^2 = p^2c^2$, or

$$p = \frac{E}{c}.$$

Since $E = hf$, the momentum of a photon is related to its wavelength by

$$p = \frac{hf}{c} = \frac{h}{\lambda}. \tag{27–4}$$

When a photon passes through matter, it interacts with the atoms and electrons. There are four important types of interactions that a photon can undergo. First, the photon can be scattered off an electron (or a nucleus) and in the process lose some energy; this is the Compton effect (Fig. 27–6). But notice that the photon is not slowed down. It still travels with speed c, but its frequency will be lower. A second type of interaction is the photoelectric effect: a photon may knock an electron out of an atom and in the process itself disappear. The third process is similar: the photon may knock an atomic electron to a higher energy state in the atom if its energy is not sufficient to knock the electron out altogether. In this process the photon also disappears, and all its energy is given to the atom. Such an atom is then said to be in an excited state, and we shall discuss this more later. Finally, a photon can actually create matter. The most common process is the production of an electron and a positron, Fig. 27–7. (A positron has the same mass as an electron, but the opposite charge, $+e$.) This is called **pair production** and the photon disappears in the process. This is an example of rest mass being created from pure energy, and it occurs in accord with Einstein's equation $E = mc^2$. Notice that a photon cannot create a single electron since electric charge would not then be conserved. The inverse of pair production also occurs: if an electron collides with a positron, the two **annihilate** each other and their energy, including their mass, appears as electromagnetic energy of photons. Because of this process, positrons usually do not last long in nature.

Pair production

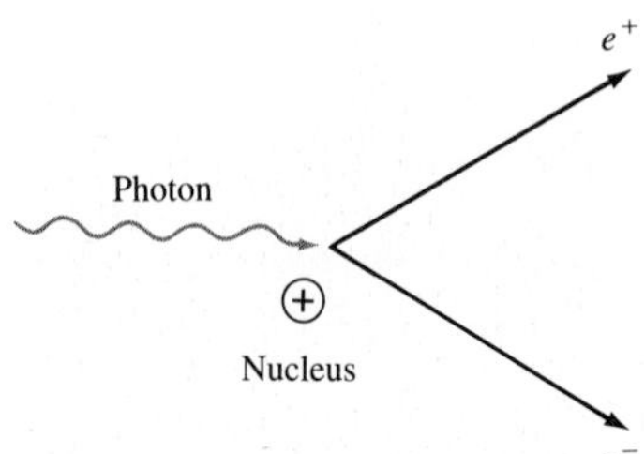

FIGURE 27–7 Pair production: a photon disappears and produces an electron and a positron.

EXAMPLE 27–4 What is the minimum energy of a photon, and its wavelength, that can produce an electron–positron pair?

SOLUTION Because $E = mc^2$, the photon must have energy $E = 2(9.1 \times 10^{-31}\text{ kg})(3.0 \times 10^8\text{ m/s})^2 = 1.64 \times 10^{-13}$ J, or 1.02 MeV. A photon with less energy cannot undergo pair production. Since $E = hf = hc/\lambda$, the wavelength of a 1.02-MeV photon is

$$\lambda = \frac{hc}{E} = \frac{(6.6 \times 10^{-34}\text{ J}\cdot\text{s})(3.0 \times 10^8\text{ m/s})}{(1.64 \times 10^{-13}\text{ J})} = 1.2 \times 10^{-12}\text{ m},$$

which is 0.0012 nm. Thus the wavelength must be very short. Such photons are in the gamma-ray (or very short X-ray) region of the electromagnetic spectrum.

Pair production cannot occur in empty space, for energy and momentum could not simultaneously be conserved. In Example 27–4, for instance, energy is conserved, but the electron–positron pair has no momentum to carry away the initial momentum of the photon. Indeed, it can be shown that at any energy, an additional massive object, such as an atomic nucleus, must take part in the interaction to carry off some of the momentum.

27–4 • Wave–Particle Duality; the Principle of Complementarity

The photoelectric effect, the Compton effect, and other experiments (see, for example, Section 28–9 on X rays) have placed the particle theory of light on a firm experimental basis. But what about the classic experiments of Young and others (Chapter 24) on interference and diffraction which showed that the wave theory of light also rests on a firm experimental basis?

We seem to be in a dilemma. Some experiments indicate that light behaves like a wave; and others indicate that it behaves like a stream of particles. These two theories seem to be incompatible, but both have been shown to have validity. Physicists have finally come to the conclusion that this duality of light must be accepted as a fact of life. It is referred to as the **wave–particle duality**. Apparently, light is a more complex phenomenon than just a simple wave or a simple beam of particles.

Wave–particle duality

To clarify the situation, the great Danish physicist Niels Bohr (1885–1962, Fig. 27–8) proposed his famous **principle of complementarity**. It states that to understand any given experiment, we must use either the wave or the photon theory, but not both. Yet we must be aware of both the wave and particle aspects of light if we are to have a full understanding of light. Therefore these two aspects of light complement one another.

Principle of complementarity

It is not possible to "visualize" this duality. We cannot picture a combination of wave and particle. Instead, we must recognize that the two aspects of light are different "faces" that light shows to experimenters.

FIGURE 27–8 Niels Bohr, walking with Enrico Fermi along the Appian Way outside Rome.

Part of the difficulty stems from how we think. Visual pictures (or models) in our minds are based on what we see in the everyday world. We apply the concepts of waves and particles to light because in the macroscopic world we see that energy is transferred from place to place by these two methods. We cannot see directly whether light is a wave or particle—so we do indirect experiments. To explain the experiments, we apply the models of waves or of particles to the nature of light. But these are abstractions of the human mind. When we try to conceive of what light really "is," we insist on a visual picture. Yet there is no reason why light should conform to these models (or visual images) taken from the macroscopic world. The "true" nature of light—if that means anything—is not possible to visualize. The best we can do is recognize that our knowledge is limited to the indirect experiments, and that in terms of everyday language and images, light reveals both wave and particle properties.

It is worth noting that Einstein's equation $E = hf$ itself links the particle and wave properties of a light beam. In this equation, E refers to the energy of a particle; and on the other side of the equation, we have the frequency f of the corresponding wave.

27–5 • Wave Nature of Matter

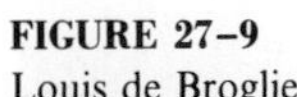

FIGURE 27–9 Louis de Broglie.

In 1923, Louis de Broglie (1892–1987; Fig. 27–9) extended the idea of the wave–particle duality. He sensed deeply the symmetry in nature and argued that if light sometimes behaves like a wave and sometimes like a particle, then perhaps those things in nature thought to be particles—such as electrons and other material objects—might also have wave properties. De Broglie proposed that the wavelength of a material particle would be related to its momentum in the same way as for a photon,[†] Eq. 27–4, $p = h/\lambda$. That is, for a particle of mass m traveling with speed v, the wavelength λ is given by

$$\lambda = \frac{h}{mv}. \tag{27–5}$$

This is sometimes called the **de Broglie wavelength** of a particle.

[†] De Broglie chose this formula (rather than, say, $E = hf$, which is not consistent with $p = h/\lambda$ for a particle with non-zero rest mass), because it allowed him to explain, or give a reason for, Bohr's model of the atom. This will be discussed in Section 27–10.

EXAMPLE 27–5 Calculate the de Broglie wavelength of a 0.20-kg ball moving with a speed of 15 m/s.

SOLUTION $\lambda = h/mv = (6.6 \times 10^{-34}\ \text{J}\cdot\text{s})/(0.20\ \text{kg})(15\ \text{m/s})$
$= 2.2 \times 10^{-34}\ \text{m}.$

This is an incredibly small wavelength. Even if the speed were extremely small, say 10^{-4} m/s, the wavelength would be about 10^{-29} m. Indeed, the wavelength of any ordinary object is much too small to be measured and detected. The problem is that the properties of waves, such as interference and diffraction, are significant only when the size of objects or slits is not much larger than the wavelength. And there are no known objects or slits to diffract waves only 10^{-30} m long, so the wave properties of ordinary objects go undetected.

But tiny elementary particles, such as electrons, are another matter. Since the mass m appears in the denominator in Eq. 27–5, a very small mass should give a much larger wavelength.

EXAMPLE 27–6 Determine the wavelength of an electron that has been accelerated through a potential difference of 100 V.

SOLUTION We assume that the speed of the electron will be much less than c, so we use nonrelativistic mechanics. (If this assumption were to come out wrong, we would have to recalculate using relativistic formulas—see Section 26–10.) The gain in KE will equal the loss in PE, so $\frac{1}{2}mv^2 = eV$ and

$$v = \sqrt{2eV/m} = \sqrt{(2)(1.6 \times 10^{-19}\ \text{C})(100\ \text{V})/(9.1 \times 10^{-31}\ \text{kg})} = 5.9 \times 10^{6}\ \text{m/s}.$$

Then

$$\lambda = \frac{h}{mv} = \frac{(6.6 \times 10^{-34}\ \text{J}\cdot\text{s})}{(9.1 \times 10^{-31}\ \text{kg})(5.9 \times 10^{6}\ \text{m/s})} = 1.2 \times 10^{-10}\ \text{m},$$

or 0.12 nm.

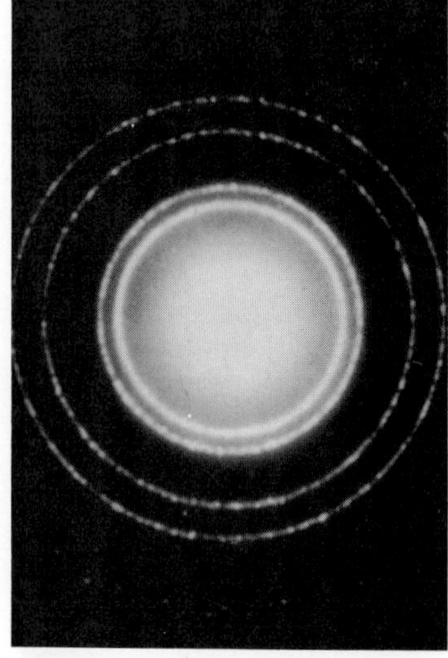

FIGURE 27–10 Diffraction pattern of electrons scattered from aluminum foil, as recorded on film.

From this example, we see that electrons can have wavelengths on the order of 10^{-10} m. Although small, this wavelength can be detected: the spacing of atoms in a crystal is on the order of 10^{-10} m and the orderly array of atoms in a crystal could be used as a type of diffraction grating, as was done earlier for X rays (see Section 25–11). C. J. Davisson and L. H. Germer performed the crucial experiment; they scattered electrons from the surface of a metal crystal and, in early 1927, observed that the electrons were scattered into a pattern of regular peaks. When they interpreted these peaks as a diffraction pattern, the wavelength of the diffracted electron wave was found to be just that predicted by de Broglie, Eq. 27–5. In the same year, G. P. Thomson (son of J. J. Thomson), using a different experimental arrangement, also detected diffraction of electrons (see Fig. 27–10). Later experiments showed that protons, neutrons, and other particles also have wave properties.

Complementarity principle applies also to matter

Thus the wave–particle duality applies to material objects as well as to light. The principle of complementarity applies to matter as well. That is, we must be aware of both the particle and wave aspects in order to have an understanding of matter, including electrons. But again we must recognize that a visual picture of a "wave–particle" is not possible.

What is an electron?

We might ask ourselves: "What is an electron?" The early experiments of J. J. Thomson (see Section 20–9) indicated a glow in a tube that moved when a magnetic field was applied. The results of these and other experiments were best interpreted as being caused by tiny negatively charged particles which we now call electrons. No one, however, has actually seen an electron directly. The drawings we sometimes make of electrons as tiny spheres with a negative charge on them are merely convenient pictures (now recognized to be inaccurate). Again we must rely on experimental results, some of which are best interpreted using the particle model and others using the wave model. These models are mere pictures that we use to extrapolate from the macroscopic world to the tiny microscopic world of the atom. And there is no reason to expect that these models somehow reflect the reality of an electron. We thus use a wave or a particle model (whichever works best in a situation) so that we can talk about what is happening. But we shouldn't be led to believe that an electron *is* a wave or a particle. Instead, we could say that an electron is the set of its properties that we can measure. Bertrand Russell said it well when he wrote that an electron is "a logical construction."

*27–6 • Electron Microscopes

The idea that electrons have wave properties led to the development of the **electron microscope**, which can produce images of much greater magnification than a light microscope. Figures 27–11 and 27–12 are diagrams of two types,

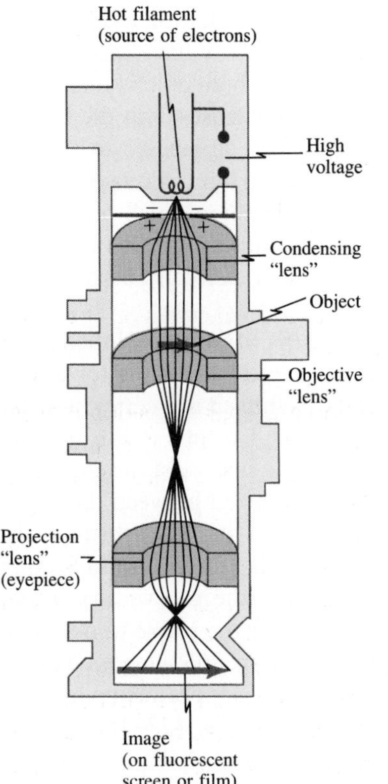

FIGURE 27–11 (left) Transmission electron microscope. The magnetic-field coils are designed to be "magnetic lenses," which bend the electron paths and bring them to a focus, as shown.

FIGURE 27–12 (below) Scanning electron microscope. Scanning coils move an electron beam back and forth across the specimen. Secondary electrons produced when the beam strikes the specimen are collected and modulate the intensity of the beam in the CRT to produce a picture.

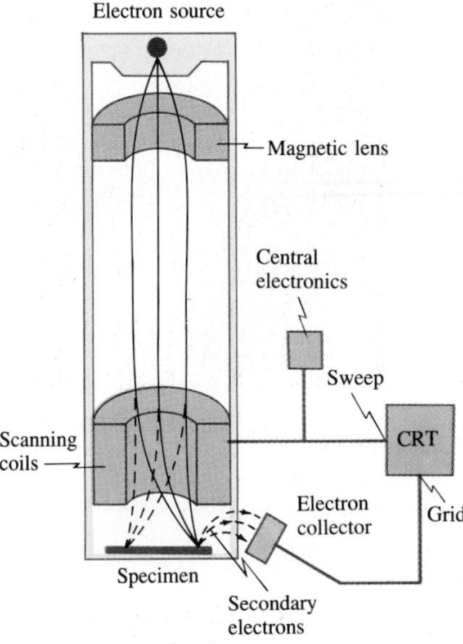

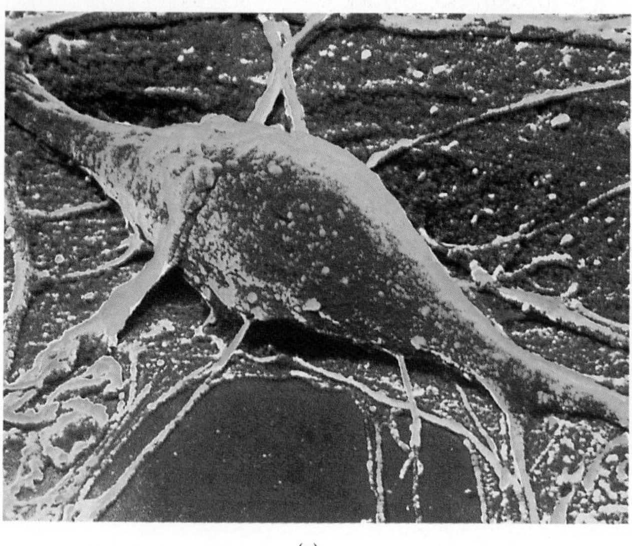

(a)

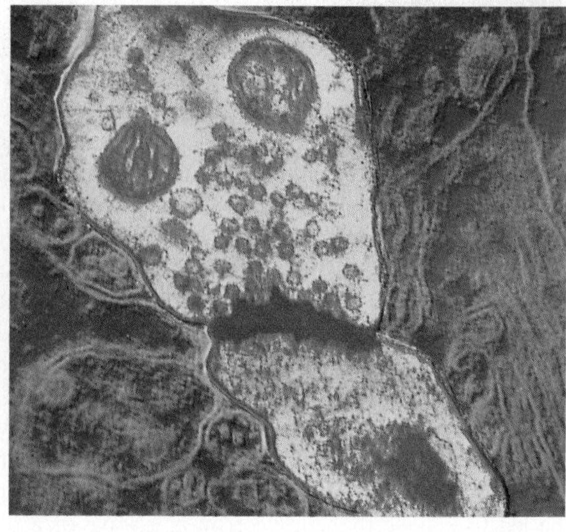

(b)

FIGURE 27–13 Electron micrographs (in false color) of neurons of the human cerebral cortex: (a) scanning electron micrograph of a single neuron ($\approx 4000\times$); (b) transmission electron micrograph of synapse (junction) between two neurons ($\approx 40{,}000\times$).

the **transmission electron microscope**, which produces a two-dimensional image, and the **scanning electron microscope**, which produces images with a three-dimensional quality. In each design, the objective and eyepiece lenses are actually magnetic fields that exert forces on the electrons to bring them to a focus. The fields are produced by carefully designed current-carrying coils of wire. Photographs using each type are shown in Fig. 27–13.

As discussed in Sections 25–7 and 25–8, the best resolution of details on an object is on the order of the wavelength of the radiation used to view it. Electrons accelerated by voltages on the order of 10^5 V have wavelengths on the order of 0.004 nm. The maximum resolution obtainable would be on this order, but in practice aberrations in the magnetic lenses limit the resolution in transmission electron microscopes to at best about 0.2 to 0.5 nm. This is still 10^3 times finer than that attainable with a light microscope, and corresponds to a useful magnification of about a million. Such magnifications are difficult to attain, and more common magnifications are 10^4 to 10^5. The maximum resolution attainable with a scanning electron microscope is somewhat less, about 5 to 10 nm at best.

FIGURE 27–14 Probe tip of scanning tunneling electron microscope moves up and down to maintain constant tunneling current, producing an image of the surface.

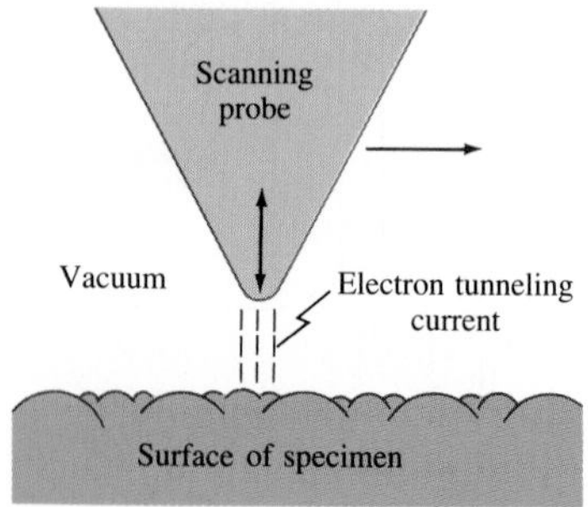

The **scanning tunneling electron microscope** (STM) was developed more recently in the 1980s. A tiny probe, whose tip may be only one (or a few) atoms wide, is moved across the specimen to be examined in a series of linear passes, like those made by the electron beam in a TV tube (CRT, Section 20–10). The tip, as it scans, remains very close to the surface of the specimen, about 1 nm above it, Fig. 27–14. A small voltage applied between the probe and the surface causes electrons to leave the surface and pass through the vacuum to the probe, by a process known as tunneling (discussed in Section 30–12). This "tunneling" current is very sensitive to the gap width, so that a feedback mechanism can be used to raise and lower the probe to maintain a constant electron current. The probe's vertical motion, following the surface of the specimen, is then plotted as a function of position, producing a three-dimensional image of the surface. (See, for example, Fig. 27–15). Surface

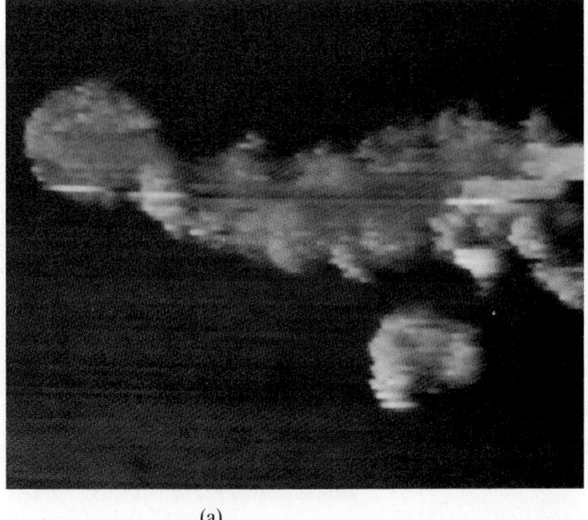

(a)

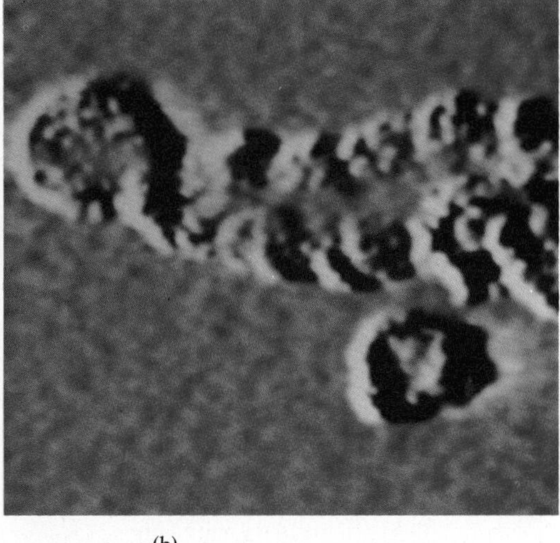

(b)

(c)

FIGURE 27–15 These images of cellular DNA, magnified one million times, were taken with a scanning tunneling microscope. They show an isolated length of DNA making a loop and crossing over on itself. Next to this length is a fragment of DNA: (a) shows the raw image as it was recorded by the STM, and (b) shows the same image after it was enhanced by a computer; (c) is a so-called "isometric" projection of the same DNA segment revealing the average distance between the coils of the helix to be about 35 angstroms.

features as fine as the size of an atom can be resolved: a resolution better than 0.1 nm laterally and 10^{-2} to 10^{-3} nm vertically. This kind of resolution was not available previously and has given a great impetus to the study of the surface structure of materials. The "topographic" image of a surface actually represents the distribution of electron charge.

27–7 • Early Models of the Atom

The idea that matter is made up of atoms was accepted by most scientists by 1900. With the discovery of the electron in the 1890s, scientists began to think of the atom itself as having a structure and that electrons were part of that structure. We now trace, in the remainder of this chapter and the next,

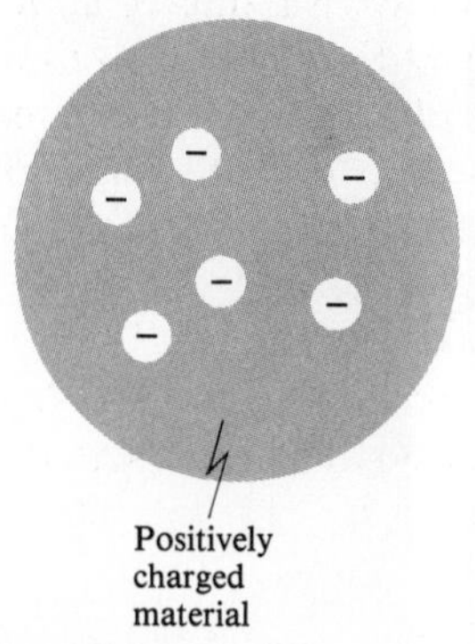

FIGURE 27–16 Plum-pudding model of the atom.

FIGURE 27–17 Ernest Rutherford, with J. J. Thomson.

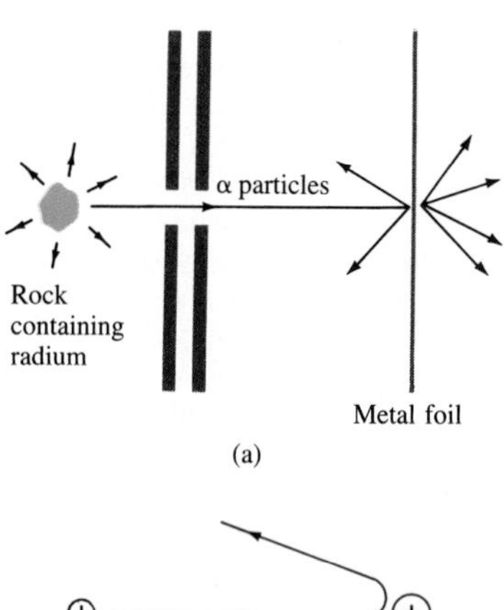

FIGURE 27–18 (a) Experimental setup for Rutherford's experiment: α particles emitted by radium strike metallic foil and some rebound backward; (b) backward rebound of α particles explained as repulsion from heavy positively charged nucleus.

FIGURE 27–19 Rutherford's model of the atom, in which electrons orbit a tiny positive nucleus (not to scale). The atom is visualized as mostly empty space.

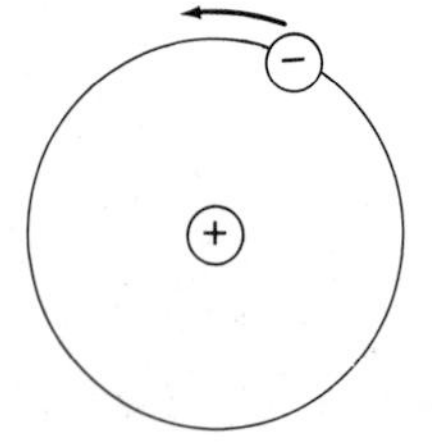

the development of our modern understanding of the atom, and of the quantum theory with which it is intertwined.†

A typical model of the atom in the 1890s visualized the atom as a homogeneous sphere of positive charge inside of which there were the negatively charged electrons, a little like plums in a pudding, Fig. 27–16. J. J. Thomson, soon after his discovery of the electron in 1897, argued that the electrons in this model should be moving.

Around 1911, Ernest Rutherford (1871–1937; Fig. 27–17) and his colleagues performed experiments whose results contradicted Thomson's model of the atom. In these experiments a beam of positively charged "alpha (α) particles" was directed at a thin sheet of metal foil such as gold, Fig. 27–18a. (These newly discovered α particles were emitted by certain radioactive materials and were soon shown to be doubly ionized helium atoms—see Chapter 30.) It was expected from Thomson's model that the alpha particles would not be deflected significantly since electrons are so much lighter than alpha particles, and the alpha particles should not have encountered any massive concentration of positive charge to strongly repel them. The experimental results completely contradicted these predictions. It was found that most of the alpha particles passed through the foil unaffected, as if the foil were mostly empty space. And of those deflected, a few were deflected at very large angles—some even nearly back in the direction from which they had come. This could happen, Rutherford reasoned, only if the positively charged alpha particles were being repelled by a massive positive charge concentrated in a very small region of space (see Fig. 27–18b). He theorized that the atom must consist of a tiny but massive positively charged nucleus, containing over 99.9 percent of the mass of the atom, surrounded by electrons some distance away. The electrons would be moving in orbits about the nucleus—much as the planets move around the sun—because if they were at rest, they would fall into the nucleus due to electrical attraction, Fig. 27–19. Ruther-

† Some readers may say: "Tell us the facts as we know them today, and don't bother us with the historical background and its outmoded theories." Such an approach would ignore the creative aspect of science and thus give a false impression of how science develops. Moreover, it is not really possible to understand today's view of the atom and the quantum without insight into the concepts that led to it.

ford's experiments suggested that the nucleus must have a radius of about 10^{-15} to 10^{-14} m. From kinetic theory, and especially Einstein's analysis of Brownian movement (see Section 13–1), the radius of atoms was estimated to be about 10^{-10} m. Thus the electrons would seem to be at a distance from the nucleus of about 10,000 to 100,000 times the radius of the nucleus itself; so an atom would be mostly empty space.

Rutherford's "planetary" model of the atom (also called the "nuclear model of the atom") was a major step toward how we view the atom today. It was not, however, a complete model and presented some major problems, as we shall see.

27–8 • Atomic Spectra: Key to the Structure of the Atom

At the beginning of this chapter we saw that heated solids (as well as liquids and dense gases) emit light with a continuous spectrum of wavelengths. This radiation is assumed to be due to oscillations of atoms and molecules, which are largely governed by the interaction of each atom or molecule with its neighbors.

Rarefied gases can also be excited to emit light. This is done by intense heating, or more commonly by applying a high voltage to a "discharge tube" containing the gas at low pressure, Fig. 27–20. The radiation from excited gases had been observed early in the nineteenth century, and it was found that the spectrum was not continuous, but *discrete*. Since excited gases emit light of only certain wavelengths, when this light is analyzed through the slit of a spectroscope or spectrometer, a **line spectrum** is seen rather than a continuous spectrum. The line spectra emitted by a number of elements in the visible region were shown in Fig. 24–23. The **emission spectrum** is characteristic of the material and can serve as a type of "fingerprint" for identification of the gas. We also saw (Chapter 24) that if a continuous spectrum passes through a gas, dark lines are observed in the spectrum, corresponding to lines normally emitted by the gas. This is called an **absorption spectrum**, and it became clear that gases absorb light at the same frequencies at which they

FIGURE 27–20 Gas-discharge tube: (a) is a diagram; (b) and (c) are photos of actual discharge tubes for hydrogen (b) and mercury gas (c).

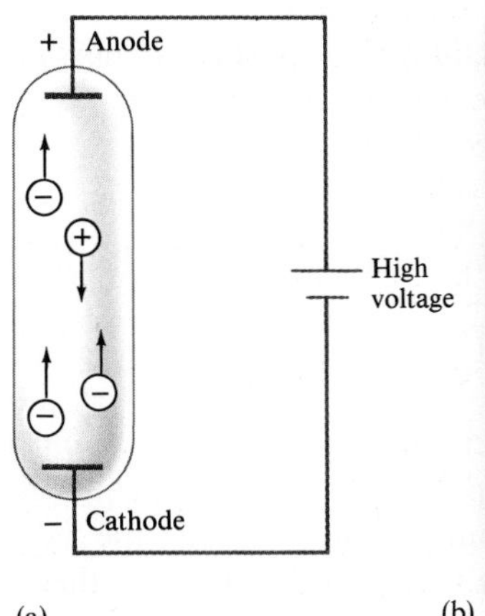

(a)

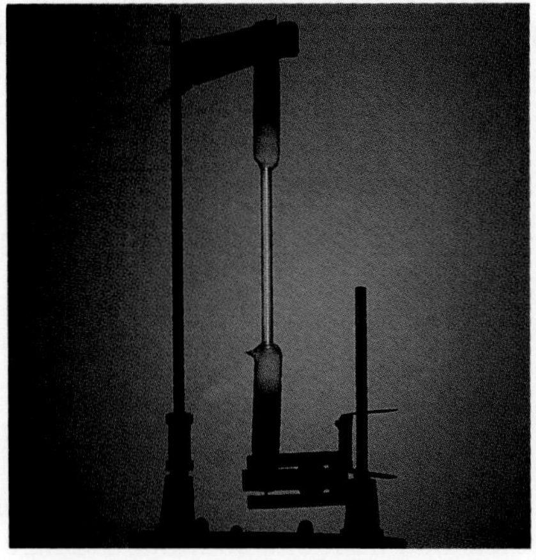

(b)

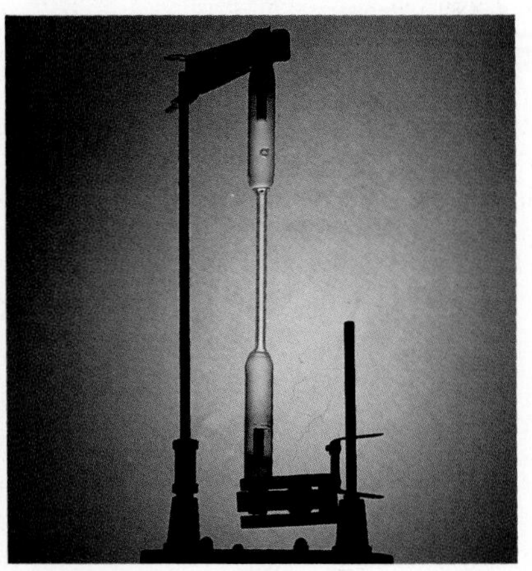

(c)

emit. With the use of film sensitive to ultraviolet and to infrared light, it was found that gases emit and absorb discrete frequencies in these regions as well as in the visible.

For our purposes here, the importance of the line spectra is that they are emitted (or absorbed) by gases that are not dense. In such thin gases, the atoms are far apart on the average and hence the light emitted or absorbed must be by *individual atoms* rather than through interactions between atoms as in a solid, liquid, or dense gas. Thus the line spectra serve as a key to the structure of the atom: any theory of atomic structure must be able to explain why atoms emit light only of discrete wavelengths, and it should be able to predict what these frequencies are.

Hydrogen is the simplest atom—it has only one electron orbiting its nucleus. It also has the simplest spectrum. The spectrum of most atoms shows little apparent regularity. But the spacing between lines in the hydrogen spectrum decreases in a regular way (Fig. 27–21). Indeed, in 1885, J. J. Balmer (1825–1898) showed that the four visible lines in the hydrogen spectrum (with measured wavelengths 656 nm, 486 nm, 434 nm, and 410 nm) would fit the following formula

Spectrum of hydrogen

Balmer series

$$\frac{1}{\lambda} = R\left(\frac{1}{2^2} - \frac{1}{n^2}\right), \qquad n = 3, 4, \cdots, \tag{27–6}$$

where n takes on the values 3, 4, 5, 6 for the four lines, and R, called the **Rydberg constant**, has the value $R = 1.097 \times 10^7\ \text{m}^{-1}$. Later it was found that this **Balmer series** of lines extended into the UV region, ending at $\lambda = 365$ nm, as shown in Fig. 27–21. Balmer's formula, Eq. 27–6, also worked for these lines with higher integer values of n. The lines near 365 nm became too close together to distinguish, but the limit of the series at 365 nm corresponds to $n = \infty$ (so $1/n^2 = 0$ in the formula).

FIGURE 27–21
Balmer series of lines for hydrogen.

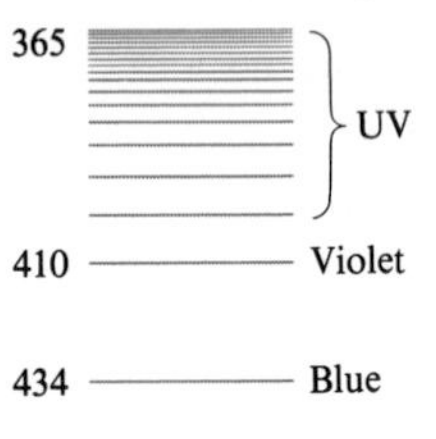

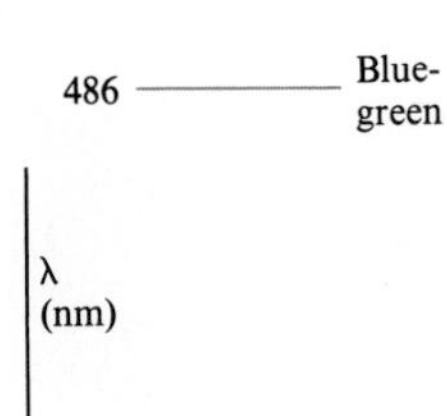

656 Red

Later experiments on hydrogen showed that there were other series of lines in the UV and IR, and each series had a pattern just like the Balmer series, but at different wavelengths. Each of these series was found to fit a formula with the same form as Eq. 27–6 but with the $1/2^2$ replaced by $1/1^2$, $1/3^2$, $1/4^2$, and so on. For example, the so-called **Lyman series** contains lines with wavelengths from 91 nm to 122 nm and fits the formula

$$\frac{1}{\lambda} = R\left(\frac{1}{1^2} - \frac{1}{n^2}\right), \qquad n = 2, 3, \cdots.$$

And the wavelengths of the **Paschen series** fit

$$\frac{1}{\lambda} = R\left(\frac{1}{3^2} - \frac{1}{n^2}\right), \qquad n = 4, 5, \cdots.$$

The Rutherford model, as it stood, was unable to explain why atoms emit line spectra. It had other difficulties as well. According to the Rutherford model, electrons orbit the nucleus, and since their paths are curved the electrons are accelerating. Hence they should give off light like any other accelerating electric charge (Chapter 22). Then, since energy is conserved, the electron's own energy must decrease to compensate. Hence electrons would be expected to spiral into the nucleus. As they spiraled inward, their frequency would increase in a short time and so too would the frequency of the light emitted. Thus the two main difficulties of the Rutherford model are these: (1) it predicts that light of a continuous range of frequencies will be emitted, whereas experiment shows line spectra; (2) it predicts that atoms are unstable—electrons should quickly spiral into the nucleus—but we know that atoms in general are stable, since the matter around us is stable.

Clearly Rutherford's model was not sufficient. Some sort of modification was needed, and it was Niels Bohr who provided it by adding an essential idea—the quantum hypothesis.

27–9 • The Bohr Model

Bohr had studied in Rutherford's laboratory for several months in 1912 and was convinced that Rutherford's planetary model of the atom had validity. But in order to make it work, he felt that the newly developing quantum theory would somehow have to be incorporated in it. The work of Planck and Einstein had shown that in heated solids, the energy of oscillating electric charges must change discontinuously—from one discrete energy state to another with the emission of a quantum of light. Perhaps, Bohr argued, the electrons in an atom also cannot lose energy continuously, but must do so in quantum "jumps." In working out his theory during the next year, Bohr postulated that electrons move about the nucleus in circular orbits, but that only certain orbits are allowed. He further postulated that an electron in each orbit would have a definite energy and would move in the orbit *without radiating energy* (even though this violated classical ideas since accelerating electric charges are supposed to emit EM waves; see Chapter 22). He thus called the possible orbits **stationary states**. Light is emitted, he hypothesized, only when an electron jumps from one stationary state to another of lower energy. When such a jump occurs, a single photon of light is emitted whose energy, since energy is conserved, is given by

Stationary state

$$hf = E_u - E_l, \tag{27–7}$$

where E_u refers to the energy of the upper state and E_l the energy of the lower state.

Bohr next set out to determine what energies these orbits would have, since the spectrum of light emitted could then be predicted from Eq. 27–7. When he became aware of the Balmer formula in early 1913, he had the key he was looking for. Bohr quickly found that his theory would be in accord with the Balmer formula if he assumed that the electron's angular momentum L is quantized and equal to an integer n times $h/2\pi$. As we saw in Chapter 8, angular momentum is given by $L = I\omega$, where I is the moment of inertia and ω is the angular velocity. For a single particle of mass m moving in a circle of radius r with speed v, $I = mr^2$ and $\omega = v/r$; hence, $L = I\omega = (mr^2)(v/r) = mvr$. Bohr's **quantum condition** then is

$$L = mvr_n = n\frac{h}{2\pi}, \qquad n = 1, 2, 3, \cdots, \tag{27–8}$$

where n is an integer and r_n is the radius of the nth possible orbit. The allowed orbits are numbered 1, 2, 3, . . . , according to the value of n, which is called the **quantum number** of the orbit.

Quantum number

Equation 27–8 did not have a firm theoretical foundation. Bohr had searched for some "quantum condition," and such tries as $E = hf$ (where E

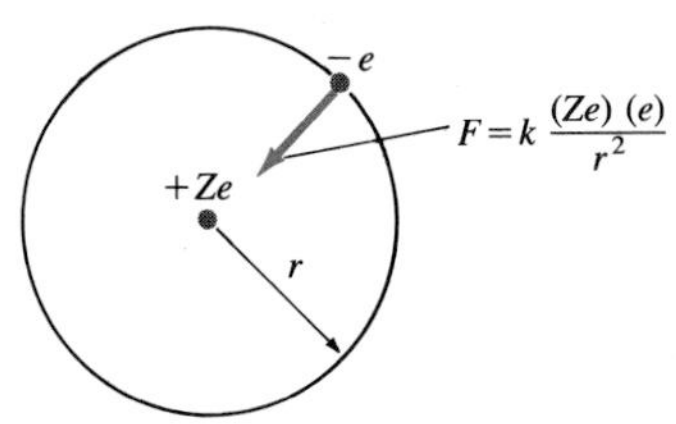

FIGURE 27–22 Electric force (Coulomb's law) keeps the negative electron in orbit around the positively charged nucleus.

represents the energy of the electron in an orbit) did not give results in accord with experiment. Bohr's reason for using Eq. 27–8 was simply that it worked; and we now look at how.

An electron in a circular orbit of radius r_n (Fig. 27–22) would have a centripetal acceleration v^2/r_n produced by the electrical force of attraction between the negative electron and the positive nucleus according to Coulomb's law,

$$F = k\frac{(Ze)(e)}{r^2}.$$

The charge on the electron is $q_1 = -e$, and that on the nucleus is $q_2 = +Ze$, where Z is the number of positive charges† (or protons). For the hydrogen atom, $Z = +1$.

In Newton's second law, $F = ma$, we substitute $a = v^2/r_n$ and Coulomb's law for F, and obtain

$$F = ma$$

$$k\frac{Ze^2}{r_n^2} = \frac{mv^2}{r_n}.$$

We solve this for r_n, and then substitute for v from Eq. 27–8 (which says $v = nh/2\pi mr_n$):

$$r_n = \frac{kZe^2}{mv^2} = \frac{kZe^2 4\pi^2 mr_n^2}{n^2h^2}.$$

We solve for r_n (it appears on both sides, so we cancel one of them) and find

$$r_n = \frac{n^2h^2}{4\pi^2 mkZe^2}. \qquad (27\text{–}9)$$

This equation gives the radii of the possible orbits. The smallest orbit is for $n = 1$ and for hydrogen ($Z = 1$) has the value

$$r_1 = \frac{(1)^2(6.626 \times 10^{-34}\ \mathrm{J \cdot s})^2}{4(3.14)^2(9.11 \times 10^{-31}\ \mathrm{kg})(9.00 \times 10^9\ \mathrm{N \cdot m^2/C^2})(1)(1.602 \times 10^{-19}\ \mathrm{C})^2}$$

Bohr radius

$$r_1 = 0.529 \times 10^{-10}\ m. \qquad (27\text{–}10)$$

The radius of the smallest orbit in hydrogen, r_1, is sometimes called the **Bohr radius**. From Eq. 27–9, we can see that the radii of the larger orbits‡ increase as n^2, so

$$r_2 = 4r_1 = 2.12 \times 10^{-10}\ \mathrm{m},$$

$$r_3 = 9r_1 = 4.76 \times 10^{-10}\ \mathrm{m},$$

and so on. The first four are shown in Fig. 27–23. Notice that, according to

FIGURE 27–23 Possible orbits in the Bohr model of hydrogen; $r_1 = 0.529 \times 10^{-10}$ m.

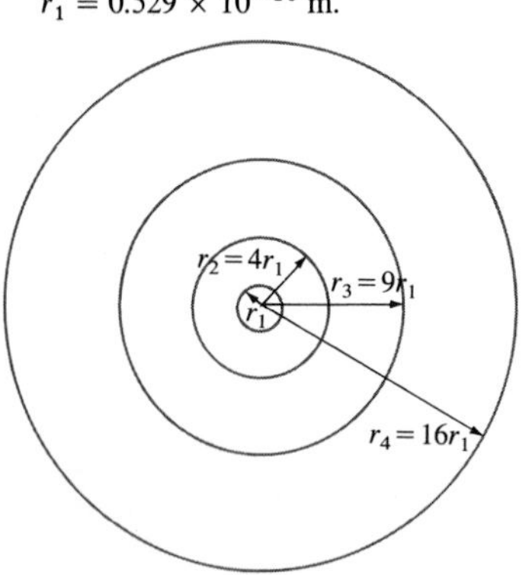

† We include Z in our derivation so that we can treat other single-electron ("hydrogenlike") atoms such as the ions He^+ ($Z = 2$) and Li^{2+} ($Z = 3$). Helium in the neutral state has two electrons; if one electron is missing, the remaining He^+ ion consists of one electron revolving around a nucleus of charge $+2e$. Similarly, doubly ionized lithium, Li^{2+}, also has a single electron, and in this case $Z = 3$.

‡ Be careful not to believe that these well-defined orbits actually exist. The Bohr model is only a model, not reality. The idea of electron orbits was rejected a few years later, and today electrons are better thought of as forming "clouds," as discussed in Chapter 28.

Bohr's model, an electron can exist only in the orbits given by Eq. 27–9; it cannot be in between.

In each of its possible orbits, the electron would have a definite energy, as the following calculation shows. The total energy equals the sum of the kinetic and potential energies. The potential energy of the electron is given by PE $= qV = -eV$, where V is the potential due to a point charge $+Ze$ as given by Eq. 17–3: $V = kQ/r = kZe/r$. So

$$\text{PE} = -eV = -k\frac{Ze^2}{r}.$$

The total energy E_n for an electron in the nth orbit of radius r_n is the sum of the kinetic and potential energies:

$$E_n = \frac{1}{2}mv^2 - \frac{kZe^2}{r_n}.$$

When we substitute v from Eq. 27–8 and r_n from Eq. 27–9 into this equation, we obtain

$$E_n = -\frac{2\pi^2 Z^2 e^4 mk^2}{h^2}\frac{1}{n^2}, \qquad n = 1, 2, 3, \cdots. \tag{27–11}$$

Energy levels in hydrogen

For hydrogen ($Z = 1$), the lowest energy level has $n = 1$, and when numbers are substituted into Eq. 27–11, we find

Ground state

$$E_1 = -2.17 \times 10^{-18}\ \text{J} = -13.6\ \text{eV},$$

where we have converted joules to electron volts, as is customary in atomic physics. Since n^2 appears in the denominator of Eq. 27–11, the energies of the larger orbits are given by

$$E_n = \frac{-13.6\ \text{eV}}{n^2}.$$

For example,

Excited states (first two)

$$E_2 = \frac{-13.6\ \text{eV}}{4} = -3.40\ \text{eV},$$

$$E_3 = \frac{-13.6\ \text{eV}}{9} = -1.51\ \text{eV}.$$

We see from Eq. 27–11 that not only are the orbit radii quantized, but so is the energy. The quantum number n that labels the orbit radii also labels the energy levels. The lowest **energy level** or **energy state** has energy E_1, and is called the **ground state**. The higher states, E_2, E_3, and so on, are called **excited states**.

Notice that although the energy for the larger orbits has a smaller numerical value, all the energies are less than zero. Thus, -3.4 eV is a greater energy than -13.6 eV. Hence the orbit closest to the nucleus (r_1) has the lowest energy. The reason the energies have negative values has to do with the way we defined the zero for potential energy; for two point charges, PE $= kq_1q_2/r$ corresponds to zero PE when the two charges are infinitely far apart. Thus, an electron with KE $= 0$ that is just barely free from the atom will have $E = 0$, corresponding to $n = \infty$ in Eq. 27–11. To remove an electron that

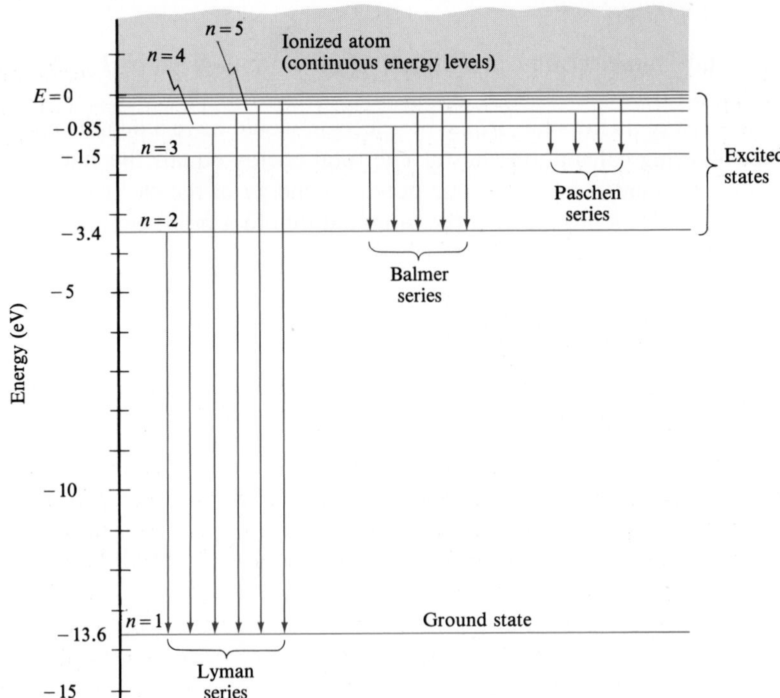

FIGURE 27–24 Energy-level diagram for the hydrogen atom, showing origin of spectral lines for the Lyman, Balmer, and Paschen series.

is part of an atom requires an energy input (otherwise atoms would not be stable); since $E \geq 0$ for a free electron, then an electron bound to an atom must have $E < 0$. That is, energy must be added to bring its energy up, from a negative value, to at least zero in order to free it.† The minimum energy required to remove an electron from the ground state of an atom is called the **binding energy** or **ionization energy**. The ionization energy for hydrogen has been measured to be 13.6 eV, and this corresponds precisely to removing an electron from the lowest state, $E_1 = -13.6$ eV, up to $E = 0$ where it will be free.

It is useful to show the various possible energy values as horizontal lines on an energy-level diagram.‡ This is shown for hydrogen in Fig. 27–24. The electron in a hydrogen atom can be in any one of these levels according to Bohr theory. But it could never be in between, say at -9.0 eV. At room temperature, nearly all H atoms will be in the ground state ($n = 1$). At higher temperatures, or during an electric discharge when there are many collisions between free electrons and atoms, many electrons can be in excited states ($n > 1$). Once in an excited state, an electron can jump down to a lower state, and give off a photon in the process. This is, according to the Bohr model, the origin of the emission spectra of excited gases. The vertical arrows in Fig. 27–24 represent the transitions or jumps that correspond to the various observed spectral lines. For example, an electron jumping from the level

† This is not unlike the case of the gravitational potential energy of a stone, PE $= mgh$; if we measure h from the top of a table, say, then when the stone is on the floor below the table, it will have negative potential energy. The atomic case is similar to that for the planets and comets (such as Halley's comet) that are bound to the sun and orbit around it. The energy of each planet is less than zero, whereas an energy equal to or greater than zero would apply to an object free of the sun and not bound to it.

‡ Note that an electron can have any energy above $E = 0$, for there it is free. Thus there is a continuum of energy states above $E = 0$, as indicated in the energy-level diagram of Fig. 27–24.

$n = 3$ to $n = 2$ would give rise to the 656-nm line in the Balmer series, and the jump from $n = 4$ to $n = 2$ would give rise to the 486-nm line (see Fig. 27–21). We can predict wavelengths of the spectral lines emitted by combining Eq. 27–7 with Eq. 27–11. Since $hf = hc/\lambda$, we have

$$\frac{1}{\lambda} = \frac{hf}{hc} = \frac{1}{hc}(E_n - E_{n'}),$$

or

$$\frac{1}{\lambda} = \frac{2\pi^2 Z^2 e^4 m k^2}{h^3 c}\left(\frac{1}{n'^2} - \frac{1}{n^2}\right), \tag{27–12}$$

where n refers to the upper state and n' to the lower state. This theoretical formula has the same form as the experimental Balmer formula, Eq. 27–6, with $n' = 2$. Thus we see that the Balmer series of lines corresponds to transitions or "jumps" that bring the electron down to the second energy level. Similarly, $n' = 1$ corresponds to the Lyman series and $n' = 3$ to the Paschen series (see Fig. 27–24). When the constant $(2\pi^2 Z^2 e^4 m k^2/h^3 c)$ in Eq. 27–12 is evaluated with $Z = 1$, it is found to have the measured value of the Rydberg constant, $R = 1.0974 \times 10^7\ \text{m}^{-1}$, in accord with experiment (see Problem 33).

The great success of Bohr's theory is that it explains why atoms emit line spectra and accurately predicts, for hydrogen, the wavelengths of emitted light. The Bohr theory also explains absorption spectra: photons of just the right wavelength can knock an electron from one energy level to a higher one. To conserve energy, only photons that have just the right energy will be absorbed. This explains why a continuous spectrum passing through a gas will have dark (absorption) lines at the same frequencies as the emission lines.

The Bohr theory also ensures the stability of atoms. It establishes stability by fiat: the ground state is the lowest state for an electron and there is no lower energy level to which it can go and emit more energy. Finally, as we saw above, the Bohr theory accurately predicts the ionization energy of 13.6 eV for hydrogen. Nonetheless, the Bohr theory was not a complete theory and has been superceded, as we shall discuss in the next chapter.

EXAMPLE 27–7 Use Fig. 27–24 to determine the wavelength of the first Lyman line, the transition from $n = 2$ to $n = 1$.

SOLUTION In this case, $hf = E_2 - E_1 = 13.6\ \text{eV} - 3.4\ \text{eV} = 10.2\ \text{eV} = 1.63 \times 10^{-18}\ \text{J}$. Since $\lambda = c/f$, we have

$$\lambda = \frac{c}{f} = \frac{hc}{E_2 - E_1} = \frac{(6.63 \times 10^{-34}\ \text{J}\cdot\text{s})(3.00 \times 10^8\ \text{m/s})}{1.63 \times 10^{-18}\ \text{J}}$$
$$= 1.22 \times 10^{-7}\ \text{m},$$

or 122 nm, which is in the UV.

EXAMPLE 27–8 Determine the wavelength of light emitted when a hydrogen atom makes a transition from the $n = 6$ to the $n = 2$ energy level according to the Bohr model.

SOLUTION We can use Eq. 27–12 or its equivalent, Eq. 27–6, with $R = 1.097 \times 10^7\ \text{m}^{-1}$. Thus

$$\frac{1}{\lambda} = (1.097 \times 10^7\ \text{m}^{-1})\left(\frac{1}{4} - \frac{1}{36}\right) = 2.44 \times 10^6\ \text{m}^{-1}.$$

So $\lambda = 1/(2.44 \times 10^{-6}\ \text{m}^{-1}) = 4.10 \times 10^{-7}$ m or 410 nm. This is the fourth line in the Balmer series, Fig. 27–21, and is violet in color.

EXAMPLE 27–9 Use the Bohr model to determine the ionization energy of the He^+ ion, which has a single electron. Also calculate the minimum wavelength a photon must have to cause ionization.

SOLUTION We want to determine the minimum energy required to lift the electron from its ground state and to barely reach the free state at $E = 0$. The ground state energy of He^+ is given by Eq. 27–11 with $n = 1$ and $Z = 2$. Since all the symbols in Eq. 27–11 are the same as for the calculation for hydrogen, except that Z is 2 instead of 1, we see that E_1 will be $Z^2 = 2^2 = 4$ times the E_1 for hydrogen. That is,

$$E_1 = 4(-13.6\ \text{eV}) = -54.4\ \text{eV}.$$

Thus, to ionize the He^+ ion should require 54.4 eV, and this value agrees with experiment. The minimum wavelength photon that can cause ionization will have energy $hf = 54.4$ eV and wavelength $\lambda = c/f = hc/hf = (6.63 \times 10^{-34}\ \text{J}\cdot\text{s})(3.00 \times 10^8\ \text{m/s})/(54.4\ \text{eV})(1.60 \times 10^{-19}\ \text{J/eV}) = 22.8$ nm. If the atom absorbed a photon of greater energy (wavelength shorter than 22.8 nm), the atom could still be ionized and the freed electron would have kinetic energy of its own.

In this last example, we saw that E_1 for the He^+ ion is four times more negative than that for hydrogen. Indeed, the energy-level diagram for He^+ looks just like that for hydrogen, Fig. 27–24, except that the numerical values for each energy level are four times larger. It is important to note, however, that we are talking here about the He^+ *ion*. Normal (neutral) helium has two electrons and its spectrum is entirely different.

EXAMPLE 27–10 Estimate the average kinetic energy of hydrogen atoms (or molecules) at room temperature, and use the result to explain why nearly ali H atoms are in the ground state at room temperature.

SOLUTION According to kinetic theory (Chapter 13), the average KE of atoms or molecules in a gas is given by Eq. 13–8:

$$\overline{\text{KE}} = \tfrac{3}{2}kT,$$

where $k = 1.38 \times 10^{-23}$ J/K is Boltzmann's constant, and T is the kelvin (absolute) temperature. Room temperature is about $T = 300$ K, so

$$\overline{\text{KE}} = \tfrac{3}{2}(1.38 \times 10^{-23}\ \text{J/K})(300\ \text{K}) = 6.2 \times 10^{-21}\ \text{J},$$

or, in electron volts:

$$\overline{\text{KE}} = \frac{6.2 \times 10^{-21}\ \text{J}}{1.6 \times 10^{-19}\ \text{J/eV}} = 0.04\ \text{eV}.$$

The average KE is thus very small compared to the energy between the ground state and the next higher energy state (13.6 eV − 3.4 eV = 10.2 eV). Any atoms in excited states emit light and eventually fall to the ground state. Once in the ground state, collisions with other atoms can transfer energy of only 0.04 eV on the average. A small fraction of atoms can have much more energy (see Section 13–9 on the distribution of molecular speeds), but even a KE that is 10 times the average is not nearly enough to excite atoms above the ground state. Thus, at room temperature, nearly all atoms are in the ground state. Atoms can be excited to upper states by very high temperatures, or by passing a current of high KE electrons through the gas, as in a discharge tube (Fig. 27–20).

We should note that Bohr made some radical assumptions that were at variance with classical ideas. He assumed that electrons in fixed orbits do not radiate light even though they are accelerating (moving in a circle), and he assumed that angular momentum is quantized. Furthermore, he was not able to say how an electron moved when it made a transition from one energy level to another. On the other hand, there is no real reason to expect that in the tiny world of the atom electrons would behave as ordinary-sized objects do. Nonetheless, he felt that where quantum theory overlaps with the macroscopic world, it should predict classical results. This is the **correspondence principle** already mentioned in regard to relativity (Section 26–13). This principle does work for Bohr's theory of the hydrogen atom. The orbit sizes and energies are quite different for $n = 1$ and $n = 2$, say. But orbits with n = 100,000,000 and 100,000,001 would be very close in size and energy (see Fig. 27–24). Indeed, jumps between such large orbits, which would approach everyday sizes, would be imperceptible. Such orbits would thus appear to be continuous, which is what we expect in the everyday world.

Correspondence principle

27–10 • De Broglie's Hypothesis

Bohr's theory was largely of an *ad hoc* nature. Assumptions were made so that theory would agree with experiment. But Bohr could give no reason why the orbits were quantized. Finally, 10 years later, a reason was found by de Broglie.

We saw in Section 27–5 that in 1923, Louis de Broglie proposed that material particles, such as electrons, have a wave nature; and that this hypothesis was confirmed by experiment several years later.

One of de Broglie's original arguments in favor of the wave nature of electrons was that it provided an explanation for Bohr's theory of the hydrogen atom. According to de Broglie, a particle of mass m moving with speed v would have a wavelength (Eq. 27–5) of

$$\lambda = \frac{h}{mv}.$$

Each electron orbit in an atom, he proposed, is actually a standing wave.

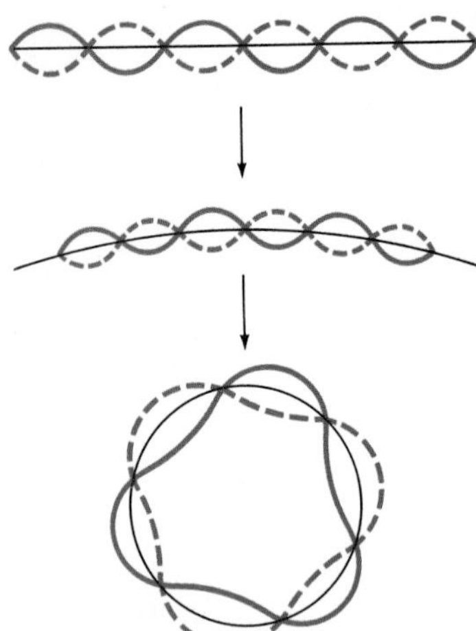

FIGURE 27–25 An ordinary standing wave compared to a circular standing wave.

FIGURE 27–26 When a wave does not close (and hence interferes destructively with itself), it rapidly dies out.

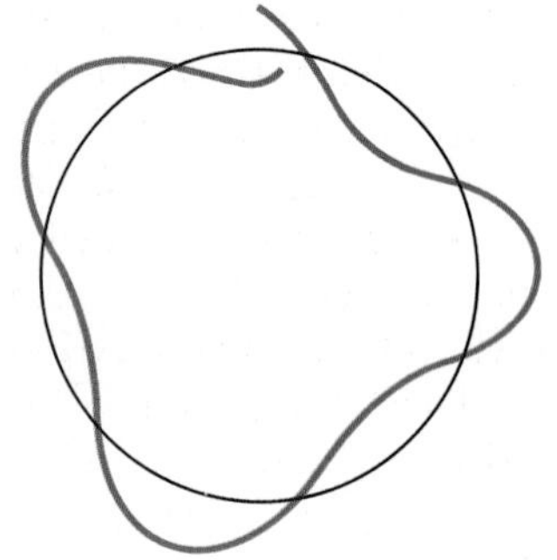

As we saw in Chapter 11, when a violin or guitar string is plucked, a vast number of wavelengths are excited. But only certain ones—those that have nodes at the ends—are sustained. These are the *resonant* modes of the string. Waves with other wavelengths interfere with themselves upon reflection and their amplitudes quickly drop to zero. Since electrons move in circles, according to Bohr's theory, de Broglie argued that the electron wave must be a *circular* standing wave that closes on itself, Fig. 27–25. If the wavelength of a wave does not close on itself, as in Fig. 27–26, destructive interference takes place as the wave travels around the loop and it quickly dies out. Thus, the only waves that persist are those for which the circumference of the circular orbit contains a whole number of wavelengths, Fig. 27–27. The circumference of a Bohr orbit of radius r_n is $2\pi r_n$ so we must have

$$2\pi r_n = n\lambda, \qquad n = 1, 2, 3, \cdots.$$

When we substitute $\lambda = h/mv$, we get

$$2\pi r_n = \frac{nh}{mv},$$

or

$$mvr_n = \frac{nh}{2\pi}.$$

This is just the *quantum condition* proposed by Bohr on an *ad hoc* basis, Eq. 27–8. And it is from this equation that the discrete orbits and energy levels were derived. Thus we have an explanation for the quantized orbits and energy states in the Bohr model: they are due to the wave nature of the electron and the fact that only resonant "standing" waves can persist. This implies that the *wave–particle duality* is at the root of atomic structure.

It should be noted in viewing the circular electron waves of Fig. 27–27 that the electron is not to be thought of as following the oscillating wave pattern. In the Bohr model of hydrogen, the electron, considered as a particle, moves in a circle. The circular wave, on the other hand, represents the *amplitude* of the electron "matter wave," and in Fig. 27–27 the wave amplitude is shown superimposed on the circular path of the particle orbit for convenience.

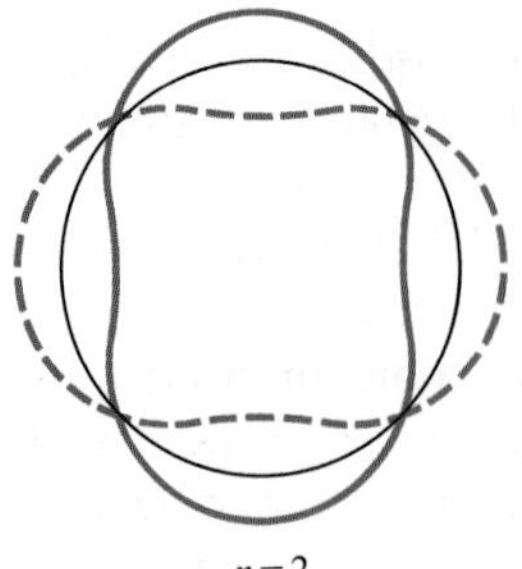

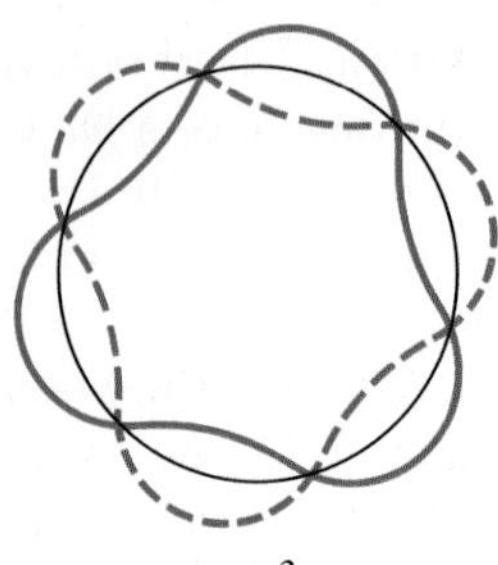

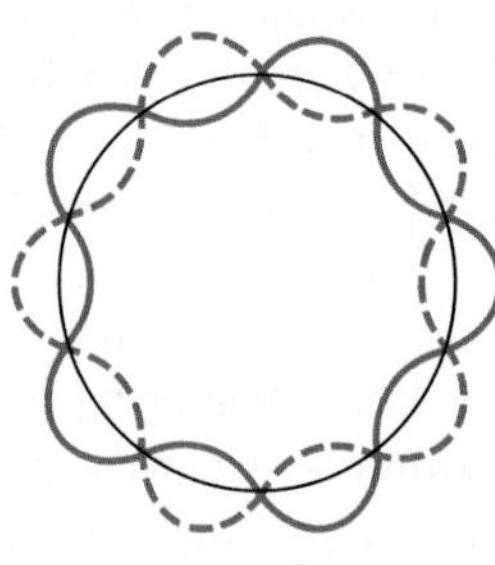

FIGURE 27–27 (right) Standing circular waves for two, three, and five wavelengths on the circumference; *n*, the number of wavelengths, is also the quantum number.

Bohr's theory worked well for hydrogen and for one-electron ions. It did not prove as successful for multielectron atoms. We will discuss this and other problems with the Bohr theory in the next chapter, and we will see how a new and radical theory, quantum mechanics, finally solved the problem of atomic structure and gave us a very different view of the atom: the idea of electrons in well-defined orbits was replaced with the idea of electron "clouds." And this new (and now generally accepted) theory of quantum mechanics has given us a wholly different view of the basic mechanisms underlying physical processes.

SUMMARY

Quantum theory has its origins in the *quantum hypothesis* that molecular oscillations are quantized: their energy E can only be integer (n) multiples of hf, where h is Planck's constant and f is the natural frequency of oscillation: $E = nhf$. This hypothesis explained the spectrum of radiation emitted by (black) bodies at high temperature.

Einstein proposed that for some purposes, light might be pictured as being emitted and absorbed as quanta (particles), which we now call *photons*, each with energy

$$E = hf.$$

He proposed the photoelectric effect as a test for the photon theory of light. In the photoelectric effect, the photon theory says that each incident photon can strike an electron in a material and eject it if it has sufficient energy. The maximum energy of ejected electrons is then linearly related to the frequency of the incident light. The photon theory is also supported by the Compton effect and the observation of electron–positron *pair production.*

The wave–particle duality refers to the idea that light and matter (such as electrons) have both wave and particle properties. The wavelength of a material object is given by

$$\lambda = \frac{h}{mv},$$

where mv is the momentum of the object. The *principle of complementarity* states that we must be aware of both the particle and wave properties of light and of matter for a complete understanding of them.

Early models of the atom include Thomson's modified plum-pudding model and Rutherford's planetary (or nuclear) model. Rutherford's model, which was created to explain the backscattering of alpha particles from thin metal foils, assumes that an atom consists of a tiny but massive positively charged nucleus surrounded (at a relatively great distance) by electrons.

To explain the line spectra emitted by atoms, as well as the stability of atoms, Bohr proposed a theory which postulated (1) that electrons bound in an atom can only occupy orbits for which the angular momentum is quantized, which results in discrete values for the radius and energy; (2) that an electron in such a *stationary state* emits no radiation; (3) that, if an electron jumps to a lower state, it emits a photon whose energy equals the difference in energy between the two states; (4) that the angular momentum L of atomic electrons is quantized by the rule $L = nh/2\pi$, where n is an integer called a *quantum number*. The $n = 1$ state in hydrogen is the *ground state*, which has an energy $E_1 = -13.6$ eV; higher values of n correspond to *excited states* and their energies are $E_n = -13.6\ \text{eV}/n^2$. Atoms are excited to these higher states by collisions with other atoms or electrons or by absorption of a photon of just the right frequency.

De Broglie's hypothesis that electrons (and other matter) have a wavelength $\lambda = h/mv$ gave an explanation for Bohr's quantized orbits by bringing in the wave–particle duality: the orbits correspond to circular standing waves in which the circumference of the orbit equals a whole number of wavelengths.

SECTION 27–10

42. (III) Suppose a particle of mass m is confined to a one-dimensional box of width L. According to quantum theory, the particle's wave (with $\lambda = h/mv$) is a standing wave with nodes at the edges of the box. (*a*) Show the possible modes of vibration on a diagram. (*b*) Show that the kinetic energy of the particle has quantized energies given by $\text{KE} = n^2h^2/8mL^2$, where n is an integer. (*c*) Calculate the ground-state energy ($n = 1$) for an electron confined to a box of width 0.50×10^{-10} m. (*d*) What is the ground-state energy for a baseball ($m = 140$ g) in a box 0.50 m wide? (*e*) An electron confined to a box has a ground-state energy of 20 eV. What is the width of the box?

GENERAL PROBLEMS

43. For what maximum kinetic energy is a collision between an electron and a hydrogen atom in its ground state definitely elastic?
44. At low temperatures, nearly all the atoms in hydrogen gas will be in the ground state. What minimum frequency photon is needed if the photoelectric effect is to be observed?
45. A beam of 70 eV electrons is scattered from a crystal, as in X-ray diffraction, and a first-order peak is observed at $\theta = 45°$. What is the spacing between planes in the diffracting crystal? (See Section 25–11.)
46. Show that the energy E (in electron volts) of a photon whose wavelength is λ (meters) is given by

$$E = 1.24 \times 10^{-6}/\lambda.$$

47. Sunlight reaching the earth has an intensity of about 1300 W/m^2. How many photons per square meter per second does this represent? Take the average wavelength to be 550 nm.
48. If a 100-W light bulb emits 3.0 percent of the input energy as visible light (average wavelength 550 nm) uniformly in all directions, estimate how many photons per second of visible light will strike the pupil (4.0 mm diameter) of the eye of an observer 10 km away.
49. By what potential difference must (*a*) a proton ($m_0 = 1.67 \times 10^{-27}$ kg), and (*b*) an electron ($m_0 = 9.11 \times 10^{-31}$ kg), be accelerated to have a wavelength $\lambda = 2.00 \times 10^{-12}$ m?
50. In certain of Rutherford's experiments (Fig. 27–18), the α particles (mass $= 6.68 \times 10^{-27}$ kg) had a kinetic energy of 4.8 MeV. How close could they get to a gold nucleus (charge $= +79e$)? Ignore the recoil motion of the nucleus.
51. By what fraction does the mass of an H atom decrease when it makes an $n = 3$ to $n = 1$ transition?
52. Calculate the ratio of the gravitational to electric force for the electron in a hydrogen atom. Can the gravitational force be safely ignored?
53. Electrons accelerated by a potential difference of 12.3 V pass through a gas of hydrogen atoms at room temperature. What wavelengths of light will be emitted?
54. Atoms can be formed in which a muon (mass = 207 times the mass of an electron) replaces one of the electrons in an atom. Calculate, using Bohr theory, the energy of the photon emitted when a muon makes a transition from $n = 2$ to $n = 1$ in a muonic $^{208}_{82}\text{Pb}$ atom (lead whose nucleus has a mass 208 times the proton mass and charge $+82e$).
55. In an X-ray tube (see Fig. 25–29 and discussion in Section 25–11), the high voltage between filament and target is V. After being accelerated through this voltage, an electron strikes the target where it is decelerated (by positively charged nuclei) and in the process one or more X-ray photons are emitted. (*a*) Show that the photon of shortest wavelength will have

$$\lambda_0 = \frac{hc}{eV}.$$

(*b*) What is the shortest wavelength of X ray emitted when accelerated electrons strike the face of a 30-kV television picture tube?
56. Show that the wavelength of a particle of mass m_0 with kinetic energy KE is given by the relativistic formula $\lambda = hc/\sqrt{(\text{KE})^2 + 2m_0c^2(\text{KE})}$.
57. What is the kinetic energy and wavelength of a "thermal" neutron (one that is in equilibrium at room temperature—see Chapter 13).
58. What is the theoretical limit of resolution for an electron microscope whose electrons are accelerated through 50 kV? (Relativistic formulas should be used.)

C H A P T E R 28

Quantum Mechanics of Atoms

Neon tubes are thin glass tubes (moldable into various shapes) filled with neon (or other) gas that glows with a particular color when a high voltage current passes through it. The atoms of the gas are excited to upper energy levels, and when their electrons jump down to lower energy levels they emit light (photons) whose frequencies (color) are characteristic of the type of gas. A tube filled with neon produces a red-orange color, helium produces pink, and a mixture of Ne, A, and Hg produces blue, for example. See Fig. 24–23. The color of the glass, if not clear, also affects the color seen.

Bohr's model of the atom gave us a first (though rough) picture of what an atom is like. It proposed an explanation for why there should be emission and absorption of light by atoms at discrete wavelengths, as well as for the stability of atoms. The wavelengths of the line spectra and the ionization energy for hydrogen (and one-electron ions) that it predicted are in excellent agreement with experiment. But the Bohr theory had important limitations. It was not able to predict the line spectra for more complex atoms—not even for the neutral helium atom, which has only two electrons. Nor could it explain why the emission lines, when viewed with great precision, actually consist of two or more very closely spaced lines (referred to as *fine structure*). The Bohr theory also didn't explain why some spectral lines were brighter than others. And it couldn't explain the bonding of atoms in molecules or in solids and liquids.

From a theoretical point of view, too, the Bohr theory was not satisfactory. For it was a strange mixture of classical and quantum ideas. And the wave–particle duality was still not really resolved.

We mention these limitations of the Bohr theory not to disparage it—for it was a landmark in the history of science. Rather, we mention them to

FIGURE 28–1 Erwin Schrödinger with Lise Meitner (see Chapter 31).

show why, in the early 1920s, it became increasingly evident that a new, more comprehensive theory was needed. It was not long in coming. Less than two years after de Broglie gave us his matter–wave hypothesis, Erwin Schrödinger (1887–1961; Fig. 28–1) and Werner Heisenberg (1901–1976; Fig. 28–2) independently developed a new comprehensive theory. Their separate approaches were quite different but were soon shown to be fully compatible.

FIGURE 28–2 Werner Heisenberg (center) on Lake Como with Wolfgang Pauli (right) and Enrico Fermi (left).

28–1 • Quantum Mechanics—A New Theory

The new theory, called **quantum mechanics**, unifies the wave–particle duality into a single consistent theory. As a theory, quantum mechanics has been extremely successful. It has successfully dealt with the spectra emitted by complex atoms, even the fine details. It explains the relative brightness of spectral lines and how atoms form molecules. It is also a much more general theory that covers all quantum phenomena from blackbody radiation to atoms and molecules. It has explained a wide range of natural phenomena and from its predictions many new practical devices have become possible. Indeed, it has been so successful that it is accepted today by nearly all physicists as the fundamental theory underlying physical processes.

Quantum mechanics deals mainly with the microscopic world of atoms and light. But in our everyday macroscopic world, we do perceive light and we accept that ordinary objects are made up of atoms. This new theory must therefore also account for the verified results of classical physics. That is, when it is applied to macroscopic phenomena, quantum mechanics must be able to produce the old classical laws. This, the **correspondence principle** (already mentioned in Section 27–9), is met fully by quantum mechanics. This doesn't mean we throw away classical theories such as Newton's laws. In the everyday world, the latter are far easier to apply and they give an accurate description. But when we deal with high speeds, close to the speed of light, we must use the theory of relativity; and when we deal with the tiny world of the atom, we use quantum mechanics.

Although we won't go into the detailed mathematics of quantum mechanics, we will discuss the main ideas and how they involve the wave and particle properties of matter to explain atomic structure and other applications.

28–2 • The Wave Function and Its Interpretation; the Double-Slit Experiment

The important properties of any wave are its wavelength, frequency, and amplitude. For an electromagnetic wave, the wavelength determines whether the light is visible or not, and if so, what color it is. We also have seen that the wavelength (or frequency) is a measure of the energy of the corresponding photon ($E = hf$). The amplitude of an electromagnetic wave at any point is the strength of the electric (or magnetic) field at that point, and is related to the intensity of the wave (the brightness of the light).

For material particles such as electrons, quantum mechanics relates the wavelength to momentum according to de Broglie's formula, $\lambda = h/mv$. But what about the amplitude of a matter wave? In quantum mechanics the amplitude of, say, an electron wave is called the **wave function** and is given the symbol Ψ (the Greek letter psi). Thus Ψ represents the amplitude, as a function of time and position, of a new kind of field which we might call a "matter" field or a matter wave.

To calculate the wave function Ψ in a given situation (say, for an electron in an atom) is one of the basic tasks of quantum mechanics. Indeed, the development of an equation to do so was Schrödinger's great contribuit. According to quantum mechanics, the position and velocity of an object cannot even be known accurately at the same time. This is expressed in the uncertainty principle, and arises because basic entities, such as electrons, are not considered simply as particles: they have wave properties as well. Quantum mechanics allows us to calculate only the probability† that, say, an electron (when thought of as a particle) will be observed at various places. Quantum mechanics says there is some inherent unpredictability in nature.

Since matter is considered to be made up of atoms, even ordinary-sized objects are expected to be governed by probability, rather than by strict determinism. For example, there is a finite (but very small) probability that when you throw a stone, its path will suddenly curve upward instead of following the downward-curved parabola of normal projectile motion. Quantum mechanics predicts with very high probability that ordinary objects will behave just as the classical laws of physics predict. But these predictions are probabilities, not certainties. The reason that macroscopic objects behave in accordance with classical laws with very high probability is due to the large number of molecules involved: when large numbers of objects are present in a statistical situation, deviations from the average are negligible. It is the average configuration of vast numbers of molecules that follows the so-called fixed laws of classical physics with such high probability, and gives rise to an apparent "determinism." Deviations from classical laws are readily observed when small numbers of molecules are dealt with. We can say, then, that although there are no precise deterministic laws in quantum mechanics, there are statistical laws based on probability.

It is important to note that there is a difference between the probability imposed by quantum mechanics and that used in the nineteenth century to understand thermodynamics and the behavior of gases in terms of molecules (Chapters 13 and 15). In thermodynamics, probability is used because there are far too many molecules to be kept track of. But the molecules were still assumed to move and interact in a deterministic way according to Newton's laws. Probability in quantum mechanics is quite different. It is seen as *inherent* in nature, and not as a limitation on our abilities to calculate.

Although a few physicists have not given up the deterministic view of nature and have refused to accept quantum mechanics as a complete theory—one was Einstein—nonetheless, the vast majority of physicists do accept quantum mechanics and the probabilistic view of nature. This view, which as presented here is the generally accepted one, is called the **Copenhagen interpretation** of quantum mechanics in honor of Niels Bohr's home, since it was largely developed there through discussions between Bohr and other prominent physicists.

Because electrons are not simply particles, they cannot be thought of as following particular paths in space and time. This suggests that a description of matter in space and time may not be completely correct. This deep and far-reaching conclusion has been a lively topic of discussion among philosophers. Perhaps the most important and influential philosopher of quan-

† Note that these probabilities can be calculated precisely, just like exact predictions at dice or playing cards, but unlike predictions of probabilities at sporting events or for natural or man-made disasters.

tum mechanics was Bohr. He argued that a space–time description of actual atoms and electrons is not possible. Yet a description of experiments on atoms or electrons must be given in terms of space and time and other concepts familiar to ordinary experience, such as waves and particles. We must not let our *descriptions* of experiments lead us into believing that atoms or electrons themselves actually exist in space and time as particles. This distinction between our interpretations of experiments and what is "really" happening in nature is crucial.

28–5 • Quantum-Mechanical View of Atoms

At the beginning of this chapter, we discussed the limitations of the Bohr theory of atomic structure. Now we examine the quantum-mechanical theory of atoms, which is far more complete than the old Bohr theory. Although the Bohr model has been discarded as an accurate description of nature, nonetheless, quantum mechanics reaffirms certain aspects of the older theory, such as that electrons in an atom exist only in discrete states of definite energy, and that a photon of light is emitted (or absorbed) when an electron makes a transition from one state to another. But quantum mechanics is a much deeper theory, and has provided us with a very different view of the atom. According to quantum mechanics, electrons do not exist in well-defined circular orbits as in the Bohr theory. Rather, the electron (because of its wave nature) is spread out in space as a "cloud" of negative charge. The size and shape of the electron cloud can be calculated for a given state of an atom. For the ground state in the hydrogen atom, the electron cloud is spherically symmetric, as shown in Fig. 28–4. The electron cloud roughly indicates the "size" of an atom; but just as a cloud may not have a distinct border, atoms do not have a precise boundary or a well-defined size. Not all electron clouds have a spherical shape, as we shall see later in this chapter.

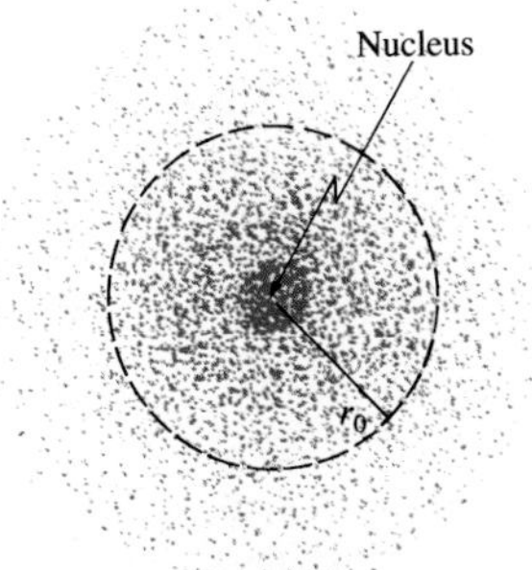

FIGURE 28–4 Electron cloud or "probability distribution" for the ground state of the hydrogen atom. The circle represents the Bohr radius.

The electron cloud can be interpreted using either the particle or the wave viewpoint. Remember that by a particle we mean something that is localized in space—it has a definite position at any given instant. By contrast, a wave is spread out in space. The electron cloud, spread out in space as in Fig. 28–4, is a result of the wave nature of electrons. Electron clouds can also be interpreted as **probability distributions** for a particle. If you were to make 500 different measurements of the position of an electron (considering it as a particle), the majority of the results would show the electron at points where the probability is high (darker areas with more dots in Fig. 28–4). Only occasionally would the electron be found where the probability is low. We cannot predict the path an electron will follow. As we saw in Section 28–3, after one measurement of its position we cannot predict exactly where it will be at a later time. We can only calculate the probability that it will be found at different points. This is clearly different from classical Newtonian physics. Indeed, as Bohr later pointed out, it is meaningless even to ask how an electron gets from one state to another when the atom emits a photon of light.

Probability distributions

28–6 • Quantum Mechanics of the Hydrogen Atom; Quantum Numbers

We now look more closely at what quantum mechanics tells us about the hydrogen atom. Much of what we say here also applies to more complex atoms, which are discussed in the next section.

Quantum mechanics predicts exactly the same energy levels (Fig. 27–24) for the hydrogen atom as does the Bohr theory. That is,

$$E_n = -\frac{13.6\text{ eV}}{n^2}, \qquad n = 1, 2, 3, \ldots,$$

where n is an integer. In the simple Bohr theory, there was only one quantum number, n. In quantum mechanics, it turns out that four different quantum numbers are needed to specify each state in the atom.

Quantum numbers

The *quantum number*, n, from Bohr theory is found also in quantum mechanics and is called the **principal quantum number**. It can have any integer value from 1 to ∞. The total energy of a state in the hydrogen atom depends on n, as we saw above.

The **orbital quantum number**, l, is related to the angular momentum of the electron; l can take on integer values from 0 to $(n - 1)$. For the ground state, $n = 1$, l can only be zero.† But for $n = 3$, say, l can be 0, 1, or 2. The actual magnitude of the angular momentum L is related to the quantum number l by

$$L = \sqrt{l(l+1)}\hbar \tag{28–3}$$

(where again $\hbar = h/2\pi$). The value of l does not affect the total energy in the hydrogen atom; only n does to any appreciable extent.‡ But in atoms with two or more electrons, the energy does depend on l as well as n, as we shall see in the next section.

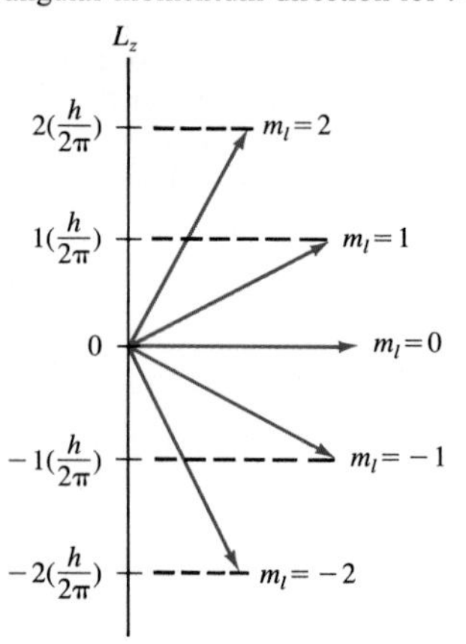

FIGURE 28–5 Quantization of angular momentum direction for $l = 2$.

The **magnetic quantum number**, m_l, is related to the direction of the electron's angular momentum, and it can take on integer values ranging from $-l$ to $+l$. For example, if $l = 2$, then m_l can be $-2, -1, 0, +1$, or $+2$. Since angular momentum is a vector, it is not surprising that both its magnitude and its direction would be quantized. For $l = 2$, the five different directions allowed can be represented by the diagram of Fig. 28–5. This limitation on the direction of L is often called **space quantization**. In quantum mechanics, the direction of the angular momentum is usually specified by giving its component along the z axis (this choice is arbitrary). Then L_z is related to m_l by the equation

$$L_z = m_l\hbar.$$

The name for m_l derives not from theory (which relates it to L_z), but from experiment. It was found that when a gas-discharge tube was placed in a magnetic field, the spectral lines were split into several very closely spaced lines. This splitting, known as the **Zeeman effect**, implies that the energy

† Contrast this with Bohr theory, which assigned $l = 1$ to the ground state (Eq. 27–8).

‡ See discussion of *fine structure* later in this section.

levels must be split (Fig. 28–6), and thus that the energy of a state depends not only on n but also on m_l when a magnetic field is applied—hence the name "magnetic quantum number." (Why the energy should depend on the direction of **L** can be seen from a semiclassical view of a moving electron as an electric current that interacts with the magnetic field—see Chapter 20.)

Finally, there is the **spin quantum number**, m_s, which can have only two values, $m_s = +\frac{1}{2}$ and $m_s = -\frac{1}{2}$. The existence of this quantum number did not come out of Schrödinger's original theory, as did n, l, and m_l. Instead, a subsequent modification by P. A. M. Dirac (1902–1984) explained its presence as a relativistic effect. The first hint that m_s was needed, however, came from experiment. A careful study of the spectral lines of hydrogen showed that each actually consisted of two (or more) very closely spaced lines even in the absence of an external magnetic field. It was at first hypothesized that this tiny splitting of energy levels, called **fine structure**, might be due to angular momentum associated with a spinning of the electron. That is, the electron might spin on its axis as well as orbit the nucleus, just as the earth spins on its axis as it orbits the sun. The interaction between the tiny current of the spinning electron could then interact with the magnetic field due to the orbiting charge and cause the small observed splitting of energy levels. (So the energy depends slightly on m_l and m_s.) This picture of the electron as spinning is, however, wholly discredited today. We cannot even view an electron as a localized object, much less a spinning one. What is important is that the electron can have two different states due to some intrinsic property and (perhaps unfortunately) we still call this property "spin." The two possible values of m_s ($+\frac{1}{2}$ and $-\frac{1}{2}$) are often said to be "spin up" and "spin down," referring to the two possible directions of the spin angular momentum.

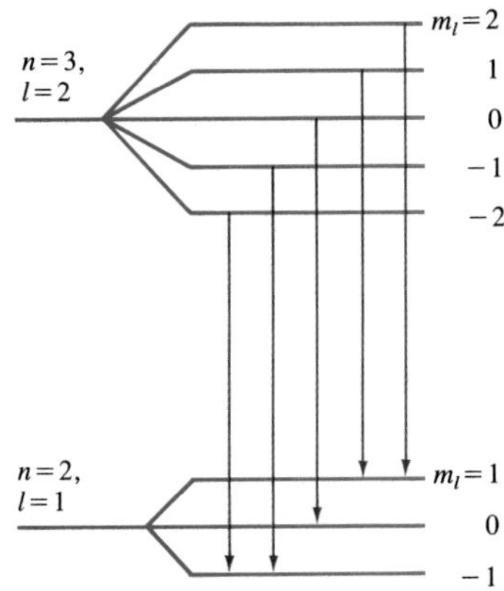

FIGURE 28–6 When a magnetic field is applied, an $n = 3$, $l = 2$ energy level is split into five separate levels, corresponding to the five values of m_l (2, 1, 0, −1, −2). An $n = 2$, $l = 1$ level is split into three levels ($m_l = 1$, 0, −1). Transitions can occur between levels (not all transitions are shown), with photons of several slightly different frequencies being given off (the Zeeman effect).

The possible values of the four quantum numbers for an electron in the hydrogen atom are summarized in Table 28–1.

TABLE 28–1
Quantum Numbers for an Electron

Name	Symbol	Possible Values
Principal	n	1, 2, 3, . . . , ∞.
Orbital	l	For a given n: l can be 0, 1, 2, . . . , $n-1$.
Magnetic	m_l	For given n and l: m_l can be l, $l-1$, . . . , 0, . . . , $-l$.
Spin	m_s	For each set of n, l, and m_l: m_s can be $+\frac{1}{2}$ or $-\frac{1}{2}$.

EXAMPLE 28–4 How many different states are possible for an electron whose principal quantum number is $n = 3$?

SOLUTION For $n = 3$, l can have the values $l = 2, 1, 0$. For $l = 2$, m_l can be 2, 1, 0, −1, −2, which is five different possibilities. For each of these, m_s can be either up or down ($+\frac{1}{2}$ or $-\frac{1}{2}$); so for $l = 2$, there are $2 \times 5 = 10$ states. For $l = 1$, m_l can be 1, 0, −1, and since m_s can be $+\frac{1}{2}$ or $-\frac{1}{2}$ for each of these, we have six more possible states. Finally, for $l = 0$, m_l can only be 0, and there are only two states corresponding to $m_s = +\frac{1}{2}$ and $-\frac{1}{2}$.

Helium, $Z = 2$

n	l	m_l	m_s
1	0	0	$\frac{1}{2}$
1	0	0	$-\frac{1}{2}$

Lithium, $Z = 3$

n	l	m_l	m_s
1	0	0	$\frac{1}{2}$
1	0	0	$-\frac{1}{2}$
2	0	0	$\frac{1}{2}$

Sodium, $Z = 11$

n	l	m_l	m_s
1	0	0	$\frac{1}{2}$
1	0	0	$-\frac{1}{2}$
2	0	0	$\frac{1}{2}$
2	0	0	$-\frac{1}{2}$
2	1	1	$\frac{1}{2}$
2	1	1	$-\frac{1}{2}$
2	1	0	$\frac{1}{2}$
2	1	0	$-\frac{1}{2}$
2	1	-1	$\frac{1}{2}$
2	1	-1	$-\frac{1}{2}$
3	0	0	$\frac{1}{2}$

basis not only for understanding complex atoms, but also for understanding molecules and bonding, and other phenomena as well.

Let us now look at the structure of some of the simpler atoms when they are in the ground state. After hydrogen, the next simplest atom is helium with two electrons. Both electrons can have $n = 1$, since one can have spin up ($m_s = +\frac{1}{2}$) and the other spin down ($m_s = -\frac{1}{2}$), thus satisfying the exclusion principle. Of course, since $n = 1$, l and m_l must be zero (Table 28–1). Thus the two electrons have the quantum numbers indicated in the table in the margin.

Lithium has three electrons, two of which can have $n = 1$. But the third cannot have $n = 1$ without violating the exclusion principle. Hence the third electron must have $n = 2$. Since it happens that the $n = 2, l = 0$ level has a lower energy than $n = 2$, $l = 1$, the electrons in the ground state have the quantum numbers indicated in the table in the margin. Of course, the quantum numbers of the third electron could also be, say, $(3, 1, -1, \frac{1}{2})$. But the atom in this case would be in an excited state since it would have greater energy. It would not be long before it jumped to the ground state with the emission of a photon. At room temperature, unless extra energy is supplied (as in a discharge tube), the vast majority of atoms are in the ground state.

We can continue in this way to describe the quantum numbers of each electron in the ground state of larger and larger atoms. That for sodium, with its eleven electrons, is shown in the table in the margin.

The ground-state configuration for all atoms is given in the **periodic table**, which is displayed inside the back cover of this book, and discussed in the next section.

*28–8 • The Periodic Table of Elements

A century ago, Dmitri Mendeleev (1834–1907) arranged the then known elements into what we now call the **periodic table** of the elements. The atoms were arranged according to increasing mass, but also so that elements with similar chemical properties would fall in the same column. Today's version is shown inside the back cover. Each square contains the atomic number Z, the symbol for the element, and the atomic mass (in atomic mass units). Finally, in the lower left corner the configuration of the ground state of the atom is given. This requires some explanation. Electrons with the same value of n are referred to as being in the same **shell**. Electrons with $n = 1$ are in one shell (called the K shell), those with $n = 2$ are in a second shell (the L shell), those with $n = 3$ are in the third (M) shell, and so on. Electrons with the same values of n and l are referred to as being in the same **subshell**. Letters are often used to specify the value of l as is shown in Table 28–2. That is, $l = 0$ is the s subshell; $l = 1$ is the p subshell; $l = 2$ is the d subshell; beginning with $l = 3$, the letters follow the alphabet, f, g, h, i, and so on. (The first letters s, p, d, and f were originally abbreviations of "sharp," "principal," "diffuse," and "fundamental," experimental terms referring to the spectra.)

Shells and subshells

The Pauli exclusion principle limits the number of electrons possible in each shell and subshell. For any value of l, there are $2l + 1$ different m_l values (m_l can be any integer from 1 to l, from -1 to $-l$, or zero), and two different m_s values. There can be, therefore, at most $2(2l + 1)$ electrons in any l subshell. For example, for $l = 2$, five m_l values are possible $(2, 1, 0, -1, -2)$, and for each of these, m_s can be $+\frac{1}{2}$ or $-\frac{1}{2}$, for a total of $2(5) = 10$ states. Table 28–2 lists the maximum number of electrons that can occupy each subshell.

Table 28–2
Values of l

Value of l	Letter symbol	Maximum number of electrons in subshell
0	*s*	2
1	*p*	6
2	*d*	10
3	*f*	14
4	*g*	18
5	*h*	22
⋮	⋮	⋮

Since the energy levels depend almost entirely on the values of n and l, it is customary to specify the electron configuration simply by giving the n value and the appropriate letter for l, with the number of electrons in each subshell given as a superscript. The ground-state configuration of sodium, for example, is written as $1s^2 2s^2 2p^6 3s^1$. This is simplified in the periodic table by specifying the configuration only of the outermost electrons and any other nonfilled subshells. The grouping of atoms in the periodic table is according to increasing mass. There is also a regularity according to chemical properties. And although this is treated in chemistry textbooks, we discuss it here briefly because it is a result of quantum mechanics.

All the noble gases (in the last column of the table) have completely filled shells or subshells. That is, their outermost subshell is completely full, and the electron distribution is spherically symmetric. With such full spherical symmetry, other electrons are not readily attracted nor are electrons readily lost (ionization energy is high). This is why the noble gases are nonreactive (more on this when we discuss molecules and bonding in Chapter 29). Column seven contains the **halogens**, which lack one electron from a filled shell. Because of the shapes of the orbits (see Section 29–1), an additional electron can be accepted from another atom, and hence these elements are quite reactive. They have a valence of -1, meaning that when an extra electron is acquired, the resulting ion has a net charge of $-1e$. At the left of the table, column I contains the **alkali metals**, all of which have a single outer s electron. This electron spends most of its time outside the inner closed shells and subshells which shield it from most of the nuclear charge; indeed, it is relatively far from the nucleus and is attracted to it by a net charge of only $+1e$, because of the shielding effect of the other electrons. Hence this outer electron is easily removed and can spend much of its time around another atom, forming a molecule. This is why the alkali metals are highly reactive and have a valence of $+1$. The other columns of the table can be treated similarly.

The presence of the transition elements in the center of the table, as well as the lanthanides (rare earths) and actinides below, is a result of incomplete inner shells. For the lowest Z elements, the subshells are filled in a simple order: first $1s$, then $2s$, followed by $2p$, $3s$, and $3p$. You might expect that $3d$ ($n = 3$, $l = 2$) would be filled next, but it isn't. Instead, the $4s$ level actually has a slightly lower energy than the $3d$, so it fills first (K and Ca). Only then does the $3d$ shell start to fill up, beginning with Sc. (The $4s$ and $3d$ levels are close, so some elements have only one $4s$ electron, such as Cr.) Most of the chemical properties are governed by the relatively loosely held $4s$ electrons and hence these **transition elements** usually have valences of $+1$ or $+2$. A similar effect is responsible for the rare earths, which are shown at the bottom of the periodic table for convenience. All have very similar chemical properties which are determined by their two outer $6s$ or $7s$ electrons, whereas the different numbers of electrons in the unfilled inner shells have little effect.

28–9 • X-Ray Spectra and Atomic Number

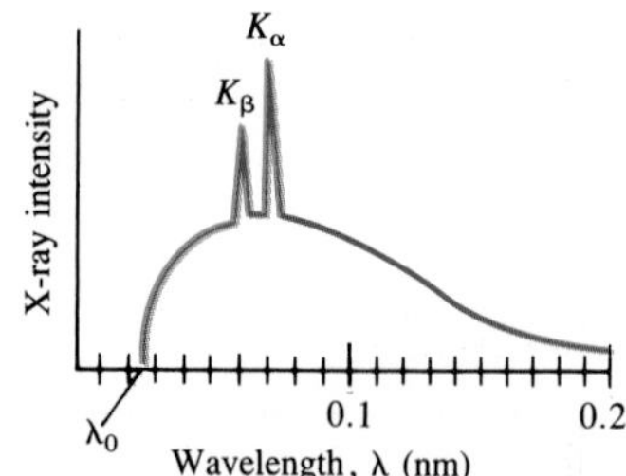

FIGURE 28–8 Spectrum of X rays emitted from a molybdenum target in an X-ray tube operated at 50 kV.

The line spectra of atoms in the visible, UV, and IR regions of the EM spectrum are mainly due to transitions between states of the outer electrons. For these electrons, much of the charge of the nucleus is shielded from them by the negative charge on the inner electrons. But the innermost electrons in the $n = 1$ shell "see" the full charge of the nucleus. Since the energy of a level is proportional to Z^2 (see Eq. 27–11), for an atom with $Z = 50$, we would expect wavelengths about $50^2 = 2500$ times shorter than those found in the Lyman series of hydrogen (around 100 nm), or 10^{-2} to 10^{-1} nm. Such short wavelengths lie in the X-ray region of the spectrum.

We saw in Section 25–11 that X rays are produced when electrons accelerated by a high voltage strike the metal target inside the X-ray tube. If we look at the spectrum of wavelengths emitted by an X-ray tube, we see that the spectrum consists of two parts: a continuous spectrum with a cutoff at some λ_0 which depends only on the voltage across the tube, and a series of peaks superimposed. A typical example is shown in Fig. 28–8. The smooth curve and the cutoff wavelength λ_0 move to the left as the voltage across the tube increases. The peaks (labeled K_α and K_β in Fig. 28–8), however, remain at the same wavelength when the voltage is changed, although they are located at different places when different target materials are used. This observation suggests that the peaks are characteristic of the material used. Indeed, we can explain them by imagining that the electrons accelerated by the high voltage of the tube can reach sufficient energies that when they collide with the atoms of the target, they can knock out one of the very tightly held inner electrons. These **characteristic X rays** (the peaks in Fig. 28–8) are photons emitted when an electron in an upper state drops down to fill the vacated lower state.

The *K* lines result from transitions *into* the *K* shell ($n = 1$). The K_α line is a transition that originates from the $n = 2$ (*L*) shell, the K_β line from the $n = 3$ (*M*) shell. An *L* line is due to a transition into the *L* shell, and so on.

Measurement of the characteristic X-ray spectra has allowed a determination of the inner energy levels of atoms. It has also allowed the determination of *Z* values for many atoms, since (as we have seen) the wavelength of the shortest X rays emitted will be inversely proportional to Z^2. Actually, for an electron jumping from, say, the $n = 2$ to the $n = 1$ level, the wavelength is inversely proportional to $(Z - 1)^2$ because the nucleus is shielded by the one electron that still remains in the 1*s* level. In 1914, H.G.J. Moseley (1887–1915) found that a plot of $\sqrt{1/\lambda}$ versus *Z* produced a straight line, Fig. 28–9. The *Z* values of a number of elements were determined by fitting them to such a **Moseley plot**. The work of Moseley put the concept of atomic number on a firm experimental basis.

FIGURE 28–9 Plot of $\sqrt{1/\lambda}$ versus *Z* for K_α X-ray lines.

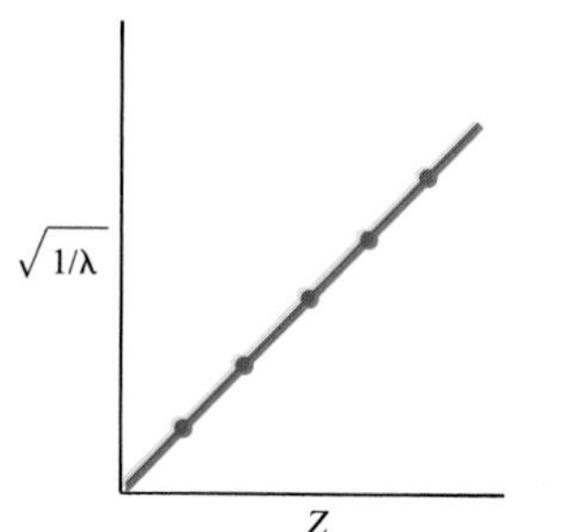

EXAMPLE 28–6 Estimate the wavelength for an $n = 2$ to $n = 1$ transition in molybdenum ($Z = 42$).

SOLUTION We use the Bohr formula, Eq. 27–12, with Z^2 replaced by $(Z - 1)^2 = (41)^2$. Or, more simply, we can use the result of Example 27–7 for the $n = 2$ to $n = 1$ transition in hydrogen ($Z = 1$). Since $\lambda \propto 1/(Z - 1)^2$, we will have

$$\lambda = (1.22 \times 10^{-7}\ \text{m})/(41)^2 = 0.073\ \text{nm}.$$

This is close to the measured value (Fig. 28–8) of 0.071 nm.

Now we briefly analyze the continuous part of an X-ray spectrum (Fig. 28–8) based on the photon theory of light. When electrons strike the target, they collide with atoms of the material and give up most of their energy as heat (about 99 percent, so X-ray tubes must be cooled). However, electrons can also give up energy by emitting a photon of light. An electron can be decelerated by interaction with atoms of the target (Fig. 28–10); but an accelerating charge can emit radiation (Chapter 22), and in this case it is called **bremsstrahlung** (German for "braking radiation"). Because energy is conserved, the energy of the emitted photon, hf, must equal the loss of kinetic energy of the electron, $\Delta\text{KE} = \text{KE} - \text{KE}'$, so

$$hf = \Delta\text{KE}.$$

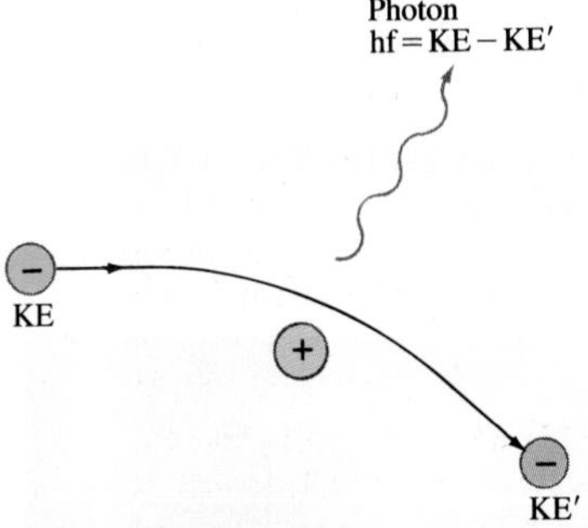

FIGURE 28–10 Bremsstrahlung photon produced by an electron decelerated by interaction with a target atom.

An electron may lose all or a part of its energy in such a collision. The continuous X-ray spectrum (Fig. 28–8) is explained as being due to such bremsstrahlung collisions in which varying amounts of energy are lost by the electrons. The shortest-wavelength X ray produced (the highest frequency) must be due to an electron that gives up all its kinetic energy to one photon in a single collision. Since the initial kinetic energy of an electron is equal to the energy given it by the accelerating voltage, V, then $\text{KE} = eV$. In a single collision in which the electron is brought to rest, we have

$$hf_0 = eV$$

or

$$\lambda_0 = \frac{hc}{eV}, \tag{28–4}$$

where $\lambda_0 = c/f_0$ is the cutoff wavelength, Fig. 28–8. This prediction for λ_0 corresponds precisely with that observed experimentally. This result is further evidence that X rays are a form of light† and that the photon theory of light is valid.

EXAMPLE 28–7 What is the shortest-wavelength X-ray photon emitted in an X-ray tube subjected to 50 kV?

SOLUTION From Eq. 28–4,

$$\lambda_0 = \frac{(6.6 \times 10^{-34}\ \text{J}\cdot\text{s})(3.0 \times 10^8\ \text{m/s})}{(1.6 \times 10^{-19}\ \text{C})(5.0 \times 10^4\ \text{V})} = 2.5 \times 10^{-11}\ \text{m},$$

or 0.025 nm. This agrees well with experiment, Fig. 28–8.

† If X rays were not photons but rather neutral particles with rest mass m_0, Eq. 28–4 would not hold.

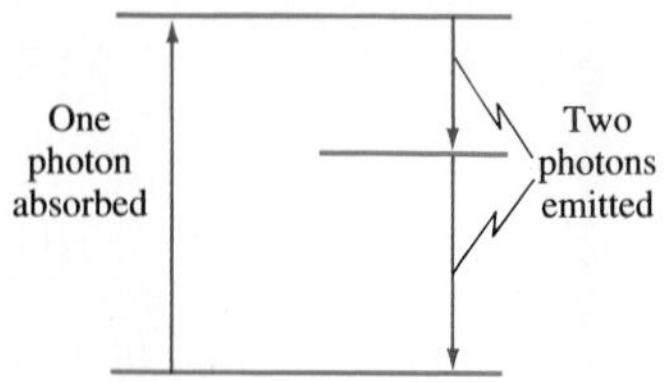

FIGURE 28–11 Fluorescence.

*28–10 • Fluorescence and Phosphorescence

When an atom is excited from one energy state to a higher one by the absorption of a photon, it may return to the lower level in a series of two (or more) jumps if there is an energy level in between (Fig. 28–11). The photons emitted will consequently have lower energy and frequency than the absorbed photon. This phenomenon is called **fluorescence**.

The wavelength for which fluorescence will occur depends on the energy levels of the particular atoms. Common fluorescent rocks and paints, for example, often emit visible light after absorbing UV light (Fig. 28–12). Because the frequencies are different for different substances, and because many substances fluoresce readily, fluorescence is a powerful tool for identification of compounds. It is also used for assaying—determining how much of a substance is present—and for following substances along a natural pathway as in plants and animals. For detection of a given compound, the stimulating light must be monochromatic, and solvents or other materials present must not fluoresce in the same region of the spectrum. Often the observation of fluorescent light being emitted is sufficient. In other cases, spectrometers are used to measure the wavelengths and intensities of the light.

FIGURE 28–12 When UV light is shown on these various "fluorescent" rocks, they fluoresce in the visible region of the spectrum.

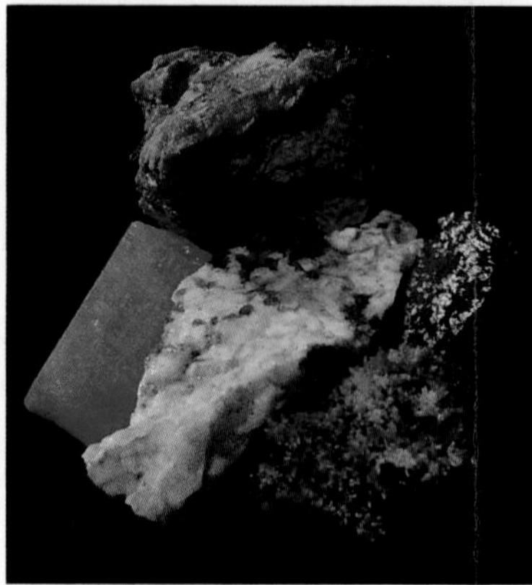

Fluorescent light bulbs work in a two-step process. The applied voltage accelerates electrons that strike atoms of the gas in the tube and cause them to be excited. When the excited atoms jump down to their normal levels, they emit UV photons which strike a fluorescent coating on the inside of the tube. The light we see is a result of this material fluorescing in response to the UV light striking it.

Materials such as those used for luminous watch dials are said to be **phosphorescent**. In a phosphorescent substance, atoms can be excited by absorption of a photon to an energy level said to be metastable. When an atom is raised to a normal excited state, it drops back down within about 10^{-8} s. **Metastable** states last much longer—even a few seconds or longer. In a collection of such atoms, many of the atoms will descend to the lower state fairly soon, but many will remain in the excited state for over an hour. Hence light will be emitted even after long periods. When you put your watch dial close to a bright lamp, it excites many atoms to metastable states, and you can see the glow a long time after.

*28–11 • Lasers and Holography

A laser is a device that can produce a very narrow intense beam of monochromatic coherent light. The emitted beam is a nearly perfect plane wave. An ordinary light source, on the other hand, emits light in all directions (so the intensity decreases rapidly with distance), and the emitted light is incoherent (the different parts of the beam are not in phase with each other). The excited atoms that emit the light in an ordinary light bulb act independently, so each photon emitted can be considered as a short wave train, typically 30 cm long and lasting 10^{-8} s; these wave trains bear no phase relation to one another.

The action of a laser is based on quantum theory. We have seen that a photon can be absorbed by an atom if (and only if) its energy hf corresponds to the energy difference between an occupied energy level of the atom and an available excited state, Fig. 28–13a. This is, in a sense, a resonant situation. If the atom is already in the excited state, it may of course jump spontaneously to the lower state with the emission of a photon. However, if a photon with this same energy strikes the excited atom, it can stimulate the atom to make the transition sooner to the lower state, Fig. 28–13b. This phenomenon is called **stimulated emission**, and it can be seen that not only do we still have the original photon, but also a second one of the same frequency as a result of the atom's transition. And these two photons are exactly *in phase*, and they are moving in the same direction. This is how coherent light is produced in a laser. Hence the name "laser," which is an acronym for **l**ight **a**mplification by **s**timulated **e**mission of **r**adiation.

Stimulated emission

Normally, most atoms are in the lower state, so incident photons will mostly be absorbed. In order to obtain the coherent light from stimulated emission, two conditions must be satisfied. First, the atoms must be excited to the higher state. That is, an **inverted population** is needed, one in which more atoms are in the upper state than in the lower one, so that *emission* of photons will dominate over absorption. And second, the higher state must be a **metastable state**—a state in which the electrons remain longer than usual so that the transition to the lower state occurs by stimulated emission rather than spontaneously. How these conditions are achieved for different lasers will be discussed shortly. For now, we assume that the atoms have been excited to an upper state. Figure 28–14 is a schematic diagram of a laser: the "lasing" material is placed in a long narrow tube at the ends of which are two mirrors, one of which is partially transparent (perhaps 1 or 2 percent). Some of the excited atoms drop down fairly soon after being excited. One of these is the atom shown on the far left in Fig. 28–14. If the emitted photon strikes another atom in the excited state, it stimulates this atom to emit a photon of the *same* frequency, moving in the *same* direction, and *in phase* with it. These two photons then move on to strike other atoms causing more stimulated emission. As the process continues, the number of photons multiplies. When the photons strike the end mirrors, most are reflected back, and as they move in the opposite direction, they continue to stimulate more atoms to emit photons. As the photons move back and forth between the mirrors, a small percentage passes through the partially transparent mirror at one end. These photons make up the narrow coherent external laser beam.

FIGURE 28–13 (a) Absorption of a photon. (b) Stimulated emission.

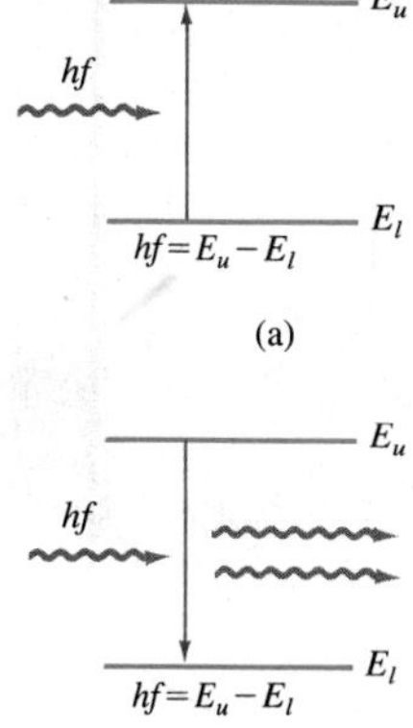

Inside the tube, some spontaneously emitted photons will be emitted at an angle to the axis, and these will merely go out the side of the tube and

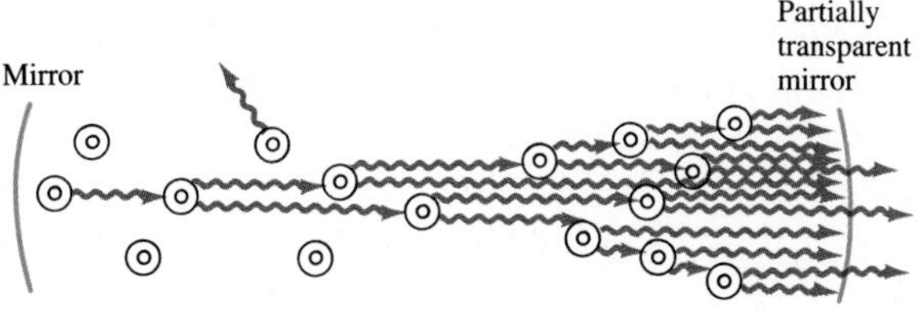

FIGURE 28–14 Laser diagram, showing excited atoms stimulated to emit light.

not contribute to the main beam. Thus the beam can be very narrow. In a well-designed laser, the spreading of the beam is limited only by diffraction, so the angular spread is $\approx \lambda/D$ (see Eq. 24–3) where D is the diameter of the end mirror. The diffraction spreading can be incredibly small. The light energy, instead of spreading out in space as it does for an ordinary light source, is directed in a pencil-thin beam.

Ruby laser

The excitation of the atoms in a laser can be done in several ways to produce the necessary inverted population. In a ruby laser, the lasing material is a ruby rod consisting of Al_2O_3 with a small percentage of aluminum (Al) atoms replaced by chromium (Cr) atoms. The Cr atoms are the ones involved in lasing. The atoms are excited by strong flashes of light of wavelength 550 nm, which corresponds to a photon energy of 2.2 eV. As shown in Fig. 28–15, the atoms are excited from state E_0 to state E_2. This process is called **optical pumping**. The atoms quickly decay either back to E_0 or to the intermediate state E_1, which is metastable with a lifetime of about 3×10^{-3} s (compared to 10^{-8} s for ordinary levels). With strong pumping action, more atoms can be forced into the E_1 state than are in the E_0 state. Thus we have the inverted population needed for lasing. As soon as a few atoms in the E_1 state jump down to E_0, they emit photons that produce stimulated emission of the other atoms and the lasing action begins. A ruby laser thus emits a beam whose photons have energy 1.8 eV and a wavelength of 694.3 nm (or "ruby-red" light).

FIGURE 28–15 Energy levels of chromium in a ruby crystal. Photons of energy 2.2 eV "pump" atoms from E_0 to E_2, which then decay to metastable state E_1. Lasing action occurs by stimulated emission of photons in transition from E_1 to E_0.

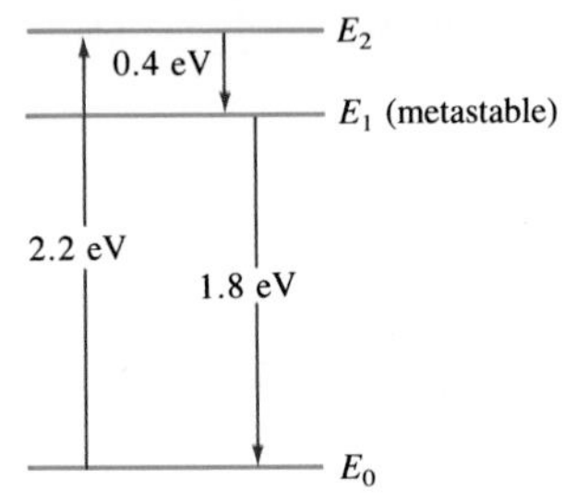

He-Ne laser

In a helium–neon (He–Ne) laser, the lasing material is a gas, a mixture of about 15 percent He and 85 percent Ne. This combination works as a lasing material because of certain compatible properties of the two gases. In this laser, the atoms are excited by applying a high voltage to the tube so that an electric discharge takes place within the gas. In the process, some of the He atoms are raised to the metastable state E_1 shown in Fig. 28–16, which corresponds to a jump of 20.61 eV. Now Ne atoms have an excited state that is almost exactly the same energy above the ground state, 20.66 eV. The He atoms do not quickly return to the ground state by spontaneous emission, but instead often give their excess energy to a Ne atom when they collide. In such a collision, the He drops to the ground state and the Ne atom is excited to the state E'_3 (the prime refers to neon states). The slight difference in energy (0.05 eV) is supplied by the kinetic energy of the moving molecules.

FIGURE 28–16 Energy levels for He and Ne. He is excited in the electric discharge to the E_1 state. This energy is transfered to the E'_3 level of the Ne by collision. E'_3 is metastable and decays to E'_2 by stimulated emission.

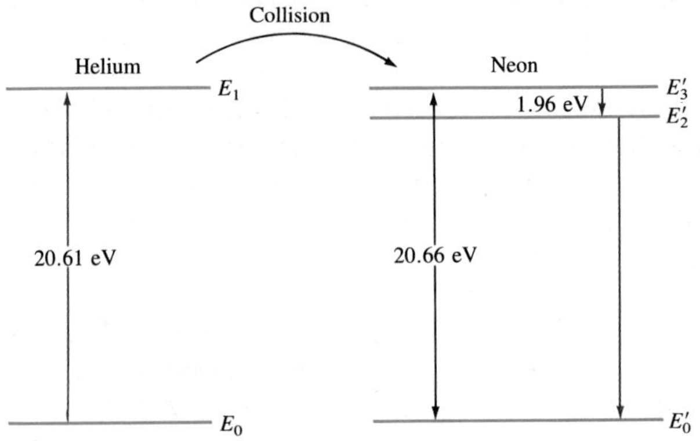

In this manner, the E'_3 state in Ne—which is metastable—becomes more populated than the E'_2 level. This inverted population between E'_3 and E'_2 is what is needed for lasing.

Other lasers

Other types of laser include: chemical lasers, in which the energy input comes from the chemical reaction of highly reactive gases; dye lasers, whose frequency is tunable; CO_2 gas lasers, capable of high power output in the infrared; rare-earth solid-state lasers such as the high-power neodymium: YAG laser; and the *pn* junction laser in which the transitions occur between the bottom of the conduction band and the upper part of the valence band (Section 29–6).

The excitation of the atoms in a laser can be done continuously or in pulses. In a **pulsed laser**, the atoms are excited by periodic inputs of energy. The multiplication of photons continues until all the atoms have been stimulated to jump down to the lower state, and the process is repeated with each pulse. In a **continuous laser**, the energy input is continuous so that as atoms are stimulated to jump down to the lower level, they are soon excited back up to the upper level. In either case, of course, the laser is not a source of energy. Energy must be put in, and the laser converts a part of this energy into an intense narrow beam. Thus a laser, like an electronic amplifier, uses energy to amplify a certain type of signal.

Medical and other uses of lasers

The unique feature of light from a laser is, as mentioned before, that it is a coherent narrow beam of a single frequency (or several distinct frequencies). Because of this feature, the laser has found many applications. Lasers are a useful surgical tool. The narrow intense beam can be used to destroy tissue in a localized area, or to break up gallstones and kidney stones. Because of the heat produced, a laser beam can be used to "weld" broken tissue, such as a detached retina. For some types of internal surgery, the laser beam can be carried by an optical fiber (Section 23–6) to the surgical point, sometimes as an additional fiber-optic path on an endoscope (also discussed in Section 23–6). An example is removal of plaques clogging human arteries. Tiny organelles within a living cell have been destroyed by researchers using lasers to study how the absence of that organelle affects the behavior of the cell. The finely focused beam of a laser has been used to destroy cancerous and precancerous cells; at the same time, the heat seals off capillaries and lymph vessels, thus "cauterizing" the wound in the process to prevent spread of the disease. The intense heat produced in a small area by a laser beam is also used for welding and machining metals and for drilling tiny holes in hard materials. The beam of a laser is narrow in itself (typically, a few mm). But because the beam is coherent, monochromatic, and essentially parallel and narrow, lenses can be used to focus the light into incredibly small areas without the usual aberration problems. The limiting factor thus becomes diffraction, and the energy crossing unit area per unit time can be very large. The precise straightness of a laser beam is also useful to surveyors for lining up equipment precisely, especially in inaccessible places.

Holography

One of the most interesting applications of laser light is the production of three-dimensional images called **holograms**. In an ordinary photograph, the film simply records the intensity of light reaching it at each point. When the photograph or transparency is viewed, light reflecting from it or passing through it gives us a two-dimensional picture. In holography, the images are

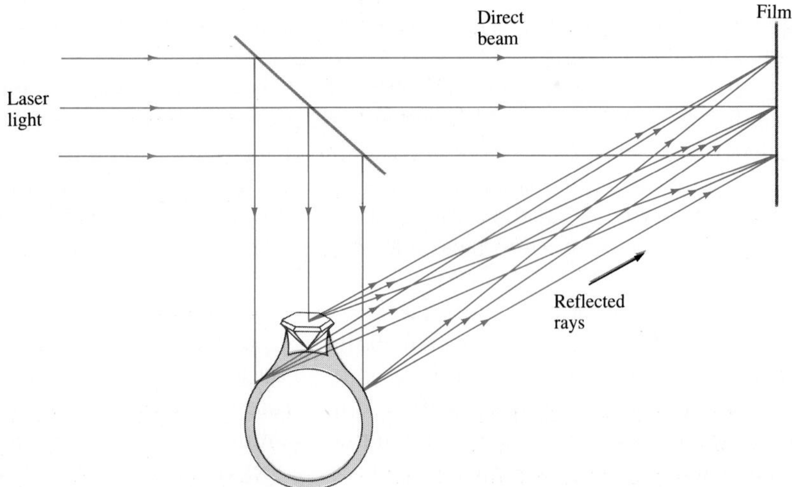

FIGURE 28–17 Making a hologram.

formed by interference, without lenses. When a laser hologram is made on film, a broadened laser beam is split into two parts by a half-silvered mirror, Fig. 28–17. One part goes directly to the film; the rest passes to the object to be photographed, from which it is reflected to the film. Light from every point on the object reaches each point on the film, and the interference of the two beams allows the film to record both the intensity and relative phase of the light at each point. After the film is developed, it is placed again in a laser beam and a three-dimensional image of the object is created. You can walk around such an image and see it from different sides as if it were the original object. Yet, if you try to touch it with your hand, there will be nothing material there.

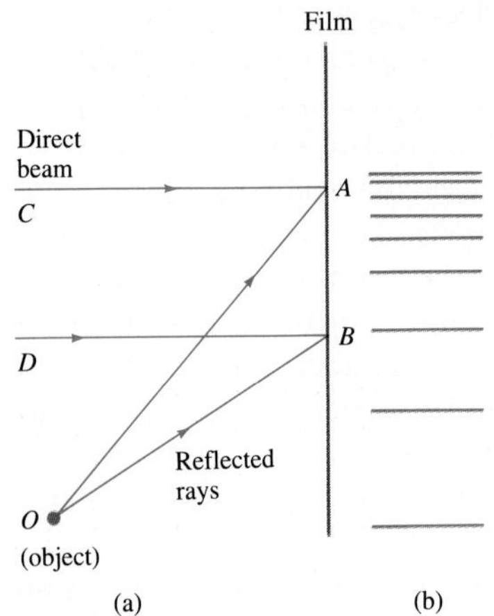

FIGURE 28–18 Light from point O on the object interferes with light of direct beam (rays CA and DB).

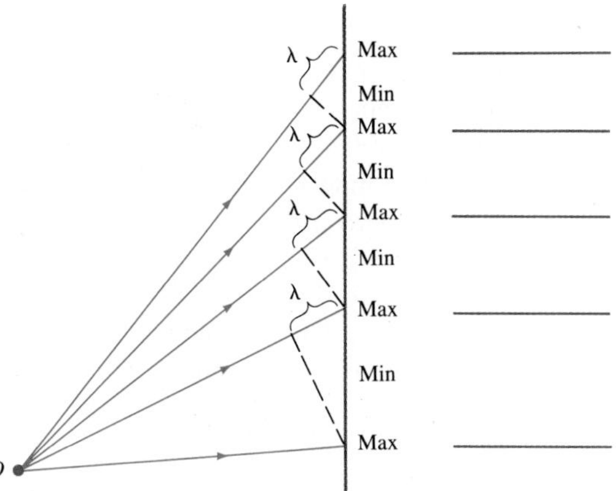

FIGURE 28–19 Each of the rays shown leaving point O on the object is one wavelength shorter than the one above it. If the top ray is in phase with the direct beam (not shown), which has the same phase at all points on the screen, all the rays shown produce constructive interference. From this diagram, it can be seen that the fringe spacing increases toward the bottom.

The details of how the image is formed are quite complicated. But we can get the basic idea by considering one single point on the object. In Fig. 28–18a, the rays OA and OB have reflected from one point on our object. The rays CA and DB come directly from the source and interfere with OA and OB at points A and B on the film. A set of interference fringes is produced as shown in Fig. 28–18b. The spacing between the fringes changes from top to bottom as shown. Why this happens is explained in Fig. 28–19. Thus the hologram of a single point object would have the pattern shown in Fig. 28–18b. The film in this case looks like a diffraction grating with variable spacing. Hence, when coherent laser light is passed back through the developed film, the diffracted rays in the first-order maxima occur at slightly different angles because the spacing changes. (Remember Eq. 24–4, $\sin\theta = m\lambda/d$; so where the spacing d is greater, the angle θ is smaller.) Hence, the rays diffracted upward (in first order) seem to diverge from a single point, Fig. 28–20. This is a virtual image of the original object, which can be seen with the eye. Rays diffracted in first order *downward* converge to make a real image, which can be seen and also photographed. (Note that the straight-through undiffracted rays are of no interest.) Of course real objects consist of many points, so a hologram will be a complex interference pattern which when laser light is incident on it, will reproduce an image of the object. Each image point will be at the correct (three-dimensional) position with respect to other points, so the image accurately represents the original object. The

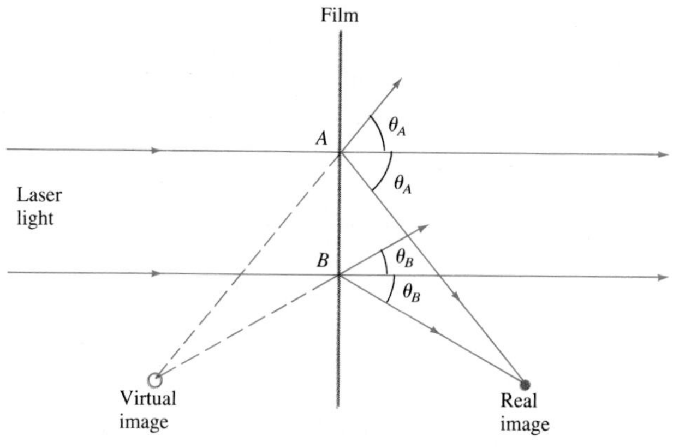

FIGURE 28–20 Reconstructing the image. Laser beam strikes film that is like a diffraction grating of variable spacing. Rays corresponding to the first diffraction maxima are shown emerging. The angle $\theta_A > \theta_B$ because the spacing at B is greater than at A ($\sin\theta = \lambda/d$). Hence real and virtual images of the point are reproduced as shown.

image can be viewed from different angles as if viewing the original object. Holograms can be made in which a viewer can walk entirely around the image (360°) and see all sides of it.

White-light holograms

So-called **volume** or **white-light holograms** do not require a laser to see the image, but can be viewed with ordinary white light (preferably a nearly point source, such as the sun or a clear bulb with a small strong filament). Such holograms must be made, however, with a laser. They are made not on thin film, but on a *thick* emulsion. The interference pattern, instead of being two-dimensional as for an ordinary hologram, is actually three-dimensional (hence the name "volume hologram"). The interference pattern in the film emulsion can be thought of as an array of bands or ribbons (consisting of the silver grains from the development process) where constructive interference occurred. This array, and the reconstruction of the image, can be compared to Bragg scattering of X rays from the atoms in a crystal (see Section 25–11 and Figs. 25–31 and 25–32). White light can reconstruct the image because the Bragg condition (Eq. 25–9, $m\lambda = 2d \sin\theta$) selects out the appropriate single wavelength. If the hologram is originally produced by lasers emitting the three additive primary colors (red, green, and blue), the three-dimensional image can be seen in full color when viewed with white light.

SUMMARY

In 1925, Schrödinger and Heisenberg separately worked out a new theory, *quantum mechanics*, which is now considered to be the basic theory at the atomic level. It is a statistical theory rather than a deterministic one.

An important aspect of quantum mechanics is the Heisenberg *uncertainty principle*. It results from the wave–particle duality and the unavoidable interaction between the observed object and the observer. One form of the uncertainty principle states that the position x and momentum p of an object cannot both be measured precisely at the same time; the products of the uncertainties, $(\Delta x)(\Delta p)$, can be no less than $\hbar$ $(=h/2\pi)$:

$$(\Delta p)(\Delta x) \gtrsim \hbar.$$

Another form states that the energy can be uncertain, or nonconserved, by an amount ΔE for a time $\Delta t \approx \hbar/\Delta E$.

According to quantum mechanics, the electrons in an atom do not have well-defined orbits, but instead exist as a "cloud." Electron clouds can be interpreted as an electron wave spread out in space, or as a *probability distribution* for electrons considered as particles.

According to quantum mechanics, the state of an electron in an atom is specified by four *quantum numbers*: n, l, m_l, and m_s. The principal quantum number, n, can take on any integer value (1, 2, 3, . . .) and corresponds to the quantum number of the old Bohr theory; l can take on values from 0 up to $n - 1$; m_l can take on integer values from $-l$ to $+l$; and m_s can be $+\frac{1}{2}$ or $-\frac{1}{2}$. The energy levels in the hydrogen atom depend on n, whereas in other atoms they depend on n and l. When an external magnetic field is applied, the spectral lines are split (the *Zeeman effect*), indicating that the energy depends also on m_l in this case. Even in the absence of a magnetic field, precise measurements of spectral lines show a tiny splitting of the lines called *fine structure*, whose explanation is that the energy depends very slightly on m_l and m_s.

The arrangement of electrons in multielectron atoms is governed by the *Pauli exclusion principle*, which states that no two electrons can occupy the same quantum state—that is, they cannot have the same set of quantum numbers n, l, m_l, and m_s. Electrons, as a result, are grouped into *shells* (according to the value of n) and *subshells* (according to l).

X rays, which are a form of electromagnetic radiation of very short wavelength, are produced when high-speed electrons strike a target. The spectrum of X rays so produced consists of two parts: a continuous spectrum produced when the electrons are decelerated by nuclei of the target, and peaks representing photons emitted by atoms of the target after being excited by collision with the high-speed electrons.

QUESTIONS

1. Compare a matter wave Ψ to (*a*) a wave on a string, and (*b*) an EM wave. Discuss similarities and differences.
2. Explain why Bohr's theory of the atom is not compatible with quantum mechanics, particularly the uncertainty principle.
3. Explain why it is that the more massive an object is, the easier it becomes to predict its future position.
4. In view of the uncertainty principle, why does a baseball seem to have a well-defined position and speed whereas an electron does not?
5. Discuss whether something analogous to the uncertainty principle operates when taking a public opinion survey. That is, do we alter what we are trying to measure when we take such a survey?
6. A cold thermometer is placed in a hot bowl of soup. Will the temperature reading of the thermometer be the same as the temperature of the hot soup before the measurement was made?
7. When you check the pressure in a tire, doesn't some air inevitably escape? Is it possible to avoid this escape of air altogether? What is the relation to the uncertainty principle?
8. It has been said that the ground-state energy in the hydrogen atom can be precisely known but the excited states have some uncertainty in their values (an "energy width"). Is this consistent with the uncertainty principle in its energy form? Explain.
9. If Planck's constant were much larger than it is, how would this affect our everyday life?
10. In what ways is Newtonian mechanics contradicted by quantum mechanics?
11. Discuss the differences between Bohr's view of the atom and the quantum-mechanical view.
12. The 589-nm yellow line in sodium is actually two very closely spaced lines. This splitting is due to an "internal" Zeeman effect. Can you explain this? [*Hint:* put yourself in the reference frame of the electron.]
13. Which of the following electron configurations are forbidden? (*a*) $1s^2 2s^2 2p^6 3s^3$; (*b*) $1s^2 2s^2 2p^4 3s^2 4p^2$; (*c*) $1s^2 2s^2 2p^8 3s^1$.
14. Give the complete electron configuration for a uranium atom.
*15. In what column of the periodic table would you expect to find the atom with each of the following configurations? (*a*) $1s^2 2s^2 2p^6 3s^2$; (*b*) $1s^2 2s^2 2p^6 3s^2 3p^6$; (*c*) $1s^2 2s^2 2p^6 3s^2 3p^6 4s^1$; (*d*) $1s^2 2s^2 2p^5$.
*16. The ionization energy for neon ($Z = 10$) is 21.6 eV and that for sodium ($Z = 11$) is 5.1 eV. Explain the large difference.
*17. Why do chlorine and iodine exhibit similar properties?
*18. Explain why potassium and sodium exhibit similar properties.
*19. Why are the chemical properties of the rare earths so similar?
20. Why do we not expect perfect agreement between measured values of X-ray line wavelengths and those calculated using Bohr theory, as in Example 28–6?
21. How would you figure out which lines in an X-ray spectrum correspond to K_α, K_β, L, etc., transitions?
22. Why do the characteristic X-ray spectra vary in a systematic way with Z, whereas the visible spectra (Fig. 24–23) do not?
23. Why do we expect electron transitions deep within an atom to produce shorter wavelengths than transitions by outer electrons?
*24. Certain dyes and other materials fluoresce by emitting visible light when UV light falls on them. Can infrared light produce fluorescence?
*25. Compare spontaneous emission to stimulated emission.
*26. How does laser light differ from ordinary light? How is it the same?
*27. Explain how a 0.0005-W laser beam, photographed at a distance, can seem much stronger than a 1000-W street lamp.

PROBLEMS

SECTION 28–2

1. (II) The neutrons in a parallel beam, each having kinetic energy $\frac{1}{40}$ eV, are directed through two slits 1.0 mm apart. How far apart will the interference peaks be on a screen 1.0 m away?
2. (II) Bullets of mass 1.0 g are fired in parallel paths with speeds of 300 m/s through a hole 5.0 mm wide. How far from the hole must you be to detect a 1.0-cm spread in the beam?

SECTION 28–3

3. (I) A proton is traveling with a speed of $(8.880 \pm 0.012) \times 10^5$ m/s. With what maximum accuracy can its position be ascertained?
4. (I) If an electron's position can be measured to an accuracy of 1.6×10^{-8} m, how accurately can its velocity be known?
5. (I) An electron remains in an excited state of an atom for typically 10^{-8} s. What is the minimum uncertainty in the energy of the state (in eV)?
6. (II) An electron and a 140-g baseball are each traveling 230 m/s measured to an accuracy of 0.075 percent. Calculate and compare the uncertainty in position of each.
7. (II) Estimate the lowest possible energy of a neutron contained in a typical nucleus of radius 1.0×10^{-15} m.

8. (II) Use the uncertainty principle to show that if an electron were present in the nucleus ($r \approx 10^{-15}$ m), its kinetic energy would be hundreds of MeV. (Since such electron energies are not observed, we conclude that electrons are not present in the nucleus.)

9. (III) The uncertainty principle can be stated in terms of angular quantities as follows:

$$\Delta L \, \Delta\phi \gtrsim \hbar.$$

Here, L stands for angular momentum along a given axis, and ϕ for the angular position measured in a plane perpendicular to that axis. (*a*) Make a plausibility argument for this relation. (*b*) Electrons in atoms have well-defined quantized values of angular momentum, with no uncertainty. What does this say about the uncertainty in angular position and the concept of electron orbits?

10. (III) In a double-slit experiment on electrons (or photons), suppose that we use indicators to determine which slit each electron passed through (Section 28–2). These indicators must tell us the y coordinate to within $a/2$, where a is the distance between slits. Use the uncertainty principle to show that the interference pattern will be destroyed. (Note: first show that the angle θ between maxima and minima of the interference pattern is given by $\lambda/2a$, Fig. 28–21.)

11. (III) How accurately can the position of a 2.50-keV electron be measured assuming its energy is known to 1.00 percent?

SECTIONS 28–6 AND 28–7

12. (I) For $n = 5$, what values can l have?

13. (I) For $n = 6$, $l = 3$, what are the possible values of m_l and m_s?

14. (I) List the quantum numbers for each electron in the ground state of carbon ($Z = 6$).

15. (I) How many different states are possible for an electron whose principal quantum number is $n = 4$? Write down the quantum numbers for each state.

16. (I) How many electrons can be in the $n = 5$, $l = 3$ subshell?

17. (I) List the quantum numbers for each electron in the ground state of aluminum ($Z = 13$).

18. (I) A hydrogen atom is known to have $l = 4$. What are the possible values for n, m_l, and m_s?

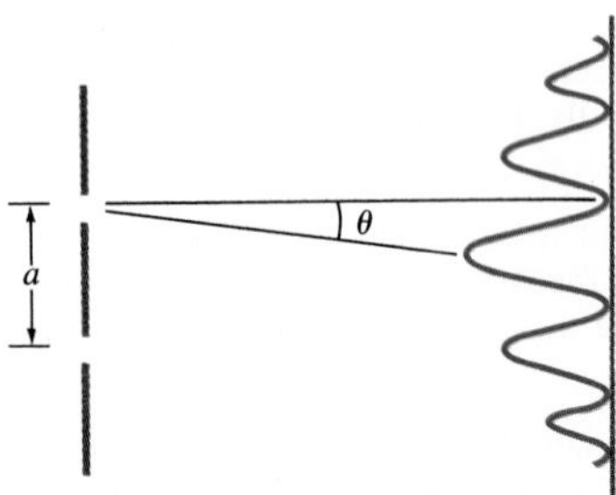

FIGURE 28–21 Problem 10.

19. (I) If a hydrogen atom has $m_l = -3$, what are the possible values of n, l, and m_s?

20. (I) Calculate the magnitude of the angular momentum of an electron in the $n = 4$, $l = 3$ state of hydrogen.

21. (II) A hydrogen atom is in the $7g$ state. Determine (*a*) the principal quantum number, (*b*) the energy of the state, (*c*) the orbital angular momentum and its quantum number l, and (*d*) the possible values for the magnetic quantum number.

22. (II) Estimate the binding energy of the third electron in lithium using Bohr theory. [*Hint:* this electron has $n = 2$ and "sees" a net charge of approximately $+1e$.] The measured value is 5.36 eV.

23. (II) The ionization (binding) energy of the outermost electron in boron is 8.26 eV. (*a*) Determine the "effective charge," Z_{eff}, seen by this electron. [*Hint:* use the Bohr model.] (*b*) Estimate the average orbital radius.

24. (II) Show that the total angular momentum is zero for a filled subshell.

25. (II) Show that the maximum number of electrons allowed in any subshell is equal to $2(2l + 1)$ where l is the angular momentum quantum number of the subshell.

SECTION 28–9

26. (I) What are the shortest-wavelength X rays emitted by electrons striking the face of a 40-kV TV picture tube? What are the longest wavelengths?

27. (I) If the shortest-wavelength bremsstrahlung X rays emitted from an X-ray tube have $\lambda = 0.019$ nm, what is the voltage across the tube?

28. (I) Show that the cutoff wavelength λ_0 is given by

$$\lambda_0 = 1240 \text{ nm}/V,$$

where V is the X-ray tube voltage in volts.

29. (II) Use the result of Example 28–6 to estimate the X-ray wavelength emitted when a Ti ($Z = 22$) atom jumps from $n = 2$ to $n = 1$.

30. (II) Estimate the wavelength for an $n = 2$ to $n = 1$ transition in iron ($Z = 26$).

31. (II) Use Bohr theory to estimate the wavelength for an $n = 3$ to $n = 1$ transition in molybdenum. The measured value is 0.063 nm. Why do we not expect perfect agreement?

32. (II) A mixture of iron and an unknown material are bombarded with electrons. The wavelength of the K_α lines are 194 pm for iron and 229 pm for the unknown. What is the unknown material?

*SECTION 28–11

*33. (II) A laser used to weld detached retinas puts out 25-ms-long pulses of 640-nm light which average 0.50-W output during a pulse. How much energy can be deposited per pulse and how many photons does each pulse contain?

*34. (II) Estimate the angular spread of a laser beam due to diffraction if the beam emerges through a 5.0-mm-diameter mirror. Assume that $\lambda = 694$ nm. How large would be the diameter of this beam if it struck a satellite 1000 km above the earth?

GENERAL PROBLEMS

35. An electron in the $n = 2$ state of hydrogen remains there on the average about 10^{-8} s before jumping to the $n = 1$ state. (*a*) Estimate the uncertainty in the energy of the $n = 2$ state. (*b*) What fraction of the transition energy is this? (*c*) What is the wavelength, and width (in nm), of this line in the spectrum of hydrogen?

36. Estimate (*a*) the quantum number l for the orbital angular momentum of the earth about the sun, and (*b*) the number of possible orientations for the plane of earth's orbit.

37. A 10-g bullet leaves a rifle at a speed of 400 m/s. (*a*) What is the wavelength of this bullet? (*b*) If the position of the bullet is known to an accuracy of 0.60 cm (radius of the barrel), what is the minimum uncertainty in its momentum? (*c*) If the accuracy of the bullet were determined only by the uncertainty principle (an unreasonable assumption), by how much might the bullet miss a pinpoint target 300 m away?

38. Using the Bohr formula for the radius of an electron orbit, estimate the average distance from the nucleus for an electron in the innermost ($n = 1$) orbit in uranium ($Z = 92$). Approximately how much energy would be required to remove this innermost electron?

39. An X-ray tube operates at 100 kV with a current of 20 mA and nearly all the electron energy goes into heat. If the specific heat capacity of the 78-g plate is 0.11 kcal/kg·C°, what will be the temperature rise per minute if no cooling water is used?

40. Use Bohr theory (especially Eq. 27–12) to show that the Moseley plot (Fig. 28–9) can be written

$$\sqrt{\frac{1}{\lambda}} = a(Z - b),$$

where $b \approx 1$, and evaluate a.

*41. (*a*) Construct a periodic table for the first 20 elements assuming that the Pauli exclusion principle holds but that the electron has no spin, so there are only three quantum numbers, n, l, m_l. (*b*) Which elements would be noble gases (i.e., with filled shells or subshells as in the real periodic table)?

42. Use the uncertainty principle to estimate the position uncertainty for the electron in the ground state of the hydrogen atom. [*Hint*: determine the momentum using the Bohr model of Section 27–9.] How does this compare to the Bohr radius?

43. In the so-called *vector model* of the atom, space quantization of angular momentum (Fig. 28–5) is illustrated as shown in Fig. 28–22. The angular momentum vector of magnitude $L = \sqrt{l(l+1)}(h/2\pi)$ is thought of as precessing around the z axis (like a spinning top or gyroscope) in such a way that the z component of angular momentum, $L_z = m_l(h/2\pi)$, also stays constant. Calculate the possible values for the angle θ between **L** and the z axis (*a*) for $l = 1$, (*b*) $l = 2$, and (*c*) $l = 3$. (*d*) Determine the minimum value of θ for $l = 1000$ and $l = 10^6$. Is this consistent with the correspondence principle?

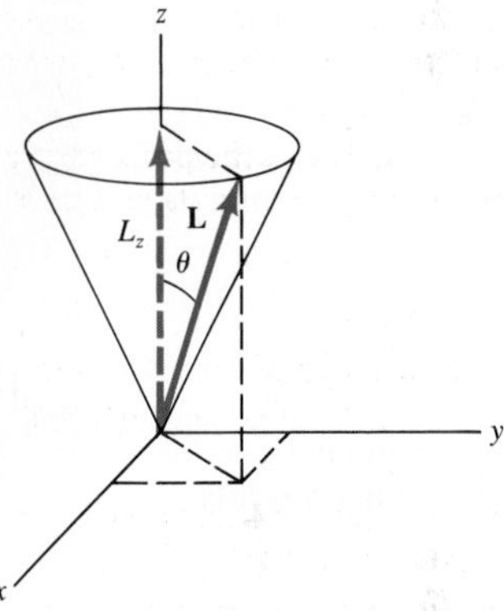

FIGURE 28–22 The vector model for orbital angular momentum. The orbital angular momentum vector **L** is imagined to precess about the z axis; L and L_z remain constant, but L_x and L_y continually change. Problem 43.

CHAPTER 29

Molecules and Solids

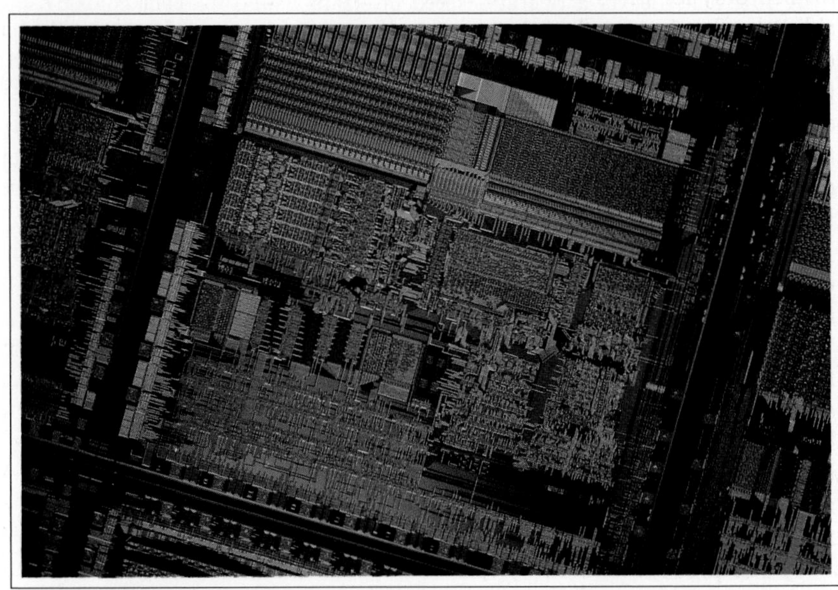

A silicon "chip" containing a complex of electronic elements.

Since its development in the 1920s, quantum mechanics has had a profound influence on our lives, both intellectually and technologically. Even the way we view the world has changed, as we saw in Chapter 28. In the present chapter, we will discuss how quantum mechanics has given us an understanding of the structure of molecules and matter in bulk, as well as a number of important applications including semiconductor devices, and applications to biology. Our discussions will necessarily be qualitative for the most part.

*29–1 • Bonding in Molecules

One of the great successes of quantum mechanics was to give scientists, at last, an understanding of the nature of chemical bonds. Since it is based in physics, and because this understanding is so important in many fields, we discuss it here.

By a molecule, we mean a group of two or more atoms that are strongly held together so as to function as a single unit. When atoms make such an attachment, we say that a chemical **bond** has been formed. There are two main types of strong chemical bond: covalent and ionic. Many bonds are actually intermediate between these two types.

Covalent Bonds. To understand how covalent bonds are formed, we take the simplest case, the bond that holds two hydrogen atoms together to form the hydrogen molecule, H_2. The mechanism is basically the same for other covalent bonds. As two H atoms approach each other, the electron clouds begin to overlap, and the electrons from each atom can "orbit" both nuclei. (This is sometimes called "sharing" electrons.) If both electrons are in the ground state ($n = 1$) of their respective atoms, there are two possibilities: their spins can be parallel (both up or both down), in which case the total spin is $S = \frac{1}{2} + \frac{1}{2} = 1$; or the spins can be opposite ($m_s = +\frac{1}{2}$ for one, $m_s = -\frac{1}{2}$ for the other), so that the total spin $S = 0$. We shall now see that a bond is formed only for the $S = 0$ state, when the spins are opposite. First we consider the $S = 1$ state, for which the spins are the same. The two electrons cannot both be in the lowest energy state and be attached to the same atom, for then they would have identical quantum numbers in violation of the exclusion principle. The exclusion principle tells us that since no two electrons can occupy the same quantum state, if two electrons have the same quantum numbers, they must be different in some other way—namely, by being in different places in space (for example, attached to different atoms). When the two atoms approach, the electrons will stay away from each other as shown by the probability distribution of Fig. 29–1. The positively charged nuclei then repel each other, and no bond is formed.

Covalent bond

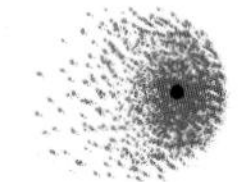

FIGURE 29–1 Electron probability distribution (electron cloud) for two H atoms when their spins are the same ($S = 1$).

For the $S = 0$ state, on the other hand, the spins are opposite and the two electrons are consequently in different quantum states. Hence they can come close together spatially. In this case, the probability distribution looks like Fig. 29–2. As can be seen, the electrons spend much of their time between the two positively charged nuclei, and the latter are attracted to the negatively charged electron cloud between them. This attraction, which holds the two atoms together to form a molecule, constitutes a *covalent bond*.

FIGURE 29–2 Electron probability distribution (cloud) around two H atoms when their spins are opposite ($S = 0$). In this case, a bond is formed because the positive nuclei are attracted to the concentration of negative charge between them. This is a hydrogen molecule, H_2.

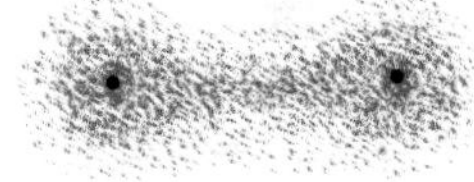

The probability distributions of Figs. 29–1 and 29–2 can perhaps be better understood on the basis of waves. What the exclusion principle requires is that when the spins are the same, there is destructive interference of the electron wave functions in the region between the two atoms. But when the spins are opposite, constructive interference occurs in the region between the two atoms, resulting in a large amount of negative charge there. Thus a covalent bond can be said to be the result of constructive interference of the electron wave functions in the space between the two atoms, and of the electrostatic attraction of the two positive nuclei for the negative charge concentration between them.

Why a bond is formed can also be understood from the energy point of view. When the two H atoms approach close to one another, if the spins of their electrons are opposite, the electrons can occupy the same space, as discussed above. This means that each electron can now move about in the space of two atoms instead of in the volume of only one. Because each electron now occupies more space, it is less well localized. From the uncertainty principle, with Δx increased, the momentum, and hence the energy, can be less. Another way of understanding this is from the point of view of de Broglie waves (Section 27–10). Because each electron has a larger "orbit," its wavelength λ can be longer, so its momentum $p = h/\lambda$ (Eq. 27–5) can be less. With less momentum, each electron has less energy when the two atoms combine than when they are separate. That is, the molecule has less energy than the two separate atoms, and so is more stable. It takes an energy input to break the H_2 molecule into two separate H atoms, so the H_2 molecule is a stable entity. This is what we mean by a *bond*. The energy required to break a bond is called the **bond energy**, the **binding energy**, or the **dissociation energy**.

Bond energy

Ionic bond

Ionic Bonds. An ionic bond is, in a sense, a special case of the covalent bond. Instead of the electrons being shared equally, they are shared unequally. For example, in sodium chloride (NaCl), the outer electron of the sodium spends nearly all its time around the chlorine (Fig. 29–3). The chlorine atom acquires a net negative charge as a result of the extra electron, whereas the sodium atom has a net positive charge. The electrostatic attraction between these two charged atoms holds them together. The resulting bond is called an *ionic bond* because it is created by the attraction between the two ions (Na^+ and Cl^-). But to understand the ionic bond, we must understand why the extra electron from the sodium spends so much of its time around the chlorine. After all, the chlorine is neutral; why should it attract another electron?

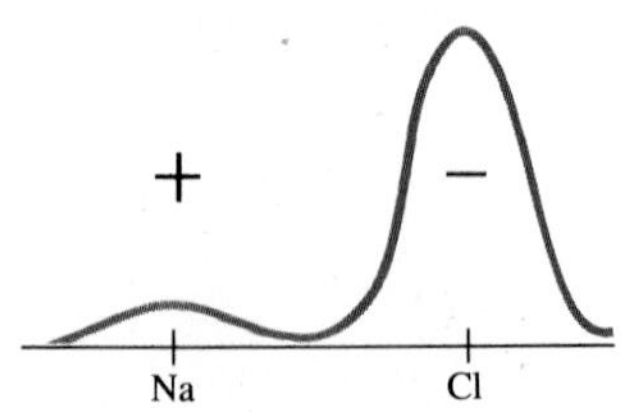

FIGURE 29–3 Probability distribution for last electron of Na in NaCl.

The answer lies in the probability distributions of the two neutral atoms. Sodium contains 11 electrons, 10 of which are in spherically symmetric closed shells (Fig. 29–4). The last electron spends most of its time beyond these closed shells. Because the closed shells have a total charge of $-10e$ and the nucleus has $+11e$, the outermost electron in sodium "feels" a net attraction due to $+1e$. It is not held very strongly. On the other hand, 12 of chlorine's 17 electrons form closed shells, or subshells (corresponding to $1s^2 2s^2 2p^6 3s^2$). These 12 form a spherically symmetric shield around the nucleus. The other five electrons are in $3p$ states whose probability distributions are not spherically symmetric and have a form similar to those for the $2p$ states in hydrogen shown in Fig. 28–7b and c. Four of these $3p$ electrons can have "donut-shaped" distributions symmetric about the z axis, as shown in Fig. 29–5. The fifth can have a "barbell-shaped" distribution (as for $m_l = 0$ in Fig. 28–7b), which in Fig. 29–5 is shown only in faint outline because it is half empty. That is, the exclusion principle allows one more electron to be

FIGURE 29–4 In a neutral sodium atom, the 10 inner electrons shield the nucleus, so the single outer electron is attracted by a net charge of $+1e$.

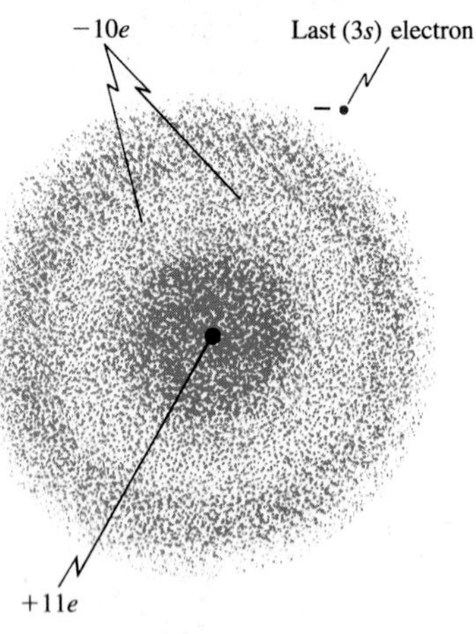

FIGURE 29–5 Neutral chlorine atom. The $+17e$ of the nucleus is shielded by the 12 electrons in the inner shells and subshells. Four of the five $3p$ electrons are shown in donut-shaped clouds, and the fifth is in the (dashed-line) cloud concentrated about the z axis. An extra electron at x will be attracted by a net charge that can be as much as $+5e$.

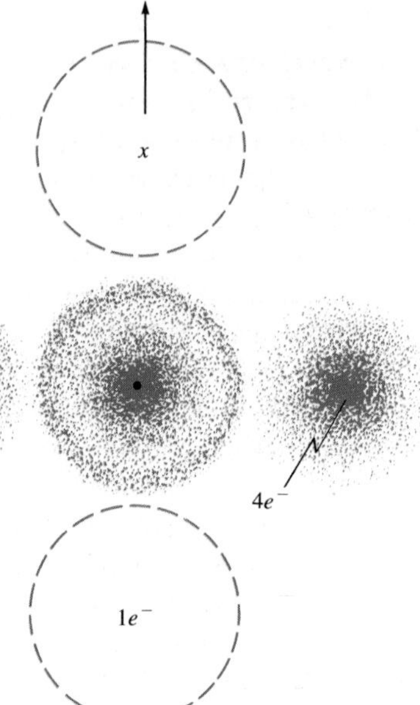

in this state (it will have spin opposite to that of the electron already there). If an extra electron—say from a Na atom—happens to be in the vicinity, it can be in this state, say at point x in Fig. 29–5. It could experience an attraction due to as much as $+5e$ because the $+17e$ of the nucleus is partly shielded at this point by the 12 inner electrons. Thus the outer electron of a Na atom will be attracted by the chlorine atom as well as by its own atom. This, combined with the strong attraction between the two ions when the extra electron stays with the Cl^-, produces the charge distribution of Fig. 29–3, and hence the ionic bond.

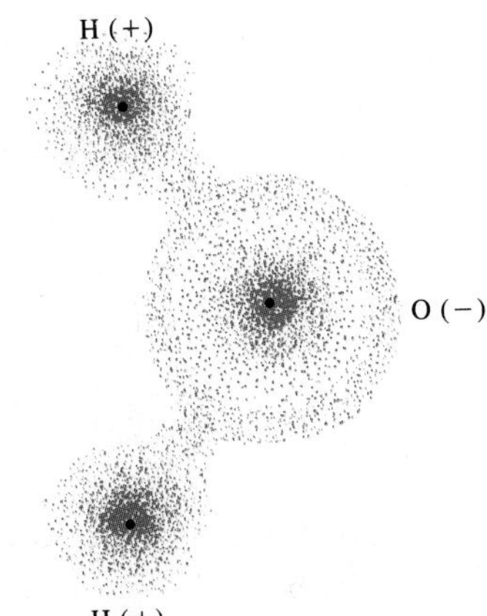

FIGURE 29–6
The water molecule is polar.

Partial Ionic Character of Covalent Bonds. A pure covalent bond in which the electrons are shared equally occurs mainly in symmetrical molecules such as H_2, O_2, and Cl_2. When the atoms involved are different from each other, it is usual to find that the shared electrons are more likely to be in the vicinity of one atom than the other. The extreme case is an ionic bond; in intermediate cases the covalent bond is said to have a *partial ionic character*. The molecules themselves are **polar**—that is, one part (or parts) of the molecule has a net positive charge and other parts a net negative charge. An example is the water molecule, H_2O (Fig. 29–6). The shared electrons are more likely to be found around the oxygen atom than around the two hydrogens. The reason is similar to that discussed above in connection with ionic bonds. Oxygen has eight electrons ($1s^2 2s^2 2p^4$), of which four form a spherically symmetric core and the other four could have, for example, a donut-shaped distribution. The barbell-shaped distribution on the z axis (like that shown dashed in Fig. 29–5) could be empty, so electrons from hydrogen atoms can be attracted by a net charge of $+4e$. They are also attracted by the H nuclei, so they actually orbit the H atoms as well as the O atom. The net effect is that there is a net positive charge on each H atom (less than $+1e$), because the electrons spend only part of their time there. And, there is a net negative charge on the O atom.

FIGURE 29–7 Potential energy as a function of separation for two point charges of (a) like sign and (b) opposite sign.

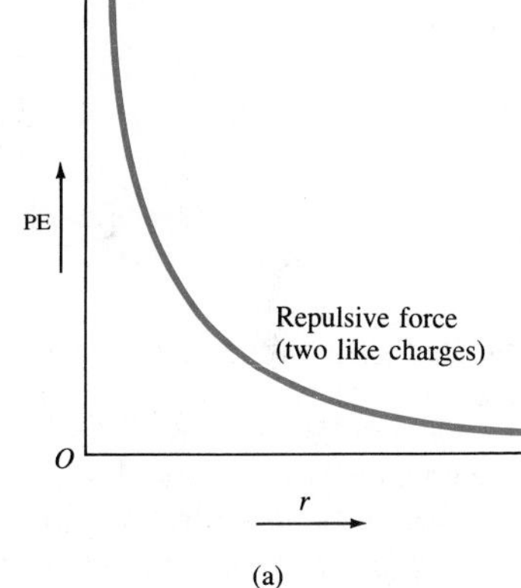

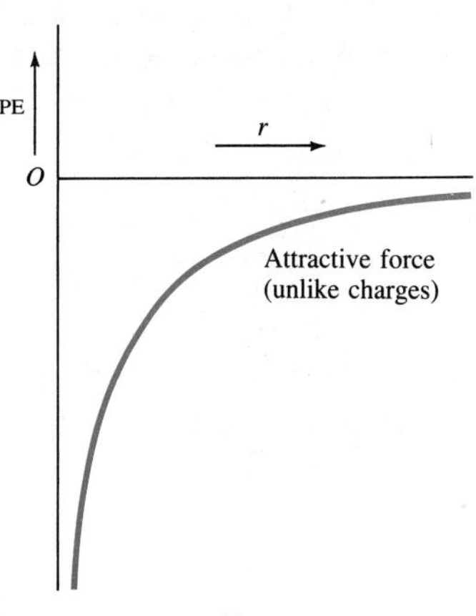

*29–2 • Potential-Energy Diagrams for Molecules

It is useful to analyze the interaction between two objects—say, between two atoms or molecules—with the use of a potential-energy diagram, a plot of the potential energy versus distance.

For the simple case of two point charges, q_1 and q_2, the PE is given by (see Chapter 17):

$$\text{PE} = \frac{1}{4\pi\varepsilon_0}\frac{q_1 q_2}{r},$$

where r is the distance between the charges, and the constant $(1/4\pi\varepsilon_0)$ is equal to $9.0 \times 10^9\ \text{N}\cdot\text{m}^2/\text{C}^2$. If the two charges have the same sign, the PE is positive for all values of r, and a graph of PE versus r in this case is shown in Fig. 29–7a. The force is repulsive (the charges have the same sign) and the curve rises as r decreases. If the two charges are of opposite sign, the PE is negative because the product $q_1 q_2$ is negative. The force is attractive in this case and the graph of PE ($\propto -1/r$) versus r looks like Fig. 29–7b. The PE becomes more *negative* as r decreases.

Now let us look at the potential-energy diagram for the formation of a covalent bond, such as for the hydrogen molecule. The potential energy of

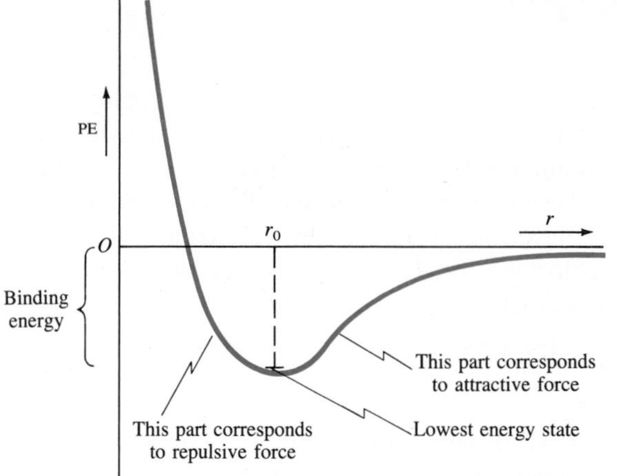

FIGURE 29–8 Potential-energy diagram for H_2 molecule; r is the separation of the two H atoms. The binding energy (the energy difference between PE = 0 and the lowest energy state near the bottom of the well) is 4.5 eV, and $r_0 = 0.074$ nm.

one H atom in the presence of the other is plotted in Fig. 29–8. The PE decreases as the atoms approach, because the electrons concentrate between the two nuclei (Fig. 29–2), so attraction occurs. However, at very short distances, the electrons would be "squeezed out"—there is no room for them between the two nuclei. Without the electrons between them, each nucleus would feel a repulsive force due to the other, so the curve rises as r decreases further. There is an optimum separation of the atoms, r_0 in Fig. 29–8, at which the energy is lowest. This is the point of greatest stability for the hydrogen molecule, and r_0 is the average separation of atoms in the H_2 molecule. The depth of this "well" is the binding energy,† as shown. This is how much energy must be put into the system to separate the two atoms to infinity, where the PE = 0. For the H_2 molecule, the binding energy is about 4.5 eV, and $r_0 = 0.074$ nm. In larger atoms, say, oxygen or nitrogen, repulsion also occurs at short distances, because the closed inner electron shells begin to overlap and the exclusion principle forbids their coming too close. The repulsive part of the curve rises even more steeply than $1/r$. A reasonable approximation to the potential energy, at least in the vicinity of r_0, is

$$\text{PE} = -\frac{A}{r^m} + \frac{B}{r^n},$$

where A and B are constants associated with the attractive and repulsive parts of the PE, and m and n are small integers. For ionic and some covalent bonds, the attractive term can often be written with $m = 1$ (Coulomb potential).

For many bonds, the potential-energy curve has the shape shown in Fig. 29–9. There is still an optimum distance r_0 at which the molecule is stable. But when the atoms approach from a large distance, the force is initially repulsive rather than attractive. The atoms thus do not interact spontaneously. Instead, some additional energy must be injected into the system to get it over the "hump" (or barrier) in the potential-energy diagram. This required energy is called the **activation energy**.

The curve of Fig. 29–9 is much more common than that of Fig. 29–8. The activation energy often reflects a need to break other bonds, before the one under discussion can be made. For example, to make water from O_2

† The binding energy corresponds not quite to the bottom of the PE curve, but to the lowest energy state, slightly above it, as shown in Fig. 29–8.

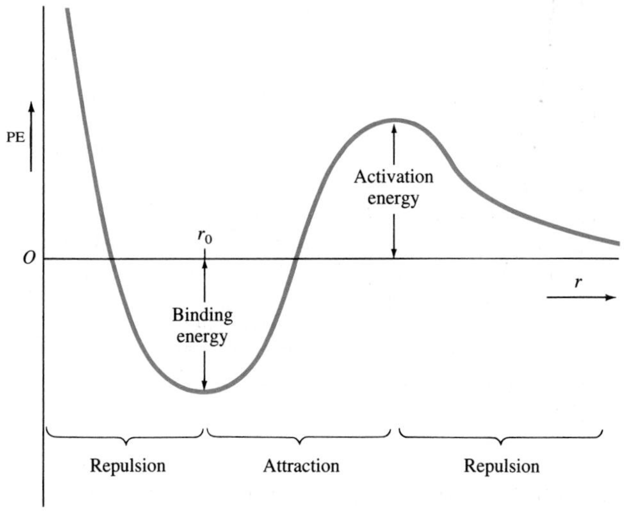

FIGURE 29–9 Potential-energy diagram for a bond requiring an activation energy.

and H_2, the H_2 and O_2 molecules must first be broken into H and O atoms; this is what the activation energy represents. Then the H and O atoms can combine to form H_2O with the release of a great deal more energy than was put in initially. The initial activation energy can be provided by applying an electric spark to a mixture of H_2 and O_2, breaking a few of these molecules into H and O atoms. The resulting explosive release of energy when these atoms combine to form H_2O quickly provides the activation energy needed for further reactions, so additional H_2 and O_2 molecules are broken up and recombined to form H_2O.

The potential-energy diagram for ionic bonds is similar in shape. In NaCl, for example, the Na^+ and Cl^- ions attract each other at distances larger than some r_0, but at shorter distances the overlapping of inner electron shells gives rise to repulsion. The two atoms thus are most stable at some intermediate separation r_0.

Sometimes the potential energy of a bond looks like that of Fig. 29–10. In this case, the energy of the bonded molecule, at a separation r_0, is greater than when there is no bond ($r = \infty$). That is, an energy *input* is required to make the bond (hence the binding energy is negative), and there is energy release when the bond is broken. Such a bond is stable only because there is the barrier of the activation energy. This type of bond is important in living cells, for it is in such bonds that energy can be stored efficiently in certain molecules, particularly ATP (adenosine triphosphate). The bond that connects the last phosphate group (designated Ⓟ in Fig. 29–10) to the rest of the molecule (ADP, meaning adenosine diphosphate, since it contains only two phosphates) is of the form shown in Fig. 29–10. Energy is actually stored in this bond. When the bond is broken (ATP → ADP + Ⓟ), energy is released and this energy can be used to make other chemical reactions "go."

FIGURE 29–10 Potential-energy diagram for formation of ATP from ADP and phosphate (Ⓟ).

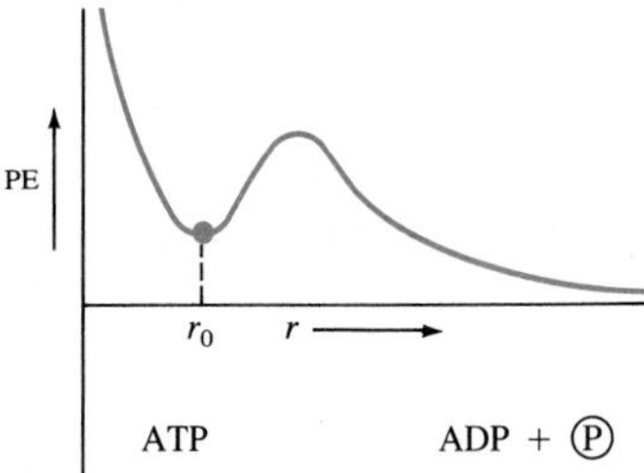

In living cells, many chemical reactions have activation energies that are often on the order of several eV. Such energy barriers are not easy to overcome in the cell. This is where enzymes come in. They act as catalysts, which means that they act to lower the activation energy so that reactions can occur that otherwise would not. Enzymes act by distorting the bonding electrons so that the initial bonds are easily broken.

*29–3 • Weak (van der Waals) Bonds

Once a bond between two atoms or ions is made, energy must normally be supplied to break the bond and separate the atoms. As mentioned in Section 29–1, this energy is called the *bond energy* or *binding energy*. The binding energy for covalent and ionic bonds is typically 2 to 5 eV. These bonds, which hold atoms together to form molecules, are often called **strong bonds** to distinguish them from so-called "weak bonds." The term **weak bond** as we use it here refers to an attachment between molecules due to simple electrostatic attraction—such as *between* polar molecules (and not *within* a polar molecule, which is a strong bond). The strength of the attachment is much less than for the strong bonds. Binding energies are typically in the range 0.04 to 0.3 eV—hence their name "weak bonds."

FIGURE 29–11 The C^+—O^- and H^+—N^- dipoles attract each other. (These dipoles may be part of, for example, thymine and adenine molecules, respectively.)

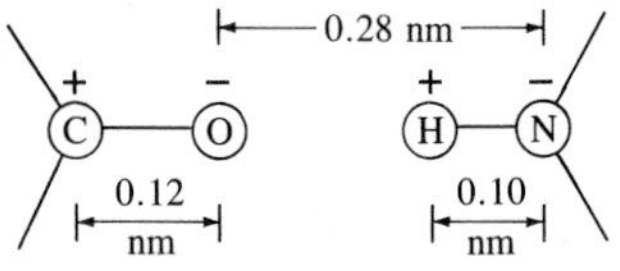

Weak bonds are generally the result of attraction between dipoles. For example, Fig. 29–11 shows two molecules that have a permanent dipole moment attracting one another. Besides such **dipole–dipole bonds**, there can also be **dipole–induced dipole bonds**, in which a polar molecule with a permanent dipole moment can induce a dipole moment in an otherwise electrically balanced (nonpolar) molecule, just as a single charge can induce a separation of charge in a nearby object (see Fig. 16–6). There can even be an attraction between two nonpolar molecules. Even though a molecule may not have a permanent dipole moment on the average, we can think of its electrons as moving about so that at any instant there may be a separation of charge. Such transient dipoles can induce a dipole moment in a nearby molecule, creating a weak attraction. All these weak bonds are referred to as **van der Waals bonds**, and the forces involved **van der Waals forces**. The potential energy has the general shape shown in Fig. 29–8, with the attractive van der Waals PE varying as $1/r^6$.

When one of the atoms in a dipole–dipole bond is hydrogen, it is often called a **hydrogen bond**. A hydrogen bond is generally the strongest of the weak bonds. This is because the hydrogen atom is the smallest atom and thus can be approached more closely. Hydrogen bonds also have a partial "covalent" character. That is, electrons between the two dipoles may be shared to a small extent.

Weak bonds are important in liquids and solids when strong bonds are absent (see Section 29–5), and they are also very important for understanding the activities of cells, such as DNA replication (see Section 16–10). The average kinetic energy of molecules in a cell is around $\frac{3}{2}kT \approx 0.04$ eV, about the magnitude of weak bonds. This means that a weak bond can readily be broken just by a molecular collision. Hence weak bonds are not very permanent—they are, instead, brief attachments. But because of this, they play an important role in the cell. On the other hand, strong bonds—those that hold molecules together—are almost never broken simply by molecular collision. Thus they are relatively permanent. They can be broken by chemical action (the making of even stronger bonds), and this usually happens in the cell with the aid of an enzyme.

*29–4 • Molecular Spectra

When atoms combine to form molecules, the probability distributions of the outer electrons overlap and this interaction alters the energy levels. None-

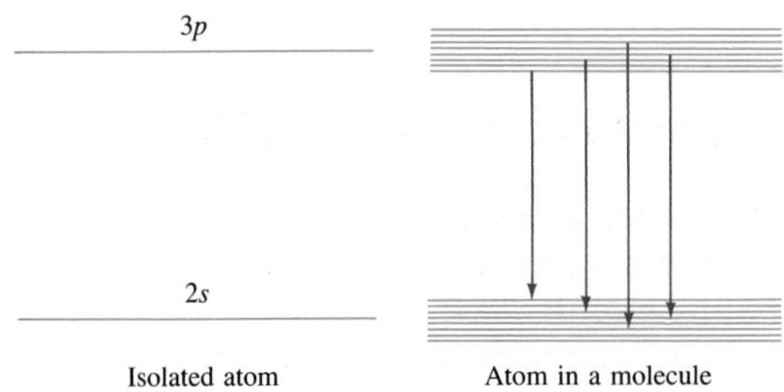

FIGURE 29–12 The individual energy levels of an isolated atom become bands of closely spaced levels in molecules, as well as in solids and liquids.

theless, molecules can undergo transitions between electron energy levels just as atoms do. For example, the H_2 molecule can absorb a photon of just the right frequency to excite one of its $1s$ electrons to a $2p$ state. The excited electron can then return to the ground state, emitting a photon. The energy of photons emitted by molecules is of the same order of magnitude as for atoms, typically 1 to 10 eV.

Additional energy levels become possible for molecules (but not for atoms) because the molecule as a whole can rotate, and the atoms of the molecule can vibrate relative to each other. The energy levels for both rotational and vibrational levels are quantized, and are generally spaced much more closely (10^{-3} to 10^{-1} eV) than the lower electronic levels. Each atomic energy level thus becomes a set of closely spaced levels corresponding to the vibrational and rotational motions, Fig. 29–12. Transitions from one level to another appear as many very closely spaced lines. In fact, the lines are not always distinguishable, and these spectra are called **band spectra**. Each type of molecule has its own characteristic spectrum, which can be used for identification and for determination of structure.

Rotational levels

Molecular Rotation. Let us now look in more detail at rotational and vibrational states in molecules. We begin with rotation, considering here only diatomic molecules, although the analysis can be extended to polyatomic molecules. When a diatomic molecule rotates about its center of mass, as shown in Fig. 29–13, its kinetic energy of rotation (see Section 8–6) is

$$E_{\text{rot}} = \frac{1}{2} I\omega^2 = \frac{(I\omega)^2}{2I},$$

where $(I\omega)$ is the angular momentum (Section 8–7). Quantum mechanics predicts quantization of angular momentum just as in atoms (see Eq. 28–3):

$$I\omega = \sqrt{L(L+1)}\hbar, \qquad L = 0, 1, 2, \ldots,$$

where L is an integer, the **rotational angular momentum quantum number**. Thus the rotational energy is quantized:

$$E_{\text{rot}} = \frac{(I\omega)^2}{2I} = L(L+1)\frac{\hbar^2}{2I}, \qquad L = 0, 1, 2, \ldots. \tag{29–1}$$

Transitions between rotational energy levels are subject to the *selection rule*:

$$\Delta L = \pm 1.$$

The energy of a photon emitted or absorbed for a transition between rotational states with angular-momentum quantum number L and $L-1$ will be

$$\Delta E_{\text{rot}} = E_L - E_{L-1} = \frac{\hbar^2}{2I} L(L+1) - \frac{\hbar^2}{2I}(L-1)(L)$$

$$= \frac{\hbar^2}{I} L. \tag{29–2}$$

FIGURE 29–13 Diatomic molecule rotating about a vertical axis.

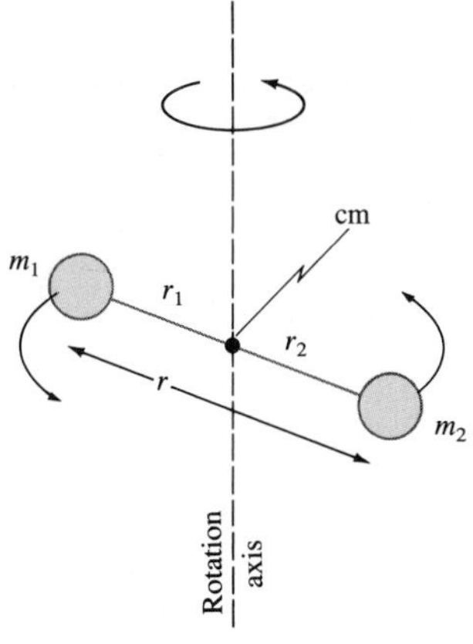

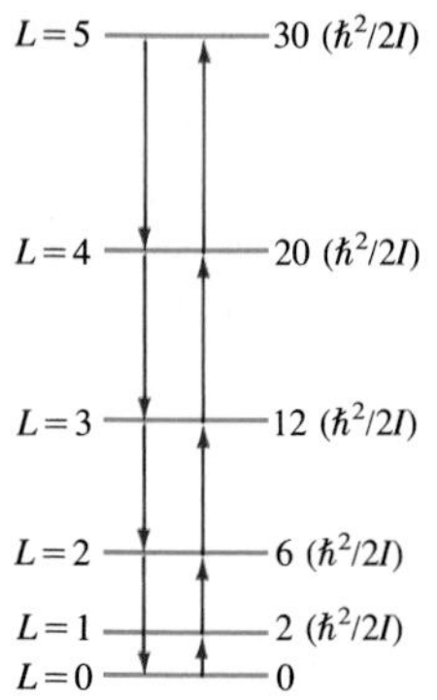

FIGURE 29–14 Rotational energy levels and allowed transitions (emission and absorption) for a diatomic molecule. Upward-pointing arrows represent absorption of a photon, and downward arrows represent emission.

We see that the transition energy increases directly with L. Figure 29–14 shows some of the allowed rotational energy levels and transitions. Measured absorption lines fall in the microwave or far-infrared regions of the spectrum, and their frequencies are generally 2, 3, 4, ... times higher than the lowest one, as predicted by Eq. 29–2 (see Example 29–1).

EXAMPLE 29–1 A rotational transition $L = 1$ to $L = 0$ for the molecule CO has a measured absorption wavelength $\lambda_1 = 2.60$ mm (microwave region). Use this to calculate (*a*) the moment of inertia of the CO molecule, and (*b*) the CO bond length, r. (*c*) Calculate the wavelength and energies of the next three rotational transitions.

SOLUTION (*a*) From Eq. 29–2, we can write

$$\frac{\hbar^2}{I}L = \Delta E = hf = \frac{hc}{\lambda_1}.$$

With $L = 1$ (the upper state) in this case, we solve for I:

$$I = \frac{\hbar^2 L}{hc}\lambda_1 = \frac{h\lambda_1}{4\pi^2 c} = \frac{(6.63 \times 10^{-34}\ \mathrm{J\cdot s})(2.60 \times 10^{-3}\ \mathrm{m})}{4\pi^2(3.00 \times 10^8\ \mathrm{m/s})}$$

$$= 1.46 \times 10^{-46}\ \mathrm{kg\cdot m^2}.$$

(*b*) We need to write the moment of inertia of the rotating molecule, which rotates about its center of mass (cm), in terms of the CO separation r. In Fig. 29–13, let m_1 be the mass of the C, which is $m_1 = 12$ u (atomic mass units), and let m_2 be the mass of the O ($m_2 = 16$ u). The distance of the cm from the C atom, which is r_1 in Fig. 29–13, is given by the cm formula, Eq. 7–8:

$$r_1 = \frac{0 + m_2 r}{m_1 + m_2} = \frac{16}{12 + 16}r = 0.57r.$$

The O atom is a distance $r_2 = r - r_1 = 0.43r$ from the cm. The moment of inertia of the CO molecule about its cm is then (see Example 8–8)

$$I = m_1 r_1^2 + m_2 r_2^2$$

$$= [(12\ \mathrm{u})(0.57r)^2 + (16\ \mathrm{u})(0.43r)^2][1.66 \times 10^{-27}\ \mathrm{kg/u}]$$

$$= (1.14 \times 10^{-26}\ \mathrm{kg})r^2.$$

We solve for r and use the result of part (a) for I:

$$r = \sqrt{\frac{1.46 \times 10^{-46}\ \mathrm{kg\cdot m^2}}{1.14 \times 10^{-26}\ \mathrm{kg}}} = 1.13 \times 10^{-10}\ \mathrm{m},$$

or 0.113 nm.

(*c*) From Eq. 29–2, $\Delta E \propto L$. Hence $\lambda = c/f = hc/\Delta E$ is proportional to $1/L$. Thus, for $L = 2$ to $L = 1$ transitions, $\lambda_2 = \frac{1}{2}\lambda_1 = 1.30$ mm. For $L = 3$ to $L = 2$, $\lambda_3 = \frac{1}{3}\lambda_1 = 0.87$ mm. And for $L = 4$ to $L = 3$, $\lambda_4 = 0.65$ mm. All are close to measured values. The energies of the photons, $hf = hc/\lambda$, are respectively 9.5×10^{-4} eV, 1.4×10^{-3} eV, and 1.9×10^{-3} eV.

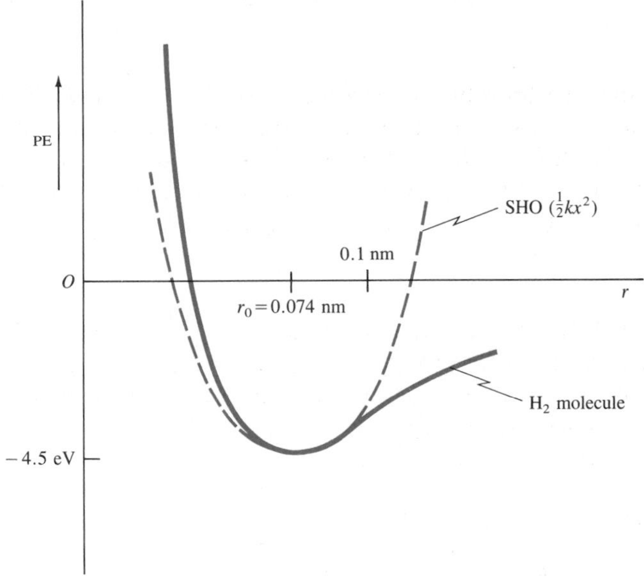

FIGURE 29–15 Potential energy for the H_2 molecule and for a simple harmonic oscillator (PE $= \frac{1}{2}kx^2$, with $|x| = |r - r_0|$).

The spacing of rotational energy levels is thus of the order 10^{-3} eV for the CO molecule. For the H_2 molecule, the wavelengths are on the order of 10^{-4} m, which is in the far-infrared region of the spectrum, and the transition energies are on the order of 10^{-2} eV (see Problem 7).

Vibrational levels

Molecular Vibration. The potential energy of the two atoms in a typical diatomic molecule has the shape shown in Fig. 29–8 or 29–9, and Fig. 29–15 again shows the PE for the H_2 molecule. We note that the PE, at least in the vicinity of the equilibrium separation r_0, closely resembles the potential energy of a harmonic oscillator, PE $= \frac{1}{2}kx^2$, which is shown superposed in dashed lines. Thus, for small displacements from r_0, each atom experiences a restoring force proportional to the displacement, and the molecule vibrates as a simple harmonic oscillator (SHO)—see Chapter 11. According to quantum mechanics, the possible energy levels are quantized according to

$$E_{\text{vib}} = (v + \tfrac{1}{2})hf, \qquad v = 0, 1, 2, \cdots, \tag{29–3}$$

where f is the classical frequency (see Chapter 11) and v is an integer called the **vibrational quantum number**. The lowest energy state ($v = 0$) is not zero (as for rotation), but has $E = \frac{1}{2}hf$. This is called the **zero-point energy**. Higher states have energy $\frac{3}{2}hf$, $\frac{5}{2}hf$, and so on, as shown in Fig. 29–16. Transitions are subject to the selection rule:

$$\Delta v = \pm 1,$$

so allowed transitions occur only between adjacent states and all give off photons of energy

$$\Delta E_{\text{vib}} = hf. \tag{29–4}$$

This is very close to experimental values for small v, but for higher energies, the PE curve (Fig. 29–15) begins to deviate from a perfect SHO curve, and this then affects the wavelengths and frequencies of the transitions. Typical transition energies are on the order of 10^{-1} eV, about 10 times larger than for rotational transitions, with wavelengths in the infrared region of the spectrum ($\approx 10^{-5}$ m).

FIGURE 29–16 Allowed vibrational energies for a diatomic molecule, where f is the fundamental frequency of vibration (see Chapter 11). The energy levels are equally spaced. Transitions are allowed only between adjacent levels ($\Delta v = \pm 1$).

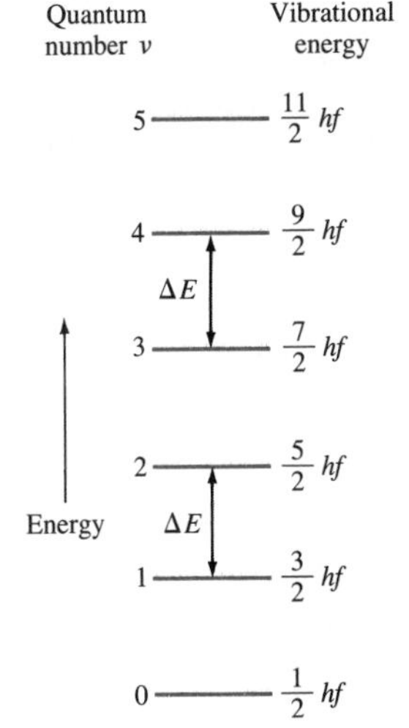

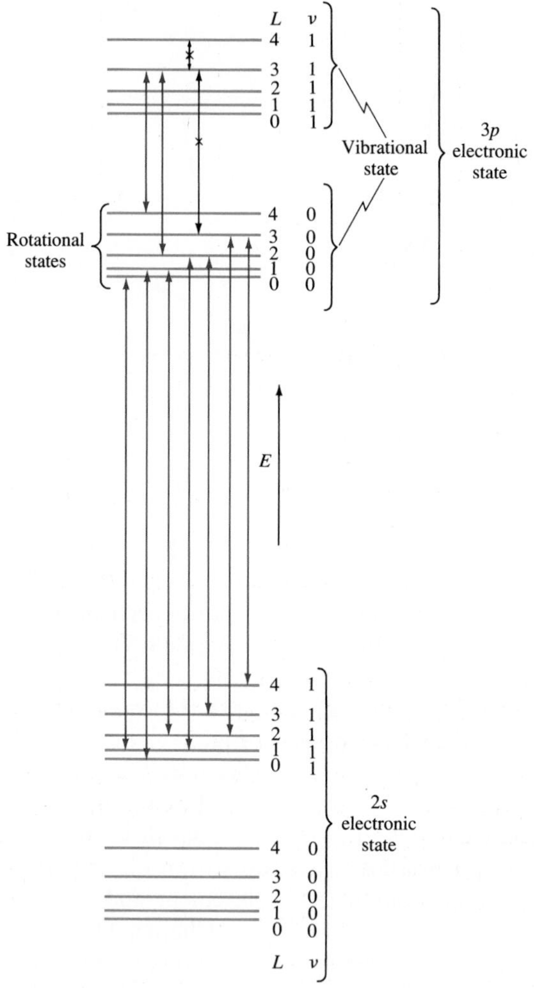

FIGURE 29–17 Combined electronic, vibrational, and rotational energy levels. Transitions marked with an × are not allowed by the selection rules.

When energy is imparted to a molecule, both the rotational and vibrational modes can be excited. Because rotational energies are an order of magnitude or so smaller than vibrational energies, which in turn are smaller than the electronic energy levels, we can represent the grouping of levels as shown in Fig. 29–17. Transitions with emission of a photon are subject to the *selection rules*:

$$\Delta v = \pm 1 \qquad \text{and} \qquad \Delta L = \pm 1.$$

Some allowed and forbidden (marked x) transitions are indicated in Fig. 29–17. Not all transitions and levels are shown, and the separation between vibrational levels, and (even more) between rotational levels, has been exaggerated. But we can clearly see the origin of the very closely spaced lines that give rise to the band spectra, as mentioned with reference to Fig. 29–12 earlier in this section.

The spectra are quite complicated, so we consider briefly only transitions within the same electronic level, such as those at the top of Fig. 29–17. A transition from a state with quantum numbers v and L, to one with quan-

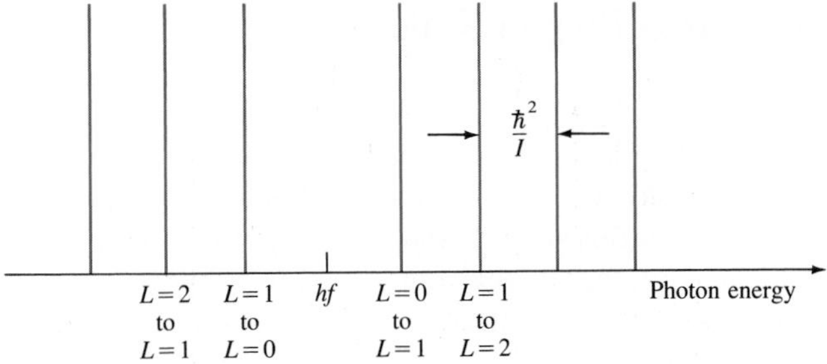

FIGURE 29–18 Expected spectrum for transitions between combined rotational and vibrational states.

tum numbers $v + 1$ and $L \pm 1$ (see the selection rules above), will absorb† a photon of energy:

$$\begin{aligned}\Delta E &= \Delta E_{\text{rot}} + \Delta E_{\text{vib}} \\ &= hf + (L+1)\frac{\hbar^2}{I} \qquad [L \to L+1], \quad L = 0, 1, 2, \cdots \\ &= hf - L\frac{\hbar^2}{I} \qquad [L \to L-1], \quad L = 1, 2, 3, \cdots,\end{aligned} \tag{29–5}$$

where we have used Eqs. 29–2 and 29–4 (see Problem 10). Note that for $L \to L-1$ transitions, L cannot be zero since there is then no state with $L-1$ ($=-1$, which doesn't exist). Equations 29–5 predict an absorption spectrum like that shown schematically in Fig. 29–18, with transitions $L \to L-1$ on the left and $L \to L+1$ on the right. Figure 29–19 shows the molecular absorption spectrum of HCl, which follows this pattern very well. (Each line in that spectrum is split into two because Cl consists of two isotopes of different mass; hence there are two kinds of HCl molecule with different moments of inertia I.)

† This is for absorption; for emission of a photon, the transition would be $v \to v-1$, $L \to L \pm 1$.

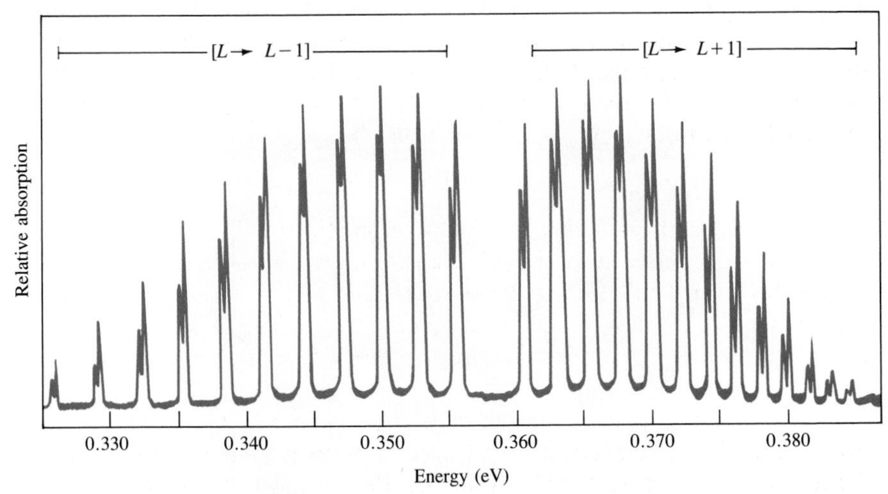

FIGURE 29–19 Absorption spectrum for HCl molecules. Lines on the left correspond to transitions where $L \to L-1$; those on the right are for $L \to L+1$. Each line has a double peak because chlorine has two isotopes of different mass and different moment of inertia.

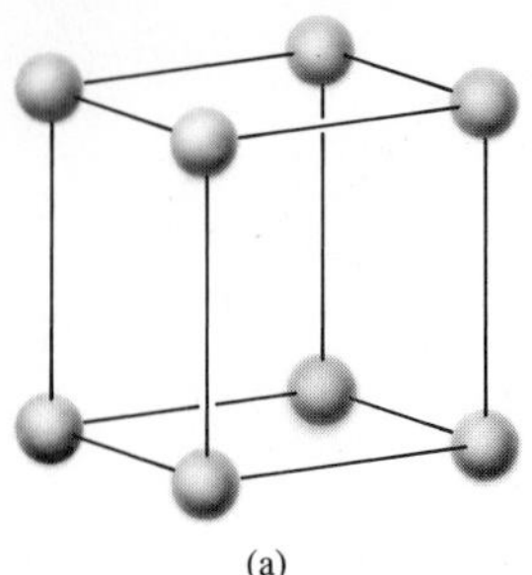

(a)

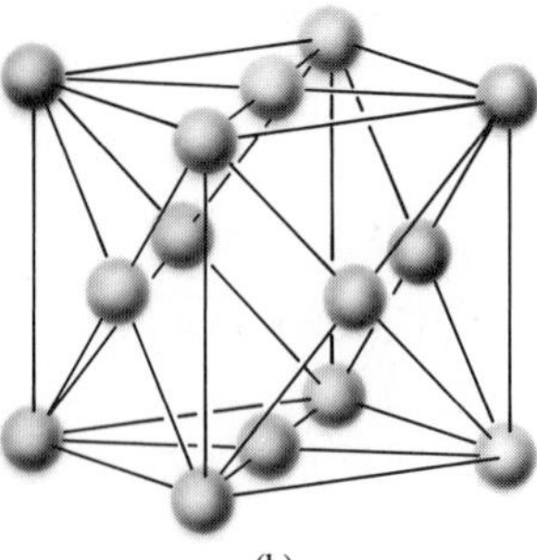

(b)

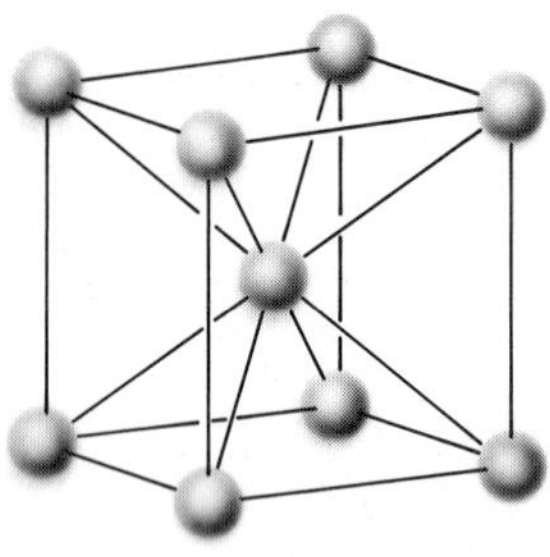

(c)

FIGURE 29–20 Arrangement of atoms in (a) a simple cubic crystal, (b) face-centered cubic crystal (note the atom at the center of each face), and (c) body-centered cubic crystal. Each diagram shows the relationship of the bonds.

*29–5 • Bonding in Solids

Quantum mechanics has been a great tool for understanding the structure of solids. This active field of research today is called **solid-state physics**, or **condensed-matter physics** so as to include liquids as well. The rest of this chapter is devoted to this subject, and we begin with a brief look at the structure of solids and the bonds that hold them together.

Although some solid materials are *amorphous* in structure, in that the atoms and molecules show no long-range order, we will be interested here in the large class of crystalline substances whose atoms, ions, or molecules form an orderly array whose geometric arrangement is known as a **lattice**. Figure 29–20 shows three of the possible arrangements of atoms in a crystal: simple cubic, face-centered cubic, and body-centered cubic. The NaCl crystal lattice is face-centered cubic, with one Na^+ ion or one Cl^- ion at each lattice point (see Fig. 29–21).

The molecules of a solid are held together in a number of ways. The most common are by *covalent* bonding (as between the carbon atoms of the diamond crystal) or *ionic* bonding (as in a NaCl crystal). Often the bonds are partially covalent and partially ionic. Our discussion of these bonds earlier in this chapter for molecules applies equally well here to solids.

Let us look for a moment at the NaCl crystal of Fig. 29–21. Each Na^+ ion feels an attractive Coulomb potential due to each of the six "nearest neighbor" Cl^- ions surrounding it. Note that one Na^+ does not "belong" exclusively to one Cl^-, so we must not think of ionic solids as consisting of individual molecules. Each Na^+ also feels a repulsive Coulomb potential due to other Na^+ ions, although this is weaker since the Na^+ ions are farther away.

A different type of bond occurs in metals. Metal atoms have relatively loosely held outer electrons. Present-day **metallic bond** theories propose that in a metallic solid, these outer electrons roam rather freely among all the metal atoms which, without their outer electrons, act like positive ions. The electrostatic attraction between the metal ions and this negative electron "gas" is at least in part responsible for holding the solid together. The binding energy of metal bonds is typically 1 to 3 eV, somewhat weaker than ionic or covalent bonds (5 to 10 eV in solids). The "free electrons" are responsible for the high electrical and thermal conductivity of metals. This theory also nicely

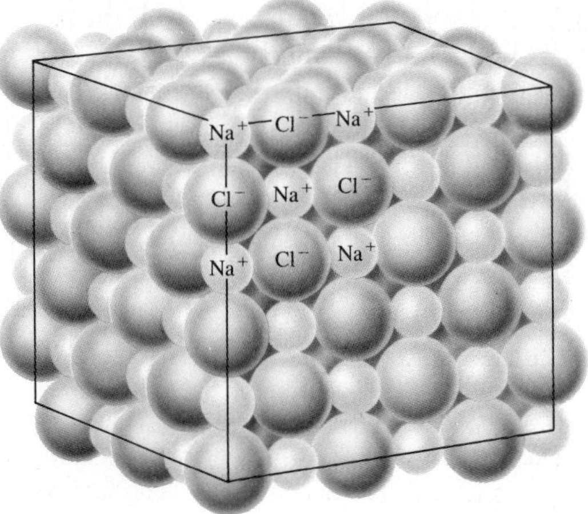

FIGURE 29–21 (right) Diagram of NaCl crystal.

accounts for the shininess of smooth metal surfaces: the free electrons can vibrate at any frequency, so when light of a range of frequencies falls on a metal, the electrons can vibrate in response and reemit light of those same frequencies. Hence, the reflected light will consist largely of the same frequencies as the incident light. Compare this to non-metallic materials that have a distinct color—the atomic electrons exist only in certain energy states, and when white light falls on them, the atoms absorb at certain frequencies, and reflect other frequencies which make up the color we see.

The atoms or molecules of some materials, such as the noble gases, can form only **weak bonds** with each other. As we saw in Section 29–3, weak bonds have very low binding energies and would not be expected to hold atoms together as a liquid or solid at room temperature. The noble gases condense only at very low temperatures, where the atomic (thermal) kinetic energy is small and the weak attraction can then hold the atoms together.

*29–6 • Band Theory of Solids

We saw in Section 29–1 that when two hydrogen atoms approach each other, the wave functions overlap, and the two 1*s* states (one for each atom) divide into two states of different energy. (As we saw, only one of these states, $S = 0$, has low enough energy to give a bound H_2 molecule.) Figure 29–22a shows this situation for 1*s* and 2*s* states for two atoms. If six atoms come together, as in Fig. 29–22b, each of the states splits into six levels. If a large number of atoms come together to form a solid, then each of the original atomic levels becomes a **band** as shown in Fig. 29–22c. The energy levels are so close together in each band that they seem essentially continuous. This is why the spectrum of heated solids (Section 27–1) appears continuous.

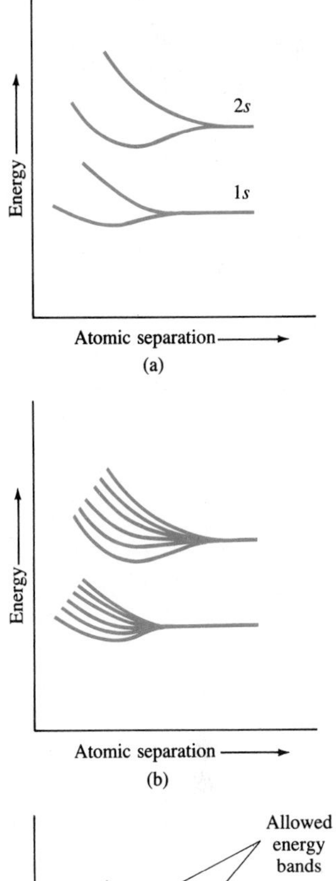

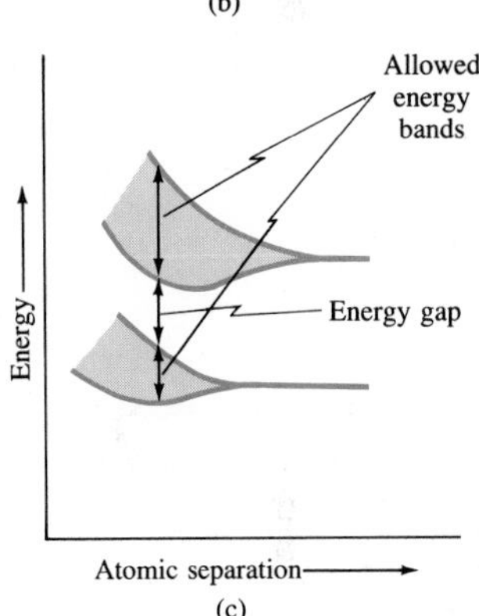

FIGURE 29–22 The splitting of 1*s* and 2*s* atomic energy levels as (a) two atoms approach each other (the atomic separation decreases, moving toward the left), (b) the same for six atoms, and (c) for many atoms when they come together to form a solid.

The crucial aspect of a good **conductor** is that the highest energy band containing electrons is only partially filled. Consider sodium, for example, whose energy bands are shown in Fig. 29–23. The 1*s*, 2*s*, and 2*p* bands are full (just as in a Na atom) and don't concern us. The 3*s* band is, however, only half full. To see why, recall that the exclusion principle stipulates that in an atom, only two electrons can be in the 3*s* state, one with spin up and one with spin down. These two states have slightly different energy. For a solid consisting of N atoms, the 3*s* band will contain $2N$ possible energy states. Now a sodium atom has a single 3*s* electron, so in a sample of sodium metal containing N atoms, there are N electrons in the 3*s* band, and N unoccupied states. When a potential difference is applied across the metal, electrons can respond by accelerating and increasing their energy, since there are plenty of unoccupied states of slightly higher energy available. Hence, a current flows readily and sodium is a good conductor. The characteristic of all good conductors is that the highest energy band is only partially filled, or two bands overlap so that unoccupied states are available. An example of the latter is magnesium, which has two 3*s* electrons, so its 3*s* band is filled. But the unfilled 3*p* band overlaps the 3*s* band in energy, so there are lots of available states for the electrons to move into. Thus magnesium, too, is a good conductor.

FIGURE 29–23 Energy bands for sodium.

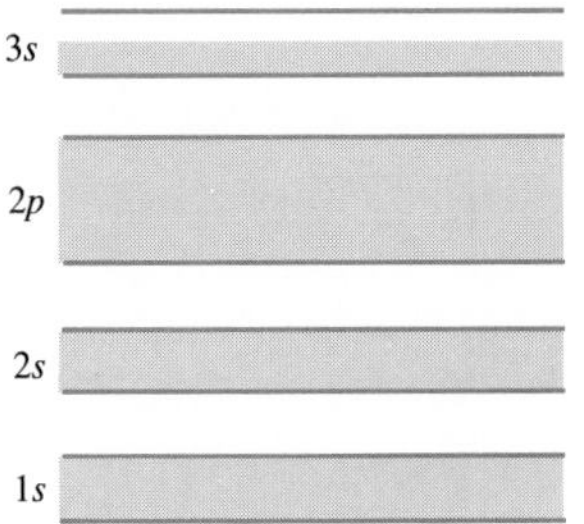

In a material that is a good **insulator**, on the other hand, the highest band containing electrons, called the **valence band**, is completely filled. The next highest energy band, called the **conduction band**, is separated from the

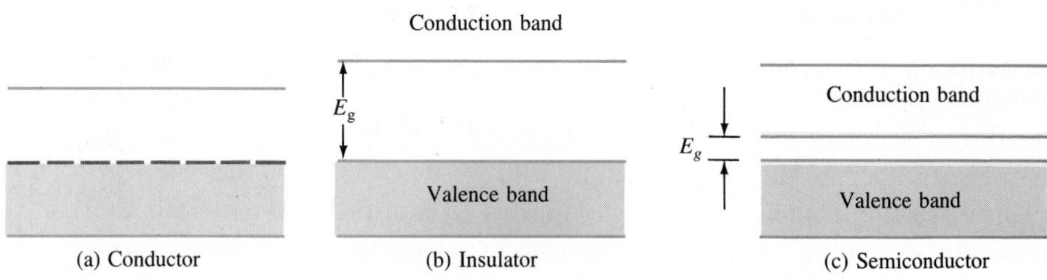

FIGURE 29–24 Energy bands for (a) a conductor, (b) an insulator, and (c) a semiconductor. Shading represents occupied states. Pale shading in (c) represents electrons that can pass from the valence band to the conduction band due to thermal agitation at room temperature (exaggerated).

valence band by a "**forbidden**" **energy gap**, E_g, of typically 5 to 10 eV. So at room temperature (300 K), where thermal energies (that is, average kinetic energy—see Chapter 13) are on the order of $\frac{3}{2}kT \approx 0.04$ eV, almost no electrons can acquire the 5 eV needed to reach the conduction band. When a potential difference is applied across the material, there are no available states accessible to the electrons, and no current flows. Hence, the material is a good insulator.

Semiconductors

Figure 29–24 compares the relevant energy bands (*a*) for conductors, (*b*) for insulators, and also (*c*) for the important class of materials known as **semiconductors**. The bands for a pure (or **intrinsic**) semiconductor, such as silicon or germanium, are like those for an insulator, except that the unfilled conduction band is separated from the filled valence band by a much smaller energy gap, E_g, typically on the order of 1 eV. At room temperature, there will be a few electrons that can acquire enough thermal energy to reach the conduction band, and so a very small current can flow when a voltage is applied. At higher temperatures, more electrons will have enough energy to jump the gap. This effect can often more than offset the effects of more frequent collisions due to increased disorder at higher temperature, so that the resistivity of semiconductors can *decrease* with increasing temperature (see Table 18–1). But this is not the whole story of semiconductor conduction. When a potential difference is applied to a semiconductor, the few electrons in the conduction band move toward the positive electrode. Electrons in the valence band try to do the same thing, and a few can because there are a small number of unoccupied states which were left empty by the electrons reaching the conduction band. Such unfilled electron states are called **holes**. Each electron in the valence band that fills a hole in this way as it moves toward the positive electrode leaves behind a hole, so that the holes migrate toward the negative electrode. As the electrons tend to accumulate at one side of the material, the holes tend to accumulate on the opposite side. We will look at this phenomenon in more detail in the next section.

Holes

*29–7 • Semiconductors and Doping

Doping

The most commonly used semiconductors in modern electronics are silicon (Si) and germanium (Ge). An atom of silicon or germanium has four outer electrons that act to hold the atoms in the regular lattice structure of the crystal, shown schematically in Fig. 29–25a. Germanium and silicon acquire useful properties for use in electronics only when a tiny amount of impurity is introduced into the crystal structure (perhaps 1 part in 10^6 or 10^7). This is called **doping** the semiconductor. Two kinds of doped semiconductor can be made, depending on the type of impurity used. If the impurity is an element whose atoms have five outer electrons, such as arsenic, we have the

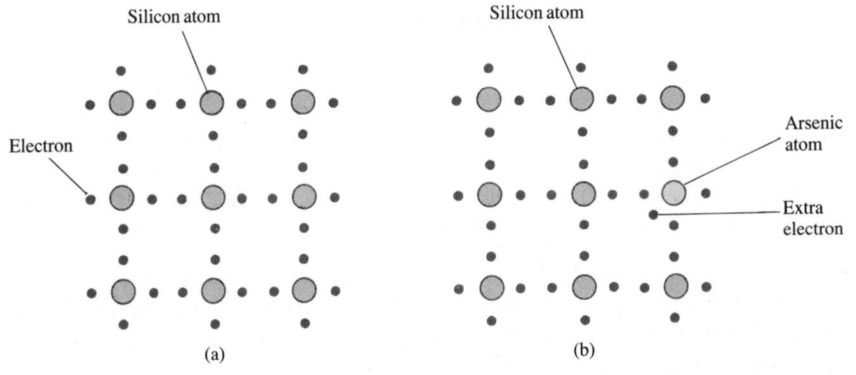

FIGURE 29–25 A silicon crystal. (a) Four (outer) electrons surround each silicon atom. (b) Silicon crystal doped with a few arsenic atoms: the extra electron doesn't fit into the crystal lattice and so is free to move about. This is an *n*-type semiconductor.

situation shown in Fig. 29–25b. Only four of arsenic's electrons fit into the bonding structure. The fifth does not fit in and can move relatively freely, somewhat like the electrons in a conductor. Because of this small number of extra electrons, a doped semiconductor becomes slightly conducting. The density of conduction electrons in an intrinsic semiconductor is about 1 per 10^{10} atoms. With an impurity concentration of 1 in 10^6 or 10^7, the impurity will dominate the conductivity and its level can be controlled with great precision. An arsenic-doped silicon crystal is called an **n-type semiconductor** because *n*egative charges (electrons) carry the electric current.

n-type

p-type

In a **p-type semiconductor**, a small amount of an element with three outer electrons—such as gallium—is added to the semiconductor. As shown in Fig. 29–26a, there is a "**hole**" in the lattice structure near a gallium atom since it has only three outer electrons. Electrons from nearby silicon atoms can jump into this hole and fill it. But this leaves a hole where that electron had previously been, Fig. 29–26b. The vast majority of atoms are silicon, so holes are almost always next to a silicon atom. Since silicon atoms require four outer electrons to be neutral, this means that there is a net positive charge at the hole. Whenever an electron moves to fill a hole, the positive hole is then at the previous position of that electron. Another electron can then fill this hole, and the hole thus moves to a new location; and so on. (It must be added that the holes, as well as electrons, are not precisely localized and must be considered as clouds.) This type of semiconductor is called *p-type* because it is the *p*ositive holes that seem to carry the electric current. Note, however, that both *p*-type and *n*-type semiconductors have *no net charge* on them.

According to the band theory (Section 29–6), in a doped semiconduc-

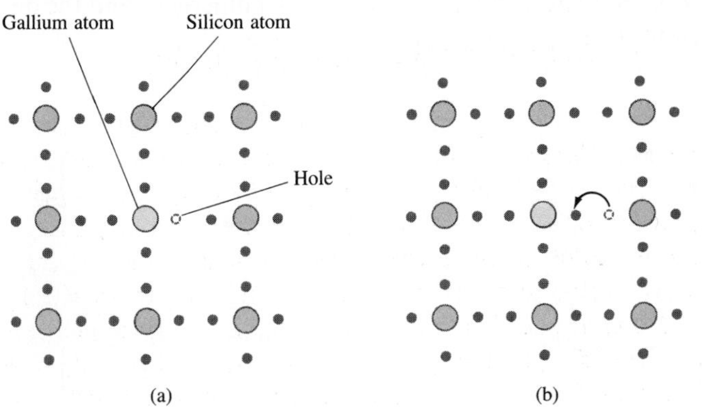

FIGURE 29–26 A *p*-type semiconductor, gallium-doped silicon. (a) Gallium has only three outer electrons, so there is an empty spot, or *hole*, in the structure. (b) Electrons from silicon atoms can jump into the hole and fill it. As a result, the hole moves to a new location (to the right in this figure), to where the electron used to be.

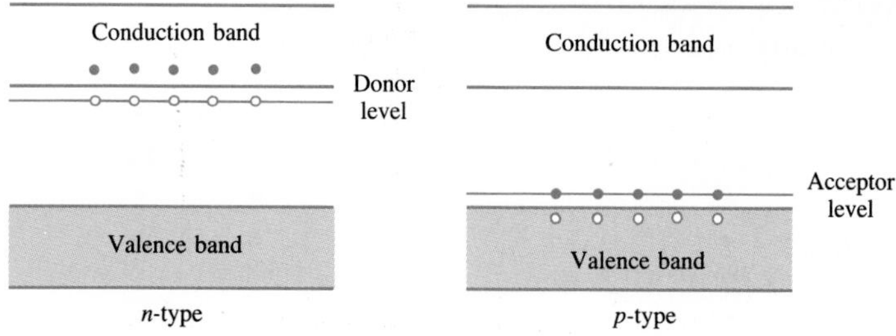

FIGURE 29–27 Impurity energy levels in doped semiconductors.

tor, the impurity provides additional energy states between the bands, as shown in Fig. 29–27. In an n-type semiconductor, the impurity energy level lies just below the conduction band. Electrons in this energy level need only about 0.05 eV (in Si; even less in Ge) of energy to reach the conduction band, so transitions occur readily at room temperature. Since this energy level supplies electrons to the conduction band, it is called a **donor** level. In p-type semiconductors, the impurity energy level is just above the valence band (Fig. 29–27). It is called an **acceptor** level because electrons from the valence band can easily jump into it. A positive hole is left behind in the valence band, and as other electrons move into this hole, the hole moves about as discussed earlier.

*29–8 • Semiconductor Diodes and Transistors

Semiconductor diodes and transistors are essential components of modern electronic devices. The miniaturization achieved today allows many thousands of diodes, transistors, resistors, and so on, to be placed on a single *chip* only a millimeter on a side. We now discuss, briefly and qualitatively, the operation of diodes and transistors.

Diodes

When an n-type semiconductor is joined to a p-type, a **p-n junction diode** is formed. Separately, the two semiconductors are electrically neutral. When joined, a few electrons near the junction diffuse from the n-type into the p-type semiconductor, where they fill a few of the holes. The n-type is left with a positive charge, and the p-type acquires a net negative charge. Thus a potential difference is established, with the n side positive relative to the p side, and this prevents further diffusion of electrons.

If a battery is connected to a diode with the positive terminal to the p side and the negative terminal to the n side as in Fig. 29–28a, the externally applied voltage opposes the internal potential difference and the diode is said to be **forward biased**. If the voltage is great enough (about 0.3 V for Ge, 0.6 V for Si at room temperature), a current will flow. The positive holes in the

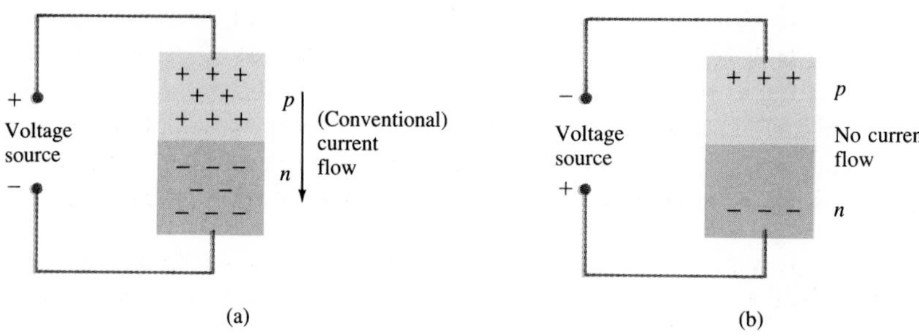

FIGURE 29–28 Schematic diagram showing how a semiconductor diode operates. Current flows when the voltage is connected in forward bias, as in (a), but not when connected in reverse bias, as in (b).

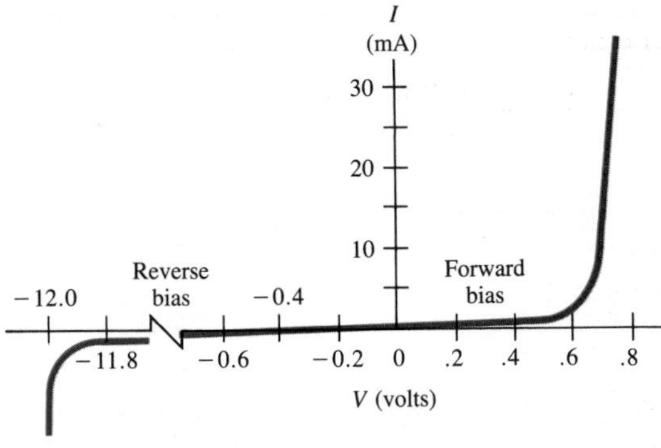

FIGURE 29–29 Current through a diode as a function of applied voltage.

p-type semiconductor are repelled by the positive terminal of the battery and the electrons in the *n*-type are repelled by the negative terminal of the battery. The holes and electrons meet at the junction, and the electrons cross over and fill the holes. A current is flowing. Meanwhile, the positive terminal of the battery is continually pulling electrons off the *p* end, forming new holes, and electrons are being supplied by the negative terminal at the *n* end. Consequently, a large current flows through the diode.

When the diode is **reverse biased**, as in Fig. 29–28b, the holes in the *p* end are attracted to the battery's negative terminal and the electrons in the *n* end are attracted to the positive terminal. The current carriers do not meet near the junction and, ideally, no current flows.

A graph of current versus voltage for a typical diode is shown in Fig. 29–29. As can be seen, a real diode does allow a small amount of reverse current to flow. For most practical purposes, this is negligible.†

EXAMPLE 29–2 The diode whose current–voltage characteristics are shown in Fig. 29–29 is connected in series with a 4.0-V battery and a resistor. If a current of 10 mA is to pass through the diode, what resistance must the resistor have?

SOLUTION In Fig. 29–29, we see that the voltage drop across the diode is about 0.7 V when the current is 10 mA. Therefore, the voltage drop across the resistor is 4.0 V − 0.7 V = 3.3 V, so $R = V/I = (3.3\ \text{V})/(1.0 \times 10^{-2}\ \text{A}) = 330\ \Omega$.

If the voltage across a diode connected in reverse bias is increased greatly, a point is reached where breakdown occurs. The electric field across the junction becomes so large that ionization of atoms results. The electrons thus pulled off their atoms contribute to a larger and larger current as breakdown continues. The voltage remains constant over a wide range of currents. This is shown on the far left in Fig. 29–29. This property of diodes can be used to accurately regulate a voltage supply. A diode designed for this purpose is called a **zener diode**. When placed across the output of an unregulated power supply, a zener diode can maintain the voltage at its own breakdown voltage as long as the supply voltage is always above this point. Zener diodes can be obtained corresponding to voltages of a few volts to hundreds of volts.

Since a *p-n* junction diode allows current to flow only in one direction (as long as the voltage is not too high), it can serve as a **rectifier**—to change

† Reverse current in Ge at room temperature is typically a few μA; for Si, a few pA. The reverse current increases rapidly with temperature, however, and may render a diode ineffective above 200°C.

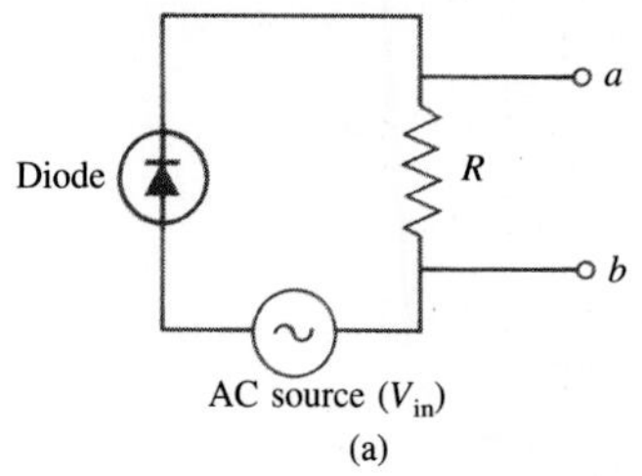

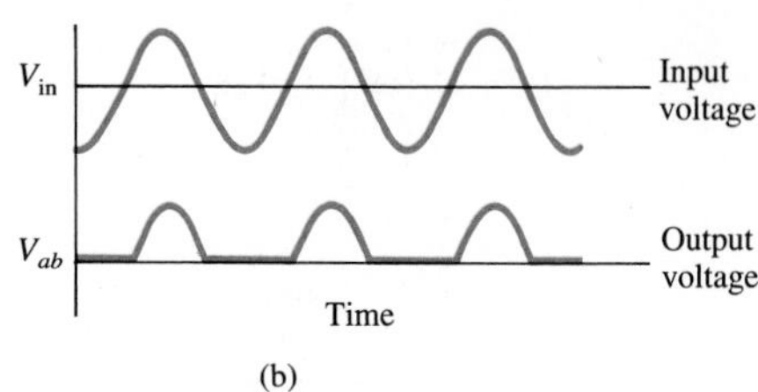

FIGURE 29–30 (a) A simple (half-wave) rectifier circuit using a semiconductor diode. (b) AC source input voltage, and output voltage across R, as functions of time.

Diodes as rectifiers

ac into dc. A simple rectifier circuit is shown in Fig. 29–30a where the arrow inside the symbol for a diode indicates the direction in which a diode conducts conventional (+) current. The ac source applies a voltage across the diode alternately positive and negative. Only during half of each cycle will a current pass through the diode; so only then is there a current through the resistor R. Hence, a graph of the voltage V_{ab} across R as a function of time looks like Fig. 29–30b. This **half-wave rectification** is not exactly dc, but it is unidirectional. More useful is a **full-wave rectifier** circuit, which uses two diodes (or sometimes four) as shown in Fig. 29–31a. At any given instant, either one diode or the other will conduct current to the right. Therefore, the output across the load resistor R will be as shown in Fig. 29–31b. Rather this is the voltage if the capacitor C were not in the circuit. The capacitor tends to store charge, and thus helps to smooth out the current as shown in Fig. 29–31c.

Rectifier circuits are important because most line voltage is ac, and most electronic devices require a dc voltage for their operation. Hence, diodes are found in nearly all electronic devices including radio and TV sets, calculators, and computers.

A diode is called a **nonlinear device** because the current is not proportional to the voltage; that is, a graph of current versus voltage (Fig. 29–29) is not a straight line as it is for a resistor (which ideally *is* linear). Transistors are also *nonlinear* devices.

Transistors

A simple **junction transistor** consists of a crystal of one type of doped semiconductor sandwiched between two crystals of the opposite type. Both *pnp* and *npn* transistors are made and they are shown schematically in Fig.

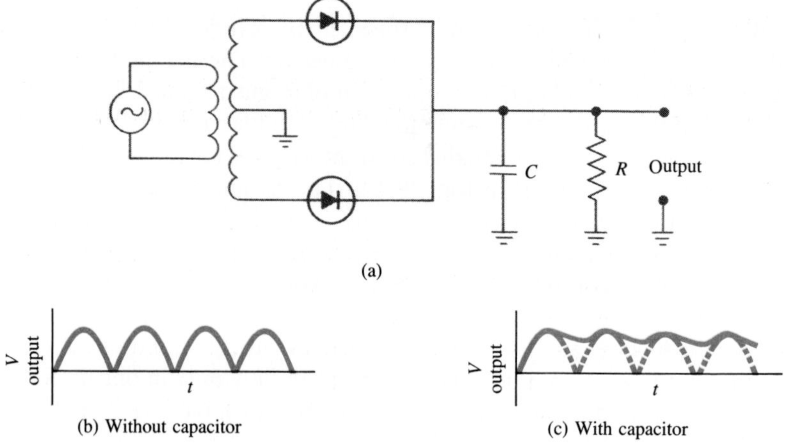

FIGURE 29–31 (a) Full-wave-rectifier circuit (including a transformer so the magnitude of the voltage can be changed). (b) Output voltage in the absence of capacitor C. (c) Output voltage with the capacitor in the circuit.

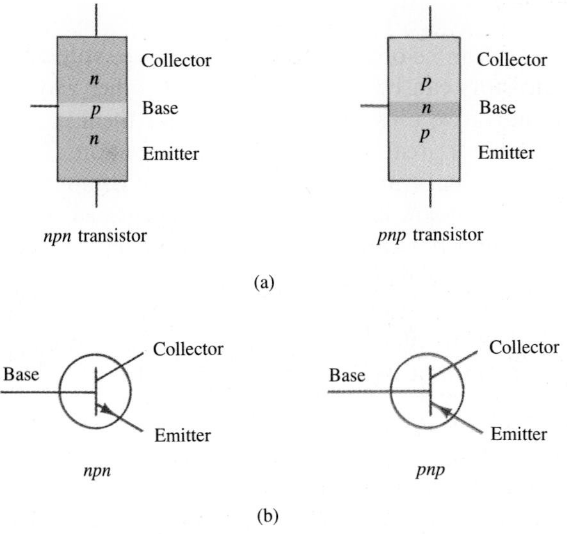

FIGURE 29–32 (a) Schematic diagram of *npn* and *pnp* transistors. (b) Symbols for *npn* and *pnp* transistors.

29–32a. The three semiconductors are given the names *collector*, *base*, and *emitter*. The symbols for *npn* and *pnp* transistors are shown in Fig. 29–32b. The arrow is always placed on the emitter and indicates the direction of (conventional) current flow in normal operation.

The operation of a transistor can be analyzed qualitatively—very briefly—as follows. Consider an *npn* transistor connected as shown in Fig. 29–33. A voltage V_{CE} is maintained between the collector and emitter by the battery $\mathscr{E}_C$. The voltage applied to the base is called the *base bias voltage*, V_{BE}. If V_{BE} is positive, conduction electrons in the emitter are attracted into the base. Since the base region is very thin (perhaps 1 μm), most of these electrons flow right across into the collector, which is maintained at a positive voltage. A large current, I_C, flows between collector and emitter and a much smaller current, I_B, through the base. A small variation in the base voltage due to an input signal causes a large change in the collector current and therefore a large change in the voltage drop across the output resistor R_C. Hence a transistor can *amplify* a small signal into a larger one. In fact, transistors are the basic elements in modern electronic **amplifiers** of all sorts.

Amplifiers

A *pnp* transistor operates in the same fashion, except that holes move instead of electrons. The collector voltage is negative, and so is the base voltage in normal operation.

In Fig. 29–33 two batteries are shown: $\mathscr{E}_C$ supplies the collector voltage and $\mathscr{E}_B$ the base bias voltage. In practice, only one source is often used, and

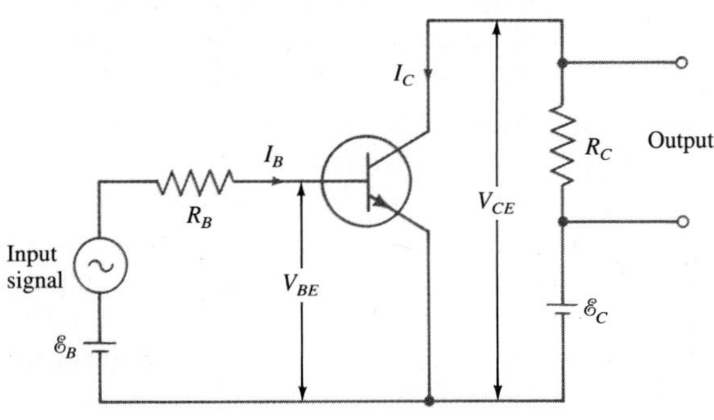

FIGURE 29–33 An *npn* transistor used as an amplifier.

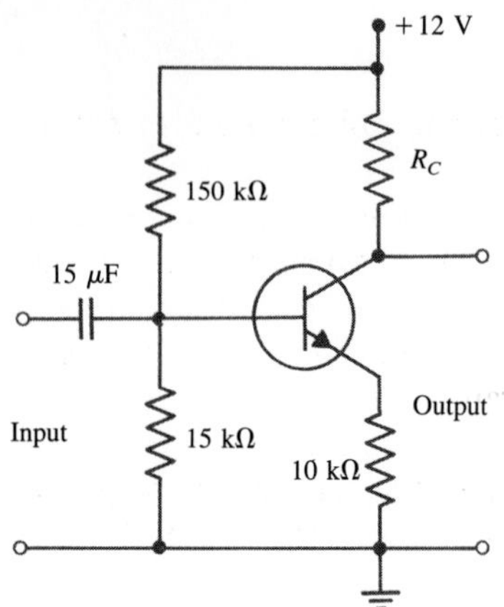

FIGURE 29–34 Typical transistor circuit involving an *npn* transistor. $\mathscr{E}_C = +12$ V and $\mathscr{E}_B$ is determined by the resistors.

the base bias voltage can be obtained using a resistance voltage divider as in Fig. 29–34. Transistors can be connected in many other ways as well, and many new and innovative uses have been found for them.

Transistors were a great advance in miniaturization of electronic circuits. Although individual transistors are very small compared to the once used vacuum tubes, they are huge compared to **integrated circuits** or **chips**. Tiny amounts of impurities can be placed at particular locations within a single silicon crystal. These can be arranged to form diodes, transistors, and resistors (undoped semiconductors). Capacitors and inductors can also be formed, although they are often connected separately. A tiny chip, only 1 mm on a side, may contain thousands of transistors and other circuit elements.

SUMMARY

Quantum mechanics explains the bonding together of atoms to form molecules. In a *covalent bond*, the electron clouds of two or more atoms overlap because of constructive interference between the electron waves. The positive nuclei are attracted to this concentration of negative charge between them, forming the bond. An *ionic bond* is an extreme case of a covalent bond in which one or more electrons from one atom spend much more time around the other atom than around their own. The atoms then act as oppositely charged ions that attract each other, forming the bond. These *strong bonds* hold molecules together, and also hold atoms and molecules together in solids. Also important are *weak bonds* (or *van der Waals bonds*), which are generally dipole attractions between molecules.

When atoms combine to form molecules, the energy levels of the outer electrons are altered because they now interact with each other. Additional energy levels also become possible because the atoms can vibrate with respect to each other, and the molecule as a whole can rotate. The energy levels for both vibrational and rotational motion are quantized, and are very close together (typically, 10^{-1} eV to 10^{-3} eV apart). Each atomic energy level thus becomes a set of closely spaced levels corresponding to the vibrational and rotational motions. Transitions from one level to another appear as many very closely spaced lines. The resulting spectra are called *band spectra*. The quantized rotational energy levels are given by

$$E_{\text{rot}} = L(L+1)\frac{\hbar^2}{2I}, \qquad L = 0, 1, 2, \cdots,$$

where I is the moment of inertia of the molecule. The energy levels for vibrational motion are given by

$$E_{\text{vib}} = (v + \tfrac{1}{2})hf, \qquad v = 0, 1, 2, \cdots,$$

where f is the classical natural frequency of vibration for the molecule. Transitions are subject to the selection rules $\Delta L = \pm 1$ and $\Delta v = \pm 1$.

Some solids are bound together by covalent and ionic bonds, just as molecules are. In metals, the electrostatic force between free electrons and positive ions helps form the *metallic bond*.

In a crystalline solid, the possible energy states for electrons are arranged in *bands*. Within each band the levels are very close together, but between the bands there may be forbidden *energy gaps*. Good conductors are characterized by the highest occupied band being only partially full, so there are many accessible states available to electrons to move about and accelerate when a voltage is applied. In a good insulator, the highest occupied energy band (the *valence band*) is completely full, and there is a large energy gap (5 to 10 eV) to the next highest band, the *conduction band*. At room temperature, molecular kinetic energy (thermal energy) available due to collisions is only about 0.04 eV, so almost no electrons can jump from the valence to the conduction band. In a semiconductor, the gap between valence and conduction bands is much smaller, on the order of 1 eV, so a few electrons can make the transition from the essentially full valence band to the nearly empty conduction band.

In a *doped* semiconductor, a small percentage of

impurity atoms with five or three valence electrons replace a few of the normal silicon atoms with their four valence electrons. A five-electron impurity produces an *n-type* semiconductor with negative electrons as carriers of current. A three-electron impurity produces a *p-type* semiconductor in which positive *holes* carry the current. The energy level of impurity atoms lies slightly below the conduction band in an *n*-type semiconductor, and acts as a *donor* from which electrons readily pass into the conduction band. The energy level of impurity atoms in a *p*-type semiconductor lies slightly above the valence band and acts as an *acceptor* level, since electrons from the valence band easily reach it, leaving holes behind to act as charge carriers.

A semiconductor *diode* consists of a *pn*-junction and allows current to flow in one direction only; it can be used as a *rectifier* to change ac to dc. Common *transistors* consist of three semiconductor sections, either as *pnp* or *npn*. Transistors can amplify electrical signals and find many other uses. An integrated circuit consists of a tiny semiconductor crystal or "chip" on which many transistors, diodes, resistors, and other circuit elements have been constructed using careful placement of impurities.

QUESTIONS

*1. What type of bond would you expect for (*a*) the N_2 molecule, (*b*) the HCl molecule, (*c*) Fe atoms in a solid?

*2. Describe how the molecule $CaCl_2$ could be formed.

*3. Does the H_2 molecule have a permanent dipole moment? Does O_2? Does H_2O? Explain.

*4. Although the molecule H_3 is not stable, the ion H_3^+ is. Explain, using the Pauli exclusion principle.

*5. The energy of a molecule can be divided into four categories. What are they?

*6. If conduction electrons are free to roam about in a metal, why don't they leave the metal entirely?

*7. A silicon semiconductor is doped with phosphorus. Will these atoms be donors or acceptors? What type of semiconductor will this be?

*8. Explain why the resistivity of metals increases with temperature whereas the resistivity of semiconductors decreases with increasing temperature.

*9. Can a diode be used to amplify a signal?

*10. Figure 29–35 shows a "bridge-type" full-wave rectifier. Explain how the current is rectified and how current flows during each half cycle.

*11. Compare the resistance of a *pn* junction diode connected in forward bias to its resistance when connected in reverse bias.

*12. Explain how a transistor could be used as a switch.

*13. If $\mathscr{E}_C$ were reversed in Fig. 29–33, how would the amplification be altered?

*14. Describe how a *pnp* transistor can operate as an amplifier.

*15. In a transistor, the base–emitter junction and the base–collector junction are essentially diodes. Are these junctions reverse-biased or forward-biased in the application shown in Fig. 29–33?

*16. What purpose does the capacitor in Fig. 29–34 serve?

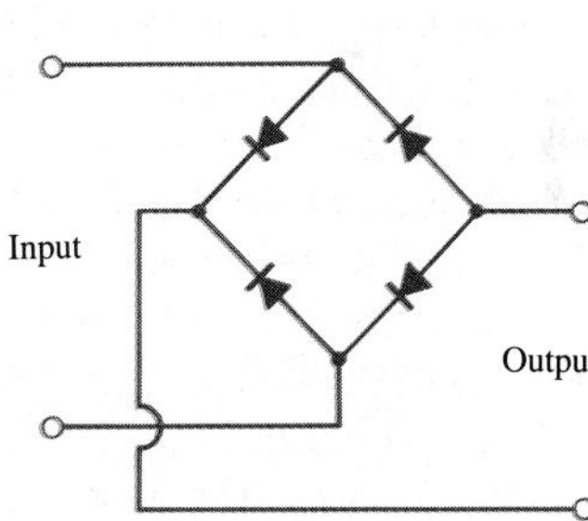

FIGURE 29–35 Question 10.

PROBLEMS

***SECTIONS 29–1 TO 29–3**

*1. (I) Estimate the binding energy of a KCl molecule by calculating the electrostatic potential energy when the K^+ and Cl^- ions are at their stable separation of 0.28 nm. Assume each has a charge of magnitude 1.0*e*.

*2. (II) The measured binding energy of KCl is 4.43 eV. From the result of Problem 1, estimate the contribution to the binding energy of the repelling electron clouds at the equilibrium distance $r_0 = 0.28$ nm.

*3. (II) Estimate the binding energy of the H_2 molecule, assuming the two H nuclei are 0.074 nm apart and the two electrons spend 33 percent of their time midway between them.

*4. (III) Apply reasoning similar to that in the text for the $S = 0$ and $S = 1$ states in the formation of the H_2 molecule to show why the molecule He_2 is *not* formed. Show also why the He_2^+ molecular ion *is* formed (with a binding energy of 3.1 eV at $r_0 = 0.11$ nm).

*SECTION 29–4

*5. (II) The so-called "characteristic rotational energy," $\hbar^2/2I$, for N_2 is 2.48×10^{-4} eV. Calculate the N_2 bond length.

*6. (II) Calculate the characteristic rotational energy, $\hbar^2/2I$, for the O_2 molecule whose bond length is 0.121 nm.

*7. (II) The equilibrium separation of H atoms in the H_2 molecule is 0.074 nm (Fig. 29–8). Calculate the energies and wavelengths of photons for the rotational transitions (*a*) $L = 1$ to $L = 0$; (*b*) $L = 2$ to $L = 1$; and (*c*) $L = 3$ to $L = 2$.

*8. (II) Calculate the bond length for the NaCl molecule given that three successive wavelengths for rotational transitions are 23.1 mm, 11.6 mm, and 7.71 mm.

*9. (II) In the absorption spectrum of the HCl molecule (Fig. 29–19), the lines are separated by 2.6×10^{-3} eV. Determine (*a*) the moment of inertia of the HCl molecule, and (*b*) the bond length. Assume that Cl has mass of 35 u.

*10. (II) Derive Eqs. 29–5.

*11. (III) (*a*) Use the curve of Fig. 29–15 to show that the stiffness constant k for the H_2 molecule is about $k \approx$ 550 N/m. (Recall that PE $= \frac{1}{2}kx^2$.) (*b*) Then estimate the fundamental wavelength for vibrational transitions using the classical formula (Chapter 11).

*SECTION 29–6

*12. (I) Explain on the basis of energy bands why the sodium chloride crystal is a good insulator. [*Hint:* consider the shells of Na^+ and Cl^- ions.]

*13. (II) We saw that there are $2N$ possible electron states in the 3*s* band of Na, where N is the total number of atoms. How many possible electron states are there in the (*a*) 2*s* band, (*b*) 2*p* band, and (*c*) 3*p* band? (*d*) State a general formula for the total number of possible states in any given electron band.

*14. (II) Calculate the longest-wavelength photon that can cause an electron in silicon ($E_g = 1.1$ eV) to jump from the valence band to the conduction band.

*15. (II) The energy gap E_g in germanium is 0.72 eV. When used as a photon detector, roughly how many electrons can be made to jump from the valence to the conduction band by the passage of a 660-keV photon that loses all its energy in this fashion?

*SECTION 29–7

*16. (II) Suppose that a silicon semiconductor is doped with phosphorus so that one silicon atom in 10^6 is replaced by a phosphorus atom. Assuming that the "extra" electron in every phosphorus atom is donated to the conduction band, by what factor is the density of conduction electrons increased? The density of silicon is 2330 kg/m^3, and the density of conduction electrons in pure silicon is about 10^{16}m^{-3} at room temperature.

*SECTION 29–8

*17. (II) A silicon diode whose current–voltage characteristics are given in Fig. 29–29 is connected in series with a battery and a 200-Ω resistor. What battery voltage is needed to produce a 10-mA current?

*18. (II) Suppose that the *current gain* of the transistor in Fig. 29–33 is $\beta = I_C/I_B = 100$. If $R_C = 3.0$ kΩ, calculate the output voltage for a time-varying input current of 2.0 μA.

*19. (II) If the current gain of the transistor amplifier in Fig. 29–33 is $\beta = I_C/I_B = 60$, what value must R_C have if a 1.0-μA base current is to produce an output voltage of 0.40 V?

*20. (II) A transistor, whose current gain $\beta = I_C/I_B = 80$, is connected as in Fig. 29–33 with $R_B = 2.0$ kΩ and $R_C =$ 6.5 kΩ. Calculate (*a*) the voltage gain, and (*b*) the power amplification.

*21. (II) An amplifier has a *voltage gain* of 90 and a 10-kΩ load (output) resistance. What is the peak output current through the load resistor if the input voltage is an ac signal with a peak of 0.080 V?

*22. (II) An ac voltage of 240 V rms is to be rectified. Estimate the average current in the output resistor R (15 kΩ) for (*a*) a half-wave rectifier (Fig. 29–30), and (*b*) a full-wave rectifier (Fig. 29–31) without capacitor.

*23. (III) Suppose that the diode of Fig. 29–29 is connected in series to a 100-Ω resistor and a 2.0-V battery. What current flows in the circuit? [*Hint:* draw a line on Fig. 29–29 representing the current in the resistor as a function of the voltage across the diode; the intersection of this line with the characteristic curve will give the answer.]

*24. (III) A silicon diode passes significant current only if the forward-bias voltage exceeds about 0.6 V. Estimate the average current in the output resistor R of (*a*) a half-wave rectifier (Fig. 29–30), and (*b*) a full-wave rectifier (Fig. 29–31) without a capacitor. Assume that $R =$ 200 Ω in each case and that the ac voltage is 10.0 V rms in each case.

GENERAL PROBLEMS

*25. Estimate the binding energy of the H_2 molecule by calculating the difference in KE of the electrons between when they are in separate atoms and when they are in the molecule, using the uncertainty principle. Take Δx for the electrons in the separated atoms to be the radius of the first Bohr orbit, 0.053 nm, and for the molecule take Δx to be the separation of the nuclei, 0.074 nm.

*26. For an arsenic donor atom in a doped silicon semiconductor, assume that the "extra" electron moves in a Bohr orbit about the arsenic ion. For this electron in the ground state, take into account the dielectric constant $K = 12$ of the Si lattice (which represents the weakening of the Coulomb force due to all the other atoms or ions in the lattice), and estimate (*a*) the binding energy, and (*b*) the orbit radius for this extra electron. [*Hint:* substitute $\varepsilon = K\varepsilon_0$ in Coulomb's law; see Section 17-8.]

***27.** A 240-V rms 60-Hz voltage is to be rectified with a full-wave rectifier as in Fig. 29–31, where $R = 15\ \text{k}\Omega$, and $C = 30\ \mu\text{F}$. (*a*) What will be the average current (approximately)? (*b*) What if $C = 0.10\ \mu\text{F}$? [*Hint:* see Section 19–7.]

***28.** A full-wave rectifier (Fig. 29–31) uses two diodes to rectify a 74-V rms ac voltage. If $R = 10\ \text{k}\Omega$ and $C = 20\ \mu\text{F}$, what will be the approximate percent variation in the output voltage? This is called *ripple voltage* (see Fig. 29–31c). [*Hint:* see Section 19–7 and assume the discharge of the capacitor is approximately linear.]

***29.** A zener-diode voltage regulator is shown in Fig. 29–36. Suppose that $R = 2.00\ \text{k}\Omega$ and that the diode breakdown voltage is 130 V; the diode is rated at a maximum current of 100 mA. (*a*) If $R_{\text{load}} = 15.0\ \text{k}\Omega$, over what range of supply voltages will the circuit maintain the voltage at 130 V? (*b*) If the supply voltage is 200 V, over what range of load resistance will the voltage be regulated?

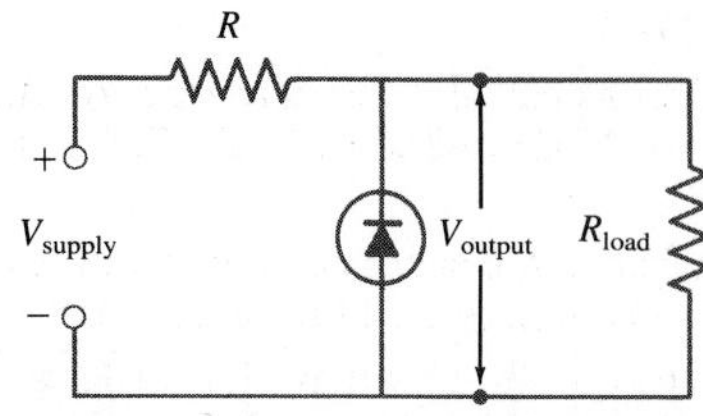

FIGURE 29–36 Problem 29.

CHAPTER 30

Nuclear Physics and Radioactivity

A geologist can search for radioactive elements (perhaps uranium) using a portable Geiger counter or scintillation counter.

In the early part of the twentieth century, Rutherford's experiments led to the idea that at the center of an atom there is a tiny but massive nucleus. At the same time that the quantum theory was being developed and scientists were attempting to understand the structure of the atom and its electrons, investigations into the nucleus itself had also begun. In this chapter and the next, we take a brief look at *nuclear physics*.

30–1 • Structure and Properties of the Nucleus

An important question to physicists in the early part of this century was whether the nucleus had a structure, and what that structure might be. It turns out that the nucleus is a complicated entity and is not fully understood even today. However, by the early 1930s, a model of the nucleus had been developed that is still useful. According to this model, a nucleus is considered as an aggregate of two types of particles: protons and neutrons. (Of course, we must remember that these "particles" also have wave properties, but for ease of visualization and language, we often refer to them simply as "particles.") A **proton** is the nucleus of the simplest atom, hydrogen. It has a positive charge ($= +e = +1.6 \times 10^{-19}$ C) and a mass

Proton

$$m_p = 1.6726 \times 10^{-27}\ \text{kg}.$$

The **neutron**, whose existence was ascertained only in 1932 by the Englishman James Chadwick (1891–1974), is electrically neutral ($q = 0$), as its name implies. Its mass, which is almost identical to that of the proton, is

Neutron

$$m_n = 1.6749 \times 10^{-27}\ \text{kg}.$$

These two constituents of a nucleus, neutrons and protons, are referred to collectively as **nucleons**.

Nucleons

Although the hydrogen nucleus consists of a single proton alone, the nuclei of all other elements consist of both neutrons and protons. The different types of nuclei are often referred to as **nuclides**. The number of protons in a nucleus (or nuclide) is called the **atomic number** and is designated by the symbol Z. The total number of nucleons, neutrons plus protons, is designated by the symbol A and is called the **atomic mass number**. This name is used since the mass of a nucleus is very closely A times the mass of one nucleon. A nuclide with 7 protons and 8 neutrons thus has $Z = 7$ and $A = 15$. The **neutron number** N is $N = A - Z$.

Z and A

To specify a given nuclide, we need give only A and Z. A special symbol is commonly used which takes the form

$${}^{A}_{Z}X,$$

where X is the chemical symbol for the element (see the periodic table inside the back cover), A is the atomic mass number, and Z is the atomic number. For example, ${}^{15}_{7}\text{N}$ means a nitrogen nucleus containing 7 protons and 8 neutrons for a total of 15 nucleons. In a neutral atom, the number of electrons orbiting the nucleus is equal to the atomic number Z (since the charge on an electron has the same magnitude but opposite sign to that of a proton). The main properties of an atom, and how it interacts with other atoms, are largely determined by the number of electrons. Hence Z determines what kind of atom it is: carbon, oxygen, gold, or whatever. It is redundant to specify both the symbol of a nucleus and its atomic number Z as described above. If the nucleus is nitrogen, for example, we know immediately that $Z = 7$. The subscript Z is thus sometimes dropped and ${}^{15}_{7}\text{N}$ is then written simply ${}^{15}\text{N}$; in words we say "nitrogen fifteen."

For a particular type of atom (say, carbon), nuclei are found to contain different numbers of neutrons, although they all have the same number of protons. For example, carbon nuclei always have 6 protons, but they may have 5, 6, 7, 8, 9, or 10 neutrons. Nuclei that contain the same number of protons but different numbers of neutrons are called **isotopes**. Thus, ${}^{11}_{6}\text{C}$, ${}^{12}_{6}\text{C}$, ${}^{13}_{6}\text{C}$, ${}^{14}_{6}\text{C}$, ${}^{15}_{6}\text{C}$, and ${}^{16}_{6}\text{C}$ are all isotopes of carbon. Of course, the isotopes of a given element are not all equally common. For example, 98.9 percent of naturally occurring carbon (on earth) is the isotope ${}^{12}_{6}\text{C}$ and about 1.1 percent is ${}^{13}_{6}\text{C}$. These percentages are referred to as the **natural abundances.**† Many isotopes that do not occur naturally can be produced in the laboratory by means of nuclear reactions (more on this later). Indeed, all

Isotopes

† The mass value for each element as given in the periodic table (inside back cover) is an average weighted according to the natural abundances of its isotopes.

nium and radium. Other radioactive elements were soon discovered as well. The radioactivity was found in every case to be unaffected by the strongest physical and chemical treatments, including strong heating or cooling and the action of strong chemical reagents. It soon became clear that the source of radioactivity must be deep within the atom, that it must emanate from the nucleus. And it became apparent that radioactivity is the result of the *disintegration* or *decay* of an unstable nucleus. Certain isotopes are not stable under the action of the nuclear force, and they decay with the emission of some type of radiation or "rays."

Many unstable isotopes occur in nature, and such radioactivity is called "natural radioactivity." Other unstable isotopes can be produced in the laboratory by nuclear reactions (Section 31–1); these are said to be produced "artificially" and to have "artificial radioactivity."

Rutherford and others began studying the nature of the rays emitted in radioactivity about 1898. They found that the rays could be classified into three distinct types according to their penetrating power. One type of radiation could barely penetrate a piece of paper. The second type could pass through as much as 3 mm of aluminum. The third was extremely penetrating: it could pass through several centimeters of lead and still be detected on the other side. They named these three types of radiation alpha (α), beta (β), and gamma (γ), respectively, after the first three letters of the Greek alphabet.

α, β, γ radiation

FIGURE 30–4 Alpha and beta rays are bent in opposite directions by a magnetic field, whereas gamma rays are not bent at all.

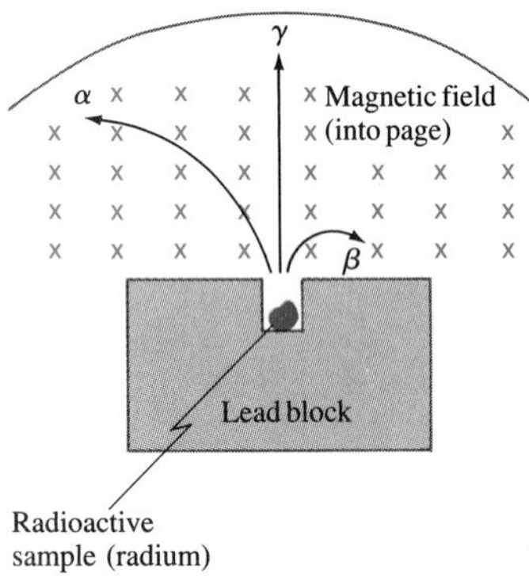

Each type of ray was found to have a different charge and hence is bent differently in a magnetic field, Fig. 30–4; α rays are positively charged, β rays are negatively charged, and γ rays are neutral. It was soon found that all three types of radiation consisted of familiar kinds of particles. Gamma rays are very high-energy photons whose energy is even higher than that of X rays. Beta rays are electrons, identical to those that orbit the nucleus (but they are created within the nucleus itself). Alpha rays (or α particles) are simply the nuclei of helium atoms, ${}^4_2\text{He}$; that is, an α ray consists of two protons and two neutrons bound together.

We now discuss each of these three types of radioactivity, or decay, in more detail.

30–4 • Alpha Decay

When a nucleus emits an α particle (${}^4_2\text{He}$), it is clear that the remaining nucleus will be different from the original: for it has lost two protons and two neutrons. Radium 226 (${}^{226}_{88}\text{Ra}$), for example, is an α emitter. It decays to a nucleus with $Z = 88 - 2 = 86$ and $A = 226 - 4 = 222$. The nucleus with $Z = 86$ is radon (Rn)—see Appendix D or the periodic table. Thus the radium decays to radon with the emission of an α particle. This is written

$${}^{226}_{88}\text{Ra} \rightarrow {}^{222}_{86}\text{Rn} + {}^4_2\text{He}.$$

Daughter nucleus
Parent nucleus
Transmutation

It is clear that when α decay occurs, a new element is formed. The **daughter** nucleus (${}^{222}_{86}\text{Rn}$ in this case) is different from the **parent** nucleus (${}^{226}_{88}\text{Ra}$ in this case). This changing of one element into another is called **transmutation**.

Alpha decay occurs because the strong nuclear force is unable to hold very large nuclei together. Because the nuclear force is a short-range force, it acts only between neighboring nucleons. But the electric force can act clear across the nucleus. For very large nuclei, the large Z means the repulsive electric force becomes very large (Coulomb's law) and acts between all protons. The strong nuclear force, since it acts only between neighboring nucleons, is overpowered and is unable to hold the nucleus together.

Why strong nuclear force cannot hold nucleus together

We can express the instability in terms of energy (or mass): the mass of the parent nucleus is greater than the mass of the daughter nucleus plus the mass of the α particle. The mass difference appears as kinetic energy, which is carried away by the α particle and the recoiling daughter nucleus. The total energy released is called the **disintegration energy**, Q, or the **Q-value** of the decay; it is defined as

$$Q = (M_P - M_D - m_\alpha)c^2, \qquad (30\text{–}2)$$

Q-value

where M_P, M_D, and m_α are the masses of the parent nucleus, daughter nucleus, and α particle, respectively. If the parent has *less* mass than the daughter plus the α particle (so $Q < 0$), the decay could not occur, for the conservation of energy law would be violated.

EXAMPLE 30–2 Calculate the disintegration energy when ${}^{232}_{92}U$ (mass = 232.037131 u) decays to ${}^{228}_{90}Th$ (228.028716 u) with the emission of an α particle. (As always, masses are for neutral atoms.)

SOLUTION Since the mass of the ${}^{4}_{2}He$ is 4.002602 u (Appendix D), the total mass in the final state is

$$228.028716\ \text{u} + 4.002602\ \text{u} = 232.031318\ \text{u}.$$

The mass lost when the ${}^{232}_{92}U$ decays is

$$232.037131\ \text{u} - 232.031318\ \text{u} = 0.005813\ \text{u}.$$

Since 1 u = 931.5 MeV, the energy Q released is

$$Q = (0.005813\ \text{u})(931.5\ \text{MeV/u}) \approx 5.4\ \text{MeV},$$

and this energy appears as KE of the α particle and the daughter nucleus. (Using conservation of momentum, it can be shown that the α particle in this decay has a KE of about 5.3 MeV. Thus, the daughter nucleus—which recoils in the opposite direction from the emitted α particle—has about 0.1 MeV of kinetic energy. See Problem 25.)

Why, you may wonder, do nuclei emit this combination of four nucleons called an α particle? Why not just four nucleons, or even one? The answer is that the α particle is very strongly bound, so that its mass is significantly less than that of four separate nucleons. As we saw in Example 30–1, two protons and two neutrons separately have a total mass of about 4.032980 u. The total mass of a ${}^{228}_{90}Th$ nucleus plus four separate nucleons is 232.061696 u, which is greater than the mass of the parent nucleus. Such a decay could not occur because it would violate the conservation of energy. Similarly, it is almost always true that the emission of a single nucleon is energetically not possible.

30–5 • Beta Decay

Transmutation of elements also occurs when a nucleus decays by β decay—that is, with the emission of an electron or β particle. The nucleus ${}^{14}_{6}C$, for example, emits an electron when it decays:

$${}^{14}_{6}C \rightarrow {}^{14}_{7}N + e^- + \text{a neutrino},$$

where e^- is the symbol for the electron. (The symbol ${}^{0}_{-1}e$ is sometimes used for the electron whose charge corresponds to $Z = -1$ and, since it is not a nucleon and has very small mass, has $A = 0$.) The particle known as the neutrino, with rest mass $m_0 = 0$ and charge $Q = 0$, was not initially detected and was only hypothesized to exist later, as we shall discuss later in this section. No nucleons are lost when an electron is emitted, and the total number of nucleons, A, is the same in the daughter nucleus as in the parent. But because an electron has been emitted, the charge on the daughter nucleus is different from that on the parent. The parent nucleus had $Z = +6$. In the decay, the nucleus loses a charge of -1, so (from charge conservation), the nucleus remaining behind must have an extra positive charge for a total of 7. So the daughter nucleus has $Z = 7$, which is nitrogen.

It must be carefully noted that the electron emitted in β decay is *not* an orbital electron. Instead, the electron is created *within the nucleus itself.* It is as if one of the neutrons changes to a proton and in the process (to conserve charge) throws off an electron. Indeed, free neutrons actually do decay in this fashion: $n \rightarrow p + e^-$ (plus a neutrino). Because of their origin in the nucleus, the electrons emitted in β decay are often referred to as "β particles," rather than as electrons, to remind us of their origin. They are, nonetheless, indistinguishable from orbital electrons.

Example 30–3 How much energy is released when ${}^{14}_{6}C$ decays to ${}^{14}_{7}N$ by β emission? Use Appendix D.

Solution Because the masses given in Appendix D are those of the neutral atom, we have to keep track of the electrons involved. Assume the parent nucleus has six orbiting electrons so it is neutral, and its mass is 14.003242 u. The daughter, in this decay ${}^{14}_{7}N$, is not neutral however since it has the same six electrons circling it but the nucleus has a charge of $+7e$. However, the mass of this daughter with its six electrons, plus the mass of the emitted electron (which makes a total of seven electrons), is just the mass of a neutral nitrogen atom. That is, the total mass in the final state is

$$(\text{mass of } {}^{14}_{7}N \text{ nucleus} + 6 \text{ electrons}) + (\text{mass of 1 electron}),$$

and this is equal to

$$\text{mass of neutral } {}^{14}_{7}N \text{ (includes 7 electrons)},$$

which, from Appendix D is a mass of 14.003074 u. (Note that the neutrino doesn't contribute to either the mass or charge balance since it has $m_0 = 0$ and $Q = 0$.) Hence the mass after decay is 14.003074 u, whereas before decay, it is 14.003242 u. So the mass difference is 0.000168 u, which corresponds to 0.156 MeV or 156 keV.

According to this example, we would expect the emitted electron to have a kinetic energy of 156 keV. (The daughter nucleus, because its mass is very much larger than that of the electron, recoils with very low velocity and hence gets very little of the kinetic energy.) Indeed, very careful measurements indicate that a few emitted β particles do have kinetic energy close to this calculated value. But the vast majority of emitted electrons have somewhat less energy. In fact, the energy of the emitted electron can be anywhere from zero up to the maximum value as calculated above. This range of electron KE was found for any β decay. It was as if the law of conservation of energy was being violated, and indeed Bohr actually considered this possibility. Careful experiments indicated that linear momentum and angular momentum also did not seem to be conserved. Physicists were troubled at the prospect of having to give up these laws, which had worked so well in all previous situations. In 1930, Wolfgang Pauli proposed an alternate solution: perhaps a new particle that was very difficult to detect was emitted during β decay in addition to the electron. This hypothesized particle could be carrying off the energy, momentum, and angular momentum required to maintain the conservation laws. This new particle was named the **neutrino**—meaning "little neutral one"—by the great Italian physicist Enrico Fermi (1901–1954; Fig. 30–5), who in 1934 worked out a detailed theory of β decay. (It was Fermi who, in this theory, postulated the existence of the fourth force in nature, which we call the weak nuclear force.) The neutrino has zero charge and seems to have zero rest mass, although there are recent suggestions that it may have a very tiny rest mass. If its rest mass is zero, it is much like a

Neutrino invented

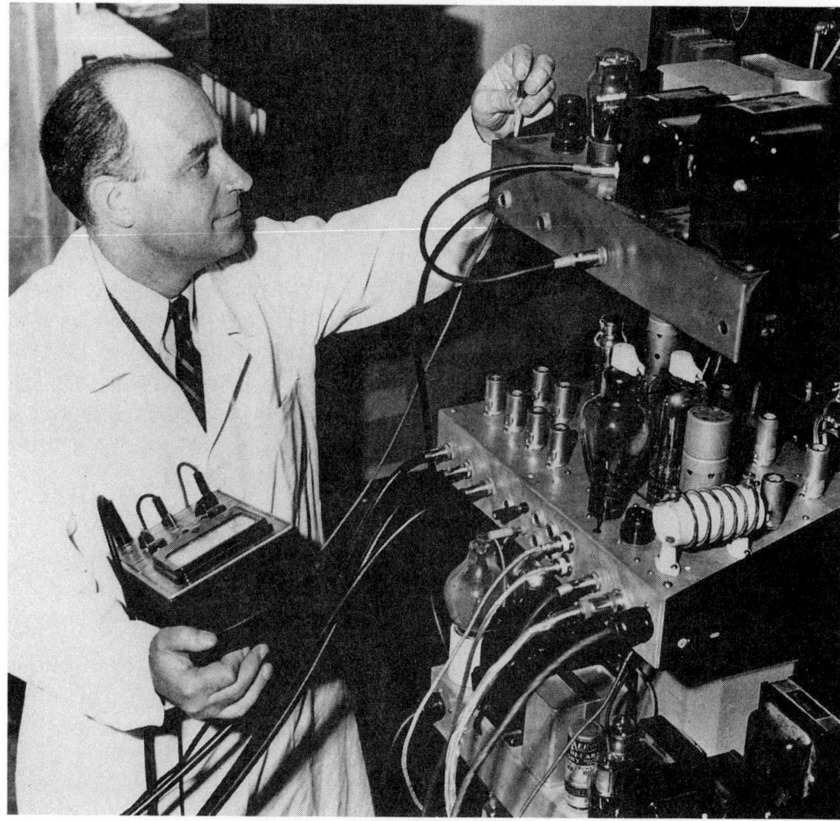

FIGURE 30–5 Enrico Fermi. Fermi contributed significantly to both theoretical and experimental physics, a feat almost unique in this century.

photon in that it is neutral and travels at the speed of light. But the neutrino is far more difficult to detect. In 1956, complex experiments produced further evidence for the existence of the neutrino; but by then, most physicists had already accepted its existence.

The symbol for the neutrino is the Greek letter nu (ν). The correct way of writing the decay of $^{14}_{6}\text{C}$ is then

$$^{14}_{6}\text{C} \rightarrow {}^{14}_{7}\text{N} + e^- + \bar{\nu}.$$

The bar ($\bar{}$) over the neutrino symbol is to indicate that it is an "antineutrino." (Why this is called an antineutrino rather than simply a neutrino need not concern us now; it is discussed in Chapter 32).

Many isotopes decay by electron emission. They are always isotopes that have too many neutrons compared to the number of protons. That is, they are isotopes that lie above the stable isotopes plotted in Fig. 30–2. But what about unstable isotopes that have too few neutrons compared to their number of protons—those that fall below the stable isotopes of Fig. 30–2?

Positron (β^+) decay

These, it turns out, decay by emitting a **positron** instead of an electron. A positron (sometimes called an e^+ or β^+ particle) has the same mass as the electron, but it has a positive charge of $+1e$. Because it is so like an electron, except for its charge, the positron is called the **antiparticle**† to the electron. An example of a β^+ decay is that of $^{19}_{10}\text{Ne}$:

$$^{19}_{10}\text{Ne} \rightarrow {}^{19}_{9}\text{F} + e^+ + \nu,$$

where e^+ (or $^{0}_{1}e$) stands for a positron. (Note that the ν emitted here is a neutrino, whereas that emitted in β^- decay is called an antineutrino. Thus an antielectron (=positron) is emitted with a neutrino, whereas an antineutrino is emitted with an electron; this is discussed in Chapter 32).

Electron capture

Besides β^- and β^+ emission, there is a third related process. This is **electron capture** and occurs when a nucleus absorbs one of its orbiting electrons. An example is $^{7}_{4}\text{Be}$, which as a result becomes $^{7}_{3}\text{Li}$. The process is written

$$^{7}_{4}\text{Be} + e^- \rightarrow {}^{7}_{3}\text{Li} + \nu.$$

Usually it is an electron in the innermost (K) shell that is captured, in which case it is called "K-capture." The electron disappears in the process and a proton in the nucleus becomes a neutron; a neutrino is emitted as a result. This process is inferred experimentally by detection of emitted X rays (due to electrons jumping down to fill the empty state) of just the proper energy.

In β decay, it is the weak nuclear force that plays the crucial role. The neutrino is unique in that it interacts with matter only via the weak force, which is why it is so hard to detect.

30–6 • Gamma Decay

Gamma rays are photons having very high energy. The decay of a nucleus by emission of a γ ray is much like emission of photons by excited atoms. Like an atom, a nucleus itself can be in an excited state. When it jumps down to a lower energy state, or to the ground state, it emits a photon. The

† Discussed in Chapter 32. Briefly, an antiparticle has the same mass as its corresponding particle, but opposite charge.

possible energy levels of a nucleus are much farther apart than those of an atom: on the order of keV or MeV, as compared to a few eV for electrons in an atom. Hence, the emitted photons have energies that can range from a few keV to several MeV. For a given decay, the γ ray always has the same energy. Since a γ ray carries no charge, there is no change in the element as a result of a γ decay.

How does a nucleus get into an excited state? It may occur because of a violent collision with another particle. More commonly, the nucleus remaining after a previous radioactive decay may be in an excited state. A typical example is shown in the energy-level diagram of Fig. 30–6. $^{12}_{5}B$ can decay by β decay directly to the ground state of $^{12}_{6}C$; or it can go by β decay to an excited state of $^{12}_{6}C$, which then decays by emission of a 4.4 MeV γ ray to the ground state.

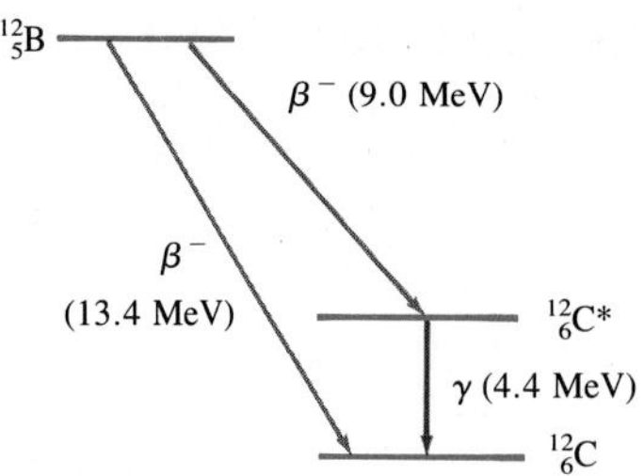

FIGURE 30–6 Energy-level diagram showing how $^{12}_{5}B$ can decay to the ground state of $^{12}_{6}C$ by β decay (total energy released = 13.4 MeV) or can β decay to an excited state of $^{12}_{6}C$ (indicated by *), which subsequently decays to its ground state by emitting a 4.4-MeV γ ray.

In some cases, a nucleus may remain in an excited state for some time before it emits a γ ray as in Fig. 30–6. The nucleus is then said to be in a **metastable state** and is called an **isomer**.

An excited nucleus can sometimes return to the ground state by another process known as **internal conversion**. In this process, the excited nucleus interacts with one of the orbital electrons and ejects this electron from the atom with the same KE that an emitted γ ray would have had.

What, you may wonder, is the difference between a γ ray and an X ray? They both are electromagnetic radiation (photons) and their range of energies overlap to some extent. The difference is not intrinsic. We use the term X ray if the photon is produced by an electron–atom interaction, and γ ray if the photon is produced in a nuclear process.

30–7 • Conservation of Nucleon Number and Other Conservation Laws

In all three types of radioactive decay, the classical conservation laws hold. Energy, linear momentum, angular momentum, and electric charge are all conserved. These quantities are the same before the decay as after. But a new conservation law is also revealed, the **law of conservation of nucleon number**. According to this law, the total number of nucleons (A) remains constant in any process, although one type can change into the other type (protons into neutrons or vice versa). This law holds in all three types of decay. Table 30–2 gives a summary of α, β, and γ decay.

TABLE 30–2
The Three Types of Radioactive Decay†

α decay:
$N(A, Z) \rightarrow N(A-4, Z-2) + {}^4_2\text{He}$
β decay:
β^-: $N(A, Z) \rightarrow N(A, Z+1) + e^- + \bar{\nu}$
β^+: $N(A, Z) \rightarrow N(A, Z-1) + e^+ + \nu$
Electron capture: $N(A, Z) + e^- \rightarrow N(A, Z-1) + \nu$
γ decay:
$N^*(A, Z) \rightarrow N(A, Z) + \gamma$

† $N(A, Z)$ means a nucleus with atomic number Z and mass number A; * indicates the excited state of a nucleus.

30–8 • Half-Life and Rate of Decay

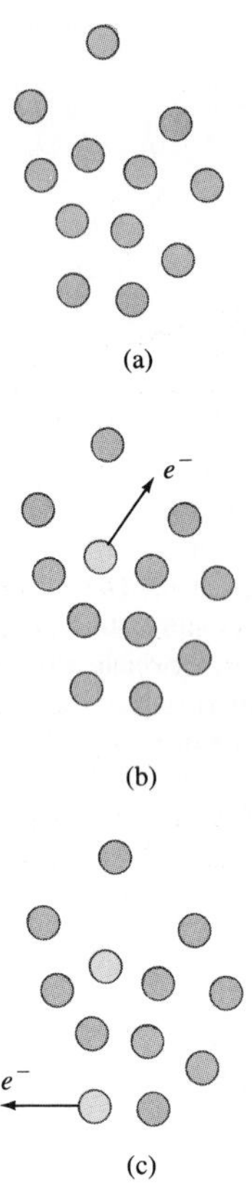

Legend

$^{14}_{6}$C atom (parent)

$^{14}_{7}$N atom (daughter)

FIGURE 30–7 Radioactive nuclei decay one by one. Hence, the number of parent nuclei in a sample is continually decreasing. When a $^{14}_{6}$C nucleus emits the electron, it becomes a $^{14}_{7}$N nucleus.

A macroscopic sample of any radioactive isotope consists of a vast number of radioactive nuclei. These nuclei do not all decay at one time. Rather, they decay one by one over a period of time. This is a random process: we can't predict exactly when a given nucleus will decay. But we can determine, on a probabilistic basis, approximately how many nuclei in a sample will decay over a given time period.

The number of decays ΔN that occur in a very short time interval Δt is found to be proportional to Δt and to the total number N of radioactive nuclei present:

$$\Delta N = -\lambda N \, \Delta t. \tag{30–3}$$

In this equation, λ is a constant of proportionality called the **decay constant**, which is different for different isotopes. The greater λ is, the greater the rate of decay and the more radioactive that isotope is said to be. The number of decays that occur in the short time interval Δt is designated ΔN because each decay that occurs corresponds to a decrease by one in the number N of nuclei present. That is, radioactive decay is a "one-shot" process, Fig. 30–7. Once a particular parent nucleus decays into its daughter, it cannot do it again. The minus sign in Eq. 30–3 is needed to indicate that N is decreasing.

Equation 30–3 can be solved for N (using calculus) and the result is

$$N = N_0 e^{-\lambda t}, \tag{30–4}$$

where N_0 is the number of nuclei present at time $t = 0$, and N the number remaining after a time t. The symbol e is the natural exponential (encountered earlier in Sections 19–7 and 21–11) whose value is $e = 2.718 \cdots$. Thus the number of parent nuclei in a sample decreases exponentially in time, as shown in Fig. 30–8a for $^{14}_{6}$C decay. Equation 30–4 is called the **radioactive decay law**.

The number of decays per second, $\Delta N/\Delta t$, is called the **activity** of the sample. Since $\Delta N/\Delta t$ is proportional to N (see Eq. 30–3), it, too, decreases exponentially in time with the same time constant (Fig. 30–8b). The activity at time t is given by

$$\frac{\Delta N}{\Delta t} = \left(\frac{\Delta N}{\Delta t}\right)_0 e^{-\lambda t}, \tag{30–5}$$

where $(\Delta N/\Delta t)_0$ is the activity at $t = 0$.

The rate of decay of any isotope is often specified by giving its "half-life" rather than the decay constant. *Half-life* The **half-life** of an isotope is defined as the time it takes for half the original amount of isotope in a given sample to decay. For example, the half-life of $^{14}_{6}$C is 5730 years. If at some time a piece of petrified wood contains, say, 1.00×10^{22} nuclei of $^{14}_{6}$C, then 5730 years later it will contain only 0.50×10^{22} of these nuclei. After another 5730 years it will contain 0.25×10^{22} nuclei, and so on. This is shown in Fig. 30–8a. Since the rate of decay $\Delta N/\Delta t$ is proportional to N, it, too, decreases by a factor of 2 every half-life (Fig. 30–8b).

The half-lives of known radioactive isotopes vary from about 10^{-22} s to 10^{28} s (about 10^{21} yr). The half-lives of many isotopes are given in Ap-

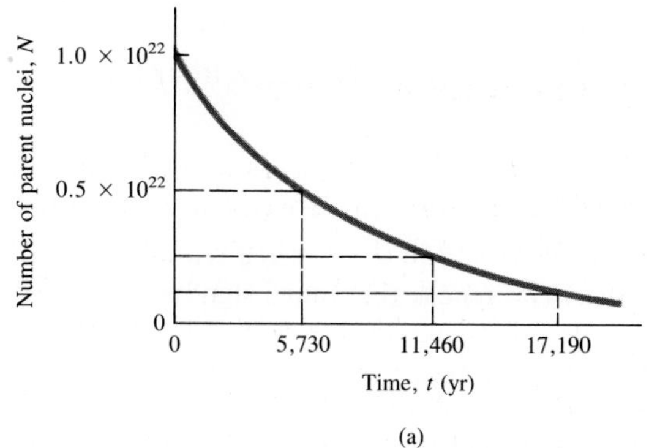

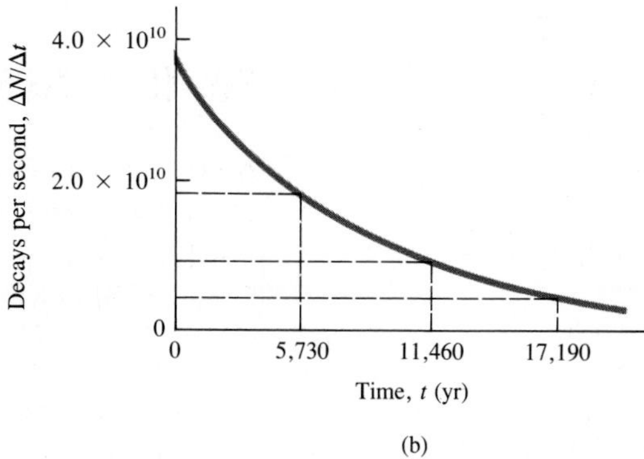

FIGURE 30–8 (a) The number N of parent nuclei in a given sample of $^{14}_{6}C$ decreases exponentially. (b) The number of decays per second also decreases exponentially. The half-life of $^{14}_{6}C$ is 5730 yr, which means that the number of parent nuclei, N, and the rate of decay, $\Delta N/\Delta t$, decrease by half every 5730 yr.

pendix D. It should be clear that the half-life (which we designate $T_{\frac{1}{2}}$) bears an inverse relationship to the decay constant. The longer the half-life of an isotope, the more slowly it decays, and hence λ is smaller. Conversely, very active isotopes (large λ) have very short half-lives. The precise relationship between half-life and decay constant is†

$$T_{\frac{1}{2}} = \frac{0.693}{\lambda}. \tag{30–6}$$

In the next section we will see how to make calculations involving $T_{\frac{1}{2}}$.

30–9 • Calculations Involving Decay Rates and Half-Life

Let us now consider examples of what we can determine about a sample of radioactive material if we know the half-life.

EXAMPLE 30–4 The isotope $^{14}_{6}C$ has a half-life of 5730 yr. If at some time a sample contains 1.00×10^{22} carbon-14 nuclei, what is the activity of the sample?

† Starting with Eq. 30–4, at $t = 0$, $N = N_0 e^0 = N_0$. At $t = T_{\frac{1}{2}}$, $N = N_0/2$ by definition of $T_{\frac{1}{2}}$ (half the parent nuclei remain). Then from Eq. 30–4,

$$\frac{N_0}{2} = N_0 e^{-\lambda T_{\frac{1}{2}}},$$

so

$$\frac{1}{2} = e^{-\lambda T_{\frac{1}{2}}},$$

or

$$e^{\lambda T_{\frac{1}{2}}} = 2.$$

We take natural logs of both sides (remember that "ln" and "e" are inverse operations) and find

$$\ln\left(e^{\lambda T_{\frac{1}{2}}}\right) = \lambda T_{\frac{1}{2}} = \ln 2,$$

so

$$T_{\frac{1}{2}} = \frac{\ln 2}{\lambda} = \frac{0.693}{\lambda}.$$

SOLUTION First we calculate the decay constant λ from Eq. 30–6, and obtain

$$\lambda = \frac{0.693}{T_{\frac{1}{2}}} = \frac{0.693}{(5730\ \text{yr})(3.156 \times 10^7\ \text{s/yr})} = 3.83 \times 10^{-12}\ \text{s}^{-1},$$

since there are $(60)(60)(24)(365\frac{1}{4}) = 3.156 \times 10^7$ s in a year. From Eq. 30–3, the activity or rate of decay (we ignore the minus sign) is

$$\frac{\Delta N}{\Delta t} = \lambda N = (3.83 \times 10^{-12}\ \text{s}^{-1})(1.00 \times 10^{22})$$

$$= 3.83 \times 10^{10}\ \text{decays/s}.$$

(The unit "decays/s" is often written simply as s^{-1} since "decays" is not a unit but refers only to the number.) Note that the graph of Fig. 30–8b starts at this value, corresponding to the original value of $N = 1.0 \times 10^{22}$ nuclei in Fig. 30–8a.

EXAMPLE 30–5 A laboratory has 1.49 μg of pure $^{13}_{7}\text{N}$, which has a half-life of 10.0 min (600 s). (*a*) How many nuclei are present initially? (*b*) What is the activity initially? (*c*) What is the activity after 1.0 h? (*d*) After approximately how long will the activity drop to less than one per second?

SOLUTION (*a*) Since the atomic mass is 13.0, then 13.0 g will contain 6.02×10^{23} nuclei (Avogadro's number). Since we have only 1.49×10^{-6} g, the number of nuclei, N_0, that we have initially is given by the ratio

$$\frac{N_0}{1.49 \times 10^{-6}\ \text{g}} = \frac{6.02 \times 10^{23}}{13.0\ \text{g}},$$

so $N_0 = 6.90 \times 10^{16}$ nuclei.

(*b*) From Eq. 30–6, $\lambda = (0.693)/(600\ \text{s}) = 1.16 \times 10^{-3}\ \text{s}^{-1}$. Then, at $t = 0$ (Eq. 30–3),

$$\left(\frac{\Delta N}{\Delta t}\right)_0 = \lambda N_0 = (1.16 \times 10^{-3}\ \text{s}^{-1})(6.90 \times 10^{16}) = 8.00 \times 10^{13}\ \text{decays/s}.$$

(*c*) Since the half-life is 10.0 min, the decay rate decreases by half every 10.0 min. We can make the following table of activity after given periods of time:

Time (min)	Activity (decays/s)
0	8.00×10^{13}
10	4.00×10^{13}
20	2.00×10^{13}
30	1.00×10^{13}
40	0.500×10^{13}
50	0.250×10^{13}
60	0.125×10^{13}

Thus, after 1.0 h, the activity is 1.25×10^{12} decays/s. A simpler way of seeing this is to note that 60 min is 6 half-lives, so the activity will decrease to $(\frac{1}{2})(\frac{1}{2})(\frac{1}{2})(\frac{1}{2})(\frac{1}{2})(\frac{1}{2}) = (\frac{1}{2})^6 = \frac{1}{64}$ of its original value, or $(8.00 \times 10^{13})/(64) = 1.25 \times 10^{12}$ per second. A third way to arrive at this answer is to use Eq. 30–5 (setting $t = 10.0\ \text{min} = 3600$ s):

$$\frac{\Delta N}{\Delta t} = \left(\frac{\Delta N}{\Delta t}\right)_0 e^{-\lambda t}$$

$$= (8.00 \times 10^{13}\ \mathrm{s}^{-1})e^{-(1.16\times 10^{-3}\ \mathrm{s}^{-1})(3600\ \mathrm{s})} = 1.23 \times 10^{12}\ \mathrm{s}^{-1}.$$

(The discrepancy in values arises because we kept only three significant figures.)

(*d*) After each hour, the activity drops by a factor of 64, so our table is extended to read as follows:

Time (h)	Activity (s^{-1})
0	8.00×10^{13}
1	1.25×10^{12}
2	1.95×10^{10}
3	3.05×10^{8}
4	4.77×10^{6}
5	7.45×10^{4}
6	1.16×10^{3}
7	$1.8 \times 10^{1} = 18$
8	<1

So the activity drops to less than one per second within 8 h. We can obtain a more precise result using the properties of the exponential and its inverse, the natural logarithm (ln). We want to determine the time t when

$$\frac{\Delta N}{\Delta t} = 1.00\ \mathrm{s}^{-1}.$$

From Eq. 30–5, we have

$$e^{-\lambda t} = \frac{(\Delta N/\Delta t)}{(\Delta N/\Delta t)_0} = \frac{1.00\ \mathrm{s}^{-1}}{8.00 \times 10^{13}\ \mathrm{s}^{-1}} = 1.25 \times 10^{-14}.$$

We take the natural log (ln) of both sides (remember $\ln e^{-\lambda t} = -\lambda t$) and divide by λ to find

$$t = -\frac{\ln(1.25 \times 10^{-14})}{\lambda} = 2.76 \times 10^4\ \mathrm{s}$$

or 7.66 h.

30–10 • Decay Series

It is often the case that one radioactive isotope decays to another isotope that is also radioactive. Sometimes this daughter decays to yet a third isotope which also is radioactive. Such successive decays are said to form a **decay series**. An important example is illustrated in Fig. 30–9. As can be seen, $^{238}_{92}\mathrm{U}$ decays by α emission to $^{234}_{90}\mathrm{Th}$, which in turn decays by β decay to $^{234}_{91}\mathrm{Pa}$. The series continues as shown, with several possible branches near the bottom. For example, $^{218}_{84}\mathrm{Po}$ can decay either by α decay to $^{214}_{82}\mathrm{Pb}$ or by β decay to $^{218}_{85}\mathrm{At}$. The series ends at the stable lead isotope $^{206}_{82}\mathrm{Pb}$. Other radioactive series also exist.

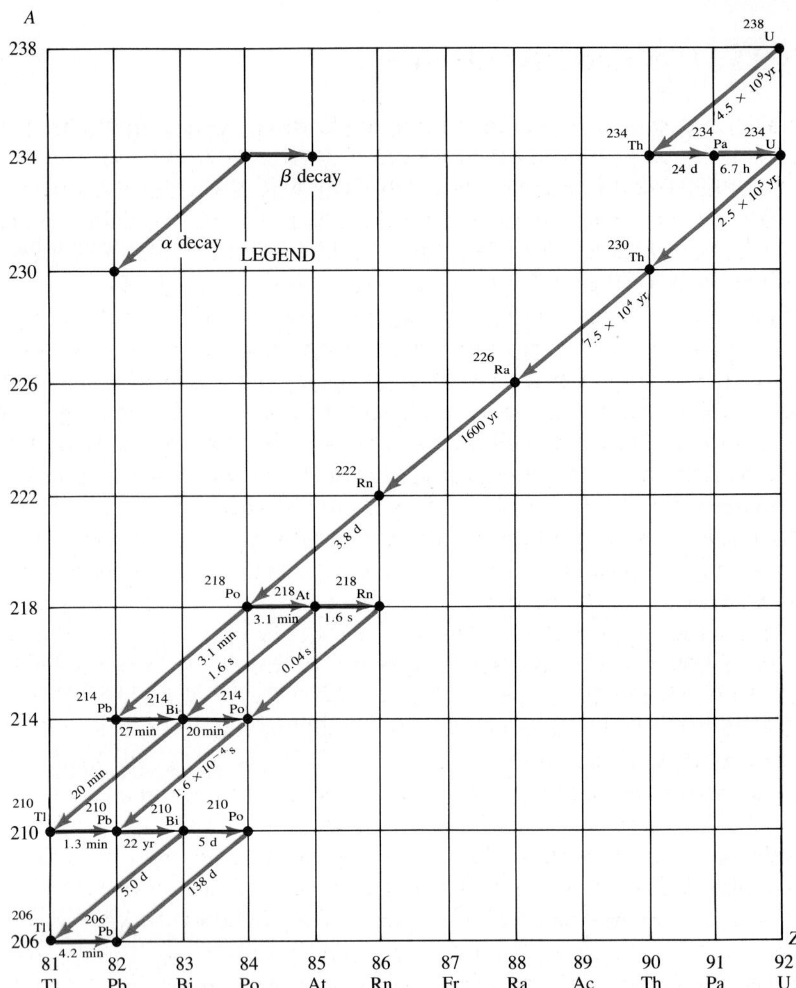

FIGURE 30–9 Decay series beginning with ${}^{238}_{92}\text{U}$. Nuclei in the series are specified by a dot representing A and Z values. Half-lives are given in seconds (s), minutes (min), hours (h), days (d), or years (yr). Note that a horizontal arrow represents β decay (A does not change), whereas a diagonal line represents α decay (A changes by 4, Z changes by 2).

Because of such decay series, certain radioactive elements are found in nature that otherwise would not be. For when the solar system acquired its present form about 5 billion years ago, it is believed that nearly all nuclides were formed (by the fusion process, Sections 31–3 and 33–2). Many isotopes with short half-lives decayed quickly and no longer exist in nature today. But long-lived isotopes, such as ${}^{238}_{92}\text{U}$ with a half-life of 4.5×10^9 yr, still do exist in nature today. Indeed, about half of the original ${}^{238}_{92}\text{U}$ still remains (assuming that the origin of the solar system was about 5×10^9 yr ago). We might expect, however, that radium (${}^{226}_{88}\text{Ra}$), with a half-life of 1600 yr, would long since have disappeared from the earth. Indeed, the original ${}^{226}_{88}\text{Ra}$ nuclei must by now have all decayed. However, because ${}^{238}_{92}\text{U}$ decays (in several steps) to ${}^{226}_{88}\text{Ra}$, the supply of ${}^{226}_{88}\text{Ra}$ is continually replenished, which is why it is still found on earth today. The same can be said for many other radioactive nuclides.

30–11 • Radioactive Dating

Radioactive decay has many interesting applications. One is the technique of *radioactive dating* by which the age of ancient materials can be determined.

The age of any object made from once-living matter, such as wood, can be determined using the natural radioactivity of ${}^{14}_{6}C$. All living plants absorb carbon dioxide (CO_2) from the air and use it to synthesize organic molecules. The vast majority of these carbon atoms are ${}^{12}_{6}C$, but a small fraction, about 1.3×10^{-12}, is the radioactive isotope ${}^{14}_{6}C$. The ratio of ${}^{14}_{6}C$ to ${}^{12}_{6}C$ in the atmosphere has remained roughly constant over many thousands of years, in spite of the fact that ${}^{14}_{6}C$ decays with a half-life of about 5730 yr. This is because neutrons in the cosmic radiation that impinges on the earth from outer space collide with atoms of the atmosphere. In particular, collisions with nitrogen nuclei produce the following nuclear transformation: $n + {}^{14}_{7}N \rightarrow {}^{14}_{6}C + p$. That is, a neutron strikes and is absorbed by a ${}^{14}_{7}N$ nucleus, and a proton is knocked out in the process. The remaining nucleus is ${}^{14}_{6}C$. This continual production of ${}^{14}_{6}C$ in the atmosphere roughly balances the loss of ${}^{14}_{6}C$ by radioactive decay. As long as a plant or tree is alive, it continually uses the carbon from carbon dioxide in the air to build new tissue and to replace old. Animals eat plants, so they too are continually receiving a fresh supply of carbon for their tissues. Organisms cannot distinguish† ${}^{14}_{6}C$ from ${}^{12}_{6}C$, and since the ratio of ${}^{14}_{6}C$ to ${}^{12}_{6}C$ in the atmosphere remains nearly constant, the ratio of the two isotopes within the living organism remains nearly constant as well. But when an organism dies, carbon dioxide is no longer absorbed and utilized. Because the ${}^{14}_{6}C$ decays radioactively, the ratio of ${}^{14}_{6}C$ to ${}^{12}_{6}C$ in a dead organism decreases in time. Since the half-life of ${}^{14}_{6}C$ is about 5730 yr, the ${}^{14}_{6}C/{}^{12}_{6}C$ ratio decreases by half every 5730 yr. If, for example, the ${}^{14}_{6}C/{}^{12}_{6}C$ ratio of an ancient wooden tool is half of what it is in living trees, then the object must have been made from a tree that was felled about 5730 yr ago. Actually, corrections must be made for the fact that the ${}^{14}_{6}C/{}^{12}_{6}C$ ratio in the atmosphere has not remained precisely constant over time. The determination of what this ratio has been over the centuries has required using techniques such as comparing the expected ratio to the actual ratio for objects whose age is known, such as very old trees whose annual rings can be counted.

Carbon dating is useful only for determining the age of objects less than about 60,000 yr old. The amount of ${}^{14}_{6}C$ remaining in older objects is usually too small to measure accurately, although new techniques are allowing detection of even smaller amounts of ${}^{14}_{6}C$, pushing the time frame further back. On the other hand, radioactive isotopes with longer half-lives can be used in certain circumstances to obtain the age of older objects. For example, the decay of ${}^{238}_{92}U$, because of its long half-life of 4.5×10^{9} years, is useful in determining the ages of rocks on a geologic time scale. When molten material solidifies into rock, the uranium present in the material becomes fixed in its position and the daughter nuclei that result from the decay of uranium will

† Organisms operate almost exclusively via chemical reactions—which involve only the outer orbital electrons of the atom; extra neutrons in the nucleus have almost no effect.

also be fixed in that position. Thus, by measuring the amount of $^{238}_{92}U$ remaining in the material relative to the amount of daughter nuclei, the time when the rock solidified can be determined.

Radioactive dating methods using $^{238}_{92}U$ and other isotopes have shown the age of the oldest earth rocks to be about 4×10^9 yr. The age of rocks in which the oldest fossilized organisms are embedded indicates that life appeared at least 3 billion years ago. The earliest fossilized remains of mammals are found in rocks 200 million years old, and the first humanlike creatures seem to have appeared about 2 million years ago. Radioactive dating has been indispensable for the reconstruction of earth's history and the evolution of its biological organisms.

*30–12 • Stability and Tunneling

We have seen that radioactive decay occurs only when the mass of the parent nucleus is greater than the sum of the masses of the daughter nucleus and all particles emitted. For example, $^{238}_{92}U$ can decay to $^{234}_{90}Th$ because the mass of $^{238}_{92}U$ is greater than the mass of the $^{234}_{90}Th$ plus the mass of the α particle. Since systems tend to go in the direction that reduces their internal or potential energy (a ball rolls downhill, a positive charge moves toward a negative charge), you may wonder why an unstable nucleus doesn't fall apart immediately. In other words, why do $^{238}_{92}U$ nuclei ($T_{\frac{1}{2}} = 4.5 \times 10^9$ yr) and other isotopes have such long half-lives? Why don't parent nuclei all decay at once?

The answer has to do with quantum theory and the nature of the forces involved. One way to view the situation is with the aid of a potential-energy diagram, as in Fig. 30–10. Let us consider the particular case of the decay $^{238}_{92}U \rightarrow {}^{234}_{90}Th + {}^{4}_{2}He$. The solid line represents the potential energy, including rest mass, where we imagine the α particle as a separate entity within the $^{238}_{92}U$ nucleus. The region labeled A in Fig. 30–10 represents the PE of the α particle when it is held within the uranium nucleus by the nuclear force (R_0 is the nuclear radius). Region C represents the PE when the α particle is free of the nucleus. The downward-curving PE (proportional to $1/r$) represents the electrical (Coulomb's law) repulsion between the positively charged α and the $^{234}_{90}Th$ nucleus. In order to get to region C, the α particle has to get by the barrier shown. Since the PE just beyond $r = R_0$ (region B) is greater

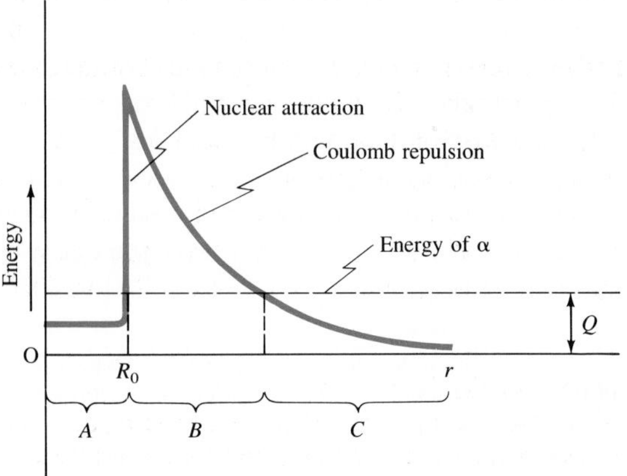

FIGURE 30–10 Potential energy for alpha particle and nucleus, showing the Coulomb barrier through which the α particle must tunnel to escape. The Q-value of the reaction is also shown.

than the energy of the alpha particle (dashed line), the α particle could not escape the nucleus if it were governed by classical physics. It could escape only if there were an input of energy equal to the height of the barrier†. Nuclei decay spontaneously, however, without any input of energy. How, then, does the α particle get from point A to point C? It actually passes through the barrier in a process known as **tunneling**. Classically, this could not happen, because an α particle in region B (within the barrier) would be violating the conservation-of-energy principle‡. The uncertainty principle, however, tells us that energy conservation can be violated by an amount ΔE for a length of time Δt given by

Tunneling

$$(\Delta E)(\Delta t) \approx \frac{h}{2\pi}.$$

We saw in Section 28–3 that this is a result of the wave–particle duality. Thus quantum mechanics allows conservation of energy to be violated for brief periods that may be long enough for an α particle to "tunnel" through the barrier. The higher and wider the barrier, the less time the α particle has to escape and the less likely it is to do so. It is therefore the height and width of this barrier that controls the rate of decay and half-life of an isotope.

*30–13 • Detection of Radiation

Individual particles such as electrons, protons, α particles, neutrons, and γ rays are not detected directly by our senses. Consequently, a variety of instruments have been developed to detect them.

One of the most common is the **Geiger counter**. As shown in Fig. 30–11, it consists of a cylindrical metal tube filled with a certain type of gas. A long wire runs down the center and is kept at a high positive voltage ($\sim 10^3$ V) with respect to the outer cylinder. The voltage is just slightly less than that required to ionize the gas atoms. When a charged particle enters through the thin "window" at one end of the tube, it ionizes a few atoms of the gas. The freed electrons are attracted toward the positive wire and as they are accelerated they strike and ionize additional atoms. An "avalanche" of electrons is quickly produced, and when it reaches the wire anode, it produces a voltage pulse. The pulse, after being amplified, can be sent to an electronic counter, which counts how many particles have been detected. Or the pulses can be sent to a loudspeaker and each detection of a particle is heard as a "click."

FIGURE 30–11 Diagram of a Geiger counter.

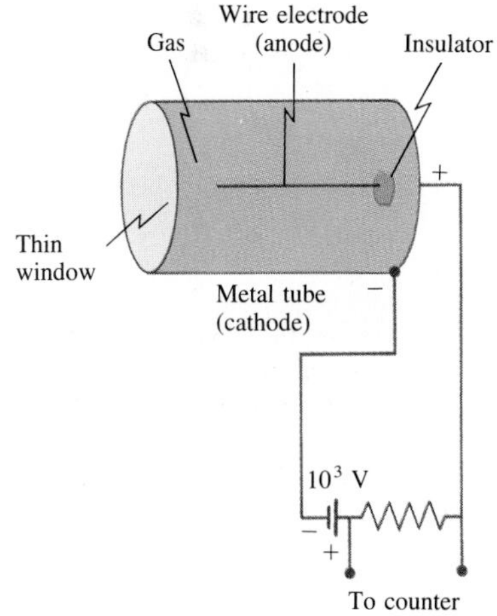

A **scintillation counter** makes use of a solid, liquid, or gas known as a **scintillator** or **phosphor**. The atoms of a scintillator are easily excited when struck by an incoming particle and emit visible light when they return to their ground states. Typical scintillators are crystals of NaI and certain plastics. One face of a solid scintillator is cemented to a photomultiplier tube, and the whole is wrapped with opaque material to keep it light-tight or is placed within a light-tight container. The **photomultiplier** (PM) **tube** converts

† Compare Section 29-2 and Fig. 29–10.

‡ Since KE $= \frac{1}{2}mv^2 \geq 0$, then $E =$ KE + PE $\geq$ PE. But in Fig. 30-10, the total E (dashed line) is less than the PE in region B, so a particle reaching region B would violate conservation of energy.

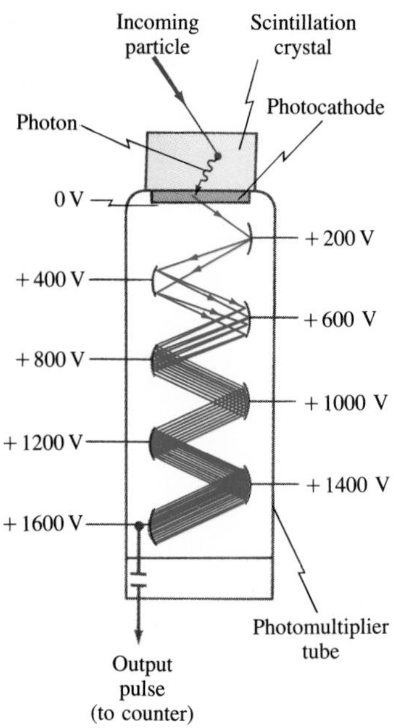

FIGURE 30–12 Scintillation counter with a photomultiplier tube.

the energy of the scintillator-emitted photon(s) into an electric signal. A PM tube is a vacuum tube containing several electrodes (typically 8 to 14), called *dynodes*, which are maintained at successively higher voltages as shown in Fig. 30–12. At its top surface is a photoelectric surface, called the *photocathode*, whose work function (Section 27–2) is low enough that electrons are easily released when struck by a photon from the scintillator. Such an electron is accelerated toward the first dynode. When it strikes the first dynode, the electron has acquired sufficient kinetic energy so that it can eject two to five more electrons. These, in turn, are accelerated to the second dynode, and a multiplication process begins. The number of electrons striking the last dynode may be 10^6 or more. Thus the passage of a particle through the scintillator results in an electric signal at the output of the PM tube that can be sent to an electronic counter just as for a Geiger tube. Because a scintillator crystal is much more dense than the gas of a Geiger counter, it is a more efficient detector—especially for γ rays, which interact less with matter than do β rays.

In tracer work and other biological experiments (Section 31–7), liquid scintillators are often used. Radioactive samples taken at different times or from different parts of an organism are placed directly in small bottles containing the liquid scintillator. This is particularly convenient for detection of β rays from ${}^{3}_{1}\mathrm{H}$ and ${}^{14}_{6}\mathrm{C}$, which have very low energies and have difficulty passing through the outer covering of a crystal scintillator or Geiger tube. A PM tube is still used to produce the electric signal.

A **semiconductor detector** consists of a reverse-biased *pn* junction diode. A particle passing through the junction can excite electrons into the conduction band, leaving holes in the valence band. The freed charges produce a short electrical pulse that can be counted just as for Geiger and scintillation counters.

The three devices discussed so far are used for counting the number of particles (or decays of a radioactive isotope). Other devices have been developed that allow the track of a charged particle to be seen. The simplest is the **photographic emulsion** (which, being small and simple and therefore portable, is now used particularly for cosmic-ray studies from balloons). A particle passing through a layer of photographic emulsion ionizes the atoms along its path. This results in a chemical change at these points, and when the emulsion is developed, the particle's path is revealed.

In a **cloud chamber**, a gas is cooled to a temperature slightly below its usual condensation point. (It is said to be "supercooled.") The gas molecules begin to condense on any ionized molecules present. Thus the ions produced when a charged particle passes through serve as centers on which tiny droplets form (Fig. 30–13). Light scatters more from these droplets than

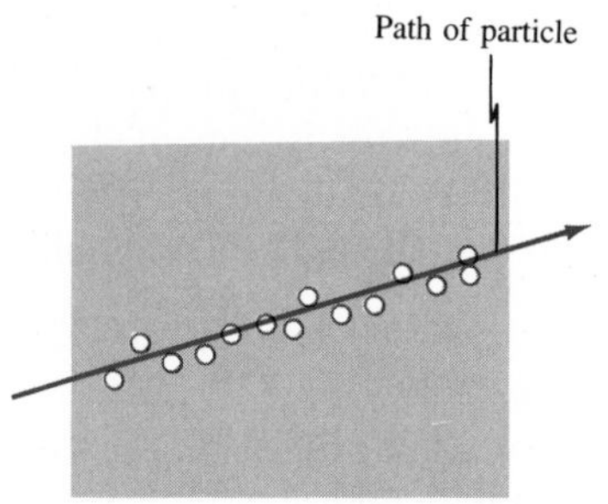

FIGURE 30–13 In a cloud or bubble chamber, droplets or bubbles are formed around ions produced by the passage of a charged particle.

(a)

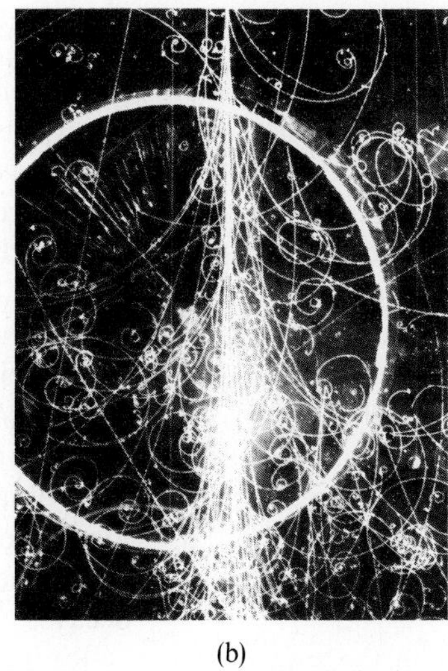

(b)

FIGURE 30–14 (a) Bubble chamber; (b) particle tracks in a bubble chamber.

from the gas background, so a photo of the cloud chamber at the right moment shows the track of the particle. An important instrument in the early days of nuclear physics, it is little used today.

The **bubble chamber**, invented in 1952 by D. A. Glaser (1926–), makes use of a superheated liquid. The liquid is kept close to its normal boiling point and the bubbles characteristic of boiling form around ions produced by the passage of a charged particle (Fig. 30–14). A photograph of the interior of the chamber thus reveals the path of particles that recently passed through. Because the bubble chamber uses a liquid—often liquid hydrogen—the density of atoms is much greater than in a cloud chamber. Hence it is a much more efficient device for observing the tracks of charged particles and their interactions with the nuclei of the liquid. Usually, a magnetic field is applied across the chamber and the momentum of the moving particles can be determined from the radius of curvature of their paths.

A **spark chamber** or **wire chamber** consists of a set of closely spaced parallel plates or fine wires. Alternate plates or planes of wires are grounded and the ones in between are kept at very high voltage. When a charged particle passes through, the ions produced in the gas between the plates or wires become an avalanche, and a large current results which produces a visible spark. Thus the path of the particle is made visible and can be photographed. Alternatively, in the case of a wire chamber, the sparks produce electric pulses whose positions can be determined by the time it takes them to reach detectors placed at the ends of the wires, and the particle's path is reconstructed electronically. In a high-voltage **streamer chamber**, there are only two plates, widely spaced, and the electric discharge closely follows the particle's path, which is seen as a glow or streamer. Spark and streamer chambers have the advantage that they recycle more quickly than bubble chambers, so more particles or events can be observed in a given time, although they can handle fewer tracks per photograph.

SUMMARY

Nuclear physics is the study of atomic nuclei. Nuclei contain *protons* and *neutrons*, which are collectively known as *nucleons*. The total number of nucleons, A, is the *atomic mass number*. The number of protons, Z, is the *atomic number*. The number of neutrons equals $A - Z$. *Isotopes* are nuclei with the same Z, but with different numbers of neutrons. For an element X, an isotope of given Z and A is represented by ${}^{A}_{Z}\mathrm{X}$.

The nuclear radius is proportional to $A^{1/3}$, indicating that all nuclei have about the same density. Nuclear masses are specified in *atomic mass units* (u), where the mass of ${}^{12}_{6}\mathrm{C}$ is defined as exactly 12.000000 u, or in terms of their energy equivalent (because $E = mc^2$), where $1\ \mathrm{u} = 931.5\ \mathrm{MeV}/c^2$.

The mass of a nucleus is less than the sum of the masses of its constituent nucleons. The difference in mass (times c^2) is the *total binding energy*. It represents the energy needed to break the nucleus into its constituent nucleons. The *binding energy per nucleon* averages about 8 MeV per nucleon, and is lowest for very light and very heavy nuclei.

Unstable nuclei undergo *radioactive decay*; they change into other nuclei with the emission of an α, β, or γ particle. An α particle is a ${}^{4}_{2}\mathrm{He}$ nucleus; a β particle is an electron or positron; and a γ ray is a high-energy photon. In β decay, a *neutrino* is also emitted. The transformation of the parent into the daughter nucleus is called *transmutation* of the elements. Radioactive decay occurs spontaneously only when the rest mass of the products is less than the mass of the parent nucleus. The loss in mass appears as kinetic energy of the products.

Nuclei are held together by the *strong nuclear force*. The *weak nuclear force* makes itself apparent in β-decay. These two forces, plus the gravitational and electromagnetic forces, are the four known types of force. Electric charge, linear and angular momentum, mass–energy, and *nucleon number* are *conserved* in all decays.

Radioactive decay is a statistical process. The number that decay (ΔN) in a time Δt is proportional to the number N of parent nuclei present: $\Delta N = -\lambda N\ \Delta t$. The proportionality constant, λ, is called the *decay constant* and is characteristic of the given nucleus. The number N of nuclei remaining after a time t decreases exponentially ($N \propto e^{-\lambda t}$), as does the *activity*, $\Delta N/\Delta t$. The *half-life*, $T_{\frac{1}{2}}$, is the time required for half the nuclei of a radioactive sample to decay. It is related to the decay constant by $T_{\frac{1}{2}} = 0.693/\lambda$.

QUESTIONS

1. What do different isotopes of a given element have in common? How are they different?

2. What are the elements represented by the X in the following: (*a*) ${}^{232}_{92}X$; (*b*) ${}^{18}_{7}X$; (*c*) ${}^{1}_{1}X$; (*d*) ${}^{82}_{38}X$; (*e*) ${}^{247}_{97}X$?

3. How many protons and how many neutrons do each of the isotopes in Question 2 have?

4. Why are the atomic masses of many elements (see the periodic table) not close to whole numbers?

5. How do we know there is such a thing as the strong nuclear force?

6. What are the similarities and the differences between the strong nuclear force and the electric force?

7. What is the experimental evidence in favor of radioactivity being a nuclear process?

8. The isotope ${}^{64}_{29}\mathrm{Cu}$ is unusual in that it can decay by γ, β^-, and β^+ emission. What is the resulting nuclide for each case?

9. A ${}^{238}_{92}\mathrm{U}$ nucleus decays to a nucleus containing how many neutrons?

10. Describe, in as many ways as possible, the differences between α, β, and γ rays.

11. What element is formed by the radioactive decay of (*a*) ${}^{24}_{11}\mathrm{Na}(\beta^-)$; (*b*) ${}^{22}_{11}\mathrm{Na}(\beta^+)$; (*c*) ${}^{210}_{84}\mathrm{Po}(\alpha)$?

12. What element is formed by the decay of (*a*) ${}^{32}_{15}\mathrm{P}(\beta^-)$; (*b*) ${}^{35}_{16}\mathrm{S}(\beta^-)$; (*c*) ${}^{211}_{83}\mathrm{Bi}(\alpha)$?

13. Fill in the missing particle or nucleus:
(*a*) ${}^{45}_{20}\mathrm{Ca} \rightarrow ? + e^- + \bar{\nu}$
(*b*) ${}^{58}_{29}\mathrm{Cu} \rightarrow ? + \gamma$
(*c*) ${}^{46}_{24}\mathrm{Cr} \rightarrow {}^{46}_{23}\mathrm{V} + ?$
(*d*) ${}^{234}_{94}\mathrm{Pu} \rightarrow ? + \alpha$
(*e*) ${}^{239}_{93}\mathrm{Np} \rightarrow {}^{239}_{92}\mathrm{U} + ?$

14. Immediately after a ${}^{238}_{92}\mathrm{U}$ nucleus decays to ${}^{234}_{90}\mathrm{Th} + {}^{4}_{2}\mathrm{He}$, the daughter thorium nucleus still has 92 electrons circling it. Since thorium normally holds only 90 electrons, what do you suppose happens to the two extra ones?

15. Do isotopes that undergo electron capture generally lie above or below the line of stability in Fig. 30–2?

16. Can hydrogen or deuterium emit an α particle?

17. Why are many artificially produced radioactive isotopes rare in nature?

18. An isotope has a half-life of one month. After two months, will a given sample of this isotope have completely decayed? If not, how much remains?

***19.** Explain the absence of β^+ emitters in the radioactive decay series of Fig. 30–9.

*20. Describe how the potential energy curve for an α particle in an α-emitting nucleus differs from that for a stable nucleus.

*21. Can $^{14}_{6}C$ dating be used to measure the age of stone walls and tablets of ancient civilizations?

PROBLEMS

SECTION 30–1

1. (I) What is the rest energy of an α particle in MeV/c^2?
2. (I) A pi meson has a mass of 139 MeV/c^2. What is this in atomic mass units?
3. (II) (*a*) What is the approximate radius of a $^{64}_{29}Cu$ nucleus? (*b*) Approximately what is the value of A for a nucleus whose radius is 3.6×10^{-15} m?
4. (II) How much energy must an α particle have to just "touch" the surface of a $^{238}_{92}U$ nucleus?
5. (II) What stable nucleus has approximately half the radius of a uranium nucleus? [*Hint:* find A and use Appendix D to get Z.]

SECTION 30–2

6. (I) Estimate the total binding energy for $^{40}_{20}Ca$, using Fig. 30–1.
7. (I) Use Fig. 30–1 to estimate the total binding energy of (*a*) $^{238}_{92}U$, and (*b*) $^{107}_{47}Ag$.
8. (II) Use Appendix D and calculate the binding energy of $^{2}_{1}H$.
9. (II) Calculate the binding energy per nucleon for a $^{12}_{6}C$ nucleus.
10. (II) Calculate the total binding energy and the binding energy per nucleon for $^{6}_{3}Li$. Use Appendix D.
11. (II) Calculate the binding energy of the last neutron in a $^{12}_{6}C$ nucleus. [*Hint:* compare the mass of $^{12}_{6}C$ with that of $^{11}_{6}C + ^{1}_{0}n$; use Appendix D.]
12. (II) (*a*) Show that the nucleus $^{8}_{4}Be$ (mass = 8.005308 u) is unstable to decay into two α particles. (*b*) Is $^{12}_{6}C$ stable against decay into three α particles? Show why or why not.

SECTIONS 30–3 TO 30–7

13. (II) $^{60}_{27}Co$ in an excited state emits a 1.33-MeV γ ray as it jumps to the ground state. What is the mass of the excited cobalt atom?
14. (II) Show that the decay $^{11}_{6}C \rightarrow ^{10}_{5}B + p$ is not possible because energy would not be conserved.
15. (II) Give the result of a calculation that shows whether or not the following decays are possible: (*a*) $^{236}_{92}U \rightarrow ^{235}_{92}U + n$; (*b*) $^{16}_{8}O \rightarrow ^{15}_{8}O + n$; (*c*) $^{23}_{11}Na \rightarrow ^{22}_{11}Na + n$.
16. (II) A $^{232}_{92}U$ nucleus emits an α particle with KE = 5.32 MeV. What is the final nucleus and what is the approximate atomic mass (in u) of the final atom?
17. (II) When $^{23}_{10}Ne$ (mass = 22.9945 u) decays to $^{23}_{11}Na$ (mass = 22.9898 u), what is the maximum kinetic energy of the emitted electron? What is its minimum energy? What is the energy of the neutrino in each case?
18. (II) The nuclide $^{32}_{15}P$ decays by emitting an electron whose maximum kinetic energy can be 1.71 MeV. (*a*) What is the daughter nucleus? (*b*) What is its atomic mass (in u)?
19. (II) The isotope $^{218}_{84}Po$ can decay by either α or β^- emission. What is the energy release in each case? The mass of $^{218}_{84}Po$ is 218.008965 u.
20. (II) How much energy is released in electron capture by beryllium: $^{7}_{4}Be + ^{0}_{-1}e \rightarrow ^{7}_{3}Li + \nu$?
21. (II) What is the energy of the α particle emitted in the decay $^{210}_{84}Po \rightarrow ^{206}_{82}Pb + \alpha$?
22. (III) The α particle emitted when $^{238}_{92}U$ decays has KE = 4.20 MeV. Calculate the recoil KE of the daughter nucleus and the Q-value of the decay.
23. (III) Show that when a nucleus decays by β^+ decay, the total energy released is equal to $(M_P - M_D - 2m_e)c^2$, where M_P and M_D are the masses of the parent and daughter atoms (neutral), and m_e is the mass of an electron or positron.
24. (III) Use the result of Problem 23 to determine the maximum kinetic energy of β^+ particles released when $^{11}_{6}C$ decays to $^{11}_{5}B$. What is the maximum energy the neutrino can have? What is its minimum energy?
25. (III) In α decay of, say, a $^{226}_{88}Ra$ nucleus, show that the nucleus carries away a fraction $1/(1 + A_D/4)$ of the total energy available, where A_D is the mass number of the daughter nucleus. [*Hint:* use conservation of momentum as well as conservation of energy.] Approximately what percentage of the energy available is thus carried off by the α particle in the case cited?

SECTIONS 30–8 TO 30–12

26. (I) A radioactive material registers 1280 counts per minute on a Geiger counter at one time, and 6 h later registers 320 counts per minute. What is its half-life?
27. (I) What is the decay constant of $^{238}_{92}U$ whose half-life is 4.468×10^9 yr?
28. (I) The decay constant of a given nucleus is 8.5×10^{-6} s^{-1}. What is its half-life?
29. (I) What is the activity of a sample of $^{14}_{6}C$ that contains 5.6×10^{20} nuclei?
30. (I) What fraction of a sample of $^{68}_{32}Ge$, whose half-life is about 9 months, will remain after 4.5 yr?
31. (I) How many nuclei of $^{238}_{92}U$ remain in a rock if the activity registers 1.8×10^4 decays per second?
32. (II) In a series of decays, the nuclide $^{235}_{92}U$ becomes $^{207}_{82}Pb$. How many α and β^- particles are emitted in this series?

33. (II) $^{124}_{55}Cs$ has a half-life of 30.8 s. (*a*) If we have 9.5 μg initially, how many nuclei are present? (*b*) How many are present 2.0 min later? (*c*) What is the activity at this time? (*d*) After how much time will the activity drop to less than about 1 per second?

34. (II) Calculate the activity of a pure 3.5-μg sample of $^{32}_{15}P$ ($T_{\frac{1}{2}} = 1.23 \times 10^6$ s).

35. (II) The activity of a sample of $^{35}_{16}S$ ($T_{\frac{1}{2}} = 7.56 \times 10^6$ s) is 6.80×10^6 decays per second. What is the mass of sample present?

36. (II) A sample of $^{233}_{92}U$ ($T_{\frac{1}{2}} = 1.59 \times 10^5$ yr) contains 7.00×10^{18} nuclei. (*a*) What is the decay constant? (*b*) Approximately how many disintegrations will occur

37. (II) The activity of a sample drops by a factor of 10 in 88 minutes. What is its half-life?

38. (II) A 60-g sample of pure carbon contains 1.3 parts in 10^{12} (atoms) of $^{14}_{6}C$. How many disintegrations occur per second?

39. (II) A radioactive nuclide registers 2880 counts per minute on a Geiger counter at one time, and 1.6 h later registers 820 counts per minute. What is the half-life of the nuclide?

40. (II) The rubidium isotope $^{87}_{37}Rb$, a β emitter with a half-life of 4.75×10^{10} yr, is used to determine the age of rocks and fossils. Rocks containing fossils of early animals contain a ratio of $^{87}_{38}Sr$ to $^{87}_{37}Rb$ of 0.0160. Assuming that there was no $^{87}_{38}Sr$ present when the rocks were formed, calculate the age of these fossils. [*Hint:* use Eq. 30–3.]

41. (II) Use Fig. 30–9 and calculate the relative decay rates for α decay of $^{218}_{84}Po$ and $^{214}_{84}Po$.

42. (III) An ancient club is found that contains 170 g of carbon and has an activity of 5.0 decays per second. Determine its age assuming that in living trees the ratio of $^{14}C/^{12}C$ atoms is about 1.3×10^{-12}.

43. (III) At $t = 0$, a pure sample of radioactive nuclei contains N_0 nuclei whose decay constant is λ. Determine a formula for the number of daughter nuclei, N_D, as a function of time; assume that $N_D = 0$ at $t = 0$.

GENERAL PROBLEMS

44. Show that the radius of the largest nuclei (say, $^{238}_{92}U$) is only about 6 times greater than that of the smallest ($^{1}_{1}H$).

45. (*a*) Determine the density of nuclear matter in kg/m^3. (*b*) What would be the radius of the earth if it had its actual mass but had the density of nuclei? (*c*) What would be the radius of a $^{238}_{92}U$ nucleus if it had the density of the earth?

46. The $^{3}_{1}H$ isotope of hydrogen, which is called *tritium* (because it contains three nucleons), has a half-life of 12.33 yr. It can be used to measure the age of objects up to about 100 yr. It is produced in the upper atmosphere by cosmic rays and brought to earth by rain. As an application, determine approximately the age of a bottle of wine whose $^{3}_{1}H$ radiation is about $\frac{1}{10}$ that present in new wine.

47. A neutron star consists of neutrons at approximately nuclear density. Estimate, for a 10-km-diameter neutron star, (*a*) its mass number, (*b*) its mass (kg), and (*c*) the acceleration of gravity at its surface.

48. Recent elementary particle theories (Section 32–11) suggest that the proton may be unstable, with a half-life $\gtrsim 10^{32}$ yr. How long would you expect to wait for one proton in your body to decay (consider that your body is all water)?

49. Decay series, such as that shown in Fig. 30–9, can be classified into four families, depending on whether the mass numbers have the form $4n$, $4n + 1$, $4n + 2$, $4n + 3$, where n is an integer. Justify this statement and show that for a nuclide in any family, all its daughters will be in the same family.

50. The nuclide $^{191}_{76}Os$ decays with β^- energy of 0.14 MeV accompanied by γ rays of energy 0.042 MeV and 0.129 MeV. (*a*) What is the daughter nucleus? (*b*) Draw an energy-level diagram showing the ground states of the parent and daughter and excited states of the daughter. To which of the daughter states does β decay of $^{191}_{76}Os$ occur?

51. Use the uncertainty principle to argue why electrons are unlikely to be found in the nucleus. Use relativity.

52. Estimate the total binding energy for copper and then estimate the energy, in joules, needed to break a 3-g copper penny into its constituent nucleons.

53. Instead of giving atomic masses for nuclides as in Appendix D, some tables give the *mass excess*, Δ, defined as $\Delta = M - A$, where A is the atomic number and M is the mass in u. Determine the mass excess, in u and in MeV/c^2, for: (*a*) $^{4}_{2}He$; (*b*) $^{12}_{6}C$; (*c*) $^{107}_{47}Ag$; (*d*) $^{235}_{92}U$. (*e*) From a glance at Appendix D, can you make a generalization about the sign of Δ as a function of Z or A?

CHAPTER 31

Nuclear Energy; Effects and Uses of Radiation

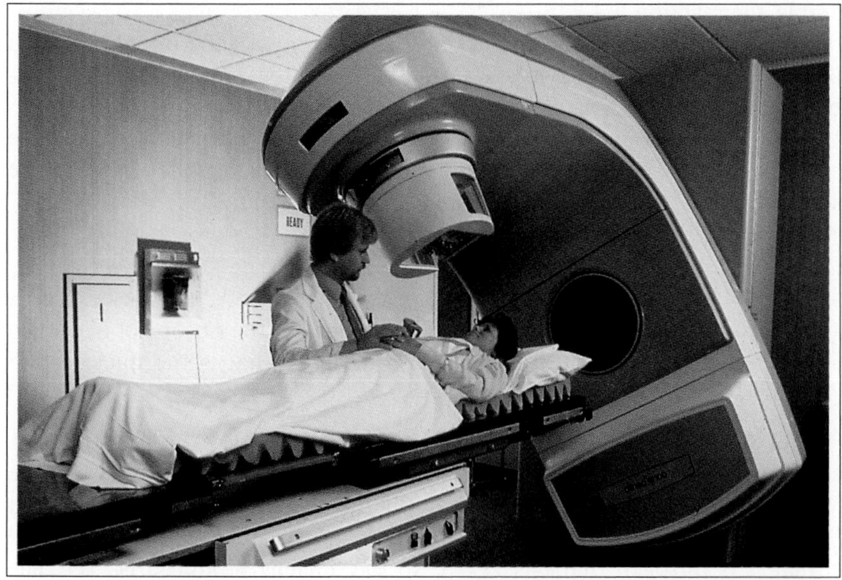

Cancer patient is being treated with radioactive $^{60}_{27}Co$, which is a γ and β emitter. The cobalt source, produced by nuclear reactions with a nuclear reactor, is placed in the well-shielded upper housing. The collimated rays pass through the body. At different rotation positions of the machine, the radiation beam passes through different parts of the body, thus minimizing the radiation dose through non-cancerous tissue, but the beam always passes through the cancerous area which is positioned on the rotation axis.

We continue our study of nuclear physics in this chapter. We begin with a discussion of nuclear reactions, after which we examine the important large energy-releasing processes of fission and fusion. The remainder of the chapter deals with the effects of nuclear radiation when it passes through matter, particularly biological matter, and how radiation is used medically for therapy and diagnosis, including recently developed imaging techniques.

31–1 • Nuclear Reactions and the Transmutation of Elements

When a nucleus undergoes α or β decay, the daughter nucleus is that of a different element from the parent. The transformation of one element into another, called *transmutation*, also occurs by means of nuclear reactions. A **nuclear reaction** is said to occur when a given nucleus is struck by another nucleus, or by a simpler particle such as a γ ray or neutron, so that an interaction takes place. Ernest Rutherford was the first to report seeing a nuclear reaction. In 1919 he observed that some of the α particles passing through nitrogen gas were absorbed and protons emitted. He concluded that nitrogen nuclei had been transformed into oxygen nuclei via the reaction

$$^{4}_{2}\text{He} + ^{14}_{7}\text{N} \longrightarrow ^{17}_{8}\text{O} + ^{1}_{1}\text{H},$$

where $^{4}_{2}\text{He}$ is an α particle, and $^{1}_{1}\text{H}$ is a proton.

Since then, a great many nuclear reactions have been observed. Indeed, many of the radioactive isotopes used in the laboratory are made by means of nuclear reactions. Nuclear reactions can be made to occur in the laboratory, but they also occur regularly in nature. In Chapter 30 we saw an example of this: $^{14}_{6}\text{C}$ is continually being made in the atmosphere via the reaction $n + ^{14}_{7}\text{N} \rightarrow ^{14}_{6}\text{C} + p$.

Nuclear reactions are sometimes written in a shortened form: for example, the reaction $n + ^{14}_{7}\text{N} \rightarrow ^{14}_{6}\text{C} + p$ is written $^{14}_{7}\text{N}(n, p)^{14}_{6}\text{C}$. The symbols outside the parentheses on the left and right represent the initial and final nuclei, respectively. The symbols inside the parentheses represent the bombarding particle (first) and the emitted small particle (second).

In any nuclear reaction, both electric charge and nucleon number are conserved. These conservation laws are often useful, as the following example shows.

EXAMPLE 31–1 A neutron is observed to strike an $^{16}_{8}\text{O}$ nucleus and a deuteron is given off. (A **deuteron**, or **deuterium**, is the isotope of hydrogen containing one proton and one neutron, $^{2}_{1}\text{H}$.) What is the nucleus that results?

SOLUTION We have the reaction $n + ^{16}_{8}\text{O} \rightarrow ? + ^{2}_{1}\text{H}$. The total number of nucleons initially is $16 + 1 = 17$, and the total charge is $8 + 0 = 8$. The same totals apply to the right side of the reaction. Hence the product nucleus must have $Z = 7$ and $A = 15$. From the periodic table, we find that it is nitrogen that has $Z = 7$, so the nucleus produced is $^{15}_{7}\text{N}$. The reaction can be written $^{16}_{8}\text{O}(n, d)^{15}_{7}\text{N}$, where d represents deuterium, $^{2}_{1}\text{H}$.

Energy (as well as momentum) is conserved in nuclear reactions, and we can use this to determine whether a given reaction can occur or not. For example, if the total mass of the products is less than the total mass of the initial particles, then energy will be released by the reaction—it will appear as kinetic energy of the outgoing particles. But if the total mass of the products is greater than the total mass of the initial reactants, the reaction requires energy. The reaction will then not occur unless the bombarding particle has sufficient kinetic energy. Consider a nuclear reaction of the general form

$$a + X \longrightarrow Y + b,$$

where a is a projectile particle (or small nucleus) that strikes nucleus X, producing nucleus Y and particle b (typically, p, n, α, γ). We define the **reaction energy**, or ***Q*-value**, in terms of the masses involved, as

Q-value

$$Q = (M_a + M_X - M_b - M_Y)c^2. \tag{31–1a}$$

Since energy is conserved, Q is equal to the change in kinetic energy (final minus initial):

$$Q = \text{KE}_b + \text{KE}_Y - \text{KE}_a - \text{KE}_X. \tag{31–1b}$$

Normally, $\text{KE}_X = 0$ since X is the target nucleus at rest (or nearly so) struck by an incoming particle a. For $Q > 0$, the reaction is said to be *exothermic* or *exoergic*; energy is released in the reaction, so the total KE is greater after the reaction than before. If Q is negative ($Q < 0$), the reaction is said to be *endothermic* or *endoergic*. In this case the final total KE is less than the initial KE, and an energy input is required to make the reaction happen; the energy input comes from the kinetic energy of the initial colliding particles (a and X).

EXAMPLE 31–2 The nuclear reaction

$$n + {}^{10}_{5}\text{B} \longrightarrow {}^{7}_{3}\text{Li} + {}^{4}_{2}\text{He}$$

is observed to occur even when very slow-moving neutrons ($M_n = 1.0087$ u) strike a boron atom at rest. For a particular reaction in which $\text{KE}_n \approx 0$, the helium ($M_{\text{He}} = 4.0026$ u) is observed to have a speed of 9.30×10^6 m/s. Determine (*a*) the KE of the lithium ($M_{\text{Li}} = 7.0160$ u), and (*b*) the Q-value of the reaction.

SOLUTION (*a*) Since the neutron and boron are both essentially at rest, the total momentum before the reaction is zero, and afterward is also zero. Therefore,

$$M_{\text{Li}}v_{\text{Li}} = M_{\text{He}}v_{\text{He}}.$$

Hence

$$\text{KE}_{\text{Li}} = \frac{1}{2}M_{\text{Li}}v_{\text{Li}}^2 = \frac{1}{2}M_{\text{Li}}\left(\frac{M_{\text{He}}v_{\text{He}}}{M_{\text{Li}}}\right)^2 = \frac{M_{\text{He}}^2v_{\text{He}}^2}{2M_{\text{Li}}}.$$

We put in numbers, changing the mass in u to kg and recalling that $1.60 \times 10^{-13}\ \text{J} = 1$ MeV:

$$\text{KE}_{\text{Li}} = \frac{(4.0026\ \text{u})^2(1.66 \times 10^{-27}\ \text{kg/u})^2(9.30 \times 10^6\ \text{m/s})^2}{2(7.0160\ \text{u})(1.66 \times 10^{-27}\ \text{kg/u})}$$

$$= 1.64 \times 10^{-13}\ \text{J} = 1.02\ \text{MeV}.$$

(*b*) We are given the data $\text{KE}_a = \text{KE}_X = 0$ in Eq. 31–1b, so $Q = \text{KE}_{\text{Li}} + \text{KE}_{\text{He}}$, where

$$\text{KE}_{\text{He}} = \tfrac{1}{2}M_{\text{He}}v_{\text{He}}^2 = \tfrac{1}{2}(4.0026\ \text{u})(1.66 \times 10^{-27}\ \text{kg/u})(9.30 \times 10^6\ \text{m/s})^2$$
$$= 2.84 \times 10^{-13}\ \text{J} = 1.78\ \text{MeV}.$$

Hence, $Q = 1.02\ \text{MeV} + 1.78\ \text{MeV} = 2.80\ \text{MeV}$.

EXAMPLE 31–3 Can the reaction ${}^{13}_{6}\text{C}(p, n){}^{13}_{7}\text{N}$ occur when ${}^{13}_{6}\text{C}$ is bombarded by 2.0-MeV protons?

SOLUTION We look up the masses of the nuclei in Appendix D. The total masses before and after the reaction are:

Before		After	
$m(^{13}_{6}\text{C})$ =	13.003355	$m(^{13}_{7}\text{N})$ =	13.005738
$m(^{1}_{1}\text{H})$ =	1.007825	$m(n)$ =	1.008665
	14.011180		14.014403

(We must use the mass of the $^{1}_{1}$H atom rather than that of the bare proton because the masses of $^{13}_{6}$C and $^{13}_{7}$N include the electrons, and we must include an equal number of electrons on each side of the equation since none are created or destroyed.) The products have an excess mass of 0.003223 u × 931.5 MeV/u = 3.00 MeV. Thus $Q = -3.00$ MeV, and the reaction is endoergic. This reaction requires energy, and the 2.0 MeV protons do not have enough to make it go. Hence this reaction won't occur.

The proton in this last example would have to have somewhat more than 3.00 MeV of KE to make this reaction go; 3.00 MeV would be enough to conserve energy, but a proton of this energy would produce the $^{13}_{7}$N and n with no KE and hence no momentum. Since an incident 3.0-MeV proton has momentum, conservation of momentum would be violated. A more complicated calculation shows that to conserve both energy and momentum, the minimum proton energy required, called the **threshold energy**, is 3.23 MeV in this case (see Problem 9).

The artificial transmutation of elements took a great leap forward in the 1930s when Enrico Fermi realized that neutrons would be the most effective projectiles for causing nuclear reactions and in particular for producing new elements. Because neutrons have no net electric charge, they are not repelled by positively charged nuclei as are protons or alpha particles. Hence the probability of a neutron reaching the nucleus and causing a reaction is much greater than for charged projectiles,[†] particularly at low energies. Between 1934 and 1936, Fermi and his co-workers in Rome produced many previously unknown isotopes by bombarding different elements with neutrons. Fermi realized that if the heaviest known element, uranium, were bombarded with neutrons, it might be possible to produce new elements whose atomic numbers were greater than that of uranium. After several years of hard work, it was suspected that two new elements had been produced, neptunium ($Z = 93$) and plutonium ($Z = 94$). The full confirmation that such "transuranic" elements could be produced came several years later at the University of California, Berkeley. The reactions are shown in Fig. 31–1.

(a) $n + {}^{238}_{92}\text{U} \rightarrow {}^{239}_{92}\text{U}$

Neutron captured by $^{238}_{92}$U

(b) ${}^{239}_{92}\text{U} \rightarrow {}^{239}_{93}\text{Np} + e^- + \bar{\nu}$

$^{239}_{92}$U decays by β decay to neptunium 239.

(c) ${}^{239}_{93}\text{Np} \rightarrow {}^{239}_{94}\text{Pu} + e^- + \bar{\nu}$

$^{239}_{93}$Np itself decays by β decay to produce plutonium 239.

FIGURE 31–1 Neptunium and plutonium are produced in this series of reactions, after bombardment of $^{238}_{92}$U by neutrons.

It was soon shown that what Fermi actually had observed when he bombarded uranium was an even stranger process—one that was destined to play an extraordinary role in the world at large.

31–2 • Nuclear Fission; Nuclear Reactors

In 1938, the German scientists Otto Hahn and Fritz Strassmann made an amazing discovery. Following up on Fermi's work, they found that uranium bombarded by neutrons sometimes produced smaller nuclei which were

[†] That is, positively charged particles. Electrons rarely cause nuclear reactions because they do not interact via the strong nuclear force.

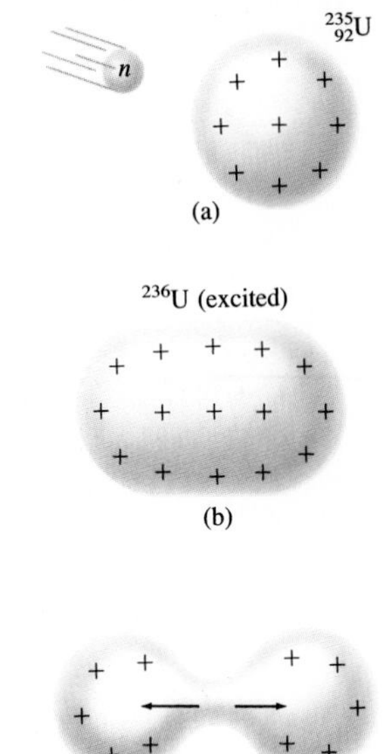

roughly half the size of the original uranium nucleus. Lise Meitner and Otto Frisch, two refugees from Nazi Germany working in Scandinavia, quickly realized what had happened: the uranium nucleus, after absorbing a neutron, actually had split into two roughly equal pieces. This was startling, for until then the known nuclear reactions involved knocking out only a tiny fragment (for example, n, p, or α) from a nucleus.

This new phenomenon was named **nuclear fission** because of its resemblance to biological fission (cell division). It occurs much more readily for $^{235}_{92}\mathrm{U}$ than for the more common $^{238}_{92}\mathrm{U}$. The process can be visualized by imagining the uranium nucleus to be like a liquid drop. According to this **liquid-drop model**, the neutron absorbed by the $^{235}_{92}\mathrm{U}$ nucleus gives the nucleus extra internal energy (like heating a drop of water). This intermediate state, or **compound nucleus**, is $^{236}_{92}\mathrm{U}$ (because of the absorbed neutron). The extra energy of this nucleus—it is in an excited state—appears as increased motion of the individual nucleons, which causes the nucleus to take on abnormal elongated shapes, Fig. 31–2. When the nucleus elongates into the shape shown in Fig. 31–2c, the attraction of the two ends via the short-range nuclear force is greatly weakened by the increased separation distance, and the electric repulsive force becomes dominant. So the nucleus splits in two. The two resulting nuclei, N_1 and N_2, are called **fission fragments**, and in the process a number of neutrons (typically two or three) are also given off. The reaction can be written

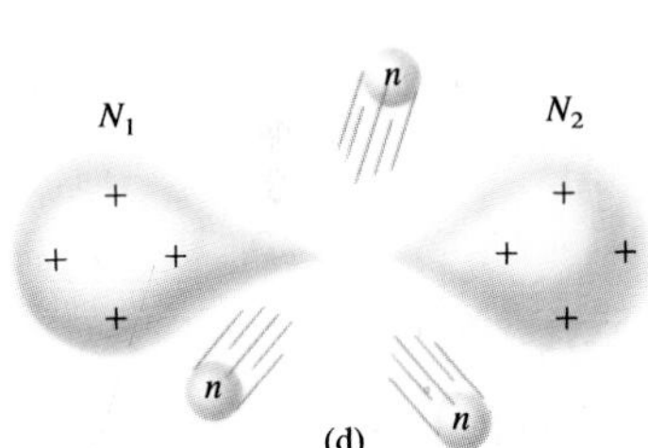

FIGURE 31–2 Fission of a $^{235}_{92}\mathrm{U}$ nucleus after capture of a neutron.

$$n + {}^{235}_{92}\mathrm{U} \longrightarrow {}^{236}_{92}\mathrm{U} \longrightarrow N_1 + N_2 + \text{neutrons.} \qquad (31\text{–}2a)$$

The compound nucleus, $^{236}_{92}\mathrm{U}$, exists for less than 10^{-12} s, so the process occurs very quickly. The two fission fragments have roughly half the mass of the uranium, although rarely are they exactly equal in mass. A typical fission reaction is

A fission reaction

$$n + {}^{235}_{92}\mathrm{U} \longrightarrow {}^{141}_{56}\mathrm{Ba} + {}^{92}_{36}\mathrm{Kr} + 3n, \qquad (31\text{–}2b)$$

although many others also occur.

A tremendous amount of energy is released in a fission reaction because the mass of $^{235}_{92}\mathrm{U}$ is considerably greater than the total mass of the fission fragments plus neutrons. This can be seen from the binding-energy-per-nucleon curve of Fig. 30–1; the binding energy per nucleon for uranium is about 7.6 MeV/nucleon, but for fission fragments that have intermediate mass (in the center portion of the graph, $Z \approx 100$), the average binding energy per nucleon is about 8.5 MeV/nucleon. Since the fission fragments are more tightly bound, they have less mass. The difference in mass (or energy) between the original uranium nucleus and the fission fragments is about $8.5 - 7.6 = 0.9$ MeV per nucleon. Since there are 236 nucleons involved in each fission, the total energy released per fission is

$$(0.9\ \mathrm{MeV/nucleon})(236\ \mathrm{nucleons}) \approx 200\ \mathrm{MeV}.$$

This is an enormous amount of energy on the nuclear scale. At a practical level, the energy from one fission is, of course, tiny. However, a great deal of energy at the macroscopic level would be available if many such fissions could occur at once. A number of physicists, including Fermi, recognized that the neutrons released in each fission (Eqs. 31–2) could be used to create a **chain reaction**. That is, one neutron initially causes one fission of a uranium nucleus; the two or three neutrons released can go on to cause additional

Chain reaction

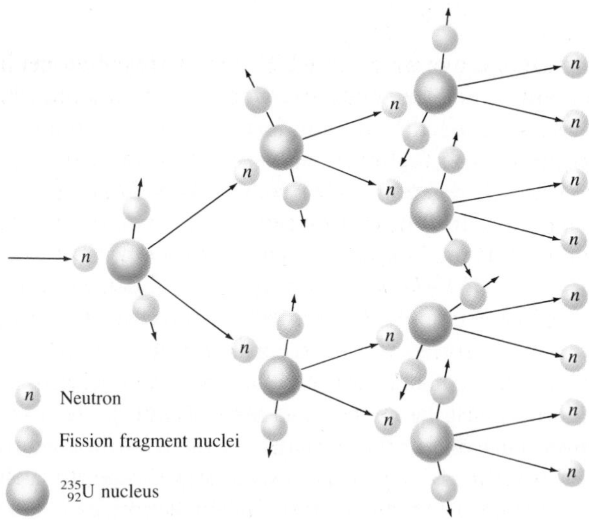

FIGURE 31–3 Chain reaction.

fissions, so the process multiplies as shown schematically in Fig. 31–3. If a **self-sustaining chain reaction** was actually possible in practice, the enormous energy available in fission could be released on a large scale. Fermi and his co-workers (at the University of Chicago) showed it was possible by constructing the first **nuclear reactor** in 1942 (Fig. 31–4).

Moderator

Several problems have to be overcome to make any nuclear reactor function. First, the probability that a $^{235}_{92}\text{U}$ nucleus will absorb a neutron is large only for slow neutrons, but the neutrons emitted during a fission, and which are needed to sustain a chain reaction, are moving very fast. A substance known as a **moderator** must be used to slow down the neutrons. The most effective moderator will consist of atoms whose mass is as close as possible to that of the neutrons. (To see why this is true, recall from Chapter 7 that a billiard ball striking an equal mass at rest can itself be stopped in one collision; but a billiard ball striking a heavy object bounces off with nearly the same speed it had.) The best moderator would thus contain $^{1}_{1}\text{H}$ atoms. Unfortunately, $^{1}_{1}\text{H}$ tends to absorb neutrons. But the isotope of hydrogen called *deuterium*, $^{2}_{1}\text{H}$, does not absorb many neutrons and is thus an almost ideal moderator. Either $^{1}_{1}\text{H}$ or $^{2}_{1}\text{H}$ can be used in the form of water.

FIGURE 31–4 Color painting of the first nuclear reactor, built by Fermi under the grandstand of Stagg Field at the University of Chicago. (There are no photographs of the original reactor because of military secrecy.) Natural uranium was used with graphite as moderator. On December 2, 1942, Fermi slowly withdrew the cadmium control rods and the reactor went critical. This first self-sustaining chain reaction was announced to Washington, by telephone, by Arthur Compton who witnessed the event and reported: "The Italian navigator has just landed in the new world."

In the latter case, it is **heavy water**, in which the hydrogen atoms have been replaced by deuterium. Another common moderator is *graphite*, which consists of ${}^{12}_{6}C$ atoms.

A second problem is that the neutrons produced in one fission may be absorbed and produce other nuclear reactions with other nuclei in the reactor, rather than produce further fissions. In a "light-water" reactor, the ${}^{1}_{1}H$ nuclei absorb neutrons, as does ${}^{238}_{92}U$ to form ${}^{239}_{92}U$ in the reaction $n + {}^{238}_{92}U \rightarrow {}^{239}_{92}U + \gamma$. Naturally occurring uranium† contains 99.3 percent ${}^{238}_{92}U$ and only 0.7 percent fissionable ${}^{235}_{92}U$. To increase the probability of fission of ${}^{235}_{92}U$ nuclei, natural uranium is often **enriched** to increase the percentage of ${}^{235}_{92}U$ using processes such as diffusion or centrifugation. (Enrichment is not usually necessary for reactors using heavy water as moderator since heavy water doesn't absorb neutrons.)

Enriched uranium

The third problem is that some neutrons will escape through the surface of the reactor core before they cause further fissions (Fig. 31–5). Thus the mass of fuel must be sufficiently large for a self-sustaining chain reaction to take place. The minimum mass of uranium needed is called the **critical mass**. The value of the critical mass depends on the moderator, the fuel (${}^{239}_{94}Pu$ may be used instead of ${}^{235}_{92}U$), and how much the fuel is enriched, if at all. Typical values are on the order of a few kilograms (that is, not grams nor thousands of kilograms).

Critical mass

To have a self-sustaining chain reaction, it is clear that on the average at least one neutron produced in each fission must go on to produce another fission. The average number of neutrons per each fission that do go on to produce further fissions is called the **multiplication factor**, f. For a self-sustaining chain reaction, we must have $f \geqslant 1$. If $f < 1$, the reactor is "subcritical". If $f > 1$, it is "supercritical." Reactors are equipped with movable **control rods** (usually of cadmium or boron), whose function is to absorb neutrons and maintain the reactor at just barely "critical," $f = 1$. The release of neutrons and subsequent fissions caused by them occurs so quickly that manipulation of the control rods to maintain $f = 1$ would not be possible if it weren't for the small percentage ($\sim 1\%$) of so-called **delayed neutrons**. They come from the decay of neutron-rich fission fragments (or their daughters) having lifetimes on the order of seconds—sufficient to allow enough reaction time to operate the control rods and maintain $f = 1$.

Control rods

Delayed neutrons

Nuclear reactors have been built for use in research and to produce electric power. Fission produces many neutrons and a "research reactor" is basically an intense source of neutrons. These neutrons can be used as pro-

Types of nuclear reactor

† ${}^{238}_{92}U$ will fission, but only with fast neutrons (${}^{238}_{92}U$ is more stable than ${}^{235}_{92}U$). The probability of absorbing a fast neutron and producing a fission is too low to produce a self-sustaining chain reaction.

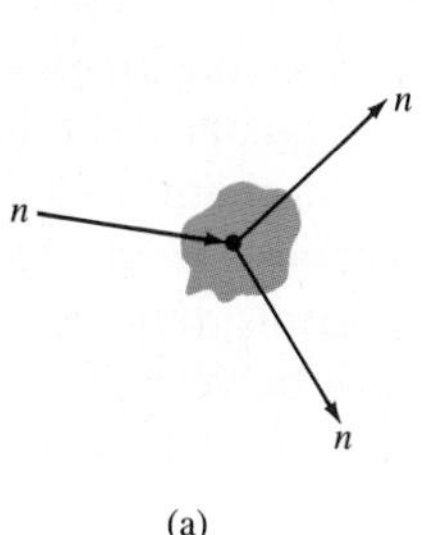

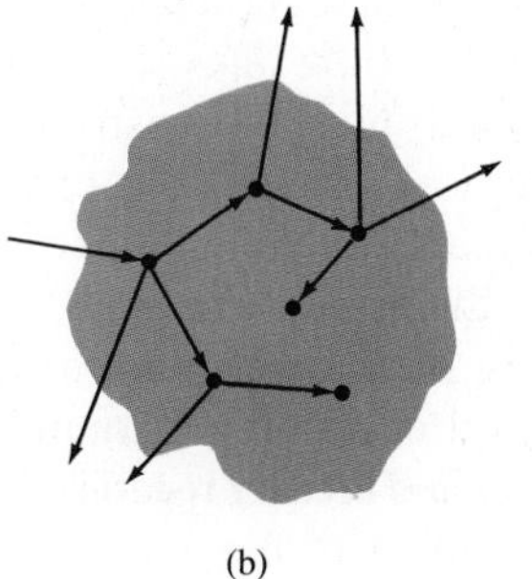

FIGURE 31–5 If the amount of uranium exceeds the critical mass, as in (b), a sustained chain reaction is possible. If the mass is less than critical, as in (a), most neutrons escape before additional fissions occur, and the chain reaction is not sustained.

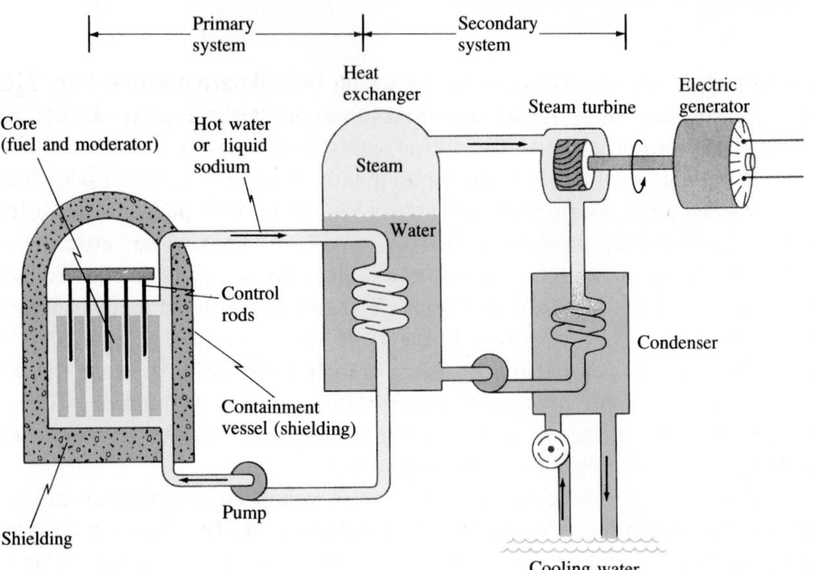

FIGURE 31–6 A nuclear reactor. The heat generated by the fission process in the fuel rods is carried off by hot water or liquid sodium and is used to boil water to steam in the heat exchanger. The steam drives a turbine to generate electricity and is then cooled in the condenser.

jectiles in nuclear reactions to produce nuclides not found in nature, including isotopes used as tracers and for medical therapy. A "power reactor" is used to produce electric power. The energy released in the fission process appears as heat, which is used to boil water and produce steam to drive a turbine connected to an electric generator (Fig. 31–6). The **core** of a nuclear reactor consists of the fuel and a moderator (water in most U.S. commercial reactors). The fuel is usually uranium enriched so that it contains 2 to 4 percent ${}^{235}_{92}U$. Water or other liquid (such as liquid sodium) is allowed to flow through the core. The thermal energy it absorbs is used to produce steam in the heat exchanger, so the fissionable fuel acts as the heat input for a heat engine (see Sections 15–5 and 15–11).

Many problems are associated with nuclear power plants. Besides the usual thermal pollution associated with any heat engine (Section 15–11), there is the problem of disposal of the radioactive fission fragments produced in the reactor, plus radioactive nuclides produced by neutrons interacting with the structural parts of the reactor. Fission fragments, like their uranium or plutonium parents, have about 50 percent more neutrons than protons. Nuclei with atomic number in the typical range for fission fragments ($Z \approx 30$ to 60) are stable only if they have more nearly equal numbers of protons and neutrons (see Fig. 30–2). Hence the highly neutron-rich fission fragments are very unstable and decay radioactively. The accidental release of highly radioactive fission fragments into the atmosphere poses a serious threat to human health (Section 31–4), as does possible leakage of the radioactive wastes when they are disposed of. The accidents at Three Mile Island and at Chernobyl have illustrated some of the dangers and have shown that nuclear plants must be constructed, maintained, and operated with great care and precision. Finally, the lifetime of nuclear power plants is limited to 30 some years, due to buildup of radioactivity and the fact that the structural materials themselves are weakened by the intense conditions inside. Decommissioning of a power plant could take a number of forms, but the cost of any method of decommissioning a large plant seems to be very great.

So-called **breeder reactors** were proposed as a solution to the problem of limited supplies of fissionable uranium. A breeder reactor is one in which some of the neutrons produced in the fission of $^{235}_{92}U$ are absorbed by $^{238}_{92}U$, and $^{239}_{94}Pu$ is produced via the set of reactions shown in Fig. 31–1. $^{239}_{94}Pu$ is fissionable with slow neutrons, so after separation it can be used as a fuel in a nuclear reactor. Thus a breeder reactor "breeds" new fuel[†] ($^{239}_{94}Pu$) from otherwise useless $^{238}_{92}U$. Since natural uranium is 99.3 percent $^{238}_{92}U$, this means that the supply of fissionable fuel could be increased by more than a factor of 100. But breeder reactors would not only have the same problems as other reactors, but in addition present other serious problems. Plutonium is a highly toxic substance that poses a serious danger to health. Furthermore, plutonium produced in a reactor can readily be used in a bomb. Thus the use of a breeder reactor, even more than a conventional uranium reactor, presents the danger of nuclear proliferation, and the possibility of theft of fuel by terrorists who could produce a bomb.

It is clear that nuclear power presents many risks. Other large-scale energy-conversion methods, such as conventional coal-burning steam plants, also present health and environmental hazards, some of which were discussed in Chapter 15, including air pollution, oil spills, and the release of CO_2 gas which may be trapping heat as in a greenhouse and raising the earth's temperature. The solution to the world's needs for energy is not only technological, but economic and political as well. Probably the best "solution" is to "conserve"—to not waste energy and use as little as possible.

Atom bomb

The first use of fission was not, however, to produce electric power. Instead, it was first used as a fission bomb (the "atomic bomb"). In early 1940, with Europe already at war, Hitler banned the sale of uranium from the Czech mines he had recently taken over. Research into the fission process suddenly was enshrouded in secrecy by Germany and by the Western powers. Physicists in the United States were alarmed. A group of them approached Einstein—a man whose name was a household word—to send a letter to President Roosevelt about the possibilities of using nuclear fission for a bomb far more powerful than any previously known, and inform him that Germany might already have begun development of such a bomb. Roosevelt responded by authorizing the program known as the Manhattan Project, to see if a bomb could be built. Work began in earnest after Fermi's demonstration in 1942 that a sustained chain reaction was possible. A new secret laboratory was developed on an isolated mesa in New Mexico known as Los Alamos. Under the direction of J. Robert Oppenheimer (1904–1967; Fig. 31–7), it became the home of famous scientists from all over Europe and the U.S.

FIGURE 31–7 Robert Oppenheimer, on the left, with General Leslie Groves who was the administrative head of Los Alamos during the war. The photograph was taken at the Trinity site in the New Mexico desert, where the first atomic bomb was exploded.

To build a bomb that was subcritical during transport but that could be made supercritical (to produce a chain reaction) at just the right moment, two pieces of uranium were used, each less than the critical mass but together greater than the critical mass. The two masses would be kept separate until the moment of detonation arrived. Then a kind of gun would force the two pieces together very quickly, a chain reaction of explosive proportions would occur, and a tremendous amount of energy would be released very suddenly. The first fission bomb was tested in the New Mexico desert in July, 1945. It was successful. In early August, a fission bomb using uranium was dropped on Hiroshima and a second, using plutonium, was dropped on Nagasaki.

[†] A breeder reactor does *not* produce more fuel than it uses, for a nearly endless supply, as popular articles sometimes imply.

year. The U.S. government specifies the recommended upper limit of allowed radiation for an individual in the general populace at about 0.5 rem per year, exclusive of natural sources. However, since even low doses of radiation are believed by many scientists to increase the chances of cancer or genetic defects, the attitude today is to keep the radiation dose as low as possible.

People who work around radiation—in hospitals, in power plants, in research—often are subjected to much higher doses than 0.5 rem/yr. The upper limit for such occupational exposures has been set somewhat higher, on the order of 5 rem/yr whole-body dose (presumably because such people know what they are getting into). To protect them from excess exposure, those people who work around radiation generally carry some type of dosimeter, typically a **radiation film badge**, which is a piece of film wrapped in light-tight material. The passage of ionizing radiation (see Section 30–13) through the film changes it so that the film is darkened upon development, and so indicates the received dose.

Large doses of radiation can cause reddening of the skin, drop in white-blood-cell count, and a large number of unpleasant symptoms such as nausea, fatigue, and loss of body hair. Such effects are sometimes referred to as **radiation sickness**. Large doses can also be fatal, although the time span of the dose is important. A short dose of 1000 rem is nearly always fatal. A 400-rem dose in a short period of time is fatal in 50 percent of the cases. However, the body possesses remarkable repair processes, so that a 400-rem dose spread over several weeks is not usually fatal. It will, nonetheless, cause considerable damage to the body.

EXAMPLE 31–6 What whole-body dose is received by a 70-kg patient exposed to a 1000-Ci $^{60}_{27}Co$ source if 2.0 percent of the γ rays reach the patient? $^{60}_{27}Co$ emits γ rays of energy 1.33 MeV and 1.17 MeV in equal amounts. Approximately 50 percent of the γ rays interact in the body and deposit all their energy. (The rest pass through.)

SOLUTION The average γ-ray energy is 1.25 MeV, so the total energy passing through the body is $(1000\text{ Ci})(3.7 \times 10^{10}\text{ decays/Ci}\cdot\text{s})\cdot(1.25\text{ MeV})(0.020) = 9.3 \times 10^{11}\text{ MeV/s}$. (The factor 0.020 represents the 2.0 percent that reach the body.) Only half this energy is deposited in the body, so the dose rate is $(\frac{1}{2})(9.3 \times 10^{11}\text{ MeV/s})/(1.6 \times 10^{-13}\text{ J/MeV}) = 7.4 \times 10^{-2}\text{ J/s}$. Since $1\text{ Gy} = 1\text{ J/kg}$, the whole-body dose rate is $(7.4 \times 10^{-2}\text{ J/s})/(1\text{ J/kg}\cdot\text{Gy})(70\text{ kg}) = 1.1 \times 10^{-3}\text{ Gy/s}$.

*31–6 • Radiation Therapy

The applications of radioactivity and radiation to human beings and other organisms is a vast field that has filled many books. This subject is often called **radiation biology**, and when applied in medicine, it is called **nuclear medicine**. Nuclear medicine involves two basic aspects: (1) radiation *therapy*—the treatment of disease—which we discuss in this section; and (2) the *diagnosis* of disease, which we discuss in the following sections of this chapter.

Radiation can cause cancer. It can also be used to treat it. Rapidly growing cancer cells are especially susceptible to destruction by radiation. Nonetheless, large doses are needed to kill the cancer cells, and some of the surrounding normal cells are inevitably killed as well. It is for this reason that cancer patients receiving radiation therapy often suffer side effects characteristic of radiation sickness. To minimize the destruction of normal cells, a narrow beam of γ or X rays is often used when the cancerous tumor is well localized. The beam is directed at the tumor, and the source (or body) is rotated so that the beam passes through various parts of the body to keep the dose at any one place as low as possible—except at the tumor and its immediate surroundings, where the beam passes at all times. (See photo at start of this chapter.) The radiation may be from a radioactive source such as $^{60}_{27}\text{Co}$; or it may be from an X-ray machine that produces photons in the range 200 keV to 5 MeV. Protons, neutrons, electrons, and pions, which are produced in particle accelerators (Section 32–2), are also being used in cancer therapy.

In some cases, a tiny radioactive source may be inserted directly inside a tumor, which will eventually kill the majority of the cells. A similar technique is used to treat cancer of the thyroid with the radioactive isotope $^{131}_{53}\text{I}$. The thyroid gland tends to concentrate any iodine present in the bloodstream; so when $^{131}_{53}\text{I}$ is injected into the blood, it becomes concentrated in the thyroid, particularly in any area where abnormal growth is taking place. The intense radioactivity emitted can then destroy the defective cells.

Although radiation can increase the lifespan of many patients, it is not always completely effective. It may not be possible to kill all the diseased cancer cells, so a recurrence of the disease is possible. Many cases, especially when the cancerous cells are not well localized in one area, are difficult to treat at all without damaging the rest of the organism.

Another application of radiation is for sterilizing bandages, surgical equipment, and even packaged foods, since bacteria and viruses can be killed or deactivated by large doses of radiation.

*31–7 • Tracers and Imaging in Research and Medicine

Radioactive isotopes are commonly used in biological and medical research as **tracers**. A given compound is artificially synthesized using a radioactive isotope such as $^{14}_{6}\text{C}$ or $^{3}_{1}\text{H}$. Such "tagged" molecules can then be traced as they move through an organism or as they undergo chemical reaction. The presence of these tagged molecules (or parts of them, if they undergo chemical change) can be detected by a Geiger or scintillation counter, which detects emitted radiation (see Section 30–13). The details of how food molecules are digested, and to what parts of the body they are diverted, can be traced in this way. Radioactive tracers have been used to determine how amino acids and other essential compounds are synthesized by organisms. The permeability of cell walls to various molecules and ions can be determined using radioactive isotopes: the tagged molecule or ion is injected into the extracellular fluid and the radioactivity present inside and outside the cells is measured as a function of time.

FIGURE 31–11 Autoradiograph of a mature leaf of the squash plant *Cucurbita melopepo* exposed for 30 s to $^{14}CO_2$. The photosynthetic (green) tissue has become radioactive; the nonphotosynthetic tissue of the veins is free of ^{14}C and therefore does not blacken the X-ray sheet. This technique is very useful in following patterns of nutrient transport in plants.

FIGURE 31–12 An autoradiograph of a fiber of chromosomal DNA isolated from the higher plant *Arabidopsis thaliana*. The dashed arrays of silver grains show the Y-shaped growing point of replicating DNA.

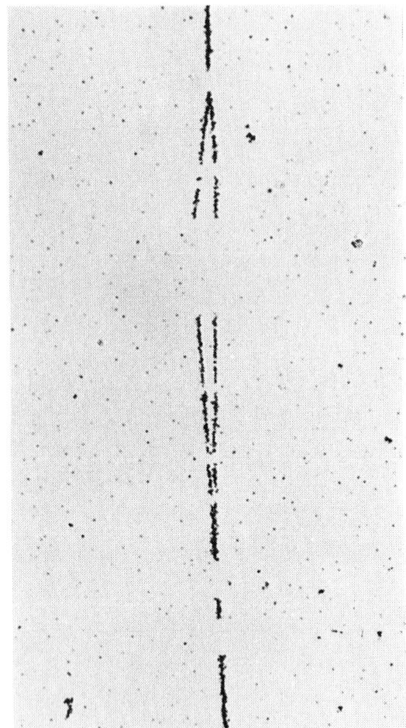

In many cases, the cells or chemicals involved are removed from the organism and placed in the Geiger or scintillation counter. To follow a process in time, different sets of organisms are treated for different periods of time before the cells or chemicals are extracted. (See Example 31–5.)

In a technique known as **autoradiography**, the position of the radioactive isotopes is detected on film. For example, the distribution of carbohydrates produced in the leaves of plants from absorbed CO_2 can be observed by keeping the plant in an atmosphere where the carbon atom in the CO_2 is $^{14}_{6}C$. After a time, a leaf is placed firmly on a photographic plate and the emitted radiation darkens the film most strongly where the isotope is most strongly concentrated (Fig. 31–11). Autoradiography using labeled nucleotides (components of DNA) has revealed much about the details of DNA replication (Fig. 31–12).

For medical diagnosis, the radionuclide commonly used today is $^{99m}_{43}Tc$, a long-lived excited state of technicium-99 (the "m" in the symbol stands for "metastable" state). It is formed when $^{99}_{42}Mo$ decays. The great usefulness of $^{99m}_{43}Tc$ derives from its convenient half-life of 6 h (short, but not too short) and the fact that it can combine with a large variety of compounds. The compound to be labeled with the radionuclide is so chosen because it concentrates in the organ or region of the anatomy to be studied. Detectors outside the body then record, or image, the distribution of the radioactively labeled compound. The detection can be done by a single detector (Fig. 31–13) which is moved across the body, measuring the intensity of radioactivity at a large number of points. The image represents the relative intensity of radioactivity at each point. The relative radioactivity is a diagnostic tool. For example, high or low radioactivity may represent overactivity or underactivity of an organ or part of an organ, or in another case may represent a lesion or tumor. More complex *gamma cameras* make use of many detectors which simultaneously record the radioactivity at many points. The measured intensities can be displayed on a CRT (TV monitor or oscilloscope screen), and allow "dynamic" studies (that is, images that change in time) to be performed.

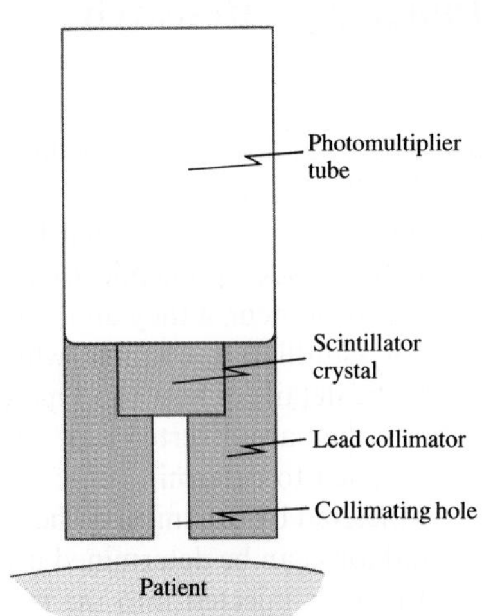

FIGURE 31–13 Collimated gamma-ray detector for scanning (moving) over patient. The collimator is necessary to select γ rays that come in a straight line from the patient. Without the collimator, γ rays from all parts of the body could strike the scintillator, producing a very poor image.

*31–8 • Emission Tomography

The images formed using the standard techniques of nuclear medicine, as briefly discussed in the previous section, are produced from radioactive tracer sources within the *volume* of the body. It is also possible to image the radioactive emissions in a single plane or slice through the body using the computed tomography techniques discussed in Section 25–12. A basic gamma camera is moved around the patient to measure the radioactive intensity from the tracer at many points and angles; the data are processed in much the same way as for X-ray CT scans (Section 25–12). This technique is referred to as **single photon emission tomography** (SPET).† *SPET*

Another important technique is **positron emission tomography** (PET), which makes use of positron emitters such as ${}^{11}_{6}\mathrm{C}$, ${}^{13}_{7}\mathrm{N}$, ${}^{15}_{8}\mathrm{O}$, and ${}^{18}_{9}\mathrm{F}$. These isotopes are incorporated into molecules that, when inhaled or injected, accumulate in the organ or region of the body to be studied. When such a nuclide β decays, the emitted positron travels at most a few millimeters before it collides with a normal electron. In this collision, the positron and electron are annihilated, producing two γ rays ($e^+e^- \rightarrow 2\gamma$), each having energy of 510 keV. The two γ rays fly off in opposite directions ($180° \pm 0.25°$) since they must have almost exactly equal and opposite momenta to conserve momentum (the momenta of the e^+ and e^- are essentially zero compared to the momenta of the γ rays). Because the photons travel along the same line in opposite directions, their detection in coincidence by rings of detectors surrounding the patient (Fig. 31–14) readily establishes the line along which the emission took place. If the difference in time of arrival of the two photons could be determined accurately, the actual position of the emitting nuclide along that line could be calculated. Present-day electronics can measure times to at best ± 300 ps, so at the γ ray's speed ($c = 3 \times 10^8$ m/s), the actual position could be determined to an accuracy on the order of about $vt \approx (3 \times 10^8\ \mathrm{m/s})\cdot(300 \times 10^{-12}\ \mathrm{s}) \approx 10$ cm, which is not very useful. Although there may be future potential for time-of-flight measurements to determine position, today computed tomography techniques are used instead, similar to those for X-ray CT, which can reconstruct PET images with a resolution on the order of 3–5 mm. One big advantage of PET is that no collimators *PET*

† Also known as SPECT, "single photon emission computed tomography".

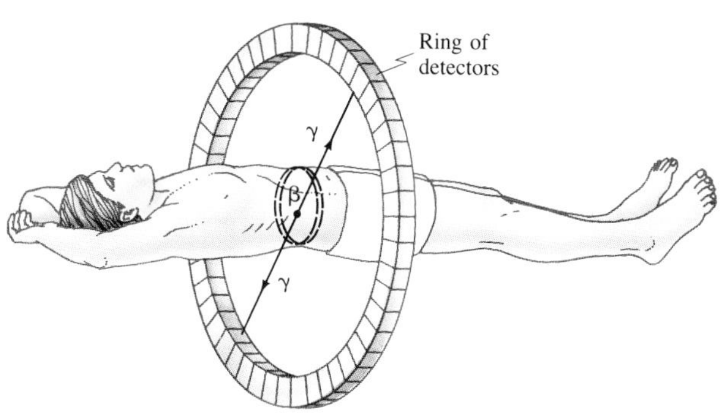

FIGURE 31–14 Positron emission tomography (PET) system showing a ring of detectors to detect the two annihilation γ rays ($e^+e^- \rightarrow 2\gamma$) emitted at 180° to each other.

are needed (as for detection of a single photon—see Fig. 31–13). Thus, fewer photons are "wasted" and lower doses can be administered to the patient with PET.

Both PET and SPET systems can give images related to biochemistry, metabolism, and function. This is to be compared to X-ray CT scans (Section 25–12), whose images reflect shape and structure—that is, the anatomy of the imaged region.

*31–9 • Nuclear Magnetic Resonance (NMR) and NMR Imaging

NMR

Nuclear magnetic resonance (NMR) is a phenomenon that soon after its discovery in 1946 became a powerful research tool in a variety of fields from physics to chemistry and biochemistry. In recent years, it has been developed into an important and widely used medical imaging technique. We first briefly discuss the phenomenon, and then we will look at its applications.

We saw in Chapter 20 (Section 20–8) that a loop of wire carrying an electric current has a torque on it when placed in a magnetic field. The loop is said to have a magnetic moment, and the torque on the loop tends to align it so that the direction of its magnetic moment (taken to be perpendicular to the plane of the coil) is parallel to the field, **B**. (The coil's plane is perpendicular to **B**.) When the coil is so aligned, it has its lowest potential energy. Then in Chapter 28 (Section 28–6), we saw that atomic electrons in their orbital motions act as tiny current loops, and furthermore that they have an intrinsic angular momentum called *spin*, which also acts like the magnetic moment of a current loop. When atoms are placed in a magnetic field, atomic energy levels split into several closely spaced levels (see Fig. 28–6).

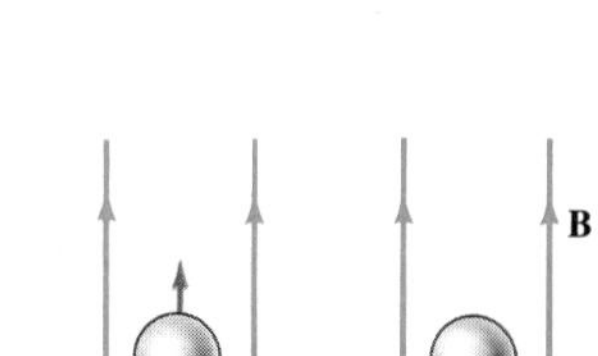

FIGURE 31–15 Schematic picture of an electron represented in a magnetic field **B** (pointing upward) with its two possible states of spin, up and down.

Nuclei, too, exhibit these magnetic properties. Many nuclides have magnetic moments, but we will examine only the simplest, the hydrogen (H) nucleus, since it is the one most used even for medical imaging. The ${}^{1}_{1}H$ nucleus consists of a single proton. Its spin angular momentum (and its magnetic moment), like that of the electron, can take on only two values when placed in a magnetic field: we call these "spin up" (parallel to the field) and "spin down" (antiparallel to the field), as suggested in Fig. 31–15. When a magnetic field is present, the energy of the nucleus splits into two levels as shown in Fig. 31–16, with the spin up (parallel to field) having the lower energy. (This is like the Zeeman effect for atomic levels—see Fig. 28–6.) The difference in energy, ΔE, between these two levels is proportional to the total magnetic field, B_T, at the nucleus:

$$\Delta E = kB_T,$$

where k is a proportionality constant that is different for different nuclides.

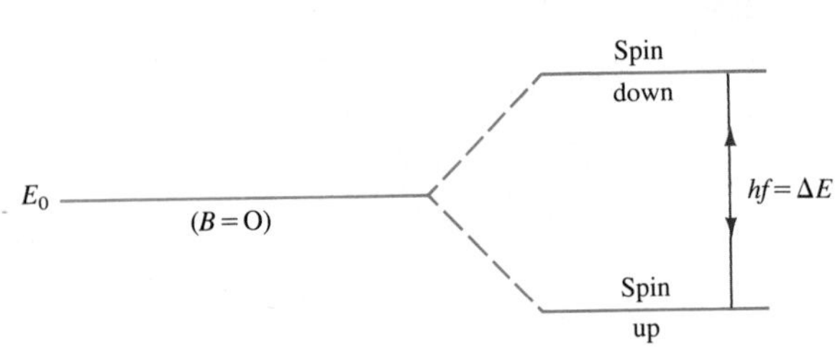

FIGURE 31–16 Energy E_0 in the absence of a magnetic field splits into two levels in the presence of a magnetic field.

In a standard nuclear magnetic resonance (NMR) setup, the sample to be examined is placed in a static magnetic field. A radiofrequency (RF) pulse of electromagnetic radiation (that is, photons) is applied to the sample. If the frequency, f, of this pulse corresponds precisely to the energy difference between the two energy levels (Fig. 31–16), so that

$$hf = \Delta E = kB_T, \tag{31-6}$$

then the photons of the RF beam will be absorbed, exciting many of the nuclei from the lower state to the upper state. This is a resonance phenomenon since there is significant absorption only if f is very near $f = kB_T/h$; hence the name "nuclear magnetic resonance." For free $^1_1$H nuclei, the frequency is 42.58 MHz for a field $B_T = 1.0$ T. If the H atoms are bound in a molecule, on the other hand, the total magnetic field B_T at the H nuclei will be the sum of the external applied field plus the magnetic field due to electrons and nuclei of neighboring atoms. Since f is proportional to B_T, the value of f will be slightly different for the bound H atoms than for free atoms in the same external field. This change in frequency, which can be measured, is called the "chemical shift." A great deal has been learned about the structure of molecules and bonds using such NMR measurements.

NMR imaging (MRI)

For producing medically useful NMR images (often called MRI, or **magnetic resonance imaging**†), the element most used is hydrogen since it is the commonest element in the human body and gives the strongest NMR signals. The experimental apparatus is shown in Fig. 31–17. The large coils set up the static magnetic field, and the RF coils produce the RF pulse of electromagnetic waves (photons) that cause the nuclei to jump from the lower state to the upper one (Fig. 31–16). These same coils (or another coil) can detect the absorption of energy or the emitted radiation (also of frequency $f = \Delta E/h$, Eq. 31-6) when the nuclei jump back down to the lower state.

The formation of a two-dimensional or three-dimensional image can be done using the techniques of computed tomography similar to those for

† This term was coined by the medical profession to avoid confusing patients who feared they were being treated with nuclear radiation.

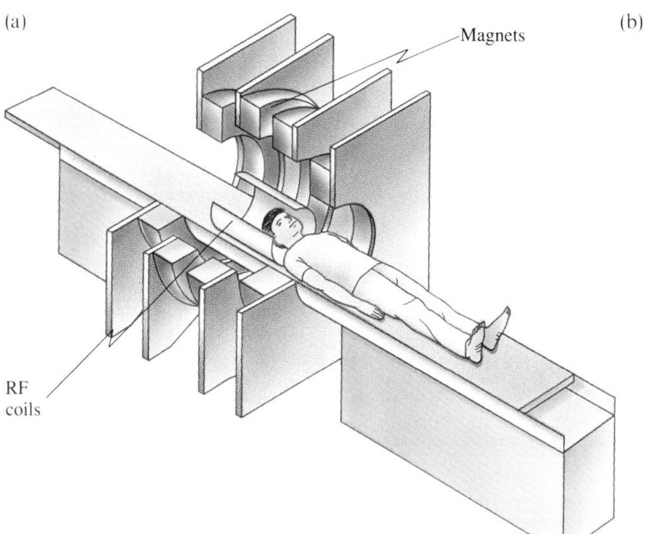

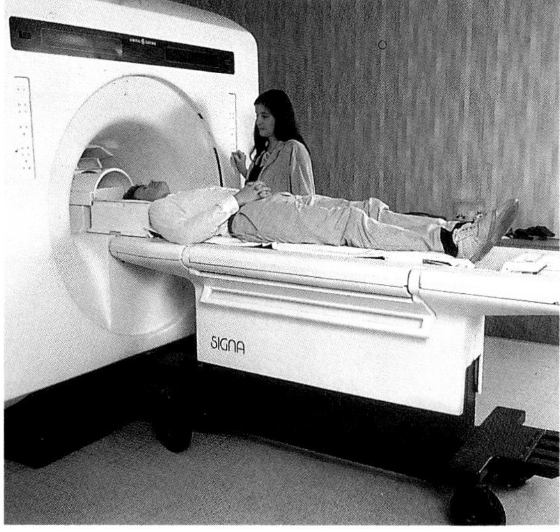

FIGURE 31–17 Typical NMR imaging setup.

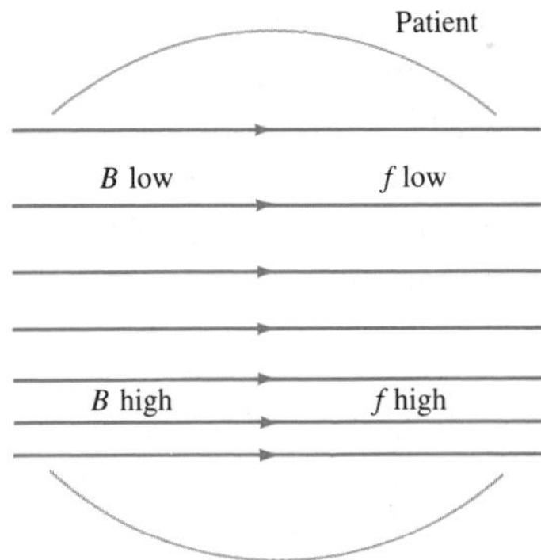

FIGURE 31–18 A static field that is stronger at the bottom than at the top. The frequency of absorbed, or emitted, light is proportional to B in NMR.

X rays discussed in Section 25–12. The simplest thing to measure for creating an image is the intensity of absorbed and/or reemitted radiation from many different points of the body, and this would be a measure of the density of H atoms at each point. But how do we determine from what part of the body a given photon comes? One technique is to give the static magnetic field a gradient; that is, instead of applying a uniform magnetic field, B_T, the field is made to vary with position across the width of the sample (or patient). Since the frequency absorbed by the H nuclei is proportional to B_T (Eq. 31–6), only one plane within the body will have the proper value of B_T to absorb photons of a particular frequency f. By varying f, absorption by different planes can be measured. Alternately, if the field gradient is applied *after* the RF pulse, the frequency of the emitted photons will be a measure of where they were emitted. See Fig. 31–18. If a magnetic field gradient in one direction is applied during excitation (absorption of photons) and photons of a single frequency are transmitted, only H nuclei in one thin slice will be excited. By applying a gradient in a different direction, perpendicular to the first, during reemission, the frequency f of the reemitted radiation will represent depth in that slice. Other ways of varying the magnetic field throughout the volume of the body can be used in order to correlate NMR frequency with position.

A reconstructed image based on the density of H atoms (that is, the intensity of absorbed or emitted radiation) is not very interesting. More useful are images based on the rate at which the nuclei decay back to the ground state, and such images can produce resolution of 1 mm or better. This NMR technique (sometimes called **spin-echo**) is producing images of great diagnostic value, both in the delineation of structure (anatomy) and in the study of metabolic processes. An NMR image is shown in Fig. 31–19.

FIGURE 31–19 False-color NMR image of a vertical section through the head showing structures in the normal brain.

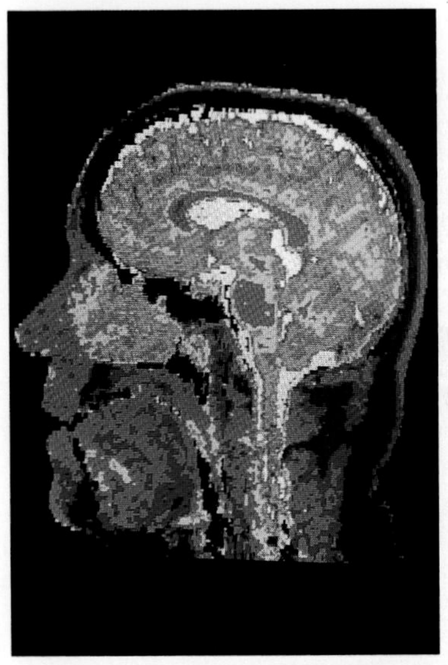

Table 31–2 lists the recently developed techniques we have discussed for imaging the interior of the body, along with the optimum resolution attainable today. Of course, resolution is only one factor that must be considered; it must be remembered that the different imaging techniques provide different types of information, useful for different types of diagnosis.

TABLE 31–2

Technique	Where Discussed In This Book	Resolution
Conventional X ray	Section 25–12	$\frac{1}{2}$ mm
CT scan, X ray	Section 25–12	$\frac{1}{2}$ mm
Nuclear medicine	Section 31–7	1 cm
SPET (single photon emission)	Section 31–8	1 cm
PET (positron emission)	Section 31–8	3–5 mm
NMR	Section 31–9	$\frac{1}{2}$–1 mm
Ultrasound	Section 12–10	2 mm

SUMMARY

A *nuclear reaction* occurs when two nuclei collide and two or more other nuclei (or particles) are produced. In this process, as in radioactivity, *transmutation* (change) of elements occurs.

In *fission*, a heavy nucleus such as uranium splits into two intermediate-sized nuclei after being struck by a neutron. $^{235}_{92}U$ is fissionable by slow neutrons, whereas some fissionable nuclei require fast neutrons.

Much energy is released in fission because the binding energy per nucleon is lower for heavy nuclei than it is for intermediate-sized nuclei, so the mass of a heavy nucleus is greater than the total mass of its fission products. The fission process releases neutrons, so that a *chain reaction* is possible. The *critical mass* is the minimum mass of fuel needed to sustain a chain reaction. In a *nuclear reactor* or nuclear bomb, a moderator is needed to slow down the released neutrons.

The *fusion* process, in which small nuclei combine to form larger ones, also releases energy. The energy from our sun is believed to originate in the fusion reactions known as the *proton–proton cycle* in which four protons fuse to form a $^4_2$He nucleus producing over 25 MeV. A working fusion reactor for power generation has not yet proved possible because of the difficulty in containing the fuel (e.g., deuterium) long enough at the high temperature required. Nonetheless, great progress has been made in confining the collection of charged ions known as a *plasma*. The two main methods are *magnetic confinement*, using a magnetic field in a device such as the toroidal-shaped *tokamak*, and *inertial confinement*, in which intense laser beams compress a fuel pellet of deuterium and tritium.

Radiation can cause damage to materials, including biological tissue. Quantifying amounts of radiation is the subject of *dosimetry*. The *curie* (Ci) and the *becquerel* (Bq) are units that measure the activity or rate of decay of a sample: 1 Ci = 3.70×10^{10} disintegrations per second, whereas 1 Bq = 1 disintegration/s. The *rad* measures the amount of energy deposited per unit mass of absorbing material: 1 rad is the amount of radiation that deposits energy at the rate of 10^{-2} J/kg of material. The SI unit of absorbed dose is the *gray*: 1 Gy = 1 J/kg. The *rem* = rad × QF, where QF is the "quality factor" of a given type of radiation; 1 rem of any type of radiation does approximately the same amount of biological damage. The average dose received per person per year is about 0.20 rem. The SI unit for effective dose is the *sievert*: 1 Sv = 10^2 rem.

QUESTIONS

1. Fill in the missing particles or nuclei: (*a*) $^{137}_{56}$Ba(n, γ)?; (*b*) $^{137}_{56}$Ba(n, ?)$^{137}_{55}$Cs; (*c*) $^2_1$H(d, ?)$^4_2$He; (*d*) $^{197}_{79}$Au(α, d)?, where *d* stands for deuterium.
2. The isotope $^{32}_{15}$P is produced by an (n, p) reaction. What must be the target nucleus?
3. When $^{22}_{11}$Na is bombarded by deuterons ($^2_1$H), an α particle is emitted. What is the resulting nuclide?
4. Why are neutrons such good projectiles for producing nuclear reactions?
5. A proton strikes a $^{20}_{10}$Ne nucleus, and an α particle is observed to emerge. What is the residual nucleus? Write down the reaction equation.
6. Are fission fragments β^+ or β^- emitters?
7. If $^{235}_{92}$U released only 1.5 neutrons per fission on the average, would a chain reaction be possible? If so, what would be different?
8. $^{235}_{92}$U releases an average of 2.5 neutrons per fission compared to 2.7 for $^{239}_{94}$Pu. Pure samples of which of these two nuclei do you think would have the smaller critical mass?
9. The energy from nuclear fission appears in the form of thermal energy—but the thermal energy of what?
10. How can a neutron, with practically no KE, excite a nucleus to the extent shown in Fig. 31–2?
11. Why would a porous block of uranium be more likely to explode if kept under water rather than in air?
12. A reactor that uses highly enriched uranium can use ordinary water (instead of heavy water) as a moderator and still have a self-sustaining chain reaction. Explain.
13. Why must the fission process release neutrons if it is to be useful?
14. Discuss the relative merits and disadvantages, including pollution and safety, of power generation by fossil fuels, nuclear fission, and nuclear fusion.
15. What is the reason for the "secondary system" in a nuclear reactor, Fig. 31–6? That is, why is the water heated by the fuel in a nuclear reactor not used directly to drive the turbines?
16. Discuss how the course of history might have been changed if, during World War II, scientists had refused to work on developing a nuclear bomb. Do you think it would have been possible to delay the building of a bomb indefinitely?
17. Research in molecular biology is moving toward the ability to perform genetic manipulations on human beings. The moral implications of future discoveries along these lines had led to a warning that this may be the molecular biologists' "Hiroshima." Discuss.
18. Does $E = mc^2$ apply in (*a*) fission, (*b*) fusion, (*c*) nuclear reactions?
19. Light energy emitted by the sun and stars comes from the fusion process. What conditions in the interior of stars makes this possible?
20. What is the basic difference between fission and fusion?
21. Why is the recommended maximum radiation dose higher for women beyond the child-bearing age than for younger women?

*22. How might radioactive tracers be used to find a leak in a pipe?

*23. How might radioactive tracers be used to measure (*a*) engine wear, (*b*) tire wear?

PROBLEMS

SECTION 31–1

1. (I) Determine whether the reaction ${}^2_1\text{H}(d, n){}^3_2\text{He}$ requires a threshold energy. (d stands for deuterium, ${}^2_1\text{H}$.)
2. (I) Is the reaction ${}^{238}_{92}\text{U}\ (n, \gamma){}^{239}_{92}\text{U}$ possible with slow neutrons? Explain.
3. (II) Does the reaction ${}^7_3\text{Li}(p, \alpha){}^4_2\text{He}$ require energy or does it release energy? How much energy?
4. (II) The reaction ${}^{18}_8\text{O}(p, n){}^{18}_9\text{F}$ requires an input of energy equal to 2.453 MeV. What is the mass of ${}^{18}_9\text{F}$?
5. (II) Calculate the energy released (or energy input required) for the reaction ${}^9_4\text{Be}(\alpha, n){}^{12}_6\text{C}$.
6. (II) (*a*) Can the reaction ${}^{24}_{12}\text{Mg}(n, d){}^{23}_{11}\text{Na}$ occur if the bombarding particles have 10.00 MeV of KE? (*b*) If so, how much energy is released?
7. (II) (*a*) Can the reaction ${}^7_3\text{Li}(p, \alpha){}^4_2\text{He}$ occur if the incident proton has KE = 2500 keV? (*b*) If so, what is the total kinetic energy of the products?
8. (II) In the reaction ${}^{14}_7\text{N}(\alpha, p){}^{17}_8\text{O}$, the incident α particles have 7.68 MeV of kinetic energy. (*a*) Can this reaction occur? (*b*) If so, what is the total kinetic energy of the products? The mass of ${}^{17}_8\text{O}$ is 16.999131 u.
9. (III) Use conservation of energy and momentum to show that a bombarding proton must have an energy of 3.23 MeV in order to make the reaction ${}^{13}_6\text{C}(p, n){}^{13}_7\text{N}$ occur. (See Example 31–3.)
10. (III) Show, using the laws of conservation of energy and momentum, that for a nuclear reaction requiring energy, the minimum kinetic energy of the bombarding particle (the *threshold energy*) is equal to $[Qm_{\text{pr}}/(m_{\text{pr}} - m_{\text{b}})]$, where Q is the energy required (difference in total mass between products and reactants), m_b is the rest mass of the bombarding particle, and m_{pr} the total rest mass of the products. Assume the target nucleus is at rest before an interaction takes place, and that all particles are nonrelativistic.

SECTION 31–2

11. (I) Calculate the energy released in the fission reaction $n + {}^{235}_{92}\text{U} \rightarrow {}^{88}_{38}\text{Sr} + {}^{136}_{54}\text{Xe} + 12n$. Use Appendix D, and assume the initial KE of the neutron is very small.
12. (I) What is the energy released in the fission reaction of Eq. 31–2b? (The masses of ${}^{141}_{56}\text{Ba}$ and ${}^{92}_{36}\text{Kr}$ are 140.91440 u and 91.92630 u, respectively.)
13. (I) How many fissions take place per second in a 25-MW reactor? Assume that 200 MeV is released per fission.
14. (II) Suppose that the average power consumption in a typical house is 300 W. What initial mass of ${}^{235}_{92}\text{U}$ would have to undergo fission to supply the electrical needs of such a house for a year? (Assume 200 MeV is released per fission.)
15. (II) What initial mass of ${}^{235}_{92}\text{U}$ is required to operate a 500-MW reactor for 1 yr? Assume 40% efficiency.
16. (II) If a 1.0-MeV neutron emitted in a fission reaction loses one-half of its KE in each collision with moderator nuclei, how many collisions must it make to reach thermal energy ($\frac{3}{2}kT = 0.040$ eV)?
17. (III) Suppose that the neutron multiplication factor is 1.0004. If the average time between successive fissions in a chain of reactions is 1.0 ms, by what factor will the reaction rate increase in 1.0 s?

SECTION 31–3

18. (I) What is the average kinetic energy of protons at the center of a star where the temperature is 10^7 K?
19. (II) Show that the energy released in the fusion reaction ${}^2_1\text{H} + {}^3_1\text{H} \rightarrow {}^4_2\text{He} + n$ is 17.59 MeV.
20. (II) Show that the energy released when two deuterium nuclei fuse to form ${}^3_2\text{He}$ with the release of a neutron is 3.27 MeV.
21. (II) Verify the energy output stated for each of the reactions of Eq. 31–3. [*Hint:* be careful with electrons.]
22. (II) Calculate the energy release per gram of fuel for the reactions of Eqs. 31–5a, b, and c. Compare to the energy release per gram of uranium in the fission process.
23. (II) If a typical house requires 300 W of electric power on average, how much deuterium fuel would have to be used in a year to supply these electrical needs? Assume the reaction of Eq. 31–5b.
24. (II) Show that the energies carried off by the ${}^4_2\text{He}$ nucleus and the neutron for the reaction of Eq. 31–5c are about 3.5 MeV and 14 MeV, respectively. Are these fixed values, independent of the plasma temperature?
25. (III) How much energy (J) is contained in 1.00 kg of water if the natural deuterium is used in the fusion reaction of Eq. 31–5a? Compare this to the energy obtained from the burning of 1.0 kg of gasoline, about 5×10^7 J.
26. (III) The energy output of massive stars is believed to be due to the *carbon cycle* (see text). (*a*) Show that no carbon is consumed in this cycle and that the net effect is the same as for the proton–proton cycle. (*b*) What is the total energy release? (*c*) Determine the energy output for each reaction and decay. (*d*) Why does the carbon cycle require a higher temperature ($\approx 2 \times 10^7$ K) than the proton–proton cycle ($\approx 1.5 \times 10^7$ K)?
27. (III) The deuterium-tritium pellet in a laser fusion device contains equal numbers of ${}^2_1\text{H}$ and ${}^3_1\text{H}$ atoms raised to a density of $200 \times 10^3\ \text{kg/m}^3$ by the laser pulses. Estimate (*a*) the density of particles in this compressed state, and (*b*) the length of time τ the pellet must be confined to meet the Lawson criterion for ignition.

SECTION 31–5

28. (I) A 0.018-μCi sample of ${}^{32}_{15}\text{P}$ is injected into an animal for tracer studies. If a Geiger counter intercepts 20 percent of the emitted β particles and is 90 percent efficient in counting them, what will be the counting rate?
29. (I) An average adult body has about 0.10 μCi of ${}^{40}_{19}\text{K}$, which comes from food. How many decays occur per second?

30. (I) A dose of 500 rem of γ rays in a short period would be lethal to about half the people subjected to it. How many rads is this?

31. (I) Fifty rads of α-particle radiation is equivalent to how many rads of X rays in terms of biological damage?

32. (I) How many rads of slow neutrons will do as much biological damage as 50 rads of fast neutrons?

33. (I) How much energy is deposited in the body of a 70-kg adult exposed to a 50-rad dose?

34. (II) A 1.0-mCi source of $^{32}_{15}P$ (in $NaHPO_4$), a β^- emitter, is implanted in an organ where it is to administer 5000 rad. The half-life of $^{32}_{15}P$ is 14.3 days and 1 mCi delivers about 1 rad/min. Approximately how long should the source remain implanted?

35. (II) About 35 eV is required to produce one ion pair in air. Show that this is consistent with the two definitions of the roentgen given in the text.

36. (II) $^{57}_{27}Co$ emits 122-keV γ rays. If a 70-kg person swallowed 2.0 μCi of $^{57}_{27}Co$, what would be the dose rate (rad/day) averaged over the whole body? Assume that 50 percent of the γ-ray energy is deposited in the body. [*Hint:* determine the rate of energy deposited in the body and use the definition of the rad.]

37. (II) What is the mass of a 1.00-μCi $^{14}_{6}C$ source?

*SECTION 31–9

*38. (II) Calculate the wavelength of photons needed to produce NMR transitions in free protons in a 1.000-T field. In what region of the spectrum does it lie?

GENERAL PROBLEMS

39. Fusion temperatures are often given in keV. Determine the conversion factor from kelvins to keV using, as is usual in this field, KE $= kT$ without the factor $\frac{3}{2}$.

40. One means of enriching uranium is by diffusion of the gas UF_6. Calculate the ratio of the speeds of molecules of this gas containing $^{235}_{92}U$ and $^{238}_{92}U$, on which this process depends.

41. (*a*) What mass of $^{235}_{92}U$ was actually fissioned in the first atomic bomb, whose energy was the equivalent of about 20 kilotons of TNT (1 kiloton of TNT releases 5×10^{12} J)? (*b*) What was the actual mass transformed to energy?

42. In a certain town the average yearly background radiation consists of 25 mrad of X rays and γ rays plus 3.0 mrad of particles having a QF of 10. How many rem will a person receive per year on the average?

43. Deuterium makes up 0.015% of natural hydrogen. Make a rough estimate of the total deuterium in the earth's oceans and estimate the total energy released if all of it were used in fusion reactors.

44. A shielded γ-ray source yields a dose rate of 0.050 rad/h at a distance of 1.0 m for an average-sized person. If workers are allowed a maximum dose rate of 5.0 rem/yr, how close to the source may they operate, assuming a 40-h work week? Assume that the intensity of radiation falls off as the square of the distance. (It actually falls off more rapidly than $1/r^2$ because of absorption in the air, so the answer above will give a better-than-permissible value.)

45. Consider a system of nuclear power plants that produce 4000 MW. (*a*) What total mass of $^{235}_{92}U$ fuel would be required to operate these plants for 1 yr, assuming that 200 MeV is released per fission? (*b*) Typically 6 percent of the $^{235}_{92}U$ nuclei that fission produce $^{90}_{38}Sr$, a β^- emitter with a half-life of 29 yr. What is the total radioactivity of the $^{90}_{38}Sr$, in curies, produced in 1 yr? (Neglect the fact that some of it decays during the 1-yr period.)

46. In the net reaction, Eq. 31–4, for the proton–proton cycle in the sun, the neutrinos escape from the sun with energy of about 0.5 MeV. The remaining energy, 26.2 MeV, is available within the sun. Use this value to calculate the "heat of combustion" per kilogram of hydrogen fuel and compare it to the heat of combustion of coal, about 3×10^7 J/kg.

47. Energy reaches the earth from the sun at a rate of about 1400 W/m^2. Calculate (*a*) the total energy output of the sun, and (*b*) the number of protons consumed per second in the reaction of Eq. 31–4, assuming that this is the source of all the sun's energy. (*c*) Assuming that the sun's mass of 2.0×10^{30} kg was originally all protons and that all could be involved in nuclear reactions in the sun's core, how long would you expect the sun to "glow" at its present rate?

CHAPTER 32

Elementary Particles

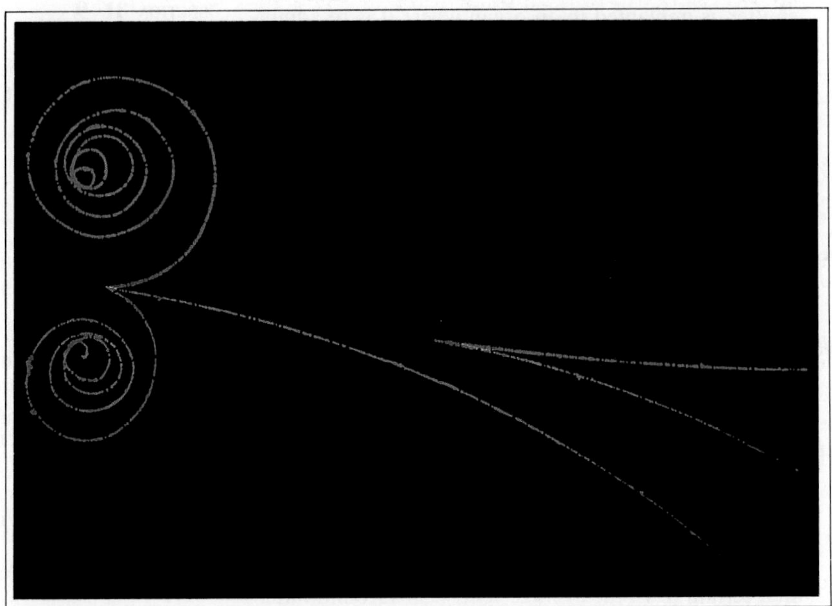

False-color bubble chamber photograph of electron (green) and positron (red) tracks. The positron is the antiparticle of the electron: it has the same mass as the electron but opposite charge. In this photo, two gamma rays enter from the left (but leave no track because they are electrically neutral) and undergo pair production (Chapter 27). In the interaction on the left, one γ ray knocks out an atomic electron (long curved green path) as well as producing an e^+e^- pair, whose paths in the magnetic field are strongly curved (Chapter 20) because they have low energy. In the interaction on the right, all the energy of a γ ray goes to the e^+e^- pair whose energies are higher and hence the curvature of their paths is less.

In this chapter we discuss the exciting subject of *elementary particle* physics, which represents the human endeavor to understand the basic building blocks of all matter.

In the years after World War II, it was found that if the incoming particle in a nuclear reaction has sufficient energy, new types of particles can be produced. In order to produce high-energy particles, various types of particle accelerators have been constructed. Most commonly they accelerate protons or electrons, although heavy ions can also be accelerated, depending on the design. These high-energy accelerators have been used to probe the nucleus more deeply, to produce and study new particles, and to give us information about the basic forces and constituents of nature.

32–1 • High-Energy Particles

Particles accelerated to high speeds are projectiles which can probe the interior of nuclei and nucleons that they strike. An important factor is that faster-moving projectiles can reveal more detail. The wavelength of projectile particles is given by de Broglie's wavelength formula:

$$\lambda = \frac{h}{mv},$$

de Broglie wavelength

showing that the greater the momentum of the bombarding particle, the shorter its wavelength. As discussed in Chapter 25, resolution of details is limited by the wavelength: the shorter the wavelength, the finer the detail that can be obtained. This is one reason why particle accelerators of higher and higher energy have been built in recent years.

EXAMPLE 32–1 What is the wavelength, and hence the expected resolution, for a beam of 1.3-GeV electrons?

SOLUTION 1.3 GeV = 1300 MeV, which is about 2500 times the mass of the electron. We are clearly dealing with relativistic speeds here, and it is easily shown that the speed of the electron is nearly $c = 3.0 \times 10^8$ m/s. (From Eq. 26–4, KE $= mc^2 - m_0c^2 \approx mc^2$; and from Eq. 26–7, $E^2 = p^2c^2 + m_0^2c^4 \approx p^2c^2$ since m_0c^2 is small compared to pc; hence, $p^2c^2 = m^2v^2c^2 \approx m^2c^4$, so $v \approx c$.) Therefore

$$\lambda = \frac{h}{mv} \approx \frac{h}{mc} = \frac{hc}{mc^2}$$

where $mc^2 = 1.3$ GeV. Hence

$$\lambda = \frac{(6.6 \times 10^{-34}\ \mathrm{J \cdot s})(3.0 \times 10^8\ \mathrm{m/s})}{(1.3 \times 10^9\ \mathrm{eV})(1.6 \times 10^{-19}\ \mathrm{J/eV})} = 0.96 \times 10^{-15}\ \mathrm{m},$$

or 0.96 fm. This is somewhat less than the size of nuclei.

Another major reason for building high-energy accelerators is that new particles of greater mass can be produced at higher energies, as we will discuss shortly. Now we look briefly at several types of particle accelerator.

32–2 • Particle Accelerators

Van de Graaff accelerator. The key component of a Van de Graaff accelerator (Fig. 32–1) is a Van de Graaff generator, invented in 1931. A large, hollow, spherical conductor is supported by an insulating column above the base. The sphere is charged to a high potential by a nonconducting moving belt in the following way. A voltage of typically 50,000 V is applied to a pointed conductor A, which "sprays" positive charge onto the moving belt. (Actually, electrons are pulled off the belt onto the electrode A.) The belt carries the positive charge into the interior of the sphere, where it is "wiped off" the belt at B and races to the outer surface of the spherical conductor.

Van de Graaff generator and accelerator

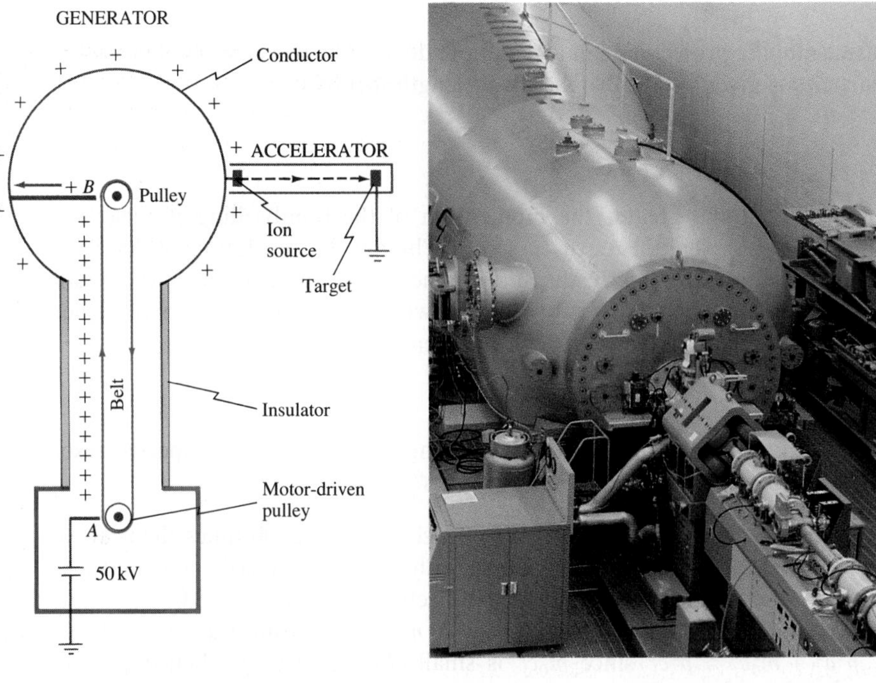

FIGURE 32–1 Van de Graff generator and accelerator: (a) diagram; (b) in use at Brookhaven National Laboratory.

(Remember that charge collects on the outer surface of any conductor, since the charges repel each other and try to get as far from each other as possible.) As more and more charge is brought upward, the sphere becomes more highly charged and reaches greater voltage. The process requires energy, of course, since the upward-moving charged belt is repelled by the charged sphere. The energy is supplied by the motor driving the belt. Very high potential differences can be obtained in this way, often giving the accelerated particles as much as 30 MeV of KE when two or more generators are used in tandem. Connected to the Van de Graaff generator is an evacuated tube which serves as the particle accelerator. A source of H or He ions (p or α) is located inside the tube, and the large positive voltage repels them so they are accelerated toward the grounded target at the far end of the tube.

Cyclotron

Cyclotron. The cyclotron was developed in 1930 by E. O. Lawrence (1901–1958; Fig. 32–2) at the University of California at Berkeley. It uses a magnetic field to maintain charged ions—usually protons—in nearly circular paths (Chapter 20). The protons move within two D-shaped cavities, as shown in Fig. 32–3. Each time they pass into the gap between the "dees," a voltage is applied that accelerates them. This increases their speed and also increases the radius of curvature of their path. After many revolutions, the protons acquire high kinetic energy and reach the outer edge of the cyclotron. They then either strike a target placed inside the cyclotron or leave the cyclotron with the help of a carefully placed "bending magnet" and are directed to an external target. The voltage applied to the dees to produce the acceleration must be alternating. When the protons are moving to the right across the gap in Fig. 32–3, the right dee must be negative and the left one positive. A half-cycle later, the protons are moving to the left, so the left dee must be negative in order to accelerate them. The frequency, f, of the applied voltage must be equal to that of the circulating protons, and this can be determined as follows. When ions of charge q (for a proton, $q = +e$) are circulating within the dees, the net force F on each is simply that due to the magnetic field B, so $F = qvB$, where v is the speed of the ion at a given moment (Eq. 20–4). (We ignore the electric force due to the applied voltage since it acts only when the ions are between the dees.) Since the ions move in circles, the acceleration is centripetal and equals v^2/r, where r is the radius of the ion's path at a given moment. We use Newton's second law, $F = ma$, and find that

$$F = ma$$

$$qvB = \frac{mv^2}{r}$$

or

$$v = \frac{qBr}{m}.$$

The time required for a complete revolution is the period T and is equal to

$$T = \frac{\text{distance}}{\text{speed}} = \frac{2\pi r}{qBr/m} = \frac{2\pi m}{qB}.$$

Hence the frequency of revolution f is

$$f = \frac{1}{T} = \frac{qB}{2\pi m}. \qquad (32\text{–}1)$$

This is known as the **cyclotron frequency**.

FIGURE 32–2 Ernest O. Lawrence holding the first cyclotron, around 1930.

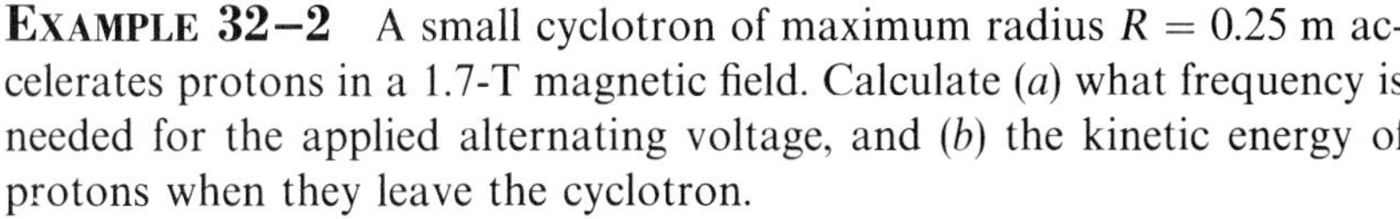

EXAMPLE 32–2 A small cyclotron of maximum radius $R = 0.25$ m accelerates protons in a 1.7-T magnetic field. Calculate (*a*) what frequency is needed for the applied alternating voltage, and (*b*) the kinetic energy of protons when they leave the cyclotron.

SOLUTION (*a*) From Eq. 32–1,

$$f = \frac{qB}{2\pi m} = \frac{(1.6 \times 10^{-19}\ \text{C})(1.7\ \text{T})}{(6.28)(1.67 \times 10^{-27}\ \text{kg})} = 2.6 \times 10^{7}\ \text{Hz}.$$

(*b*) The protons leave the cyclotron at $r = R = 0.25$ m. Then, since $v = qBr/m$,

$$\text{KE} = \frac{1}{2}mv^2 = \frac{1}{2}m\frac{q^2B^2R^2}{m^2} = \frac{q^2B^2R^2}{2m}$$

$$= \frac{(1.6 \times 10^{-19}\ \text{C})^2(1.7\ \text{T})^2(0.25\ \text{m})^2}{(2)(1.67 \times 10^{-27}\ \text{kg})} = 1.4 \times 10^{-12}\ \text{J},$$

or 8.8 MeV. Note that the magnitude of the voltage applied to the dees does not affect the final energy. But the higher this voltage, the fewer revolutions are required to bring the protons to full energy.

FIGURE 32–3 Diagram of a cyclotron. The magnetic field, applied by a large electromagnet, points into the page. A is the ion source. The field lines shown are for the electric field in the gap.

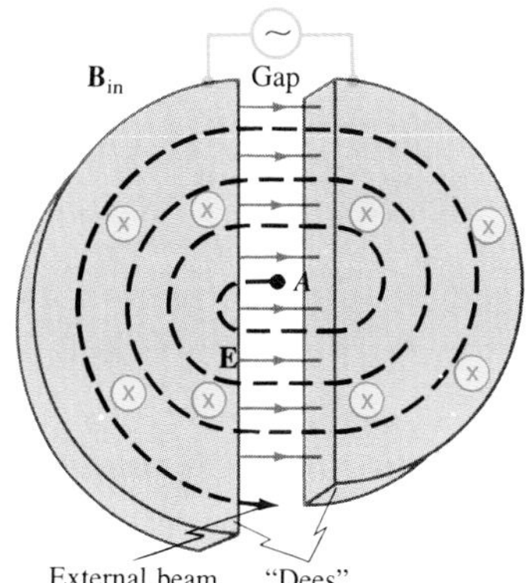

An interesting aspect of the cyclotron is that the frequency of the applied voltage, as given by Eq. 32–1, does not depend on the radius r. That is, the frequency does not have to be changed as the ions start from the source and are accelerated to paths of larger and larger radii. Unfortunately, this is only true at nonrelativistic energies. For at higher speeds, the mass of the ions will increase according to Einstein's formula, $m = m_0/\sqrt{1 - v^2/c^2}$, where m_0 is the rest mass. As can be seen from Eq. 32–1, as the mass increases, the frequency of the applied voltage must be reduced. To achieve large energies, complex electronics is needed to decrease the frequency as a packet of protons increases in speed and reaches larger orbits. Such a modified cyclotron is called a **synchrocyclotron**.

Synchrocyclotron

Synchrotron. Another way to deal with the increase in mass with speed is to increase the magnetic field B as the particles speed up. Such a device is called a *synchrotron*, and today they can be enormous. The Fermi National Accelerator Laboratory (Fermilab) at Batavia, Illinois, has a radius of 1.0 km and that at CERN (European Center for Nuclear Research) in Geneva, Switzerland, is 1.1 km in radius. These machines can accelerate protons to energies of 500 GeV. The new *Tevatron* at Fermilab can accelerate protons to 1 TeV (hence its name: 1 TeV = 10^{12} eV) using superconducting magnets. These large synchrotrons do not use enormous magnets 1 km in radius. Instead, a narrow ring of magnets is used (see Fig. 32–4) which are all placed at the same radius from the center of the circle. The magnets are interrupted

Synchrotron

(a)

(b)

FIGURE 32–4 (a) Aerial view of Fermilab at Batavia, Illinois; the accelerator is a circular ring 1.0 km in radius. (b) The interior of the tunnel of the main accelerator at Fermilab. The upper (rectangular-shaped) ring of magnets is for the 500-GeV accelerator. Below it is the ring of superconducting magnets for the Tevatron.

by gaps where high voltage accelerates the particles. After the particles are injected, they must then move in a circle of constant radius. This is accomplished by giving them considerable energy initially in a much smaller accelerator, and then slowly increasing the magnetic field as they speed up in the large synchrotron.

One problem of any accelerator is that accelerating electric charges radiate electromagnetic energy (see Chapter 22). Since ions or electrons are accelerated in an accelerator, we can expect considerable energy to be lost by radiation. The effect increases with speed and is especially important in circular machines, where centripetal acceleration is present, particularly in synchrotrons, and hence is called **synchrotron radiation**. Synchrotron radiation can actually be useful, however, when intense beams of photons are needed, and it is usually obtained from an electron synchrotron.

Linac

Linear accelerators. A Van de Graaff accelerator is essentially a linear accelerator since the ions move in a linear path. But the name *linear accelerator* is usually reserved for a more complex arrangement in which particles are accelerated many, many times along a straight-line path. Figure 32–5 is a diagram of a simple "linac." The ions pass through a series of tubular conductors. The voltage applied to the tubes must be alternating so that when the ions reach a gap, the tube in front of them is negative and the one they just left is positive. This assures that they are accelerated at each gap. As the ions increase in speed, they cover more distance in the same amount of time. Consequently, the tubes must be longer the farther they are from the source. The Heavy Ion Linear Accelerator (Hilac) at Berkeley, California, can accelerate ions from hydrogen through uranium, imparting to them as much as 8.5 MeV per nucleon. Linear accelerators are of particular importance for accelerating electrons. Because of their small mass, electrons reach high speeds very quickly. Indeed, an electron linac such as the one shown in Fig. 32–5 would have tubes nearly equal in length, since the electrons would be traveling close to $c = 3.0 \times 10^8$ m/s for almost the entire distance. The amount of energy radiated away by electrons in a linear machine is much less than for a circular machine (see mention of synchrotron radiation above). The largest electron linear accelerator is that at Stanford, called SLAC (Stanford Linear Accelerator Center). It is over 3 km (2 miles) long and can accelerate electrons to 50 GeV (recall 1 GeV = 10^9 eV). Many hospitals have 10-MeV electron linacs that produce photons to irradiate tumors.

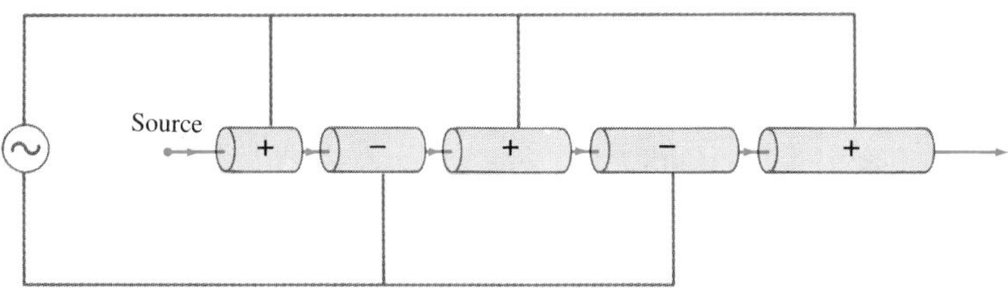

FIGURE 32–5 Simple linear accelerator.

Colliding Beams. One way in which **high-energy-physics** experiments (as this field in which high-energy accelerators are used is called) can be carried out is to allow the beam of particles from an accelerator to strike a stationary target. An alternative method which increases the energy of a collision, is the recent development of *colliding beams*—that is, the target particles are

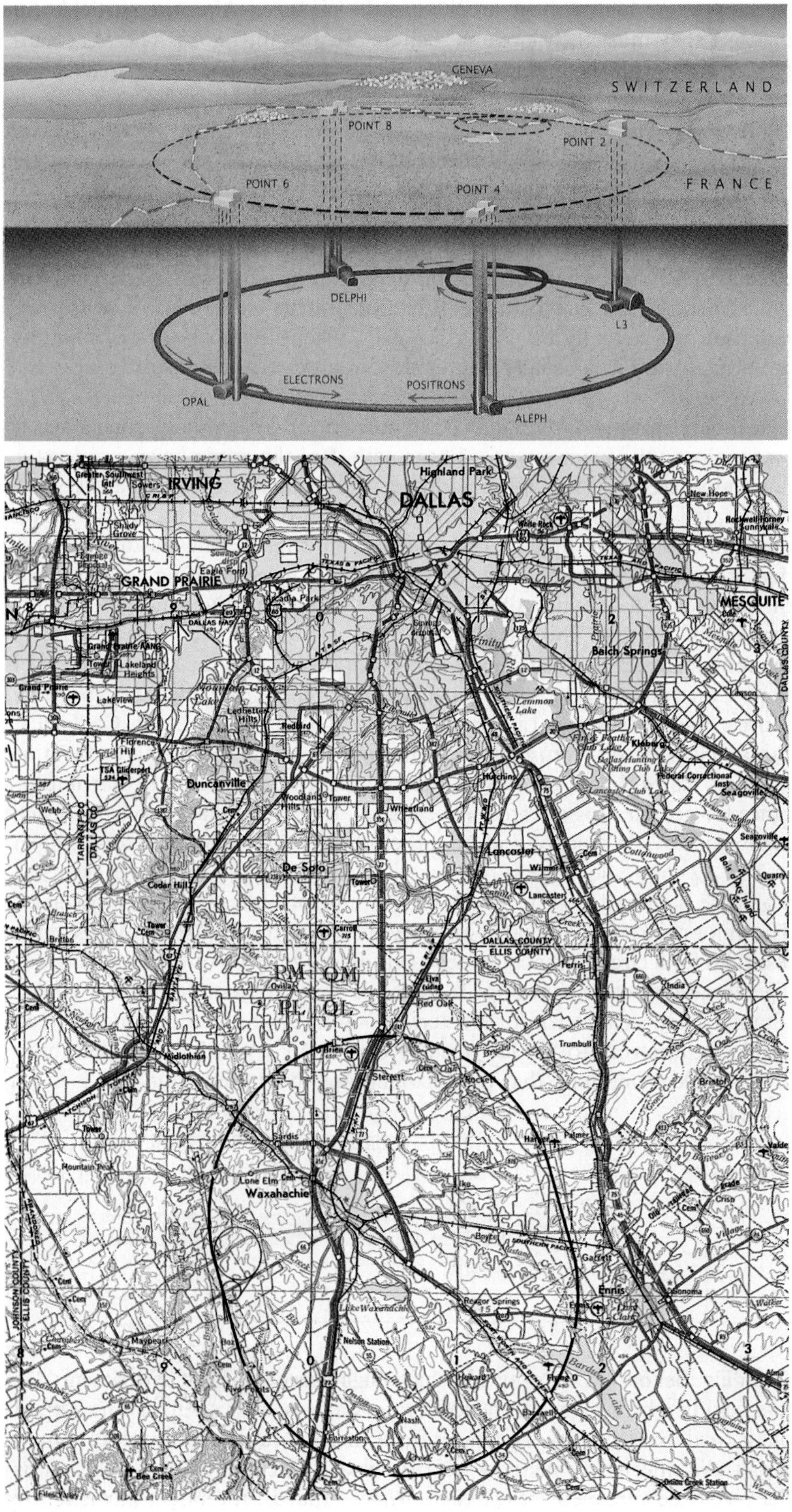

FIGURE 32–6 (a) The 8.5 km diameter LEP collider, about 100 m below ground, straddles the French-Swiss border near Geneva. (b) Diagram of the SSC (superconducting supercollider) shown superimposed on a map of Texas, to give an idea of its size.

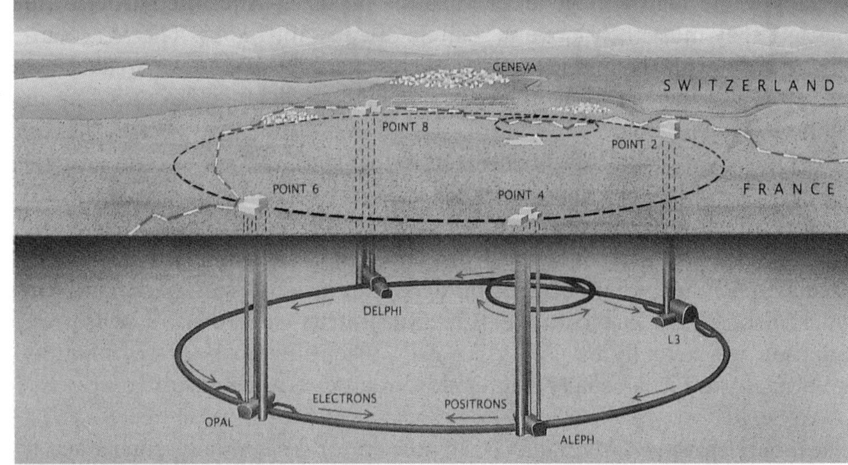

(a)

(b)

moving as well as the projectile particles. This can be accomplished with only one accelerator through the use of **storage rings**. The accelerator accelerates one type of particle (say electrons or protons) to a maximum energy and then magnets are used to steer these particles into one circular storage ring where the particles can continue to circulate for many hours. It then accelerates a second type of particle (say positrons), or another group of the first type (such as protons), and these are sent to a second storage ring. The two storage rings overlap at several places and the two beams, circulating in opposite directions, collide head-on with each other at these points of intersection. Storage rings for colliding-beam experiments are in use at a number of facilities around the world, including CERN, Fermilab, and at the Stanford Linear Collider (SLC), and they have played an important role in recent advances in elementary particle physics. The largest collider-accelerator today is the truly enormous Large Electron-Positron (LEP) Collider at CERN, which has a circumference of 26.7 km (Fig. 32–6a). Inaugurated in 1989, it produces oppositely revolving beams of e^+ and e^-, each of energy 50 GeV, for a total interaction energy of 100 GeV. The even larger proposed **Superconducting Super Collider** (SSC) in the United States is planned to accelerate protons to 20 TeV in each beam, for a total energy of 40 TeV (see Fig.32–6b).

32–3 • Beginnings of Elementary Particle Physics—the Yukawa Particle

By the mid-1930s, it was recognized that all atoms can be considered to be made up of neutrons, protons, and electrons. The basic constituents of the universe were no longer considered to be atoms but rather the proton, neutron, and electron. Besides these three *elementary particles*, as they could be called, several others were also known: the positron (a positive electron), the neutrino, and the γ particle (or photon), for a total of six elementary particles.

Looking back, things seemed fairly simple in 1935. But in the decades that followed, hundreds of other subnuclear particles were discovered. The properties and interactions of these particles, and which ones should be considered as fundamental or "elementary," became the substance of research in the field of **elementary particle physics**.

Elementary particle physics, as it exists today, can be said to have begun in 1935 when the Japanese physicist Hideki Yukawa (1907–1981) predicted the existence of a new particle that would in some way mediate the strong nuclear force. To understand Yukawa's idea, we first look at the electromagnetic force. When we first discussed electricity, we saw that the electric force acts over a distance, without contact. To better perceive how a force can act over a distance, we saw that Faraday introduced the idea of a *field*. The force that one charged particle exerts on a second can be said to be due to the electric field set up by the first. Similarly, the magnetic field can be said to carry the magnetic force. Later (Chapter 22), we saw that electromagnetic fields can travel through space as waves. Finally, in Chapter 27, we saw that electromagnetic radiation (light) can be considered as either a wave or as a collection of particles called photons. Because of this wave–particle duality, it is possible to imagine that the electromagnetic force between charged particles is due (1) to the EM field set up by one and felt by

(a) Repulsive force (children throwing pillows)

(b) Attractive force (children grabbing pillows from each other's hands)

FIGURE 32–7 Forces equivalent to particle exchange. (a) Replusive force (children throwing pillows at each other). (b) Attractive force (children grabbing pillows from each other's hands).

FIGURE 32–8 Feynman diagram showing a photon acting as the carrier of the electromagnetic force between two electrons.

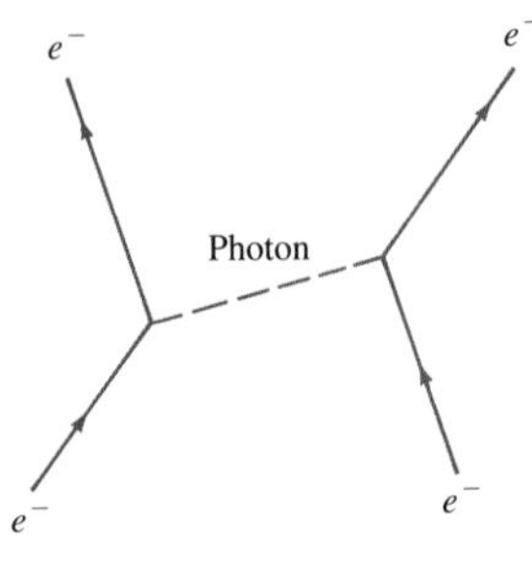

FIGURE 32–9 (below) Meson exchange when a proton and neutron interact via the strong nuclear force.

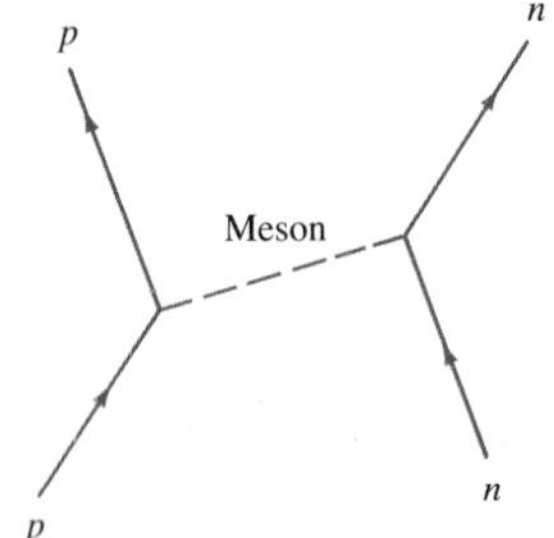

the other, or (2) to an exchange of photons or γ particles between them. It is (2) that we want to concentrate on here, and a crude analogy for how an exchange of particles could give rise to a force is suggested in Fig. 32–7. In part (a), two children start throwing pillows at each other; each catch results in the child being pushed backward by the impulse. This is the equivalent of a repulsive force. On the other hand, if the two children exchange pillows by grabbing them out of the other person's hand, they will be pulled toward each other, as when an attractive force acts.

For the electromagnetic force between two charged particles, it is photons that are exchanged between the two particles that give rise to the force. A simple diagram describing this photon exchange is shown in Fig. 32–8. Such a diagram, called a **Feynman diagram** (after its inventor, the American physicist, Richard Feynman (1918–1988), is based on the theory called **quantum electrodynamics** (QED). The case shown is the simplest, in which a single photon is exchanged. One of the charged particles emits the photon and recoils somewhat as a result; and the second particle absorbs the photon. In any such collision or *interaction*, energy and momentum are transferred from one particle to the other, carried by the photon. Because the photon is absorbed by the second particle very shortly after it is emitted by the first, it is not observable, and is referred to as a *virtual* photon, in contrast to one that is free and can be detected by instruments. The photon is said to *mediate*, or *carry*, the electromagnetic force.

Now to Yukawa's prediction. By analogy with photon exchange that mediates the electromagnetic force, Yukawa argued that there ought to be a particle that mediates the strong nuclear force—the force that holds nucleons together in the nucleus. Just as the photon is called the quantum of the electromagnetic field or force, so the Yukawa particle would represent the quantum of the strong nuclear force.

Yukawa predicted that this new particle would have a mass intermediate between that of the electron and the proton. Hence it was called a **meson**, meaning "in the middle," and Fig. 32–9 is a Feynman diagram of meson exchange. We can make a rough approximation of the mass of the meson as follows. Suppose the proton on the left in Fig. 32–9 is at rest. For it to emit a meson would require energy (to make the mass) that would have to come from nowhere; such a process would violate conservation of energy. But the uncertainty principle allows nonconservation of energy by an amount ΔE if it occurs only for a time Δt given by $(\Delta E)(\Delta t) \approx h/2\pi$. We set ΔE equal to the energy needed to create the mass m of the meson: $\Delta E = mc^2$. Now conservation of energy is violated only as long as the meson exists, which is the time Δt required for the meson to pass from one nucleon to the other. If we assume the meson travels at relativistic speed, close to the speed of light c, then Δt need be at most about $\Delta t = d/c$, where d is the maximum distance that can separate the interacting nucleons. Thus we have

$$\Delta E\,\Delta t \approx \frac{h}{2\pi}$$

$$mc^2\left(\frac{d}{c}\right) \approx \frac{h}{2\pi}$$

or

$$mc^2 \approx \frac{hc}{2\pi d}. \tag{32–2}$$

The range of the strong nuclear force is about $d \approx 1.5 \times 10^{-15}$ m (this is the maximum distance away it can be felt), so from Eq. 32–2,

$$mc^2 \approx \frac{(6.6 \times 10^{-34}\ \text{J}\cdot\text{s})(3.0 \times 10^8\ \text{m/s})}{(6.28)(1.5 \times 10^{-15}\ \text{m})} \approx 2.2 \times 10^{-11}\ \text{J} = 130\ \text{MeV}.$$

The mass of the predicted meson is thus very roughly 130 MeV/c^2 or about 250 times the electron mass of 0.51 MeV/c^2. (Note, incidentally, that since the electromagnetic force has infinite range ($d = \infty$), Eq. 32–2 tells us that the exchanged particle for the electromagnetic force, the photon, will have zero rest mass.)

Just as photons can be observed as free particles, as well as acting in an exchange, so it was expected that mesons might be observed directly. Such a meson was searched for in the cosmic radiation that enters the earth's atmosphere from the sun and other sources in the universe. In 1937 a new particle was discovered whose mass was 106 MeV (207 times the electron mass). This is close to the mass predicted. But it turned out that this new particle, called the **muon** (or *mu meson*), did not interact strongly with matter. *Muon* It could hardly mediate the strong nuclear force if it didn't interact via the strong nuclear force. Thus the muon, which can have either a positive or a negative charge and seems to be nothing more than a very massive electron, is not the Yukawa particle.

The particle predicted by Yukawa was finally found in 1947. It is called the "π" or pi meson, or simply the **pion**. It comes in three charge states: +, *Pion* −, or 0. The π^+ and π^- have mass of 139.6 MeV/c^2 and the π^0 a mass of 135.0 MeV/c^2. All three interact strongly with matter. Soon after their discovery in cosmic rays, pions were produced in the laboratory using a particle accelerator. Reactions observed include

$$\begin{aligned} p + p &\rightarrow p + p + \pi^0, \\ p + p &\rightarrow p + n + \pi^+. \end{aligned} \qquad (32\text{–}3)$$

The incident proton from the accelerator must have sufficient energy to produce the additional mass of the pion. A number of other mesons were discovered in subsequent years which were also considered to mediate the strong nuclear force. The recent theory of quantum chromodynamics, however, which involves quarks, has replaced mesons with *gluons* as the basic carriers of the strong force, as we shall discuss in Section 32–10.

So far we have discussed the particles that mediate the electromagnetic and strong nuclear forces. But there are four known types of force—or interaction—in nature. What about the other two: the weak nuclear force, and gravity? Theorists believe that these are also mediated by particles. The particles presumed to transmit the weak force are referred to as the W^+, W^-, and Z^0. In 1983, after extensive searches, the long-awaited discovery of the W and Z^0 particles was announced by Carlo Rubbia, the physicist behind the building of the proton–antiproton collider at the very high-energy accelerator at CERN that produced the results, and leader of a large group of scientists (over 100) who worked on the project. See Fig. 32–10. The quantum of the gravitational force, called the **graviton**, has not yet been identified. A comparison of the four forces is given in Table 32–1, where they are listed according to their (approximate) relative strengths. Notice that although gravity may be the most obvious force in daily life (because of the huge mass

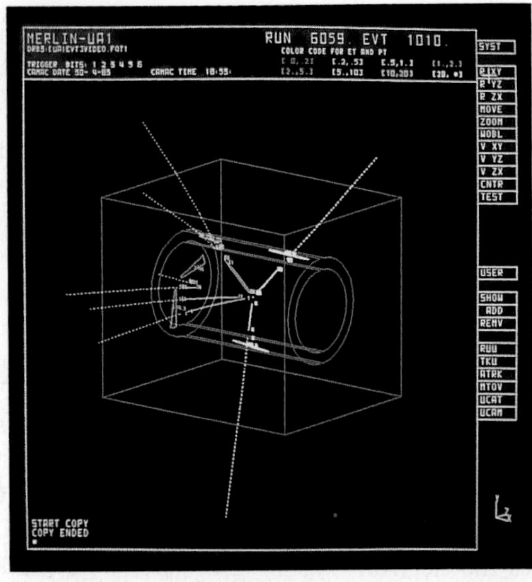

(a)

(b)

FIGURE 32–10 (a) Computer reconstruction of a Z-particle decay into an electron and a positron ($Z^0 \rightarrow e^+e^-$) whose tracks are shown in white, which took place in the UA1 detector at CERN. (b) Photo of the UA1 detector at CERN as it was being built, showing a wall of photomultiplier tubes at right.

of the earth), on a nuclear scale, it is much the weakest of the four forces and its effect at the nuclear level can nearly always be ignored.

32–4 • Particles and Antiparticles

The positron, as we have seen, is basically a positive electron. That is, many of its properties are the same as for the electron, such as mass, but it has the opposite charge. The positron is said to be the **antiparticle** to the electron. After the positron was discovered in 1932, it was predicted that other particles also ought to have antiparticles. In 1955 the antiparticle to the proton was found, the **antiproton** ($\bar{p}$); see Fig. 32–11. (The bar over the p is used to indicate antiparticle.) A large amount of energy was needed to produce this massive particle (mass = proton's mass). Its discovery by Emilio Segré (1905–1989) and Owen Chamberlain (1920–   ) was made only after the completion of the large accelerator (the Bevatron) at the University of California at Berkeley. Soon after, the antineutron ($\bar{n}$) was found. Most other particles also have antiparticles. But the photon, the π^0, and a few other particles do not have distinct antiparticles—or we say that they are their own antiparticles.

Table 32–1
The Four Forces in Nature

Type	Relative Strength (approx.)	Field Particle
Strong nuclear	1	Mesons/gluons†
Electromagnetic	10^{-2}	Photon
Weak nuclear	10^{-13}	$W^{\pm}$ and Z^0
Gravitational	10^{-40}	Graviton (?)

† See Section 32–10.

Antiparticles are produced in nuclear reactions when there is sufficient energy available, and they do not live very long in the presence of matter. For example, when a positron encounters an electron, the two annihilate each other. The energy of their vanished mass, plus any kinetic energy they possessed, is converted to the energy of γ rays or of other particles. Annihilation occurs for all other particle–antiparticle pairs.

32–5 • Particle Interactions and Conservation Laws

One of the important uses of high-energy accelerators is to study the interactions of elementary particles with each other. As a means of ordering this subnuclear world, the conservation laws are indispensable. The laws of conservation of energy, of momentum, of angular momentum, and of electric charge are found to hold precisely in all particle interactions. (Although the uncertainty principle allows, for example, nonconservation of energy for times $\Delta t \approx h/2\pi\Delta E$, as we have seen, the times involved are much too short for experimental observation and violation has never been detected.

A study of particle interactions has revealed a number of new conservation laws, some of which we now discuss. These new conservation laws (just like the old ones) are ordering principles: they help to explain why some reactions occur and others do not. For example, the following reaction has never been found to occur:

$$p + n \nrightarrow p + p + \bar{p}$$

even though charge, energy, and so on, are conserved ($\bar{p}$ means an antiproton and $\nrightarrow$ means the reaction does not occur). To understand why such a reaction doesn't occur, physicists hypothesized a new conservation law, the conservation of **baryon number**. (Baryon number is the same as nucleon number—which we saw earlier is conserved in nuclear reactions.) An important addition to this law is the proposal that whereas all nucleons have baryon number $B = +1$, all antinucleons (antiprotons, antineutrons) have $B = -1$. The reaction above does not conserve baryon number since on the left side we have $B = (+1) + (+1) = +2$, and on the right $B = (+1) + (+1) + (-1) = +1$. On the other hand, the following reaction does conserve B and *does* occur if the incoming proton has sufficient energy:

$$p + n \rightarrow p + n + \bar{p} + p,$$

$$B = +1 + 1 = +1 + 1 - 1 + 1.$$

As indicated, $B = +2$ on both sides of this equation. From these and other reactions, the conservation of baryon number has been established as a basic law of physics.

Also useful are the conservation laws for the three **lepton numbers**, associated with weak interactions including decays. In ordinary β decay, an electron or positron is emitted along with a neutrino or antineutrino. In a similar type of decay, a muon is emitted instead of an electron. The neutrino (ν_e) that accompanies an emitted electron is found to be different from the neutrino (ν_μ) that accompanies an emitted muon. Each of these neutrinos has an antiparticle: $\bar{\nu}_e$ and $\bar{\nu}_\mu$. In ordinary β decay we have, for example,

$$n \rightarrow p + e^- + \bar{\nu}_e$$

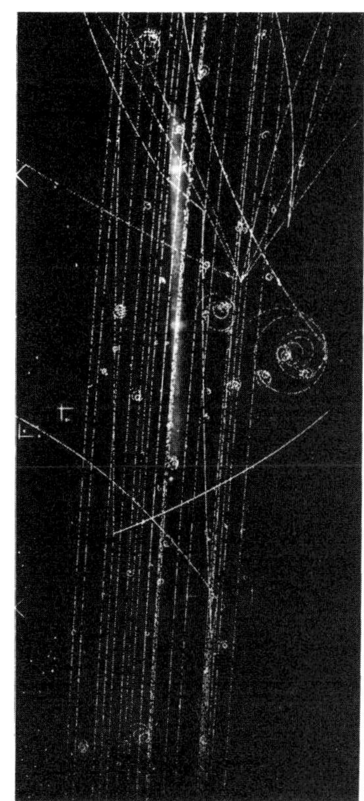

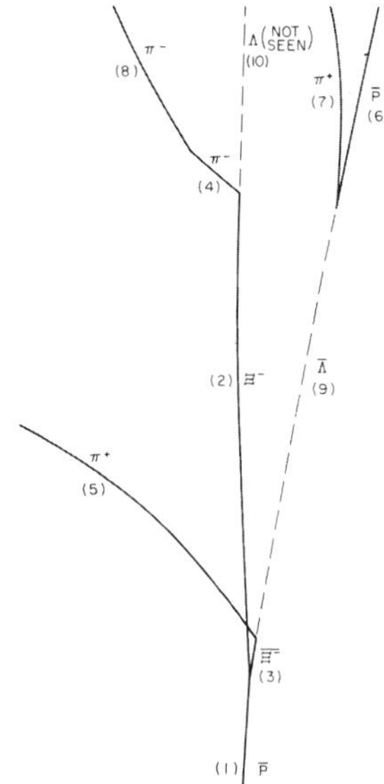

FIGURE 32–11 Liquid-hydrogen bubble-chamber photograph of an antiproton ($\bar{p}$) colliding with a proton, producing a hyperon pair ($\bar{p} + p \rightarrow \Xi^- + \bar{\Xi}^-$) that subsequently decay into other particles. The drawing indicates the assignment of particles to each track. Neutral particle paths are shown by dashed lines since neutral particles produce no bubbles and hence no tracks.

but never $n \rightarrow p + e^- + \bar{\nu}_\mu$ or $n \rightarrow p + e^- + \bar{\nu}_e + \nu_e$. To explain why these do not occur, the concept of electron lepton number, L_e, was invented. If the electron (e^-) and the electron neutrino (ν_e) are given $L_e = +1$, and e^+ and $\bar{\nu}_e$ are given $L_e = -1$, whereas all other particles have $L_e = 0$, then all observed decays conserve L_e. For example, in $n \rightarrow p + e^- + \bar{\nu}_e$, $L_e = 0$ initially, and $L_e = 0 + (+1) + (-1) = 0$ after the decay. Decays that do not conserve L_e but would obey the other conservation laws are not observed to occur. Hence it is believed that L_e is conserved in all interactions.

In a decay involving muons, such as

$$\pi^+ \rightarrow \mu^+ + \nu_\mu,$$

a second quantum number, muon lepton number (L_μ), is conserved. The μ^- and ν_μ are assigned $L_\mu = +1$, and μ^+ and $\bar{\nu}_\mu$ have $L_\mu = -1$, whereas other particles have $L_\mu = 0$. It is believed that L_μ is also conserved in all interactions or decays. Similar assignments can be made for a third lepton number, L_τ, associated with the recently discovered τ lepton and its neutrino, ν_τ.

Antiparticles have opposite Q, B, Ls

Keep in mind that antiparticles have not only opposite electric charge from their particles, but also opposite B, L_e, L_μ, and L_τ.

EXAMPLE 32–3 Which of the following decay schemes is possible for muon decay: (*a*) $\mu^- \rightarrow e^- + \bar{\nu}_e$; (*b*) $\mu^- \rightarrow e^- + \bar{\nu}_e + \nu_\mu$; (*c*) $\mu^- \rightarrow e^- + \nu_e$? All of these particles have $L_\tau = 0$.

SOLUTION A μ^- has $L_\mu = +1$ and $L_e = 0$. This is the initial state, and the final state (after decay) must also have $L_\mu = +1$, $L_e = 0$. In (*a*), the final state has $L_\mu = 0 + 0 = 0$, and $L_e = +1 - 1 = 0$; L_μ would not be conserved and indeed this decay is not observed to occur. The final state of (*b*) has $L_\mu = 0 + 0 + 1 = +1$ and $L_e = +1 - 1 + 0 = 0$, so both L_μ and L_e are conserved. This is in fact the most common decay mode of the μ^-. Finally, (*c*) does not occur because L_e ($= +2$ in final state) is not conserved, nor is L_μ.

Recent theoretical work suggests that the conservation laws of baryon and lepton numbers may be only approximate, rather than exact, as we shall discuss in Section 32–11.

32–6 • Particle Classification

In the decades following the discovery of the π meson in the late 1940s, a great many other subnuclear particles were discovered. Today they number in the hundreds. Much theoretical and experimental work has been done to try to understand this multitude of particles. An important aid to understanding is to arrange the particles in categories according to their properties. One way of doing this is according to their interactions. Since not all particles interact by means of all four of the forces known in nature, this is used as a classification scheme. Table 32–2 lists many of the known particles classified in this way along with many of their properties. The particles listed are those that are stable, and many that are unstable (namely, those whose decay does not occur via the strong interaction—see the next section and

Table 32–2
Elementary Particles (stable under strong decay)†

Category	Particle Name	Symbol	Antiparticle	Rest Mass (MeV/c^2)	B	L_e	L_μ	L_τ	S	Lifetime (s)	Principal Decay Modes
Gauge bosons	Photon	γ	Self	0	0	0	0	0	0	Stable	
	W	W^+	W^-	80.6×10^3	0	0	0	0	0	3×10^{-25}	$e\nu_e$, $\mu\nu_\mu$, $\tau\nu_\tau$, hadrons
	Z	Z^0	Self	91.2×10^3	0	0	0	0	0	3×10^{-25}	e^+e^-, $\mu^+\mu^-$, $\tau^+\tau^-$, hadrons
Leptons	Electron	e^-	e^+	0.511	0	+1	0	0	0	Stable	
	Neutrino (e)	ν_e	$\bar{\nu}_e$	0(?)§	0	+1	0	0	0	Stable	
	Muon	μ^-	μ^+	105.7	0	0	+1	0	0	2.20×10^{-6}	$e^-\bar{\nu}_e\nu_\mu$
	Neutrino (μ)	ν_μ	$\bar{\nu}_\mu$	0(?)§	0	0	+1	0	0	Stable	
	Tau	τ^-	τ^+	1784.	0	0	0	−1	0	3×10^{-13}	$\mu^-\bar{\nu}_\mu\nu_\tau$, $e^-\bar{\nu}_e\nu_\tau$, hadrons + ν_τ
	Neutrino (τ)	ν_τ	$\bar{\nu}_\tau$	0(?)§	0	0	0	−1	0	Stable	
Hadrons											
Mesons	Pion	π^+	π^-	139.6	0	0	0	0	0	2.60×10^{-8}	$\mu^+\nu_\mu$
		π^0	Self	135.0	0	0	0	0	0	0.84×10^{-16}	2γ
	Kaon	K^+	K^-	493.6	0	0	0	0	+1	1.24×10^{-8}	$\mu^+\nu_\mu$, $\pi^+\pi^0$
		K_S^0	$\bar{K}_S^0$	497.7	0	0	0	0	+1	0.89×10^{-10}	$\pi^+\pi^-$, $2\pi^0$
		K_L^0	$\bar{K}_L^0$	497.7	0	0	0	0	+1	5.2×10^{-8}	$\pi^\pm e^\mp \overset{(-)}{\nu}_e$, $\pi^\pm\mu^\mp \overset{(-)}{\nu}_\mu$, 3π
	Eta	η^0	Self	548.8	0	0	0	0	0	5×10^{-19}	2γ, $3\pi^0$, $\pi^+\pi^-\pi^0$
Baryons	Proton	p	$\bar{p}$	938.3	+1	0	0	0	0	Stable‡	
	Neutron	n	$\bar{n}$	939.6	+1	0	0	0	0	900	$pe^-\bar{\nu}_e$
	Lambda	Λ^0	$\bar{\Lambda}^0$	1115.6	+1	0	0	0	−1	2.6×10^{-10}	$p\pi^-$, $n\pi^0$
	Sigma	Σ^+	$\bar{\Sigma}^-$	1189.4	+1	0	0	0	−1	0.80×10^{-10}	$p\pi^0$, $n\pi^+$
		Σ^0	$\bar{\Sigma}^0$	1192.5	+1	0	0	0	−1	7×10^{-20}	$\Lambda^0\gamma$
		Σ^-	$\bar{\Sigma}^+$	1197.4	+1	0	0	0	−1	1.5×10^{-10}	$n\pi^-$
	Xi	Ξ^0	$\bar{\Xi}^0$	1315	+1	0	0	0	−2	2.9×10^{-10}	$\Lambda^0\pi^0$
		Ξ^-	Ξ^+	1321	+1	0	0	0	−2	1.64×10^{-10}	$\Lambda^0\pi^-$
	Omega	Ω^-	Ω^+	1672	+1	0	0	0	−3	0.82×10^{-10}	$\Xi^0\pi^0$, Λ^0K^-

†See also Table 32–4 for particles with charm and bottomness.
‡Upper limits on neutrino masses: $m_{\nu_e} \lesssim 17\ eV/c^2$; $m_{\nu_\mu} \lesssim 270\ keV/c^2$; $m_{\nu_\tau} \lesssim 35\ MeV/c^2$.
§ $>4.5 \times 10^{32}$ yr for $p \to \pi^0 e^+$.

also Table 32–4). At the top of the table are the **gauge bosons** (so-named† after the theory that describes them, "gauge theory"), which include the *photon* and the *W* and *Z* particles that carry the electromagnetic and weak interactions, respectively.

Gauge bosons

Next in Table 32–2 are the **leptons**, which are particles that do not interact via the strong force but do interact via the weak nuclear force (as well as the much weaker gravitational force); those that carry electric charge also interact via the electromagnetic force. The leptons include the electron, the muon, and the tau (or τ, discovered in 1976 and more than 3000 times heavier than the electron), and three types of neutrino: the electron neutrino (ν_e), the muon neutrino (ν_μ), and the tau neutrino (ν_τ). They each have antiparticles, as indicated in Table 32–2.

Leptons

The third category of particle in Table 32–2 is the **hadron**. Hadrons are those particles that can interact via the strong nuclear force. Hence they are said to be **strongly interacting particles**. They also interact via the other

Hadrons

† Bosons are particles that are not governed by the Pauli exclusion principle (Section 28–7).

forces, but the strong force predominates at short distances. The hadrons include nucleons, pions, and a large number of other particles. They are divided into two subgroups:† **baryons**, which are those particles that have baryon number $+1$ (or -1 in the case of their antiparticles); and **mesons**, which have baryon number $= 0$.

Baryons

Mesons

Notice that the baryons Λ, Σ, Ξ, and Ω all decay to lighter-mass baryons, and eventually to a proton or neutron. All these processes conserve baryon number. Since there is no lighter particle than the proton with $B = +1$, if baryon number is strictly conserved, the proton itself cannot decay and is stable (but see Section 32–11).

32–7 • Particle Stability and Resonances

Lifetime depends on which force is acting

Many of the particles listed in Table 32–2 are unstable. The lifetime of an unstable particle depends on which force is most active in causing the decay. When we say the strong nuclear force is stronger than the electromagnetic, we mean that two particles will interact more strongly and more quickly if this force is acting. When a stronger force influences a decay, that decay occurs more quickly. Decays caused by the weak force typically have lifetimes of 10^{-13} s or longer. Particles that decay via the electromagnetic force have much shorter lifetimes, typically about 10^{-16} to 10^{-19} s. (Exceptions to this scheme are the W and Z particles, which decay via the weak interaction but have very short lifetimes due to their special nature as exchange particles.) The unstable particles listed in Table 32–2 decay either via the weak or the electromagnetic interaction. Decays that involve a γ (photon) are electromagnetic (such as $\pi^0 \rightarrow 2\gamma$). The other decays shown take place via the weak interaction, often accompanied by a neutrino which interacts only via the weak interaction: examples are $\pi^+ \rightarrow \mu^+\nu_\mu$ and $\Sigma^- \rightarrow n\pi^-$.

A great many particles have been found that decay via the strong interaction, and these are not listed in Table 32–2. Such particles decay into other strongly interacting particles (say, n, p, π, but not involving γ, e, ν, and so on) and their lifetimes are very short, typically 10^{-23} s. In fact their lifetimes are so short that they do not travel far enough to be detected before decaying. Their decay products can be detected, however, and it is from them that the existence of such short-lived particles is inferred. To see how this is done, let us consider the first such particle discovered (by Fermi). Fermi used a beam of π^+ directed through a hydrogen target (protons) with varying amounts of energy. A graph of the number of interactions (π^+ scattered) versus the pion's kinetic energy is shown in Fig. 32–12. The large peak around 200 MeV was much higher than expected and certainly much higher than the number of interactions at neighboring energies. This led Fermi to conclude that the π^+ and proton combined momentarily to form a short-lived particle before coming apart again, or at least that they resonated

† Originally, particles were divided according to their mass into leptons (meaning light particles), baryons (meaning "heavy"), and those of intermediate mass, the mesons (meaning "middle"). The newer classification according to their interactions is not always consistent with this. The muon, for example, is now called a lepton (it doesn't interact strongly), although it was once called the mu meson because of its mass. Other exceptions are the ψ particles, which are very heavy but have $B = 0$ and so are classified as mesons (Section 32–9), and the τ, which is a lepton although it is heavier than some baryons.

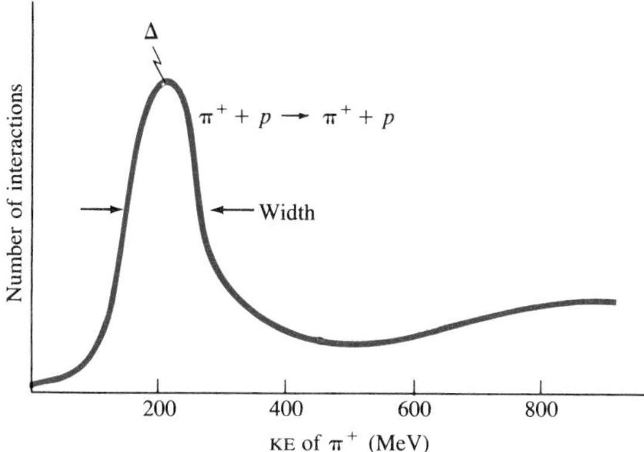

FIGURE 32–12
Number of π^+ particles scattered by a proton target as a function of the incident π^+ kinetic energy.

together for a short time. Indeed, the large peak in Fig. 32–12 resembles a resonance curve (see Figs. 11–12 and 21–31) and this new "particle"—now called the Δ—is referred to as a **resonance**. Hundreds of other resonances have been found in a similar way. Many resonances are regarded as excited states of other particles such as the nucleon (proton or neutron).

The width of a resonance—in Fig. 32–12 the width of the Δ peak is on the order of 100 MeV—is an interesting application of the uncertainty principle. If a particle lives only 10^{-23} s, then its mass (that is, its rest energy) will be uncertain by an amount $\Delta E \approx h/2\pi\Delta t \approx (6.6 \times 10^{-34}\ \mathrm{J\cdot s})/(6)(10^{-23}\ \mathrm{s}) \approx 10^{-11}\ \mathrm{J} \approx 100$ MeV, which is what is observed. Actually, the lifetimes of $\approx 10^{-23}$ s for such resonances are inferred by the reverse process: from the measured width being ≈ 100 MeV.

32–8 • Strange Particles

In the early 1950s, certain of the newly found particles, namely, the K, Λ, and Σ, were found to behave rather strangely in two ways. First, they were always produced in pairs. For example, the reaction

$$\pi^- + p \rightarrow K^0 + \Lambda^0$$

occurred with high probability, but the reaction $\pi^- + p \rightarrow K^0 + n$ was never observed to occur. This seemed strange, since no known conservation law would have been violated in the unobserved reaction and there was plenty of energy available. The second feature of these *strange particles* (as they came to be called) was that although they were clearly produced via the strong interaction (that is, at a high rate), they did not decay at a rate characteristic of the strong interaction even though they decayed into strongly interacting particles (for example, $K \rightarrow 2\pi$, $\Sigma^+ \rightarrow p + \pi^0$). Instead of lifetimes of 10^{-23} s as expected for strongly interacting particles, strange particles have lifetimes of 10^{-10} to 10^{-8} s, which are characteristic of the weak interaction.

Strangeness and its conservation

To explain these observations, a new quantum number, **strangeness**, and a new conservation law, conservation of strangeness, were introduced. By assigning the strangeness numbers (S) indicated in Table 32–2, the production of strange particles in pairs was readily explained. Antiparticles were assigned opposite strangeness from their particles: one of each pair was assigned $S = +1$ and the other $S = -1$ (see Table 32–2). For example, in the reaction $\pi^- + p \rightarrow K^0 + \Lambda^0$, the initial state has strangeness $S = 0 + 0 = 0$, and the final state has $S = +1 - 1 = 0$, so strangeness is conserved. But for $\pi^- + p \rightarrow K^0 + n$, the initial state has $S = 0$ and the final state has $S = +1 + 0 = +1$, so strangeness would not be conserved; and this reaction isn't observed.

To explain the decay of strange particles, it is assumed that strangeness is conserved in the strong interaction but is *not* conserved in the weak interaction. Thus, although strange particles were forbidden by strangeness conservation to decay to lower-mass nonstrange particles via the strong interaction, they could undergo such decay by means of the weak interaction. This would occur much more slowly, of course, which accounts for their longer lifetimes of 10^{-10} to 10^{-8} s.

The conservation of strangeness was the first example of a "partially conserved" quantity. In this case, the quantity strangeness is conserved by strong interactions but not by weak.

32–9 • Quarks and Charm

All observed particles fall into two categories: **fermions** (which obey the Pauli exclusion principle) and **bosons** (which do not). (See Section 28–7.) The fermions can be further subdivided into two groups: leptons and hadrons (Table 32–2). The principal difference between these two groups is that the hadrons interact via the strong interaction, whereas the leptons do not. Another important difference that physicists had to deal with in the 1960s was that there were only four known leptons (e^-, μ^-, ν_e, ν_μ; the τ and ν_τ were not yet discovered), but there are well over a hundred hadrons.

The leptons are considered to be truly elementary particles since they do not seem to break down into smaller entities, do not show any internal structure, and have no measurable size. (Attempts to determine the size of leptons have put an upper limit of about 10^{-18} m.)

The hadrons, on the other hand, are more complex. Experiments indicate they do have an internal structure. And the fact that there are so many of them suggests that they can't all be elementary. To deal with this problem, M. Gell-Mann and G. Zweig in 1963 independently proposed that none of the hadrons so far observed, not even the proton and neutron, was elementary. Instead, they proposed that the hadrons are made up of combinations of three, more fundamental, pointlike entities called **quarks**.[†] Quarks, then, would be considered truly elementary particles, like leptons. The three quarks were labeled *u*, *d*, *s*, and given the names *up*, *down*, and *sideways* (or, more commonly now, *strange*). They were assumed to have fractional charge ($\frac{1}{3}$ or $\frac{2}{3}$ the charge on the electron—that is, less than the previously thought smallest charge). Other properties of quarks and antiquarks are indicated in Table 32–3. All hadrons known at the time could be constructed in theory from these three types of quark. Mesons would consist of a quark–antiquark

Quarks

[†] Gell-Mann chose the word from a phrase in James Joyce's *Finnegan's Wake*.

TABLE 32–3
Properties of Quarks and Antiquarks

				Quarks				
Name	**Symbol**	**Spin**	**Charge**	**Baryon Number**	**Strangeness**	**Charm**	**Bottomness**	**Topness**
Up	u	$\frac{1}{2}$	$+\frac{2}{3}e$	$\frac{1}{3}$	0	0	0	0
Down	d	$\frac{1}{2}$	$-\frac{1}{3}e$	$\frac{1}{3}$	0	0	0	0
Strange	s	$\frac{1}{2}$	$-\frac{1}{3}e$	$\frac{1}{3}$	-1	0	0	0
Charmed	c	$\frac{1}{2}$	$+\frac{2}{3}e$	$\frac{1}{3}$	0	$+1$	0	0
Bottom	b	$\frac{1}{2}$	$-\frac{1}{3}e$	$\frac{1}{3}$	0	0	-1	0
Top	t	$\frac{1}{2}$	$+\frac{2}{3}e$	$\frac{1}{3}$	0	0	0	$+1$
				Antiquarks				
Name	**Symbol**	**Spin**	**Charge**	**Baryon Number**	**Strangeness**	**Charm**	**Bottomness**	**Topness**
Up	$\bar{u}$	$\frac{1}{2}$	$-\frac{2}{3}e$	$-\frac{1}{3}$	0	0	0	0
Down	$\bar{d}$	$\frac{1}{2}$	$+\frac{1}{3}e$	$-\frac{1}{3}$	0	0	0	0
Strange	$\bar{s}$	$\frac{1}{2}$	$+\frac{1}{3}e$	$-\frac{1}{3}$	$+1$	0	0	0
Charmed	$\bar{c}$	$\frac{1}{2}$	$-\frac{2}{3}e$	$-\frac{1}{3}$	0	-1	0	0
Bottom	$\bar{b}$	$\frac{1}{2}$	$+\frac{1}{3}e$	$-\frac{1}{3}$	0	0	$+1$	0
Top	$\bar{t}$	$\frac{1}{2}$	$-\frac{2}{3}e$	$-\frac{1}{3}$	0	0	0	-1

pair. For example, a π^+ meson is considered a $u\bar{d}$ pair (note that for the $u\bar{d}$ pair, $Q = \frac{2}{3}e + \frac{1}{3}e = +1e$, $B = \frac{1}{3} - \frac{1}{3} = 0$, $S = 0 + 0 = 0$, as they must for a π^+); and a $K^+ = u\bar{s}$, with $Q = +1$, $B = 0$, $S = +1$. Baryons, on the other hand, would consist of three quarks. For example, a neutron is $n = ddu$, whereas an antiproton is $\bar{p} = \bar{u}\bar{u}\bar{d}$.

Soon after the quark theory was proposed, physicists began looking for these fractionally charged particles. Although there is indirect experimental evidence in favor of their existence, direct detection of them remains elusive. Indeed, there are suggestions that quarks are so tightly bound together that they may not ever exist in the free state.

In 1964, several physicists proposed that there ought to be a fourth quark. Their argument was based on the expectation that there exists a deep symmetry in nature, including a connection between quarks and leptons. If there are four leptons (as was thought in the 1960s), then symmetry in nature would suggest there should also be four quarks. The fourth quark was said to be *charmed*. Its charge would be $+\frac{2}{3}e$ and it would have another property to distinguish it from the other three quarks. This new property, or quantum number, was called **charm** (see Table 32–3). Charm was assumed to be like strangeness: it would be conserved in strong and electromagnetic interactions, but would not be conserved by the weak. The new charmed quark would have charm $C = +1$ and its antiquark $C = -1$.

Experimentally, however, there seemed to be no need for a charmed quark. Before 1974, all known hadrons could be explained as combinations of the three original quarks. In fact, hadrons corresponding to all three quark, and quark–antiquark, combinations had been found. Furthermore, the Ω^- baryon had been predicted by the three-quark theory ($\Omega^- = sss$) and was discovered soon after, Fig. 32–13. But in 1974, a new heavy meson was

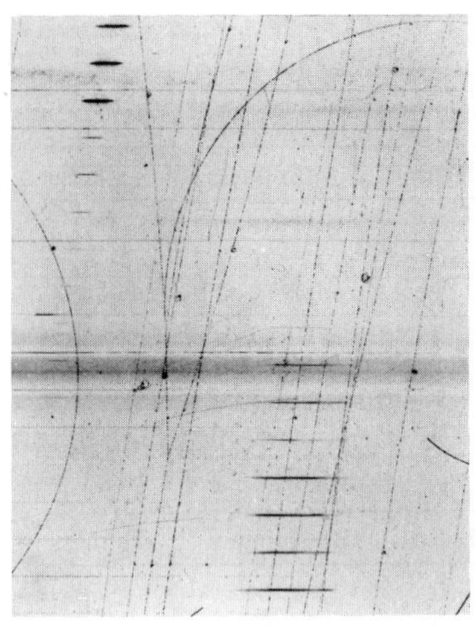

discovered simultaneously by two different groups of experimenters. This new meson, called the J/ψ (often simply the ψ), did not fit the old three-quark scheme. The J/ψ, whose mass is 3100 MeV/c^2, far higher than for other known mesons, could also not be an excited state of a smaller-mass meson (into which it would decay) because its lifetime would have to be about 10^{-23} s. In fact, its lifetime was found to be 1000 times greater than this, about 10^{-20} s. It soon became clear that the existence of the J/ψ could be accounted for on the basis of the charmed quark: a J/ψ would be a combination of a charmed quark and its antiquark ($c\bar{c}$). The charm of the J/ψ itself is zero ($C = +1 - 1 = 0$), so it can decay strongly into hadrons (such as several pions, $p\bar{p}$, $\Lambda\bar{\Lambda}$, etc.), which is observed. Why it lives 1000 times longer than other strongly decaying particles is theorized as due to the c and $\bar{c}$ quarks that make it up having each to be converted into noncharmed quarks that become the hadrons seen in the decay. This process would inhibit the rate at which the decay could occur, so the charmed-quark model yielded a useful explanation.

Soon after the J/ψ was discovered, a similar meson, called the ψ' (mass about 3685 MeV/c^2), was also found. Many other ψ-like mesons were also found, and these are all believed to be bound states of a $c\bar{c}$ pair.

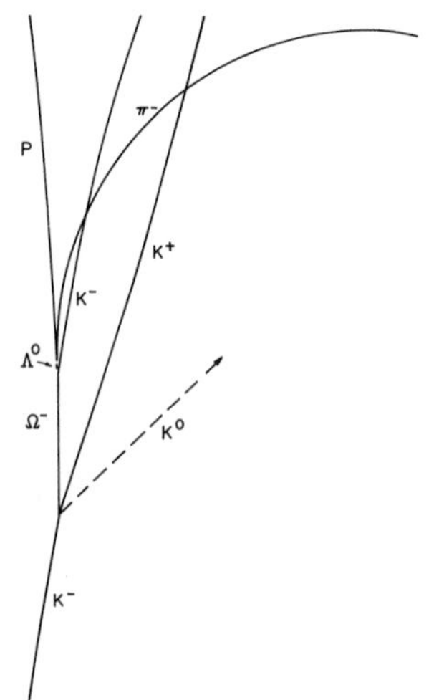

FIGURE 32–13 Liquid-hydrogen bubble-chamber photograph showing the production of a negatively charged omega baryon (Ω^-) by the interaction of a negative K meson (K^-) with a proton (a hydrogen nucleus in the bubble chamber).

Although the J/ψ meson and its relatives have no net charm themselves, it stands to reason that in the decay of one of the larger-mass $c\bar{c}$ combinations, there ought to be mesons that do have charm. That is, the c and $\bar{c}$ quarks ought to appear in separate particles and lend a charm to them of $+1$ or -1. In such a decay process, other quarks would also be produced. For example, the decay

$$\psi \rightarrow D^+ + D^-$$

might be written as

$$\psi = c\bar{c} \rightarrow c\bar{c}d\bar{d} \rightarrow c\bar{d} + \bar{c}d.$$

The $d\bar{d}$ quarks in the third step are produced from energy. Since the d and $\bar{d}$ quarks have opposite quantum numbers and charge, no conservation laws are violated—only energy is required to produce their mass. The $c\bar{d}$ combination, which has $Q = +1$, $B = 0$, $S = 0$, and $C = +1$, has been dubbed the D^+ meson, and $\bar{c}d$ is its antiparticle, D^-. This meson, along with its neutral counterpart, D^0, was found in 1977 with a mass around 1870 MeV/c^2. More recent experiments indicate that charmed baryons also exist. Some of the new particles are listed in Table 32–4.

About the same time, strong evidence appeared for the tau (τ) lepton, with a mass of 1784 MeV/c^2. This lepton, like the electron and muon, presumably has a neutrino associated with it. Thus, the family of leptons is at present believed to have six members. This would upset the balance between leptons and quarks, the presumed basic building blocks of matter, unless two new quarks also exist. Indeed, theoretical physicists postulated the existence of a fifth and sixth quark. These were named **top** and **bottom** quarks, since they are thought to resemble the "up" and "down" quarks. (Some physicists prefer the names *truth* and *beauty* for these t and b quarks.) The names apply also to the new properties (quantum numbers) that distinguish the new quarks from the old quarks, and which (like strangeness) are conserved in strong, but not weak, interactions. (These are included in Table 32–3.) Indeed, another new meson, Υ (upsilon), considered to be a $b\bar{b}$ combination, was detected at about 9400 MeV/c^2. And in 1983, the B meson was observed which

t and b quarks

TABLE 32–4
Partial List of Hadrons Associated with Charm and Bottomness ($L_e = L_\mu = L_\tau = 0$)

Category	Particle	Anti-Particle	Rest Mass (MeV/c^2)	Baryon Number	Strange-ness	Charm	Bottom-ness	Lifetime (s)	Principal Decay Modes
Mesons	D^+	D^-	1869	0	0	+1	0	10.6×10^{-13}	K + others, e + others
	D^0	$\bar{D}^0$	1865	0	0	+1	0	4.2×10^{-13}	K + others
	D_S^+	D_S^-	1969	0	+1	+1	0	4.5×10^{-13}	$K\bar{K}\pi$, $K\bar{K}3\pi$, $\eta\pi$, $\eta 3\pi$
	J/ψ (3097)	Self	3097	0	0	0	0	1.0×10^{-20}	Hadrons, e^+e^-, $\mu^+\mu^-$
	ψ' (3685)	Self	3686	0	0	0	0	3×10^{-21}	$J/\psi\pi\pi$, hadrons, e^+e^-, $\mu^+\mu^-$
	Υ	Self	9460	0	0	0	0	1×10^{-20}	Hadrons, $\mu^+\mu^-$, e^+e^-, $\tau^+\tau^-$
	B^-	B^+	5278	0	0	0	−1	1.2×10^{-12}	D^0 + others
	B^0	$\bar{B}^0$	5280	0	0	0	−1	1×10^{-12}	D^0 + others
Baryons	Λ_C^+	Λ_C^-	2285	+1	0	+1	0	1.9×10^{-13}	Hadrons (e.g., Λ + others)
	Σ_C^{++}	Σ_C^{--}	2453	+1	0	+1	0	?	$\Lambda_C^+\pi^+$
	Σ_C^+	Σ_C^-	2453	+1	0	+1	0	?	$\Lambda_C^+\pi^0$
	Σ_C^0	$\bar{\Sigma}_C^0$	2453	+1	0	+1	0	?	$\Lambda_C^+\pi^-$

has "bottomness", consisting of only one b quark plus a non-b quark. There is still no firm experimental evidence (as of 1990) for the top quark.

Today, the "truly" elementary particles are considered to be the six quarks, the six leptons, and the gauge bosons that carry the fundamental forces. See Table 32–5.

32–10 • The "Standard Model": Quantum Chromodynamics (QCD) and the Electroweak Theory

Not long after the quark theory was proposed, it was suggested that quarks have another property (or quality) called **color**. The distinction between the six quarks (u, d, s, c, b, t) was referred to as **flavor**. According to theory, each of the flavors of quark can have three colors, usually designated red, green, and blue. (Note that the names "color" and "flavor" have nothing to do with our senses, but are purely whimsical—as are other names, like charm, in this new field.) The antiquarks are colored antired, antigreen, and antiblue. Baryons are made up of three quarks, one of each color. Mesons consist of a quark–antiquark pair of a particular color and its anticolor. Thus baryons and mesons are white or colorless.

Originally, the idea of quark color was proposed to preserve the Pauli exclusion principle (Section 28–7), which applies to fermions such as electrons and nucleons, all of which have spin $\frac{1}{2}$ (or any half-integral spin, like $\frac{3}{2}$, $\frac{5}{2}$, and so on). Quarks, too, are fermions (they have spin $\frac{1}{2}$) and therefore should obey the exclusion principle. Yet for three particular baryons (uuu, ddd, and sss), all three quarks would have the same quantum numbers, and at least two of them have their spin in the same direction (since there are only two choices, spin up [$m_s = +\frac{1}{2}$] or spin down [$m_s = -\frac{1}{2}$]). This would seem to violate the exclusion principle, but if quarks have an additional quantum number (color), which could be different for each quark, it would serve to distinguish them and the exclusion principle would hold. Al-

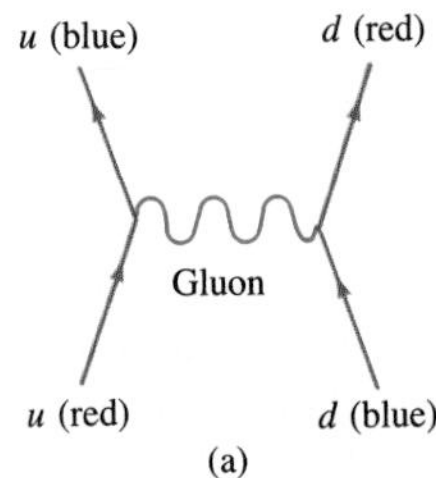

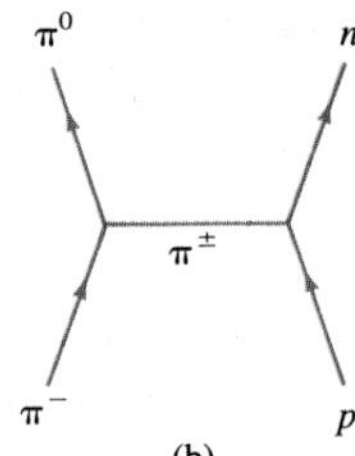

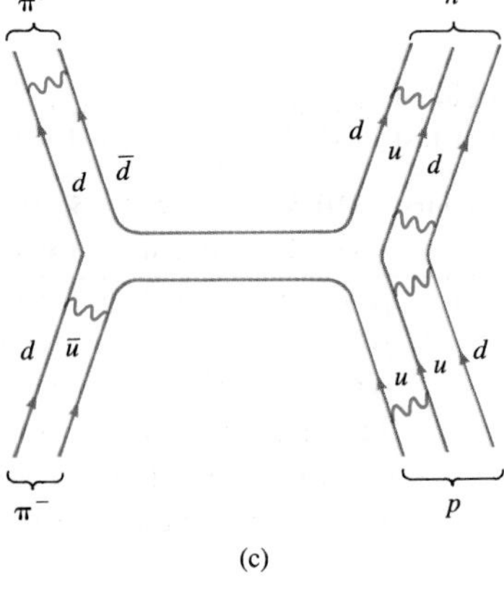

FIGURE 32–14 (a) The force between two quarks holding them together as part of a proton, for example, is carried by a gluon, which in this case involves a change in color. (b) Strong interaction $\pi^- p \rightarrow \pi^0 n$ with the exchange of a charged π meson (+ or −, depending on whether it is considered moving to the left or to the right). (c) Quark representation of the same interaction $\pi^- p \rightarrow \pi^0 n$; the intermediate pion ($d\bar{u}$ or $\bar{d}u$) can be interpreted as the annihilation of a $u\bar{u}$ pair with production of a $d\bar{d}$ pair. The wavy lines between quarks represent gluon exchanges holding the hadrons together.

though quark color, and the resulting threefold increase in the number of quarks, was thus originally an *ad hoc* idea, it also served to bring the theory into better agreement with experiment, such as predicting the correct lifetime of the π^0 meson. The idea of color soon became, in addition, a central feature of the theory as determining the force binding quarks together in a hadron. Each quark is assumed to carry a *color charge*, analogous to electric charge, and the strong force between quarks is often referred to as the **color force**. This new theory of the strong force is called **quantum chromodynamics** (*chrome* = color in Greek), or **QCD**, to indicate that the force acts between color charges (and not between, say, electric charges). The strong force between two hadrons[†] is considered to be a force between the quarks that make them up, as suggested in Fig. 32–14. The particles that transmit the color force (analogous to photons for the EM force) are called **gluons** (a play on "glue"). They are included in Table 32–5. There are eight gluons, according to the theory, all massless, and six of them have color charge.[‡] Thus gluons have replaced mesons (Table 32–1) as the particles carrying the strong (color) force.

The color force has the interesting property that its strength increases with increasing distance; and as two quarks approach each other very closely (equivalently, have high energy), the force between them approaches zero; this aspect is referred to as **asymptotic freedom**.

The weak force, as we have seen, is thought to be mediated by the W^+, W^-, and Z^0 particles. It acts between the "weak charges" that each particle has. Each elementary particle can thus have electric charge, weak charge, color charge, and gravitational mass, although one or more of these could

[†] The strong force between hadrons appears feeble, however, in comparison to the force directly between quarks within hadrons.

[‡] Compare to the EM interaction, where the photon has no electric charge. Because gluons have color charge, they could attract each other and form composite particles (photons cannot). Such "glueballs" are being searched for.

TABLE 32–5
The Elementary Particles[†] as Seen Today

Quarks	u, d	s, c,	b, (t)
Leptons	e, ν_e,	μ, ν_μ,	τ, (ν_τ)
Gauge bosons	γ (photon),	$W^\pm$, Z^0,	gluons

[†] Note that the quarks and leptons are arranged into three families each, and the gauge particles are arranged in groups for the forces they mediate.

be zero. For example, all leptons have color charge of zero, so they do not interact via the strong force.

To summarize, the latest theories consider the truly elementary particles to be the leptons, the quarks, and the gauge bosons (photon, W and Z, and the gluons), and perhaps other bosons. The photon and the leptons are observed in experiments, and finally so were the W^+, W^-, and Z^0. But so far only combinations of quarks (baryons, mesons) have been observed, and it seems likely that free quarks and gluons are unobservable. On the other hand, some physicists believe that leptons and quarks are not fundamental, but are composites of still more fundamental objects. Only the future can tell.

One important aspect of new theoretical work is the attempt to find a unified basis for the different forces in nature. This was a long-held hope of Einstein, which he was never able to fulfill. A so-called "gauge" theory that unifies the weak and electromagnetic interactions was put forward in the 1960s by S. Weinberg, S. Glashow, and A. Salam. In this **electroweak theory**, the weak and electromagnetic forces are seen as two different manifestations of a single, more fundamental, electroweak interaction. The electroweak theory has had many successes, including the prediction of the $W^{\pm}$ particles as carriers of the weak force, with masses of 81 ± 2 GeV/c^2 in excellent agreement with the measured values of 80.7 ± 1.3 GeV/c^2 (and similar accuracy for the Z^0). The electroweak theory and QCD for the strong interaction are often referred to today as the **standard model**.

Electroweak theory and the standard model

32–11 • Grand Unified Theories

With the success of the unified electroweak theory, attempts have recently been made also to incorporate it and QCD for the strong (color) force into a so-called **grand unified theory** (GUT). One type of such a grand unified theory of the electromagnetic, weak, and strong forces has been worked out in which there is only one class of particle—leptons and quarks belong to the same family and are able to change freely from one type to the other—and the three forces are different aspects of a single underlying force. The unity is predicted to occur, however, only on a scale of less than about 10^{-30} m. If two elementary particles (leptons or quarks) approach each other to within this **unification scale**, the apparently fundamental distinction between them would not exist at this level, and a quark could readily change to a lepton, or vice versa. Baryon and lepton numbers would not be conserved. The weak, electromagnetic, and strong (color) force would blend to a force of a single strength.

GUT

Unification of forces

How could a lepton become a quark, or vice versa? The theory predicts the existence of particles that can be exchanged between a quark and a lepton allowing one to change into the other, just as the charged pion exchanged between the π^- and p in Fig. 32–14b allows the proton to become a neutron. The mass of the new exchange particles, consistent with the uncertainty principle as applied earlier in this chapter (see Eq. 32–2 with $d \approx 10^{-30}$ m), would be about 10^{14} GeV/c^2, or 10^{14} times the proton mass. With such an incredibly large mass, there is little hope of seeing them in the laboratory. It is also this huge mass that is thought to keep baryon and lepton numbers conserved in observed reactions, since the likelihood of producing such a massive particle, even as a virtual exchange particle, is extremely small at even the highest laboratory energies.

Symmetry breaking

What happens between the unification distance of 10^{-30} m and more normal (larger) distances is referred to as **symmetry breaking**. As an analogy, consider an atom in a crystal. Deep within the atom, there is much symmetry—in the innermost regions the electron cloud is spherically symmetric. Further out, this symmetry breaks down—the electron clouds are distributed preferentially along the lines (bonds) joining the atoms in the crystal. In a similar way, at 10^{-30} m the force between elementary particles is theorized to be a single force—it is symmetrical and does not single out one type of "charge" over another. But at larger distances, that symmetry is broken and we see three distinct forces. (In the "standard model" of electroweak interactions, Section 32–10, the symmetry breaking between the electromagnetic and the weak interactions occurs at about 10^{-18} m.)

Proton decay?

Since unification occurs at such tiny distances and huge energies, the theory is difficult to test experimentally. But it is not completely impossible. One possibly testable prediction is the basis for the idea, suggested at the end of Section 32–5, that the proton might decay (via, for example, $p \rightarrow \pi^0 e^+$) and violate conservation of baryon number. This could happen if two quarks approached to within 10^{-30} m of each other, which would require energies on the order of 10^{14} GeV (an average energy at a temperature of $T = 10^{27}$ K). At such distances and energies, quarks and electrons could change freely into one another, as mentioned above. At normal temperatures, there is a very tiny probability that such an event could occur, and if it does, it implies that a proton could decay (its constituent quarks becoming electrons in the decay). But it is very unlikely at normal temperature and energy, so the decay of a proton can only be an unlikely process. In the simplest form of GUT, the theoretical estimate of the proton lifetime for the decay mode $p \rightarrow \pi^0 e^+$ is $\approx 10^{31}$ yr, and this has just come within the realm of testability. Proton decays have still not been seen and experiments put the lower limit on the proton lifetime for the above mode to be about 4.5×10^{32} yr, an order of magnitude greater than the prediction. This may seem a disappointment, but on the other hand, it presents a challenge. Indeed, more complex GUTs are not affected by this result.

Connection with cosmology

Another interesting prediction of unified theories relates to cosmology (see Chapter 33). It is thought that during the first 10^{-35} s after the theorized big bang that created the universe, the temperature was so extremely high that particles had energies corresponding to the unification scale. Baryon number would not have been conserved then, and this could account for the observed predominance of matter ($B > 0$) over antimatter ($B < 0$) in the universe.

This last example is interesting, for it illustrates a deep connection between investigations at either end of the size scale: theories about the tiniest objects (elementary particles) have a strong bearing on the understanding of the universe as a whole. We will look at this more in the next chapter.

Even more ambitious than grand unified theories are attempts to also incorporate gravity, and thus unify all four forces in nature into a single theory. A new theory, developed in the mid-1980s, that seems to successfully unite the four forces is called **superstring** theory, in which the elementary particles (Table 32–5) are imagined not as points but as one-dimensional strings about 10^{-35} m long. It is very difficult, however, to test the theory experimentally. The world of elementary particles is opening new vistas. What happens in the near future is bound to be exciting.

SUMMARY

Particle accelerators are used to accelerate charged particles such as electrons and protons to very high energy. High-energy particles have short wavelength and so can be used to probe the structure of matter at very small distances in great detail. High kinetic energy also allows the creation of new particles through collision (via $E = mc^2$). Van de Graaff and linear accelerators use high voltage to accelerate particles along a line. Cyclotrons and synchrotrons use a magnetic field to keep the particles in a circular path and accelerate them at intervals by high voltage.

An *antiparticle* has the same mass as a particle but opposite charge. Certain other properties may also be opposite: for example, the antiproton has *baryon number* (nucleon number) opposite to that for the proton. In all nuclear and particle reactions, the following conservation laws hold: momentum, mass–energy, angular momentum, electric charge, baryon number, and the three *lepton numbers*. Certain particles have a property, called *strangeness*, which is conserved by the strong force but not by the weak force. The more recently noted properties, *charm*, *bottomness*, and (presumably) *topness*, also are believed to be conserved by the strong force but not by the weak.

Just as the electromagnetic force can be said to be due to an exchange of photons, the strong nuclear force is thought to be carried by *mesons* that have rest mass, or, according to more recent theory, by massless *gluons*. The W and Z particles carry the weak force. These fundamental force carriers (photon, W and Z, gluons) are called *gauge bosons*.

Other particles can be classified as either *leptons* or *hadrons*. Leptons participate in the weak and electromagnetic interactions. Hadrons participate in the strong interaction as well. The hadrons can be classified as *mesons*, with baryon number zero, and *baryons*, with nonzero baryon number.

All particles, except for the photon, electron, neutrinos, and (so far at least) proton, decay with measurable half-lives varying from 10^{-25} s to 10^3 s. The half-life depends on which force is predominant in the decay. Weak decays usually have half-lives greater than about 10^{-13} s. Electromagnetic decays have half-lives on the order of 10^{-16} to 10^{-19} s. The shortest lived particles, called *resonances*, decay via the strong interaction and live typically for only about 10^{-23} s.

The latest theories of elementary particle physics postulate the existence of *quarks* as the basic building blocks of the hadrons. Initially, three quarks were proposed (called *up*, *down*, and *strange*). Recent evidence suggests that three additional quarks are needed, called the *charmed*, *bottom*, and *top* quarks. It is expected that there are the same number of quarks as leptons (six of each), and that quarks and leptons are the truly elementary particles along with the gauge bosons (γ, W, Z, gluons). Quarks are said to have *color*, and, according to *quantum chromodynamics* (QCD), the strong color force acts between their color charges and is transmitted by *gluons*. *Electroweak theory* views the weak and electromagnetic forces as two aspects of a single underlying interaction. QCD plus the electroweak theory are referred to as the *standard model*.

Grand unified theories of forces suggest that at very short distances (10^{-30} m) and very high energy, the weak, electromagnetic, and strong forces appear as a single force, and the fundamental difference between quarks and leptons disappears.

QUESTIONS

1. What limits the maximum energy attainable for protons in an ordinary cyclotron? How is this limitation overcome in a synchrotron?
2. A proton in a synchrotron has a speed of $0.99c$. What must be done to increase its energy?
3. Give a reaction between two nucleons, similar to Eq. 32–3, that could produce a π^-.
4. If a proton is moving at very high speed, so that its KE is much greater than its rest energy (m_0c^2), can it then decay via $p \rightarrow n + \pi^+$?
5. What would an "antiatom," made up of antiparticles to constituents of normal atoms, consist of? What might happen if *antimatter*, made of such antiatoms, came in contact with our normal world of matter?

6. Does the presence of a neutrino among the decay products of a particle necessarily mean that the decay occurs via the weak interaction? Do all decays via the weak interaction produce a neutrino?
7. What particle in a decay signals the electromagnetic interaction?
8. Why is it that a neutron decays via the weak interaction even though the neutron and one of its decay products (proton) are strongly interacting?
9. Which of the four interactions (strong, electromagnetic, weak, gravitational) does an electron take part in? A neutrino? A proton?
10. Check that charge and baryon number are conserved in each of the decays in Table 32–2.
11. Which of the particle decays in Table 32–2 occur via the electromagnetic interaction?
12. Which of the particle decays in Table 32–2 occur by the weak interaction?
13. By what interaction, and why, does $\Sigma^{\pm}$ decay to Λ^0? What about Σ^0 decaying to Λ^0?
14. The Δ baryon has spin $\frac{3}{2}$, baryon number 1, and charge $Q = +2, +1, 0,$ or -1. Why is there no charge state $Q = -2$?
15. Which of the particle decays in Table 32–4 occur via the electromagnetic interaction?
16. Which of the particle decays in Table 32–4 occur by the weak interaction?

PROBLEMS

SECTIONS 32–1 AND 32–2

1. (I) What is the total energy of a proton whose kinetic energy is 15.0 GeV?
2. (I) Calculate the wavelength of 30-GeV electrons.
3. (I) What strength of magnetic field is used in a cyclotron in which protons make 1.9×10^7 revolutions per second?
4. (I) The voltage across the dees of a cyclotron is 50 kV. How many revolutions do protons make to reach a kinetic energy of 15 MeV?
5. (I) What is the total energy of a proton whose kinetic energy is 35 GeV? What is its wavelength?
6. (I) If α particles are accelerated by the cyclotron of Example 32–2, what must be the frequency of the voltage applied to the dees?
7. (I) What is the time for one complete revolution for a very high-energy proton in the 1.0-km-radius Fermilab accelerator?
8. (II) (*a*) If the cyclotron of Example 32–2 accelerated α particles, what maximum energy could they attain? What would their speed be? (*b*) Repeat for deuterons ($^2_1$H). (*c*) In each case, what frequency of voltage is required?
9. (II) What is the wavelength, and maximum resolving power attainable, using 400-GeV protons at Fermilab?
10. (II) Protons are injected into the 1.0-km-radius Fermilab synchrotron with an energy of 8.0 GeV. If they are accelerated by 2.5 MV each revolution, how far do they travel and approximately how long does it take for them to reach 400 GeV?
11. (III) What magnetic field intensity is needed at the 1.0-km-radius Fermilab synchrotron for 400-GeV protons? Use the relativistic mass.
12. (III) Show that the energy of a particle (charge *e*) in a synchrotron, in the relativistic limit ($v \approx c$), is given by E (in eV) $= Brc$, where B is magnetic field strength and r the radius of the orbit (SI units).

SECTIONS 32–3 TO 32–6

13. (I) Draw a Feynman diagram for $n + p \rightarrow n + p + \pi^0$.
14. (I) Draw a Feynman diagram for $p\bar{p}$ annihilation with the production of two pions.
15. (I) Draw a Feynman diagram for the photoelectric effect.
16. (I) Two protons are heading toward each other with equal speeds. What minimum kinetic energy must each have if a π^0 meson is to be created in the process? (See Table 32–2.)
17. (I) How much energy is released in the decay $\pi^+ \rightarrow \mu^+ + \nu_\mu$?
18. (I) About how much energy is released when a Λ^0 decays to $n\pi^0$? (See Table 32–2.)
19. (I) How much energy is required to produce a neutron–antineutron pair?
20. (II) Which of the following decays are possible? For those that are forbidden, explain which laws are violated.
(*a*) $\Xi^0 \rightarrow \Sigma^+ + \pi^-$
(*b*) $\Omega^- \rightarrow \Sigma^0 + \pi^- + \nu$
(*c*) $\Sigma \rightarrow \Lambda + \gamma + \gamma$
21. (II) What are the wavelengths of the two photons produced when a proton–antiproton pair at rest annihilate?
22. (II) What would be the wavelengths of the two photons produced when an electron–positron pair, each with 300 keV of KE, annihilate?
23. (II) In the rare decay $\pi^+ \rightarrow e^+ + \nu_e$, what is the kinetic energy of the positron?
24. (II) What minimum kinetic energy must a neutron and proton each have if they are traveling at the same speed toward each other, collide, and produce a K^+K^- pair in addition to themselves? (See Table 32–2.)
25. (II) Calculate the KE of each of the two products in the decay $\Xi^- \rightarrow \Lambda^0\pi^-$.
26. (III) Could a π^+ meson be produced if a 100-MeV proton struck a proton at rest? What minimum KE must it have?
27. (III) For the reaction $p + p \rightarrow 3p + \bar{p}$, where one of the initial protons is at rest, use relativistic formulas to show that the threshold energy is $6m_pc^2$, equal to three times the Q-value of the reaction, where m_p is the proton mass.

SECTIONS 32–7 TO 32–11

28. (I) The measured width of the ψ meson is 63 keV. Estimate its lifetime.
29. (I) The measured width of the ψ' meson is 215 keV. Estimate its lifetime.
30. (I) What is the energy width (or uncertainty) of (*a*) η, (*b*) Σ^0?
31. (I) Use Fig. 32–12 to estimate the energy width and then the lifetime of the Δ resonance.
32. (I) The B^- meson is presumed to be a $b\bar{u}$ quark combination. (*a*) Show that this is consistent for all quantum numbers. (*b*) What are the quark combinations for B^+, B^0, $\bar{B}^0$?
33. (II) What are the quark combinations that can form (*a*) a neutron, (*b*) an antineutron, (*c*) a Λ^0, (*d*) a $\bar{\Xi}^0$?
34. (II) What particles do the following quark combinations produce? (*a*) uud, (*b*) $\bar{u}\bar{u}\bar{s}$, (*c*) $\bar{u}s$, (*d*) $d\bar{u}$, (*e*) $\bar{c}s$.
35. (II) What is the quark combination needed to produce a D^0 meson ($Q = B = S = 0$, $C = +1$)?
36. (II) The D_S^+ meson has $S = C = +1$, $B = 0$. What quark combination would produce it?
37. (II) (*a*) Show that the so-called unification distance of 10^{-30} m in recent grand unified theory is equivalent to an energy of about 10^{14} GeV. Use either the uncertainty principle or de Broglie's wavelength formula and explain how they apply. (*b*) Calculate what temperature this corresponds to.
38. (II) Draw possible Feynman diagrams using quarks (as in Fig. 32–14c) for the reactions (*a*) $pn \rightarrow pn$, (*b*) $\bar{p}p \rightarrow \pi^+\pi^-$.
39. (II) Draw a possible quark Feynman diagram (see Fig. 32–14c) for the reaction $K^-p \rightarrow K^-p$.

GENERAL PROBLEMS

40. (*a*) How much energy is released when an electron and a positron annihilate each other? (*b*) How much energy is released when a proton and an antiproton annihilate each other?
41. Which of the following reactions are possible, and by what interaction could they occur? For those forbidden, explain why.
    (*a*) $\pi^-p \rightarrow K^+\Sigma^-$
    (*b*) $\pi^+p \rightarrow K^+\Sigma^+$
    (*c*) $\pi^-p \rightarrow \Lambda^0K^0\pi^0$
    (*d*) $\pi^+p \rightarrow \Sigma^0\pi^0$
    (*e*) $\pi^-p \rightarrow pe^-\bar{\nu}_e$
    (*f*) $\pi^-p \rightarrow K^0p\pi^0$
    (*g*) $K^-p \rightarrow \Lambda^0\pi^0$
    (*h*) $K^+n \rightarrow \Sigma^+\pi^0\gamma$
    (*i*) $K^+ \rightarrow \pi^0\pi^0\pi^+$
    (*j*) $\pi^+ \rightarrow e^+\nu_e$
42. For the decay $\Lambda^0 \rightarrow p\pi^-$, calculate (*a*) the Q-value (energy released), and (*b*) the KE of the p and π^-. (Use relativistic formulas.)
43. Symmetry breaking occurs in the electroweak theory at about 10^{-18} m. Show that this corresponds to an energy that is on the order of the mass of the $W^\pm$.
44. The mass of a π^0 can be measured by observing the reaction $\pi^- + p \rightarrow \pi^0 + n$ at very low incident π^- kinetic energy (assume it is zero). The neutron is observed to be emitted with a KE of 0.60 MeV. Use conservation of energy and momentum to determine the π^0 mass.
45. Calculate the Q value for each of the reactions, Eq. 32–3, for producing a pion.
46. Calculate the maximum KE of the electron in the decay $\mu^- \rightarrow e^- + \bar{\nu}_e + \nu_\mu$. [*Hint:* in what direction do the two neutrinos move relative to the electron in order to give the latter the maximum KE? Both energy and momentum are conserved; use relativistic formulas.]

CHAPTER 33

Astrophysics and Cosmology

The Andromeda galaxy, which is the nearest spiral galaxy to us and the one most easily seen. It is 2.2 million light-years from us and contains over 300 billion stars. This photo has been computer enhanced to better reveal differences in brightness. Our Galaxy is also a spiral galaxy. (What would it take to get a photo of our Galaxy like this, seen from the outside?)

In the previous chapter, we studied the tiniest objects in the universe—the elementary particles. Now we leap to the largest—stars and galaxies. These two extreme realms, elementary particles and the cosmos, are among the most intriguing and exciting subjects in science. And, surprising though it may seem, these distant realms are related in a fundamental way, as already hinted in Chapter 32.

Use of the techniques and ideas of physics to study the heavens is often referred to as **astrophysics**. At the base of our present theoretical understanding of the universe is Einstein's *general theory of relativity* and its theory of gravitation—for in the large-scale structure of the universe, gravity is the dominant force. General relativity serves also as the foundation for modern **cosmology**, which is the study of the universe as a whole. Cosmology deals especially with the search for a theoretical framework to understand the observed universe, its origin, and its future. The questions posed by cosmology are complex and difficult; the possible answers are often unimaginable. They are questions like "Has the universe always existed, or did it have a beginning in time?" Either alternative is difficult to imagine: time going back indefinitely into the past, or an actual moment when the universe began (but, then, what was there before?). And what about the size of the universe? Is it infinite in size? (It is hard to imagine infinity.) Or is it finite in size? (This is also hard to imagine, for if the universe is finite, it does not make sense to ask what is beyond it, because the universe is all there is.)

Our survey of astrophysics and cosmology will be necessarily brief and qualitative, but we will nonetheless touch on the major ideas. We begin with a look at what can be seen beyond the earth.

33–1 • Stars and Galaxies

According to the ancients, the stars, except for the few that seemed to move (the planets), were fixed on a sphere beyond the last planet. The universe was neatly self-contained, and we on earth were at or near its center. But in the centuries following Galileo's first telescopic observations of the heavens in 1610, our view of the universe has changed dramatically. We no longer place ourselves at the center, and we view the universe as vastly larger. The distances involved are so great that we specify them in terms of the time it takes light to travel the given distance: for example, 1 light-second = $(3.0 \times 10^8 \text{ m/s})(1.0 \text{ s}) = 3.0 \times 10^8 \text{ m} = 3.0 \times 10^5 \text{ km}$; 1 light-minute = 18×10^6 km; and 1 **light-year** (ly) is

$$1 \text{ ly} = (2.998 \times 10^8 \text{ m/s})(3.156 \times 10^7 \text{ s/yr}) = 9.46 \times 10^{15} \text{ m} \approx 10^{13} \text{ km}.$$

Light-year

For specifying distances to the sun and moon, we usually use meters or kilometers, but we could specify them in terms of light. The earth–moon distance is 384,000 km, which is 1.28 light-seconds. The earth–sun distance is 1.50×10^{11} m, or 150,000,000 km; this is equal to 8.3 light-minutes. The nearest star to us, other than the sun, is Proxima Centauri, about 4.3 ly away.

On a clear moonless night, thousands of stars of varying degrees of brightness can be seen, as well as the elongated cloudy stripe known as the Milky Way (Fig. 33–1). It was Galileo who first observed, about 1610, that the Milky Way is comprised of countless individual stars. A century and a half later (about 1750), Thomas Wright suggested that the Milky Way was a flat disc of stars extending to great distances in a plane, which we call the **Galaxy** (Greek for "milky way").

Our Galaxy has a diameter of almost 100,000 light-years and a thickness of about 6000 light-years. It has a bulging central nucleus and spiral arms (Figure 33–2). Our sun, which seems to be just another star, is located more than halfway from the center to the edge, about 28,000 ly from the center. Our Galaxy contains about 10^{11} stars. The sun orbits the galactic center approximately once every 200 million years, so its speed is about 250 km/s relative to the center of the Galaxy. The total mass of all the stars in our Galaxy is about 3×10^{41} kg.

FIGURE 33–1 A section of the Milky Way. The thin line is the trail of an artificial earth satellite.

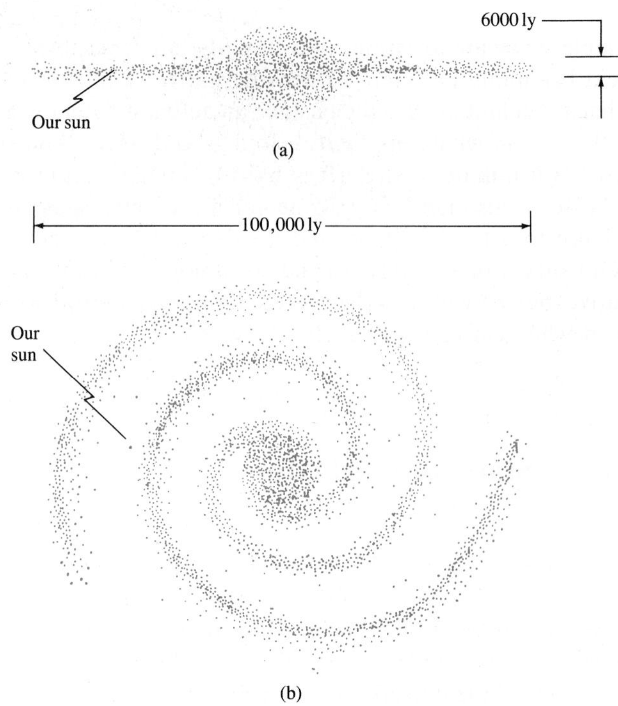

FIGURE 33–2 Our Galaxy, as it would appear from the outside: (a) "end view," in the plane of the disc; (b) "top view," looking down on the disc. (If only we could see it like this—from the outside!)

FIGURE 33–3 This globular star cluster is located in the constellation Hercules.

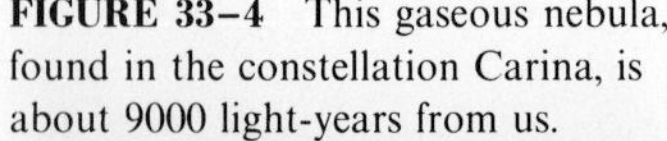

FIGURE 33–4 This gaseous nebula, found in the constellation Carina, is about 9000 light-years from us.

We can see by telescope, in addition to stars both within and outside the Milky Way, many faint cloudy patches in the sky which were all referred to once as "nebulae" (Latin for "clouds"). A few of these, such as those in the constellations Andromeda and Orion, can actually be discerned with the naked eye on a clear night. Some are **star clusters** (Fig. 33-3), groups of stars that are so numerous they appear to be a cloud. Others are glowing clouds of gas or dust (Fig. 33–4), and it is for these that we now mainly reserve the word **nebula**. Most fascinating are those that belong to a third category: they often have fairly regular elliptical shapes and seem to be a great distance beyond our Galaxy. Immanuel Kant (about 1755) seems to have been the first to suggest that these latter might be circular discs like our Galaxy: they appear elliptical because we see them at an angle, and are faint because they are so distant. At first it was not universally accepted that these objects were *extragalactic* (that is, outside our Galaxy). The very large telescopes constructed in this century revealed that individual stars could be resolved within these extragalactic objects and that many contained spiral arms. Edwin Hubble (1889–1953), who did much of this observational work in the 1920s using the 2.5-m (100-inch) telescope† on Mt. Wilson near Los Angeles, California, was also able to demonstrate that these objects were indeed extragalactic because of their great distances. (The distance to the Andromeda nebula (a galaxy) is over 2 million light-years, a distance 20 times greater than the diameter of our Galaxy.) Thus it was determined that these nebulae are *galaxies* similar to ours. Today, the largest telescopes can see about 10^{11} galaxies. See Fig. 33–5. (Note that it is usual to capitalize the word galaxy only when it refers to our own.)

† 2.5 m (= 100 inches) refers to the diameter of the curved objective mirror. The bigger the mirror, the more light it collects and the less diffraction there is, so more and fainter stars can be seen. See Chapter 25.

(a)

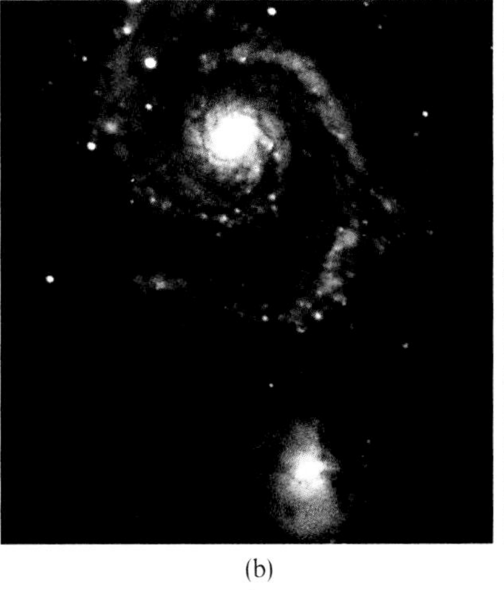

(b)

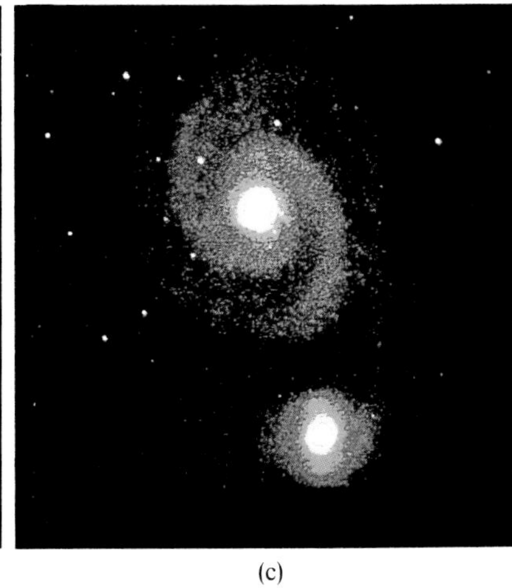

(c)

FIGURE 33–5 Photographs of galaxies. (a) Spiral galaxy in the constellation Hydra. (b) Two galaxies; the larger and more dramatic one is known as the Whirlpool galaxy. (c) The same galaxies as in (b), but this is a false-color infrared image which shows the arms of the spiral as being more regular than in the visible light photo (b); the different colors correspond to different light intensities.

Galaxies tend to be grouped in **galaxy clusters**, with only a few to many thousands of galaxies in each cluster. Furthermore, clusters themselves seem to be organized into even larger aggregates: clusters of clusters of galaxies, or **superclusters**. The galaxies nearest us are about 2 million light-years away. The farthest galaxies visible with optical telescopes are more than a thousand times farther away, about 3×10^9 ly. (See Table 33–1.) Recent evidence suggests that galaxies and galaxy clusters are not distributed randomly in space but lie in bubblelike patterns around vast empty regions of diameter on the order of 10^8 ly.

From this perspective of a vast universe, we living creatures on earth may feel very small and anonymous—a far cry from the view held only a few centuries ago that placed us at the very center of the universe. If this modern view of the universe seems too stark, however, we can take a philosophical approach: all of these observations and interpretations have been made here on earth, and the ideas were created here on earth. Who is to say that we are in any way unimportant in the universe, even though we may not hold a geometrically central position?

Besides the usual stars, clusters of stars, galaxies, and clusters and superclusters of galaxies, the universe contains a number of other interesting objects. Among these are strange types of stars known as *red giants*, *white dwarfs*, *neutron stars*, *black holes* (at least theoretically), and exploding stars called *novae* and *supernovae*. In addition there are *quasars* (quasistellar objects), which, if we judge their distance correctly, are galaxies thousands of times brighter than ordinary galaxies. Furthermore, there is radiation that reaches the earth but does not emanate from the bright pointlike objects we

Table 33–1
Heavenly Distances

Object	Approx. Distance from Earth (ly)
Moon	4×10^{-8}
Sun	1.5×10^{-5}
Other planets (out to Pluto)	6×10^{-4}
Nearest star (Proxima-Centauri)	4
Center of our Galaxy	3×10^4
Nearest galaxy	2×10^6
Farthest galaxy seen optically	3×10^9

SOLUTION The luminosity L is the same for both stars, so the apparent brightness depends only on their relative distances. We use the inverse square law as stated above: the star appears dimmer by a factor

$$\frac{l_1}{l_2} = \frac{d_2^2}{d_1^2} = \frac{(1.5 \times 10^8 \text{ km})^2}{(10 \text{ pc})^2(3.26 \text{ ly/pc})^2(10^{13} \text{ km/ly})^2} \approx 10^{-13},$$

where d_1 and d_2 are the distances from earth to the star and to the sun, respectively.

Because of the wide range in apparent brightnesses, astronomers use a logarithmic scale to specify brightness. The **apparent magnitude**, m, is defined so that for two objects whose apparent brightnesses, l_1 and l_2, differ by a factor of 100, their apparent magnitudes, m_1 and m_2, differ by 5. That is (see Appendix A for a review of logarithms),

$$\frac{l_1}{l_2} = 100^{(m_2 - m_1)/5}, \qquad (33\text{–}1a)$$

or

$$\frac{l_1}{l_2} = 10^{2(m_2 - m_1)/5} = 10^{(m_2 - m_1)/2.5}.$$

We take logs of both sides and obtain

Apparent magnitude

$$m_2 - m_1 = 2.5 \log_{10} (l_1/l_2). \qquad (33\text{–}1b)$$

The zero point was chosen, historically, so that the brightest stars (other than the sun) have m on the order of 1. The precise definition of the zero point today is $l(m = 0) = 2.52 \times 10^{-8}$ W/m^2. Hence we can also write

$$m = 2.5 \log_{10} \left(\frac{2.52 \times 10^{-8} \text{ W/m}^2}{l} \right) \qquad (33\text{–}1c)$$

for the apparent magnitude of a star of brightness l. Visible stars (other than the sun) fall in the range $m \approx -1$ (the brightest) to $m = 6$ (just barely visible to the naked eye). Stars with apparent magnitude greater than 6 can be seen only with a telescope. The sun, on this scale, has apparent magnitude of about -27 (note the minus sign).

The **absolute magnitude**, M, of an object is defined as the apparent magnitude that object would have at a distance of 10 pc. It is easy to show (see Problem 15) that

Absolute magnitude

$$M = m - 5 \log_{10} (d/10), \qquad (33\text{–}2)$$

where d is the distance in parsecs to the object. The sun has $M \approx 4.7$, typical for a star, whereas bright galaxies have $M \approx -22$.

EXAMPLE 33–3 Estimate the number of stars in a typical bright galaxy with $M \approx -22$, assuming our sun with $M \approx 4.7$ is typical for a star.

SOLUTION The number of stars, N, should be roughly equal to the total luminosity of the galaxy divided by the luminosity of a typical star:

$$N \approx \frac{L_{\text{gal}}}{L_{\text{star}}}.$$

The absolute luminosities (L) are proportional to the apparent brightnesses

(l) at a given distance, say 10 pc; and at this distance of $d = 10$ pc, Eq. 33–2 gives $M = m$ (because $\log 1 = 0$). Thus

$$N \approx \frac{L_{\text{gal}}}{L_{\text{star}}} = \frac{l_{\text{gal}}\,(10\text{ pc})}{l_{\text{star}}\,(10\text{ pc})},$$

and using Eq. 33–1a we get

$$N \approx 100^{(m_{\text{star}} - m_{\text{gal}})/5} = 100^{(4.7+22)/5} \approx 5 \times 10^{10}.$$

Careful study of nearby stars has shown that for most stars, the absolute luminosity depends on the mass: *the more massive the star, the greater its luminosity*. Another important parameter of a star is its surface temperature, which can be determined from the spectrum of electromagnetic frequencies it emits, just as for a blackbody (Section 27–1). As we saw in Chapter 27, the spectrum of hotter and hotter bodies shifts from predominately lower frequencies (and longer wavelengths, such as red) to higher frequencies (and shorter wavelengths such as blue). Quantitatively, the relation is given by Wien's displacement law (Section 27–1): The peak wavelength λ_{p} in the spectrum of light emitted by a blackbody (and stars are fairly good approximations to blackbodies) is inversely proportional to its kelvin temperature T; that is, $\lambda_{\text{p}} T = 2.90 \times 10^{-3}\ \text{m}\cdot\text{K}$. The surface temperatures of stars typically range from about 3500 K (reddish) to perhaps 50,000 K (bluish).

An important astronomical discovery, made near the turn of the century, was that for most stars, the color is related to the absolute luminosity and therefore to the mass. A useful way to present this relationship is by the so-called Hertzsprung–Russell (H–R) diagram. On the H–R diagram, one axis represents the absolute magnitude and the other represents the temperature, and each star is represented by a point on the diagram, Fig. 33–7. Most stars fall along the diagonal band termed the **main sequence**. Starting at the lower right we find the coolest stars, reddish in color; they are the least luminous and therefore must be of low mass. Farther up toward the left we find hotter and more luminous stars that are yellowish-white, like our sun.

Main sequence stars

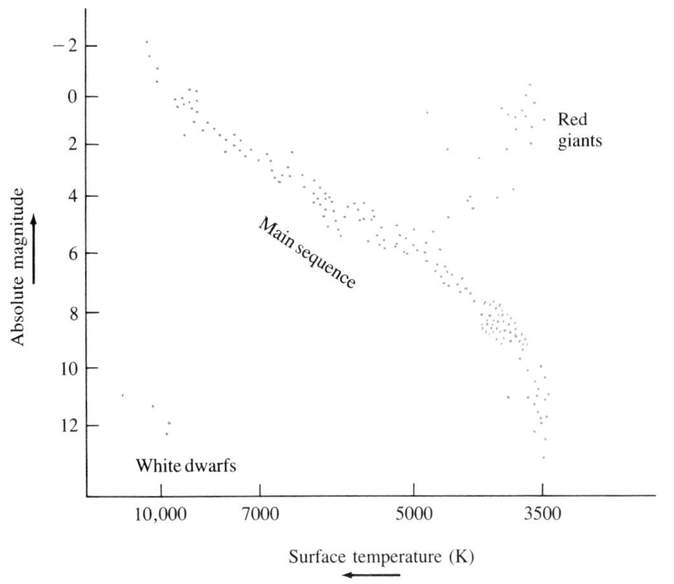

FIGURE 33–7 Hertzsprung–Russell (H–R) diagram.

Still farther up we find still more massive and more luminous stars, bluish in color. Stars that fall on this diagonal band are called *main-sequence stars.* There are also stars that fall outside the main sequence. Above and to the right we find extremely large stars, with high luminosities but with low (reddish) color temperature; these are called *red giants.* At the lower left, there are a few stars of low luminosity but with high temperature; these are the *white dwarfs.*

Red giants

White dwarfs

Before considering the different types of stars, let us first look at a couple of Examples that show how physical principles can be used to gain information about stars.

EXAMPLE 33–4 Suppose that the distances to two nearby stars can be reasonably estimated and this data, together with their measured apparent magnitudes, suggests that the two stars have about the same absolute luminosity, L. The spectrum of one of the stars peaks at about 700 nm (so it is reddish); the spectrum of the other peaks at about 350 nm (bluish). Use Wien's displacement law (Section 27–1) and the Stefan-Boltzmann law (Section 14–9) to determine (*a*) the surface temperature of each star, and (*b*) how much larger one star is than the other.

SOLUTION (*a*) Wien's displacement law (Section 27–1) states that $\lambda_p T = 2.90 \times 10^{-3}\ \mathrm{m \cdot K}$. So the temperature of the reddish star is

$$T_r = \frac{2.90 \times 10^{-3}\ \mathrm{m \cdot K}}{700 \times 10^{-9}\ \mathrm{m}} = 4140\ \mathrm{K}.$$

The temperature of the bluish star will be double this since its peak wavelength is half (350 nm vs 700 nm); just to check:

$$T_b = \frac{2.90 \times 10^{-3}\ \mathrm{m \cdot K}}{350 \times 10^{-9}\ \mathrm{m}} = 8280\ \mathrm{K}.$$

(*b*) The Stefan-Boltzmann law, which we discussed in Chapter 14 (see Eq. 14–4), states that the power radiated *per unit area* of surface from a body is proportional to the fourth power of the kelvin temperature, T^4. Now the temperature of the bluish star is double that of the reddish star, so the bluish one must radiate $(2)^4 = 16$ times as much energy per unit area. But we are given that they have the same luminosity (the same total power output), so the surface area of the blue star must be $\frac{1}{16}$ that of the red one. Since the surface area of a sphere is $4\pi r^2$, we conclude that the radius of the reddish star is $\sqrt{16} = 4$ times larger than that of the bluish star (or $4^3 = 64$ times the volume).

EXAMPLE 33–5 Detailed study of a star suggests that it most likely fits on the main sequence of an H–R diagram. Its measured apparent magnitude is $m = 11$, and the peak wavelength of its spectrum is $\lambda_p \approx 600$ nm. Estimate its distance from us.

SOLUTION The star's temperature, from Wien's law, is

$$T \approx \frac{2.90 \times 10^{-3}\ \mathrm{m \cdot K}}{600 \times 10^{-9}\ \mathrm{m}} \approx 4800\ \mathrm{K}.$$

A star on the main sequence of an H–R diagram at this temperature has absolute magnitude of about $M \approx 6$, read off of Fig. 33–7. Equation 33–2 relates the apparent and absolute magnitudes, m and M, which we now know, to the star's distance from us. We rearrange Eq. 33–2:

$$\log_{10}\left(\frac{d}{10\text{ pc}}\right) = \frac{m - M}{5};$$

then we raise both sides to the power 10:

$$\frac{d}{10\text{ pc}} = 10^{(m-M)/5},$$

which in our case gives

$$\frac{d}{10\text{ pc}} = 10^{(11-6)/5} = 10^1 = 10.$$

So the star is about $d \approx (10)(10\text{ pc}) = 100$ pc from us.

Why are there different types of stars, such as red giants and white dwarfs as well as main-sequence stars? Were they all born this way, in the beginning? Or might each different type represent a different age in the life cycle of a star? Astronomers and astrophysicists today believe the latter is largely true. Note, however, that we cannot actually follow any but the tiniest part of the life cycle of any given star since they live for ages vastly greater than ours, on the order of millions or billions of years. Nonetheless, let us follow the process of **stellar evolution** from the birth to the death of a star, as astrophysicists have reconstructed it today.

Birth of a star

Stars are born, it is believed, when gaseous clouds (mostly hydrogen) contract due to the pull of gravity. A huge gas cloud might fragment into numerous contracting masses, each mass centered in an area where the density was only slightly greater than that at nearby points. Once such "globules" formed, gravity would inexorably pull each one in toward its center of mass. As the particles of such a *protostar* accelerate inward, their kinetic energy increases. When the kinetic energy is sufficiently high, the Coulomb repulsion that keeps the hydrogen nuclei apart can be overcome and nuclear fusion can take place. In a star like our sun, the "burning" of hydrogen[†] occurs via the proton–proton cycle (Section 31–3, Eqs. 31–3), in which four protons fuse to form a ${}^4_2\text{He}$ nucleus with the release of γ rays and neutrinos. These reactions require a temperature of about 10^7 K, corresponding to an average KE (kT) of about 1 keV. (In more massive stars, the carbon cycle produces the same net effect: four ${}^1_1\text{H}$ produce a ${}^4_2\text{He}$—see Section 31–3.) The fusion reactions take place primarily in the core of a star, where T is sufficiently high. (The surface temperature is, of course, much lower—on the order of a few thousand kelvins.) The tremendous release of energy in these fusion reactions produces a pressure sufficient to halt the gravitational contraction, and our protostar, now really a young *star*, stabilizes on the main sequence. Exactly where the star falls along the main

Reaching the main sequence

[†] The word "burn" is put in quotation marks because these high temperature fusion reactions occur via a *nuclear* process, and must not be confused with ordinary burning (of, say, paper, wood, or coal) in air which are chemical reactions at the *atomic* level (and at a much lower, though still hot, temperature).

sequence depends on its mass. The more massive the star, the farther up (and to the left) it falls on the diagram. To reach the main sequence requires perhaps 30 million years, if it is a star like our sun, and it will remain there† about 10 billion years (10^{10} yr). Although most stars are billions of years old, there is evidence that stars are actually being born at this moment.

As hydrogen "burns"—that is, fuses to form helium—the helium that is formed is denser and tends to accumulate in the central core where it was formed. As the core of helium grows, hydrogen continues to "burn" in a shell around it. When much of the hydrogen within the core has been consumed, the production of energy decreases and is no longer sufficient to prevent the huge gravitational forces from once again causing the core to contract and heat up. The hydrogen in the shell around the core then "burns" even more fiercely because of this rise in temperature, causing the outer envelope of the star to expand and to cool. The surface temperature, thus reduced, produces a spectrum of light that peaks at longer wavelength. By this time the star has left the main sequence. It has become redder, and as it has grown in size, has become more luminous. So it has moved to the right and upward on the H–R diagram, as shown in Fig. 33–8. As it moves upward, it enters the **red giant** stage. Our sun, for example, has been on the main sequence for about $4\frac{1}{2}$ billion years. It will probably remain there another 4 or 5 billion years. (We can take comfort in that!) When our sun leaves the main sequence, it will grow in size (as a red giant) until it occupies all the volume out to perhaps the orbit of Venus or even earth.

Red giants

As a star's envelope expands, the core is shrinking and heating up. When the temperature reaches about 10^8 K, helium nuclei, with their greater

† More massive stars, since they are hotter and the Coulomb repulsion is more easily overcome, "burn" much more quickly and so use up their fuel faster, resulting in shorter lives. A star 10 times more massive than our sun, for example, will remain on the main sequence only for about 10^7 years. Stars less massive than our sun live much longer than our sun's 10^{10} yr.

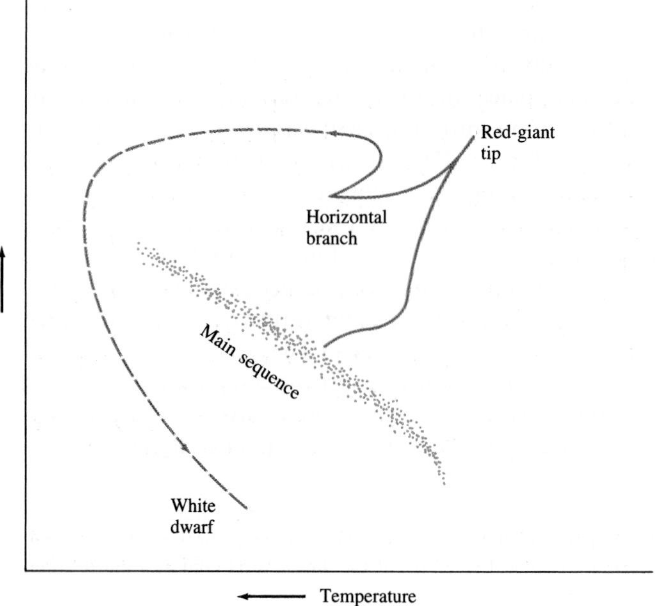

FIGURE 33–8
Evolution of a star like our sun represented on an H–R diagram.

charge and hence greater electrical repulsion, can then undergo fusion. The reactions are

Nucleosynthesis

$$\begin{aligned} {}^4_2\text{He} + {}^4_2\text{He} &\longrightarrow {}^8_4\text{Be} + \gamma, \\ {}^4_2\text{He} + {}^8_4\text{Be} &\longrightarrow {}^{12}_6\text{C} + \gamma. \end{aligned} \tag{33–3}$$

These two reactions occur in quick succession, and the net effect is

$$3\,{}^4_2\text{He} \longrightarrow {}^{12}_6\text{C}.$$

The burning of helium causes a rapid and major change in the star, and the star moves rapidly to the "horizontal branch" on the H–R diagram (see Fig. 33–8). As carbon accumulates, but before all the ${}^4_2\text{He}$ is used up, other reactions are possible such as

$${}^4_2\text{He} + {}^{12}_6\text{C} \longrightarrow {}^{16}_8\text{O} + \gamma, \tag{33–4a}$$

and, to a lesser extent,

$${}^4_2\text{He} + {}^{16}_8\text{O} \longrightarrow {}^{20}_{10}\text{Ne} + \gamma, \tag{33–4b}$$

$${}^4_2\text{He} + {}^{20}_{10}\text{Ne} \longrightarrow {}^{24}_{12}\text{Mg} + \gamma. \tag{33–4c}$$

Hydrogen and helium are by far the most common elements in the universe, with H making up about $\frac{3}{4}$ of the mass of the elements and He about $\frac{1}{4}$. The next most common elements found in the universe are C and O, as is suggested by their production in the above reactions.

Once the "burning" of He is complete, the core cannot contract sufficiently to enter another stage of "burning" (of C and O) if the star's mass is less than about 0.7 solar masses. If the mass is greater than this, the core can contract and heat up further, allowing fusion reactions between C nuclei and between O nuclei at around $6–7 \times 10^8$ K, and 1×10^9 K, respectively. Among the more likely of these reactions are

$$\begin{aligned} {}^{12}_6\text{C} + {}^{12}_6\text{C} &\longrightarrow {}^{24}_{12}\text{Mg} + \gamma \\ {}^{16}_8\text{O} + {}^{16}_8\text{O} &\longrightarrow {}^{28}_{14}\text{Si} + {}^4_2\text{He} \\ &\longrightarrow {}^{31}_{16}\text{S} + n. \end{aligned}$$

In the range $T = 2.5–5 \times 10^9$ K, nuclei as heavy as ${}^{56}_{26}\text{Fe}$ and ${}^{56}_{28}\text{Ni}$ can be made. But here the process of **nucleosynthesis**, the formation of heavy nuclei from lighter ones by fusion, ends. As we saw in Fig. 30–1, the average binding energy per nucleon begins to decrease for A greater than about 60. Further fusions would require energy, rather than release it. (Elements heavier than Ni are probably formed mainly by neutron capture, in supernova explosions, as we shall discuss shortly.)

What happens next depends on the mass of the star. If the star has a residual mass less than about 1.4 solar masses[†] (known as the *Chandrasekhar limit*), no further fusion energy can be obtained and the star collapses under the action of gravity. As it shrinks, the star cools and typically follows the route shown in Fig. 33–8, descending from the upper left downward, becoming a **white dwarf**. A white dwarf with a mass equal to that of the sun

White dwarf

[†] This cutoff refers to the mass of the star remaining *after* mass loss from the red giant stage. A star whose residual mass is 1.4 solar masses had an original mass of perhaps 6 to 8 solar masses.

would be about the size of the earth. What determines its size is that it collapses to the point at which the electron clouds of the atoms start to overlap; at this point it collapses no further, being restricted by the Pauli exclusion principle, in that no two electrons can be in the same quantum state. A white dwarf continues to lose internal energy, decreasing in temperature and becoming dimmer and dimmer until its light goes out. It has then become a *black dwarf*, a dark cold chunk of ash.

More massive stars (residual mass greater than 1.4 solar masses) follow a quite different scenario. A star with this great a mass can contract under gravity and heat up even further, reaching extremely high temperatures. The KE of the nuclei is then so high that fusion of elements heavier than iron can occur even though the reactions are energy-requiring. Furthermore the high-energy collisions can also cause the breaking apart of iron and nickel nuclei into He nuclei, and eventually into protons and neutrons:

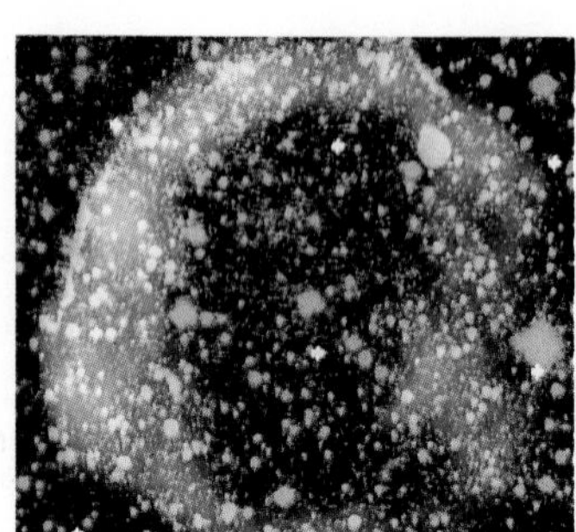

FIGURE 33–9 Computer-enhanced false-color image of the shell of material expanding outward from "Tycho's supernova" (seen by Tycho Brahe in 1572).

$$^{56}_{26}\mathrm{Fe} \longrightarrow 13\,^{4}_{2}\mathrm{He} + 4n,$$

$$^{4}_{2}\mathrm{He} \longrightarrow 2p + 2n.$$

These are energy-requiring (endoergic) reactions, but at such extremely high temperature and pressure there is plenty of energy available, enough even to force electrons and protons together to form neutrons in inverse β decay:

$$e^- + p \longrightarrow n + \nu.$$

As the core contracts under the huge gravitational forces, the tremendous mass becomes essentially an enormous nucleus made up almost exclusively of neutrons. The size of the star is no longer limited by the exclusion principle applied to electrons, but rather applied to neutrons, and the star contracts rapidly to form an enormously dense **neutron star**. The contraction of the core signals a great reduction in gravitational potential energy. Somehow this energy must be released. Indeed, it was suggested in the 1930s that core collapse may be accompanied by a catastrophic explosion whose tremendous energy could form virtually all elements of the periodic table and blow away the entire outer envelope of the star (Fig. 33–9), spreading its contents into interstellar space. Such explosions, now believed to produce one type of observed **supernova** (there may be other mechanisms as well), are thus thought to be the origin of heavy elements in our solar system, the stuff of which we and our environment are made. Indeed, the presence of heavy elements on earth suggests that the earth is composed of debris from at least one supernova.

Neutron stars

Supernovae

In a supernova explosion, a star's brightness is observed to suddenly increase billions of times in a period of just a few days and then fade away over the next few months. In February, 1987, such a supernova occurred in an appendage to our Galaxy known as the Large Magellanic Cloud, only 170,000 ly away, and was visible to the naked eye in the southern hemisphere (Fig. 33–10). Supernovae are seen relatively often in distant galaxies, but this recent one, called SN1987a, is the first to have occurred close enough to be visible to the naked eye since the famous one of 1604 observed by Kepler and Galileo (and thus SN1987a is the first since the invention of the telescope). A supernova visible to the naked eye was observed by Chinese astronomers in A.D. 1054, and its remains are still visible in the sky (in the Crab Nebula), in the midst of which there is now a **pulsar**. Pulsars are as-

Pulsars

FIGURE 33–10
SN1987a, the supernova that appeared in February 1987, is the very bright star on the right. The bright nebulous object is the Tarantula nebula.

tronomical objects that emit sharp pulses of radiation at regular intervals, on the order of a second. They are now believed to be neutron stars which, because of conservation of angular momentum, increase greatly in rotational speed as their moment of inertia decreases as they contract. The intense magnetic field of such a rapidly rotating star (perhaps as high as 10^6 to 10^{10} T) could trap and accelerate charged particles, which then give off radiation in a beam that rotates with the star. The discovery of pulsars in 1967 has lent support to the theory that neutron stars are the result of supernova explosions.

The core of a neutron star contracts to the point at which all neutrons are as close together as they are in a nucleus. That is, the density of a neutron star is on the order of 10^{14} times that of normal solids and liquids on earth. The mass of a neutron star 1.5 times that of our sun would have a diameter of only about 10 km.

If the mass of a neutron star is less than about two or three solar masses, its subsequent evolution is similar to that of a white dwarf. If the mass is greater than this, the gravitational force is so strong that the star contracts to an even smaller diameter and an even greater density, overcoming, in a sense, even the neutron exclusion principle. Gravity is so strong that even light emitted from it cannot escape—it is pulled back in by the force of gravity. In other words, the escape velocity from such a star is greater than c, the speed of light. Since no radiation can escape from such a star, we cannot see it—it is black. A body may pass by it and be deflected by its gravitational field, but if it comes too close, it will be swallowed up, never to escape. This is a **black hole**. Black holes are predicted by theory to exist. Although observation has not yet unambiguously confirmed their existence, strong evidence suggests there may be a giant black hole at the center of our Galaxy—its mass is estimated at several million times that of the sun—and it seems possible that many or all galaxies have black holes at their centers.

Black holes

33–3 • General Relativity: Gravity and the Curvature of Space

We have seen that the force of gravity plays a dominant role in the processes that occur in stars and, in general, in the evolution of the universe as a whole. The reasons gravity, and not one of the other of the four forces in nature, plays the dominant role in the universe are (1) it is long-range and (2) it is always attractive. The strong and weak nuclear forces act over very short distances only, on the order of the size of a nucleus; hence they do not act over astronomical distances (although of course they act between nuclei and nucleons in stars to produce nuclear reactions). The electromagnetic force, like gravity, acts over great distances; but it can be either attractive or repulsive. And since the universe does not seem to contain large areas of net electric charge, a large net force does not occur. But gravity acts as an attractive force between *all* masses, and there are large accumulations in the universe of only the one type of mass (not + and − as with electric charge).

However, the force of gravity as Newton described it in his law of universal gravitation shows discrepancies on a cosmological scale. Einstein, in his general theory of relativity, developed a theory of gravity that resolves these problems and forms the basis of cosmological dynamics.

In the *special theory of relativity* (Chapter 26), Einstein concluded that there is no way for an observer to determine whether a given frame of reference is at rest or is moving at constant velocity in a straight line. Thus, Einstein said, the laws of physics must be the same in different inertial reference frames. (As we saw, the actual *paths* of objects are usually different in different reference frames.) To consider only uniformly moving reference frames is somewhat restricting. What about the general case of motion, where reference frames can be *accelerating*?

It is in the **general theory of relativity** that Einstein tackled the problem of accelerating reference frames and developed a theory of gravity. The mathematics of general relativity is very complex, so our discussion will be mainly qualitative.

We begin with Einstein's famous **principle of equivalence**, which states that

Principle of equivalence

No observer can determine by experiment if he or she is accelerating or is rather in a gravitational field.

If some observers sensed that they were accelerating (as in a vehicle speeding around a sharp curve), they could not prove by any experiment that in fact they weren't simply experiencing the pull of a gravitational field. Conversely, we might think we are being pulled by gravity when in fact we are undergoing an "inertial" acceleration having nothing to do with gravity. For example, pilots making a steeply banked turn in fog often have this experience, and cannot tell in which direction the earth lies.

As a thought experiment, consider a person in a freely falling elevator near the earth's surface. If our observer held out a book and let go of it, what would happen? Gravity would pull it downward toward the earth, but at the same rate ($g = 9.8\ \text{m/s}^2$) at which the person and elevator were falling. So the book would hover right next to the person's hand (Fig. 33–11). The effect is exactly the same as if this reference frame were at rest and *no* forces were acting. On the other hand, suppose the elevator were far out in space where there is no gravitational field. If the person released the book, it would float, just as it did in Fig. 33–11b. If, instead, the elevator were accelerating upward at an acceleration of $9.8\ m/s^2$, the book as seen by our observer would fall to the floor with an acceleration of $9.8\ m/s^2$, just as if it were falling because of gravity. According to the principle of equivalence, the observer could not be sure whether the book fell because the elevator was accelerating upward at $a = 9.8\ m/s^2$ in the absence of gravity, or because a gravitational field with $g = 9.8\ m/s^2$ was acting downward and he was at rest (say on the earth). The two descriptions are equivalent.

The principle of equivalence is related to the concept of mass and to the idea that there are two types of mass. For any force, Newton's second law says that $F = ma$, where m is the mass—or more precisely, the **inertial mass**. The more inertial mass a body has, the less it is affected by a given force (the less acceleration it undergoes). You might say that inertial mass represents resistance to any type of force whatever. The second type of mass is **gravitational mass**. When one body attracts another by the gravitational force (Newton's law of universal gravitation, $F \propto mm'/d^2$, Chapter 5), the strength of the force is proportional to the product of the gravitational masses of the two bodies. This is much like the electric force between two bodies that is proportional to the product of their electric charges. Now the property of a body termed its electric charge is clearly not related to its inertial mass; so why should we expect that the gravitational mass (call it gravitational charge if you like) of a body is related to its inertial mass? We have, up to now, assumed they were the same. Why? Because no experiment—not even high-precision experiments—has been able to discern any measurable difference between them. This, then, is another way to state the equivalence principle: *gravitational mass is equivalent to inertial mass.*

The principle of equivalence can be used to show that light ought to be deflected due to the gravitational force of a massive body. We will not do the complex calculation, but will consider a thought experiment to give a rough idea. Consider an elevator in free space where no gravity acts. If there is a hole in the side of the elevator and a beam of light enters from outside, the beam travels straight across the elevator and makes a spot on the opposite side, if the elevator is at rest (Fig. 33–12a). If the elevator is accelerating up-

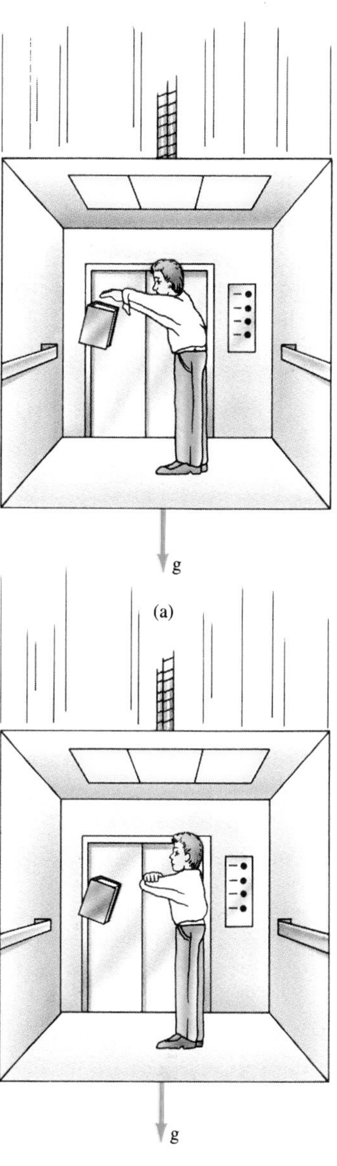

FIGURE 33–11 (above)
A freely falling elevator. The released book hovers next to the owner's hand.

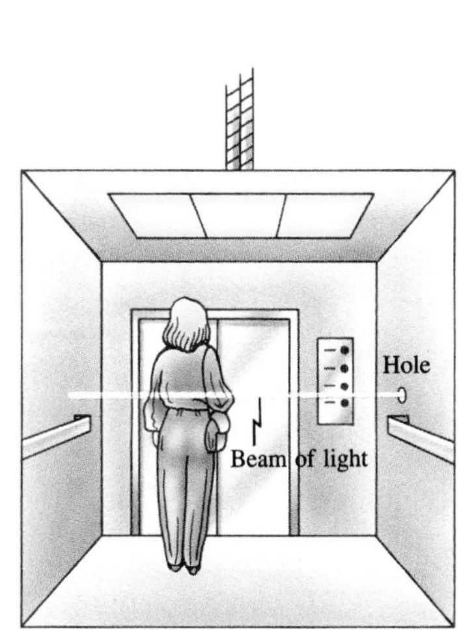

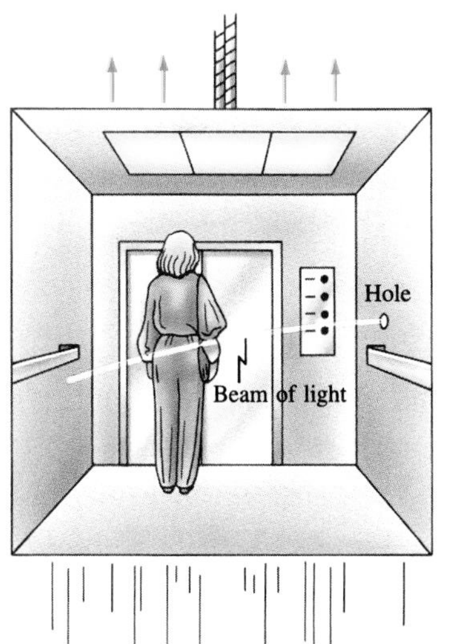

FIGURE 33–12 (left)
The bending of a light beam (exaggerated) in an elevator (b) accelerating in an upward direction. In (a), there is no acceleration.

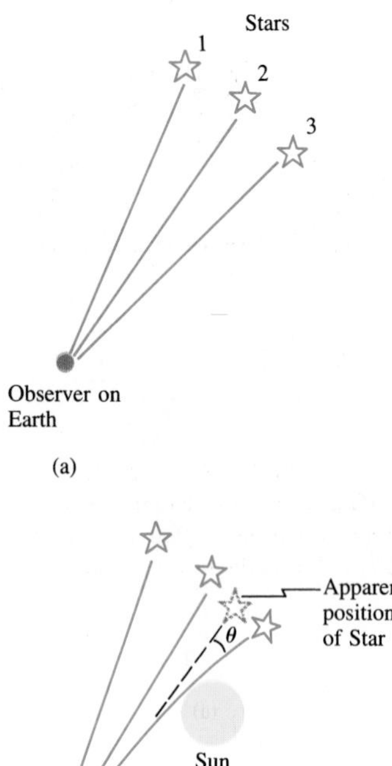

FIGURE 33–13 (a) Three stars in the sky. (b) If the light from one of these stars passes very near the sun, whose gravity bends the rays, the star will appear higher than it actually is.

ward as in Fig. 33–12b, the light beam still travels straight across in a reference frame at rest. In the upwardly accelerating elevator, however, the beam is observed to curve downward. Why? Because during the time the light travels from one side of the elevator to the other, the elevator is moving upward at ever-increasing speed. Now, according to the equivalence principle, an upwardly accelerating reference frame is equivalent to a downward gravitational field. Hence, we can picture the curved light path in Fig. 33–12b as being the effect of a gravitational field. Thus we expect gravity to exert a force on a beam of light and to bend it out of a straight-line path! This is an important prediction of Einstein's general theory of relativity. And it can be tested. The amount a light beam would be deflected from a straight-line path must be small even when passing a massive body. (For example, light near the earth's surface after traveling 1 km would drop only about 10^{-10} m, which is equal to the diameter of a small atom and not detectable.) The most massive body near us is the sun, and it was calculated that light from a distant star would be deflected by 1.75″ of arc (tiny but detectable) as it passed near the sun (Fig. 33–13). However, such a measurement could only be made during a total eclipse of the sun, so the sun's tremendous brightness would not overwhelm the starlight passing nearby. An opportune eclipse occurred in 1919 and scientists journeyed to the South Atlantic to observe it. Their photos of stars around the sun revealed shifts in accordance with Einstein's prediction. One newspaper flaunted the headline "Light Caught Bending." Newton's universal law of gravitation, coupled with the equivalence principle, also predicts a gravitational deflection of light, considered as photons of mass $m = E/c^2$ (see Problem 44), but the predicted deflection is half that predicted by Einstein's general relativity; and it is the latter that is observed.

Now a light beam must travel by the shortest, most direct, path between two points. If it didn't, some other object could travel between the two points in a shorter time and thus have a greater speed than the speed of light—a clear contradiction of the special theory of relativity. If a light beam can follow a curved path (as discussed above), then this curved path must be the shortest distance between the two points—which suggests that *space itself is curved* and that it is the gravitational field that causes the curvature. Indeed, the curvature of space—or rather, of four-dimensional space-time—is a basic aspect of general relativity.

FIGURE 33–14 On a two-dimensional curved surface, the sum of the angles of a triangle may not be 180°.

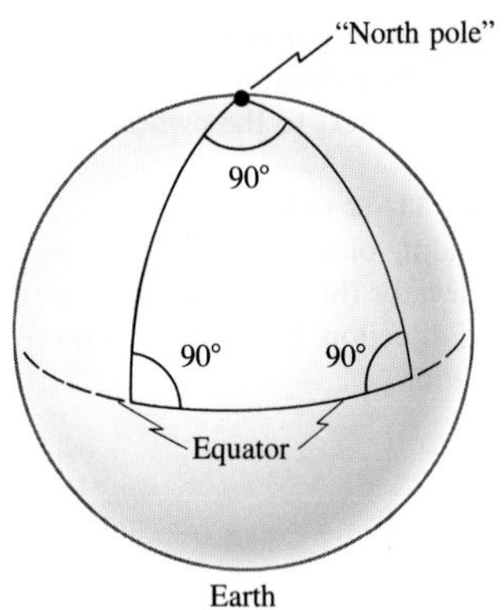

What is meant by *curved space*? To understand, let us recall that our normal method of viewing the world is via Euclidean plane geometry. In Euclidean geometry, there are many axioms and theorems we take for granted, such as that the sum of the angles of any triangle is 180°. Other geometries, non-Euclidean which involve curved space, have also been imagined by mathematicians. Now it is hard enough to imagine three-dimensional curved space, much less curved four-dimensional space-time. So let us explain the idea of curved space by using two-dimensional surfaces.

Consider, for example, the two-dimensional surface of a sphere. It is clearly curved, Fig. 33–14, at least to us who view it from the outside—from our three-dimensional world. But how would hypothetical two-dimensional creatures determine whether their two-dimensional space were flat (a plane) or curved? One way would be to measure the sum of the angles of a triangle. If the surface is a plane, the sum of the angles is 180°. But if the space is curved, and a sufficiently large triangle is constructed, the sum of the angles will *not* be 180°. To construct a triangle on a curved surface, say a sphere, we must use the equivalent of a straight line: that is, the shortest distance between two points, which is called a **geodesic**. On a sphere, a geodesic is an arc of a great circle (an arc contained in a plane passing through the center of the sphere) such as the earth's equator and the earth's longitude lines. Consider, for example, the large triangle of Fig. 33–14 whose sides are two "longitude" lines passing from the "north pole" to the equator, a part of which forms the third side. The two longitude lines make 90° angles with the equator (look at a world globe to see this more clearly); if they make, say, a 90° angle with each other at the north pole, the sum of these angles is $90° + 90° + 90° = 270°$. This is clearly *not* a Euclidean space. (Note, however, that if the triangle is small in comparison to the radius of the sphere, the angles will add up to nearly 180°, and the triangle will seem flat.)

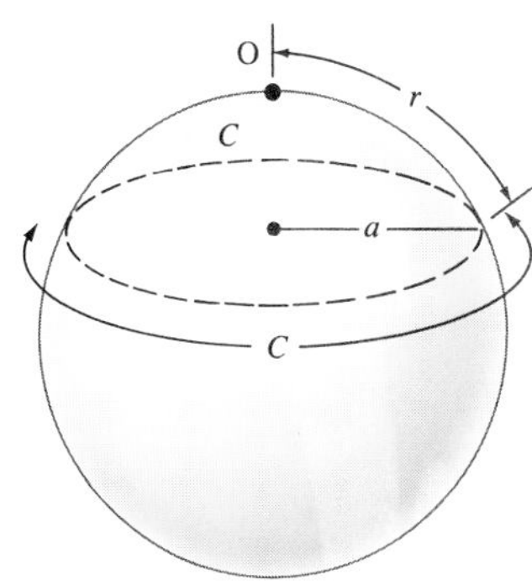

FIGURE 33–15 On a spherical surface, a circle of circumference C is drawn about point O as the center. The radius is the distance r along the surface. (Note that in our three-dimensional view, we can tell that $2\pi a = C$; since $r > a$, then $2\pi r > C$.)

Another way to test the curvature of space is to measure the radius r and circumference C of a large circle. On a plane surface, $C = 2\pi r$. But on a two-dimensional spherical surface, C is *less* than $2\pi r$, as can be seen in Fig. 33–15. The proportionality constant between C and r is *less* than 2π. Such a surface is said to have *positive curvature*. On the saddlelike surface of Fig. 33–16, the circumference of a circle is greater than $2\pi r$, and the sum of the angles of a triangle is less than 180°. Such a surface is said to have a *negative curvature.*

FIGURE 33–16 Example of a two-dimensional surface with negative curvature.

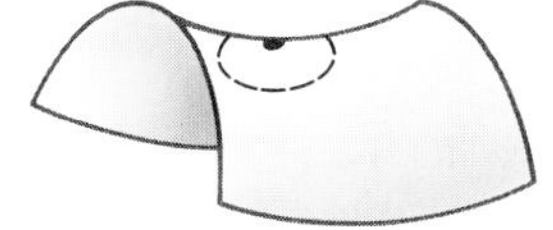

Now, what about our universe? On a large scale (not just near a large mass), what is the overall curvature of the universe? Does it have positive curvature, negative curvature, or is it flat (zero curvature)? In the nineteenth century, Carl Friedrich Gauss (1777–1855) tried to determine whether our natural three-dimensional space deviated from Euclidean space by measuring the angles of a triangle formed by three mountain peaks using light rays as sides of the triangle. He was unable to detect any deviation from 180°, nor are experiments today precise enough to detect any deviation.

The universe: open or closed?

Nonetheless, the question of the curvature of space in the real world is an important one in cosmology. And the answer is still not known. If the universe has a positive curvature, then the universe is *finite*, or *closed*. This does not mean that in such a universe the stars and galaxies would extend out to a certain boundary, beyond which there is empty space. Rather, galaxies would be spread throughout the space, and the space would fold back and "close on itself." There is no boundary or edge in such a universe. If a particle were to move in a straight line in a particular direction, it would eventually return to the starting point—albeit eons of time later. To ask "What is beyond such a closed universe?" is futile, since the space of the universe is all there is. If the curvature of space is zero or negative, on the other hand, the universe would be *open*. It would just go on and on and never fold back on itself. An open universe would be *infinite*. Whether the universe is open or closed depends, in part, on how much total mass there is in the universe, as we will discuss in Section 33–7. If the mass is great enough, it bends space into a positively curved, closed, and finite space.

According to Einstein's theory, space–time is curved, especially near massive bodies. To visualize this, we might think of space as being like a thin rubber sheet; if a heavy weight is hung from it, it curves as shown in

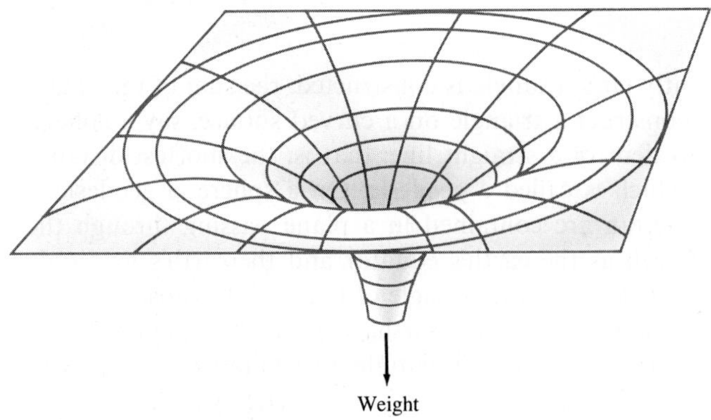

FIGURE 33–17
Rubber-sheet analogy for space–time curved by matter.

Gravity as curvature of space–time

Fig. 33–17. The weight corresponds to a huge mass that causes space (space itself!) to curve. Thus, in Einstein's theory† we do not speak of the "force" of gravity acting on bodies. Instead we say that bodies and light rays move along geodesics (the equivalent of straight lines in plane geometry) in curved space–time. Thus, a body at rest or moving slowly near the great mass of Fig. 33–17 would follow a geodesic toward that body.

Black holes

The extreme curvature of space–time shown in Fig. 33–17 could be produced by a **black hole**. A black hole, as we saw in the previous section, is so dense that even light cannot escape from it. To become a black hole, a body of mass M must undergo **gravitational collapse**, contracting by gravitational self-attraction to within a radius called the **Schwarzschild radius**:

$$R = \frac{2GM}{c^2},$$

where G is the gravitational constant and c is the speed of light.

How might we observe black holes? We cannot see them because no light can escape from them. They would be black objects against a black sky. But they do exert a gravitational force on nearby bodies. The suspected black hole at the center of our Galaxy was discovered by examining the motion of matter in its vicinity. Another technique is to examine stars which appear to be rotating as if they were members of a *binary system* (two stars rotating about their common center of mass), although the companion is invisible. If the unseen star is a black hole, it might be expected to pull off gaseous material from its visible companion (Fig. 33–18). As this matter approached the black hole, it would be highly accelerated and should emit X rays of a characteristic type. At present there is fairly reasonable evidence for a black hole in the binary star Cygnus X-1; whether it actually is or not must be determined by further observations.

† Alexander Pope (1688–1744) wrote an epitaph for Newton:

"Nature and Nature's laws lay hid in night.
God said, Let Newton be! and all was light."

Sir John Squire (1884–1958), perhaps uncomfortable with Einstein's profound thoughts, added:

"It did not last; the Devil howling *Ho*
Let Einstein be, restored the status quo."

(Thanks to D. Edgell for drawing my attention to the second quote.)

FIGURE 33–18 Artist's conception of how a black hole, which is one star of a binary pair, might pull matter from the other (more normal) star.

33–4 • The Expanding Universe

We discussed in Section 33–2 how individual stars evolve from their birth to their death as white dwarfs, neutron stars, and black holes. But what about the universe as a whole: is it static or does it evolve? The evolution of stars suggests that the universe as a whole evolves. Let us look at the evidence.

In his theory of the universe, Newton assumed the universe was *static*: no large-scale changes occur over time. Newton recognized the difficulties in imagining a universe either as finite or as infinite. If it is finite and has a boundary, then we naturally ask, "What is beyond the boundary?" Yet, if there were anything beyond, why wouldn't such a region also be part of the universe? Furthermore, a static finite universe would seem to be unstable, since why wouldn't gravity pull everything into one glob at the center? An infinite universe is also hard to imagine. It too would be confronted by problems, one of which later came to be known as **Olbers' paradox**: why is it that *the sky at night is dark*? It was argued that if the universe were infinite, with stars uniformly distributed throughout, then any line of sight should eventually fall on the surface of a star. Hence the night sky would appear bright everywhere. This seems to imply that the universe could not be infinite—a conclusion that itself causes problems, as we've seen. Furthermore, stars are not points, but have a finite size—they subtend a finite (though minute) angle. Thus the night sky could be bright even if the universe were finite.

Why is the night sky dark?

There are possible answers to Olbers' paradox. For example, the universe might be infinite in extent but not infinitely old, so light from distant stars would not yet have reached us. It is commonly believed that this idea, together with reduced light energy from the redshift in the expansion of the universe (which we are about to discuss) provides a solution to Olbers' paradox. Yet evidence for these two ideas—that the universe did not always exist and that it is not static—was not found until well into the twentieth century. Even Einstein, while working on solutions to his general relativity equations in 1917, assumed the universe was static.

One of the most important scientific ideas of this century proposes that distant galaxies are racing away from us, and that the farther they are from us, the faster they are moving away. How astronomers reached this astonishing conclusion, and what it means for the past history of the universe as well as its future, will occupy us for the remainder of the book.

Hubble and the redshift

The idea that the universe is expanding was first put forth by Hubble in 1929. It was based on observations of the Doppler shift of light emitted by stars. In Chapter 12 we discussed how the frequency and wavelength of sound are altered if the source is moving toward or away from an observer. If the source moves toward us, the frequency is higher and the wavelength is shorter. If the source moves away from us, the frequency is lower and the wavelength is longer. The Doppler effect occurs also for light, but the shifted wavelength or frequency is given by a formula slightly different from that for sound because for light (according to special relativity), we can make no distinction between motion of the source and motion of the observer. (Recall that sound travels in a medium, such as air, but light does not; according to relativity, there is no ether.) According to special relativity, the formula is (see Problem 31)

$$\lambda' = \lambda\sqrt{\frac{1 + v/c}{1 - v/c}}, \tag{33–5}$$

where λ is the emitted wavelength as seen in the source's reference frame, and λ' is the wavelength measured in a frame moving with velocity v away from the source along the line of sight. Note that this relation depends only on the relative velocity v. (For relative motion toward each other, $v < 0$ in this formula.) Thus, when a source emits light of a particular wavelength and the source is moving away from us, the wavelength appears longer to us: the color of the light (if it is visible) is shifted toward the red end of the visible spectrum, an effect known as a **redshift**. If the source moves toward us, the color shifts toward the blue (shorter wavelength) end of the spectrum. The amount of shift depends on the velocity of the source (Eq. 33–5). For speeds not too close to the speed of light, it is easy to show (Problem 30) that the fractional change in wavelength is proportional to the speed of the source to or away from us (as was the case for sound):

$$\frac{\Delta\lambda}{\lambda} = \frac{\lambda' - \lambda}{\lambda} \approx \frac{v}{c}. \qquad [v \ll c] \tag{33–6}$$

In the spectra of stars and galaxies, lines are observed that correspond to lines in the known spectra of particular atoms (see Section 27–8). What Hubble found was that the lines seen in the spectra of galaxies were generally *redshifted*, and that the amount of shift seemed to be proportional to the distance of the galaxy from us. That is, the velocity, v, of a galaxy moving away from us is proportional to its distance, d, from us:

Hubble's law

$$v = Hd. \tag{33–7}$$

This is **Hubble's law**, one of the most important astronomical discoveries of this century. The constant H is called *Hubble's constant* (although it may not be constant). Hubble's law does not work well for nearby galaxies—in fact some are actually moving toward us ("blue-shifted"); but this is believed to merely represent random motion of the galaxies. For more distant galaxies, the velocity of recession (Hubble's law) is much greater than that of random motion, and so is dominant.

The value of H is not known very precisely. It is generally taken to be about

Hubble's constant

$$H \approx 50 \text{ km/s/Mpc}$$

(that is, 50 km/s per megaparsec of distance). If we use light years for distance, $H \approx 15$ km/s per million light years of distance. However, H could be as low as 40 km/s/Mpc or as high as about 100 km/s/Mpc (12 to 30 km/s/Mly). This rather large uncertainty arises mainly from the uncertainty in distance measurement for remote galaxies.

What does it mean that distant galaxies are all moving away from us, and with ever greater speed the farther they are from us? It is as if there had been a great explosion at some distant time in the past. And at first sight we seem to be in the middle of it all. But we aren't, necessarily. The expansion appears the same from any other point in the universe. To understand why, see Fig. 33–19. In Fig. 33–19a we have the view from earth (or

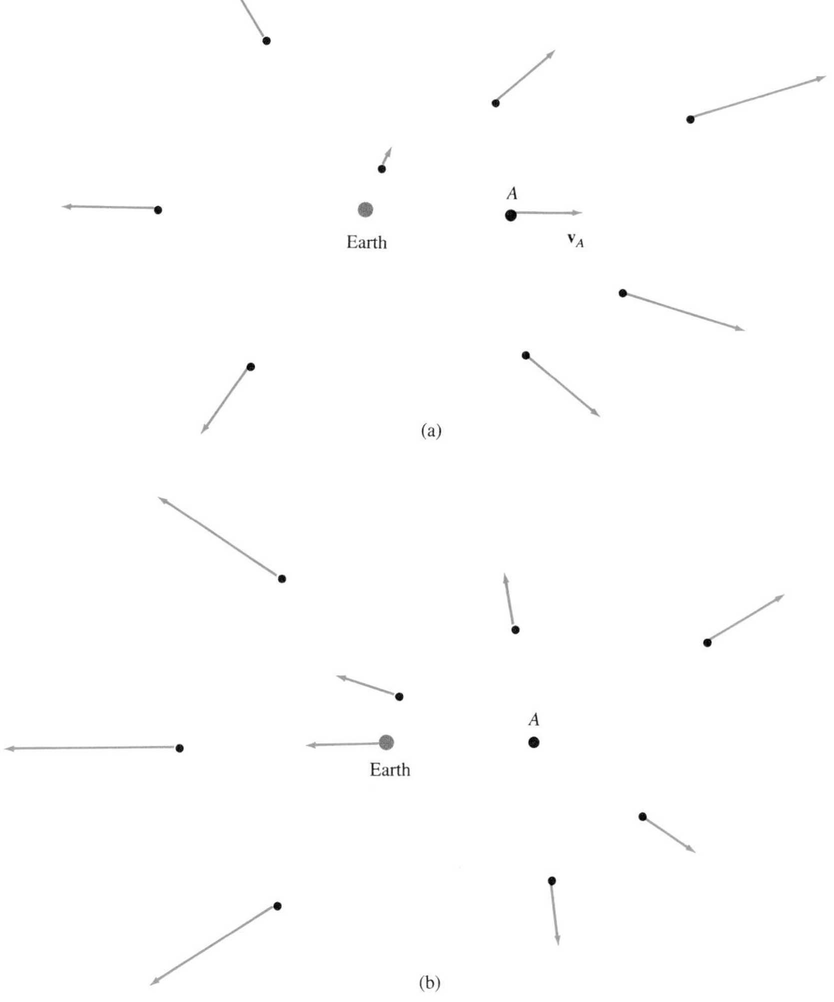

FIGURE 33–19 Expansion of the universe looks the same from any point in the universe.

our Galaxy). The velocities of surrounding galaxies are indicated by arrows, pointing away from us, and greater for galaxies more distant from us. Now what if we were on the galaxy labeled A in Fig. 33–19a. From earth, galaxy A appears to be moving to the right at a velocity, call it $\mathbf{v}_A$, represented by the arrow pointing to the right. If we were *on* galaxy A, earth would appear to be moving to the left at velocity $-\mathbf{v}_A$. To determine the velocities of other galaxies relative to A, we vectorially add the velocity vector, $-\mathbf{v}_A$, to all the velocity arrows shown in Fig. 33–19a. This yields Fig. 33–19b, where we see clearly that the universe is expanding away from galaxy A as well; and the velocities of galaxies receding from A are proportional to their distance from A.

Thus the expansion of the universe can be stated as follows: all galaxies are racing away from *each other* at an average rate of about 50 km/s per megaparsec of distance between them (or 15 km/s per million light years of separation). The ramifications of this profound discovery are enormous, and we discuss them in a moment.

Quasars

At this point we must point out that there is a class of objects called "quasistellar objects," or **quasars**, that do not seem to conform to Hubble's law. Quasars are as bright as nearby stars but display very large redshifts. If quasars are normal participants in the general expansion of the universe according to Hubble's law, their large redshifts suggest they are very distant. If they are so far away, they must be incredibly bright—sometimes thousands of times brighter than normal galaxies. On the other hand, it has been suggested that some quasars, at least, are not of abnormal brightness because they are nearer than their redshifts suggest. In the first case we would have an unresolved brightness problem, in the second case an unresolved redshift problem. An interesting piece of evidence is that the population density of quasars seems to increase with distance from us. If they are at great distances from us—as their redshifts suggest—this would merely mean they were more common in the early universe than they are now. But if instead they are much closer to us—as their brightness suggests—it would seem that *we* are in a special place in the universe, the place where quasars are least populous. Most astronomers are unwilling to accept this, for it would violate a basic assumption of uniformity, the so-called *cosmological principle*, which we discuss shortly. A recent suggestion that could resolve the quasar question is that these mysterious objects may be cores of galaxies that emit enormous amounts of energy when they draw in nearby material, for example when galaxies "collide." If the center of a galaxy is a black hole, a large amount of matter accelerated toward it would give off a prodigious amount of energy. It seems possible, then, that quasars are powered by black holes.

Cosmological principle

A basic assumption in cosmology has been that on a large scale, the universe looks the same to observers at different places at the same time. In other words, the universe is both isotropic (looks the same in all directions) and homogeneous (would look the same if we were located elsewhere, say in another galaxy). This assumption is called the **cosmological principle**. On a local scale, say in our solar system or within our Galaxy, it clearly does not apply (the sky looks different in different directions). But it has long been thought to be valid if we look on a large enough scale, so that the average population density of galaxies and clusters of galaxies ought to be the same in different areas of the sky. The expansion of the universe is consistent with the cosmological principle (Fig. 33–19) and the uniformity of the cosmic microwave background radiation (discussed in the next section) supports it. But matter (i.e., galaxies), even on the largest scale, seems not to be homogeneous but tends to clump, as already discussed. Although astronomers are reluctant to give up the cosmological principle, there is now some doubt about its validity. One possible resolution might be that over 90 percent of the universe may be nonluminous dark matter, which might be uniformly distributed. In any case, the cosmological principle has aided us in treating the universe as a single evolving entity and not as a random collection of material bodies. Another way of stating the cosmological principle is this: there is nothing special about the earth on a cosmological scale; our large-scale observations are no different from those that might be made elsewhere in the universe. What a change has occurred since that time, only a few hundred years ago, when we saw ourselves at the center of the universe!

The expansion of the universe, as described by Hubble's law, strongly suggests that galaxies must have been closer together in the past than they are now. Further, Hubble's law is consistent with all galaxies having been quite close together at some time in the past. This is, in fact, the basis of the *Big Bang* theory of the origin of the universe (discussed in the next section) which pictures the beginning of the universe as a great explosion. Let us see what can then be said about the age of the universe.

Age of the universe

If we use the value of the Hubble constant, $H \approx 15$ km/s per million light years, then the time required for the galaxies to arrive at their present separations would have been 10^6 light-years divided by 15 km/s:

$$t_0 = \frac{d}{v} \approx \frac{(10^6\ \text{ly})(10^{13}\ \text{km/ly})}{(15\ \text{km/s})(3 \times 10^7\ \text{s/y})} \approx 20 \times 10^9\ \text{y},$$

or 20 billion years. The age of the universe calculated in this way, which we could call the *characteristic expansion time* or "Hubble age", is equal to the reciprocal of the Hubble constant (see Eq. 33–7), $t_0 = 1/H$, and is probably an overestimate. The actual age should be less, as galaxies have not been moving with fixed velocities but have been slowing down under the action of their mutual gravitational attraction. Perhaps 10 to 15 billion years would be a better estimate.

There are two other independent checks on the age of the universe. The first is determination of the age of the earth (and solar system) from radioactivity, primarily using uranium, which places the age of the solar system at about $4\frac{1}{2}$ billion years. Second, by using the theory of stellar evolution, the ages of stars have been estimated to be about 10 to 15 billion years. These independent and unrelated determinations are consistent with a Big Bang occurring 10 to 15 billion years ago. The lower value determined from radioactivity is consistent since we would expect the origin of the earth (and the solidification of rocks) to have occurred somewhat after the origin of the universe as a whole.

Steady-state model

Before discussing the Big Bang theory in more detail, we briefly discuss one of the alternatives to the Big Bang—the **steady-state model**—which assumes that the universe is infinitely old and on the average looks the same now as it always has.† Thus, according to the steady-state model, no large-

† That the universe should be uniform in time as well as in space is essentially an extension of the cosmological principle and is sometimes referred to as the *perfect cosmological principle*.

scale changes have taken place in the universe as a whole, particularly no Big Bang. To maintain this view in the face of the recession of galaxies away from each other, mass–energy conservation must be violated. That is, matter is assumed to be created continuously, keeping the density of the universe constant. The rate of mass creation required is very small—about one nucleon per cubic meter every 10^9 years. This is much too small to be detected, and thus cannot be tested.

The steady-state model provided the Big Bang model with healthy competition in the 1940s and 50s. However, the discovery of the cosmic microwave background radiation (next section) pushed the Big Bang model to the forefront.

33–5 • The Big Bang and the Cosmic Microwave Background

The expansion of the universe seems to suggest, as we have seen, that the matter of the universe was once much closer together than it is now. This is the basis for the idea that the universe began 10 to 15 billion years ago with a huge explosion known affectionately as the **Big Bang**.

The Big Bang

If there was a Big Bang, it must have occurred simultaneously at all points in the universe. If the universe is *finite*, the explosion would have taken place in a tiny volume approaching a point. However, this point of extremely dense matter is not to be thought of as a concentrated mass in the midst of a much larger space around it. Rather, the initial dense point *was* the universe—the entire universe. There wouldn't have been anything else. If, on the other hand, the universe is *infinite*, then the explosion would have occurred at *all* points in the universe since an infinite universe, even if smaller at an earlier time, would still have been infinite. In either case, when we say the universe was once smaller than it is now, we mean that the average separation between galaxies was less. Thus, it is the *size of the universe itself* that has increased since the Big Bang.

What is the evidence supporting the Big Bang? First, the age of the universe as calculated from the Hubble expansion, from stellar evolution, and from radioactivity, all point to a consistent time of origin for the universe, as we saw in the last section. Another, and crucial, piece of evidence was the discovery in the 1960s of the "cosmic microwave background radiation," which came about as follows.

The 2.7 K cosmic microwave background radiation

In 1964, Arno Penzias and Robert Wilson were experiencing difficulty with what they assumed to be background noise, or "static," in their radio telescope (a large antenna device for detecting radio waves from the heavens). Eventually, they were convinced that it was real and that it was coming from outside our Galaxy. They made precise measurements at a wavelength $\lambda =$ 7.35 cm, which is in the microwave region of the electromagnetic spectrum. (See Fig. 22–10; this radiation is called "microwave" because the wavelength, though much greater than that of visible light, is somewhat smaller than wavelengths for ordinary radio waves, which are typically meters or hundreds of meters.) The intensity of this radiation was found not to vary by day or night or time of year. Neither did it depend on direction; it came from all directions in the universe with equal intensity (within less than one

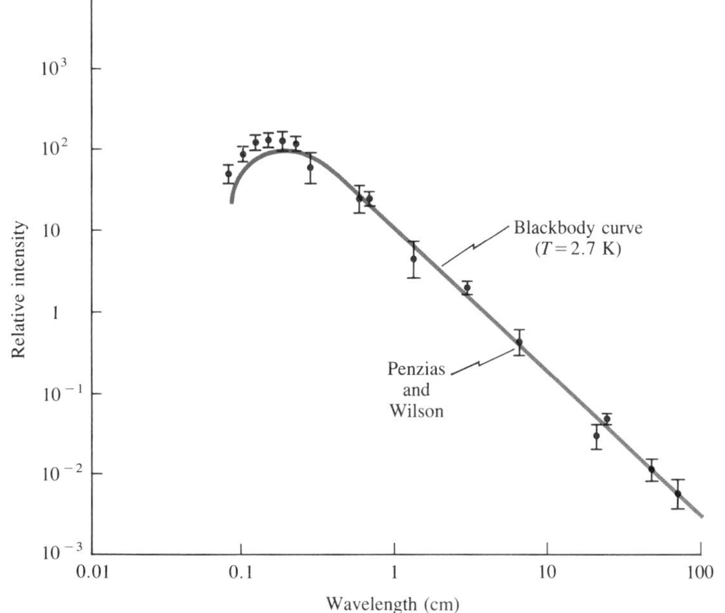

FIGURE 33–20 Spectrum of cosmic microwave background radiation, showing blackbody curve and experimental measurements, including that of Penzias and Wilson. (The vertical bars represent the experimental uncertainty in each measurement.)

part per thousand)†. It could only be concluded that the radiation came from beyond our Galaxy, from the universe as a whole.

The intensity of this radiation at $\lambda = 7.35$ cm corresponds to blackbody radiation (see Section 27–1) at a temperature of about 3 K, now measured more precisely to be 2.7 ± 0.1 K. When radiation at other wavelengths was measured, the intensities were found to fall on a blackbody curve, as shown in Fig. 33–20 (see also Fig. 27–1), with a peak near 0.1 cm corresponding to a temperature of 2.7 K.

Importance of CMBR: the Big Bang

The discovery of this **cosmic microwave background radiation** at a temperature of 2.7 K ranks as one of the two most significant cosmological discoveries of this century. (The other was Hubble's expanding universe.) It is highly significant because it provides strong evidence in support of the Big Bang, and it gives us some idea of conditions in the very early universe. In fact, in the late 1940s, George Gamow and his collaborators calculated that a Big Bang origin of the universe should have generated just such a microwave background radiation.

To understand this, let us look at what a Big Bang might have been like. There must have been a tremendous release of concentrated energy. The temperature must have been extremely high, so high that there could not have been any atoms in the very early stages of the universe. Instead, the universe must have consisted solely of radiation (photons) and elementary particles. The universe would have been opaque—the photons in a sense "trapped," since as soon as they were emitted they would have been scattered or absorbed, primarily by electrons. Indeed, the microwave background radiation is strong evidence that matter and radiation were once in equilibrium at a very high temperature. As the universe expanded, the energy

† The remarkable uniformity of the cosmic microwave background radiation is in accordance with the cosmological principle. A detected nonuniformity of less than 0.1% probably reflects the motion of our Galaxy (at about 600 km/s) relative to the overall distribution of matter and radiation in the universe.

would have spread out over an increasingly larger volume and the temperature would have dropped. Only when the temperature had reached about 3000 K could nuclei and electrons have stayed together as atoms. With the disappearance of free electrons, the radiation would have been freed—"decoupled" from matter, we could say—to spread throughout the universe. As the universe expanded, so too the wavelengths of the radiation expanded, redshifting to longer wavelengths that correspond to lower temperature (recall Wien's law, $\lambda_{max} T =$ constant, Section 27–1), until they would have reached the 2.7 K background radiation we observe today.

Although the total energy associated with the cosmic microwave background radiation is much larger than that from other radiation sources (such as the light emitted by stars), it is small compared to the energy, and mass (remember $E = mc^2$), associated with matter. In fact today, radiation is believed to make up less than $\frac{1}{1000}$ of the energy of the universe: today the universe is *matter-dominated*. But it was not always so. The cosmic microwave background radiation strongly suggests that early in its history the universe was *radiation-dominated*†. But, as we shall see in the next section, that period lasted less than $\frac{1}{10,000}$ of the history of the universe (thus far).

33–6 • The Standard Cosmological Model: The Early History of the Universe

It is now almost generally agreed that the evolution of the universe must have been determined in the first few moments of the Big Bang. In the last decade or two, a convincing theory of the origin and evolution of the universe has developed, now known as the **standard model**. Although a few cosmologists hold other views, most favor the standard model. Much of this theory is based on recent theoretical and experimental advances in elementary particle physics. Indeed, in the last few years cosmology and elementary particle physics have cross-fertilized to a surprising extent.

Let us go back now to the earliest of times—as close as possible to the Big Bang—and follow a standard model scenario of events as the universe expanded and cooled after the Big Bang. Initially, we will be talking of extremely small time intervals, as well as extremely high temperatures, far beyond anything in the universe today. Figure 33–21 is a graphical representation of the events, and it may be helpful to consult it as we go along.

The standard-model "scenario" of the history of the universe after the Big Bang

We begin at a time only a minuscule fraction of a second after the Big Bang, 10^{-43} s. Although this is an unimaginably short time, predictions as early as this can be made based on present theory, albeit somewhat speculatively. Earlier than this instant, we can say nothing since we do not yet have a theory of quantum effects on gravity which would be needed for the

† If there had not been such intense radiation in the first few minutes of the universe, nuclear reactions might have produced a much larger percentage of heavy nuclei than we see. Instead, nearly 75 percent of visible matter is hydrogen, presumably because the intense radiation immediately blasted apart any heavy nuclei that formed into their constituent protons and neutrons.

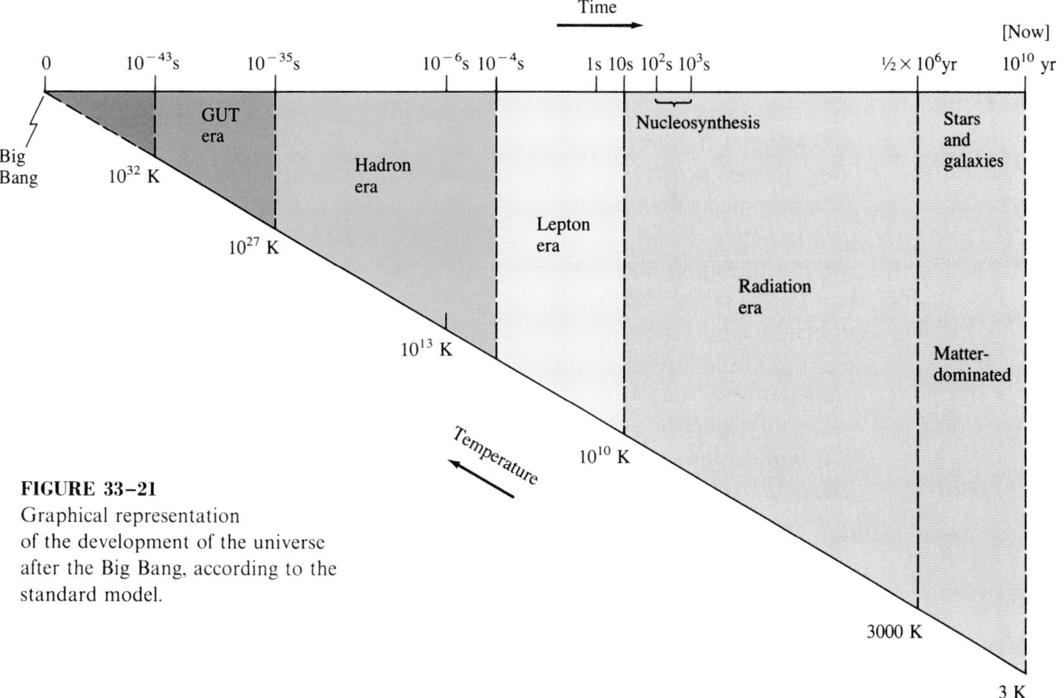

FIGURE 33–21 Graphical representation of the development of the universe after the Big Bang, according to the standard model.

incredibly high densities and temperatures then. It is imagined, however, that prior to 10^{-43} s, the four forces in nature were unified—there was only one force. The temperature would have been about 10^{32} K, corresponding to particles moving about every which way with an average kinetic energy of 10^{19} GeV (see Eq. 13–8):

$$\text{KE} \approx kT \approx \frac{(1.4 \times 10^{-23}\ \text{J/K})(10^{32}\ \text{K})}{1.6 \times 10^{-19}\ \text{J/eV}} \approx 10^{28}\ \text{eV} = 10^{19}\ \text{GeV}.$$

(Note that the factor $\frac{3}{2}$ in Eq. 13–8 is usually ignored in such calculations.) At $t = 10^{-43}$ s, a kind of "phase transition" is believed to have occurred during which the gravitational force, in effect, "condensed out" as a separate force. This, and subsequent phase transitions, are somewhat analogous to the phase transitions water undergoes as it cools from a gas, condenses into a liquid, and with further cooling freezes into ice. The symmetry of the four forces was broken, but the strong, weak, and electromagnetic forces were still unified. At this point, the universe entered the so-called **grand unified era** (after grand unified theory—see Chapter 32). There was no distinction between quarks and leptons; baryon and lepton numbers were not conserved. Very shortly thereafter, as the universe expanded considerably and the temperature had dropped to about 10^{27} K, there was another phase transition during which the strong force condensed out. This probably occurred about 10^{-35} s after the Big Bang. Now the universe was filled with a soup of leptons and quarks. The leptons included electrons, muons, taus, neutrinos, and all their antiparticles. The quarks were initially free (something we have not seen in our present universe), but soon they began to "condense" into more normal particles: nucleons and the other hadrons and their antiparticles. With this **confinement of quarks**, the universe entered the **hadron era**.

Quark confinement (hadron era)

We can think of this "soup" as a grand mixture of particles and antiparticles, as well as photons—all in roughly equal numbers—colliding with one another frequently and exchanging energy.

By the time the universe was only about a microsecond (10^{-6} s) old, it had cooled to about 10^{13} K, corresponding to an average KE of 1 GeV, and the vast majority of hadrons disappeared. To see why, let us focus on the most familiar hadrons: nucleons and their antiparticles. When the average kinetic energy of particles was somewhat higher than 1 GeV, protons, neutrons, and their antiparticles were continually being created out of the energies of collisions involving photons and other particles, such as

Most hadrons disappear

$$\text{photons} \longrightarrow p + \bar{p},\ \ n + \bar{n}.$$

But just as quickly, the particles and antiparticles would annihilate: for example

$$p + \bar{p} \longrightarrow \text{photons or leptons.}$$

So the processes of creation and annihilation of nucleons were in equilibrium. The numbers of nucleons and antinucleons were high—roughly as many as there were electrons, positrons, or photons. But as the universe cooled and the average kinetic energy of particles dropped below about 1 GeV, which is the minimum energy needed to create nucleons and antinucleons (about 940 MeV each) in a typical collision, the process of nucleon creation could not continue. The process of annihilation could continue, however, until there were almost no nucleons left. But not quite zero. To explain our present world, which consists mostly of matter with very little antimatter in sight, we must suppose that earlier in the universe, perhaps around 10^{-35} s after the Big Bang, a slight excess of quarks over antiquarks was formed.† This would have resulted in a slight excess of nucleons over antinucleons. And it is these "leftover" nucleons that we are made of today. The excess of nucleons over antinucleons was about one part in 10^9. Earlier, during the hadron era, there were about as many nucleons as photons. After it ended, the "leftover" nucleons thus numbered only about one nucleon per 10^9 photons, and this ratio has persisted to this day. Protons, neutrons, and all other heavier particles were thus tremendously reduced in number by about 10^{-6} s after the Big Bang. The lightest hadrons, the pions, disappeared as the nucleons had; because they are the lightest-mass hadrons (140 MeV), they were the last hadrons to go, around 10^{-4} s after the Big Bang. Lighter particles, including electrons, positrons, neutrinos, photons—in roughly equal numbers—dominated, and the universe entered the **lepton era**.

Lepton era

By the time the first full second had passed (clearly the most eventful second in history!), the universe had cooled to about 10 billion degrees, 10^{10} K. The average KE was about 1 MeV. This was still sufficient energy to create electrons and positrons and balance their annihilation reactions, since their masses correspond to about 0.5 MeV. So there were about as many e^+ and e^- as there were photons. But within a few more seconds, the temperature had dropped sufficiently so that e^+ and e^- could no longer be formed. Annihilation ($e^+ + e^- \rightarrow$ photons) continued. And, like nucleons

† An alternative possibility is that there was perfect symmetry between quarks and antiquarks, matter and antimatter, but that the universe somehow separated into domains, some containing only matter, others only antimatter. If this were true, we would expect antiparticles from such distant domains to reach us, at least occasionally, in cosmic rays; but none has ever been detected.

before them, electrons and positrons all but disappeared from the universe—except for a slight excess of electrons over positrons (eventually to join with nuclei to form atoms). Thus, about $t = 10$ s after the Big Bang, the universe entered the **radiation era**. Its major constituents were now photons and neutrinos. But the neutrinos, partaking only in the weak force, rarely interacted. So the universe, until then experiencing an energy balance between matter and radiation, became **radiation-dominated**: much more energy was contained in radiation than in matter, a situation that would last almost a million years.

Radiation-dominated universe

Meanwhile, during the next few minutes, crucial events were taking place. Beginning about 2 or 3 minutes after the Big Bang, nuclear fusion began to occur. The temperature had dropped to about 10^9 K, corresponding to $\overline{\text{KE}} \approx 100$ keV, where nucleons could strike each other and be able to fuse (Section 31–3), but not so high that the newly formed nuclei would be immediately broken apart by subsequent collisions. Deuterium, helium, and very tiny amounts of lithium were probably made. But the universe was cooling too quickly, and larger nuclei were not made. After only a few minutes, probably not even a quarter of an hour, the temperature dropped far enough that nucleosynthesis stopped, not to start again for millions of years (in stars). Thus, after the first hour or so of the universe, matter consisted mainly of bare nuclei of hydrogen (about 75%) and helium (about 25%)† and electrons. But radiation (photons) continued to dominate.

Our story is almost complete. The next important event occurred a few hundred thousand years later. The universe had expanded to about $\frac{1}{1000}$ of its present size, and the temperature had cooled to about 3000 K. The average KE of nuclei, electrons, and photons was less than an electron volt. Since ionization energies of atoms are on the order of eV, then as the temperature dropped below this point, electrons could orbit the bare nuclei and remain there (without being ejected by collisions), thus forming atoms. With the birth of atoms, the photons which were continually scattering from the free electrons became much freer to spread throughout the universe. The total energy contained in radiation had been decreasing (redshifting as the universe expanded), until at this point it was about equal to the total energy contained in matter. As the universe continued to expand, the radiation cooled further (to 2.7 K today, forming the cosmic microwave background radiation we detect from everywhere in the universe), and lost energy. But the mass of material particles did not decrease, so beginning at about this point the energy of the universe became increasingly concentrated in matter rather than in radiation: the universe became **matter-dominated**, as it remains today.‡

Birth of stable atoms

Matter-dominated universe

Shortly after the birth of atoms, stars and galaxies formed—probably by self-gravitation around mass concentrations (inhomogeneities). This tran-

† This standard model prediction of a 25% primordial production of helium is fully in accord with what we observe today—the universe *does* contain about 25% He—and it is strong evidence in support of the standard Big Bang model. Furthermore, the theory says that 25% He abundance is fully consistent with there being three neutrino types, which is the number we observe so far. And it sets an upper limit of four to the maximum number of possible neutrino types. Actually, the fourth could be another type of low-mass particle, a *photino* or a *gravitino*, for example. Here we have a situation where cosmology actually makes a specific prediction about fundamental physics.

‡ Although today matter contains more of the energy of the universe than does radiation, there are many more photons (perhaps 10^9 times more) than atoms, nuclei, and electrons. But each photon (at $T \approx 2.7$ K) has very little energy.

spired about a million years after the Big Bang. The universe continued to evolve (Section 33–2) until today, some 15×10^9 years later.

This scenario is by no means "proven" in any sense. Nor does it answer all questions. But it does provide a tentative picture, for the first time, of how the universe began and evolved. It does have some problems, however, some of which have been resolved by a modification first proposed in the early 1980s known as the **inflationary scenario**. According to the inflationary scenario, at the earliest stages of cosmological evolution, before about 10^{-35} s after the Big Bang, the universe underwent a very rapid exponential expansion associated with the phase transition (symmetry breaking) that separated the strong force from the electroweak (as discussed earlier in this section). After the brief inflationary period, the scenario settles back to the standard expansion already discussed. The appeal of the inflationary scenario is that it provides natural explanations for a number of problems not resolved by the standard model, such as why the universe is as close to being as flat as it is, and why the cosmic microwave background radiation is so uniform.

Inflationary scenario

There are, however, some questions we haven't yet treated such as: Was there a stage before the Big Bang, or did time just begin with the Big Bang? And what of the future of the universe? We'll look at these next, in the final section of this book.

33–7 • The Future of the Universe?

According to the standard Big Bang model, the universe is evolving and changing. Individual stars are evolving and dying as white dwarfs, neutron stars, black holes. At the same time, the universe as a whole is expanding. One important question is whether the universe will continue to expand forever. This question is connected to the curvature of space–time (Section 33–3) and to whether the universe is open (and infinite) or closed (and finite). There are three possibilities as shown in Fig. 33–22. If the curvature of the universe is *negative*, the expansion of the universe will never stop, although it should decrease due to the gravitational attraction of its parts. Such a

An open or closed universe?

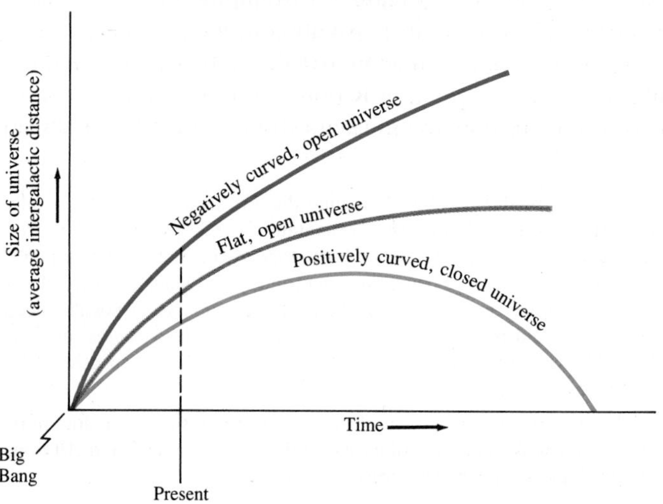

FIGURE 33–22 Three future possibilities for the universe.

universe would be *open* and infinite. If the universe is *flat* (no curvature), it will still be open and infinite but its expansion will slowly approach a zero rate. Finally, if the universe has *positive* curvature, it will be *closed* and finite. The effect of gravity in this case would be strong enough so that the expansion would eventually stop and the universe would then begin to contract. All matter eventually would collapse back onto itself in a **big crunch**. If, in this last case, the maximum expansion of the universe corresponded to, say, an intergalactic separation twice what it is now, the maximum expansion would occur about 30 or 40 billion years from now. Then, as the universe began to contract, the big crunch would occur about 100 billion years after the Big Bang.

Whether we live in an open and continually expanding universe, or a closed one that eventually will contract, is a basic question in cosmology. But we don't know the answer. How might we find out? One way is to determine the average mass density in the universe. If the average mass density is above a critical value known as the **critical density**, estimated to be about

Critical density of the universe

$$\rho_c \approx 10^{-26}\ \mathrm{kg/m^3}$$

(i.e., a few nucleons/m^3 on average throughout the universe), then gravity will prevent expansion from continuing forever, and will eventually pull the universe back into a big crunch. To say it another way, there would be sufficient mass that gravity would give space–time a zero or positive curvature. If the actual density is equal to the critical density, the universe will be flat and open. If the actual density is less than the critical density, the universe will have negative curvature and be open, expanding forever.

Great efforts have gone into measuring the actual density of the universe. Estimates of the amount of visible matter in the universe put the actual density an order of magnitude less than the critical density, thereby suggesting an open universe. However, there is evidence for a significant amount of nonluminous matter in the universe, often referred to as the **missing mass** or **dark matter**, enough to bring the density to almost exactly ρ_c. For example, observations of rotating galaxies suggest that they rotate as if they had considerably more mass than we can see. And observations of the motion of galaxies within clusters also suggest they have considerably more mass than can be seen. Furthermore, the highly regarded inflationary theory offers a strong argument in favor of ρ being precisely equal to ρ_c, and thus that space on a grand scale is Euclidean. If there is nonluminous matter in the universe, what might it be? One suggestion is that the missing mass consists of previously unknown weakly interacting massive particles ("WIMPS") or perhaps small primordial black holes made in the early stages of the universe. Or perhaps the missing mass is so-called *brown matter*—huge amounts of cool nonglowing hydrogen gas or small faint stars ("brown dwarfs"). Another proposal is that neutrinos, once believed to be massless, may actually have nonzero rest mass. Since the universe probably contains roughly as many neutrinos of each type as it does photons (that is, about 10^9 times the number of nucleons, although this neutrino background has yet to be detected), neutrino masses of only about 30 to 50 eV would be sufficient to bring the actual density of the universe up to the critical density. The supernova of 1987 offered a unique opportunity to measure the neutrino mass, at least that for the electron neutrino. If neutrinos have nonzero rest mass, then their

"Missing" mass (dark matter)

velocities are less than the speed of light ($v < c$). High-energy neutrinos emitted from the supernova should in this case have arrived at earth earlier than lower-energy (and therefore slower) neutrinos emitted at the same instant. Since they traveled a distance of about 170,000 ly from SN1987a, the time difference ought to be measurable. To get an idea of the size of the effect, let us consider the following example.

EXAMPLE 33–6 [A challenging Example.] Suppose two neutrinos from SN1987a were detected on earth (via the reaction $\bar{\nu}_e + p \to n + e^+$) 10 seconds apart, with measured kinetic energies of about 20 MeV and 10 MeV. (*a*) Estimate the rest mass of the neutrino, m_0, using this data, assuming both neutrinos were emitted at the same time. (*b*) Theoretical models of supernova explosions suggest that the neutrinos are emitted in a burst that lasts from a second or two up to perhaps 10 s. If we assume the neutrinos are not emitted simultaneously but rather at any time over a 10-s interval, what then can we say about the neutrino mass based on the data given above?

SOLUTION (*a*) We expect the neutrino mass to be less than 100 eV (from laboratory measurements). Since our neutrinos have KE of 20 MeV and 10 MeV, we can make the approximation $m_0c^2 \ll E$, where E (the total energy) is essentially equal to the kinetic energy. From Eqs. 26–3 and 26–5, we have

$$E = mc^2 = \frac{m_0c^2}{\sqrt{1 - v^2/c^2}}.$$

We solve this for v, the velocity of a neutrino with energy E:

$$v = c\left(1 - \frac{m_0^2c^4}{E^2}\right)^{\frac{1}{2}} = c\left(1 - \frac{m_0^2c^4}{2E^2} + \cdots\right),$$

where we have used the binomial expansion $(1 + x)^{\frac{1}{2}} = 1 + \frac{1}{2}x + \cdots$ (see Appendix A), and we ignore higher-order terms since $m_0^2c^4 \ll E^2$. The time t for a neutrino to travel a distance d ($=170{,}000$ ly) is

$$t = \frac{d}{v} = \frac{d}{c\left(1 - \dfrac{m_0^2c^4}{2E^2}\right)} \approx \frac{d}{c}\left(1 + \frac{m_0^2c^4}{2E^2}\right),$$

where again we used the binomial expansion $[(1 + x)^{-1} = 1 - x + \cdots]$. The difference in arrival times for our two neutrinos of energies $E_1 =$ 20 MeV and $E_2 = 10$ MeV is

$$t_2 - t_1 = \frac{d}{c}\frac{m_0^2c^4}{2}\left(\frac{1}{E_2^2} - \frac{1}{E_1^2}\right).$$

We solve this for m_0c^2 and set $t_2 - t_1 = 10$ s:

$$\begin{aligned} m_0c^2 &= \left[\frac{2c(t_2 - t_1)}{d}\frac{E_1^2E_2^2}{E_1^2 - E_2^2}\right]^{\frac{1}{2}} \\ &= \left[\frac{2(3 \times 10^8\ \text{m/s})(10\ \text{s})}{(1.7 \times 10^5\ \text{ly})(10^{16}\ \text{m/ly})}\frac{(400\ \text{MeV}^2)(100\ \text{MeV}^2)}{(400\ \text{MeV}^2 - 100\ \text{MeV}^2)}\right]^{\frac{1}{2}} \\ &= 22 \times 10^{-6}\ \text{MeV} = 22\ \text{eV}. \end{aligned}$$

We thus estimate the mass of the neutrino to be 22 eV, but there would of course be experimental uncertainties.

(*b*) If the two neutrinos of energy $E_1 = 20$ MeV and $E_2 = 10$ MeV were emitted at unknown times over a 10-s interval, then the 10-s difference in their arrival times could be due to a 10-s difference in their emission time. In this case our data are consistent with zero rest mass and it puts an approximate *upper limit* on the neutrino mass of 22 eV.

In the actual experiments, the most sensitive detector consisted of several thousand tons of water in an underground chamber. It detected 11 events in 12 seconds, probably via the reaction $\bar{\nu}_e + p \rightarrow n + e^+$ and (less likely) $\nu + e^- \rightarrow \nu + e^-$. There was not a clear correlation between energy and time of arrival. Nonetheless, a careful analysis of this experiment, plus other careful measurements, has set a rough upper limit on the electron anti-neutrino mass of about 12 eV. The upper limits for the masses of the other neutrinos are much higher (see Table 32–2), so we still cannot rule out neutrinos as being able to "close" the universe.

Another factor that could provide the answer to the question of an open or closed universe, if we could measure it accurately enough, is the so-called **deceleration parameter**. It is a measure of the rate at which the expansion of the universe is slowing. But to measure this rate requires looking far back in time, to the galaxies farthest away, whose light we receive now was emitted at a time closer to the beginning of the universe. At that time, the rate of expansion was much faster than today. Unfortunately, we would have to know the distance to these galaxies more precisely than is now possible. So this method does not at present yield an answer to whether the universe is open or closed. A recent related analysis of the number of galaxies as a function of redshift (i.e., distance) gives a result consistent with $\rho = \rho_c$, although the experimental uncertainty allows fairly wide limits on ρ. Nonetheless, the suggestion of the simplest universe of all, with no overall curvature, is appealing.

Deceleration parameter

If the universe is open, how would it evolve in the future? According to the latest theories, which rely to a large extent on elementary particle theory, after about 10^{18} years, galaxies would have much of their matter knocked away and scattered throughout the universe by collisions with other stars. The remaining matter would eventually condense into massive "galactic black holes." Clusters of these would then coalesce into extremely massive "supergalactic black holes." Finally, the black holes themselves would "evaporate"—the matter within them, through the slow quantum-mechanical process of tunneling (see Section 30–12), would "leak out." This process is so slow that it would take on the order of 10^{100} years. The universe would then be mainly a thin gas of electrons, positrons, neutrinos, and photons.

On the other hand, if the universe is closed, it might turn around and begin to contract even before all the stars have burnt out. As the universe contracts, the background radiation would increase in energy and temperature. The universe might simply retrace its steps, if it weren't for black holes. As density increases and the universe rushed toward its inevitable end in the big crunch, black holes might gobble up more and more matter until the entire universe coalesced into a single supermassive black hole—which would then be the universe.

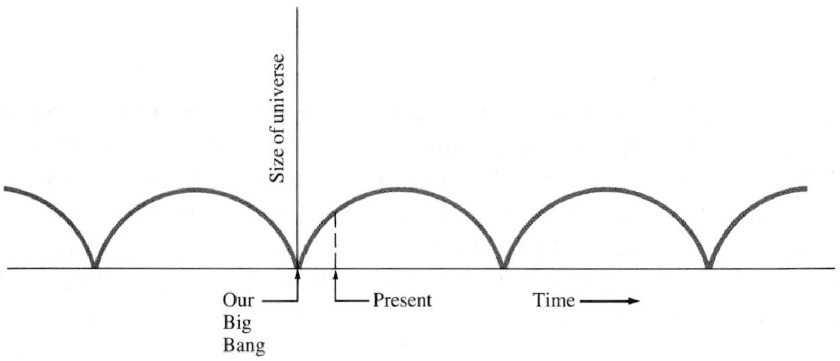

Size of universe

Our Big Bang — Present — Time →

FIGURE 33–23 Cyclic model of the universe. Although the cycles shown here are the same, they could have different periods and different expansion rates.

If the universe is closed, what happens after the big crunch? We don't know, of course. What is possible, though, is a "bounce." That is, the dense fiery nucleus of the big crunch might explode again, resulting once more in an expanding universe. Thus the universe might be cyclic as shown in Fig. 33–23. Such a **cyclic** or **pulsating** universe proposes a possible answer to one of our favorite "unanswerable" questions: What happened before the Big Bang? In this model there was simply a previous cycle. But left unanswered would be a number of other questions such as "When did it all begin?"

The questions raised by cosmology can seem absurd at times, they are so removed from everyday "reality." We can always say: the sun is shining, it's going to burn on for an unimaginably long time, all is well. Nonetheless, the questions of cosmology are deep ones that fascinate the human intellect. One aspect that is especially intriguing is this: calculations on the formation and evolution of the universe have been performed that deliberately varied the values—just slightly—of certain fundamental physical constants. The result? A universe in which life as we know it could not exist. This has given rise to the so-called **Anthropic principle**, which says that if the universe were even a little different than it is, we couldn't be here. It's as if the universe were exquisitely tuned, almost as if to accommodate us.

SUMMARY

The night sky contains myriads of stars including those in the Milky Way, which is a "side view" of our *Galaxy* looking along the plane of the disc. Our Galaxy includes about 10^{11} stars. Beyond our Galaxy are billions of other galaxies. Astronomical distances are measured in *light-years* (1 ly $\approx 10^{13}$ km). The nearest star is about 4 ly away and the nearest other galaxy is 2 million ly away. Our galactic disc has a diameter of about 100,000 ly. Distances are often specified in *parsecs*, where 1 parsec = 3.26 ly.

Stars are believed to begin life as collapsing masses of hydrogen gas (protostars). As they contract, they heat up (PE is transformed to KE). When the temperature reaches 10 million degrees, nuclear fusion begins and forms heavier elements (*nucleosynthesis*), mainly helium at first. The energy released during these reactions balances the gravitational force, and the young star stabilizes as a *main-sequence* star. The tremendous luminosity of stars comes from the energy released during these thermonuclear reactions. After billions of years, as helium is collected in the core and hydrogen is used up, the core contracts and heats further. The envelope expands and cools, and the star becomes a *red giant* (larger diameter, redder color). The next stage of stellar evolution depends on the mass of the star. Stars of residual mass less than about 1.4 solar masses cool further and become *white dwarfs*, eventually fading and going out altogether. Heavier stars contract further due to their greater gravity: the density approaches nuclear density, the huge pressure forces electrons to combine with protons to form neutrons, and the star becomes essentially a huge nucleus of neutrons. This is a *neutron star*, and the energy released from its final core contraction is believed to produce *supernovae.* If the star's residual mass is greater than two or three solar masses, it may contract even further and form a *black hole*, which is so dense that no matter or light can escape from it.

In the *general theory of relativity*, the *equivalence principle* states that an observer cannot distinguish acceleration from a gravitational field. Said another way, gravitational and inertial mass are the same. The theory predicts gravitational bending of light rays to a degree consistent with experiment. Gravity is treated as a curvature in space and time, the curvature being greater near massive bodies. The universe as a whole may be curved. If there is sufficient mass, the curvature of the universe is positive, and the universe is *closed* and *finite*; otherwise, it is *open* and *infinite*.

Distant galaxies display a *redshift* of spectral lines, interpreted as a Doppler shift. The universe seems to be expanding, its galaxies racing away from each other at speeds (v) proportional to the distance (d) between them:

$$v = Hd,$$

which is known as *Hubble's law* (H is Hubble's constant). This expansion of the universe suggests an explosive origin, the *Big Bang*, which probably occurred 10 to 15 billion years ago. *Quasars* are objects with a large redshift (suggesting great distance) and high luminosity (suggesting closeness or, more likely, extraordinary energy output). The *cosmological principle* assumes the universe, on a large scale, is homogeneous and isotropic.

Important evidence for the *Big Bang* model of the universe was the discovery of the *cosmic microwave background radiation*, which conforms to a blackbody radiation curve at a temperature of 2.7 K. The *standard model* of the Big Bang provides a possible scenario as to how the universe developed as it expanded and cooled after the Big Bang. Starting at 10^{-43} s after the Big Bang, according to this model, there was a series of *phase transitions* during which previously unified forces of nature "condensed out" one by one. The *inflationary scenario* assumes that during one of these phase transitions, the universe underwent a brief but rapid exponential expansion. Until about 10^{-35} s, there was no distinction between quarks and leptons. Shortly thereafter, quarks were *confined* into hadrons (the *hadron era*). About 10^{-6} s after the Big Bang, the majority of hadrons disappeared, introducing the *lepton era*. By the time the universe was about 10 s old, the electrons too had mostly disappeared and the universe became *radiation-dominated*. A couple of minutes later, nucleosynthesis began, but lasted only a few minutes. It was then several hundred thousand years before the universe was cool enough for electrons to combine with nuclei and form atoms. Also about this time, the background radiation had expanded and cooled so much that its total energy equaled the energy in matter. As the radiation cooled further, losing energy, the universe became *matter-dominated* (not in numbers, but in energy). Shortly thereafter stars and galaxies formed, producing a universe not much different than it is today—10 or 15 billion years later.

If the universe is *open*, it will continue to expand indefinitely. If it is *closed*, gravity is sufficiently strong to halt expansion and the universe will eventually begin to collapse back on itself, ending in a *big crunch*. Whether the universe is open or closed depends on whether its average mass density is above or below a critical density. If the universe is closed, it may rebound from the big crunch and perhaps reexpand in a *cyclic* manner.

QUESTIONS

1. The Milky Way was once thought to be "cloudy," but no longer is considered so. Explain.
2. Give an explanation for why some galaxies have arms.
3. If you were measuring star parallaxes from the moon instead of earth, what corrections would you have to make? What changes would occur if you were measuring parallaxes from Mars?
4. A star is in equilibrium when it radiates at its surface all the energy generated at its core. What happens when it begins to generate more energy than it radiates? Less energy? Explain.
5. Describe a red giant star. List some of its properties.
6. Select a point on the H–R diagram. Mark several directions away from this point. Now describe the changes that would take place in a star moving in each of these directions.
7. Does the H–R diagram reveal anything about the core of a star?
8. Why do some stars end up as white dwarfs, and others as neutron stars or black holes?
9. What is a geodesic? What is its role in general relativity?
10. State Olber's paradox in your own words. How is it resolved?
11. If it were discovered that the redshift of spectral lines of galaxies was due to something other than expansion, how might our view of the universe change? Would there be conflicting evidence? Discuss.

12. All galaxies appear to be moving away from us. Are we therefore at the center of the universe? Explain.

13. If you were located in a galaxy near the boundary of our observable universe, would galaxies in the direction of the Milky Way appear to be approaching you or receding from you? Explain.

14. What is the difference between the Hubble age of the universe and the actual age? Which is greater?

15. Compare an explosion on earth to the Big Bang. Consider such questions as: Would the debris spread at a higher speed for more distant particles, as in the Big Bang? Would the debris come to rest? What type of universe would this correspond to, open or closed?

16. When the primordial nucleus exploded, thus creating the universe, into what did it expand? Discuss.

17. Explain what the 2.7 K cosmic microwave background radiation is. Where does it come from? Why is its temperature now so low?

18. The birth of atoms—that is, the combination of electrons with nuclei about which they orbit—occurred when the universe had cooled to about 3000 K and is generally called *recombination*. Why is this term misleading?

19. Why were atoms unable to exist until almost a million years after the Big Bang?

20. Explain why today the universe is said to be matter-dominated, yet there are probably 10^9 times as many photons as there are massive particles.

21. If the universe is open, what will eventually happen to the cosmic microwave background radiation? If it is closed, what will happen to it?

22. Under what circumstances would the universe eventually collapse in on itself?

PROBLEMS

SECTIONS 33–1 AND 33–2

1. (I) A star exhibits a parallax of 0.33 seconds of arc. How far away is it?

2. (I) The parallax angle of a star is 0.00014°. How far away is the star?

3. (I) A star is 30 parsecs away. What is its parallax angle? State (*a*) in seconds of arc, and (*b*) in degrees.

4. (I) What is the parallax angle for a star that is 30 light-years away? How many parsecs is this?

5. (I) A star is 60 parsecs away. How long does it take for its light to reach us?

6. (I) If one star is twice as far away from us as a second star, will the parallax angle of the first star be greater or less than that of the second star? By what factor?

7. (II) Show that the absolute magnitude of the sun is about 4.7, typical for a star.

8. (II) Calculate the Q-values for the He burning reactions of Eqs. 33–3. (The mass of ${}^{8}_{4}\text{Be}$ is 8.005304u.)

9. (II) Calculate the Q-values for the fusion reactions of Eqs. 33–4.

10. (II) When our sun becomes a red giant, what will be its average density if it expands out to the orbit of Venus (1.1×10^{11} m from the sun)?

11. (II) Calculate the density of a white dwarf whose mass is equal to the sun's and whose radius is equal to the earth's. How many times larger than earth's density is this?

12. (II) A neutron star whose mass is 1.5 solar masses has a radius of about 11 km. Calculate its average density and compare to that for a white dwarf (Problem 11) and to that of nuclear matter.

13. (II) If a star has absolute magnitude M, what is its absolute luminosity L?

14. (II) The largest telescopes can just detect objects whose apparent magnitude is 22.7. How far away could our sun be and still be detectable telescopically?

15. (II) Derive Eq. 33–2 for the absolute magnitude of a star or galaxy, starting from Eq. 33–1.

16. (II) Show that the absolute luminosities and absolute magnitudes of two stars are related by

$$M_2 - M_1 = 2.5 \log_{10} \frac{L_1}{L_2}.$$

17. (II) Suppose two stars of the same apparent magnitude m are also believed to be the same size. The spectrum of one star peaks at 800 nm whereas that of the other peaks at 400 nm. Use Wien's displacement law (Section 27–1) and the Stefan-Boltzmann law (Eq. 14–4) to estimate their relative distances from us. [*Hint:* see Examples 33–4 and 33–5.]

18. (III) Stars located in a certain cluster are assumed to be about the same distance from us. Two such stars have apparent magnitudes $m_1 = 7.0$ and $m_2 = 9.6$, and spectra that peak at $\lambda_1 = 500$ nm and $\lambda_2 = 700$ nm. Estimate their relative sizes (give ratio of their diameters). [*Hint:* use the Stefan-Boltzmann law, Eq. 14–4.]

SECTION 33–3

19. (I) Describe a triangle, drawn on the surface of a sphere, for which the sum of the angles is (*a*) 360°, and (*b*) 180°.

20. (I) Show that the Schwarzschild radius for a star with mass equal to that: (*a*) of our sun is 2.95 km; (*b*) of earth is 8.9 mm.

21. (II) What is the Schwarzschild radius for a typical galaxy?

22. (II) What is the maximum sum-of-the-angles for a triangle on a sphere?

SECTION 33–4

23. (I) The redshift of a galaxy indicates a velocity 2500 km/s. How far away is it?

24. (I) If a galaxy is traveling away from us at 1.0% of the speed of light, roughly how far away is it?

25. (I) Estimate the speed of a galaxy (relative to us) that is near the "edge" of the universe, say 10 billion light-years away.

26. (I) Make an approximate calculation for the age of the universe using Hubble's constant. Assume (*a*) $H = 15$ km/s/Mly and (*b*) $H = 30$ km/s/Mly.

27. (II) Estimate the observed wavelength for the 656-nm line in the Balmer series of hydrogen in the spectrum of a galaxy whose distance from us is (*a*) 1.0×10^6 ly, (*b*) 1.0×10^8 ly, (*c*) 1.0×10^{10} ly.

28. (II) Estimate the speed of a galaxy, and its distance from us, if the wavelength for the hydrogen line at 434 nm is measured on earth as being 610 nm.

29. (II) Starting from Eq. 33–5, show that the Doppler shift in wavelength is $\Delta\lambda/\lambda \approx v/c$ (Eq. 33–6) for $v \ll c$. [*Hint:* use the binomial expansion–see Appendix A–5.]

30. (III) Derive the Doppler shift formula for light (Eq. 33–5) using the special theory of relativity. [*Hint:* proceed as in Section 12–8 for sound, but invoke special relativity including time dilation.]

SECTIONS 33–5 TO 33–7

31. (I) Calculate the wavelength at the peak of the blackbody radiation distribution at 2.7 K using the Wien displacement law.

32. (II) Calculate what the critical density, $\rho_c \approx 10^{-26}$ kg/m^3, would be in average number of nucleons per cubic meter.

33. (II) The size of the universe (the average distance between galaxies) at any one moment is believed to have been inversely proportional to the absolute temperature. Estimate the size of the universe, compared to today, at (*a*) $t = 10^6$ yr, (*b*) $t = 1$ s, (*c*) $t = 10^{-6}$ s, and (*d*) $t = 10^{-35}$ s.

34. (II) At approximately what time had the universe cooled below the threshold temperature for producing (*a*) kaons ($M \approx 500$ MeV/c^2), (*b*) Υ ($M \approx 9500$ MeV/c^2), and (*c*) muons ($M \approx 100$ MeV/c^2)?

GENERAL PROBLEMS

35. Suppose that three main-sequence stars could undergo the three changes represented by the three arrows, *A*, *B*, and *C*, in the H–R diagram of Fig. 33–24. For each case, describe the changes in temperature, luminosity, and size.

36. Assume that the nearest stars to us have an intrinsic luminosity about the same as the sun's. Their apparent brightness, however, is about 10^{11} times fainter than the sun. From this, estimate the distance to the nearest stars. (Newton did this calculation, although he made a numerical error of a factor of 100.)

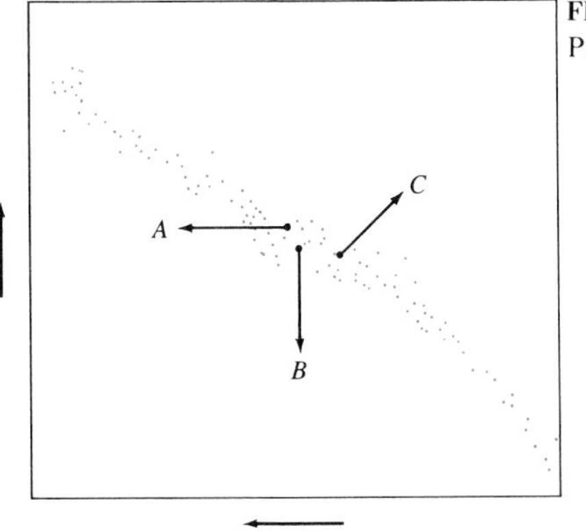

FIGURE 33–24 Problem 36.

37. Use conservation of angular momentum to estimate the angular velocity of a neutron star which has collapsed to a diameter of 10 km, from a star whose radius was equal to that of our sun (7×10^8 m), of mass 1.5 times that of the sun, and which rotated (like our sun) about once a month.

38. By what factor does the rotational KE change when the star in Problem 37 collapses to a neutron star?

39. A certain pulsar, believed to be a neutron star of mass 1.5 times that of the sun, with diameter 10 km, is observed to have a rotation speed of 1.0 rev/s. If it loses rotational KE at the rate of 1 part in 10^9 per day, which is all transformed into radiation, what is the power output of the star?

40. Estimate the rate at which hydrogen atoms would have to be created, according to the steady-state model, to maintain the present density of the universe of about 10^{-27} kg/m^3, assuming the universe is expanding with Hubble constant $H = 50$ km/s/Mpc.

41. Estimate what neutrino rest mass (in eV) would provide the critical density to close the universe. Assume the neutrino density is, like photons, about 10^9 times that of nucleons, and that nucleons make up only (*a*) 2% of the mass needed, or (*b*) 5% of the mass needed.

42. Two stars, whose spectra peak at 600 nm and 400 nm, respectively, both lie on the main sequence. Use Wien's law, the Stefan-Boltzmann law, and the H–R diagram (Fig. 33–7) to estimate the ratio of their diameters. [*Hint:* see Problem 16 and Examples 33–4 and 33–5.]

APPENDIX A

Mathematical Review

A–1 • Relationships, Proportionality, and Equations

One of the important aspects of physics is the search for relationships between different quantities—that is, determining how one quantity affects another. For example, how does temperature affect the air pressure in a tire? Or how does the net force on an object affect its acceleration? Sometimes a given quantity is affected by two or more quantities; for instance, the acceleration of an object is related to both its mass and the applied force. If it is suspected that a relationship exists between two or more quantities, one can try to determine the precise nature of this relationship. This is done by varying one of the quantities and measuring how the other varies as a result. If it is likely that a particular quantity will be affected by more than one factor or quantity, only one of the latter is varied at a time, while the others are held constant.[†]

As a simple example, the ancients found that if one circle has twice the diameter of a second circle, the first also has twice the circumference. If the diameter is three times as large, the circumference is also three times as large. In other words, an increase in the diameter results in a proportional increase in the circumference. We say that the circumference is *directly proportional to* the diameter. This can be written in symbols as $C \propto D$, where "$\propto$" means "is proportional to," and C and D refer to the circumference and diameter of a circle, respectively. The next step is to change this proportionality to an equation, which will make it possible to link the two quantities numerically. This merely entails inserting a proportionality constant, which in many cases is determined by measurement. (In some cases it can be chosen arbitrarily, as in Section 4–4.) The ancients found that the ratio of the circumference to the diameter of any circle was 3.1416 (to keep only the first few decimal places). This number is designated by the Greek letter π. It is the constant of proportionality for the relationship $C \propto D$, and to obtain an equation we insert it into the proportion and change the $\propto$ to $=$. Thus, $C = \pi D$.

Direct proportion

Other kinds of proportionality occur as well. For example, the area of a circle is proportional to the *square* of its radius. That is, if the radius is doubled, the area becomes four times as large; and so on. In this case we

[†] When one quantity affects another, we often use the expression "is a function of" to indicate this dependence; for example, we say that the pressure in a tire is a function of the temperature.

can write $A \propto r^2$, where A stands for the area and r for the radius of the circle.

Sometimes two quantities are related in such a way that an increase in one leads to a proportional *decrease* in the other. This is called *inverse proportion*. For example, the time required to travel a given distance is inversely proportional to the speed of travel. The greater the speed, the less time it takes. We can write this inverse proportion as: time $\propto$ 1/speed. The larger the denominator of a fraction, the smaller the fraction is as a whole. For example, $\frac{1}{4}$ is smaller than $\frac{1}{2}$. Thus, if the speed is doubled, the time is halved, which is what we want to express by this inverse proportionality relationship.

Inverse proportion

Whatever kind of proportion is found to hold, it can be changed to an equality by insertion of the proper proportionality constant. Quantitative statements or predictions about the physical world can then be made with the equation.

A–2 • Exponents

When we write 10^4 we mean that you multiply 10 times itself four times: $10^4 = 10 \times 10 \times 10 \times 10 = 10{,}000$. The superscript 4 is called an *exponent*, and 10 is said to be raised to the fourth power. Any number or symbol can be raised to a power; special names are used when the exponent is 2 (a^2 is "a squared") or 3 (a^3 is "a cubed"). For any other power, we say a^n is "a to the nth power." If the exponent is 1, it is usually dropped: $a^1 = a$, since no multiplication is involved.

The rules for multiplying numbers expressed as powers are as follows:

$$(a^n)(a^m) = a^{n+m}. \qquad \text{(A–1)}$$

That is, the exponents are added. To see why, consider the result of the multiplication of 3^3 by 3^4:

$$(3^3)(3^4) = (3)(3)(3) \times (3)(3)(3)(3) = (3)^7.$$

Here the sum of the exponents is $3 + 4 = 7$, so rule A–1 works. Notice that this rule works only if the base numbers (a in Eq. A–1) are the same. Thus we *cannot* use the rule of summing exponents for $(6^3)(5^2)$; these numbers would have to be written out. However, if the base numbers are different but the exponents are the same, we can write a second rule:

$$(a^n)(b^n) = (ab)^n. \qquad \text{(A–2)}$$

For example, $(5^3)(6^3) = (30)^3$ since

$$(5)(5)(5)(6)(6)(6) = (30)(30)(30).$$

The third rule involves a power raised to another power: $(a^3)^2$ means $(a^3)(a^3)$, which is equal to $a^{3+3} = a^6$. The general rule is then

$$(a^n)^m = a^{nm}. \qquad \text{(A–3)}$$

In this case, the exponents are multiplied.

Negative exponents are used for reciprocals. Thus,

$$\frac{1}{a} = a^{-1}, \qquad \frac{1}{a^3} = a^{-3},$$

and so on. The reason for using negative exponents is that it allows us to use the multiplication rules given above. For example, $(a^5)(a^{-3})$ means

$$\frac{(a)(a)(a)(a)(a)}{(a)(a)(a)} = a^2.$$

Rule A–1 gives us the same result:

$$(a^5)(a^{-3}) = a^{5-3} = a^2.$$

What does an exponent of zero mean? That is, what is a^0? Any number raised to the zeroth power is defined as being equal to 1:

$$a^0 = 1.$$

This definition is used because it follows from the rules for adding exponents. For example,

$$a^3a^{-3} = a^{3-3} = a^0 = 1.$$

But *does* a^3a^{-3} actually equal 1? Yes, because

$$a^3a^{-3} = \frac{a^3}{a^3} = 1.$$

Fractional exponents are used to represent *roots*. For example, $a^{\frac{1}{2}}$ means the square root of a; that is $a^{\frac{1}{2}} = \sqrt{a}$. Similarly, $a^{\frac{1}{3}}$ means the cube root of a, and so on. The fourth root of a means that if you multiply the fourth root of a by itself four times, you again get a:

$$(a^{\frac{1}{4}})^4 = a.$$

This is consistent with rule A–3 since $(a^{\frac{1}{4}})^4 = a^{\frac{4}{4}} = a^1 = a$.

A–3 • Powers of 10, or Exponential Notation

Writing out very large and very small numbers such as the distance of Neptune from the sun, 4,500,000,000 km, or the diameter of a typical atom, 0.00000001 cm, is inconvenient and prone to error. It also leaves in question (see Section 1–4) the number of significant figures. (How many of the zeros are significant in the number 4,500,000,000 km?) We therefore make use of the "powers of 10," or exponential notation. The distance from Neptune to the sun is then expressed as 4.50×10^9 km (assuming that the value is significant to three digits) and the diameter of an atom 1.0×10^{-8} cm. This way of writing numbers is based on the use of exponents, where a^n signifies a multiplied by itself n times. For example, $10^4 = 10 \times 10 \times 10 \times 10 = 10{,}000$. Thus, $4.50 \times 10^9 = 4.50 \times 1{,}000{,}000{,}000 = 4{,}500{,}000{,}000$. Notice that the exponent (9 in this case) is just the number of places the decimal point is moved to the right to obtain the fully written out number (4.500,000,000.)

When two numbers are multiplied (or divided), you first multiply (divide) the simple parts and then the powers of 10. Thus, 2.0×10^3 multiplied by 5.5×10^4 equals $(2.0 \times 5.5) \times (10^3 \times 10^4) = 11 \times 10^7$, where we have used the rule for adding exponents (Appendix A–2). Similarly, 8.2×10^5 divided by 2.0×10^2 equals

$$\frac{8.2 \times 10^5}{2.0 \times 10^2} = \frac{8.2}{2.0} \times \frac{10^5}{10^2} = 4.1 \times 10^3.$$

For numbers less than 1, say 0.01, the exponent or power of 10 is written with a negative sign: $0.01 = 1/100 = 1/10^2 = 1 \times 10^{-2}$. Similarly, $0.002 = 2 \times 10^{-3}$. The decimal point has again been moved the number of places expressed in the exponent. The negative exponent allows calculations to come out correctly. Thus, $0.020 \times 3600 = 72$; in exponential notation $(2.0 \times 10^{-2}) \times (3.6 \times 10^3) = 7.2 \times 10^1 = 72$.

Notice also that $10^1 \times 10^{-1} = 10 \times 0.1 = 1$, and by the law of exponents, $10^1 \times 10^{-1} = 10^0$. Therefore, $10^0 = 1$.

When writing a number in exponential notation, it is usual to make the simple number be between 1 and 10. Thus it is conventional to write 4.5×10^9 rather than 45×10^8, although they are the same number.† This notation also allows the number of *significant figures* to be clearly expressed. We write 4.50×10^9 if this value is accurate to three significant figures, but 4.5×10^9 if it is accurate to only two.

A–4 • Algebra

Physical relationships between quantities can be represented as equations involving symbols (usually letters of the alphabet) that represent the quantities. The manipulation of such equations is the field of algebra, and is used a great deal in physics. An equation involves an equals sign, which tells us that the quantities on either side of the equals sign have the same value. Examples of equations are

$$3 + 8 = 11$$

$$2x + 7 = 15$$

$$a^2b + c = 6.$$

The first equation involves only numbers, so is called an arithmetic equation. The other two equations are algebraic since they involve symbols. In the third equation, the quantity a^2b means the product of a times a times b: $a^2b = a \times a \times b$.

Solving for an Unknown

Often we wish to solve for one (or more) symbols, and we treat it as an *unknown*. For example, in the equation $2x + 7 = 15$, x is the unknown; this equation is true, however, only when $x = 4$. Determining what value (or

† Another convention used, particularly with computers, is that the simple number be between 0.1 and 1. Thus we would write 4,500,000,000 as 0.450×10^{10}.

values) the unknown(s) can have to satisfy the equation(s) is called *solving the equation.* To solve an equation, the following rule can be used:

> *An equation will remain true if any operation performed on one side is also performed on the other side:* for example, (*a*) addition or subtraction of a number or symbol; (*b*) multiplication or division by a number or symbol; (*c*) raising each side of the equation to the same power, or taking the same root (such as square root).

EXAMPLE A–1 Solve for x in the equation

$$2x + 7 = 15.$$

SOLUTION We first subtract 7 from both sides:

$$2x + 7 - 7 = 15 - 7$$

$$2x = 8.$$

Then we divide both sides by 2 to get

$$\frac{2x}{2} = \frac{8}{2}$$

$$x = 4,$$

and this solves the equation.

EXAMPLE A–2 (*a*) Solve the equation

$$a^2b + c = 24$$

for the unknown a in terms of b and c. (*b*) Solve for a assuming that $b = 2$ and $c = 6$.

SOLUTION (*a*) We are trying to solve for a, so we first subtract c from both sides:

$$a^2b = 24 - c,$$

then divide by b:

$$a^2 = \frac{24 - c}{b},$$

and finally take square roots:

$$a = \sqrt{\frac{24 - c}{b}}.$$

(*b*) If we are given that $b = 2$ and $c = 6$, then

$$a = \sqrt{\frac{24 - 6}{2}} = 3.$$

To check a solution, we put it back into the original equation (this is really a check that we did all the manipulations correctly). In the equation

$$a^2b + c = 24,$$

we put in $a = 3$, $b = 2$, $c = 6$ and find

$$(3)^2(2) + (6) \stackrel{?}{=} 24$$

$$24 = 24,$$

which checks.

Two or More Unknowns

If we have two or more unknowns, one equation is not sufficient to find them. In general, if there are n unknowns, n independent equations are needed. For example, if there are two unknowns, we need two equations. If the unknowns are called x and y, a typical procedure is to solve one equation for x in terms of y, and substitute this into the second equation.

Example A–3 Solve the following pair of equations for x and y:

$$3x - 2y = 19$$

$$x + 4y = -3.$$

SOLUTION We solve the second equation for x in terms of y by subtracting $4y$ from both sides:

$$x = -3 - 4y.$$

We substitute this expression for x into the first equation, and simplify:

$$3(-3 - 4y) - 2y = 19$$

$$-9 - 12y - 2y = 19 \qquad \text{(carried out the multiplication by 3)}$$

$$-14y = 28 \qquad \text{(added 9 to both sides)}$$

$$y = -2 \qquad \text{(divided both sides by } -14\text{).}$$

Now that we know $y = -2$, we substitute this into the expression for x:

$$x = -3 - 4y$$

$$= -3 - 4(-2) = -3 + 8 = 5.$$

Our solution is $x = 5$, $y = -2$. We check this solution by putting these values back into the original equations:

$$3x - 2y \stackrel{?}{=} 19$$

$$3(5) - 2(-2) \stackrel{?}{=} 19$$

$$15 + 4 \stackrel{?}{=} 19$$

$$19 = 19 \qquad \text{(it checks)}$$

and

$$x + 4y \stackrel{?}{=} -3$$

$$5 + 4(-2) \stackrel{?}{=} -3$$

$$-3 = -3 \qquad \text{(it checks).}$$

Other methods for solving two or more equations, such as the method of determinants, can be found in an algebra textbook.

The Quadratic Formula

We sometimes encounter equations that involve an unknown, say x, that appears not only to the first power, but squared as well. Such a *quadratic equation* can be written in the form

$$ax^2 + bx + c = 0.$$

The quantities a, b, and c are typically given numbers or constants.[†] The general solutions to such an equation are given by the *quadratic formula*:

Quadratic formula

$$x = \frac{-b \pm \sqrt{b^2 - 4ac}}{2a}.$$

The $\pm$ sign indicates that there are two solutions for x: one where the plus sign is used, the other where the minus sign is used.[‡]

EXAMPLE A–4 Find the solutions for x in the equation

$$3x^2 - 5x = 2.$$

SOLUTION First we write this equation in the standard form

$$ax^2 + bx + c = 0$$

by subtracting 2 from both sides:

$$3x^2 - 5x - 2 = 0.$$

In this case, a, b, and c in the standard formula take the values $a = 3$, $b = -5$, and $c = -2$. The two solutions for x are

$$x = \frac{+5 + \sqrt{25 - (4)(3)(-2)}}{(2)(3)} = \frac{5 + 7}{6} = 2$$

and

$$x = \frac{+5 - \sqrt{25 - (4)(3)(-2)}}{(2)(3)} = \frac{5 - 7}{6} = -\frac{1}{3}.$$

In this example, the two solutions are $x = 2$ and $x = -\frac{1}{3}$. In physics problems, it sometimes happens that only one of the solutions corresponds to a real-life solution; in this case, the other solution is discarded. In other cases, both solutions may correspond to physical reality.

Notice, incidentally, that b^2 must be greater than $4ac$, so that $\sqrt{b^2 - 4ac}$ yields a real number. If $(b^2 - 4ac)$ is less than zero (negative), there is no real solution. The square root of a negative number is called *imaginary*.

[†] Or one or more of them could be variables, in which case additional equations are needed.

[‡] A second-order equation—one in which the highest power of x is 2—has two solutions; a third-order equation—involving x^3—has three solutions, and so on.

A–5 • The Binomial Expansion

Sometimes we end up with a quantity of the form $(1 + x)^n$: that is, the quantity $(1 + x)$ is raised to the nth power. This can be written as an infinite sum of terms, known as a *series expansion*, as follows:

$$(1 + x)^n = 1 + nx + \frac{n(n-1)}{2!}x^2 + \cdots.$$

This formula is useful for us mainly when x is very small compared to one ($x \ll 1$). In this case, each successive term is much smaller than the preceding term. For example, if $x = 0.01$, and $n = 2$, say, then whereas the first term equals 1, the second term is $nx = (2)(0.01) = 0.02$, and the third term is $[(2)(1)/2](0.01)^2 = 0.0001$, and so on. Thus, when x is small, we can ignore all but the first two (or three) terms and can write

$$(1 + x)^n \approx 1 + nx.$$

This approximation often allows us to easily solve an equation that otherwise might be very difficult. Some examples are

$$(1 + x)^2 \approx 1 + 2x$$

$$\frac{1}{1+x} = (1 + x)^{-1} \approx 1 - x$$

$$\sqrt{1 + x} = (1 + x)^{\frac{1}{2}} \approx 1 + \tfrac{1}{2}x$$

$$\frac{1}{\sqrt{1+x}} = (1 + x)^{-\frac{1}{2}} \approx 1 - \tfrac{1}{2}x$$

where $x \ll 1$.

As a numerical example, let us evaluate $\sqrt{1.02}$ using the binomial theorem since $x = 0.02$ is much smaller than 1:

$$\sqrt{1.02} = (1.02)^{\frac{1}{2}} = (1 + 0.02)^{\frac{1}{2}} \approx 1 + \tfrac{1}{2}(0.02) = 1.01.$$

You can check with a calculator (and maybe not even more quickly) that $\sqrt{1.02} \approx 1.01$.

A–6 • Plane Geometry

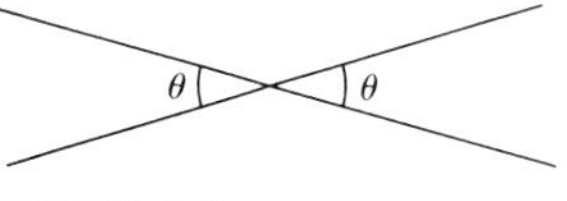

FIGURE A–1

We review here a number of theorems involving angles and triangles that are useful in physics.

1. *Equal angles.* Two angles are equal if any of the following conditions are true:
   (*a*) They are vertical angles (Fig. A–1); *or*
   (*b*) the left side of one is parallel to the left side of the other, and the right side of one is parallel to the right side of the other (the left and right sides are as seen from the vertex, where the two sides meet; Fig. A–2); *or*

FIGURE A–2

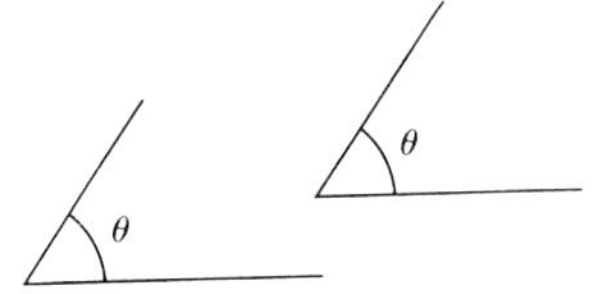

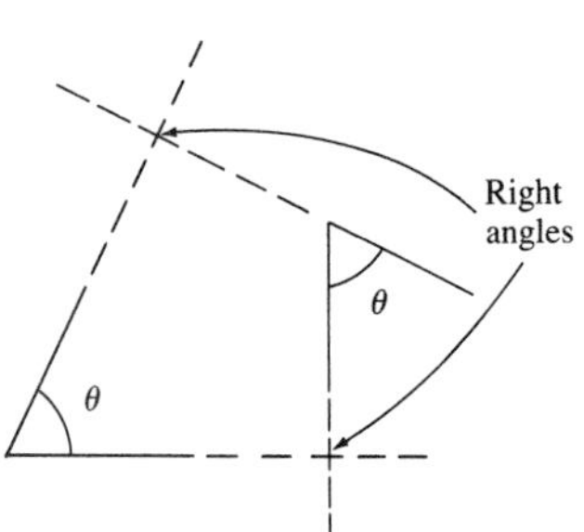

FIGURE A–3

FIGURE A–4

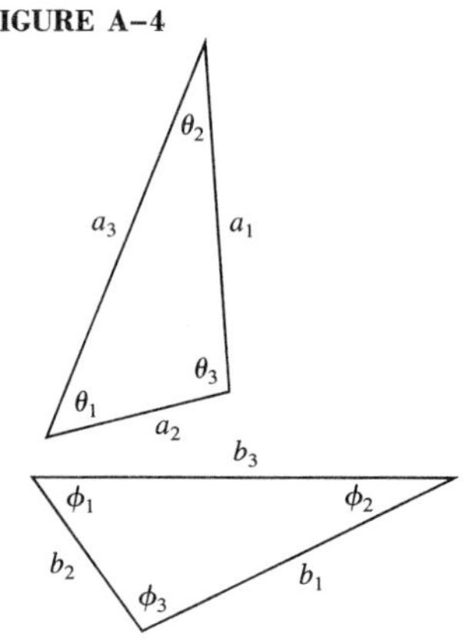

FIGURE A–5

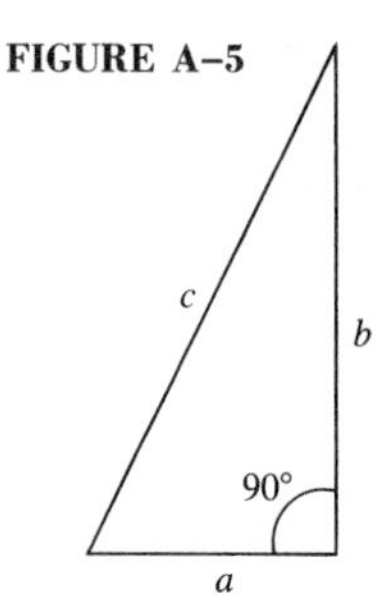

(*c*) the left side of one is perpendicular to the left side of the other, and the right sides are likewise perpendicular (Fig. A–3).

2. *The sum of the angles* in any plane triangle is 180°.
3. *Similar triangles.* Two triangles are said to be similar if all three of their angles are equal (in Fig. A–4, $\theta_1 = \phi_1$, $\theta_2 = \phi_2$, $\theta_3 = \phi_3$). Similar triangles thus have the same basic shape but may be different sizes and have different orientations. Two useful theorems about similar triangles are:
   (*a*) Two triangles are similar if any two of their angles are equal. (This follows because the third angles must also be equal since the sum of the angles of a triangle is 180°.)
   (*b*) The ratio of corresponding sides of two similar triangles are equal. That is (Fig. A–4),
$$\frac{a_1}{b_1} = \frac{a_2}{b_2} = \frac{a_3}{b_3}.$$
4. *Congruent triangles.* Two triangles are congruent if one can be placed precisely on top of the other. That is, they are similar triangles and they have the same size. Two triangles are congruent if any of the following holds:
   (*a*) The three corresponding sides are equal.
   (*b*) Two sides and the enclosed angle are equal ("side-angle-side").
   (*c*) Two angles and the enclosed side are equal ("angle-side-angle").
5. *Right triangles.* A right triangle has one angle that is 90° (a *right angle*); that is, the two sides that meet at the right angle are perpendicular. The two other (acute) angles in the right triangle add up to 90°.
6. *Pythagorean theorem.* In any right triangle, the square of the length of the hypotenuse (the side opposite the right angle) is equal to the sum of the squares of the lengths of the other two sides. In Fig. A–5,
$$c^2 = a^2 + b^2.$$

A–7 • Areas and Volumes

Object	Surface Area	Volume
Circle, radius r	πr^2	—
Sphere, radius r	$4\pi r^2$	$\frac{4}{3}\pi r^3$
Right circular cylinder, radius r, height h	$2\pi r^2 + 2\pi rh$	$\pi r^2 h$

A–8 • Trigonometric Functions and Identities

(See diagram)

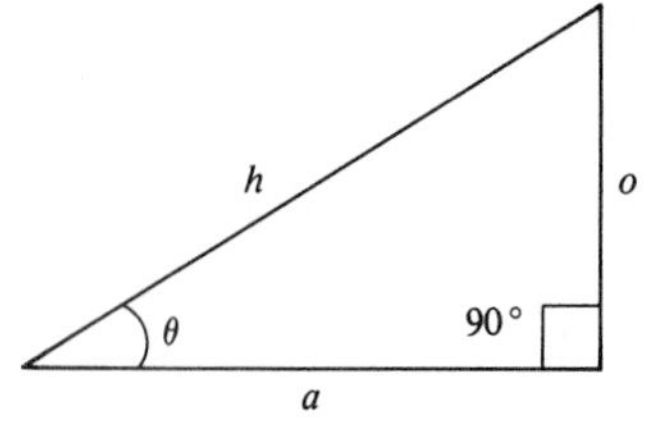

$$\sin\theta = \frac{o}{h} \qquad \csc\theta = \frac{1}{\sin\theta} = \frac{h}{o}$$

$$\cos\theta = \frac{a}{h} \qquad \sec\theta = \frac{1}{\cos\theta} = \frac{h}{a}$$

$$\tan\theta = \frac{o}{a} = \frac{\sin\theta}{\cos\theta} \qquad \cot\theta = \frac{1}{\tan\theta} = \frac{a}{o}$$

$$a^2 + o^2 = h^2 \qquad \text{(Pythagorean theorem)}$$

The following are some useful identities among the trigonometric functions:

$$\sin^2\theta + \cos^2\theta = 1, \qquad \sec^2\theta - \tan^2\theta = 1, \qquad \csc^2\theta - \cot^2\theta = 1$$

$$\sin 2\theta = 2\sin\theta\cos\theta$$

$$\cos 2\theta = \cos^2\theta - \sin^2\theta = 2\cos^2\theta - 1 = 1 - 2\sin^2\theta$$

$$\tan 2\theta = \frac{2\tan\theta}{1-\tan^2\theta}$$

$$\sin(A \pm B) = \sin A\cos B \pm \cos A\sin B$$

$$\cos(A \pm B) = \cos A\cos B \mp \sin A\sin B$$

$$\tan(A \pm B) = \frac{\tan A \pm \tan B}{1 \mp \tan A\tan B}$$

$$\sin\tfrac{1}{2}\theta = \sqrt{\frac{1-\cos\theta}{2}}, \qquad \cos\tfrac{1}{2}\theta = \sqrt{\frac{1+\cos\theta}{2}}, \qquad \tan\tfrac{1}{2}\theta = \sqrt{\frac{1-\cos\theta}{1+\cos\theta}}$$

$$\sin A \pm \sin B = 2\sin\left(\frac{A \pm B}{2}\right)\cos\left(\frac{A \mp B}{2}\right)$$

For any triangle (see diagram):

$$\frac{\sin\alpha}{a} = \frac{\sin\beta}{b} = \frac{\sin\gamma}{c} \qquad \text{(Law of sines)}$$

$$c^2 = a^2 + b^2 - 2ab\cos\gamma \qquad \text{(Law of cosines)}$$

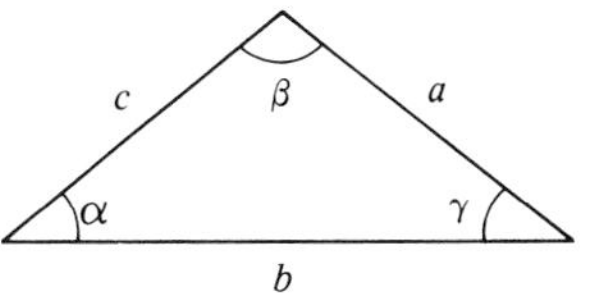

A–9 • Logarithms

Logarithms are defined in the following way:

$$\text{if } y = A^x, \quad \text{then } x = \log_A y.$$

That is, the logarithm of a number y to the base A is that number which, as the exponent of A, gives back the number y. For *common logarithms*, the base is 10, so

$$\text{if } y = 10^x, \quad \text{then } x = \log y.$$

The subscript 10 on $\log_{10}$ is usually omitted when dealing with common logs. Another base sometimes used is the natural number $e = 2.718\cdots$.[†] Such logarithms are called *natural logarithms* and are written ln. Thus,

$$\text{if } y = e^x, \quad \text{then } x = \ln y.$$

[†] The natural number e can be written as an infinite series:

$$e = 1 + \frac{1}{1} + \frac{1}{1\cdot 2} + \frac{1}{1\cdot 2\cdot 3} + \frac{1}{1\cdot 2\cdot 3\cdot 4} + \cdots.$$

For any number y, the two types of logarithm are related by

$$\ln y = 2.3026 \log y.$$

Some simple rules for logarithms are as follows:

$$\log (ab) = \log a + \log b. \tag{A–4}$$

This is true because if $a = 10^n$ and $b = 10^m$, then $ab = 10^{n+m}$. From the definition of logarithm, $\log a = n$, $\log b = m$, and $\log (ab) = n + m$; hence, $\log (ab) = n + m = \log a + \log b$. In a similar way, we can show that

$$\log \left(\frac{a}{b}\right) = \log a - \log b \tag{A–5}$$

and

$$\log a^n = n \log a. \tag{A–6}$$

These three rules apply not only to common logs but to natural or any other kind of logarithm.

Logs were once used as a technique for simplifying certain types of calculation. Because of the advent of electronic calculators and computers, they are not often used any more for this purpose. However, logs do appear in certain physical equations, so it is helpful to know how to deal with them. If you do not have a calculator that calculates logs, you can easily use a *log table*, such as the small one shown here (Table A–1). The number N is given to two digits (some tables give N to three or more digits); the first digit is in the vertical column to the left, the second digit is in the horizontal row across the top. For example, the table tells us that $\log 1.0 = 0.000$, $\log 1.1 = 0.041$, and $\log 4.1 = 0.613$. Note that the table does not include the decimal point—it is understood. The table gives logs for numbers between 1.0 and 9.9; for larger or smaller numbers we use rule A–4:

$$\log (ab) = \log a + \log b.$$

For example,

$$\log (380) = \log (3.8 \times 10^2) = \log (3.8) + \log (10^2).$$

From the table, $\log 3.8 = 0.580$; and from rule A–6, $\log (10^2) = 2 \log (10) = 2$

Table A–1
Short Table of Common Logarithms

N	0.0	0.1	0.2	0.3	0.4	0.5	0.6	0.7	0.8	0.9
1	000	041	079	114	146	176	204	230	255	279
2	301	322	342	362	380	398	415	431	447	462
3	477	491	505	519	531	544	556	568	580	591
4	602	613	624	633	643	653	663	672	681	690
5	699	708	716	724	732	740	748	756	763	771
6	778	785	792	799	806	813	820	826	833	839
7	845	851	857	863	869	875	881	887	892	898
8	903	908	914	919	924	929	935	940	944	949
9	954	959	964	968	973	978	982	987	991	996

since $\log(10) = 1$. [This follows from the definition of the logarithm: if $10 = 10^1$, then $1 = \log(10)$.] Thus,

$$\begin{aligned}\log(380) &= \log(3.8) + \log(10^2)\\ &= 0.580 + 2\\ &= 2.580.\end{aligned}$$

Similarly,

$$\begin{aligned}\log(0.081) &= \log(8.1) + \log(10^{-2})\\ &= 0.908 - 2 = -1.092.\end{aligned}$$

Sometimes we need to do the reverse process: find the number N whose log is, say, 2.670. This is called "taking the antilogarithm." To do so, we separate our number 2.670 into two parts, making the separation at the decimal point:

$$\begin{aligned}\log N = 2.670 &= 2 + 0.670\\ &= \log 10^2 + 0.670.\end{aligned}$$

We now look in the table to see what number has its log equal to 0.670; none does, so we must *interpolate*: we see that $\log 4.6 = 0.663$ and $\log 4.7 = 0.672$. So the number we want is between 4.6 and 4.7, and closer to the latter by 7/9. Approximately we can say that $\log 4.68 = 0.670$. Thus

$$\begin{aligned}\log N &= 2 + 0.670\\ &= \log(10^2) + \log(4.68) = \log(4.68 \times 10^2),\end{aligned}$$

so $N = 4.68 \times 10^2 = 468$. If the given logarithm is negative, say, -2.180, we proceed as follows:

$$\begin{aligned}\log N = -2.180 &= -3 + 0.820\\ &= \log 10^{-3} + \log 6.6 = \log 6.6 \times 10^{-3},\end{aligned}$$

so $N = 6.6 \times 10^{-3}$. Notice that what we did was to add to our given logarithm the next largest integer (3 in this case) so that we have an integer, plus a decimal number between 0 and 1.0 whose antilogarithm can be looked up in the table.

APPENDIX B

Dimensional Analysis

When we speak of the *dimensions* of a quantity, we are referring to the type of units that must be used. The dimensions of area, for example, are always length squared (abbreviated $[L^2]$, using square brackets) and the units can be square meters, square feet, and so on. Velocity, on the other hand, can be measured in units of km/h, m/s, and mi/h, but the dimensions are always a length $[L]$ divided by a time $[T]$, that is, $[L/T]$. The formula for a quantity may be different in different cases, but the dimensions remain the same. For example, the area of a triangle of base b and height h is $A = \frac{1}{2}bh$, whereas the area of a circle of radius r is $A = \pi r^2$. The formulas are different in the two cases, but the dimensions in both cases are the same: $[L^2]$.

When we specify the dimensions of a quantity, we usually do so in terms of base quantities, not derived quantities (see Section 1–5). The four common base quantities are length $[L]$, time $[T]$, mass $[M]$, and electric current $[I]$. (There are also three other, less used, basic quantities—see Section 1–5). Thus, force, which by Newton's second law has the same units as mass $[M]$ times acceleration $[L/T^2]$, has dimensions of $[ML/T^2]$.

Dimensions can be used as a help in working out relationships, and such a procedure is referred to as *dimensional analysis.* One useful technique is the use of dimensions to check a relationship for correctness. Two simple rules apply here. First, we can add or subtract quantities only if they have the same dimensions (we do not add centimeters and pounds); second, the quantities on each side of an equals sign must have the same dimensions.

For example, suppose that you derived the equation $v = v_0 + \frac{1}{2}at^2$, where v is the speed of an object after a time t, when it starts with an initial speed v_0 and undergoes an acceleration a. Let us do a dimensional check to see if this equation can be correct. We write a dimensional equation as follows, remembering that the dimensions of speed are $[L/T]$ and of acceleration are $[L/T^2]$:

$$\left[\frac{L}{T}\right] \stackrel{?}{=} \left[\frac{L}{T}\right] + \left[\frac{L}{T^2}\right][T^2]$$

$$\stackrel{?}{=} \left[\frac{L}{T}\right] + [L].$$

The dimensions are incorrect: on the right side, we have the sum of two quantities whose dimensions are not the same. Thus, we conclude that an error was made in the derivation of the original equation.

If such a dimensional check does come out correct, it does not prove that the equation is correct; for example, a dimensionless numerical factor (such as $\frac{1}{2}$ or 2π) could be wrong. Thus a dimensional check can only tell you when a relationship is wrong; it cannot tell you if it is completely right.

Another use of dimensional analysis is for a quick check on an equation you are not sure about. For example, suppose that you cannot remember whether the equation for the period T of an oscillating mass m on the end of a spring with spring constant k is $T = 2\pi\sqrt{k/m}$ or is $T = 2\pi\sqrt{m/k}$. A dimensional check can tell you. The dimensions of k, since from Hooke's law k = force/distance are $[mL/T^2/L] = [M/T^2]$. Thus the formula $T = 2\pi\sqrt{m/k}$ is correct:

$$[T] = \sqrt{\frac{[M]}{[M/T^2]}},$$

whereas the formula $T = 2\pi\sqrt{k/m}$ is not:

$$[T] \neq \sqrt{\frac{[M/T^2]}{[M]}}.$$

Finally, an important use of dimensional analysis, but one with which much care must be taken, is to obtain the *form* of an equation. That is, we may want to determine how one quantity depends on others. To take a concrete example, let us try to find an expression for the period T of a simple pendulum. First, we try to figure out what T could depend on, and make a list of these variables. It might depend on its length l, on the mass m of the bob, on the angle of swing θ, and on the acceleration due to gravity, g. It might also depend on air resistance (we would use the viscosity of air), the gravitational pull of the moon, and so on; but everyday experience suggests that the earth's gravity is the major force involved, so we ignore the other possible forces. So let us assume that T is a function of l, m, θ, and g, and that each of these factors is present to some power:

$$T = Cl^w m^x \theta^y g^z.$$

C is a dimensionless constant, and w, x, y, and z are exponents we want to solve for. We now write down the dimensional equation for this relationship:

$$[T] = [L]^w[M]^x[L/T^2]^z;$$

because θ has no dimensions (a radian is a length over a length—see Eq. 8–1), it does not appear. We simplify and obtain

$$[T] = [L]^{w+z}[M]^x[T]^{-2z}.$$

To have dimensional consistency, we must have

$$1 = -2z$$

$$0 = w + z$$

$$0 = x.$$

We solve these equations and find that $z = -\frac{1}{2}$, $w = \frac{1}{2}$, and $x = 0$. Thus our desired equation must be

$$T = C\sqrt{l/g}\, f(\theta), \qquad \text{(B–1)}$$

where $f(\theta)$ is some function of θ that we cannot determine using this technique. Nor can we determine in this way the dimensionless constant C. (Of course, to obtain C and f, we would have to do an analysis such as that in Chapter 11 using Newton's laws, which reveals that $C = 2\pi$ and $f \approx 1$ for small θ). But look what we *have* found, using only dimensional consistency. We obtained the form of the expression that relates the period of a simple pendulum to the major variables of the situation, l and g (see Eq. 11–8) and saw that it indeed does not depend on the mass m.

How did we do it? And how useful is this technique? Basically, we had to use our intuition as to which variables were important and which were not. This is not always easy, and often requires a lot of insight. As to usefulness, the final result in our example could have been obtained from Newton's laws, as in Chapter 11. But in many physical situations, such a derivation from other laws cannot be done. In those situations, dimensional analysis can be a powerful tool.

In the end, any expression derived by the use of dimensional analysis (or by any other means, for that matter) must be checked against experiment. For example, in our derivation of Eq. B–1, we can compare the periods of two pendula of different lengths, l_1 and l_2, whose amplitudes (θ) are the same. For, using Eq. B–1, we would have

$$\frac{T_1}{T_2} = \frac{C\sqrt{l_1/g}\,f(\theta)}{C\sqrt{l_2/g}\,f(\theta)} = \sqrt{\frac{l_1}{l_2}}.$$

Because C and $f(\theta)$ are the same for both pendula, they cancel out, so we can experimentally determine if the ratio of the periods varies as the ratio of the square roots of the lengths. This comparison to experiment checks our deviation, at least in part. C and $f(\theta)$ could be determined by further experiment.

APPENDIX C

Gauss's Law

An important relation in electricity is Gauss's law, developed by the great mathematician Karl Friedrich Gauss (1777–1855). It relates electric charge and electric field, and is a more general and elegant version of Coulomb's law.

Gauss's law involves the concept of *electric flux*, which (just as in the case of magnetic flux, Section 21–2) refers to the electric field passing through a given area. For a uniform electric field **E** passing through an area A, as shown in Fig. C–1a, the electric flux Φ_E is defined as

$$\Phi_E = EA\cos\theta, \quad \text{(C–1)}$$

where θ is the angle between the electric-field direction and a line drawn perpendicular to the area, as shown. The flux can be written equivalently as

$$\Phi_E = E_\perp A = EA_\perp,$$

where $E_\perp = E\cos\theta$ is the component of **E** perpendicular to the area (Fig. C–1b), and, similarly, $A_\perp = A\cos\theta$ is the projection of the area A perpendicular to the field **E** (Fig. C–1c).

Electric flux has a simple intuitive interpretation in terms of field lines. We saw in Section 16–8 that field lines can always be drawn so that the number (N) passing through unit area perpendicular to the field ($A_\perp$) is proportional to the magnitude of the field (E): that is, $E \propto N/A_\perp$. Hence,

$$N \propto EA_\perp = \Phi_E, \quad \text{(C–2)}$$

so the flux through an area is proportional to the number of lines passing through that area.

Gauss's law involves the *total* flux through a closed surface. For any such surface, such as that shown in Fig. C–2, we divide the surface up into

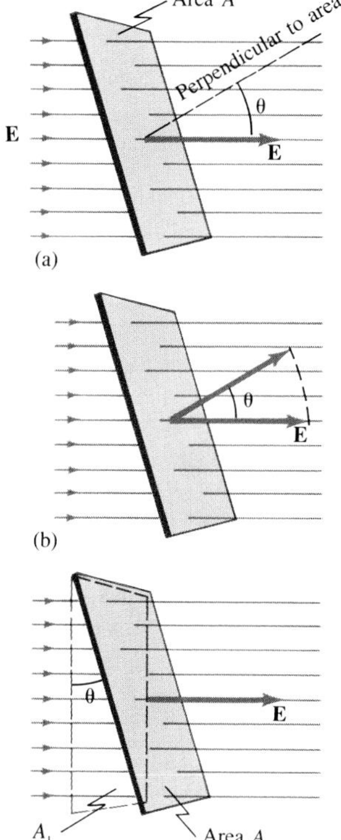

FIGURE C–1 (a) A uniform electric field **E** passing through a flat area A. (b) $E_\perp = E\cos\theta$ is the component of **E** perpendicular to the plane of the area A. (c) $A_\perp = A\cos\theta$ is the projection (shown dashed) of the area A perpendicular to the field **E**.

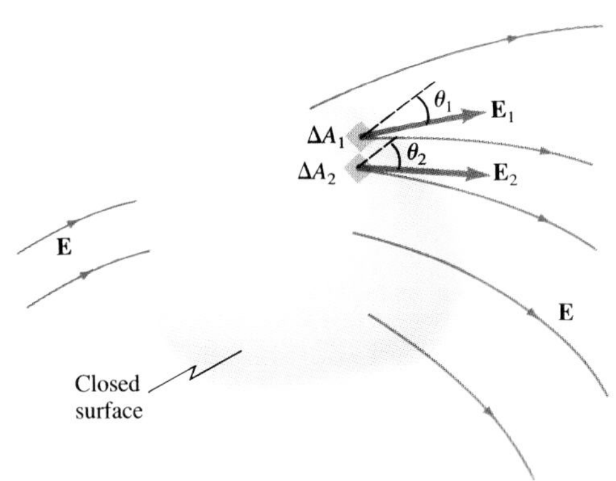

FIGURE C–2 Electric-field lines passing through a closed surface. The surface is divided up into many tiny areas, ΔA_1, ΔA_2, . . . , and so on, of which only two are shown.

many tiny areas, ΔA_1, ΔA_2, $\Delta A_3, \ldots$, and so on. We make the division so that each ΔA is small enough that it can be considered flat and so that the electric field can be considered constant within each ΔA. Then the *total* flux through the entire surface is the sum over all the individual fluxes through each of the tiny areas:

$$\begin{aligned}\Phi_E &= E_1\,\Delta A_1 \cos\theta_1 + E_2\,\Delta A_2 \cos\theta_2 + \cdots \\ &= \sum E\,\Delta A \cos\theta = \sum E_\perp\,\Delta A,\end{aligned}$$

where the symbol $\sum$ means "sum of." We saw in Section 16–8 that the number of field lines starting on a positive charge or ending on a negative charge is proportional to the magnitude of the charge. Hence, the *net* number of lines N pointing out of any closed surface (number of lines pointing out minus the number pointing in) must be proportional to the *net* enclosed charge, Q. But from Eq. C–2, we have that the net number of lines N is proportional to the total flux Φ_E. Therefore,

$$\Phi_E = \sum_{\substack{\text{closed}\\ \text{surface}}} E_\perp\,\Delta A \propto Q.$$

The constant of proportionality is $1/\varepsilon_0$, as we shall see, so we have

Gauss's law

$$\sum_{\substack{\text{closed}\\ \text{surface}}} E_\perp\,\Delta A = \frac{Q}{\varepsilon_0}, \qquad \text{(C–3)}$$

where the sum ($\sum$) is over any closed surface, and Q is the net charge enclosed within that surface. This is **Gauss's law**.

This argument to justify Gauss's law may seem rather abstract. So let us take a particular example, that of a single point charge, and show that Gauss's law is equivalent to Coulomb's law, and at the same time show that the proportionality constant is indeed $1/\varepsilon_0$. In Fig. C–3, we have a single charge Q, and for our "gaussian surface," we choose an imaginary sphere of radius r centered on the charge. Since Gauss's law is supposed to be valid for any surface, we have chosen one that will make our calculation easy, namely a sphere centered on the charge. Since the charge is at the center of the sphere, the magnitude of the electric field E at all points on the sphere is the same, namely

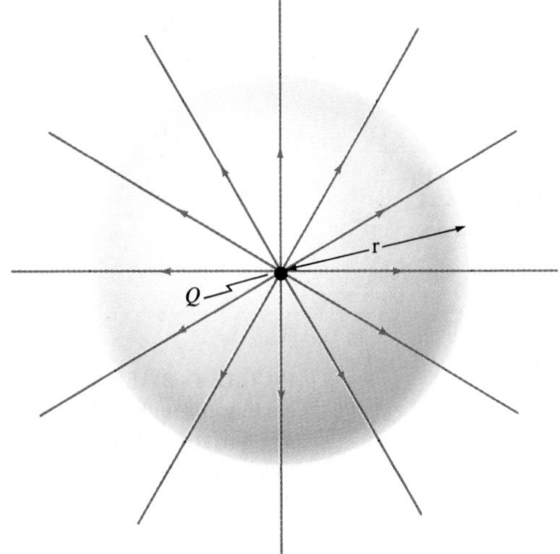

FIGURE C–3 A single point charge Q at the center of an imaginary sphere of radius r (our "gaussian surface"—that is, the closed surface we choose to use for applying Gauss's law in this case).

$$E = \frac{1}{4\pi\varepsilon_0}\frac{Q}{r^2},$$

from Coulomb's law (Eq. 16–4b). The direction of **E** at all points is perpendicular to the sphere, so $E_\perp = E$. Hence, the total flux is

$$\sum_{\substack{\text{closed}\\ \text{surface}}} E_\perp \,\Delta A = E \sum \Delta A = \frac{1}{4\pi\varepsilon_0}\frac{Q}{r^2}\sum \Delta A.$$

Since E is the same through each of the areas ΔA, we were able to factor it out of the sum. Now the sum over the entire area of the spherical surface, $\sum \Delta A$, is simply the surface area of a sphere, $4\pi r^2$. Hence,

$$\sum E_\perp \,\Delta A = \frac{1}{4\pi\varepsilon_0}\frac{Q}{r^2}(4\pi r^2) = \frac{Q}{\varepsilon_0},$$

which is just Gauss's law.

We can argue that Gauss's law will hold for any closed surface other than a sphere by noting that the number of lines N passing through it would be the same as through our sphere (Fig. C–4). Since the flux through the surface is proportional to N, then $\sum E_\perp \,\Delta A = Q/\varepsilon_0$ is valid for any surface surrounding a point charge.

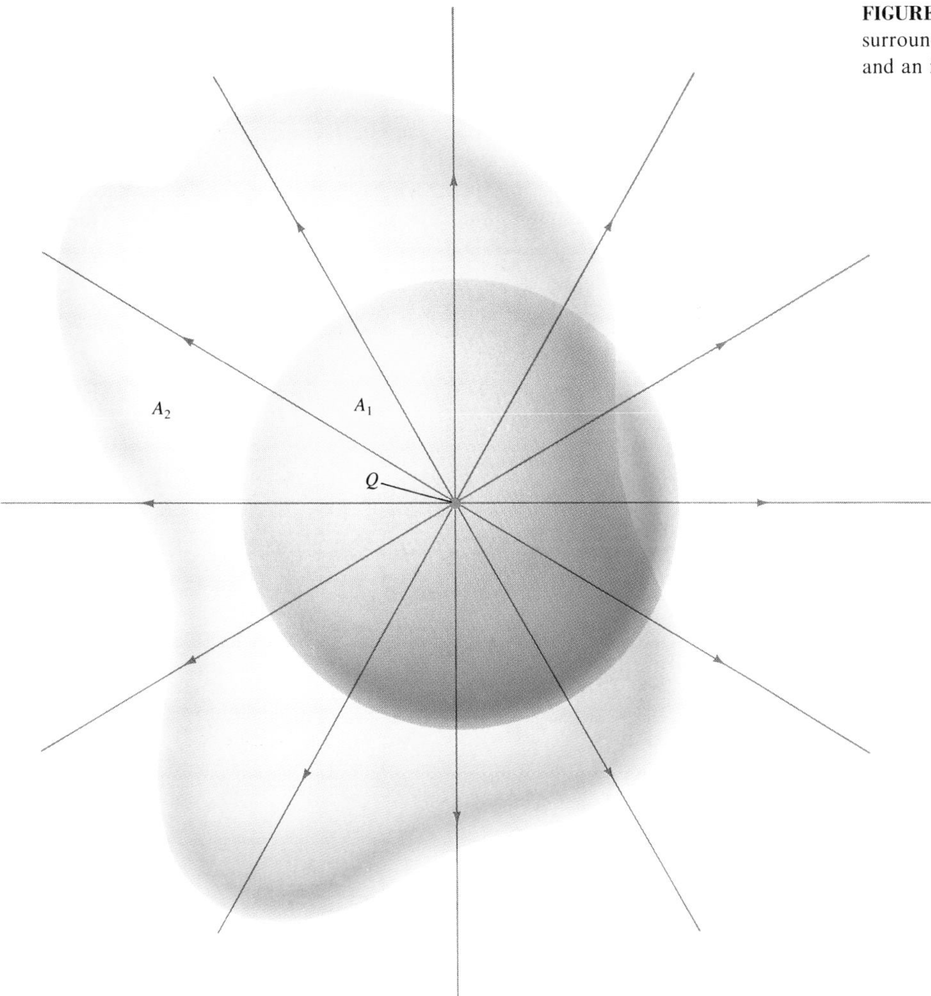

FIGURE C–4 A single point charge Q, surrounded by a spherical surface A_1 and an irregular surface A_2.

If we have many charges, $Q_1, Q_2, \ldots$, then for each charge individually,

$$\sum E_{i\perp}\,\Delta A = \frac{Q_i}{\varepsilon_0}$$

where the subscript i refers to any one of the charges. The sum of all the charges is the total charge $Q\ (=\sum Q_i)$. And, by the superposition principle, the total electric field E is equal to the sum of the fields due to each separate charge: $E = \sum E_i$. Hence

$$\sum E_{\perp}\,\Delta A = \frac{\sum Q_i}{\varepsilon_0} = \frac{Q}{\varepsilon_0}.$$

So Gauss's law is consistent with Coulomb's law for any number of charges enclosed within a closed surface of any shape.

Coulomb's law and Gauss's law are equivalent for electric fields produced by static charges. However, electric fields can also be produced by (changing) magnetic fields, as discussed in Chapter 21. Coulomb's law does not apply for such electric fields, but Gauss's law does apply. Hence, Gauss's law is considered a more general law than Coulomb's law.

Coulomb's law and Gauss's law can be used to determine the electric field due to a given (static) charge distribution. Gauss's law is useful when the charge distribution is simple and symmetrical. However, we must choose the "gaussian" surface very carefully so we can determine **E**. We normally try to think of a surface that has just the symmetry needed so E will be constant on all or on parts of its surface.

EXAMPLE C–1 *Spherical shell.* A thin spherical shell of radius r_0 possesses a total net charge Q that is uniformly distributed on it, Fig. C–5. Determine the electric field at points (*a*) outside the shell, and (b) inside the shell.

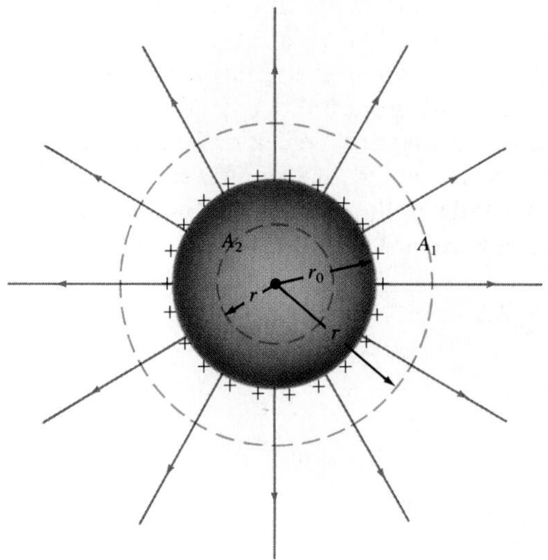

FIGURE C–5 Uniform distribution of charge on a thin spherical shell of radius r_0. Example C–1.

SOLUTION (*a*) Because the charge is distributed symmetrically, the electric field must also be symmetric. Thus the field must be directed radially outward (inward if $Q < 0$) and must depend only on r. The electric field will thus have the same magnitude at all points on an imaginary gaussian surface, which we draw as a sphere of radius r, concentric with the shell, and labeled A_1 in Fig. C–5. Since **E** is perpendicular to this surface, Gauss's law gives

$$\sum E_\perp \, \Delta A = E \sum \Delta A = E(4\pi r^2) = \frac{Q}{\varepsilon_0},$$

or

$$E = \frac{1}{4\pi\varepsilon_0} \frac{Q}{r^2}. \qquad [r > r_0]$$

Thus the field outside a uniformly charged spherical shell is the same as if all the charge were concentrated at the center as a point charge.

(*b*) Inside the shell, the field must also be symmetric. So E must again have the same value at all points on a spherical gaussian surface (A_2 in Fig C–5) concentric with the shell. Thus, E can be factored out of the sum and we have

$$\sum E_\perp \, \Delta A = E \sum \Delta A = E(4\pi r^2) = \frac{Q}{\varepsilon_0} = 0$$

since the charge Q enclosed by the shell is zero. Hence

$$E = 0 \qquad [r < r_0]$$

inside a uniform spherical shell of charge.

EXAMPLE C–2 *Long uniform line of charge.* A very long straight wire possesses a uniform charge per unit length, λ. Calculate the electric field at points near (but outside) the wire, far from the ends.

SOLUTION Because of the symmetry, we expect the field to be directed radially outward (if $Q > 0$) and to depend only on the perpendicular distance, r, from the wire. Because of this cylindrical symmetry, the field will be the same at all points on a gaussian surface that is a cylinder with the wire along its axis, Fig. C–6. **E** is perpendicular to this surface at all points. But for Gauss's law, we need a closed surface, so we must include the flat ends of the cylinder. Since **E** is parallel to the ends, there is no flux through the ends (the $\cos\theta$ in Gauss's law is $\cos 90° = 0$). So Gauss's law tells us

$$\sum E_\perp \, \Delta A = E \sum \Delta A = E(2\pi r l) = \frac{Q}{\varepsilon_0} = \frac{\lambda l}{\varepsilon_0},$$

where l is the length of our chosen gaussian surface ($l \ll$ length of wire), and $2\pi r$ is its circumference. Hence

$$E = \frac{1}{2\pi\varepsilon_0} \frac{\lambda}{r}.$$

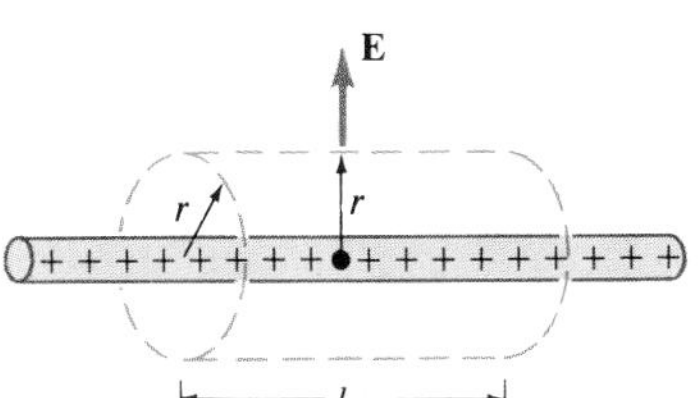

FIGURE C–6 Calculation of **E** due to a very long line of charge. Example C–2.

(1) Atomic Number Z	(2) Element	(3) Symbol	(4) Mass Number, A	(5) Atomic Mass[‡]	(6) % Abundance (or Radioactive Decay Mode)	(7) Half-life (if radioactive)
12	Magnesium	Mg	24	23.985042	78.99%	
13	Aluminum	Al	27	26.981538	100%	
14	Silicon	Si	28	27.976927	92.23%	
			31	30.975362	β^-, γ	157.3 min
15	Phosphorus	P	31	30.973762	100%	
			32	31.973908	β^-	14.262 days
16	Sulfur	S	32	31.972071	95.02%	
			35	34.969033	β^-	87.51 days
17	Chlorine	Cl	35	34.968853	75.77%	
			37	36.965903	24.23%	
18	Argon	Ar	40	39.962384	99.600%	
19	Potassium	K	39	38.963708	93.2581%	
			40	39.964000	0.0117%; β^-, EC, γ, β^+	1.227×10^9 yr
20	Calcium	Ca	40	39.962591	96.941%	
21	Scandium	Sc	45	44.955911	100%	
22	Titanium	Ti	48	47.947947	73.8%	
23	Vanadium	V	51	50.943962	99.750%	
24	Chromium	Cr	52	51.940511	83.79%	
25	Manganese	Mn	55	54.938048	100%	
26	Iron	Fe	56	55.934940	91.72%	
27	Cobalt	Co	59	58.933198	100%	
			60	59.933820	β^-, γ	5.2714 yr
28	Nickel	Ni	58	57.935346	68.077%	
			60	59.930789	26.223%	
29	Copper	Cu	63	62.929599	69.17%	
			65	64.927791	30.83%	
30	Zinc	Zn	64	63.929144	48.6%	
			66	65.926035	27.9%	
31	Gallium	Ga	69	68.925580	60.108%	
32	Germanium	Ge	72	71.922079	27.66%	
			74	73.921177	35.94%	
33	Arsenic	As	75	74.921594	100%	
34	Selenium	Se	80	79.916519	49.61%	
35	Bromine	Br	79	78.918336	50.69%	
36	Krypton	Kr	84	83.911508	57.0%	
37	Rubidium	Rb	85	84.911793	72.17%	
38	Strontium	Sr	86	85.909266	9.86%	
			88	87.905618	82.58%	
			90	89.907737	β^-	29.1 yr
39	Yttrium	Y	89	88.905847	100%	
40	Zirconium	Zr	90	89.904702	51.45%	
41	Niobium	Nb	93	92.906376	100%	
42	Molybdenum	Mo	98	97.905407	24.13%	

(1) Atomic Number Z	(2) Element	(3) Symbol	(4) Mass Number, A	(5) Atomic Mass[‡]	(6) % Abundance (or Radioactive Decay Mode)	(7) Half-life (if radioactive)
43	Technetium	Tc	98	97.907215	β^-, γ	4.2×10^6 yr
44	Ruthenium	Ru	102	101.904348	31.6%	
45	Rhodium	Rh	103	102.905502	100%	
46	Palladium	Pd	106	105.903481	27.33%	
47	Silver	Ag	107	106.905091	51.839%	
			109	108.904754	48.161%	
48	Cadmium	Cd	114	113.903359	28.73%	
49	Indium	In	115	114.903876	95.7%; β^-, γ	4.41×10^{14} yr
50	Tin	Sn	120	119.902197	32.59%	
51	Antimony	Sb	121	120.903820	57.36%	
52	Tellurium	Te	130	129.906228	33.87%	$< 1.25 \times 10^{21}$ yr
53	Iodine	I	127	126.904474	100%	
			131	130.906111	β^-, γ	8.04 days
54	Xenon	Xe	132	131.904141	26.9%	
			136	135.90721	8.9%	$\geqslant 2.36 \times 10^{21}$ yr
55	Cesium	Cs	133	132.905436	100%	
56	Barium	Ba	137	136.905816	11.23%	
			138	137.905236	71.70%	
57	Lanthanum	La	139	138.906346	99.9098%	
58	Cerium	Ce	140	139.905434	88.43%	
59	Praseodymium	Pr	141	140.907647	100%	
60	Neodymium	Nd	142	141.907718	27.13%	
61	Promethium	Pm	145	144.912745	EC, γ, α	17.7 yr
62	Samarium	Sm	152	151.919728	26.7%	
63	Europium	Eu	153	152.921226	52.2%	
64	Gadolinium	Gd	158	157.924099	24.84%	
65	Terbium	Tb	159	158.925344	100%	
66	Dysprosium	Dy	164	163.929172	28.2%	
67	Holmium	Ho	165	164.930320	100%	
68	Erbium	Er	166	165.930292	33.6%	
69	Thulium	Tm	169	168.934213	100%	
70	Ytterbium	Yb	174	173.938861	31.8%	
71	Lutecium	Lu	175	174.940772	97.41%	
72	Hafnium	Hf	180	179.946547	35.100%	
73	Tantalum	Ta	181	180.947993	99.988%	
74	Tungsten (wolfram)	W	184	183.950929	30.7%	$> 3 \times 10^{17}$ yr
75	Rhenium	Re	187	186.955746	62.60%; β^-	4.35×10^{10} yr
76	Osmium	Os	191	190.960922	β^-, γ	15.4 days
			192	191.961468	41.0%	
77	Iridium	Ir	191	190.960585	37.3%	
			193	192.962916	62.7%	
78	Platinum	Pt	195	194.964765	33.8%	
79	Gold	Au	197	196.966543	100%	

(1) Atomic Number Z	(2) Element	(3) Symbol	(4) Mass Number, A	(5) Atomic Mass‡	(6) % Abundance (or Radioactive Decay Mode)	(7) Half-life (if radioactive)
80	Mercury	Hg	199	198.968253	16.87%	
			202	201.970617	29.86%	
81	Thallium	Tl	205	204.974400	70.476%	
82	Lead	Pb	206	205.974440	24.1%	
			207	206.975871	22.1%	
			208	207.976627	52.4%	
			210	209.984163	β^-, γ, α	22.3 yr
			211	210.988734	β^-, γ	36.1 min
			212	211.991872	β^-, γ	10.64 h
			214	213.999798	β^-, γ	26.8 min
83	Bismuth	Bi	209	208.980374	100%	
			211	210.987254	α, γ, β^-	2.14 min
84	Polonium	Po	210	209.982848	α, γ	138.376 days
			214	213.995177	α, γ	164.3 μs
85	Astatine	At	218	218.00868	α, β^-	1.6s
86	Radon	Rn	222	222.017571	α, γ	3.8235 days
87	Francium	Fr	223	223.019733	β^-, γ, α	21.8 min
88	Radium	Ra	226	226.025402	α, γ	1600 yr
89	Actinium	Ac	227	227.027749	β^-, γ, α	21.773 yr
90	Thorium	Th	228	228.028716	α, γ	1.9131 yr
			232	232.038051	100%; α, γ	1.405×10^{10} yr
91	Protactinium	Pa	231	231.035880	α, γ	3.276×10^4 yr
92	Uranium	U	232	232.037131	α, γ	68.9 yr
			233	233.039630	α, γ	1.592×10^5 yr
			235	235.043924	0.720%; α, γ	7.038×10^8 yr
			236	236.045562	α, γ	2.3415×10^7 yr
			238	238.050784	99.2745%; α, γ	4.468×10^9 yr
			239	239.054289	β^-, γ	23.50 min
93	Neptunium	Np	239	239.052932	β^-, γ	2.355 days
94	Plutonium	Pu	239	239.052157	α, γ	24,119 yr
95	Americium	Am	243	243.061373	α, γ	7380 yr
96	Curium	Cm	245	245.065484	α, γ	8500 yr
97	Berkelium	Bk	247	247.07030	α, γ	1380 yr
98	Californium	Cf	249	249.074844	α, γ	351 yr
99	Einsteinium	Es	254	254.08802	α, γ, β^-	275.7 days
100	Fermium	Fm	253	253.085174	EC, α, γ	3.00 days
101	Mendelevium	Md	255	255.09107	EC, α, γ	27 min
102	Nobelium	No	255	255.09324	α, γ, EC	3.1 min
103	Lawrencium	Lr	257	257.0995	α, EC	0.646 s
104	Rutherfordium (?)	Rf	261	261.1086	α	65 s
105	Hahnium (?)	Ha	262	262.1138	α, fission	34 s
106			263	263.1182	α, fission	0.8 s
107			262	262.1231	α	102 ms
108			264	264.1285	α	0.08 ms
109			266	266.1378	α	3.4 ms

Answers to Odd-Numbered Problems

CHAPTER 1

1. (*a*) 3; (*b*) 4; (*c*) 3; (*d*) 1; (*e*) 3; (*f*) 2; (*g*) 2, 3, or 4
3. (*a*) 36,000; (*b*) 100; (*c*) 0.55; (*d*) 31,600,000; (*e*) 0.0000862
5. 0.2%
7. 5%
9. (*a*) 0.0356 m; (*b*) 0.000025 V; (*c*) 0.00025 kg; (*d*) 0.000000000500 s; (*e*) 0.0000000000000025 m; (*f*) 25,000,000,000 V
13. 3.9×10^{-9} in
15. (*a*) 9.46×10^{15} m; (*b*) 6.3×10^4 AU; (*c*) 7.19 AU/h
17. 2×10^9
19. (*a*) 2000
21. (*a*) 3.156×10^7 s; (*b*) 3.156×10^{16} ns; (*c*) 3.17×10^{-8} yr
23. 3.5×10^6 m
25. 240

CHAPTER 2

1. 97.8 km/h
3. 0.73 h
5. (*a*) 89 km/h; (*b*) 25 m/s; (*c*) 81 ft/s
7. (*a*) 3.97 m/s; (*b*) 0 m/s
9. 913 km/h
11. 6.73 m/s
13. 2.5 s
15. (*a*) 0.44 m/s, 1.4 m/s, etc; (*b*) 3.8 m/s^2, 4.0 m/s^2, etc
17. (*a*) 2.6 m/s^2; (*b*) 93 m
19. 2.1 m/s^2
21. 80 m/s
23. 20 *g*'s
25. (*a*) 71 m; (*b*) 40 m
27. She should brake.
29. 28 *g*'s
31. (*a*) 3.19 s; (*b*) 31.3 m/s
33. 1.5 s
37. 5.60 s
41. (*a*) ±9.30 m/s; (*b*) 1.09 s, 2.99 s; (*c*) stone passes same height once while ascending and again while falling.
43. 41 m
45. (*a*) 4.8 s; (*b*) 37 m/s; (*c*) 75 m
47. (*a*) 0.28 m/s; (*b*) 1.1 m/s; (*c*) 0.28 m/s; (*d*) 1.5 m/s; (*e*) −0.9 m/s
49. (*a*) 50 s; (*b*) 90 s to 107 s; (*c*) when curve is straight: 0 to 15 s, 65 s to 75 s, 90 s to 107 s; (*d*) $t \simeq 70$ s
51. (*a*) 5 m/s^2; (*b*) 2.5 m/s^2; (*c*) 0.35 m/s^2; (*d*) 1.6 m/s^2
57. 6
59. 243 km/h
61. 12 s
63. (*a*) 52 min; (*b*) 31 min
65. 1.9 m

CHAPTER 3

1. 103 km; 12° south of west
5. (*a*) 2.8 in *x* direction; (*b*) 9.8 in *x* direction; (*c*) 9.8 in negative *x* direction
7. 24 m, 31° north of east
9. (*a*) $v_N = 704$ km/h, $v_W = 560$ km/h; (*b*) 2110 km north, 1680 km west
11. (*a*) (−9, −4, −5); (*b*) (−3, 4, −11)
13. 64.6; 53.1°
15. (*a*) 62.7, −31.1°; (*b*) 77.4, 71.9°
17. ±79.0; ±65.9°
19. 8.4 m/s; 4.8 m/s
21. (*a*) 3.26 m/s, 72.1° to shore; (*b*) 13.0 m, 72.1° to shore
23. (*a*) 337 km/h, 6.0° E of N; (*b*) 25 km NE of destination (or 17.7 km perpendicular distance from correct course)
25. (*a*) 1.95 m/s; (*b*) 2.73 m/s
27. (*a*) 133 m; (*b*) 167 s
29. 72 km/h, 34° from ⊥; 72 km/h, 56° from ⊥
31. 0.444 m/s^2
33. 108 km/h
35. 7.0 m
37. 6.4 m
39. 2.23 s
41. 22 m
45. 6.9° above horizontal
47. (*a*) 14.3 s; (*b*) 1140 m; (*c*) $v_X = 79.9$ m/s, $v_Y = -79.9$ m/s; (*d*) 11.3 m/s; (*e*) −45.0°
49. $\tan \theta = -gt/v_0$
51. 18 m north, 25 m east, 36 m up; 47 m
53. 28.3 km/h, from 41.7° east of south
55. 0.89 s; 0.97 m
57. 35 m/s
59. 46.9° below horizontal

CHAPTER 4

1. 720 N
3. 60.7 kg
5. 1370 N
7. −4200 N
9. 3.2 m/s^2 downward
11. 210 N
13. The child must *accelerate* down in such a way that the tension in the rope is no more than the *weight* it can support (30 kg) *g*; that is, $a = 0.25g$

15. $3.84\ m/s^2$
17. (a) 9.9 m/s downward; (b) 4.0×10^3 N upward
19. (a) 40 N; (b) 10 N; (c) 0
21. (a) 19 N, $1.9\ m/s^2$, at 230°; (b) 14 N, $1.4\ m/s^2$, at 41°
23. (a) $13\ m/s^2$; (b) 5.1 m/s
27. (a) 370 N; (b) $0.98\ m/s^2$
29. (a) $a = F/(m_1 + m_2 + m_3)$; (b) $F_1 = m_1F/(m_1 + m_2 + m_3)$ $F_2 = m_2F/(m_1 + m_2 + m_3)$ $F_3 = m_3F/(m_1 + m_2 + m_3)$; (c) $F_{12} = (m_2 + m_3)F/(m_1 + m_2 + m_3)$ $F_{23} = m_3F/(m_1 + m_2 + m_3)$; (d) $a = 3.3\ m/s^2$ $F_1 = F_2 = F_3 = 33$ N $F_{12} = 67$ N $F_{23} = 33$ N
31. $1.7\ m/s^2$; 20.9 N, 22.6 N
33. 59 N; 0
35. 38°
37. 10.0 kg
39. (a) $1.5\ m/s^2$; (b) 380 N
41. $6.4\ m/s^2$
43. (a) $3.4\ m/s^2$; (b) 9.0 m/s
45. (a) $1.5\ m/s^2$, 6.0 m/s; (b) 2.4 m, 2.7 s
47. 90.7 N; 0.613
49. (a) $a = (m_1 \sin\theta - m_2)g/(m_1 + m_2)$; (b) if $m_1 \sin\theta > m_2$, a is down plane; if $m_1 \sin\theta < m_2$, a is up plane
51. 0.58
53. 0.020 N
55. 0.20
57. ≈ 35 m/s ≈ 130 km/h
59. 12 m/s
61. yes; 13 m/s
63. (a) 9.1 N·h/km; (b) 130 N

CHAPTER 5

1. 6.38 g's
3. $5.95 \times 10^{-3}\ m/s^2$; 3.56×10^{22} N
5. 22 m/s; yes
7. 0.26
11. (a) 0.880 N; (b) 7.45 N
13. $T_1 = (2\pi f)^2(m_1r_1 + m_2r_2)$ $T_2 = (2\pi f)^2 m_2r_2$
15. 0.39
17. $F_t = 3200$ N; $F_c = 1800$ N
19. (a) 1.45 m/s; (b) 3.66 m/s
21. 3.2×10^{-9} N; 3.6×10^{-6} N
23. $1.7\ m/s^2$
25. $25\ m/s^2$
27. 3.2×10^7 m
29. increase by factor of 1.26
31. 3.3×10^{-8} N toward diagonally opposite sphere
33. 2.01×10^{30} kg
35. 6.45×10^3 m/s tangentially
37. (a) 530 N; (b) 530 N; (c) 700 N; (d) 350 N; (e) 0
39. (a) 18 N toward moon; (b) 220 N away from moon
43. 1.41 h
45. 164 yr
47. $\approx 2 \times 10^8$ yr
49. $r_1 = 4.22 \times 10^8$ m; $r_E = 6.71 \times 10^8$ m; $r_G = 10.71 \times 10^8$ m; $r_C = 18.84 \times 10^8$ m
51. 2.64×10^3 km
53. $(rg)^{1/2}$
55. 3.46×10^8 m from center of earth
57. (a) 640 m; (b) 5500 N
59. $\theta = \tan^{-1}(M_M R_E^2 / M_E D_M^2)$
61. 1.41 h

CHAPTER 6

1. 2.7×10^3 J
3. 5.9×10^6 J
5. 6.63 m
7. 29 J
9. (a) 180 N; (b) −800 J; (c) −4800 J; (d) 5600 J; (e) 0
11. 4900 J
13. 0.070 J
15. 1.5×10^{10} J
17. 6.83×10^3 m/s
19. 4
21. 3.86×10^5 J
23. 420 N
25. (a) -1.0×10^5 J; (b) 9.1×10^4 N; (c) -1.9×10^5 J
27. (a) 2.03×10^3 N; (b) 6.09×10^3 J; (c) 4.67×10^4 J; (d) -4.06×10^4 J; (e) 8.22 m/s
29. 55 J
31. (a) 7.6×10^5 J (b) 7.6×10^5 J; (c) yes, thermal energy
33. 1.8 m; no
35. 2.4×10^6 J
37. 6.5 m/s
39. 5.2 m/s
41. 18.1 m/s
43. (a) 21.1 m/s; (b) 23.1 m
45. (a) 7.64×10^4 J; (b) 717 N
47. 400 N
51. 5.4 MJ
53. 1.86 W
55. 24°
57. (a) $20\ m/s^2$; (b) −1.9 N
59. 0.54
61. $E = \frac{1}{2}kx_0^2 = \frac{1}{2}mv^2 + \frac{1}{2}kx^2$
63. (a) 1.8×10^5 J; (b) 22 m/s; (c) 3.2 m
65. $12\ (Mg/h)$
67. (a) 48.5 m/s; (b) 7.95×10^5 W

CHAPTER 7

1. 0.18 kg·m/s
3. 0.425 m/s (opposite direction of package)
5. 13.5 m/s
7. 640 m/s
9. 6.7×10^3 m/s
11. (a) $v_1 = 4.98 \times 10^3$ m/s, $v_2 = 7.43$ m/s in original direction; (b) 5.25×10^8 J
13. 130 N; no
15. (a) 450 kg·m/s; (b) 450 kg·m/s; (c) 450 N
17. (a) 4.5 N·s; (b) 75 m/s
19. $v_1 = 6.00$ m/s, $v_2 = 1.50$ m/s in original direction
21. $v_1 = 3.00$ m/s, $v_2 = 2.00$ m/s
25. 10.6×10^{-23} kg·m/s; 144°
27. (a) −30°; (b) both $v/\sqrt{3}$; (c) $\frac{2}{3}$
29. 2.4
31. 0.63 m
33. less massive, 2700 J; more massive, 1800 J
35. (b) $e = (h'/h)^{1/2}$
37. 141°
39. 0.48×10^{-10} m
43. From elbow: (8.4 cm, 8.5 cm)
45. $6.67r_0$ from center of smallest sphere
47. $0.27R$ to left of center of plate
49. (a) 4.7 m from girl; (b) 3.2 m; (c) 3.3 m

51. $mv/(m + M)$, upward; the balloon stops
53. 14.2 m/s
57. $m/2$
59. 1:2
61. (*a*) −5.04 m/s, 3.36 m/s; (*b*) 2.6 m along plane

CHAPTER 8

1. (*a*) $\pi/6 = 0.52$; (*b*) $\pi/2 = 1.6$; (*c*) $7\pi/3 = 7.3$
3. 6.8 km diameter
5. 21 m/s
7. -8.4 rad/s^2
9. 3.4×10^6 m
11. (*a*) 464 m/s; (*b*) 232 m/s
13. (*a*) 3.5 rad/s^2; (*b*) $a_c = 230$ m/s^2, $a_T = 1.05$ m/s^2
15. 0.38 rad/s^2
17. (*a*) −54 rad/s^2; (*b*) 190 rev
21. (*a*) −2.0 rad/s^2; (*b*) 14 s
23. 106 N·m
25. 1.2 N·m clockwise
27. 5.72 kg·m^2
29. 0.85×10^{-10} m
31. (*a*) 6.88×10^{-3} kg·m^2; (*b*) 0.230 N·m
33. 17 kg
35. 1760 rev; 21.1 s
37. (*a*) 152 rad/s^2; (*b*) 1130 N
39. (*a*) 3.34 rad/s^2; (*b*) 7.01 m/s^2; (*c*) 352 m/s^2; (*d*) 2.57×10^3 N; (*e*) 1.14°
41. 12.0 m/s
43. 2.49×10^5 J
45. 2.95 m/s
47. 1.89 kg·m^2/s
49. 0.38 rev/s
51. (*a*) 3.18 kg·m^2/s; (*b*) 0.398 N·m
53. 4.8 rev/s
55. 0.32 rad/s
57. $\tan\theta = r\omega^2/g$
59. (*a*) ball passes ahead of woman; (*b*) $d = v\omega t^2$; $a_c = 2\omega v$
61. north, south
63. $L/2$; $L/2$
65. 8.21×10^{-6}
67. $M/2$
69. (*a*) 8.1 m; (*b*) 6.2 s
71. $F = Mg(2Rh - h^2)^{1/2}/(R - h)$
73. (*c*) 157 hp
75. $h = 2.7(R_0 - r)$

CHAPTER 9

1. 1.44 N
3. 5.37 kg
5. near end, 2350 N; far end, 1370 N
7. left, 240 N; right, 210 N
9. $F_1 = -2.94 \times 10^3$ N; $F_2 = 1.47 \times 10^4$ N
11. $F_1 = 1470$ N down; $F_2 = 1960$ N up
13. 6.47 kg
15. $F_T = 140$ N; $F_w = 140$ N, $\theta = 45°$
17. (*a*) 290 N; (*b*) 2.8×10^3 N
19. $\theta_{min} = \tan^{-1}(1/2\mu)$
21. $y = 2.2$ m
23. 94 N
25. 990 N
27. 1790 N
29. 2700 N
31. (*a*) $F_M = 1500$ N, $F_{Jx} = 520$ N, $F_{Jy} = 2400$ N; (*b*) $F_M = 1700$ N, $F_{Jx} = 580$ N, $F_{Jy} = 2400$ N
33. 446 N
37. (*a*) 3.9; (*b*) 1.5
39.

$T_f:T_r$	IMA	d
52:13	0.13	8.5 m
52:15	0.15	7.4 m
52:17	0.17	6.5 m
52:20	0.20	5.5 m
52:24	0.24	4.6 m
52:28	0.28	4.0 m
42:13	0.16	6.9 m
42:15	0.19	6.0 m
42:17	0.21	5.3 m
42:20	0.25	4.5 m
42:24	0.30	3.8 m
42:28	0.35	3.2 m

41. $0.85mg$
43. (*b*) yes; (*c*) $\sum_{i=1}^{i=N} (1/2i)$; (*d*) 32
45. 2.0×10^{-3} cm
47. left, 0.15%; right, 0.13%
49. 9×10^7 N/m^2 = 900 atm
51. 1.7×10^{-2} J
53. 5.1×10^4 N
55. 0.33 cm^2
57. 1.02 cm
59. 12.0 m
61. 1.0×10^9 N·m; will not topple
63. 40°; same
65. (*a*) 5.8×10^8 N/m^2 on each leg; (*b*) will break; (*c*) 1.3×10^7 N/m^2 on each, will not break
67. $T_1 = 4.54mg$; $T_2 = 4.96mg$; $h = 158$ m
69. (*a*) $mg/2 \tan\theta$; (*b*) $mg/2 \sin\theta$; (*c*) a: horizontal, b: at $\measuredangle\theta$ to horizontal
71. 3.82
73. 7.2 m vertically
75. 2.7 m; neglect energy loss as sound, heat on impact

CHAPTER 10

1. 2.7×10^{11} kg
3. 0.8563
5. 7.4×10^7 N/m^2 = 730 atm
7. 1.75×10^4 N/m^2
9. 1.16×10^7 N, 1.21×10^5 N/m^2; same
11. 3.8×10^9 N
13. 13.1 m
17. (*a*) 0.34 kg; (*b*) 1.5×10^4 N
19. 987 kg/m^3
21. 0.199
23. (*a*) SG = 1.004; (*b*) zero (it floats)
25. 954 kg/m^3
27. 0.105
31. 5.0 m/s
33. 2.0×10^5 N/m^2
35. 7.1×10^5 N
37. 2.2 m/s; 1.06 atm
39. 0.074 Pa·s
41. 3200; turbulent
43. 3.7×10^{-3} Pa·s
45. 13 cm
47. $\approx 2 \times 10^{-4}$ m^3/s
49. 3.1 atm
51. 3.7 yr
55. 0.13 m
57. 2.0×10^{-3} N, down
59. (*a*) $\gamma = (F - mg)/2\pi r$; (*b*) 0.048 N/m
61. 0.29 μm (diameter)
65. (*a*) 8×10^{-4} N; (*b*) 0.19 N

67. 0.63 N
69. 140 N
71. 1.1 m
73. 2.2×10^7 kg (sea water)
75. (*a*) 153 m/s; (*b*) 0.85 atm
77. 33 N; no

CHAPTER 11

1. 165 N/m
3. $4A$
7. 2.9 Hz
9. 0.56 kg
11. (*a*) 0.32 m; (*b*) 1.18 Hz; (*c*) 1.4 J; (*d*) KE = 0.48 J; PE = 0.92 J
13. 10.3 m/s
17. (*a*) $x = 0.285 \sin 20.5t$; (*b*) at $t = 0.077$ s and every 0.153 s thereafter
19. $A_1 = \sqrt{10}A_2$
23. (*a*) 1.36 s; (*b*) 0.73 Hz
25. (*a*) 1.8 s; (*b*) ∞
27. (*a*) 0.83 Hz; (*b*) 0.49 m/s
29. 2.4 m/s
31. 545 m to 188 m; 3.41 m to 2.78 m
33. 1.41 times greater in rod with lower density
35. 0.36 m
37. (*a*) 1200 km; (*b*) no, could be anywhere on circle of this radius
39. (*a*) 0.08 m; (*b*) increase by factor of $2\frac{1}{4}$
41. (*a*) 1.0×10^{10} J/m²·s; (*b*) 5.0×10^{10} W
45. 49° to shelf
47. (*a*) $\theta_{im} = \sin^{-1}(v_1/v_2)$; (*b*) $\theta \geq 54°$
49. 587 Hz
51. 0.417 m
53. 300 Hz; 600 Hz, 900 Hz
57. (*b*) 1.8%; (*c*) yes
59. 0.044 m; 1.7 m
61. (*a*) 1.02 Hz; (*b*) 16.5 J
63. (*a*) 784 Hz, 1176 Hz, 880 Hz, 1320 Hz; (*b*) 1.26; (*c*) 1.12; (*d*) 0.794
65. $T = 12\frac{1}{2}$ h; $f = 2.2 \times 10^{-5}$ Hz; $v = 450$ m/s
67. 0.69 cm
69. 0.334
71. slow; 16.4 s
73. (*d*) $k = C^4/D^3$; (*e*) $T = 2\pi\sqrt{mD^3/C^4}$

CHAPTER 12

1. 206 m
3. 470 m
5. 10^{-6} W/m²
7. 1.6×10^6
9. 1.6×10^{-7} W
11. 86.2 dB
13. (*a*) 9; (*b*) 9.54 dB
15. 0.089 mm
17. (*a*) 132 dB; (*b*) 97 dB; (*c*) 55 dB
19. 500 Hz; 10,000 Hz
21. 24.6 N
23. 0.324 m
25. (*a*) 123 Hz, 368 Hz, 613 Hz, 858 Hz; (*b*) 245 Hz, 490 Hz, 735 Hz, 980 Hz
29. 4.3 m, open
31. (*a*) 115; (*b*) 116
33. 6 Hz
35. 15 Hz, inaudible; 3.75 Hz, inaudible
37. (*a*) $131\frac{1}{3}$ Hz or $132\frac{2}{3}$ Hz; (*b*) 1.02%
39. (*a*) 343 Hz; (*b*) 1029 Hz; 1715 Hz
43. (*a*) 849 Hz; (*b*) 951 Hz
45. 31 Hz
47. 0.205 m/s
49. 220 m/s
53. (*a*) 3.08; (*b*) 19°
55. 41 m
57. 3400 Hz; yes
59. (*a*) 280 m/s, 40 N; (*b*) 19 cm; (*c*) 880 Hz, 1320 Hz
61. 635 Hz
63. 2.7 Hz
65. 32.7 m/s
67. $\Delta d = \lambda[(l/2)^2 + d^2]^{1/2}/4d$
69. (*a*) 650 Hz; (*b*) 650 Hz; (*c*) 650 Hz; (*d*) 650 Hz; (*e*) 677 Hz; (*f*) 678 Hz

CHAPTER 13

1. 7.3 kg
3. (*a*) 38.4°C; (*b*) 68.2°C
5. −40°
7. 5/9
11. 1.8 mL
13. −70°C
15. (*b*) 0.0052
17. −0.0014
19. (*a*) 6.6×10^7 N/m²; (*b*) no; (*c*) 6.6×10^6 N/m²; yes
21. (*a*) 310 K; (*b*) 300 K; (*c*) 77 K; (*d*) 5773 K
23. (*a*) 4.3×10^3 K, 1.5×10^7 K; (*b*) 6%, 0.002%
25. 1.43 kg/m³
27. 2.29 atm
29. 1590 atm
31. 3.88 atm
33. 4.04 cm³
35. 2.69×10^{25} molecules/m³
37. 6×10^3 m/s
39. (*a*) 3.83; (*b*) 4.53
41. $\sqrt{2}$
43. (*a*) 294 m/s; (*b*) 12 m/s
47. (*a*) solid, vapor; (*b*) 5.11 to 73 atm, −56.6°C to 31°C
49. vapor
51. 0.47 atm
53. 2.22×10^3 N/m²
55. 6.4 kg
57. 120°C
59. 29%
61. 0.16 s, ≪ thermal speed
63. (*b*) N_2, by 6.9%
65. $\frac{1}{8}$
67. ≈300 molecules/cm³
69. 11 L; not advisable
71. $5\frac{3}{4}$ min faster
75. 420 m/s
77. 3.34×10^{-9} m
79. (*a*) lower; (*b*) 0.4%

CHAPTER 14

1. 1.7×10^6 J
3. 11°C
5. 188 kg
7. 840 kcal
9. 215 kg
11. 0.223 kg
13. 20°C
15. 24.2°C
17. 0.057 kcal/kg·°C

19. 0.334 kg
21. (a) 6.6 h (b) 50 h
23. 1.12×10^4 J/kg
25. 16 kW
27. 0.75 mm
29. 10°C
31. $\Delta Q/\Delta t = (k_1A_1/l_1 + k_2A_2/l_2)\,\Delta T$
33. 17 h
35. 7800 kg
37. 2.8°C
39. (a) $C = cm$ (b) 1 kcal/C° (c) 35 kcal/C°
41. (a) 49 kcal/h; (b) 8000 kcal/h
43. ≈ 20 W; only 10%
45. (a) $U = k/l$; (b) no; thickness (c) $\Delta Q/\Delta t = [A_1U_1 + A_2/(1/U_2 + 1/U_3 + 1/U_4)]\,\Delta T$ (d) 1.33 cal/s·m²·C°
47. (a) 360 C°, assuming no melting (b) 4.2 g melts
49. 4.4 g/h

CHAPTER 15

3. (a) -7.52×10^5 J (b) 0
5. (a) 0 (b) -1860 J (c) fell
7. (a) 25 J (b) 63 J (c) -95J (d) -120 J (e) -35 J
9. 1.19×10^7 J
11. 2.8 kg
13. 530°C
15. 4.3×10^{10} J/h
17. 722°C
19. 2.7×10^8 kg/h
21. -5.9 kcal/K
23. 0.22 kcal/K
25. 3.6 cal/K
27. (a) $\frac{1}{36}$ (b) $\frac{1}{6}$
29. (a) 2.47×10^{-23} J/K (b) -9.21×10^{-22} J/K (c) thermodynamic systems have many more particles, hence much greater ΔS
31. (a) 210 m² (b) yes
33. (a) 7.8×10^6 kWh (b) 0.39 MW
35. 1×10^{12} kcal
37. (a) 1.83×10^5 J (b) -1.35×10^5 J
39. 21%
41. 0.15 J/K
43. (a) 4.4×10^4 J (b) 3.5 min
45. (a) 5.6 C° (b) 83 J/kg·K

CHAPTER 16

1. 31 N
3. 2.7×10^{-3} N
5. 68 N
7. $F_1 = -144$ N (left)
$F_2 = 529$ N (right)
$F_3 = -385$ N (left)
9. 7.7×10^5 N away from center of square
11. both $= Q_T/2$; 0, Q_T
13. $Q_1 = 55.2\ \mu C$, $Q_2 = 34.8\ \mu C$; $Q_1 = 107.8\ \mu C$, $Q_2 = -17.8\ \mu C$
15. -1.2×10^6 N/C, up
17. 2.52×10^7 N/C, up
19. A: 2.6×10^6 N/C, up;
B: 6.7×10^6 N/C, 56° above
21. horizontal to the right
23. 1.7×10^6 N/C, away from center (0.289l, 0.5l)
27. 3.46×10^5 km
29. (a) 4×10^{-10} N (b) 7×10^{-10} N (c) $\approx 10^{-4}$ N
31. 6.1×10^{-10}
33. 6.8×10^5 C; negative
35. 1.3×10^6 N/C, parallel to 41 side
37. 1.4×10^7
39. $0.402Q_0$, $0.366l$ From $-Q_0$ on line joining the charges
41. (a) 1.9×10^{-3} (b) 3.2×10^{-9} s

CHAPTER 17

1. -6.0×10^{-4} J
3. 2600 V; B
5. 2.8×10^4 V/m
7. 19 kV
9. 1.83×10^7 m/s
11. 2.08×10^5 V
13. 0.045 J
15. (a) 7300 V; (b) 7.1×10^4 N/C, 57° north of east
17. (a) 8.5×10^{-32} C·m; (b) zero
19. (a) -0.088 V; (b) 0.6%
23. 3.25 pF
25. 90 μC
27. 220 pF
29. 2.97 μC
31. 8.25 μF
33. 1.62×10^{-4} J
35. Doubled
37. (a) 216 μJ; (b) 81 μJ; (c) -135 μJ; (d) yes
39. (a) 26 MJ; (b) 61 kg
41. Yes; 1.4×10^{-11} V
43. 9.4 MV
45. 0.50 μC
47. (a) 8.9 cm beyond -2.0 μC charge; (b) 4.0 cm beyond -2.0 μC charge; also between them, 0.80 m from -2.0 μC.
49. $Q_1 = C_1Q_0/(C_1 + C_2)$;
$Q_2 = C_2Q_0/(C_1 + C_2)$;
$V_1 = V_2 = Q_0/(C_1 + C_2)$

CHAPTER 18

1. 1.1×10^5 C
3. 4.0×10^{-11} A
5. 34 Ω
7. 4.7×10^{20}
9. 2×10^{-5} m/s
11. 0.33 mm
13. 3.7 mm diameter
15. -136°C
17. (a) 5.0×10^{-4} Ω; (b) 2.0×10^{-3} Ω (c) 4.5×10^{-3} Ω
19. $R_C = 1.24 \times 10^3$ Ω; $R_N = 1.56 \times 10^3$ Ω
21. 3.33 A
23. 8.7 V
25. 0.975 kWh
27. 1.7×10^6 J
29. 71%
31. 262 kg/h
33. 0.11 A

F

G

H

I

J

K

L

S

T

U

V

Periodic Table of the Elements§

Symbol — Cl 17 — Atomic number
Atomic mass§ — 35.453
$3p^5$ — Electron configuration

Group I	Group II	Transition elements										Group III	Group IV	Group V	Group VI	Group VII	Group 0
H 1 1.0079 $1s^1$																	He 2 4.00260 $1s^2$
Li 3 6.94 $2s^1$	Be 4 9.01218 $2s^2$											B 5 10.81 $2p^1$	C 6 12.011 $2p^2$	N 7 14.0067 $2p^3$	O 8 15.9994 $2p^4$	F 9 18.9984 $2p^5$	Ne 10 20.18 $2p^6$
Na 11 22.9898 $3s^1$	Mg 12 24.305 $3s^2$											Al 13 26.9815 $3p^1$	Si 14 28.0855 $3p^2$	P 15 30.974 $3p^3$	S 16 32.06 $3p^4$	Cl 17 35.453 $3p^5$	Ar 18 39.948 $3p^6$
K 19 39.0983 $4s^1$	Ca 20 40.08 $4s^2$	Sc 21 44.9559 $3d^14s^2$	Ti 22 47.9 $3d^24s^2$	V 23 50.9415 $3d^34s^2$	Cr 24 51.996 $3d^54s^1$	Mn 25 54.938 $3d^54s^2$	Fe 26 55.847 $3d^64s^2$	Co 27 58.9332 $3d^74s^2$	Ni 28 58.7 $3d^84s^2$	Cu 29 63.546 $3d^{10}4s^1$	Zn 30 65.39 $3d^{10}4s^2$	Ga 31 69.73 $4p^1$	Ge 32 72.6 $4p^2$	As 33 74.9216 $4p^3$	Se 34 78.96 $4p^4$	Br 35 79.904 $4p^5$	Kr 36 83.80 $4p^6$
Rb 37 85.47 $5s^1$	Sr 38 87.62 $5s^2$	Y 39 88.9059 $4d^15s^2$	Zr 40 91.22 $4d^25s^2$	Nb 41 92.9064 $4d^45s^1$	Mo 42 95.94 $4d^55s^1$	Tc 43 (98) $4d^55s^2$	Ru 44 101.07 $4d^75s^1$	Rh 45 102.906 $4d^85s^1$	Pd 46 106.4 $4d^{10}5s^0$	Ag 47 107.868 $4d^{10}5s^1$	Cd 48 112.41 $4d^{10}5s^2$	In 49 114.82 $5p^1$	Sn 50 118.7 $5p^2$	Sb 51 121.75 $5p^3$	Te 52 127.60 $5p^4$	I 53 126.90 $5p^5$	Xe 54 131.3 $5p^6$
Cs 55 132.905 $6s^1$	Ba 56 137.33 $6s^2$	57-71†	Hf 72 178.49 $5d^26s^2$	Ta 73 180.95 $5d^36s^2$	W 74 183.85 $5d^46s^2$	Re 75 186.207 $5d^56s^2$	Os 76 190.2 $5d^66s^2$	Ir 77 192.22 $5d^76s^2$	Pt 78 195.08 $5d^96s^1$	Au 79 196.97 $5d^{10}6s^1$	Hg 80 200.59 $5d^{10}6s^2$	Tl 81 204.38 $6p^1$	Pb 82 207.2 $6p^2$	Bi 83 208.980 $6p^3$	Po 84 (209) $6p^4$	At 85 (210) $6p^5$	Rn 86 (222) $6p^6$
Fr 87 (223) $7s^1$	Ra 88 226.025 $7s^2$	89-103‡	Rf 104 (261) $6d^27s^2$	Ha 105 (262) $6d^37s^2$	106 (263)	107 (262)	108 (265)	109 (266)									

† Lanthanide series

La 57	Ce 58	Pr 59	Nd 60	Pm 61	Sm 62	Eu 63	Gd 64	Tb 65	Dy 66	Ho 67	Er 68	Tm 69	Yb 70	Lu 71
138.906	140.12	140.908	144.24	(145)	150.4	151.96	157.25	158.925	162.50	164.930	167.26	168.934	173.04	174.967
$5d^16s^2$	$5d^14f^16s^2$	$4f^36s^2$	$4f^46s^2$	$4f^56s^2$	$4f^66s^2$	$4f^76s^2$	$5d^14f^76s^2$	$5d^14f^86s^2$	$4f^{10}6s^2$	$4f^{11}6s^2$	$4f^{12}6s^2$	$4f^{13}6s^2$	$4f^{14}6s^2$	$5d^14f^{14}6s^2$

‡ Actinide series

Ac 89	Th 90	Pa 91	U 92	Np 93	Pu 94	Am 95	Cm 96	Bk 97	Cf 98	Es 99	Fm 100	Md 101	No 102	Lr 103
(227)	232.038	231.036	238.029	237.048	(244)	(243)	(247)	(247)	(251)	(252)	(257)	(258)	(259)	(260)
$6d^17s^2$	$6d^27s^2$	$5f^26d^17s^2$	$5f^36d^17s^2$	$5f^46d^17s^2$	$5f^66d^07s^2$	$5f^76d^07s^2$	$5f^76d^17s^2$	$5f^96d^07s^2$	$5f^{10}6d^07s^2$	$5f^{11}6d^07s^2$	$5f^{12}6d^07s^2$	$5f^{13}6d^07s^2$	$6d^07s^2$	$6d^17s^2$

§ Atomic mass values averaged over isotopes in percentages they occur on earth's surface. For many unstable elements, mass number of the most stable known isotope is given in parentheses.